INTRODUCTION TO THERMAL AND FLUID ENGINEERING

HEAT TRANSFER

A Series of Reference Books and Textbooks

Series Editor

Afshin J. Ghajar

Regents Professor
School of Mechanical and Aerospace Engineering
Oklahoma State University

Engineering Heat Transfer: Third Edition, *William S. Janna*

Conjugate Problems in Convective Heat Transfer, *Abram S. Dorfman*

Introduction to Thermal and Fluid Engineering,
Allan D. Kraus, James R. Welty, Abdul Aziz

Upcoming titles include:

Thermal Measurements and Inverse Techniques,
Helcio R.B. Orlande, Olivier Fudym, Denis Maillet, and Renato M. Cotta

INTRODUCTION TO THERMAL AND FLUID ENGINEERING

Allan D. Kraus
James R. Welty
Abdul Aziz

CRC Press
Taylor & Francis Group
Boca Raton London New York

CRC Press is an imprint of the
Taylor & Francis Group, an **informa** business

CRC Press
Taylor & Francis Group
6000 Broken Sound Parkway NW, Suite 300
Boca Raton, FL 33487-2742

First issued in paperback 2019

ISBN-13: 978-1-4200-8808-3 (hbk)
ISBN-13: 978-0-367-38242-1 (pbk)

Library of Congress Cataloging-in-Publication Data

Kraus, Allan D.
Introduction to thermal and fluid engineering / Allan D. Kraus, James R. Welty, Abdul Aziz.
p. cm. -- (Heat transfer ; 3)
"A CRC title."
Includes bibliographical references and index.
ISBN 978-1-4200-8808-3 (hardcover : alk. paper)
1. Thermodynamics. 2. Fluid dynamics. 3. Heat--Transmission. 4. Fluid mechanics. I. Kraus, Allan D. II. Welty, James R. III. Aziz, Abdul S. IV. Title. V. Series.

TJ265.K73 2010
621.402'1--dc22 2010011767

Visit the Taylor & Francis Web site at
http://www.taylorandfrancis.com

and the CRC Press Web site at
http://www.crcpress.com

Dedicated to Fred Landis, a fine friend,

a valued colleague, and a superb servant of our profession for which he

never received the credit that he deserved.

Contents

Preface

This text treats the disciplines of thermodynamics, fluid mechanics, and heat transfer, in that order, as comprising what are generally referred to as the **thermal/fluid sciences**. The study of these separate and independent disciplines has been a standard part of the mechanical and chemical engineering curricula for decades. Other engineering majors have commonly taken one or more of these subjects but, generally, not all three.

The first component, classical **thermodynamics**, involves the interaction of work, heat, and the change in the energy level of a system as it undergoes a change between equilibrium states. The laws of thermodynamics form a framework by which these state changes are evaluated and related to measurable properties, most notably temperature. Knowledge of the thermodynamic limits of processes is essential to the evaluation of energetic systems and all engineers need such knowledge upon occasion.

The second component, **fluid mechanics**, treats the change of mass, energy, and momentum associated with the movement of fluids. Mass flow into and out of an open system is an important part of the overall energy balance for the system. Large, multicomponent systems such as municipal water supplies, petroleum refineries, and manufacturing facilities involve numerous pipes, ducts, and other passageways through which fluids are transported and fluid flow analyses are important for describing the rates at which energetic processes take place.

The third component of the thermal/fluid sciences is **heat transfer**. Heat transfer, as one knows from a study of introductory physics is accomplished by conduction, convection, and radiation. Each of these modes enables one to evaluate the rate at which heat is transported between sites that are at different temperatures. Convection heat transfer is intimately involved with fluid motion and is, therefore, directly coupled to a knowledge of fluid mechanics.

As mentioned, thermodynamic analysis will determine the limiting equilibrium states that a process may experience. The project design of an energetic process requires, in addition to a listing of the appropriate thermodynamic limits, knowledge of the rates at which the process progresses from its initial to its final states. For example, in the case of a heat exchanger, the cross-sectional area of the unit is evaluated by the rate of mass throughput as determined by using fluid mechanics while the length of the unit is determined from the rate of heat transfer.

The foregoing comments speak directly to the interdependence of the three disciplines that comprise the thermal/fluid sciences. This text introduces these subjects to students who, most likely, will not take further course work in this area. We anticipate that the text will be used in a rather intensive one-semester course most likely in the junior year of an engineering curriculum. Knowledge of basic physics and mathematics through ordinary differential equations is assumed.

Each chapter includes several example problems, worked in detail, that illustrate the application of the material being introduced. In addition, numerous problem sets are included at the end of each chapter.

A uniform, nonintegrated, treatment of the subject matter, using consistent units and nomenclature is one of the features of this text. The authors believe that a fundamental knowledge of energetic processes and the ability to quantify them is essential for all engineering graduates. We hope that the text will induce more students to acquire these important capabilities in their education and subsequent careers. The means to this end is **motivation**.

The authors have managed to garner diplomas in mechanical, chemical, and electrical engineering, as well as applied mathematics. All have been involved in teaching this material on the undergraduate and graduate levels as well as the "survey course" that you may be taking or are about to take. With regard to this survey course, we have found a rather unusual lack of motivation among students, probably because of the lack of interest in the subject matter, resentment at being required to take such a course, and/or the pressure of other course work that is time consuming and considered more important.

As we wrote this book, we anticipated a lack of motivation on the part of the student/reader and have written the book with the student in mind. Accordingly, you may find that the material covered is extremely fascinating. Indeed, by the end of Chapter 5, you will be extremely adroit at the bookkeeping required to keep track of the forms of energy necessary to solve detailed problems involving closed and open thermodynamic systems.

We do not lose sight of the difference between system synthesis and system analysis and close to a score of "design problems" have been included. These design problems range from a simple first law "heat balance" to a detailed design of a heat sink to be employed in the cooling of a package of electronics.

You will learn how automobile and aircraft engines work, how steam power plants and refrigeration systems are put together, and you will become well versed in such diverse topics as fluid statics, buoyancy, stability, the flow of fluids in pipes and fluid machinery. Finally, the thermal control of electronic components pervade the entire heat transfer section of the book.

The authors have tried to anticipate problems that you will encounter in your learning process and this has been their motivation. It is a book that has not been easy to produce. Your motivation is to give the subject matter contained here half a chance and to recognize that while you are sitting in the classroom, after paying a substantial tuition, if you listen, you might just learn something.

We have not provided a disk of thermal, fluid, and thermodynamic properties because all of the properties that you will need may be found in the 22 tabulations in Appendix A. Neither have we provided a disk containing any computer programs as we see little point in providing computer codes for a small portion of the problems in the text.

The disk that is attached is a comprehensive study guide and is approximately 600 pages. It is intended to help you, to motivate you, and to provide you with ready reference to the material contained herein.

We acknowledge the assistance of Jonathan Plant, our editor and an old friend B. J. Clark, who is our editorial consultant.

Allan Kraus
James Welty
Abdul Aziz
September 2010

1

The Thermal/Fluid Sciences: Introductory Concepts

Chapter Objectives

- To introduce the component disciplines that comprise the thermal/fluid sciences—thermodynamics, fluid mechanics, and heat transfer.
- To discuss the physical laws upon which the thermal sciences are based.
- To present a list of products and/or processes whose analysis and design rely upon thermal/fluid science principles.

1.1 Introduction

The subject areas of thermodynamics, fluid mechanics, and heat transfer comprise what are generally referred to as the **thermal/fluid sciences.** These subjects are often studied separately but are sufficiently interrelated that they are frequently taken sequentially by engineering students and, as with this text, are often treated in a single book.

In this book, these subjects will be treated in a relatively fundamental fashion. Our goal is to provide a basic understanding of the physical laws and processes that involve energy utilization in ways that accomplish useful tasks.

Thermal/fluid systems involve energy that can be stored, transferred, or converted within the system. Energy storage manifests itself in different forms such as gravitational or potential and kinetic energy as well as energy that is contained in the matter that constitutes the system.

Energy can be transferred between a system and its surroundings via the flow of heat, work, and the flow of hot and/or cold fluid streams of matter. It may be converted from one form to another such as the conversion of the energy contained within fuels to electrical energy as in a power plant, or to mechanical energy, which is used to propel your automobile. A steam power plant (Figure 1.1) converts the chemical energy contained in a fossil fuel to electrical power, while the automobile and the turbojet engines contained in a fuel are converted to propulsive power. The familiar household refrigerator provides a conditioned low temperature environment by employing electrical energy to drive a heat transfer process that involves the use of a refrigerant.

Additional applications include the pressure cooker, the human cardiovascular system (Figure 1.2), and the solar panel (Figure 1.3). A study of thermodynamics (in Chapter 4) tells us that the boiling point of water decreases as the surrounding pressure decreases. The pressure cooker affords an opportunity to raise the cooking temperature by means of a regulated pressure. The cardiovascular system is a rather complex combination of

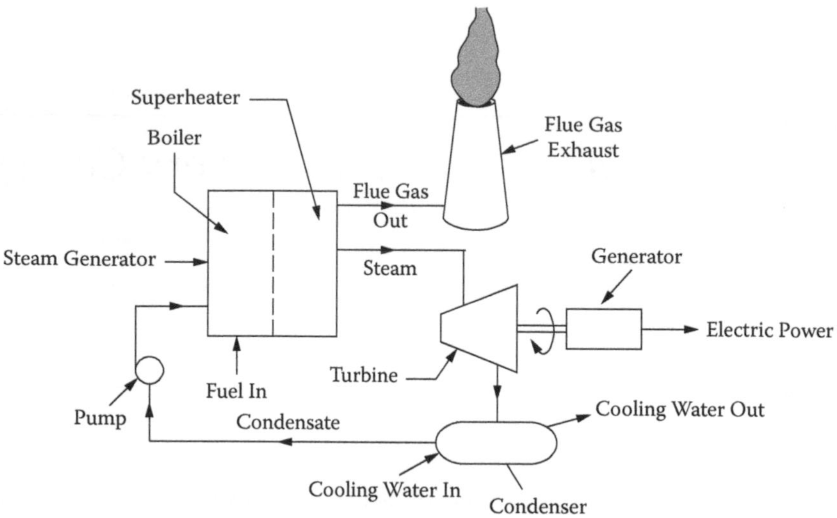

FIGURE 1.1
Artist's rendering of a steam power plant.

heat transfer and fluid flow components that regulate the flow of air and blood within the relatively narrow range necessary to maintain life. In the solar collector pump and storage system displayed in Figure 1.3, the solar panel provides an inexpensive supply of hot water.

Thermal/fluid systems are all around us. Some occur naturally while others are man-made. All are amenable to analysis using a small number of physical laws. These same laws provide the basic building blocks for engineering design.

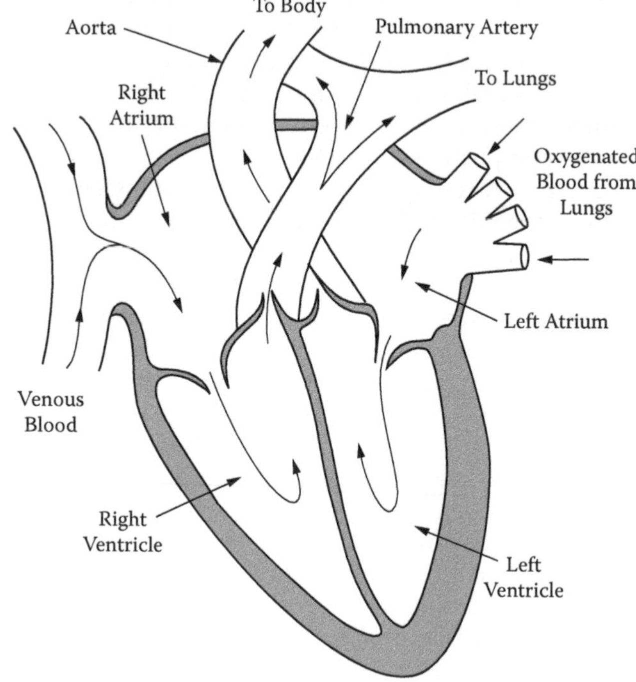

FIGURE 1.2
Artist's rendering of the human cardiovascular system.

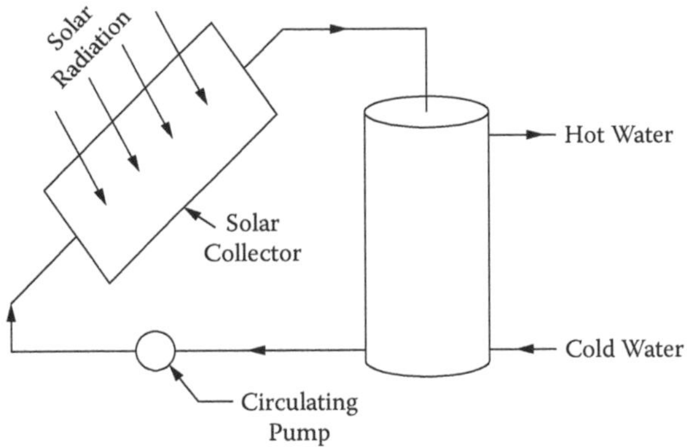

FIGURE 1.3
Artist's rendering of a solar collector system.

In the remaining portion of this introductory chapter, we will attempt to set the stage by discussing the disciplines of thermodynamics, fluid mechanics, and heat transfer in a general way. More formal statements of these subjects and the laws that they involve will be treated in subsequent chapters.

Several examples of engineering systems and manufactured products will be cited. We hope to make it clear that the thermal/fluid sciences comprise an important part of all engineers' understanding and are essential components for analyzing and/or designing processes that utilize energy to enhance our lives.

1.2 Thermodynamics

Thermodynamics can be generally defined as the study of the relationships between **heat, work,** and the **properties** of systems. System properties are those that define energy content. An important feature of thermodynamic analysis is that it relates heat transfer and work done on or by a system as it experiences changes between equilibrium states.

The building blocks of thermodynamic analysis include the following:

- The **zeroth law of thermodynamics** is, essentially, a necessary concept for defining the property we know as **temperature.**
- The **first law of thermodynamics** is a statement that energy must be conserved. This precept is one of the most powerful tools available for the analysis and design of products, processes, and systems.
- The **second law of thermodynamics**, which pertains to energy transport being restricted such that processes, while still in compliance with the first law, must obey certain constraints. The system property, **entropy**, is useful in second law analysis.

To illustrate the concepts involved in the foregoing statements, consider a thermodynamic **system** consisting of a cup of hot coffee sitting on a kitchen counter as indicated in Figure 1.4. As we know, the heat transfer between the coffee and room air will be directly related to the magnitude of its temperature change. The first law of thermodynamics requires that, in the absence of any work being done on or by the coffee in the cup, the heat exchange

FIGURE 1.4
A cup of coffee resting on a kitchen counter.

with the environment must equal the amount by which the coffee and the cup change in energy content. The first law requirement could be satisfied by either the coffee becoming hotter or cooler. The second law of thermodynamics does not allow heat to transfer from a cooler to a warmer temperatures of its own accord. Hence, the coffee must decrease in temperature.

1.3 Fluid Mechanics

Fluid mechanics is defined as the branch of applied mechanics that involves fluid behavior, either at rest or in motion.

All fluid mechanics analyses are based on one or more of the following laws:

- The **first law of thermodynamics** involves the conservation of energy. Energy conservation is a major consideration when evaluating the effects of fluid flow. Most often, we work with **open systems** which, by definition, have mass entering and leaving via fluid flow.
- The **conservation of mass** is a straightforward and powerful tool and is an absolute requirement for all engineering analysis. Mass conservation equations are often referred to as **continuity** expressions.
- **Newton's second law of motion** relates force to the time rate of change of **momentum**. The term **momentum equation** is often used to refer to a statement of Newton's second law.
- The **second law of thermodynamics** has important consequences when flow is compressible, involving gases and vapors under certain conditions. When flow may be considered to be incompressible, the second law is not a necessary consideration.

Most fluid properties are temperature dependent. For compressible fluids such as gases, pressure is also an important factor. A less familiar property, which is important for describing fluid flow, is the **viscosity**.

1.4 Heat Transfer

We consider **heat transfer** to be the rate of energy exchange between systems or objects as a result of a temperature difference.

Heat transfer occurs in three ways:

- **Conduction heat transfer** occurs as higher-energy-level molecules (as indicated by higher temperature) transfer thermal energy to lower-energy-level molecules (clearly at lower temperature). Conduction can occur in the steady state (independent of time) or it can be transient (time dependent). Conduction occurs through direct contact between high and low temperature materials.
- **Convection heat transfer** is associated with energy exchanged between a surface and an adjacent fluid. The distinction is made between **forced** and **free** or **natural** convection. Forced convection occurs when the fluid flow past the surface is produced by an external agent such as a fan, pump, or blower. Free or natural convection occurs when the fluid motion is due to a warmer (or cooler) fluid adjacent to a surface. This flow occurs because of the density differences resulting from the temperature variations that result from the heat exchange.
- **Radiation heat transfer** is associated with energy that is exchanged by electromagnetic waves that lie in the thermal band of the electromagnetic spectrum. Electromagnetic wave/lengths vary from cosmic waves that have extremely small wave/lengths (known as the low end of the electromagnetic spectrum) to radio waves at the high end. The thermal band is generally considered to encompass the range between 0.38 and 0.76 μm (micrometers).

  In contrast to both conduction and convection, radiation heat transfer requires no medium for its propagation. Indeed, radiant energy exchange between surfaces is at a maximum when the surfaces are separated by a perfect vacuum.

1.5 Engineered Systems and Products

An exhaustive compilation of familiar engineering systems that influence our lives would encompass many pages. The list that follows in Table 1.1 is organized along the lines of general technical sectors, with some subtopics in each sector. These subtopics identify more specific products and processes.

Three specific cases will now be considered.

1.5.1 Case I: A Commercial Airliner

Figure 1.5 shows a photograph of a commercial airliner that most of us have seen either personally or in the media.

Any commercial aircraft is a complicated conglomeration of systems, each of which requires detailed consideration in order to achieve optimal operation. Subsystems of such an aircraft include

- Engine/propulsion
- Climate control
- Flight instrumentation

TABLE 1.1

Examples of Engineered Systems and/or Products

Aerospace
• Commercial aircraft
• Space systems
Environmental control
• Greenhouses
• Clean rooms
• Heat pumps
Transportation
• Automobiles and trucks
• Gas turbines
Electronics
• Ink jet printers
• Chip cooling
Public works
• Hydroelectricity
• Municipal water supplies
Manufacturing
• Chemical plants
• Silicon wafers
Power plants
• IC engines
• Steam and gas turbines
• Fuel cells
• Nuclear fission

- Flight controls
- Fuel storage and delivery
- Engine monitoring and control
- Hydraulics

FIGURE 1.5
A commercial jet airliner.

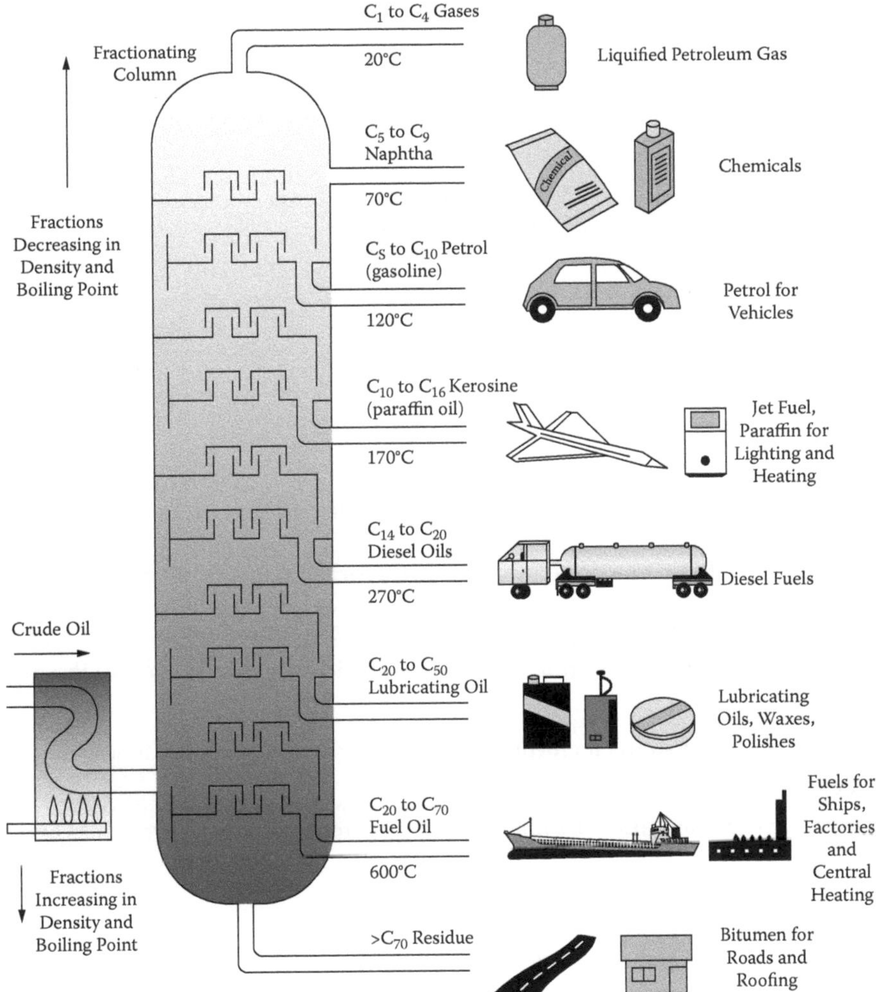

FIGURE 1.6
A distillation tower in a petroleum refinery.

- Communications
- Navigation systems

The design, manufacture and operation of such a complex system involves a huge array of engineering talents and efforts. Thermal/fluid sciences play roles in every item listed.

1.5.2 Case II: A Chemical Plant Distillation Column

The manufacture of chemical products is a major endeavor of contemporary society. Industries that rely on the production and/or conversion of chemicals include pharmaceuticals, agriculture, pulp and paper, plastics, petroleum, and many others. In addition to the manufacture of products from feedstocks, engineers in the chemical industry must meet stringent environmental standards regarding effluents in solid, liquid, and gaseous forms.

A component of a petroleum refinery is shown in Figure 1.6. This component is a distillation tower that converts an inlet stream of crude oil to a variety of hydrocarbon products

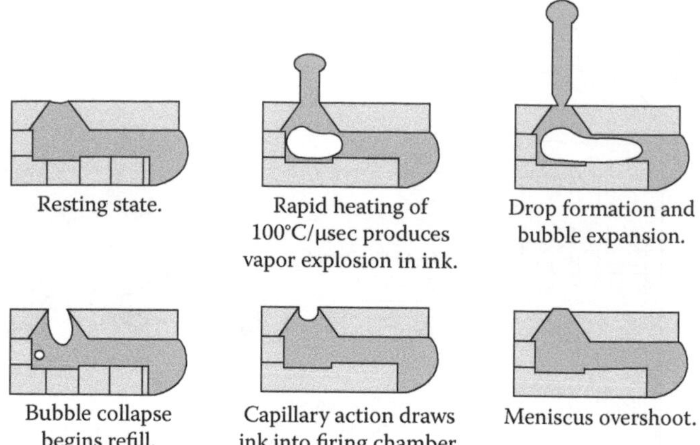
Resting state.
Rapid heating of 100°C/μsec produces vapor explosion in ink.
Drop formation and bubble expansion.
Bubble collapse begins refill.
Capillary action draws ink into firing chamber.
Meniscus overshoot.

FIGURE 1.7
Drop formation in a thermal ink jet printer. (Adapted Courtesy of the Hewlett-Packard Co., by permission.)

from methane to paraffins. In a petroleum plant, the primary products are fuels for internal combustion engines.

The design and operation of such a column involves many components which, in turn, rely on several of the principles within the thermal/fluid sciences. Subsets of the distillation column include

- Extensive piping systems including pumps, valves, and expansion joints
- Heat exchangers
- Extensive instrumentation to monitor system temperatures, pressure, and flow rates
- Controls to adjust operating conditions

1.5.3 Case III: The Thermal Ink Jet Printer

Printers, as auxiliaries to personal computers, have become commonplace. In Figure 1.7, we show a representative device that employs a thermal ink jet to deliver ink to a page.

The objective of a printer is simple; it is to provide a hard copy image that is clear and informative. The medium employed to impart an image to the paper is ink that may be black or in various colors. The jet of ink is driven by a pressure pulse which, in the thermal process, is provided by the well-controlled and extremely fast formation and collapse of a vapor bubble. The controlled formation and collapse produces a series of ink droplets which, in an aggregate, produce written text, tables, illustrations, and even pictures. The device that accomplishes all of this at an affordable price is a marvelous achievement. The thermal/fluid sciences are vital building blocks to the design, fabrication, and successful operation of these ubiquitous devices.

The subsystems of an ink jet printer include:

- Storage and maintenance of an ink at the proper operating conditions
- Paper feed operation and control
- Thermal driver operation
- Ink delivery and purging systems
- Fuel storage and delivery
- Cooling for electronic controls

1.6 Historical Development

1.6.1 Thermodynamics

Thermodynamics can be traced back almost 2500 years to about 400 BC when Democritus wrote that all matter consists of tiny material bits called atoms. However, developments in thermodynamics were sparse until the beginning of the eighteenth century, although Leonardo da Vinci (1452–1519) stated (about 1500) that air contained two gases; Galileo (1564–1642) approached the concept of temperature (1638) through the use of his thermoscope; Grand Duke Ferdinand of Tuscany invented (1640) the sealed-stem alcohol thermometer; and G. W. Leibnitz (1646–1716) anticipated the first law of thermodynamics by demonstrating (1695) that the sum of kinetic and potential energies remains constant in an isolated mechanical system.

The **caloric theory** held that heat was a subtle elastic fluid possessing mass whose particles repelled each other. It was attracted by ordinary matter in varying degrees and it was indestructible and uncreatable. Moreover, it was either sensible or latent and, in its latent form, it could combine with solids to form liquids and combine with liquids to form gases. The caloric theory was apparently introduced by Joseph Black in 1770, who provided ideas on specific heats and latent heats of phase change and, later, by Black's student, William Cleghorn, who fortified the caloric theory with several basic postulates.

The indestructibility of caloric caused some difficulty. For example, in 1772, John Locke noted that axles were heated by friction at the contact point to the wheels and even Black acknowledged that friction could occur. The caloric theory was further challenged by Count Rumford (Benjamin Thompson, 1753–1814) in his 1798 experiments pertaining to the boring of cannon and Sir Humphrey Davy (1778–1829) who, in 1799, made observations of frictional heating, which indicated that, if indeed there was a caloric fluid, it surely could be created.

In 1824, Carnot (1796–1832) formulated the second law of thermodynamics, which appeared to strengthen the caloric theory. His reasoning was that a certain amount of caloric passed through a heat engine in the same manner as water flowed through a turbine and, in doing so, degraded the potential for doing further work.

However, it was the quantitative experimental work of Joule (1818–1899), between 1843 and 1849, that led to the law of conservation of energy (or, the first law of thermodynamics). But, in spite of Joule's work, the caloric theory lasted for a few more years while the argument raged over which of the two laws, Joule's or Carnot's, was correct. Indeed, Joule's theory contradicted Carnot's, which had been formulated on the basis of caloric theory. The argument persisted until it was carried forward by Helmholtz (1821–1894) in 1847 and by Lord Kelvin (1824–1907) in 1848 and was finally resolved by Clausius (1822–1888) in 1850. Clausius showed that Joule's work did not require that heat and work be mutually interchangeable under all circumstances and that, because the first and second laws are independent of each other, Carnot's second law was correct, despite his use of the erroneous caloric theory.

The subsequent organizational work of Clausius, Planck (1858–1947), and Poincare (1854–1912) and extensions of the subject by the American, Gibbs (1839–1903), essentially completed the structure of classical thermodynamics by the beginning of the twentieth century.

1.6.2 Fluid Mechanics

Interest in fluid behavior began in the ancient Greek and Roman civilizations. Archimedes (287–212 BC) was the first to express the principles of buoyancy and flotation and Frontinus

(40–103 AD) was apparently the first to provide details on the elaborate water supply systems that were built by the Romans.

While there was considerable interest in the design of water conduits, the interest was haphazard until Leonardo da Vinci derived the equation for conservation of mass in one-dimensional flow. Coupled with Newton's (1642–1727) three laws and his law of viscosity pertaining to linear fluids, several of the great mathematicians such as Bernoulli[1] (1700–1782), Euler (1707–1783), d'Alembert (1717–1783), Lagrange (1736–1813), and Laplace (1749–1827) provided solutions to the flow problems that arose from the assumption of frictionless flow.

Because most flows are particularly vulnerable to the effects of viscosity, engineers rejected the frictionless flow assumption and developed the science of **hydraulics**, which relied substantially on experimental evidence.

The link between theoretical **hydrodynamics** and experimental hydraulics was established at the end of the nineteenth century. William Froude (1810–1879) and his son Robert (1846–1924) developed the conversion of wave resistance between a model and its prototype. Sir Osborne Reynolds (1842–1912) published the results of his classic pipe experiment (1883), which demonstrated the importance of the dimensionless parameter now known as the Reynolds number. Lord Rayleigh (John W. Strutt, 1842–1919) investigated wave motion, jet instability, and laminar flow analogies.

It was the German engineer, Ludwig Prandtl (1875–1953), who introduced the concept of a fluid boundary layer (1904). This concept provided a foundation for the unification of the experimental and theoretical aspects of fluid mechanics. Prandtl proposed that for flow adjacent to a solid boundary, a thin boundary layer, in which frictional effects are very important, is developed. However, outside of the boundary layer, the fluid behaves very much like a frictionless fluid and subsequent investigations by von Karman (1881–1963) and Taylor (1886–1975) provide the basis for the present state of the art in fluid mechanics at the end of the twentieth century.

1.6.3 Heat Transfer

Historically, conduction, convection, and radiation, which are generally considered as the three modes of heat transfer, all begin in different places and at different times.

Fourier (1768–1830) began working on the problem of heat flow by conduction while he was in Egypt during Napoleon's invasion and, in 1805, with the help of Biot (1774–1862), he was ready to publish the results of his work. The final publication of his "The Analytical Theory of Heat" was some 20 years later in 1822, 1824, and 1826. These included another of Fourier's major contributions, the representation of a periodic function in terms of a trigonometric series.

In the meantime, Navier (1785–1836) and Poisson (1781–1840) completed the formulation of the viscous field equations that eventually were employed to predict heat convection. Following Lord Kelvin (1824–1907), Stokes (1819–1903) developed a more independent formulation of these equations and these became known as "the Navier-Stokes equations." It was Maxwell (1831–1879) who set the foundation for the kinetic theory of gases thereby eliminating any further consideration of the caloric theory. Then, Lord Rayleigh (1842–1919) did a considerable amount of work in fluid mechanics that would soon serve all areas of convection heat transfer.

While it is observed that convection heat transfer had been driven by developments in fluid mechanics, heat transfer in general and convection heat transfer in particular became a separate discipline as fundamental "real-world" problems began to evolve.

---

[1] These mathematicians are listed chronologically according to date of birth.

In 1904, Prandtl presented his paper proposing the boundary layer and this idea was carried forward by his students, Blasius (1883–1970) and von Karman (1881–1963). Nusselt (1882–1959) was another contributor to the subject of analytical convective heat transfer. His paper in 1909 employed dimensional analysis to prove concepts in natural convection *before* the basic theory of dimensional analysis by Buckingham (1867–1940) appeared in 1914.

In 1884, Boltzmann (1844–1906), who had been working in the area of the kinetic theory of gases to provide a greater capability for the prediction of transport phenomena, turned his attention to thermal radiation. The relationship between emittance and absorptance had been established by Kirchhoff (1824–1887) and, in 1879, Stefan (1835–1893) showed, experimentally, that the heat radiated from a thermally active hot blackbody, should be a function of the fourth power of the absolute temperature. Boltzmann used a well-conceived heat engine argument to show that Stefan's contention was true and the "fourth power law" is now known as the Stefan-Boltzmann law.

It was Wien (1864–1928), a German physicist, who, in 1896, provided his distribution law. This was confirmed by Lord Rayleigh who, in 1900, used classical statistical mechanics to obtain the wavelength distribution of thermal radiation. Finally, Planck (1858–1947), whose explanation of thermal radiation followed in 1901, showed that radiation could only occur in discrete energy levels.

1.7 The Thermal/Fluid Sciences and the Environment

It has been noted that the thermal/fluid sciences form the underpinning for almost all energy transfer and conversion processes. Indeed, they are responsible for many aspects of the quality of life that we all enjoy. However, their utilization is accompanied by an impact on the environment that has become more and more difficult to ignore.

The effect of the thermal/fluid sciences on the environment is noticeable in three significant areas. These are **global warming** due to thermal radiation, the formation of **smog** due to lift and drag of solid particles, and **acid rain**, which involves the combustion process. These effects are now briefly considered.

1.7.1 The Ozone Layer

The general stratum of the earth's upper atmosphere in which there is an appreciable **ozone**, O_3 concentration is called the **ozone layer** or **ozonosphere**. In this layer (Figure 1.8), the ozone plays an important part in the radiation balance of the atmosphere. The

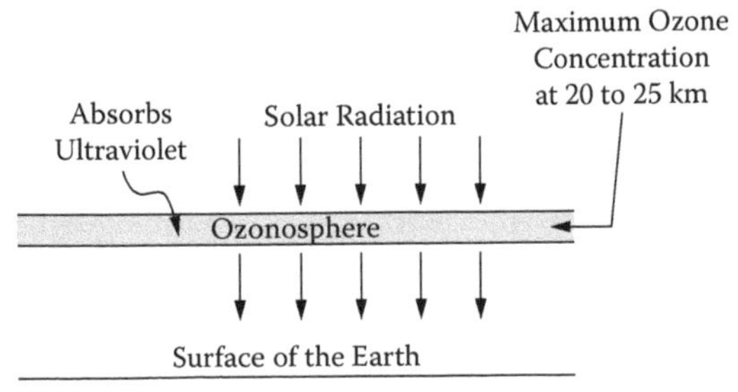

FIGURE 1.8
The ozonosphere.

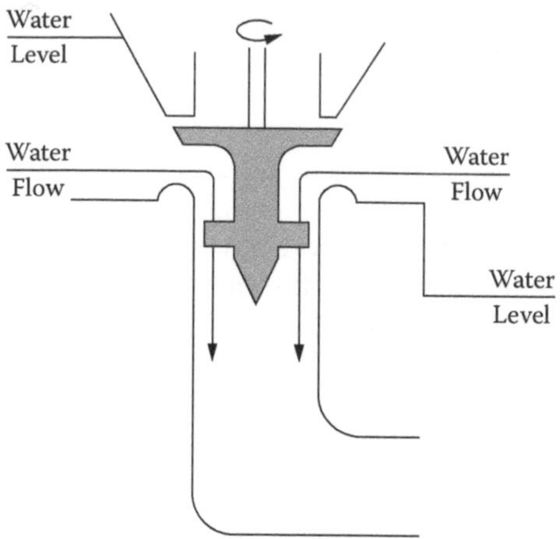

FIGURE 1.9
Artist's conception of a hydroelectric trubine employed in a hydroelectric power plant.

layer lies roughly between 10 and 50 km above the surface of the earth with maximum ozone concentration at about 20 to 25 km.

The ozone layer transmits most of the oncoming solar radiation incident upon it but, thankfully, provides a barrier to harmful ultraviolet radiation. The temperature at the surface of the earth increases during daylight hours and the earth cools down at night by radiating part of the energy to deep space via infrared radiation. Water vapor, carbon dioxide and traces of methane, and the oxides of nitrogen inhibit the nocturnal radiation. These are called **greenhouse gases**, which lead to a phenomenon called the **greenhouse effect**.

The greenhouse effect keeps the earth warm and makes life, as we know it, possible. However, excessive amounts of these greenhouse gases causes some of the heat reradiated by the earth to be trapped by these gases. This causes the average temperature of the earth to increase and the climate at some localities to undergo an undesirable change. The phenomenon is usually referred to as **global warming** and can cause severe changes in weather patterns and the melting of ice at the poles.

The seriousness of global warning has resulted the United Nations forming a committee to deal with climate changes due to global warming and world summits were held in Rio de Janiero (1992) and Kyoto (1997) to set goals for the reduction of CO_2 and other greenhouse gases. These goals can be assisted by resorting to the use of hydroelectric (Figure 1.9), wind (Figure 1.10), and geothermal energy sources.

1.7.2 Smog

Smog is a brown and/or yellow haze that exists as a stagnant air mass and surrounds congested and populated areas on warm or hot summer days. It is composed of carbon monoxide, volatile organic compounds such as butane and other hydrocarbons, dust and soot in particulate form, and ground level ozone.

Carbon monoxide is an odorless and colorless poisonous gas that is emitted by motor vehicles. It can combine with red blood cells that actually carry oxygen, decreasing the supply of oxygen to the brain and other organs, thereby slowing body reactions and reflexes. At high concentrations, an excess of carbon monoxide in the environment can be fatal.

On the other hand, ground level ozone, which should not be confused with the ozone layer that exists in the atmosphere, is a pollutant that has significant adverse health effects. It can cause smartness and irritation to the eyes, shortness of breath, general fatigue, headaches,

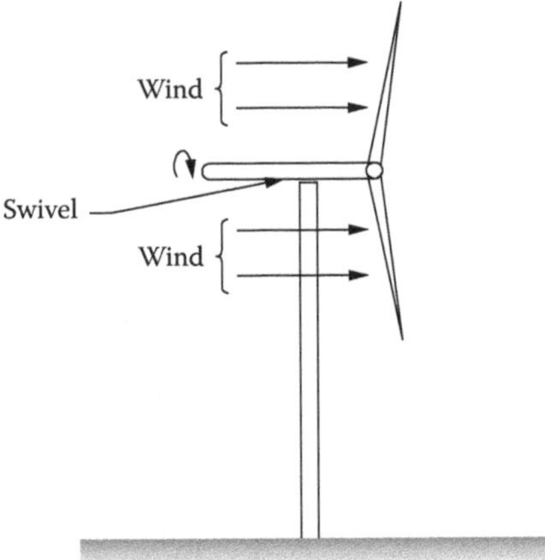

FIGURE 1.10
Artist's conception of the use of wind power.

and nausea. It damages the air sacs in the lungs causing temporary problems such as asthma. It is formed when hydrocarbons and the oxides of nitrogen react in the presence of sunlight.

Although ground level smog and ozone form in congested urban areas with heavy traffic (the Los Angeles freeways, for example) prevailing winds can transport them to other locations. This makes the production of smog a global rather than a local problem.

1.7.3 Acid Rain

Fossil fuels such as coal, gasoline, and diesel oil contain an amount of sulfur that varies from just a trace to an appreciable amount. During the combustion process in motor vehicles and power plants where nitrogen is a major ingredient of the air used, oxides of sulfur, and nitrogen are produced (Figure 1.11).

In the presence of water vapor, which is also contained in the products of combustion, the oxides of sulfur and nitrogen form droplets of sulfuric and nitric acid. These droplets are washed from the atmosphere and form what is known as **acid** or **acid rain**. The soil is capable of neutralizing certain amounts of this acid rain but the amounts produced by power plants using high-sulfur-laden fuels may exceed this capability.

The result may be the deterioration and eventual destruction of forests because the leaves and root systems of the trees and shrubs absorb these acids. Indeed, lakes and rivers may become too acidic for fish to spawn and exist.

$$S + O_2 \longrightarrow SO_2$$
$$SO_2 + H_2O \longrightarrow H_2SO_3 \quad \text{(Sulfurous acid)}$$
$$2H_2SO_3 + O_2 \longrightarrow 2H_2SO_4 \quad \text{(Sulfuric acid)}$$

$$N + O_2 \longrightarrow NO_2$$
$$NO_2 + H_2O \longrightarrow H_2NO_3 \quad \text{(Nitrous acid)}$$
$$2H_2NO_3 + O_2 \longrightarrow 2H_2NO_4 \quad \text{(Nitric acid)}$$

FIGURE 1.11
The combustion of sulfur and nitrogen in air.

The problem of acid rain has been recognized and legislation that prohibits the use of high-sulfur-laden coal now exists.

1.8 Summary

In this chapter, we have examined the component disciplines that comprise the thermal/fluid sciences. These components are thermodynamics, fluid mechanics, and heat transfer. The fundamental laws, which are the building blocks of these disciplines, have been introduced and discussed in general terms. Important new properties have also been introduced.

In this chapter, we have also included examples of engineered systems whose development and operation involve the thermal/fluid sciences to a significant degree.

The chapters to follow will address thermodynamics, fluid mechanics, and heat transfer in considerable detail. Numerous illustrative examples will be employed to illustrate techniques for applying the basic laws.

2

Thermodynamics: Preliminary Concepts and Definitions

Chapter Objectives

- To briefly introduce the subject of thermodynamics.
- To provide precise definitions of some of the working terms used in a study of thermodynamics.
- To consider the dimensions and units that pertain to thermodynamics.
- To examine density and its related properties.
- To define pressure and consider how it is measured.
- To define temperature and to present the zeroth law of thermodynamics.
- To outline a problem-solving methodology.

2.1 The Study of Thermodynamics

When most people think about thermodynamics, they think about the transfer of energy and the utilization of such energy transfer for the useful production of work. This often leads many engineering students in fields such as computer science and electrical or civil engineering to wonder why this particular subject is relevant to them. In reality, thermodynamics deals with much more than the study of heat or energy transfer and the development of work. Indeed, it deals with virtually all aspects of our lives, from the combustion processes that run our automobiles and produce our electric power in power plants to the refrigeration cycles that cool our beer, from the cryogenic pumping of liquids and gases in space to the distillation processes used to produce the gasoline that runs our automobiles.

Thermodynamics is important to electrical engineers so that they can better understand that the limiting factor in the microminiaturization of electronic components is the rejection of heat. It is important to civil engineers because a knowledge of thermal expansion and thermal stresses is requisite to the design of structures and to the computer scientists who need to thoroughly understand the systems that they are trying to model and develop.

Many engineering students have heard the rumors about thermodynamics that say it is a difficult subject, it is a great leveler in that taking a course in it will destroy one's grade point average and that it is something to be avoided until it becomes a matter of taking it or not graduating. Be assured that the authors have been there. All of this is, of course, not true and if one keeps an open mind, a study of thermal and fluid sciences in general and thermodynamics in particular can be quite a rewarding experience. Common terms like **work, reversibility**, and **system** all have very well-defined meanings in thermodynamics and we will study them. Moreover, we will also study a whole host of terms that are completely unfamiliar to the student who is not pursuing a curriculum in mechanical or chemical engineering. Fortunately, however, everyone knows a little something about thermodynamics, which has been learned through everyday experience, that a ball rolls downhill, that heat moves from a region at high temperature to one at low temperature, and that wind-up toys will just not run on their own.

Thermodynamics is with us in our everyday lives and, through its study, we can better relate to the world around us. The first law of thermodynamics can help us understand the need for conservation of our natural resources and the second law of thermodynamics can give us an understanding of why it is so difficult to separate gases after they mix and why your sock drawer is always such a mess. And, to be sure, the property of entropy can help us better understand the vast extent and operation of systems and devices in our universe.

The harnessing of available energy and the transformation of this energy to a usable form has long been a goal of society. Indeed, energy has driven society, in a little more than a century, from muscular effort and the horse and buggy to the modern day automobile, supersonic transport, modern tools and appliances, heating, air conditioning, and the exploration of space.

Thermodynamics provides the basic principles of energy transfer. However, because the development of particular sources of energy is becoming increasingly expensive, thermodynamics is also concerned with the efficiency of energy utilization. Moreover, thermodynamics is concerned with the environmental impact of various energy conversion alternatives. These are the reasons why we study thermodynamics.

Our first step will be the definition of some of the working terms.

2.2 Some Definitions

2.2.1 Systems

A **system** is any portion of the universe that is chosen for thermodynamic analysis. It can be real, imaginary, fixed, or movable and it is separated from its **surroundings** or **environment** by its **boundaries** (Figure 2.1). If neither mass nor energy crosses its boundaries, it is said

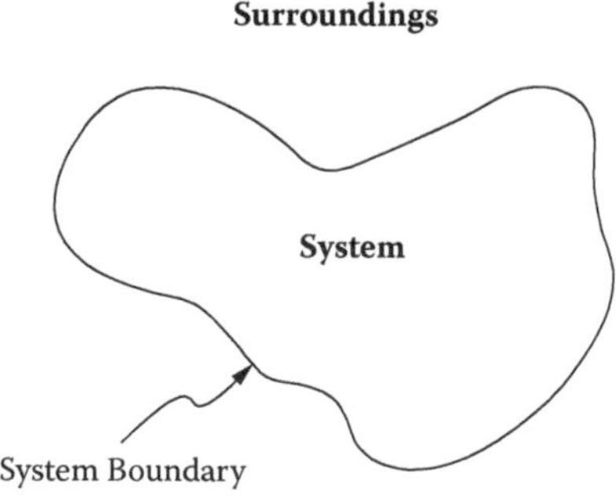

FIGURE 2.1
A system is separated from its surroundings by its boundaries.

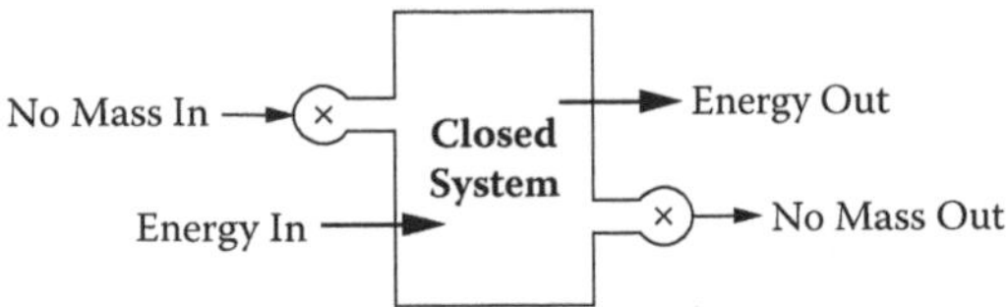

FIGURE 2.2
A closed system or control mass. Energy may flow in or out but mass may not.

to be an **isolated system**. In a **closed system** or **control mass**, only energy (and no mass) may be transferred across the system boundaries (Figure 2.2). If both mass and energy are transmitted across the system boundaries, the system is said to be an **open system** or **control volume** (Figure 2.3). If the **properties** of the system (to be discussed shortly) are everywhere the same, the system is considered **uniform** and if the properties do not change with time, the system is said to be at **steady state**.

2.2.2 Fluids

A **fluid** is a substance that deforms continuously under the action or the influence of a shear force. This characteristic allows it to assume the shape of its container. Both **liquids** and **gases** are fluids.

2.2.3 Substances

A **substance** is a tangible material that occurs in macroscopic amounts and a **pure substance** is one that is uniform and homogeneous in chemical structure. A pure substance may exist in a **single phase** (solid, liquid, or vapor), as a **two-phase** (solid-liquid, liquid-vapor, or solid-vapor) mixture, or as a **three-phase** (solid-liquid-vapor) mixture. In a two-phase or three-phase mixture, the phases are separated by **phase boundaries** (Figure 2.4). We may note that the vapor phase is frequently referred to as the **gaseous phase**, which brings us to the difference between a gas and a vapor.

We find that the terms **gas** and **vapor** are often used interchangeably. However, they may be distinguished from one another by recognizing that a substance which, at standard conditions of temperature and pressure, normally exists in the gaseous phase (oxygen and nitrogen, for example) is normally called a gas. On the other hand, if a substance at standard conditions of temperature and pressure normally exists as a liquid (water, for example), it can produce a vapor.

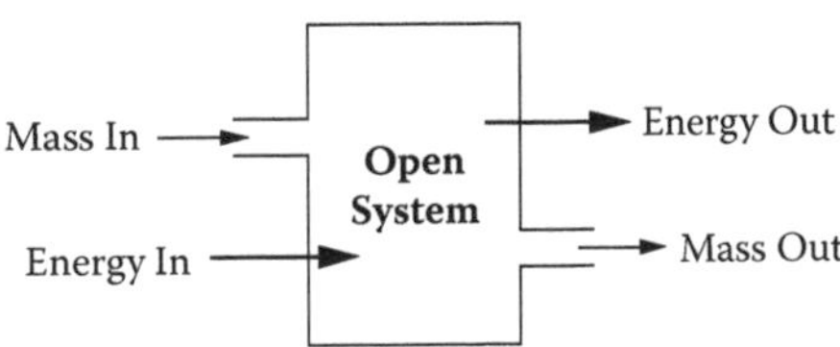

FIGURE 2.3
An open system or control volume. Both energy and mass may flow in or out and the mass flows may contain energy.

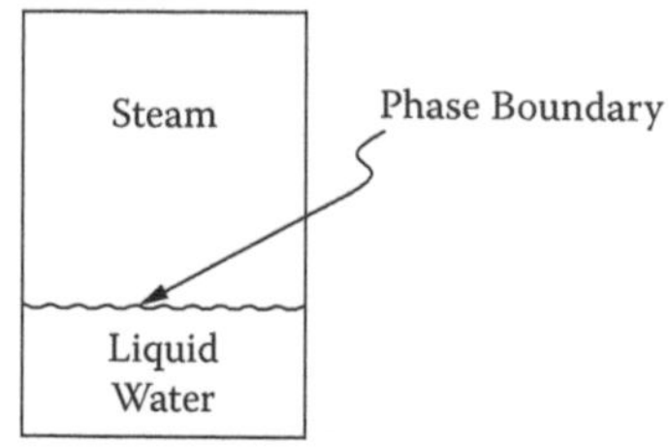

FIGURE 2.4
A two-phase system containing steam or vapor and liquid water. The two phases are separated by the phase boundary.

2.2.4 State and Properties

The **state** of a system is the condition in which the system exists. A thermodynamic **property** is any observable characteristic that describes the state of the system. Properties that depend on the mass of the system and hence the extent of the system, such as volume (designated by V) and energy (designated by E), are called **extensive properties**. Properties that do not depend on the extent of the mass in the system such as pressure (designated by P) and temperature (designated by T) are called **intensive properties.** We note that both extensive and intensive properties are designated by capital letters. Extensive properties, when given on a per unit mass basis are called **specific properties** and are designated by a lower case letter (Figure 2.5). Examples of specific extensive properties are specific volume, $v = V/m$, and specific energy, $e = E/m$.

2.2.5 Processes

A **simple substance** is a substance whose state is determined by two independent intensive properties that are independently variable (Figure 2.6). We observe that the density, $\rho = m/V$ is *not* an independent intensive property because it is the reciprocal of the specific volume, v. We also see that, because the state of a system is determined by its properties, whenever any property changes, its state changes, and the system is said to have undergone a **process**. A sequence of processes that returns a system to its initial state is called **cycle** (Figure 2.7) and we note that the change in the values of a property depends only on the **end states** of the process and is independent of the process itself. Therefore, the properties of a system that undergoes a cycle are the same at the initial and final states.

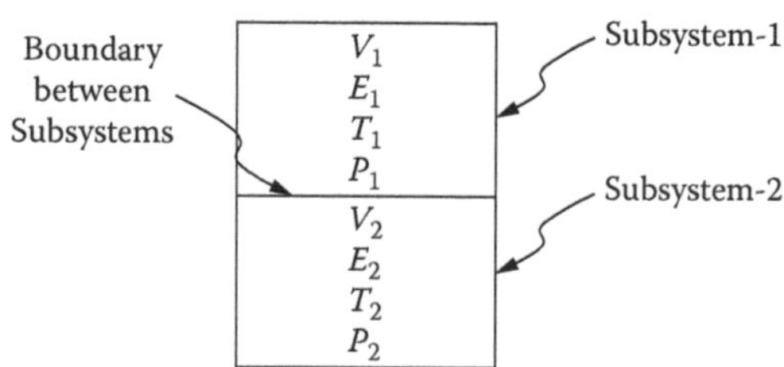

FIGURE 2.5
Two subsystems separated by a boundary. The subsystems contain substances having the extensive properties of volume, V, and energy, E, and the intensive properties of temperature, T, and pressure, P.

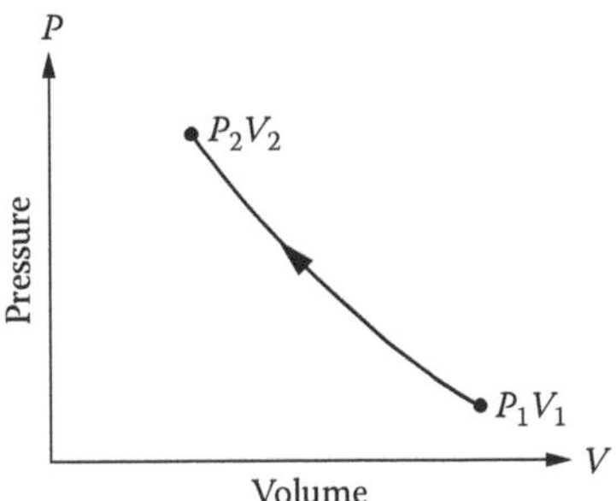

FIGURE 2.6
A P-V diagram of a process in which the state of the system changes because the system properties change.

2.2.6 Thermodynamic Equilibrium

2.2.6.1 Requirements for Thermodynamic Equilibrium

A system is said to be in **thermodynamic equilibrium** when it cannot change its state without some outside influence, that is, when there is no tendency of the system to change spontaneously. In order to achieve thermodynamic equilibrium, **mechanical, thermal, chemical**, and **phase equilibrium** must all exist individually and simultaneously:

- **Mechanical equilibrium** requires a condition of balance between all opposing forces, moments, and other influences.
- **Thermal equilibrium** requires all parts of the system to be at the same temperature (Figure 2.8).
- **Chemical equilibrium** requires that the system show no tendency to undergo further chemical reaction.
- **Phase equilibrium** requires that the system be either in a single phase or in a number of phases that show no tendency to change their condition when the system is isolated from its surroundings.

A system that is in equilibrium (Figure 2.8) is said to be in an **equilibrium state.**

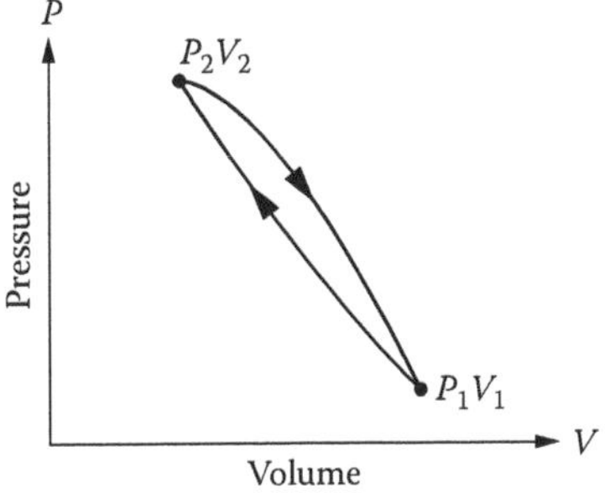

FIGURE 2.7
A P-V diagram of a cycle indicating the properties of the system at both the initial and the final states. Both the initial and final states are indicated by P_1 and V_1. Here, P_2 and V_2 are at some intermediate state.

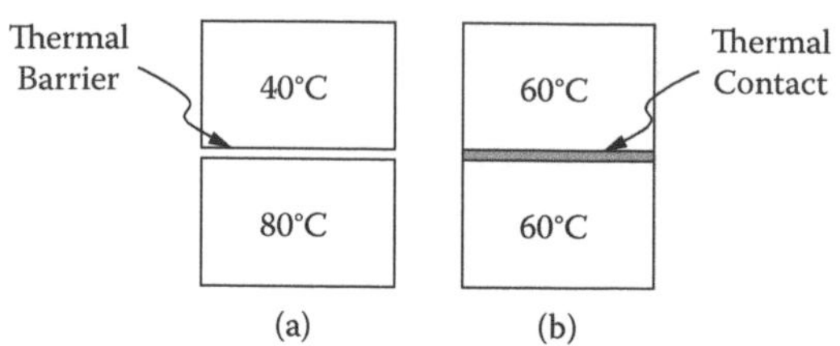

FIGURE 2.8
Two systems (a) not in thermal equilibrium and (b) in thermal equilibrium.

2.2.6.2 Quasi-Equilibrium or Quasi-Static Processes

Regardless of the process, a system can proceed from one equilibrium state to another only if the system is perturbed, that is, only if some external force or other influence acts upon the system. At the end of the process, another, but different, equilibrium state (which also isolates the system from its surroundings) is obtained. If the process occurs at a finite rate, deviations from equilibrium conditions may become significant. However, the process may be thought of as a series of infinitesimal processes, each of which achieves an equilbrium state at its conclusion. Under these circumstances, we see that the entire process is a sum of a series of **quasi-equilibrium** or **quasi-static** processes (Figure 2.9) in which the deviation from thermodynamic equilibrium is not only negligible but infinitesimal.

2.2.7 Statistical and Classical Thermodynamics

The **microscopic** approach to the study of thermodynamics involves individual molecules and their internal structure which may include individual atoms, electrons, and nuclei. Because the number of these particles is huge, statistical techniques using the laws of probability can be employed to determine the most probable distribution of particles between momentum and energy states. When these statistical techniques are employed, the microscopic viewpoint is then described by **statistical thermodynamics** or **statistical dynamics**.

The **macroscopic** approach to the study of thermodynamics, which we use in this book, involves quantities of matter that are finite in size. The detailed molecular or atomic structure of matter is ignored and the macroscopic approach is adopted in the study of **classical thermodynamics**.

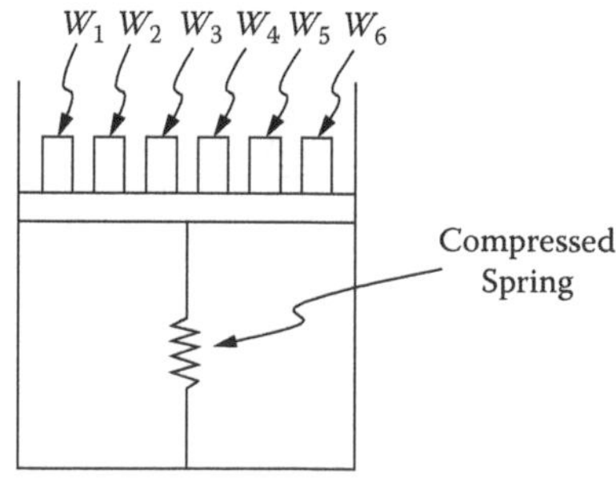

FIGURE 2.9
A platform with several weights held in place by a compressible spring. As each weight is removed, the platform rises in a series of quasi-equilibrium or quasi-static processes.

TABLE 2.1

Primary Dimensions and Units for the SI System

Dimension or Quantity	SI Unit and Symbol
Mass	kilogram (kg)
Length	meter (m)
Time	second (s)
Temperature	Kelvin (K)
Amount of substance	mole (mol)
Luminous intensity	candela (Cd)
Electric current	ampere (A)

2.3 Dimensions and Units

2.3.1 Introduction

Primary dimensions refer to a rather small group of physical quantities such as mass, time, length, and temperature that may be employed to describe the nature of a physical system. **Secondary dimensions** such as pressure, velocity, and area are measured in terms of the primary dimensions.

A system of **units** is used to provide arbitrary magnitudes and give names that are assigned to the dimensions. The units are adopted as standards of measurement and are referred to as **base units.**

2.3.2 The SI (Système Internationale D'unités) System

The SI system of units is an extension of the metric system and has been adopted in many countries as the only legally acceptable system. SI units are divided into three classes that are the base units, the derived units, and some special units. The primary dimensions used in a study of the thermal and fluid sciences along with their base units and symbols embrace the first five entries in Table 2.1.

There are "standards" for all of the units shown in Table 2.1. For example, the standard for the second is the duration of 9,192,631,770 periods corresponding to the transition states between two levels of the ground state of the cesium-133 atom.

Examples of derived units expressed in terms of the base units are given in Table 2.2 and examples of derived units with special names are given in Table 2.3.

TABLE 2.2

Examples of SI Derived Units Expressed in Terms of the Base Units

Dimesion or Quantity	SI Unit Name and Symbol
Area	square meter (m^2)
Volume	cubic meter (m^3)
Velocity	meter per second (m/s)
Acceleration	meter per second squared (m/s^2)
Density	kilogram per cubic meter (kg/m^3)
Specific volume	cubic meter per kilogram (m^3/kg)
Current density	ampere per square meter (A/m^2)

TABLE 2.3

Examples of SI Derived Units with Special Names

Dimension or Quantity	Name and Symbol	Expression in Terms of Other Units	Expression in Terms of Base Units
Force	newton, N	m-kg/s^2	
Pressure	pascal, Pa	N/m^2	kg/m-s^2
Energy, heat, work	joule, J	m^2-kg/s^2	
Work	watt, W	J/s	m^2-kg/s^3
Electrical charge	coulomb, C	A-s	A-s
Electrical potential	volt, V	W/A	m^2-kg/s^3-A
Electrical resistance	ohm, Ω	V/A	m^2-kg/s^3-A^2
Magnetic flux	weber, Wb	V/s	m^2-kg/s^2-A
Electrical inductance	henry, L	Wb/A	m^2-kg/s^2-A^2

The mole is defined as the molecular weight of a substance expressed in the appropriate mass unit. For example, a gram-mol (gmol) of hydrogen contains 2.016 grams (g) and a kilogram-mol (kgmol) of hydrogen contains 2.016 kilograms (kg). The number of moles of a substance, n, is related to its mass, m and the molecular weight, M, by the simple expression

$$n = \frac{m}{M} \tag{2.1}$$

In the SI system, force is a **secondary dimension** because the unit of force is a derived unit. Table 2.3 shows that the unit of force is the **newton**, N, which can be obtained from Newton's second law, $F \propto ma$ as

$$1\,\text{N} = (1\,\text{kg})(1\,\text{m/s}^2) = 1\,\text{kg-m/s}^2$$

and we see that the constant of proportionality is unity or

$$1\,\frac{\text{N-s}^2}{\text{kg-m}} = 1 \tag{2.2}$$

This allows Newton's law to be written as $F = ma$.

Because pressure is force per unit area, $P = F/A$, the unit of pressure, the **pascal**, can be expressed in terms of

$$1\,\text{Pa} = 1\,\text{N/m}^2 = (1\,\text{kg-m/s}^2)(1\,/\text{m}^2) = 1\,\text{kg/m-s}^2 \tag{2.3}$$

Classically, a work interaction or **work**, W, is defined as the **dot** or **scalar** product of a force vector, $\boldsymbol{F}$, and a displacement vector, $d\boldsymbol{x}$, as indicated in Figure 2.10.

$$\delta W = \boldsymbol{F} \cdot d\boldsymbol{x} = F dx \cos\theta$$

where θ is the angle between the line of action of the force, $\boldsymbol{F}$, and the displacement direction, $\boldsymbol{x}$. Here δW represents an inexact differential which, along with the direction of the work interaction, is discussed in detail in Chapter 3. In this book, a work interaction or work will be represented by

$$\delta W = F x$$

and we recognize that the unit of work is the **joule** (J) defined as

$$1\,\text{J} = 1\,\text{N-m} \tag{2.4}$$

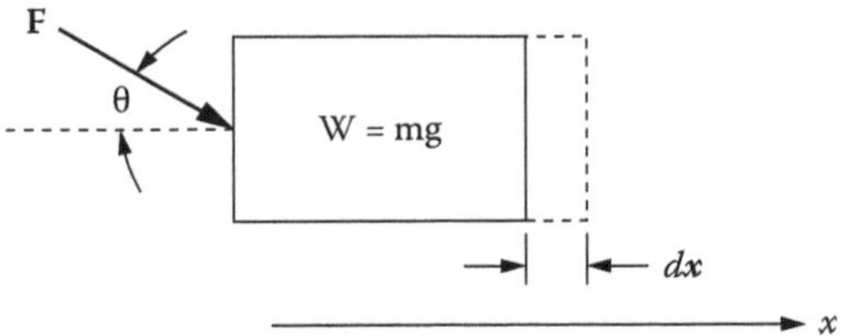

FIGURE 2.10
A force applied at an angle θ with the horizontal. The weight moves in the horizontal direction indicated by the length coordinate, x, and the differential amount of work done is given by the dot or scalar product of the vector force, $\boldsymbol{F}$. and the displacement, x, or $\delta W = \boldsymbol{F} \cdot dx$.

and because power, P, is the rate of doing work, the unit of power is the **watt** (W)

$$P = \frac{\delta W}{dt} = \dot{W} = 1\,\text{J/s} = 1\,\text{N-m/s} \tag{2.5}$$

Finally, we note that the **weight** of a body is equal to the force of gravity on the body. Hence, weight always refers to a force and, in the SI system, this force is always in newtons. The mass of the body is related to the weight via

$$W = mg \tag{2.6}$$

where g is the local gravitational acceleration, which has a mean sea level value of

$$g = 9.807\,\text{m/s}^2$$

and is a function of location. This shows that the weight of a body may vary while the mass of the body is always the same (Figure 2.11).

2.3.3 The English Engineering System

The **English Engineering System** of units is often used in the United States. This system adds the force in lb_f as a primary dimension as indicated in Table 2.4. Here we find the

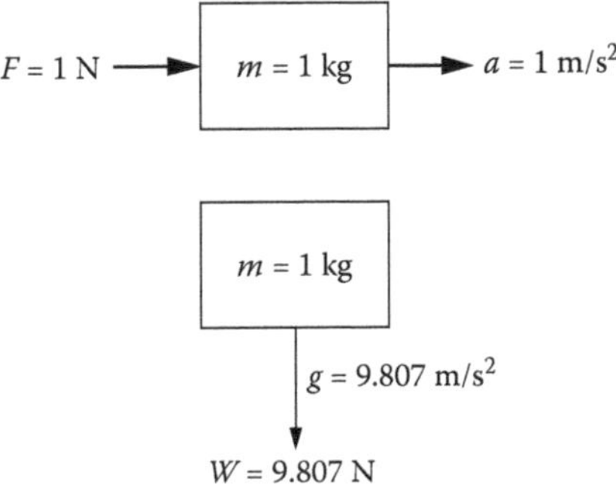

FIGURE 2.11
The weight of a body may vary but the mass of the body is invariant. Here, in the SI system of units, $g_c = 9.807\,\text{m/s}^2$.

TABLE 2.4

Primary Dimensions and Units for the English Engineering System

Dimension or Quantity	Unit and Symbol
Mass	pound-mass (lb_m)
Length	foot (ft)
Time	second (s)
Temperature	degree Rankine (°R)
Amount of substance	mole (mol)
Luminous intensity	candle
Electric current	ampere (A)
Force	pound-force (lb_f)

pound used as *both* the unit of mass (lb_m) and the unit of force (lb_f). This leads to more than just a little confusion when this system is employed for two reasons:

- The use of the symbol, lb, is not permitted and use must always be made of the subscript to designate lb_m or lb_f.
- Because there are six primary dimensions, Newton's law $F \propto ma$ must be written with a proportionality constant (Figure 2.12)

$$F = \frac{ma}{g_c} \tag{2.7}$$

where g_c is a proportionality constant. The force required to accelerate a mass of 1 lb_m at a rate of 32.174 ft/s^2 is designated as 1 lb_f. Thus,

$$1\,lb_f = \frac{(1\,lb_m)(32.174\,ft/s^2)}{g_c}$$

or

$$g_c = 32.174\,\frac{lb_m\text{-}ft}{lb_f\text{-}s^2} \tag{2.8}$$

Thus, Newton's law must be written as

$$F = \frac{ma}{32.174} \tag{2.9}$$

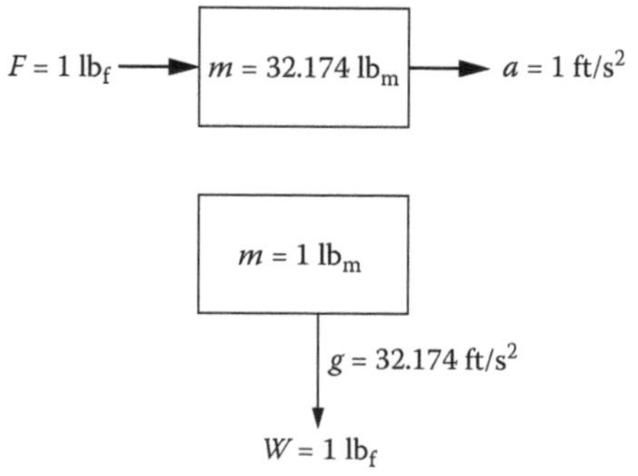

$F = 1\ lb_f$ → $m = 32.174\ lb_m$ → $a = 1\ ft/s^2$

$m = 1\ lb_m$

$g = 32.174\ ft/s^2$

$W = 1\ lb_f$

FIGURE 2.12

Weight and mass in the English system of units require the use of a proportionality constant, $g_c = 32.174\,m/s^2$ for their conversion.

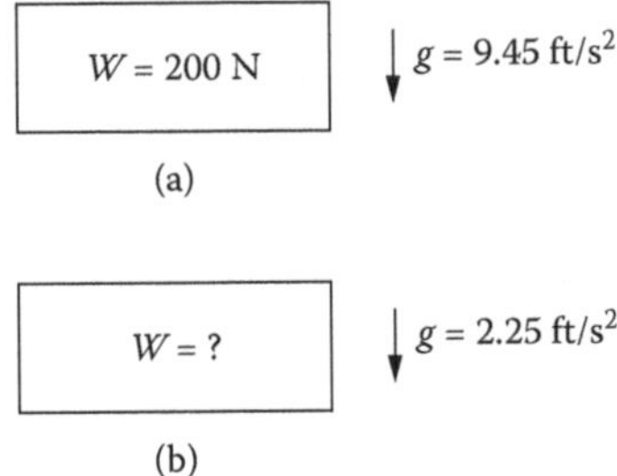

FIGURE 2.13
Data for Example 2.1.

It is important to remember that the SI system of units requires this conversion factor between force and mass, g_c but, because its value is unity, it is often omitted. In this book, we will employ the SI system exclusively.

Example 2.1 A metal ingot weighs (Figure 2.13) 200 N at a location where the acceleration of gravity is $9.45\,\text{m/s}^2$. Determine its mass and weight when the acceleration of gravity is $2.25\,\text{m/s}^2$.

Solution

Assumptions and Specifications

1. The metal ingot is the system.
2. Two values of the acceleration of gravity are specified.

Here, the mass can be computed using Equation 2.6

$$m = \frac{W}{g} = \frac{200\,\text{N}}{9.45\,\text{m/s}^2} = 21.16\,\text{N-s}^2/\text{m}$$

and because $1\,\text{kg} = 1\,\text{N-s}^2/\text{m}$, the mass is

$$m = 21.16\,\text{kg} \Longleftarrow$$

The mass of the ingot is invariant but its weight changes as the acceleration of gravity changes. Hence, when $g = 2.25\,\text{m/s}^2$

$$W = mg = (21.16\,\text{kg})(2.25\,\text{m/s}^2) = 47.61\,\text{kg-m/s}^2 = 47.61\,\text{N} \Longleftarrow$$

2.3.3.1 The British Gravitational System

In Section 2.3.3, we observed that the lb_m is the preferred unit of mass in the English Engineering System. Another unit of mass is the **slug**, which is defined as the quantity of mass that could be accelerated at $1\,\text{m/s}^2$ when it is acted upon by a force of $1\,\text{lb}_\text{f}$.

With the slug designated as the preferred mass unit, we have the **British Gravitational System** with all of the entries in Table 2.4 remaining the same except for the unit of mass that is changed from the lb_f to the slug. In this case, Newton's second law reads

$$1\,\text{lb}_\text{f} = (1\,\text{slug})(1\,\text{m/s}^2) = 1\,\text{slug-ft/s}^2$$

which shows that the relationship between the slug and the pound mass is

$$1\,\text{slug} = 32.174\,\text{lb}_\text{m}$$

TABLE 2.5
Standard SI Unit Prefixes

Factor	Prefix	Symbol	Factor	Prefix	Symbol
10^{-12}	pico	p	10^2	hecto	h
10^{-9}	nano	n	10^3	kilo	k
10^{-6}	micro	μ	10^6	mega	M
10^{-3}	milli	m	10^9	giga	G
10^{-2}	centi	c	10^{12}	tera	T

2.3.4 SI Unit Prefixes

Because it is often necessary to work with very large or very small values of the quantities in the SI system, standard prefixes have been provided to simplify the usage. These are listed in Table 2.5. With these prefixes in hand, 1,000,000 watts become 1 MW, 1000 grams become 1 kg and 0.001 m becomes 1 mm.

2.4 Density and Related Properties

The **density**, ρ, of a substance is defined as its mass per unit volume

$$\rho = \frac{m}{V} \tag{2.10}$$

and the reciprocal of the density is the **specific volume,** v

$$v = \frac{1}{\rho} = \frac{V}{m} \tag{2.11}$$

The **specific gravity** of a substance is the ratio of its density to the density of water at some reference temperature

$$\text{SG} = \frac{\rho}{\rho_{H_2O}} \tag{2.12}$$

and we note that, because specific gravity is a ratio, it has no units or dimensions. The density of water at 25°C can be taken as

$$\rho_{H_2O} = 997.11\,\text{kg/m}^3$$

and, because the density of mercury is 13,560 kg/m^3, its specific gravity can be taken as 13.6.

The **specific weight,** γ, of a substance is its weight per unit volume

$$\gamma = \frac{W}{V} = \frac{mg}{V} = \rho g \tag{2.13}$$

It is sometimes necessary to express the specific volume and other specific quantities on a **per mole** basis. Thus, for the specific volume,

$$\bar{v} = Mv \tag{2.14}$$

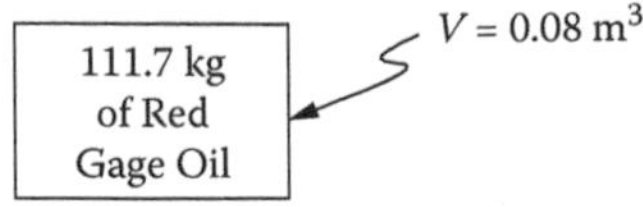

FIGURE 2.14
A quantity of gage oil that occupies a certain volume.

where the **molal specific volume** in $m^3/kgmol$ is the quantity carrying the overbar and where M is the molecular weight of the substance in kg/kgmol. Values for the molecular weight for various substances are listed in Table A.1 in Appendix A.

Example 2.2 As shown in Figure 2.14, 111.7 kg of red gage oil occupies a volume of $0.08\,m^3$. Determine its density, specific volume, specific gravity, and its specific weight at sea level.

Solution

Assumptions and Specifications

1. The red gage oil is the system.
2. The mass and volume are specified.

The density is given by Equation 2.10

$$\rho = \frac{m}{V} = \frac{111.7\,\mathrm{kg}}{0.08\,\mathrm{m}^3} = 1396.3\,\mathrm{kg/m}^3 \Longleftarrow$$

and the specific volume is the reciprocal of the density

$$v = \frac{1}{\rho} = \frac{1}{1396.3\,\mathrm{kg/m}^3} = 7.16 \times 10^{-4}\,\mathrm{m}^3/\mathrm{kg} \Longleftarrow$$

With the density of water taken as $997.11\,kg/m^3$, the specific gravity is computed using Equation 2.12

$$\mathrm{SG} = \frac{\rho}{\rho_{\mathrm{H_2O}}} = \frac{1396.3\,\mathrm{kg/m}^3}{997.11\,\mathrm{kg/m}^3} = 1.40 \Longleftarrow$$

Finally, we find that because the specific weight is related to the density by the acceleration of gravity, which at sea level is $g = 9.807\,kg/m^3$

$$\gamma = (1396.3\,\mathrm{kg/m}^3)(9.807\,\mathrm{m/s}^2) = 13,693.5\,\mathrm{kg/m}^2\text{-}\mathrm{s}^2$$

Because $1\,kg = 1\,N\text{-}s^2/m$, then $1\,kg/m^2\text{-}s^2 = 1\,N/m.^3$ Thus,

$$\gamma = 13,693.5\,\mathrm{N/m}^3 \Longleftarrow$$

2.5 Pressure

2.5.1 Gage, Absolute, and Vacuum Pressure

Pressure is a type of stress that is exerted uniformly in all directions. The pressure, P, is defined as the normal force per unit area exerted on the confining surface of a fluid system

$$P = \frac{dF_n}{dA} \tag{2.15}$$

and is created by bombardment of the surface by the fluid molecules. Here, the differential area, dA, is the smallest area for which the effects are the same as for a continuous medium or **continuum**.

The unit of pressure is N/m^2 and is called **pascal**. Because the pascal is a small quantity, kPa $= 10^3$ Pa and MPa $= 10^6$ Pa are often used in thermodynamics.

If we consider zero pressure at a datum level, the actual pressure at a given position in the fluid system can be considered as its **absolute pressure**. However, most pressure measuring devices indicate a **gage pressure** that is the difference between the absolute pressure and the atmospheric pressure taken as

$$1 \text{ standard atmosphere} = \begin{cases} 101,325\,\text{Pa} \\ 760\,\text{mm of Hg at } 20^\circ\,\text{C} \end{cases}$$

We also note that a frequently used unit of pressure is the **bar**

$$1\,\text{bar} = 10^5\,\text{Pa} = 100\,\text{Pa}$$

The **gage pressure** is the difference between the absolute pressure in the system and atmospheric pressure

$$P_{\text{gage}} = P_{\text{abs}} - P_{\text{atm}} \tag{2.16a}$$

For pressures below atmospheric, the term **vacuum** indicates the difference between the atmospheric pressure and the absolute pressure

$$P_{\text{vac}} = P_{\text{atm}} - P_{\text{abs}} \tag{2.16b}$$

These relationships are summarized in Figure 2.15 where we see that the datum for absolute zero pressure is a perfect vacuum.

2.5.2 Measuring Pressure

Pressure may be measured by **Bourdon tube gages, pressure transducers, strain gages, piezoelectric sensors**, or by **manometers** (Figure 2.16), which are devices that measure pressure differences in terms of the height or length of a column of liquid such as water, mercury or, what is commonly known as **gage oil**. A manometer that measures atmospheric pressure is called a **barometer** (Figure 2.17), although not all barometers are manometers. Manometers are discussed in detail in Chapter 12.

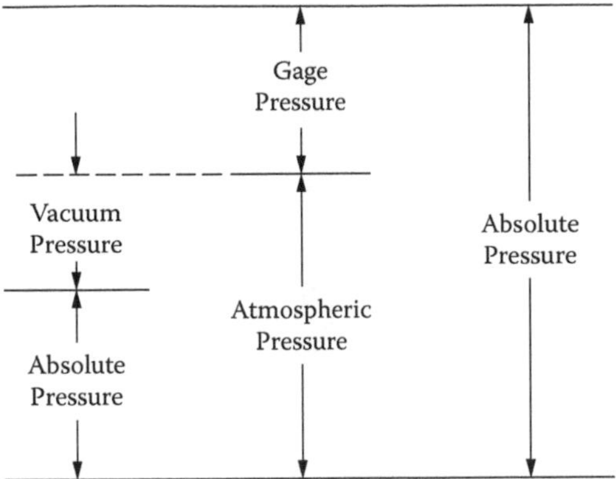

FIGURE 2.15
The relationship between absolute, gage, atmospheric, and vacuum pressure.

2.6 Temperature and the Zeroth Law of Thermodynamics

Temperature is that property of a substance or object that determines the direction of heat flow when the object is placed into contact with another object. Unfortunately, we find that the words "property" and "direction of heat flow" are insufficient for the precise evaluation of temperature that is required in our study of thermodynamics and, as a result, it is often advantageous to think of temperature as a measure of molecular activity.

Consider body *A* at temperature T_A, which is brought into intimate physical contact with body *B* at temperature T_B. If both bodies are isolated from the surroundings with $T_A > T_B$, heat will flow from body *A* to body *B* and after a period of time of sufficient duration has elapsed, there will be no further heat flow from body *A* to body *B*. At this time, both bodies will have reached a state of thermal equilibrium (Figure 2.18) and both may be presumed to be at the same temperature. This serves to define **equality of temperature** and leads to the statement of the **zeroth law of thermodynamics**.

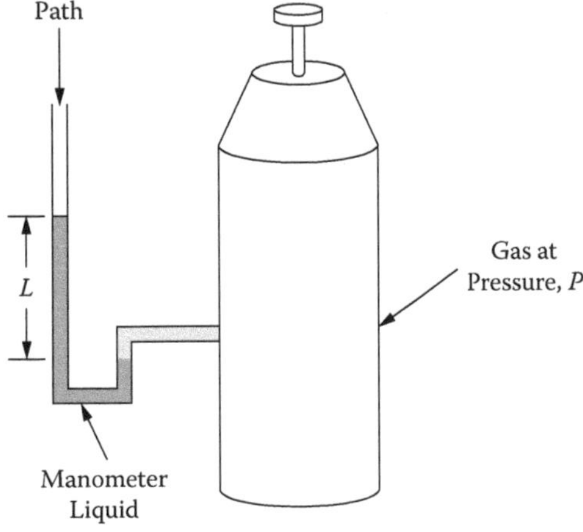

FIGURE 2.16
Manometer.

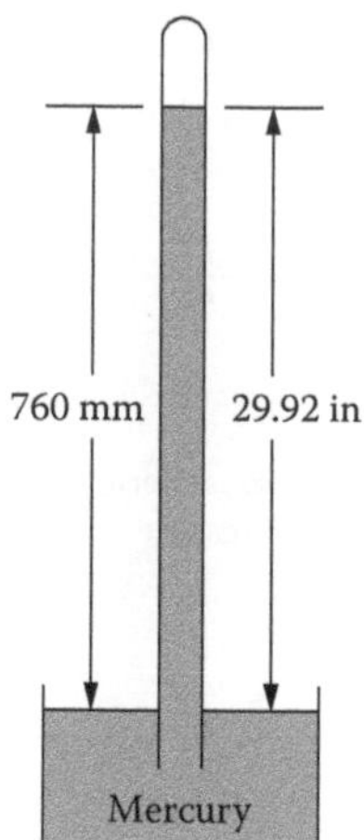

FIGURE 2.17
A mercury barometer at the standard atmospheric pressure of 29.92 in (760 mm) of mercury.

> If body A is in thermal equilibrium with body C and if body B is also in thermal equilibrium with body C, then body A and body B must be in thermal equilibrium with each other.

The zeroth law of thermodynamics forms the basis of temperature measurement or thermometry. A **thermometer** is a device that has at least one property that varies with temperature. The property itself is referred to as a **thermometric property** and the substance that displays this property is called a **thermometric substance**. In the common "mercury-in-glass" thermometer (Figure 2.19), the mercury is the thermometric substance and the coefficient of expansion that causes a change in volume of the mercury is the thermometric property.

Thermometers are not the only temperature measuring devices. **Thermometry**, which is the art of temperature measurement also makes use of changes of pressure in a gas confined at constant volume, changes in electrical resistivity (the **thermistor**), changes in electrical potential (the **thermocouple**), and optical changes (the **radiometer**). We note that the third body considered in the zeroth law may be employed as a thermometer that can be calibrated by bringing it into thermal equilibrium with a body at known temperature such as a mass of melting ice.

The absolute temperature scale in the SI system is the **Kelvin scale** and the unit of temperature is kelvin (K without the degree symbol). Its reference state value is the **triple point** of water, which is the point where a state of phase equilibrium exists among the three

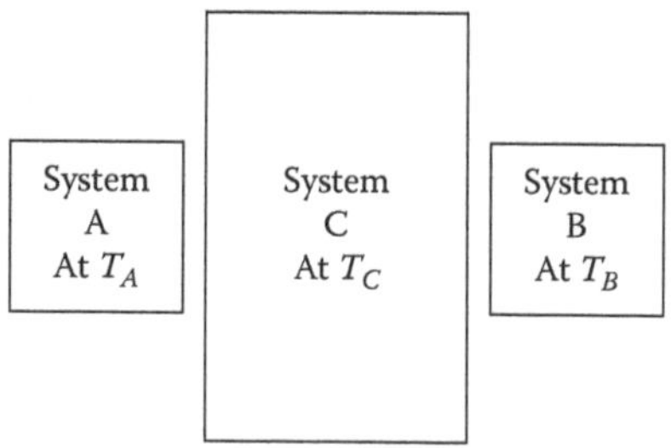

FIGURE 2.18
Three systems in close proximity. If $T_A = T_B = T_C$, the systems are said to be in thermal equilibrium.

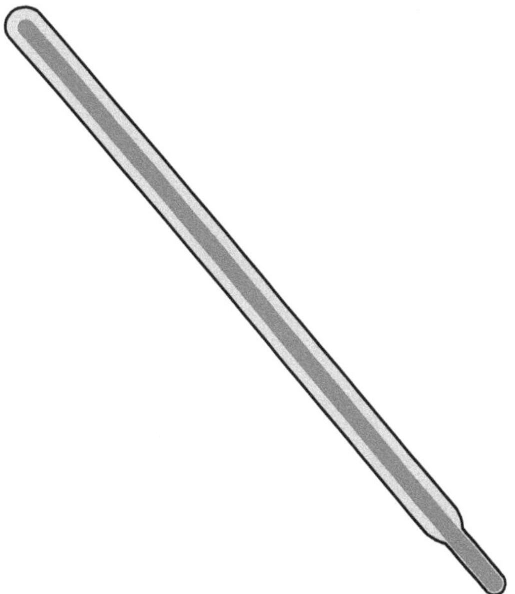

FIGURE 2.19
Mercury-in-glass thermometer.

phases: ice (solid), water (liquid), and steam (vapor). This point occurs at a pressure of one atmosphere and a temperature of 273.16 K. The freezing point of water, called the **ice point,** is the state of phase equilibrium between ice and water at one atmosphere. It is set at a point 0.01 K lower at 273.15 K. The **Celsius** temperature scale (formerly known as the **centigrade** temperature scale) has 0°C (note the use of the degree symbol) as the ice point. Hence,

$$^{\circ}\text{C} = \text{K} - 273.15^{\circ}$$

and we will find enough accuracy in the relationship

$$^{\circ}\text{C} = \text{K} - 273^{\circ} \tag{2.17}$$

The boiling point of water, which is the state of equilibrium between steam and liquid water at 1 atmosphere, is set 100 K higher at 373.15 K. The difference between the ice point and the boiling point of water is 100°C.

The temperature scales used in the English system of units are the **Rankine** and the **Fahrenheit** scales. We know that in the Fahrenheit scale, the ice point is taken as 32°F and the difference between the ice point and the boiling point is 180°F. Following the proportion, $180/100 = 1.80$ between the Fahrenheit and the Celsius scales, the Rankine temperature (°R) is defined as 1.80 times the temperature in kelvin

$$^{\circ}\text{R} = 1.80\,\text{K} \tag{2.18}$$

which puts the triple point of water at 491.69°R and the ice point of water at 459.67°F. The four temperature scales are shown in Figure 2.20.

From the foregoing, we can list the conversions between the Celsius and Fahrenheit scales as

$$^{\circ}\text{C} = \frac{5}{9}(^{\circ}\text{F} - 32^{\circ}\text{F}) \tag{2.19a}$$

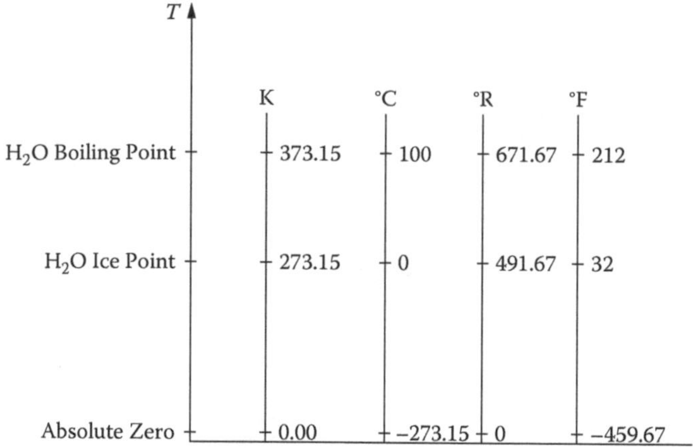

FIGURE 2.20
Kelvin, Celsius, Rankine, and Fahrenheit temperature scales and the relation between them.

and

$$°F = \frac{9}{5}°C + 32°C \tag{2.19b}$$

We can incorporate both of these into the single expression

$$\frac{°C}{5} = \frac{°F - 32°F}{9} \tag{2.19c}$$

2.7 Problem-Solving Methodology

Simplification of a problem with provision of appropriate fundamental principles is called **modeling** and the result is in the form of a **mathematical model** or merely, a model. If the model does not require a computer solution, certain steps are needed in order to come to an expeditious solution involving the following steps:

- Determine the purpose of the problem.
- State what is given with the realization that some of the information may eventually prove to be unnecessary.
- Ascertain what is to be found.
- Make and list any assumptions that you **may** need in order to simplify the problem.
- If required, draw and label a sketch identifying all known values.
- Consider what fundamental principles may be needed.
- Carefully consider the strategy for the solution and check alternative approaches to determine the most expeditious solution procedure.
- Label each step of the solution.
- Be sure that each answer is properly identified and has the proper units.
- Make sure that the solution of a "real-world" problem is a "real-world" solution.
- Provide any comments deemed to be necessary to support your answer.

Observe that the foregoing list is presented as a guide and some judgment as to what to include may be required.

2.8 Summary

In this chapter we have presented a discussion of the working terminology that is germane to a study of thermodynamics. This discussion was based on the definitions of systems, fluids, substances, states and properties, and processes. In addition, the various types of equilibrium were listed and quasi-equilibrium and quasi-static processes were considered.

The difference between a control mass and a control volume is fundamental and these definitions have been provided in Section 2.2.1.

- A **closed system**, often referred to as a **control mass**, is a system in which energy may be transferred across its boundaries but there is no transfer of mass across the boundaries from or to the surroundings.
- An **open system** or **control volume**, is a system in which both energy and mass may cross the system boundaries.

The difference between the control mass and control volume system formulations is in the transfer of mass across the system boundaries. The mass that is transferred may contain energy.

A discussion of the SI system of units was provided and it was shown that the primary dimensions having a bearing on thermodynamics were mass, length, time, the amount of substance, and temperature. Force was shown to be a secondary dimension because the unit of force is a derived unit.

The English engineering system was then considered and it was shown that, in this system, *both* mass *and* force are primary dimensions. The proportionality constant, g_c, was then developed and it was shown that in the SI system

$$g_c = 1\,\frac{\text{kg-m}}{\text{N-s}^2}$$

and in the English engineering system

$$g_c = 32.174\,\frac{\text{lb}_\text{m}\text{-ft}}{\text{lb}_\text{f}\text{-s}^2}$$

Density, specific volume, specific volume, specific gravity, specific weight and gage, absolute and vacuum pressure were then discussed.

After an exposition of the zeroth law of thermodynamics including the definition of temperature scales and the conversion between them, the chapter concluded with some thoughts on the methodology of problem solving.

2.9 Problems

Weight, Mass, and Newton's Law

2.1: A force of 80 N is applied to a mass of 9 kg. Determine the acceleration of the mass.

2.2: Determine the weight of a 40-kg mass in a location where $g = 8.25\,\text{m/s.}^2$

2.3: An object weighs 400 N on the surface of the earth where $g = 9.81\,\text{m/s.}^2$ Determine its weight on a planet where the acceleration of gravity is $1.25\,\text{m/s.}^2$

2.4: A system has a mass of 8 kg and is subjected to an external vertical force of 160 N. If the local gravitational acceleration is $9.47\,\text{m/s}^2$ and frictional effects may be neglected, determine the acceleration of the mass if the external force is acting (a) upward and (b) downward.

2.5: An inhabitant of another planet weighs 150 N on a spring-type scale in his planet atmosphere where $g = 1.85\,\text{m/s.}^2$ If he appears at a location on earth where the local gravitational acceleration is $9.68\,\text{m/s,}\ ^2$ determine (a) his mass on his planet, (b) his mass on earth, and (c) his weight on a spring scale.

2.6: Given that the acceleration of gravity decreases at a rate of $3.318 \times 10^{-6}\,/\text{s}^2$ for every meter of elevation, take the acceleration of gravity at sea level as $9.807\,\text{m/s}^2$ and determine the weight of a 100-kg mass in Denver, Colorado where the altitude is 1609.3 m.

2.7: Given that the acceleration of gravity decreases at a rate of $3.318 \times 10^{-6}\,/\text{s}^2$ for every meter of elevation, take the acceleration of gravity at sea level as $9.807\,\text{m/s}^2$ and determine the altitude where the weight of an object is decreased by 5%.

2.8: A 2-kg mass is "weighed" with a beam balance at a location where g is $9.75\,\text{m/s}^2$. Determine (a) its weight and (b) its weight as determined by a spring scale that reads correctly for a standard gravitational acceleration of $9.81\,\text{m/s.}^2$

2.9: An astronaut who weighs 1962 N at Cape Canaveral where $g = 9.81\,\text{m/s}^2$ goes to a planet where $g = 5.25\,\text{m/s}^2$. While on the planet, he gathers a bag of rocks weighing 105 N to be brought back as samples. Upon his return to earth, it is found that the rocks and the astronaut together weigh 2011 N. How much weight did the astronaut lose during the mission.

2.10: An object that occupies a volume of $0.625\,\text{m}^3$ weighs 3920 N where the acceleration of gravity is $9.80\,\text{m/s}^2$. Determine the mass of the object and its density.

Density and Related Properties

2.11: A cylindrical drum that is 60 cm in diameter and 85 cm high is completely filled with a fluid whose density is $1040\,\text{kg/m}^3$. Determine (a) the total volume of the fluid, (b) the total mass of the fluid, and (c) the specific volume of the fluid.

2.12: A fluid mass of 13.5 kg completely fills a 12 L container. Given the acceleration of gravity as $9.81\,\text{m/s}^2$, determine (a) its density, (b) its specific volume, (c) its specific weight, and (d) its specific gravity if the density of water is $1000\,\text{kg/m}^3$.

2.13: A pump discharges 75 gal/min of water whose density is $998\,\text{kg/m}^3$. Find (a) the mass flow rate of the water in kg/s and (b) the time required to fill a cylindrical vat that is 6 m in diameter and 5 m high.

2.14: A fluid mass of 20 kg has a density of $2\,\text{kg/m}^3$. Determine (a) its volume, (b) its specific volume, (c) its specific weight, and (d) its weight.

2.15: The mass of the earth has been estimated as 5.981×10^{24} kg and its mean radius is 6.38×10^6 m. Estimate its apparent density.

2.16: Suppose that $2\,\text{m}^3$ of liquid A with a specific gravity of 1.24 and $3\,\text{m}^3$ of liquid B with a density of $880\,\text{kg/m}^3$ are completely mixed. Determine (a) the density of the mixture and (b) the weight of the mixture contained in a volume of $7.38\,\text{m}^3$.

2.17: A 150-kg mass has a uniform density of 4250 kg/m^3. Determine (a) its weight, (b) its volume, and (c) its specific volume.

2.18: A liquid has a specific volume of 8×10^{-4} m^3/kg. Taking the acceleration of gravity as 9.80 m/s^2 and the density of water as 1000 kg/m^3, determine (a) its density, (b) its specific weight, and (c) its specific gravity.

2.19: A glass tube with an inner diameter of 1 cm contains 8 g of water ($\rho = 1000$ kg/m^3) and 14 g of gage oil with a specific gravity of 1.40. Taking the acceleration of gravity as 9.807 m/s^2, determine the height of the liquid in the tube.

2.20: A cylindrical container with a diameter of 2 m contains three fluids:

Fluid	ρ, kg/m^3	SG	γ, N/m^3	Weight, N
Water	1000			27,500
Oil			1420	2000
Mercury		13.6		40,000

Determine the total height of the fluids in the cylinder.

Pressure

2.21: A pipe with an inner diameter of 3 cm and a wall thickness of 2.286 mm contains air at a pressure of 27.6 kPa. Determine the tensile stress in the pipe wall.

2.22: A penstock, 1.75 m in diameter is built of longitudinal wood staves held together by circumferential steel hoops spaced 10 cm center-to-center apart. The water pressure is 56 kPa and the allowable tensile stress in the hoops is 130 MPa. The hoops are threaded for nuts. Determine their root diameter.

2.23: A wood stave pipe, 75 cm in diameter, is to operate under a water pressure of 350 kPa. The pipe will be wound spirally with 0.635 cm diameter steel wire at a tensile strength of 168 MPa. What spacing of wire is required?

2.24: A composite vertical fluid column that is open to the atmosphere is composed of 8 cm of mercury with a specific gravity of 13.6, 12 cm of water with a density of $\rho =$ 1000 kg/m^3, and 14 cm of oil with a specific gravity of 0.831. Taking the acceleration of gravity as 9.81 m/s^2, determine (a) the pressure at the base of the column, (b) the pressure at the oil-water interface, and (c) the pressure at the water-mercury interface.

2.25: Can you determine the density of mercury from the observation that a pressure of 1.01325 bar supports a column of mercury that is 760 mm high?

2.26: Determine the absolute pressure exerted by a liquid having a depth of 4 m and a density of 875 kg/m^3 when the atmospheric pressure is 1 bar and the acceleration of gravity is 9.804 m/s^2.

2.27: The pressure of a system is indicated by a pressure gage reading a vacuum pressure of 600 mbar when the atmospheric pressure is 1.025 bar. Determine the absolute pressure of the system.

2.28: Figure P2.28 shows a piston whose mass is 160 kg, which is raised on the inside of a 40 cm diameter vertical cylinder. The lower end of the cylinder is held in a pool of water whose density is 1000 kg/m^3. The pool of water is exposed to the atmosphere at 1.034 bar and the acceleration of gravity is 9.76 m/s^2. The water rises to a height of 6 m as shown and friction is to be neglected. Find (a) the pull on the piston and (b) the pressure exerted by the water on the piston.

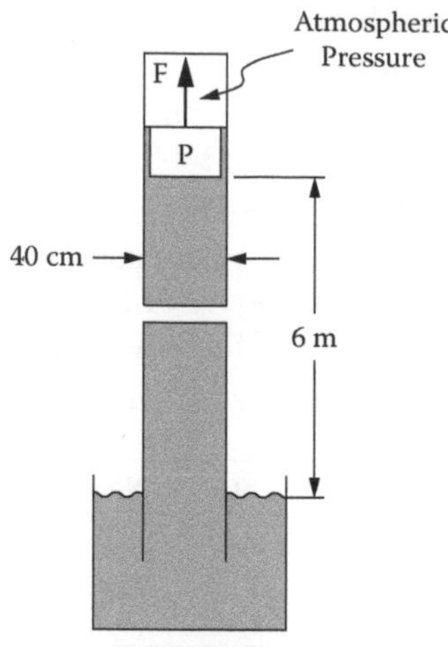

FIGURE P2.28

2.29: A vertical tube contains 86 cm of mercury having a specific gravity of 13.6 and a layer of water at a density of $875\,g/m^3$. Taking the acceleration of gravity as $9.81\,m/s^2$, determine the height of the water layer if the tube is open to the atmosphere and the pressure at the bottom of the mercury layer is 1.225 bar.

2.30: Determine the height of a column of water at a density of $1000\,kg/m^3$ that can be supported by the atmosphere at 1.0135 Pa with the acceleration of gravity equal to $9.807\,m/s^2$.

Temperature

2.31: At what point are the Celsius and Fahrenheit temperatures identical?

2.32: Fahrenheit and Celsius thermometers are immersed in a fluid and give readings such that the Fahrenheit reading is 2.2 times the Celsius reading. Determine the two readings.

2.33: Convert a reading of 77°F to K.

2.34: Convert a reading of 86°C to °R.

2.35: A new proposed absolute temperature scale is known as the **Beaver scale** (°B). In the Beaver scale, water turns to ice at 6279°B. Determine (a) the boiling point of water and (b) the temperature in °C at 10, 810°B.

2.36: Define a new linear temperature scale, say °Z, where the freezing and boiling points of water are 200°Z and 500°Z. Determine a correlation between the Z scale and the Celsius scale.

2.37: An element boils at 480°C. Using the Z scale developed in Problem 2.36, determine the boiling point in °Z.

2.38: If a substance boils at 800°Z (Problem 2.36), determine the boiling point in °C.

2.39: Determine the Celsius equivalent of 248°F.

2.40: Determine the Fahrenheit equivalent of 380°C.

3

Energy and the First Law of Thermodynamics

Chapter Objectives

- To describe the forms of energy and to define what is meant by kinetic, potential, internal, and total energy.
- To define work and to show that it is not a property but a path function that depends on the path between two state points.
- To define heat transfer and to show that it is not a property but a path function that depends on the path between two state points.
- To present the concept of conservation of energy and to link all of the energy quantities considered into the first law of thermodynamics.
- To consider thermodynamic cycles.
- To develop the ideal gas model.
- To consider enthalpy and specific heats for ideal gases.
- To present the equations that govern three of the five fundamental processes of the ideal gas.

3.1 Introduction

Energy may be defined as the capability or capacity to produce work. It is contained in all **matter** and while it exists in many different forms, these forms, however, are well defined. Because **matter** is anything that **possesses mass** and **occupies space**, energy is related to mass. Moreover, we may note that Einstein's theory of relativity suggests that mass, m, may be converted to energy, E, (and energy may be converted to mass) via

$$E = mc^2$$

where $c = 2.9997 \times 10^8$ m/s is the speed of light. However, for all energy-mass interactions other than nuclear reactions, the amount of mass converted to energy is extremely small and can be neglected. Thus, in this study, we state the **conservation of mass principle** that is often quoted in subsequent discussions as

> Mass can neither be created nor destroyed and its composition cannot be altered from one form to another unless it undergoes a chemical change.

FORMS OF ENERGY

- **Kinetic Energy** is the energy that a body possesses by virtue of its motion.
- **Potential Energy** is the energy that a body possesses by virtue of its position.
- **Internal Energy** is a characteristic property of the state of a thermodynamic system including the intrinsic energies of individual molecules, kinetic energies of internal motions, and contributions from interations between individual molecules.

FIGURE 3.1
Forms of energy. The internal energy is treated as a macroscopic quantity even though it is composed of many microscopic energies resulting from the atomic and molecular structure.

Having thus observed that mass is a conserved property, we next focus our attention on the forms of energy and a general **conservation of energy principle** known as the **first law of thermodynamics**. The forms of energy are summarized in Figure 3.1.

3.2 Kinetic, Potential, and Internal Energy

Macroscopic **kinetic energy, gravitational potential energy** (most often referred to merely as "potential energy"), and microscopic **internal energy** are scalar properties and we examine them in some detail in the subsections that follow.

3.2.1 Kinetic Energy

In Newton's second law, $F = ma$, the acceleration can be represented by

$$a = \frac{d\hat{V}}{dt} = \frac{d\hat{V}}{dx}\frac{dx}{dt} = \hat{V}\frac{d\hat{V}}{dx}$$

where x is the displacement and $\hat{V}$ is the velocity in the direction of the applied force, F. If this form of the acceleration is substituted into $F = ma$ where the applied force is constant, the result is

$$F = m\hat{V}\frac{d\hat{V}}{dx}$$

or, after separation of the variables

$$F\,dx = m\hat{V}d\hat{V}$$

Integration between x_1, where the velocity is $\hat{V}_1$, and x_2, where the velocity is $\hat{V}_2$, gives

$$Fs = \frac{m}{2}(\hat{V}_2^2 - \hat{V}_1^2)$$

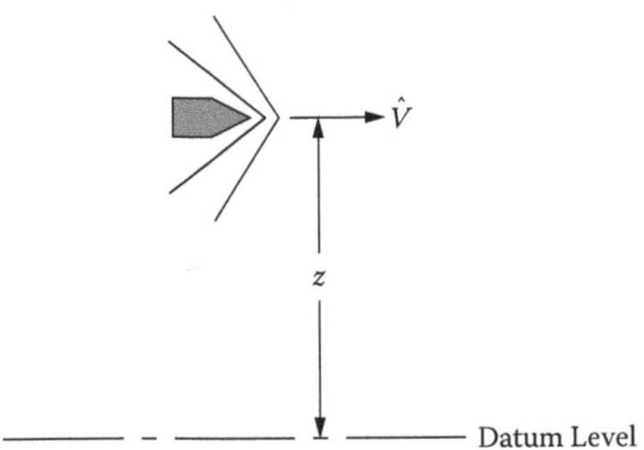

FIGURE 3.2
A projectile that possesses kinetic energy by virtue of its velocity and potential energy by virtue of its elevation, z, above the datum level.

where $s = x_2 - x_1$. The product of m and $\hat{V}^2/2$ leads to the **kinetic energy** (Figure 3.2) defined as

$$\text{KE} = \frac{m\hat{V}^2}{2} \qquad \text{(N-m or joules)} \tag{3.1}$$

and the kinetic energy may also be written as a specific intensive property

$$\text{ke} = \frac{\text{KE}}{m} = \frac{\hat{V}^2}{2} \qquad \text{(N-m/kg or joules/kg)} \tag{3.2}$$

3.2.2 Potential Energy

Potential energy is energy that is associated with a system by virtue of its position in a force field and gravitational potential energy is associated with the gravitational force field. Potential energy is stored in a system because of its elevation above some arbitrary **datum** (Figure 3.3) and is potentially available for conversion to work. The change in potential energy, $d(\text{PE})$, is equal to

$$d(\text{PE}) = Wdz = mgdz$$

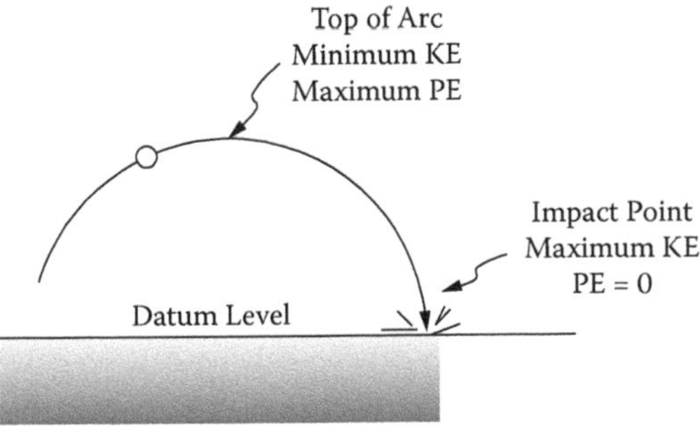

FIGURE 3.3
A projectile in a semicircular arc.

where W is the weight of the system and z is its elevation above the datum. An integration in a constant force field gives

$$\Delta(\text{PE}) = mg(z_2 - z_1)$$

where z_1 and z_2 represent the initial and final elevations. Hence, potential energy is given by

$$\text{PE} = mgz \qquad (\text{N-m or joules}) \tag{3.3}$$

Potential energy is also a scalar property and may be written as a specific intensive property

$$\text{pe} = \frac{\text{PE}}{m} = gz \qquad (\text{N-m/kg or joules/kg}) \tag{3.4}$$

We note that potential energy does not have an absolute value and that only changes in potential energy with respect to some arbitrary datum are of interest.

3.2.3 Internal Energy

The **internal energy** of a substance is energy that is associated with the motion and configuration of its molecules, atoms, and subatomic particles such as nuclei and protons. It is a property consisting of the combined molecular kinetic and potential energies due to the random motion, vibration, rotation, and the intermolecular forces between the molecules and can be determined from other properties such as temperature and pressure. The internal energy is designated by the letter U and because it is a unit of energy, it is expressed in N-m or joules. Specific internal energy, which is an intensive property, is

$$u = \frac{U}{m} \qquad (\text{N-m/kg or joules/kg}) \tag{3.5}$$

3.2.4 The Total Energy

The total energy possessed by a substance is the sum of the kinetic, potential, and internal energies

$$E = \text{KE} + \text{PE} + U \tag{3.6}$$

The total energy is a system property, and we illustrate this idea in Example 3.1.

Example 3.1 The datum for a moving mass of 50 kg (Figure 3.4) is taken at sea level (the surface of the earth) where the acceleration of gravity is $9.81\,\text{m/s}^2$. The mass is at an elevation of 175 m and has a total energy of 2068 kJ with a specific internal energy of $u = 28.44\,\text{kJ/kg}$. Determine the velocity of the mass.

Solution

Assumptions and Specifications

(1) The 50-kg mass forms the system.
(2) The system is moving with uniform velocity.
(3) The local acceleration of gravity is $9.81\,\text{m/s}^2$.
(4) The elevation, total energy, and specific internal energy are all specified.

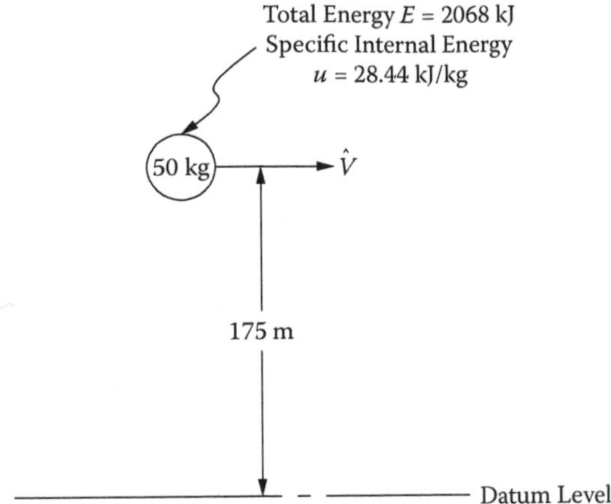

FIGURE 3.4
The moving mass in Example 3.1.

The strategy here is to compute the potential energy and then use the total and internal energies to find the kinetic energy. We employ a modification of Equation 3.6

$$\mathrm{KE} = E - \mathrm{PE} - mu$$

and with

$$\begin{aligned}\mathrm{PE} &= mgz \\ &= (50\,\mathrm{kg})(9.81\,\mathrm{m/s^2})(175\,\mathrm{m}) \\ &= 85.84 \times 10^3\,\mathrm{kg\text{-}m^2/s^2} \\ &= 85.84\,\mathrm{kJ}\end{aligned}$$

Then with $E = 2068\,\mathrm{kJ}$, we have

$$\begin{aligned}\mathrm{KE} &= 2068\,\mathrm{kJ} - 85.84\,\mathrm{kJ} - (50\,\mathrm{kg})(28.44\,\mathrm{kJ/kg}) \\ &= 2068\,\mathrm{kJ} - 85.84\,\mathrm{kJ} - 1422\,\mathrm{kJ} \\ &= 560.16\,\mathrm{kJ}\end{aligned}$$

The velocity can be obtained from this value of kinetic energy

$$\mathrm{KE} = \frac{1}{2}m\hat{V}^2 = \frac{(50\,\mathrm{kg})\hat{V}^2}{2(1\,\mathrm{kg\text{-}m/N\text{-}s^2})} = 560.16\,\mathrm{kJ}$$

Thus,

$$\hat{V}^2 = \frac{2(560.16 \times 10^3\,\mathrm{N\text{-}m})(1\,\mathrm{kg\text{-}m/N\text{-}s^2})}{50\,\mathrm{kg}} = 22,406\,\mathrm{m^2/s^2}$$

and

$$\hat{V} = 149.7\,\mathrm{m/s} \Longleftarrow$$

3.3 Work

Work is defined as the energy that is expended when a force acts through a displacement

$$W = \int \boldsymbol{F} \cdot d\boldsymbol{x} = \int F \cos\theta \, dx \tag{3.7}$$

where θ is the angle between the line of action of the force and the direction of the displacement (Figure 3.5). We note that in $\boldsymbol{F} \cdot d\boldsymbol{x}$, $\boldsymbol{F}$, and $d\boldsymbol{x}$ are vectors and, because the scalar or dot product yields a scalar, work is a scalar quantity. Equation 3.7 defines work in the macroscopic mechanical terms of force and displacement and, because the force may not be constant, we see that the indicated integration may require a knowledge of how the force varies with the displacement.

We note that the thermodynamic view of work is that it is a form of **energy transfer** between a system and its surroundings. Thus, it is identified at the **system boundary** where its sole external effect is equivalent to the raising or lowering of a weight. Of course, in the real world, the raising or lowering of a weight may never be observed and the test is really whether or not the sole effort on the surroundings "could be" the raising or lowering of a weight.

The foregoing considerations for work show that it is energy that crosses a system boundary. Thus, we see that it is a transfer phenomenon and we will designate work as being positive if it is done *by* the system and negative if it is done *on* the system

$$\text{Work} \longrightarrow \begin{cases} \text{done } by \text{ a system is positive} \\ \text{done } on \text{ a system is negative} \end{cases}$$

Consider Figure 3.6a, which shows a compressible gas within a piston-cylinder assembly. The gas at state 1 is compressed to state 2 and the path is shown in pressure and volume coordinates that we refer to as the P-V plane in Figure 3.6b. With the differential amount of work designated as δW,

$$\delta W = -F dx$$

where the force on the piston, $F = PA$, is in the opposite direction of the displacement, dx. Hence,

$$\delta W = -PA\, dx$$

and because the differential volume is $dV = A dx$, we see that

$$\delta W = -P dV$$

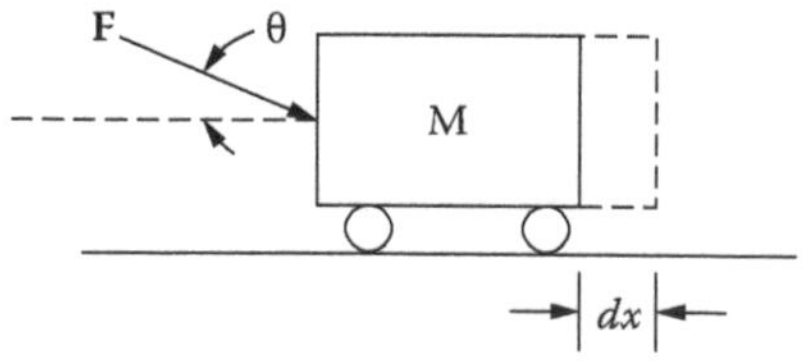

FIGURE 3.5
A vector force $\boldsymbol{F}$ on a moving mass, m, such that the displacement to the right is dx.

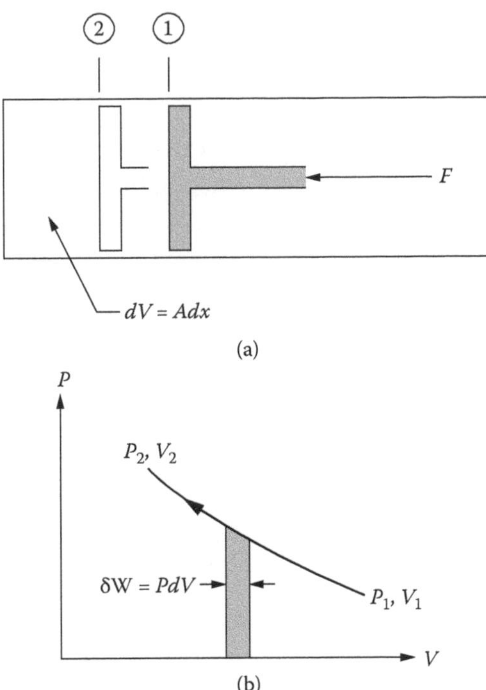

FIGURE 3.6
Work in a quasi-equilibrium compression process: (a) the piston-cylinder arrangement and (b) the P-V diagram.

and

$$W = -\int_{V_1}^{V_2} PdV \tag{3.8}$$

If a functional relationship between P and V is known, Equation 3.8 can be employed to evaluate the work in this quasi-equilibrium process. Reference to Figure 3.6b shows δW as the shaded strip and the integral indicates that the entire area under the curve between V_1 and V_2 is equal to the work done on the system by the surroundings. We note that had the process proceeded in the opposite direction, the magnitude of the work would have remained the same but the work would have been positive indicating that work was done by the gas on the surroundings. Indeed, a different functional relationship between P and V would have led to a different path between the two end states.

We are led by the foregoing development to the conclusion that work depends on the process between the end states of a system and that systems may possess energy but they cannot possess work. Thus, work at a system boundary is *not* a system property and different P-V relationships will yield different amounts of work. Because the magnitude of the work depends on the path between the initial and final states, work is considered to be a path function, which has an **inexact differential** that we have designated as δW.

Example 3.2 The initial and final volumes of a gas are 8 m^3 and 16 m^3. If the initial pressure is 20 N/m^2, determine the work done if (a) the pressure during expansion is constant and (b) the pressure during the expansion process is inversely proportional to the volume, that is, $PV = C$ (a constant).

Solution

Assumptions and Specifications

(1) The gas in the enclosure is taken as a closed system.
(2) Kinetic and potential energy changes are negligible.
(3) The displacement of the moving boundary represents the only work mode.
(4) The expansion is governed by a series of quasi-static equilibrium states.

(a) With P constant

$$W = \int_{V_1}^{V_2} P\,dV = P\int_{V_1}^{V_2} dV$$

$$= (20\,\text{N/m}^2)V\Big|_{8\,\text{m}^3}^{16\,\text{m}^3}$$

$$= (20\,\text{N/m}^2)(16\,\text{m}^3 - 8\,\text{m}^3)$$

$$= 160\,\text{N-m} \qquad (160\,\text{J}) \Longleftarrow$$

(b) With $P = C/V$ or $PV = C$

$$W = \int_{V_1}^{V_2} P\,dV = C\int_{V_1}^{V_2} \frac{dV}{V}$$

$$= C\ln V\Big|_{V_1}^{V_2} = C\ln\frac{16\,\text{m}^3}{8\,\text{m}^3}$$

$$= C\ln 2$$

However, $PV = P_1V_1 = C$ so that

$$C = (20\,\text{N/m}^2)(8\,\text{m}^3) = 160\,\text{N-m}$$

and with $\ln 2 = 0.6931$

$$W = 0.6931(160\,\text{N-m}) = 110.9\,\text{N-m} \qquad (110.9\,\text{J}) \Longleftarrow$$

In both cases (a) and (b), an expansion of the gas is indicated (Figure 3.7). Under this circumstance, the work is done by the system and is positive. Figure 3.7 illustrates what the example shows in a straightforward manner; that the work done *by* the system varies as the functional relationship between the P and V values. Moreover, work is a path function and not a system property.

The work in an expansion or compression quasi-equilibrium process is but one example of a work interaction between a system and its surroundings. Other examples include electrical and magnetic work, extension, and compression of a spring, work in stretching a film or membrane and work transmitted by a rotating shaft. We will discuss **shaft** or **paddle wheel work, electrical work,** and the work involved in making a bar elongate and the stretching of a spring in Section 3.6.2.

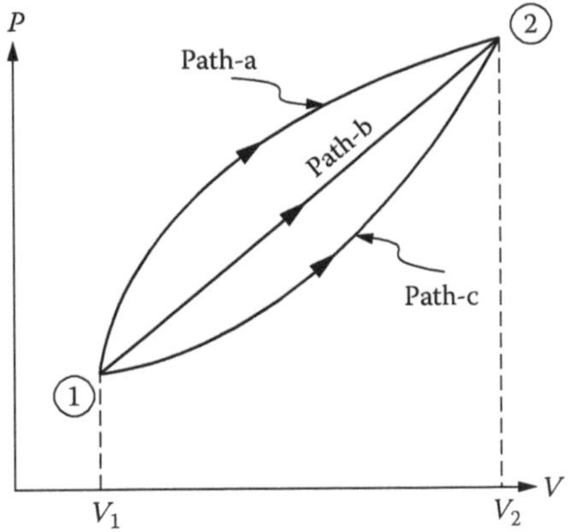

FIGURE 3.7
Work is a path function and not a system property. These three paths enclose different areas in the P-V plane and all yield different amounts of work.

3.3.1 Power

Power is the rate of doing work

$$\dot{W} = \frac{dW}{dt} \tag{3.9}$$

and the unit of power is the **watt** (W)

$$1\,\text{W} = 1\,\text{J/s} \qquad \text{or} \qquad 1\,\text{N-m/s}$$

Power in a translational system can be expressed as

$$\dot{W}_{\text{trans}} = \frac{dW}{dt} = \frac{d(Fx)}{dt} = F\frac{dx}{dt} = F\hat{V} \quad \text{(watts)} \tag{3.10a}$$

and in a rotational system where the work is the product of the torque, τ, in joules and the angular displacement, θ, in radians

$$\dot{W}_{\text{rot}} = \frac{dW}{dt} = \frac{d(\tau\theta)}{dt} = \tau\frac{d\theta}{dt} = \tau\omega \qquad \text{(watts)} \tag{3.10b}$$

3.4 Heat

All energy transfer, other than work, across the boundaries of a system is accounted for by **heat**,[1] Q. This heat is also a form of energy and its flow is induced by a temperature difference between the system and the surroundings and it will flow in the direction of

[1] There is a huge difference between heat and heat transfer. Here, we consider heat as energy, and in Chapter 5, where we deal with open systems, we consider the rate of heat transfer, $\dot{Q}$.

decreasing temperature. The transfer of heat is taken as positive when it is transferred **to** the system and negative when it is transferred **from** the system

$$\text{Heat} \longrightarrow \begin{cases} \text{transferred } \textit{to} \text{ a system is positive} \\ \text{transferred } \textit{from} \text{ a system is negative} \end{cases}$$

From this, we note that the sign convection pertaining to heat transfer is exactly opposite to the sign convention adopted for the work.

Like the work interaction, the heat interaction depends on the path between the end states of a process and is not a property. The **rate of heat transfer**, already designated as $\dot{Q}$ in watts or J/s, is related to the heat energy, Q, by

$$\dot{Q} = \frac{\delta Q}{dt} \qquad \text{watts} \tag{3.11}$$

where δQ, like δW is an inexact differential. The **heat flux**, $\dot{q}$, is the rate of heat transfer per unit surface area on the system boundary and, just as Q is related to $\dot{Q}$ by

$$Q = \int \dot{Q}\, dt \qquad \text{(joules)}$$

$\dot{Q}$ is related to $\dot{q}$ by

$$\dot{Q} = \int \dot{q}\, dS \qquad \text{(watts)}$$

where S is the surface area on the system boundary through which the heat passes.

3.5 The First Law of Thermodynamics

We now consider a process in which both heat and work interactions (Q and W) occur between a system and its surroundings. For the closed system or control mass shown in Figure 3.8, we modify Equation 3.6 to include both heat and work, as follows:

$$E_2 - E_1 = Q - W \tag{3.12}$$

This is the **first law of thermodynamics** that can be written in a differential form

$$dE = \delta Q - \delta W \tag{3.13}$$

Both Equations 3.12 and 3.13 indicate that energy within a closed system can be changed only by means of a transfer of energy by heat or work between the system and its surroundings. As a word statement, the first law of thermodynamics is

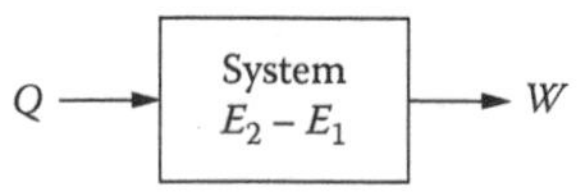

FIGURE 3.8
A system with both heat and work interaction between the system and its surroundings.

> The change of total energy (kinetic, potential, and internal) is equal to the net heat transferred to the control mass minus the work done by the control mass.

We may also observe that either of Equations 3.12 and 3.13 demonstrate the **conservation of energy** principle:

> Energy may neither be created nor destroyed but may be converted from one form to another.

3.6 The Energy Balance for Closed Systems

3.6.1 Processes

Equation 3.12

$$E_2 - E_1 = Q - W \tag{3.12}$$

indicates that **over a time interval**

$$\left\{\begin{array}{c}\text{the change in}\\\text{energy content}\\\text{within a system}\end{array}\right\} = \left\{\begin{array}{c}\text{the net amount of}\\\text{energy transferred}\\\text{into the system}\\\text{boundary as heat}\end{array}\right\} - \left\{\begin{array}{c}\text{the net amount of}\\\text{energy transferred}\\\text{out of the system}\\\text{boundary as work}\end{array}\right\} \tag{3.13}$$

Equation (3.12) may be written using total or specific quantities. We will find it convienient to use a form that places the properties at state 1 and the heat transfer on the left-hand side and the properties at state 2 and the work interaction on the right-hand side. This is consistent with the conventions that we have adopted for the direction of the heat flow and whether work is done on or by the system. Thus, in terms of total energy, we have

$$\frac{1}{2}m\hat{V}_1^2 + mgz_1 + U_1 + Q = \frac{1}{2}m\hat{V}_2^2 + mgz_2 + U_2 + W \qquad \text{(J)} \tag{3.14a}$$

or in terms of specific energy,

$$\frac{1}{2}\hat{V}_1^2 + gz_1 + u_1 + \frac{Q}{m} = \frac{1}{2}\hat{V}_2^2 + gz_2 + u_2 + \frac{W}{m} \qquad \text{(J/kg)} \tag{3.14b}$$

We may write Equation 3.13 as

$$\Delta E = \Delta Q - \Delta W$$

Then, over some time interval, this becomes

$$\frac{\Delta E}{\Delta t} = \frac{\Delta Q}{\Delta t} - \frac{\Delta W}{\Delta t}$$

and in the limit as $\Delta t \longrightarrow 0$

$$\frac{dE}{dt} = \dot{Q} - \dot{W} \tag{3.15}$$

With

$$\frac{dE}{dt} = \frac{d\text{KE}}{dt} + \frac{d\text{PE}}{dt} + \frac{dU}{dt}$$

we see that Equation 3.15 can be written as

$$\frac{d\text{KE}}{dt} + \frac{d\text{PE}}{dt} + \frac{dU}{dt} = \dot{Q} - \dot{W} \tag{3.16}$$

Example 3.3 During the execution of a certain process, the work done per degree of temperature increase is 60 J/K and the internal energy is given as a function of temperature, $U = 25(1+0.16T)$ J/kg. Determine the heat transferred in joules if the temperature changes from 325 K to 400 K.

Solution

Assumptions and Specifications

(1) We consider the working medium as the closed system.
(2) Kinetic and potential energy changes are negligible.
(3) The process is represented by a series of quasi-equilibrium processes.
(4) The internal energy is a specified function of the temperature.

Here, from Equation 3.14a we have

$$U_1 + Q = U_2 + W$$

and because W and $\Delta U = U_2 - U_1$ can be evaluated

$$Q = W + \Delta U = W + U_2 - U_1$$

With $dW/dT = 60\,\text{J/K}$

$$W = \int_{T_1}^{T_2} (60\,\text{J/K})dT = (60\,\text{J/K}) \int_{325\,\text{K}}^{400\,\text{K}} dT = (60\,\text{J/K})(75\,\text{K}) = 4500\,\text{J}$$

Moreover,

$$\begin{aligned}\Delta U &= 25(1 + 0.16T_2)\,\text{J/K} - 25(1 + 0.16T_1)\,\text{J/K} \\ &= (4.0\,\text{J/K})(400\,\text{K} - 325\text{K}) \\ &= (4.0\,\text{J/K})(75\,\text{K}) \\ &= 300\,\text{J}\end{aligned}$$

Hence,

$$\begin{aligned}Q &= W + U_2 - U_1 \\ &= 4500\,\text{J} + 300\,\text{J} \\ &= 4800\,\text{J} \Longleftarrow\end{aligned}$$

and this positive value indicates that this heat is transferred **to** the system.

Example 3.4 An elevator weighs 4750 N and is moving downward at a velocity of 1.92 m/s (Figure 3.9). The elevator has a 3250-N counterweight and an associated braking system. Determine the frictional energy absorbed by the braking system when the elevator is brought to a stop in a distance of 2 m.

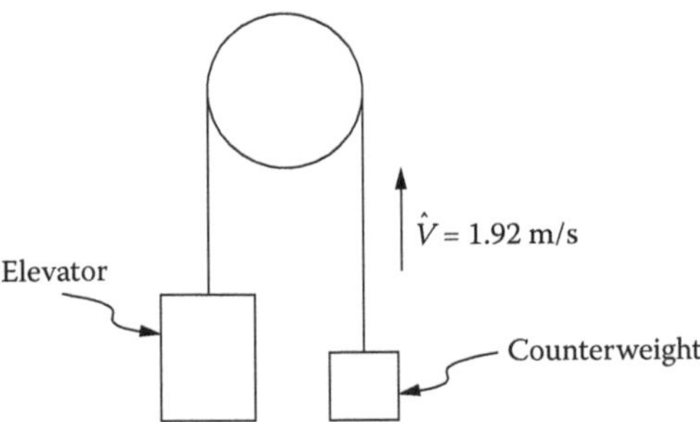

FIGURE 3.9
Elevator, counterweight, and rotating pulley for Example 3.4.

Solution

Assumptions and Specifications

(1) Both the elevator and the counterweight are taken as the system.
(2) Rotating kinetic and potential energies in all cables and pulleys in the system are negligible.
(3) The local acceleration of gravity is taken as $g = 9.81\,\text{m/s}^2$.

With the subscripts E, CW, and B designating the elevator, counterweight, and braking system respectively, conservation of energy dictates that

$$\text{KE}_\text{E} + \text{PE}_\text{E} = \text{KE}_\text{CW} + \text{PE}_\text{CW} + W_\text{B}$$

With $\hat{V} = 1.92\,\text{m/s}$ for both the elevator and the counterweight

$$\text{KE}_\text{E} = \frac{1}{2}m\hat{V}^2 = \frac{1}{2}\left(\frac{4750\,\text{N}}{9.81\,\text{m/s}^2}\right)(1.92\,\text{m/s})^2 = 892.5\,\text{J}$$

and for a change in elevation of 2 m

$$\text{PE}_\text{E} = mgz = \left(\frac{4750\,\text{N}}{9.81\,\text{m/s}^2}\right)(9.81\,\text{m/s}^2)(2\,\text{m}) = 9500\,\text{J}$$

For the counterweight by straight proportion

$$\text{KE}_\text{CW} = \left(\frac{3250\,\text{N}}{4750\,\text{N}}\right)(892.5\,\text{J}) = 610.7\,\text{J}$$

and

$$\text{PE}_\text{CW} = \left(\frac{3250\,\text{N}}{4750\,\text{N}}\right)(9500\,\text{J}) = 6500\,\text{J}$$

We may now solve for the braking energy

$$\begin{aligned} W_\text{B} &= \text{KE}_\text{E} - \text{KE}_\text{CW} + \text{PE}_\text{E} - \text{PE}_\text{CW} \\ &= 892.5\,\text{J} - 610.7\,\text{J} + 9500\,\text{J} - 6500\,\text{J} \\ &= 3281.8\,\text{J} \Longleftarrow \end{aligned}$$

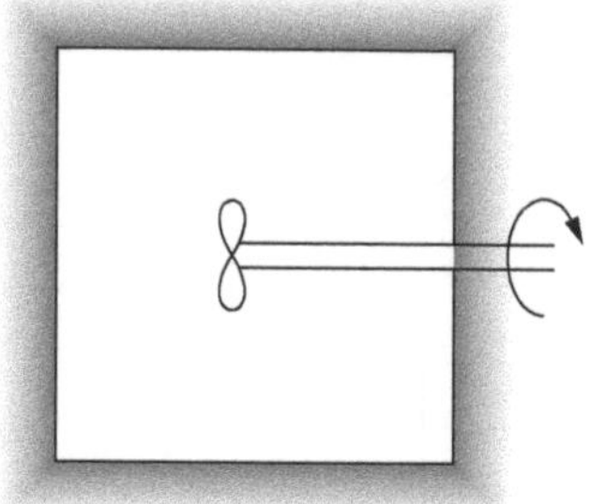

FIGURE 3.10
Rotating paddle wheel and shaft inside an enclosure.

3.6.2 Other Forms of Work

3.6.2.1 *Shaft or Paddle Wheel Work*

Consider Figure 3.10, which shows a closed, rigid, and insulated container holding a mass of gas, m, at a temperature, T_1. A shaft passes through the container wall and the shaft is rotated to spin the paddle wheel (or vane or rotor).

We may write the first law energy balance

$$E_2 - E_1 = Q - W \tag{3.12}$$

and we note that because the container is closed, rigid, and insulated, $Q = W = 0$. Under these circumstances $E_1 = E_2$ or $U_1 = U_2$. Because, as indicated in Section 3.2.3, the internal energy is a function of temperature, this further infers that $\Delta T = 0$.

A little thought leads us to the conclusion that energy is delivered to the mass of gas inside of the container by the rotation of the shaft. This energy is called **shaft** or **paddle wheel work**

$$W_p = \tau\theta$$

where τ is the average torque applied to the shaft and θ is the angular displacement.

Clearly, the first law must be generalized to accommodate other kinds of work in addition to the boundary work (the energy that flows into or out of the system across the system boundaries). Moreover, the power due to the paddle wheel will be

$$\dot{W}_p = \tau\omega$$

where ω is the angular velocity of the shaft.

3.6.2.2 *Electrical Work*

A study of electrostatics will show that work is done on or by an electric charge as it is moved through a potential difference ΔV. Thus,

$$W_e = \Delta V q = (V_2 - V_1)q$$

and if $V_1 = 0$, we have with $V_2 = V$

$$W_e = Vq$$

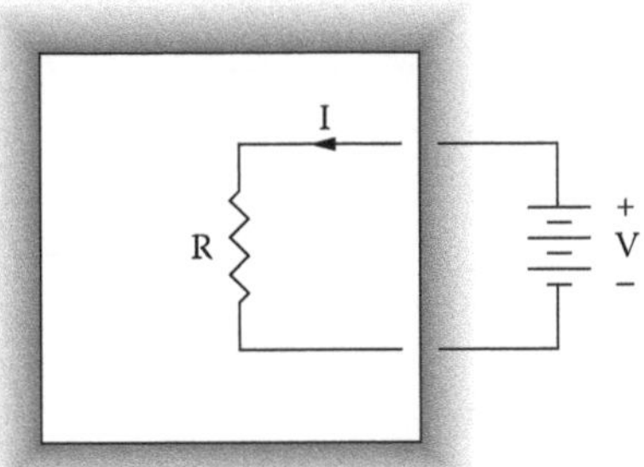

FIGURE 3.11
Current flow in a resistor.

Because instantaneous current, in amperes, is the rate of flow of charge

$$i \equiv \frac{dq}{dt}$$

and with v taken as the instaneous voltage we have the instantaneous electric power

$$\dot{W}_e = v\frac{dq}{dt} = vi$$

If the system is time invariant (Figure 3.11), the electrical work over a time interval, Δt, will be

$$\dot{W}_e = VI$$

3.6.2.3 Spring Work

A spring with a spring constant, k_s, in N/m can be stretched (Figure 3.12a) from an original length, x_0, to $x = x_1 - x_0$, so that the tension force will be

$$F_t = -k_s(x_1 - x_0)$$

where $k_s > 0$. If the spring is compressed (Figure 3.12b) from its original length, x_0, to $x = x_2 + x_0$, the compression force is

$$F_c = k_s(x_2 - x_0)$$

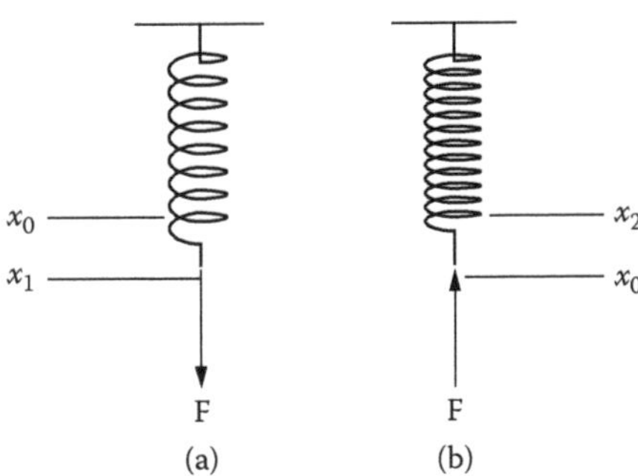

FIGURE 3.12
A spring (a) stretched and (b) compressed from an original length, x_o.

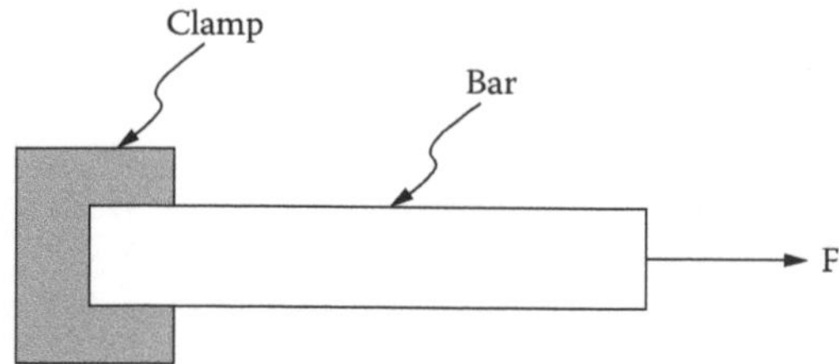

FIGURE 3.13
A solid bar under tension.

Here

$$x = x_1 - x_0 - (x_2 - x_0) = x_1 - x_2$$

and the work done on the spring in changing its length from x_1 to x_2 will then be

$$W = \int_{x_1}^{x_2} F x\, dx = k_s \int_{x_1}^{x_2} x\, dx = \frac{k_s}{2}(x_2^2 - x_1^2) \qquad \text{(W)}$$

3.6.2.4 *Two Additional Examples*

Two other examples of the work concept are the extension of a solid bar and the stretching of a liquid film.

3.6.2.4.1 The Stretching of a Solid Bar In Figure 3.13, one end of the solid bar is under tension and the other end is fixed at $x = x_o$. When a force is applied at a point where the unstressed length is x_1, the bar will elongate to a length, x_2. With σ_n taken as the normal stress in the bar and A the cross-sectional area, the work done will be

$$W = -\int_{x_1}^{x_2} F\, dx = -\int_{x_1}^{x_2} \sigma_n A dx = (x_1 - x_2)\sigma_n A$$

where the minus sign is needed because work is done *on* the bar when dx is positive.

3.6.2.4.2 The Stretching of a Liquid Film A liquid film suspended in a rigid wire frame is shown in Figure 3.14. Here, the microscopic forces between the molecules near the liquid-air interfaces support the thin liquid layer within the frame by the effect of **surface tension**. These microscopic forces establish a measurable macroscopic force that is perpendicular to any line in the liquid film and the force per unit length in such a line is the surface tension, σ.

Thus, the force, F, shown in Figure 3.14 can be written as

$$F = 2\sigma L$$

because there are two film surfaces that act on the wire.

If the movable wire is displaced a distance, x, the work is given by

$$W = -\int_{x_1}^{x_2} F\, dx = 2L\sigma(x_1 - x_2)$$

where the minus sign is required because the work is done on the system when dx is positive.

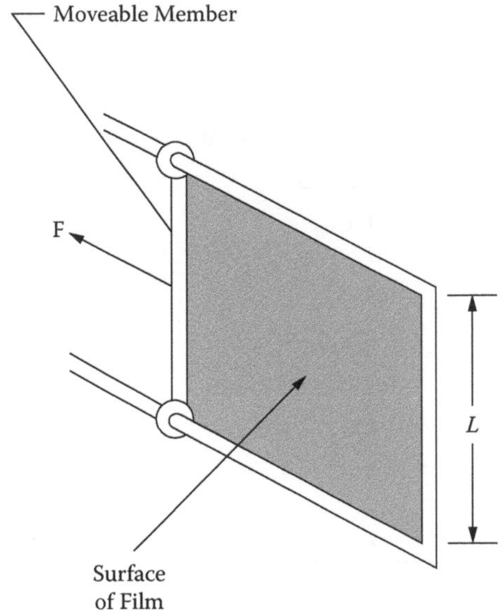

FIGURE 3.14
A liquid film suspended in a rigid wire frame.

3.6.3 Cycles

The study of devices that operate on a cycle is an important application of thermodynamics. Because the working substance returns to its initial state, $\Delta E = 0$ and Equation 3.12 may be written as

$$Q - W = 0$$

or with emphasis on the application to a cycle

$$Q_{\text{cycle}} = W_{\text{cycle}} \tag{3.17}$$

Figure 3.15 shows two types of cycles and the directions of all heat and work interactions are indicated by arrows. We note that the heat interactions derive from an external body at a "hot side" temperature, T_H, and an external body at a "cold side" temperature, T_L. For the **power cycle** shown in Figure 3.15a, the net work is

$$W_{\text{cycle}} = Q_{\text{in}} - Q_{\text{out}}$$

and we observe that in order to have work done by the system in a power cycle, Q_{in} must exceed Q_{out}. In this case, the **cycle efficiency** is defined as the ratio of the work done by the cycle to the heat input

$$\eta = \frac{W_{\text{cycle}}}{Q_{\text{in}}} = \frac{Q_{\text{in}} - Q_{\text{out}}}{Q_{\text{in}}} = 1 - \frac{Q_{\text{out}}}{Q_{\text{in}}} \tag{3.18}$$

and is less than unity.

Notice that, because a cycle begins and ends at the same state point, when dealing with cycles where only W_{cycle}, Q_{out}, and Q_{in} are involved, each of these quantities can be represented by a positive number. In this event, any negative signs in Equation 3.17 will automatically take care of any subtraction. The use of positive numbers to represent Q_{out}, and Q_{in} will be followed in subsequent chapters when cycles (with $E_1 = E_2$) are considered.

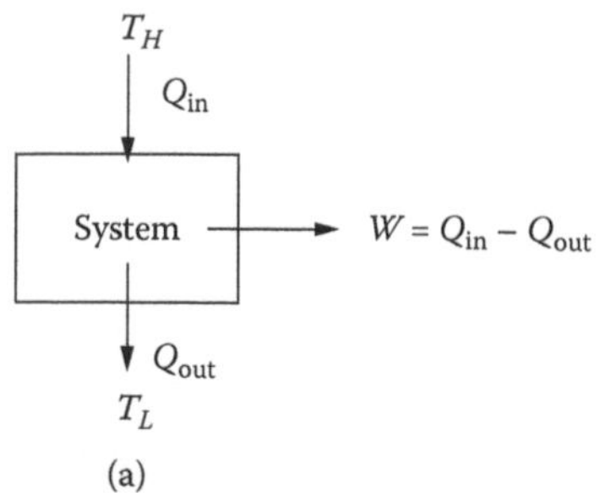

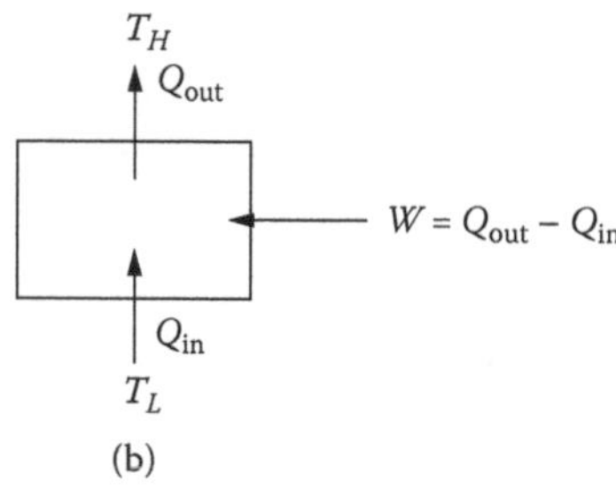

FIGURE 3.15
Sketches of two cycles: (a) power cycle and (b) refrigeration or heat pump cycle.

The cycle in Figure 3.15b is either a **refrigeration** or a **heat pump cycle**. If it is to represent a refrigerator, its purpose is to remove heat from a body or system (such as a refrigerated space) at T_L. If it is to represent a heat pump, its purpose is to provide heat to a body or system (such as a dwelling) at T_H. In the case of either the refrigerator or the heat pump

$$W_{cycle} = Q_{out} - Q_{in}$$

and Q_{out} must exceed Q_{in}. In both, the **coefficient of performance** is a measure of the effectiveness of the device or system. For the refrigerator

$$\beta = \frac{Q_{in}}{W_{cycle}} = \frac{Q_{in}}{Q_{out} - Q_{in}} \tag{3.19}$$

and for the heat pump

$$\gamma = \frac{Q_{out}}{W_{cycle}} = \frac{Q_{out}}{Q_{out} - Q_{in}} \tag{3.20}$$

Example 3.5 A thermal cycle contains an engine that delivers 24 horsepower (HP) to a load and uses a small pump, which absorbs 1.85 KW to circulate the fluid in the system (Figure 3.16). If the heat supplied in a heat exchanger is 90 KW, determine (a) the net work, (b) the heat rejected, and (c) the efficiency of the cycle.

Solution

Assumptions and Specifications

(1) The working medium is considered as the system.
(2) The system operates in the steady state.

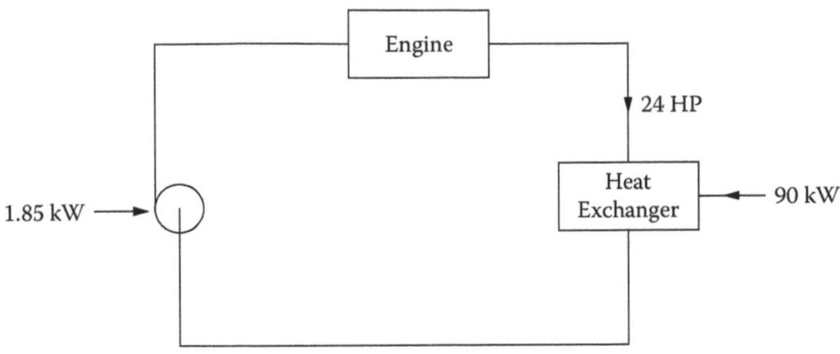

FIGURE 3.16
Thermal cycle for Example 3.5.

(a) The engine delivers

$$W_{out} = (24\,\text{HP})(0.746\,\text{kW/HP}) = 17.90\,\text{kW}$$

and the net work, which will be W_{cycle}, is

$$W_{cycle} = W_{out} - W_{pump} = 17.90\,\text{kW} - 1.85\,\text{kW} = 16.05\,\text{kW} \Longleftarrow$$

(b) The heat rejection is obtained from

$$W_{cycle} = Q_{in} - Q_{out}$$

so that

$$Q_{out} = Q_{in} - W_{cycle} = 90\,\text{kW} - 16.05\,\text{kW} = 73.95\,\text{kW} \Longleftarrow$$

(c) The efficiency is

$$\eta = \frac{W_{cycle}}{Q_{in}} = \frac{16.05\,\text{kW}}{90\,\text{kW}} = 0.178 \quad \text{or} \quad 17.8\,\% \Longleftarrow$$

3.7 The Ideal Gas Model

From our study of physics, we may recall **Charles law,**[2] which derived from a consideration of the relationships between pressure and volume for a gas. These relationships are idealizations and give very close approximations to the behavior of real gases at low pressures and moderate temperatures. We may consider the **ideal gas model** by referring to the alternate forms of Charles' law:

> At constant pressure, the volume of a fixed mass of a gas varies directly with its absolute temperature.

[2] J. A. C. Charles (1746–1823) was a French chemist, physicist, and inventor.

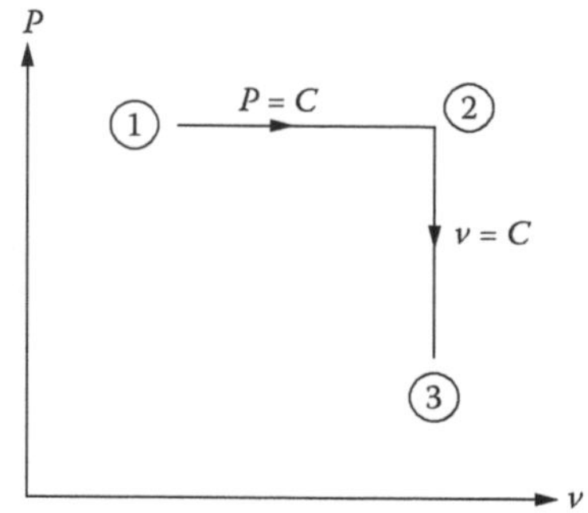

FIGURE 3.17
The P-v plane with a constant pressure process followed by a constant volume process.

and

> At constant volume, the pressure of a fixed mass of a gas varies directly with its absolute temperature.

Consider Figure 3.17, which shows the two alternative forms of Charles' law in the P-v (pressure-specific volume) plane. We may note that the process from state 1 to state 2 is at constant pressure and that the process from state 2 to state 3 is at constant specific volume. From state 1 to state 2, Charles' law states that

$$\frac{v_1}{v_2} = \frac{T_1}{T_2} \quad \text{so that} \quad T_2 = T_1 \frac{v_2}{v_1}$$

and from state 2 to state 3,

$$\frac{P_2}{P_3} = \frac{T_2}{T_3} \quad \text{so that} \quad T_2 = T_3 \frac{P_2}{P_3}$$

Hence,

$$T_1 \frac{v_2}{v_1} = T_3 \frac{P_2}{P_3}$$

and because $P_1 = P_2$ and $v_2 = v_3$

$$T_1 \frac{v_3}{v_1} = T_3 \frac{P_1}{P_3}$$

or

$$\frac{P_1 v_1}{T_1} = \frac{P_3 v_3}{T_3} = \frac{PVC}{T} \tag{3.21}$$

Because P, v, and T are all independent variables, each side of Equation 3.21 must be uniquely equal to a constant, R (called the **gas constant**), so that

$$PV = RT \tag{3.22}$$

and with $v = V/m$

$$PV = mRT \tag{3.23}$$

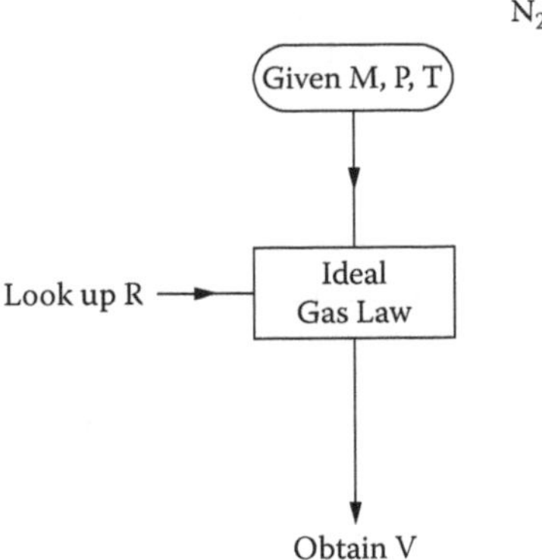

FIGURE 3.18
The procedure for finding the volume of nitrogen in a closed, rigid container in Example 3.6.

Equations 3.22 and 3.23 are alternative expressions of the **equation of state** for the ideal gas, which is also referred to as the **ideal gas law**. In order to make this equation of state universal, the gas constant is related to another gas constant that holds for all gases via

$$R = \frac{\bar{R}}{M} \tag{3.24}$$

where M is the molecular weight of the gas. Here, $\bar{R}$ is called the **universal gas constant**. Because the mass is related to the number of moles, n, by the molecular weight

$$n = \frac{m}{M}$$

we can provide another representation for the equation of state

$$PVC = m\frac{\bar{R}}{M}T = n\bar{R}T \tag{3.25}$$

Moreover, because the density of the gas is $\rho = m/V$, the equation of state can also be written as

$$P = \rho RT \tag{3.26}$$

The value of the universal gas constant, $\bar{R}$, is based on **Avogadro's law**[3] which states that one mole of a gas at standard conditions of temperature and pressure contains 6.023×10^{23} molecules

$$\bar{R} = 8314\,\text{Pa-m}^3/\text{kgmol-K} \quad \text{or} \quad \bar{R} = 8314\,\text{J/kgmol-K} \tag{3.27}$$

Values of the gas constant, R, for several gases are listed in Table A.1 in Appendix A, which also lists values of the molecular weight and **specific heats** that are considered in the next section.

[3] Amadeo Avogadro (1776–1856) was an Italian physicist.

Example 3.6 32 kg of nitrogen at 1.2 MPa and 177°C are in a closed, rigid container. Use the procedure provided in Figure 3.18 to determine the volume of the container.

Solution

Assumptions and Specifications

(1) We assume that the nitrogen is the system.
(2) The nitrogen is an ideal gas.
(3) Kinetic and potential energy changes are negligible.

From Table A.1, we read $R = 297\,\text{J/kg-K}$. Then with $T = 117°\text{C} + 273°\text{C} = 390\,\text{K}$, the volume of the container will be

$$V = \frac{mRT}{P}$$

$$= \frac{(32\,\text{kg})(297\,\text{N-m/kg-K})(390\,\text{K})}{1.2 \times 10^6\,\text{N/m}^2}$$

$$= 3.089\,\text{m}^3 \Longleftarrow$$

3.8 Ideal Gas Enthalpy and Specific Heats

Because the internal energy, U, and the product, PV, occur so frequently in so many thermodynamic considerations, their sum is given the name **enthalpy**, designated by H

$$H = U + PV \qquad (\text{J}) \tag{3.28a}$$

or **specific enthalpy** designated by h

$$h = \frac{H}{m} = u + Pv \qquad (\text{J/kg}) \tag{3.28b}$$

The **specific heat** of a substance may be defined as

> The quantity of energy that is required to raise a unit mass of the substance by one degree.

For a closed system containing a gas and with no kinetic or potential energy change (a stationary system), Equation 3.12 tells us that in a constant volume process where no work is done by or on the gas

$$\delta Q = dU$$

Here, δQ is the energy transferred into the system so that we may define the specific heat at constant volume on a unit mass basis as

$$c_v(T) = \left.\frac{du}{dT}\right|_{v=\text{constant}} \qquad (\text{J/kg-K}) \tag{3.29a}$$

and we note that the specific heat, c_v, is a function of temperature.

Similar reasoning leads to the constant pressure process where we find that from Equation 3.28a

$$dH = dU + d(PV) = dU + \delta W$$

and from Equation 3.12

$$\delta Q = dH$$

so that the **specific heat at constant pressure,** c_p, also defined on a unit mass basis is

$$c_p(T) = \left.\frac{dh}{dT}\right|_{P=\text{constant}} \qquad (\text{J/kg-K}) \tag{3.29b}$$

which is also a function of temperature.

Now we know that u is a function of temperature (Section 3.2.3) and we see from Equation 3.28b that, because R is a constant,

$$h = u(T) + RT = h(T)$$

From Equation 3.29a, we may write

$$du = c_v(T)dT \tag{3.30a}$$

and from Equation 3.29b

$$dh = c_p(T)dT \tag{3.30b}$$

Equations 3.30 may be integrated yielding

$$u_2 - u_1 = \int_{T_1}^{T_2} c_v(T)dT \tag{3.31a}$$

and

$$h_2 - h_1 = \int_{T_1}^{T_2} c_p(T)dT \tag{3.31b}$$

If the specific heats are constant, the result of the integrations in Equation 3.31 will be

$$u_2 - u_1 = c_v(T_2 - T_1) \tag{3.31c}$$

and

$$h_2 - h_1 = c_p(T_2 - T_1) \tag{3.31d}$$

Specific heats at constant volume and constant pressure for ideal gases may be taken as constant, determined from a table of specific heat values such as Table A.2 or evaluated from an equation such as the polynomial equation for the molar constant pressure specific heat

$$\frac{\bar{c}_p}{\bar{R}} = a + bT + cT^2 + dT^3 + eT^4$$

We also note that the ideal gas law permits the representation

$$h = u + Pv \quad \longrightarrow \quad h = u + RT$$

and because

$$\frac{dh}{dT} = \frac{du}{dT} + R$$

we see that

$$c_p(T) = c_v(T) + R \tag{3.32}$$

or

$$R = c_p(T) - c_v(T) \tag{3.33}$$

Equation 3.33 shows that, because R is greater than 0, $c_p(T)$ must exceed $c_v(T)$. In subsequent sections, we will encounter the ratio of specific heats, c_p/c_v, designated as k. Some useful relationships are

$$k = \frac{c_p(T)}{c_v(T)} > 1 \tag{3.34}$$

then

$$R = c_v(T)\left[\frac{c_p(T)}{c_v(T)} - 1\right] = c_v(T)(k-1)$$

so that

$$c_v(T) = \frac{R}{k-1} \tag{3.35a}$$

and

$$c_p(T) = \frac{kR}{k-1} \tag{3.35b}$$

Table A.2 provides specific heats and their ratio, k, as a function of temperature for air and five other gases.

Example 3.7 Air at one atmosphere (101.3 kPa) and 25°C is contained in a piston-cylinder assembly (Figure 3.19) where the volume is initially at 1.60×10^{-3} m^3. Heat is added in a manner such that the movement of the piston holds the air at constant pressure until the air reaches 275°C. Determine (a) the work done by the gas, (b) the heat added to the gas, and (c) the change in internal energy.

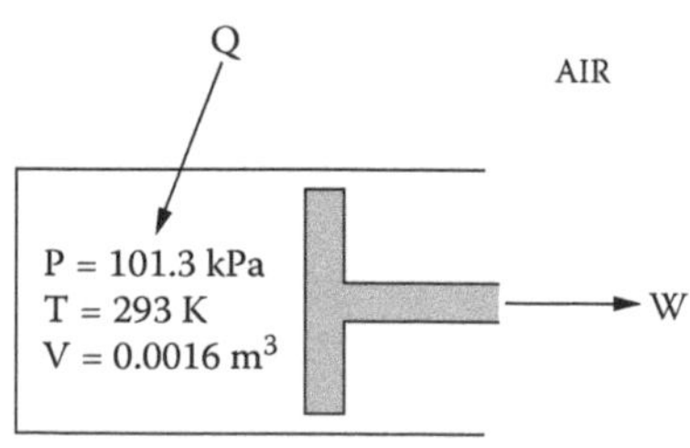

FIGURE 3.19
Air inside of a piston-cylinder assembly for Example 3.7.

Solution

Assumptions and Specifications

(1) The air contained in the piston-cylinder assembly represents a closed system.
(2) The expansion process is quasi-static.
(3) Kinetic and potential energy changes are negligible.
(4) The air may be considered as an ideal gas.

(a) The work in a constant pressure process can be evaluated using Equation 3.8

$$W = \int_{V_1}^{V_2} P dV = P(V_2 - V_1)$$

and because $mR = PV/T$, the volume at state 2 can be obtained from conditions at state 1

$$mR = \frac{P_1 V_1}{T_1} = \frac{P_2 V_2}{T_2}$$

and with $P_1 = P_2$

$$V_2 = \frac{T_2}{T_1} V_1$$

Here, $T_1 = 25°\text{C} + 273°\text{C} = 298\,\text{K}$, $T_2 = 275°\text{C} + 273°\text{C} = 548\,\text{K}$ and the average temperature is

$$T_{\text{avg}} = \frac{298\,\text{K} + 548\,\text{K}}{2} = 423\,\text{K}$$

Then

$$V_2 = \left(\frac{548\,\text{K}}{298\,\text{K}}\right)(1.6 \times 10^{-3}\,\text{m}^3) = 2.942 \times 10^{-3}\,\text{m}^3$$

and

$$\begin{aligned} W &= P(V_2 - V_1) \\ &= (1.013 \times 10^5\,\text{N/m}^2)(2.942 \times 10^{-3}\,\text{m}^3 - 1.600 \times 10^{-3}\,\text{m}^3) \\ &= (1.013 \times 10^5\,\text{N/m}^2)(1.342 \times 10^{-3}\,\text{m}^3) \\ &= 136.0\,\text{J} \Longleftarrow \end{aligned}$$

Because the work is a positive quantity, we conclude that the work is done by the system (the gas in the cylinder) on the surroundings.

(b) Because kinetic and potential energy changes are negligible, we can rearrange Equation 3.12 to show that

$$Q = m(u_2 - u_1) + W$$

For this constant pressure process

$$Q = m(u_2 - u_1) + mP(v_2 - v_1) = m(h_2 - h_1)$$

and we can compute the heat added to the gas from the change in enthalpy using Equation 3.31d

$$h_2 - h_1 = c_p(T_2 - T_1) \tag{3.31d}$$

First, however, we will need the amount of air present. We read from Table A.1 for air at 298 K ≈ 300 K that $R = 287\,\mathrm{J/kg\text{-}K}$ and we calculate m from conditions at point 1 using the ideal gas law.

$$m = \frac{PV}{RT} = \frac{(1.013 \times 10^5\,\mathrm{N/m^2})(1.60 \times 10^{-3}\,\mathrm{m^3})}{(287\mathrm{J/kg\text{-}K})(298\,\mathrm{K})} = 1.895 \times 10^{-3}\,\mathrm{kg}$$

Then with $T_2 - T_1 = 548\,\mathrm{K} - 298\,\mathrm{K} = 250\,\mathrm{K}$ and with $T_{avg} = 423\,\mathrm{K}$, Table A.2 reveals that $c_p = 1.016\,\mathrm{kJ/kg\text{-}K}$ and $c_v = 0.729\,\mathrm{kJ/kg\text{-}K}$ so that

$$\begin{aligned} Q &= m(h_2 - h_1) \\ &= mc_p(T_2 - T_1) \\ &= (1.895 \times 10^{-3}\,\mathrm{kg})(1016\,\mathrm{J/kg\text{-}K})(250\,\mathrm{K}) \\ &= 481.4\,\mathrm{J} \Longleftarrow \end{aligned}$$

(c) The change in internal energy can be computed from the energy equation of Equation 3.12 or from Equation 3.31a. First, from the energy equation

$$\begin{aligned} U_2 - U_1 &= Q - W \\ &= 481.3\,\mathrm{J} - 136.0\,\mathrm{J} \\ &= 345.3\,J \Longleftarrow \end{aligned}$$

and alternatively from Equation 3.31c with $c_v = 0.729\,\mathrm{kJ/kg\text{-}K}$ from Table A.2 at 423 K

$$\begin{aligned} U_2 - U_1 &= mc_v(T_2 - T_1) \\ &= (1.895 \times 10^{-3}\,\mathrm{kg})(729\,\mathrm{J/kg\text{-}K})(250\,\mathrm{K}) \\ &= 345.3\,\mathrm{J} \end{aligned}$$

3.9 Processes of an Ideal Gas

3.9.1 Introduction

We will find it extremely useful to characterize quasi-static processes of an ideal gas by the relationship

$$PV^n = C \tag{3.36}$$

where the exponent, n, may take on different values that are dictated by the particular process. We consider five such processes of the ideal gas, three of which are:

- The **isometric** or constant volume process where $n = \infty$

- The **isobaric** or constant pressure process where $n = 0$
- The **isothermal** or constant temperature process where $n = 1$

The other two, which are discussed in subsequent chapters:

- The **isentropic** or constant entropy process where n is equal to $k = c_p/c_v$
- The **polytropic** process where n is unspecified and can take on any value between 0 and ∞

At this point in our study, we are able to treat the first three of these ideal gas processes with modest detail.

3.9.2 The Constant Volume Process

The constant volume process is characterized by $n = \infty$ so that Equation 3.36 takes the form

$$V = C$$

and the pressure-temperature relationship is

$$\frac{T_2}{T_1} = \frac{P_2}{P_1} \tag{3.37}$$

Because $V_2 = V_1$, Equation 3.8 is evaluated as

$$W = \int_{V_1}^{V_2} P dV = 0$$

which tells us that no work is done on or by the system in a constant volume process. Thus, in the absence of kinetic and potential energy changes, Equation 3.12 becomes

$$U_2 - U_1 = Q$$

and via Equation 3.31a

$$Q = U_1 - U_1 = m \int_{T_1}^{T_2} c_v(T) dT \tag{3.38a}$$

and for small temperature excursions where $c_v(T)$ can be taken as constant

$$Q = mc_v(T_2 - T_1) \tag{3.38b}$$

3.9.3 The Constant Pressure Process

The constant pressure process is characterized by $n = 0$ so that Equation 3.36 takes the form

$$P = C$$

and the volume-temperature relationship is

$$\frac{T_2}{T_1} = \frac{V_2}{V_1} \tag{3.39}$$

Here, by Equation 3.8, the work is

$$W = \int_{V_1}^{V_2} P dV = P \int_{V_1}^{V_2} dV = P(V_2 - V_1) \tag{3.40}$$

and if $V_2 > V_1$, the work is done by the system on the surroundings.

Finally, by Equation 3.12, we see that

$$\begin{aligned} U_2 - U_1 &= Q - W \\ &= Q - P(V_2 - V_1) \\ Q &= U_2 - U_1 + (PV_2 - PV_1) = H_2 - H_1 \end{aligned}$$

or via Equation 3.31b

$$Q = H_2 - H_1 = m \int_{T_1}^{T_2} c_p(T) dT \tag{3.41a}$$

and for small temperature excursions where $c_p(T)$ can be taken as constant

$$Q = H_2 - H_1 = mc_p(T_2 - T_1) \tag{3.41b}$$

3.9.4 The Isothermal Process

The constant temperature process or isothermal process is characterized by $n = 1$ so that Equation 3.36 takes the form

$$PV = C$$

and the pressure-volume relationship is

$$\frac{P_1}{P_2} = \frac{V_2}{V_1} \tag{3.42}$$

Here, by Equation 3.8, the work is

$$\int_{V_1}^{V_2} P dV = \int_{V_1}^{V_2} \frac{C}{V} dV = C \ln \frac{V_2}{V_1}$$

But $C = P_1 V_1 = P_2 V_2$ so that we have

$$W = P_1 V_1 \ln \frac{V_2}{V_1} = P_2 V_2 \ln \frac{V_2}{V_1} \tag{3.43}$$

and we note that if $V_2 > V_1$ work is done by the system on the surroundings. Moreover, we note that because $T_2 = T_1$

$$Q = U_2 - U_1 = 0$$

and

$$Q = W$$

Thus, we see that in an isothermal process, work is done *on* the surroundings and must equal the heat transferred *to* the system because there is no change in internal energy in the system.

Example 3.8 A closed and rigid cylinder has a volume of $0.34\,m^3$ and contains helium initially at 25°C and 6.0×10^5 Pa. More helium is added until the temperature reaches 75°C and the pressure reaches 20×10^5 Pa. The contents of the cylinder are then allowed to cool to the original temperature. Determine (a) the mass of helium added, (b) the final pressure, and (c) the heat lost during the cooling process.

Solution

Assumptions and Specifications

(1) The helium is the system and is an ideal gas.
(2) "Closed and rigid" implies a constant volume process.
(3) Kinetic and potential energy changes are negligible.
(4) The helium addition and cooling processes occur under quasi-equilibrium conditions.
(5) The specific heat may be taken from Table A.1 at 300 K.

(a) We call the initial condition state 1 and with $V_1 = 0.34\,m^3$, $P_1 = 6.5 \times 10^5\,N/m^2$ and $T_1 = 25°C + 273°C = 298\,K$, we use Table A.1 to obtain for helium

$$R = 2078\,J/kg\text{-}K \qquad \text{and} \qquad c_v = 3.146\,kJ/kg\text{-}K$$

The original mass of helium in the tank can be found from the ideal gas law

$$m_1 = \frac{P_1 V_1}{RT_1} = \frac{(6 \times 10^5\,N/m^2)(0.34\,m^3)}{(2078\,J/kg\text{-}K)(298\,K)} = 0.329\,kg$$

At state 2, $V_2 = 0.34\,m^3$ and with $P_2 = 20 \times 10^5\,N/m^2$ and $T_2 = 75°C + 273°C = 348\,K$, the mass of the helium at state 2 will be

$$m_2 = \frac{P_2 V_2}{RT_2} = \frac{(20 \times 10^5\,N/m^2)(0.34\,m^3)}{(2078\,J/kg\text{-}K)(348\,K)} = 0.940\,kg$$

This means that the mass of helium added between state 1 and state 2 is

$$\Delta m = m_2 - m_1 = 0.940\,kg - 0.329\,kg = 0.611\,kg \Longleftarrow$$

(b) We designate the final condition as state 3 where $m_3 = 0.940\,kg$, $T_3 = 298\,K$ and $V_3 = 0.34\,m^3$. Thus,

$$P_3 = \frac{m_3 R T_3}{V_3} = \frac{(0.940\,kg)(2078\,J/kg\text{-}K)(298\,K)}{0.34\,m^3} = 1.712 \times 10^6\,Pa \Longleftarrow$$

(c) For this constant volume process with $W = 0$, Equations 3.12 and 3.31c give with $T_3 - T_2 = 298\,K - 348\,K = -50\,K$

$$\begin{aligned} Q &= m(u_3 - u_2) = mc_v(T_3 - T_2) \\ &= (0.940\,kg)(3.146\,kJ/kg\text{-}K)(-50\,K) = -147.9\,kJ \Longleftarrow \end{aligned}$$

The minus sign shows that heat is rejected to the surroundings.

3.10 Summary

The conservation of mass principle can be stated as:

> Mass can neither be created nor destroyed and its composition cannot be altered from one form to another unless it undergoes a chemical change.

There are three forms of energy

- Kinetic energy

$$\mathrm{KE} = \frac{m\hat{V}^2}{2} \qquad \text{(N-m or joules)}$$

- Potential energy

$$\mathrm{PE} = mgz \qquad \text{(N-m or joules)}$$

- Internal energy, U, represents all energy changes in a system other than kinetic or potential energy. As specific internal energy, it is written as

$$u = \frac{U}{m} \qquad \text{(N-m/kg or J/kg)}$$

Total energy is the sum of kinetic, potential, and internal energy

$$E = \mathrm{KE} + \mathrm{PE} + U$$

Work is defined as the energy that is expended when a force acts through a displacement. The thermodynamic view of work is that it is a form of energy transfer between a system and its surroundings.

$$\text{Work} \longrightarrow \begin{cases} \text{done } \textbf{by} \text{ a system is positive} \\ \text{done } \textbf{on} \text{ a system is negative} \end{cases}$$

and with regard to a system

$$W = \int_{V_1}^{V_2} P\,dV$$

Other forms of work such as paddle wheel work, electrical work, and work required to elongate or compress a spring must often be considered.

All energy transfer, other than work, across the boundaries of a system is accounted for by the heat transfer, Q. This heat transfer is induced by a temperature difference between the system and the surroundings and flows in the direction of decreasing temperature. Heat transfer is taken as positive when it is transferred **to** the system and negative when it is transferred **from** the system

$$\text{Heat} \longrightarrow \begin{cases} \text{transferred } \textbf{to} \text{ a system is positive} \\ \text{transferred } \textbf{from} \text{ a system is negative} \end{cases}$$

From this, we note that the sign convection pertaining to heat transfer is exactly opposite to the sign convention adopted for the work.

The first law of thermodynamics states that

> The change of total energy (kinetic, potential, and internal) is equal to the net heat transferred to the control mass minus the work done by the control mass.

It may be considered as an alternative statement to the conservation of energy principle:

> Energy may neither be created nor destroyed but may be converted from one form to another.

For a closed system where there is a change of kinetic or potential energy, the first law may be summarized by Equation 3.12

$$E_2 - E_1 = Q - W \tag{3.12}$$

The efficiency of a power cycle is

$$\eta = \frac{W_{\text{cycle}}}{Q_{\text{in}}} = \frac{Q_{\text{in}} - Q_{\text{out}}}{Q_{\text{in}}} = 1 - \frac{Q_{\text{out}}}{Q_{\text{in}}} \tag{3.18}$$

and for a refrigeration cycle, the coefficient of performance takes two forms. For the refrigerator

$$\beta = \frac{Q_{\text{in}}}{W_{\text{cycle}}} = \frac{Q_{\text{in}}}{Q_{\text{out}} - Q_{\text{in}}} \tag{3.19}$$

and for the heat pump

$$\gamma = \frac{Q_{\text{out}}}{W_{\text{cycle}}} = \frac{Q_{\text{out}}}{Q_{\text{out}} - Q_{\text{in}}} \tag{3.20}$$

In the ideal gas law

$$PV = mRT \qquad \text{or} \qquad Pv = RT$$

the gas constant, R, depends on the universal gas constant

$$R = \frac{\bar{R}}{M}$$

where $\bar{R} = 8314\,\text{J/kgmol}$ and M is the molecular weight of the gas.

The enthalpy of a substance is

$$H = U + PV \qquad \text{or} \qquad h = u + Pv$$

and the specific heats are defined by

$$c_v = \left.\frac{du}{dT}\right|_{v=\text{constant}}$$

for constant volume and

$$c_p = \left.\frac{dh}{dT}\right|_{P=\text{constant}}$$

for constant pressure. For gases, the ratio of the specific heats is

$$k = \frac{c_p}{c_v}$$

and the specific heats are related by

$$c_p = c_v + R$$

Processes of an ideal gas depend on

$$PV^n = C \qquad \text{(a constant)}$$

where the exponent, n, can vary. Three processes of an ideal gas are:

- The **isometric** or constant volume process where $n = \infty$
- The **isobaric** or constant pressure process where $n = 0$
- The **isothermal** or constant temperature process where $n = 1$

3.11 Problems

Forms of Energy and Work

3.1: A 16-kg mass has a potential energy of 4500 J with respect to a certain datum and $g = 9.807\,\text{m/s}^2$. Determine the height relative to the datum.

3.2: A 200-N pile driver hammer is released 4 m above the top of a piling. For the pile driver hammer, determine, at the instant of impact, (a) the maximum change in potential energy and (b) the velocity if friction is negligible and $g = 9.81\,\text{m/s.}^2$

3.3: It takes 40 HP to drive an elevator system that has a mass of 3000 kg. Consider that the system as ideal, that is, with negligible losses. With $g = 9.81\ \text{m/s,}^2$ determine (a) the upward uniform velocity and (b) the kinetic energy.

3.4: Consider an elevator weighing 6000 N and a counterweight weighing 4800 N in motion with a velocity of $\hat{V} = 4.25\,\text{m/s}$. The system also contains a braking pulley and the kinetic energy of all cables and rotating parts is negligible. Determine the energy absorbed by the braking pulley if the elevator is stopped in 1.625 m and the gravitational acceleration is $9.81\,\text{m/s.}^2$

3.5: A mass of 1500 kg is accelerated uniformly along a horizontal surface from 10.8 km/h to 10.8 km/h in 4 seconds. Determine (a) the change in kinetic energy, (b) the force required, and (c) the work done by this force over the time interval.

3.6: A 6-gm mass is dropped from rest from a height of 12 m above the earth's surface. What will be its speed upon impact?

3.7: An elevator cab of weight $W = 981\,\text{N}$ moves from street level to the top of a tower, 412 m above the surface. Determine the change in potential energy.

3.8: Determine the energy expended by a climber weighing 600 N in climbing Mount Everest whose altitude is 8851 m.

3.9: The rolling resistance of a railroad car is 10 N per 6250 N of weight of car. Starting from rest, the car rolls down an incline 120 m long on a 5% grade. It then continues along a horizontal stretch. How far will the car roll on the horizontal stretch before coming to rest?

3.10: A mass of 4 kg moves at a velocity of 5 m/s at a height of 40 m above sea level. Determine (a) its weight, (b) its kinetic energy, and (c) its potential energy.

3.11: A disgruntled New York Mets fan drops a baseball with a mass of 143 g from a window in the Empire State Building at an elevation of 305 m. The ball reaches a terminal velocity of 40 m/s at the instant that it strikes the ground. Determine the increase in internal energy.

3.12: An object having a mass of 12 kg with an initial downward velocity of 35 m/s falls from a height of 125 m. The acceleration of gravity is $9.78\,\text{m/s}^2$ and the air resistance is negligible. Determine the velocity at impact with the ground.

3.13: A 2-kg piece of metal is moving horizontally at 8 m/s. If $g = 9.81\,\text{m/s}^2$, determine (a) the change in velocity necessary for an increase of kinetic energy of 12 N-m and (b) the change in elevation for an increase of potential energy of 16 N-m.

3.14: The acceleration of gravity varies as a function of the elevation above sea level in accordance with

$$g = 9.807\,\text{m/s}^2 - 3.32 \times 10^{-6}\,\text{s}^{-2} z$$

where z is in meters. A satellite with a mass of 300 kg is boosted to an altitude of 450 km above the surface of the earth. Determine the work required.

3.15: A mass of 10 kg is 200 m above a given datum where $g = 9.75\,\text{m/s}^2$. Determine the gravitational force in newtons and the potential energy in joules.

3.16: The work required to stretch a spring 8.59 cm from its free length is 362 J. The elemental equation that pertains to the spring is

$$F = k_s y$$

where k_s is the **spring constant** in N/m that relates the perturbing force, F in N to the stretch or compression of the spring, y in meters. Determine the value of the spring constant.

3.17: A weight of 480 N is attached on to the end of a rope 6 m below a frictionless pivot. The weight is pushed so that the rope makes an angle of 36.87° from the vertical. Determine (a) the gain in potential energy and (b) the mass if $g = 9.81\,\text{m/s}^2$.

3.18: A quantity of gas at a constant pressure of 16 Pa is compressed from $6\,\text{m}^3$ to $1.5\,\text{m}^3$. Determine the work done.

3.19: A sample of gas is expanded in a process at a constant pressure, P, from a volume of $1\,\text{m}^3$ to a volume of $4\,\text{m}^3$. The amount of work done is 432 J. Determine the pressure.

3.20: Work is being done by a substance in a non-flow process in accordance with

$$P = \frac{100}{V} \qquad (\text{N/m}^2)$$

where V is in m^3. Determine the work done on or by the substance as the pressure increases from 0.8 to 8 bar.

The Energy Balance for Closed Systems

3.21: Consider a nonflow process from state point 1 to state point 2 where 100 J of work is done on the system and where 50 J of heat is transferred from the system. The initial internal energy is 50 J and potential and kinetic energy changes are negligible. Determine the final internal energy.

3.22: Consider a nonflow process from state point 1 to state point 2 where 100 J of work is done by the system and where 100 J of heat is transferred from the system. There is no change in internal energy and the increase in potential energy is 100 J. Determine the change in kinetic energy.

3.23: Consider a nonflow process from state point 1 to state point 2 where 100 J of work is done on the system and where 100 J of heat is transferred to the system. The initial internal energy is 80 J, the potential energy change is negligible and the kinetic energy increases by 40 J. Determine the final internal energy.

3.24: Consider a nonflow process from state point 1 to state point 2 where 100 J of heat is transferred to the environment, the internal energy increases by 20 J, the potential energy increases by 40 J, and the kinetic energy decreases by 40 J. Determine the magnitude and direction of the work done.

3.25: Consider a nonflow process from state point 1 to state point 2 where 20 J of work is done by the system and where 100 J of heat is transferred from the system. The final internal energy is 50 J, the potential energy increases by 40 J and the kinetic energy increases by 20 J. Determine the initial internal energy.

3.26: Consider a nonflow process from state point 1 to state point 2 where 50 J of work is done by the system. The change in internal energy is 100 J, the potential energy increases by 40 J and the change in kinetic energy is negligible. Determine the heat transferred.

3.27: Consider a nonflow process from state point 1 to state point 2 where 40 J of work is done by the system and where 40 J of heat is transferred to the system. The initial internal energy is 80 J and there is no increase in either potential or kinetic energy. Determine the final internal energy.

3.28: Consider a nonflow process from state point 1 to state point 2 where 40 J of work is done by the system and where there 100 J of heat is transferred from the system. The change in initial internal energy is 40 J and the kinetic energy increases by 60 J. Determine the change in potential energy.

3.29: Consider a nonflow process from state point 1 to state point 2 where 100 J of work is transferred to the environment, the internal energy increases by 50 J, the potential energy increase is negligible and no heat is transferred from or to the environment. Determine the change in kinetic energy.

3.30: Consider a nonflow process from state point 1 to state 2 where 200 J of heat is transferred to the system. Determine the work done if the internal energy decreases by 50 J, the potential energy increases by 50 J, and the kinetic energy increases by 50 J.

3.31: Consider a nonflow process from state point 1 to state point 2 where 40 kJ of work is done by the system and where 25 kJ of heat is transferred from the system. The system is returned to state 1, and the heat transferred to the system is 20 J. Determine the work done in the return process.

3.32: Consider a nonflow process from state point 1 to state point 2 where 50 kJ of work is done on the system and 100 kJ of heat is transferred to the system. The system is returned to state 1 and the heat transferred from the system is 20 J. Determine the work done in the return process.

3.33: Consider a nonflow process from state 1 to state 2 where 100 kJ of work is done by the system and no heat is transferred from or to the system. The system is returned to state 1 and the work done on the system is 80 kJ. Determine the heat transferred in the return process.

3.34: Consider a nonflow process from state 1 to state 2 where 12 kJ of work is done by the system and where 40 kJ of heat is transferred to the system. The process continues to state 3 with a heat transfer of 18 kJ to the system. During the return to state 1, 24 kJ of work is done by the system and the change of energy is $\Delta E_{31} = 12$ kJ. Determine (a) Q_{31}, (b) W_{23}, (c) ΔE_{12}, and (d) ΔE_{23}.

3.35: Consider a nonflow process from state 1 to state 2 where 50 kJ of work is done on the system and 80 kJ of heat is transferred to the system. The process continues to state 3 with a heat transfer of 25 kJ from the system. During the return to state 1, 10 kJ of work is done on the system and the heat transferred to the system is 35 kJ. Find (a) ΔE_{12}, (b) ΔE_{23}, (c) ΔE_{31}, and (d) W_{23}.

The Ideal Gas

3.36: In a constant volume nonflow process, 2 kg of oxygen is heated from 30°C until the pressure is tripled. Treating oxygen as an ideal gas, determine (a) its final temperature and (b) the work done by the oxygen.

3.37: In a nonflow process 10 kg of air at 25°C and 1 bar are heated at constant pressure until the volume is 18 m^3. Treating air as an ideal gas and assuming constant specific heats, find (a) the change in internal energy, (b) the change in enthalpy, (c) the work done, and (d) the heat added.

3.38: A tank contains air at 50 bar and 25°C. The air may be assumed as an ideal gas and an amount measuring 20 m^3 at 1 bar and 20°C is removed. The pressure in the tank is then found to be 20 bar when the temperature is 25°C. Determine the volume of the tank.

3.39: The temperature of an ideal gas remains constant while the pressure increases from 1 to 4 bar. The initial volume is 2.88 m^3. Determine (a) the final volume and (b) the work done.

3.40: An automobile tire contains a volume of air at a gage pressure of 2 bar and 24°C. The barometric pressure is 29.75 in of mercury. Under running conditions, the temperature of the air in the tire increases to 76°C. Determine the gage pressure.

3.41: An automobile tire has a volume of 0.052 m^3 and contains air at a pressure of 2 bar and 25°C. Determine the mass of air in the tire.

3.42: For the automobile tire in Problem 3.41, determine the pressure in the tire if the automobile is left out so that the temperature falls to 0°C.

3.43: For the automobile tire in Problem 3.42, how many kilograms of air must be added to bring the pressure to 2.18 bar when the air temperature is 0°C?

3.44: Determine the density of oxygen at atmospheric pressure and 35°C.

3.45: At 37°C, 20 kg of propane occupies a volume of 3.125 m^3. Determine (a) the pressure and (b) the mass of propane that must be added to bring the pressure to 1.20 MPa.

3.46: Ethane contained in a 0.625 m^3 tank at an absolute pressure of 225 kPa and 77°C begins to escape: 0.25 kg of the ethane leaks out of the tank and the temperature falls to 27°C. Determine the gage pressure of the ethane remaining in the tank if the barometer is at 99 kPa.

3.47: A balloon is filled with helium at 25°C and 30.12 in of mercury until the volume becomes 1000 m^3. Determine (a) the mass of helium required and (b) the volume if the balloon rises to an altitude where $P = 28$ in of mercury and the temperature is 0°C.

3.48: Suppose that hydrogen is substituted for helium in Problem 3.47. All other conditions remain the same. Determine (a) the mass of hydrogen required and (b) the volume if the balloon rises to an altitude where $P = 28$ in of mercury and the temperature is 0°C.

3.49: A closed rigid container with a volume of 4 m^3 contains a gas having a molecular weight of 32 kg/kgmol at 1 MPa and 57°C. The container develops a leak and gas is lost leaving a pressure of 0.4 MPa and a temperature of 27°C. Determine the volume of gas lost if it is reckoned at a pressure of 1 bar and a temperature of 25°C.

3.50: A rigid container weighing 200 N contains propane at 1.25 MPa and 27°C. After some propane has been used, the contents of the container are at 5.125 bar and 0°C and the container was then found to weigh 140 N. Determine (a) the mass of the container, (b) the internal volume of the propane container, and (c) the kilograms of propane used.

4

Properties of Pure, Simple Compressible Substances

Chapter Objectives

- To introduce and describe the state postulate.
- To present several relationships that relate the pressure, volume, and temperature of a pure substance.
- To develop thermodynamic property data including some hints for phase determination.
- To investigate real gas behavior.
- To consider the principle of corresponding states.
- To examine equations of state.
- To provide data for the polytropic process of an ideal gas.

4.1 The State Postulate

We recall that a pure substance is one that is uniform and homogeneous in its chemical structure. It may exist in a single phase, or, if each phase is separated by a phase boundary, it may exist as a two- or three-phase mixture. However, a mixture of liquid water and an air-water vapor mixture is not a pure substance even though all its components remain in the gaseous phase (Figure 4.1).

When a system undergoes a change of state, a large number of properties may change and, because **intrinsic properties** are those that are functions of molecular behavior, we expect that all intrinsic properties are functionally related. For example, suppose that the dependent intrinsic property, z_o, is a function of p independent intrinsic properties, $z_1, z_2, z_3, \ldots, z_p$. Then

$$z_o = f(z_1, z_2, z_3, \ldots, z_p)$$

and once the values of $z_1, z_2, z_3, \ldots, z_p$ are specified, the value of z_o becomes fixed. However, we must be able to answer the question, "what is the value of p?"

A **simple compressible substance** is defined as any pure substance where nonrelevant effects such as surface tension, velocity or motion, electrical and thermal conductivity, modulus of elasticity, and viscosity can be neglected. The words "simple compressible" imply that the only possible work interaction between the substance acting as the system and the surroundings that can change the state of the system is Pdv work. We note that, in

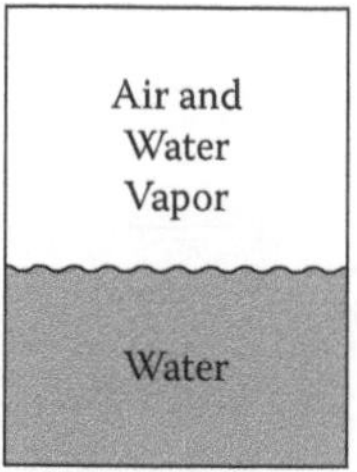

FIGURE 4.1
A mixture of water and air containing water is not a pure substance.

general, for this case, the use of the specific volume and other specific quantities, permits the extent of the system to be removed from consideration.

For a pure, simple compressible, substance, repeated observations have shown that two independent properties are necessary and sufficient to provide the equilibrium state of a system. While this observation cannot be derived from the basic postulates of the classical approach to thermodynamics, no violations have been observed. Thus, in the ideal gas model for a particular gas with a specified value of R, two properties specify a third

$$P = f_1(v, T), \quad v = f_2(P, T), \quad \text{and} \quad T = f_3(P, v)$$

The **state postulate** supplies the number of variables that are required to completely specify the state of a single-phase or homogeneous system. In general

> If there are n relevant work interactions between a system and its surroundings, the number of independent intensive and intrinsic properties required to specify the state of the system is $n + 1$.

In particular, in the case of a simple compressible substance where Pdv is the only relevant work interaction, $n = 1$ and it is necessary to fix two independent properties to completely specify the equilibrium state.

4.2 *P*-*v*-*T* Relationships

4.2.1 Introduction

The state postulate asserts that if P, v, and T are the properties of a simple compressible substance, then, for a single phase system, the pressure, P can be considered a function of the specific volume, v, and the temperature, T

$$P = f(v, T) \tag{4.1}$$

The graph of such a function is called a **surface** and, in particular, the functional relationship of Equation 4.1 is a P-v-T surface.

4.2.2 The *P*-*v*-*T* Surface

Figure 4.2 shows a P-v-T surface for a substance that expands upon freezing. Water is a typical example of such a substance. However, most substances contract upon freezing and the P-v-T surface for these substances is shown in Figure 4.3. We note that the coordinates of the P-v-T surface are pressure, specific volume, and temperature.

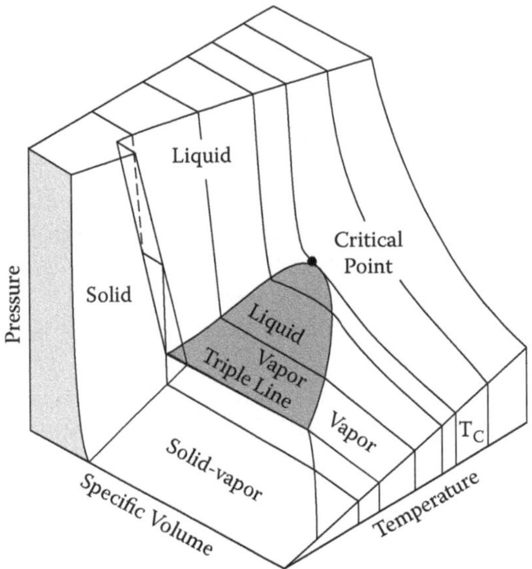

FIGURE 4.2
The pressure-specific volume-temperature surface for a substance that expands upon freezing.

In Figures 4.2 and 4.3, we can see that there are three regions that contain a single phase. These are the **solid, liquid**, and **vapor phases**. Three two-phase regions are located between the single-phase regions and these are called the **solid-liquid**, the **liquid-vapor**, and the **solid-vapor regions**.

4.2.3 Single-Phase Regions

In the single-phase regions of Figures 4.2 and 4.3, the state is fixed by any two of the properties, P, v, or T because all of these are independent.

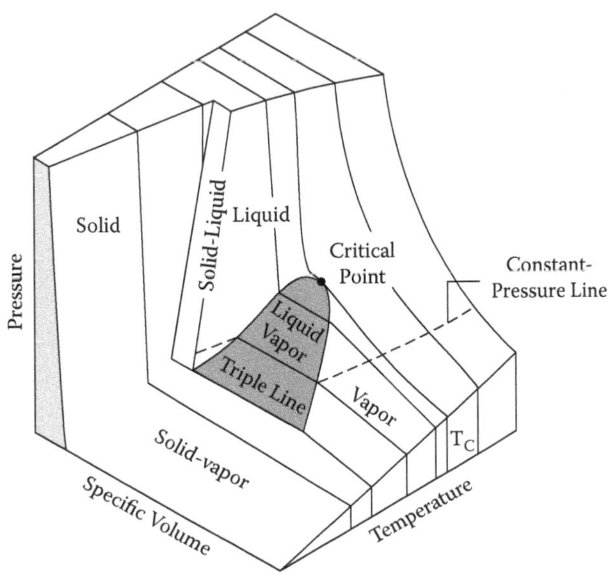

FIGURE 4.3
The pressure-specific volume-temperature surface for a substance that contracts upon freezing.

4.2.4 Two-Phase Regions

In the two-phase regions of Figures 4.2 and 4.3, two phases coexist at equilibrium. These are the solid-liquid, the liquid-vapor and the solid-vapor regions. In these two-phase regions, pressure and temperature are *not* independent and the state is fixed by specific volume *and* pressure *or* temperature.

The change of phase from solid to liquid is referred to as **melting** and the change from liquid to solid is called **freezing** or **solidification**. The change from liquid to vapor is called **vaporization** or **boiling** and the change from vapor to liquid is called **condensation**. The change of phase from solid to vapor is known as **sublimation.**

The quantity of heat required to change a solid to a liquid is called the **latent heat of fusion** or merely the **heat of fusion**. Similarly, the quantity of heat required to change a liquid to a vapor is called the **latent heat of vaporization** or merely the **heat of vaporization.** The quantity of heat required to change a solid to a vapor is called the **heat of sublimation.**

The magnitudes of the latent heats depends upon the temperature at which the phase change occurs. For water at 0°C (273 K), the latent heat of fusion is

$$\lambda_f = 333.7\,\text{kJ/kg}$$

and at 100°C (373 K), the latent heat of vaporization is

$$\lambda_v = 2257.1\,\text{kJ/kg}$$

4.2.5 Three-Phase Regions

Three phases can coexist in equilibrium along a line shown in Figures 4.2 and 4.3, which is called the **triple line** and represents the **triple state**. For water, this state is found at the triple point where the pressure is 611.2 Pa and the temperature is 273.16 K (0.16°C). We may recall that the triple point of water was used to establish the Kelvin temperature scale.

4.2.6 The *P*-*T* Surface

The projection of the P-v-T surface normal to the v-axis is called the ***P*-*T* surface** or **phase diagram**. The phase diagram for a substance that expands upon freezing is shown in Figure 4.4. The triple state is represented by a line on the P-v-T surface which is, in reality, a

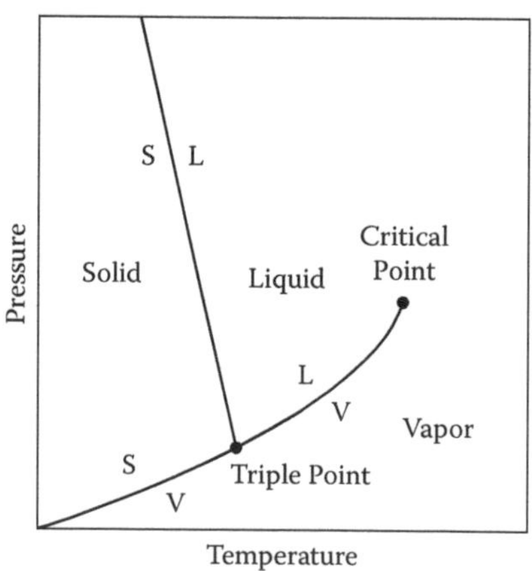

FIGURE 4.4
Phase diagram for water (not to scale).

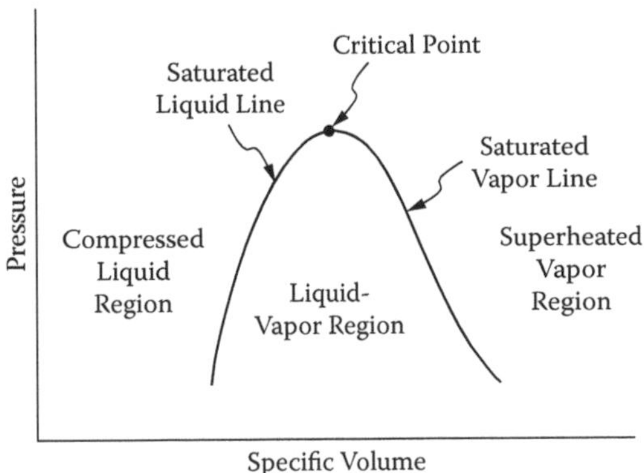

FIGURE 4.5
The P-v surface for water showing the saturated liquid and saturated vapor lines and the critical point.

line that proceeds into the plane of the paper and leaves at a point on the P-T diagram. This is the reason for designating the **triple point** as the point where all of the lines intersect. The lines between the phases are called **saturation lines.**

We may note, in particular, the saturation line between the liquid and vapor planes that begins at the triple point and ends at a point called the **critical point** having a pressure designated as P_c and a temperature designated as T_c. The horizontal and vertical lines from this point are called the **critical pressure isobars** and the **critical temperature isotherms,** respectively.

4.2.7 The *P*-*v* Surface

Water, a substance that expands upon freezing, can be used to provide a typical illustration of the P-v surface shown in Figure 4.5. Here, we see two saturation lines. The line that divides the liquid region, usually referred to as the **compressed liquid** or **subcooled liquid** region and the liquid-vapor region, is called the **saturated liquid line**. A compressed liquid is a liquid *at a particular temperature* whose pressure is higher than the saturation pressure at that temperature or liquid *at a particular pressure* whose temperature is lower than the saturation temperature at that pressure. Properties associated with the saturated liquid line are given the subscript f (f for "fluid") and typical examples are the specific volume, v_f, the specific internal energy, u_f and the specific enthalpy, h_f.

The line that divides the liquid-vapor region and the vapor region, called the **superheated vapor region,** is called the **saturated vapor line**. Properties associated with the saturated vapor are given the subscript g (g for "gas") and typical examples are the specific volume, v_g, the specific internal energy, u_g and the specific enthalpy, h_g. The difference in properties between the saturated liquid and saturated vapor lines are given the subscript fg (for "the difference between gas and fluid") and typical examples are u_{fg} and h_{fg}.

In Figure 4.5, we can observe that the saturated liquid and saturated vapor lines intersect at the critical point and form a **vapor dome**. The area under the vapor dome represents the so-called **wet region** because the liquid-vapor region contains both liquid and vapor to varying degrees. Thermodynamic properties under the vapor dome (with the saturated liquid and saturated vapor lines as extremes) depend on a quantity known as the **quality** that is the mass fraction of the vapor present.

$$x = \frac{m_{\text{vap}}}{m_{\text{tot}}} = \frac{m_g}{m_f + m_g} \tag{4.2}$$

where the subscripts represent the amounts of liquid and vapor present. Sometimes, y, the amount of liquid or **moisture** present, is employed

$$y = \frac{m_{\text{liq}}}{m_{\text{tot}}} = \frac{m_f}{m_f + m_g} \tag{4.3}$$

Of course, there is a relationship between x and y

$$x + y = 1$$

4.2.8 Some Hints For Phase Determination

The location of the phase of a substance on the P-v surface can be facilitated by noting that the saturation pressure and the saturation temperature are coupled.

- To be in the superheated vapor region, the temperature must be greater than the saturation temperature.
- To be in the liquid-vapor region, the temperature must be equal to the saturation temperature and the pressure must be equal to the saturation pressure.
- To be in the subcooled or compressed liquid region, the temperature must be less than the saturation temperature.

4.3 Thermodynamic Property Data

4.3.1 Table Arrangement

Thermodynamic property data are available in listed equations, graphs, computer codes, or tables such as those published by Keenan et al. (1969) and ASME (1967). In the steam tables, extractions of which are included here in Tables A.3 to A.5 in Appendix A, the data are divided into convenient increments and linear interpolation may be employed to provide results to an acceptable accuracy. Tables for Refrigerant R-134a (a relatively new non-CFC refrigerant) extracted from ASHRAE (1994) are given in the same format as the steam tables in Tables A.6 to A.8 in Appendix A.

The tables are categorized into:

- Saturation tables (Tables A.3 and A.4 for water-steam and Tables A.6 and A.7 for refrigerant R-134a),
- Superheated vapor tables (Table A.5 for steam and Table A.8 for refrigerant R-134a), and
- Compressed liquid tables.

Details for the saturation and superheated vapor tables will be considered in Sections 4.3.2 and 4.3.3, which now follow, and an alternative method for the evaluation of compressed liquid properties is provided in Section 4.3.4.

4.3.2 The Saturation Tables

The saturation tables provide values of specific volume, v, specific internal energy, u, specific enthalpy, h, and specific entropy, s, a property that will be introduced to our study in

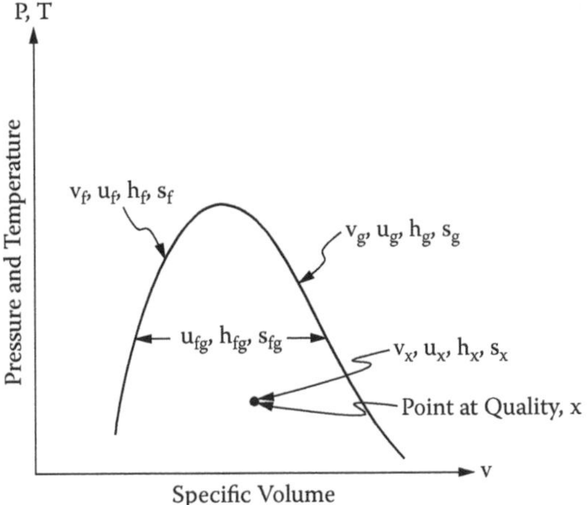

FIGURE 4.6
The basis for the saturation tables.

Chapter 7. The properties are listed for both the saturated liquid and saturated vapor states that form the boundaries of the vapor dome and are to be used for the liquid-vapor region where either the pressure, P, or the temperature, T, must be provided. Table A.3 is the **temperature table** where the saturation pressure is also given to the immediate right of the saturation temperature. Table A.4 is the **pressure table** where the saturation temperature is listed to the right of the saturation pressure.

We recall that the subscripts f and g, respectively, designate a property of a saturated liquid and a saturated vapor. The values that designate the difference between saturated vapor and saturated liquid are designated by the subscript fg.

At points under the vapor dome that are neither saturated liquid nor saturated vapor, the property value is obtained by adding the contributions of the two phases. For example, with the quality and moisture fraction defined by Equations 4.2 and 4.3, we find that the volume in the liquid-vapor region having a quality, x, will be

$$V = V_{\text{liq}} + V_{\text{vap}}$$

and with the specific volume at point x designated as v_x

$$v_x = \frac{V_{\text{liq}}}{m} + \frac{V_{\text{vap}}}{m}$$

and then

$$mv_x = m_f v_f + m_g v_g$$

Because $m_f = m - m_g$

$$mv_x = (m - m_g)v_f + m_g v_g$$

and then a division by the total mass provides

$$v_x = \left(1 - \frac{m_g}{m}\right) v_f + \frac{m_g}{m} v_g = (1 - x)v_f + xv_g \qquad (4.4a)$$

If the difference between v_g and v_f is

$$v_{fg} = v_g - v_f$$

then Equation 4.4a can be rearranged to

$$v_x = v_f + xv_{fg} \tag{4.4b}$$

In those cases where values of v_{fg} are not listed, Equation 4.4a may be employed.

In the cases of specific internal energy, specific enthalpy and specific entropy, which is introduced in Chapter 7,

$$u_x = (1 - x)u_f + xu_g \tag{4.5a}$$

or

$$u_x = u_f + xu_{fg} \tag{4.5b}$$

$$h_x = (1 - x)h_f + xh_g \tag{4.6a}$$

or

$$h_x = h_f + xh_{fg} \tag{4.6b}$$

and

$$s_x = (1 - x)s_f + xs_g \tag{4.7a}$$

or

$$s_x = s_f + xs_{fg} \tag{4.7b}$$

If the property of a substance in the liquid-vapor region is given, the quality can be found via

$$x = \frac{v_x - v_f}{v_g - v_f} = \frac{u_x - u_f}{u_{fg}} = \frac{h_x - h_f}{h_{fg}} = \frac{s_x - s_f}{s_{fg}} \tag{4.8}$$

Here, v_x, u_x, h_x, or s_x is the given property.

Example 4.1 At 160°C, 5 kg of water are enclosed in a rigid container having a volume of 0.5 m^3 (Figure 4.7). Determine (a) the pressure, (b) the enthalpy, and (c) the mass and volume of the vapor.

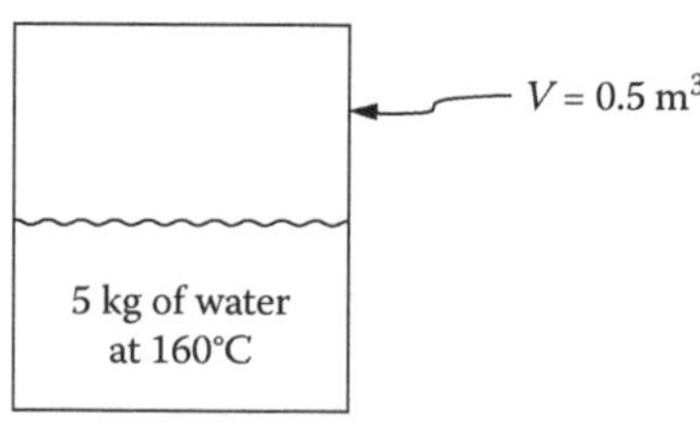

FIGURE 4.7
The control volume for Example 4.1.

Solution

Assumptions and Specifications

(1) The 5 kg of water is the system.
(2) There is vapor present (this will require checking).
(3) The given conditions of V and T represent an equilibrium state.

The specific volume of the mixture will be

$$v_{\text{mix}} = \frac{V}{m} = \frac{0.50\,\text{m}^3}{5\,\text{kg}} = 0.10\,\text{m}^3/\text{kg}$$

Because the temperature is given, we can find from Table A.3 that

$$v_f = 1.102 \times 10^{-3}\,\text{m}^3/\text{kg} \qquad \text{and} \qquad v_g = 0.3068\,\text{m}^3/\text{kg}$$

We note that the mixture specific volume is between v_f and v_g. Thus, the mixture exists in the liquid-vapor phase and assumption 2 has been verified.

(a) The pressure is the saturation pressure and we can read from Table A.3 at 160°C.

$$P_{\text{sat}} = 0.618\,\text{MPa} \qquad (618\,\text{kPa}) \Longleftarrow$$

(b) The enthalpy value depends on the quality that we can find from the specific volume data. From Equation 4.8

$$\begin{aligned} x &= \frac{v_x - v_f}{v_g - v_f} \\ &= \frac{0.1000\,\text{m}^3/\text{kg} - 1.102 \times 10^{-3}\,\text{m}^3/\text{kg}}{0.3068\,\text{m}^3/\text{kg} - 1.102 \times 10^{-3}\,\text{m}^3/\text{kg}} \\ &= \frac{0.0989\,\text{m}^3/\text{kg}}{0.3057\,\text{m}^3/\text{kg}} = 0.324 \end{aligned}$$

Then, with h_f and h_{fg} from Table A.3

$$h_f = 675.2\,\text{kJ/kg} \qquad \text{and} \qquad h_{fg} = 2081.7\,\text{kJ/kg}$$

the value of h_x is found from Equation 4.6b

$$\begin{aligned} h_x &= h_f + x h_{fg} \\ &= 675.2\,\text{kJ/kg} + (0.324)(2081.7\,\text{kJ/kg}) \\ &= 675.2\,\text{kJ/kg} + 674.5\,\text{kJ/kg} \\ &= 1349.7\,\text{kJ/kg} \Longleftarrow \end{aligned}$$

(c) The mass of vapor, m_g, depends on the quality

$$m_g = x m_{\text{mix}} = (0.324)(5\,\text{kg}) = 1.620\,\text{kg} \Longleftarrow$$

and the volume of the vapor will be

$$V_g = m_g v_g = (1.620\,\text{kg})(0.3068\,\text{m}^3/\text{kg}) = 0.497\,\text{m}^3 \Longleftarrow$$

We note that the vapor occupies almost the entire container.

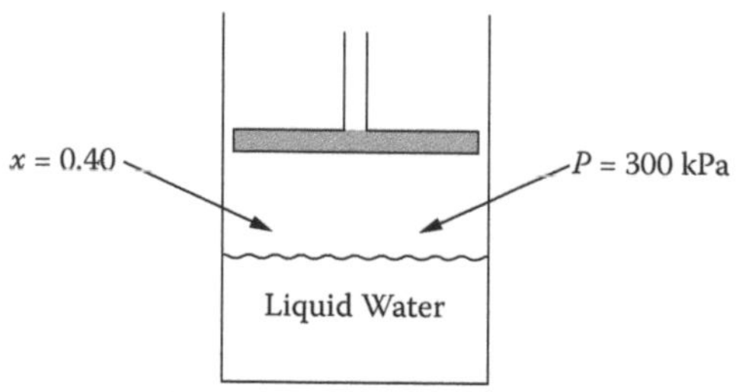

FIGURE 4.8
The control volume for Example 4.2.

Example 4.2 A mass of saturated liquid water is enclosed in a container where the pressure is held at 300 kPa (Figure 4.8). Heat is added to the water until the quality is 40%. Determine (a) the initial temperature, (b) the final temperature, (c) the final specific volume, and (d) the final internal energy.

Solution

Assumptions and Specifications

(1) The water and any vapor produced by the heat addition constitutes the system.
(2) Both the initial state and the state after the heat addition represent equilibrium states.

(a) At 300 kPa. Table A.4 gives a saturation temperature of

$$133.5°\text{C} \Longleftarrow$$

(b) A vaporization of some of the water to form a mixture at a certain quality merely shows that the mixture has moved at constant pressure and temperature from the saturated liquid line to a point somewhere under the vapor dome. Thus, the temperature remains at the saturation value

$$133.5°\text{C} \Longleftarrow$$

(c) With v_f and v_g taken from Table A.4 at 300 kPa

$$v_f = 1.073 \times 10^{-3}\,\text{m}^3/\text{kg} \qquad \text{and} \qquad v_g = 0.6056\,\text{m}^3/\text{kg}$$

the final specific volume at a quality of 40% can be found from Equation 4.4a

$$\begin{aligned} v_x &= (1-x)v_f + xv_g \\ &= (1-0.40)(1.073 \times 10^{-3}\,\text{m}^3/\text{kg}) + (0.40)(0.6056\,\text{m}^3/\text{kg}) \\ &= 6.438 \times 10^{-4}\,\text{m}^3/\text{kg} + 0.2422\,\text{m}^3/\text{kg} = 0.2429\,\text{m}^3/\text{kg} \Longleftarrow \end{aligned}$$

(d) With u_f and u_{fg} from Table A.4

$$u_f = 560.83\,\text{kJ/kg} \qquad \text{and} \qquad u_{fg} = 1982.1\,\text{kJ/kg}$$

the value of u_x is found from Equation 4.5b

$$\begin{aligned} u_x &= u_f + xu_{fg} \\ &= 560.83\,\text{kJ/kg} + (0.40)(1982.1\,\text{kJ/kg}) \\ &= 560.83\,\text{kJ/kg} + 792.84\,\text{kJ/kg} = 1353.7\,\text{kJ/kg} \Longleftarrow \end{aligned}$$

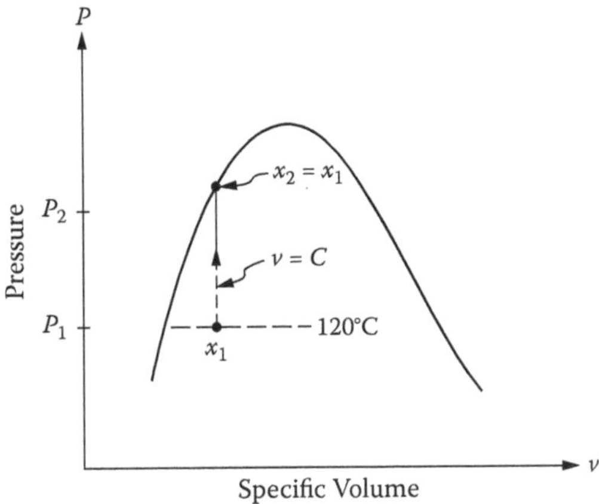

FIGURE 4.9
Pressure-volume diagram for Example 4.3.

The following example shows that, under certain conditions, the addition of heat can cause a liquid-vapor mixture to become saturated liquid.

Example 4.3 A closed rigid vessel contains water in the form of a liquid-vapor mixture at 120°C. The specific volume of the mixture is 0.002213 m³/kg. Heat is added to the mixture until all of the vapor condenses and the mixture turns into a saturated liquid. Determine (a) the initial pressure of the mixture, (b) the initial quality of the mixture, (c) the final pressure, (d) the final temperature, and (e) the heat added during the process.

Solution

Assumptions and Specifications

(1) The liquid vapor mixture constitutes the system.

(2) Both the initial state and the state after the heat addition represent equilibrium states.

(a) Because the initial state is a mixture of liquid and vapor, the initial pressure must be the saturation pressure corresponding to 120°C. Table A.3 gives the pressure as

$$P_1 = 198.55\,\text{kPa} \Longleftarrow$$

(b) With v_f and v_g taken from Table A.3 at 120°C

$$v_f = 1.060 \times 10^{-3}\,\text{m}^3/\text{kg} \qquad \text{and} \qquad v_g = 0.8921\,\text{m}^3/\text{kg}$$

we calculate the quality of the mixture at point 1 using Equation 4.8

$$\begin{aligned} x &= \frac{v_x - v_f}{v_g - v_f} \\ &= \frac{2.213 \times 10^{-3}\,\text{m}^3/\text{kg} - 1.060 \times 10^{-3}\,\text{m}^3/\text{kg}}{0.8921\,\text{m}^3/\text{kg} - 1.060 \times 10^{-3}\,\text{m}^3/\text{kg}} \\ &= \frac{1.153 \times 10^{-3}\,\text{m}^3/\text{kg}}{0.8910\,\text{m}^3/\text{kg}} = 0.00129 \Longleftarrow \end{aligned}$$

This shows that at point 1, just a trace of vapor is present.

(c) Because the container is closed and rigid, the final specific volume must equal the initial specific volume and we find that at point 2,

$$v_2 = 2.213 \times 10^{-3}\,\text{m}^3/\text{kg}$$

The final pressure is found from Table A.4 with interpolation

$$P_2 = 21.012\,\text{MPa} \Longleftarrow$$

(d) Noting that the final state lies on the saturated liquid line, the final temperature is read from Table A.4 with interpolation as

$$T_2 = 369.8°\text{C} \Longleftarrow$$

(e) Because no work is done in a constant volume process, the first law of thermodynamics gives

$$\frac{Q}{m} = u_2 - u_1$$

The initial internal energy is found from Table A.3 at 120°C. With

$$u_f = 503.28\,\text{kJ/kg} \qquad \text{and} \qquad u_{fg} = 2025.4\,\text{kJ/kg}$$

the value of u_1 is found from Equation 4.5b

$$\begin{aligned} u_1 &= u_f + x u_{fg} \\ &= 503.28\,\text{kJ/kg} + (0.00129)(2025.4\,\text{kJ/kg}) \\ &= 503.28\,\text{kJ/kg} + 2.61\,\text{kJ/kg} = 505.89\,\text{kJ/kg} \Longleftarrow \end{aligned}$$

and with u_2 read from Table A.4 at 369.8°C

$$u_2 = u_f = 1839.7\,\text{kJ/kg}$$

Hence

$$\begin{aligned} \frac{\dot{Q}}{m} &= u_2 - u_1 \\ &= 1839.7\,\text{kJ/kg} - 505.89\,\text{kJ/kg} \\ &= 1333.8\,\text{kJ/kg} \Longleftarrow \end{aligned}$$

The process is shown on a P-v diagram in Figure 4.10. It is interesting to note that if the specific volume at state 1 is less than v_c ($v_1 < v_c$), the constant volume heat addition results in a saturated liquid (state 2). However, if the initial specific volume at state 1′ is greated than v_c ($v_{1'} > v_c$), the constant volume heat addition results in a saturated vapor (state 2′).

4.3.3 The Superheated Vapor Tables

In the single-phase superheat region, two properties are needed to specify the equilibrium state. The variables chosen are the temperature, T, and the pressure, P. Table A.5 gives values of v, u, h, and s at several locations that form a grid of values based on temperature

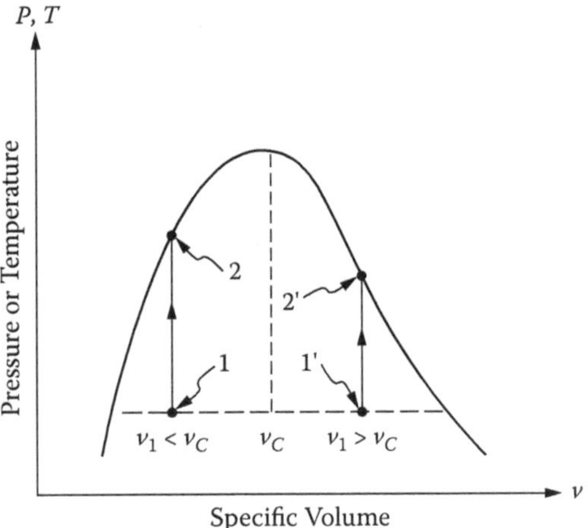

FIGURE 4.10
Constant volume processes in a liquid-vapor mixture.

and pressure. For points that do not fall within the grid, linear interpolation provides reasonable and accurate values.

Example 4.4 Determine (a) the internal energy of steam at 1.00 MPa and 340°C and (b) the enthalpy of steam at 3.00 MPa and 264°C.

Solution

Assumptions and Specifications

(1) The steam constitutes a closed system.

(2) The states represent equilibrium states.

(a) We note in Table A.5, that a point for 1.00 MPa and 340°C is not provided. Thus, we must interpolate using values at 1.00 MPa and 300°C and 1.00 MPa and 350°C. From Table A.5 at 1.00 MPa and 350°C, read

$$u = 2874.7\,\text{kJ/kg}$$

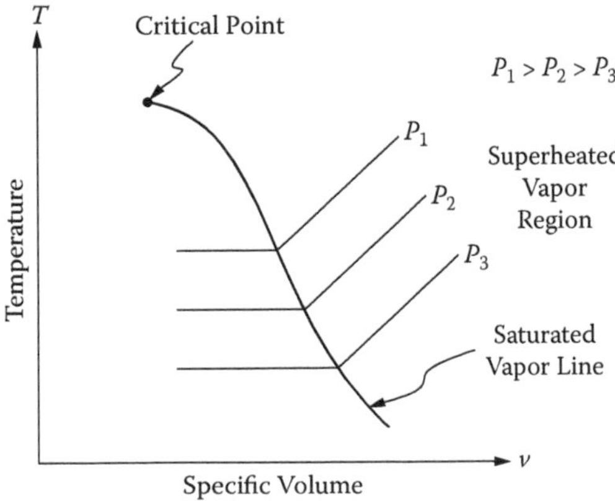

FIGURE 4.11
The superheated vapor region.

and at 1.00 MPa and 300°C, read

$$u = 2793.1\,\text{kJ/kg}$$

The value of the internal energy at 1.00 MPa and 340°C can now be found by linear interpolation. We establish an interpolation fraction, F, as

$$F = \frac{340°\text{C} - 300°\text{C}}{350°\text{C} - 300°\text{C}} = 0.80$$

and the value of u will be

$$\begin{aligned} u &= 2793.1\,\text{kJ/kg} + F(2874.7\,\text{kJ/kg} - 2793.1\,\text{kJ/kg}) \\ &= 2793.1\,\text{kJ/kg} + (0.80)(81.6\,\text{kJ/kg}) \\ &= 2793.1\,\text{kJ/kg} + 65.3\,\text{kJ/kg} = 2858.4\,\text{kJ/kg} \Longleftarrow \end{aligned}$$

(b) This time, we note that in Table A.5, that a point for 3.00 MPa and 264°C is not provided. Thus, we must interpolate using values at 3.00 MPa and 250°C and 3.00 MPa and 300°C. From Table A.5 at 3.00 MPa and 300°C, read

$$h = 2994.1\,\text{kJ/kg}$$

and at 3.00 MPa and 250°C, read

$$h = 2854.9\,\text{kJ/kg}$$

Here, the interpolation fraction is

$$F = \frac{264°\text{C} - 250°\text{C}}{300°\text{C} - 250°\text{C}} = 0.28$$

and the value of the enthalpy at 3.00 MPa and 264°C will be

$$\begin{aligned} h &= 2854.9\,\text{kJ/kg} + F(2994.1\,\text{kJ/kg} - 2854.9\,\text{kJ/kg}) \\ &= 2854.9\,\text{kJ/kg} + (0.28)(139.2\,\text{kJ/kg}) \\ &= 2854.9\,\text{kJ/kg} + 39.0\,\text{kJ/kg} = 2893.9\,\text{kJ/kg} \Longleftarrow \end{aligned}$$

4.3.4 The Compressed or Subcooled Liquid Region

Although a compressed liquid table is usually provided in compendia of thermodynamic properties, we provide a close approximation. This approximation is based upon the fact that that there is little variation in the values of v and u at pressures above saturation values. Thus,

$$v(P, T) \approx v(T) \tag{4.9a}$$

and

$$u(P, T) \approx u(T) \tag{4.9b}$$

The approximate value of the enthalpy, h, in the compressed (subcooled) liquid region can be obtained for a given temperature, T, having an associated saturation pressure, P_{sat},

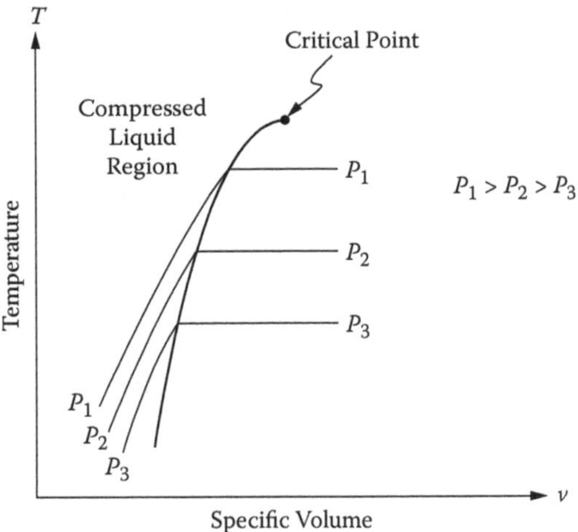

FIGURE 4.12
The compressed liquid region.

by noting that

$$h(P, T) = u(P, T) + Pv(P, T)$$

Then, with the approximations expressed in Equation 4.9, values of the enthalpy in the compressed liquid region may be evaluated according to the deviation from compressed liquid

$$\begin{aligned} dh &= du + d(Pv) \\ h - h_f &= u - u_f + (Pv) - (Pv)_f \\ &= 0 + v_f(P - P_f) \end{aligned}$$

or

$$h(P, T) \approx h_f(T) + v_f(P - P_{\text{sat}}) \tag{4.10}$$

where $h(P, T)$ is the value of the enthalpy sought at the compressed liquid pressure, P.

Liquid Water
at 100°C
and 10 MPa

FIGURE 4.13
The control volume for Example 4.5.

Example 4.5 Determine the enthalpy of liquid water at a temperature of 100°C and a pressure of 10 MPa.

Solution

Assumptions and Specifications

(1) The liquid constitutes a closed system.
(2) The liquid is in equilibrium state.

We can use Table A.3 at 100°C to find

$$P_{sat} = 0.1013\,\text{MPa}, \quad h_f = 418.9\,\text{kJ/kg} \quad \text{and} \quad v_f = 1.043 \times 10^{-3}\,\text{m}^3/\text{kJ}$$

Then with

$$P - P_{sat} = 10\,\text{MPa} - 0.1013\,\text{MPa} = 9.8987\,\text{MPa}$$

so that

$$v_f(P - P_{sat}) = (1.043 \times 10^{-3}\,\text{m}^3/\text{kg})(9.8987\,\text{MPa}) = 10.32\,\text{kJ/kg}$$

and Equation 4.10 provides

$$h \approx h_f + v_f(P - P_{sat})$$
$$\approx 418.9\,\text{kJ/kg} + 10.32\text{kJ/kg} \approx 429.2\text{kJ/kg} \Longleftarrow$$

Example 4.6 Refrigerant R-134a which is at 8°C and a quality of $x = 0.60$ (Figure 4.14) is enclosed in a closed and rigid container having a volume of 0.115 m^3. Heat is added until the pressure is 800 kPa. Determine the quantity of heat added.

Solution

Assumptions and Specifications

(1) The refrigerant and any vapor produced by the heat addition constitutes the system.
(2) Both the initial state and the state after the heat addition represent equilibrium states.
(3) Kinetic and potential energy changes are negligible.
(4) A closed and rigid container implies a constant volume heat addition.

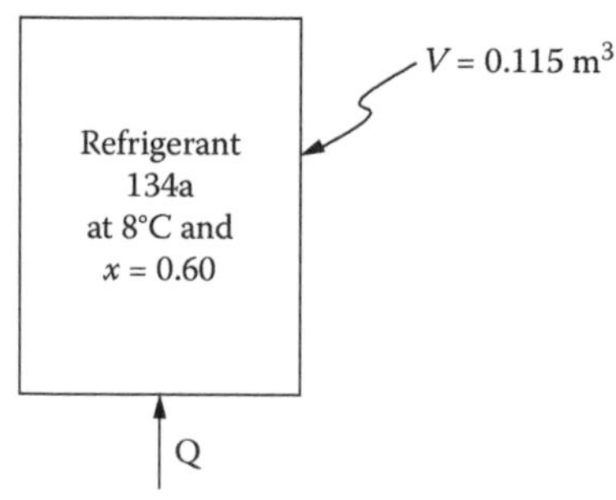

FIGURE 4.14
The control volume for Example 4.6.

Here, because no work is done on or by the system (constant volume process)

$$Q = \Delta U - W = \Delta U = m(u_2 - u_1)$$

At the initial state designated by point 1, Table A.6 at 8°C gives

$$v_f = 7.884 \times 10^{-4}\,\text{m}^3/\text{kg} \quad \text{and} \quad v_g = 0.0525\,\text{m}^3/\text{kg}$$

and

$$u_f = 60.43\,\text{kJ/kg} \quad \text{and} \quad u_g = 231.46\,\text{kJ/kg}$$

At $x = 0.60$ using Equation 4.4a

$$\begin{aligned} v_1 &= (1 - x)v_f + xv_g \\ &= (1 - 0.60)(7.884 \times 10^{-4}\,\text{m}^3/\text{kg}) + (0.60)(0.0525\,\text{m}^3/\text{kg}) \\ &= 3.154 \times 10^{-4}\,\text{m}^3/\text{kg} + 0.0315\,\text{m}^3/\text{kg} = 0.0318\,\text{m}^3/\text{kg} \end{aligned}$$

and with u_1 as the internal energy at point 1

$$\begin{aligned} u_1 &= (1 - x)u_f + xu_g \\ &= (0.40)(60.43\,\text{kJ/kg}) + (0.60)(231.46\,\text{kJ/kg}) \\ &= 24.17\,\text{kJ/kg} + 138.88\,\text{kJ/kg} = 163.05\,\text{kJ/kg} \end{aligned}$$

With $v = v_1$, the mass of refrigerant R-134a will be

$$m = \frac{V}{v} = \frac{0.115\,\text{m}^3}{0.0318\,\text{m}^3/\text{kg}} = 3.616\,\text{kg}$$

We next observe in Table A.7 that, at 800 kPa or 0.8 MPa, the saturated vapor specific volume is

$$v_g = 0.0255\,\text{m}^3/\text{kg}$$

so that the final state at point 2 is seen to be in the superheated vapor region. Thus, Table A.8 can be employed to provide at $v_2 = v_1 = 0.0318\,\text{m}^3/\text{kg}$ with interpolation

$$T_2 = 73.7^\circ\text{C}$$

The interpolation process also yields the internal energy at point 2

$$u_2 = 283.74\,\text{kJ/kg}$$

The amount of heat added is

$$Q = \Delta U - W = \Delta U$$

or

$$\begin{aligned} Q &= m(u_2 - u_1) \\ &= (3.616\,\text{kg})(283.74\,\text{kJ/kg} - 163.05\,\text{kJ/kg}) \\ &= (3.616\,\text{kg})(120.69\,\text{kJ/kg}) = 436.4\,\text{kJ} \Longleftarrow \end{aligned}$$

4.4 The *T*-*s* and *h*-*s* Diagrams

To this point, we have relied on the P-v diagram to show thermodynamic properties for water and refigerant R-134a. Indeed, the P-v diagram (the P-v surface considered in Section 4.2.7 and Figure 4.5) has been employed to show the compressed liquid, the liquid-vapor, and the superheated vapor regions.

The thermodynamic property of entropy is included in thermodynamic property data and while this property in introduced in Chapter 7, it is well, at this juncture, to describe two additional and extremely useful property diagrams containing specific entropy.

Figure 4.15a displays the temperature-specific entropy (T-s) diagram where the critical point is located at the apex of the vapor dome and where the saturated liquid and saturated vapor lines flow downward to the left and right of the critical point. In the T-s diagram, constant pressure lines are horizontal inside of the vapor dome and constant specific enthalpy

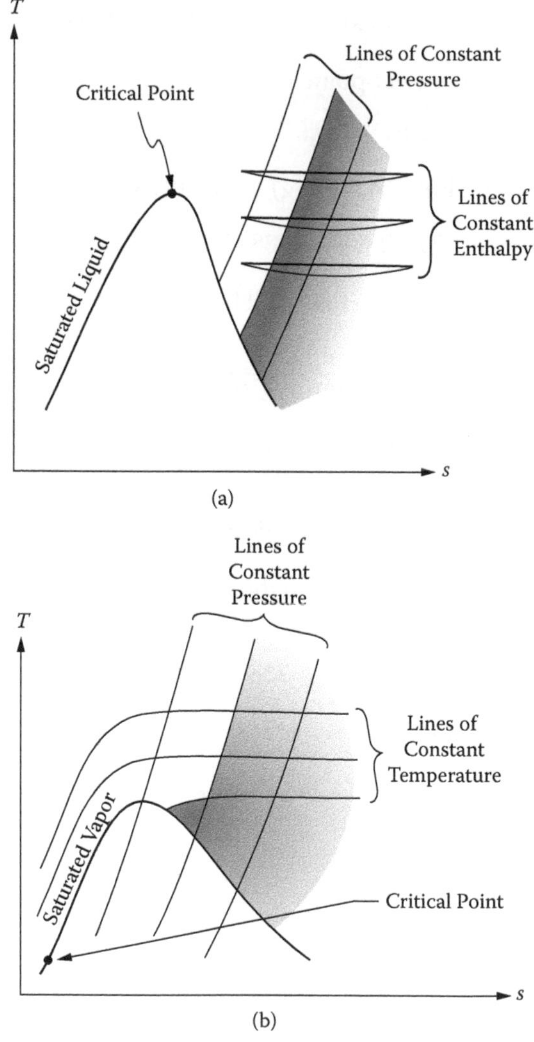

FIGURE 4.15
(a) The temperature-specific entropy (T-s) diagram and (b) specific enthalpy-specific entropy (h-s) diagram.

lines become nearly horizontal in the superheated vapor region as the pressure is reduced. This is indicated in the shaded area in Figure 14.15a and for pressures and temperatures outside of the shaded area, both temperature and pressure are needed to evaluate the specific enthalpy.

The specific enthalpy-specific entropy (h-s) diagram is shown in Figure 4.15b. Known as the **Mollier diagram** or **Mollier chart**, the diagram does not provide any liquid data. The critical point is located in the lower left-hand corner and the saturated vapor line may be noted. In the superheated vapor region, constant temperature lines become nearly horizontal as the pressure is reduced and this is indicated by the shaded area.

4.5 Real Gas Behavior

4.5.1 Introduction

On a macroscopic scale, a gas appears as if it were a homogeneous system described by its pressure, volume, and temperature. However, an examination of a gas on a microscopic basis requires the consideration of a huge number of molecules in random motion and with considerable intermolecular space between them.

For example, the pressure that a gas exerts on a surface is related to the number of molecules striking the surface. As the pressure increases or the temperature decreases, the **mean free path** between adjacent molecules decreases. The density will increase (or the specific volume will decrease) and the ideal gas relationship given in Section 3.7 by Equation 3.22 and repeated here as

$$Pv = RT \tag{4.11}$$

will no longer be valid at very high pressures. We illustrate this in the next example.

Example 4.7 Determine the percent error between the specific volume to be obtained from the steam table and the value obtained from the ideal gas law for steam at (a) 15 MPa and 400°C and (b) 15 MPa and 600°C.

Solution

Assumptions and Specifications

(1) The steam is a closed system.
(2) The steam is in the equilibrium state.

(a) At 15 MPa and 400°C (673 K), Table A.5 gives

$$v = 0.01566\,\text{m}^3/\text{kg}$$

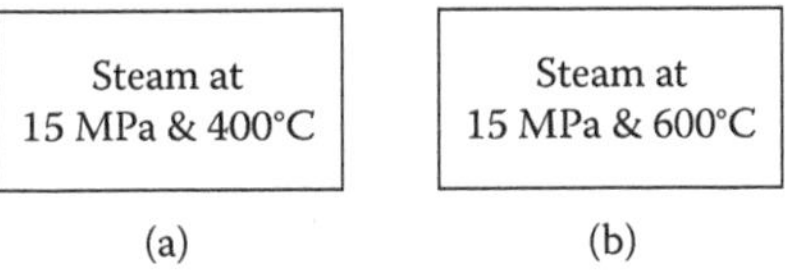

FIGURE 4.16
Two alternative control masses for Example 4.7.

For water vapor with a molecular weight of $M = 18.02\,\text{kg/kgmol}$, we obtain the value of R from

$$R = \frac{\bar{R}}{M} = \frac{8.314\,\text{kJ/kgmol-K}}{18.02\,\text{kg/kgmol}} = 0.4614\,\text{kJ/kg-K}$$

so that we can obtain v from Equation 4.11 with $T = 673\,\text{K}$

$$v = \frac{RT}{P} = \frac{(0.4614\,\text{kN-m/kg-K})(673\,\text{K})}{15 \times 10^3\,\text{kN/m}^2} = 0.02070\,\text{m}^3/\text{kg}$$

To establish the percent error, we take the steam table value as the base value because it is based on experimental evidence. Hence,

$$\begin{aligned}
\%\,\text{error} &= 100\left[\frac{0.02070\,\text{m}^3/\text{kg} - 0.01566\,\text{m}^3/\text{kg}}{0.01566\,\text{m}^3/\text{kg}}\right] \\
&= 100\left[\frac{0.00504\,\text{m}^3/\text{kg}}{0.01566\,\text{m}^3/\text{kg}}\right] \\
&= 32.19\,\% \Longleftarrow
\end{aligned}$$

(b) At 15 MPa and 600°C (873 K), Table A.3 gives

$$v = 0.02488\,\text{m}^3/\text{kg}$$

Then from Equation 4.11, we obtain

$$v = \frac{RT}{P} = \frac{(0.4614\,\text{kPa-m}^3/\text{kg-K})(873\,\text{K})}{15 \times 10^3\,\text{kPa}} = 0.02685\,\text{M}^3/\text{kg}$$

This time, the percent error is

$$\begin{aligned}
\%\,\text{error} &= 100\left[\frac{0.02685\,\text{m}^3/\text{kg} - 0.02448\,\text{m}^3/\text{kg}}{0.02448\,\text{m}^3/\text{kg}}\right] \\
&= 100\left[\frac{0.00237\,\text{m}^3/\text{kg}}{0.02448\,\text{m}^3/\text{kg}}\right] \\
&= 9.70\,\% \Longleftarrow
\end{aligned}$$

It appears that the error is greater as the saturated vapor line is approached. Moreover, we can see very clearly that there is a danger in considering water vapor as an ideal gas.

4.5.2 The Compressibility Factor

Gases do not follow the ideal gas equation unless the pressure is relatively low and the temperature is relatively high. We therefore need a method for correcting the ideal gas equation to allow for real gas behavior and still maintain reasonable accuracy. The ideal gas law of Equation 4.11 may be used to define a **compressibility factor**, Z

$$Z = \frac{Pv}{RT} \tag{4.12}$$

and this indicates that a gas can be taken as an ideal gas only if

$$Z = \frac{Pv}{RT} = 1 \tag{4.13}$$

The use of the compressibility factor, Z, permits the prediction of P-v-T data when a gas is no longer at low pressure and moderate temperatures. Values of Z for a particular gas can be developed on the basis of Equation 4.13 for a wide range of equilibrium states and may be plotted for a particular gas as functions of pressure and temperature.

When the pressure and temperature are normalized with respect to their values at the **critical state**, all gases can be seen to fit approximately on a single compressibility chart. Critical state properties are provided in Table A.9 and the normalizations involve three **reduced properties**. The **reduced pressure** is

$$P_r \equiv \frac{P}{P_c} \tag{4.14}$$

where both P and P_c must be in consistent units to make P_r dimensionless. The **reduced temperature** is

$$T_r \equiv \frac{T}{T_c} \tag{4.15}$$

with both temperatures in absolute units. Because the use of the reduced specific volume presents computational difficulties, it has been replaced by the **pseudo-reduced specific volume**

$$v_r' \equiv \frac{v}{RT_c/P_c} \tag{4.16}$$

Because the ideal gas specific volume is

$$v_{\text{id}} = \frac{RT}{P}$$

we may consider Z as the ratio of the actual gas specific volume to the ideal gas specific volume

$$Z = \frac{v}{v_{\text{id}}}$$

4.5.3 The Principle of Corresponding States

The **principle of corresponding states** asserts that

> The compressibility factor for all gases is approximately the same when gases have identical reduced pressure and temperature.

Thus, the compressibility factor for any gas is a function of the reduced pressure and temperature

$$Z = f(P_r, T_r) \tag{4.17}$$

and the **generalized compressibility charts** with values of P_r plotted as the abscissa and Z plotted as the ordinate are displayed in Figures A.1 through A.3 in Appendix A.

Example 4.8 In Example 4.7, the specific volume of steam, obtained from the ideal gas equation, was seen to differ markedly from the steam table value. Repeat Example 4.7 using the principle of corresponding states.

Solution

Assumptions and Specifications

(1) The steam is a closed system.
(2) The steam is in the equilibrium state.

Example 4.7 gives $R = 0.4641\,\text{kJ/kg-K}$ and, from Table A.9, we find for water

$$P_c = 22.09\,\text{MPa} \qquad \text{and} \qquad T_c = 647.3\,\text{K}$$

Then, for $P = 15\,\text{MPa}$

$$P_r = \frac{P}{P_c} = \frac{15\,\text{MPa}}{22.09\,\text{MPa}} = 0.679$$

(a) For $T = 673\,\text{K}$

$$T_r = \frac{T}{T_c} = \frac{673\,\text{K}}{647.3\,\text{K}} = 1.040$$

Figure A.3 gives

$$Z = 0.755$$

so that

$$v = \frac{ZRT}{P} = \frac{(0.755)(0.4641\,\text{kN-m}^3/\text{kg-K})(673\,\text{K})}{15 \times 10^3\,\text{kPa}} = 0.01572\,\text{m}^3/\text{kg}$$

With the steam table value of $0.01566\,\text{m}^3/\text{kg}$, we see that the error is now almost negligible.

(b) For $T = 873\,\text{K}$

$$T_r = \frac{T}{T_c} = \frac{873\,\text{K}}{647.3\,\text{K}} = 1.349$$

Figure A.3 then gives

$$Z = 0.91$$

so that

$$v = \frac{ZRT}{P} = \frac{(0.91)(0.4641\,\text{Pa-m}^3/\text{kg-K})(873\,\text{K})}{15 \times 10^3\,\text{kPa}} = 0.02458\,\text{m}^3/\text{kg}$$

With the steam table value of $0.02588\,\text{m}^3/\text{kg}$, we find that the error here is also negligible.

4.6 Equations of State

We are aware that the ideal gas law of Equation 4.11 may not always provide an accurate representation among pressure, specific volume, and temperature. More accurate relationships among these variables can be obtained by developing analytical formulations that are called **equations of state**.

The **generalized equations of state** are usually equations that contain two constants and are not very complicated. These include

- The **van der Waals equation**
- The **Berthold equation**
- The **Redlich-Kwong equation**

These equations have a limited range of application because of their simplicity and have been developed from generalized behavior at the critical state.

The **empirical equations of state** include

- The **Beattie-Bridgeman equation**, which has five constants
- The **Benedict-Webb-Rubin equation**, which has eight constants

These equations are more complex than the generalized equations of state. The constants have been determined from experimental data.

A **theoretical equation of state** is the **virial equation of state** in which the product of the pressure and the molal specific volume is represented by an infinite series. The constants in the infinite series, called **virial constants**, are based on principles derived from the kinetic theory of gases.

We will provide some details for the van der Waals and the Redlich-Kwong equations of state.

4.6.1 The Van der Waals Equation

The van der Waals equation of state, proposed in 1873, is

$$P = \frac{\bar{R}T}{\bar{v} - b} - \frac{a}{\bar{v}^2} \tag{4.18}$$

where

$$a = \frac{27}{64}\frac{\bar{R}^2T_c^2}{P_c} \tag{4.19a}$$

attempts to correct for intermolecular forces and

$$b = \frac{\bar{R}T_c}{8P_c} \tag{4.19b}$$

which is intended to account for the volume of the gas molecules. The term, $\bar{v}$, is the molal specific volume for the gas.

The van der Waals equation is only accurate over a limited range with an accuracy that decreases at high densities. The van der Waal constants, a and b for seventeen gases are given in Table A.10.

4.6.2 The Redlich-Kwong Equation of State

The Redlich-Kwong equation of state

$$P = \frac{\bar{R}T}{\bar{v} - b} - \frac{a}{T^{1/2}\bar{v}(\bar{v} + b)} \tag{4.20}$$

proposed in 1949, provides more accuracy than the van der Waals equation of state. In Equation 4.20

$$a = \frac{0.4275\bar{R}^2T_c^{2.50}}{P_c} \tag{4.21a}$$

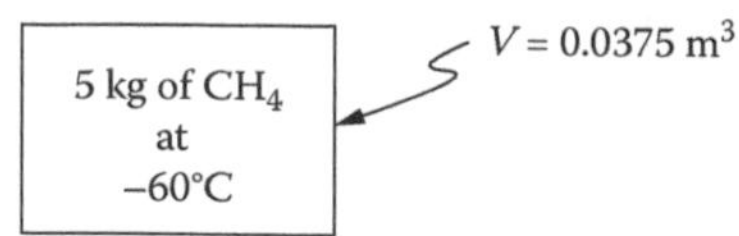

FIGURE 4.17
The control volume for Example 4.9.

and

$$b = \frac{0.0867\bar{R}T_c}{P_c} \tag{4.21b}$$

The Redlich-Kwong constants, a and b for eight gases are given in Table A.11.

Example 4.9 As indicated in Figure 4.17, consider 5 kg of methane at −60°C in a closed and rigid container having a volume of 0.0375 m^3. Determine the pressure in bars exerted by the methane using (a) the ideal gas equation of state, (b) the principle of corresponding states, (c) the van der Waals equation of state, and (d) the Redlich-Kwong equation of state.

Solution

Assumptions and Specifications

(1) The methane in the chamber is a closed system.
(2) Equilibrium exists.

With $\bar{R} = 8314$ N-m/kgmol-K and the molecular weight of methane, $M = 16.04$ kg/kgmol taken from Table A.1, we have

$$\bar{v} = Mv = M\left(\frac{V}{m}\right) = (16.04\,\text{kg/kgmol})\left(\frac{0.0375\,\text{m}^3}{5\,\text{kg}}\right) = 0.1203\,\text{m}^3/\text{kgmol}$$

(a) By the ideal gas equation of state at $T = -60$°C or 213 K

$$P = \frac{\bar{R}T}{\bar{v}} = \frac{(8314\,\text{J/kgmol-K})(213\,\text{K})}{0.1203\,\text{m}^3/\text{kgmol}} = 147.21\,\text{bars} \Longleftarrow$$

(b) By the method of corresponding states, Table A.9 gives for methane

$$T_c = 191\,\text{K} \quad \text{and} \quad P_c = 46.4\,\text{bar}$$

Then,

$$T_r = \frac{T}{T_c} = \frac{213\,\text{K}}{191\,\text{K}} = 1.115$$

and because P_r is to be found, we need, v_r' as the second parameter in Figure A.2. Thus,

$$v_r' = \frac{vP_c}{RT_c} = \frac{\bar{v}P_c}{\bar{R}T_c} = \frac{(0.1203\,\text{m}^3/\text{kgmol})(46.4\times 10^5\,\text{N/m}^2)}{(8314\,\text{N-m/kgmol-K})(191\,\text{K})} = 0.352$$

Then we read Figure A.2 to find at $T_r = 1.115$ and $v_r' = 0.352$ either

$$Z = 0.508 \quad \text{or} \quad P_r = 1.59$$

From Z we have

$$P = \frac{Z\bar{R}T}{\bar{v}} = \frac{(0.508)(8314\,\text{N-m/kgmol-K})(213\,\text{K})}{0.1203\,\text{m}^3/\text{kgmol}} = 74.78\,\text{bar}$$

and from P_r we have

$$P = P_r P_c = (1.59)(46.4\,\text{bars}) = 73.77\,\text{bar} \Longleftarrow$$

The discrepancy here is due to the lack of precision in reading Figure A.2.

(c) The constants in the van der Waals equation are obtained for methane from Table A.10

$$a = 2.293\,\text{bar-(m/kgmol)}^2 \quad \text{and} \quad b = 0.0428\,\text{m}^3/\text{kgmol}$$

Then

$$\frac{a}{\bar{v}^2} = \frac{2.293\,\text{bar-(m/kgmol)}^2}{(0.1203\,\text{m}^3/\text{kgmol})^2} = 158.44\,\text{bar}$$

and

$$\frac{\bar{R}T}{\bar{v} - b} = \frac{(8314\,\text{N-m/kgmol-K})(213\,\text{K})}{0.1203\,\text{m}^3/\text{kgmol} - 0.0428\,\text{m}^3/\text{kgmol}} = 228.50\,\text{bar}$$

Thus, by Equation 4.18 we have

$$P = \frac{\bar{R}T}{\bar{v} - b} - \frac{\bar{a}}{\bar{v}^2} = 228.50\,\text{bar} - 158.44\,\text{bar} = 70.06\,\text{bar} \Longleftarrow$$

(d) The constants in the Redlich-Kwong equation are obtained for methane from Table A.11

$$a = 32.19\,\text{bar-(m/kgmol)}^2\text{-K}^{0.50} \quad \text{and} \quad b = 0.02969\,\text{m}^3/\text{kgmol}$$

Then with

$$\bar{v}(\bar{v} + b) = 0.1203\,\text{m}^3/\text{kgmol})(0.1203\,\text{m}^3/\text{kgmol} + 0.0297\,\text{m}^3/\text{kgmol})$$

or

$$\bar{v}(\bar{v} + b) = 0.0180\,(\text{m}^3/\text{kgmol})^2$$

we have

$$\frac{a}{T^{1/2}\bar{v}(\bar{v} + b)} = \frac{32.19\,\text{bar-(m}^3/\text{kgmol)}^2\text{-K}^{0.50}}{(213\,\text{K})^{0.50}(0.0180\,\text{m}^3/\text{kgmol}^2)} = 122.53\,\text{bar}$$

and

$$\frac{\bar{R}T}{\bar{v} - b} = \frac{(8314\,\text{N-m/kgmol-K})(213\,\text{K})}{0.1203\,\text{m}^3/\text{kgmol} - 0.02969\,\text{m}^3/\text{kgmol}} = 195.44\,\text{bar}$$

Thus, by Equation 4.20 we have

$$P = \frac{\bar{R}T}{\bar{v} - b} - \frac{a}{T^{1/2}\bar{v}(\bar{v} + b)} = 195.44\,\text{bar} - 122.53\,\text{bar} = 72.91\,\text{bar} \Longleftarrow$$

We may compare the foregoing values:

Method	Pressure, bars
Ideal gas law	147.21
Corresponding states	74.78
van der Waals equation	70.06
Redlich-Kwong equation	73.91

We observe that the method of corresponding states and the Redlich-Kwong equation provide almost the same value of the pressure. The ideal gas equation is, clearly, inaccurate at this relatively high pressure. The Redlich-Kwong equation of state of 1949 is more accurate than the van der Waals equation of state developed in 1873. Hence, we are led to the conclusion that the pressure is approximately 73.50 bar.

4.7 The Polytropic Process for an Ideal Gas

In Section 3.9, we suggested that processes involving ideal gases can be characterized by Equation 3.36

$$PV^n = C \tag{3.36}$$

where the exponent, n, may have values that are dictated by the particular process. Pertinent relationships were then developed for the constant volume process ($n = \infty$), the constant pressure process ($n = 0$), and the the constant temperature process ($n = 1$). For the polytropic process, n may take on any value from $-\infty$ to ∞ ($\infty < n < \infty$).

4.7.1 *P-V-T* Relationships

In the polytropic process beginning at state 1 and ending at state 2

$$P_1 V_1^n = P_2 V_2^n = C$$

n may range between $-\infty < n < \infty$. The P-V-T relationship is

$$\frac{P_2}{P_1} = \left(\frac{V_1}{V_2}\right)^n \tag{4.22}$$

and because $P_1 V_1/T_1 = P_2 V_2/T_2$, we see that

$$\frac{T_2}{T_1} = \left(\frac{P_2}{P_1}\right)^{(n-1)/n} \tag{4.23}$$

and

$$\frac{T_2}{T_1} = \left(\frac{V_1}{V_2}\right)^{n-1} \tag{4.24}$$

4.7.2 Work

With $PV^n = \text{constant}$, the work done will be

$$W = C\int_{V_1}^{V_2} \frac{dV}{V^n} = \frac{P_2V_2 - P_1V_1}{1-n} = \frac{mR(T_2 - T_1)}{1-n} \qquad (n \neq 1) \tag{4.25}$$

If $n = 1$, we see that the polytropic process reverts to the isothermal process with

$$W = P_1V_1 \ln\frac{V_2}{V_1} = mRT_1 \ln\frac{V_2}{V_1} \tag{3.43}$$

4.7.3 Internal Energy and Enthalpy

As long as the gas is an ideal gas with specific heats c_p and c_v assumed to be constant, the change in internal energy and enthalpy will always be

$$U_2 - U_1 = m(u_2 - u_1) = mc_v(T_2 - T_1) \tag{4.26}$$

and

$$H_2 - H_1 = m(h_2 - h_1) = mc_p(T_2 - T_1) \tag{4.27}$$

4.7.4 Heat Transfer

For any process involving a closed system with negligible change in kinetic and potential energy, the first law energy equation is

$$Q = \Delta U + W$$

For a polytropic process of an ideal gas beginning at state 1 and ending at state 2 with $n \neq 1$ Equations 4.25 and 4.26 give

$$\begin{aligned} Q &= \Delta U + W \\ &= mc_v(T_2 - T_1) + \frac{mR(T_2 - T_1)}{1-n} \\ &= m\left(\frac{c_v + R - nc_v}{1-n}\right)(T_2 - T_1) \end{aligned}$$

But

$$c_p = c_v + R \qquad \text{and} \qquad c_p = kc_v$$

so that

$$Q = mc_v\left(\frac{k-n}{1-n}\right)(T_2 - T_1)$$

or

$$Q = mc_n(T_2 - T_1) \tag{4.28}$$

where c_n is the **polytropic specific heat**

$$c_n = c_v\left(\frac{k-n}{1-n}\right) \tag{4.29}$$

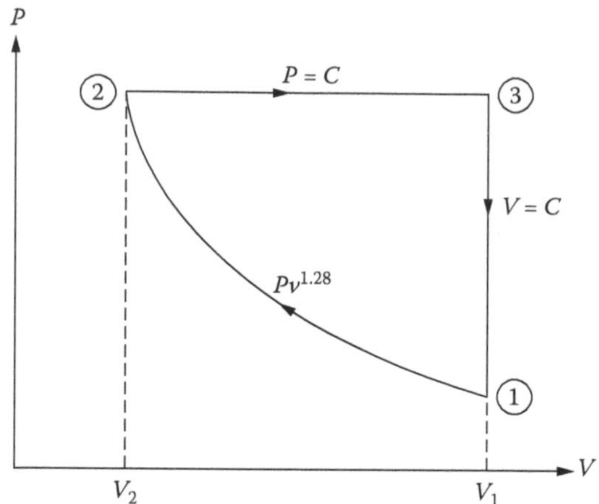

FIGURE 4.18
P-v diagram for quasistatic processes of Example 4.10. Here $V_1 = 4V_2$.

We note that for $n = 0$, which represents the constant pressure process

$$c_n = c_v k = c_p$$

and if $n = \pm\infty$, which represents the constant volume process

$$c_n = c_v\left(\frac{k-n}{1-n}\right) = \lim_{n\longrightarrow\infty} c_v \frac{n}{n}\left(\frac{\frac{k}{n}-1}{\frac{1}{n}-1}\right) = c_v$$

If $n = k$, the process is an adiabatic process in which $c_n = 0$ and $Q = 0$. The adiabatic process is the fifth process of the ideal gas that we will consider in Chapter 7.

Example 4.10 As illustrated in Figure 4.18, 8 g of air undergoes the following quasistatic processes in a piston-cylinder assembly:

state 1 to state 2	a polytropic compression with $n = 1.28$ to one-quarter of the initial volume
state 2 to state 3	a constant pressure expansion to the initial volume
state 3 to state 1	a constant volume heat rejection

If the initial pressure is 5 bar and the initial temperature is 212°C (485 K), determine the net work done on the gas.

Solution

Assumptions and Specifications

(1) The air in the piston-cylinder assembly is a closed system.
(2) The air obeys the ideal gas law.
(3) Kinetic and potential energy changes are negligible.
(4) Quasi-static processes are specified.

The strategy here is to note that no work is done at constant volume from state 3 to state 1 and that

$$W = W_{12} + W_{23}$$

For air, Table A.1 gives $R = 0.287\,\text{kJ/kg-K}$. Then at state 1 via the ideal gas law

$$\begin{aligned} V_1 &= \frac{mRT_1}{P_1} \\ &= \frac{(0.008\,\text{kg})(287\,\text{N-m/kg-K})(485\,\text{K})}{5 \times 10^5\,\text{N/m}^2} \\ &= 2.227 \times 10^{-3}\,\text{m}^3 \end{aligned}$$

and with $V_2 = V_1/4$

$$V_2 = \frac{2.227 \times 10^{-3}\,\text{m}^3}{4} = 5.568 \times 10^{-4}\,\text{m}^3$$

we can obtain P_2 from Equation 4.22

$$\begin{aligned} P_2 &= P_1 \left(\frac{V_1}{V_2}\right)^n \\ &= (5 \times 10^5\,\text{N/m}^3)(4)^{1.28} \\ &= 2.949 \times 10^6\,\text{N/m}^2 \end{aligned}$$

Because process 2-3 is a constant pressure process that returns the system to the original volume, we see that $P_3 = P_2$ and $V_3 = V_1$

$$P_3 = 2.949 \times 10^6\,\text{N/m}^2 \quad \text{and} \quad V_3 = 2.227 \times 10^{-3}\,\text{m}^3$$

All of the data required to determine the work done on the gas is at hand. From state 1 to state 2, Equation 4.25 gives

$$W_{12} = \frac{P_2V_2 - P_1V_1}{1 - n}$$

With

$$P_2V_2 = (2.949 \times 10^6\,\text{N/m}^2)(5.568 \times 10^{-4}\,\text{m}^3) = 1642.0\,\text{N-m}$$

and

$$P_1V_1 = (0.500 \times 10^6\,\text{N/m}^2)(2.227 \times 10^{-3}\,\text{m}^3) = 1113.5\,\text{N-m}$$

we have

$$W_{12} = \frac{1642.0\,\text{N-m} - 1113.5\,\text{N-m}}{1 - 1.28} = \frac{528.5\,\text{N-m}}{-0.28} = -1887.5\,\text{N-m}$$

From state 2 to state 3, $n = 0$ and Equation 4.25 gives

$$
\begin{aligned}
W_{23} &= P_3V_3 - P_2V_2 = P_3(V_3 - V_2) \\
&= (2.949 \times 10^6\,\text{N/m}^2)(2.227 \times 10^{-3}\,\text{m}^3 - 5.568 \times 10^{-4}\,\text{m}^3) \\
&= (2.949 \times 10^6\,\text{N/m}^2)(1.670 \times 10^{-3}\,\text{m}^3) = 4925.4\,\text{N-m}
\end{aligned}
$$

The work done is

$$
\begin{aligned}
W &= W_{12} + W_{13} \\
&= -1887.5\,\text{N-m} + 4925.4\,\text{N-m} \\
&= +3037.9\ \text{N-m} \Longleftarrow
\end{aligned}
$$

The positive sign indicates that the work is done by the air in the piston-cylinder assembly *on* the surroundings.

4.8 Summary

The state postulate supplies the number of variables that are required to completely specify the state of a single phase or homogeneous system. In the case of a simple, compressible, substance where Pdv is the only relevant work interaction, two independent properties are sufficient to completely specify the equilibrium state.

The steam tables are categorized into

- Saturation tables
- Superheated vapor tables
- Compressed liquid tables

The refrigerant-134a tables are categorized into

- Saturation tables
- Superheated vapor tables

Details for each of these tables have been provided.

For an ideal gas, the equation of state is

$$Pv = RT$$

Several methods are available for determining the properties of a nonideal gas. They are

- The use of the method of corresponding states in which the compressibility factor

  $$Z = \frac{Pv}{RT}$$

  is defined and charts are available to obtain values of Z.
- The use of equations of state proposed by various investigators.

Two of the equations of state that are cited here are those of van der Waals

$$\left(P + \frac{a}{\bar{v}^2}\right)(\bar{v} - b) = \bar{R}T$$

and Redlich-Kwong

$$P = \frac{\bar{R}T}{\bar{v} - b} - \frac{a}{T^{1/2}\bar{v}(\bar{v} + b)}$$

Polytropic processes of an ideal gas are decribed by

$$PV^n = C \qquad \text{(a constant)}$$

where the exponent, n, can vary. Three processes of an ideal gas were considered in Chapter 3 and, in this chapter, the polytropic process, where n can take on any value, is considered. In particular, a polytropic specific heat is

$$c_n = c_v\left(\frac{k - n}{1 - n}\right)$$

where $k = c_p/c_v$.

4.9 Problems

Use of the Steam Tables

4.1: Given a water-steam system where $T = 200°\text{C}$ with a quality of $x = 0.18$. Use the steam tables to determine the specific volume.

4.2: Given a water-steam system where $P = 500\,\text{kPa}$ with a quality of $x = 0.36$. Use the steam tables to determine the specific volume.

4.3: Given a water-steam system where $v = 1.148 \times 10^{-3}\,\text{m}^3/\text{kg}$. Use the steam tables to determine T, P, x, and h.

4.4: Given a water-steam system where $v = 0.0502\,\text{m}^3/\text{kg}$. Use the steam tables to determine T, P, x, and u.

4.5: Given a water-steam system where $T = 300°\text{C}$ and $x = 0.60$. Use the steam tables to determine P, u, and s.

4.6: Given a water-steam system where $P = 50\,\text{kPa}$ and $h = 1034.3\,\text{kJ/kg}$. Use the steam tables to determine T, x, and u.

4.7: Given a water-steam system where $T = 200°\text{C}$ and $P = 6\,\text{MPa}$. Use the steam tables to determine h and u.

4.8: Given a water-steam system where $T = 400°\text{C}$ and $P = 1.5\,\text{MPa}$. Use the steam tables to determine u, h, and s.

4.9: Given a water-steam system where $P = 1\,\text{MPa}$ and $h = 2872.6\,\text{kJ/kg}$. Use the steam tables to determine T, u, and s.

4.10: Given a water-steam system where $P = 1.25\,\text{MPa}$ and $T = 275°\text{C}$. Use the steam tables to determine u.

Systems Containing Water and/or Steam

4.11: Two steam boilers discharge equal amounts of steam into the same line. In boiler-1, the pressure is 800 kPa and the temperature is 600°C. In boiler-2, the pressure is 800 kPa and the quality is 92%. Determine (a) the enthalpy at the quilibrium condition and (b) the temperature at the equilibrium condition.

4.12: A 0.12-m^3 drum contains a mixture of water and steam at 300°C. Determine (a) the volume occupied by each substance if their masses are equal and (b) the mass of each substance if their volumes are equal.

4.13: Heat is extracted from a saturated water-vapor mixture in a closed rigid container at 4 MPa. Determine the quality of the water-vapor mixture when it reaches 125°C.

4.14: Three kilograms of steam at 2 MPa and 400°C are expanded at constant volume to a pressure of 200 kPa. Determine (a) the volume of the container, (b) the final temperature, and (c) the final quality.

4.15: Determine the volume occupied by 0.525 kg of steam at 15 MPa and 88% quality.

4.16: One kilogram of wet steam at 200°C and with an enthalpy of 2250 kJ/kg is confined in a rigid container. Heat is applied until the steam becomes saturated. Determine (a) the initial pressure, (b) the initial quality, (c) the initial specific volume, and (d) the final pressure.

4.17: Two kilograms of steam at 1.5 MPa and a quality of $x = 0.50$ are expanded at constant pressure to an unknown final state. If 240 kJ of work are done by the steam, determine (a) the final temperature and (b) the heat added to the steam.

4.18: Two kilograms of steam at 1.6 MPa and a quality of $x = 0.50$ are expanded at constant pressure to an unknown final state. If 128 kJ of work are done by the steam, determine (a) the final temperature and (b) the heat added to the steam.

4.19: Five kilograms of steam at 2.5 MPa and 300°C undergo a constant pressure process until the quality becomes 50%. Determine (a) Δh, (b) Δu, (c) Δs, (d) W, and (e) Q.

4.20: State point 1 for a water-steam mixture inside of a weighted piston-cylinder assembly is at 3 MPa and 250°C. Heat is added until the temperature reaches state point 2 at 350°C. Determine (a) the work done to raise the piston and (b) the heat transferred.

Use of Refrigerant R-134a Tables

4.21: Given a liquid-vapor refrigerant R-134a system where $T = 48°C$ with a quality of $x = 0.22$. Determine the specific volume.

4.22: Given a liquid-vapor refrigerant R-134a system where $P = 600$ kPa with a quality of $x = 0.42$. Determine the specific volume.

4.23: Given a refrigerant R-134a system where $v = 0.0409\,m^3/kg$. Determine T, P, x, and h.

4.24: Given a refrigerant R-134a system where $v = 9.308 \times 10^{-4}\,m^3/kg$. Determine T, P, x, and u.

4.25: Given a liquid-vapor refrigerant R-134a system where $T = 30°C$ and $x = 0.40$. Use the refrigerant R-134a tables to determine P, h, and s.

4.26: Given a refrigerant R-134a system where $P = 8.6247$ bars and $h = 163$ kJ/kg. Determine T, x, and u.

4.27: Given a refrigerant R-134a system where $T = -10°C$ and $P = 100$ kPa. Determine u, h, and s.

4.28: Given a refrigerant R-134a system where $P = 400\,\text{kPa}$ and $h = 277.00\,\text{kJ/kg}$. Use the refrigerant R-134a tables to determine T, u, and s.

4.29: Refrigerant R-134a at 2 bars and 40°C is cooled at constant temperature until the final specific volume is $0.0165\,\text{m}^3/\text{kg}$. Determine the change in specific enthalpy.

4.30: A refrigerant R-134a system at 600 kPa and $h = 120\,\text{kJ/kg}$ is expanded at constant enthalpy to a pressure of 100 kPa. Determine the change in entropy.

4.31: Refrigerant R-134a is confined in a closed-rigid container having a volume of $0.125\,\text{m}^3$ at 200 kPa and a quality of 0.488. Heat is added until the pressure reaches 400 kPa. Determine (a) the mass of refrigerant in the container and (b) the heat added to the refrigerant.

4.32: Refrigerant R-134a at 100 kPa and a specific volume of $0.160\ \text{m}^3/\text{kg}$ undergoes a constant pressure nonflow process until the temperature is 0°C. For a mass of 1 kg, determine (a) Δh, (b) Δv, (c) Δs, (d) Δu, (e) W, and (f) Q.

4.33: Two kilograms of saturated refrigerant R-134a vapor at 100 kPa are compressed in an adiabatic nonflow process at constant entropy until the pressure is 1 MPa. Determine (a) the final temperature and (b) the work done.

4.34: Determine the heat rejected from 2.5 kg of refrigerant R-134a in a constant pressure process from 1.2 MPa and 60°C to a quality of 20%.

4.35: One half kilogram of refrigerant R-134a with a quality of $x = 0.625$ is expanded isothermally to a pressure of 600 kPa and 30°C. The measured work output during the expansion is 65 kJ/kg. Determine the heat transferred.

Polytropic Processes of an Ideal Gas

4.36: Three kilograms of air are expanded polytropically ($PV^{1.28} = \text{constant}$) from $P_1 = 600\,\text{kPa}$ and $T_1 = 90°\text{C}$ to $P_2 = 200\,\text{kPa}$. Determine (a) V_2, (b) T_2, (c) ΔH, (d) ΔU, (e) W, and (f) Q.

4.37: Four kilograms of carbon dioxide are compressed polytropically ($PV^{1.25} = \text{constant}$) from $P_1 = 100\,\text{kPa}$ and $T_1 = 60°\text{C}$ to $T_2 = 227°\text{C}$. Determine (a) P_2, (b) W, and (c) Q.

4.38: Nitrogen expands in a cylinder from 1500 kPa and 267°C to 100 kPa and 27°C. Determine the value of n for a polytropic ($PV^n = \text{constant}$) process.

4.39: Two kilograms of air at 25°C are expanded in a polytropic process ($PV^{1.26} = \text{constant}$) until the pressure is halved. Determine (a) ΔU, (b) ΔH, (c) W, and (d) Q.

4.40: Two kilograms of hydrogen undergo a polytropic compression from $P_1 = 6\,\text{MPa}$ and $v_1 = 0.1875\,\text{m}^3/\text{kg}$ until the pressure is doubled. Assume constant specific heats with the values given in Table A.1 and determine (a) v_1, (b) T_1 and T_2, (c) ΔH, (d) ΔU, (e) W, and (f) Q.

4.41: In a polytropic process, hydrogen expands from $P_1 = 1\,\text{MPa}$, $T_1 = 60°\text{C}$ and $V_1 = 0.275\,\text{m}^3$ to $P_2 = 100\,\text{kPa}$. The polytropic exponent is 1.3. Using the specific heats given in Table A.1 as constant, determine (a) the mass of hydrogen present, (b) T_2, (c) W, (d) Q, and (e) ΔH.

4.42: Six kilograms of methane are compressed in a polytropic process ($PV^{1.18} = \text{constant}$) from $P_1 = 100\,\text{kPa}$ and $T_1 = 57°\text{C}$ to $T_2 = 257°\text{C}$. Assuming that the methane is an ideal gas and that the specific heats listed in Table A.1 are constant, determine (a) P_2, (b) W, and (c) Q.

4.43: A quantity of propane expands polytropically ($PV^n = \text{constant}$) from 1 MPa and $0.20\,\text{m}^3$ to 10 MPa and $1.262\,\text{m}^3$. Determine (a) the value of n and (b) the work done.

4.44: Two kilograms of helium undergoes a polytropic process (PV^n = constant) from $V_1 = 4.092\,\text{m}^3$ and $P_1 = 650\,\text{kPa}$ to $P_2 = 100\,\text{kPa}$ and $T_2 = 367°\text{C}$. Determine (a) the value of n and (b) the work done.

4.45: One kilogram of a gas, which may be presumed to be ideal, has a value of $R = 0.136\,\text{kJ/kg-K}$ and $k = 1.352$. The gas undergoes a polytropic process (PV^n = constant) from $P_1 = 100\,\text{kPa}$ and $T_1 = 32°\text{C}$ to $P_2 = 500\,\text{kPa}$ and $V_2 = 0.105\,\text{m}^3$. Determine (a) the value of n, (b) ΔU, and (c) ΔH.

The Principle of Corresponding States

4.46: Determine the specific volume of superheated steam at 20 MPa and 500°C using (a) the steam tables, (b) the ideal gas equation, and (c) the principle of corresponding states.

4.47: Use the principle of corresponding states to obtain the density of oxygen at 4 MPa and −100°C.

4.48: Use the principle of corresponding states to determine the pressure at which the specific volume of oxygen at −100°C is $3.75 \times 10^{-3}\ \text{m}^3/\text{kg}$.

4.49: Propane at $P_1 = 2.5\,\text{MPa}$ and $v_1 = 0.01875\,\text{m}^3/\text{kg}$ is heated at constant volume until the pressure is $P_2 = 4\,\text{MPa}$. Use the principle of corresponding states to estimate the change in temperature.

4.50: Steam at 1 MPa and 327°C is heated at constant volume until the pressure is doubled. Use the principle of corresponding states to find the final temperature.

4.51: Steam at 10 MPa and 427°C is expanded isothermally until the volume is doubled. Use the principle of corresponding states to find the final pressure.

4.52: Steam at 8 MPa and 227°C is heated at constant pressure until the temperature is doubled. Use the principle of corresponding states to find the final specific volume.

4.53: Air at 1 MPa and 227°C flows through a 5.08 cm diameter tube at a velocity of $\hat{V} = 40\,\text{m/s}$. Determine the mass flow (a) by considering air to be an ideal gas and (b) by using the principle of corresponding states.

4.54: Consider refrigerant R-134a at 1.2 MPa and 63.6°C and determine its specific volume by (a) the ideal gas law and (b) the principle of corresponding states.

4.55: Refrigerant R-134a is heated at a constant pressure of 2 MPa from 82.3°C to 157.1°C. Use the principle of corresponding states to determine the work done.

Equations of State

4.56: Two kilograms of propane at 20°C occupy a closed and rigid container having a volume of $0.05\,\text{m}^3$. Determine the pressure using (a) the ideal gas law, (b) the van der Waals equation of state, and (c) the Redlich-Kwong equation of state.

4.57: The specific volume of superheated steam at 1 MPa and 200°C is $0.2059\,\text{m}^3/\text{kg}$. Determine the specific volume of the steam using (a) the ideal gas law, (b) the principle of corresponding states, and (c) the van der Waals equation of state. Then use the steam table value as the basis for determining the percent error in each.

4.58: One kilogram of nitrogen at −110°C is contained in a closed and rigid vessel having a volume of $0.1\,\text{m}^3$. Determine the pressure exerted by the nitrogen using the van der Waals equation of state.

4.59: Use the Redlich-Kwong equation of state to find the pressure of carbon monoxide gas with a temperature of 310 K and a specific volume of $0.025\,\text{m}^3/\text{kg}$ and compare the value with that predicted by the ideal gas equation.

4.60: The pressure of R-134a vapor is 500 kPa at a temperature of 40°C. Determine the specific volume of the vapor by means of the van der Waals equation of state and compare the result with the value read from Table A.8.

4.61: Determine the compressibility factor for

(a) Argon at 200 K and $P = 5$ MPa

(b) Carbon dioxide at 500 K and $P = 10$ MPa

(c) Methane at 382.4 K and $P = 2.32$ MPa

(d) Helium at $T = 15.9$ K and $P = 0.46$ MPa

4.62: Oxygen gas in a container is at 210 K and has a specific volume of 0.002 m^3/kg. Use the Redlich-Kwong equation of state to determine the pressure of the oxygen and then compare your result with the value obtained from the ideal gas equation.

4.63: For oxygen at 80 K with a specific volume of $v = 0.0125$ m^3/kg, determine the pressure via (a) the ideal gas law, (b) the van der Waals equation of state, and (c) the Redlich-Kwong equation of state.

4.64: Hydrogen gas at 100 K has a specific volume of 1.98 m^3/kg. Use the Beattie-Bridgeman equation of state

$$P = \frac{\bar{R}T}{\bar{v}^2}\left(1 - \frac{c}{\bar{v}T^3}\right)(\bar{v} + B) - \frac{A}{\bar{v}^2} \qquad \text{(kPa)}$$

where

$$A = A_o\left(1 - \frac{a}{\bar{v}}\right) \quad \text{and} \quad B = B_o\left(1 - \frac{b}{\bar{v}}\right)$$

with

$$A_o = 20.0117 \qquad a = -0.00506$$
$$B_o = 0.02096 \qquad b = -0.04359$$
$$c = 0.0504$$

to find the pressure. All of the constants given will provide P in kPa as long as $\bar{v}$ is in m^3/kgmol.

5

Control Volume Mass and Energy Analysis

Chapter Objectives

- To discuss what is meant by an open system, or a control volume.
- To define what is meant by the conservation of mass with regard to a control volume.
- To develop a relationship between the mass rate of flow and the volumetric rate of flow.
- To discuss the conservation of energy and to show how energy equations, in accordance with the first law of thermodynamics, may be written for a control volume.
- To develop the concept of flow work.
- To consider, in detail and on an energy and first law of thermodynamics basis, the analysis and performance of several of the devices that may make up a thermal system.
- To give the definitions of synthesis and analysis and to provide a design example involving a "first law" heat balance.

5.1 Introduction

Several mechanical "real-world" devices lend themselves to control volume analysis. Examples include nozzles, diffusers, turbines, compressors, pumps, mixing chambers, and heat exchangers. This chapter gives illustrations of control volume analysis as it pertains to these devices.

We begin this development with a restatement of what is meant by closed and open systems. In Section 2.1, we defined a **closed system** or **control mass** as a system in which only energy (and not mass) may cross the system boundaries. An **open system** or **control volume** is a system in which **both** mass and energy can be transmitted across the system boundaries. The boundaries of a control volume are referred to as **control surfaces**. Flow through a control volume can be **steady** or **transient**. Steady flow implies no change with respect to time and transient flow occurs when the flow is not steady. The term **uniform flow** refers to flow that does not change with respect to location within a specified region.

5.2 The Control Volume

As we have indicated, the difference between the control mass and the control volume formulation lies in the movement of mass across system boundaries. The mass transferred will contain energy. We will examine the conservation of mass and the conservation of energy for a control volume providing several illustrations.

We turn next to a discussion of the conservation of mass (Section 5.3) and the conservation of energy (Section 5.4).

5.3 Conservation of Mass

5.3.1 The Mass Balance

We considered the conservation of mass principle in Section 3.1 where we noted that

> Mass can neither be created nor destroyed and, except in chemical processes, its composition cannot be altered from one form to another.

Thus, in the absence of nuclear reactions, mass is a **conserved property**, and the conservation of mass principle applied to the "multiple input-multiple output" control volume (designated by cv) shown in Figure 5.1 may be stated as

$$\left\{\begin{array}{c}\text{the net change}\\ \text{in the mass within}\\ \text{a control volume}\\ \text{during a time period}\end{array}\right\} =$$

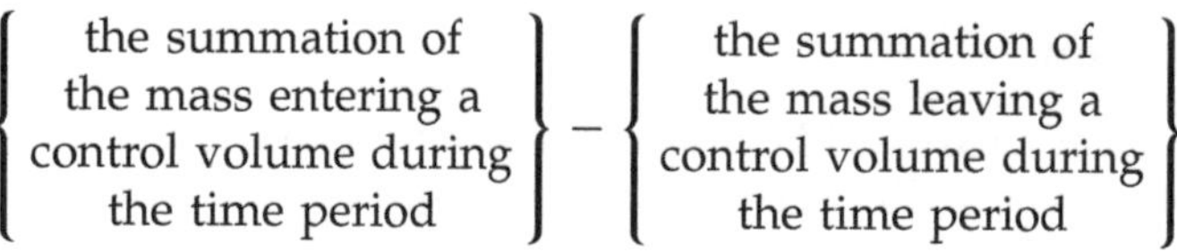

the summation of the mass entering a control volume during the time period
−
the summation of the mass leaving a control volume during the time period

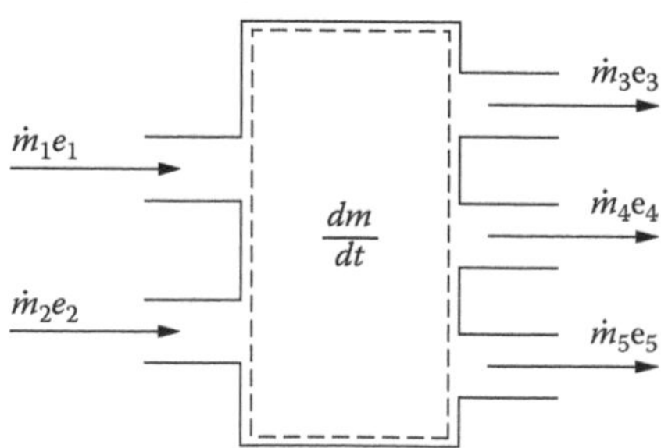

FIGURE 5.1
A control volume with multiple inputs and outputs.

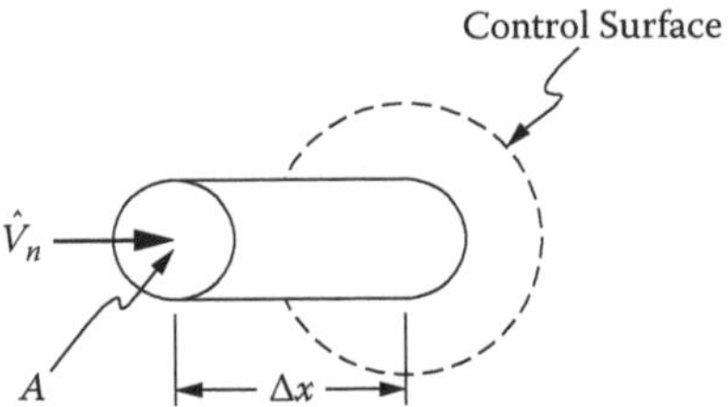

FIGURE 5.2
A fluid element of area A and length dx, located just upstream of a control volume.

On a rate basis, this may be written as

$$\begin{Bmatrix} \text{the rate of change} \\ \text{of the mass within} \\ \text{a control volume} \end{Bmatrix} = \begin{Bmatrix} \text{the rate of} \\ \text{mass entering} \\ \text{the control volume} \end{Bmatrix} - \begin{Bmatrix} \text{the rate of} \\ \text{mass leaving} \\ \text{the control volume} \end{Bmatrix} \tag{5.1}$$

or

$$\frac{dm_{\text{cv}}}{dt} = \sum_i \dot{m}_i - \sum_e \dot{m}_e \tag{5.2}$$

where $\dot{m}$ is defined as the mass flow rate, m_{cv} is the mass residing inside the control volume, and the subscripts, i and e, refer, respectively, to *all* inlets and exits. Because Equation 5.2 represents the conservation of mass, it is usually called the **mass balance** and is a form of what is referred to as a **continuity relationship** or merely as **continuity**.

In the case of **steady flow**, where no mass accumulates within the control volume, $dm_{\text{cv}}/dt = 0$ and Equation 5.2 reduces to

$$\sum_i \dot{m}_i = \sum_e \dot{m}_e \tag{5.3}$$

and for a "single inlet-single outlet" control volume with state point 1 at the inlet and state point 2 at the outlet

$$\dot{m}_1 = \dot{m}_2 \tag{5.4}$$

5.3.2 The Volumetric Flow Rate

The volumetric flow rate, typically in units of m^3/s, can be related to the mass flow rate, $\dot{m}$, in kg/s, in terms of the local fluid density (or specific volume) and flow area. A fluid element that is located just "upstream" of a control surface is shown in Figure 5.2. With a cross-sectional area, A, we see that the volume of the fluid element is

$$V = A\Delta x$$

where Δx is the distance that the fluid element traverses in time Δt. The average volumetric flow rate will be

$$\dot{V} = \lim_{\Delta t \to 0} A\frac{\Delta x}{\Delta t}$$

and because the normal velocity to the area dA is

$$\hat{V}_n = \lim_{\Delta t \to 0} \frac{\Delta x}{\Delta t}$$

we have

$$\dot{V} = \hat{V}_n A$$

Uniform flow applies to a flow where all of the measurable fluid properties are uniform throughout any particular inlet or exit area. If the flow is not uniform over the area, dA, then we must integrate over the area

$$\dot{V} = \int_A \hat{V}_n dA$$

and in this case, the mass flow rate will be

$$\dot{m} = \int_A \rho \hat{V}_n dA \tag{5.5}$$

If the flow is uniform over the entire cross-sectional area, Equation 5.5 reduces to

$$\dot{m} = \rho A \hat{V} \tag{5.5a}$$

or, because $v = 1/\rho$

$$\dot{m} = \frac{A\hat{V}}{v} \tag{5.5b}$$

where $\hat{V}$ denotes the uniform value of the velocity at the inlet or outlet being considered.

We call the product of the area, A, and the velocity $\hat{V}$

$$\dot{V} = A\hat{V} \tag{5.6}$$

the **volumetric flow rate**.

Example 5.1 A pipe with an inner diameter of 30 cm carries 10,000 L/min of water and supplies a pipe tee with two branches. The smaller branch has a diameter of 10 cm and carries 2000 L/min and the larger branch has a diameter of 17.5 cm. Determine the velocities in each pipe.

Solution

Assumptions and Specifications

(1) The control volume involves the pipe tee with the contained fluid.
(2) The water is incompressible.
(3) Steady and uniform flow prevails.
(4) There are no heat and work interactions with the environment.
(5) The potential energy change is negligible.

A sketch of the control volume with pertinent physical and flow data is shown in Figure 5.3. Note that the water flows are in liters per minute.

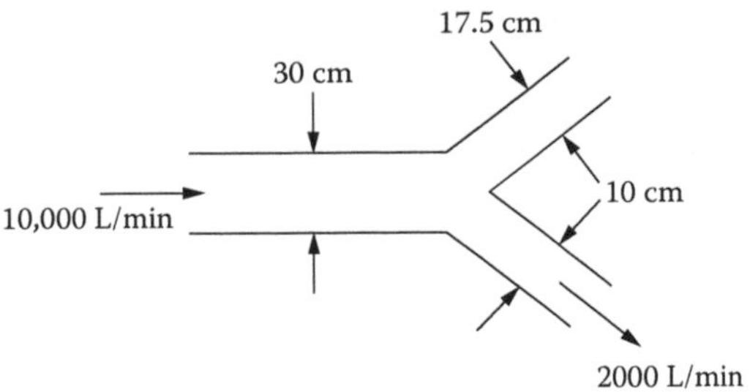

FIGURE 5.3
The control volume for Example 5.1.

The velocity at point 1 will be

$$\hat{V}_1 = \frac{\dot{V}_1}{A_1} = \frac{(10,000\,\text{L/min})(10^{-3}\,\text{m}^3/\text{L})}{\frac{\pi}{4}(0.30\,\text{m})^2} = 141.5\,\text{m/min}$$

or

$$\hat{V}_1 = 2.36\,\text{m/s} \Longleftarrow$$

At the pipe intersection continuity gives

$$\dot{V}_1 = \dot{V}_2 + \dot{V}_3$$

$$10,000\,\text{L/min} = \hat{V}_2 A_2 + 2,000\,\text{L/min}$$

The velocity at point 2 is therefore,

$$\hat{V}_2 = \frac{(10,000\,\text{L/min}^3 - 2000\,\text{L/min})(10^{-3}\,\text{m}^3/\text{L})}{\frac{\pi}{4}(0.175\text{ m})^2} = 332.6\,\text{m/min}$$

or

$$\hat{V}_2 = 5.54\,\text{m/s} \Longleftarrow$$

Finally, the velocity of the water at point 3 is

$$\hat{V}_3 = \frac{(2000\,\text{L/min})(10^{-3}\,\text{m}^3/\text{L})}{\frac{\pi}{4}(0.10\text{ m})^2} = 254.7\,\text{m/min}$$

or

$$\hat{V}_3 = 4.24\,\text{m/s} \Longleftarrow$$

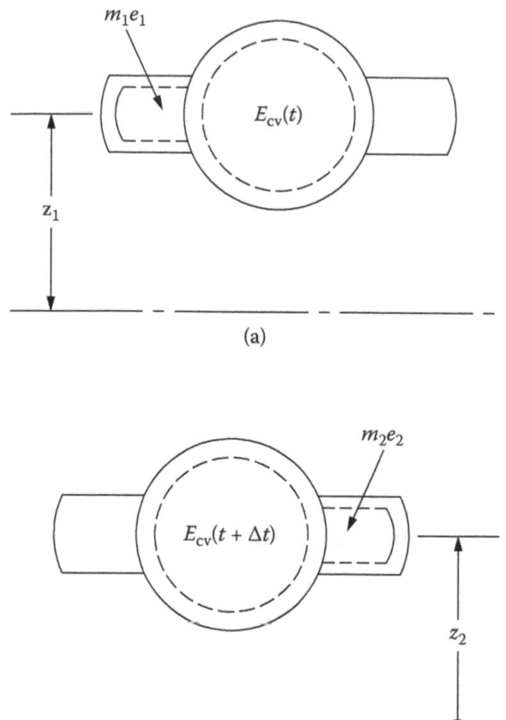

FIGURE 5.4
Control volume showing a parcel of mass located (a) just "upstream" of the control volume entrance and (b) just "downstream" of the control volume exit.

5.4 Conservation of Energy for a Control Volume

5.4.1 Introduction

We have noted that the significant difference between control volume and control mass analysis is that, in a control volume analysis, mass may flow across the system boundaries. Here, we will apply a form of the first law of thermodynamics to the control volume:

> Energy, a conservative property, can neither be created nor destroyed but can be changed or converted from one form to another.

We will then see that, because of mass flow, $\dot{m}$, we will need to replace the energy balance used in control mass analysis by an **energy rate balance** where the individual energy terms are in joules or kilojoules per second, which translate to watts or kilowatts.

The control volumes to be employed in our development are shown in Figure 5.4. For each basic control volume, the surfaces are indicated by dashed lines. In addition, however, each of these control volumes shows a small mass. In Figure 5.4a, this mass is designated by the subscript 1 and its location is at the inlet to the control volume at state point i. In Figure 5.4b, the small mass is located at the outlet of the control volume at state point e and is designated by the subscript 2.

This discussion is based on the fact that in the time interval, Δt, the entire mass m_1 passes into the control volume and at the end of this time period, a quantity of mass, originally within the control volume has passed out to the exit region as indicated in Figure 5.4b. This allows us to identify the mass within the control volume with the subscript cv.

We also specify that, consistent with the control mass analysis in Chapter 3, the conventions regarding the direction of the heat and work interactions through the control surfaces will be maintained. However, they will be energy rate quantities

$$\dot{W}_{cv} \longrightarrow \begin{cases} \text{done } \textbf{by} \text{ a system is positive} \\ \text{done } \textbf{on} \text{ a system is negative} \end{cases}$$

and

$$\dot{Q}_{cv} \longrightarrow \begin{cases} \text{transferred } \textbf{to} \text{ a system is positive} \\ \text{transferred } \textbf{from} \text{ a system is negative} \end{cases}$$

5.4.2 The Energy Rate Balance

We begin our study of the energy rate balance with the conservation of energy principle for the control mass that occupies the entire interior of the control volume with a modification of Equation 3.12

$$\Delta E = E_2 - E_1 = Q - W \tag{3.12}$$

which, on a rate basis, can be represented as

$$\frac{dE}{dt} = \dot{Q} - \dot{W}$$

Because this equation was written for a closed system, it cannot possibly account for any energy transfer due to the mass transport at the inlet and exit from the control volume.

However, we can write equations that provide the energy associated with the control mass at both times, $t + \Delta t$ and t

$$E\Big|_{t+\Delta t} = E_{cv}\Big|_{t+\Delta t} + E_2$$

and

$$E\Big|_{t} = E_{cv}\Big|_{t} + E_1$$

so that the change in energy within the control mass over the time interval, Δt, will be

$$E\Big|_{t+\Delta t} - E\Big|_{t} = E_{cv}\Big|_{t+\Delta t} - E_{cv}\Big|_{t} + E_2 - E_1$$

But

$$E_1 = m_1 e_1 \quad \text{and} \quad E_2 = m_2 e_2$$

so that after we divide throughout by Δt we obtain

$$\frac{E_{t+\Delta t} - E_t}{\Delta t} = \frac{E_{cv,t+\Delta t} - E_{cv,t}}{\Delta t} + \frac{m_2 e_2}{\Delta t} - \frac{m_1 e_1}{\Delta t}$$

or in the limit as $\Delta t \longrightarrow 0$

$$\frac{dE}{dt} = \frac{dE_{cv}}{dt} + \dot{m}_2 e_2 - \dot{m}_1 e_1$$

With this in Equation 3.12 we obtain

$$\frac{dE_{cv}}{dt} + \dot{m}_e e_e - \dot{m}_i e_i = \dot{Q}_{cv} - \dot{W} \tag{5.7}$$

where we have changed the subscripts so that $\dot{m}_1 \rightarrow \dot{m}_i$ and $\dot{m}_2 \rightarrow \dot{m}_e$ and we have added the subscript cv to the heat transmitted across the system boundaries. The reason for holding off on changing $\dot{W}$ to $\dot{W}_{cv}$ will be apparent in the next section.

At this point, we must make the distinction between work interactions at control surfaces where flow does and does not occur

$$\dot{W} = \dot{W}_{flow} + \dot{W}_{nonflow}$$

where these terms are work rates (or power) and are expressed in watts or kilowatts. The $\dot{W}_{nonflow}$ term represents the contribution of the work interaction across the control volume surfaces where no flow occurs. It is often referred to as Pdv work and we will refer to it as $\dot{W}_{cv}$

$$\dot{W}_{nonflow} = \dot{W}_{PdV} = \dot{W}_{cv}$$

5.4.3 Flow Work

Because control volume analysis can (and does) involve mass flow across its boundaries, we need to recognize the effects of the work required to push a mass of fluid into or out of the control volume. This work is called **flow work** and is represented as $\dot{W}_{flow}$.

In Figure 5.5 we see a portion of a control volume, its control surface and a single inlet located just "upstream" of the control volume. The inlet duct has a cross-sectional area, A, and, because of the pressure P that exists in the duct, there will be a force $F = PA$ on the element of mass m. Because work is the product of force and distance, we have the flow work

$$d\dot{W}_{flow} = PA\,dx$$

and, in this case, the flow work rate will be

$$\dot{W}_{flow} = \int PA\,dx = PA\hat{V}$$

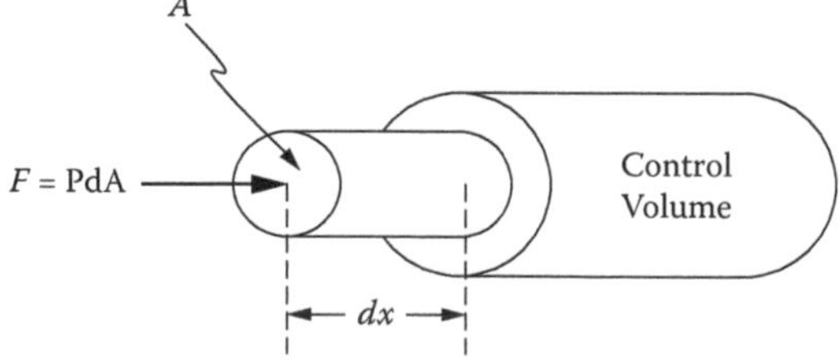

FIGURE 5.5
A control volume, its control surface, and a single inlet located just upstream of the control volume.

and because

$$\dot{m} = \frac{A\hat{V}}{v}$$

we have

$$\dot{W}_{\text{flow}} = \dot{m}(Pv) \tag{5.8}$$

For a single input-single output control volume with one-dimensional flow, we may write

$$\dot{W} = \dot{W}_{\text{nonflow}} + \dot{W}_{\text{flow}}$$

as

$$\dot{W} = \dot{W}_{\text{cv}} + \dot{m}_i P_i v_i - \dot{m}_e P_e v_e \tag{5.9}$$

and we are ready, at last, to write the first law energy equation for the control volume.

5.4.4 The Control Volume Energy Equation

The first law energy balance for the control volume will be

$$\left\{\begin{array}{c}\text{the net rate of}\\\text{change in the}\\\text{energy within}\\\text{the control volume}\end{array}\right\} = \left\{\begin{array}{c}\text{the rate at which}\\\text{energy crosses the}\\\text{control surfaces as}\\\text{heat and/or work}\end{array}\right\}$$

$$+\left\{\begin{array}{c}\text{the rate at which}\\\text{energy enters the control}\\\text{volume by mass flow}\end{array}\right\} - \left\{\begin{array}{c}\text{the rate at which}\\\text{energy exits the control}\\\text{volume by mass flow}\end{array}\right\}$$

With

$$e_i = u_i + \frac{\hat{V}_i^2}{2} + gz_i$$

and

$$e_e = u_e + \frac{\hat{V}_e^2}{2} + gz_e$$

this becomes, for a multiple input-multiple outlet system,

$$\frac{dE_{\text{cv}}}{dt} = \dot{Q}_{\text{cv}} - \dot{W}_{\text{cv}} + \sum_i \dot{m}_i \left(u_i + P_i v_i + \frac{\hat{V}_i^2}{2} + gz_i\right)$$

$$- \sum_e \dot{m}_e \left(u_e + P_e v_e + \frac{\hat{V}_e^2}{2} + gz_e\right) \tag{5.10}$$

The specific enthalpy was introduced in Section 3.8

$$h = u + Pv \tag{3.28b}$$

and may be employed here

$$\frac{dE_{cv}}{dt} = \dot{Q}_{cv} - \dot{W}_{cv} + \sum_i \dot{m}_i \left(h_i + \frac{\hat{V}_i^2}{2} + gz_i\right) - \sum_e \dot{m}_e \left(h_e + \frac{\hat{V}_e^2}{2} + gz_e\right) \tag{5.11}$$

If the flow is steady, $dE_{cv}/dt = 0$, and Equation 5.11 becomes

$$0 = \dot{Q}_{cv} - \dot{W}_{cv} + \sum_i \dot{m}_i \left(h_i + \frac{\hat{V}_i^2}{2} + gz_i\right) - \sum_e \dot{m}_e \left(h_e + \frac{\hat{V}_e^2}{2} + gz_e\right) \tag{5.12}$$

and for a single input-single output system in steady flow

$$0 = \dot{Q}_{cv} - \dot{W}_{cv} + \dot{m}\left(h_i + \frac{\hat{V}_i^2}{2} + gz_i\right) - \dot{m}\left(h_e + \frac{\hat{V}_e^2}{2} + gz_e\right) \tag{5.13}$$

5.5 Specific Heats of Incompressible Substances

An **incompressible substance** is one whose specific volume is assumed to be constant and whose specific internal energy is assumed to vary only with temperature. Because the change in specific volume is zero, it can be shown that the constant volume and constant pressure specific heats for an incompressible substance are equal. Thus, c_v and c_p may be replaced by c without a subscript

$$c_v = c_p = c \tag{5.14}$$

This indicates that the change in internal energy will be given by

$$u_2 - u_1 = c(T_2 - T_1) \tag{5.15}$$

and that the change in enthalpy will be

$$h_2 - h_1 = u_2 - u_1 + v_2 P_2 - v_1 P_1$$

Moreover, with constant specific volume, $v_1 = v_2 = v$ and

$$h_2 - h_1 = u_2 - u_1 + v(P_2 - P_1) \tag{5.16a}$$

or

$$h_2 - h_1 = c(T_2 - T_1) + v(P_2 - P_1) \tag{5.16b}$$

5.6 Applications of Control Volume Energy Analysis

5.6.1 Introduction

Control volume energy analysis is fundamental in the estimation of the performance of equipment of all types. Seven types will be reviewed here and in each case, a fairly comprehensive example will be provided. The types of equipment include

- The nozzle and the diffuser
- The turbine
- The compressor
- The pump
- The open type of heat exchanger
- The closed type of heat exchanger
- The throttling valve

5.6.2 The Nozzle and the Diffuser

A **nozzle** is a device that has a varying cross-sectional area that takes a fluid at high pressure and low velocity at its inlet and provides a high velocity stream at its outlet. The **diffuser** operates in exactly the opposite manner. It converts a high velocity low pressure fluid at the inlet to fluid at low velocity and a higher pressure at the outlet. We consider the nozzle in an example.

Figure 5.6 shows an artist's conception of both a nozzle and a diffuser. A combination of the two may be found in the sketch of the wind tunnel shown in Figure 5.7.

The control volume analysis of the nozzle, with its control volume as indicated in Figure 5.8, is based on the assumptions that

1. The flow is steady.
2. The rate of heat transfer across the walls of the nozzle is negligible.
3. There is no work interaction between the system represented by the control volume and the environment.
4. Potential energy changes between inlet and outlet are negligible.

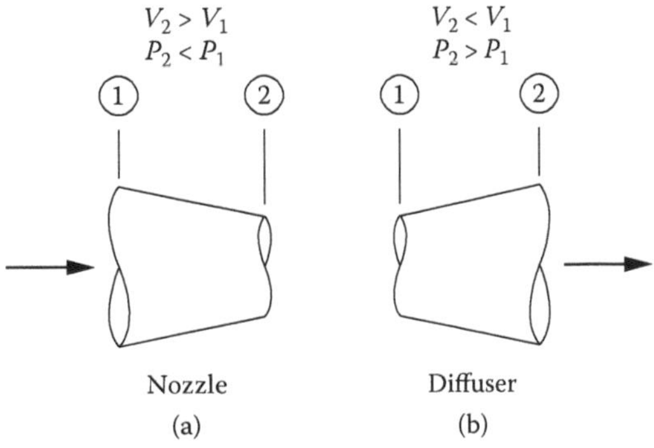

FIGURE 5.6
(a) A nozzle and (b) a diffuser.

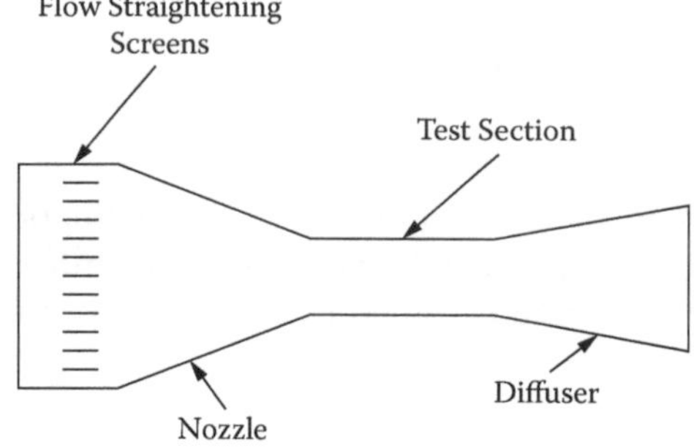

FIGURE 5.7
A wind tunnel may contain both a nozzle and diffuser.

With the foregoing assumptions in hand, we see that Equation 5.13 applies for this single input-single output system ($\dot{m}_1 = \dot{m}_2$) and that it may be written as

$$h_1 + \frac{\hat{V}_1^2}{2} = h_2 + \frac{\hat{V}_2^2}{2} \tag{5.17}$$

Example 5.2 A nozzle operates with steam. At the nozzle inlet (state point 1), $P_1 = 5\,\text{MPa}$, $T_1 = 600°\text{C}$ and $\hat{V}_1 = 12\,\text{m/s}$ and at the nozzle exit (state point 2), $P_2 = 1.5\,\text{MPa}$ and $\hat{V}_2 = 800\,\text{m/s}$. If the mass flow rate through the nozzle is 2.25 kg/s, determine the nozzle exit area.

Solution

Assumptions and Specifications

(1) The system is the steam in the nozzle in steady flow.
(2) The control volume is the nozzle.
(3) There are no heat and/or work interactions between the system and the surroundings.
(4) Potential energy changes are negligible.

The control volume, with the specified conditions, is shown in Figure 5.9. The strategy is to find the nozzle outlet area from

$$A_2 = \frac{\dot{m}_2 v_2}{\hat{V}_2}$$

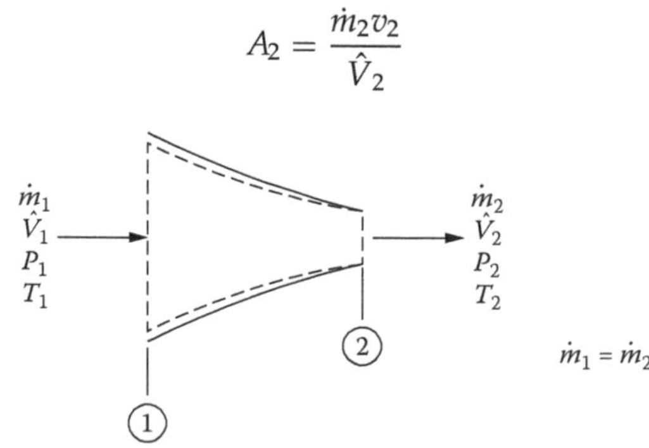

FIGURE 5.8
The control volume for the nozzle.

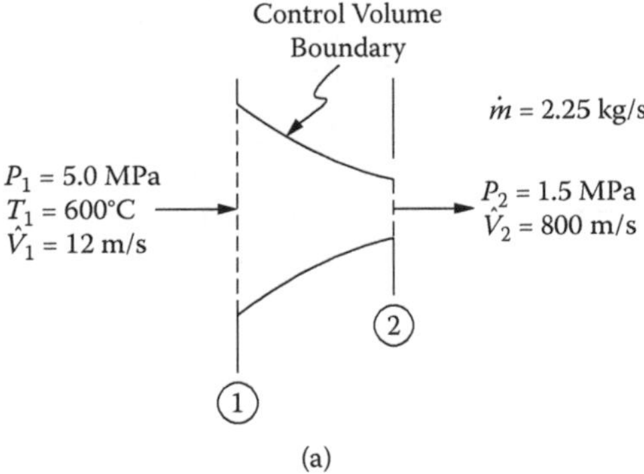

(a)

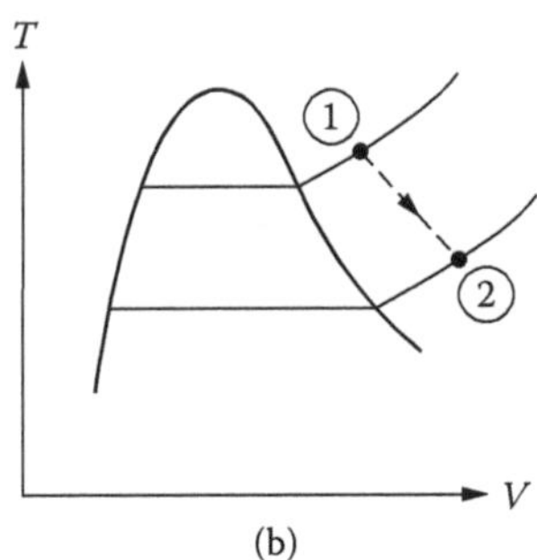

(b)

FIGURE 5.9
(a) The control volume for the nozzle in Example 5.2 and (b) the T-v diagram.

where the specific volume at the outlet, v_2, must be determined from P_2 and some other independent intensive property.

In this case, Equation 5.17 applies

$$h_1 + \frac{\hat{V}_1^2}{2} = h_2 + \frac{\hat{V}_2^2}{2} \tag{5.17}$$

and we may solve for h_2, which is the other independent intensive property

$$h_2 = h_1 + \frac{\hat{V}_1^2 - \hat{V}_2^2}{2}$$

Table A.5 in Appendix A may be used to find h_1 at $P_1 = 5\,\text{MPa}$ and $T_1 = 600°\text{C}$

$$h_1 = 3662.3\,\text{kJ/kg}$$

and with

$$\frac{\hat{V}_1^2 - \hat{V}_2^2}{2} = \frac{(12\,\text{m/s})^2 - (800\,\text{m/s})^2}{2}$$

$$= \frac{144\,\text{m}^2/\text{s}^2 - 6.40 \times 10^5\,\text{m}^2/\text{s}^2}{2}$$

$$= -3.199 \times 10^5\,\text{m}^2/\text{s}^2 = -319.93\,\text{kJ/kg}$$

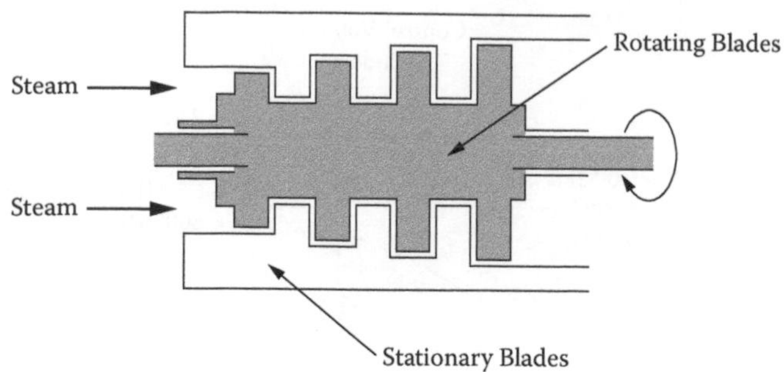

FIGURE 5.10
Cutaway view of an axial flow turbine.

the enthalpy at the outlet will be

$$\begin{aligned} h_2 &= h_1 + \frac{\hat{V}_1^2 - \hat{V}_2^2}{2} \\ &= 3662.3\,\text{kJ/kg} - 319.93\,\text{kJ/kg} \\ &= 3342.4\,\text{kJ/kg} \end{aligned}$$

The two independent intensive properties needed to establish v_2 are P_2 and h_2. Table A.5 with interpolation yields

$$T_2 = 440.6°\text{C} \quad \text{and} \quad v = 216.1 \times 10^{-3}\text{m}^3/\text{kg}$$

Thus,

$$A_2 = \frac{\dot{m}v_2}{\hat{V}_2} = \frac{(2.25\,\text{kg/s})(216.1 \times 10^{-3}\text{m}^3/\text{kg})}{800\,\text{m/s}} = 6.078 \times 10^{-4}\,\text{m}^2$$

or

$$A_2 = 6.078\,\text{cm}^2 \Longleftarrow$$

5.6.3 The Turbine

A reaction turbine is a device that possesses a set of blades that are mounted on a rotating shaft. Useful work is obtained from the **working fluid** as it expands between the inlet and outlet pressures. The gas turbine is treated in Chapter 8, the steam turbine is considered in Chapter 9, and impulse turbines are discussed in Chapters 14 and 18. A cutaway view of an axial flow turbine is shown in Figure 5.10. The example selected for this section is for a steam turbine.

The control volume analysis of a turbine is based on the assumptions that

1. The flow is steady.
2. Potential energy changes between inlet and outlet are negligible.

With the foregoing assumptions in hand, we see that Equation 5.13 applies for this single input-single output system and that it may be rewritten as

$$\dot{Q}_{\text{cv}} = \dot{W}_{\text{cv}} + \dot{m}\left(h_2 + \frac{\hat{V}_2^2}{2}\right) - \dot{m}\left(h_1 + \frac{\hat{V}_1^2}{2}\right) \tag{5.18}$$

Example 5.3 Steam enters a turbine at a flow rate of 1.75 kg/s and develops a shaft power of 1020 kW. Conditions at the inlet are $P_1 = 80\,\text{bar}$, $T_1 = 400°\text{C}$ and $\hat{V}_1 = 12\,\text{m/s}$. At the turbine outlet, $P_2 = 10\,\text{kPa}$, $\hat{V}_2 = 48\,\text{m/s}$ and the quality is 0.96 (96 %). The turbine operates at steady state and potential energy changes are negligible. Determine the heat loss through the turbine casing.

Solution

Assumptions and Specifications

(1) The control volume is the turbine.
(2) The steam in the turbine is the system.
(3) The system operates in the steady state.
(4) Changes in potential energy are negligible.

The configuration with the given data is shown in Figure 5.11 and in this case, Equation 5.18 applies

$$\dot{Q}_{cv} = \dot{W}_{cv} + \dot{m}\left(h_2 + \frac{\hat{V}_2^2}{2}\right) - \dot{m}\left(h_1 + \frac{\hat{V}_1^2}{2}\right)$$

or

$$\dot{Q}_{cv} = \dot{W}_{cv} + \dot{m}\left[(h_2 - h_1) + \left(\frac{\hat{V}_2^2 - \hat{V}_1^2}{2}\right)\right]$$

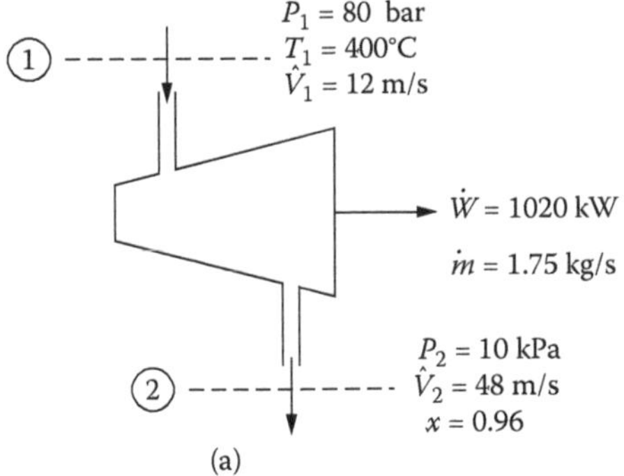

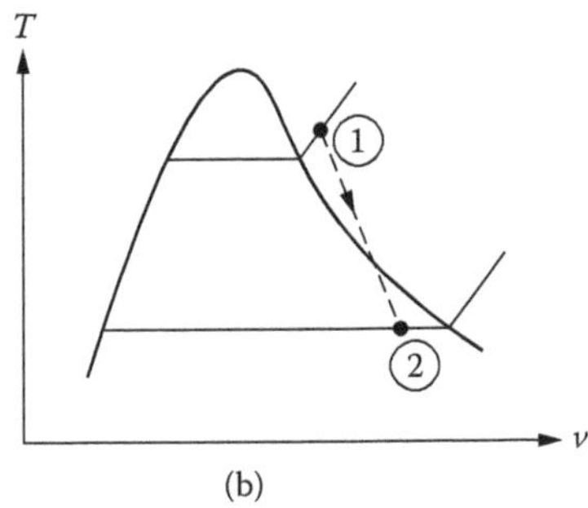

FIGURE 5.11
(a) The control volume for the turbine in Example 5.3 and (b) the T-v diagram.

The enthalpy at the inlet to the turbine is found from Table A.5 at 8 MPa and 400°C as

$$h_1 = 3140.3\,\text{kJ/kg}$$

and at $P_2 = 10\,\text{kPa}$, we find from Table A.4 that

$$h_f = 191.7\,\text{kJ/kg} \quad \text{and} \quad h_{fg} = 2393.3\,\text{kJ/kg}$$

Equation 4.6b then gives at state point 2

$$\begin{aligned} h_2 &= h_f + xh_{fg} \\ &= 191.7\,\text{kJ/kg} + (0.96)(2393.3\,\text{kJ/kg}) \\ &= 191.7\,\text{kJ/kg} + 2297.6\,\text{kJ/kg} \\ &= 2489.3\,\text{kJ/kg} \end{aligned}$$

Thus,

$$h_2 - h_1 = 2489.3\,\text{kJ/kg} - 3140.3\,\text{kJ/kg} = -651.0\,\text{kJ/kg}$$

and

$$\frac{\hat{V}_2^2 - \hat{V}_1^2}{2} = \frac{(48\,\text{m/s})^2 - (12\,\text{m/s})^2}{2} = 1080\,\text{m}^2/\text{s}^2 = 1.08\,\text{kJ/kg}$$

Thus,

$$\begin{aligned} \dot{Q}_{cv} &= \dot{W}_{cv} + \dot{m}\left[(h_2 - h_1) + \left(\frac{\hat{V}_2^2 - \hat{V}_1^2}{2}\right)\right] \\ &= 1020\,\text{kW} + (1.75\,\text{kg/s})(-651.0\,\text{kJ/kg} + 1.08\,\text{kJ/kg}) \\ &= 1020\,\text{kW} + (1.75\,\text{kg/s})(-649.9\,\text{kJ/kg}) \\ &= 1020\,\text{kW} - 1137.4\,\text{kW} = -117.4\,\text{kW} \Longleftarrow \end{aligned}$$

We note that this negative value means that heat is flowing from the turbine (the control volume) through the turbine casing *out* into the surrounding environment. We also note that the effect of the change in kinetic energy is much smaller than the effect of the specific enthalpy change.

5.6.4 The Compressor

A **compressor** is a device whose purpose is to convert the work expended to drive it into an increase in pressure of the **gas** passing through it. Compressors may be of the reciprocating type, the axial flow type, or the centrifugal type.

The control volume analysis of a compressor is based on the assumptions that

1. The flow is steady.
2. Kinetic and potential energy changes between inlet and outlet are negligible.

With the foregoing assumptions in hand, we see that Equation 5.13 applies for this single input-single output system and that it may be rewritten as

$$\dot{Q}_{cv} - \dot{W}_{cv} = \dot{m}(h_2 - h_1) \tag{5.19}$$

Example 5.4 Air enters a water-jacketed air compressor at $P_1 = 1\,\text{bar}$, $T_1 = 285\,\text{K}$ and a velocity of $\hat{V}_1 = 6\,\text{m/s}$ through an inlet area of $A_1 = 0.09375\,\text{m}^2$. At the exit, $P_2 = 8\,\text{bar}$, $T_2 = 480\,\text{K}$ and the velocity is $\hat{V}_2 = 1.5\,\text{m/s}$. Heat enters the water jacket at the rate of 2.75 kW. Use the ideal gas model and determine the power required to drive the compressor.

Solution

Assumptions and Specifications

(1) The control volume is the compressor.
(2) The air in the compressor is the system.
(3) The flow is steady.
(4) Kinetic and potential energy changes are negligible.
(5) The air may be treated as an ideal gas.

The configuration with the given data is shown in Figure 5.12. Here, the use of the ideal gas model is specified and we use Equation 5.19, which may be modified to give the work required to drive the compressor.

$$\dot{W}_{cv} = \dot{Q}_{cv} + \dot{m}(h_1 - h_2)$$

Because we are dealing with an ideal gas, the mass flow rate of the air can be found from the known conditions at the compressor inlet

$$\dot{m} = \frac{A_1 \hat{V}_1}{v_1} = \frac{A_1 \hat{V}_1 P_1}{RT_1}$$

and with R and c_p (which will be needed) obtained from Table A.1

$$R = 287\,\text{N-m/kg-K} \quad \text{and} \quad c_p = 1.005\,\text{kJ/kg-K}$$

we find

$$\dot{m} = \frac{(0.09375\,\text{m}^2)(6.0\,\text{m/s})(100\,\text{kPa})}{(287\,\text{N-m/kg-K})(285\,\text{K})} = 0.688\,\text{kg/s}$$

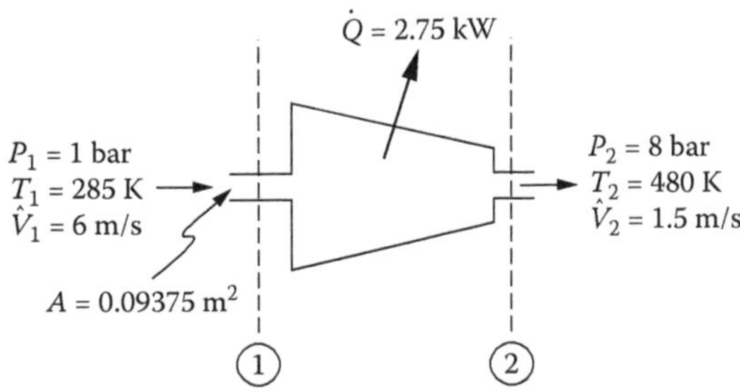

FIGURE 5.12
Control volume for the compressor in Example 5.4.

The change in specific enthalpy will be

$$\begin{aligned} h_1 - h_2 &= c_p(T_1 - T_2) \\ &= (1.005\,\text{kJ/kg-K})(285\,\text{K} - 480\,\text{K}) \\ &= (1.005\,\text{kJ/kg-K})(-195\,\text{K}) \\ &= -195.8\,\text{kJ/kg} \end{aligned}$$

Because the heat flow is out of the control volume, $\dot{Q}_{cv}$ is negative

$$\begin{aligned} \dot{W}_{cv} &= \dot{Q}_{cv} + \dot{m}(h_1 - h_2) \\ &= -2.75\,\text{kW} + (0.688\,\text{kg/s})(-195.8\,\text{kJ/kg}) \\ &= -2.75\,\text{kW} - 134.85\,\text{kJ/s} \\ &= -137.6\,\text{kW} \Longleftarrow \end{aligned}$$

Our sign convention for work indicates it to be *into* the control volume and is, thus, the power required to drive the compressor.

The horsepower required to drive the compressor will be

$$\text{HP} = (137.6\,\text{kW})(1.34\,\text{HP/kW}) = 184.4\,\text{HP} \Longleftarrow$$

5.6.5 Pumps

A **pump** is a device that changes the state of a liquid that passes through it as the working fluid. Usually, the work expended to drive the pump is converted to a change in **head**, either in the form of a change in pressure or a change in elevation. Pumps are treated in detail in Chapter 18.

The control volume analysis of a pump is based on the assumptions that

1. The flow is steady.
2. The heat flow across the casing of the pump is negligible.

With the foregoing assumptions in hand, we see that Equation 5.13 applies for this single input-single output system and that it may be rewritten as

$$\dot{W}_{cv} = \dot{m}_i\left(h_i + \frac{\hat{V}_i^2}{2} + gz_i\right) - \dot{m}_e\left(h_e + \frac{\hat{V}_e^2}{2} + gz_e\right)$$

or with $\dot{m}_i = \dot{m}_e = \dot{m}$ and the inlet and outlet designated by the subscripts 1 and 2

$$\dot{W}_{cv} = \dot{m}\left[(h_1 - h_2) + \left(\frac{\hat{V}_1^2 - \hat{V}_2^2}{2}\right) + g(z_1 - z_2)\right] \tag{5.20}$$

Example 5.5 Determine the required horsepower for a pump that carries water at a constant temperature and a flow rate of 8 kg/s through a piping system in which the outlet is 40 m higher than the inlet. The temperature in the pump and piping system is a constant 27°C. The pump conditions at the inlet are $P_1 = 1$ bar and $\hat{V}_1 = 3$ m/s and the conditions at the outlet are $P_2 = 2$ bar and $\hat{V}_2 = 12$ m/s. The pump is at a location where the acceleration of gravity may be taken as 9.81 m/s^2.

Solution

Assumptions and Specifications

(1) The pump is the control volume.
(2) The water at a constant temperature of 27°C is the system.
(3) The system operates in steady flow.
(4) There is no heat transfer to or from the surroundings.
(5) The acceleration of gravity may be taken as $9.81\,\text{m/s}^2$.

The configuration with the given data is shown in Figure 5.13 and we note that the acceleration of gravity at $9.81\,\text{m/s}^2$ is specified. Under these conditions, Equation 5.20 is the energy rate balance

$$\dot{W}_{\text{cv}} = \dot{m}\left[(h_1 - h_2) + \left(\frac{\hat{V}_1^2 - \hat{V}_2^2}{2}\right) + g(z_1 - z_2)\right] \tag{5.20}$$

We evaluate the kinetic and potential energy changes first.

$$\frac{\hat{V}_1^2 - \hat{V}_2^2}{2} = \frac{(3\,\text{m/s})^2 - (12\,\text{m/s})^2}{2} = -67.5\,\text{m}^2/\text{s}^2 = -0.0675\,\text{kJ/kg}$$

Then with $z_1 - z_2 = -40\,\text{m}$

$$g(z_1 - z_2) = (9.81\,\text{m/s}^2)(-40\,\text{m}) = -392.4\,\text{m}^2/\text{s}^2 = -0.392\,\text{kJ/kg}$$

Now

$$h_2 - h_1 = \left[h_f + v_f(P - P_{\text{sat}})\right]_2 - \left[h_f + v_f(P - P_{\text{sat}})\right]_1$$

and because $T_1 = T_2$, the P_{sat} and h_f terms cancel, leaving

$$h_2 - h_1 = v_f(P_2 - P_1)$$

The value of v_f is obtained from Table A.3 at $T_1 = T_2 = 27°\text{C}$ (with interpolation).

$$v_f = 1.0034 \times 10^{-3}\,\text{m}^3/\text{kg}$$

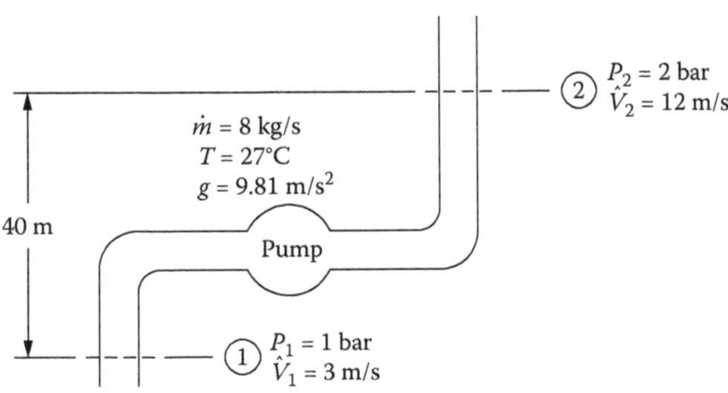

FIGURE 5.13
The control volume for the pump in Example 5.5.

so that

$$
\begin{aligned}
h_1 - h_2 &= v_f(P_1 - P_2) \\
&= (1.0034 \times 10^{-3}\,\text{m}^3/\text{kg})(1\,\text{bar} - 2\,\text{bar}) \\
&= (1.0034 \times 10^{-3}\,\text{m}^3/\text{kg})(-10^5\,\text{N}/\text{m}^2) \\
&= -100.3\,\text{N-m}/\text{kg} \quad \text{or} \quad -0.1003\,\text{kJ}/\text{kg}
\end{aligned}
$$

Then

$$
\dot{W}_{\text{cv}} = \dot{m}\left[(h_1 - h_2) + \left(\frac{\hat{V}_1^2 - \hat{V}_2^2}{2}\right) + g(z_1 - z_2)\right]
$$

and with a flow rate of 8 kg/s, we have

$$
\begin{aligned}
\dot{W}_{\text{cv}} &= (8\,\text{kg/s})(-0.100\,\text{kJ/kg} - 0.0675\,\text{kJ/kg} - 0.392\,\text{kJ/kg}) \\
&= (8\,\text{kg/s})(-0.560\,\text{kJ/kg}) = -4.48\,\text{kJ/s} \quad \text{or} \quad -4.48\,\text{kW}
\end{aligned}
$$

This negative value for the power interaction shows that work is done on the system. The horsepower required to drive the pump will be

$$
\text{HP} = (4.48\,\text{kW})(1.34\,\text{HP/kW}) = 6.00\,\text{HP} \Longleftarrow
$$

5.6.6 The Mixing Chamber

A mixing chamber may have more than one inlet and outlet and because of this, conservation of mass requires a mass balance. For steady flow, where no mass accumulates inside the control volume, Equation 5.3 pertains

$$
\sum_i \dot{m}_i = \sum_e \dot{m}_e \tag{5.2}
$$

After all mass flow rates are established we then write an energy rate balance based on

1. The flow is steady.
2. Heat flow across the control volume boundary is negligible.
3. There is no work interaction between the control volume and the surroundings.
4. All kinetic and potential energy changes are negligible.

With these assumptions, we employ a modification of Equation 5.12, which applies to a multiple input-multiple output system under steady flow conditions

$$
\sum_i \dot{m}_i h_i = \sum_e \dot{m}_e h_e \tag{5.21}
$$

Example 5.6 A water mixing chamber has two inlets and one outlet. At inlet 1, $P_1 = 8\,\text{bar}$, $T_1 = 200°\text{C}$, and $\dot{m}_1 = 50\,\text{kg/s}$. At inlet 2, $P_2 = 8\,\text{bar}$, $T_2 = 60°\text{C}$, and $A_2 = 22.5\,\text{cm}^2$. At the outlet, $P_3 = 8\,\text{bar}$ and $\dot{V}_3 = 0.075\,\text{m}^3/\text{s}$. Determine (a) the mass flow rate at the outlet, (b) the mass flow rate at inlet 2, and (c) the velocity at inlet 2.

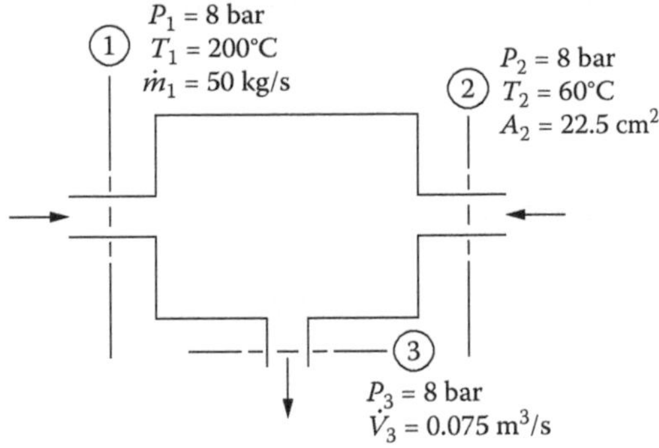

FIGURE 5.14
The control volume for the mixing chamber in Example 5.6.

Solution

Assumptions and Specifications

(1) The mixing chamber is taken as the control volume.
(2) The fluid within the chamber is taken as the system.
(3) Operation is in steady flow.
(4) There are negligible heat transfer and work interactions between the control volume and the surroundings.
(5) There are negligible kinetic and potential energy changes.

The configuration with the given data is shown in Figure 5.14.

(a) The mass flow rate at the outlet can be obtained from a combination of Equations 5.5b and 5.6

$$\dot{m}_3 = \frac{\dot{V}_3}{v_3}$$

where, v_3 is the specific volume for the saturated liquid at 0.80 MPa and Table A.4 reveals that

$$v_3 = v_f = 1.115 \times 10^{-3}\,\text{m}^3/\text{kg}$$

and

$$\dot{m}_3 = \frac{\dot{V}_3}{v_3} = \frac{0.075\,\text{m}^3/\text{s}}{1.115 \times 10^{-3}\,\text{m}^3/\text{kg}} = 67.27\,\text{kg/s} \Longleftarrow$$

(b) For this control volume with two inlets and one outlet, we may use Equation 5.3 to obtain

$$\dot{m}_1 + \dot{m}_2 = \dot{m}_3$$

so that

$$\dot{m}_2 = \dot{m}_3 - \dot{m}_1 = 67.27\,\text{kg/s} - 50\,\text{kg/s} = 17.27\,\text{kg/s} \Longleftarrow$$

(c) We find the velocity at inlet 2 from a modification of Equation 5.5b.

$$\hat{V}_2 = \frac{\dot{m}_2 v_2}{A_2}$$

and because state point 2 is a compressed liquid

$$v_2 \approx v_f(T_2)$$

Table A.3 at 60°C gives

$$v_f = 1.017 \times 10^{-3}\,\mathrm{m^3/kg}$$

so that

$$\hat{V}_2 = \frac{\dot{m}_2 v_2}{A_2} = \frac{(17.27\ \mathrm{kg/s})(1.017 \times 10^{-3}\,\mathrm{m^3/kg})}{22.5 \times 10^{-4}\,\mathrm{m^2}} = 7.81\,\mathrm{m/s} \Longleftarrow$$

5.6.7 Heat Exchangers

The primary purpose of a **heat exchanger** is to transfer energy between two fluids. They are treated in detail in Chapter 25. Here, we note that they are usually classified as **regenerators, open-type exchangers**, and **closed-type exchangers** or **recuperators.**

Regenerators are heat exchangers in which the hot and cold fluids flow alternately through the same space with as little mixing between the two streams as possible. The quantity of energy transferred is dependent upon the fluid and flow properties of the fluid streams as well as the geometry and thermal properties of the surface.

Open-type heat exchangers, as the name implies, are devices in which actual mixing of the two fluid streams occurs. The hot and cold fluids enter the open-type exchanger and leave as a single stream. We note that the mixing chamber considered in Section 5.6.6 and in Example 5.6 is an open-type heat exchanger.

In the closed type of heat exchanger or recuperator, the fluid streams do not come into direct contact with one another and are separated by a tube wall or surface. The heat transfer is by convection from the hotter fluid to the surface, conduction through the tube wall or surface by conduction and then by convection to the cooler fluid. Typical examples of this type of heat exchanger are indicated in Figure 5.15.

The energy rate balance for the open-type heat exchanger is much like that for the closed-type of exchanger or mixing chamber considered in Section 5.6.6. It is based on the following assumptions:

1. The flow is steady.
2. Because insulation can be employed to minimize the heat loss from the heat exchanger to the surroundings, the heat flow across the control volume boundary is assumed to be negligible.
3. Other than the energy carried by the fluid streams, there is no work interaction between the control volume and the surroundings.
4. All kinetic and potential energy changes are negligible.

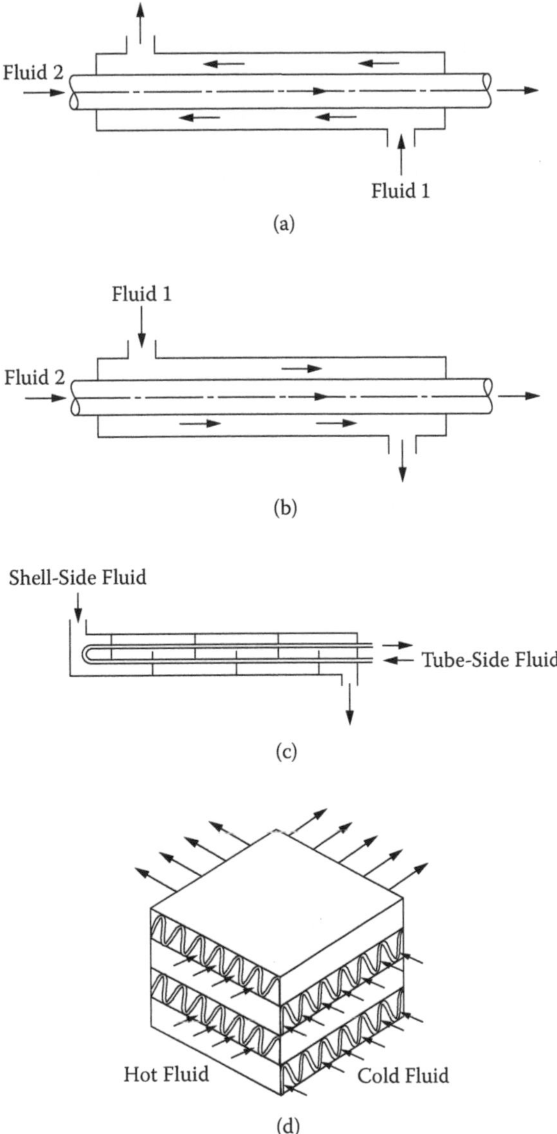

FIGURE 5.15
Examples of closed-type heat exchangers or recuperators: (a) the counterflow double-pipe exchanger, (b) the cocurrent or parallel flow double-pipe exchanger, (c) the shell-and-tube heat exchanger with one shell pass and two tube passes, and (d) the compact cross-flow exchanger.

With these assumptions, we employ Equations 5.3 and 5.12 that apply to the multiple input-multiple output system. Under steady-state conditions, Equation 5.3 is the mass balance and Equation 5.12 is the energy rate balance, which can be modified to

$$\sum_i \dot{m}_i h_i = \sum_e \dot{m}_e h_e \tag{5.22}$$

While we observe that this is the same equation used for the mixing chamber, we note that, for a two-fluid exchanger, there are *two* inlets and *two* outlets.

Example 5.7 A steam power plant condenser takes steam at its inlet at 8 kPa and a quality of 0.96 (96%) and the condensate leaves as a saturated liquid at 8 kPa. Cooling water is supplied at atmospheric pressure by a river at 25°C and leaves at 35°C. The condensate flow rate is 2.75 kg/s and there is no pressure drop between the cooling water inlet and outlet. Determine (a) the ratio of the mass flow rate of the cooling water to the mass flow rate of the condensate, (b) the mass flow rate of the cooling water, and (c) the heat transferred from the condensate to the cooling water.

Solution

Assumptions and Specifications

(1) The overall condenser is divided into two control volumes.
(2) The two flowing streams constitute the system.
(3) Steady-state conditions apply.
(4) There are no heat and/or work interactions between the overall control volume and the surroundings.
(5) Kinetic and potential energy changes are negligible.
(6) The cooling water is at constant pressure and its flow may be considered to be incompressible.

Figure 5.16a shows the control volume for the entire steam condenser with the pertinent inlet and outlet conditions; Figure 5.16b is the control volume for just the steam side of the condenser. We note, in Figure 5.16b, that there is heat flow from the condensing steam to the cooling water but that this heat flow does not appear in the overall control volume shown in Figure 5.16a. We also note that the cooling water side is at constant pressure.

(a) To find the ratio of the mass flow rate of the cooling water to the mass flow rate of the condensate, we first note that a mass balance for both the condensate and cooling water sides of the condenser will yield

$$\dot{m}_1 = \dot{m}_2 \quad \text{and} \quad \dot{m}_3 = \dot{m}_4$$

Then, for the control volume in Figure 5.18a, Equation 5.22 gives

$$\dot{m}_1 h_1 + \dot{m}_3 h_3 = \dot{m}_2 h_2 + \dot{m}_4 h_4$$

or

$$\dot{m}_1 (h_1 - h_2) = \dot{m}_3 (h_4 - h_3)$$

The mass flow rate ratio will be

$$\frac{\dot{m}_3}{\dot{m}_1} = \frac{h_1 - h_2}{h_4 - h_3}$$

and the problem becomes one of finding four enthalpy values.

We can pinpoint h_2 as h_f at $P_2 = 8\,\text{kPa}$ and we find h_f and h_{fg} at 8 kPa from Table A.4 at $T_{sat,2} = 41.5°\text{C}$ as

$$h_2 = h_f = 173.8\,\text{kJ/kg} \quad \text{and} \quad h_{fg} = 2403.6\,\text{kJ/kg}$$

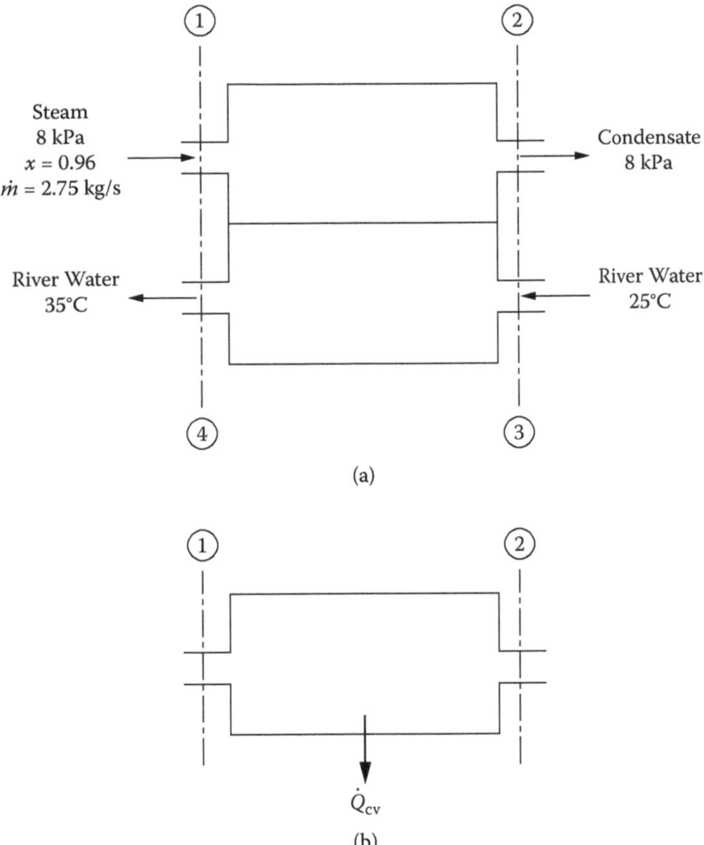

FIGURE 5.16
(a) The overall control volume for the entire steam condenser and (b) the control volume for the steam side of the condenser.

Then, use of Equation 4.6b gives h_1

$$\begin{aligned} h_1 &= h_f + xh_{fg} \\ &= 173.8\,\text{kJ/kg} + (0.96)(2403.6\,\text{kJ/kg}) \\ &= 173.8\,\text{kJ/kg} + 2307.5\,\text{kJ/kg} \\ &= 2481.3\,\text{kJ/kg} \end{aligned}$$

and

$$h_1 - h_2 = 2481.3\,\text{kJ/kg} - 173.8\,\text{kJ/kg} = 2307.5\,\text{kJ/kg}$$

For the cooling water, the actual pressure is much greater than the saturation pressure at both T_3 and T_4. This puts h_3 and h_4 into the compressed liquid region and, as indicated in Chapter 4, for compressed liquids at small temperature differences,

$$h_4 - h_3 = h_{f,4} - h_{f,3}$$

Thus, we read Table A.3 at $T_3 = 25°\text{C}$ and $T_4 = 35°\text{C}$ to obtain

$$h_4 = 146.5\,\text{kJ/kg} \quad \text{and} \quad h_3 = 104.8\,\text{kJ/kg}$$

so that

$$h_4 - h_3 = 146.5\,\text{kJ/kg} - 104.8\,\text{kJ/kg} = 41.7\,\text{kJ/kg}$$

Therefore,

$$\begin{aligned}\frac{\dot{m}_3}{\dot{m}_1} &= \frac{h_1 - h_2}{h_4 - h_3} \\ &= \frac{2307.5\,\text{kJ/kg}}{41.7\,\text{kJ/kg}} \\ &= 55.4 \Longleftarrow\end{aligned}$$

(b) We establish the heat transferred between the condensing steam and the cooling water by employing the control volume of Figure 5.13b and we write the simple energy rate balance

$$\dot{Q}_{cv} = \dot{m}(h_2 - h_1)$$

Hence,

$$\begin{aligned}\dot{Q}_{cv} &= \dot{m}(h_2 - h_1) \\ &= (2.75\,\text{kg/s})(173.8\,\text{kJ/kg} - 2481.3\,\text{kJ/kg}) \\ &= (2.75\,\text{kg/s})(-2307.5,\,\text{kJ/kg}) \\ &= -6345.6\,\text{kJ/s} \quad \text{or} \quad -6.345\,\text{MW} \Longleftarrow\end{aligned}$$

The negative sign indicates that heat is flowing *out* of the control volume that contains the condensing steam.

(c) The cooling water flow rate is merely

$$\dot{m}_3 = 55.4\dot{m}_1 = (55.4)(2.75\,\text{kg/s}) = 152.35\,\text{kg/s} \Longleftarrow$$

5.6.8 The Throttling Valve

A **throttling process** is a steady flow process that occurs across a flow restriction with a subsequent reduction in the pressure of the flowing fluid. **Throttling devices,** as indicated in Figure 5.17, are usually in the form of a **partially opened valve** or a **porous plug. Capillary tubes** are also employed.

A typical example of a throttling device is the **throttling calorimeter** (Figure 5.18), which measures the quality in a liquid vapor mixture.

The analysis of the throttling device will be governed by the assumptions:

1. The flow is steady.
2. The heat flow across the control volume boundary is negligible.
3. There is no work interaction between the control volume and the surroundings.
4. All kinetic and potential energy changes are negligible.

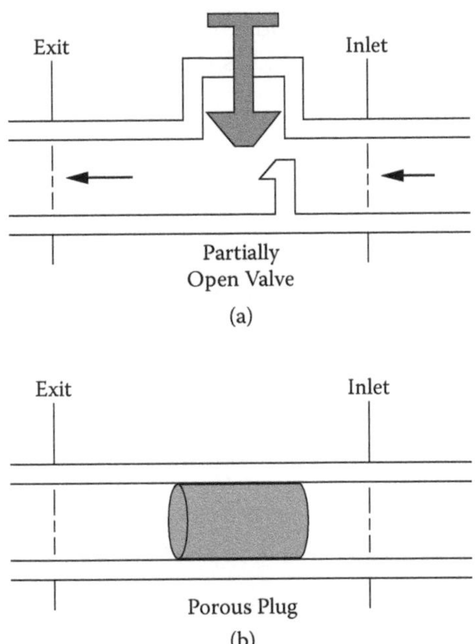

FIGURE 5.17
Two throttling devices.

With regard to Figure 5.18, these assumptions show that the mass balance of Equation 5.3 for this single input-single output system gives $\dot{m}_1 = \dot{m}_2 = \dot{m}$ and that the energy rate balance of Equation 5.13 reduces to

$$\dot{m}_1 h_1 = \dot{m}_2 h_2$$

or because $\dot{m}_1 = \dot{m}_2$

$$h_2 = h_1 \tag{5.23}$$

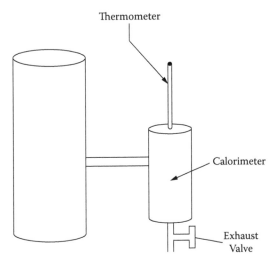

FIGURE 5.18
Artist's conception of the throttling calorimeter showing its control volume.

Example 5.8 Steam in the form of a liquid-vapor mixture at 2 MPa passes through a throttling calorimeter (throttling valve) and is discharged at 1 bar and 150°C. Determine the quality of the steam in the supply line.

Solution

Assumptions and Specifications

(1) The control volume is the calorimeter shown in Figure 5.19.
(2) The steam in the calorimeter is the system.
(3) Steady-state conditions apply.
(4) There are no heat and/or work interactions between the control volume and the surroundings.
(5) Kinetic energy and potential energy changes are negligible.

Figure 5.19 shows the control volume for the device and the foregoing assumptions and the fact that this is specified as a throttling valve allows us to use Equation 5.23

$$h_1 = h_2$$

Here, state point 2 is in the superheated region and we may read Table A.5 at 0.1 MPa and 150°C to obtain

$$h_2 = 2776.8\,\text{kJ/kg}$$

Then, with $h_1 = h_2 = 2776.8\,\text{kJ/kg}$, we use Table A.4 at 2 MPa to obtain

$$h_f = 908.3\,\text{kJ/kg} \quad \text{and} \quad h_{fg} = 1888.7\,\text{kJ/kg}$$

and, from Equation 4.8, we find that

$$\begin{aligned} x &= \frac{h_1 - h_f}{h_{fg}} \\ &= \frac{2776.8\,\text{kJ/kg} - 908.3\,\text{kJ/kg}}{1888.7\,\text{kJ/kg}} \\ &= \frac{1868.5\,\text{kJ/kg}}{1888.7\,\text{kJ/kg}} \\ &= 0.989 \quad \text{or} \quad (98.9\%) \end{aligned}$$

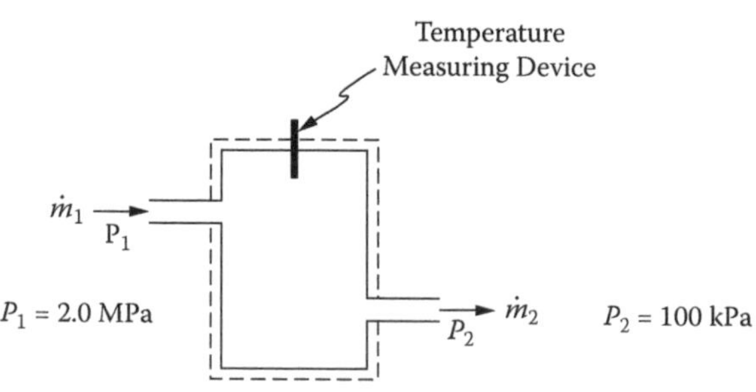

FIGURE 5.19
The control volume for the throttling calorimeter of Example 5.8.

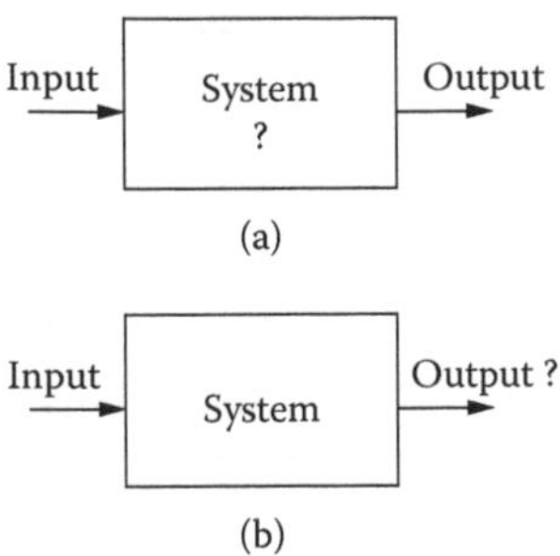

FIGURE 5.20
Two systems. In (a) the input and output are both specified and the system is to be designed or **synthesized**. In (b) the system is specified and the output is to be found for a given input. This is an example of an **analysis** problem.

5.7 Synthesis or Analysis?

Synthesis refers to a procedure where a system is developed to achieve a desired result. On the other hand, **analysis** refers to the determination of the response of a specified system to a prescribed input. These may be cast into the system input-output framework of Figure 5.20 where synthesis and analysis are both described.

As a somewhat trivial example, consider a system consisting of a single resistor shown in Figure 5.21. The input to this system is an applied voltage of 110 V and it is desired to obtain 5.5 A as the output. A simple synthesis procedure may be employed to obtain

$$R = \frac{V}{I} = \frac{110\,\text{V}}{5.5\,\text{A}} = 20\,\Omega$$

which shows that the system consists of a single 20 Ω resistor.

If the resistance (system) is set at 40 Ω, then the solution to the analysis problem for the same system input of 110 V will be

$$I - \frac{V}{R} = \frac{110\,\text{V}}{40\,\Omega} = 2.75\,\text{A}$$

Almost all of the problems encountered in the thermal fluid sciences are not as simple as the foregoing "resistor" problem. In such cases, the system must be **synthesized by analysis** in which a system is selected and then subjected to analysis to determine whether the results are satisfactory. If not, the system is adjusted and the procedure is repeated until the specified result is achieved. This procedure may require several iterations and it is easy to see that the synthesis procedure may often be a monumental trial-and-error analysis problem.

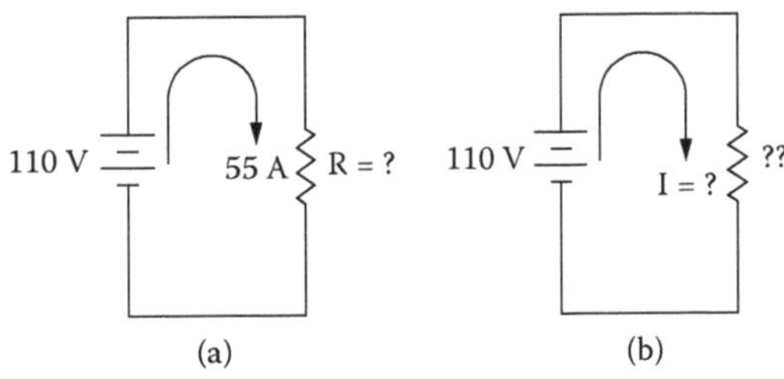

FIGURE 5.21
The simple system illustrating (a) the synthesis problem and (b) the analysis problem.

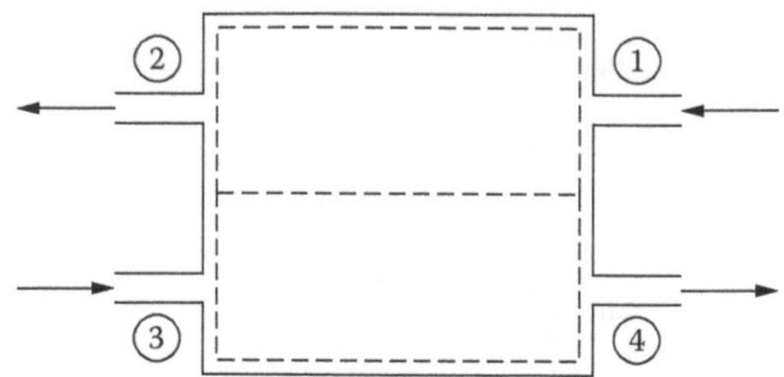

FIGURE 5.22
Control volume for a two-fluid heat exchanger.

5.8 The First Law Heat Balance

It was observed in Section 5.6.7 that the energy rate balance for a two-fluid heat exchanger (Figure 5.22) is given by Equation 5.22

$$\dot{Q} = \sum_i \dot{m}_i h_i = \sum_e \dot{m}_e h_e$$

and for the case of a surface dissipating heat to a single fluid (Figure 5.23), this can be further modified to

$$\dot{Q} = \dot{m}(h_e - h_i)$$

Almost every problem involving the flow of heat between two fluids or between a surface and a flowing fluid requires a first law energy rate balance or **heat balance**.

Indeed, in Figure 5.22 for a liquid ($\dot{m}_a$) flowing between two temperatures, T_1 and T_2 and another liquid ($\dot{m}_b$) flowing between two temperatures, T_3 and T_4, we have

$$\dot{Q} = \dot{m}_a(T_1 - T_2) = \dot{m}_b(T_4 - T_3)$$

This is the heat balance showing that liquid a is the hotter fluid and liquid b is the cooler fluid.

In Figure 5.23, with a single fluid flowing at a mass flow rate of $\dot{m}$ between two temperatures T_1 and T_2, the heat balance for a surface dissipation of $\dot{Q}$ will be

$$\dot{Q} = \dot{m}c(T_1 - T_2)$$

A firm grasp on this fundamental problem, the so-called **heat lost equals heat gained problem** is required. Otherwise, the analyst will be shooting at a moving target and disappointment will be inevitable.

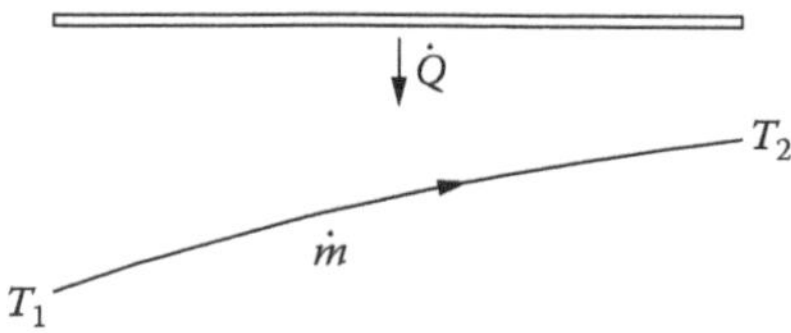

FIGURE 5.23
A surface dissipating heat to a single fluid.

In Chapter 22, attention will be directed to the liquid cooling of an electronic component having a high heat dissipation. The determination of the component surface temperature requires an input from at least two of the disciplines treated in this book:

- It begins with a "first law" consideration in which the range of liquid flows to accommodate the specified heat dissipation are established.
- The surface temperature is then found using procedures developed in Chapter 22.

Our attention now turns to the "first law" portion of the problem.

5.9 Design Example 1

A liquid cooled traveling wave tube collector is to be cooled in an application where the **uniform** dissipation in the collector is 1500 W. Coolanol-45 (Monsanto Chemical Co.), which enters the collector at 77°C, is to be used. The collector material is copper ($k = 385$ W/m-K) and in order to prevent "outgassing" of the copper that will destroy the vacuum, the walls of the collector cavity must be held to a maximum temperature of 150°C. The details of the collector cavity and the properties of coolanol-45 are provided in the second part of this design problem in Chapter 22.

Coolanol-45 is but one of several liquids that are available and they all meet the requirements of high dielectric strength, chemical inertness, thermal decomposition and impurities, effects of moisture, pour and flash points, flammability, toxicity, surface tension, and vapor pressure. Note that water is not suitable because of its low dielectric strength.

Solution

Assumptions and Specifications

(1) Steady-state conditions exist.
(2) The collector cavity walls are isothermal.
(3) The heat flow is radially outward from the collector cavity to the coolant passage.
(4) There is no heat flow longitudinally toward the ends of the collector.
(5) Thermal properties vary with temperature in a prescribed manner.
(6) The specific heat of the coolanol-45 varies in accordance with

$$c = 3.40 \times 10^{-3}T + 1.818 \text{ kJ/kg-K}$$

where T is in °C.

The calculations for the heat balance are displayed in Table 5.1. Observe that the liquid exit temperature is assumed and a specific heat is obtained. The coolant temperature increases are then calculated and the resulting exit temperature is then compared to the value assumed. It is almost always necessary to repeat the procedure until the exit temperatures match.

This concludes the first law portion of the problem. The second portion of the problem involves forced convection heat transfer and the problem will be completed in Chapter 22.

Assume, $\dot{m}$, kg/s	0.100	0.125	0.150	0.175	0.200
T_1, °C	77.0	77.0	77.0	77.0	77.0
Assume T_2, °C	84.6	83.0	81.8	81.0	80.6
T_b, °C	80.8	80.0	79.4	79.0	78.8
c at T_b, J/kg-K	2093	2090	2088	2087	2086
$\Delta T = \dot{Q}/\dot{m}c$, °C	7.2	5.7	4.8	4.1	3.6
T_2, °C	84.2	82.7	81.8	81.1	80.6
T_2 OK?	no	no	yes	yes	yes
T_b, °C	80.6	79.9			
c at T_b, J/kg-K	2092	2089			
$\Delta T = \dot{Q}/\dot{m}c$, °C	7.2	5.7			
T_2, °C	84.2	82.7			
T_2 OK?	yes	yes			

5.10 Summary

A control volume represents an open system in which both mass and energy may be transferred across the system boundaries. By contrast, in Section 2.2.1, a **closed system,** often referred to as a **control mass,** was defined as a system in which energy may be transferred across its boundaries but there is no transfer of mass across the boundaries from or to the surroundings.

The volumetric flow rate, designated as $\dot{V}$, is related to the fluid velocity and the area of the flow cross section

$$\dot{V} = A\hat{V} = \frac{\dot{m}}{v}$$

The conservation of energy principle leads to the energy rate balance for a control volume, which is

$$\frac{dE_{cv}}{dt} = \dot{Q}_{cv} - \dot{W}_{cv} + \sum_i \dot{m}_i \left(h_i + \frac{\hat{V}_i^2}{2} + gz_i\right) - \sum_e \dot{m}_e \left(h_e + \frac{\hat{V}_e^2}{2} + gz_e\right) \tag{5.11}$$

For steady flow, this becomes

$$0 = \dot{Q}_{cv} - \dot{W}_{cv} + \sum_i \dot{m}_i \left(h_i + \frac{\hat{V}_i^2}{2} + gz_i\right) - \sum_e \dot{m}_e \left(h_e + \frac{\hat{V}_e^2}{2} + gz_e\right)$$

and for a single input-single output system, we have

$$0 = \dot{Q}_{cv} - \dot{W}_{cv} + \dot{m} \left(h_i + \frac{\hat{V}_i^2}{2} + gz_i\right) - \dot{m} \left(h_e + \frac{\hat{V}_e^2}{2} + gz_e\right) \tag{5.13}$$

For incompressible substances, no distinction is made between specific heats at constant volume and constant pressure. Thus, $c_p = c_v = c$ and

$$u_2 - u_1 = c(T_2 - T_1) \quad \text{and} \quad h_2 - h_1 = c(T_2 - T_1) + v(P_2 - P_1)$$

Control volume energy analysis is fundamental in the estimation of the performance of several types of equipment.

- For the control volume analysis of the nozzle and diffuser

$$h_1 + \frac{\hat{V}_1^2}{2} = h_2 + \frac{\hat{V}_2^2}{2} \tag{5.17}$$

- For the turbine

$$\dot{Q}_{cv} = \dot{W}_{cv} + \dot{m}\left(h_2 + \frac{\hat{V}_2^2}{2}\right) - \dot{m}\left(h_1 + \frac{\hat{V}_1^2}{2}\right) \tag{5.18}$$

- For the compressor

$$\dot{Q}_{cv} - \dot{W}_{cv} = \dot{m}(h_2 - h_1) \tag{5.19}$$

- For the pump

$$\dot{W}_{cv} = \dot{m}\left[(h_1 - h_2) + \left(\frac{\hat{V}_1^2 - \hat{V}_2^2}{2}\right) + g(z_1 - z_2)\right] \tag{5.20}$$

- For the mixing chamber

$$\sum_i \dot{m}_i h_i = \sum_e \dot{m}_e h_e \tag{5.21}$$

- For the heat exchanger

$$\sum_i \dot{m}_i h_i = \sum_e \dot{m}_e h_e \tag{5.22}$$

- For the throttling valve

$$h_2 = h_1 \tag{5.23}$$

5.11 Problems

Energy Equations

5.1: A steady-flow system has a mass flow rate of 4 kg/s of fluid entering the system at $P_1 = 0.6875$ MPa, $\rho_1 = 3\,\text{kg/m}^3$, $\hat{V}_1 = 40$ m/s, and $u_1 = 1600$ kJ/kg. Conditions at the outlet of the system are $P_2 = 0.125$ MPa, $\rho_2 = 0.75\,\text{kg/m}^3$, $\hat{V}_2 = 200$ m/s, and $u_2 = 1480$ kJ/kg. Each kilogram rejects 12 kJ of heat during its passage through the system. Determine the work done.

5.2: A steady-flow system takes in 0.75 kg/s of a fluid and discharges it at a point 27.5 m above the inlet. The fluid enters at a velocity of $\hat{V}_1 = 60$ m/s and leaves at $\hat{V}_2 = 20$ m/s.

During the process, 2 kW of heat are supplied from an external source and the increase in enthalpy is 1 kJ/kg. Determine the work done by the system.

5.3: A steam-generating unit receives 144,000 kg/h of water as saturated liquid at 280°C. The water leaves as saturated steam at 4 MPa. The heating value of the coal is 30,000 kJ/kg and air to support the combustion is supplied in the ratio of 13.5 kg of air/kg of coal. The air enters with an enthalpy of 50 kJ/kg and leaves with an enthalpy of 300 kJ/kg. Determine the amount of coal supplied.

5.4: A cylindrical tank with a diameter of 1.50 m has a hole in the bottom that has a diameter of 1.25 cm; a steady stream of water flows from the hole. Determine the quantity of flow, $\dot{V}$, that must be added to the contents of the tank to maintain the water level at 2.5 m.

5.5: Water at 10 bar (gage) and 25°C enters a hydraulic turbine with a volumetric flow rate of 40 m^3/s at a velocity of 1.25 m/s at a height of 125 m above a given datum. The water leaves the turbine at 2 bar (gage) and 25.2°C at the datum level at a velocity of 10 m/s. Assume that the water is incompressible and that the turbine loses 4 J/kg of water flowing through it and determine (a) Δu, (b) Δ ke, (c) Δ pe, (d) Δh, and (e) $\dot{W}_{cv}$.

5.6: Assuming that water is incompressible and that no heat is transferred, determine the power required to pump 20 kg/s of water at 101 kPa and 75°C from a tank whose outlet is 5 m below ground level if the pipe diameter is 10 cm.

5.7: Refrigerant R-134a flows through a horizontal pipe reducer with the following inlet (point 1) and outlet (point 2) conditions:

Inlet		Outlet
10	P, bar	5
40	T°C	70
6.25	$\hat{V}$, m/s	
7.52	d, cm	5.08

Determine (a) the mass flow and (b) the exit velocity.

5.8: An insulated container is large enough to contain 4 kg of water and 1 kg of ice at 0°C with heat of fusion of 333 kJ/kg. A copper sphere having a mass of 16 kg at 98°C is dropped without splashing into the water-ice mixture. Determine the equilibrium temperature of the water-ice-copper system.

5.9: Determine how long it will take to heat a mountain cabin measuring 8 m × 4 m × 3 m containing air at 0°C to a comfort level of 21°C using a 2-kW heater.

5.10: A compressible fluid is flowing in a device with a single input and a single output. The data for the fluid and the device are as follows:

Inlet (point 1)		Outlet (point 2)
6.8	P, bar abs	4.0
6.25	ρ, kg/m^3	0.80
58	$\hat{V}$, m/s	680
290	u, kJ/kg	125
0.720	c_v, kJ/kg-K	0.720
0.280	R, kJ/kg-K	0.280
12.70	d, cm	10.37
0	z, m	4

Determine between inlet and outlet (a) Δ ke, (b) Δh, (c) Δ pe, (d) the mass flow rate, (e) the total heat transferred if no work is done between inlet and outlet, and (f) the specific heat at constant pressure if an ideal gas is assumed.

5.11: Water flows at the rate of 750 L/min through the venturi shown in Figure P5.11. Assume that $\dot{Q}_{cv} = \dot{W}_{cv} = \Delta u = 0$ and determine the pressure drop between points 1 and 2.

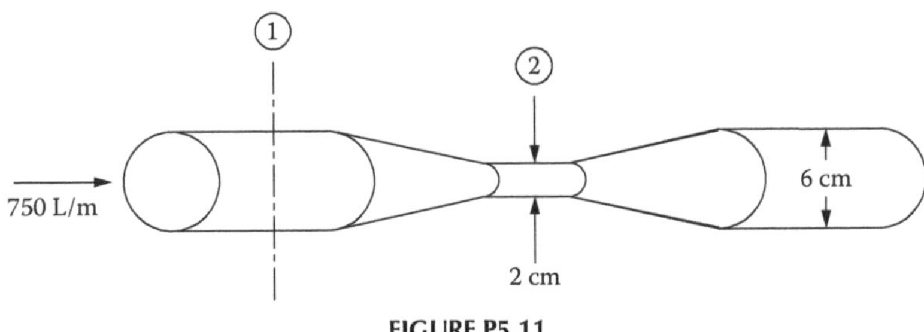

FIGURE P5.11

5.12: A compressor takes in 30 m³/min of air having a density of 1.266 kg/m³. The device discharges the air with a density of 4.870 kg/m³. At the inlet, $P_1 = 100$ kPa and at the outlet, $P_2 = 550$ kPa. The increase in internal energy is 15 kJ/kg and the heat rejected by the air in passing through the compressor is 6 kJ/kg. Neglecting changes in kinetic and potential energy, determine the work required to drive the compressor.

Nozzles and Diffusers

5.13: Steam at 20 bar and 800°C at a velocity of 75 m/s is provided to the inlet (point 1) of a nozzle. Determine the exit velocity (point 2) if the exit conditions are a liquid-vapor mixture at 5 bar and 92% quality.

5.14: Air at 4 bar and 67°C enters a nozzle with a mass flow rate of 0.625 kg/s. The exit area is 0.0025 m² and the exit pressure and velocity are 1.20 bar and 240 m/s, respectively. Determine (a) the exit temperature assuming that the air is an ideal gas and that constant specific heats may be used, (b) the change in specific kinetic energy, and (c) the inlet velocity.

5.15: Steam at 4 MPa and 400°C at a mass flow rate of 0.75 kg/s is introduced at negligible inlet velocity to a horizontal nozzle. The steam exits the nozzle at 1.5 MPa at a velocity of 750 m/s. Determine (a) the exit condition (temperature or quality) and (b) the exit area.

5.16: Air, which may be treated as an ideal gas, enters a converging nozzle at 400 kPa and 247°C with a velocity of 40 m/s through an inlet area of 125 cm². The air leaves the nozzle at 100 kPa and a velocity of 200 m/s. During its passage through the nozzle, the air loses 40 kJ/kg of air flow. Determine (a) the mass flow rate of the air, (b) its outlet temperature, and (c) the inlet area of the nozzle.

5.17: Steam enters a horizontal and adiabatic nozzle at 4 MPa and 600°C at a velocity of 240 m/s through an area of 60 cm². The steam leaves the nozzle at 1.5 MPa and 300°C. Determine (a) the mass flow rate of the steam, (b) the exit velocity of the steam, and (c) the exit area of the nozzle.

5.18: Refrigerant R-134a at a steady flow rate of 2 kg/s enters an adiabatic nozzle at 500 kPa and 100°C with a velocity of 50 m/s. The fluid leaves the nozzle at 200 kPa and a velocity of 320 m/s. Determine (a) the exit temperature and (b) the inlet/outlet area ratio.

5.19: Air in steady flow enters an adiabatic diffuser through an inlet flow area of 100 cm^2 at 100 kPa, 147°C, and 250 m/s and leaves at 150 kPa and 20 m/s. Consider the air as an ideal gas and determine (a) the exit temperature and (b) the exit area.

5.20: Saturated steam at 250°C at a steady flow rate enters a diffuser at a velocity of 600 m/s through an inlet flow area of 125 cm^2. The steam is discharged at 2.5 MPa and 280°C. Determine (a) the mass flow rate through the diffuser, (b) the exit velocity, and (c) the exit area.

5.21: Water, which may be taken as incompressible, enters an adiabatic diffuser in steady flow at 1 bar, 25°C, and a velocity of 90 m/s through an inlet flow area of 250 cm^2. The water leaves the diffuser at 5 bar and an exit area of 1500 cm^2. Determine (a) the exit velocity and (b) the outlet temperature.

5.22: Steam enters a diffuser as saturated vapor at 125°C with a velocity of 275 m/s. The exit pressure and temperature are 300 kPa and 150°C; and the exit area is 62.5 cm^2. Determine (a) exit velocity and (b) the mass flow rate.

Turbines

5.23: Steam at a mass flow rate of 15 kg/s at 10 MPa and 500°C enters a well-insulated turbine at a velocity of 100 m/s. At the exit, the steam is at 1 MPa and 96% quality with a velocity of 40 m/s. Determine (a) the power output and (b) the inlet area.

5.24: An adiabatic turbine delivers 10 MW using steam at a steady-flow rate of 20 kg/s. Inlet conditions are 10 MPa and 500°C and the outlet pressure is 1 MPa. Potential and kinetic energy changes are negligible. Determine the condition of the outlet steam by giving either a temperature or a quality.

5.25: An adiabatic turbine uses steam to deliver 10 MW at a steady-flow rate of 10 kg/s. Inlet conditions are 10 MPa and 500°C and the outlet pressure is 1 MPa. Potential and kinetic energy changes are negligible. Determine the condition of the outlet steam by giving either a temperature or a quality.

5.26: Nitrogen, which may be taken as an ideal gas, expands in an adiabatic turbine from 1 MPa and 475°C to a pressure of 200 kPa. Inlet and exit velocities are 100 m/s and 180 m/s, respectively. The power delivered by the turbine is 280 kW. The inlet area is 62.5 cm^2. Determine the outlet temperature.

5.27: A steam turbine is designed to have a power output of 10 MW when the steady-flow mass flow rate is 16 kg/s and the steam is supplied at 4 MPa and 375°C, at a velocity of 200 m/s. The steam is discharged in a dry saturated condition at 20 kPa at a velocity of 60 m/s. Determine the heat transfer through the turbine casing.

5.28: Determine the mass flow rate of steam required to produce 750 kW output in a turbine where potential and kinetic energy changes are negligible. The steam enters at 1 MPa and 250°C and leaves at 200 kPa and a quality of 92%.

5.29: A steam turbine operates with inlet conditions of 10 MPa, 500°C, and a velocity of 80 m/s. Dry saturated steam leaves the outlet at 0.1 MPa at a velocity of 160 m/s. The datum level is taken at the outlet and the elevation of the inlet is at 2 m. The heat loss through the turbine casing is 40 kW and the mass flow rate is 2 kg/s. Determine (a) the power delivered by the turbine and (b) the power delivered by the turbine if potential and kinetic energy changes are negligible.

5.30: Determine the mass flow rate of steam required to produce 750 kW output in a turbine where potential and kinetic energy changes are negligible. The steam enters at 1 MPa and 250°C and leaves at 200 kPa and a quality of 92%.

5.31: Air, which may be taken as an ideal gas, enters a turbine in a steady-flow process at 750 kPa and 487°C and 140 m/s. The air leaves the turbine at 150 kPa, 187°C, and 240 m/s. The inlet area is 5.625 cm^2 and during passage through the turbine, the air loses 12.5 kJ/kg per kg/s of flow through the casing. Determine the power output.

5.32: Refrigerant R-134a enters a steady-flow turbine at 800 kPa, 80°C, and 60 m/s. Exit conditions are 0.60 bar, −10°C, and 120 m/s. The inlet area is 27.5 cm^2 and the power developed is 280 kW. Determine (a) the mass flow rate and (b) the heat transferred through the turbine casing.

Compressors and Pumps

5.33: Air enters a compressor at 100 kPa and 20°C with a velocity of 2 m/s and leaves at 600 kPa. The power input to the compressor is 500 kW of which 120 kW is lost to the surroundings in the form of heat. The area at the compressor inlet is 0.25 m^2. Assuming that the air behaves as an ideal gas and neglecting kinetic and potential energy changes, determine the temperature of the compressed air leaving the compressor.

5.34: In a vapor compression refrigerator, refrigerant R-134a enters the compressor as a saturated vapor at −4°C. The refrigerant leaves the compressor as a superheated vapor at 1 MPa and 50°C and the mass flow of the refrigerant is 0.02 kg/s. The compressor requires a work input of 750 W and kinetic and potential energy changes are negligible. Determine the heat lost to the surroundings.

5.35: In a water-cooled air compressor, the inlet pressure and temperature are 100 kPa and 298 K, respectively. Air leaves the compressor at a pressure of 500 kPa. The compression process follows the law $Pv^{1.3} = \text{constant}$ and heat rejected to the cooling water amounts to 35 kW/kg of air. Assuming a mass flow of 2 kg/s and neglecting kinetic and potential energy change, determine the power input to the compressor.

5.36: An adiabatic compressor draws carbon dioxide gas at 95 kPa and 298 K and delivers it at 300 kPa and 450 K. The volumetric flow rate of carbon dioxide is 0.6 m^3/kg. Kinetic and potential energy changes during the compression process are negligible and the carbon dioxide can be treated as an ideal gas. For a flow rate of 2 kg/s, determine (a) the volumetric flow rate at the exit of the compressor and (b) the power required to drive the compressor.

5.37: Air enters a compressor with a mass flow rate of 1.25 kg/s at 105 kPa and 27°C and exits at 675 kPa and a specific volume of 0.1575 m^3/kg. The work required to drive the compressor is 200 kW. The heat lost in this process is absorbed by an insulated water jacket where the temperature rises to 294 K from an initial value of 282 K with no change in pressure. Any heat loss from the water jacket may be ignored and all kinetic and potential energy changes are negligible. Determine the mass flow of the cooling water in the water jacket.

5.38: A steady-state compression of nitrogen takes place in a compressor between 1 bar and 25°C at a mass flow rate of 0.875 kg/s and a velocity of 20 m/s to 2.5 bar and 375°C at a velocity of 100 m/s. Potential energy change and heat exchange with the environment are negligible. Determine the power required to drive the compressor.

5.39: The feedwater pump in a steam power plant takes in saturated liquid water at 20 kPa with a velocity of 1.5 m/s and delivers it at 1 MPa with a velocity of 10 m/s. The difference in elevation between the inlet and the outlet of the pump is 15 m and the water temperature remains constant at 60.1°C. The acceleration of gravity is taken as 9.81 m/s^2 and the water is to be delivered at 25 kg/s. Determine the power required to operate the pump.

5.40: In a 200-horsepower pump installed in a long pipeline the increase in kinetic energy of the oil is 8.4 kJ/s-kg but the potential energy change is negligible. If the temperature of the oil increases by 8°C and the oil may be considered as incompressible with a density of 860 kg/m^3 with a specific heat of 1.909 kJ/kg-K. The difference in pressure between the inlet and the outlet is 12 kPa. Determine the mass flow rate of the oil.

5.41: A pump is used to deliver water from an underground tank to an overhead tank at a rate of 20 kg/s. The flow velocities in the inlet and outlet pipes are 1 m/s and 5 m/s, respectively. Assuming no change in water pressure and the water temperature constant at 300 K, determine the pumping power required if the water experiences a change in elevation of 15 m.

5.42: Water is delivered in a steady flow by a pump at a volumetric flow rate of 6.25×10^{-3} m^3/s through a pipe having an inner diameter (ID) of 18 cm. The outlet pipe is 80 m above the inlet and has an ID of 16 cm. Both the inlet and outlet pressures (1.25 bar) and temperatures (27°C) are nearly equal. The acceleration of gravity is 9.81 m/s^2. Determine the power required.

Mixing Chambers

5.43: Two streams of steam enter a mixing chamber and exit as a single stream. Stream 1 has $P_1 = 500$ kPa, $x_1 = 0.50$, and $\dot{m}_1 = 2$ kg/s. For stream 2, $P_1 = 500$ kPa, $T_2 = 350$°C, and $\dot{m}_2 = 4$ kg/s and for the exit (stream 3), $P_3 = 500$ kPa, and $\dot{m}_3 = 6$ kg/s. Assuming that the mixing chamber is adiabatic, determine the enthalpy, quality, and the temperature of the outlet stream.

5.44: Water at 200 kPa and 110°C flows into a mixing device at the rate of 5 kg/s. The water is heated by mixing it with superheated steam at 200 kPa and 500°C. The outlet stream consists of saturated liquid water at 200 kPa. Assume that there is no heat loss to the surroundings and determine the mass flow of the superheated steam.

5.45: Two streams of refrigerant R-134a enter a mixing chamber and exit as a single stream. Stream 1 is saturated liquid with $T_1 = 8$°C and $\dot{m}_1 = 1.5$ kg/s. Stream 2 is saturated vapor with $T_2 = 8$°C and $\dot{m}_2 = 2.2$ kg/s. Heat is gained by the mixing chamber at the rate of 20 kW. Determine (a) the quality of the outlet stream and (b) the outlet area if the velocity at the outlet is 3 m/s.

5.46: In an adiabatic mixing chamber, two streams of refrigerant R-134a mix to form a single stream. Stream 1 is compressed liquid with $P_1 = 0.20$ MPa, $T_1 = -20$°C, and $\dot{m}_1 = 0.6$ kg/s while stream 2 is superheated vapor with $P_2 = 0.20$ MPa and $T_2 = 30$°C. Determine the mass flow of stream 2 if the outlet tream is to have a pressure of 0.20 MPa and a quality of 0.75.

5.47: The pressure in an adiabatic mixing chamber is 1 MPa. Steam enters at inlet 1 at a rate of 3.0 kg/s at a quality of 90%. Steam enters inlet 2 at 4.5 kg/s and 300°C. The exit velocity (at point 3) is 9 m/s. Determine (a) the exit temperature and (b) the diameter of the exit pipe.

5.48: Hot water (inlet 1) enters an adiabatic mixing chamber at 300 kPa and 75°C at a flow rate of 0.625 kg/s. It is mixed with a stream of cold water (inlet 2) at 300 kPa and 15°C. If the mixture is to leave the chamber at 300 kPa and 50°C, determine the flow rate of the cold water stream.

5.49: A hot stream of refrigerant R-134a enters an adiabatic mixing chamber at 8 bar and 80°C. A cold stream enters the chamber at 8 bar as a saturated liquid. The ratio of the mass flow rates of the hot to cold streams is 2.75:1. Determine the condition of the stream leaving at 8 bar (either a temperature if superheated vapor or a quality if a liquid-vapor mixture).

5.50: At point 1, refrigerant R-134a enters an adiabatic mixing chamber as saturated vapor at 6 bar. At point 2 liquid enters at 6 bar as saturated liquid. The flow rates of each stream are identical. Determine the quality of the exiting stream.

Heat Exchangers

5.51: In an air-water heat exchanger, air enters the heat exchanger at 125 kPa and 60°C and leaves at 37°C at a flow rate of 2 kg/s. The water enters at 20°C with a mass flow rate of 0.4 kg/s, the heat loss to the surroundings is estimated to be 5 kW and the air may be assumed to be an ideal gas. Determine the outlet temperature of the water.

5.52: Water at 1 MPa enters a boiler as saturated liquid and leaves as saturated vapor at 0.8 MPa. The stream enters a superheater where its temperature is raised to 400°C without further pressure loss. For a steam mass flow rate of 48 kg/s, determine the rate of heat supplied in (a) the boiler and (b) the superheater.

5.53: The condenser of a refrigerator receives R-134a at 400 kPa and 20°C and cools it to saturated liquid at 360 kPa. Determine the heat rejected in the condenser per kilogram of refrigerant.

5.54: In a cross-flow heat exchanger, air enters at 27°C and flows over a bank of tubes with a mass flow rate of 0.7 kg/s. The air is heated by the products of combustion flowing inside the tubes, which enter at a rate of 0.30 kg/s at 300°C and leave at 260°C. Both the air and the products of combustion with $c_p = 1.041$ kJ/kg-K may be assumed to behave like ideal gases. Determine the temperature of the air leaving the heat exchanger.

5.55: In a steam power plant, steam, flowing at the rate of 160,000 kg/h, enters the condenser as a mixture at 20 kPa with a quality of 0.92 and is condensed into saturated liquid with no pressure loss. Cooling water drawn from a river enters the condenser tubes at 20°C. If the maximum temperature rise of the cooling water is not to exceed 8°C, determine the mass flow rate of water circulating through the condenser tubes.

5.56: Refrigerant R-134a is contained in the cooling unit of an air-conditioning system. It enters at a volumetric flow rate of 36 m^3/s at 1.15 bar and 40°C and exits at 1.12 bar and 20°C. The refrigerant enters the unit at 8°C with a quality of 36% and leaves as saturated vapor at 8°C. Neglecting kinetic and potential energy changes and any heat transfer across the enclosure of the cooling unit, determine (a) the heat transferred between the air and the refrigerant and (b) the mass flow rate of the refrigerant.

5.57: In an air preheater, flue gases from a boiler are used to preheat the incoming combustion air. The flue gases, at a flow rate of 1.8 kg/s, enter at a pressure of 102 kPa and a temperature of 525°C. Outside air at a mass flow rate of 1.7 m/s enters at 20°C and leaves at 265°C. The flue gases may be treated an ideal gas having the properties of air. Neglect kinetic and potential energy changes and any heat transfer across the enclosure of the preheater. Determine (a) the heat transferred in the air heater and (b) the exiting flue gas temperature.

5.58: In a steady-flow heat exchanger, steam enters at 200 kPa and 300°C and leaves at 175 kPa with a quality of 72%. This cooling is accomplished by 48 kg/min of air entering at 32°C and leaving at 48°C. Neglecting kinetic and potential energy changes and any heat transfer across the enclosure of the exchanger, determine the mass flow of the steam in kg/min.

5.59: In a heat exchanger, air is heated by passing it over tubes containing steam which enters the tubes at 300 kPa and 250°C and leaves at 200 kPa and 200°C at a flow rate of 0.15 kg/s. The air enters at 101.3 kPa and 17°C and leaves at 37°C. The tubing has an inside diameter of 2 cm and the steam velocity is 4.8 m/s. Neglecting kinetic and potential energy changes and any heat transfer across the boundaries of the enclosure, determine (a) the heat transferred to the air, (b) the mass flow rate of the air, and (c) the number of tubes in the exchanger.

5.60: In the evaporator coil of an air-conditioning machine, dry air, which may be taken as an ideal gas at 30°C and atmospheric pressure, enters at a mass flow rate of 1.25 m/s. Refrigerant R-134a at −12°C and a quality of 22% is used in the coil. The refrigerant leaves the coil as saturated vapor and 24 kW are absorbed by the refrigerant. Assuming that both the heating and cooling processes take place at constant pressure with negligible kinetic and potential energy changes and no heat transfer across the boundaries of the system, determine (a) the flow rate of the refrigerant and (b) the exit temperature of the air.

Throttling Devices

5.61: Saturated liquid water at 80°C passes through a throttling valve where its pressure drops to 35 kPa and the water becomes a liquid-vapor mixture. For the liquid-vapor mixture, determine (a) the temperature, (b) the quality, (c) the specific volume, and (d) the specific internal energy.

5.62: The properties of R-134a after throttling are $P_2 = 60$ kPa and $x_2 = 0.32$. Assuming the refrigerant at the inlet to the throttling device is in the saturated liquid state, determine the pressure and temperature before throttling.

5.63: A porous plug is placed in a steam supply line carrying saturated vapor at 1 MPa. As the steam flows through the porous plug, its pressure drops to 0.2 MPa. Assuming that the flow through the plug is an adiabatic throttling process, determine the temperature of the steam after its passage through the porous plug.

5.64: At the inlet of a throttling valve, the state of refrigerant R-134a is $P_1 = 700$ kPa, and $T_1 = 20$°C. At the outlet, the refrigerant is a liquid-vapor mixture at 100 kPa. Assume the process to be adiabatic with negligible changes in kinetic and potential energies and determine at the outlet, (a) the quality, (b) the specific volume, and (c) the specific internal energy.

5.65: Consider a throttling process where steam at 280°C and a quality of 96% is throttled to $P_2 = 200$ kPa. At point 2, determine the quality (if a liquid-vapor mixture) or the temperature (if a superheated vapor).

5.66: Steam is throttled adiabatically in a steady flow from 1500 kPa and 350°C to 200 kPa at a flow rate of 0.01 kg/s. The inlet and outlet areas to and from the throttling device are 4 cm^2, and 24 cm^2, respectively. Determine (a) the inlet velocity, (b) the outlet velocity, and (c) the volumetric flow rate at the inlet.

5.67: Steam at 400 kPa is in a chamber equipped with a throttling calorimeter. A small quantity of steam is led to the calorimeter and is exhausted from the calorimeter (101.32 kPa) at a temperature of 150°C. Determine the quality of the steam in the chamber.

6

The Second Law of Thermodynamics

Chapter Objectives

- To consider the idea of **quality** of energy.
- To introduce the two statements of the **second law of thermodynamics** (the Kelvin-Planck statement and the Clausius statement) and to establish their equivalence.
- To define what is meant by a reversible process and identify the causes of irreversibility.
- To describe the Carnot cycle and its use as a heat engine, a refrigerator, and a heat pump.
- To discuss the Carnot cycle with external irreversibilities.
- To develop the Kelvin and Rankine absolute temperature scales.

6.1 Introduction

In Chapter 3, heat and work were identified as two forms of energy that are of primary interest in the study of thermodynamics and it was shown that the first law of thermodynamics relates the heat and work interactions at the boundary of a system to the change of energy within the system. The first law views heat and work as **quantitatively equivalent** to each other but places no restrictions on the conversion of heat into work or work into heat. In other words, the first law places no constraints on the direction of conversion.

In practice, however, there is a **preferred direction** in which processes occur. For example, a sugar cube dissolves readily in a cup of coffee but the reverse, if it happens at all, would be considered a miracle. Similarly, objects fall off a table quite readily but do not rise up against gravity on their own. Heat flows from a higher temperature to a lower temperature on its own but the reverse process requires an expenditure of work. Gases expand, liquids mix, and fuels burn quite readily but a reversal of these processes is difficult to comprehend. Indeed, conversion of work into heat (friction) is commonplace but the conversion of heat into work is not so readily accomplished.

The preferential direction for the occurrences of a process is intimately linked to the idea of the **quality of energy**. A process occurs in a direction such that **high-quality** energy is degraded into **low-quality** energy. This is why work that is high-quality energy can readily be converted into heat that is low-quality energy but the reverse process will never be complete. This is the concept of the degradation of energy that is the essence of the second law of thermodynamics and the two classical statements of the second law of thermodynamics are the Kelvin-Planck and the Clausius statements.

6.2 The Kelvin-Planck Statement and Heat Engines

The conversion of heat into work is accomplished in a device called a **heat engine**. Usually, heat engines are designed to operate in a thermodynamic cycle, which means that the initial and final states of the working fluid are the same. The cycle consists of a number (usually four) of processes such as constant volume, constant pressure, and isothermal processes. We discussed some of these processes of an ideal gas in Chapters 3 and 4 and another, the **isentropic process** will be considered in Chapter 7.

For the present discussion, we represent the heat engine by the diagram shown in Figure 6.1. The engine is shown operating between two **thermal reservoirs**, a high-temperature reservoir or a heat source at T_H and a low-temperature reservoir or a heat sink at T_L. Each reservoir is assumed to have an infinite thermal capacity so that its temperature, T_H or T_L, is unaffected by the withdrawal or deposition of heat. In practice, these reservoirs have finite thermal capacities. Consequently, to maintain a constant reservoir temperature, the heat withdrawn from the source reservoir must be replenished by an equivalent amount of heat flow to it. For example, in a steam power plant, the heat supplied by the boiler to generate the steam is continuously replenished by the burning of the fuel. Similarly, the heat rejected into the sink must be removed if the sink temperature is to be maintained at a constant value. In a steam power plant, the heat sink may be a river that eventually receives the heat that is picked up by the cooling water as it circulates through the condenser tubes. This rejected heat is carried downstream by the river currents.

The heat engine in Figure 6.1 receives Q_H from a source at a high temperature, T_H, and rejects heat, Q_L at a low temperature, T_L. The difference, $Q_H - Q_L$, is converted to work so that via the first law for the engine cycle

$$W_{\text{net}} = Q_H - Q_L \tag{6.1}$$

and the thermal efficiency of the engine cycle is

$$\eta = \frac{W_{\text{net}}}{Q_H} = \frac{Q_H - Q_L}{Q_H} = 1 - \frac{Q_L}{Q_H} \tag{6.2}$$

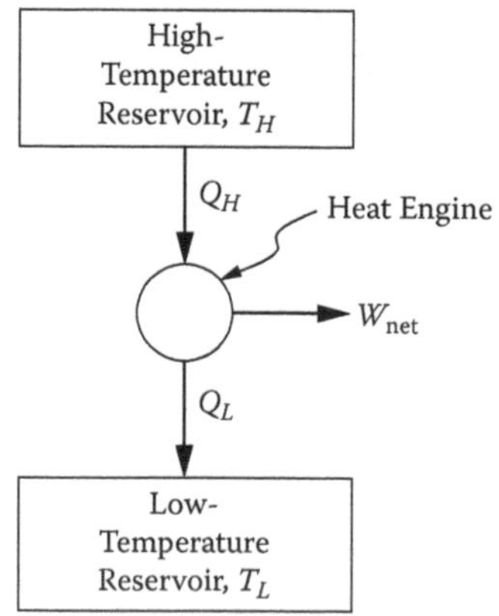

FIGURE 6.1
A heat engine taking heat, Q_H, from a high-temperature reservoir at T_H and rejecting heat, Q_L, to a low-temperature reservoir at T_L. The work done is W_{net}.

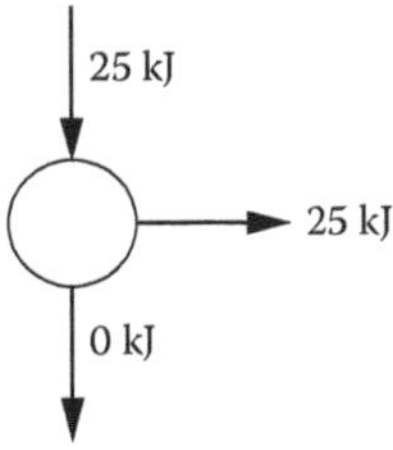

FIGURE 6.2
Proposed heat engine for Example 6.1a.

The Kelvin[1]-Planck[2] statement of the second law of thermodynamics prohibits Q_L from being zero and if Q_L is not equal to zero, η must be less than unity. A 100% efficient heat engine is consequently forbidden by the second law of thermodynamics and the promised classical statement that asserts this fact is

The Kelvin-Planck Statement

It is impossible to construct a heat engine that operates in a thermodynamic cycle and produces work while exchanging heat with a single thermal reservoir.

If an attempt is made to operate a steam power plant without a condenser, thereby eliminating Q_L, the work required to pump the turbine exhaust back into the boiler would exceed the turbine work output and the net work would be negative. When a condenser is employed, $Q_L \neq 0$. The work required to pump the condensate to the boiler is a small fraction of the turbine work output and the net work is positive.

Example 6.1 The following data for Q_H, Q_L, and W_{net} apply to a heat engine. For each case, determine if the heat engine is feasible in light of both the first and second laws of thermodynamics. The conditions are (a) $Q_H = 25\,\text{kJ}$, $Q_L = 0\,\text{kJ}$, and $W_{net} = 25\,\text{kJ}$ (Figure 6.2), (b) $Q_H = 50\,\text{kJ}$, $Q_L = 20\,\text{kJ}$, and $W_{net} = 30\,\text{kJ}$ (Figure 6.3), (c) $Q_H = 30\,\text{kJ}$, $Q_L = 15\,\text{kJ}$ (Figure 6.4), and $W_{net} = 20\,\text{kJ}$, and (d) $Q_H = 40\,\text{kJ}$, $Q_L = 40\,\text{kJ}$, and $W_{net} = 0\,\text{kJ}$ (Figure 6.5).

Solution

Assumptions and Specifications

(1) Steady-state conditions exist.

(2) In each case, the system is the heat engine.

(a) The energy balance of the first law of thermodynamics is a modification of Equation 6.1

$$Q_H = Q_L + W_{net} = 0\,\text{kJ} + 25\,\text{kJ} = 25\,\text{kJ}$$

The engine satisfies the first law. However, $Q_L = 0\,\text{kJ}$ means that all of the heat supplied is converted into work, which is impossible according to the Kelvin-Planck statement of the second law. Hence,

The engine in part (a) is *not* feasible $\Longleftarrow$

---

[1] William Thompson (Lord Kelvin) (1824–1907) was a British mathematician and physicist who invented the Kelvin balance and contributed to the science of thermodynamics.

[2] Max Planck (1858–1947) was a German physicist who presented the quantum theory and introduced Planck's constant.

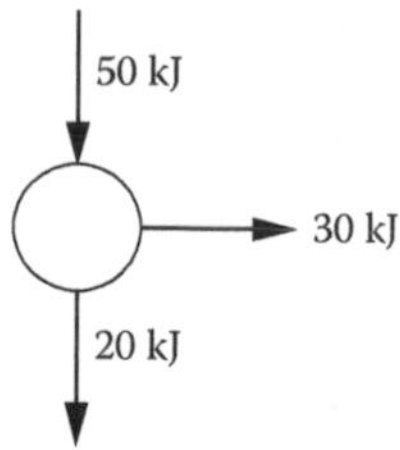

FIGURE 6.3
Proposed heat engine for Example 6.1b.

(b) The energy balance of the first law of thermodynamics is a modification of Equation 6.1

$$Q_H = Q_L + W_{\text{net}} = 20\,\text{kJ} + 30\,\text{kJ} = 50\,\text{kJ}$$

The engine satisfies the first law and because Q_L is not zero, it does not violate the second law. However, the thermal efficiency

$$\eta = 1 - \frac{Q_L}{Q_H} = 1 - \frac{20\,\text{kJ}}{50\,\text{kJ}} = 1 - 0.40 = 0.60$$

is on the high side. As we will see in Chapter 8, thermal efficiencies of spark ignition engines lie between 25% and 30% and those of compression ignition engines are between 35% and 40%.

The engine in part (b) *is* feasible $\Longleftarrow$

(c) The energy balance of the first law of thermodynamics is a modification of Equation 6.1

$$Q_H = Q_L + W_{\text{net}} = 15\,\text{kJ} + 20\,\text{kJ} = 35\,\text{kJ} \neq 30\,\text{kJ}$$

These numbers do not add up and the first law is violated. This makes any second law consideration a "don't care."

The engine in part (c) is *not* feasible $\Longleftarrow$

(d) The energy balance of the first law of thermodynamics is a modification of Equation 6.1

$$Q_H = Q_L + W_{\text{net}} = 40\,\text{kJ} + 0\,\text{kJ} = 40\,\text{kJ}$$

The engine satisfies the first law. Because, $Q_L \neq 0\,\text{kJ}$, the second law is also satisfied. However, the net work is zero, which implies that the device is not a heat engine. The data

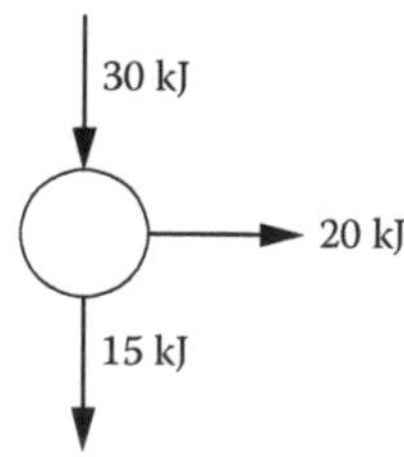

FIGURE 6.4
Proposed heat engine for Example 6.1c.

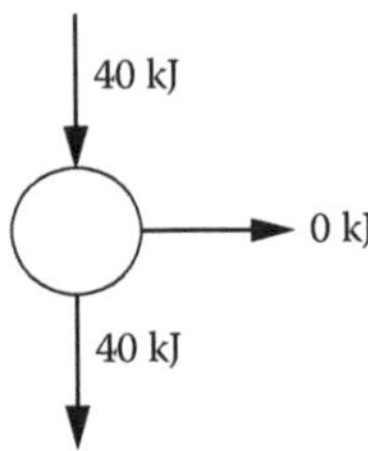

FIGURE 6.5
Proposed heat engine for Example 6.1d.

indicates a heat transfer process in which 40 kJ is transferred between a hot reservoir to a cold reservoir.

$$\text{The system in part (d) } \textit{is not} \text{ a heat engine } \Longleftarrow$$

Example 6.2 A high-temperature reservoir supplies heat, Q_{H1}, to heat engine 1. The heat rejected by this engine, Q_{L1}, is supplied to heat engine 2 ($Q_{L1} = Q_{H2}$). Heat engine 2 rejects heat, Q_{L2}, to a low-temperature reservoir. If the thermal efficiencies of heat engine 1 and heat engine 2 are η_1 and η_2, respectively (Figure 6.6), show that the thermal efficiency, η, of a heat engine receiving heat, Q_{H1}, and rejecting heat, Q_{L2}, is given by

$$\eta = \eta_1 + \eta_2 - \eta_1\eta_2$$

Solution

Assumptions and Specifications

(1) Steady-state conditions exist.
(2) The system is the two heat engines.

The thermal efficiency of heat engine 1 is given by

$$\eta_1 = 1 - \frac{Q_{L1}}{Q_{H1}} \tag{a}$$

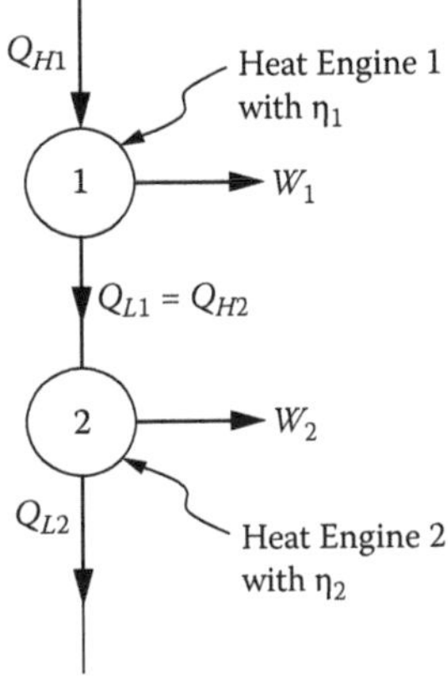

FIGURE 6.6
Two heat engines in cascade for Example 6.2.

Similarly, the thermal efficiency of heat engine 2 is given by

$$\eta_2 = 1 - \frac{Q_{L2}}{Q_{H2}} \tag{b}$$

The thermal efficiency of the overall heat engine receiving heat, Q_{H1}, and rejecting heat, Q_{L2}, is

$$\eta = 1 - \frac{Q_{L2}}{Q_{H1}} \tag{c}$$

Because $Q_{L1} = Q_{H2}$, Equation (a) may be written as

$$\eta_1 = 1 - \frac{Q_{H2}}{Q_{H1}}$$

and this may be adjusted to

$$Q_{H2} = Q_{H1}(1 - \eta_1) \tag{d}$$

With Equation (d) in Equation (b), we have

$$\eta_2 = 1 - \frac{Q_{L2}}{Q_{H1}(1 - \eta_1)} \tag{e}$$

and from Equation (c)

$$\frac{Q_{L2}}{Q_{H1}} = 1 - \eta \tag{f}$$

With Equation (f) in Equation (e)

$$\eta_2 = 1 - \frac{1 - \eta}{1 - \eta_1}$$

and when this is solved for η, we obtain the required result

$$\eta = \eta_1 + \eta_2 - \eta_1\eta_2 \Longleftarrow$$

6.3 The Clausius Statement: Refrigerators and Heat Pumps

It is a common observation that heat flows of its own accord from regions at higher temperature to regions at lower temperature. However, the direction of heat flow cannot be reversed without the expenditure of work. This idea is embodied in the Clausius[3] statement of the second law:

The Clausius Statement

Without a supply of work, it is impossible to construct a refrigerator or a heat pump that operates in a thermodynamic cycle and transfers heat from a low-temperature reservoir to a high-temperature reservoir.

[3] R. J. E. Clausius (1822–1888) was a German physicist who was a founder of thermodynamics.

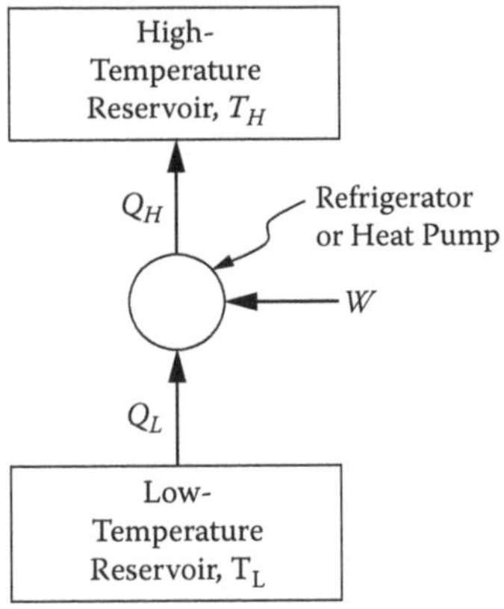

FIGURE 6.7
A refrigerator or heat pump taking heat, Q_L, from a low-temperature reservoir at T_L and rejecting heat, Q_H, to a high-temperature reservoir at T_H. The work required to drive the device is W.

A modest discussion of the refrigeration or heat pump cycle was provided in Section 3.6.3 and those remarks will now be amplified. A representation of a refrigerator or a heat pump in which work, W, is supplied to drive the flow of heat from a low-temperature reservoir at T_L to a higher-temperature reservoir at T_H is shown in Figure 6.7. For the refrigerator operating on an ideal vapor compression cycle, the refrigerant enters the compressor as a saturated vapor and is compressed to a higher pressure and temperature. The superheated vapor is then circulated through a condenser coil and rejects heat, Q_H, to the surroundings. The refrigerant leaves the condenser as a saturated liquid and passes through a throttling valve experiencing a pressure drop and emerging as a mixture of liquid and vapor. This mixture enters the evaporator coil located inside the refrigerator, picks up heat from the refrigerated space, and is converted to saturated vapor ready to go through another cycle.

Application of the first law of thermodynamics to the system in Figure 6.7 gives

$$Q_L + W = Q_H \qquad \text{or} \qquad W = Q_H - Q_L \tag{6.3}$$

and, as we have indicated in Section 3.6.3, the performance of the refrigerator is measured in terms of the **coefficient of performance** (COP), which is the ratio of the heat removed from the refrigerated space to the work supplied

$$\beta = \frac{Q_L}{W} = \frac{Q_L}{Q_H - Q_L} = \frac{1}{\dfrac{Q_H}{Q_L} - 1} \tag{6.4}$$

If the system is a heat pump, the appropriate coefficient of performance is the ratio of the heat supplied to a heated space, Q_H, to the work supplied

$$\gamma = \frac{Q_H}{W} = \frac{Q_H}{Q_H - Q_L} = \frac{1}{1 - \dfrac{Q_L}{Q_H}} \tag{6.5}$$

From Equations 6.4 and 6.5 we have

$$W = \frac{Q_L}{\beta} = \frac{Q_H}{\gamma}$$

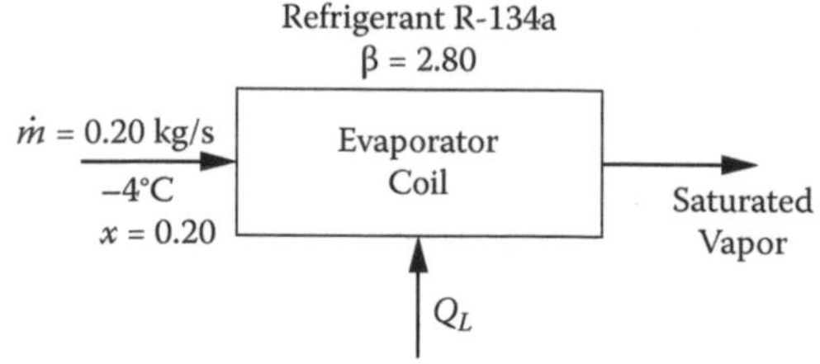

FIGURE 6.8
Confguration for the evaporator coil of Example 6.3.

and solving for γ gives

$$\gamma = \frac{Q_H}{Q_L}\beta = \left(\frac{W + Q_L}{Q_L}\right)\beta = \left(\frac{W}{Q_L} + 1\right)\beta = \left(\frac{1}{\beta} + 1\right)\beta$$

or

$$\gamma = \beta + 1 \tag{6.6}$$

Example 6.3 Refrigerant R-134a enters the evaporator coil (Figure 6.8) of a refrigeration system at a temperature of −4°C and a quality of 0.20 and emerges as a saturated vapor at the same temperature. The coefficient of performance of the system is 2.80. If the refrigerant flow rate is 0.20 kg/s, determine the power required to drive the system.

Solution

Assumptions and Specifications

(1) Steady-state and steady-flow conditions exist.
(2) The system is the evaporator coil.
(3) Kinetic and potential energy changes are negligible.

The change in enthalpy across the evaporator is given by

$$\Delta h = h_g - (h_f + x h_{fg}) = h_g - h_f - x h_{fg} = (1 - x) h_{fg}$$

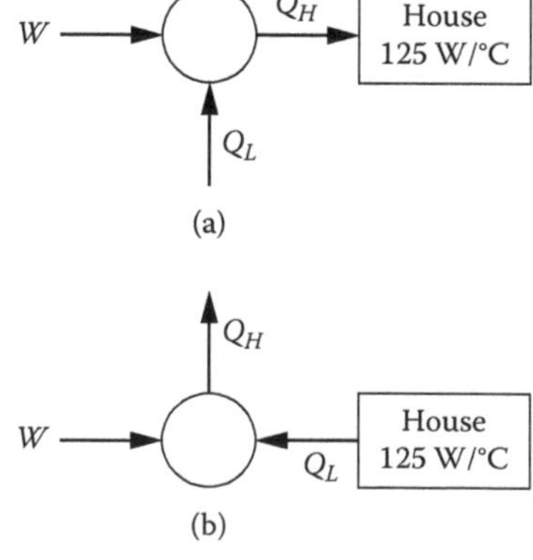

FIGURE 6.9
Dual-purpose heating/cooling unit for Example 6.4.

At $-4°C$, Table A.6 in Appendix A gives $h_{fg} = 200.15\,kJ/kg$. Thus,

$$\Delta h = (1 - 0.20)(200.15\,kJ/kg) = 160.12\,kJ/kg$$

and the rate of heat removal, $\dot{Q}_L$ is

$$\dot{Q}_L = \dot{m}\Delta h = (0.20\,kg/s)(160.12\,kJ/kg) = 32.02\,kW$$

Equation 6.4 may be expressed in rate form as

$$\beta = \frac{\dot{Q}_L}{\dot{W}}$$

or

$$\dot{W} = \frac{\dot{Q}_L}{\beta} = \frac{32.02\,kW}{2.80} = 11.44\,kW \Longleftarrow$$

Example 6.4 A dual-purpose heating/cooling unit (Figure 6.9) is designed to maintain a house at 20°C throughout the year. The power input to the unit is 1.20 kW and the heat loss or gain is 125 W per unit temperature difference between the inside and the outside. During the winter, the unit operates as a heat pump with a COP of $\gamma = 3.80$ while during the summer it serves as an air conditioner with a COP of $\beta = 3.0$. Determine (a) the maximum temperature and (b) the minimum temperature for which the unit meets the duty.

Solution

Assumptions and Specifications

(1) Steady-state conditions exist.
(2) The system is the dual purpose heat/cooling unit.

(a) Let T_{max} denote the maximum outside temperature during the summer when the unit functions as an air conditioner (refrigerator). The rate at which the unit can extract heat from the house is, according to Equation 6.4

$$\dot{Q}_L = \beta\dot{W} = 3.00(1.20\,kW) = 3.60\,kW$$

The maximum heat gain of the house is

$$\dot{Q}_{max} = (125\,W/°C)(T_{max} - 20°C)$$

which may be equated to $\dot{Q}_L$ and then solved for T_{max}

$$(125\,W/°C)(T_{max} - 20°C) = 3600\,W$$

or

$$T_{max} = 20°C + \frac{3600\,W}{125\,W/°C} = 20°C + 28.8°C = 48.8°C \Longleftarrow$$

(b) Let T_{min} denote the minimum outside temperature during the winter when the unit functions as a heat pump. According to Equation 6.5, the rate at which the unit can supply heat to the house is

$$\dot{Q}_H = \gamma\dot{W} = 3.80(1.20\,kW) = 4.56\,kW$$

The maximum heat loss of the house is

$$\dot{Q}_{max} = (125\,\text{W}/^\circ\text{C})(20^\circ\text{C} - T_{min})$$

which may be equated to $\dot{Q}_H$ and then solved for T_{min}

$$(125\,\text{W}/^\circ\text{C})(20^\circ\text{C} - T_{min}) = 4560\,\text{W}$$

or

$$T_{min} = 20^\circ\text{C} - \frac{4560\,\text{W}}{125\,\text{W}/^\circ\text{C}} = 20^\circ\text{C} - 36.5^\circ\text{C} = -16.5^\circ\text{C} \Longleftarrow$$

6.4 The Equivalence of the Kelvin-Planck and Clausius Statements

For the second law of thermodynamics to be absolutely general, the Kelvin-Planck and Clausius statements must be equivalent. This means that a violation of the Kelvin-Planck statement implies a violation of the Clausius statement and that a violation of the Clausius statement leads to a violation of Kelvin-Planck.

For example, consider Figure 6.10 and suppose that it would be possible to transfer heat from the low-temperature reservoir at T_L to the high-temperature reservoir at T_H as indicated in Figure 6.10a. This is clearly in violation of the Clausius statement of the second law. If we add the engine, shown in Figure 6.10b, which is not in violation of the second law, the work done will be $W = Q_H - Q_L$. For identical values of Q_H, the combination of both systems is shown in Figure 6.10c, which implies that no heat is being exchanged with the hot reservoir. Such an engine is excluded by the Kelvin-Planck statement that no engine may operate in a cycle while exchanging heat with a single reservoir. Thus, the violation of the Clausius statement has led to a violation of the Kelvin-Planck statement.

On the other hand, the assumption that the heat supplied, Q_H, from a high-temperature reservoir at T_H can be completely converted to work, W, is a violation of the Kelvin-Planck

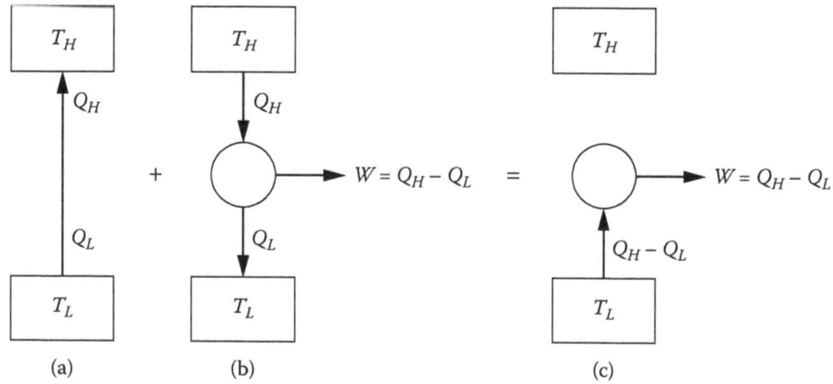

FIGURE 6.10
A violation of the Clausius statement implies a violation of the Kelvin-Planck statement. (a) A transfer of heat from T_L to T_H without work, which is a violation of the Clausius statement. (b) A heat engine not in violation of the Clausius or the Kelvin-Planck statement. (c) A superposition of (a) and (b) that violates the Kelvin-Planck statement.

statement. Under these circumstances, the work, W, can be converted to heat Q by means of friction. With this heat, the temperature of the body can be raised to a temperature $T > T_H$. The net result is the transfer of heat from a body at a lower temperature, T_H, to a body at higher temperature, T. This violates the Clausius statement.

6.5 Reversible and Irreversible Processes

According to the Kelvin-Planck statement of the second law, it is impossible to construct a heat engine with 100% thermal efficiency. Given the reservoir temperatures, T_H and T_L, the paramount question concerns the maximum efficiency that a heat engine operating between these temperatures can attain. The answer to this question is rooted in the concept of a reversible process that may be defined as:

> A reversible process is a process that can be reversed without any residual change in the system or its surroundings.

Both the system and the surroundings must return to their original states as if the forward and reverse processes never occurred. A reversible process is an ideal process that cannot be achieved in practice. Nonetheless, the reversible process provides a yardstick for measuring the irreversibility in an actual process.

The four main effects that cause a process to be irreversible are friction, heat transfer through a finite temperature difference, mixing, and combustion.

- Friction causes energy (work) to be dissipated into heat (Figure 6.11) but because heat, according to the Kelvin-Planck statement of the second law, cannot be completely converted to work, the presence of friction makes a process irreversible.
- When heat transfer from a high temperature to a low temperature body occurs (Figure 6.12), the process cannot be reversed unless the surroundings supply some work as demanded by the Clausius statement. Because the surroundings are not restored to their original state, heat transfer through a finite temperature difference is an irreversible process.
- When two or more gases are allowed to mix (Figure 6.13), they do so spontaneously. However, if the mixture is to be separated into its original components, work must be done. Consequently, mixing is an irreversible process.
- The combustion of a fuel (Figure 6.14) produces a mixture of gases. The impossibility of the return of these gases to the form of the original fuel shows that combustion is an irreversible process.

Other effects that cause irreversibilities include the conversion of electrical power into heat (the I^2R loss in conductors), unrestrained expansion of fluids, hysteresis, and inelastic deformation.

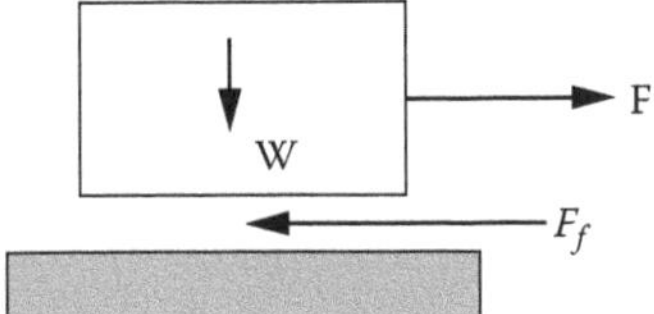

FIGURE 6.11
One cause of an irreversibility is friction.

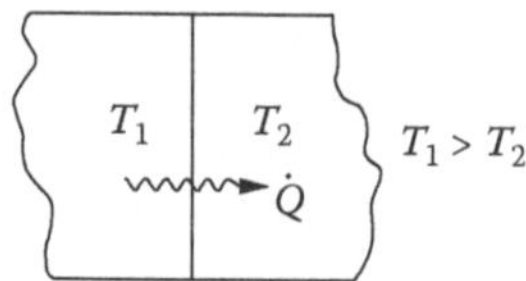

FIGURE 6.12
One cause of an irreversibility is heat transfer through a finite temperature difference.

A heat engine that operates on a thermodynamic cycle consisting only of reversible processes is a **reversible heat engine** and provides the maximum thermal efficiency. One such cycle is the **Carnot cycle**, which is considered next.

6.6 The Carnot Cycle

The Carnot cycle[4] is shown, for an ideal gas in P-V coordinates, in Figure 6.15. The cycle consists of four reversible processes.

- *Process 1-2*: A reversible isothermal process in which the working fluid receives heat, Q_H, from a high-temperature reservoir at temperature, T_H.
- *Process 2-3*: A reversible adiabatic expansion process during which the temperature of the working fluid decreases from T_H to T_L where T_L is the temperature of the low-temperature reservoir.
- *Process 3-4*: A reversible isothermal process in which the working fluid rejects heat, Q_L, to the low-temperature reservoir at temperature, T_L.
- *Process 4-1*: A reversible adiabatic compression process during which the temperature of the working fluid increases from T_L to T_H.

The foregoing reversible processes are displayed in Figure 6.15.

The thermal efficiency of the Carnot cycle may be derived by assuming, for convenience, that the working fluid is an ideal gas.

- *Process 1-2*: The first law energy balance of Equation 3.12 can be written as

$$Q_{12} - W_{12} = U_2 - U_1 = mc_v(T_2 - T_1) \tag{6.7}$$

However, because the process is isothermal, $T_1 = T_2 = T_H$, we find that $Q_{12} = Q_H = W_{12}$ and Equation 3.43, which gives the work done in an isothermal process, produces

$$Q_H = W_{12} = P_1 V_1 \ln \frac{V_2}{V_1} = mRT_H \ln \frac{V_2}{V_1} \tag{6.8}$$

- *Process 2-3*: The first law energy balance of Equation 3.12 can be written as

$$Q_{23} - W_{23} = U_3 - U_2 = mc_v(T_3 - T_2)$$

but because the process is adiabatic, $Q_{23} = 0$, $T_2 = T_H$ and $T_3 = T_L$. Hence,

$$W_{23} = mc_v(T_H - T_L) \tag{6.9}$$

[4] Sadi Carnot (1796–1832) was a French physicist who formulated Carnot's theorems in thermodynamics.

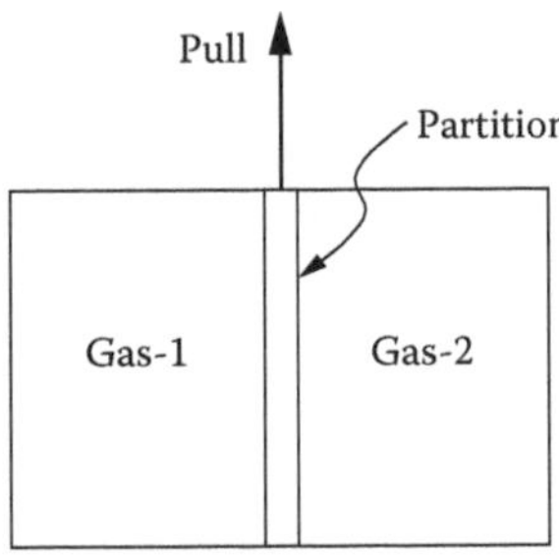

FIGURE 6.13
One cause of an irreversibility is in the mixing of gases.

- *Process 3-4*: The first law energy balance of Equation 3.12 can be written as

$$Q_{34} - W_{34} = U_4 - U_3 = mc_v(T_4 - T_3)$$

This is another isothermal process where $T_3 = T_4 = T_L$ and $Q_{34} = Q_L = W_{34}$. This time Equation 3.43 gives

$$Q_L = W_{34} = P_3V_3 \ln\frac{V_4}{V_3} = mRT_L \ln\frac{V_4}{V_3} \tag{6.10}$$

- *Process 4-1*: Here, the first law energy balance of Equation 3.12 can be written as

$$Q_{41} - W_{41} = U_1 - U_4 = mc_v(T_1 - T_4)$$

but because the process is adiabatic, $Q_{41} = 0$, $T_1 = T_H$ and $T_4 = T_L$, and

$$W_{41} = mc_v(T_L - T_H) \tag{6.11}$$

The net work of the cycle is

$$W_{\text{net}} = W_{12} + W_{23} + W_{34} + W_{41}$$

and use of Equations 6.8 through 6.11 shows that

$$W_{\text{net}} = mRT_H \ln\frac{V_2}{V_1} + mc_v(T_H - T_L) + mRT_L \ln\frac{V_4}{V_3} + mc_v(T_L - T_H)$$

or

$$W_{\text{net}} = mRT_H \ln\frac{V_2}{V_1} + mRT_L \ln\frac{V_4}{V_3} \tag{6.12}$$

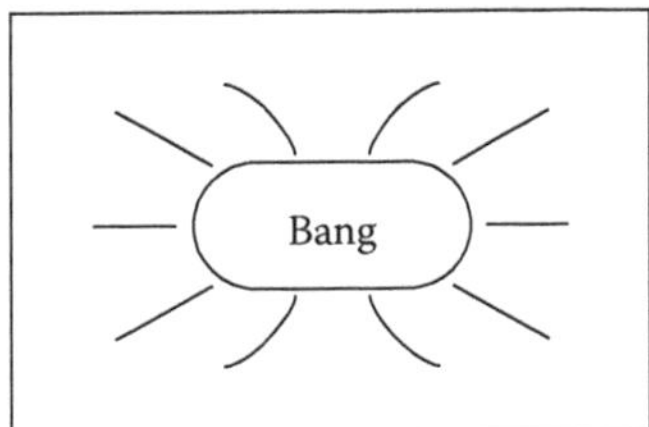

FIGURE 6.14
One cause of an irreversibility is in combustion.

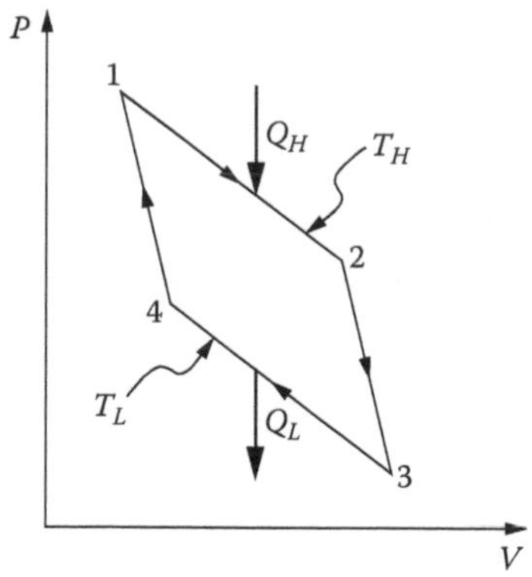

FIGURE 6.15
Four reversible processes in a Carnot cycle.

The heat into the cycle is given by Equation 6.8

$$Q_H = mRT_H \ln \frac{V_2}{V_1} \tag{6.8}$$

The thermal efficiency of the Carnot cycle, η_C, will be

$$\eta_C = \frac{W_{\text{net}}}{Q_H} = \frac{mRT_H \ln \dfrac{V_2}{V_1} + mRT_L \ln \dfrac{V_4}{V_3}}{mRT_H \ln \dfrac{V_2}{V_1}}$$

or

$$\eta_C = 1 + \frac{T_L}{T_H} \frac{\ln \dfrac{V_4}{V_3}}{\ln \dfrac{V_2}{V_1}} \tag{6.13}$$

It will be shown in the next chapter that for an ideal gas executing a reversible adiabatic process,

$$TV^{k-1} = C \qquad \text{(a constant)} \tag{6.14}$$

and when we apply this to processes 2-3 and 4-1, which are reversible adiabatic processes, we obtain

$$\left(\frac{V_2}{V_3}\right)^{k-1} = \frac{T_L}{T_H} \tag{6.15}$$

and

$$\left(\frac{V_1}{V_4}\right)^{k-1} = \frac{T_L}{T_H} \tag{6.16}$$

From these, it follows that

$$\frac{V_2}{V_1} = \frac{V_3}{V_4} = \left(\frac{V_4}{V_3}\right)^{-1} \tag{6.17}$$

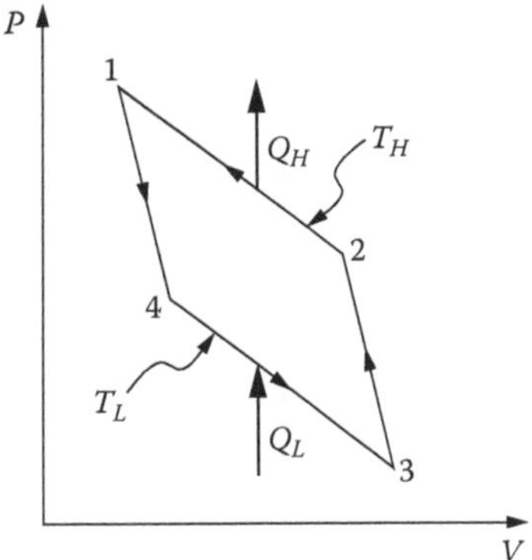

FIGURE 6.16
The P-V diagram for the reversed Carnot cycle showing four reversible processes.

Then, after substitution of Equation 6.17 into Equation 6.13, the efficiency of the Carnot cycle is seen to be

$$\eta_C = 1 - \frac{T_L}{T_H} \tag{6.18}$$

Even though this efficiency has been derived for an ideal gas, it applies to every reversible engine operating between the temperature limits of T_H and T_L.

When we compare Equations 6.2 and 6.18, we find that for a Carnot cycle and indeed for any reversible cycle that

$$\frac{Q_L}{Q_H} = \frac{T_L}{T_H} \tag{6.19}$$

Equation 6.19 will be employed in the next chapter when we identify a new property known as **entropy**.

The reversed Carnot cycle (shown in Figure 6.16 with the directions of Q_L and Q_H reversed) can either represent a refrigeration or a heat pump cycle. The performance of a refrigerator is measured in terms of the **coefficient of performance** and for the refrigerator, the coefficient of performance was defined by Equations 3.19 and 6.4. For the heat pump, the coefficient of performance was defined by Equations 3.20 and 6.5. With Q_L/Q_H replaced by T_L/T_H in accordance with Equation 6.19 in Equations 6.4 and 6.5, we have for the Carnot refrigerator

$$\beta_C = \frac{1}{\dfrac{T_H}{T_L} - 1} \tag{6.20}$$

and for the Carnot heat pump

$$\gamma_C = \frac{1}{1 - \dfrac{T_L}{T_H}} \tag{6.21}$$

Example 6.5 A Carnot heat pump (Figure 6.17) supplies heat to a steam generator where water enters as saturated liquid at 100 kPa and exits as saturated vapor at the same pressure. The heat pump consumes 80 kW of power and draws heat from the environment at 10°C. Determine the rate of steam generation.

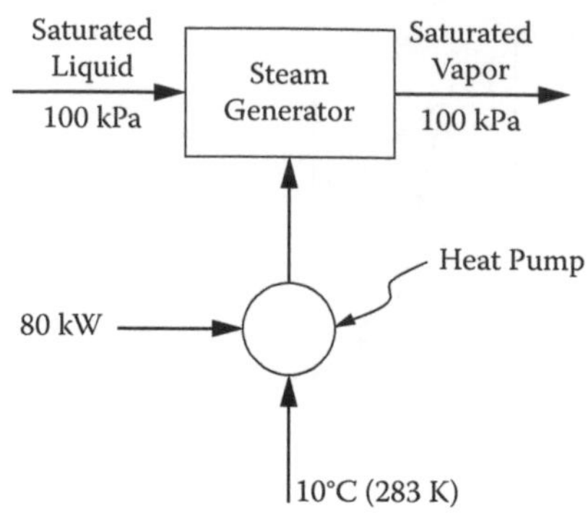

FIGURE 6.17
Configuration for Example 6.5.

Solution

Assumptions and Specifications

(1) All processes in the cycle are internally reversible.
(2) The control volume is the steam generator.
(3) The system is the steam in the steam generator.
(4) There are no potential and kinetic energy changes.
(5) The system operates in the steady state.

From Table A.4 at 100 kPa, read

$$T_{\text{sat}} = 99.6^\circ\text{C} \quad \text{and} \quad h_{\text{fg}} = 2258.0\,\text{kJ/kg}$$

Then with

$$T_H = 99.6^\circ\text{C} + 273^\circ\text{C} = 372.6\,\text{K} \quad \text{and} \quad T_L = 10^\circ\text{C} + 273^\circ\text{C} = 283\,\text{K}$$

Equation 6.21 gives

$$\gamma_C = \frac{1}{1 - \dfrac{T_L}{T_H}} = \frac{1}{1 - \dfrac{283\,\text{K}}{372.6\,\text{K}}} = \frac{1}{1 - 0.760} = 4.16$$

For $\dot{W} = 80\,\text{kW}$

$$\dot{Q}_H = \gamma_C \dot{W} = (4.16)(80\,\text{kW}) = 332.7\,\text{kW}$$

and

$$\dot{m} = \frac{\dot{Q}_H}{h_{\text{fg}}} = \frac{332.7\,\text{kW}}{2258.0\,\text{kJ/kg}} = 0.147\,\text{kg/s} \Longleftarrow$$

Example 6.6 A three-process power cycle employing an ideal gas with constant specific heats is shown in Figure 6.18. It consists of the following:

- *Process 1-2*: A constant volume heat addition
- *Process 2-3*: An adiabatic expansion with $k = c_p/c_v = 1.40$
- *Process 3-1*: A constant pressure heat rejection with $V_3/V_1 = 2$

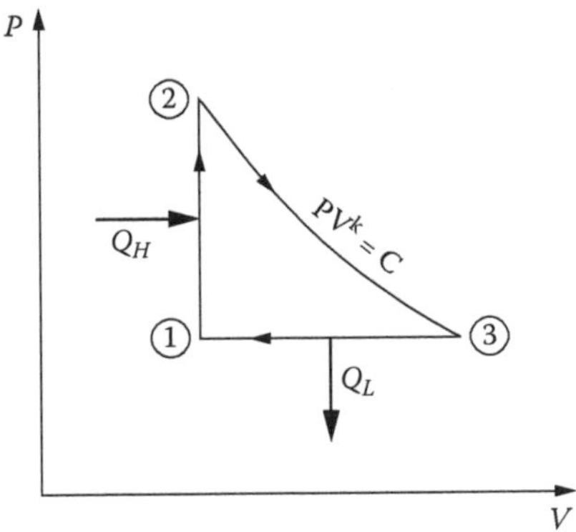

FIGURE 6.18
Three processes for Example 6.6.

(a) Show that the thermal efficiency of the cycle is given by

$$\eta_C = \frac{W_{net}}{Q_H} = 1 - k\frac{T_1}{T_2}\left[\frac{\dfrac{V_3}{V_1} - 1}{1 - \dfrac{T_1}{T_2}}\right]$$

and then compute, with $T_H = 900\,\text{K}$ and $T_L = 300\,\text{K}$, (b) the cycle efficiency and (c) the Carnot efficiency.

Solution

Assumptions and Specifications

(1) All processes in the cycle are internally reversible.
(2) An ideal gas with constant specific heats is specified.
(3) Steady flow and steady state applies.

(a) Application of the first law stated by Equation 3.12 to process 1-2 gives

$$Q_{12} - W_{12} = U_2 - U_1 = mc_v(T_2 - T_1)$$

and because for this constant volume process, $W_{12} = 0$

$$Q_{12} = Q_H = mc_v(T_2 - T_1)$$

For process 2-3, the first law gives

$$Q_{23} - W_{23} = U_3 - U_2 = mc_v(T_3 - T_2)$$

and because for this adiabatic process, $Q_{23} = 0$

$$W_{23} = mc_v(T_2 - T_3)$$

For process 3-1, the first law provides

$$Q_{31} - W_{31} = U_1 - U_3 = mc_v(T_1 - T_3)$$

and for a constant pressure process where $W_{31} = P_1V_1 - P_3V_3$, we may write

$$Q_{31} = mc_p(T_1 - T_3) + P_1V_1 - P_3V_3 = H_1 - H_3$$

or

$$Q_L = -Q_{31} = mc_p(T_3 - T_1)$$

The thermal efficiency of the cycle may now be expressed as

$$\eta = 1 - \frac{Q_L}{Q_H} = 1 - \frac{mc_p(T_3 - T_1)}{mc_v(T_2 - T_1)} = 1 - k\frac{T_1\left(\frac{T_3}{T_1} - 1\right)}{T_2\left(1 - \frac{T_1}{T_2}\right)}$$

Because the ideal gas law requires that, at constant pressure, $T_3/T_1 = V_3/V_1$, we have

$$\eta_C = 1 - k\frac{T_1}{T_2}\left(\frac{\frac{V_3}{V_1} - 1}{1 - \frac{T_1}{T_2}}\right) \Longleftarrow$$

and this is the required result.

(b) We note that $T_1 = T_L = 300\,\text{K}$ and $T_2 = T_H = 900\,\text{K}$. Then with $V_3/V_1 = 2.0$,

$$\eta_C = 1 - k\frac{T_1}{T_2}\left(\frac{\frac{V_3}{V_1} - 1}{1 - \frac{T_1}{T_2}}\right) = 1 - (1.40)\left(\frac{300\,\text{K}}{900\,\text{K}}\right)\left(\frac{2 - 1}{1 - \frac{300\,\text{K}}{900\,\text{K}}}\right) = 0.30 \Longleftarrow$$

(c) For a Carnot engine

$$\eta_C = 1 - \frac{T_L}{T_H} = 1 - \frac{300\,\text{K}}{900\,\text{K}} = 0.667 \qquad (66.7\%) \Longleftarrow$$

6.7 The Carnot Cycle with External Irreversibilities

Isothermal heat addition and heat rejection processes are not achievable in practice. A more feasible approach is to assume that the heat is supplied from a reservoir at a source temperature, T_{source}, through a heat exchanger of surface area, S_H, with an overall heat transfer coefficient, U_H, such that

$$\dot{Q}_H = U_H S_H(T_{\text{source}} - T_H) \tag{6.22}$$

Because the heat supply involves a finite temperature difference, $T_{\text{source}} - T_H$, the process is externally irreversible. If the heat rejection, $\dot{Q}_L$, is assumed to be isothermal at a temperature, T_L, then the power output, W_{net} will be

$$\dot{W}_{\text{net}} = \eta_C\dot{Q}_H = \left(1 - \frac{T_L}{T_H}\right)U_H S_H(T_{\text{source}} - T_H) \tag{6.23}$$

and $\dot{W}_{\text{net}}$ can be optimized if

$$\frac{d\dot{W}_{\text{net}}}{dT_H} = 0 \tag{6.24}$$

The result of the indicated differentiation, upon simplification, will be

$$T_H = (T_{\text{source}} T_L)^{1/2} \tag{6.25}$$

and the efficiency of the cycle then becomes

$$\eta_C = 1 - \frac{T_L}{T_H} = 1 - \frac{T_L}{(T_{\text{source}} T_L)^{1/2}} = 1 - \left(\frac{T_L}{T_{\text{source}}}\right)^{1/2} \tag{6.26}$$

On the other hand, if the heat addition is isothermal but the heat rejection is to a heat sink at temperature, T_{sink}, through a heat exchanger of surface area, S_L, via a heat transfer coefficient of U_L, then

$$\dot{Q}_L = U_L S_L (T_L - T_{\text{sink}}) \tag{6.27}$$

and

$$\dot{W}_{\text{net}} = \eta_C \dot{Q}_H = \eta_C \dot{Q}_L \frac{T_H}{T_L} = \left(1 - \frac{T_L}{T_H}\right)\left(\frac{T_H}{T_L}\right) U_H S_H (T_L - T_{\text{sink}}) \tag{6.28}$$

Now the optimization procedure involving $\dot{W}_{\text{net}}$ and T_L will require setting $d\dot{W}_{\text{net}}/dT_L = 0$ and the result of the indicated differentiation, upon simplification, will be

$$T_L = (T_{\text{sink}} T_H)^{1/2} \tag{6.29}$$

and the efficiency of the cycle becomes

$$\eta_C = 1 - \frac{T_L}{T_H} = 1 - \frac{(T_{\text{sink}} T_H)^{1/2}}{T_H} = 1 - \left(\frac{T_{\text{sink}}}{T_H}\right)^{1/2} \tag{6.30}$$

When both the heat addition and the heat rejection involve heat transfers through finite temperature differences, $T_{\text{source}} - T_H$ and $T_L - T_{\text{sink}}$ respectively, then it can be shown that $\dot{W}_{\text{net}}$ is at its optimum value when

$$\frac{T_L}{T_H} = \left(\frac{T_{\text{sink}}}{T_{\text{source}}}\right)^{1/2} \tag{6.31}$$

in which case, the cycle efficiency will be

$$\eta_C = 1 - \left(\frac{T_{\text{sink}}}{T_{\text{source}}}\right)^{1/2} \tag{6.32}$$

Example 6.7 A Carnot engine operates between a heat source at 1200 K and a heat sink at 300 K. The heat supply rate to the engine is 100 kW. Determine the temperatures T_H and T_L, the engine efficiency, the net power, and the heat rejection rate for (a) a heat supply through a finite temperature difference but an isothermal heat rejection, (b) an isothermal heat supply but heat rejection through a finite temperature difference, (c) both the heat supply and the heat rejection occurring through a finite temperature difference, and (d) the engine operates on a Carnot cycle with no external irreversibilities.

Solution

Assumptions and Specifications

(1) W_{net} is optimized in each case.
(2) $T_H = 1200\,\text{K}$ and $T_L = 300\,\text{K}$ are specified.
(3) The control volume is a Carnot engine.

(a) For a heat supply through a finite temperature difference but an isothermal heat rejection, the temperature, T_L equals the sink temperature

$$T_{\text{sink}} = T_L = 300\ \text{K} \Longleftarrow$$

The temperature, T_H is given by Equation 6.25 with $T_{\text{source}} = 1200\,\text{K}$

$$T_H = (T_{\text{source}} T_L)^{1/2} = [(1200\,\text{K})(300\,\text{K})]^{1/2} = 600\,\text{K} \Longleftarrow$$

Equation 6.26 gives the thermal efficiency

$$\eta_C = 1 - \left(\frac{T_L}{T_{\text{source}}}\right)^{1/2} = 1 - \left(\frac{300\,\text{K}}{1200\,\text{K}}\right)^{1/2} = 0.50 \quad (50\%) \Longleftarrow$$

The net engine power will be

$$\dot{W}_{\text{net}} = \eta_C \dot{Q}_H = (0.50)(100\,\text{kW}) = 50\,\text{kW} \Longleftarrow$$

and the heat rejected derives from the first law

$$\dot{Q}_L = \dot{Q}_H - \dot{W}_{\text{net}} = 100\,\text{kW} - 50\,\text{kW} = 50\,\text{kW} \Longleftarrow$$

(b) For an isothermal heat supply but heat rejection through a finite temperature difference, the temperature, T_H equals the source temperature

$$T_{\text{source}} = T_H = 1200\,\text{K} \Longleftarrow$$

The temperature, T_L is given by Equation 6.29 with $T_{\text{sink}} = 300\,\text{K}$

$$T_L = (T_{\text{sink}} T_H)^{1/2} = [(300\,\text{K})(1200\,\text{K})]^{1/2} = 600\,\text{K} \Longleftarrow$$

Equation 6.30 gives the thermal efficiency

$$\eta_C = 1 - \left(\frac{T_{\text{sink}}}{T_H}\right)^{1/2} = 1 - \left(\frac{300\,\text{K}}{1200\,\text{K}}\right)^{1/2} = 0.50 \quad (50\%) \Longleftarrow$$

The net engine power will be

$$\dot{W}_{\text{net}} = \eta_C \dot{Q}_H = (0.50)(100\,\text{kW}) = 50\,\text{kW} \Longleftarrow$$

and the heat rejected derives from the first law

$$\dot{Q}_L = \dot{Q}_H - \dot{W}_{\text{net}} = 100\,\text{kW} - 50\,\text{kW} = 50\,\text{kW} \Longleftarrow$$

(c) For both the heat supply and the heat rejection occurring through a finite temperature difference, we have $T_{\text{source}} = 1200\,\text{K}$ and $T_{\text{sink}} = 300\,\text{K}$. We cannot find T_H and T_L unless $U_H S_H$ or $U_C S_C$ are known. However, their ratio may be obtained from Equation 6.31

$$\frac{T_L}{T_H} = \left(\frac{T_{\text{sink}}}{T_{\text{source}}}\right)^{1/2} = \left(\frac{300\ \text{K}}{1200\,\text{K}}\right)^{1/2} = \frac{1}{2}$$

The thermal efficiency may be found from Equation 6.32

$$\eta_C = 1 - \left(\frac{T_{\text{sink}}}{T_{\text{source}}}\right)^{1/2} = 1 - \left(\frac{300\,\text{K}}{1200\,\text{K}}\right)^{1/2} = 0.50 \quad (50\%) \Longleftarrow$$

The net engine power will be

$$\dot{W}_{\text{net}} = \eta_C \dot{Q}_H = (0.50)(100\,\text{kW}) = 50\,\text{kW} \Longleftarrow$$

and the heat rejected derives from the first law

$$\dot{Q}_L = \dot{Q}_H - \dot{W}_{\text{net}} = 100\,\text{kW} - 50\,\text{kW} = 50\,\text{kW} \Longleftarrow$$

(d) For a Carnot cycle with no external irreversibilities, $T_H = T_{\text{source}} = 1200\,\text{K}$ and $T_L = T_{\text{sink}} = 300\,\text{K}$. The thermal efficiency is given by

$$\eta_C = 1 - \frac{T_L}{T_H} = 1 - \frac{300\,\text{K}}{1200\,\text{K}} = 0.75 \quad (75\%)$$

The net engine power will be

$$\dot{W}_{\text{net}} = \eta_C \dot{Q}_H = (0.75)(100\,\text{kW}) = 75\,\text{kW} \Longleftarrow$$

and the heat rejected derives from the first law

$$\dot{Q}_L = \dot{Q}_H - \dot{W}_{\text{net}} = 100\,\text{kW} - 75\,\text{kW} = 25\,\text{kW} \Longleftarrow$$

6.8 The Absolute Temperature Scales

Equation 6.19 that applies to every reversible heat engine provides a basis for the absolute Kelvin temperature scale. Because Equation 6.19 defines only a ratio of temperatures, a reference state must be assigned. The reference state selected is the triple point of water that is assigned a value of 273.16 K. If a reversible heat engine receives heat, Q, from a reservoir at the reference temperature and rejects heat, Q_{ref}, to a reservoir at the reference temperature, T_{ref}, then

$$T = T_{\text{ref}}\frac{Q}{Q_{\text{ref}}} = 273.16\frac{Q}{Q_{\text{ref}}} \tag{6.33}$$

Equation 6.33 defines the absolute Kelvin temperature scale. The scale is independent of the working fluid used in a reversible engine.

The Rankine temperature scale is defined in terms of the Kelvin temperature scale. The relationship between the two scales is linear and is given by

$$T(^\circ\text{R}) = 1.8T(K) \tag{6.34}$$

6.9 Summary

The preferential direction for the occurrence of a thermodynamic process is intimately linked to the idea of the quality of energy and this is the essence of the second law of thermodynamics.

Two classical statements of the second law of thermodynamics are the Kelvin-Planck statement

> It is impossible to construct a heat engine that operates in a thermodynamic cycle and produces work while exchanging heat with a single thermal reservoir.

and the Clausius statement

> Without a supply of work, it is impossible to construct a refrigerator or a heat pump that operates in a thermodynamic cycle and transfers heat from a low-temperature reservoir to a high-temperature reservoir.

These statements are equivalent and a violation of one of them infers a violation of the other.

The performance of both refrigeration and heat pump cycles is measured by the coefficient of performance, abbreviated COP. For the refrigerator, the coefficient of performance is

$$\beta = \frac{1}{\dfrac{Q_H}{Q_L} - 1}$$

and for the heat pump, the coefficient of performance is

$$\gamma = \frac{1}{1 - \dfrac{Q_L}{Q_H}}$$

The relationship between β and γ is

$$\gamma = \beta + 1$$

A reversible process is a process that can be reversed without any residual change in the system or its surroundings. Examples of irreversible processes include the following:

- Motion with friction
- Unrestrained expansion
- Mixing and diffusion
- Spontaneous chemical reactions
- Heat transfer through a finite temperature difference
- Current flow through an electrical resistor

The Carnot cycle is composed of four reversible processes working between the temperatures, T_H, and, T_L. The efficiency of a Carnot engine is

$$\eta_C = 1 - \frac{T_L}{T_H}$$

The coefficient of performance of the Carnot refrigerator is

$$\beta_C = \frac{1}{\dfrac{T_H}{T_L} - 1}$$

and the coefficient of performance of the Carnot heat pump is

$$\gamma_C = \frac{1}{1 - \dfrac{T_L}{T_H}}$$

6.10 Problems

Heat Engines

6.1: A steam power plant generates 5 MW of electrical power while burning natural gas at a rate of 420 kg/s in the boiler. The plant rejects heat to a condenser in which the cooling water enters at 27°C and leaves at 35°C. Assume the heating value of the natural gas to be 30,000 kJ/kg and the specific heat of water to be $c = 4.184$ kJ/kg-K. Determine (a) the thermal efficiency and (b) the mass flow rate of the cooling water required.

6.2: The following data for Q_H, Q_L, and W_{net} apply to a heat engine. For the heat engine to be feasible, both the first and second laws of thermodynamics must be satisfied. For each case, determine whether the heat engine is feasible. The cases are (a) $Q_H = 200$ kJ, $Q_L = 90$ kJ, and $W_{net} = 110$ kJ, (b) $Q_H = 50$ kJ, $Q_L = 0$, and $W_{net} = 35$ kJ, and (c) $Q_H = 18$ kJ, $Q_L = 18$ kJ, and $W_{net} = 0$.

6.3: Two heat engines operate in series. Engine 1 receives heat at the rate of 20 kW and produces a power of 8 kW. The heat rejected by engine 1 is used to drive engine 2, which produces 4 kW. Determine (a) the thermal efficiency of engine 1, (b) the thermal efficiency of engine 2, and (c) the thermal efficiency of the combined system.

6.4: Figure P6.4 shows two heat engines, one operating between reservoirs at absolute temperatures T_1 and T_2 with efficiency η_1 and the other operating between reservoirs at absolute temperatures T_2 and T_3 with efficiency η_2. Show that (a)

$$\frac{Q_3}{Q_1} = (1 - \eta_1)(1 - \eta_2)$$

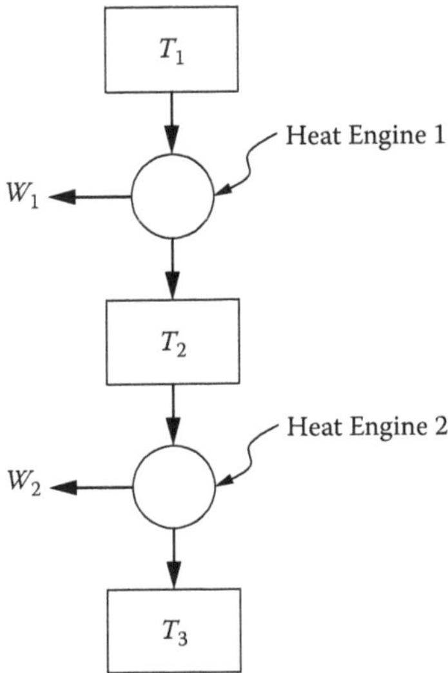

FIGURE P6.4

and (b)

$$W_1 + W_2 = (\eta_1 + \eta_2 - \eta_1\eta_2)Q_1$$

6.5: A gasoline engine delivers a power output of 20 kW and has a thermal efficiency of 28% based on a the heating value of gasoline of 44,000 kJ/kg. Determine (a) the rate of fuel consumption in the engine and (b) the power lost through the exhaust and the radiator.

6.6: An engine uses fuel oil at the rate of 0.1 kg/s and produces 850 W. The heating value of the oil is 41,850 kJ/kg. Determine (a) the heat rejection from the engine and (b) its thermal efficiency.

6.7: A power cycle uses an ideal gas with constant specific heat as the working medium. The cycle consists of four processes:

Process 1-2: an isothermal compression

Process 2-3: a constant volume heat addition

Process 3-4: an isothermal expansion

Process 4-1: a constant volume heat rejection

Draw a P-V diagram and develop an expression for the thermal efficiency of the cycle in terms of R, c_v, T_1, T_3, V_1, and V_3.

6.8: Two reversible engines are connected between a heat source and a heat sink, as indicated in Figure P6.4. In addition, $\eta_1 = \eta_2$, $T_1 = 580\,\text{K}$ and $T_3 = 220\,\text{K}$ and the heat entering engine 1 is 120 W. Determine (a) T_2, (b) W_1, (c) W_2, and (d) the heat rejected by engine 2.

6.9: The thermal efficiency of a particular engine operating on an ideal cycle is 35%. Determine (a) the heat supplied per 1.5 kW of work developed, (b) the ratio of the heat supplied to the heat rejected, and (c) the ratio of the work developed to the heat rejected.

6.10: A thermodynamic cycle has a steam generator to which 72 kW are added. The working fluid drives an engine that produces 27 kW. The pump that circulates the working medium absorbs 2 kW. Determine (a) the net work, (b) the heat rejected, and (c) the thermal efficiency.

6.11: A heat engine draws in heat at the rate of 100 kW from a high-temperature source. If the thermal efficiency of the heat engine is 28%, determine the rate at which heat is rejected to the low-temperature sink.

6.12: A coal burning power plant produces 30 MW of power. Each kilogram of coal yields 6000 kJ of energy when burned, and the plant has an efficiency of 30%. Determine the coal consumption of the plant in kg/h.

6.13: Determine the unknown quantity in each case in the following table:

Case	Q_H, kJ	Q_L, kJ	η,%	W_{net}, kJ
(1)	100	(a)	(b)	35
(2)	250	200	(c)	(d)
(3)	(e)	(f)	28	60
(4)	1000	(g)	32	(h)

6.14: The following data for Q_H, Q_L, and η apply to a heat engine. For each case, determine whether the heat engine is feasible and, if feasible, determine the heat output.

Case	$\dot{Q}_H$, kW	$\dot{Q}_L$, kW	η,%
(1)	300	150	60
(2)	500	300	40
(3)	800	400	65
(4)	200	150	25

6.15: Three heat engines operate in series. Engine 1 receives heat at the rate of 100 kW and produces 25 kW of power. The heat rejected by engine 1 is received by engine 2, which produces a power of 20 kW. The heat rejected by engine 2 is received by engine 3, which produces a power of 15 kW. Determine (a) the thermal efficiency of each engine, (b) the heat rejected by each engine, and (c) the thermal efficiency of the combined system.

6.16: For the three-engine arrangement shown in Figure P6.16, show that the thermal efficiency, η, is

$$\eta = (1 - \eta_1)(1 - \eta_2)(1 - \eta_3)$$

where η_1, η_2, and η_3 are the respective thermal efficiencies of the individual engines

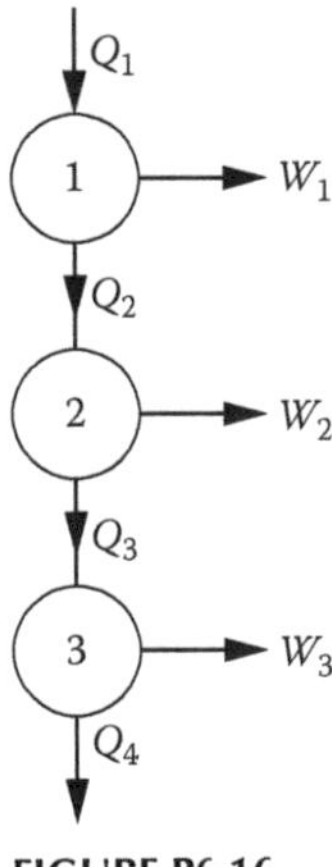

FIGURE P6.16

6.17: A steam power plant having a thermal efficiency of 35% generates 25 MW of power. The plant discharges heat in a condenser that draws cooling water at the rate of 820 kg/s from a river at 25°C. Determine the temperature of the water after it circulates through the condenser.

6.18: Two heat engines operate in parallel between the same source and sink as shown in Figure P6.18. Determine (a) the work outputs, W_1 and W_2, (b) the thermal efficiencies, η_1 and η_2, and (c) the thermal efficiency of the combined system.

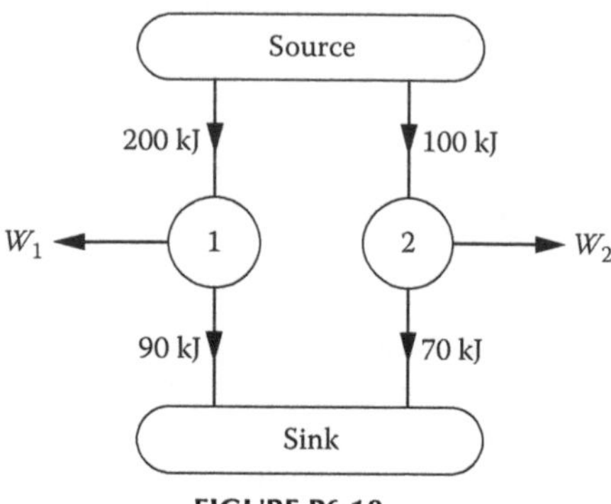

FIGURE P6.18

6.19: Consider the arrangement shown in Figure P6.19 and prove that the thermal efficiency of the combined arrangement, η, is given by

$$\eta = \frac{\eta_1 + \eta_2}{2}$$

where η_1 and η_2 are the respective thermal efficiencies of engines 1 and 2 and $Q_{H1} = Q_{H_2} = Q_H$.

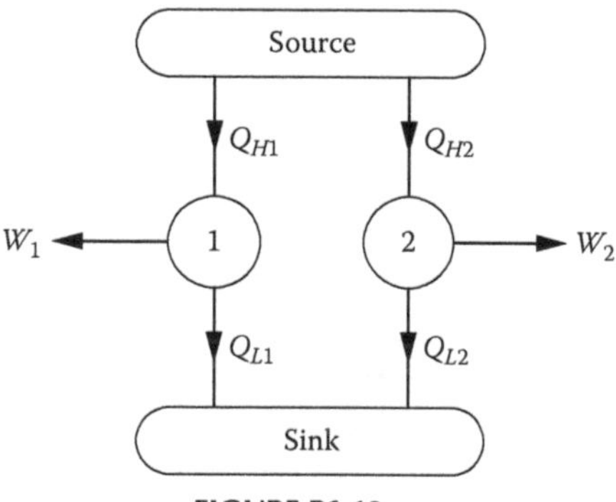

FIGURE P6.19

6.20: A diesel engine with a thermal efficiency of 26% provides a power output of 220 kW. If the engine consumes 80 kg/h of diesel fuel, determine (a) the heating value of the fuel in kJ/kg and (b) the power lost to the environment.

Refrigerators and Heat Pumps

6.21: In a vapor compression refrigeration system using refrigerant R-134a, heat is extracted from the refrigerated space as the refrigerant circulates through the evaporator and is rejected to the surroundings when the refrigerant flows through the condenser. In one such system, the R-134a enters the evaporator at −8°C with a quality of 0.30 and leaves the evaporator as saturated vapor at −8°C. The refrigerant enters the condenser at 0.8 MPa and 40°C and leaves the condenser as saturated liquid at 0.8 MPa. Determine (a) the coefficient of performance of the system and (b) the work input to the system.

6.22: A 1-ton window air conditioner draws energy from an electric outlet at the rate of 1.20 kW. The cost of electricity is 0.25/kW-h and 1 ton of refrigeration is equivalent to 211 kJ/min. Determine (a) the coefficient of performance of the air conditioner, (b) the heat rejected to the surroundings, and (c) the cost to operate the unit over 30 days with operation for 6 h/day.

6.23: A refrigerant enters the evaporator coil of a refrigeration system with a quality of 0.20 and leaves as saturated vapor without experiencing any pressure loss. The specific enthalpies of saturated liquid and saturated vapor are 157.31 kJ/kg and 1436.7 kJ/kg, respectively. The system incorporates a compressor that requires 300 kJ/kg of work. Determine the coefficient of performance of the system.

6.24: A household refrigerator with a coefficient of performance of 2.60 is able to cool its contents from a temperature of 20 to 10°C in 2 h. The mass of the contents may be assumed to be 10 kg and their average specific heat is taken as 3.10 kJ/kg-K. Determine (a) the rate at which heat is rejected by the refrigerator to the surrounding air and (b) the power input to the compressor of the refrigerator.

6.25: Figure P6.25 shows two refrigerators, one operating between reservoirs at absolute temperatures T_1 and T_2 with a coefficient of performance of β_1 and the other operating between reservoirs at absolute temperatures T_2 and T_3 with a coefficient of performance of β_2. Show that the coefficient of performance between reservoirs at absolute temperatures T_1 and T_3 is given by

$$\beta = \frac{\beta_1\beta_2}{1+\beta_1+\beta_2}$$

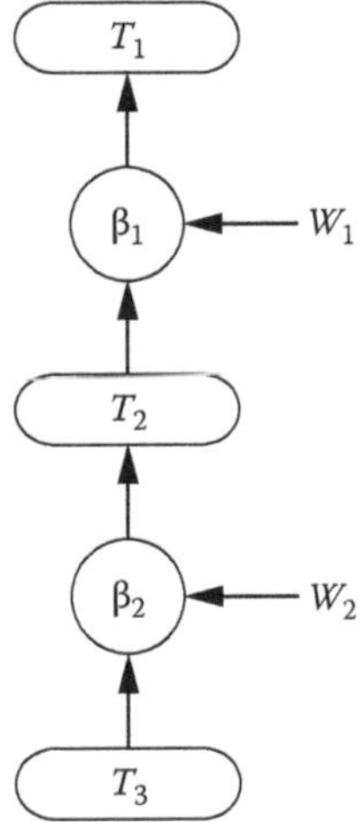

FIGURE P6.25

6.26: A heat pump draws heat from the outside air and delivers it to the inside of a house that is maintained at 20°C. The heat loss from the house is estimated to be 40,000 kJ/h. The heat pump has a coefficient of performance of 3.50. Determine (a) the power required to drive the heat pump, (b) the rate at which heat is extracted from the outside air, (c) the cost of operating the heat pump for 8 h if the cost of electricity is 20 cents per kW-h, and (d) the cost of heating the house for 8 h if resistance heating is used instead of the heat pump.

6.27: Figure P6.27 shows two heat pumps, one operating between reservoirs at absolute temperatures T_1 and T_2 with a coefficient of performance of γ_1 and the other operating between reservoirs at absolute temperatures T_2 and T_3 with a coefficient of performance of γ_2. Show that the coefficient of performance between reservoirs at absolute temperatures T_1 and T_3 is given by

$$\gamma = \frac{\gamma_1\gamma_2}{\gamma_1+\gamma_2-1}$$

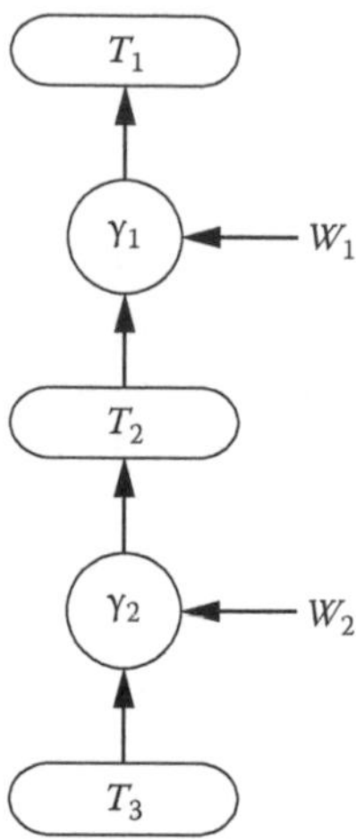

FIGURE P6.27

6.28: For a cooling load of 2.6 kW, determine (a) the power required to drive a refrigeration unit with a coefficient of performance of 3.50 and (b) the coefficient of performance if the unit is used as a heat pump.

6.29: A refrigeration unit is required to handle a cooling load of 15,000 kJ/h. The unit is driven by a 1.2-kW electric motor. Determine its coefficient of performance.

6.30: Refrigerant R-134a enters the evaporator of a refrigerator at −8°C with a quality of $x = 0.22$ and leaves the evaporator as saturated vapor at −8°C. The volumetric flow rate of the refrigerant is 4.23×10^{-3} m^3/s at the entrance of the evaporator. The refrigerator consumes 12 kW of power. Determine its coefficient of performance.

6.31: The refrigeration unit of a food storage room consumes 3.2 kW and operates with a coefficient of performance of 2.96. Determine (a) the rate of heat removal from the room and (b) the heat rejected to the environment in which the unit is located.

6.32: An air-conditioning unit consumes 10 kW of power and removes 860 kJ/m of heat from the air-conditioned space. Determine (a) the coefficient of performance of the unit, (b) the heat rejected to the environment in kilojoules over a period of 2 h, and (c) the daily cost of electricity to run the unit if the unit operates for 8 h and the cost of electricity is $0.20 per kW-h.

6.33: The contents of a household refrigerator have a mass of 12 kg with an average specific heat of 3.20 kJ/kg-K. When the refrigerator is turned on, the contents are initially at 25°C. The refrigerator has a coefficient of performance of 2.75 and draws 600 W of power. Determine how long it will take to cool the contents from 25 to 12°C.

6.34: An air-conditioning system with a coefficient of performance of 3.20 is used to maintain the temperature of a dwelling at 22°C. Heat leakage into the dwelling from the surroundings is at the rate of 5 kW. Heat generated owing to the occupants and several electrical appliances located in the dwelling amounts to 2 kW. Determine the power consumption of the system.

6.35: A dual-purpose heating/cooling unit is designed to maintain a building at 20°C throughout the year. The unit consumes 20,000 kJ/h and the heat loss/gain to or from the building is 1800 kJ/h per unit of temperature difference between the inside and outside of the building. During the summer months, the unit operates as an air conditioner with a coefficient of performance of 3.20; during the winter months, it operates as a heat pump with a coefficient of performance of 4.00. Determine the maximum and minimum outside temperatures for satisfactory operation of the unit.

6.36: For the two-refrigerator arrangement shown in Figure P6.36, determine a relationship between COP_1, COP_2, and the combined COP for the entire system. Here COP_1 is the coefficient of performance of refrigerator 1 and COP_2 is the coefficient of performance of refrigerator 2.

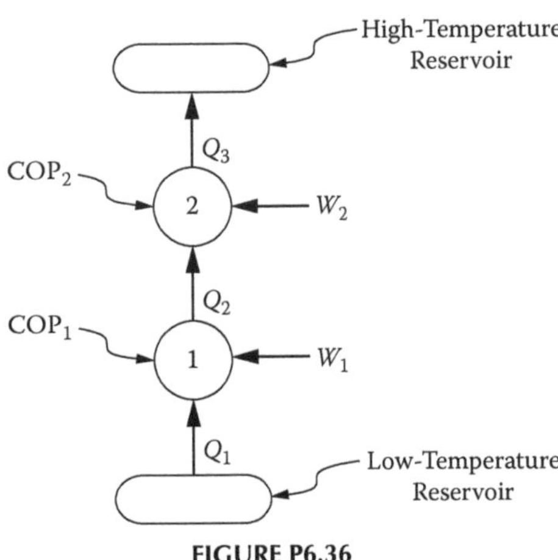

FIGURE P6.36

6.37: Determine the unknown quantity for each refrigeration unit in the following table:

Case	$\dot{Q}_H$, kW	$\dot{Q}_L$, kW	COP	$\dot{W}$ kW
(1)	120	(a)	3.10	(b)
(2)	220	160	(c)	(d)
(3)	(e)	200	(f)	80
(4)	(g)	(h)	3.20	60

6.38: Determine the unknown quantity for each of the cases in Problem 6.37 if the data are for a heat pump.

6.39: Reconsider Problem 6.36 and assume that the arrangement pertains to two heat pumps. Determine a relationship between COP_1, COP_2, and the combined COP for the entire system. Here COP_1 is the coefficient of performance of refrigerator 1 and COP_2 is the coefficient of performance of refrigerator 2.

6.40: A heat pump with a coefficient of performance of 3.8 is used to heat a house in winter and maintain a temperature of 20°C for comfort. The heat loss from the house to the outside environment occurs at the rate of 23,600 kJ/h. Determine the power required (in kW) to drive the heat pump.

6.41: For the two-heat pump arrangement shown in Figure P6.41, show that

$$COP_{1\text{-}2} = \frac{2COP_1COP_2}{COP_1COP_2}$$

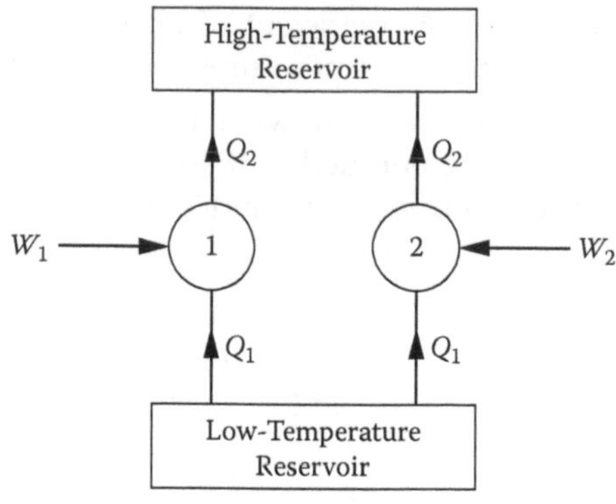

FIGURE P6.41

The Carnot Cycle

6.42: A Carnot engine receives heat at a rate of 50 kW and rejects waste heat to a sink at 300 K. The efficiency of the engine is 0.55. Determine (a) the temperature of the source and (b) the power to the engine.

6.43: The hot reservoir of a Carnot engine is a combustion chamber that burns fuel at the rate of 2 kg/s with products of combustion at 1200 K. The exhaust from the engine is at 400 K and is rejected to the atmosphere. If the heating value of the fuel is 30,000 kJ/kg, find (a) the maximum power output of the system and (b) the maximum thermal efficiency attainable.

6.44: Figure P.6.44 shows a reversible heat pump drawing heat from a reservoir at absolute temperature, T_2 and delivering it to a reservoir at absolute temperature, T_1. The heat pump is driven by a reversible heat engine that operates between two reservoirs at absolute temperatures T_3 and T_4. Show that the ratio of heat delivered by the heat pump, Q_1, to the heat drawn by the heat engine, Q_3, is given by

$$\frac{Q_1}{Q_3} = \frac{1 - (T_4/T_3)}{1 - (T_2/T_1)}$$

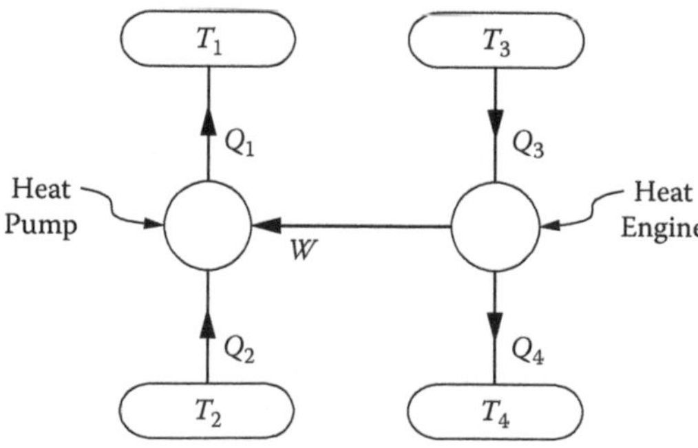

FIGURE P6.44

6.45: Consider the two heat pumps shown in Figure P6.45 with $T_1 = 293\,\text{K}$, $T_2 = 263\,\text{K}$, and $\dot{Q}_1 = 10\,\text{kW}$. The heat pumps are reversible with identical coefficients of performance. Determine (a) the coefficient of performance of each heat pump, (b) the temperature, T_2, of the intermediate reservoir, (c) $\dot{Q}_2$, and (d) $\dot{Q}_3$.

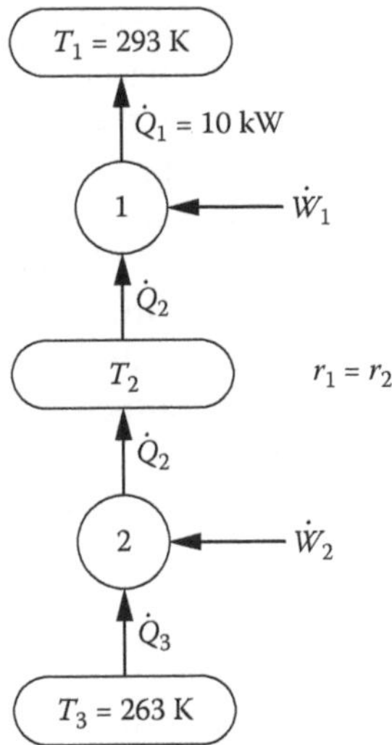

FIGURE P6.45

6.46: Consider the two heat engines shown in Figure P6.46. Assume the heat engines to be reversible and to have identical efficiencies with $T_1 = 800$ K and $T_3 = 300$ K. Determine (a) the thermal efficiency of each engine, (b) the thermal efficiency of the two engines working together, and (c) the work output of the two engines if $Q_1 = 500$ kJ.

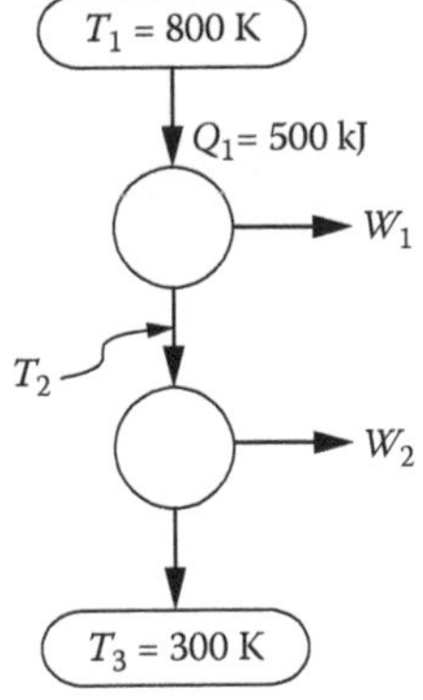

FIGURE P6.46

6.47: A Carnot engine draws heat Q_1 from a source at absolute temperature T_1 and rejects heat to two sinks, Q_2 to a sink at absolute temperature T_2 and Q_3 to a sink at absolute temperature T_3. Show that for $Q_2 = Q_3/2$, the thermal efficiency of the engine is given by

$$\eta = 1 - \frac{3T_2T_3}{T_1(2T_2 + T_3)}$$

6.48: A Carnot heat pump draws heat from two reservoirs, Q_2 from a source at absolute temperature, T_2, and Q_3 from a source at absolute temperature, T_3. The heat pump delivers heat, Q_1, to a reservoir at absolute temperature, T_1. Show that for $Q_2 = Q_3/3$, the coefficient of performance of the heat pump is given by

$$\gamma = 1 - \frac{1}{1 - \dfrac{4T_2T_3}{T_1(3T_2 + T_3)}}$$

6.49: A Carnot engine draws one-third of its heat supply from a source at 1000 K and two-thirds from a source at 800 K. The engine rejects half of its heat to a sink at 400 K and the other half to a sink at 300 K. Determine the thermal efficiency of the engine.

6.50: A Carnot refrigerator extracts heat from the interior of a refrigerated space at −10°C and discharges it to a room at 20°C. The working fluid is refrigerant R-134a, which circulates through the system at the rate of 0.40 kg/s. The refrigerant experiences a change in enthalpy of 180 kJ/kg as it flows through the evaporator coil. Determine (a) the coefficient of performance of the refrigerator, (b) the power input to the refrigerator, (c) the rate of heat rejection to the surroundings, and (d) the pressure and temperature of the refrigerant entering the condenser.

6.51: The following data pertain to four heat engines. Determine whether each engine is reversible, irreversible, or impossible:

Case	Q_H, kJ	Q_L, kJ	T_H K	T_C K	η
(1)	500	300	1000	600	
(2)	700	600	1200	700	
(3)	800	400	2600	1200	
(4)			800	300	0.72

6.52: A reversible heat engine, with a thermal efficiency of 70%, rejects heat to a sink at 400 K. Determine (a) the temperature of the source and (b) the work output of the engine.

6.53: A heat pump uses R-134a as the working fluid and provides heating at a rate of 16 kW. Saturated vapor at 2 bar leaves the evaporator and a liquid-vapor mixture leaves with a quality of $x = 0.20$ leaves the condenser at 8 bar. Determine (a) the coefficient of performance and (b) the power required to drive the system.

6.54: A reversible heat engine operates between reservoirs at temperatures, T_H and T_S as indicated in Figure P6.54. The engine drives a reversible refrigerator that operates between temperatures, T_C and T_S. Show that

$$\frac{Q_C}{Q_H} = \frac{1}{T_s}\left(\frac{1}{T_C} - \frac{1}{T_S}\right)\left(1 - \frac{T_S}{T_H}\right)$$

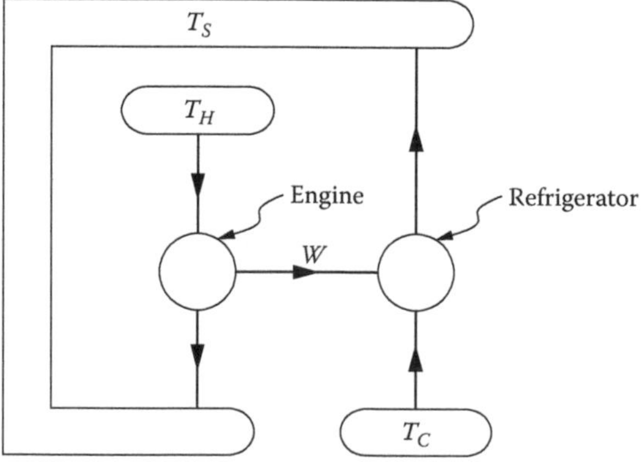

Both the Engine and the Refrigerator Are Reversible

FIGURE P6.54

6.55: A refrigerator extracts heat, Q_L, from a reservoir at 268 K and rejects heat, Q_H, to a reservoir at 300 K. For each of the following, determine whether the refrigerator operates reversibly, irreversibly, or is impossible:

(1) $Q_L = 670\,\text{kJ}$ and $W = 80\,\text{kJ}$.
(2) Coefficient of performance (COP) = 9.
(3) $Q_H = 1200\,\text{kJ}$ and $Q_L = 800\,\text{kJ}$.
(4) $Q_H = 1000\,\text{kJ}$ and $W = 200\,\text{kJ}$.

7

Entropy

Chapter Objectives

- To introduce entropy from both statistical and classical thermodynamic perspectives.
- To derive the Clausius inequality.
- To develop and illustrate the use of the temperature-entropy diagram.
- To consider the Gibbs or TdS relations for a simple compressible substance.
- To present equations for the entropy change of solids, liquids, and gases.
- To establish the isentropic relations for an ideal gas.
- To define the isentropic efficiencies for turbines, compressors, pumps, and nozzles.
- To develop entropy balance equations for closed and open systems.

7.1 Introduction

Chapter 6 emphasized that the preferred direction for the occurrence of a process is one in which **high-quality energy** degrades into **low-quality energy**. This principle is akin to the idea of **molecular disorder**. The idea of disorder is a familiar one as things always seem to become disordered spontaneously but never return to an orderly arrangement on their own.

In 1868, Rudolf Clausius[1] suggested that the property **entropy** be used as a measure of molecular disorder within a system and Ludwig Boltzmann[2] related the concept of entropy to the probability or number of ways in which the state of a system can be formulated. Thus, a substance has a lower value of entropy when it is in the solid phase because the number of ways in which the molecules can arrange themselves in a solid is limited. On the other hand, a substance in a gaseous phase has a higher entropy value because there are a very large number of ways in which the molecules can arrange themselves. So we are led to the fact that, as the number of ways that a state may arise increases, the more probable the state and the more disordered the system in that state.

---

[1] Rudolf Julius Emmanuel Clausius (1822–1888) was a German physicist and a founder of thermodynamics.

[2] L. E. Boltzmann (1844–1906) was an Austrian physicist who was a pioneer in the development of the kinetic theory of gases.

Based on the foregoing rationale, the entropy, S, of a system has been defined in statistical thermodynamics via the **Boltzmann equation**

$$S = k \ln P \tag{7.1}$$

where P is the thermodynamic probability (or the total number of quantum states available to the system) and k is the Boltzmann constant. Although in this text, our approach to thermodynamics is classical, we have briefly introduced the statistical approach in order to provide the reader with a more fundamental insight.

7.2 The Classical Definition of Entropy

The thermodynamic analysis of Carnot cycles, presented in Section 6.6, led to Equation 6.19 that may be rearranged to yield

$$\frac{Q_H}{T_H} - \frac{Q_L}{T_L} = 0 \tag{7.2}$$

We note that Q_H and Q_L are reversible heat transfers and Equation 7.2 reveals that, in a Carnot cycle, the algebraic sum of the reversible heat transfers divided by their corresponding absolute temperatures is zero. Because the initial and final states in a cycle are the same, the net (algebraic) change in every property in the system undergoing the cycle is zero. It follows that the quantity, Q_{rev}/T, whose algebraic sum is zero, as reflected by Equation 7.2, must be a property of the system. This is the property entropy, S, whose definition from a statistical perspective was given by Equation 7.1.

From a classical point of view, the differential change in entropy, dS, is defined as

$$dS = \left(\frac{\delta Q}{T}\right)_{rev} \tag{7.3}$$

and the change in entropy, $S_2 - S_1$, may be obtained by integrating Equation 7.3 between points 1 and 2 giving

$$S_2 - S_1 = \int_1^2 \left(\frac{\delta Q}{T}\right)_{rev} \tag{7.4}$$

We observe that S is an extensive property whose units are kJ/K. The changes in specific entropy, s, with units kJ/kg-K are

$$ds = \left(\frac{\delta q}{T}\right)_{rev} \tag{7.5}$$

and

$$s_2 - s_1 = \int_1^2 \left(\frac{\delta q}{T}\right)_{rev} \tag{7.6}$$

where δq is the specific reversible heat transfer in kJ/kg.

Each of Equations 7.3 through 7.6 shows that, for a reversible adiabatic process, (i.e., for $\delta q = 0$), the change in entropy is zero. Such a process is called an **isentropic process** and it may be recalled that two such processes (processes 2-3 and 4-1 in Figure 6.15) were used in constructing the Carnot cycle.

It is important to recognize that entropy is a **property** and the change in entropy between state 1 and state 2 is **independent** of the path between the two states. However, the calculation of $S_2 - S_1$ using Equation 7.4 or $s_2 - s_1$, using Equation 7.6, requires the computation of Q or q along a **reversible** path between the states.

The classical entropy relationships given by Equations 7.3 to 7.6 are in terms of the change in entropy. The absolute values of S or s may be found by using the reference $S = s = 0$ when $T = 0$ from Equation 7.1. At a temperature of absolute zero, only a single state is available to the system, $P = 1$, and consequently the reference of zero entropy at absolute zero is used to tabulate the values of specific entropy for gases. However, in the steam tables, the entropy of saturated liquid water at 0.01°C is assigned a zero value (see Tables A.3 and A.4 in Appendix A). For refrigerant R-134a, the entropy of saturated liquid at −40°C is taken as zero (Tables A.6 and A.7). As we are concerned with changes in entropy, the assignment of a reference state, where $s = 0$, is arbitrary.

From a mathematical standpoint, it is interesting to note that in Equation 7.3, the factor, $1/T$, converts the inexact differential, δQ, into an exact differential, dS, for a reversible process.

7.3 The Clausius Inequality

Equation 7.2, which resulted from a Carnot cycle analysis, is true for any reversible cycle that is designed to receive heat Q_H from a reservoir at T_H and reject heat Q_L to a reservoir at T_L. We may express Equation 7.2 in a cyclic integral form

$$\oint \left(\frac{\delta Q}{T}\right)_{\text{rev}} = 0 \tag{7.7}$$

for a reversible cycle.

Now consider the irreversible cycle operating between the same reservoirs and assume that from the reservoir at T_H, it receives the same heat, Q_H, as the reversible cycle. Because the irreversible cycle is less efficient than the reversible cycle, it must reject more heat, that is, $(Q_L)_{\text{irrev}} > (Q_L)_{\text{rev}}$. This means that

$$\frac{Q_H}{T_H} - \left(\frac{Q_L}{T_L}\right)_{\text{irrev}} < 0$$

Consequently,

$$\oint \left(\frac{\delta Q}{T}\right) < 0 \tag{7.8}$$

for an irreversible process. Equations 7.7 and 7.8 may be combined into a single equation giving the **Clausius inequality**

$$\oint \left(\frac{\delta Q}{T}\right) \leq 0 \tag{7.9}$$

In Equation 7.9, the equality applies for a reversible cycle and the inequality applies to an irreversible cycle. The Clausius inequality is often stated as a corollary of the second law of thermodynamics:

> When a system operates in a cycle and the differential heat, δQ, added or removed at a point is divided by the absolute temperature, T, at that point, the algebraic sum of all $\delta Q/T$ values is zero for a reversible cycle and less than zero for an irreversible cycle.

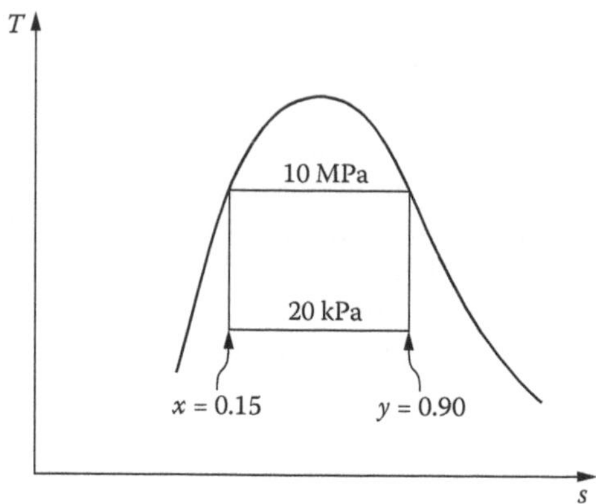

FIGURE 7.1
T-s diagram for Example 7.1.

Although Equation 7.9 was developed for a reversible heat engine cycle, it also holds for the refrigeration and heat pump cycles.

Example 7.1 In a steam power plant, heat is supplied to the working fluid in a boiler and rejected in a condenser. Saturated liquid enters the boiler at 10 MPa and exits as saturated vapor at the same pressure. The steam exhausted by the turbine enters the condenser at 20 kPa with a quality of 0.90 and leaves the condenser with a quality of 0.15 at the same pressure. The mass flow rate is 10 kg/s and the pump and turbine processes are adiabatic. Evaluate the cyclic integral $\oint(\dot{Q}/T)$ and show that the Clausius inequality is satisfied.

Solution

Assumptions and Specifications

(1) The pump and turbine processes are adiabatic.
(2) The system is the steam in the steam generator.
(3) There are no potential and kinetic energy changes.
(4) The system operates in the steady state.
(5) The mass flow rate and the qualities in the condenser are specified.

A plot of temperature against entropy is shown in Figure 7.1. The rate of heat supply to the boiler is

$$\dot{Q} = \dot{m}(\Delta h) = \dot{m}h_{fg}$$

From Table A.4 at 10 MPa, read

$$h_{fg} = 1319.5\,\text{kJ/kg} \quad \text{and} \quad T_H = T_{sat} = 311°\text{C} + 273°\text{C} = 584\,\text{K}$$

Thus,

$$\dot{Q}_H = \dot{m}h_{fg} = (10\,\text{kg/s})(1319.5\,\text{kJ/kg}) = 13,195\,\text{kW}$$

At 20 kPa, Table A.4 also gives

$$h_{fg} = 2358.7\,\text{kJ/kg} \quad \text{and} \quad T_L = T_{sat} = 60.1°\text{C} + 273°\text{C} = 333.1\,\text{K}$$

The rate of heat rejection in the condenser is

$$\dot{Q}_L = \dot{m}\Delta x h_{fg} = (10\,\text{kg/s})(0.90 - 0.15)(2358.7\,\text{kJ/kg}) = 17,690\,\text{kW}$$

The cyclic integral may also be expressed in terms of the rate of heat transfer

$$\oint\left(\frac{\delta\dot{Q}}{T}\right) \le 0$$

and

$$\oint\left(\frac{\delta Q}{T}\right) = \frac{\dot{Q}_H}{T_H} - \frac{\dot{Q}_L}{T_L} = \frac{13{,}195\ \text{kW}}{584\,\text{K}} - \frac{17{,}690\,\text{kW}}{333.1\,\text{K}} = -30.51\,\text{kW/K} \Longleftarrow$$

which satisfies the Clausius inequality.

Because the numerical value of the cyclic integral is negative, the cycle is irreversible. We observe that both the heat supply and the heat rejection processes are isothermal. If either of them was not isothermal, the evaluation of the cyclic integral would have been more difficult.

Example 7.2 Determine the value of the cyclic integral, $\oint(\dot{Q}/T)$, if the steam in Example 7.1 enters the condenser at 20 kPa with a quality of 0.6758 and leaves the condenser with a quality of 0.3572 at the same pressure.

We make the same assumptions as we made in Example 7.1. Figure 7.1 also shows that the specified values of the quality in the condenser have changed.

In this case, the rate at which heat is rejected in the condenser is given by

$$\dot{Q}_L = \dot{m}\Delta x h_{fg} = (10\,\text{kg/s})(0.6758 - 0.3572)(2358.7\,\text{kJ/kg}) = 7514.8\,\text{kW}$$

and

$$\oint\left(\frac{\delta Q}{T}\right) = \frac{\dot{Q}_H}{T_H} - \frac{\dot{Q}_L}{T_L} = \frac{13,195\ \text{kW}}{584\,\text{K}} - \frac{7514.8\,\text{kW}}{333.1\,\text{K}} \approx 0 \Longleftarrow$$

Because the cyclic integral is zero, the cycle is reversible. Once again, note that both the heat supply and the heat rejection processes are isothermal.

7.4 The Temperature-Entropy Diagram

The temperature-specific entropy (T-s) and specific enthalpy-specific entropy (h-s) diagrams were briefly introduced in Section 4.4. Now that entropy has been identified as a property, we can fill in further details pertaining to the T-s diagram which we will refer to as the temperature-entropy diagram.

Figure 7.2, the temperature-entropy diagram for water (steam) very much resembles the P-v diagram of Figure 4.4. We observe that the saturated liquid and saturated vapor lines intersect at the critical point and form a vapor dome. The region under the vapor dome is the liquid-vapor region. The region to the left of the saturated liquid line is the **compressed liquid** or **subcooled liquid** region and that to the right of the saturated vapor line is the **superheated vapor** region.

Figure 7.2 also shows four constant pressure lines marked P_1, P_2, P_3, and P_c, with $P_c > P_3 > P_2 > P_1$. The diagram also identifies the specific entropies of saturated liquid, s_f,

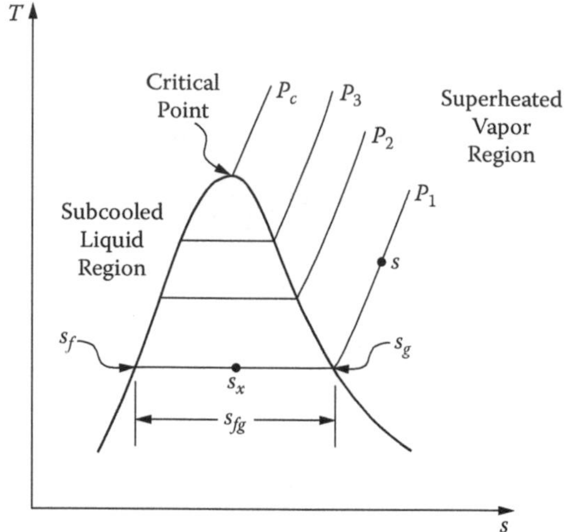

FIGURE 7.2
T-s diagram showing four pressures an pertinent entropy values.

saturated vapor, s_g, superheated vapor, s, and the specific entropy at a particular quality, s_x, all corresponding to a pressure, P_1. Tables A.3 to A.5 provide data for s_f, s_g, $s_{fg} = s_g - s_f$ and s. The specific entropy of a liquid-vapor mixture, s_x, depends on the quality, x, and can be evaluated using Equation 4.7a or 4.7b.

Example 7.3 Consider the cycles described in Examples 7.1 and 7.2 and show them on a temperature-entropy diagram (Figure 7.3).

Solution

We make the same assumptions that were made in Example 7.1

The cycle for Example 7.1 is shown in Figure 7.2 as 1-2-3-4-1. In process 1-2, a mixture of liquid and vapor at 20 kPa and a quality of $x = 0.15$ is compressed to saturated liquid

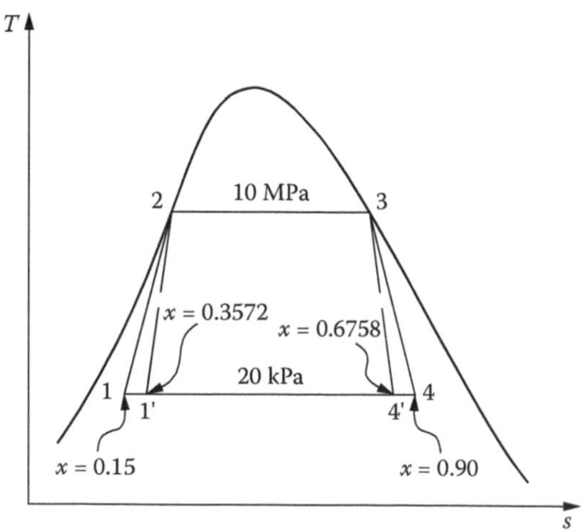

FIGURE 7.3
T-s diagram for Example 7.3.

water at 10 MPa. We read Table A.4 to find at 20 kPa

$$s_f = 0.8314\,\text{kJ/kg-K} \quad \text{and} \quad s_{fg} = 7.0779\,\text{kJ/kg-K}$$

and at 10 MPa

$$s_f = 3.3580\,\text{kJ/kg-K} \quad \text{and} \quad s_g = 5.6196\,\text{kJ/kg-K}$$

Then by Equation 4.7b at 20 kPa

$$s_1 = s_f + xs_{fg} = 0.8314\,\text{kJ/kg-K} + (0.15)(7.0779\,\text{kJ/kg-K}) = 1.8931\,\text{kJ/kg-K}$$

and at 10 MPa

$$s_2 = s_f = 3.3580\,\text{kJ/kg-K}$$

Clearly, the entropy increases in process 1-2 and makes the process irreversible.

In process 2-3, which occurs in the boiler, the heat is supplied isothermally at 10 MPa and changes the saturated liquid to saturated vapor. Since we have determined that the saturated vapor entropy value at 10 MPa is $s_g = 5.6196\,\text{kJ/kg-K}$, we can write

$$s_3 = s_g = 5.6196\,\text{kJ/kg-K}$$

Although there is an increase in entropy from state 2 to state 3, this increase is associated with a reversible heat supply process.

Process 3-4 occurs in the turbine where saturated vapor at 10 MPa expands to yield a mixture of liquid and vapor with a quality of $x = 0.90$. The entropy may be found via Equation 4.7b using values that have already been determined at 20 kPa

$$s_4 = s_f + xs_{fg} = 0.8314\,\text{kJ/kg-K} + (0.90)(7.0779\,\text{kJ/kg-K}) = 7.2015\,\text{kJ/kg-K}$$

Because $s_4 > s_3$, the process of expansion in the turbine is irreversible.

Process 4-1 occurs in the condenser where the heat is rejected isothermally at 333.1 K and the quality of the mixture of liquid and vapor decreases from $x = 0.90$ to $x = 0.15$. Observe that the decrease in entropy from state 4 to state 1 is due to a reversible heat rejection process. It may be noted that the net change in entropy over the cycle is zero and that because processes 1-2 and 3-4 are irreversible, the cycle is irreversible.

The cycle for Example 7.2 is shown in Figure 7.2 as 1′-2-3-4′-1′ and the entropies at state 1′ and 4′ may be determined as follows:

$$s_1' = s_f + xs_{fg} = 0.8314\,\text{kJ/kg-K} + (0.3572)(7.0779\,\text{kJ/kg-K}) = 3.3596\,\text{kJ/kg-K}$$

and

$$s_4' = s_f + xs_{fg} = 0.8314\,\text{kJ/kg-K} + (0.6758)(7.0779\,\text{kJ/kg-K}) = 5.6146\,\text{kJ/kg-K}$$

We note that $s_1' \approx s_2$ and $s_4' \approx s_3$ indicating that the compression from state 1′ to state 2 and the expansion from state-3 to state-4′ are isentropic (reversible adiabatic) processes. The cycle 1′-2-3-4′-1′ is therefore reversible. Because entropy is a property of the system, its net change is zero.

Example 7.4 The compressor of a refrigeration unit receives saturated R-134a vapor at a temperature of −20°C. The vapor is compressed and delivered to a condenser as a superheated vapor at 1.2 MPa (Figure 7.4). Assuming the compression process to be isentropic, determine (a) the temperature of the superheated vapor at the compressor outlet, (b) the

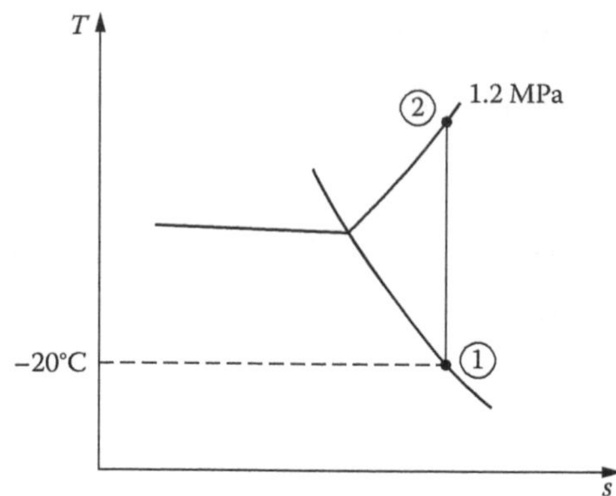

FIGURE 7.4
T-*s* diagram for the compressor in Example 7.4.

work input into the compressor, (c) the change in the specific internal energy of the refrigerant, and (d) the specific volume of the refrigerant at the compressor outlet.

Solution

Assumptions and Specifications

(1) The compression process is isentropic.
(2) The system is the refrigerant in the compressor.
(3) There are no potential and kinetic energy changes.
(4) The system operates in the steady state.

(a) The specific entropy of saturated R-134a vapor at −20°C (state 1) may be read from Table A.6

$$s_1 = s_g = 0.9332\,\text{kJ/kg-K}$$

Because the compression process is isentropic from state 1 to state 2 (superheated vapor at 1.2 MPa), there is no change in entropy so that

$$s_2 = s_1 = 0.9332\,\text{kJ/kg-K}$$

When we consult Table A.8 at $P = 1.2\,\text{MPa}$, we find that the value, $s_2 = 0.9332\,\text{kJ/kg-K}$, lies between the tabulated values of $s = 0.9164\,\text{kJ/kg-K}$ at 50°C and $s = 0.9527\,\text{kJ/kg-K}$ at 60°C. A linear interpolation with an interpolation factor, F,

$$F = \frac{0.9332\,\text{kJ/kg-K} - 0.9164\,\text{kJ/kg-K}}{0.9527\,\text{kJ/kg-K} - 0.9164\,\text{kJ/kg-K}} = 0.46$$

gives

$$T = 50^\circ\text{C} + (0.46)(60^\circ\text{C} - 50^\circ\text{C}) = 54.6^\circ\text{C} \Longleftarrow$$

(b) For the compressor as a control volume with isentropic compression, the rate of heat transfer, $\dot{Q}_{cv} = 0$, and Equation 5.19 that ignores the changes in kinetic and potential energies gives the specific compressor work, w_c as

$$w_c = \frac{\dot{W}_{cv}}{\dot{m}} = h_1 - h_2$$

Noting that h_1 is equal to h_g at −20°C, we have from Table A.6

$$h_1 = h_g = 235.31\,\text{kJ/kg}$$

To find h_2 we interpolate using the data for h from Table A.8. Then

$$h_2 = 275.52\,\text{kJ/kg} + (0.46)(287.44\,\text{kJ/kg} - 275.52\,\text{kJ/kg}) = 281.00\,\text{kJ/kg}$$

and

$$w_c = h_1 - h_2 = 235.31\,\text{kJ/kg} - 281.00\,\text{kJ/kg} = -45.69\,\text{kJ/kg} \Longleftarrow$$

This negative value shows that this is the work required to drive the compressor.

(c) The change in specific internal energy may be obtained in similar fashion. Noting that u_1 is equal to u_g at −20°C, we have from Table A.6

$$u_1 = u_g = 215.84\,\text{kJ/kg}$$

To find u_2 we interpolate using the data for u from Table A.8.

$$u_2 = 254.98\,\text{kJ/kg} + (0.46)(265.42\,\text{kJ/kg} - 254.98\,\text{kJ/kg}) = 259.78\,\text{kJ/kg}$$

Then the change in specific internal energy is

$$u_2 - u_1 = 259.78\,\text{kJ/kg} - 215.84\,\text{kJ/kg} = 43.94\,\text{kJ/kg} \Longleftarrow$$

(d) The specific volume, v_2 can be found by interpolating the data for v from Table A.8

$$v_2 = 0.01712\,\text{m}^3/\text{kg} + (0.46)(0.01835\,\text{m}^3/\text{kg} - 0.01712\,\text{m}^3/\text{kg})$$

or

$$v_2 = 0.01769\,\text{m}^3/\text{kg} \Longleftarrow$$

7.5 The Gibbs Property Relations

Because entropy is not directly measurable, its numerical values appearing in Tables A.3 through A.8 and A.12 in Appendix A are calculated from a knowledge of measurable properties such as pressure, temperature, and specific volume and utilizing the thermodynamic relations that relate these properties to entropy. Two such relations are the **Gibbs equations** also called the *TdS* equations.

Consider a closed stationary system composed of a simple compressible substance and experience heat and work interactions at its boundary. The differential form of the first law of thermodynamics for this system is

$$\delta Q - \delta W = dU \tag{7.10}$$

If the heat and work transfers are reversible, then Equation 7.10 may be written as

$$\delta Q_{\text{rev}} - \delta W_{\text{rev}} = dU \tag{7.11}$$

From Equation 7.3 it follows that

$$\delta Q_{\text{rev}} = TdS \tag{7.12}$$

and, for a closed system, the only form of work that is reversible is the flow work (or PdV work)

$$\delta W_{rev} = PdV \tag{7.13}$$

When we substitute Equations 7.12 and 7.13 into Equation 7.11 we obtain, after a little rearrangement, the first Gibbs or TdS equation

$$TdS = dU + PdV \tag{7.14}$$

or

$$Tds = du + Pdv \tag{7.15}$$

on a unit mass basis.

Although Equations 7.14 and 7.15 are derived on the assumption of a reversible process, they are relations involving properties and, because changes in properties are independent of any particular process, must apply to irreversible processes as well.

The second Gibbs or TdS relationship may be derived by introducing the expression for the enthalpy

$$H = U + PV \tag{7.16}$$

Differentiation of Equation 7.16 provides

$$dH = dU + PdV + VdP$$

or

$$dU + PdV = dH - VdP \tag{7.17}$$

When we replace $dU + PdV$ on the right-hand side of Equation 7.14 with $dH - VdP$, Equation 7.14 may be expressed as

$$TdS = dH - VdP \tag{7.18}$$

or

$$Tds = dh - vdP \tag{7.19}$$

on a unit mass basis. Equation 7.18 is the second Gibbs or TdS equation.

Equations 7.15 and 7.19 are employed in the next section to develop expressions for the change in entropy of solids, liquids, and ideal gases in terms of pressure, temperature, and specific volume.

7.6 Entropy Change for Solids, Liquids, and Ideal Gases

The entropy change for a solid or liquid can be expressed in terms of a single property but the entropy change for an ideal gas requires a knowledge of two properties.

7.6.1 Entropy Change for Solids and Liquids

Consider Equation 7.15. For a solid or a liquid, the change in specific volume with temperature is rather negligible. Hence, with $du = c\,dT$

$$ds = \frac{du}{T} = c\frac{dT}{T} \tag{7.20}$$

If $\bar{c}$ is the average value of c between the temperatures, T_1 and T_2, Equation 7.20 may be integrated to give

$$s_2 - s_1 = \bar{c} \ln \frac{T_2}{T_1} \tag{7.21}$$

Equation 7.21 shows that the change in entropy for a solid or liquid may be found from just the knowledge of the temperatures.

Example 7.5 Two aluminum blocks, one at a temperature of 80°C and the other at 40°C interact thermally until an equilibrium temperature, T_e, is reached. Each block has a mass of 2 kg and an average specific heat of 903J/kg-K. Determine (a) the equilibrium temperature and (b) the change in entropy.

Solution

Assumptions and Specifications

(1) The system consists of the two blocks.
(2) There is no heat interaction between the blocks and the surroundings.

(a) Let m and c be the mass and average specific heat of each block with T_1 and T_2 taken as the initial temperatures of the hot and cold blocks. Because there are no heat or work interactions between the blocks and the surroundings, the first law energy balance is

$$mc(T_1 - T_e) = mc(T_e - T_2)$$

or

$$T_e = \frac{1}{2}(T_1 + T_2) = \frac{1}{2}(80°\text{C} + 40°\text{C}) = 60°\text{C} \quad (333\,\text{K}) \Longleftarrow$$

(b) With $T_1 = 80°\text{C}$ (353 K) and $T_2 = 40°\text{C}$ (313 K) we can use Equation 7.21 to provide the change in entropy for each block. For block 1 (the hot block)

$$\begin{aligned}\Delta S_1 &= m\bar{c} \ln \frac{T_e}{T_1} \\ &= (2\,\text{kg})(903\text{J/kg-K}) \ln \frac{333\text{ K}}{353\,\text{K}} \\ &= (1806\,\text{J/K}) \ln(0.9433) = -105.34\,\text{J/K}\end{aligned}$$

Similarly, for block 2 (the cold block)

$$\begin{aligned}\Delta S_2 &= m\bar{c} \ln \frac{T_e}{T_2} \\ &= (2\,\text{kg})(903\,\text{J/kg-K}) \ln \frac{333\,\text{K}}{313\,\text{K}} \\ &= (1806\,\text{J/K}) \ln(1.0639) = 111.86\,\text{J/K}\end{aligned}$$

The two blocks represent the system and the change in entropy will be

$$\Delta S = \Delta S_1 + \Delta S_2 = -105.34\,\mathrm{J/K} + 111.86\,\mathrm{J/K}) = 6.52\,\mathrm{J/K} \Longleftarrow$$

7.6.2 Entropy Change for an Ideal Gas

For an ideal gas, the change in specific volume cannot be neglected. Further, the specific heat is a function of temperature. The change in specific internal energy is

$$du = c_v(T)dT \tag{3.30a}$$

Equation 3.30a, along with the ideal gas law

$$P = \frac{RT}{v}$$

allows us to put Equation 7.15 into the form

$$ds = c_v(T)\frac{dT}{T} + R\frac{dv}{v} \tag{7.22}$$

If the average value of c_v between the temperatures T_1 and T_2 is designated as $\bar{c}_v$, then Equation 7.22 may be integrated to yield

$$s_2 - s_1 = \bar{c}_v \ln\frac{T_2}{T_1} + R\ln\frac{v_2}{v_1} \tag{7.23}$$

Alternatively, we may employ Equation 7.19 to develop an expression for the change in entropy using the specific heat at constant pressure. Noting that

$$dh = c_p(T)dT \tag{3.30b}$$

Equation 3.30b, along with the ideal gas law, allows us to put Equation 7.19 into the form

$$ds = c_p(T)\frac{dT}{T} - R\frac{dP}{P} \tag{7.24}$$

If the average value of c_p between the temperatures T_1 and T_2 is designated as $\bar{c}_p$, then an integration of Equation 7.24 yields

$$s_2 - s_1 = \bar{c}_p \ln\frac{T_2}{T_1} - R\ln\frac{P_2}{P_1} \tag{7.25}$$

We see from Equations 7.23 and 7.25 that, unlike a solid or a liquid, the change in entropy of an ideal gas requires a knowledge of two properties.

Example 7.6 Carbon dioxide, initially at a pressure of 100 kPa and a temperature of 300 K, is compressed to a pressure of 300 kPa and a temperature of 400 K. The mass of carbon dioxide is 0.20 kg and the compression work is 200 kJ. Determine (a) the heat transferred and (b) the change in entropy for this process.

Solution

Assumptions and Specifications

(1) The carbon dioxide is an ideal gas and is the system.
(2) The entire process operates under quasi-equilibrium conditions.
(3) Kinetic and potential energy changes are negligible.
(4) The work of compression is specified.

Table A.1 gives $R = 0.189\,\text{kJ/kg-K}$ and Table A.2 may be employed to provide the average specific heats at 350 K

$$\bar{c}_v = 0.706\,\text{kJ/kg-K} \quad \text{and} \quad \bar{c}_p = 0.895\,\text{kJ/kg-K}$$

(a) The first law expression for this process is

$$Q - W = U_2 - U_1 = m\bar{c}_v(T_2 - T_1)$$

and we note that the work of compression specified as 200 kJ is negative because it is done on the carbon dioxide. Hence,

$$\begin{aligned} Q &= W + m\bar{c}_v(T_2 - T_1) \\ &= -200\,\text{kJ} + (0.20\,\text{kg})(0.706\,\text{kJ/kg-K})(400\,\text{K} - 300\,\text{K}) \\ &= -200\,\text{kJ} + 14.12\,\text{kJ} = -185.88\,\text{kJ} \Longleftarrow \end{aligned}$$

The negative sign shows that heat is lost to the surroundings.

(b) The change in entropy is evaluated using Equation 7.25

$$\begin{aligned} s_2 - s_1 &= \bar{c}_p \ln \frac{T_2}{T_1} - R \ln \frac{P_2}{P_1} \\ &= (0.895\,\text{kJ/kg-K}) \ln \frac{400\,\text{K}}{300\,\text{K}} - (0.189\,\text{kJ/kg-K}) \frac{300\,\text{kPa}}{100\,\text{kPa}} \\ &= 0.2575\,\text{kJ/kg-K} - 0.2076\,\text{kJ/kg-K} = 0.0499\,\text{kJ/kg-K} \end{aligned}$$

Then

$$\Delta S = m(s_2 - s_1) = (0.20\,\text{kg})(0.0499\,\text{kJ/kg-K}) = 0.0100\,\text{kJ/K} \Longleftarrow$$

7.7 The Isentropic Process for an Ideal Gas

In an isentropic process, there is no change in entropy ($\Delta S = 0$). Thus, for an ideal gas in an isentropic process, Equation 7.23 gives

$$s_2 - s_1 = \bar{c}_v \ln \frac{T_2}{T_1} + R \ln \frac{v_2}{v_1} = 0$$

or

$$\ln \frac{T_2}{T_1} = -\frac{R}{\bar{c}_v} \ln \frac{V_2}{V_1} = \frac{R}{\bar{c}_v} \ln \frac{V_1}{V_2} = \ln \left(\frac{V_1}{V_2}\right)^{R/\bar{c}_v}$$

and it then follows that

$$\frac{T_2}{T_1} = \left(\frac{V_1}{V_2}\right)^{R/\bar{c}_v}$$

We note from Equation 3.35a that

$$\frac{R}{\bar{c}_v} = k - 1$$

which allows us to write

$$\frac{T_2}{T_1} = \left(\frac{V_1}{V_2}\right)^{k-1} \tag{7.26}$$

If we put $s_2 - s_1 = 0$ in Equation 7.25, a similar development brings us to

$$\ln \frac{T_2}{T_1} = \frac{R}{\bar{c}_p} \ln \frac{P_2}{P_1}$$

It then follows that

$$\frac{T_2}{T_1} = \left(\frac{P_2}{P_1}\right)^{R/\bar{c}_p}$$

and we note from Equation 3.35b that

$$\frac{R}{\bar{c}_p} = \frac{k-1}{k}$$

and hence,

$$\frac{T_2}{T_1} = \left(\frac{P_2}{P_1}\right)^{(k-1)/k} \tag{7.27}$$

When we equate Equations 7.26 and 7.27 we obtain

$$\frac{T_2}{T_1} = \left(\frac{V_1}{V_2}\right)^{k-1} = \left(\frac{P_2}{P_1}\right)^{(k-1)/k}$$

which gives

$$\frac{P_2}{P_1} = \left(\frac{V_1}{V_2}\right)^{k} \tag{7.28}$$

Equations 7.26 through 7.28 are the P-V-T relationships for an ideal gas undergoing an isentropic process. We may write Equation 7.28 as

$$P_1 V_1^k = P_2 V_2^k \tag{7.29}$$

or more generally

$$PV^k = C \qquad \text{(a constant)} \tag{7.30}$$

The work in an isentropic process will be

$$W = \int_{V_1}^{V_2} P dV = \frac{P_2 V_2 - P_1 V_1}{1-k} = \frac{mR(T_2 - T_1)}{1-k} \tag{7.31}$$

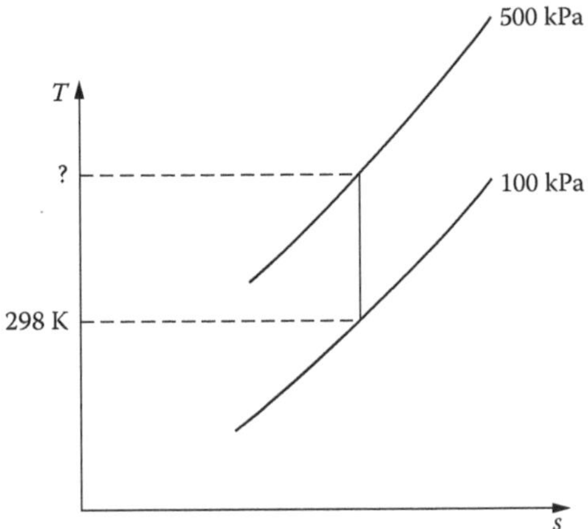

FIGURE 7.5
T-s diagram for the compressor in Example 7.7.

Example 7.7 An air compressor draws air at 100 kPa and 298 K and delivers it at a pressure of 500 kPa (Figure 7.5). Assuming the compression process to be isentropic with $k = 1.4$ and specific heats constant at 300 K, determine (a) the final specific volume, (b) the final temperature of the air, and (c) the work of compression.

Solution

Assumptions and Specifications

(1) The compressor is the control volume.
(2) The air is an ideal gas and is the system.
(3) The entire process operates under quasi-equilibrium conditions.
(4) Kinetic and potential energy changes are negligible.

(a) The initial specific volume of the air may be determined from the ideal gas law. With $R = 0.287\,\text{kJ/kg-K}$ and $c_p = 1.005\,\text{kJ/kg-K}$ from Table A.1, we can write

$$v_1 = \frac{RT_1}{P_1} = \frac{(0.287\,\text{kJ/kg-K})(298\,\text{K})}{100\,\text{kPa}} = 0.8553\,\text{m}^3/\text{kg}$$

The final specific volume may now be found from a rearrangement of Equation 7.28

$$v_2 = v_1 \left(\frac{P_1}{P_2}\right)^{1/k} = (0.8553\,\text{m}^3/\text{kg}) \left(\frac{100\,\text{kPa}}{500\,\text{kPa}}\right)^{1/1.4} = 0.2709\,\text{m}^3/\text{kg} \Longleftarrow$$

(b) To find the final temperature, we use Equation 7.27

$$T_2 = T_1 \left(\frac{P_2}{P_1}\right)^{(k-1)/k} = (298\,\text{K}) \left(\frac{500\,\text{kPa}}{100\,\text{kPa}}\right)^{0.4/1.4} = 472\text{ K} \Longleftarrow$$

(c) The first law of thermodynamics for a control volume, in the absence of kinetic and potential energy changes, on a unit mass basis is

$$q_{\text{cv}} - w_{\text{cv}} = h_2 - h_1$$

For an isentropic process, $q_{cv} = 0$, which then gives

$$w_{cv} = h_1 - h_2$$

For an ideal gas with constant specific heats, $\Delta h = c_p \Delta T$. Thus,

$$\begin{aligned} w_{cv} &= h_1 - h_2 = c_p(T_1 - T_2) \\ &= (1.005\,\text{kJ/kg-K})(298\,\text{K} - 472\,\text{K}) = -174.87\,\text{kJ/kg} \Longleftarrow \end{aligned}$$

We note that the negative sign indicates that work is being done on the system (the air).

7.8 Isentropic Efficiencies of Steady Flow Devices

From a power production or power consumption perspective, it is highly desirable to consider the use of steady flow devices such as turbines, pumps, and nozzles in an isentropic or reversible adiabatic process. In actual practice, however, a reversible adiabatic process cannot be realized because of the presence of irreversibilities such as friction and heat transfer through a finite temperature difference. The concept of **isentropic efficiency** is used to compare the actual performance of a steady flow device with its ideal performance under isentropic conditions. It is defined in such a manner that its numerical value is always less than unity.

For a steam or gas turbine, we define the isentropic efficiency as the ratio of the actual enthalpy change, $h_1 - h_2$, to the enthalpy change in an isentropic process, $h_1 - h_{2s}$

$$\eta_s = \frac{h_1 - h_2}{h_1 - h_{2s}} \tag{7.32}$$

We see that, neglecting kinetic and potential energy changes, the isentropic efficiency is the ratio of the actual specific work output to the isentropic work output and because $h_1 - h_2 < h_1 - h_{2s}$, the isentropic efficiency is always less than unity. The actual and

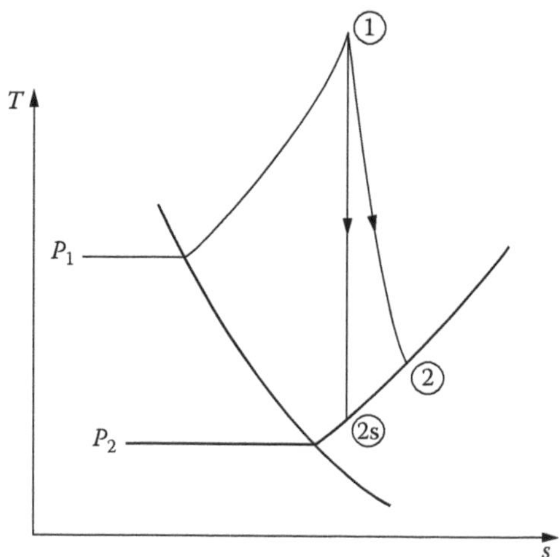

FIGURE 7.6
T-s diagram for a turbine with isentropic operation between points 1 and 2s and actual operation between points 1 and 2.

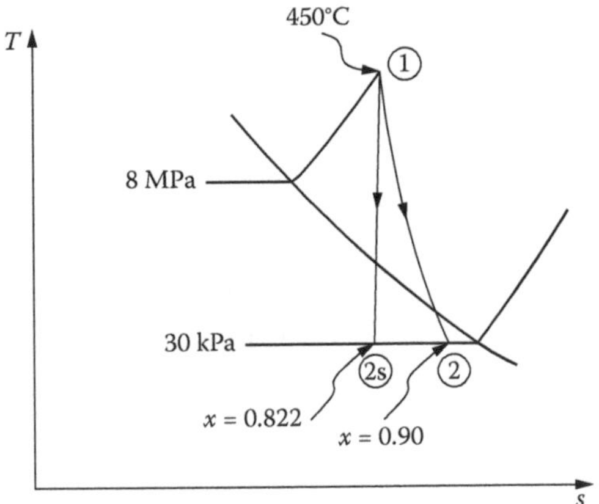

FIGURE 7.7
T-s diagram for the turbine in Example 7.8. Point 2 is at a quality of 0.900 and it is shown in the example that point 2s is at 0.822.

isentropic processes for a steam turbine and gas turbine are shown in Figure 7.6.

Example 7.8 Steam at 8 MPa and 450°C expands in a steam turbine to a pressure of 30 kPa with a quality of $x = 0.90$. The mass flow rate of steam is 25 kg/s. Determine (a) the power output and (b) the isentropic efficiency of the turbine.

Solution

Assumptions and Specifications

(1) The water-steam mixture in the turbine forms the system.
(2) The turbine is the control volume.
(3) Steady flow exists.
(4) Kinetic and potential energy changes are negligible.

(a) With reference to Figure 7.7, state 1 corresponds to 8 MPa and 450°C and Table A.5 reveals that

$$h_1 = 3273.5\,\text{kJ/kg} \quad \text{and} \quad s_1 = 6.5574\,\text{kJ/kg-K}$$

State 2 is at 30 kPa and data from Table A.4 provides

$$h_f = 289.10\,\text{kJ/kg} \quad \text{and} \quad h_{fg} = 2336.3\,\text{kJ/kg}$$

Then, after noting that the expansion from state 1 to state 2 is not isentropic, we find that

$$\begin{aligned} h_2 &= h_f + xh_{fg} \\ &= 289.10\,\text{kJ/kg} + (0.90)(2336.3\,\text{kJ/kg}) \\ &= 289.10\,\text{kJ/kg} + 2102.7\,\text{kJ/kg} = 2391.8\,\text{kJ/kg} \end{aligned}$$

and

$$\begin{aligned} \dot{W}_{turb} &= \dot{m}(h_1 - h_2) \\ &= (25\,\text{kg/s})(3273.5\,\text{kJ/kg} - 2391.8\,\text{kJ/kg}) = 22.04\,\text{MW} \Longleftarrow \end{aligned}$$

(b) To find the enthalpy at state 2s (h_{2s}), we first find the quality after an isentropic expansion. With $s_1 = s_{2s} = 6.5574\,\text{kJ/kg-K}$, we obtain from Table A.4 at 30 kPa

$$s_f = 0.9433\,\text{kJ/kg-K} \quad \text{and} \quad s_{fg} = 6.8260\,\text{kJ/kg-K}$$

so that via Equation 4.8

$$x_{2s} = \frac{s_1 - s_f}{s_{fg}} = \frac{6.5574\,\text{kJ/kg-K} - 0.9433\,\text{kJ/kg-K}}{6.8260\,\text{kJ/kg-K}} = 0.823$$

and then

$$\begin{aligned} h_{2s} &= h_f + x_{2s} h_{fg} \\ &= 289.10\,\text{kJ/kg} + (0.823)(2336.3\,\text{kJ/kg}) \\ &= 289.10\,\text{kJ/kg} + 1922.8\,\text{kJ/kg} = 2211.9\,\text{kJ/kg} \end{aligned}$$

The isentropic efficiency may be found using Equation 7.32

$$\begin{aligned} \eta_s &= \frac{h_1 - h_2}{h_1 - h_{2s}} = \frac{3273.5\,\text{kJ/kg} - 2391.8\,\text{kJ/kg}}{3273.5\,\text{kJ/kg} - 2211.9\,\text{kJ/kg}} \\ &= \frac{881.7\,\text{kJ/kg}}{1061.6\,\text{kJ/kg}} = 0.831 \qquad (83.1\%) \Longleftarrow \end{aligned}$$

For a compressor, a pump or a diffuser operating between state 1 and state 2, the isentropic efficiency is

$$\eta_s = \frac{h_{2s} - h_1}{h_2 - h_1} \tag{7.33}$$

and the actual and isentropic processes for a pump and a gas compressor are shown in Figure 7.8.

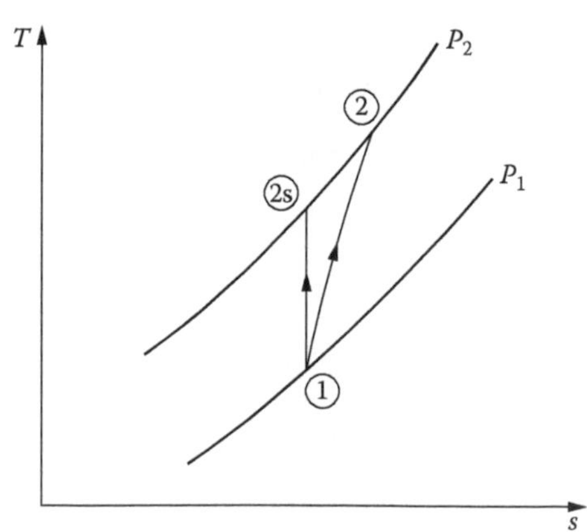

FIGURE 7.8
T-*s* diagram for a pump, compressor, or diffuser with ideal and nonideal entropy changes.

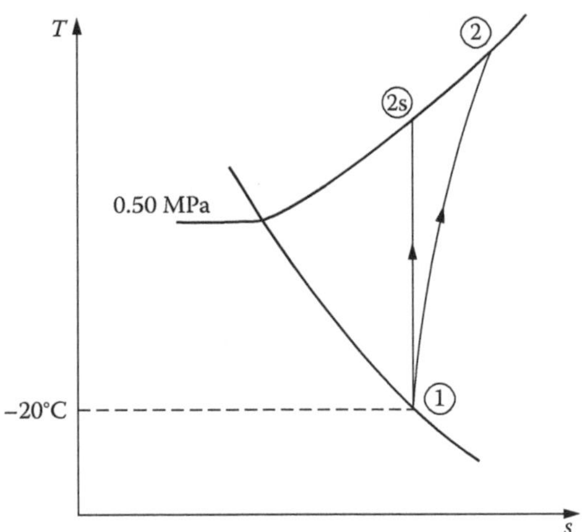

FIGURE 7.9
T-*s* diagram for the compressor in Example 7.9.

Example 7.9 A refrigeration compressor takes saturated R-134a vapor at −20°C and compresses it to a pressure of 0.50 MPa (Figure 7.9). If the isentropic efficiency is 0.85, determine (a) the specific work required to drive the compressor and (b) the temperature of the refrigerant after compression.

Solution

Assumptions and Specifications

(1) The compressor is the control volume.
(2) The refrigerant is the system.
(3) Steady flow exists.
(4) The system operates in the steady state.
(5) Kinetic and potential energy changes are negligible.

(a) Table A.6 provides at −20°C

$$h_1 = h_g = 235.31\,\text{kJ/kg} \quad \text{and} \quad s_1 = s_g = 0.9332\,\text{kJ/kg-K}$$

Table A.8 shows that at 0.50 MPa, $s = 0.9264\,\text{kJ/kg-K}$ at 20°C and $s = 0.9597\,\text{kJ/kg-K}$ at 30°C. With $s_{2s} = s_1 = 0.9332\,\text{kJ/kg-K}$, we find an interpolation fraction

$$F = \frac{0.9332\,\text{kJ/kg-K} - 0.9264\,\text{kJ/kg-K}}{0.9597\,\text{kJ/kg-K} - 0.9264\,\text{kJ/kg-K}} = \frac{0.0068\,\text{kJ/kg-K}}{0.0333\,\text{kJ/kg-K}} = 0.204$$

so that

$$h_{2s} = 260.34 + (0.204)(270.28\,\text{kJ/kg} - 260.34\,\text{kJ/kg})$$
$$= 260.34\,\text{kJ/kg} + 2.03\,\text{kJ/kg} = 262.37\,\text{kJ/kg}$$

We may solve Equation 7.33 for h_2

$$h_2 = h_1 + \frac{h_{2s} - h_1}{\eta_s}$$

$$= 235.31\,\text{kJ/kg} + \frac{262.37\,\text{kJ/kg} - 235.31\,\text{kJ/kg}}{0.85}$$

$$= 235.31 + 31.84\,\text{kJ/kg} = 267.15\,\text{kJ/kg}$$

Then, the specific work to drive the compressor will be

$$w_{\text{comp}} = h_2 - h_1 = 267.15\,\text{kJ/kg} - 235.31\,\text{kJ/kg} = 31.84\,\text{kJ/kg} \Longleftarrow$$

(b) To find the final temperature, we use the data of Table A.8 with interpolation

$$T_2 = 20°\text{C} + (0.204)(30°\text{C} - 20°\text{C}) = 20°\text{C} + 2.0°\text{C} = 22°\text{C} \Longleftarrow$$

For a water- or air-cooled compressor, the ideal compression process is a reversible isothermal process and, therefore, a more appropriate measure of the efficiency is the ratio of the isothermal specific work to the actual specific work. The isothermal specific work may be obtained from a consideration of the steady flow energy equation which, in differential form (neglecting kinetic and potential energy changes), may be written for a reversible process as

$$\delta q_{\text{rev}} - \delta w_{\text{rev}} = dh \tag{7.34}$$

Use of Equations 7.12 and 7.19 permits the representation

$$\delta q_{\text{rev}} = T ds = dh - v dP$$

so that Equation 7.34 becomes

$$dh - v dP - \delta w_{\text{rev}} = dh$$

or

$$\delta w_{\text{rev}} = -v dP \tag{7.35}$$

Now, for an ideal gas, $v = RT/P$, so that

$$\delta w_{\text{rev}} = -RT \frac{dP}{P} \tag{7.36}$$

Because T is constant during an isothermal compression, $T_1 = T_2 = T$, and an integration of Equation 7.36 will reveal

$$w_{T=C} = RT \ln \frac{P_2}{P_1} \tag{7.37}$$

where the minus sign may be omitted when it is agreed that W is the specific work input to the compressor.

Thus, the efficiency of the compressor will be

$$\eta_{\text{comp}} = \frac{w_{T=C}}{h_2 - h_1} = \frac{RT_1 \ln P_2/P_1}{\bar{c}_p (T_2 - T_1)} \tag{7.38}$$

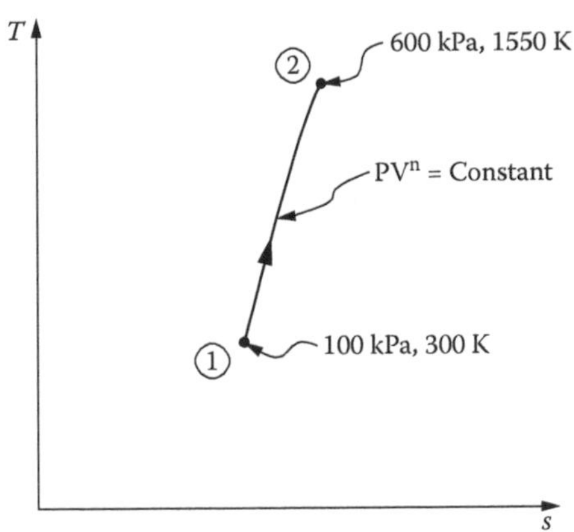

FIGURE 7.10
T-s diagram for the compressor in Example 7.10.

Example 7.10 A compressor for a gas turbine (Figure 7.10) draws air at 100 kPa and 300 K and delivers it at a pressure of 600 kPa and 550 K. Assuming that the compression is in accordance with PV^n = a constant, determine (a) the isothermal efficiency and (b) the polytropic exponent, n.

Solution

Assumptions and Specifications

(1) The compressor is the control volume.
(2) The air is an ideal gas and is the system.
(3) The system operates in the steady state.
(4) Steady flow exists.
(5) Kinetic and potential energy changes are negligible.

Table A.1 gives $R = 0.287\,\text{kJ/kg-K}$ and with an average temperature of

$$T_{\text{avg}} = \frac{1}{2}(300\,\text{K} + 550\,\text{K}) = 425\,\text{K}$$

We find from Table A.2 (with interpolation) that $\bar{c}_p = 1.0165\,\text{kJ/kg-K}$

(a) The thermal efficiency of the compressor may be found directly from Equation 7.38. With $P_2/P_1 = 600\,\text{kPa}/100\,\text{kPa} = 6$,

$$\eta_{\text{comp}} = \frac{RT_1 \ln P_2/P_1}{\bar{c}_p(T_2 - T_1)} = \frac{(0.287\,\text{kJ/kg-K})(300\,\text{K}) \ln 6}{(1.0165\,\text{kJ/kg-K})(550\,\text{K} - 300\,\text{K})} = 0.607 \Longleftarrow$$

(b) The ideal gas law gives

$$v_1 = \frac{RT_1}{P_1} = \frac{(0.287\,\text{kJ/kg-K})(300\,\text{K})}{100\,\text{kN/m}^2} = 0.861\,\text{m}^3/\text{kg}$$

and

$$v_2 = \frac{RT_2}{P_2} = \frac{(0.287\,\text{kJ/kg-K})(550\,\text{K})}{600\,\text{kN/m}^2} = 0.263\,\text{m}^3/\text{kg}$$

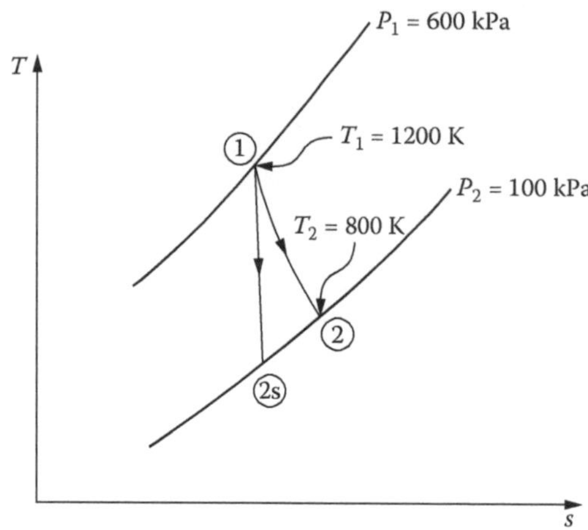

FIGURE 7.11
T-s diagram for the nozzle in Example 7.11.

and the polytropic exponent derives from $P_1 v_1^n = P_2 v_2^n$ or

$$n = \frac{\ln P_2/P_1}{\ln v_1/v_2} = \frac{\ln(600\,\text{kPa}/100\,\text{kPa})}{\ln(0.861\,\text{m}^3/\text{kg}/0.263\,\text{m}^3/\text{kg})} = 1.51 \Longleftarrow$$

We note that because $n \neq k$, the process is not isentropic.

Finally, we consider the isentropic efficiency of a nozzle that may be defined as the ratio of the actual specific kinetic energy at the nozzle exit to the specific kinetic energy at the exit of the nozzle obtained under isentropic expansion

$$\eta_s = \frac{\hat{V}^2_{\text{act}}}{\hat{V}^2_{S=C}} \tag{7.39}$$

Example 7.11 The nozzle of a jet engine (Figure 7.11) is fed with hot products of combustion at 600 kPa and 1200 K. The products exit the nozzle at 100 kPa and 800 K. Assuming the products of combustion to behave as an ideal gas with $\bar{c}_p = 1.060\,\text{kJ/kg-K}$ and $k = 1.37$, determine (a) the actual nozzle exit velocity if the expansion is adiabatic, (b) the nozzle exit velocity if the expansion is isentropic, and (c) the isentropic efficiency of the nozzle.

Solution

Assumptions and Specifications

(1) The nozzle is the control volume.
(2) The products of combustion are an ideal gas and comprise the system.
(3) The system operates in the steady state.
(4) Steady flow exists.
(5) There are no heat or work interactions between the nozzle and the environment.
(6) Kinetic energy at the inlet is negligible.
(7) Potential energy changes are negligible.
(8) The average specific heat is specified.

(a) For the specified conditions, with subscripts 1 and 2 designating the inlet and outlet respectively, the energy equation becomes

$$h_1 = h_2 + \frac{\hat{V}_2^2}{2}$$

or

$$\hat{V}_2 = [2(h_1 - h_2)]^{1/2} = [2\bar{c}_p(T_1 - T_2)]^{1/2}$$

Hence,

$$\hat{V}_2 = [2(1.060\,\text{kJ/kg-K})(1000\,\text{J/kJ})(1200\,\text{K} - 800\,\text{K})]^{1/2} = 920.9\,\text{m/s} \Longleftarrow$$

(b) For isentropic flow, the temperature, T_{2s} at the nozzle exit is given by Equation 7.27

$$\begin{aligned} T_{2s} &= T_1 \left(\frac{P_2}{P_1}\right)^{(k-1)/k} = (1200\,\text{K}) \left(\frac{100\,\text{kPa}}{600\,\text{kPa}}\right)^{0.37/1.37} \\ &= (1200\,\text{K})(0.1667)^{0.270} = (1200\,\text{K})(0.6164) = 739.6\,\text{K} \end{aligned}$$

The velocity, $\hat{V}_{2s}$ is then given by

$$\begin{aligned} \hat{V}_{2s} &= [2\bar{c}_p(T_1 - T_{2s})]^{1/2} \\ &= [2(1.06\,\text{kJ/kg-K})(1000\,\text{J/kJ})(1200\,\text{K} - 739.6\,\text{K})]^{1/2} = 987.9\,\text{m/s} \Longleftarrow \end{aligned}$$

(c) The use of Equation 7.39 then gives the nozzle efficiency

$$\eta_{\text{noz}} = \frac{\hat{V}_2^2}{\hat{V}_{2s}^2} = \frac{(920.9\,\text{m/s})^2}{(987.9\,\text{m/s})^2} = 0.869 \qquad (86.9\%) \Longleftarrow$$

7.9 The Entropy Balance Equation

Besides the mass and energy conservation equations that were treated in detail in Chapters 3 and 5, another equation of fundamental importance in thermodynamics is the entropy balance equation. In this section, we develop two steady-state entropy balance equations, one for the closed system or control mass and one for the open system or control volume.

7.9.1 The Entropy Balance for Closed Systems

Figure 7.12 shows a closed system with heat and work interactions at its boundary. The system is assumed to be at the absolute temperature, T, and we will let the absolute temperature of the surroundings be fixed at T_o. Because of the heat transfer, Q, the closed system experiences an increase in entropy of Q/T. The internal irreversibilities within the closed system, if any, would give rise to entropy generation, S_{gen}, and because work transfer, W, is entropy free, we may write the entropy balance equation as

$$\Delta S_{\text{sys}} = \frac{Q}{T} + S_{\text{gen}} \tag{7.40}$$

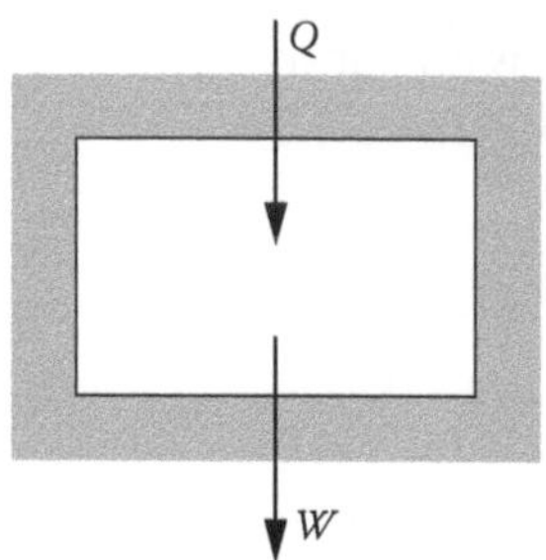

FIGURE 7.12
A closed system with heat and work interactions at its boundary.

Now consider the surroundings that have experienced a decrease in entropy, ΔS_o, given by

$$\Delta S_o = -\frac{Q}{T_o} \tag{7.41}$$

The change in entropy of the universe (the system plus the surroundings), designated as ΔS_u, is then seen to be

$$\Delta S_u = Q\left(\frac{1}{T} - \frac{1}{T_o}\right) + S_{\text{gen}} \tag{7.42}$$

The temperature of the surroundings, T_o, must be greater than the boundary temperature, T. For the heat transfer to occur from the surroundings to the system, and, because $S_{\text{gen}} \geq 0$ (zero only if internal irreversibilities are absent), it follows from Equation 7.42 that

$$\Delta S_u \geq 0 \tag{7.43}$$

where the equality applies if Q is reversible, that is, if $T = T_o$ and $S_{\text{gen}} = 0$.

Equation 7.43 establishes the principle of increase of entropy, which states that the entropy of the universe always increases. For an adiabatic closed system, $Q = 0$, which makes $\Delta S_o = 0$, it follows from Equations 7.40 and 7.41 that

$$\Delta S_{\text{sys}} = \Delta S_u = S_{\text{gen}} \tag{7.44}$$

Example 7.12 Saturated steam at 200 kPa is contained in a closed, stationary vessel and is cooled by water surrounding the vessel. The mass of the vapor is 2 kg and during the period in which the steam condenses and is converted to saturated liquid water, the cooling water temperature rises from 25 to 35°C (Figure 7.13). Determine the net change in entropy for (a) the steam and (b) the cooling water.

Solution

Assumptions and Specifications

(1) The 2 kg of steam-water mixture is the system.
(2) The system operates in the steady state.
(3) There is enough cooling water to allow all of the steam to condense.
(4) A stationary, closed, and rigid vessel is specified.
(5) The temperature of the surroundings may be taken as the average between 35°C and 25°C or 30°C.

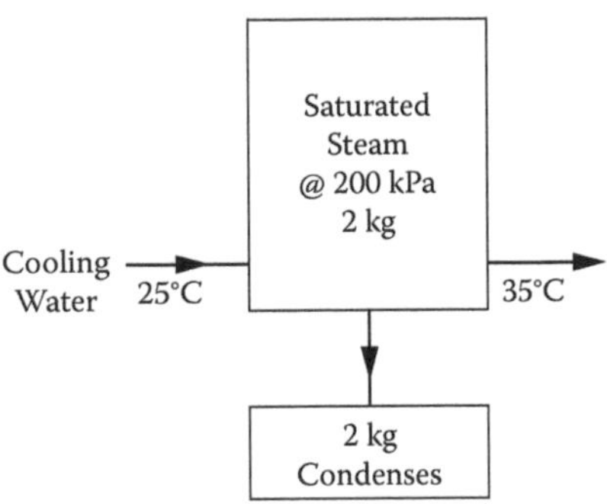

FIGURE 7.13
Configuration for Example 7.12.

The strategy here is to first determine the heat transferred across the boundaries of system. A stationary, closed, and rigid vessel is specified and for this closed system, the first law of thermodynamics in the absence of work is

$$Q = U_2 - U_1 = m\Delta u$$

Table A.4 may be read at 200 kPa to reveal that $u_{fg} = 2024.7\,\text{kJ/kg}$, which then gives

$$Q = mu_{fg} = (2\,\text{kg})(2024.7\,\text{kJ/kg}) = 4049.4\,\text{kJ}$$

The temperature of the system (the steam) is the saturation temperature corresponding to 200 kPa. Table A.4 provides $T = 120.2°\text{C}$ (393.2 K).

(a) The decrease in the entropy of the steam is

$$\Delta S = -\frac{Q}{T} = -\frac{4049.4\,\text{kJ}}{393.2\,\text{K}} = -10.30\,\text{kJ/K} \Longleftarrow$$

(b) The increase in the entropy of the cooling water at 30°C (303 K) is

$$\Delta S = \frac{Q}{T} = \frac{4049.4\,\text{kJ}}{303\,\text{K}} = 13.36\,\text{kJ/K} \Longleftarrow$$

We note that the net change in entropy is the algebraic sum of the entropy change of the water and the steam

$$\Delta S = -10.30\,\text{kJ/K} + 13.36\,\text{kJ/K} = 3.06\,\text{kJ/K} \Longleftarrow$$

7.9.2 The Entropy Rate Balance for an Open System (Control Volume)

The increase of entropy principle can be extended to an open system or control volume. Figure 7.14 shows a multiple input-multiple output control volume where a typical stream, i, with mass flow, $\dot{m}_i$, and specific entropy, s_i, is carrying entropy into the control volume at a rate of $\dot{m}_i s_i$. In a similar fashion, we may consider an exiting stream, e, with mass flow, $\dot{m}_e$, and specific entropy, s_e, carrying entropy out of the control volume at a rate of $\dot{m}_e s_e$. Summations over all inlet and outlet streams will then give the total flow of entropy into and out of the control volume

$$\sum_i \dot{m}_i s_i \quad \text{and} \quad \sum_e \dot{m}_e s_e$$

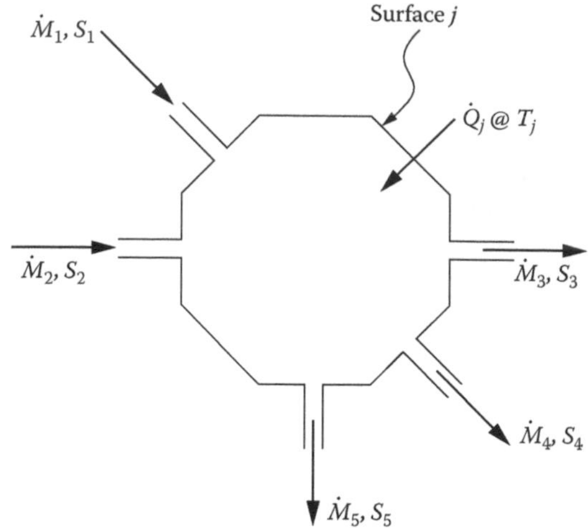

FIGURE 7.14
Multiple input-multiple output control volume.

Let the control volume consist of a number of isothermal surfaces. If surface, j, at an absolute temperature, T_j, receives heat at the rate of $\dot{Q}_j$, then the rate at which entropy enters the control volume through surface-j is $\dot{Q}_j/T_j$. The rate at which entropy enters the control volume over all of its surfaces is then

$$\sum_j \frac{\dot{Q}_j}{T_j}$$

If the control volume generates entropy at the rate of $\dot{S}_{gen}$ due to internal irreversibilities, the entropy rate balance for steady state and steady flow may be written as

$$\sum_i \dot{m}_i s_i + \sum_j \frac{\dot{Q}_j}{T_j} + \dot{S}_{gen} = \sum_e \dot{m}_e s_e \tag{7.45}$$

We see that with regard to a control volume, the rate at which entropy enters due to mass transfer at all of the inlet ports plus the rate at which entropy enters due to heat transfers at its confining surfaces plus any entropy generated due to internal irreversibilities is equal to the rate at which entropy leaves due to mass transfer at all of the outlet ports.

Example 7.13 Superheated steam enters a nozzle at 5 MPa and 300°C and exits the nozzle at 200 kPa and a quality of 0.95 (Figure 7.15). The rate of heat loss from the nozzle to the surroundings is 10 kW. Assume that the nozzle surface is at the average between the inlet and outlet steam temperatures. The mass flow rate is 2 kg/s. Determine the rate of entropy generation in the nozzle.

Solution

Assumptions and Specifications

(1) The 2 kg of steam-water mixture is the system.
(2) The nozzle is the control volume.

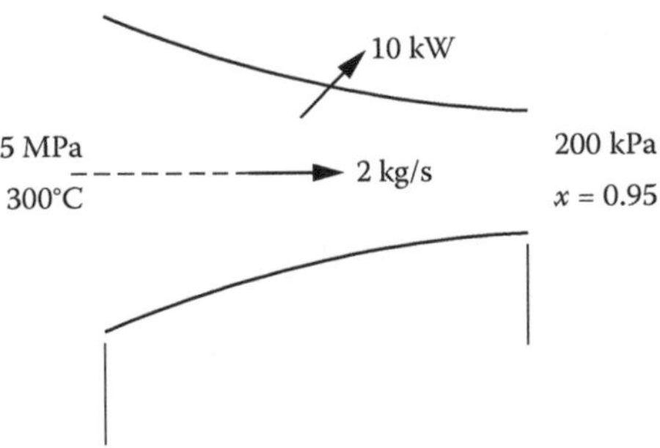

FIGURE 7.15
Steam flow through a nozzle for Example 7.13.

(3) The system operates in the steady state with steady flow.
(4) The temperature of the nozzle wall may be taken as the average of the inlet and outlet stream temperatures.

This is a single input-single output steady-state steady flow system with a single heat transfer from a uniform temperature surface. When we observe that the heat transferred is out of the system, we may write Equation 7.45 as

$$\dot{m}s_i - \frac{\dot{Q}}{T} + \dot{S}_{\text{gen}} = \dot{m}s_e$$

or

$$\dot{m}s_i + \dot{S}_{\text{gen}} = \frac{\dot{Q}}{T} + \dot{m}s_e$$

where $\dot{m}$ is the mass flow rate of the steam.

At 5 MPa and 300°C, Table A.5 provides

$$s_i = 6.2085 \text{ kJ/kg-K}$$

and at the outlet of the nozzle at 200 kPa, Table A.4 gives

$$s_f = 1.5295 \text{ kJ/kg-K} \quad \text{and} \quad s_{fg} = 5.5974 \text{ kJ/kg-K}$$

The entropy at the exit will then be

$$\begin{aligned} s_e &= s_f + x s_{fg} \\ &= 1.5295 \text{ kJ/kg-K} + (0.95)(5.5974 \text{ kJ/kg-K}) \\ &= 1.5295 \text{ kJ/kg-K} + 5.3175 \text{ kJ/kg-K} = 6.8470 \text{ kJ/kg-K} \end{aligned}$$

At the exit of the nozzle where $P = 200$ kPa, Table A.4 reveals that $T_{\text{sat}} = 120.2°\text{C}$ so that the average temperature of the nozzle walls will be

$$T = \frac{1}{2}(300°\text{C} + 120.2°\text{C}) = 210.1°\text{C} \quad (483.1 \text{ K})$$

We may now solve for S_{gen}

$$
\begin{aligned}
S_{gen} &= \frac{\dot{Q}}{T} + \dot{m}(s_e - s_i) \\
&= \frac{10\,\text{kW}}{483.1\,\text{K}} + (2\,\text{kg/s})(6.8470\,\text{kJ/kg-K} - 6.2085\,\text{kJ/kg-K}) \\
&= 0.0207\,\text{kW/K} + (2\,\text{kg/s})(0.6385\,\text{kJ/kg-K}) = 1.2977\,\text{kW/K}
\end{aligned}
$$

7.10 Summary

Entropy is a measure of molecular disorder. For our purposes, classical thermodynamics defines the difference in specific entropy as

$$s_2 - s_1 = \int_1^2 \left(\frac{\delta q}{T}\right)_{\text{rev}}$$

Entropy is a property and, in any process, is independent of the path. Absolute values are tabulated based on various substance standards.

The Clausius inequality states that

> When a system operates in a cycle and the differential heat, δQ, added or removed at a point is divided by the absolute temperature, T, at that point, the algebraic sum of all $\delta Q/T$ values is zero for a reversible cycle and less than zero for an irreversible cycle.

The Gibbs property relations, or TdS equations, are

$$TdS = dU + PdV \tag{7.14}$$

$$Tds = du + Pdv \tag{7.15}$$

$$TdS = dH - VdP \tag{7.18}$$

and

$$Tds = dh - vdP \tag{7.19}$$

For solids and liquids, the specific energy change is given by

$$s_2 - s_1 = \bar{c} \ln \frac{T_2}{T_1} \tag{7.21}$$

where $\bar{c}$ is the average specific heat. For ideal gases

$$s_2 - s_1 = \bar{c}_v \ln \frac{T_2}{T_1} + R \ln \frac{v_2}{v_1} \tag{7.23}$$

and

$$s_2 - s_1 = \bar{c}_p \ln \frac{T_2}{T_1} - R \ln \frac{P_2}{P_1} \tag{7.24}$$

For an isentropic process of an ideal gas, the P-V-T relations are

$$\frac{T_2}{T_1} = \left(\frac{V_1}{V_2}\right)^{k-1} \quad \frac{T_2}{T_1} = \left(\frac{P_2}{P_1}\right)^{(k-1)/k} \quad \frac{P_2}{P_1} = \left(\frac{V_1}{V_2}\right)^{k}$$

and the work will be

$$W = \int_{V_1}^{V_2} P dV = \frac{P_2 V_2 - P_1 V_1}{1-k} = \frac{mR(T_2 - T_1)}{1-k} \tag{7.31}$$

Isentropic efficiencies of steady flow devices include the turbine

$$\eta_s = \frac{h_1 - h_2}{h_1 - h_{2s}} \tag{7.32}$$

the compressor or pump

$$\eta_s = \frac{h_{2s} - h_1}{h_2 - h_1} \tag{7.33}$$

and the nozzle

$$\eta_s = \frac{\hat{V}^2_{\text{act}}}{\hat{V}^2_{S=C}} \tag{7.39}$$

The entropy balance equation for the closed system is

$$\Delta S_{\text{sys}} = \frac{Q}{T} + S_{\text{gen}} \tag{7.40}$$

and for the open system or control volume

$$\sum_i \dot{m}_i s_i + \sum_j \frac{\dot{Q}_j}{T_j} + \dot{S}_{\text{gen}} = \sum_e \dot{m}_e s_e \tag{7.45}$$

7.11 Problems

The Clausius Inequality

7.1: The following heat engines operate between a hot reservoir at 1000 K and a cold reservoir at 300 K. Verify the validity of the Clausius inequality in (a) $Q_H = 500$ kJ and $Q_L = 200$ kJ, (b) $Q_H = 600$ kJ and $Q_L = 180$ kJ, and (c) $Q_H = 900$ kJ and $Q_L = 250$ kJ.

7.2: A heat engine draws heat from a flame at 800 K at the rate of 20 W. The engine rejects heat to an ocean at 290 K and the engine efficiency is 32%. Determine the cyclic integral $\oint \delta\dot{Q}/T$ and comment on whether the engine satisfies the Clausius inequality.

7.3: A hot reservoir at 1500 K supplies 120 kJ to heat engine 1, which has a thermal efficiency of 28%. The heat rejected by engine 1 is supplied to engine 2, which operates with a thermal efficiency of 35%. Heat engine 2 rejects heat to a sink at 400 K and another

12 kJ to a sink at 300 K. Apply the Clausius inequality to determine whether engine 2 is reversible.

7.4: A heat engine receives 200 kJ from one source at 1000 K and 400 kJ from another source at 800 K. The engine rejects 120 KJ to a sink at 400 K and 120 kJ to another sink at 300 K. Apply the Clausius inequality to determine whether or not the engine is reversible.

7.5: Saturated liquid water at 5 MPa enters a boiler and is heated to saturated vapor at the same pressure. The vapor expands in an adiabatic turbine to 15 kPa with a quality of 0.85. The exhaust from the turbine enters a condenser where heat is extracted from the mixture and its quality is reduced to 0.20. The mixture than enters an adiabatic pump, which compresses it to saturated liquid at 5 MPa. Determine the value of the cyclic integral $\oint \delta \dot{Q}/T$.

7.6: A heat engine draws 800 kJ from a hot reservoir at 600 K and rejects heat to a cold reservoir at 300 K. The work output from the engine is used to drive a heat pump, which draws 300 kJ from a cold reservoir at 280 K and delivers it to a hot reservoir at 310 K. The thermal efficiency of the engine is 25% and the coefficient of performance of the heat pump is 3.5. Determine whether the combination of the engine and the heat pump satisfies the Clausius inequality.

7.7: A vapor compression refrigeration cycle using R-134a executes four processes:

Process 1-2 An isentropic compression, $T_1 = -8°\text{C}$ and

$$s_1 = 0.9066\,\text{kJ/kg-K. to } T_2 = 31.33°\text{C and}$$
$$s_2 = 0.9066\,\text{kJ/kg-K.}$$

Process 2-3 A constant pressure heat rejection to saturated liquid at

$$P_3 = 0.8\,\text{MPa.}$$

Process 3-4 An adiabatic throttling process to a mixture of liquid and

$$\text{vapor at } P_4 = 0.21704\,\text{MPa.}$$

Process 4-1 A constant pressure heat extraction from the evaporator.

Evaluate the cyclic integral $\oint \delta \dot{Q}/T$ and comment on whether the system satisfies the Clausius inequality.

Temperature-Entropy Diagrams

7.8: Determine the entropy of water for the following states and indicate the states on a T-s diagram, (a) $T = 40°\text{C}$ and $P = 5\,\text{MPa}$, (b) $T = 45.80°\text{C}$ and $P = 10\,\text{kPa}$, (c) $T = 70°\text{C}$ and $v = 4\,\text{m}^3/\text{kg}$, (d) $T = 200°\text{C}$, and (e) $T = 325°\text{C}$ and $P = 2.50\,\text{MPa}$.

7.9: Show the expansion of steam with $T_1 = 342.4°\text{C}$ and $P_1 = 15\,\text{MPa}$ (saturated vapor) to $T_2 = 150°\text{C}$ and $P_2 = 200\,\text{kPa}$ on a T-s diagram utilizing the entropy data for the three intermediate points, (1) $T = 300°\text{C}$ and $P = 8\,\text{MPa}$, (2) $T = 250°\text{C}$ and $P = 1.5\,\text{MPa}$, and (3) $T = 200°\text{C}$ and $P = 600\,\text{kPa}$.

7.10: Draw a T-s plot showing the throttling of saturated liquid water at 0.8 to 0.01 MPa. Indicate on your plot, the quality of the mixture corresponding to $P = 0.60\,\text{MPa}$, $P = 0.40\,\text{MPa}$, $P = 0.20\,\text{MPa}$, and $P = 0.0.01\,\text{MPa}$.

7.11: Determine the entropy of refrigerant R-134a for the following states and indicate the states on a T-S diagram: (a) $T = 40°\text{C}$ and $P = 0.09135\,\text{MPa}$, (b) $T = -12°\text{C}$ and $v = 0.08\,\text{m}^3/\text{kg}$, (c) $P = -0.2\,\text{MPa}$ and $u = 170\,\text{kJ/kg}$, (d) $P = 0.5\,\text{MPa}$ and $h = 275\,\text{kJ/kg}$, and (e) $T = 40°\text{C}$ and $v = 0.04633\,\text{m}^3/\text{kg}$.

7.12: Refrigerant R-134a is compressed from $T_1 = -4°C$ (saturated vapor) to $P_2 = 1.0\,MPa$ and $T_2 = 80°C$. Show the path of compression on a T-s diagram indicating any three intermediate state points giving the values of T, s, and h.

7.13: Draw the cycle 1-2-3-4-1 on a T-s diagram with the state points in accordance with the following:

State-1: $T_1 = -12°C,\ x_1 = 0.90$

State-2: $T_2 = 80°C,\ P_2 = 1.0\,MPa$

State-3: $T_2 = 35°C,\ P_3 = 1.00\,MPa$

State-4: $T_1 = -12°C,\ x_4 = 0.20$

7.14: A throttling of saturated refrigerant R-134a occurs from 6 to 0.6 bar. Draw a T-s plot marking the curve with quality values corresponding to $P = 6, 4, 2, 1$, and 0.6 bar.

Entropy Changes for Solids, Liquids, and Ideal Gases

7.15: Two steel blocks, one at a temperature of 200°C and the other at 100°C, are brought into thermal contact with one another until thermal equilibrium is established. The mass of each block is 5 kg and the average specific heat of steel is 0.434 kJ/kg-K. Assuming no interaction with the surroundings, determine (a) the equilibrium temperature and (b) the change in entropy of the two blocks together as the system.

7.16: A 2-kg aluminum ball is removed from a furnace at 120°C. The ball is allowed to cool in an environment at 20°C until equilibrium is established. Assume the specific heat for aluminum to be 0.90 kJ/kg-K and determine (a) the change in entropy for the ball, (b) the change in entropy of the environment, and (c) the change of entropy of the ball and the environment taken together.

7.17: An ice cube of dimensions 2.5 cm × 2.5 cm × 2 cm at 0°C is placed in an insulated cup containing 0.4 kg of water at 16°C. Assuming that the final temperature is 0°C, determine (a) the amount of ice melted and (b) the change in entropy of the ice and water taken together.

7.18: A 1.5-kg copper block at 200°C is quenched in a water bath containing 6 kg of water at 20°C. Assume that the specific heats of copper and water are respectively 385J/kg-K and 4179J/kg-K with no heat loss to the surroundings and determine (a) the final equilibrium temperature, (b) the entropy change of the copper block, and (c) the entropy change of the water.

7.19: A hot block of mass, m_1, specific heat, c_1, and a temperature of T_1 is brought into thermal contact with a cold block of mass, m_2, specific heat, c_2, and a temperature of T_2. If the heat loss to the surroundings is Q, show that the equilibrium temperature, T_e, is given by

$$T_e = \frac{m_1c_1T_1 + m_2c_2T_2 - Q}{m_1c_1 + m_2c_2}$$

7.20: In a refrigeration process, 10 kg of saturated liquid water at 0°C is converted to ice at −5°C. Assume the latent heat of water to be 335 kJ/kg and c for ice to be 2040J/kg-K and determine the decrease of entropy of the water during the process.

7.21: Determine the change in entropy that occurs when 3 kg of oil (c = 1909J/kg-K) are heated from 20 to 50°C.

7.22: A hot reservoir at 500 K is used to heat 12 kg of water from 25 to 85°C. Determine the change of entropy of the water and the hot reservoir using (a) $c = 4178\,\text{J/kg-K}$ for water and (b) using the entropy data from Table A.3.

7.23: Using c_p and c_v values at an average temperature of 450 K, determine the change in entropy when 3 kg of air is heated at constant volume from $P_1 = 200\,\text{kPa}$ and $T_1 = 300\,\text{K}$ to $P_2 = 1000\,\text{kPa}$ and $T_2 = 600\,\text{K}$.

7.24: A kilogram of nitrogen at 1200 kPa and 700 K expands to 200 kPa and 400 K. The work in the expansion process is 400 kJ. Use the specific heat values from Table A.1 to determine (a) the heat transfer and (b) the change in entropy of the nitrogen.

7.25: One kilogram of air at 900 kPa and 500 K expands to 100 kPa and 300 K. Determine the change in entropy of the air if (a) the air is cooled at 900 kPa from 600 to 300 K followed by an isothermal expansion from 900 to 100 kPa and (b) if the air is expanded to its final volume followed by a constant volume cooling to 300 K.

7.26: Consider the flow of 2 kg of helium at 300 kPa and 450 K through a throttling valve. The pressure of the helium after throttling is 100 kPa. Using specific heat values from Table A.1 and assuming them to be constant, determine the change in entropy.

7.27: Three kilograms of argon experience an increase in entropy of 0.65 kJ/K while executing an internally irreversible process. The initial temperature of the argon is 300 K. Using specific heat values from Table A.1 and assuming them to be constant, evaluate the final temperature of the argon for (a) a constant pressure process and (b) a constant volume process.

7.28: One kilogram of air executes a sequence of processes in a steady-flow system.

Process 1-2	A polytropic compression ($n = 1.2$) from $P_1 = 100$ kPa and $T_1 = 300$ K to 600 kPa
Process 2-3	A constant pressure heat addition from T_2 to $T_3 = 1200\,\text{K}$
Process 3-4	A polytropic expansion ($n = 1.2$) from P_3 and T_3 to $P_4 = 100\,\text{kPa}$
Process 4-1	A constant pressure heat rejection to $T_4 = 300\,\text{K}$

Assume that Table A.1 may be used for R and the specific heats and determine (a) the temperatures T_3 and T_4 and (b) the change in entropy for each of the processes.

7.29: Assuming 1 kg of an ideal gas with constant specific heats, determine the change in entropy between the initial and final states in each of the following cases:
(a) Air; $P_1 = 120\,\text{kPa}$ and $T_1 = 300\,\text{K}$; $P_2 = 480\,\text{kPa}$ and $T_2 = 550\,\text{K}$
(b) Argon; $P_1 = 100\,\text{kPa}$ and $T_1 = 298\,\text{K}$; $P_2 = 350\,\text{kPa}$ and $v_2 = 0.05\,\text{m}^3/\text{kg}$
(c) Carbon dioxide; $P_1 = 110\,\text{kPa}$ and $T_1 = 310\,\text{K}$; $P_2 = 420\,\text{kPa}$ and $v_2 = 0.06\,\text{m}^3/\text{kg}$
(d) Oxygen; $P_1 = 500\,\text{kPa}$ and $T_1 = 500\,\text{K}$; $P_2 = 100\,\text{kPa}$ and $T_2 = 325\,\text{K}$
(e) Nitrogen; $P_1 = 800\,\text{kPa}$ and $T_1 = 600\,\text{K}$; $P_2 = 200\,\text{kPa}$ and $T_2 = 400\,\text{K}$

7.30: One kilogram of ethane at $P_1 = 100\,\text{kPa}$ and $T_1 = 300\,\text{K}$ is compressed to $P_2 = 600\,\text{kPa}$ with $s_2 - s_1 = -0.25\,\text{kJ/kg-K}$. Assuming ethane to be an ideal gas with properties given in Table A.1, determine (a) T_2 and (b) v_2.

7.31: Nitrogen at 100 kPa and 300 K is contained in a piston-cylinder assembly. The mass is 0.6 kg and 250 kJ of heat is supplied to the nitrogen, The specific heats are constant as given in Table A.1. Determine for the nitrogen (a) the final temperature, (b) the final volume, and (c) the change in entropy.

7.32: One kilogram of helium executes the following cycle in a closed system:

Process 1-2 A constant volume compression from $P_1 = 100\,\text{kPa}$ and $T_1 = 300\,\text{K}$ to $P_2 = 300\,\text{kPa}$

Process 2-3 An isothermal expansion from $P_2 = 300\,\text{kPa}$ to $P_3 = 100\,\text{kPa}$

Process 3-1 A constant pressure heat rejection to $T_1 = 300\,\text{K}$

Sketch the cycle on a P-V diagram and, assuming that helium behaves as an ideal gas with constant specific heats given in Table A.1, determine (a) T_2, (b) v_3, (c) the heat transferred, (d) the work done, and (e) the change in entropy between state 3 and state 1.

7.33: A power cycle consists of three internally reversible processes:

Process 1-2 A polytropic compression ($n = 1.3$) from $P_1 = 100\,\text{kPa}$ and $T_1 = 300\,\text{K}$ to $P_2 = 650\,\text{kPa}$

Process 2-3 An isothermal expansion from $P_2 = 650\,\text{kPa}$ to $P_3 = 100\,\text{kPa}$

Process 3-1 A constant pressure compression back to $P_1 = 100\,\text{kPa}$

Show the cycle on a P-v diagram. Then, assume that the working fluid is 1 kg of nitrogen that may be taken as an ideal gas with specific heats taken as constant from Table A.1 and determine (a) T_2 and v_2, (b) T_3 and v_3, and (c) the changes in entropy in each of the three processes.

Isentropic Relations for an Ideal Gas

7.34: Air is compressed isentropically from $P_1 = 100\,\text{Pa}$ and $T_1 = 300\,\text{K}$ to $P_2 = 500\,\text{kPa}$. Assuming air to be an ideal gas with constant specific heats, determine (a) T_2 and (b) V_1/V_2.

7.35: An Otto cycle consists of four processes that are executed in a closed system:

Process 1-2 An isentropic compression from T_1 and v_1 to T_2 and v_2

Process 2-3 A constant volume heat addition from T_2 and v_2 to T_3 and $v_3 = v_2$

Process 3-4 An isentropic expansion from T_3 and v_3 to T_4 and $v_4 = v_1$

Process 4-1 A constant volume heat rejection from T_4 and v_4 to T_1 and v_1

Assume that the air is an ideal gas with constant specific heats and draw the cycle on both P-v and T-s coordinates and show that the increase in entropy in process 2-3 is equal to the decrease in entropy in process 4-1.

7.36: For Problem 7.35, prove that

$$\frac{T_1}{T_2} = \frac{T_4}{T_3} = \left(\frac{v_2}{v_1}\right)^{k-1}$$

7.37: Oxygen is compressed isentropically from $P_1 = 100\,\text{kPa}$ and $T_1 = 398\,\text{K}$ to $P_2 = 600\,\text{kPa}$. Assuming that oxygen is a perfect gas with constant specific heats, determine the specific work done on the oxygen when (a) the oxygen is compressed in a piston-cylinder arrangement and (b) when the oxygen is compressed in a steady-flow compressor.

7.38: One kilogram of air executes a sequence of processes in a steady-flow system.

Process 1-2 An isentropic compression from $P_1 = 100\,\text{kPa}$ and $T_1 = 300\,\text{K}$ to $P_2 = 600\,\text{kPa}$

Process 2-3 A constant pressure heat addition from P_2 and T_2 to $T_3 = 1300\,\text{K}$

Process 3-4 An isentropic expansion from P_3 and T_3 to $P_4 = P_1$ and T_4

Process 4-1 A constant pressure heat rejection from P_3 and T_3 to P_4 and T_4

Assume that air is an ideal gas with constant specific heats and determine (a) the temperatures T_3 and T_4 and (b) the changes in specific entropy, $s_3 - s_2$ and $s_1 - s_4$.

7.39: One kilogram of methane executes a cycle shown in Figure P7.39. If $P_1 = 95\,\text{kPa}$, $T_1 = 298\,\text{K}$, and $V_1/V_2 = 4$, assume that methane is an ideal gas with constant specific heats and determine (a) the pressure and temperature at state 2 and, if the pressure, P_3 is 350 kPa, (b) the changes in specific entropy, $s_3 - s_2$ and $s_1 - s_3$.

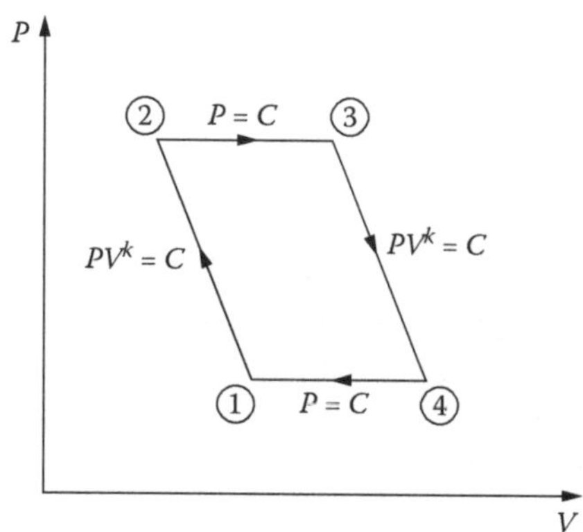

FIGURE 7P.39

7.40: Carbon monoxide undergoes a change if state from an initial temperature of 47°C and a pressure of 200 kPa to a final state of 107°C and a pressure of 475 kPa. Determine (a) the heat transferred and (b) the change in specific entropy.

Isentropic Efficiencies of Steady-Flow Devices

7.41: A steam turbine receives steam at 6 MPa and 400°C and exhausts it to a condenser that operates at a pressure of 40 kPa. Kinetic and potential energy changes are negligible and the turbine has an isentropic efficiency of 0.80. Determine (a) the enthalpy and quality of the exhaust steam and (b) the specific work output from the turbine.

7.42: Superheated steam at 6 MPa and 800°C with a velocity of 15 m/s enters a steam turbine. The steam is exhausted at a pressure of 15 kPa with a velocity of 60 m/s and a quality of 0.95. The turbine produces 800 kW with a mass flow rate of 5000 kg/h. Determine (a) the rate of heat flow from the turbine to the surroundings and (b) the isentropic efficiency of the turbine.

7.43: Steam at 10 MPa and 650°C enters a turbine and leaves as saturated vapor at 10 kPa. Kinetic and potential energy changes are negligible. Determine (a) the specific work output of the turbine, (b) the change in specific entropy of the steam, and (c) the isentropic efficiency of the turbine.

7.44: A well-insulated steam turbine operating at steady state develops 8 MW of power for a steam flow of 10 kg/s. The steam enters at 4 MPa and leaves at 8 kPa with a quality of 0.92. Kinetic and potential energy changes are negligible. Determine the inlet temperature of the steam.

7.45: A gas turbine receives hot gases at 600 kPa and 1400 K and exhausts them to the atmosphere at 100 kPa. The isentropic efficiency is 0.88. Assume the hot gases to behave like an ideal gas with $c_p = 0.9861\,\text{kJ/kg-K}$ and $c_v = 0.7472\,\text{kJ/kg-K}$ and determine the temperature at the turbine exhaust.

7.46: A compressor draws in nitrogen at 100 kPa and 300 K and delivers it at 300 kPa and 500 K. Determine the isentropic efficiency assuming that nitrogen is an ideal gas with constant specific heats.

7.47: Air at 100 kPa and 27°C enters a compressor at the rate of 2 kg/s and leaves at 600 kPa. Assuming air to be an ideal gas with constant specific heats, plot (a) the temperature of the air leaving the compressor and (b) the power input to the compressor as a function of the isentropic efficiency of the compressor for values ranging from 0.50 to 1.00 in increments of 0.10. What conclusions can you draw from the plot?

7.48: A compressor for a gas turbine system draws air at 100 kPa and 298 K and delivers it at 700 kPa and 600 K. Determine (a) the polytropic exponent, n, if the compression follows $PV^n = \text{constant}$ and (b) the efficiency of the compressor.

7.49: A feedwater pump draws saturated liquid condensate at 10 kPa and delivers it to a boiler at 7.5 MPa. The isentropic efficiency of the pump is 0.87, and kinetic and potential energy changes are negligible. Determine the outlet enthalpy.

7.50: A compressor draws in refrigerant R-134a at 0.60 bar and −20°C and delivers it at 10 bar. Neglecting any heat interaction with the environment and any kinetic and potential energy changes, determine the exit temperature of the refrigerant and the specific work input to the compressor for (a) isentropic compression and (b) an 83% isentropic efficiency of the compressor.

7.51: Steam at 8 MPa and 550°C is expanded in a well-insulated and horizontal nozzle to saturated vapor at 90 kPa. Determine the nozzle exit velocity and the exit temperature for (a) an isentropic expansion and (b) a 76% isentropic efficiency of the nozzle.

7.52: Steam at 4 MPa and 450°C flows through a well-insulated and horizontal nozzle. The exit velocity is 800 m/s. The isentropic efficiency of the nozzle is 0.85. Determine the actual exit velocity if the inlet kinetic energy of the steam is negligible.

The Entropy Balance Equation

7.53 Saturated liquid water at 150 kPa is contained in a piston-cylinder device where 950 kJ/kg is supplied to the water from a source at 200°C. Determine (a) the quality of the liquid-vapor mixture that is produced as a result of the heat supply, (b) the change in specific entropy of the water, and (c) the change in the specific entropy of the universe (water plus the source).

7.54: A piston-cylinder device contains a mixture of water and steam with a quality of 0.20 at 200 kPa. The energy supplied from the surroundings consists of 100 kJ/kg of heat from a source at 180°C and an unknown amount of specific work (kJ/kg). As a result of the energy supplied, the mixture is converted into saturated vapor. Determine (a) the amount of work supplied and (b) the entropy generated.

7.55: A closed stationary vessel contains 1.5 kg of saturated water vapor at 300 kPa. The vessel loses heat to a large body of coolant at 20°C. As a result of the heat loss, the vapor condenses and becomes a mixture with a quality of 0.40. Determine the change in entropy of the steam and the coolant.

7.56: One kilogram of refrigerant R-134a with a quality of 0.85 at 1.6 bar is contained in a piston-cylinder arrangement. Because of a heat loss of 300 kJ/kg to the surroundings, the quality of the mixture changes to 0.20. Determine the entropy generated in the cooling process.

7.57: One kilogram of nitrogen is compressed adiabatically from 100 kPa and 300 K to 600 kPa. Because of internal irreversibilities, the work required for compression is 35% more than the work for isentropic compression. Assuming that the nitrogen is an ideal gas with constant specific heats, determine (a) the actual work required for compression, (b) the temperature of the nitrogen after compression, and (c) the entropy generated.

7.58: The inside and outside temperatures of a single pane window are 20°C and −5°C. The window glass is 4 mm thick with a surface area of 0.80 m^2 and has a thermal conductivity of 1.4 W/m-K. Determine (a) the rate of heat conduction through the window and (b) the entropy generated in the window.

7.59: A plane wall of dimensions 1 m × 0.5 m × 0.01 m generates heat internally at a rate of 50 kW/m^3. The left face of the wall receives heat at the rate of 1.5 kW and is maintained at 40°C. The temperature on the right face of the wall is 80°C. Determine the rate of entropy generation in the wall.

7.60: Steam at 8 MPa and 450°C expands in a turbine to 50 kPa and a quality of 0.90. The outer surface of the turbine, which is at 85°C, loses heat to the surroundings at 85°C at the rate of 25 kW and the mass flow rate of the steam is 50 kg/s. Assuming that kinetic and energy changes are negligible, determine the rate of entropy generation in the turbine.

7.61: Refrigerant R-134a in the form of a saturated liquid at 8 bar flows at 1.2 kg/s through a throttling valve and expands to a pressure of 1.8 bar. The throttling valve receives 100 W from the surroundings at 25°C through its surface, which is at 18°C. Assuming that kinetic and energy changes are negligible, determine the rate of entropy generation during the process.

7.62: Nitrogen gas at 100 kPa and 300 K flows through a steady-flow compressor at the rate of 10 kg/s. The power input to the compressor is 500 kW but 100 kW is lost in the form of heat transfer through the casing to the surroundings. The pressure of the compressed nitrogen is 300 kPa. Assuming that nitrogen is an ideal as with $c_p = 1.044$ kJ/kg-K and $c_v = 0.747$ kJ/kg-K as the average values and 60°C as the average surface temperature of the compressor, determine the rate of entropy generation during the process.

7.63: Refrigerant R-134a enters a horizontal pipe of 15 cm inside diameter as a mixture of liquid and vapor with a quality of 0.20 at a pressure of 4 bar and a velocity of 8 m/s. The refrigerant leaves the pipe as saturated vapor at 2 bar. During the process, heat is transferred from the surroundings to the pipe, which is at an average temperature of −2°C. For steady flow, determine (a) the mass flow rate, (b) the exit velocity, (c) the rate of heat transfer from the surroundings, and (d) the rate of entropy generation in the pipe.

7.64: In a double-pipe heat exchanger, air at 3 kg/s enters the inner tube at 120 kPa and 320 K and leaves at 100 kPa and 270 K. The outer tube carries refrigerant R-134a, which enters as saturated liquid at 140 kPa and exits as saturated vapor at the same pressure. Heat transfer between the heat exchanger and the surroundings is negligible. For steady operation and neglecting kinetic and potential energy changes, determine (a) the mass flow rate of R-134a and (b) the rate of entropy production in the heat exchanger.

8

Gas Power Systems

Chapter Objectives

- To establish the basis for the consideration of gas power systems.
- To describe the operation of the spark-ignition (Otto)[1] and compression-ignition (Diesel)[2] engines, the gas turbine or Brayton engine, and the jet engine.
- To define what is meant by the air standard and cold air standard cycles.
- To consider both the air standard and cold air standard performance of the ideal Otto, Diesel, Brayton or gas turbine, and jet engine cycles.
- To discuss the compression ratio, its limitations, and its effect on the performance of the Otto and Diesel cycles.
- To show the effect of regeneration on the ideal Brayton cycle.

8.1 Introduction

In this chapter, we focus our attention on gas power systems that always use a gas as a working fluid. The gas power systems that we will consider are the reciprocating **internal combustion engine** (both spark-ignition or the Otto cycle and compression-ignition or the Diesel cycle), the rotating gas turbine, and the aircraft jet engine. These power systems are open systems because the working fluid is exhausted at the end of each cycle.

Section 3.6.3 was devoted to an introduction to cycle analysis. There we saw that the work done by a power cycle is equal to the heat input less the heat rejected

$$W_{\text{cycle}} = Q_{\text{in}} - Q_{\text{out}}$$

and that the thermal efficiency of the cycle is

$$\eta = \frac{W_{\text{cycle}}}{Q_{\text{in}}} = \frac{Q_{\text{in}} - Q_{\text{out}}}{Q_{\text{in}}} = 1 - \frac{Q_{\text{out}}}{Q_{\text{in}}} \tag{3.18}$$

---

[1] Nikolaus Otto (1832–1891) was a German inventor who built the first four-stroke cycle internal combustion engine.

[2] Rudolf Diesel (1858–1913) was a German inventor who designed and built the Diesel engine.

These equations are applied to the **ideal** internal combustion and gas turbine cycles that are characterized by all of the following:

1. Operation is frictionless with reversible compressions and expansions.
2. All processes are quasi-static or quasi-equilibrium processes.
3. There is no heat transfer in the conduits connecting the component parts of the system.
4. The working substance is air that is treated as an ideal gas.

Because air is assumed as the working fluid, in the analysis of gas power systems, considerable simplification is evident. **Air standard analysis** considers that

1. All processes are internally reversible.
2. A fixed amount of air is used.
3. The combustion process is represented as a heat addition from an external source.
4. The exhaust process is represented as a heat rejection to an external sink.

In air standard analysis, variable specific heats are employed and we obtain specific internal energy and enthalpy values from the "air tables," which are provided in Table A.12 in Appendix A. Because both the compression and expansion processes in the cycle are isentropic, use of the reference volume

$$\frac{v_{ri}}{v_{rj}} \equiv \frac{V_i}{V_j}$$

and the reference pressure

$$\frac{p_{ri}}{p_{rj}} \equiv \frac{P_i}{P_j}$$

are required. They will be considered in further detail as we proceed. If a **cold air standard analysis** is to be conducted, then the specific heats are taken as constants.

8.2 The Internal Combustion Engine

A single piston-cylinder assembly of an internal combustion engine is shown in Figure 8.1(a). The inside diameter of the cylinder is called its **bore** and the distance that the piston travels from **bottom dead center** to **top dead center** is called the **stroke**. The volume in the cylinder when the piston is at top dead center is called the **clearance volume** and the volume that is displaced by the piston in one stroke as it moves from bottom dead center to top dead center is called the **displacement volume**. The ratio of the cylinder volume at bottom dead center to the volume at top dead center taken respectively as V_1 and V_2 is called the **compression ratio,** r

$$r \equiv \frac{V_1}{V_2}$$

Internal combustion engines can operate on either the **four-stroke** or **two-stroke** cycle. The four-stroke cycle consists of four separate and distinct strokes and we employ Figure 8.1(b) to describe this cycle.

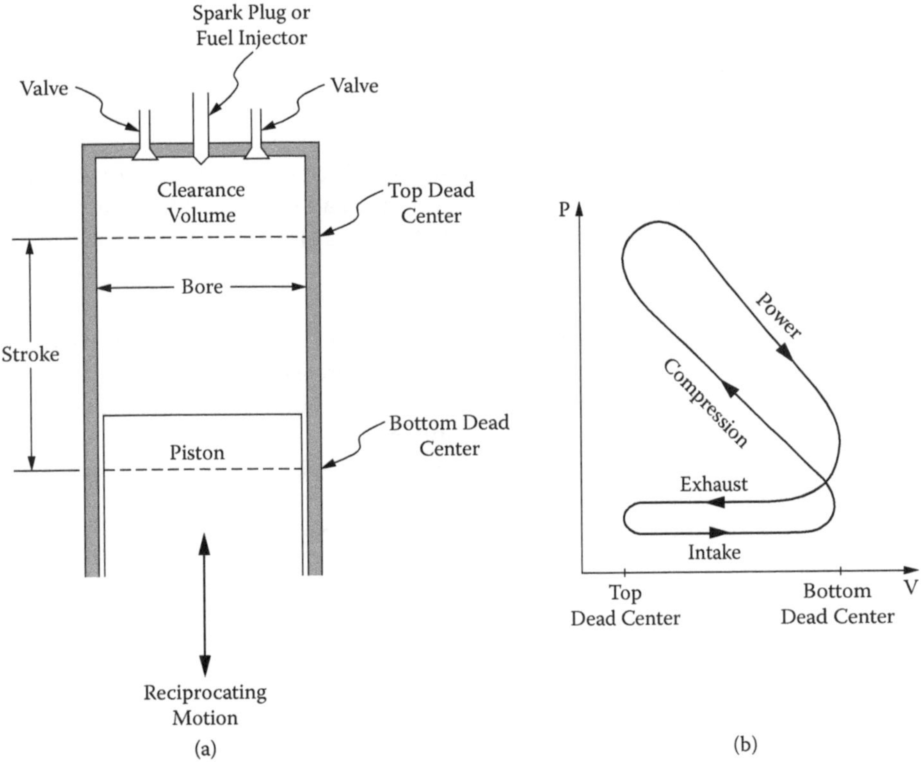

FIGURE 8.1
(a) A piston-cylinder assembly in an internal combustion engine and (b) its operation.

1. The **intake stroke** begins with the piston at top dead center. The intake valve opens and for the spark-ignition engine, a **charge** of air *and* fuel (such as gasoline) is drawn into the cylinder. In the compression-ignition engine, the charge is air without the fuel.
2. The **compression stroke** begins with the piston at bottom dead center and ends with the piston at top dead center. During the compression stroke both intake and exhaust (exit) valves are closed, and during this stroke, the pressure and temperature of the charge are markedly increased.

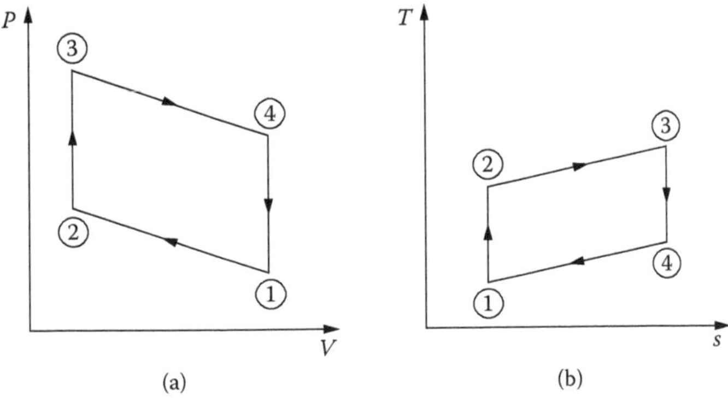

FIGURE 8.2
The ideal air standard gasoline engine or Otto cycle, which operates on a four-stroke cycle: (a) the P-v plane and (b) the T-s plane.

3. The **power stroke** begins with the piston at top dead center. This is the stroke in which a combustion process occurs with the valves shut. The combustion process is induced by a spark in the spark-ignition engine and by the injection of the fuel in the compression-ignition engine; the power is derived from the expansion of the air fuel mixture as the piston moves from top dead center to bottom dead center.
4. In the **exhaust stroke**, in which the exhaust valve is open, the spent air fuel mixture is discharged as the piston moves from bottom dead center to top dead center. At the conclusion of the exhaust stroke, the piston-cylinder assembly is ready to begin a new cycle.

We note that two revolutions of the **crankshaft** are needed to execute a four-stroke cycle. The two-stroke cycle combines the intake and the compression strokes and the power and exhaust strokes by using intake and exit **ports**. Only one revolution is required to execute the two-stroke cycle and, because of the unavoidable mixing of the incoming air-fuel mixture and the exhaust gases, an engine using the two-stroke cycle is generally less efficient than one that employs the four-stroke cycle.

We now turn our attention to the air standard Otto cycle.

8.3 The Air Standard Otto Cycle

8.3.1 Performance

The performance of the automobile engine may be modeled by the *ideal* air standard Otto cycle with the T-s and P-v representations shown in Figure 8.2. The four strokes in the cycle may be summarized by

- *Process 1-2:* Isentropic compression
- *Process 2-3:* Constant volume heat addition
- *Process 3-4:* Isentropic expansion
- *Process 4-1:* Constant volume heat rejection

We note that the compression and expansion processes are both isentropic and that the heat addition and exhaust processes occur at constant volume, and in accordance with its definition, the compression ratio is

$$r \equiv \frac{V_1}{V_2} \equiv \frac{V_4}{V_3} \tag{8.1}$$

If all of the work and heat interactions are considered as *positive quantities*, we can represent the work of compression and expansion by the closed-system expressions

$$W_{12} = m(u_2 - u_1) \quad \text{and} \quad W_{34} = m(u_3 - u_4) \tag{8.2}$$

respectively and the heat transfer during the power stroke by

$$Q_{23} = m(u_3 - u_2) \tag{8.3a}$$

and the exhaust stroke by

$$Q_{41} = m(u_4 - u_1) \tag{8.3b}$$

We are quick to note that, because of the convention adopted here that all of the work and heat interactions are to be considered positive, W_{12} is the work **input** during compression and Q_{41} is the heat **rejected** by the system. Both of these quantities are represented by a positive number.

The net work and heat of the cycle are evaluated as

$$W_{\text{cycle}} = W_{34} - W_{12} = m[(u_3 - u_4) - (u_2 - u_1)] \tag{8.4a}$$

and

$$Q_{\text{cycle}} = Q_{23} - Q_{41} = m[(u_3 - u_2) - (u_4 - u_1)] \tag{8.4b}$$

The cycle efficiency is the net work of the cycle divided by the heat input

$$\eta = \frac{W_{\text{cycle}}}{Q_{\text{in}}} = \frac{m[(u_3 - u_4) - (u_2 - u_1)]}{m(u_3 - u_2)} = 1 - \frac{u_4 - u_1}{u_3 - u_2} \tag{8.5}$$

The specific internal energy values required by Equations 8.4 and 8.5 are functions of temperature and, in air standard analysis, use is made of the **reference volumes** that apply to the isentropic processes in the cycle

$$v_{r2} = v_{r1}\left(\frac{V_2}{V_1}\right) = \frac{v_{r1}}{r} \tag{8.6a}$$

and

$$v_{r4} = v_{r3}\left(\frac{V_4}{V_3}\right) = r v_{r3} \tag{8.6b}$$

These reference volumes are functions of temperature and may be found in the "air tables" provided.

If the analysis is based on the cold air standard where the specific heat at constant volume, c_v, is taken as constant, then

$$W_{\text{cycle}} = mc_v[(T_3 - T_4) - (T_2 - T_1)] \tag{8.7a}$$

$$Q_{\text{cycle}} = mc_v[(T_3 - T_2) - (T_4 - T_1)] \tag{8.7b}$$

and

$$\eta = \frac{W_{\text{cycle}}}{Q_{\text{in}}} = \frac{mc_v[(T_3 - T_4) - (T_2 - T_1)]}{mc_v(T_3 - T_2)} = 1 - \frac{T_4 - T_1}{T_3 - T_2} \tag{8.7c}$$

In the cold air standard cycle, the temperature-volume relationships for the isentropic compression and expansion were developed in Chapter 7. We have seen that these relationships employ the specific heat ratio which, for air, is taken as $k = c_p/c_v = 1.40$. Hence, as indicated in Chapter 7

$$\frac{T_2}{T_1} = \left(\frac{V_1}{V_2}\right)^{k-1} = r^{k-1} \tag{8.8a}$$

and

$$\frac{T_4}{T_3} = \left(\frac{V_3}{V_4}\right)^{k-1} = \frac{1}{r^{k-1}} \tag{8.8b}$$

and Equations 8.8a and 8.8b show that

$$\frac{T_2}{T_1} = \frac{T_3}{T_4} = r^{k-1} \tag{8.8c}$$

The efficiency relationship of Equation 8.7c can be modified to

$$\eta = 1 - \frac{T_4 - T_1}{T_3 - T_2} = 1 - \frac{T_1}{T_2}\left[\frac{(T_4/T_1) - 1}{(T_3/T_2) - 1}\right]$$

and using Equation 8.8c, we see that this can be adjusted to

$$\eta = 1 - \frac{1}{r^{k-1}} \tag{8.9}$$

The **mean effective pressure** (mep) is defined as the ratio of the net work of the cycle to the displacement volume, $V_1 - V_2$

$$\text{mep} = \frac{W_{\text{cycle}}}{V_1 - V_2} = \frac{r\,W_{\text{cycle}}}{V_1(r-1)} \tag{8.10}$$

If the pressure on the piston during the entire power stroke is constant and equal to the mean effective pressure, then

$$W_{\text{cycle}} = (\text{mep})(\text{displacement volume})$$

or

$$W_{\text{cycle}} = (\text{mep})(\text{stroke})((\text{piston area})$$

This shows that the mean effective pressure is a yardstick that provides a comparison of engines running at the same speed.

We now turn to two examples concerning the Otto cycle. It is the intent of these examples to consider both the cold air standard performance and the air standard cycle performance, showing the differences and discrepancies between the pressures and temperatures at various points in the cycle as well as the efficiencies.

Example 8.1 In an Otto cycle to be analyzed on the cold air standard, the initial pressure is 0.92 bar, the initial temperature is 27°C and the bottom dead center volume of the piston-cylinder assemblies is 555 cm^3. The compression ratio is 7.75 and the maximum temperature during the cycle is not to exceed 2100 K. Determine (a) the pressure and temperature at each of the four points (as shown in Figure 8.3) of the cycle, (b) the thermal efficiency, and (c) the mean effective pressure.

Solution

Assumptions and Specifications

(1) The air contained in the system is an ideal gas.
(2) The cycle consists of two constant volume and two isentropic processes.
(3) All processes are quasi-static.

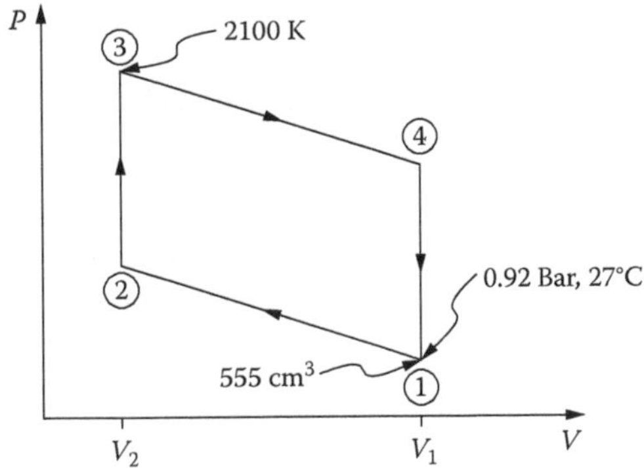

FIGURE 8.3
P-V diagram for the Otto cycle in Example 8.1.

(4) There are no kinetic and potential energy changes.
(5) A cold air standard analysis is specified.

(a) For the conditions at the end of the isentropic compression (point 2 in Figure 8.3), we use Equation 8.8a with $k = 1.40$ and $T_1 = 300\,\text{K}$

$$\frac{T_2}{T_1} = \left(\frac{V_1}{V_2}\right)^{k-1} = r^{k-1}$$

$$T_2 = T_1 r^{k-1} = (300\,\text{K})(7.75)^{0.40} = 680.5\,\text{K} \Longleftarrow$$

Then, because for an ideal gas, $P_1V_1/T_1 = P_2V_2/T_2$

$$P_2 = P_1\left(\frac{V_1}{V_2}\right)\left(\frac{T_2}{T_1}\right) = P_1 r\left(\frac{T_2}{T_1}\right)$$

we have

$$P_2 = (0.92\,\text{bar})(7.75)\left(\frac{680.5\,\text{K}}{300\,\text{K}}\right) = 16.17\,\text{bar} \Longleftarrow$$

For the conditions at the end of the constant volume heat addition (point 3 in Figure 8.3) the maximum temperature allowed in the cycle is specified

$$T_3 = 2100\,\text{K}$$

and the pressure will be in accordance with Equation 3.37

$$P_3 = P_2\left(\frac{T_3}{T_2}\right) = (16.17\ \text{bar})\left(\frac{2100\,\text{K}}{680.5\,\text{K}}\right) = 49.91\ \text{bar} \Longleftarrow$$

For conditions at the end of isentropic expansion (point 4 in Figure 8.3) use Equation 8.8b

$$\frac{T_4}{T_3} = \left(\frac{V_3}{V_4}\right)^{k-1} = \left(\frac{1}{r}\right)^{k-1}$$

so that with $T_3 = 2100\,\text{K}$

$$T_4 = (2100\,\text{K})\left(\frac{1}{7.75}\right)^{0.40} = (2100\,\text{K})(0.441) = 925.8\,\text{K} \Longleftarrow$$

Then, because $P_4V_4/T_4 = P_3V_3/T_3$

$$P_4 = P_3\left(\frac{V_3}{V_4}\right)\left(\frac{T_4}{T_3}\right) = \frac{P_3}{r}\left(\frac{T_4}{T_3}\right)$$

we have

$$P_4 = \frac{49.91\text{bar}}{7.75}\left(\frac{925.8\,\text{K}}{2100\,\text{K}}\right) = 2.84\,\text{bar} \Longleftarrow$$

(b) We obtain the thermal efficiency from Equation 8.9

$$\eta = 1 - \frac{1}{r^{k-1}} = 1 - \frac{1}{(7.75)^{0.40}} = 1 - 0.441 = 0.559 \Longleftarrow$$

(c) One of the inputs to Equation 8.10 for the mean effective pressure is the net work of the cycle, which, in turn, requires that we evaluate the mass in the closed system. For air, Table A.1 provides

$$R = 287\,\text{J/kg-K} \quad \text{and} \quad c_v = 0.718\,\text{kJ/kg-K}$$

and the ideal gas law may then be employed to obtain the mass

$$m = \frac{P_1V_1}{RT_1} = \frac{(0.92 \times 10^5\,\text{N/m}^2)(5.55 \times 10^{-4}\,\text{m}^3)}{(287\,\text{J/kg-K})(300\,\text{K})} = 5.930 \times 10^{-4}\,\text{kg}$$

We then use Equation 8.7a to determine the net work

$$\begin{aligned} W_{\text{cycle}} &= mc_v[(T_3 - T_4) - (T_2 - T_1)] \\ &= (5.930 \times 10^{-4}\,\text{kg})(0.718\,\text{kJ/kg-K}) \\ &\qquad [(2100\,\text{K} - 925.8\,\text{K}) - (680.5\,\text{K} - 300\,\text{K})] \\ &= (4.258 \times 10^{-4}\,\text{kJ/K})(1174.2\,\text{K} - 380.5\,\text{K}) \\ &= (4.258 \times 10^{-4}\,\text{kJ/K})(793.7\,\text{K}) = 0.338\,\text{kJ} \end{aligned}$$

Now, Equation 8.10 gives the mean effective pressure

$$\text{mep} = \frac{rW_{\text{cycle}}}{V_1(r-1)}$$

or

$$\text{mep} = \frac{(7.75)(0.338\,\text{kJ})}{(555 \times 10^{-6}\,\text{m}^3)(6.75)} = 699.1\,\text{kN/m}^2 \quad \text{or} \quad 6.99\,\text{bar} \Longleftarrow$$

Example 8.2 Repeat Example 8.1 using an air standard analysis.

Solution

We make the same assumptions as were made in Example 8.1 except that an air standard analysis is specified.

At 300 K, Table A.12 gives

$$v_{r1} = 621.2 \quad \text{and} \quad u_1 = 214.07\,\text{J/kg}$$

Then, to find the conditions at the end of the isentropic compression (point 2 in Figure 8.3) we use the reference value ratio $v_{r2}/v_{r1} = V_2/V_1$ so that

$$v_{r2} = v_{r1}\left(\frac{V_2}{V_1}\right) = \frac{v_{r1}}{r} = \frac{621.2}{7.75} = 80.16$$

We then use Table A.12 once again but this time with interpolation to find T_2 and u_2

$$T_2 = 665.3\,\text{K} \Longleftarrow \quad \text{and} \quad u_2 = 484.70\,\text{kJ/kg}$$

Because for the ideal gas, $P_2V_2/T_2 = P_1V_1/T_1$ we see that

$$P_2 = P_1\left(\frac{V_1}{V_2}\right)\left(\frac{T_2}{T_1}\right) = P_1 r\left(\frac{T_2}{T_1}\right)$$

or

$$P_2 = (0.92\,\text{bar})(7.75)\left(\frac{665.3\,\text{K}}{300\,\text{K}}\right) = 15.81\,\text{bar} \Longleftarrow$$

At the end of the combustion process (the constant volume heat addition) at point 3 in Figure 8.3, T_3 is the maximum temperature permitted in the cycle and is specified at

$$T_3 = 2100\,\text{K} \Longleftarrow$$

Table A.12 then gives

$$v_{r3} = 2.356 \quad \text{and} \quad u_3 = 1775.3\,\text{kJ/kg}$$

and because the combustion process takes place at constant volume, Equation 3.37 gives us

$$P_3 = P_2\left(\frac{T_3}{T_2}\right) = (15.81\,\text{bar})\left(\frac{2100\,\text{K}}{665.3\,\text{K}}\right) = 49.90\,\text{bar} \Longleftarrow$$

For conditions at point 4 (see Figure 8.3 at the end of the isentropic expansion)

$$\frac{v_{r4}}{v_{r3}} = \frac{V_4}{V_3} = r$$

so that

$$v_{r4} = v_{r3}r = (2.356)(7.75) = 18.26$$

Then interpolation of Table A.12 gives

$$T_4 = 1112.6\,\text{K} \Longleftarrow \quad \text{and} \quad u_4 = 856.33\,\text{kJ/kg}$$

Finally, because we are working with an ideal gas, $P_4V_4/T_4 = P_3V_3/T_3$ and the value of P_4 will be

$$P_4 = P_3\left(\frac{V_3}{V_4}\right)\left(\frac{T_4}{T_3}\right) = \frac{P_3}{r}\left(\frac{T_4}{T_3}\right) = \frac{49.90\,\text{bar}}{7.75}\left(\frac{1112.6\,\text{K}}{2100\,\text{K}}\right) = 3.41\,\text{bar} \Longleftarrow$$

Equation 8.9 applies only to the cold air standard. Instead, use Equation 8.5

$$\begin{aligned}\eta &= 1 - \frac{u_4 - u_1}{u_3 - u_2}\\ &= 1 - \frac{856.33\,\text{kg/s} - 214.07\,\text{kg/s}}{1775.3\,\text{kg/s} - 484.70\,\text{kg/s}}\\ &= 1 - \frac{642.26\,\text{kg/s}}{1290.6\,\text{kg/s}}\\ &= 1 - 0.498 = 0.502 \Longleftarrow\end{aligned}$$

(c) The mean effective pressure is obtained from Equation 8.10. The value of the mass can be obtained from the specified conditions at point 1 and will be equal to the value obtained in Example 8.1. Hence,

$$m = 5.930 \times 10^{-4}\,\text{kg}$$

and we may find W_{cycle}, which Equation 8.10 requires, from Equation 8.4a.

$$\begin{aligned}W_{cycle} &= m[(u_3 - u_4) - (u_2 - u_1)]\\ &= (5.930 \times 10^{-4}\,\text{m})[(1775.3\,\text{kJ/kg} - 856.33\,\text{kJ/kg}\\ &\quad - (484.70\,\text{kJ/kg} - 214.07\,\text{kJ/kg})]\\ &= (5.930 \times 10^{-4}\,\text{m})(918.97\,\text{kJ/kg} - 270.63\,\text{kJ/kg})\\ &= (5.930 \times 10^{-4}\,\text{m})(648.34\,\text{kJ/kg}) = 0.384\,\text{kJ}\end{aligned}$$

Now Equation 8.10 gives us

$$\begin{aligned}\text{mep} &= \frac{r\,W_{cycle}}{V_1(r-1)}\\ &= \frac{(7.75)(0.384\,\text{kJ})}{(555 \times 10^{-6}\,\text{m}^3)(6.75)}\\ &= 794.4\,\text{kN/m}^2 \quad \text{or} \quad 7.94\,\text{bar} \Longleftarrow\end{aligned}$$

The goal of Examples 8.1 and 8.2 was to provide a means for the assessment of any discrepancies in the results of the two analysis procedures. The examples were contrived to provide comparison of the temperatures and pressures at each point in the cycle, the efficiencies and the mean effective pressures.

The pressures and temperatures at each point in the cycle are tabulated in Table 8.1 for both the cold air standard and the air standard analyses. With regard to both the temperature and pressure, we note a slight discrepancy at point 2, no discrepancy at points 1 and 3 (point 1 is the specified cycle starting point), and a rather marked discrepancy at point 4. The error involved at point 4, using the more correct air standard value as a basis, amounts to about 17% in both temperature and pressure.

We also see a discrepancy between the two values of the efficiency and the mean effective pressure. For the efficiency, the comparison is 0.559 for the cold air standard and 0.502 for the air standard. This represents an error of about 10%. For the mean effective pressure, the value for the cold air standard is 6.99 bar and for the air standard, it is 7.94 bar, an error of about 13%.

TABLE 8.1

A Comparison of Temperatures and Pressures between Examples 8.1 and 8.2. The Cycle Points Correspond to Figure 8.3

Cycle Point	Cold Air Standard, T K	Air Standard, T K	Cold Air Standard, P bar	Air Standard, P bar
1	300	300	0.92	0.92
2	680.5	665.3	16.17	15.81
3	2100	2100	49.91	49.90
4	925.8	1112.6	2.84	3.41

Because the specific heats for air do indeed vary with temperature, the air standard analysis is presumed to give more accurate results than its cold air standard counterpart. Use of the cold air standard temperatures and pressures can have an impact on the actual design of an engine working on the Otto cycle and the cold air standard efficiency and mean effective pressure can have a bearing on performance comparisons with other engines.

8.3.2 The Compression Ratio and Its Effect on Performance

We may refer again to Equation 8.9

$$\eta = 1 - \frac{1}{r^{k-1}} \tag{8.9}$$

which is plotted in Figure 8.4 and which shows that, on a cold air standard basis, the thermal efficiency is directly proportional to the compression ratio. We find it tempting to think that the thermal efficiency of the Otto cycle has an upper limit of 100% as the compression ratio increases without bound. However, it is a fact that limitations on the compression ratio are made by the fuel itself.

In the spark-ignition engine, the upper limit on the compression ratio is dictated by the ignition temperature of the fuel that must be below the temperature at the end of the isentropic compression. Otherwise a high-velocity, high-pressure movement of the flame front in the cylinder will propagate *before* the spark ignition. This causes **autoignition,** which manifests itself in **knocking** or **pinging**. These annoying phenomena are more apt to occur with low octane fuels and at compression ratios greater than 10.

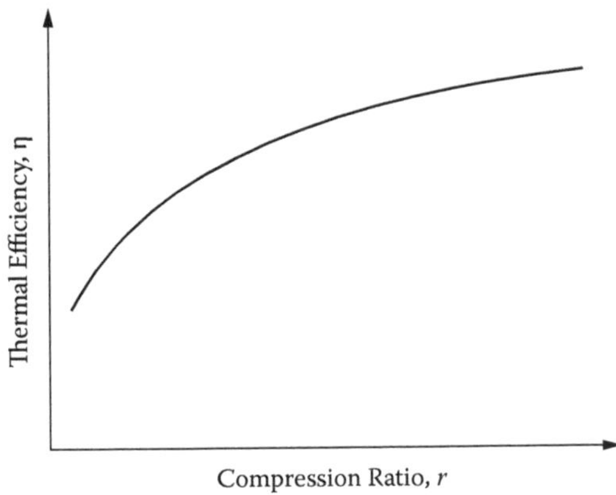

FIGURE 8.4
The cold air standard thermal efficiency of the Otto cycle as a function of the compression ratio.

We turn now to Design Example 2, which involves the establishment of the bore and stroke of an internal combustion engine to operate on the Otto cycle.

8.4 Design Example 2

Determine the bore and stroke in an Otto engine to operate on the air standard in an ideal Otto cycle. The intake pressure is 0.96 bar at 27°C with an air mass of 3.75×10^{-4} kg. The compression ratio is limited to 8.00 in order to circumvent any fuel detonation problems and the maximum temperature in the cycle is limited to 2250 K because of metallurgical reasons. The thermal efficiency of the cycle is to be $\eta = 0.500$ and the length of the stroke is to be four times the bore.

Solution

Assumptions and Specifications

(1) The air contained in the system is an ideal gas.
(2) The cycle consists of two constant volume and two isentropic processes.
(3) All processes are quasi-static.
(4) There are no kinetic and potential energy changes.
(5) An air standard analysis is specified.

Figure 8.5 shows the P-v diagram for the cycle.

A trial-and-error procedure will be necessary and Table 8.2 summarizes the pertinent calculations at assumed compression ratios of 7.00, 7.50, 7.75, and 8.00. A plot of the cycle efficiency as a function of the compression ratio is provided in Figure 8.6 where it is observed that an efficiency of 0.500 is achieved when the compression ratio, r, is 7.83.

The volume at point 1 in Figure 8.5 may be determined from the ideal gas law. We read from Table A.1

$$R = 287\,\text{J/kg-K}$$

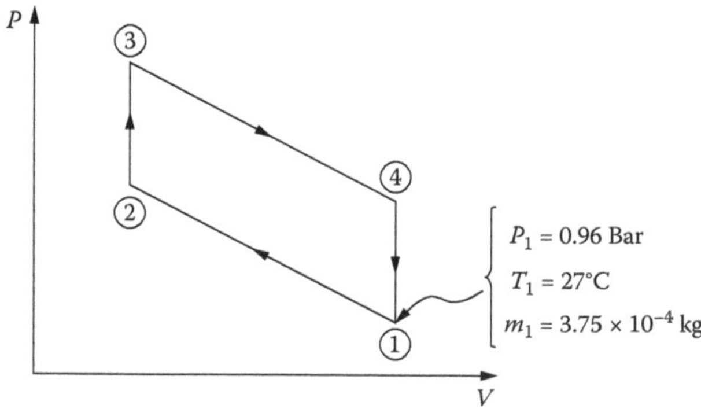

FIGURE 8.5
P-v diagram for the Otto cycle in Design Example 2.

TABLE 8.2

Computations for Design Example 2

Assume r	7.00	7.50	7.75	8.00
v_{r1} at 300 K	621.2	621.2	621.2	621.2
u_1 at 300 K, kJ/kg	214.07	214.07	214.07	214.07
$v_{r2} = v_{r1}/r$	88.743	82.837	80.155	77.650
T_2 at v_{r2} K	640.4	657.3	665.3	673.1
u_2 at T_2, K	466.07	478.91	485.14	491.28
T_2/T_1	2.135	2.191	2.218	2.244
$P_2 = P_1 r(T_2/T_1)$, MPa	14.347	15.775	16.502	17.231
T_3, K	2250	2250	2250	2250
v_{r3} at T_3	1.864	1.864	1.864	1.864
u_3 at T_3, kJ/kg	1921.3	1921.3	1921.3	1921.3
T_3/T_2	3.512	3.423	3.382	3.342
$P_3 = P_2(T_3/T_2)$, MPa	50.39	54.00	55.81	57.60
$v_{r4} = r v_{r3}$	13.048	13.980	14.440	14.912
T_4, K	1240.7	1213.2	1200.7	1188.5
u_4 at v_{r4}, kJ/kg	969.54	945.37	933.95	923.01
$u_4 - u_1$, kJ/kg	755.47	731.30	719.88	708.94
$u_3 - u_2$, kJ/kg	1455.23	1442.39	1436.16	1430.02
$\phi = (u_4 - u_1)/(u_3 - u_2)$	0.519	0.507	0.501	0.496
$\eta = 1 - \phi$	0.481	0.493	0.499	0.504

so that at $T = 27°\text{C}$ (300 K)

$$V_1 = \frac{\dot{m}RT_1}{P_1} = \frac{(3.75 \times 10^{-4}\,\text{kg})(287\text{J/kg-K})(300\,\text{K})}{0.96 \times 10^5\,\text{Pa}} = 3.363 \times 10^{-4}\text{m}^3$$

Then with

$$V_2 = \frac{V_1}{r} = \frac{3.363 \times 10^{-4}\text{m}^3}{7.83} = 4.295 \times 10^{-5}\text{m}^3$$

we have

$$LA^2 = V_1 - V_2 = 3.363 \times 10^{-4}\text{m}^3 - 4.295 \times 10^{-5}\text{m}^3 = 2.934 \times 10^{-4}\text{m}^3$$

where L is the length of the stroke and with d as the bore of the cylinder. The specification calls for

$$L = 4d$$

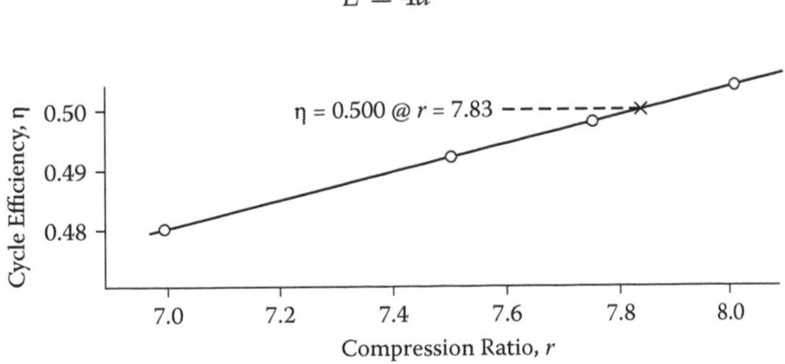

FIGURE 8.6
Plot of cycle efficiency as a function of the compression ratio for Design Example 2.

Hence,

$$LA^2 = (4d)\left(\frac{\pi}{4}d^2\right) = \pi d^3 = 2.934 \times 10^{-4}\text{m}^3$$

$$d = \left(\frac{2.934 \times 10^{-4}\text{m}^3}{\pi}\right)^{1/3} = 0.0454\,\text{m} \quad (4.54\,\text{cm}) \Longleftarrow$$

and

$$L = 4d = 4(0.0454\,\text{m}) = 0.1815\,\text{m} \quad (18.15\,\text{cm}) \Longleftarrow$$

8.5 The Air Standard Diesel Cycle

Two representations of the ideal Diesel cycle are displayed in Figure 8.7. The four strokes in the cycle may be summarized as follows:

- *Process 1-2:* Isentropic compression
- *Process 2-3:* Constant pressure heat addition
- *Process 3-4:* Isentropic expansion
- *Process 4-1:* Constant volume heat rejection

As in the case of the Otto cycle, we treat all work and heat interactions as positive quantities. For the air standard cycle, the heat transfer during the constant pressure power stroke will be

$$Q_{\text{in}} = Q_{23} = mc_p(T_3 - T_2) = m(h_3 - h_2) \tag{8.11a}$$

and the heat rejected during the constant volume exhaust stroke is

$$Q_{\text{out}} = Q_{41} = mc_v(T_4 - T_1) = m(u_4 - u_1) \tag{8.11b}$$

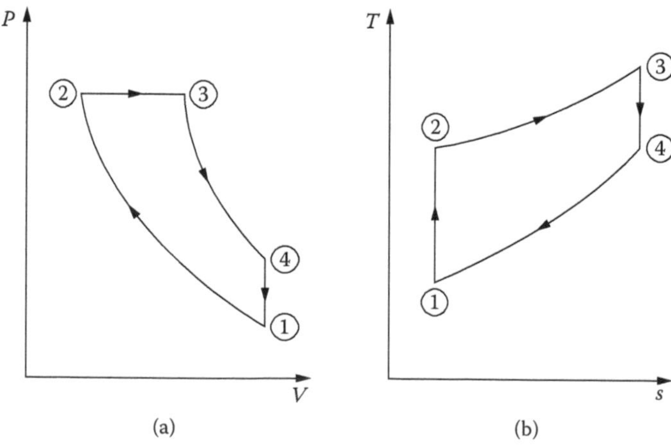

FIGURE 8.7
The ideal air standard Diesel cycle. (a) the P-V plane and (b) the T-s plane.

Consequently, the work for the cycle is

$$W_{\text{cycle}} = Q_{\text{in}} - Q_{\text{out}} = m(h_3 - h_2) - m(u_4 - u_1) \tag{8.11c}$$

and the cycle thermal efficiency will be

$$\eta = \frac{W_{\text{cycle}}}{Q_{\text{in}}} = \frac{m(h_3 - h_2) - m(u_4 - u_1)}{m(h_3 - h_2)} = 1 - \frac{u_4 - u_1}{h_3 - h_2} \tag{8.12}$$

On the cold air standard basis, we take the specific heats and $k = c_p/c_v$ as constant. The efficiency will be

$$\eta = \frac{mc_p(T_3 - T_2) - mc_v(T_4 - T_1)}{mc_p(T_3 - T_2)} = 1 - \frac{T_4 - T_1}{k(T_3 - T_2)}$$

As in the case of the Otto cycle, the compression ratio is the ratio of the cylinder displacement volume to the cylinder clearance volume. In the case of the Diesel cycle, we define the **cutoff ratio** as the ratio of the cylinder volume after the combustion is completed to the cylinder volume before combustion begins. Hence, the compression and cutoff ratios are defined by

$$r \equiv \frac{V_1}{V_2} \quad \text{and} \quad r_c \equiv \frac{V_3}{V_2}$$

The cold air standard efficiency may be put into a form that is a function of only the specific heat ratio, k, and the compression and cutoff ratios, r and r_c. In order to accomplish this, we must put all of the temperatures into terms of T_1. First, for the isentropic compression from point 1 to point 2 in Figure 8.7, Equation 8.8a gives

$$\frac{T_2}{T_1} = \left(\frac{V_1}{V_2}\right)^{k-1} = r^{k-1} \quad \text{or} \quad T_2 = T_1 r^{k-1}$$

and for the constant pressure process between point 2 and point 3 (Figure 8.5)

$$\frac{T_3}{T_2} = \frac{V_3}{V_2} = r_c \quad \text{or} \quad T_3 = T_2 r_c = T_1 r_c r^{k-1}$$

Then, for the isentropic expansion from point 3 to point 4 (Figure 8.7), Equation 8.8b gives

$$\frac{T_4}{T_3} = \left(\frac{V_3}{V_4}\right)^{k-1} = \left(\frac{r_c}{r}\right)^{k-1} \quad \text{or} \quad T_4 = T_1 r_c^k$$

Now, the cold air standard efficiency can be written as

$$\begin{aligned}
\eta &= 1 - \frac{T_4 - T_1}{k(T_3 - T_2)} \\
&= 1 - \frac{T_1 r_c^k - T_1}{k(T_1 r_c r^{k-1} - T_1 r^{k-1})} \\
&= 1 - \frac{r_c^k - 1}{k(r_c r^{k-1} - r^{k-1})}
\end{aligned}$$

or

$$\eta = 1 - \frac{r_c^k - 1}{kr^{k-1}(r_c - 1)} = 1 - \frac{1}{r^{k-1}}\left[\frac{r_c^k - 1}{k(r_c - 1)}\right] \tag{8.13}$$

We now illustrate the interplay of the foregoing relationships in a rather lengthy example.

Example 8.3 A Diesel cycle with a compression ratio of 16 and a cutoff ratio of 2 takes in air at an initial pressure of 1 bar and an initial temperature of 27°C. For an air standard analysis, determine (a) the pressure and temperature at each if the four points in the cycle, (b) the thermal efficiency, (c) the mean effective pressure, and (d) the cold air standard thermal efficiency.

Solution

Assumptions and Specifications

(1) The air contained in the piston-cylinder assemblies is an ideal gas and forms a closed system.
(2) Kinetic and potential energy changes are negligible.
(3) This is an ideal cycle with one constant pressure, one constant volume and two isentropic processes.
(4) All of the processes that comprise the cycle are quasi-static.
(5) An air standard analysis is specified.

(a) At $T_1 = 27°\text{C}$ or 300 K, Table A.12 gives

$$v_{r1} = 621.2, \quad p_{r1} = 1.386 \quad \text{and} \quad u_1 = 214.07\,\text{kJ/kg}$$

Then, to establish conditions at the end of the isentropic compression stroke (point 2 in Figure 8.8), we use the reference volume ratio

$$\frac{v_{r2}}{v_{r1}} = \frac{V_2}{V_1} \quad \text{or} \quad v_{r2} = v_{r1}\left(\frac{V_2}{V_1}\right) = \frac{v_{r1}}{r}$$

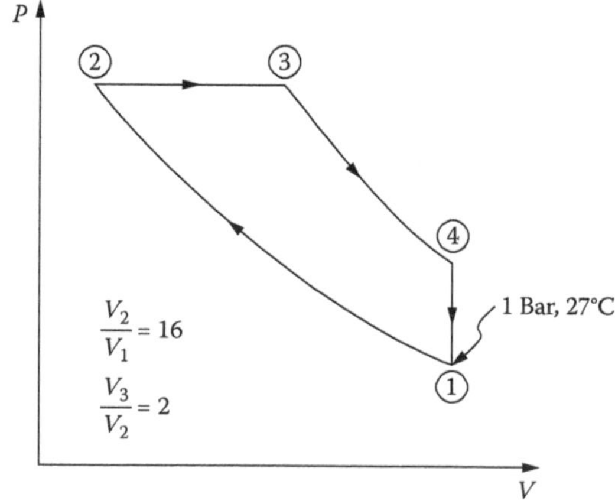

FIGURE 8.8
P-v diagram for the Example 8.3.

so that

$$v_{r2} = \frac{621.2}{16} = 38.83$$

and we may use Table A.12 to obtain (with interpolation)

$$T_2 = 862.3\,\text{K} \Longleftarrow, \quad h_2 = 890.95 \text{ kJ/kg}, \quad \text{and} \quad p_{r2} = 63.77$$

Then, P_2 may be found from

$$P_2 = P_1\left(\frac{p_{r2}}{p_{r1}}\right) = (1 \text{ bar})\left(\frac{63.77}{1.386}\right) = 46.01\,\text{bar} \quad \text{or} \quad 4.60\,\text{MPa} \Longleftarrow$$

At the end of the constant pressure heat addition at point 3 in Figure 8.8

$$P_3 = P_2 = 4.60\,\text{MPa} \Longleftarrow$$

and

$$T_3 = \left(\frac{V_3}{V_2}\right)T_2 = r_c T_2 = 2(862.3\,\text{K}) = 1724.6\,\text{K} \Longleftarrow$$

From Table A.12 with interpolation

$$v_{r3} = 4.541 \quad \text{and} \quad h_3 = 1910.2\,\text{kJ/kg}$$

and for isentropic compression with $V_4 = V_1$

$$v_{r4} = \left(\frac{V_4}{V_3}\right)v_{r3} = \left(\frac{V_1}{V_2}\right)\left(\frac{V_2}{V_3}\right)v_{r3} = \frac{r}{r_c}v_{r3}$$

so that

$$v_{r4} = \left(\frac{16}{2}\right)(4.541) = 36.32$$

Interpolation of Table A.12 gives

$$T_4 = 882.5\,\text{K} \Longleftarrow \quad \text{and} \quad u_4 = 660.05\,\text{kJ/kg}$$

Finally, because $P_4V_4/T_4 = P_3V_3/T_3$

$$P_4 = P_3\left(\frac{V_3}{V_4}\right)\left(\frac{T_4}{T_3}\right) = P_3\left(\frac{r_c}{r}\right)\left(\frac{T_4}{T_3}\right)$$

so that

$$P_4 = (4.60\,\text{MPa})\left(\frac{2}{16}\right)\left(\frac{882.5\,\text{K}}{1724.6\,\text{K}}\right) = 0.294\,\text{bar} \Longleftarrow$$

(b) The thermal efficiency is given by Equation 8.12

$$\eta = 1 - \frac{u_4 - u_1}{h_3 - h_2}$$

$$= 1 - \frac{660.05\,\text{kJ/kg} - 214.07\,\text{kJ/kg}}{1910.2\,\text{kJ/kg} - 890.95\,\text{kJ/kg}}$$

$$= 1 - \frac{445.98\,\text{kJ/kg}}{1019.25\,\text{kJ/kg}}$$

$$= 1 - 0.438 = 0.562 \Longleftarrow$$

(c) We note that neither the mass of air or the initial volume of the system has been specified and while the mean effective pressure is ordinarily obtained from Equation 8.10, we will need to make a modification that will involve the specific volume. The net work in the cycle is given by Equation 8.11c

$$W_{\text{cycle}} = Q_{\text{in}} - Q_{\text{out}} = m(h_3 - h_2) - m(u_4 - u_1) \tag{8.11c}$$

and a division by the mass, m gives $w_{\text{cycle}} = W_{\text{cycle}}/m$ or

$$w_{\text{cycle}} = (h_3 - h_2) - (u_4 - u_1)$$

However, $h_3 - h_2$ and $u_4 - u_1$ were obtained in the efficiency computation in part (b) so that

$$w_{\text{cycle}} = 1019.25\,\text{kJ/kg} - 445.98\,\text{kJ/kg} = 573.27\,\text{kJ/kg}$$

Table A.1 gives $R = 287\,\text{J/kg-K}$ for air so that the inlet specific volume at $T_1 = 300\,\text{K}$ and $P_1 = 1\,\text{bar}$ will be

$$v_1 = \frac{RT_1}{P_1} = \frac{(287\,\text{N-m/kg-K})(300\,\text{K})}{10^5\,\text{N/m}^2} = 0.861\,\text{m}^3/\text{kg}$$

Now we can put Equation 8.10 on a per kg basis

$$\text{mep} = \frac{r\,W_{\text{cycle}}}{V_1(r-1)} = \frac{r\,w_{\text{cycle}}}{v_1(r-1)}$$

so that the mean effective pressure will be

$$\text{mep} = \frac{rw_{\text{cycle}}}{v_1(r-1)}$$

$$= \frac{(16)(573.27\,\text{kJ/kg})}{(0.861\,\text{m}^3/\text{kg})(15)}$$

$$= = 710.2\,\text{kN/m}^2 \quad \text{or} \quad 0.710\,\text{MPa} \Longleftarrow$$

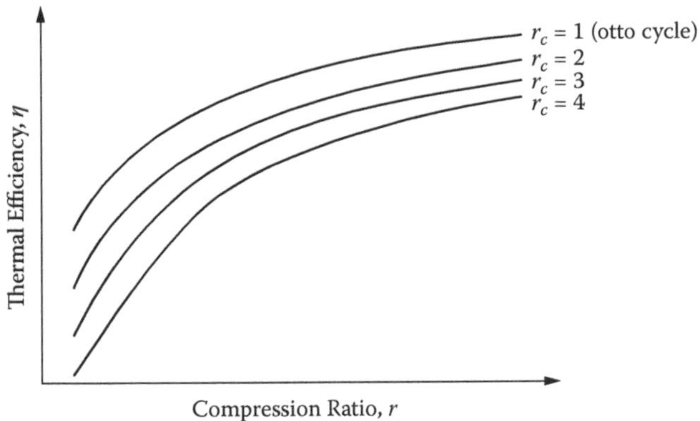

FIGURE 8.9
The cold air standard thermal efficiency of the Diesel cycle as a function of the compression and cutoff ratios ($k =$ 1.40).

(d) The cold air standard efficiency is given by Equation 8.13

$$\eta = 1 - \frac{1}{r^{k-1}}\left[\frac{r_c^k - 1}{k(r_c - 1)}\right]$$

$$= 1 - \frac{1}{16^{0.40}}\left[\frac{2^{1.40} - 1}{(1.40)(2 - 1)}\right]$$

$$= 1 - \frac{1}{3.031}\left[\frac{2.639 - 1}{1.40}\right]$$

$$= 1 - 0.386 = 0.614 \Longleftarrow$$

We note that the error involved in the use of the cold air standard amounts to about 9%.

A comparison of Equations 8.9 and 8.13 shows that the difference between the thermal efficiency of the Otto and Diesel cycles differs only in the bracketed term that appears in Equation 8.13. Because the bracketed term is always greater than unity, the thermal efficiency of the Otto cycle always exceeds that of the Diesel cycle.

As indicated in Figure 8.9, the virtue of the Diesel cycle is in its ability to operate at higher compression ratios. Because $V_2 = V_3$ in the Otto cycle, the curve in Figure 8.7 for $r_c = V_3/V_2 = 1$ represents the Otto cycle.

8.6 The Gas Turbine

8.6.1 Introduction

Figure 8.10a shows the component parts of a gas turbine power system that operates in an open cycle. Work is produced in the **turbine**, which is on the same shaft as the **compressor**. Atmospheric air is drawn into the compressor (point 1 in Figure 8.10a) where it is

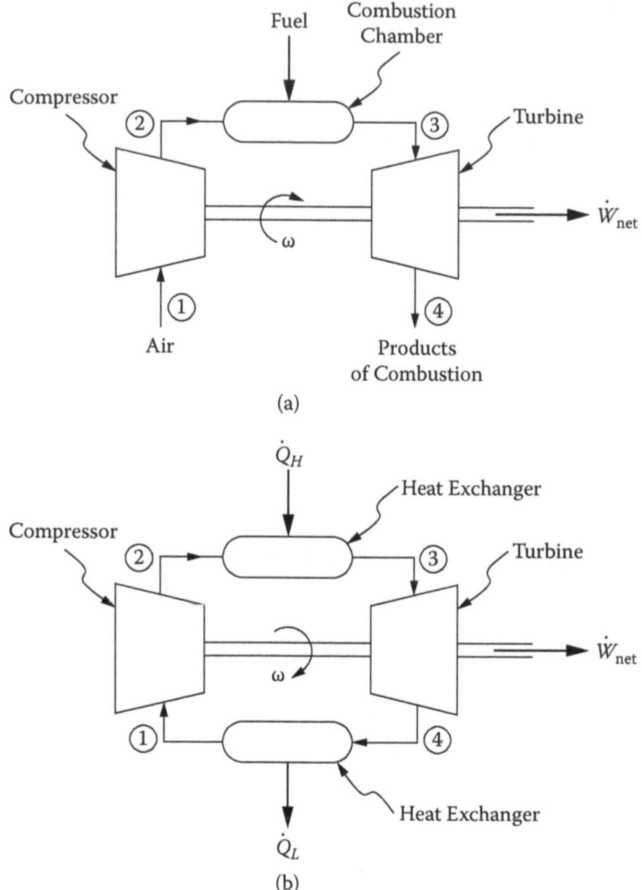

FIGURE 8.10
Component parts of a gas turbine power system (a) the open system and (b) the closed system.

compressed to high pressure and temperature. The air is discharged from the compressor into a **combustion chamber** (point 2) where fuel is introduced and where combustion occurs to yield combustion products at high temperature. These products of combustion are then led to the turbine (point 3) where they expand to low temperature and pressure and are discharged to the atmosphere at point 4.

In order to conduct an air standard analysis of the gas turbine, it is treated as a closed system as indicated in Figure 8.10b. In the air standard analysis, the temperature rise during the combustion process is accounted for by a heat addition from an external source and the return of the working fluid to its initial state is accomplished by a heat rejection to an external sink in the heat exchanger. As in other gas power cycles, the working substance is air, which we take as an ideal gas.

8.6.2 The Ideal Gas Turbine or Brayton Cycle

T-s and P-V representations of the ideal air standard **Brayton cycle** are displayed in Figure 8.11. The cycle contains four processes that may be summarized by:

- *Process 1-2:* Isentropic compression
- *Process 2-3:* Constant pressure heat addition

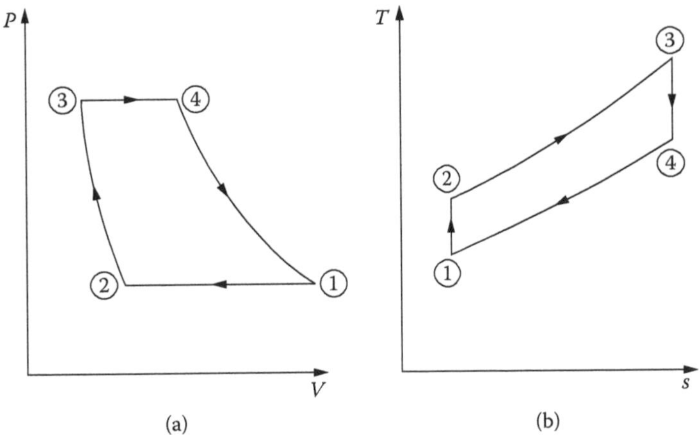

FIGURE 8.11
The air standard gas turbine or Brayton cycle (a) the P-v and (b) the T-s plane.

- *Process 3-4:* Isentropic expansion
- *Process 4-1:* Constant pressure heat rejection

We note that, because the compressor is on the same shaft as the turbine, in order to obtain the net work of the cycle, the work required to drive the compressor must be subtracted from the work developed by the turbine. For an air standard analysis on an *energy rate* or power basis, the power developed by the turbine is

$$\dot{W}_{turb} = \dot{W}_{34} = \dot{m}(h_3 - h_4) \tag{8.14a}$$

and the power required to drive the compressor will be

$$\dot{W}_{comp} = \dot{W}_{21} = \dot{m}(h_2 - h_1) \tag{8.14b}$$

Therefore, the power of the cycle is

$$\dot{W}_{cycle} = \dot{W}_{34} - \dot{W}_{21} = \dot{m}[(h_3 - h_4) - (h_2 - h_1)] \tag{8.14c}$$

The heat added in the combustion chamber is

$$\dot{Q}_{23} = \dot{m}(h_3 - h_2) \tag{8.15}$$

and the thermal efficiency is

$$\eta = \frac{\dot{W}_{cycle}}{\dot{Q}_{in}} = \frac{\dot{m}[(h_3 - h_4) - (h_2 - h_1)]}{\dot{m}(h_3 - h_2)}$$

which can be rearranged to

$$\eta = 1 - \frac{h_4 - h_1}{h_3 - h_2} \tag{8.16}$$

Because a relatively large part of the power developed by the turbine is required to drive the compressor, a useful quantity is the **back work ratio** (bwr).

$$\text{bwr} = \frac{\dot{W}_{comp}}{\dot{W}_{turb}} = \frac{h_2 - h_1}{h_3 - h_4} \tag{8.17}$$

On a cold air standard basis where the specific heats are constant, the efficiency given by Equation 8.16 can be written in terms of the temperatures

$$\eta = 1 - \frac{T_4 - T_1}{T_3 - T_2} = 1 - \frac{T_1}{T_2}\left[\frac{(T_4/T_1) - 1}{(T_3/T_2) - 1}\right]$$

and because, as shown in Chapter 7, the temperatures and pressures in an isentropic process are related by

$$\frac{T_2}{T_1} = \left(\frac{P_2}{P_1}\right)^{(k-1)/k} \quad \text{and} \quad \frac{T_3}{T_4} = \left(\frac{P_3}{P_4}\right)^{(k-1)/k}$$

Moreover, $P_1 = P_2$ and $P_3 = P_4$, and we see that

$$\frac{T_2}{T_1} = \frac{T_3}{T_4} \quad \text{and} \quad \frac{T_3}{T_2} = \frac{T_4}{T_1} = r_p^{(k-1)/k}$$

so that the cold air standard efficiency can be written as

$$\eta = 1 - \frac{1}{r_p^{(k-1)/k}} \tag{8.18}$$

where r_p is the pressure ratio that applies to both the turbine and compressor.

$$r_p \equiv \frac{P_2}{P_1} \equiv \frac{P_3}{P_4} \tag{8.19}$$

Example 8.4 An ideal Brayton cycle takes in air at a volumetric flow rate of 300,000 L/min at 1 bar and 27°C. The pressure ratio in the isentropic compression and expansion is 6 and the products of combustion leave the combustion chamber at 927°C (Figure 8.12). For the air standard analysis, determine (a) the net power developed by the turbine, (b) the back work ratio, and (c) the thermal efficiency. Then determine (d) the cold air standard thermal efficiency.

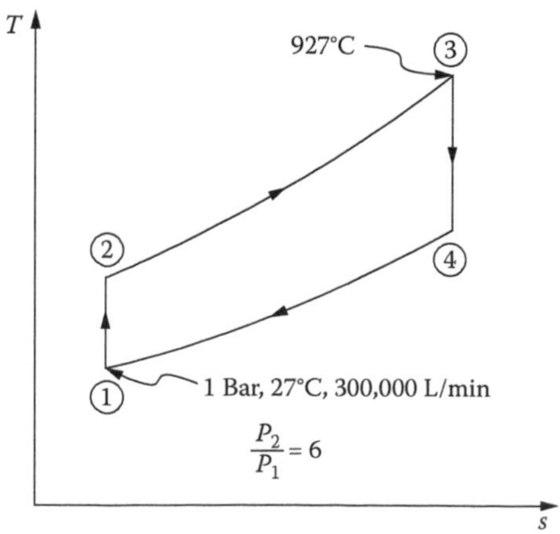

FIGURE 8.12
T-s diagram for Example 8.4.

Solution

Assumptions and Specifications

(1) The air is an ideal gas and the system is an open system.
(2) Kinetic and potential energy changes throughout the cycle are negligible.
(3) The cycle consists of two isentropic and two constant pressure processes.
(4) An air standard analysis is required.

We determine the required temperatures and enthalpies from Table A.12 using Figure 8.10. At $T_1 = 27°\text{C}$ (300 K), we read

$$p_{r1} = 1.386 \quad \text{and} \quad h_1 = 300.19\,\text{kJ/kg}$$

and then because

$$\frac{p_{r2}}{p_{r1}} = \frac{P_2}{P_1} = r_p = 6$$

we see immediately that the reference pressure at the end of the isentropic compression (point 2) is

$$p_{r2} = p_{r1} r_p = (1.386)(6) = 8.316$$

Then to establish the conditions at the end of the isentropic compression, we use Table A.12 (with interpolation)

$$T_2 = 498.4\,\text{K} \quad \text{and} \quad h_2 = 501.27\,\text{kJ/kg}$$

The temperature of the products of combustion at the end of the constant pressure heat addition (point 3) is specified as 1200 K. Table A.12 gives

$$p_{r3} = 238.0 \quad \text{and} \quad h_3 = 1277.79\,\text{kJ/kg}$$

Then because

$$\frac{p_{r4}}{p_{r3}} = \frac{P_4}{P_3} = \frac{1}{r_p} = \frac{1}{6}$$

we find that the reference pressure at point 4 will be

$$p_{r4} = \frac{p_{r3}}{r_p} = \frac{238.0}{6} = 39.67$$

With interpolation, Table A.12 provides

$$T_4 = 761.9\,\text{K} \quad \text{and} \quad h_4 = 780.32\,\text{kJ/kg}$$

(a) Table A.1 gives us $R = 287\,\text{J/kg-K}$ and because $\dot{m} = \dot{V}/v$ where $\dot{V}$ is the volumetric flow rate

$$\dot{V} = (300{,}000\,\text{L/min})(10^{-3}\,\text{m}^3/\text{L})\left(\frac{1}{60\,\text{s/min}}\right) = 5\,\text{m}^3/\text{s}$$

Hence, using conditions at point 1 gives

$$\dot{m} = \frac{\dot{V}P_1}{RT_1}$$

$$= \frac{(5\,\text{m}^3/\text{s})(1 \times 10^5\,\text{N/m}^2)}{(287\,\text{N-m/kg-K})(300\,\text{K})}$$

$$= 5.807\,\text{kg/s}$$

On an air standard basis, the net power developed by the turbine is found from Equation 8.14a

$$\dot{W}_{\text{turb}} = \dot{m}(h_3 - h_4)$$

$$= (5.807\,\text{kg/s}(1277.79\,\text{kJ/kg} - 780.32\,\text{kJ/kg})$$

$$= (5.807\,\text{kg/s})(497.47\,\text{kJ/kg}) = 2888.8\,\text{kW} \Longleftarrow$$

(b) The back work ratio given by Equation 8.17 requires a computation of the work chargeable to the compressor given by Equation 8.14b

$$\dot{W}_{\text{comp}} = \dot{m}(h_2 - h_1)$$

$$= (5.807\,\text{kg/s})(501.27\,\text{kJ/kg} - 300.19\,\text{kJ/kg})$$

$$= (5.807\,\text{kg/s})(201.08\,\text{kJ/kg}) = 1167.67\,\text{kW}$$

The back work ratio is

$$\text{bwr} = \frac{\dot{W}_{\text{comp}}}{\dot{W}_{\text{turb}}} = \frac{1167.67\,\text{kW}}{2888.8\,\text{kW}} = 0.404 \Longleftarrow$$

(c) The evaluation of the thermal efficiency via Equation 8.16 requires that the net work of the cycle

$$\dot{W}_{\text{cycle}} = \dot{W}_{\text{turb}} - \dot{W}_{\text{comp}}$$

$$= 2888.8\,\text{kW} - 1167.67\,\text{kW}$$

$$= 1721.1\,\text{kW}$$

be divided by the heat input given by Equation 8.15

$$\dot{Q}_{\text{in}} = \dot{m}(h_3 - h_2)$$

$$= (5.807\,\text{kg/s})(1277.79\,\text{kJ/kg} - 501.27\,\text{kJ/kg})$$

$$= (5.807\,\text{kg/s})(776.52\,\text{kJ/kg}) = 4509.3\,\text{kW}$$

Hence,

$$\eta = \frac{\dot{W}_{\text{cycle}}}{\dot{Q}_{\text{in}}} = \frac{1721.1\,\text{kW}}{4509.3\,\text{kW}} = 0.382 \quad \text{or} \quad 38.2\% \Longleftarrow$$

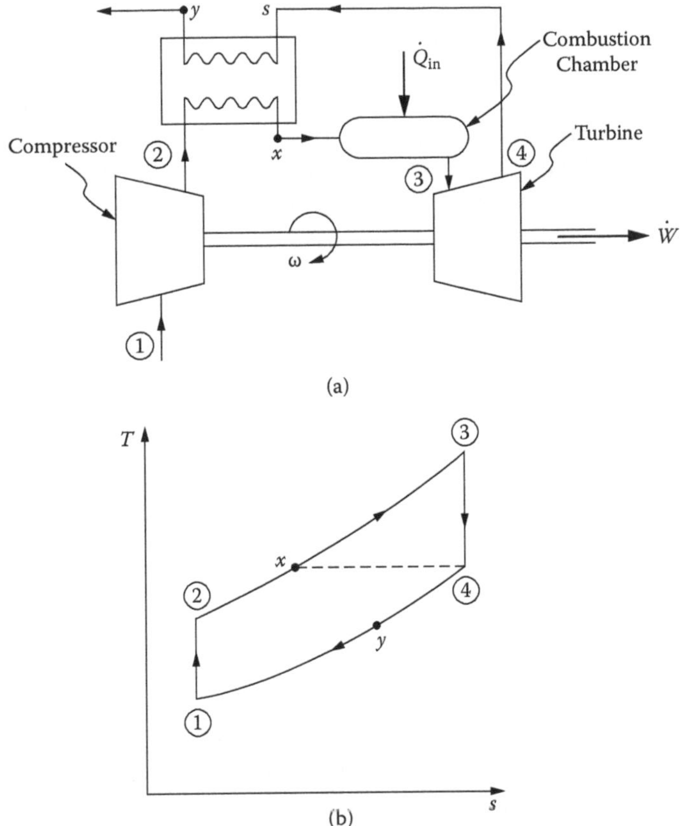

FIGURE 8.13
The ideal gas turbine or Brayton cycle with regeneration (a) the system and (b) the T-s diagram.

(d) The cold air standard efficiency is given by Equation 8.18. With $k = 1.40$

$$\eta = 1 - \frac{1}{r_p^{(k-1)/k}} = 1 - \frac{1}{6^{0.40/1.40}} = 1 - \frac{1}{1.669} = 1 - 0.599 = 0.401 \Longleftarrow$$

We note in this particular example, that the error involved in using the cold air standard efficiency is about 5%.

8.6.3 The Ideal Brayton Cycle with Regeneration

In the Brayton cycle, the products of combustion are exhausted from the turbine at a relatively high temperature. It is possible to reduce the consumption of fuel in the combustion chamber by preheating the air leaving the compressor with energy obtained from the turbine exhaust. This process takes place in a heat exchanger, known as a **recuperator**, and a system with regeneration is shown in Figure 8.13a.[3] We see that the working fluid enters the **cold side** of the recuperator at point 2 and passes through the recuperator to enter the

[3] The process of regeneration applies to the heating of the air leaving the compressor. This heating is done in a heat exchanger called a recuperator.

combustion chamber at point x. The spent air-fuel mixture, which still leaves the turbine at point 4, passes through the **hot side** of the recuperator, and leaves the recuperator at point y. The T-s diagram for this cycle is shown in Figure 8.13b.

The recuperator itself is an internally reversible heat exchanger where the fluid, in passing through it, experiences no pressure loss due to friction. We may note in Figure 8.13 that the maximum possible heat transfer in the recuperator would occur if gas entering the combustion chamber were at the exhaust temperature from the turbine:

$$\dot{Q}_{max} = \dot{m}(h_4 - h_2) \tag{8.20a}$$

In actual practice, however, because achievement of the turbine exhaust temperature would require a recuperator of infinite size, point x is at a temperature that is lower than the temperature at point 4. The actual heat transferred in the recuperator is therefore,

$$\dot{Q}_{act} = \dot{m}(h_x - h_2) \tag{8.20b}$$

Equations 8.20 can be employed to define a recuperator **effectiveness**

$$\epsilon = \frac{\dot{Q}_{act}}{\dot{Q}_{max}} = \frac{\dot{m}(h_x - h_2)}{\dot{m}(h_4 - h_2)} = \frac{h_x - h_2}{h_4 - h_2} \tag{8.21a}$$

and if the recuperator effectiveness is specified, we can determine the enthalpy at point x via

$$h_x = h_2 + \epsilon(h_4 - h_2) \tag{8.21b}$$

An effectiveness value of 1.00 (or 100%) means that the temperature of the air stream entering the combustion chamber is equal to the temperature of the combustion products leaving the turbine. In practice, however, recuperator effectiveness values range from 0.60 to 0.80 (60% to 80%) because, as already stated, higher effectiveness values are only attainable in rather impractical large sizes.

The net power developed by the cycle remains unchanged when the recuperator is employed and is still given by Equation 8.14c

$$\dot{W}_{cycle} = \dot{m}[(h_3 - h_4) - (h_2 - h_1)] \tag{8.14c}$$

However, when regeneration is employed, the heat input is

$$\dot{Q}_{in} = \dot{m}(h_3 - h_x) \tag{8.22}$$

and the thermal efficiency of the cycle is

$$\eta = \frac{\dot{W}_{cycle}}{\dot{Q}_{in}} = \frac{(h_3 - h_4) - (h_2 - h_1)}{h_3 - h_x} \tag{8.23}$$

It can be shown (Problem 8.39 asks for a derivation) that when the compressor and turbine inlet temperatures are fixed at T_1 and T_3, the net work is a maximum when

$$T_2 = (T_1 T_3)^{1/2} \tag{8.24}$$

and the pressure ratio for this condition is given by

$$\frac{P_2}{P_1} = \left(\frac{T_2}{T_1}\right)^{k/2(k-1)} \tag{8.25}$$

We now use Example 8.5 to show the improvement in thermal efficiency when a recuperator is employed.

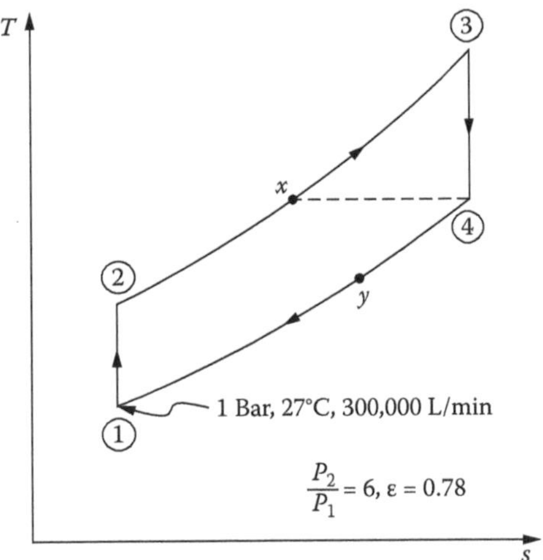

FIGURE 8.14
T-*s* diagram for Example 8.5.

Example 8.5 Once again, consider the ideal Brayton cycle of Example 8.4 and suppose that a recuperator with an effectiveness of $\epsilon = 0.78$ is placed in the system as shown in Figure 8.14. For an air standard analysis determine the thermal efficiency and compare it with the value obtained in Example 8.4.

Solution

The same assumptions made in Example 8.4 will be made here. The enthalpy values at the four points in the cycle that were found in Example 8.4 are summarized as

$$h_1 = 300.19\,\text{kJ/kg} \qquad h_2 = 501.27\,\text{kJ/kg}$$
$$h_3 = 1277.79\,\text{kJ/kg} \qquad h_4 = 780.32\,\text{kJ/kg}$$

Point x is at the exit of the cold air-side of the recuperator and its enthalpy can be found from Equation 8.21b

$$\begin{aligned} h_x &= h_2 + \epsilon(h_4 - h_2) \\ &= 501.27\,\text{kJ/kg} + (0.78)(780.32\,\text{kJ/kg} - 501.27\,\text{kJ/kg}) \\ &= 501.27\,\text{kJ/kg} + (0.78)(279.05\,\text{kJ/kg}) \\ &= 501.27\,\text{kJ/kg} + 217.66\,\text{kJ/kg} = 718.93\,\text{kJ/kg} \end{aligned}$$

The net power developed by the cycle is the same as the value determined in Example 8.4

$$\dot{W}_{\text{cycle}} = 1721.1\,\text{kW}$$

Here, however, the heat input will be

$$\begin{aligned} \dot{Q}_{\text{in}} &= \dot{m}(h_3 - h_x) \\ &= (5.807\,\text{kg/s})(1277.79\,\text{kJ/kg} - 718.93\,\text{kJ/kg}) \\ &= (5.807\,\text{kg/s})(558.86\,\text{kJ/kg}) = 3245.3\,\text{kW} \end{aligned}$$

and the thermal efficiency will be

$$\eta = \frac{\dot{W}_{cycle}}{\dot{Q}_{in}} = \frac{1721.1\,\text{kW}}{3245.31\,\text{kW}} = 0.530 \quad \text{or} \quad 53.0\% \Longleftarrow$$

The marked improvement in the thermal efficiency from 0.382 to 0.530 may be noted.

8.7 The Jet Engine

Because of its excellent power-to-weight characteristics, the gas turbine has been adapted for use in aircraft propulsion. As indicated in the cutaway view of the turbojet engine of Figure 8.15a, the power developed by the turbine is just sufficient to drive the compressor and other auxiliary components.

The components of a turbojet engine are indicated in Figure 8.15a and the T-s diagram for the ideal turbojet engine is shown in Figure 8.15b. The engine itself is composed of three main sections: the compressor, the combustion chamber or **combustor**, and the turbine. Air, treated as an ideal gas, is the working medium and enters the diffuser section (located just upstream of the compressor) at atmospheric conditions. The purpose of the diffuser is to

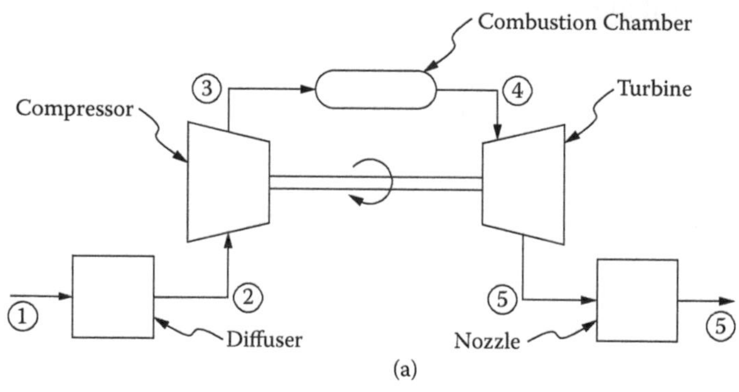

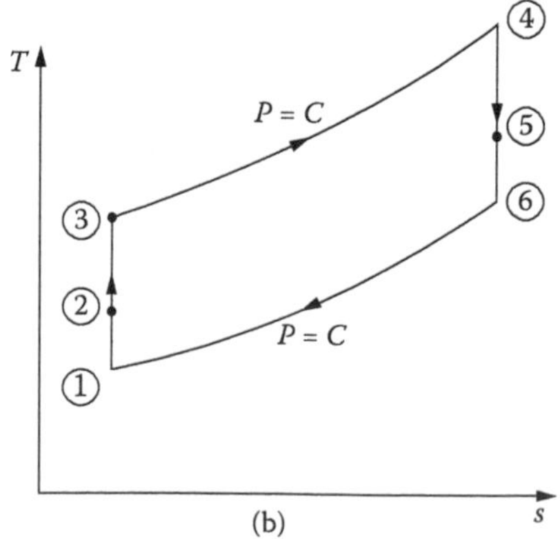

FIGURE 8.15
A turbojet engine (a) the cycle and (b) the T-s plane.

decelerate the air entering the engine and energy absorbed during this deceleration leads to a pressure rise, which is referred to as the **ram effect**.

The compressor, combustor, and turbine serve the same function as they do in the gas turbine. The air-fuel mixture leaving the turbine at a pressure that is significantly higher than atmospheric is led to the nozzle where an expansion to low pressure and high velocity takes place. It is this change in velocity between point 1 and point 6 (see Figure 8.15b) that gives rise to the **thrust** of the engine.

There are six points on the T-s diagram of Figure 8.18b that represent the ideal cycle and define the six processes that comprise the cycle. These processes are described by

- *Process 1-2:* This is an isentropic compression in the diffuser where the velocity of the incoming air is reduced.
- *Process 2-3:* This is also an isentropic compression. It occurs in the compressor where the pressure is increased through the pressure ratio,

$$r_p = \frac{P_3}{P_2}$$

- *Process 3-4:* This is a constant pressure heat addition in the combustor or burner section. Here fuel is injected and the products of combustion leaving at point 4 are at high pressure and temperature.
- *Process 4-5:* This is an isentropic expansion in the turbine where the pressure and temperature of the products of combustion are reduced. However, we have noted that the pressure at the exhaust from the turbine is significantly greater than atmospheric and this permits the nozzle to provide a high velocity at its exhaust.
- *Process 5-6:* This is an isentropic process in the nozzle where the products of combustion leave at high velocity and are discharged to the surroundings.
- *Process 6-1:* This is a constant pressure heat rejection that returns the cycle to its starting point.

We begin the analysis of the turbojet engine by noting that the thrust developed by the engine is

$$F_T = \dot{m}(\hat{V}_6 - \hat{V}_1) \qquad (\text{N}) \tag{8.26}$$

where $\dot{m}$ is the air flow rate. Here, the velocity leaving the nozzle may be developed from a first law energy balance between points 5 and 6 in Figure 8.15. In the absence of work done on or by the air, heat transferred, potential energy changes, and neglible velocity of the air at point 5, the energy rate balance will be

$$h_5 = h_6 + \frac{\hat{V}_6^2}{2}$$

so that

$$\hat{V}_6 = [2(h_5 - h_6)]^{1/2} \quad (\text{m/s}) \tag{8.27}$$

The **propulsive power** developed by the engine is the product of this thrust and the velocity of the aircraft

$$\dot{W}_P = F_T \hat{V}_{\text{ac}} = \dot{m}(\hat{V}_6 - \hat{V}_1)\hat{V}_{\text{ac}} \quad (\text{W}) \tag{8.28}$$

The heat supplied in the combustor is

$$\dot{Q}_{\text{in}} = \dot{m}(h_4 - h_3) \qquad (\text{W}) \tag{8.29}$$

and, because the net work developed by the engine is zero, the **propulsion efficiency** is defined as the ratio of the propulsive power to the energy released by the fuel

$$\eta_P = \frac{\dot{W}_P}{\dot{Q}_{in}} = \frac{(\hat{V}_6 - \hat{V}_1)\hat{V}_{ac}}{h_4 - h_3} \tag{8.30}$$

We now turn to a rather comprehensive example where we resort to the first law energy equation for a control volume to help us generate the solution.

Example 8.6 A jet aircraft is flying at a velocity of 1000 km/h at an altitude where the air temperature is −40°C and the pressure is 0.40 bar. The compressor has a pressure ratio of 10 and the temperature at the entrance to the turbine is 727°C. The inlet to the engine has an area of 0.579 m^3. Using an air standard analysis and an ideal Brayton cycle, determine (a) the temperature of the air leaving the diffuser, (b) the pressure of the air leaving the compressor, (c) the temperature of the air leaving the nozzle, (d) the velocity of the air leaving the nozzle, (e) the thrust produced by the engine, and (f) the propulsion efficiency of the engine.

Solution

Assumptions and Specifications

(1) The air is an ideal gas and forms a closed system.
(2) The cycle is an ideal cycle consisting of two isentropic processes and two constant pressure processes.
(3) Each component of the system can be analyzed as a control volume at steady state.
(4) Potential energy changes are negligible.
(5) The diffuser and the compressor undergo isentropic compressions and the turbine and the nozzle undergo isentropic expansions.
(6) All processes are quasi-static.
(7) The turbine produces just enough power to operate the compressor.
(8) An air standard analysis has been specified.

(a) For the diffuser, the inlet pressure is $P_1 = 0.40$ bar and the inlet temperature is 233 K (−40°C). Table A.12 (with interpolation) gives

$$p_{r1} = 0.5740 \quad \text{and} \quad h_1 = 233.02\,\text{kJ/kg}$$

The evaluation of the temperature leaving the diffuser (point 2 in Figure 8.16) requires consideration of the diffuser as a control volume. An energy rate balance with no heat or work interactions across the control volume boundaries and no change in either kinetic or potential energies and with negligible exit velocity produces

$$h_2 = h_1 + \frac{\hat{V}_1^2}{2}$$

and with

$$\hat{V}_1 = \frac{10^6\,\text{m/h}}{3600\,\text{s/h}} = 277.78\,\text{m/s}$$

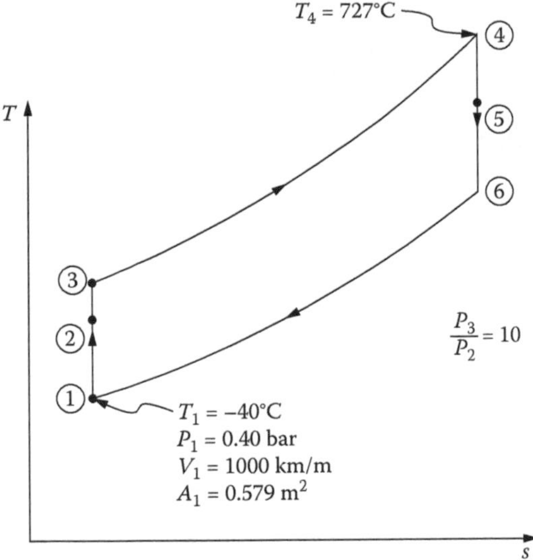

FIGURE 8.16
T-s diagram for Example 8.6.

the value of h_2 will be

$$h_2 = h_1 + \frac{\hat{V}_1^{\,2}}{2}$$

$$= 233.02\,\text{kJ/kg} + \frac{(277.78\,\text{m/s})^2}{2(1000\,\text{J/kJ})}(1\,\text{N-s}^2/\text{kg-m})$$

$$= 233.02\,\text{J/kg} + 38.58\,\text{kJ/kg} = 271.60\,\text{kJ/kg}$$

We can then interpolate Table A.12 at $h_2 = 271.60\,\text{kJ/kg}$ to obtain

$$T_2 = 271.5\,\text{K} \Longleftarrow \quad \text{and} \quad p_{r2} = 0.9783$$

and then

$$P_2 = P_1\left(\frac{p_{r2}}{p_{r1}}\right) = (0.40\,\text{bar})\left(\frac{0.9783}{0.5740}\right) = 0.6818\,\text{bar}$$

We are also able to obtain the mass flow of the air and select R from Table A.1 as 287 J/kg-K. Then using conditions at point 1, we have

$$\dot{m} = \rho_1 A_1 \hat{V}_1 = \left(\frac{P_1}{RT_1}\right) A_1 \hat{V}_1$$

Thus,

$$\dot{m} = \left[\frac{4 \times 10^4\,\text{N/m}^2}{(287\text{J/kg-K})(233\,\text{K})}\right](0.579\,\text{m}^2)(277.78\,\text{m/s}) = 96.21\,\text{kg/s}$$

(b) In seeking the compressor exit temperature at point 3 in Figure 8.16, we begin by noting that $P_3/P_2 = 10$ so that

$$P_3 = 10P_2 = 10(0.6818\,\text{bar}) = 6.818\,\text{bar} \Longleftarrow$$

and

$$p_{r3} = p_{r2}\left(\frac{P_3}{P_2}\right) = (0.9783)(10) = 9.783$$

The temperature and enthalpy at point 3 are obtained from Table A.12 with interpolation at $p_{r3} = 9.783$

$$T_3 = 521.4\,\text{K} \quad \text{and} \quad h_3 = 525.12\,\text{K}$$

(c) At point 4 (the inlet to the turbine), the temperature is specified as 1000 K (727°C) and from Table A.12

$$p_{r4} = 114.0 \quad \text{and} \quad h_4 = 1046.04\,\text{kJ/kg}$$

and because the power developed by the turbine is just sufficient to run the compressor, we have

$$\dot{W}_{\text{turb}} = \dot{W}_{\text{comp}}$$

$$\dot{m}(h_4 - h_5) = \dot{m}(h_3 - h_2)$$

Solving for h_5

$$\begin{aligned} h_5 &= h_4 - (h_3 - h_2) \\ &= 1046.04\,\text{kJ/kg} - (525.12\,\text{kJ/kg} - 271.60\,\text{kJ/kg}) \\ &= 1046.04\,\text{kJ/kg} - 253.52\,\text{kJ/kg} = 792.51\,\text{kJ/kg} \end{aligned}$$

Interpolation in Table A.12 then yields the values of T_5 and p_{r5}

$$T_5 = 773.1\,\text{kJ/kg} \Longleftarrow \quad \text{and} \quad p_{r5} = 41.95$$

and because $P_3 = P_4 = 6.818\,\text{bar}$

$$P_5 = P_4\left(\frac{p_{r5}}{p_{r4}}\right) = (6.818\ \text{bar})\left(\frac{41.95}{114.0}\right) = 2.509\,\text{bar}$$

The expansion through the nozzle is isentropic to atmospheric pressure. Thus, at point 6

$$p_{r6} = p_{r5}\left(\frac{P_6}{P_5}\right) = (41.95)\left(\frac{0.40\,\text{bar}}{2.509\,\text{bar}}\right) = 6.687$$

and this time, Table A.12 gives, with interpolation

$$T_6 = 468.9\,\text{K} \Longleftarrow \quad \text{and} \quad h_6 = 471.11\,\text{kJ/kg}$$

(d) With just the nozzle as a control volume, the energy rate balance between points 5 and 6 in Figure 8.16 can be written in the absence of heat and work interactions across the boundaries of the control volume, the absence of potential energy change and negligible inlet velocity

$$\dot{m}\left(h_6 + \frac{\hat{V}_6^2}{2}\right) = \dot{m}h_5$$

and the exit velocity is seen to be

$$\hat{V}_6 = [2(h_5 - h_6)]^{1/2}$$

and with

$$h_5 - h_6 = 792.51\,\text{kJ/kg} - 471.11\,\text{kJ/kg} = 321.40\,\text{kJ/kg}$$

then

$$\begin{aligned}\hat{V}_6 &= [2(321,400\,\text{N-m/kg})(1\,\text{kg-m/N-s}^2)]^{1/2}\\ &= [642.80\,\text{m}^2/\text{s}^2]^{1/2}\\ &= 801.75\,\text{m/s} \Longleftarrow\end{aligned}$$

(e) The thrust is given by Equation 8.26

$$\begin{aligned}F_T &= \dot{m}(\hat{V}_6 - \hat{V}_1)\\ &= (96.21\,\text{kg/s})(801.75\,\text{m/s} - 277.78\,\text{m/s})\\ &= (96.21\,\text{kg/s})(523.97\,\text{m/s})\\ &= 50{,}410\,\text{N} \Longleftarrow\end{aligned}$$

(f) The propulsive power is given by Equation 8.28

$$\begin{aligned}\dot{W}_P &= \dot{m}(\hat{V}_6 - \hat{V}_1)\hat{V}_{\text{ac}} = F_T\hat{V}_{\text{ac}}\\ &= (50,410\,\text{N})(277.78\,\text{m/s})\\ &= 14.00\,\text{MW}\end{aligned}$$

and the heat added in the combustor is given by Equation 8.29

$$\begin{aligned}\dot{Q}_{\text{in}} &= \dot{m}(h_4 - h_3)\\ &= (96.21\,\text{kg/s})(1046.04\,\text{kJ/kg} - 525.12\,\text{kJ/kg})\\ &= (96.21\,\text{kg/s})(520.92\,\text{kJ/kg}) = 5.012 \times 10^4\,\text{kW}\end{aligned}$$

Then, the propulsion efficiency, given by Equation 8.29, will be

$$\eta_P = \frac{\dot{W}_{\text{cycle}}}{\dot{Q}_{\text{in}}} = \frac{1.400 \times 10^7\,\text{W}}{5.011 \times 10^7\,\text{W}} = 0.279 \quad \text{or} \quad 27.9\% \Longleftarrow$$

8.8 Summary

Air standard analysis of gas power systems involves air as the working medium where both the ideal gas law and variable specific heats are employed. Internal energy and enthalpy values are obtained from the "air tables" (given here as Table A.12) and are based on the use of the reference volume and reference pressure

$$\frac{v_{ri}}{v_{rj}} = \frac{V_i}{V_j} \quad \text{and} \quad \frac{p_{ri}}{p_{rj}} = \frac{P_i}{P_j}$$

Cold air standard analysis of gas power systems involves air as the working medium where the ideal gas law and constant specific heats are employed. In this case, the specific heats are taken from Table A.1.

In the ideal air standard Otto or gasoline engine cycle, there are four processes that are shown in Figure 8.2.

The compression ratio is

$$r = \frac{V_1}{V_2} = \frac{V_4}{V_3}$$

the heat added is

$$Q_{in} = m(u_3 - u_2) \tag{8.3a}$$

the heat rejected is

$$Q_{out} = m(u_4 - u_1) \tag{8.3b}$$

the net work of the cycle is

$$W_{cycle} = m[(u_3 - u_4) - (u_2 - u_1)] \tag{8.4a}$$

and the efficiency is

$$\eta = 1 - \frac{u_4 - u_1}{u_3 - u_2} \tag{8.5}$$

The cold air standard efficiency is

$$\eta = 1 - \frac{1}{r^{k-1}} \tag{8.9}$$

In the ideal air standard diesel engine cycle, there are four processes as shown in Figure 8.7.

The heat input is given by

$$Q_{in} = Q_{23} = mc_p(T_3 - T_2) = m(h_3 - h_2) \tag{8.11a}$$

and the heat rejected is

$$Q_{out} = Q_{41} = mc_v(T_4 - T_1) = m(u_4 - u_1) \tag{8.11b}$$

The work done is

$$W_{cycle} = Q_{in} - Q_{out} = m(h_3 - h_2) - m(u_4 - u_1) \tag{8.11c}$$

and the thermal efficiency will be

$$\eta = 1 - \frac{u_4 - u_1}{h_3 - h_2} \tag{8.12}$$

On the cold air standard basis, the efficiency will be

$$\eta = 1 - \left[\frac{1}{r^{k-1}} \frac{r_c^k - 1}{k(r_c - 1)}\right] \tag{8.13}$$

In the ideal air standard Brayton or gas turbine cycle, there are four processes as shown in Figure 8.10a.

The heat added in the combustion chamber is

$$\dot{Q}_{in} = \dot{m}(h_3 - h_2) \tag{8.15}$$

The power of the cycle is

$$\dot{W}_{\text{cycle}} = \dot{m}[(h_3 - h_4) - (h_2 - h_1)] \tag{8.14c}$$

and the thermal efficiency is

$$\eta = 1 - \frac{h_4 - h_1}{h_3 - h_2} \tag{8.16}$$

The back work ratio is

$$\text{bwr} = \frac{h_2 - h_1}{h_3 - h_4} \tag{8.17}$$

The cold air standard efficiency is

$$\eta = 1 - \frac{1}{r_p^{(k-1)/k}} \tag{8.18}$$

where r_p is the pressure ratio

$$r_p = \frac{P_2}{P_1} = \frac{P_3}{P_4} \tag{8.19}$$

In the ideal air standard Brayton with regeneration, the cycle contains a recuperator as shown in Figure 8.13, when the recuperator effectiveness is specified, the enthalpy at point x can be determined from

$$h_x = h_2 + \epsilon(h_4 - h_2) \tag{8.21b}$$

The net power developed by the cycle is given by

$$\dot{W}_{\text{cycle}} = \dot{m}[(h_3 - h_4) - (h_2 - h_1)] \tag{8.14c}$$

the heat input is

$$\dot{Q}_{\text{in}} = \dot{m}(h_3 - h_x) \tag{8.22}$$

and the thermal efficiency of the cycle is

$$\eta = \frac{(h_3 - h_4) - (h_2 - h_1)}{h_3 - h_x} \tag{8.23}$$

In the ideal air standard jet engine cycle shown, there are six processes as shown in Figure 8.15.

The thrust developed by the engine is

$$F_T = \dot{m}(\hat{V}_6 - \hat{V}_1) \qquad \text{(N)} \tag{8.26}$$

The velocity at the exit from the nozzle will be

$$\hat{V}_6 = [2(h_5 - h_6)]^{1/2} \tag{8.27}$$

The **propulsive power** developed by the engine is

$$\dot{W}_P = F_T \hat{V}_{ac} = \dot{m}(\hat{V}_6 - \hat{V}_1)\hat{V}_{ac} \tag{8.28}$$

The heat supplied in the combustor will be

$$Q_{in} = \dot{m}(h_4 - h_3) \tag{8.29}$$

and the propulsion efficiency is

$$\eta_P = \frac{(\hat{V}_6 - \hat{V}_1)\hat{V}_{ac}}{h_4 - h_3} \tag{8.30}$$

8.9 Problems

Use of the Air Tables

8.1: Air at 37°C and 120 kPa is compressed isentropically to a pressure of 1.50 MPa. Determine the final temperature and enthalpy for (a) an air standard process and (b) a cold air standard process.

8.2: Air at 42°C and 125 kPa is compressed isentropically through $V_1/V_2 = 8$. Determine the final temperature and pressure for (a) an air standard process and (b) a cold air standard process.

8.3: In a constant volume process, 600 kJ of heat is added to air at 700 K and 16 bar. Determine the final temperature and pressure for (a) an air standard process and (b) a cold air standard process.

8.4: In a constant pressure process, 650 kJ of heat is added to air at 600 K and a volume of 0.40 m^3. Determine the final temperature and volume for (a) an air standard process and (b) a cold air standard process.

8.5: An air standard cycle (specific heats are variable) is composed of four processes. Process 1-2 is an isentropic compression from $P_1 = 100$ kPa and $T_1 = 300$ K to $P_2 = 850$ kPa, process 2-3 is a constant volume heat addition to $T_3 = 1750$ K, process 3-4 is an isentropic expansion to 100 kPa, and process 4-1 is a constant pressure heat rejection. Determine (a) the heat added, (b) the heat rejected, (c) the net work, and (d) the thermal efficiency.

8.6: Rework Problem 8.5 using the cold air standard (constant specific heats taken at 300 K).

8.7: An air standard cycle (specific heats are variable) is composed of three processes. Process 1-2 is a constant volume heat addition to $T_1 = 22$°C and 98 kPa to 392 kPa, process 2-3 is an isentropic expansion to 98 kPa and process 3-1 is a constant pressure heat rejection to the initial state. Assume that 0.005 kg of air is present and determine (a) the heat added, (b) the heat rejected, (c) the net work, and (d) the thermal efficiency.

8.8: Rework Problem 8.7 using a cold air standard cycle (constant specific heats taken at 300 K).

The Ideal Otto Cycle

8.9: An air standard Otto cycle has a compression ratio of 8.75 with inlet conditions of 100 kPa and 27°C and the heat added is 1325 kJ/kg. For a mass of 1 kg, determine (a) the heat rejection, (b) the net work, (c) the thermal efficiency, and (d) the mean effective pressure.

8.10: Rework Problem 8.9 using a cold air standard cycle (constant specific heats taken at 300 K).

8.11: Intake conditions of an air standard Otto cycle are $P_1 = 100\,\text{kPa}$, $V_1 = 420\,\text{cm}^3$, and $T_1 = 22°\text{C}$. If the maximum temperature in the cycle is to be limited to 2250 K and the compression ratio is 8, determine (a) the mass of air present, (b) the heat added, (c) the heat rejected, (d) the net work, (e) the thermal efficiency, and (f) the mean effective pressure.

8.12: Rework Problem 8.11 using a cold air standard cycle (constant specific heats taken at 300 K) to find the temperature at each point in the cycle and the thermal efficiency.

8.13: An air standard Otto cycle has a temperature of 500 K and a pressure of 600 kPa at the end of the isentropic expansion. For a compression ratio of 7.5 and 160 kJ/kg of heat rejected from the cycle, determine (a) the heat input, (b) the net work done, and (c) the thermal efficiency.

8.14: Rework Problem 8.13 using a cold air standard cycle (constant specific heats taken at 300 K) to find the temperature at each point in the cycle and the thermal efficiency.

8.15: An ideal Otto cycle has a compression ratio of 8 and the inlet conditions are $T_1 = 32°\text{C}$ and $P_1 = 98\,\text{kPa}$. The cycle heat addition is 725 kJ/kg. Use an air standard analysis to determine (a) the maximum pressure in the cycle, (b) the heat rejected, (c) the net work, (d) the thermal efficiency, and (e) the mean effective pressure.

8.16: Rework Problem 8.15 using a cold air standard analysis (constant specific heats taken at 300 K).

8.17: In an Otto cycle where the temperature at the end of the isentropic expansion is 820 K, the compression ratio is 8 and the inlet conditions are $P_1 = 100\,\text{kPa}$, $V_1 = 625\,\text{cm}^3$, and $T_1 = 27°\text{C}$. Use an air standard analysis to determine (a) the highest temperature and pressure in the cycle, (b) the heat added, (c) the heat rejected, (d) the net work, (e) the thermal efficiency, and (f) the mean effective pressure.

8.18: Use the conditions of Problem 8.17 and use a cold air standard analysis (constant specific heats taken at 300 K) to determine the maximum temperature and pressure in the cycle.

8.19: Consider an ideal Otto cycle operating on the air standard with a compression ratio of 8.6. The minimum and maximum temperatures that occur in the cycle are 37°C and 827°C. The pressure at the inlet is 100 kPa and 0.640 kJ are added to the charge during each cycle. Determine (a) the heat rejected, (b) the net work done, and (c) the thermal efficiency.

8.20: Rework Problem 8.19 using a cold air standard cycle (constant specific heats taken at 300 K).

8.21: In an air standard Otto cycle with a compression ratio of 8.75, the pressure during the constant volume heat addition is tripled. If the inlet conditions are $P_1 = 100\,\text{kPa}$ and $T_1 = 27°\text{C}$, determine (a) the temperature at points 1 and 4 shown in Figure 8.2 and (b) the thermal efficiency.

8.22: Rework Problem 8.21 using a cold air standard cycle (constant specific heats taken at 300 K).

The Ideal Diesel Cycle

8.23: The pressure and temperature at the beginning of the isentropic compression in an air standard Diesel cycle are 96 kPa and 17°C. At the end of compression, the pressure is 6.45 MPa and the temperature is 2000 K. Determine (a) the compression ratio, (b) the cutoff ratio, (c) the thermal efficiency, and (d) the mean effective pressure.

8.24: Rework Problem 8.23 using a cold air standard cycle (constant specific heats taken at 300 K).

8.25: An air standard Diesel cycle has a compression ratio of 16. Determine the thermal efficiencies for cutoff ratios of 1 (Otto cycle), 2, 3, 4, and 5.

8.26: The pressure and temperature at the beginning of the isentropic compression in an air standard Diesel cycle are 185 kPa and 102°C. The compression ratio is 20 and the heat addition is 908 kJ/kg. Determine (a) the maximum temperature and pressure in the cycle, (b) the cutoff ratio, (c) the net work in kJ/kg, and (d) the thermal efficiency.

8.27: Rework Problem 8.26 using a cold air standard cycle (constant specific heats taken at 300 K).

8.28: An air standard Diesel cycle has a compression ratio of 16 and a cutoff ratio of 2. Conditions at the beginning of the isentropic compression are $P_1 = 96\,\text{kPa}$, $T_1 = 287\,\text{K}$, and $V_1 = 0.045\,\text{m}^3$. Determine (a) the heat added in kJ, (b) the heat rejected in kJ, (c) the thermal efficiency, and (d) the mean effective pressure.

8.29: Rework Problem 8.28 using a cold air standard cycle (constant specific heats taken at 300 K).

8.30: A Diesel engine with a compression ratio of 18 and a cutoff ratio of 2.5 draws air at 100 kPa and 300 K. Using a cold air standard analysis, determine (a) the pressures and temperatures at states 2, 3, and 4 in Figure 8.7, (b) the thermal efficiency, and (c) the mean effective pressure.

8.31: Rework Problem 8.30 using an air standard analysis.

8.32: A Diesel engine is to be designed to produce a cold air standard efficiency of 55% when operating with a compression ratio of 15. Determine the cutoff ratio for the engine.

8.33: Use L'Hospital's rule to show that the efficiency of a cold air standard Diesel cycle as given by Equation 8.13 reduces to that of the Otto cycle in the limit when $r_c \longrightarrow 1$.

8.34: In an ideal air standard Diesel cycle, the compression ratio is 15 and the heat addition per cycle is 10.8 kJ. At the beginning of the isentropic compression, $P_1 = 97.5\,\text{kPa}$, $T_1 = 20°\text{C}$, and $V_1 = 0.015\,\text{m}^3$. Determine (a) the mass of air, (b) the maximum temperature in the cycle, (c) the net work, and (d) the thermal efficiency.

8.35: Rework Problem 8.34 using a cold air standard cycle.

8.36: A cold air standard Diesel cycle with a compression ratio of 15 has a maximum temperature of 1600 K and a minimum temperature of 300 K. The thermal efficiency of the cycle is 80% of that of a Carnot engine operating between the same temperature limits. Determine the cutoff ratio.

The Ideal Gas Turbine Cycle without Regeneration

8.37: A gas turbine operating on a Brayton cycle takes in air at 290 K and operates with a pressure ratio of 8. The mass flow through the turbine is 35 kg/s and the turbine inlet temperature is 1250 K. Assume operation on the air standard and determine (a) the power required to drive the compressor, (b) the back work ratio, and (c) the thermal efficiency.

8.38: Rework Problem 8.37 using the cold air standard.

8.39: Show that the net work is at a maximum in the Brayton cycle with fixed compressor and turbine inlet temperatures (T_1 and T_3) when

$$T_2 = (T_1 T_3)^{1/2}$$

and the pressure ratio for this condition is given by

$$\frac{P_2}{P_1} = \left(\frac{T_3}{T_1}\right)^{k/2(k-1)}$$

8.40: The minimum and maximum temperatures in a Brayton cycle are 300 K and 1400 K, respectively. Using the cold air standard, determine (a) the pressure ratio that produces the maximum net work, (b) the compressor work, (c) the turbine work, (d) the heat supplied in the combustion chamber, and (e) the thermal efficiency.

8.41: A gas turbine operates on a Brayton cycle with a pressure ratio of 7 for both the compressor and the turbine. The compressor inlet conditions are 100 kPa and 300 K and the turbine inlet temperature is 1500 K. The isentropic efficiencies of the compressor and the turbine are 0.85 and 0.95, respectively. Using the cold air standard and a mass flow of 25 kg/s, determine (a) the compressor power input, (b) the rate of heat supply in the combustion chamber, (c) the turbine output power, and (d) the thermal efficiency.

8.42: Rework Problem 8.41 using the air standard.

8.43: The compressor pressure ratio of an ideal air standard Brayton cycle is 10. If air enters the compressor at 101.3 kPa and 295 K with a mass flow rate of 11.50 kg/s, determine (a) the net power developed and (b) the thermal efficiency.

8.44: Rework Problem 8.43 using the cold air standard.

8.45: The compressor pressure ratio of an ideal air standard Brayton cycle is 12.5 and the net power developed is 7350 kW. If the maximum and minimum temperatures in the cycle are 1560 K and 290 K, respectively, determine (a) the mass flow rate of the air and (b) the thermal efficiency.

8.46: Rework Problem 8.45 using the cold air standard.

8.47: In an ideal air standard Brayton cycle, air enters the compressor at 101.3 kPa and 25°C with a volumetric flow rate of 4.8 m^3/s. The turbine inlet is at 1425 K and the compressor pressure ratio is 7.5. Determine (a) the thermal efficiency, (b) the back work ratio, and (c) the net power developed in kW.

8.48: Rework Problem 8.47 with a compressor pressure ratio of 10.

8.49: Rework Problem 8.47 with a compressor pressure ratio of 15.

8.50: Rework Problem 8.47 with a turbine inlet temperature of 1550 K.

8.51: Rework Problem 8.47 with a turbine inlet temperature of 1600 K.

8.52: Rework Problem 8.43 with compressor and turbine isentropic efficiencies of 0.85.

The Ideal Gas Turbine Cycle with Regeneration

8.53: Suppose that a recuperator with an effectiveness of 78% is incorporated into the Brayton cycle of Problem 8.37. Determine (a) the net power developed in kW, (b) the back work ratio, and (c) the thermal efficiency.

8.54: Rework Problem 8.55 with a recuperator effectiveness of 0.84.

8.55: Rework Problem 8.54 with a recuperator effectiveness of 0.90.

8.56: Show that for an ideal Brayton cycle with a one hundred percent effectiveness recuperator, the thermal efficiency is given by

$$\eta = 1 - \frac{T_1}{T_3}(r_p)^{(k-1)/k}$$

where r_p is the pressure ratio, T_1 is the compressor inlet temperature, and T_3 is the turbine inlet temperature.

8.57: Data for a gas turbine engine operating on the Brayton cycle equipped with a recuperator is

Pressure and temperature at the compressor inlet: 100 kPa and 298 K

Pressure ratio: $r_p = 10$

Mass flow rate: 10 kg/s

Turbine inlet temperature: 1600 K

Recuperator effectiveness: $\epsilon = 0.70$

Using the cold air standard, determine (a) the rate of heat supply in the combustion chamber and (b) the thermal efficiency of the engine.

8.58: Rework Problem 8.53 with an isentropic compressor efficiency of 0.90 with turbine operation remaining isentropic.

8.59: Rework Problem 8.53 with an isentropic turbine efficiency of 0.90 with compressor operation remaining isentropic.

8.60: Rework Problem 8.53 with isentropic compressor and turbine efficiencies of 0.90.

8.61: A gas turbine with a recuperator operates in an air standard Brayton cycle. Air enters the compressor at 96.5 bar and 300 K and is compressed to 482.4 bar. After the air passes through the regenerator its temperature is 600 K. Both the compressor and turbine each have isentropic efficiencies of 0.875 and the air enters the turbine at a mass flow rate of 35 kg/s. Determine (a) the thermal efficiency of the cycle, (b) the recuperator effectiveness, and (c) the volumetric flow rate of the air entering the compressor in m^3/s.

8.62: Rework Problem 8.61 for an air temperature at the outlet of the regenerator of 640 K.

The Jet Engine

8.63: A turbojet flying at 330 m/s and 238 K draws air at 40 kPa into its diffuser section. The air leaving the diffuser section enters a compressor that provides a pressure ratio of 6.5. The temperature of the air entering the turbine is 1200 K and the mass flow rate through the engine is 70 kg/s. Assuming an ideal jet engine cycle as described in Section 8.7 and using the cold air standard, determine (a) the temperature and pressure of the air leaving the diffuser, (b) the temperature and pressure of the air at the compressor outlet, (c) the rate of heat supply in the combustion chamber, (d) the temperature and velocity of the air at the nozzle exit, (e) the thrust developed by the engine, and (f) the propulsion efficiency of the engine.

8.64: A turbojet flying at 350 m/s draws air at 50 kPa and 237 K into its diffuser section. The air leaves the diffuser with negligible velocity and enters a compressor that provides a pressure ratio of 8.75. The temperature of the air entering the turbine is 1200 K and the mass flow rate through the engine is 72 kg/s. Assuming an ideal jet engine as described in Section 8.7 and using the cold air standard, determine (a) the temperature and pressure of the air leaving the diffuser, (b) the temperature and pressure of the air at the compressor inlet, and (c) the rate of heat supply in the combustion chamber.

8.65: Consider the ideal jet engine without the diffuser (that is, with no ram effect). The compressor draws 50 kg/s of air at 75 kPa and 280 K and compresses it to 525 kPa. The heat suppled to the combustion chamber, which burns a fuel having a heating value of 41,800 kJ/kg, raises the temperature of the air to 1500 K. Assume the cold air standard and determine (a) the temperature and pressure of the air at the exit from the turbine, (b) the mass flow rate of the fuel required, and (c) the velocity at exit from the nozzle if the exit pressure is 70 kPa.

8.66: Rework Problem 8.65 using the air standard.

8.67: Air at 20 kPa and −23°C enters a turbojet engine with a velocity of 275 m/s. The pressure ratio across the compressor is 12.5, the turbine inlet temperature is 1200 K, and the pressure at the exit from the nozzle is 20 kPa. Assuming an ideal jet engine and using an air standard cycle with a nominal flow rate of 1 kg/s, determine (a) the temperature and pressure at each point in the cycle, (b) the exit velocity from the nozzle, and (c) the thrust produced.

9

Vapor Power and Refrigeration Cycles

Chapter Objectives

- To describe the operation of the steam power plant and its component parts.
- To introduce the ideal Rankine cycle.
- To show how the performance of the Rankine cycle can be enhanced through the use of superheat, reheat, and regeneration.
- To examine the effects of irreversibilities.
- To describe the operation of the ideal vapor compression refrigeration cycle and its many component parts.
- To indicate the operational differences between a refrigerator and a heat pump.
- To define what is meant by a ton of refrigeration and to present the basis for refrigerant selection.

9.1 Introduction

In this chapter, we devote our attention to vapor power and refrigeration cycles. The power cycle that we consider is the Rankine or steam power plant cycle, which accounts for most of the electric power generation worldwide, far outweighing the uses of hydroelectric, nuclear, and solar power. In the Rankine cycle, the working medium is usually water, which is alternately evaporated and condensed. This is in contrast to the gas power systems discussed in Chapter 8 where the working medium is a gas that never changes phase.

We also consider the vapor compression refrigeration cycle that may be found in many commercial, industrial, and household refrigeration and air-conditioning units. In the vapor compression refrigeration cycle, the working medium is the refrigerant that is alternately evaporated and condensed.

We first turn to a description of the steam power plant.

9.2 The Steam Power Plant

Figure 9.1 shows the principal components of a steam power plant. The prime mover is the **turbine**, which takes high-pressure and high-temperature steam at the outlet of the **steam generator** (point 3) and discharges steam to the condenser at low pressure (point 4). The

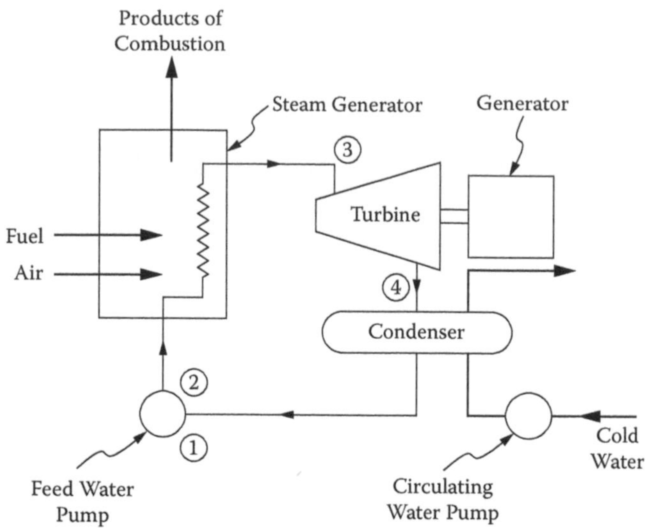

FIGURE 9.1
The basic components of a steam power plant.

work done by the turbine which, as we know, is a rotating device consisting of many stages, is delivered by the turbine shaft and is used to drive the **generator** that provides the electric power. The condenser pressure is made as low as possible because the thermal efficiency of the entire power plant is inversely proportional to the condenser pressure.

The **feedwater pump**, operating between points 1 and 2, is used to raise the pressure of the water in the **sump** of the condenser to the steam generator pressure. The water leaving the pump is in the subcooled or compressed liquid region.

A detail of a modern central station steam generator is displayed in Figure 9.2. In a nonnuclear power plant, the steam generator may be **fired** by coal (in most cases **pulverized**), fuel oil, or natural gas. The steam generator consists of a **furnace** where the steam is produced by radiation from the high-temperature products of combustion in what is frequently referred to as the **waterwall**. Essentially, the waterwall acts as a **boiler** although in some steam generators, there is a **boiler bank** consisting of tubes that are exposed to the products of combustion and operate in the convection mode. Often, there is an **economizer** consisting of a bank of finned tubes whose purpose is to extract some of the energy contained in the **flue gases** to heat the water entering the steam generator.

The outlet of the steam generator provides steam to the turbine at point 3 in the cycle (Figure 9.1). The steam may be saturated or contain liquid water. However, because turbines using saturated or wet steam have operational problems, such as turbine blade erosion, in most cases, the steam generator contains a **superheater** to put the steam at point 3 in the superheated region. In many cases, there are two **stages** in the turbine and after the steam expands through the first stage, it is led back to the steam generator so that it can be heated once again in the **reheater**.

The condenser receives the steam at the exhaust of the turbine at point 4 in Figure 9.1 and discharges the heat contained to the **ultimate sink** of circulating cooling water taken from a river or a lake. Sometimes, because of municipal restrictions, air is employed as the ultimate sink in which case the device operating between point 4 and point 1 is called a **cooling tower**. The pressure on the steam side of the condenser is set by the temperature of the cooling water and this temperature is consistent with the temperature difference maintained between the steam and the cooling water.

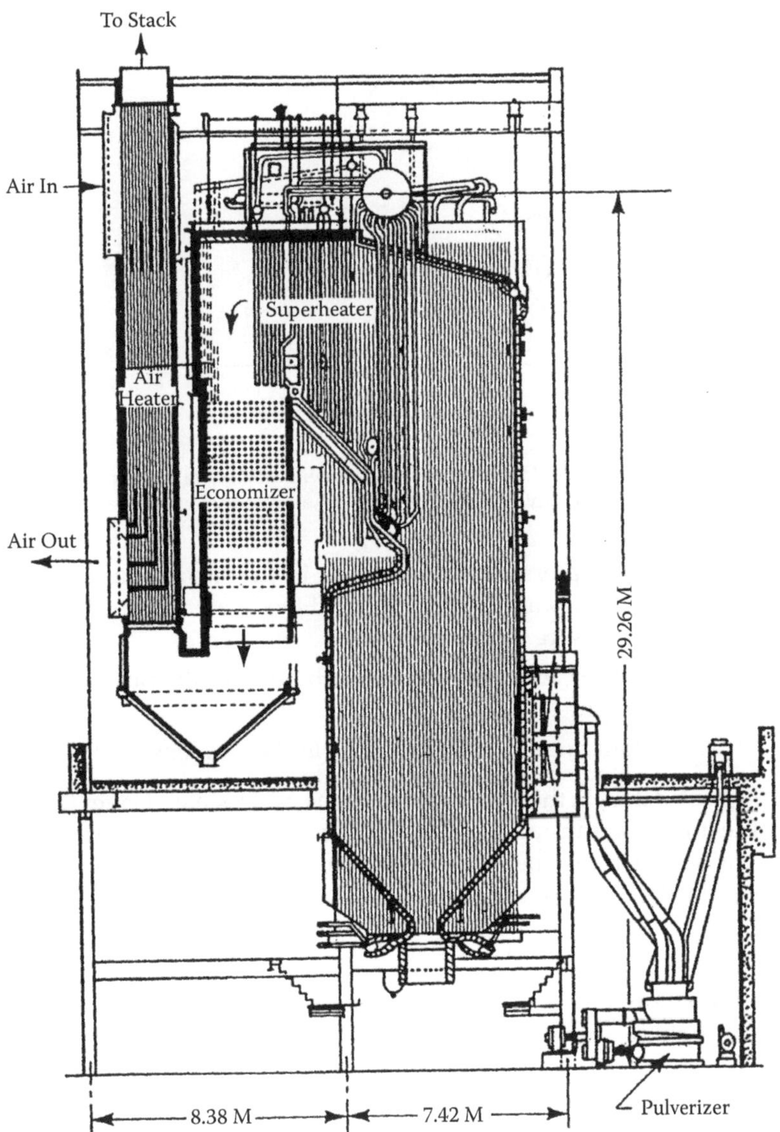

FIGURE 9.2
A typical central station steam generator.

9.3 The Ideal Rankine Cycle

9.3.1 The Ideal Rankine Cycle

The T-s diagram for the ideal Rankine cycle is indicated in Figure 9.3 and we observe that the cycle is composed of four ideal processes:

- *Process 1-2:* An isentropic compression in the pump from the condenser pressure to the boiler pressure.

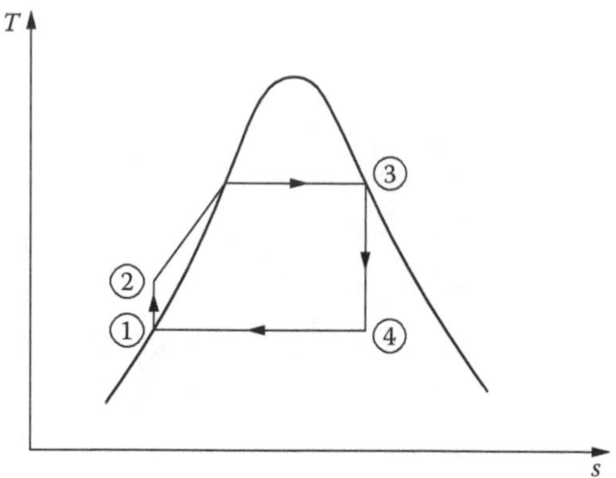

FIGURE 9.3
The T-s diagram for the Rankine cycle. In some cases, a steam engine may take the place of the steam turbine between points 3 and 4 because of the adverse effect of moist or saturated steam on the turbine blades.

- *Process 2-3:* A constant pressure heat addition in the steam generator (and its associated components such as economizers and feedwater heaters) until saturated steam at the boiler pressure is obtained.
- *Process 3-4:* An isentropic expansion in the turbine from the pump outlet pressure to the condenser pressure.
- *Process 4-1:* A constant pressure heat rejection in the condenser.

We note in Figure 9.3 that, in the basic ideal Rankine cycle, steam leaves the steam generator and enters the turbine as saturated steam. If all work and heat interactions are considered as positive quantities, then the energy rate balance for a control volume representing just the turbine operating between state points 3 and 4 will be

$$\dot{W}_{\text{turb}} = \dot{m}(h_3 - h_4) \tag{9.1}$$

where $\dot{m}$ is the mass flow rate. Equation 9.1 presumes that there is no heat transfer to the surroundings through the **casing** of the turbine and that there are no kinetic and potential energy changes.

With the pump taken as a control volume and with operation between state points 1 and 2 in Figure 9.3, a similar analysis, again with no heat transfer and negligible kinetic and potential energy changes, will yield

$$\dot{W}_{\text{pump}} = \dot{m}(h_2 - h_1) \tag{9.2}$$

The net work in the cycle is therefore

$$\dot{W}_{\text{cycle}} = \dot{W}_{\text{turb}} - \dot{W}_{\text{pump}} = \dot{m}[(h_3 - h_4) - (h_2 - h_1)] \tag{9.3}$$

Next, with the steam generator taken as a control volume, we note that there is no work interaction between it and its surroundings and with negligible kinetic and potential energy changes, an energy rate balance yields

$$\dot{Q}_{\text{in}} = \dot{m}(h_3 - h_2) \tag{9.4}$$

The condenser has a condensing side and a cooling water side and we may apply a control volume to just the condensing side. Again, with no work interaction and with negligible kinetic and potential energy changes, the heat rejected by the condenser will be

$$\dot{Q}_{out} = \dot{m}(h_4 - h_1) \tag{9.5}$$

and because the net work of the cycle is equal to the difference between the heat added and the heat rejected

$$\dot{W}_{cycle} = \dot{Q}_{in} - \dot{Q}_{out} = \dot{m}[(h_3 - h_2) - (h_4 - h_1)]$$

or, after a slight modification

$$\dot{W}_{cycle} = \dot{m}[(h_3 - h_4) - (h_2 - h_1)]$$

which confirms Equation 9.3.

The thermal efficiency of the ideal Rankine cycle is

$$\eta = \frac{\dot{W}_{cycle}}{\dot{Q}_{in}} = \frac{\dot{m}[(h_3 - h_4) - (h_2 - h_1)]}{\dot{m}(h_3 - h_2)}$$

or

$$\eta = 1 - \frac{h_4 - h_1}{h_3 - h_2} \tag{9.6}$$

As in the case of the gas turbine, a useful parameter is the back work ratio, defined as the ratio of the power required to drive the pump to the power developed by the turbine:

$$\text{bwr} = \frac{\dot{W}_{pump}}{\dot{W}_{turb}} = \frac{h_2 - h_1}{h_3 - h_4} \tag{9.7}$$

We now turn to an example that is intended to illustrate the foregoing principles. The example involves saturated steam at the outlet of the steam generator and, because saturated steam at the inlet to a power plant turbine leads to operational difficulties, the example employs a reciprocating steam engine to drive the generator. The use of the steam engine does not alter, in any manner, the equations that have been developed for the steam turbine.

Example 9.1 A steam power plant operates with saturated steam at 6 MPa at the steam generator outlet and has a condenser pressure of 8 kPa. The plant employs a steam engine that can operate with saturated steam at its inlet. The mass flow rate to the plant is 25 kg/s and the cooling water for the condenser enters at 23°C and leaves at 31°C. For the ideal Rankine cycle, determine (a) the power developed by the engine, (b) the power required to drive the pump, (c) the thermal efficiency of the cycle, (d) the back work ratio, (e) the heat rejected by the condenser, and (f) the mass flow rate of the cooling water.

Solution

Assumptions and Specifications

(1) The water-steam mixture contained in the component parts and the interconnecting piping forms a closed system.
(2) The system operates in the steady state.
(3) Each component part of the system can be treated as a control volume.

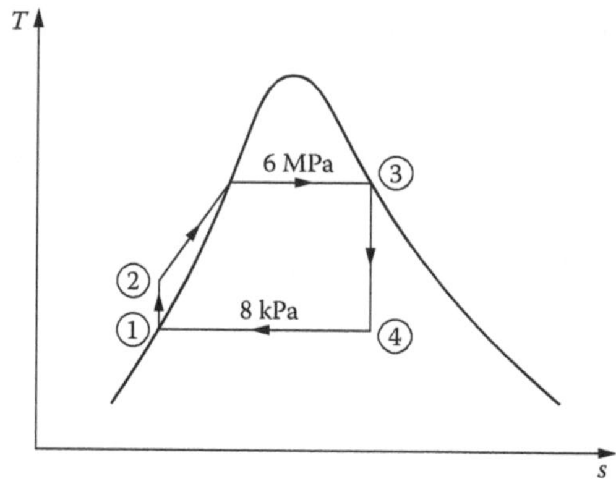

FIGURE 9.4
The T-s diagram for Example 9.1.

(4) This is an ideal Rankine cycle with two isentropic and two constant pressure processes.
(5) Kinetic and potential energy changes are negligible.
(6) The equations that involve work apply to a steam engine.
(7) The condition of saturated steam at the inlet to the engine is specified.

We will need to apply Equations 9.2 and 9.4 through 9.7 to satisfy the requirements of this problem. Thus, we will need specific enthalpy values at points 1 through 4 in Figure 9.4. At point 1 at $P_1 = 8\,\text{kPa}$, Table A.4 in Appendix A gives

$$v_1 = 1.008 \times 10^{-3}\,\text{m}^3/\text{kg} \quad \text{and} \quad h_1 = 173.8\,\text{kJ/kg}$$

Then, by Equation 4.10 with

$$P_2 - P_1 = 6 \times 10^6\,\text{N/m}^2 - 8 \times 10^3\,\text{N/m}^2 = 5.992 \times 10^6\,\text{N/m}^2$$

we find h_2

$$\begin{aligned} h_2 &= h_1 + v_1(P_2 - P_1) \\ &= 173.8\,\text{kJ/kg} + (1.008 \times 10^{-3}\,\text{m}^3/\text{kg})(5.992 \times 10^3\,\text{kN/m}^2) \\ &= 173.8\,\text{kJ/kg} + 6.04\,\text{kJ/kg} = 179.84\,\text{kJ/kg} \end{aligned}$$

At point 3, saturated vapor at 6 MPa has specific enthalpy and entropy values that we may find in Table A.4

$$h_3 = 2790.5\,\text{kJ/kg} \quad \text{and} \quad s_3 = 5.8906\,\text{kJ/kg-K}$$

Then, at point 4, because of the isentropic expansion through the turbine, $s_3 = s_4 = 5.8906\,\text{kJ/kg-K}$, Table A.4 gives at 8 kPa

$$s_f = 0.5924\,\text{kJ/kg-K} \quad \text{and} \quad s_{fg} = 7.6372\,\text{kJ/kg-K}$$

and

$$h_f = 173.8\,\text{kJ/kg} \quad \text{and} \quad h_{fg} = 2403.6\,\text{kJ/kg}$$

Then the quality at point 4, via Equation 4.8, will be

$$x_4 = \frac{s_4 - s_f}{s_{fg}} = \frac{5.8906\,\text{kJ/kg-K} - 0.5924\,\text{kJ/kg-K}}{7.6372\,\text{kJ/kg-K}} = 0.6937$$

Now we can find h_4 using Equation 4.6b

$$h_4 = h_f + xh_{fg} = 173.8\,\text{kJ/kg} + (0.6937)(2403.6\,\text{kJ/kg})$$
$$= 173.8\,\text{kJ/kg} + 1667.4\,\text{kJ/kg} = 1841.2\,\text{kJ/kg}$$

We now proceed to the determination of the required quantities.

(a) The power developed by the engine is given by Equation 9.1. With $\dot{m} = 25\,\text{kg/s}$

$$\dot{W}_{engr} = \dot{m}(h_3 - h_4)$$
$$= (25\,\text{kg/s})(2790.5\,\text{kJ/kg} - 1841.2\,\text{kJ/kg})$$
$$= (25\,\text{kg/s})(949.3\,\text{kJ/kg}) = 23{,}732\,\text{kW} \Longleftarrow$$

(b) The power required to drive the pump is given by Equation 9.2.

$$\dot{W}_{pump} = \dot{m}(h_2 - h_1)$$
$$= (25\,\text{kg/s})(179.84\,\text{kJ/kg} - 173.8\,\text{kJ/kg})$$
$$= (25\,\text{kg/s})(6.04\,\text{kJ/kg}) = 151\,\text{kW} \Longleftarrow$$

(c) The thermal efficiency of the cycle is given by Equation 9.6.

$$\eta = 1 - \frac{h_4 - h_1}{h_3 - h_2} = 1 - \frac{1841.2\,\text{kJ/kg} - 173.8\,\text{kJ/kg}}{2790.5\,\text{kJ/kg} - 179.84\,\text{kJ/kg}}$$
$$= 1 - \frac{1667.4\,\text{kJ/kg}}{2610.7\,\text{kJ/kg}} = 1 - 0.639 = 0.361 \Longleftarrow$$

(d) The back work ratio is given by Equation 9.7 with $\dot{W}_{eng}$ substituted for $\dot{W}_{turb}$.

$$\text{bwr} = \frac{\dot{W}_{pump}}{\dot{W}_{eng}} = \frac{151\,\text{kW}}{23{,}732\,\text{kW}} = 0.0064 \quad \text{or} \quad 0.64\% \Longleftarrow$$

(e) We find the heat rejected in the condenser from Equation 9.5.

$$\dot{Q}_{out} = \dot{m}(h_4 - h_1)$$
$$= (25\,\text{kg/s})(1841.2\,\text{kJ/kg} - 173.8\,\text{kJ/kg})$$
$$= (25\,\text{kg/s})(1667.4\,\text{kJ/kg}) = 41,685\,\text{kW} \Longleftarrow$$

(f) If the water side of the condenser is considered as a control volume, we see that, in the absence of any work interactions with the surroundings and negligible kinetic and potential energy changes, an energy rate balance gives

$$\dot{Q}_{out} = \dot{m}_w(h_{2w} - h_{1w})$$

where the subscript, w, refers to the cooling water. Table A.3 may be used to find, with interpolation, at $T_1 = 23°\text{C}$ and $T_2 = 31°\text{C}$, respectively

$$h_{1w} = 96.44\,\text{kJ/kg} \quad \text{and} \quad h_{2w} = 129.78\,\text{kJ/kg-K}$$

With the heat transferred to the cooling water from part (e)

$$\dot{m}_w = \frac{\dot{Q}_{\text{out}}}{h_{2w} - h_{1w}} = \frac{41{,}685\,\text{kW}}{129.78\,\text{kJ/kg-K} - 96.44\,\text{kJ/kg-K}}$$

$$= \frac{41{,}685\,\text{kW}}{33.34\,\text{kJ/kg-K}} = 1250.3\,\text{kg/s} \Longleftarrow$$

9.3.2 Increasing the Efficiency of the Ideal Rankine Cycle

In Example 9.1, we observed that the efficiency of the ideal Rankine cycle was a rather dismal 0.361. This leads us to the conclusion that the increase in the efficiency of the cycle is a subject of more than academic interest. We will find it extremely helpful if we can develop an expression that is similar in form to the equation for the thermal efficiency of the Carnot cycle given in Chapter 6 by

$$\eta = 1 - \frac{T_L}{T_H}$$

In Figure 9.5, we know that heat is added between points 2 and 3 and is represented, except for the nonisothermal heat supply in the compressed liquid region, by the integral

$$q_{\text{in}} = \int_{s_2}^{s_3} T\,ds$$

Because T is a function of s, we can obtain its average value, $\bar{T}_H$, from

$$\bar{T}_H = \frac{1}{s_3 - s_2} \int_{s_2}^{s_3} T\,ds$$

and thus,

$$q_{\text{in}} = \bar{T}_H(s_3 - s_2)$$

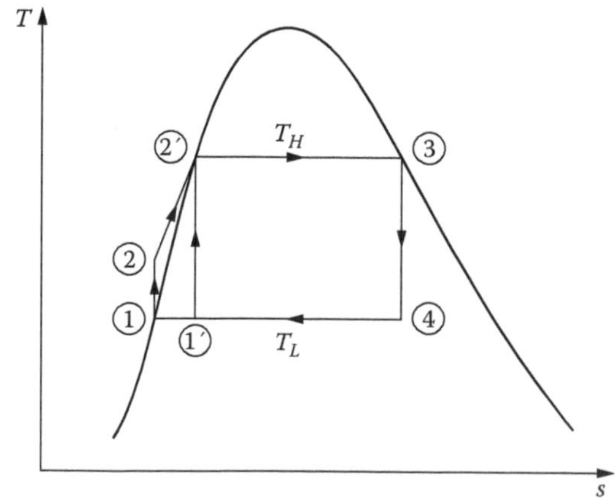

FIGURE 9.5
A T-s diagram showing the comparison between the ideal Rankine cycle and the Carnot cycle.

When we proceed in identical fashion, we can show that

$$q_{\text{out}} = \int_{s_1}^{s_4} T ds$$

with an average value such that

$$q_{\text{out}} = \bar{T}_L(s_4 - s_1)$$

Hence, with $s_1 = s_2$ and $s_3 = s_4$, the work of the cycle

$$w_{\text{cycle}} = q_{\text{in}} - q_{\text{out}}$$

can be written as

$$w_{\text{cycle}} = \bar{T}_H(s_3 - s_2) - \bar{T}_L(s_3 - s_2)$$

so that the efficiency will be

$$\eta = 1 - \frac{\bar{T}_L}{\bar{T}_H} \tag{9.8}$$

Equation 9.8 suggests that the efficiency of the ideal Rankine cycle is improved as $\bar{T}_L$ is minimized or as $\bar{T}_H$ is maximized—or both. Except that average rather than fixed hot and cold temperatures are employed, the form of Equation 9.8 is similar to the form of the equation for the Carnot cycle efficiency.

It is a fact, which we can verify by an inspection of the steam tables, that as the saturation pressure is increased, the saturation temperature is increased. Thus, we find that higher ideal Rankine cycle efficiencies occur at higher steam generator pressures. By similar observations, we know that higher ideal Rankine cycle efficiencies occur at lower condenser pressures.

There are, of course, limitations to these conclusions. The convecting surfaces in the steam generator and condenser are heat transfer devices that require a temperature difference to yield a flow of heat. The size of the steam generator is inversely proportional to the temperature difference between the flue gases and the steam. Moreover, as we proceed through the steam generator, we find that the maximum temperature of the steam at a point in the cycle cannot exceed the lowest flue gas temperature at that point. At the condenser end, the temperature is not only limited by the temperature of the cooling medium at the exit but also by the desire to operate over a temperature difference that will allow for a condenser of reasonable size. This entire subject is treated in Chapter 25.

9.4 The Ideal Rankine Cycle with Superheat

A **superheater** is essentially a heat exchanger that can be placed in the furnace of the steam generator so that it can absorb radiant heat from the flue gases. Alternatively, the superheater may consist of a bank of tubes placed in such a manner that the flue gases pass across it and transfer heat to it by convection. In either application, the purpose of the superheater is to raise the temperature of the steam leaving the steam generator and entering the turbine. A T-s diagram for the ideal Rankine cycle with superheat is shown in Figure 9.6.

Because the temperature at point 3 in the cycle of Figure 9.6 is elevated above the saturation temperature, the average temperature, T_H, has increased. In accordance with

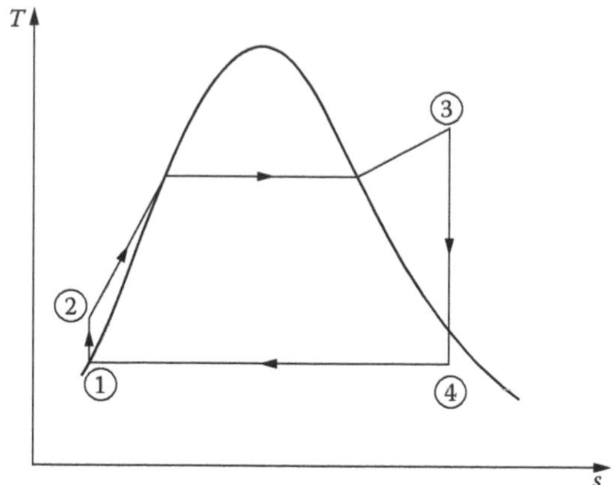

FIGURE 9.6
The T-s diagram for the ideal Rankine cycle with superheat.

Equation 9.8, we should see an increase in efficiency. Moreover, we can see that the use of the superheater begins to alleviate the problem of low-quality steam at the turbine exhaust.

Here too, state points 1 through 4 are used for the ideal Rankine cycle with superheat. Thus, Equation 9.1 through 9.7 may be used when the superheater is employed. However, we will find in Example 9.2, which now follows, that the values of specific enthalpy and entropy will change. We will also see in Example 9.2 how the performance changes when the superheater is in the cycle.

Example 9.2 A steam power plant operates with superheated steam at 6 MPa and a temperature of 600°C at the steam generator outlet and has a condenser pressure of 8 kPa (Figure 9.7). The mass flow rate in the plant is 25 kg/s and the cooling water for the condenser enters at 23°C and leaves at 31°C. For the ideal Rankine cycle, determine (a) the power developed by the turbine, (b) the power required to drive the pump, (c) the thermal efficiency of the cycle, (d) the back work ratio, (e) the heat rejected by the condenser, and (f) the mass flow rate of the cooling water.

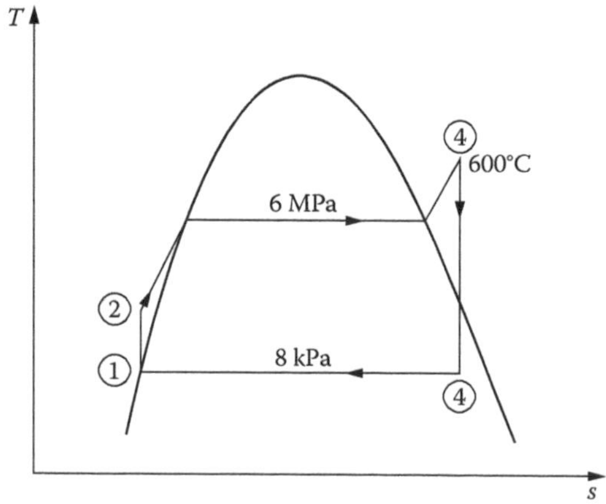

FIGURE 9.7
The T-s diagram for Example 9.2.

Solution

We make the same assumptions as in Example 9.1 except that a turbine is now involved with its inlet taking superheated steam and conditions at state points 1 and 2 remain the same.

Conditions at state points 1 and 2 are the same as those in Example 9.1.

$$h_1 = 173.8\,\text{kJ/kg} \quad \text{and} \quad h_2 = 179.84\,\text{kJ/kg}$$

At point 3, superheated vapor at 6 MPa and 600°C has specific enthalpy and entropy values that we may find in Table A.5

$$h_3 = 3654.2\,\text{kJ/kg} \quad \text{and} \quad s_3 = 7.1627\,\text{kJ/kg-K}$$

Then, at point 4, because of the isentropic expansion through the turbine, $s_3 = s_4 = 7.1627\,\text{kJ/kg-K}$, Table A.4 gives at 8 kPa

$$s_f = 0.5924\,\text{kJ/kg-K} \quad \text{and} \quad s_{fg} = 7.6372\,\text{kJ/kg-K}$$

and

$$h_f = 173.8\,\text{kJ/kg} \quad \text{and} \quad h_{fg} = 2403.6\,\text{kJ/kg}$$

We find the quality at point 4 via Equation 4.8:

$$x_4 = \frac{s_4 - s_f}{s_{fg}} = \frac{7.1627\,\text{kJ/kg-K} - 0.5924\,\text{kJ/kg-K}}{7.6372\,\text{kJ/kg-K}}$$

$$= \frac{6.5703\,\text{kJ/kg-K}}{7.6372\,\text{kJ/kg-K}} = 0.8603$$

and then by Equation 4.6b

$$h_4 = h_f + xh_{fg} = 173.8\,\text{kJ/kg} + (0.8603)(2403.6\,\text{kJ/kg})$$

$$= 173.8\,\text{kJ/kg} + 2067.8\,\text{kJ/kg} = 2241.6\,\text{kJ/kg}$$

We now proceed to the determination of the required quantities.

(a) The power developed by the turbine is given by Equation 9.1. With $\dot{m} = 25\,\text{kg/s}$

$$\dot{W}_{\text{turb}} = \dot{m}(h_3 - h_4)$$

$$= (25\,\text{kg/s})(3654.2\,\text{kJ/kg} - 2241.6\,\text{kJ/kg})$$

$$= (25\,\text{kg/s})(1412.6\,\text{kJ/kg}) = 35{,}315\,\text{kW} \Longleftarrow$$

(b) The power required to drive the pump is given by Equation 9.2 and has the same value as in Example 9.1

$$151\,\text{kW} \Longleftarrow$$

(c) The thermal efficiency of the cycle is given by Equation 9.6

$$\eta = 1 - \frac{h_4 - h_1}{h_3 - h_2} = 1 - \frac{2241.6\,\text{kJ/kg} - 173.8\,\text{kJ/kg}}{3654.2\,\text{kJ/kg} - 179.84\,\text{kJ/kg}}$$

$$= 1 - \frac{2067.8\,\text{kJ/kg}}{3474.4\,\text{kJ/kg}} = 1 - 0.595 = 0.405 \Longleftarrow$$

(d) The back work ratio is given by Equation 9.7

$$\text{bwr} = \frac{\dot{W}_{\text{pump}}}{\dot{W}_{\text{turb}}} = \frac{151\,\text{kW}}{35{,}315\,\text{kW}} = 0.0043 \quad \text{or} \quad 0.43\% \Longleftarrow$$

(e) We find the heat rejected in the condenser from Equation 9.5

$$\begin{aligned} \dot{Q}_{\text{out}} &= \dot{m}(h_4 - h_1) \\ &= (25\,\text{kg/s})(2241.6\,\text{kJ/kg} - 173.8\,\text{kJ/kg}) \\ &= (25\,\text{kg/s})(2067.8\,\text{kJ/kg}) = 51,695\,\text{kW} \Longleftarrow \end{aligned}$$

(f) The enthalpy values for the cooling water remain the same as in Example 9.1. We find that at $T_1 = 23°\text{C}$ and $T_2 = 31°\text{C}$, respectively

$$h_{1w} = 96.44\,\text{kJ/kg} \quad \text{and} \quad h_{2w} = 129.78\,\text{kJ/kg-K}$$

With the heat transferred to the cooling water from part (e)

$$\dot{Q}_{\text{out}} = \dot{m}_w(h_{2w} - h_{2w})$$

and

$$\begin{aligned} \dot{m}_w &= \frac{\dot{Q}_{\text{out}}}{h_{2w} - h_{1w}} = \frac{51,695\,\text{kW}}{129.78\,\text{kJ/kg-K} - 96.44\,\text{kJ/kg-K}} \\ &= \frac{51{,}695\,\text{kW}}{16.66\,\text{kJ/kg-K}} = 1550.5\,\text{kg/s} \Longleftarrow \end{aligned}$$

This example shows a modest increase in efficiency over the value obtained in Example 9.1. The primary reason for superheat is to increase the quality of the steam-water mixture at the low-pressure end of the turbine.

9.5 The Effect of Irreversibilities

Several irreversibilities conspire to reduce the thermal efficiency of the Rankine cycle. The most significant of these is the departure from the isentropic compression in the pump and the isentropic expansion in the turbine. As shown in Figure 9.8, these compressions and expansions are not isentropic and the specific enthalpies at the pump and turbine outlets are at higher values than in the ideal cycle.

We have shown in Chapter 7 that irreversibilities in the pump can be handled by a parameter known as the isentropic pump efficiency

$$\eta_s = \frac{h_{2s} - h_1}{h_2 - h_1} \tag{7.33}$$

where h_{2s} is the specific enthalpy of the water at the pump outlet when the compression in the pump is isentropic and h_1 is the specific enthalpy of the water at the pump inlet. If the isentropic pump efficiency and specific enthalpy values (including h_{2s}) for the ideal

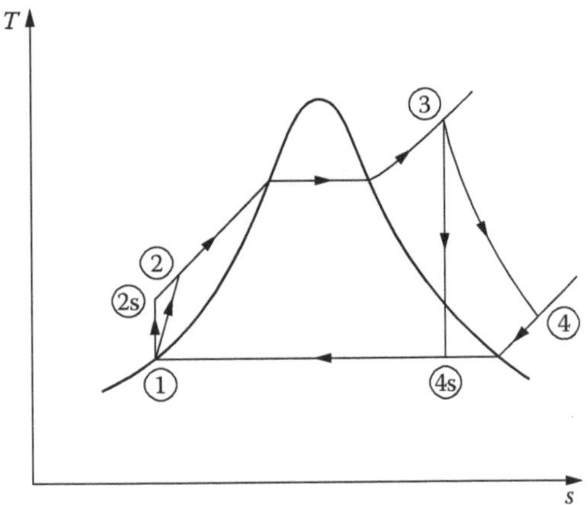

FIGURE 9.8
The T-s diagram for the ideal Rankine cycle modified to show the effect of irreversibilities in the pump and turbine.

Rankine cycle are known or obtained from the steam tables, the specific enthalpy, h_2 may be computed from

$$h_2 = h_1 + \frac{h_{2s} - h_1}{\eta_s} \tag{9.9}$$

An identical procedure may be followed for the turbine. In Chapter 7, we showed that irreversibilities in the turbine are accounted for by the isentropic turbine efficiency.[1]

$$\eta_s = \frac{h_3 - h_4}{h_3 - h_{4s}} \tag{7.32}$$

where h_{4s} is the specific enthalpy of the steam (or steam-water mixture) at the turbine outlet when the turbine irreversibility is excluded and h_3 represents the specific enthalpy of the steam at the turbine inlet. If the isentropic turbine efficiency and the specific enthalpy values (including h_{4s}) for the ideal Rankine cycle are known or obtained from the steam tables, the specific enthalpy, h_4 may be calculated from

$$h_4 = h_3 - \eta_s(h_3 - h_{4s}) \tag{9.10}$$

Other irreversibilities manifest themselves in pressure losses in the tubing in the steam generator and condenser and the connecting pipes between the power plant components. There is also an unwanted heat transfer through the casing of the various power plant components to the environment. However, these additional irreversibilities will not be considered here.

We now reconsider Example 9.2 and show how the inclusion of effects of pump and turbine irreversibilities influence thermal efficiency.

Example 9.3 Consider the Rankine cycle of Example 9.2 and suppose that both the pump and the turbine have isentropic efficiencies of 86% (0.860). All other conditions are the same as in Example 9.2. Determine the thermal efficiency of the cycle that includes both the pump and turbine irreversibilities (Figure 9.9).

[1] The subscripts have been adjusted in keeping with Figure 9.7.

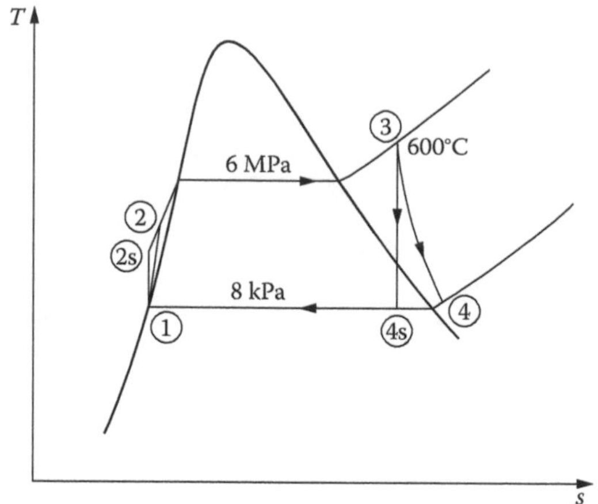

FIGURE 9.9
The T-s diagram for Example 9.3.

Solution

We make the same assumptions as in Example 9.2 except that the pump and turbine do not include isentropic processes. Specific enthalpy data at each state point in the ideal Rankine cycle may be taken from Example 9.2 but we must designate the specific enthalpies at points 2 and 4 as h_{2s} and h_{4s}, which are the constant entropy values

$$h_1 = 173.8\,\text{kJ/kg} \qquad h_3 = 3654.2\,\text{kJ/kg}$$
$$h_{2s} = 179.84\,\text{kJ/kg} \qquad h_{4s} = 2241.6\,\text{kJ/kg}$$

Now we use Equations 9.9 and 9.10 to determine the actual values of h_2 and h_4. For h_2, Equation 9.9 gives

$$\begin{aligned} h_2 = h_1 + \frac{h_{2s} - h_1}{\eta_{\text{pump}}} &= 173.8\,\text{kJ/kg} + \frac{179.84\,\text{kJ/kg} - 173.8\,\text{kJ/kg}}{0.860} \\ &= 173.8\,\text{kJ/kg} + \frac{6.04\ \text{kJ/kg}}{0.860} \\ &= 173.8\,\text{kJ/kg} + 7.02\,\text{kJ/kg} = 180.82\,\text{kJ/kg} \end{aligned}$$

and for h_4, Equation 9.10 provides

$$\begin{aligned} h_4 &= h_3 - \eta_{\text{turb}}(h_3 - h_{4s}) \\ &= 3654.2\,\text{kJ/kg} - (0.860)(3654.2\,\text{kJ/kg} - 2241.6\,\text{kJ/kg}) \\ &= 3654.2\,\text{kJ/kg} - (0.860)(1412.6\,\text{kJ/kg}) \\ &= 3654.2\,\text{kJ/kg} - 1214.84\,\text{kJ/kg} = 2439.4\,\text{kJ/kg} \end{aligned}$$

We now have a different set of specific enthalpies some of which account for the irreversibilities

$$h_1 = 173.8\,\text{kJ/kg} \qquad h_3 = 3654.2\,\text{kJ/kg}$$
$$h_2 = 180.82\,\text{kJ/kg} \qquad h_4 = 2439.4\,\text{kJ/kg}$$

The thermal efficency of the cycle is now

$$\eta = 1 - \frac{h_4 - h_1}{h_3 - h_2} = 1 - \frac{2439.4\,\text{kJ/kg}) - 173.8\,\text{kJ/kg}}{3654.2\,\text{kJ/kg}) - 180.82\,\text{kJ/kg}}$$

$$= 1 - \frac{2265.6\,\text{kJ/kg}}{3473.4\,\text{kJ/kg}} = 1 - 0.652 = 0.348 \Longleftarrow$$

It is apparent that a lack of attention to the irreversibilities in the pump and turbine (and particularly the turbine) can lead to an enormous disappointment.

9.6 The Rankine Cycle with Superheat and Reheat

In the ideal Rankine cycle with both **reheat** and superheat (Figure 9.10a), the turbine has two stages. After an expansion in stage 1, the **high-pressure turbine**, the steam is rerouted to the steam generator where it is preheated at constant pressure in a **reheater**. The steam

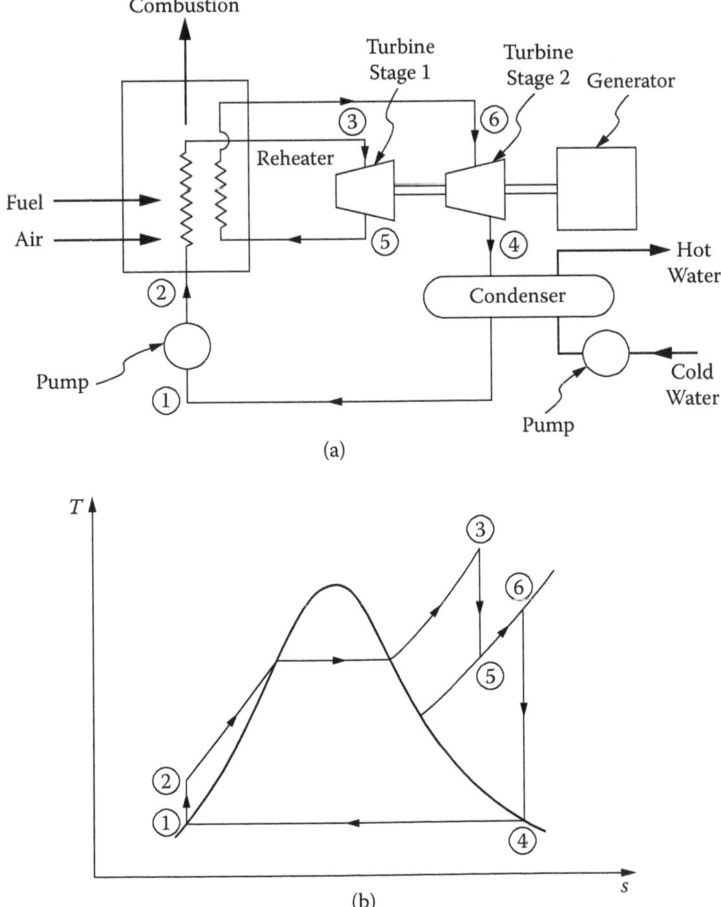

FIGURE 9.10
(a) A steam power plant with both superheat and reheat. (b) The T-s diagram for the ideal Rankine cycle.

is then led to stage 2, called **low-pressure turbine** where it is expanded to the condenser pressure.

In Figure 9.10b, the T-s diagram for the ideal Rankine cycle with both superheat and reheat, we see that there are six state points. Points 1 through 4 are in their customary location. However, point 5 is the point where the steam is taken from the high-pressure turbine and led back to the reheater inside of the steam generator. Point 6 is located at the end of the constant pressure reheat process and is at the inlet to the low-pressure turbine. At point 4, where the steam is exhausted from the low-pressure turbine, we see that the steam may still be superheated or possess a high quality. This is a significant advantage of the reheat cycle.

Whether the overall thermal efficiency of the power plant operating with reheat is improved depends on the selection of the reheat pressure and the temperature at the end of the reheat process. State 5 is usually permitted to come close to the saturated vapor line and the temperature at the end of the reheat process, T_6, is equal to or somewhat less than the total steam temperature, T_3. Use of the reheat cycle with

$$0.200 < \frac{P_5}{P_3} < 0.333$$

will usually provide an increase in the cycle thermal efficiency.

The energy rate balance for the control volume representing *both* turbines operating between state points 3 and 5 and points 6 and 4 in Figure 9.10a is

$$\dot{W}_{\text{turb}} = \dot{m}[(h_3 - h_5) + (h_6 - h_4)] \tag{9.11}$$

Equation 9.11 presumes that there is no heat transfer to the surroundings and that there are negligible kinetic and potential energy changes.

With the pump taken as a control volume and with operation between state points 1 and 2 in Figure 9.10b, a similar analysis, again with no heat transfer and negligible kinetic and potential energy changes, will yield Equation 9.2

$$\dot{W}_{\text{pump}} = \dot{m}(h_2 - h_1) \tag{9.2}$$

The net work in the cycle is therefore,

$$\begin{aligned}\dot{W}_{\text{cycle}} &= \dot{W}_{\text{turb}} - \dot{W}_{\text{pump}} \\ &= \dot{m}[(h_3 - h_5) + (h_6 - h_4) - (h_2 - h_1)]\end{aligned} \tag{9.12}$$

Then with the steam generator taken as a control volume, we note that there is no work interaction. Thus, with negligible kinetic and potential energy changes, an energy rate balance provides

$$\dot{Q}_{\text{in}} = \dot{m}[(h_3 - h_2) + (h_6 - h_5)] \tag{9.13}$$

The condenser has a condensing side and a cooling water side and we may apply a control volume to just the condensing side. Again, with no work interaction and with negligible kinetic and potential energy changes, the heat rejected by the condenser will be given by Equation 9.5

$$\dot{Q}_{\text{out}} = \dot{m}(h_4 - h_1) \tag{9.5}$$

The thermal efficiency of the ideal Rankine cycle with reheat is

$$\eta = \frac{\dot{W}_{\text{cycle}}}{\dot{Q}_{\text{in}}} = \frac{\dot{m}[(h_3 - h_5) + (h_6 - h_4) - (h_2 - h_1)]}{\dot{m}[(h_3 - h_2) + (h_6 - h_5)]} \tag{9.14}$$

and the back work ratio will be

$$\text{bwr} = \frac{\dot{W}_{\text{pump}}}{\dot{W}_{\text{turb}}} = \frac{h_2 - h_1}{(h_3 - h_5) + (h_6 - h_4)} \tag{9.15}$$

Example 9.4 A steam power plant operates with superheated steam at 6 MPa and a temperature of 600°C at the steam generator outlet (Figure 9.11). The steam leaves the first stage of the turbine at 1.5 MPa, is reheated to 600°C, and is then led to the input of the second stage of the turbine. The condenser pressure is 8 kPa, the mass flow rate in the plant is 25 kg/s and the cooling water for the condenser enters at 23°C and leaves at 31°C. For the ideal Rankine cycle, determine (a) the power developed by the turbine, (b) the power required to drive the pump, (c) the thermal efficiency of the cycle, (d) the back work ratio, (e) the heat rejected by the condenser, and (f) the mass flow rate of the cooling water.

Solution

We make the same assumptions as in Example 9.2 except that a reheater is now involved with a heat addition from state points 5 to 6.

Conditions at state points 1 and 2 are the same as those in Example 9.2.

$$h_1 = 173.8\,\text{kJ/kg} \quad \text{and} \quad h_2 = 179.84\,\text{kJ/kg}$$

and at point 3, the superheated vapor at 6 MPa and 600°C has the same specific enthalpy and entropy values as in Example 9.2

$$h_3 = 3654.2\,\text{kJ/kg} \quad \text{and} \quad s_3 = 7.1627\,\text{kJ/kg-K}$$

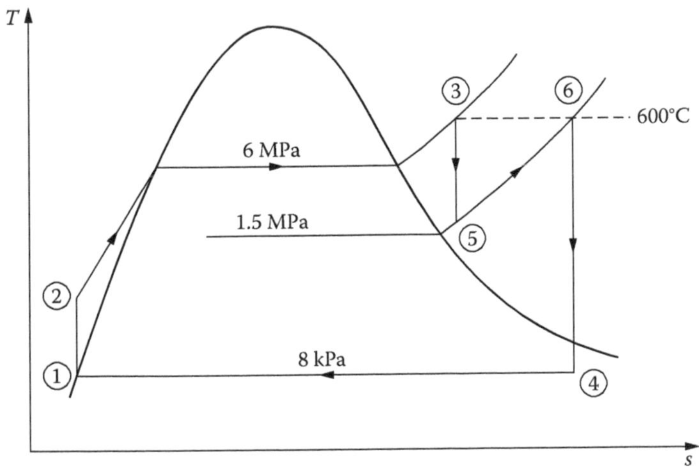

FIGURE 9.11
T-s diagram for Example 9.4.

Then, at point 5, due to the isentropic expansion through the turbine, $s_3 = s_5 = 7.1627$ kJ/kg-K, Table A.5 gives at $P_5 = 1.5\,\text{MPa}$ with interpolation

$$T_5 = 368.4°\text{C} \quad \text{and} \quad h_5 = 3186.9\,\text{kJ/kg}$$

At point 6, superheated vapor at 1.5 MPa and 600°C has specific enthalpy and entropy values given in Table A.5

$$h_6 = 3689.9\,\text{kJ/kg} \quad \text{and} \quad s_3 = 7.8331\,\text{kJ/kg-K}$$

We note that specific enthalpy and entropy data at point 4 is the same as in Example 9.2

$$s_f = 0.5924\,\text{kJ/kg-K} \quad \text{and} \quad s_{fg} = 7.6372\,\text{kJ/kg-K}$$

and

$$h_f = 173.8\,\text{kJ/kg} \quad \text{and} \quad h_{fg} = 2403.6\,\text{kJ/kg}$$

Then, $s_4 = s_6$, and we find the quality at point 4 via Equation 4.8

$$x_4 = \frac{s_4 - s_f}{s_{fg}} = \frac{7.8331\,\text{kJ/kg-K} - 0.5924\,\text{kJ/kg-K}}{7.6372\,\text{kJ/kg-K}}$$

$$= \frac{7.2407\,\text{kJ/kg-K}}{7.6372\,\text{kJ/kg-K}} = 0.9481$$

and then by Equation 4.6b

$$h_4 = h_f + xh_{fg} = 173.8\,\text{kJ/kg} + (0.9481)(2403.6\,\text{kJ/kg})$$

$$= 173.8\,\text{kJ/kg} + 2278.8\,\text{kJ/kg} = 2452.6\,\text{kJ/kg}$$

(a) The power developed by the turbine is given by Equation 9.11. With

$$h_3 - h_5 = 3654.2\,\text{kJ/kg} - 3186.9\,\text{kJ/kg} = 467.3\,\text{kJ/kg}$$

and

$$h_6 - h_4 = 3689.9\,\text{kJ/kg} - 2452.6\,\text{kJ/kg} = 1237.3\,\text{kJ/kg}$$

and with $\dot{m} = 25\,\text{kg/s}$

$$\dot{W}_{\text{turb}} = \dot{m}[(h_3 - h_5) + (h_6 - h_4)]$$

$$= (25\,\text{kg/s})(467.3\,\text{kJ/kg} + 1237.3\,\text{kJ/kg})$$

$$= (25\,\text{kg/s})(1704.6\,\text{kJ/kg}) = 42{,}615\,\text{kW} \Longleftarrow$$

(b) The power required to drive the pump is given by Equation 9.2 and has the same value as in Example 9.2

$$151\,\text{kW} \Longleftarrow$$

(c) The net work of the cycle is given by Equation 9.12

$$\dot{W}_{\text{cycle}} = \dot{W}_{\text{turb}} - \dot{W}_{\text{pump}}$$

$$= 42{,}615\,\text{kW} - 151\,\text{kW} = 42{,}464\,\text{kW}$$

and with

$$h_3 - h_2 = 3654.2\,\text{kJ/kg} - 179.84\,\text{kJ/kg} = 3474.4\,\text{kJ/kg}$$

and

$$h_6 - h_5 = 3689.9\,\text{kJ/kg} - 3186.9\,\text{kJ/kg} = 503.0\,\text{kJ/kg}$$

the heat input to the cycle, given by Equation 9.13 will be

$$\begin{aligned}\dot{Q}_{\text{in}} &= \dot{m}[(h_3 - h_2) + (h_6 - h_5)] \\ &= (25\,\text{kg/s})(3474.4\,\text{kJ/kg} + 503.0\,\text{kJ/kg}) \\ &= (25\,\text{kg/s})(3977.4\,\text{kJ/kg}) = 99{,}434\,\text{kW}\end{aligned}$$

The thermal efficiency of the cycle is given by Equation 9.14.

$$\eta = \frac{\dot{W}_{\text{cycle}}}{\dot{Q}_{\text{in}}} = \frac{42,464\,\text{kW}}{99{,}434\,\text{kW}} = 0.427 \quad \text{or} \quad (42.7\%) \Longleftarrow$$

(d) The back work ratio is given by Equation 9.15

$$\text{bwr} = \frac{\dot{W}_{\text{pump}}}{\dot{W}_{\text{turb}}} = \frac{151\,\text{kW}}{42,615\,\text{kW}} = 0.0035 \quad \text{or} \quad 0.35\% \Longleftarrow$$

(e) We find the heat rejected in the condenser from Equation 9.5.

$$\begin{aligned}\dot{Q}_{\text{out}} &= \dot{m}(h_4 - h_1) \\ &= (25\,\text{kg/s})(2452.6\,\text{kJ/kg} - 173.8\,\text{kJ/kg}) \\ &= (25\,\text{kg/s})(2278.8\,\text{kJ/kg}) = 56{,}970\,\text{kW} \Longleftarrow\end{aligned}$$

(f) The specific enthalpies of the cooling water are the same as in Example 9.2. At $T_1 =$ 23°C and $T_2 = 31$°C, respectively

$$h_{1w} = 96.44\,\text{kJ/kg} \quad \text{and} \quad h_{2w} = 129.78\,\text{kJ/kg-K}$$

and with the heat transferred to the cooling water from part (e)

$$\begin{aligned}\dot{m}_w &= \frac{\dot{Q}_{\text{out}}}{h_{2w} - h_{1w}} = \frac{56,970\,\text{kW}}{129.78\,\text{kJ/kg-K} - 96.44\,\text{kJ/kg-K}} \\ &= \frac{56{,}970\,\text{kW}}{33.34\,\text{kJ/kg-K}} = 1708.8\,\text{kg/s} \Longleftarrow\end{aligned}$$

We turn now to Design Example 3, which involves the establishment of the temperature at the entrance to a reheater to assure that the quality of the steam leaving the second stage of a turbine is equal to or greater than unity.

9.7 Design Example 3

A steam power plant on an ideal Rankine cycle operates with superheated steam at 10 MPa and a temperature of 700°C at the entrance to the first stage of the turbine (Figure 9.12). The steam leaves the first stage of the turbine at 1 MPa and is reheated and then led to the second stage of the turbine, which exhausts to a condenser at 10 kPa. Determine (a) the temperature at the exit from the reheater necessary to make the quality of the steam at the turbine exhaust equal to or greater than unity and (b) the thermal efficiency of the cycle.

Solution

Assumptions and Specifications

(1) The water-steam mixture contained in the component parts and the interconnecting piping forms a closed system.
(2) The system operates in the steady state.
(3) Each component part of the system can be treated as a control volume.
(4) This is an ideal Rankine cycle with three isentropic and three constant pressure processes.
(5) Kinetic and potential energy changes are negligible.
(6) The condition of saturated steam at the inlet to the condenser is specified.

Figure 9.12 shows a T-s diagram for the cycle.

(a) We note that the quality of the steam at point 4 in Figure 9.12 is $x \geq 1.00$ and we are able to read from Table A.4 at 10 kPa that

$$s_4 = 8.1511\,\text{kJ/kg-K}$$

Then, for an isentropic drop between points 6 and 4

$$s_6 = s_4 = 8.1511\,\text{kJ/kg-K}$$

and at 1 MPa, Table A.5 (with interpolation) will yield the temperature at the entrance to the second stage of the turbine

$$T_6 = 651.6°\text{C} \Longleftarrow$$

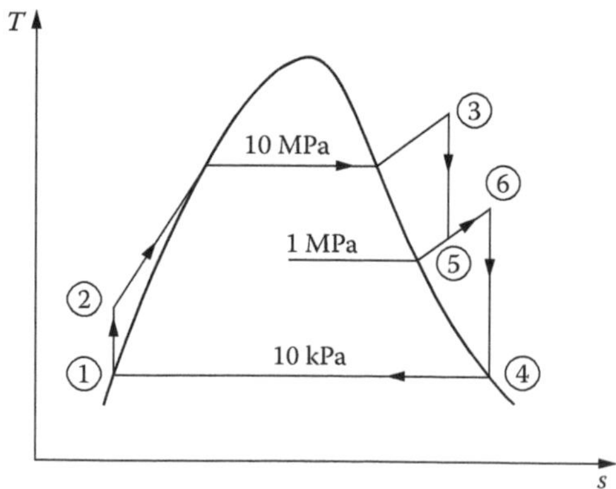

FIGURE 9.12
T-s diagram for Design Example 3.

(b) The thermal efficiency is given by Equation 9.14 and to evaluate the efficiency, the enthalpies of six points will be needed. At point 1 where $P_1 = 10\,\text{kPa}$, Table A.4 provides

$$v_1 = v_f = 1.010 \times 10^{-3}\,\text{m}^3/\text{kg} \quad \text{and} \quad h_1 = 191.7\,\text{kJ/kg}$$

Then

$$P_2 - P_1 = 10,000\,\text{kPa} - 10\,\text{kPa} = 9990\,\text{kPa}$$

and we find via Equation 4.10 that

$$\begin{aligned} h_2 &= h_1 + v_1(P_2 - P_1) \\ &= 191.7\,\text{kJ/kg} + (1.010 \times 10^{-3})\,\text{m}^3/\text{kg})(9990\,\text{kPa}) \\ &= 191.7\,\text{kJ/kg} + 10.0\,\text{kJ/kg} = 201.7\,\text{kJ/kg} \end{aligned}$$

At point 3, the superheated vapor at 10 MPa and 700°C has specific enthalpy and entropy values that can be obtained from Table A.5.

$$h_3 = 3862.5\,\text{kJ/kg} \quad \text{and} \quad s_3 = 7.1601\,\text{kJ/kg-K}$$

Then, at point 5, because of the isentropic expansion through the first stage of the turbine, $s_3 = s_5 = 7.1601\,\text{kJ/kg-K}$, Table A.5 gives at $P_5 = 1\,\text{MPa}$ (with interpolation)

$$T_5 = 310.5^\circ\text{C} \quad \text{and} \quad h_5 = 3073.3\,\text{kJ/kg}$$

The specific entropies at points 4 and 6 have been obtained.

$$s_4 = s_6 = 8.1511\,\text{kJ/kg-K}$$

and interpolation of Table A.5 yields

$$h_6 = 3808.6\,\text{K}$$

Finally at point 4, Table A.4 at 10 kPa may be employed to find

$$h_4 = h_g = 2585.0\,\text{kJ/kg}$$

Equation 9.14 gives the thermal efficiency as

$$\eta = \frac{\dot{W}_{\text{cycle}}}{\dot{Q}_{\text{in}}} = \frac{\dot{m}[(h_3 - h_5) + (h_6 - h_4) - (h_2 - h_1)]}{\dot{m}[(h_3 - h_2) + (h_6 - h_5)]} \tag{9.14}$$

and with

$$\begin{aligned} h_3 - h_5 &= 3862.5\,\text{kJ/kg} - 3073.3\,\text{kJ/kg} = 789.2\,\text{kJ/kg} \\ h_6 - h_4 &= 3808.6\,\text{kJ/kg} - 2585.0\,\text{kJ/kg} = 1223.6\,\text{kJ/kg} \\ h_2 - h_1 &= 201.7\,\text{kJ/kg} - 191.7\,\text{kJ/kg} = 10.0\,\text{kJ/kg} \\ h_3 - h_2 &= 3862.5\,\text{kJ/kg} - 201.7\,\text{kJ/kg} = 3660.8\,\text{kJ/kg} \\ h_6 - h_5 &= 3808.6\,\text{kJ/kg} - 3073.3\,\text{kJ/kg} = 735.3\,\text{kJ/kg} \end{aligned}$$

we have

$$\eta = \frac{789.2\,\text{kJ/kg} + 1223.6\,\text{kJ/kg} - 10.0\,\text{kJ/kg}}{3660.8\,\text{kJ/kg} + 735.3\,\text{kJ/kg}} = \frac{2002.8\,\text{kJ/kg}}{4396.1\,\text{kJ/kg}} = 0.456 \Longleftarrow$$

9.8 The Ideal Rankine Cycle with Regeneration

In any of the Rankine cycles considered thus far, a substantial amount of heat is added to the feedwater between the pump outlet and the point where the water reaches the saturated liquid line. The average temperature of this heat addition is considerably below the temperature of the water-steam mixture as it is vaporized. Equation 9.8

$$\eta = 1 - \frac{\bar{T}_L}{\bar{T}_H} \tag{9.8}$$

indicates that the cycle efficiency can be increased if, during this heat addition process, $\bar{T}_H$ could be increased. One way of increasing $\bar{T}_H$ during the heat addition process is through the use of **regenerative feedwater heating.**

Figure 9.13a shows that steam is **extracted** or **bled** from the turbine at state point 5 and routed to an **open feedwater heater** that is essentially an **adiabatic mixing chamber.** The T-s diagram for this important variation of the ideal Rankine cycle is displayed in Figure 9.13b. Modern steam power plants may employ several extractions and an equal number of feedwater heaters.

For a control volume embracing both stages of the turbine in Figure 9.13a, we write a mass balance

$$\dot{m}_3 = \dot{m}_4 + \dot{m}_5$$

where

$\dot{m}_3$-mass flow rate entering stage 1 of the turbine
$\dot{m}_4$-mass flow rate entering stage 2 of the turbine
$\dot{m}_5$-mass flow rate that is extracted

A division by $\dot{m}_3$ produces

$$\frac{\dot{m}_5}{\dot{m}_3} + \frac{\dot{m}_4}{\dot{m}_3} = 1$$

which defines the **extraction fraction** as

$$y \equiv \frac{\dot{m}_5}{\dot{m}_3} \tag{9.16}$$

and shows that

$$\frac{\dot{m}_4}{\dot{m}_3} = 1 - y$$

With a control volume surrounding just the feedwater heater, continuity, in the absence of kinetic and potential energy changes and external work interactions, yields with $\dot{m}_1 = \dot{m}_6$

$$-1h_6 + yh_5 + (1-y)h_2 = 0$$

or

$$y = \frac{h_6 - h_2}{h_5 - h_2} \tag{9.17}$$

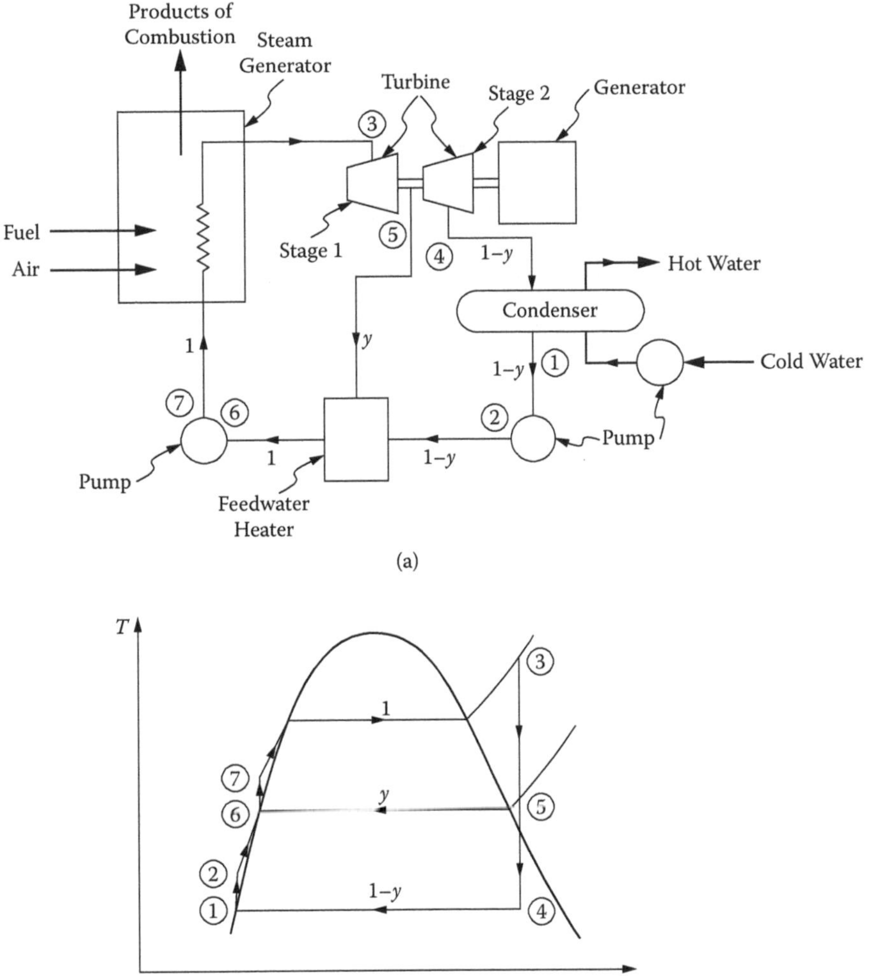

FIGURE 9.13
(a) A steam power plant with superheat and an open feedwater heater and (b) The T-s diagram for the ideal Rankine cycle.

The power developed by the turbine and the power required to drive the *two* pumps are both functions of the mass flow entering stage 1 of the turbine

$$\frac{\dot{W}_{turb}}{\dot{m}_3} = h_3 - h_5 + (1 - y)(h_5 - h_4) \tag{9.18a}$$

and

$$\frac{\dot{W}_{pump}}{\dot{m}_3} = h_7 - h_6 + (1 - y)(h_2 - h_1) \tag{9.18b}$$

The net power for the cycle will be the difference between Equations 9.18a and 9.18b

$$\frac{\dot{W}_{cycle}}{\dot{m}_3} = h_3 - h_5 + (1 - y)(h_5 - h_4) - (1 - y)(h_2 - h_1) - (h_7 - h_6)$$

or

$$\frac{\dot{W}_{cycle}}{\dot{m}_3} = (h_3 - h_5) - (h_7 - h_6) + (1 - y)[(h_5 - h_4) - (h_2 - h_1)] \tag{9.19}$$

The heat added to the steam-water mixture will be

$$\frac{\dot{Q}_{in}}{\dot{m}_3} = h_3 - h_7 \tag{9.20}$$

and the heat discharged in the condenser is

$$\frac{\dot{Q}_{out}}{\dot{m}_3} = (1 - y)(h_4 - h_1) \tag{9.21}$$

The thermal efficiency of the cycle is then

$$\eta = \frac{\dot{W}_{cycle}/\dot{m}_3}{\dot{Q}_{in}/\dot{m}_3} = \frac{\dot{W}_{cycle}}{\dot{Q}_{in}}$$

or

$$\eta = \frac{(h_3 - h_5) - (h_7 - h_6) + (1 - y)[(h_5 - h_4) - (h_2 - h_1)]}{h_3 - h_7} \tag{9.22}$$

We now focus our attention on a rework of Example 9.2 but with a steam extraction to an open feedwater heater instead of a reheater.

Example 9.5 A steam power plant operates with superheated steam at 6 MPa and a temperature of 600°C at the steam generator outlet (Figure 9.14). When the steam in the first stage of the turbine reaches 600 kPa, a portion is extracted and led to an open feedwater heater. The balance of the steam is led to the input of the second stage of the turbine. The condenser pressure is 8 kPa and the mass flow rate in the plant is 25 kg/s. For the ideal Rankine cycle, determine (a) the temperature at the extraction point, (b) the extraction fraction, and (c) the thermal efficiency of the cycle.

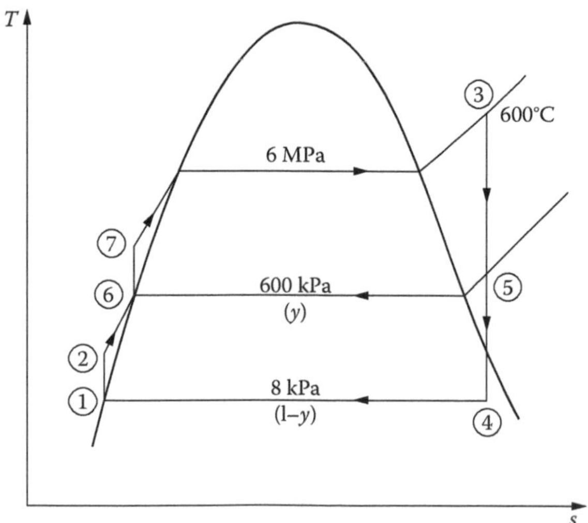

FIGURE 9.14
T-s diagram for Example 9.5.

Solution

We make the same assumptions as in Example 9.2 except that an extraction of some of the steam in the turbine is led to an open feedwater heater.

Conditions at state points 1 through 4 are the same as those in Example 9.2.

$$h_1 = 173.8\,\text{kJ/kg} \qquad h_3 = 3654.2\,\text{kJ/kg}$$

$$h_2 = 179.84\,\text{kJ/kg} \qquad h_4 = 2241.6\,\text{kJ/kg}$$

(a) For the temperature at the extraction point, we also note that

$$s_3 = 7.1627\,\text{kJ/kg-K}$$

and at point 5, because of the isentropic expansion through stage 1 of the turbine, $s_5 = s_3 = 7.1627\,\text{kJ/kg-K}$. With interpolation of Table A.5 at $P_5 = 600\,\text{kPa}$, we find that

$$T_5 = 245.5^\circ\text{C} \Longleftarrow \quad \text{and} \quad h_5 = 2947.6\,\text{kJ/kg}$$

At point 6 at $P_6 = 600\,\text{kPa}$, Table A.4 gives

$$v_6 = 1.101 \times 10^{-3}\,\text{m}^3/\text{kg} \quad \text{and} \quad h_6 = 670.1\,\text{kJ/kg}$$

Then, by Equation 4.10 with

$$P_7 - P_6 = 6 \times 10^6\,\text{N/m}^2 - 6 \times 10^5\,\text{N/m}^2 = 5.40 \times 10^6\,\text{N/m}^2$$

we find h_7

$$\begin{aligned} h_7 &= h_6 + v_6(P_7 - P_6) \\ &= 670.1\,\text{kJ/kg} + (1.101 \times 10^{-3}\,\text{m}^3\text{kg})(5.40 \times 10^3\,\text{kN/m}^2) \\ &= 670.1\,\text{kJ/kg} + 5.95\,\text{kJ/kg} = 676.1\,\text{kJ/kg} \end{aligned}$$

(b) The extraction fraction can be determined from Equation 9.17

$$\begin{aligned} y &= \frac{h_6 - h_2}{h_5 - h_2} = \frac{670.1\,\text{kJ/kg} - 179.84\,\text{kJ/kg}}{2947.6\,\text{kJ/kg} - 179.84\,\text{kJ/kg}} \\ &= \frac{490.3\,\text{kJ/kg}}{2767.8\,\text{kJ/kg}} = 0.177 \quad \text{or} \quad (17.7\%) \Longleftarrow \end{aligned}$$

(c) With

$$h_3 - h_5 = 3654.2\,\text{kJ/kg} - 2947.6\,\text{kJ/kg} = 706.6\,\text{kJ/kg}$$

$$h_7 - h_6 = 676.1\,\text{kJ/kg} - 670.1\,\text{kJ/kg} = 6.0\,\text{kJ/kg}$$

$$h_5 - h_4 = 2947.6\,\text{kJ/kg} - 2241.6\,\text{kJ/kg} = 706.0\,\text{kJ/kg}$$

$$h_2 - h_1 = 179.84\,\text{kJ/kg} - 173.8\,\text{kJ/kg} = 6.04\,\text{kJ/kg}$$

and with $\dot{m}_3 = 25$ kg/s, we obtain the power developed from a modification of Equation 9.19

$$\begin{aligned}\dot{W}_{\text{cycle}} &= \dot{m}_3\{(h_3 - h_5) - (h_7 - h_6) \\ &\quad + (1 - y)[(h_5 - h_4) - (h_2 - h_1)]\} \\ &= (25\,\text{kg/s})\{(706.6\,\text{kJ/kg} - 6.0\,\text{kJ/kg}) \\ &\quad (1 - 0.177))[706.0\,\text{kJ/kg} - 6.04\,\text{kJ/kg}]\} \\ &= (25\,\text{kg/s})[700.6\,\text{kJ/kg} + 0.823(700.0\,\text{kJ/kg})] \\ &= (25\,\text{kg/s})(1276.7\,\text{kJ/kg}) = 31,918\,\text{kW}\end{aligned}$$

The heat input to the cycle is determined from a modification of Equation 9.20

$$\begin{aligned}\dot{Q}_{\text{in}} &= \dot{m}_3(h_3 - h_7) \\ &= (25\,\text{kg/s})(3654.2\,\text{kJ/kg} - 676.1\,\text{kJ/kg}) \\ &= (25\,\text{kg/s})(2978.1\,\text{kJ/kg}) = 74{,}453\,\text{kW}\end{aligned}$$

The efficiency is

$$\eta = \frac{\dot{W}_{\text{cycle}}}{\dot{Q}_{\text{in}}} = \frac{31{,}918\,\text{kW}}{74{,}453\,\text{kW}} = 0.429 \quad \text{or} \quad (42.9\%) \Longleftarrow$$

We see that, in this case, the use of regeneration improves the plant efficiency from 0.405, in Example 9.2, to 0.429 here in Example 9.5.

Examples 9.1, 9.2, 9.4, and 9.5 have been based on the same conditions of temperature and pressure at the steam generator inlet and outlet. The only difference among these examples is in the use of different techniques and components. In order to assess the value of these techniques, we provide the following tabulation:

Example	Basis	Thermal Efficiency, η
9.1	Saturated steam	0.361
9.2	Superheated steam	0.405
9.4	Superheat and reheat	0.427
9.5	Superheat and extraction	0.429

9.9 The Ideal Refrigeration Cycle

9.9.1 Physical Description

Figure 9.15a is a diagram of an ideal vapor compression refrigeration/heat pump cycle with four principal components. These points are indicated in the T-s diagram of the ideal cycle shown in Figure 9.15b. The prime mover is the **compressor** that takes low pressure and low-temperature refrigerant at the outlet of the **evaporator** (point 1) and discharges the refrigerant to the **condenser** at higher pressure and temperature (point 2).

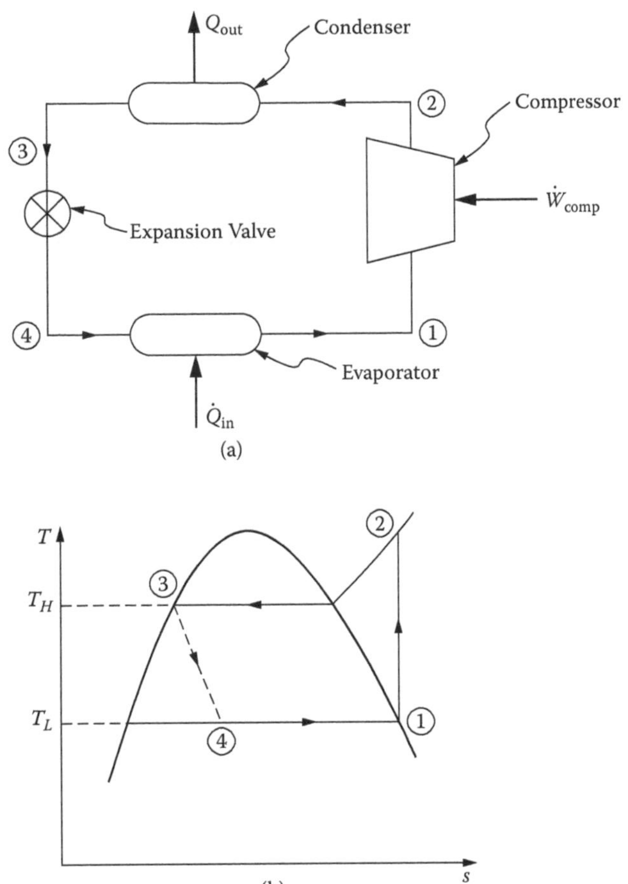

FIGURE 9.15
(a) The four basic components of a vapor compression refrigeration or heat pump cycle. (b) the T-s diagram for the ideal cycle.

The work necessary to drive the system, usually in the form of electric power, is applied to the compressor. We note that, in the ideal cycle, the refrigerant is compressed isentropically in the compressor and leaves the compressor as a superheated vapor.

At point 2, the refrigerant leaves the compressor and enters the **condenser** where it is cooled at constant pressure as a result of the heat transfer to a region at temperature, T_H. During its journey through the condenser, the refrigerant is cooled resulting in a change in phase from a superheated vapor to a saturated liquid at point 3. The temperature, T_H, is the temperature of the environment that varies with location, or, if the cycle is used as a heat pump, T_H is the temperature of a heated space such as a dwelling.

An **expansion valve** is employed to **throttle** the refrigerant from higher pressure and temperature at point 3 to a condition of lower pressure and temperature at point 4. This **throttling process** takes the saturated liquid refrigerant at the outlet of the condenser (point 3) to the inlet of the **evaporator** at point 4 at constant enthalpy ($h_3 = h_4$).

The refrigerant enters the **evaporator** at point 4 and, via a constant pressure heat exchange process, emerges as saturated vapor at point 1. Heat is absorbed by the refrigerant in this lower pressure and lower temperature process from the region at temperature, T_C, which is either the temperature of the refrigerated space or the temperature at the input side of a heat pump.

We observe that the cycles for a refrigerator and for a heat pump, while serving a different purpose (one cooling and the other heating), are basically the same.

9.9.2 Refrigerants

Refrigerants of the halogenated hydrocarbon type such as the freons are currently being phased out because of the concern about their effects on the earth's ozone layer. Instead, because of international agreements, tetrafluoromethane, commonly known as refrigerant R-134a, is being used. Tables of the thermodynamic properties of Refrigerant R-134a may be found in Tables A.6, A.7, and A.8. Like the steam tables, these tables give the properties of specific volume, internal energy, enthalpy, and entropy as functions of saturation temperature, saturation pressure, and in the superheat region where the properties are given as a function of both temperature and pressure.

Refrigerants are selected on the basis of pressure and temperature of application, chemical stability, corrosiveness, toxicity, type of compressor, and cost. The temperatures at both the evaporator and compressor ends of the cycle are fixed by the temperatures, T_H and T_C and these, in turn, yield the pressure in both the evaporator and condenser.

9.10 The Ideal Vapor Compression Refrigeration Cycle

From the T-s diagram for the ideal vapor compression refrigeration cycle displayed earlier (Figure 9.15b), we observe that the cycle is composed of four ideal processes:

- *Process 1-2:* An isentropic compression from the condenser pressure to the evaporator pressure
- *Process 2-3:* A constant pressure heat rejection in the condenser until saturated liquid is obtained at the condenser outlet
- *Process 3-4:* A constant enthalpy throttling process in which the refrigerant is reduced from higher to lower pressure and temperature
- *Process 4-1:* A constant pressure heat addition to a condition of saturated vapor in the evaporator

We note in both parts of Figure 9.15 that, in the basic ideal vapor refrigeration cycle, refrigerant enters the compressor as saturated liquid and leaves as superheated vapor. Because the work of compression is a positive quantity,

$$\dot{W}_{\text{comp}} = \dot{m}(h_2 - h_1) \tag{9.23}$$

where $\dot{m}$ is the mass flow rate of refrigerant. Equation 9.23 presumes that there is no heat transfer to the surroundings from the compressor and that there are no kinetic and potential energy changes.

The heat rejected in the condenser is

$$Q_{\text{out}} = \dot{m}(h_2 - h_3) \tag{9.24}$$

With the evaporator taken as a control volume and with operation between state points 4 and 1 in Figure 9.15b, a similar analysis, again with negligible kinetic and potential energy changes, will yield the heat added

$$Q_{\text{in}} = \dot{m}(h_1 - h_4) \tag{9.25}$$

It is important to note that in the throttling process from points 3 to 4,

$$h_3 = h_4 \tag{9.26}$$

We recall that the **coefficient of performance** is employed to measure the effectiveness of a refrigerating system. With the work required taken as

$$\dot{W}_{comp} = Q_{out} - Q_{in}$$

we may refer to Section 3.6.3 to show that for a refrigerator

$$\beta = \frac{Q_{in}}{\dot{W}_{comp}} = \frac{Q_{in}}{Q_{out} - Q_{in}} \tag{3.19}$$

and, through the use of Equation 9.23 and 9.25, we obtain

$$\beta = \frac{\dot{m}(h_1 - h_4)}{\dot{m}(h_2 - h_1)} = \frac{h_1 - h_4}{h_2 - h_1} \tag{9.27}$$

Alternatively, for a heat pump

$$\gamma = \frac{Q_{out}}{\dot{W}_{comp}} = \frac{Q_{out}}{Q_{out} - Q_{in}} \tag{4.20}$$

and using Equations 9.23 and 9.24 we obtain

$$\gamma = \frac{\dot{m}(h_2 - h_3)}{\dot{m}(h_2 - h_1)} = \frac{h_2 - h_3}{h_2 - h_1} \tag{9.28}$$

From Equations 4.19 and 4.20 we have

$$\dot{W}_{comp} = \frac{Q_{in}}{\beta} = \frac{Q_{out}}{\gamma}$$

and in solving for γ we find that

$$\gamma = \frac{Q_{out}}{Q_{in}}\beta = \left(\frac{\dot{W}_{comp} + Q_{in}}{Q_{in}}\right)\beta = \left(\frac{\dot{W}_{comp}}{Q_{in}} + 1\right)\beta = \left(\frac{1}{\beta} + 1\right)\beta$$

or

$$\gamma = \beta + 1 \tag{9.29}$$

The cooling capacity of a refrigeration system is often specified in terms of **tons of refrigeration**. The ton of refrigeration represents the quantity of refrigeration necessary to freeze one ton of water at 32°F into ice at 32°F in one day. In English units with the latent heat of fusion at 144 Btu/lb_m, we have

$$1\text{ ton} = (2000\,\text{lb}_\text{m}/\text{day})(144\,\text{Btu}/\text{lb}_\text{m}) = 288{,}000\,\text{Btu}/\text{day}$$

or

$$1\text{ ton} = 12{,}000\,\text{Btu}/\text{h} = 200\,\text{Btu}/\text{min}$$

In SI units, a ton of refrigeration is equivalent to

$$1\text{ ton} = 211\,\text{kJ}/\text{min} = 3.5167\,\text{kW}$$

We now turn to an example of an *ideal* vapor compression unit acting as a refrigerator.

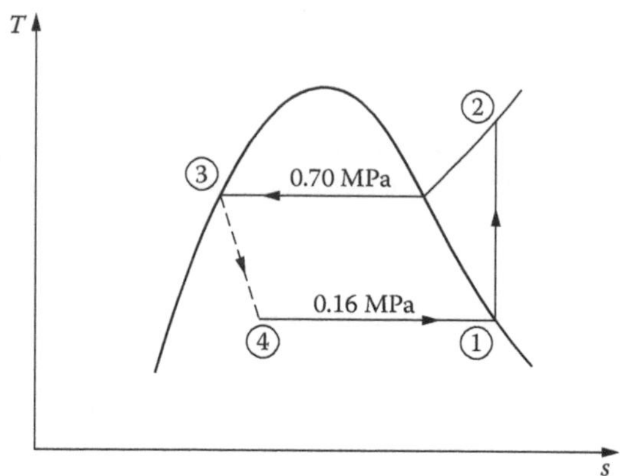

FIGURE 9.16
T-*s* diagram for Example 9.6.

Example 9.6 An ideal refrigeration system uses refrigerant R-134a as the working fluid. Saturated vapor enters the compressor at 0.16 MPa and saturated liquid leaves the condenser at 0.70 MPa. The capacity of the unit is 3.5 tons. Determine (a) the mass flow rate of refrigerant, (b) the power necessary to drive the compressor, and (c) the coefficient of performance of the unit.

Solution

Assumptions and Specifications

(1) The refrigerant-vapor mixture contained in the component parts and the interconnecting piping forms a closed system.
(2) The system operates at steady state.
(3) Each component part of the system can be treated as a control volume.
(4) The refrigeration cycle is idealized to include one isentropic, two constant pressure, and one throttling process.
(5) Kinetic and potential energy changes are negligible.
(6) There are no pressure losses in the system components or in the interconnecting piping.

The T-s diagram for this cycle is showed in Figure 9.16.
For point 1 at 0.16 MPa, Table A.7 gives for saturated vapor

$$h_1 = 237.97\,\text{kJ/kg} \quad \text{and} \quad s_1 = 0.9295\ \text{kJ/kg-K}$$

and for point 2 at 0.70 MPa with $s_2 = s_1 = 0.9295\,\text{kJ/kg-K}$, Table A.8 gives (with interpolation)

$$T_2 = 32.8°\text{C} \quad \text{and} \quad h_2 = 268.40\,\text{kJ/kg}$$

Point 3 is at 0.70 MPa and is saturated liquid. Table A.7 gives

$$h_3 = 86.78\,\text{kJ/kg}$$

and, because of the throttling process between points 3 and 4,

$$h_4 = 86.78\,\text{kJ/kg}$$

(a) The refrigerant flow rate is determined from a form of Equation 9.24.

$$\dot{m} = \frac{Q_{in}}{h_1 - h_4} = \frac{(3.5\,\text{tons})(3.5167\ \text{kW/ton})}{237.97\,\text{kJ/kg} - 86.78\,\text{kJ/kg}}$$

$$= \frac{12.309\ \text{kW}}{151.19\,\text{kJ/kg}} = 0.0814\,\text{kg/s} \Longleftarrow$$

(b) The power required to drive the compressor is given by Equation 9.23.

$$\dot{W}_{comp} = \dot{m}(h_2 - h_1)$$

$$= (0.0814\,\text{kg/s})(268.40\,\text{kJ/kg} - 237.97\,\text{kJ/kg})$$

$$= (0.0814\,\text{kg/s})(30.43\,\text{kJ/kg}) = 2.477\ \text{kW} \Longleftarrow$$

(c) The coefficient of performance is given by Equation 9.27.

$$\beta = \frac{h_1 - h_4}{h_2 - h_1} = \frac{237.97\,\text{kJ/kg} - 86.78\,\text{kJ/kg}}{268.40\,\text{kJ/kg} - 237.97\,\text{kJ/kg}} = \frac{151.19\,\text{kJ/kg}}{30.43\,\text{kJ/kg}} = 4.97 \Longleftarrow$$

This result can be checked

$$\beta = \frac{Q_{in}}{\dot{W}_{comp}} = \frac{12.309\ \text{kW}}{2.477\ \text{kW}} = 4.97$$

Example 9.7 Suppose that the refrigeration system in Example 9.6 is employed as a heat pump. Determine (a) the temperature at the outlet of the condenser and (b) the coefficient of performance of the unit.

Solution

We make the same assumptions that we made in Example 9.6 and we use the data of Example 9.6.

(a) The temperature at the outlet of the condenser is the saturation temperature of refrigerant R-134a corresponding to 0.70 MPa

$$T_3 = 26.72°\text{C} \Longleftarrow$$

(b) The coefficient of performance is given by Equation 9.28.

$$\gamma = \frac{h_2 - h_3}{h_2 - h_1} = \frac{268.40\,\text{kJ/kg} - 86.78\,\text{kJ/kg}}{268.40\,\text{kJ/kg} - 237.97\,\text{kJ/kg}} = \frac{181.62\,\text{kJ/kg}}{30.43\,\text{kJ/kg}} = 5.97 \Longleftarrow$$

Notice that this satisifies Equation 9.29. Another check is via Equation 9.29 using β from Example 9.6

$$\gamma = \beta + 1 = 4.97 + 1 = 5.97$$

9.11 Departures from the Ideal Refrigeration Cycle

Figure 9.17 shows departures from the ideal refrigeration cycle. Here, we notice points 2 and 2s at the outlet of the compressor and points 3a and 3 at the outlet of the condenser.

Heat transfer from the compressor to the environment creates a departure from the isentropic compression of the ideal refrigeration cycle. This departure is from point 2s that now marks the outlet of the compressor for the isentropic case to point 2, which is at the outlet of the compressor in the "real-world" nonideal case. The enthalpies of the two points are related by the **isentropic compressor efficiency**

$$\eta_s = \frac{h_{2s} - h_1}{h_2 - h_1} \tag{9.30}$$

Observe that if there is no change at point 3 in Figure 9.17, the shift from point 2s to point 2 results in an increase of the work input to the compressor while there is no change in the refrigeration capacity. This, of course, results in a degraded coefficient of performance.

In the actual case, pressure losses cause the refrigerant to leave the condenser as a subcooled liquid. This results in a change from point 3a (the original exit from the condenser in the ideal case) to point 3. This, in fact, is not a serious concern because it ensures that the refrigerant leaving the condenser is a liquid and, depending on the location of point 3, may even result in an increased refrigeration effect.

Example 9.8 Suppose that in Example 9.6, the compressor operates with an isentropic efficiency of 84% and that the temperature at the outlet of the condenser is 24°C. Determine the coefficient of performance for these conditions.

Solution

We make the same assumptions that we made in Example 9.6. In addition, we note that there is some pressure loss in the condenser and that there is a heat transfer through the walls of the compressor.

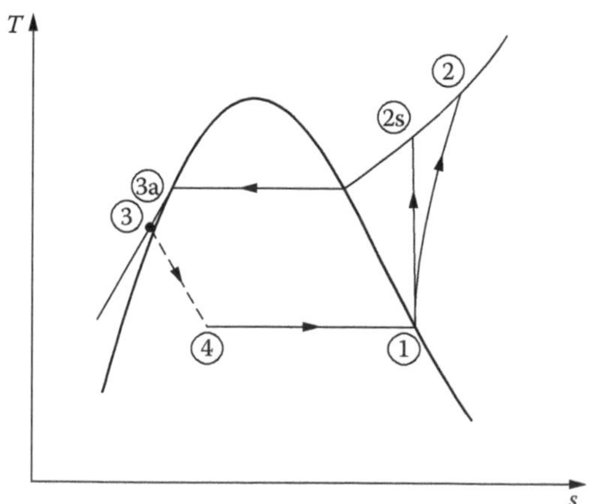

FIGURE 9.17
T-s diagram showing departures from the ideal refrigeration cycle.

Here, we designate h_{2s} as enthalpy at the outlet of the compressor for the isentropic compression. From Example 9.6

$$h_1 = 237.97\,\text{kJ/kg} \quad \text{and} \quad h_{2s} = 268.40\ \text{kJ/kg}$$

Equation 9.30 allows for the calculation of the actual h_2

$$\begin{aligned} h_2 &= h_1 + \frac{h_{2s} - h_1}{\eta_s} \\ &= 237.97\,\text{kJ/kg} + \frac{268.40\,\text{kJ/kg} - 237.97\,\text{kJ/kg}}{0.84} \\ &= 237.97\,\text{kJ/kg} + 36.23\,\text{kJ/kg} = 274.20\,\text{kJ/kg} \end{aligned}$$

Then, from Table A.6 at 24°C, we obtain $v_f = 8.257 \times 10^{-4}\,\text{m}^3/\text{kg}$, $h_f = 82.90\,\text{kJ/kg}$ and $P_{\text{sat}} = 0.64566\,\text{MPa}$. We then determine h_3 as

$$\begin{aligned} h_3 &= h_f + v_f(P_3 - P_{\text{sat}}) \\ &= 82.90\,\text{kJ/kg} + (8.257 \times 10^{-4}\,\text{m}^3/\text{kg})(700\,\text{kPa} - 645.66\,\text{kPa}) \\ &= 82.90\,\text{kJ/kg} + 0.05\,\text{kJ/kg} = 82.95\,\text{kJ/kg} \end{aligned}$$

In summary with $h_4 = h_3$

$$h_1 = 237.97\,\text{kJ/kg} \qquad h_2 = 274.20\,\text{kJ/kg}$$

$$h_2 = 82.95\,\text{kJ/kg} \qquad h_4 = 82.95\,\text{kJ/kg}$$

and

$$\beta = \frac{h_1 - h_4}{h_2 - h_1} = \frac{237.90\,\text{kJ/kg} - 82.95\,\text{kJ/kg}}{274.20\,\text{kJ/kg} - 237.97\,\text{kJ/kg}} = \frac{154.95\,\text{kJ/kg}}{36.23\,\text{kJ/kg}} = 4.28 \Longleftarrow$$

9.12 Summary

The ideal Rankine cycle possesses two isentropic processes and two constant pressure processes. Work is done isentropically in a prime mover such as a turbine or steam engine and the work required to drive the feedwater pump, also accomplished isentropically, is charged to the net power developed in the cycle. Heat is added in the steam generator and rejected in the condenser at constant pressure.

The T-s diagram for the basic ideal Rankine cycle is shown in Figure 9.2. The net work is given by the difference between the turbine work

$$\dot{W}_{\text{turb}} = \dot{m}(h_3 - h_4) \tag{9.1}$$

and the pump work

$$\dot{W}_{\text{pump}} = \dot{m}(h_2 - h_1) \tag{9.2}$$

and is

$$\dot{W}_{\text{cycle}} = \dot{W}_{\text{turb}} - \dot{W}_{\text{pump}} = \dot{m}[(h_3 - h_4) - (h_2 - h_1)] \tag{9.3}$$

The heat input is

$$\dot{Q}_{in} = \dot{m}(h_3 - h_2) \tag{9.4}$$

and the thermal efficiency of the ideal Rankine cycle is

$$\eta = 1 - \frac{h_4 - h_1}{h_3 - h_2} \tag{9.6}$$

The back work ratio is

$$\text{bwr} = \frac{\dot{W}_{pump}}{\dot{W}_{turb}} = \frac{h_2 - h_1}{h_3 - h_4} \tag{9.7}$$

When a superheater is added to the steam generator, the T-s diagram is as indicated in Figure 9.6 and Equations 9.1 through 9.7 may be used to predict the Rankine cycle performance. However, when a reheater is added to the steam generator, two additional state points are added to the T-s diagram as shown in Figure 9.9. This revision leads to a modification of Equations 9.1 through 9.9. The net work of the cycle is the difference between the work done by the turbine

$$\dot{W}_{turb} = \dot{m}[(h_3 - h_5) - (h_6 - h_4)] \tag{9.11}$$

and the work required to drive the pump

$$\dot{W}_{pump} = \dot{m}(h_2 - h_1) \tag{9.2}$$

The net work in the cycle is therefore,

$$\dot{W}_{cycle} = \dot{W}_{turb} - \dot{W}_{pump} = \dot{m}[(h_3 - h_5) + (h_6 - h_4) - (h_2 - h_1)] \tag{9.12}$$

The heat input is

$$\dot{Q}_{in} = \dot{m}[(h_3 - h_2) + (h_6 - h_5)] \tag{9.13}$$

and the thermal efficiency is

$$\eta = \frac{\dot{W}_{cycle}}{\dot{Q}_{in}} = \frac{\dot{m}[(h_3 - h_5) - (h_6 - h_4) - (h_2 - h_1)]}{\dot{m}[(h_3 - h_2) - (h_6 - h_5)]} \tag{9.14}$$

In this case, the back work ratio will be

$$\text{bwr} = \frac{\dot{W}_{pump}}{\dot{W}_{turb}} = \frac{h_2 - h_1}{(h_3 - h_5) - (h_6 - h_4)} \tag{9.15}$$

If regenerative feed water heating, represented by the T-s diagram of Figure 9.13a is employed, the extraction fraction is given by

$$y \equiv \frac{\dot{m}_5}{\dot{m}_3} \tag{9.16}$$

or

$$y = \frac{h_6 - h_2}{h_5 - h_2} \tag{9.17}$$

The power developed by turbine and the power required to drive the *two* pumps are both functions of the mass flow entering stage 1 of the turbine

$$\frac{\dot{W}_{turb}}{\dot{m}_3} = h_3 - h_5 + (1 - y)(h_5 - h_4) \tag{9.18a}$$

and

$$\frac{\dot{W}_{pump}}{\dot{m}_3} = h_7 - h_6 + (1 - y)(h_2 - h_1) \tag{9.18b}$$

The net power for the cycle will be the difference between Equations 9.18a and 9.18b

$$\frac{\dot{W}_{cycle}}{\dot{m}_3} = (h_3 - h_5) - (h_7 - h_6) + (1 - y)[(h_5 - h_4) - (h_2 - h_1)] \tag{9.19}$$

The heat input is

$$\frac{\dot{Q}_{in}}{\dot{m}_3} = h_3 - h_7 \tag{9.20}$$

and the thermal efficiency of the cycle is

$$\eta = \frac{(h_3 - h_5) - (h_7 - h_6) + (1 - y)[(h_5 - h_4) - (h_2 - h_1)]}{h_3 - h_7} \tag{9.22}$$

The efficiency of the Rankine cycle is vulnerable to the irreversibilities in the pump and turbine described by the isentropic efficiencies

$$\eta_{pump} = \frac{h_{2s} - h_1}{h_2 - h_1}$$

where h_{2s} is the specific enthalpy of the water at the pump outlet when the pump irreversibility is included and

$$\eta_{turb} = \frac{h_3 - h_4}{h_3 - h_{4s}}$$

where h_{4s} is the specific enthalpy of the steam (or steam-water mixture) at the turbine outlet.

The T-s diagram for an ideal vapor compression refrigeration cycle is shown in Figure 9.15b and the cycle is seen to consist of an isentropic compression, a constant pressure heat rejection, a throttling process, and a constant pressure heat addition.

The work required to drive the compressor is

$$\dot{W}_{comp} = \dot{m}(h_2 - h_1) \tag{9.23}$$

The heat rejected is

$$Q_{out} = \dot{m}(h_2 - h_3) \tag{9.24}$$

and this may be used in a heat pump application for maintaining a heated space at a given temperature. The heat added is given by

$$Q_{in} = \dot{m}(h_1 - h_4) \tag{9.25}$$

and this may be employed in a refrigeration application to maintain a refrigerated space at a prescribed temperature.

We recall that the **coefficient of performance** and *not* the efficiency is employed to measure the effectiveness of a refrigerating system. For a refrigerator

$$\beta = \frac{\dot{m}(h_1 - h_4)}{\dot{m}(h_2 - h_1)} = \frac{h_1 - h_4}{h_2 - h_1} \tag{9.27}$$

and for a heat pump

$$\gamma = \frac{\dot{m}(h_2 - h_3)}{\dot{m}(h_2 - h_1)} = \frac{h_2 - h_3}{h_2 - h_1} \tag{9.28}$$

and these coefficients of performance are related by

$$\gamma = \beta + 1 \tag{9.29}$$

The conversion between a ton of refrigeration and power in kW is

$$1\,\text{ton} = 211\,\text{kJ/min} = 3.5167\,\text{kW}$$

and the isentropic compressor efficiency is given by

$$\eta_s = \frac{h_{2s} - h_1}{h_2 - h_1} \tag{9.30}$$

9.13 Problems

The Rankine Cycle with Superheat

9.1: In an ideal Rankine cycle, steam enters the turbine at 4 MPa and 400°C. The turbine delivers 12 MW and the condenser operates at 50 kPa. Determine (a) the thermal efficiency and (b) the mass flow rate of the steam.

9.2: Suppose that in Problem 9.1, the conditions at the turbine inlet are changed to 5 MPa and 500°C. All other conditions remain the same. Determine (a) the thermal efficiency and (b) the mass flow rate of the steam.

9.3: Suppose that in Problem 9.1, the conditions at the turbine inlet are changed to 6 MPa and 600°C. All other conditions remain the same. Determine (a) the thermal efficiency and (b) the mass flow rate of the steam.

9.4: Suppose that in Problem 9.1, the conditions at the turbine inlet are changed to 8 MPa and 750°C. All other conditions remain the same. Determine (a) the thermal efficiency and (b) the mass flow rate of the steam.

9.5: In an ideal Rankine cycle, steam enters the turbine at 8 MPa and 750°C. The turbine delivers 12 MW and the condenser operates at 40 kPa. Determine (a) the thermal efficiency and (b) the mass flow rate of the steam.

9.6: Suppose that in Problem 9.5, the turbine discharges steam to the condenser at a pressure of 30 kPa. All other conditions remain the same. Determine (a) the thermal efficiency and (b) the mass flow rate of the steam.

9.7: Suppose that in Problem 9.5, the turbine discharges steam to the condenser at a pressure of 20 kPa. All other conditions remain the same. Determine (a) the thermal efficiency and (b) the mass flow rate of the steam.

9.8: Suppose that in Problem 9.5, the turbine discharges steam to the condenser at a pressure of 10 kPa. All other conditions remain the same. Determine (a) the thermal efficiency and (b) the mass flow rate of the steam.

9.9: In an ideal Rankine cycle, steam enters the turbine at 1.5 MPa and 500°C. The Rankine steam rate (the steam flow) is 10,000 kg/s, and the condenser operates at 8 kPa. Determine (a) the thermal efficiency and (b) the power output of the cycle.

9.10: In an ideal Rankine cycle, steam enters the turbine at 10 MPa and 850°C. The turbine delivers 300 MW and the condenser operates at 7.5 kPa. Determine (a) the thermal efficiency, (b) the mass flow rate of the steam output, and (c) the heat rejected by the condenser.

9.11: In an ideal Rankine cycle, steam enters the turbine at 8 MPa and 650°C. The turbine delivers 300 MW and the condenser operates at 5 kPa. Determine (a) the thermal efficiency, and (b) the mass flow rate of the steam, and (c) the mass flow rate of the condenser cooling water ($c = 4.177$ kJ/kg-K) needed to limit its temperature rise to 9.6°C.

9.12: The power output of an ideal Rankine cycle is limited by the condenser that can handle 200 MW. Steam enters the turbine at 5 MPa and 800°C and leaves at 20 kPa. Determine the cycle power output.

9.13: In an ideal Rankine cycle, steam enters the turbine at 5 MPa and the condenser operates at 10 kPa. Determine the temperature at the turbine inlet if the quality of the steam entering the condenser is limited to $x = 0.90$.

9.14: In an ideal Rankine cycle, steam at a flow rate of 125 kg/s, enters the turbine at 5 MPa and 750°C. The condenser operates at 10 kPa and the steam generator uses coal with a heating value of 29,800 kJ/kg. Determine the quantity of coal used.

9.15: In an ideal Rankine cycle, steam enters the turbine at 4 MPa and 600°C. If the condenser operates at 10 kPa, determine the back work ratio.

9.16: In an ideal Rankine cycle, steam enters the turbine at 10 MPa and 800°C with a steam flow of 120 kg/s. The condenser operates at 50 kPa. Determine (a) the quality at the turbine exhaust, (b) the thermal efficiency, and (c) the cycle power output.

The Rankine Cycle with Reheat

9.17: In an ideal Rankine cycle, steam enters the turbine at 10 MPa and 600°C with a steam flow of 120 kg/s. When the pressure in the turbine reaches 3 MPa, the steam is reheated to 600°C. The condenser operates at 50 kPa. Determine (a) the quality of the steam at the turbine exhaust, (b) the thermal efficiency, and (c) the power at the turbine shaft.

9.18: Suppose that in the ideal Rankine cycle of Problem 9.17, the steam enters the turbine at 700°C. All other conditions remain the same. Determine (a) the quality of the steam at the turbine exhaust, (b) the thermal efficiency, and (c) the power at the turbine shaft.

9.19: Suppose that in the ideal Rankine cycle of Problem 9.17, the steam enters the turbine at 800°C. All other conditions remain the same. Determine (a) the quality of the steam at the turbine exhaust, (b) the thermal efficiency, and (c) the power at the turbine shaft.

9.20: Suppose that in the ideal Rankine cycle of Problem 9.19, the steam enters the reheater at 700°C. All other conditions remain the same. Determine (a) the quality of the steam at the turbine exhaust, (b) the thermal efficiency, and (c) the power at the turbine shaft.

9.21: Suppose that in the ideal Rankine cycle of Problem 9.19, the steam enters the reheater at 800°C. All other conditions remain the same. Determine (a) the quality of the steam at the turbine exhaust, (b) the thermal efficiency, and (c) the power at the turbine shaft.

9.22: In an ideal Rankine cycle, steam enters the turbine at 10 MPa and 600°C with a steam flow of 120 kg/s. When the pressure in the turbine reaches 2.5 MPa, the steam is reheated to 600°C. The condenser operates at 50 kPa. Determine (a) the quality of the steam at the turbine exhaust, (b) the thermal efficiency, and (c) the power at the turbine shaft.

9.23: Suppose that in the ideal Rankine cycle of Problem 9.22, the steam enters the turbine at 700°C. All other conditions remain the same. Determine (a) the quality of the steam at the turbine exhaust, (b) the thermal efficiency, and (c) the power at the turbine shaft.

9.24: Suppose that in the ideal Rankine cycle of Problem 9.22, the steam leaves the turbine at 800°C. All other conditions remain the same. Determine (a) the quality of the steam at the turbine exhaust, (b) the thermal efficiency, and (c) the power at the turbine shaft.

9.25: Suppose that in the ideal Rankine cycle of Problem 9.24, the steam leaves the reheater at 700°C. All other conditions remain the same. Determine (a) the quality of the steam at the turbine exhaust, (b) the thermal efficiency, and (c) the power at the turbine shaft.

9.26: Suppose that in the ideal Rankine cycle of Problem 9.24, the steam enters the reheater at 800°C. All other conditions remain the same. Determine (a) the quality of the steam at the turbine exhaust, (b) the thermal efficiency, and (c) the power at the turbine shaft.

9.27: Suppose that in the ideal Rankine cycle of Problem 9.22, steam enters the condenser at 5 kPa. All other conditions remain the same. Determine (a) the quality of the steam at the turbine exhaust, (b) the thermal efficiency, and (c) the power at the turbine shaft.

9.28: Suppose that in the ideal Rankine cycle of Problem 9.22, steam enters the condenser at 10 kPa. All other conditions remain the same. Determine (a) the quality of the steam at the turbine exhaust, (b) the thermal efficiency, and (c) the power at the turbine shaft.

9.29: Suppose that in the ideal Rankine cycle of Problem 9.22, steam enters the condenser at 15 kPa. All other conditions remain the same. Determine (a) the quality of the steam at the turbine exhaust, (b) the thermal efficiency, and (c) the power at the turbine shaft.

9.30: Suppose that in the ideal Rankine cycle of Problem 9.22, steam enters the condenser at 20 kPa. All other conditions remain the same. Determine (a) the quality of the steam at the turbine exhaust, (b) the thermal efficiency, and (c) the power at the turbine shaft.

9.31: Suppose that in the ideal Rankine cycle of Problem 9.22, steam enters the condenser at 25 kPa. All other conditions remain the same. Determine (a) the quality of the steam at the turbine exhaust, (b) the thermal efficiency, and (c) the power at the turbine shaft.

9.32: Suppose that in the ideal Rankine cycle of Problem 9.22, steam enters the condenser at 40 kPa. All other conditions remain the same. Determine (a) the quality of the steam at the turbine exhaust, (b) the thermal efficiency, and (c) the power at the turbine shaft.

The Effect of Irreversibilities

9.33: Superheated steam enters a turbine at 10 MPa and 800°C and is exhausted at 10 kPa and 150°C. Determine the isentropic efficiency of the turbine.

9.34: Steam enters an adiabatic turbine at 10 MPa and 550°C with a mass flow rate of 3.25 kg/s and leaves at 10 kPa. The isentropic efficiency of the turbine is 0.875. Determine (a) the steam temperature at the turbine exit and (b) the power output of the turbine.

9.35: Superheated steam at 5 MPa and 600°C, and 70 m/s enters an adiabatic turbine and leaves at 50 kPa, 100°C, and 150 m/s. The turbine delivers 5.25 MW. Determine (a) the mass flow rate of the steam and (b) the isentropic efficiency of the turbine.

9.36: Air (with variable specific heats) enters an adiabatic nozzle at 400 kPa and 500 K at a rather low velocity. The isentropic efficiency of the nozzle is 0.880 and the exit velocity is 250 m/s. Determine (a) the outlet temperature and (b) the outlet pressure.

9.37: Nitrogen, which may be assumed as an ideal gas with constant specific heats, enters an adiabatic turbine at 1.5 MPa and 250°C at a flow rate of 90 kg/min and leaves at 200 kPa. The turbine develops 600 kW. Determine the isentropic efficiency.

9.38: An adiabatic gas turbine is to handle combustion gases that enter at 847°C and 800 kPa and leave at 490 kPa. The combustion gases may be treated as air (ideal gas) with variable specific heats. Changes in kinetic and potential energy are negligible and the isentropic efficiency of the turbine is taken as 0.875. For a mass flow rate through the turbine at 2 kg/s, determine the power delivered by the turbine.

9.39: Air is compressed in a adiabatic compressor from 100 kPa and 27°C to 650 kPa and 287°C. Neglect kinetic and potential energy changes and assume variable specific heats to determine (a) the isentropic efficiency of the compressor and (b) the exit temperature of the air after an isentropic compression.

9.40: An adiabatic compressor operates with refrigerant R-134a. Saturated vapor enters at 1.2 bar at a volumetric flow rate of 0.0052 m^3/s and leaves at 12 bar. The isentropic efficiency of the compressor is 0.825. Determine (a) the exit temperature of the refrigerant and (b) the power required to drive the compressor.

The Ideal Rankine Cycle with Regeneration

9.41: A Rankine steam power plant incorporates an open feedwater heater that operates steadily at 1 MPa. The extracted steam enters the heater at 1 MPa and a quality of 0.92 and the condensate enters the heater at 1 MPa and 115°C. The water leaves the heater at 1 MPa and has a flow rate of 32 kg/s. Determine the mass flow of the extracted steam.

9.42: The boiler in a Rankine cycle steam power plant is designed to generate 35 kg/s of steam at 7 MPa and 700°C. The condenser pressure is 10 kPa and rejects heat to cooling water causing its temperature to rise by 8°C. The plant incorporates an open feedwater heater that receives saturated steam at 2 MPa. The condensate leaves the feedwater heater as saturated liquid at 2 MPa. Determine (a) the mass flow rate of extracted steam, (b) the mass flow rate of cooling water, (c) the back work ratio, and (d) the thermal efficiency.

9.43: Determine the thermal efficiency of a regenerative Rankine cycle with steam delivered to the turbine at 7 MPa and 650°C and the condenser operating at 50 kPa. Steam is extracted at 400 kPa and is led to an open feedwater heater that supplies saturated liquid to a second pump.

9.44: Rework Problem 9.42 with the isentropic efficiency of each pump taken as 0.80. All other conditions remain the same.

9.45: Rework Problem 9.42 with the isentropic efficiency of the turbine from the point of extraction to the exhaust into the condenser is 0.85. All other conditions remain the same.

9.46: An ideal Rankine cycle with regeneration has a boiler that generates steam at 10 MPa and 800°C. The condenser operates at a pressure of 20 kPa. A fraction of the steam flowing through the turbine is extracted and diverted to an open feedwater heater that delivers condensate as saturated liquid at the extracted pressure. Determine the thermal efficiency of the cycle for an extraction pressure of 6 MPa.

9.47: An ideal Rankine cycle with regeneration has a boiler that generates steam at 10 MPa and 800°C. The condenser operates at a pressure of 20 kPa. A fraction of the steam flowing through the turbine is extracted and diverted to an open feedwater heater that delivers condensate as saturated liquid at the extracted pressure. Determine the thermal efficiency of the cycle for an extraction pressure of 5 MPa.

9.48: An ideal Rankine cycle with regeneration has a boiler that generates steam at 10 MPa and 800°C. The condenser operates at a pressure of 20 kPa. A fraction of the steam flowing through the turbine is extracted and diverted to an open feedwater heater that delivers condensate as saturated liquid at the extracted pressure. Determine the thermal efficiency of the cycle for an extraction pressure of 4 MPa.

9.49: An ideal Rankine cycle with regeneration has a boiler that generates steam at 10 MPa and 800°C. The condenser operates at a pressure of 20 kPa. A fraction of the steam flowing through the turbine is extracted and diverted to an open feedwater heater that delivers condensate as saturated liquid at the extracted pressure. Determine the thermal efficiency of the cycle for an extraction pressure of 3 MPa.

The Vapor Compression Refrigeration Cycle

9.50: On an as yet undiscovered planet (somewhere between Earth and Venus), the average environmental temperature is 82°C. A vapor compression refrigerating system is to be designed using water as the refrigerant with 35°C as the evaporator temperature and with saturated vapor at 100°C leaving the compressor. Saturated liquid enters the expansion valve. For a cooling capacity of 30 tons in an ideal cycle, determine (a) the mass flow rate of refrigerant, (b) the required power, and (c) the coefficient of performance.

9.51: Modify the isentropic compressor in Problem 9.50 so that its isentropic efficiency is 75%. Then find the new value of the coefficient of performance.

9.52: Modify Problem 9.50 for the condition that compressed liquid at 95°C enters the expansion valve. Then find the new value of the coefficient of performance.

9.53: Refrigerant R-134a enters the compressor in an ideal vapor compression refrigeration cycle as saturated vapor at 0.8 bar and leaves the compressor as saturated vapor at 6 bars. The expansion valve receives the refrigerant as saturated liquid. The flow rate of the refrigerant is 0.135 kg/s. For the ideal cycle, determine (a) the refrigeration capacity in tons, (b) the power input to the compressor, (c) the coefficient of performance, and (d) the coefficient of performance if the system is operated as a heat pump.

9.54: Modify the isentropic compressor in Problem 9.53 so that its isentropic efficiency is 80%. Then find the new value of the coefficient of performance.

9.55: Modify Problem 9.53 for the condition that compressed liquid at 16°C enters the expansion valve. Then find the new value of the coefficient of performance.

9.56: Refrigerant R-134a is used in an ideal vapor compression refrigeration cycle operating between 0.16 MPa and 1.0 MPa. The refrigerant is a saturated vapor entering the compressor. The refrigerating capacity is 10 tons. Determine (a) the vapor temperature at exit from the compressor, (b) the work required to drive the compressor, (c) the coefficient of performance, and (d) the coefficient of performance if the system is operated as a heat pump.

9.57: Modify the isentropic compressor in Problem 9.56 so that its isentropic efficiency is 78%. Then determine (a) the temperature leaving the compressor, (b) the power input to the compressor, and (c) the coefficient of performance.

9.58: Refrigerant R-134a is used in an ideal vapor compression refrigeration cycle operating between 0.16 and 0.8 MPa. The refrigerant is a saturated vapor entering the compressor. The refrigerating capacity is 20 tons. Determine (a) the vapor temperature at exit from the compressor, (b) the mass flow rate of the refrigerant, (c) the work required to drive the compressor, (d) the coefficient of performance, and (e) the coefficient of performance if the system is operated as a heat pump.

9.59: Modify the isentropic compressor in Problem 9.58 so that its isentropic efficiency is 80%. Then determine (a) the temperature leaving the compressor, (b) the coefficient of performance, and (c) the power input to the compressor.

9.60: Refrigerant R-134a enters the compressor of an ideal vapor compression refrigeration system as a saturated vapor at −16°C at a volumetric flow rate of 1.8 m^3/min. The refrigerant leaves the condenser is a saturated liquid at 0.90 MPa. Determine (a) the refrigerating capacity in tons, (b) the work required to drive the compressor, (c) the coefficient of performance, and (d) the coefficient of performance if the system is operated as a heat pump.

10

Mixtures of Gases, Vapors, and Combustion Products

Chapter Objectives

- To indicate how to convert from a gravimetric analysis to a volumetric analysis (and the reverse) of a mixture of ideal gases.
- To show how the properties of a mixture of ideal gases are determined.
- To consider the properties of an ideal gas and vapor mixture including a consideration of humidity.
- To illustrate the use of the psychrometric chart.
- To provide the essentials for preliminary combustion calculations.

10.1 Introduction

In this chapter, we devote our attention to mixtures of ideal gases and vapors. We discuss what is meant by a molal or volumetric analysis and by a weight or gravimetric analysis, the Dalton law of partial pressures, and the apparent molecular weight and gas constant of a mixture of ideal gases. Properties of ideal gases are considered, specific and relative humidity are discussed, and the use of the psychrometric chart is demonstrated. Finally, attention is focused on fuels, combustion, and the analysis of the products of combustion.

10.2 Mixtures of Ideal Gases

10.2.1 Development

We begin by considering a closed system composed of a mixture of two or more ideal gas components. Because the whole is equal to the sum of the parts, the composition of the mixture (with the total mixture designated by the subscript m) can be described either on a mass basis

$$m_m = m_1 + m_2 + m_3 + \cdots + m_k = \sum_{i=1}^{k} m_i \tag{10.1}$$

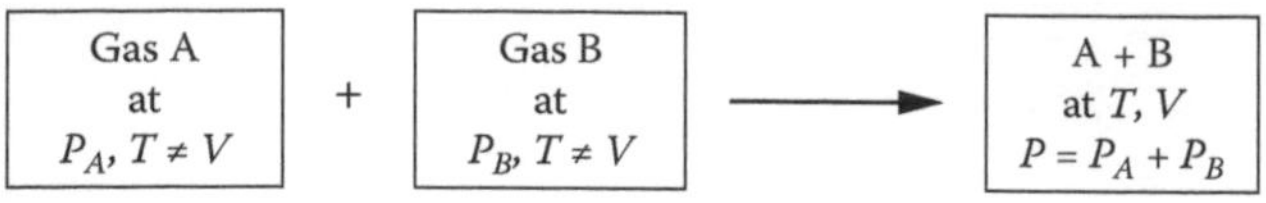

Gas A at P_A, $T \neq V$	+	Gas B at P_B, $T \neq V$	→	A + B at T, V $P = P_A + P_B$

FIGURE 10.1
Two gases used to illustrate Dalton's law of partial pressures.

or on a mole basis

$$n_m = n_1 + n_2 + n_3 + \cdots + n_k = \sum_{i=1}^{k} n_i \tag{10.2}$$

The relationship between the mass and the number of moles of each component is

$$m_i = n_i M_i \qquad \text{or} \qquad n_i = \frac{m_i}{M_i} \tag{10.3}$$

and we may define the **mass fraction** (or **gravimetric fraction**) as

$$G_i = \frac{m_i}{\sum_{i=1}^{k} m_i} = \frac{m_i}{m_m} \tag{10.4}$$

In similar fashion, we may define the **mole fraction** as

$$Y_i = \frac{n_i}{\sum_{i=1}^{k} n_i} = \frac{n_i}{n_m} \tag{10.5}$$

From these relationships it is clear that

$$\sum_{i=1}^{k} G_i = 1 \tag{10.6}$$

and

$$\sum_{i=1}^{k} Y_i = 1 \tag{10.7}$$

Equations 10.6 and 10.7 show that the sum of the gravimetric and mole fractions of a mixture are both equal to unity.

10.2.2 Dalton's Law of Partial Pressures

Two gases, A and B, are shown in Figure 10.1. The Dalton law of partial pressures pertains to n gases.

> The total pressure exerted by a mixture of gases is the sum of the pressures, P_i, that each gas would exert if it existed alone at the mixture temperature and volume.

Thus, for k-components, if

$$T_1 = T_2 = T_3 = \cdots = T_k = T_m$$

and

$$V_1 = V_2 = V_2 = \cdots = V_k = V_m$$

Dalton's law states that

$$P_m = P_1 + P_2 + P_3 + \cdots + P_k = \sum_{i=1}^{k} P_i \tag{10.8}$$

Here, the P_i's are the component pressures of each component and these are also known as the **partial pressures.** Because $T_i = T_m$ and $v_i = v_m$, the ratio of each partial pressure to the total pressure of the mixture will be

$$\frac{P_i}{P_m} = \frac{n_i \bar{R} T_i / V_i}{n_m \bar{R} T_m / V_m} = \frac{n_i T_m / V_m}{n_m T_m / V_m} = \frac{n_i}{n_m} = Y_i$$

and this shows that the ratio of the partial pressure of the ith component to the total pressure is equal to the mole or volumetric fraction

$$P_i = Y_i P_m \tag{10.9}$$

Note that Equation 10.9 is only valid for ideal gas mixtures.

10.2.3 Gravimetric and Volumetric Analyses

For an ideal gas mixture of k-components, a listing of the mass fractions constitutes a **gravimetric analysis.** Indeed, this may be extended to a listing of the mole or volumetric fractions, which represents a **volumetric analysis**. From the development in Section 10.2.1, we see that, if we are given a volumetric analysis, the mass fractions may be obtained as

$$G_i = \frac{Y_i M_i}{\sum Y_i M_i} \qquad (i = 1, 2, 3, \ldots, k) \tag{10.10}$$

If a gravimetric analysis is given, the mole fractions are

$$Y_i = \frac{G_i / M_i}{\sum G_i / M_i} \qquad (i = 1, 2, 3, \ldots, k) \tag{10.11}$$

10.2.4 The Apparent Molecular Weight and Gas Constant

The apparent molecular weight of a mixture, M_m, will be

$$M_m = \frac{m_m}{n_m}$$

and with substitution of Equations 10.3 and 10.10, we find that

$$M_m = \frac{m_m}{n_m} = \frac{\sum_{i=1}^{k} n_i M_i}{n_m} = \sum_{i=1}^{k} Y_i M_i \tag{10.12}$$

which shows that the apparent or average molecular weight of a mixture is the sum of the products of all of the mole fractions (or volumetric fractions) and the individual molecular weights. Moreover, the mixture gas constant will be equal to

$$R_m = \frac{\bar{R}}{M_m} = \frac{8314\,\text{J/kgmol-K}}{M_m} \tag{10.13a}$$

and with $M = m/n$ we have the alternate definition

$$R_m = \frac{\bar{R}}{M_m} = \sum_{i=1}^{k} G_i R_i \tag{10.13b}$$

Example 10.1 Consider air as an ideal gas. It is a fact that, neglecting trace substances, the air possesses a gravimetric analysis of approximately 23.1% oxygen and 76.9% nitrogen. Determine (a) the volumetric analysis, (b) the apparent molecular weight, and (c) the apparent gas constant.

Solution

Assumptions and Specifications

(1) The oxygen and nitrogen may be treated as ideal gases.

(2) The gravimetric fractions of the two gases are specified.

(a) The volumetric analysis can be obtained from Equation 10.11 in an easy-to-use tabular format. Molecular weights are taken from Table A.l in Appendix A.

i	component	$\% G_i$	M_i	$\% G_i/M_i$	$\% Y_i$
1	O_2	23.1	32.00	0.7219	20.82 $\Longleftarrow$
2	N_2	76.9	28.02	2.7445	79.18 $\Longleftarrow$
Σ		100.0		3.4663	

(b) The apparent molecular weight of the mixture may be obtained from Equation 10.12

$$\begin{aligned} M_m &= Y_1 M_1 + Y_2 M_2 \\ &= (0.2082)(32.00\,\text{kg/kgmol}) + (0.7918)(28.02\,\text{kg/kgmol}) \\ &= 6.662\,\text{kg/kgmol} + 22.186\,\text{kg/kgmol} = 28.85\,\text{kg/kgmol} \Longleftarrow \end{aligned}$$

(c) The apparent gas constant of the mixture is obtained from Equation 10.13a.

$$R = \frac{8.314\,\text{kJ/kgmol-K}}{28.85\,\text{kg/kgmol}} = 0.288\,\text{kJ/kg-K} \Longleftarrow$$

Comment: Considering that the given gravimetric analysis does not include the traces of carbon dioxide, argon, other gases and water vapor that are present, the results in (b) and (c) compare favorably with $M_m = 28.97\,\text{kg/kgmol}$ and $R = 0.287\,\text{kJ/kg-K}$ listed in Table A.1.

10.2.5 Properties of Ideal Gas Mixtures

The specific heats, internal energy, and enthalpy and entropy of an ideal gas mixture can be determined by adding the contribution of each component of the mixture. Thus, for the constant pressure and constant volume specific heats

$$c_{p,m} = \sum_{i=1}^{k} G_i c_{p,i} \tag{10.14}$$

and

$$c_{v,m} = \sum_{i=1}^{k} G_i c_{v,i} \tag{10.15}$$

for the internal energy

$$u_m = \sum_{i=1}^{k} G_i u_i \tag{10.16}$$

for the enthalpy

$$h_m = \sum_{i=1}^{k} G_i h_i \tag{10.17}$$

but because the entropy of an ideal gas is a function of *both* pressure and temperature, we find that the change in entropy for the ith component of an ideal gas between state 1 and state 2 will be

$$\Delta s_i = c_{p,i} \ln \frac{T_2}{T_1} - R_i \ln \frac{P_{i2}}{P_{i1}} \tag{10.18}$$

Example 10.2 A two-component ideal gas mixture consisting of 4 kg of carbon monoxide and 2 kg of methane are in a 0.75 m^3 drum at 300 K. Determine (a) the pressure of the mixture, (b) the volumetric analysis, (c) the partial pressures of each component, and (d) the heat required to raise the temperature of the mixture 10°C.

Solution

Assumptions and Specifications

(1) The carbon monoxide and methane may be treated as ideal gases.
(2) The masses of the two gases are specified.
(3) The specific heats of the gases remain constant.

(a) The volumetric analysis can be obtained from Equation 10.11 in an easy-to-use tabular format. Molecular weights are taken from Table A.l.

i	**component**	m**, kg**	$\% G_i$	M_i	$\% G_i/M_i$	$\% Y_i$
1	CO	4	66.67	28.01	2.3803	53.39 $\Longleftarrow$
2	CH_4	2	33.33	16.04	2.0779	46.61 $\Longleftarrow$
Σ			100.0		4.4582	

(b) The apparent molecular weight of the mixture may be obtained from Equation 10.12

$$\begin{aligned} M_m &= Y_1 M_1 + Y_2 M_2 \\ &= (0.5339)(28.01\,\text{kg/kgmol}) + (0.4661)(16.04\,\text{kg/kgmol}) \\ &= 14.95\,\text{kg/kgmol} + 7.48\,\text{kg/kgmol} + = 22.43\,\text{kg/kgmol} \Longleftarrow \end{aligned}$$

and the apparent gas constant of the mixture is obtained from Equation (10.13a)

$$R = \frac{8.314\,\text{kJ/kgmol-K}}{22.43\,\text{kg/kgmol}} = 0.371\,\text{kJ/kg-K} \Longleftarrow$$

Then, the ideal gas law gives

$$P_m = \frac{m_m R_m T_m}{V_m} = \frac{(6\,\text{kg})(0.371\,\text{kJ/kg-K})(300\,\text{K})}{0.75\,\text{m}^3} = 8.904\,\text{bar} \Longleftarrow$$

(c) The partial pressures are proportional to the volumetric fractions. From Equation 10.9 for the carbon dioxide

$$P_1 = Y_1 P_m = (0.5339)(8.904\,\text{bar}) = 4.754\,\text{bar} \Longleftarrow$$

and for the methane

$$P_2 = Y_2 P_m = (0.4661)(8.904\,\text{bar}) = 4.150\,\text{bar} \Longleftarrow$$

Table A.1 gives at 300 K

$$c_{v,1} = 0.745\ \text{kJ/kg-K} \quad \text{and} \quad c_{v,2} = 1.735\,\text{kJ/kg-K}$$

Then, Equation 10.15 provides

$$\begin{aligned} c_{v,m} &= G_1 c_{v,1} + G_2 c_{v,2} \\ &= (0.6667)(0.745\,\text{kJ/kg-K}) + (0.3333)(1.735\,\text{kJ/kg-K}) \\ &= 0.497\,\text{kJ/kg-K} + 0.578\,\text{kJ/kg-K} = 1.075\,\text{kJ/kg-K} \end{aligned}$$

(d) Hence, the heat added to the mixture for a 10°C temperature rise will be

$$Q_m = m_m c_{v,m} \Delta T_m = (6\,\text{kg})(1.075\ \text{kJ/kg-K})(10^\circ\text{C}) = 64.5\,\text{kJ} \Longleftarrow$$

10.3 Psychrometrics

10.3.1 Introduction

If one of the components in a two-component mixture is close to its saturation state, the condensable component is called a **vapor** and the mixture is referred to as a **gas-vapor mixture.** The temperature of this gas-vapor mixture, as measured by a conventional thermometer, is called the **dry bulb temperature**, T_{db}. A dry air-water vapor mixture, which is saturated with water vapor, is frequently referred to as **saturated air**. The study of the behavior of these two-component mixtures is called **psychrometrics.**

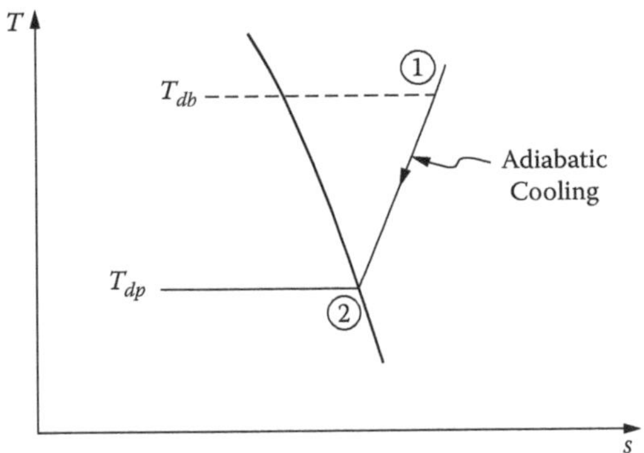

FIGURE 10.2
The T-s plane showing superheated vapor at state 1 with adiabatic cooling to state 2.

The T-s plane for the vapor in an air-vapor mixture is shown in Figure 10.2. When a vapor, at state 1, is superheated and if it is cooled adiabatically at constant pressure, it becomes saturated at point 2 and further cooling will cause some of the vapor to condense. State 2 is called the **dew point** and the temperature at point 2 is called the **dew point temperature,** T_{dp}.

10.3.2 Specific and Relative Humidity

The composition of a given moist air sample can be measured by the **humidity ratio** (also known as the **specific humidity**), ω, or the **relative humidity**, ϕ. With the moisture and dry air designated respectively with the subscripts v and da, the humidity ratio is defined as

$$\omega = \frac{m_v}{m_{\text{da}}} \qquad (\text{kg vapor/kg dry air}) \tag{10.19}$$

Use of the ideal gas law yields

$$\omega = \frac{m_v}{m_{\text{da}}} = \frac{M_v P_v V/\bar{R}T}{M_{\text{da}} P_{\text{da}} V/\bar{R}T} = \frac{M_v P_v}{M_{\text{da}} P_{\text{da}}}$$

With $M_v/M_{\text{da}} = 18.016/28.97 = 0.622$ and Dalton's law, which gives $P_v = P_m - P_{\text{da}}$, we obtain a relationship for the specific humidity in terms of the total pressure and the partial pressure of the water vapor as

$$\omega = 0.622\frac{P_v}{P_m - P_v} \tag{10.20}$$

The composition of a moist air sample can also be expressed in terms of the **relative humidity,** ϕ, which is defined as the ratio of the mole faction of the water vapor, Y_{v}, in the sample to the mole fraction, Y_{g}, in a saturated moist air sample at the same temperature and pressure.

$$\phi = \frac{Y_{\text{v}}}{Y_{\text{g}}}$$

Because Dalton's law tells us that $P_v = Y_{\text{v}} P_m$ and $P_{\text{g}} = Y_{\text{g}} P_m$, this may be written as

$$\phi = \frac{P_v}{P_{\text{g}}} = \frac{\rho_{\text{v}}}{\rho_{\text{g}}} = \frac{v_{\text{g}}}{v_{\text{v}}} \tag{10.21}$$

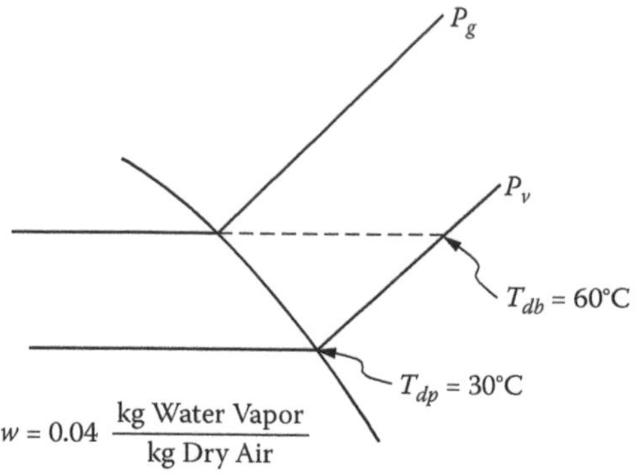

FIGURE 10.3
The T-s plane for Example 10.3.

Example 10.3 A low-pressure air-water vapor mixture at 60°C has a dew point of 30°C. The mass ratio of water vapor to air is 0.04. Determine (a) the humidity ratio, (b) the relative humidity, (c) the mixture pressure, (d) the volumetric fraction of the water vapor, (e) the density of the water vapor, and (f) the volume occupied by the dry air in 1 kg of water vapor (Figure 10.3).

Solution

Assumptions and Specifications

(1) The air and water vapor may be treated as ideal gases.
(2) The mixture dew point is specified.
(3) The water to air mass ratio is specified.

(a) The mass ratio is just another way of specifiying the humidity ratio. Hence,

$$\omega = 0.04\,\text{kg vapor/kg dry air} \Longleftarrow$$

(b) From the dew point temperature of 30°C, we read $P_v = 4.24\,\text{kPa}$ from Table A.3. We then read from Table A.3 at 60°C, a saturation pressure of 19.92 kPa and employ Equation 10.21 to obtain

$$\phi = \frac{P_v}{P_g} = \frac{4.24\,\text{kPa}}{19.92\,\text{kPa}} = 0.213 \qquad (21.3\%) \Longleftarrow$$

(c) The partial pressure of the dry air can be found from a modification of Equation 10.20

$$P_{\text{da}} = \frac{0.622P_v}{\omega} = \frac{0.622(4.24\,\text{kPa})}{0.04\,\text{kg vapor/kg dry air}} = 65.93\,\text{kPa}$$

Then by Dalton's law

$$P_m = P_v + P_{\text{da}} = 4.24\,\text{kPa} + 65.93\,\text{kPa} = 70.17\,\text{kPa} \Longleftarrow$$

(d) The volumetric fraction of the water vapor can be found from Equation 10.9

$$Y_v = \frac{P_{v1}}{P_m} = \frac{4.24\,\text{kPa}}{70.17\,\text{kPa}} = 0.0604 \Longleftarrow$$

(e) With $v_g = 7.6786\,\text{m}^3/\text{kg}$ at 60°C taken from Table A.3, we use Equation 10.21 to obtain

$$\rho_{v1} = \frac{\phi}{v_g} = \frac{0.213}{7.6786\ \text{m}^3/\text{kg}} = 0.0277\,\text{kg/m}^3 \Longleftarrow$$

(f) The total volume of dry air required for 1 kg of water vapor will be

$$V_m = \frac{1\,\text{kg}}{\rho_v} = \frac{1\,\text{kg}}{0.0277\,\text{kg/m}^3} = 36.07\text{m}^3 \Longleftarrow$$

10.3.3 Other Properties

Values of the internal energy, enthalpy, and entropy, for moist air can be obtained from an addition of the contributions of the air and the water vapor. For example, for the enthalpy

$$H_m = m_{\text{da}} h_{\text{da}} + m_v h_v$$

where h_v may be taken as the saturated vapor value, h_g, at the mixture temperature. Division by the mass of air gives the enthalpy value per unit mass of dry air

$$\frac{H_m}{m_{\text{da}}} = h_{\text{da}} + \frac{m_v}{m_{\text{da}}} h_v = h_{\text{da}} + \omega h_g \tag{10.22}$$

Because $m_m = m_v + m_{\text{da}}$, a division by the total volume, $V_m = V_v = V_{\text{da}}$ yields

$$\frac{m_m}{V_m} = \frac{m_v}{V_v} + \frac{m_{\text{da}}}{V_{\text{da}}}$$

or

$$\rho_m = \rho_v + \rho_{\text{da}} \tag{10.23}$$

10.3.4 The Wet Bulb Temperature

Consider the **adiabatic saturator** illustrated in Figure 10.4 and notice that the enclosure is insulated so that no heat can pass between the device and the surrounding environment. There are two inlets and one outlet. In steady flow:

1. At the inlet at point 1, the air-water vapor mixture flows with $\dot{m}_{a1}$ and $\dot{m}_{v1}$ at T_1 and P_1 with an unknown specific humidity, ω_1.

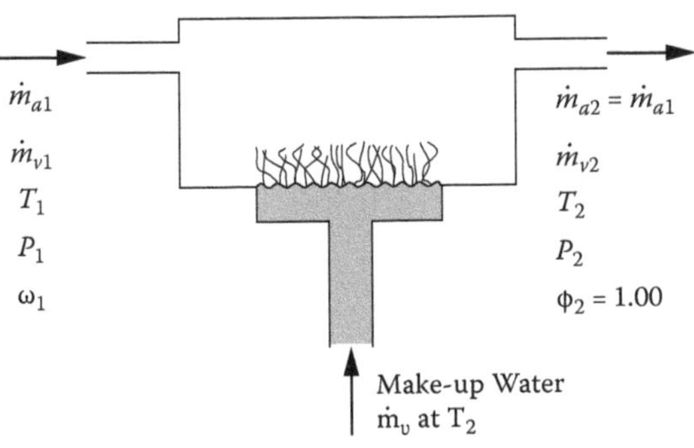

FIGURE 10.4
Adiabatic saturation process.

2. At the outlet at point 2, the mixture is presumed to be **saturated** with $\phi = 100\%$ at flow rates of $\dot{m}_{a2} = \dot{m}_{a1}$ and $\dot{m}_{v2}$ at T_2 and P_2.
3. At point 3, continuity of mass demands that enough water, $\dot{m}_\ell$, be added at the saturation temperature, T_2, to **make-up** for the moisture picked up by the air stream as it flows from point 1 to point 2

$$\dot{m}_\ell = \dot{m}_{v2} - \dot{m}_{v1}$$

We immediately observe that the heat required to evaporate some of the water in the pool must come from the sensible heat contained in the flowing air stream. Moreover, when the device operates at constant pressure, $P_m = P_1 = P_2$, because of the heat absorbed by the pool of water

$$T_2 < T_1$$

It was specified that saturation conditions exist at point 2 and if the device is long enough to allow the air to achieve this saturated condition, we can write

$$T_{as} = T_2$$

which defines the **adiabatic saturation temperature**, a temperature that is a unique function of both ω_1 and T_1 and the enthalpies of the air and the water vapor. Indeed, because the saturation condition is reached, the process is known as an **adiabatic saturation process**. It can be shown via an energy balance that use of the temperature T_{as} will yield the entering humidity ratio

$$\omega_1 = \frac{h_a T_{as} - h_a(T) + \omega_2[h_g(T_{as}) - h_f(T_{as})]}{h_g(T) - h_f(T_{as})} \tag{10.24}$$

An instrument called a **psychrometer** (Figure 10.5) contains two thermometers. In the **sling psychrometer**, one of the thermometers measures the actual temperature of the air, which we know as the dry bulb temperature, T_{db}. The bulb of the other thermometer

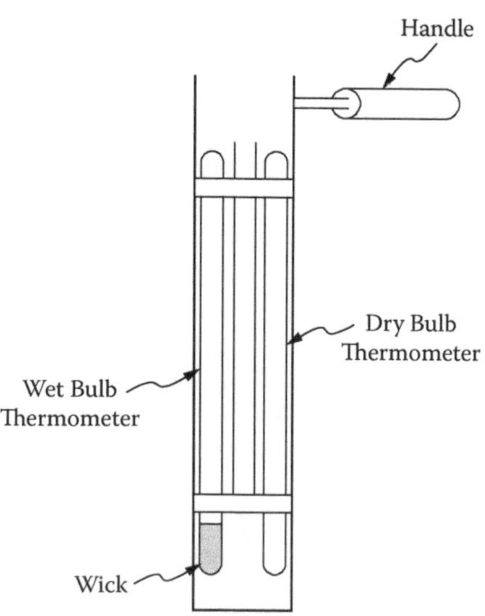

FIGURE 10.5
A sling psychrometer.

is completely covered by a wetted wick. The wetted bulb is called the **wet bulb** and it measures the **wet bulb temperature**, T_{wb}. When the sling psychrometer is whirled so that surrounding air flows across both the wet and dry bulb at a velocity between 100 and 200 m/s, the sensible heat transfer from the wet bulb to the surrounding air is exactly equal to the latent heat required for vaporization.

The wet bulb temperature closely approximates the adiabatic saturation temperature and T_{wb} can be used in Equation 10.24 in place of T_{as}. If we wish to make use of the specific heat of air as well as the wet bulb temperature, we define the **wet bulb depression** as

$$\psi = T_{db} - T_{wb} \tag{10.25}$$

and write

$$\omega_1 = \frac{\omega_2 h_{fg,wb} - c_{p,a}\psi}{h_{g,db} - h_{f,wb}} \tag{10.26}$$

An alternative is Carrier's equation, which relates the partial pressure of the vapor to the wet and dry bulb temperatures and the saturation pressure of the water vapor at the wet bulb temperature $P_{sat,wb}$.

$$P_v = P_{sat,wb} - \frac{(P_m - P_{sat,wb})\psi}{1555.6°C - 0.7222T_{wb}} \tag{10.27}$$

Example 10.4 Moist air is present in a room at 1 bar at dry bulb and wet bulb temperatures of $T_{db} = 20°C$ and $T_{wb} = 15°C$. Determine (a) the humidity ratio and the (b) the relative humidity.

Solution

Assumptions and Specifications

(1) The air and the water vapor may be treated as ideal gases.
(2) The dry bulb and wet bulb temperatures are specified.
(3) The wet bulb temperature may be employed in place of the adiabatic saturation temperature.
(4) The relative humidity at the wet bulb condition is 100%.

Here

$$\psi = 20°C - 15°C = 5°C$$

and at 20°C, Table A.3 provides

$$P = 2.34\,\text{kPa} \quad \text{or} \quad P = 0.0234\,\text{bar}$$

(a) Before using Equation 10.26, we will need to find ω_2. With $P_v = P_{sat,wb} = 1.71$ kPa (at $T_{wb} = 15°C$) from Table A.3, Equation 10.20 is used to determine

$$\omega_2 = 0.622\frac{P_v}{P_m - P_v} = \frac{0.622(0.0171\,\text{bar})}{1.00\,\text{bar} - 0.0171\,\text{bar}} = 0.0108\,\text{kg water/kg dry air}$$

Table A.1 provides $c_{p,a}$ at 300 K and Table A.3 gives us the steam properties for use in Equation 10.26

$$c_{p,a} = 1.005\,\text{kJ/kg-K} \qquad h_{fg,wb} = 2466.8\,\text{kJ/kg water}$$

$$h_{g,db} = 2538.9\,\text{kJ/kg water} \qquad h_{f,wb} = 63.0\,\text{kJ/kg water}$$

so that Equation 10.26 may be used to provide

$$\omega_1 = \frac{\omega_2 h_{\text{fg,wb}} - c_{p,a}\psi}{h_{\text{g,db}} - h_{\text{f,wb}}}$$

$$= \frac{(0.0108\,\text{kg water/kg dry air})(2466.8\,\text{kJ/kg water}) - (1.005\,\text{kJ/kg-K})(5^\circ\text{C})}{2538.9\,\text{kJ/kg water} - 63.0\,\text{kJ/kg water}}$$

$$= \frac{21.62\,\text{kJ/kg water}}{2475.9\,\text{kg water/kg dry air}} = 0.0088\,\text{kg water/kg dry air} \Longleftarrow$$

(b) The relative humidity can be found from Equation 10.21 after the water vapor pressure at the dry bulb temperature is found via a modification of Equation 10.20.

$$P_{\text{v,db}} = \frac{\omega_1 P_m}{0.622 + \omega_1}$$

$$= \frac{(0.0088\,\text{kg water/kg dry air})(1.00\,\text{bar})}{0.622\,\text{kg water/kg dry air} + 0.0088\,\text{kg water/kg dry air}}$$

$$= 0.0139\,\text{bar}$$

and then using Equation 10.21

$$\phi = \frac{P_v}{P_g} = \frac{0.0139\,\text{bar}}{0.0234\,\text{bar}} = 0.593 \qquad (59.3\%) \Longleftarrow$$

Comment: We can check the vapor pressure at the dry bulb temperature by Carrier's equation. With $p_{\text{sat,wb}} = 1.71\,\text{kPa}$, we have

$$P_v = p_{\text{sat,wb}} - \frac{(P_m - P_{\text{sat,wb}})\psi}{1555.6^\circ\text{C} - 0.7222T_{\text{wb}}}$$

$$= 1.71\,\text{bar} - \frac{(100\,\text{kPa} - 1.71\,\text{kPa})(5^\circ\text{C})}{1555.6^\circ\text{C} - 0.7222(15^\circ\text{C})}$$

$$= 1.71\,\text{kPa} - 0.32\,\text{kPa} = 1.39\,\text{kPa} \qquad (0.0139\,\text{bar})$$

10.4 The Psychrometric Chart

In the previous section, we considered several of the fundamental quantities employed in a study of psychrometrics along with some of the necessary equations required for their evaluation. These quantities can be presented in chart form and such a chart is referred to as a **psychrometric chart**. A chart, *constructed for atmospheric pressure (101.325 kPa)*, is provided here as Figure A.4 and is also shown as Figure 10.6.

We observe that in Figure 10.6, the ordinate (shown at the right) is the humidity ratio, ω, and the abscissa is the dry bulb temperature, T_{db}. Lines of constant wet bulb temperature run from the upper left to the lower right of the chart and lines of constant relative humidity

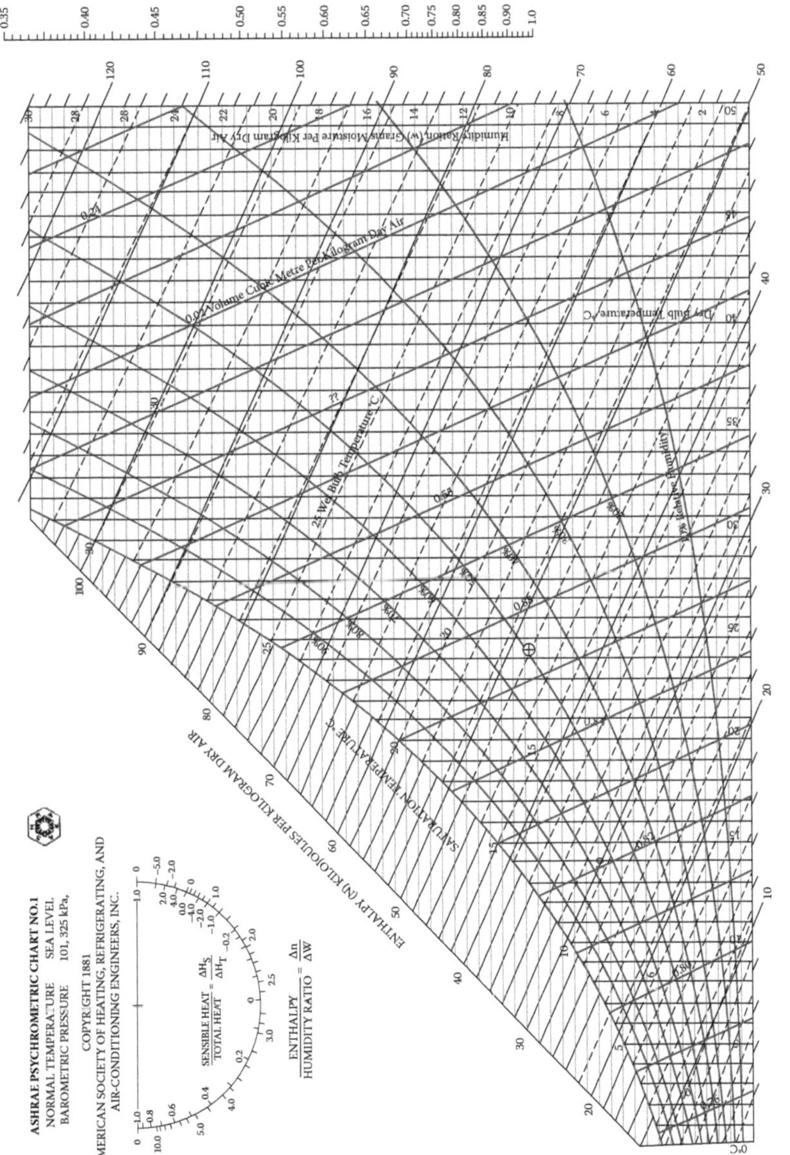

FIGURE 10.6
Psychrometric chart. (From the American Society of Heating, Refrigerating, and Air-Conditioning Engineers, used with permission.)

are indicated by curves that run concave upward from the lower left to the upper right. The **saturation curve** is the locus of points of 100% relative humidity and contain the wet bulb temperature and dew point scales. Specific enthalpy for the air-water vapor mixture is shown on a scale slanted upward from the lower left to the upper right and specific volume lines are plotted from upper left to lower right.

Example 10.5 illustrates the use of the psychrometric chart.

Example 10.5 Air at $T_{db} = 25°C$ and $T_{wb} = 20°C$ is at atmospheric pressure. Use the psychrometric chart to estimate (a) the humidity ratio, (b) the relative humidity, (c) the dew point, (d) the enthalpy of the mixture, and (e) the specific volume of the mixture.

Solution

Assumptions and Specifications

(1) The air and the water vapor may be treated as ideal gases.

(2) The dry bulb and wet bulb temperatures are specified.

In Figures A.4 or 10.6, locate a point p at the intersection of the $T_{db} = 25°C$ and $T_{wb} = 20°C$ lines

(a) Move horizontally from point p to the right and read

$$\omega = 0.0126 \text{ kg water/kg dry air} \Longleftarrow$$

(b) Point p lies between two relative humidity lines that are concave upward. Interpolate to obtain

$$\phi = 63\% \Longleftarrow$$

(c) Move horizontally from point p to the left and read at the saturation curve

$$T_{dp} = 16°C$$

(d) The mixture enthalpy is read by following the wet bulb line from point p

$$h = 57.1 \text{ kJ/kg} \Longleftarrow$$

(e) The specific volume is read by interpolating between the $v = 0.85 \text{ m}^3/\text{kg}$ and $v = 0.87 \text{ m}^3/\text{kg}$ lines

$$v = 0.862 \text{ m}^3/\text{kg} \Longleftarrow$$

10.5 The Products of Combustion

10.5.1 Fuels

Fuels are combustible substances that may come in many different forms and **combustion** refers to a chemical reaction among a fuel and an oxidizing agent that results in a release of energy. The emphasis in this section is on **hydrocarbon fuels**; this form includes gaseous hydrocarbons, liquid hydrocarbons, and coal.

- Coal is the natural, rock-like, derivative of forest-type plant material, usually accumulated in peat beds and progressively compressed and subjected to biochemical conversion until it is finally altered into graphite or graphite-like material. The composition and characteristics of coal vary with location. The three common types of coal are **anthracite, bituminous,** and **lignite**.

TABLE 10.1

Fuel Families and Their Chemical Formulas

Family	Chemical Formula
Aromatic	
Benzene	C_nH_{2n-6}
Napthalene	C_nH_{2n-12}
Diolefin	C_nH_{2n-2}
Olefin	C_nH_{2n}
Ethylene	C_2H_4
Propylene	C_3H_6
Paraffin	C_nH_{2n+2}
Methane	CH_4
Ethane	C_2H_6
Propane	C_3H_8
Butane	C_4H_{10}
Pentane	C_5H_{12}
Hexane	C_6H_{14}
Heptane	C_7H_{16}
Octane	C_8H_{18}
Decane	$C_{10}H_{22}$

A sample of coal can be subjected to a **proximate analysis** or **ultimate analysis**. Proximate analysis specifies the components on a mass basis giving the amounts of fixed carbon, volatile matter, ash, and moisture. Ultimate analysis also specifies the components on a mass basis giving the amounts of carbon, sulfur, hydrogen, nitrogen, oxygen, and ash. The moisture content is often specified and the ultimate analysis is usually given either on an "as fired" or "dry" basis.

- Liquid hydrocarbon fuel may be a mixture of many hydrocarbons. Gasoline, for example, consists of a mixture of about 35 hydrocarbon compounds. The general chemical formula of several hydrocarbon families is indicated in Table 10.1.

  Alcohols are frequently used as fuels as well. In the alcohol family, one of the H atoms in the foregoing taxonomy is replaced by an OH. For example, ethane is C_2H_6 and ethyl alcohol is C_2H_5OH. Finally, for the purpose of analyis, gasoline is usually taken as octane, C_8H_{18}, and diesel oil is taken as dodecane, $C_{12}H_{26}$. Both of these are in the paraffin family.
- Gaseous hydrocarbon fuels are obtained from natural gas wells or from certain chemical processes. While they normally consist of several hydrocarbons, the major constituent is methane, also a paraffin, CH_4.

10.5.2 The Combustion Process

When an oxidizing agent is combined with a fuel, combustion occurs and energy is released. Three examples are

$$2H_2 + O_2 \rightarrow 2H_2O$$

$$S + O_2 \rightarrow SO_2$$

$$C + O_2 \rightarrow CO_2$$

The reaction of carbon with oxygen shown represents a **complete combustion**. However,

$$2C + O_2 \rightarrow 2CO$$

represents an **incomplete combustion**. Incomplete combustion is wasteful and occurs when not enough oxidizing agent is provided.

We observe that, from the foregoing, we are dealing with reactants that include the fuel and an oxidizing agent. The process, in general, is characterized as

$$\text{fuel} + \text{oxidizing agent} \rightarrow \text{products of combustion}$$

and when we examine any of the three combustion equations provided, say for the combustion of hydrogen

$$2H_2 + O_2 \rightarrow 2H_2O \tag{10.28}$$

we note that hydrogen is the fuel and that oxygen is the oxidizing agent. Moreover, the product of the combustion is water. The numerical coefficients before the hydrogen, the 2, and oxygen, the 1, which is understood represent the moles of each reacting substance. Note that in the SI system, the number of kgmol on both sides of the equation are different. Equation 10.28 means that

$$2\,\text{kgmol } H_2 + 1\,\text{kgmol } O_2 \rightarrow 2\ \text{kgmol } H_2O$$

These coefficients are referred to as **stoichiometric coefficients.** But, because mass must be conserved, the mass of the reactants must equal the mass of the products.

10.5.3 Combustion Air

When trace substances such as carbon dioxide and argon are not considered, combustion processes involving air concern a mixture composed of 21% oxygen and 79% nitrogen by volume. This leads to the conclusion that for each kgmol of oxygen, there are 79/21 = 3.76 kgmol of nitrogen and that for the complete combustion of hydrogen to form water, Equation 10.28 becomes

$$\begin{aligned} 2\,\text{kgmol } H_2 \ + \ 1\,\text{kgmol } O_2 + 3.76\,\text{kgmol } N_2 \\ \rightarrow 2\,\text{kgmol } H_2O + 3.76\,\text{kgmol } N_2 \end{aligned} \tag{10.29}$$

and we observe that the nitrogen in this process is along for the ride. Moreover, Equation 10.29 represents the combustion of hydrogen with the **theoretical** amount of air.

Example 10.6 Determine the general relationship for the combustion of a hydrocarbon in the paraffin family, C_nH_{2n+2}, with the theoretical amount of air.

Solution

Assumptions and Specifications

(1) All constituents in the combustion process are ideal gases.
(2) The theoretical amount of air is specified.
(3) The combustion is complete.

Write the combustion equation in terms of unknown kgmol values

$$C_nH_{2n+2} + a\,(O_2 + 3.76\,N_2) \rightarrow b\,CO_2 + c\,H_2O + d\,N_2$$

and then, by an examination of each constituent, solve for a, b, c, and d.

$$\begin{aligned} \mathrm{C}:&\quad b = n \\ \mathrm{H}:&\quad 2c = 2n + 2 \\ \mathrm{O}:&\quad 2a = 2b + c \\ \mathrm{N}:&\quad 2d = 7.52a \end{aligned}$$

and hence,

$$a = \frac{3n+1}{2}, \quad b = n, \quad c = n + 1, \quad \text{and} \quad d = 3.76a$$

Then

$$C_nH_{2n+2} + \frac{3n+1}{2}(O_2 + 3.76\,N_2) \rightarrow n\,CO_2 + (n+1)\,H_2O + \frac{3n+1}{2}(3.76)\,N_2$$

The final expression in Example 10.6 is a balanced equation that yields the products of combustion for any hydrocarbon in the paraffin family. It is repeated here for ready reference and given an equation number

$$C_nH_{2n+2} + \frac{3n+1}{2}(O_2 + 3.76\,N_2) \rightarrow n\,CO_2 + (n+1)\,H_2O + \frac{3n+1}{2}(3.76)\,N_2 \tag{10.30}$$

If x-percent excess air is present, the combustion will be carried out with only a portion of the oxygen provided. For example, the complete combustion of carbon with x% excess air will be represented by

$$C + \left(1 + \frac{x}{100}\right) O_2 + \left(1 + \frac{x}{100}\right)(3.76)N_2 \rightarrow CO_2 + \frac{x}{100}O_2 + \left(1 + \frac{x}{100}\right)(3.76)N_2$$

and for a hydrocarbon in the paraffin family

$$C_nH_{2n+2} + \left(1 + \frac{x}{100}\right)\left(\frac{3n+1}{2}\right)(O_2 + 3.76\,N_2) \rightarrow nCO_2 + (n+1)H_2O + \left(1 + \frac{x}{100}\right) O_2 + \left(1 + \frac{x}{100}\right)\left(\frac{3n+1}{2}\right)(3.76)\,N_2 \tag{10.31}$$

10.5.4 The Air-Fuel Ratio

The **air-fuel ratio** is a parameter used to quantify the amount of fuel and oxidizing agent. Designated by the symbol, AFR, it is simply the ratio of the amount of air in a combustion reaction to the amount of fuel. It can be written on a mass basis

$$\text{AFR} = \frac{\text{mass of air}}{\text{mass of fuel}} \tag{10.32a}$$

or a molar basis

$$\overline{\text{AFR}} = \frac{\text{kgmol of air}}{\text{kgmol of fuel}} \tag{10.32b}$$

and, consistent with the conversion from gravimetric (mass) to volumetric (molar) fractions,

$$\text{AFR} = \overline{\text{AFR}}\left(\frac{M_{\text{air}}}{M_{\text{fuel}}}\right) \tag{10.32c}$$

where M_{air} and M_{fuel} are the molecular weights.

The amount of air actually supplied to a combustion process is commonly expressed in terms of the **percent of theoretical air**, Υ

$$\Upsilon = 100\frac{\overline{\text{AFR}}}{(\overline{\text{AFR}})_{\text{th}}} \tag{10.33}$$

where the subscript th indicates the theoretical or stoichiometric condition. A number for Υ greater than 100 indicates an excess of air and a number less than 100 shows a deficiency. For example, $\Upsilon = 118$ indicates a **rich** mixture that is equivalent to 18% excess air, and $\Upsilon = 82$ shows a **lean** mixture with an air deficiency of 18%.

Example 10.7 Propane is burned with dry air and the volumetric analysis of the products of combustion are CO_2, 8.31%, $O_2 = 8.31\%$, and $N_2 = 83.38\%$. Determine (a) the balanced combustion equation for one mole of fuel, (b) the two airflow ratios, AFR and $\overline{\text{AFR}}$, (c) the percent of theoretical air, and (d) the excess air.

Solution

Assumptions and Specifications

(1) All constituents in the combustion process are ideal gases.

(2) The combustion is complete.

Consider 100 kgmol of dry products of combustion and observe that, in this analysis, provision must be made for any water (water vapor) that is produced in the combustion process. We write the combustion equation in terms of unknown kgmol values

$$a\,C_3H_8 + b\,O_2 + c\,N_2$$
$$\rightarrow 8.31\,CO_2 + 8.31\,O_2 + 83.38N_2 + d\,H_20$$

and then, by an examination of each constituent, solve for a, b, c, and d.

The nitrogen balance yields

$$c = 83.38$$

and with

$$\frac{c}{b} = \frac{3.76}{1.00} = 3.76$$

it is seen that

$$b = \frac{83.38}{3.76} = 22.18$$

From the carbon balance

$$3a = 8.31 \quad \text{or} \quad a = 2.77$$

Finally, from the hydrogen balance

$$2d = 8a$$

$$d = 4a = 4(2.77) = 11.08$$

These values provide the balanced combustion equation

$$2.77\,C_3H_8 + 22.18\,(O_2 + 3.76\,N_2)$$
$$\rightarrow 8.31\,CO_2 + 8.31\,O_2 + (22.18)(3.76)N_2 + 11.08\,H_2O$$

(a) On the basis of one unit mole of propane:

$$C_3H_8 + 8O_2 + 30.11\,N_2$$
$$\rightarrow 3\,CO_2 + 3\,O_2 + 30.11N_2 + 4\,H_2O \Longleftarrow$$

(b) The airflow ratio on a molar basis is

$$\overline{AFR} = \frac{8.00 + 30.11}{1.00} = 38.11\,\frac{\text{kgmol air}}{\text{kgmol propane}} \Longleftarrow$$

and on a mass basis (with the molecular weights of air and propane taken from Table A.1), Equation 10.32c gives

$$AFR = 38.11\left(\frac{28.97}{44.09}\right) = 25.04\,\frac{\text{kg air}}{\text{kg propane}} \Longleftarrow$$

(c) The theoretical air comes from the chemical reaction with no excess air and a complete combustion to carbon dioxide and water vapor

$$C_3H_8 + 5O_2 + 5(3.76)N_2 \rightarrow 3CO_2 + 4H_2O + 5(3.76)N_2$$

so that

$$\overline{AFR})_{th} = \frac{5(4.76)\text{ kgmol air}}{1\text{ kgmol fuel}} = 23.80\frac{\text{kgmol air}}{\text{kgmol propane}}$$

Hence, the percent theoretical air will be

$$\Upsilon = 100\frac{\overline{AFR}}{(\overline{AFR})_{th}} = 100\left(\frac{38.11}{23.8}\right) = 1.60 \Longleftarrow$$

(d) The actual percent excess air derives from the percent theoretical air

$$x = 160 - 100 = 60\% \Longleftarrow$$

10.6 Summary

The gravimetric fraction of a particular gas in a mixture of ideal gases is

$$G_i = \frac{m_i}{\sum_{i=1}^{k} m_i} = \frac{m_i}{m_m} \tag{10.4}$$

and the mole fraction will be

$$Y_i = \frac{n_i}{\sum_{i=1}^{k} n_i} = \frac{n_i}{n_m} \tag{10.5}$$

Dalton's law of partial pressures relates the partial pressure of a constituent ideal gas to the total mixture pressure via the volumetric fraction

$$P_i = Y_i P_m \tag{10.9}$$

The gravimetric analysis of an ideal gas mixture can be obtained from the volumetric analyis by

$$G_i = \frac{Y_i M_i}{\sum Y_i M_i} \qquad (i = 1, 2, 3, \ldots, k) \tag{10.10}$$

and if the gravimetric analysis is given, the volumetric analysis can be determined from

$$Y_i = \frac{G_i/M_i}{\sum G_i/M_i} \qquad (i = 1, 2, 3, \ldots, k) \tag{10.11}$$

The apparent molecular weight of a mixture, M_m, can be obtained from the volumetric fractions

$$M_m = \frac{m_m}{n_m} = \frac{\sum_{i=1}^{k} n_i M_i}{n_m} = \sum_{i=1}^{k} Y_i M_i \tag{10.12}$$

and the mixture gas constant will be equal to

$$R_m = \frac{\bar{R}}{M_m} = \sum_{i=1}^{k} G_i R_i \tag{10.13b}$$

The composition of a given moist air sample can be measured by the specific humidity

$$\omega = \frac{m_v}{m_{da}} \qquad \text{(kg vapor/kg dry air)} \tag{10.19}$$

or

$$\omega = 0.622 \frac{P_v}{P_m - P_v} \tag{10.20}$$

or in terms of the relative humidity

$$\phi = \frac{Y_v}{Y_g} = \frac{P_v}{P_g} = \frac{\rho_v}{\rho_g} = \frac{v_g}{v_v} \tag{10.21}$$

The wet bulb depression is defined as

$$\psi = T_{db} - T_{wb} \tag{10.25}$$

and a relationship between the specific humidity or humidity ratio is

$$\omega_1 = \frac{\omega_2 h_{fg,wb} - c_{p,a}\psi}{h_{g,db} - h_{f,wb}} \tag{10.26}$$

where ω_2 represents the specific humidity for an adiabatic saturation process. An alternative is Carrier's equation, which relates the partial pressure of the vapor to the wet and dry bulb temperatures and the saturation pressure of the water vapor at the wet bulb temperature $P_{sat,wb}$.

$$P_v = P_{sat,wb} - \frac{(P_m - P_{sat,wb})\psi}{1555.6°\text{C} - 0.7222T_{wb}} \tag{10.27}$$

Several of the fundamental quantities employed in a study of psychrometrics can be presented in chart form and such a chart is referred to as a **psychrometric chart**. A psychrometric chart, constructed for atmospheric pressure (101.325 kPa) is provided as Figure 10.4.

A discussion of the types of fuels, the combustion process and the air required for combustion has been presented.

The air-fuel ratio on both a molar basis

$$\overline{\text{AFR}} = \frac{\text{kgmol of air}}{\text{kgmol of fuel}}$$

and, consistent with the conversion from gravimetric (mass) to volumetric (molar) fractions,

$$\text{AFR} = \overline{\text{AFR}}\left(\frac{M_{air}}{M_{fuel}}\right) \tag{10.32c}$$

where M_{air} and M_{fuel} are the molecular weights.

The percent theoretical air, Υ, is defined as

$$\Upsilon = 100\frac{\overline{\text{AFR}}}{(\overline{\text{AFR}})_{th}} \tag{10.33}$$

10.7 Problems

Mixtures of Ideal Gases

10.1: One kilogram of air and an unknown mass of carbon dioxide occupy a tank at 2 MPa. Determine the mass of CO_2 that has a partial pressure of 0.35 MPa.

10.2: One kgmol of a gaseous mixture has a gravimetric analysis of $O_2 = 0.316$, $CO_2 = 0.442$, and $N_2 = 0.242$. Determine (a) the number of kgmoles of each constituent, (b) the apparent molecular weight of the mixture, (c) the mass of each constituent, and (d) the apparent gas constant of the mixture.

10.3: A mixture of 58 kg of N_2 and 142 kg of CO_2 is at 2 MPa. Determine the partial pressures of both the N_2 and the CO_2.

10.4: When 4 kg of O_2 are mixed with 6 kg of an unknown gas, the resulting mixture is contained in a cube that is 15 cm on a side at 3 bar an 65°C. Determine for the unknown gas (a) the apparent gas constant, (b) the apparent molecular weight, and (c) the partial pressure.

10.5: A drum with a volume of 0.625m^3 contains a gaseous mixture of CO_2 and CH_4 each of which is 50% by mass. The mixture is at 8 bar and 40°C. Two kg of O_2 are added to the drum with the mixture temperature remaining at 40°C. For the final mixture, determine (a) the gravimetric analysis, (b) the volumetric analysis, and (c) the total pressure.

10.6: A gaseous mixture is composed of 1.75 kgmol of N_2, 2 kgmol of O_2, and 1.25 kgmol of CO_2. Determine (a) the volumetric analysis, (b) the mass of the mixture, and (c) the apparent molecular weight of the mixture.

10.7: In a closed and rigid tank having a volume of 0.125m^3, 4 kg of a mixture containing equal parts by volume of CO_2 and N_2 exist at 3 bar and 65°C. The tank receives an additional 0.50 kg of N_2 with the temperature remaining at 65°C. Determine the (a) the final gravimetric analysis, (b) the volumetric analysis, (c) the apparent molecular weight, and (d) the mixture pressure.

10.8: A gaseous mixture of CH_4, N_2, CO, and O_2 occupies a vessel at the respective partial pressures of 1.0, 1.2, 1.5, and 1.8 bar. Determine (a) the volumetric analysis, (b) the gravimetric analysis, (c) the apparent molecular weight, (d) the apparent gas constant, and (e) the volume occupied by 100 kg of the mixture at 27°C.

10.9: A gaseous mixture has a volumetric analysis of 22% N_2, 29% CO_2, and 49% CH_4. Determine (a) the gravimetric analysis, (b) the apparent molecular weight of the mixture, and (c) the partial pressure of the N_2 if that of the CH_4 is 0.60 bar.

10.10: Given 25 m^3 of a gaseous mixture whose gravimetric analysis is 13% CO_2, 9% O_2, and 78% N_2 at 1 bar and 150°C. Determine (a) the volumetric analysis, (b) the respective partial pressures, (c) the apparent molecular weight and gas constant, and (d) the kgmol of each component.

10.11: A mixture of O_2 and N_2 occupies a certain volume at 2 kPa and 95°C. The partial pressure of the O_2 is twice that of the N_2. Determine (a) the volumetric analysis, (b) the gravimetric analysis, (c) the apparent gas constant of the mixture, (d) the density of the mixture, and (e) the final mixture pressure if 40 kJ/kg of heat are added at constant volume.

10.12: A certain volume is divided into two compartments designated as A and B. Here, $V_A = V_B = 0.125\text{m}^3$, O_2 is in compartment A, CO is in compartment B and each gas is at 4 bar and 65°C. When the partition between the compartments is removed, the gases mix thoroughly. Determine (a) the mass of each constituent and (b) the change in entropy after mixing.

10.13: A certain volume is divided into two compartments designated as A and B. Compartment A contains 0.75 kg of O_2 at 4 bar and 37°C. Compartment B contains 0.15m^3 of a gaseous mixture with a volumetric composition of 12% CO_2 and 88% N_2 at 1 bar and 27°C. When the partition between the compartments is removed, the gases mix. For equilibrium conditions after mixing, determine (a) the gravimetric analysis, (b) the volumetric analysis, (c) the apparent molecular weight and gas constant of the mixture, (d) the temperature, (e) the pressure, and (f) the change of entropy.

10.14: In a closed rigid container at 3.5 MPa and 37°C, there is 0.05m^3 of air. The vessel connects with another in which there are 1.25 kg of helium at 6 bar. When the valve separating the two vessels is opened, the resulting equilibrium mixture is at 1.8 MPa and 30°C. Determine (a) the initial temperature and the volume of the helium before the valve between the tanks is opened and (b) the partial pressure of the air and helium after equilibrium is attained.

10.15: A pipeline carries 100 m^3/min of CO at 1.5 bar and 80°C. Another pipeline carries 160 m^3/min of CH_4 at 1.5 bar and 15°C. The two lines merge and kinetic and potential energy changes may be considered negligible. Determine (a) the gravimetric analysis, (b) the volumetric analysis, (c) the apparent molecular weight of the mixture, and (d) the gas constant of the mixture.

10.16: Ten kilogram of a gas having a volumetric analysis of 76% N_2, 15% O_2, and 9% CO_2 enters a heat exchanger at point 1 where $P_1 = 1.5$ bar and $T_1 = 540$°C and leaves at point 2 where $P_2 = 1.5$ bar and $T_2 = 85$°C. Using constant specific heats, determine (a) the heat transferred and (b) the change in entropy.

Psychrometrics

10.17: A barometer reads 29.35 in of mercury, $T_{db} = 35$°C and $T_{wb} = 25$°C. Determine (a) the relative humidity, (b) the humidity ratio, (c) the dew point temperature, and (d) the density.

10.18: A barometer reads 29.45 in of mercury, $T_{db} = 30$°C, and the relative humidity is 49%. Determine (a) the partial pressure of the water vapor, (b) the humidity ratio, and (c) the dew point temperature.

10.19: A low-pressure air-steam mixture at 65°C has a dew point temperature of 25°C. The humidity ratio is 0.04. Determine (a) the relative humidity, (b) the pressure of the mixture, (c) the density of the mixture, and (d) the volume occupied by 1 kg of dry air.

10.20: The humidity ratio of atmospheric air at 25°C is 0.016-kg water vapor per kg dry air. Determine (a) the relative humidity, (b) the pressure of the mixture, (c) the dew point temperature, (d) the density of the mixture, and (e) the mass of water occupied by 50 m^3 of dry air.

10.21: Air has leaked into a steam power plant condenser where the temperature is 35°C, resulting in 0.06 kg dry air per kg of water vapor. The barometer reads 29.92 in of mercury. Determine (a) the condenser pressure, (b) the dew point temperature, and (c) the humidity ratio.

10.22: The humidity ratio of atmospheric air at 25°C is 0.015-kg water vapor per kg dry air. Determine (a) the vapor pressure, (b) the relative humidity, (c) the dew point temperature, (d) the density of the mixture (e) the volumetric and gravimetric fractions of the water vapor, and (f) the mass of water vapor in 25m^3 of the atmospheric air.

10.23: The gravimetric percentage of water in atmospheric air is 3.04%. Given T_{db} = 35°C and a mixture pressure of 1 bar, determine (a) the specific humidity, (b) the partial pressure of the vapor, (c) the relative humidity, and (d) the dew point temperature.

10.24: Atmospheric air at 28.50 in of mercury has a 10°C wet bulb depression from a dry bulb temperature of 30°C. Determine the humidity ratio from Carrier's equation.

10.25: Atmospheric air at 29.75 in of mercury, 35°C and a relative humidity of 50% is heated at constant pressure to 45°C. Determine (a) the relative humidity after heating, (b) the change in mixture enthalpy, (c) the dew point of the original air and the heated air, and (d) the amount of heat required for $60\text{m}^3/\text{min}$ of atmospheric air.

10.26: Atmospheric air at 29.92 inches of mercury, 35°C and 60% relative humidity is to be cooled at constant pressure to 10°C. For an air flow at the initial state of $1200\ \text{m}^3/\text{min}$, determine (a) the specific humidity, (b) the dew point temperature, and (c) the heat removed.

10.27: For 2.5 m^3 of an air-steam mixture that is initially saturated at 1.25 bar and 15°C is heated at constant volume, with a final temperature is 65°C, determine (a) the masses of air and water vapor and (b) the partial pressure at the final state.

10.28: Consider two streams of moist air that mix adiabatically. Stream 1 has a mass flow of $\dot{m}_1$, a temperature of T_1, and a specific humidity of ω_1. Stream 2 has similar parameters designated as $\dot{m}_2$, T_2, and ω_2. Stream 3, which is a result of the mixing, has parameters $\dot{m}_3$, T_3, and ω_3. Show that

$$h_3 = \frac{\dot{m}_1(1+\omega_1)h_1 + \dot{m}_2(1+\omega_2)h_2}{\dot{m}_1(1+\omega_1) + \dot{m}_2(1+\omega_2)}$$

10.29: A cubic meter of an air-steam mixture, initially at 1 bar, a dry bulb temperature of 35°C, and a relative humidity of 60% is cooled at constant volume to 5°C. Determine (a) the initial humidity ratio, (b) the temperature at which condensation begins to occur, (c) the total amount of vapor condensed, and (d) the heat withdrawn.

10.30: $240\text{m}^3/\text{min}$ of saturated air at 100 kPa and 10°C are mixed with $200\text{m}^3/\text{min}$ of air at 100 kPa, 35°C, and a relative humidity of 60%. For the mixture, determine (a) the specific humidity, (b) the dry bulb temperature, (c) the relative humidity, and (d) the dew point.

10.31: A low-pressure air-water vapor mixture at 65°C has a dew point of 25°C. The specific humidity is 0.04 kg water vapor/kg dry air. Determine (a) the relative humidity, (b) the pressure of the mixture, (c) the density of the mixture, and (d) the volumetric fraction of the water vapor.

10.32: An adiabatic chamber is divided into two parts that are separated by a partition.

A(dry air)	← Part →	*B*(saturated vapor)
12	m, kg	0.50
105.2	P, kPa	105.2
101	T, °C	101

The partition is removed and mixing occurs until equilibrium is reached. For the final state, determine (a) the partial pressure of the water vapor and (b) the relative humidity.

The Psychrometric Chart

10.33: Use the psychrometric chart with conditions of a total pressure of 1 atm, $T_{db} = 32°\text{C}$, and $\phi = 60\%$ to find (a) the humidity ratio, (b) the wet bulb temperature, (c) the dew point, (d) the enthalpy, and (e) the vapor pressure.

10.34: Use the psychrometric chart with conditions of a total pressure of 1 atm, $T_{db} = 32°C$, and $\omega = 0.006$ kg of water vapor/kg dry air to find (a) the wet bulb temperature, (b) the relative humidity, (c) the dew point, (d) the enthalpy, and (e) the vapor pressure.

10.35: Use the psychrometric chart with conditions of a total pressure of 1 atm, $T_{db} = 24°C$, and $T_{dp} = 14°C$ to find (a) the humidity ratio, (b) the wet bulb temperature, (c) the relative humidity, (d) the enthalpy, and (e) the vapor pressure.

10.36: Use the psychrometric chart with conditions of a total pressure of 1 atm, $T_{db} = 24°C$, and an enthalpy of 48 kJ/kg to find (a) the humidity ratio, (b) the wet bulb temperature, (c) the relative humidity, (d) the dew point, and (e) the vapor pressure.

10.37: Use the psychrometric chart with conditions of a total pressure of 1 atm, $T_{db} = 24°C$, and $v = 0.86\,m^3/kg$ to find (a) the humidity ratio, (b) the wet bulb temperature, (c) the relative humidity, (d) the dew point, and (e) the enthalpy.

10.38: Use the psychrometric chart with conditions of a total pressure of 1 atm, $T_{wb} = 20°C$, and $\phi = 50\%$ to find (a) the humidity ratio, (b) the dry bulb temperature, (c) the dew point, (d) the enthalpy, and (e) the vapor pressure.

10.39: Use the psychrometric chart with conditions of a total pressure of 1 atm, $T_{db} = 80°C$, and $\phi = 50\%$ to find (a) the specific humidity, (b) the wet bulb temperature, (c) the dew point, (d) the enthalpy, and (e) the vapor pressure.

10.40: Use the psychrometric chart with conditions of a total pressure of 1 atm, $T_{db} = 30°C$, and $T_{dp} = 12°C$ to find (a) the humidity ratio, (b) the wet bulb temperature, (c) the relative humidity, and (d) the enthalpy.

10.41: Use the psychrometric chart with conditions of total pressure of 1 atm, $T_{db} = 30°C$, and $T_{wb} = 18°C$ to find (a) the humidity ratio, (b) the relative humidity, (c) the dew point, (d) the enthalpy, and (e) the vapor pressure.

10.42: Use the psychrometric chart with conditions of a total pressure of 1 atm, an enthalpy of 50 kJ/kg and $T_{dp} = 10°C$ to find (a) the humidity ratio, (b) the dry bulb temperature, (c) the wet bulb temperature, (d) the relative humidity, and (e) the vapor pressure.

10.43: An air-water vapor mixture at atmospheric pressure has a dry bulb temperature of 35°C and a mixture humidity of 50%. The mixture is heated until the dry bulb temperature is 45°C. Determine (a) the relative humidity after heating, (b) the change in enthalpy, (c) the dew point point of both the original air and heated air, and (d) the heat added to the mixture.

10.44: An air-water vapor mixture at atmospheric pressure has a dry bulb temperature of 35°C and a mixture humidity of 60%. The mixture is cooled until the dry bulb temperature is 10°C. Determine (a) the relative humidity after the cooling process, (b) the dew point temperature at both the beginning and end of the heating process, heated air, and (c) the heat extracted from the mixture.

10.45: Cold air at 10°C and a relative humidity of 30% is heated to 31°C and then passed through an evaporative cooler until the temperature reaches 22°C. Determine (a) the heat input, (b) the mass of water, and (c) the relative humidity at the conclusion of the process.

10.46: Dry air ($\phi = 0.00$) at 28°C is sprayed with water until it becomes saturated. Determine (a) the final temperature and (b) the amount of water used.

10.47: For 2.5 m^3 of an air-steam mixture initially saturated at 1.25 bar and 15°C and heated at constant pressure, the final temperature is 65°C. Determine (a) the masses of air and water vapor, (b) the partial pressure at the final state, (c) the relative humidity at the final state, and (d) the heat added.

10.48: For 2.5 m^3 of an air-steam mixture initially saturated at 1.25 bar and 30°C and heated at constant volume, the final temperature is 65°C. Determine (a) the masses of air and water vapor, (b) the partial pressure at the final state, (c) the relative humidity at the final state, and (d) the heat added.

The Products of Combustion

10.49: Develop the combustion equation for acetylene, C_2H_2 for (a) theoretical air and (b) x% excess air.

10.50: Determine the air-fuel ratio, both on a mass and molar basis, when propylene is burned in 25% excess air.

10.51: Hexane is burned in 60% excess air. Determine the percent theoretical air.

10.52: Decane is burned with 150% of theoretical air. Determine the gravimetric analysis of the products of combustion.

10.53: A mixture of 3 kgmol of ethane and 5 kgmol of propane is burned in stoichiometric air. Determine the air-fuel radio on a mass basis.

10.54: In a combustion process with dodecane, $C_{12}H_{26}$, the dry product mole fractions are 83.21% N_2, 5.07% O_2, 10.44% CO_2, and 1.28% CO. Determine the percent theoretical air of the reactants.

10.55: Determine the percent of theoretical air in the combustion of pentane if the ratio of the mass of air to the mass of pentane is 12.5.

10.56: One kilogram of pentane is burned in 20 kg of air. Determine the percent excess air used in this combustion process.

10.57: Determine the gravimetric and volumetric analysis of the products of combustion when decane is burned with 40% excess air.

10.58: Ethyl alcohol, C_2H_5OH, burns completely in the theoretical amount of air. Determine the air-fuel ratio on both a mass and molar basis.

10.59: Butane reacts with 78% of theoretical air to form products of combustion that include CO_2, CO, N_2, and H_2O. Determine (a) the balanced combustion reaction equation, (b) the air-fuel ratio on a mass basis, and (c) the gravimetric analysis of the products.

11

Introduction to Fluid Mechanics

Chapter Objectives

- To define a **fluid**.
- To distinguish between fluid properties and flow properties.
- To describe the variation of properties from point-to-point in a **fluid continuum**.
- To develop the concepts of fluid stresses and their orientation.
- To define the fluid property, **viscosity**.

11.1 The Definition of a Fluid

Fluid mechanics deals with fluids at rest or in motion. We begin by defining a fluid as a substance that deforms continuously when subjected to shearing stresses. The key words in this definition are **deform** and **continuously**.

By now, the reader is acquainted with the relationships of stress, strain, and deformation as related to solid bodies. In Figure 11.1, we illustrate how a solid object, such as a chunk of hard rubber, will react to a shearing stress. If the solid is deformed within its **elastic** range, it will return to its original shape when the stress is removed. If stressed into the **plastic** range, it will take on some permanent deformation and will not return to its original shape when the stress is removed.

A fluid, on the other hand, will deform **continuously** when subjected to a shearing stress. In order for it to come to rest, another shearing stress must be applied in the opposite direction. This process is shown in Figure 11.2 for a fluid confined between two parallel surfaces. The lower surface, as indicated, is stationary and the upper one is moving to the right with velocity, $\hat{V}_x$. Also shown is the velocity profile of the fluid, that is, the velocity of the fluid at a varying distance, y, from the lower plate. The fluid velocity is zero at the lower stationary plate and is equal to the velocity, $\hat{V}_x$, at the upper moving surface. This is because fluid particles in contact with a solid surface remain attached to it. If the surface is stationary, so is the adjacent fluid layer and if the surface moves with velocity, $\hat{V}_x$, this will also be the rate at which the attached fluid particles will move. This is a statement of the **no-slip** condition that holds for most fluid-solid contact.

Suppose that the linear variation of velocity across the fluid layer shown in Figure 11.2 is achieved a considerable period of time after the fluid motion has occurred. If no further changes occur with additional time, the fluid is said to be in the **steady-state** condition where there is no further variation with time, $\hat{V}_x(y) \neq \hat{V}_x(y, t)$.

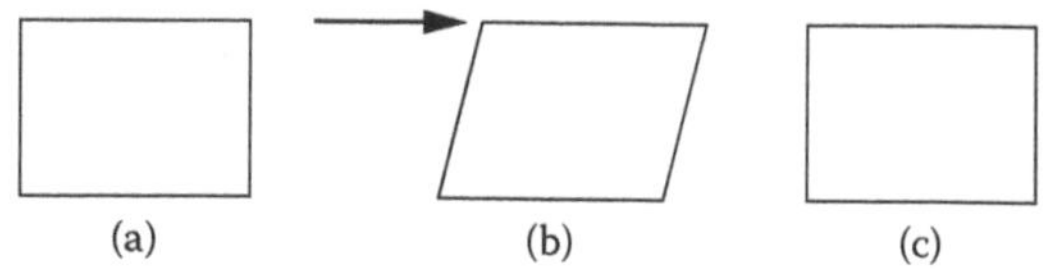

FIGURE 11.1
Deformation of a solid object under the influence of a shear stress: (a) solid body, (b) body under shear, and (c) shear stress released.

If we were to examine velocity profiles at each instant of time shortly after motion was imparted to the upper surface, we would see a different pattern at each time, $\hat{V}_x(y) = \hat{V}_x(y, t)$ and there *is* a variation with time. Such cases are termed **unsteady state** or **transient** processes and whenever there is a change in fluid motion, a transient process will occur. Steady-state conditions will then be approached after some extended time period. In Figure 11.3, the variation in $\hat{V}_x(y, t)$ is depicted at different times between $\hat{V}_x(y, 0)$ and steady state.

This discussion has followed from the defining statement for a fluid, that is, its tendency to deform continuously under a shearing stress. Clearly, if a fluid is at rest or is **static**, it is not exposed to a shear stress. This is a fundamental precept of **fluid statics** that will be considered in depth in Chapter 12.

11.2 Fluid Properties and Flow Properties

In order to describe fluid behavior, it is necessary to have a clear understanding of certain important characteristics called **properties**. In Chapter 2, we defined a thermodynamic property as any observable characteristic that describes the state of a system. Here, we extend this idea to a fluid by pointing out that the property is an observable characteristic that is useful in describing substance or material behavior. In the case of fluid mechanics, we will be concerned with both the properties of the fluid and the properties of the flow. The distinction between **fluid properties** and **flow properties** can be subtle and the reader is hereby alerted to this distinction.

Fluid properties of interest include a number of quantities that we have already considered. One example is the density. Others, such as the viscosity, will perhaps be new. Fluid properties of importance are generally tabulated in handbooks or other references, usually as functions of temperature and pressure. In some instances, properties are related by a mathematical expression termed an **equation of state.** The most familiar equation of state is that for the ideal gas, which was considered in Chapter 3. It has the form given by Equation 3.25

$$PV = n\bar{R}T \tag{11.1}$$

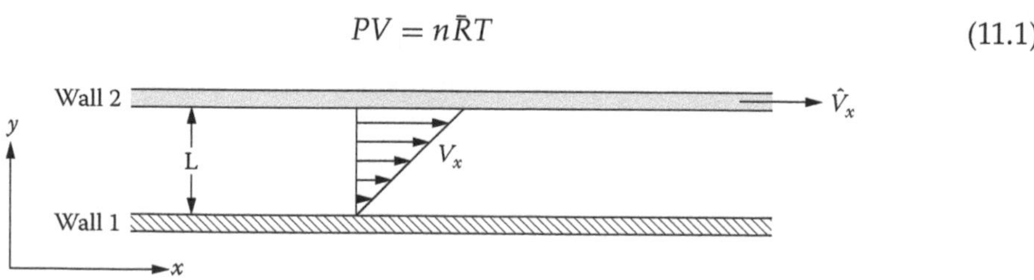

FIGURE 11.2
Fluid motion between two parallel plates.

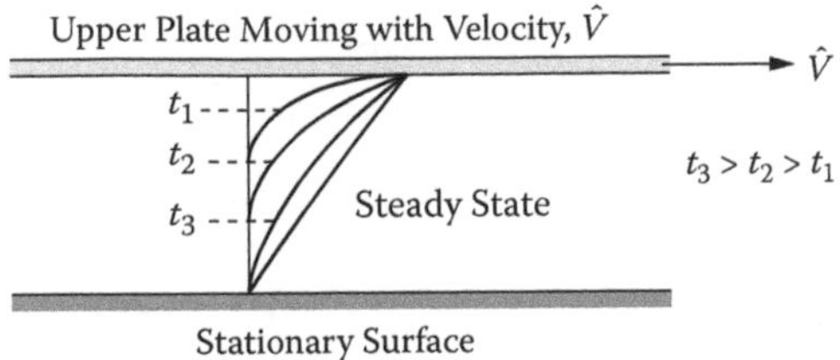

FIGURE 11.3
Velocity profiles between parallel plates as a function of time between initial motion and steady state.

If this is divided by the molecular weight of the gas (or gas mixture) in question, the equation takes the form

$$Pv = \frac{\bar{R}}{M}T = RT \tag{11.2}$$

The form of this equation of state that we will find most useful is

$$P = \rho RT \tag{11.3}$$

where ρ is the mass density of the gas in kg/m^3.

Fluid properties of air and water are provided in Tables A.13 and A.14 in Appendix A.

The concept of flow properties is not as clear as that of fluid properties. The general idea is that certain fluid behavior is a characteristic of the flow condition rather than being specific for a given fluid. An example of a flow property is velocity. Some others will be considered shortly. However, certain fundamental concepts must bc discussed before these additional ideas can be introduced.

11.3 The Variation of Properties in a Fluid

If we consider the property, pressure, which we have designated as P, and its variation with position in a two-dimensional framework, we may write

$$P = P(x, y) \tag{11.4}$$

A familiar example of pressure variation is a weather map, which is commonly found in your daily newspaper. The lines on such a map (Figure 11.4) are lines of constant pressure called **isobars**. The prefix iso- means **constant** and examples of its use are in isothermal for constant temperature, isometric for constant measure, and isotropic for constant properties in all directions.

In Figure 11.5, it is of interest to evaluate the change in pressure as we move from one position at $P_1(x_1, y_1)$ to another at $P_2(x_2, y_2)$. The change in P can be determined by expressing its differential in the form

$$dP = \frac{\partial P}{\partial x}dx + \frac{\partial P}{\partial y}dy \tag{11.5}$$

Integration of this expression between the two points of interest will allow the pressure difference, $P_2 - P_1$, to be evaluated. To do this, we must know the partial derivatives $\partial P/\partial x$ and $\partial P/\partial y$.

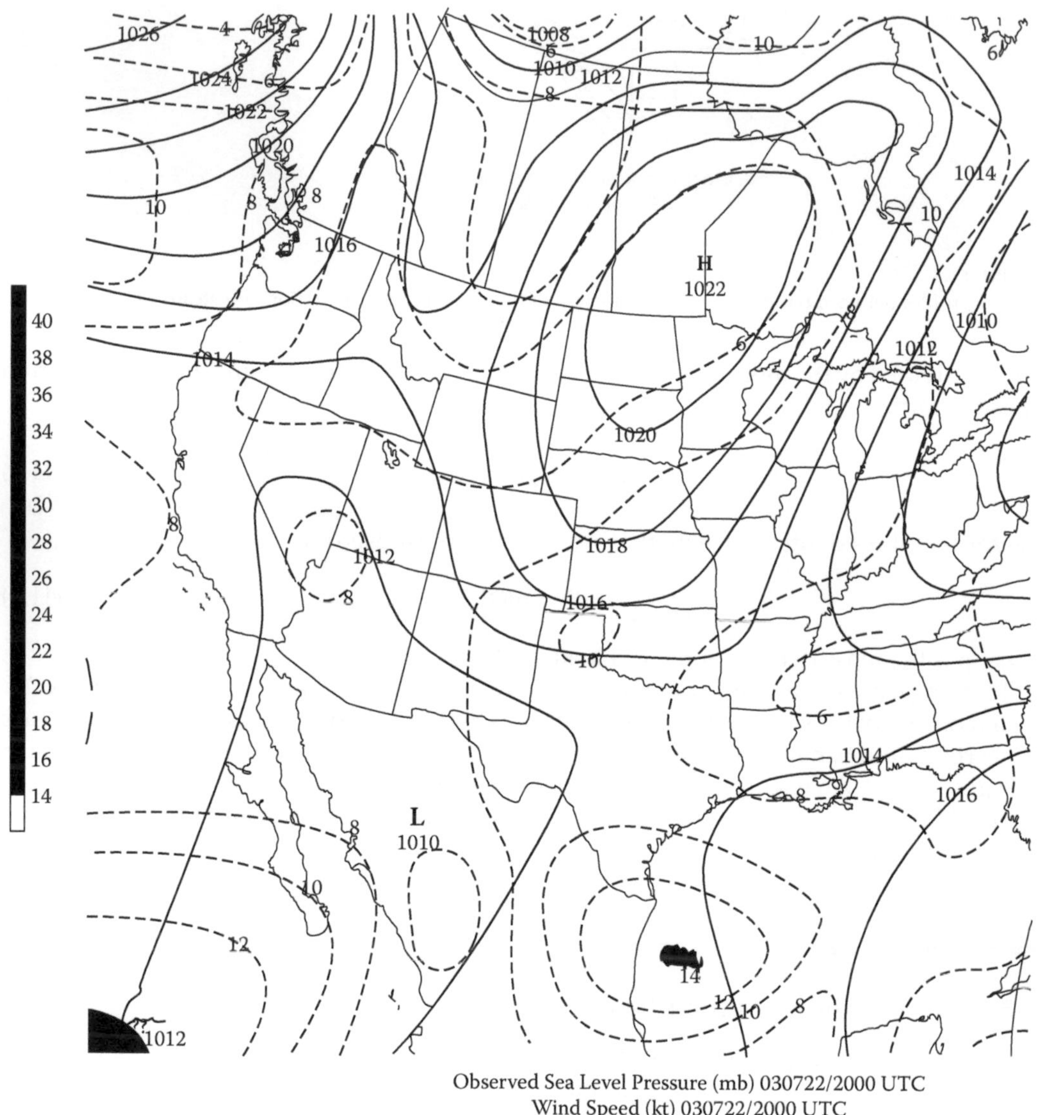

FIGURE 11.4
A weather map showing isobars. (From Ohio State University Weather Service, http:/aspl.ohio-state.edu).

It will also be of interest to evaluate the **rate** of pressure change along a given path. Denoting a general path or direction as s, we express the **directional derivative** as

$$\frac{dP}{ds} = \frac{\partial P}{\partial x}\frac{dx}{ds} + \frac{\partial P}{\partial y}\frac{dy}{ds} \tag{11.6}$$

There are, of course, an infinite number of paths that might be taken in moving away from a particular point. Two such paths are of particular interest. One is the path along which dP/ds is a minimum. Obviously this is $dP/ds = 0$ leading to $P = C$ (a constant) and is the path along an isobar. The other is the path along which dP/ds is a maximum.

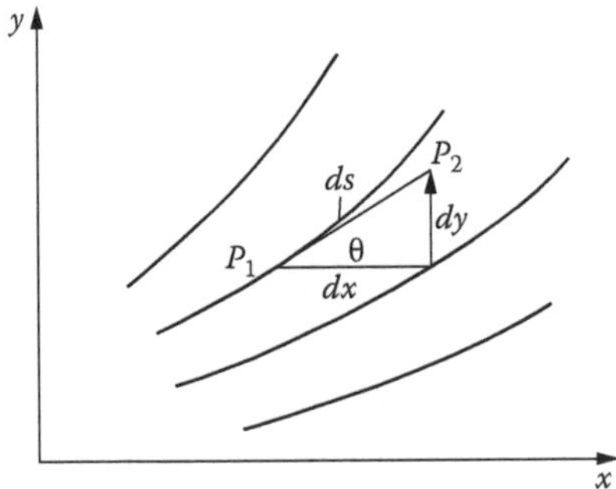

FIGURE 11.5
Isobars in the x-y plane.

The directional derivative along the path of maximum rate of change is termed the **gradient** designated by grad or ∇. The gradient of the pressure in a two-dimensional frame is written as

$$\nabla P = \frac{\partial P}{\partial x}\boldsymbol{x} + \frac{\partial P}{\partial y}\boldsymbol{y} \tag{11.7}$$

where $\boldsymbol{x}$ and $\boldsymbol{y}$ are **unit vectors** in the x- and y-directions, respectively.

The definition of the gradient can be extended to three dimensions. In rectangular coordinates with $P = P(x, y, z)$, we have

$$\nabla P = \frac{\partial P}{\partial x}\boldsymbol{x} + \frac{\partial P}{\partial y}\boldsymbol{y} + \frac{\partial P}{\partial z}\boldsymbol{z} \tag{11.8}$$

The expressions for ∇P in cylindrical and spherical coordinates are given in Appendix B. The reader will observe that, while these expressions are different, the geometric meaning of the gradient, that is, the vector having direction and magnitude of maximum rate of change of the dependent variable, remains the same.

Example 11.1 For a scalar $\phi = xy^3z^2$, determine grad ϕ and evaluate it at the point $x = 1,\ y = 1/2,\ z = 1/4$.

Solution

Here with $\phi = xy^3z^2$

$$\frac{\partial \phi}{\partial x} = y^3z^2$$

$$\frac{\partial \phi}{\partial y} = 3xy^2z^2$$

$$\frac{\partial \phi}{\partial z} = 2xy^3z$$

Thus,

$$\nabla\phi = y^3z^2\boldsymbol{x} + 3xy^2z^2\,\boldsymbol{y} + 2xy^3z\boldsymbol{z} \Longleftarrow$$

and at the point ($x = 1,\ y = 1/2,\ z = 1/4$)

$$\nabla\phi = \left(\frac{1}{2}\right)^3\left(\frac{1}{4}\right)^2 x + 3(1)\left(\frac{1}{2}\right)^2\left(\frac{1}{4}\right)^2 y + 2(1)\left(\frac{1}{2}\right)^3\left(\frac{1}{4}\right) z$$

$$= \left(\frac{1}{128}\right)x + \left(\frac{3}{64}\right)y + \left(\frac{1}{16}\right)z$$

$$= \frac{1}{128}(x + 6y + 8z) \Longleftarrow$$

The path of maximum variation in pressure is perpendicular to the path of minimum or zero variation. We thus deduce that, to move in a direction of maximum rate of pressure change, we should proceed along a path that is at right angles to an isobar. This relationship between isolines and directions of maximum rate of change is general.

11.4 The Continuum Concept

In our discussion thus far, it has been presumed that the pressure varies **continuously** in any direction. Another way of stating this idea is that $\partial P/\partial x$ and $\partial P/\partial y$ have values at the point $(x,\ y)$. Such variation requires the existence of a **continuum**. To develop this concept, we will consider the density, ρ. We already know that density is the mass per unit volume. Beyond that, of course, we must consider just what we mean when we refer to the density at a point.

If we consider the mass of air in, say, a cubic meter of space, we may evaluate the density according to

$$\rho = \frac{\Delta m}{\Delta V} \tag{11.9}$$

where ΔV is the volume (1 m^3 in this case) and Δm is the mass contained in ΔV.

We next consider a volume ΔV that is considerably smaller than 1 m^3, say a cube measuring 1 cm on a side. In this case, the density expression would be the same as Equation 11.9 and we would expect the same result for ρ.

We now continue the procedure, evaluating $\Delta m/\Delta V$ for ever smaller values of ΔV. A plot of this process is shown in Figure 11.6. Eventually, the volume becomes sufficiently

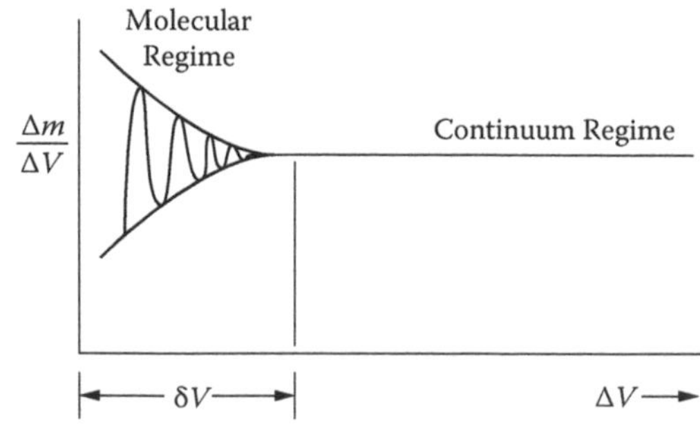

FIGURE 11.6
The density at a point.

small so that it will contain different numbers of molecules at different times of observation. From this point, as ΔV becomes smaller, the variation in $\Delta m/\Delta V$ oscillates between ever larger extremes. Clearly, the extreme values on this ratio become very large in the limit as $\Delta V \longrightarrow 0$.

For ΔV greater than a limiting value, δV, ρ is constant and we call this the **continuum regime**. When ΔV is less than ΔV, the density, ρ, will vary and this is the **molecular regime**. When we consider the density at a point to behave continuously, that is, just as though continuum conditions prevail, we are not being precise. However, such operations are still physically realistic because ΔV, for all practical purposes, is extremely small. Operationally, properties will be assumed to be continuous quantities. The density at a point will actually be considered to hold for a small region, ΔV, near a point such that

$$\rho = \lim_{\Delta V \longrightarrow \delta V} \frac{\Delta m}{\Delta V} \tag{11.10}$$

The concept of a continuum will be considered to hold throughout this book. Point-to-point continuous behavior will thus be presumed for all fluid and flow properties.

11.5 Laminar and Turbulent Flow

The terms **laminar** and **turbulent** flow are familiar to most of us and, intuitively, we have an impression as to the meaning of these concepts. A familiar situation is shown in Figure 11.7 where the characteristics of laminar and turbulent flows can be clearly seen. In Figure 11.7, smoke rises from a source of combustion in an apparently smooth, regular fashion and then suddenly changes to a much less regular and chaotic flow.

Smooth, orderly flow is called **laminar**. The term literally means layer-like and it implies that fluid flows in distinct layers, or "lamina," which retain their identity. In such a condition, fluid particles in a given layer remain in that layer as long as laminar flow continues. The only migration of fluid between adjacent layers is at the molecular level.

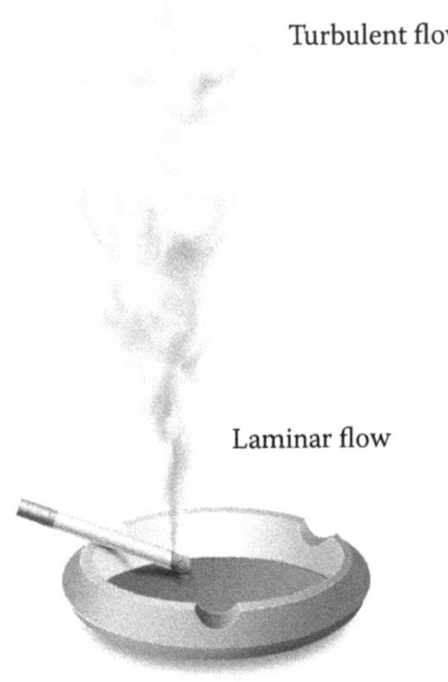

FIGURE 11.7
A common example of the transition from laminar to turbulent flow.

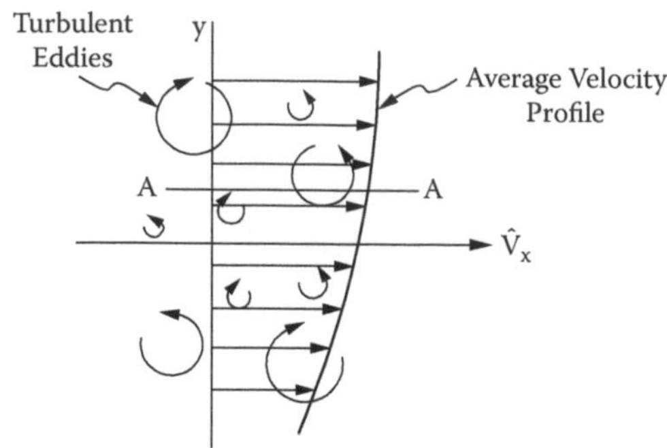

FIGURE 11.8
The structure of turbulent flow.

In contrast to the regularity of laminar flow, fluids in **turbulent** flow are characterized by apparently random motions of large numbers of fluid particles in directions other than that of the mean flow. Fluid particles that start at one position in a turbulently flowing stream may be carried, along with many neighbors, into another location by a small whirlpool-like **eddy**. The presence of eddies is a characteristic of turbulent flow and, even though the main flow is in the x-direction in Figure 11.8, turbulent eddies will propel fluid particles randomly between different y locations.

As we might expect, laminar flow is much easier to describe quantitively than is turbulent flow. We will consider both types of flow in subsequent chapters.

The region where the flow changes from laminar to turbulent, as seen in Figure 11.7, is designated as the **transition** regime.

11.6 Fluid Stress Conventions and Concepts

We defined a fluid in Section 11.1 in terms of the response to an imposed shearing stress. In a large quantity of fluid, an individual element may have imposed on it a number of influences, in addition to the shear stress.

Forces exerted on a medium are of two types. **Body forces** are those developed without physical contact and they arise from the presence of the fluid in a force field; examples would be gravitational and electromagnetic fields. The gravitational force is the most common type of body force and will be the only such force considered in this book.

Quantitatively, Newton's second law of motion describes the gravitational force as

$$\boldsymbol{F} = \boldsymbol{W} = \rho V \boldsymbol{g} \tag{11.11}$$

where ρ is the mass density of the fluid, V is the volume under consideration and $\boldsymbol{g}$ is the acceleration of gravity. We commonly call such a force the weight, W, of the fluid volume. Various forms of Equation 11.11 give rise to other terms; W/V, the weight per unit volume is the specific weight, γ, and $\boldsymbol{g}$ may be thought of as the gravitational body force per unit mass. It should be noted that Equation 11.11 is a vector expression and that the direction of the body force, $\boldsymbol{F}$, must be the same as that of, $\boldsymbol{g}$, the gravitational acceleration. If gravity is oriented downward (in the minus y-direction), we may write

$$F_x\boldsymbol{x} + F_y\boldsymbol{y} + F_z\boldsymbol{z} = \rho V(-g\boldsymbol{y})$$

giving

$$F_x = F_z = 0 \tag{11.12a}$$

and

$$F_y = W = -\rho V g \tag{11.12b}$$

Expressions for the weight will generally have obvious directional characteristics so we will not, henceforth, evaluate W with this much formality.

Surface forces are those exerted by direct contact between a fluid medium and a boundary and the result of forces acting on a fluid medium is to put a general element of the fluid into a state of stress. The variation of the stress throughout the fluid is termed the **stress field**.

A piece of the boundary of a general fluid element is shown in Figure 11.9. We note that a small portion of the surface, ΔS, has a surface force, $\Delta \boldsymbol{F}$ acting on it. The vector, $\boldsymbol{n}$, is the outwardly directed unit vector normal to ΔS. The force, $\Delta \boldsymbol{F}$, can thus be resolved into its normal and tangential components, ΔF_n and ΔF_t, as indicated. We then define the normal and shear stresses at ΔS, consistent with continuum behavior as

$$\sigma = \lim_{\Delta S \longrightarrow 0} \frac{\Delta F_n}{\Delta S} \tag{11.13}$$

and

$$\tau = \lim_{\Delta S \longrightarrow 0} \frac{\Delta F_t}{\Delta S} \tag{11.14}$$

respectively. These stress components, σ and τ, will have varying character depending on the orientation of the surface, ΔS, used in their definitions.

Stress is, in general, a **tensor** quantity of second order: this means that its complete specification involves nine components. This is in contrast to a vector quantity (a first-order tensor) having three components and a scalar (a zeroth-order tensor), which has only one.

In Figure 11.10, we show two elements in a fluid continuum. The configuration of the element in Figure 11.10a is consistent with Cartesian or rectangular coordinates; its dimensions are Δx, Δy, and Δz in the directions corresponding to the coordinate directions.

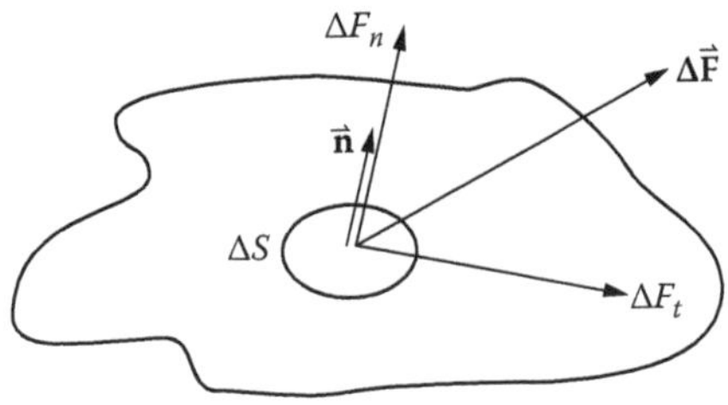

FIGURE 11.9
A surface force acting on an element of surface ΔS.

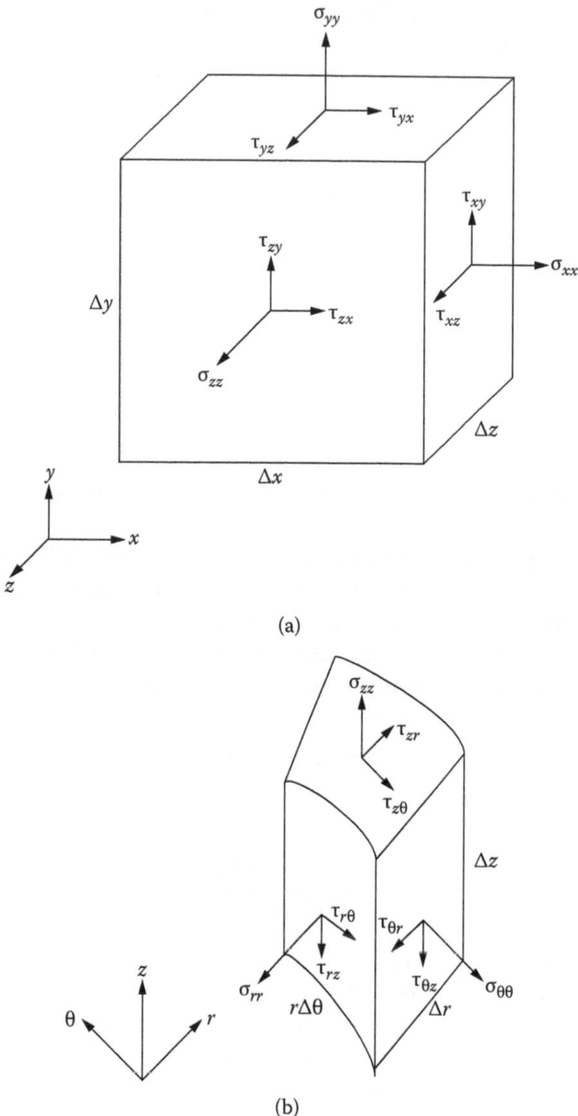

FIGURE 11.10
Stress in fluid elements: (a) rectangular coordinates and (b) cylindrical coordinates.

In Figure 11.10b, we show an element consistent with a cylindrical coordinate system where we note that its sides have lengths Δr, $r\,\Delta\theta$, and Δz.

On each face, the stress can be resolved into three components as shown. By conventions, shear stresses, parallel to the face, are labeled τ and normal stresses, perpendicular to the face, are labeled σ. Because each such element has six faces, there are six sets of stress components and the three sets on the visible faces are shown in the figure. As mentioned earlier in this section, these nine components are conveniently expressed as a matrix

$$\begin{bmatrix} \sigma_{xx} & \tau_{xy} & \tau_{xz} \\ \tau_{yx} & \sigma_{yy} & \tau_{yz} \\ \tau_{zx} & \tau_{zy} & \sigma_{zz} \end{bmatrix}$$

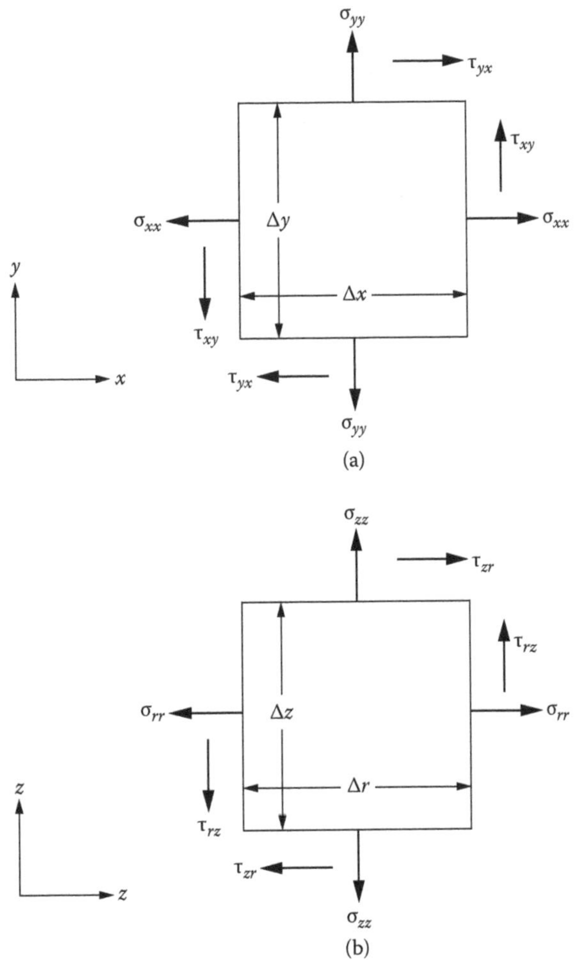

FIGURE 11.11
Two-dimensional views showing stress components: (a) the x-y plane and (b) the r-z plane.

where the diagonal elements, σ_{ii}, are the **normal** stress components having both subscripts the same. The off-diagonal elements, τ_{ij}, are the **shear** stress components and we note that the subscripts on τ are mixed, that is, they are not the same.

The convention for the labeling of σ_{ii} and τ_{ij} merits our closest attention. The first subscript identifies the plane on which the stress component acts and the second identifies the direction of action relative to the coordinate system. The plane, or face, of the element is specified by the direction that the outwardly directed normal vector to that face will point. Thus, the three faces shown in Figure 11.10a are the positive x, y, and z faces. In Figure 11.10b, these are the θ, r, and z faces. All components are conventionally drawn and considered in their positive sense and positive components are those having both subscripts with the same sign, either positive or negative.

Two-dimensional representations of these elements are shown in Figure 11.11. The x-y plane is shown in Figure 11.11a and the r-z plane in Figure 11.11b. We will use such elements frequently in the chapters to follow.

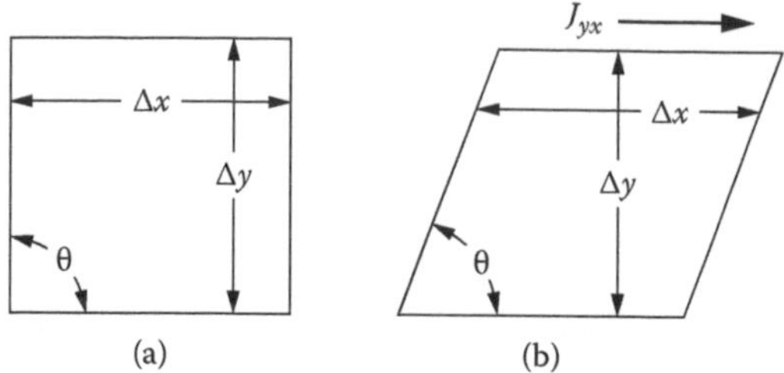

FIGURE 11.12
The deformation of a fluid element (a) at time t and (b) at time $t + \Delta t$.

11.7 Viscosity, a Fluid Property

We are now able to give meaning to a very important fluid property, the **viscosity**, symbolized by μ (the Greek lower case "mu"), which is defined as

$$\text{viscosity} = \frac{\text{shear stress}}{\text{rate of shear strain}} \tag{11.15}$$

Figure 11.12 shows a two-dimensional view of a fluid element considered at two instants in time and shows the change of the configuration, that is, the **deformation**, resulting from the presence of a shearing stress. The "word equation" of Equation 11.15 may be written in terms of the quantities shown in Figure 11.12

$$\mu = \frac{\tau_{yx}}{-d\theta/dt} \tag{11.16}$$

We recall that shear strain is related to the change in the angle θ. The **rate** of shear strain would thus be the time derivative of θ. Upon examining the relationship between cause (the shear stress) and effect (the angular deformation), we notice that a positive shear will cause a decrease in θ and a negative shear will cause θ to increase. This is the reason for the negative sign in Equation 11.16.

We may evaluate $d\theta/dt$ in terms of the flow condition causing the deformation shown. Clearly, for θ to increase in the time period, Δt, the velocity, $\hat{V}_x$, must be greater at $y + \Delta y$ than at y. The rate of shearing stress is $-d\theta/dt$ and we proceed to express it as

$$-\frac{d\theta}{dt} = \lim_{\Delta x,\, \Delta y,\, \Delta t \longrightarrow 0} \frac{\theta|_{t+\Delta t} - \theta|_t}{\Delta t}$$

Here

$$\theta_t = \frac{\pi}{2}$$

and

$$\theta|_{t+\Delta t} = \frac{\pi}{2} - \arctan \frac{(\hat{V}_x|_{y+\Delta y} - \hat{V}_x|_y)\Delta t}{\Delta y}$$

so that

$$-\frac{d\theta}{dt} = \lim_{\Delta t \longrightarrow 0} \frac{\dfrac{\pi}{2} - \arctan \dfrac{(\hat{V}_x|_{y+\Delta y} - \hat{V}_x|_y)\Delta t}{\Delta y} - \dfrac{\pi}{2}}{\Delta t}$$

Observe that the two $\pi/2$ terms cancel, which allows the Δt's to be cancelled as well. We then have

$$-\frac{d\theta}{dt} = \lim_{\Delta y \longrightarrow 0} \frac{(\hat{V}_x|_{y+\Delta y} - \hat{V}_x|_y)}{\Delta y} = \frac{d\hat{V}_x}{dy} \tag{11.17}$$

The minus sign is the result of a positive shear stress, so a consistent representation for the viscosity will be

$$\mu = \frac{\tau_{yx}}{d\hat{V}_x/dy} \tag{11.18}$$

and the common form of Equation 11.18 is written as

$$\tau_{yx} = \mu \frac{d\hat{V}_x}{dy} \tag{11.19}$$

which is referred to as **Newton's viscosity relationship**.

Many authors refer to Equation 11.19 as Newton's "law" of viscosity. Clearly, it is not a law in a fundamental sense so we will avoid such a designation. A fluid, having a viscosity that does not vary as a function of strain rate, is termed **Newtonian**. Water, all gases, and a number of fluids are Newtonian. Some other fluids, such as starch suspensions and paints, display varying values of μ depending on the rate of deformation. These fluids are designated **non-Newtonian**.

Equations 11.18 and 11.19 represent viscosity in the case of one-dimensional flow. A more general designation for velocity variation in all three coordinate directions would yield the expressions

$$\tau_{yx} = \tau_{xy} = \mu \left(\frac{\partial \hat{V}_x}{\partial y} - \frac{\partial \hat{V}_y}{\partial x} \right) \tag{11.20a}$$

$$\tau_{zy} = \tau_{yz} = \mu \left(\frac{\partial \hat{V}_y}{\partial z} - \frac{\partial \hat{V}_z}{\partial y} \right) \tag{11.20b}$$

and

$$\tau_{xz} = \tau_{zx} = \mu \left(\frac{\partial \hat{V}_z}{\partial x} - \frac{\partial \hat{V}_x}{\partial z} \right) \tag{11.20c}$$

in rectangular coordinate form.

Viscosity is generally a function of temperature. In gases, the viscosity increases with temperature as a result of increased molecular activity. In contrast, the much more closely packed molecules in liquids experience decreased intermolecular forces as the temperature rises. Hence, the viscosities of liquids decrease with temperature. Viscosity is not affected by pressure except for extremes at both the high and low ends. Thus, we generally consider viscosity to be independent of pressure. Viscosity is tabulated, as a function of temperature, for a number of important fluids, along with other properties in Table A.15 for gases and A.16 for liquids.

Another property, tabulated along with μ, is the **kinematic viscosity**, designated by ν (the Greek lower-case "nu"), which is defined as

$$\nu \equiv \frac{\mu}{\rho} \tag{11.21}$$

Because density varies with both temperature and pressure, the kinematic viscosity will also vary with both properties.

The units of viscosity can be determined from Equation 11.18. We have

$$\mu = \frac{\text{shear stress}}{\text{velocity/length}} = \frac{\text{N/m}^2}{\text{m/s/m}} = \frac{\text{N-s}}{\text{m}^2} = \text{Pa-s} \tag{11.22}$$

or, representing F as ma, an alternate form is

$$\mu = \frac{\text{N-s}}{\text{m}^2} = \frac{\text{kg-m}}{\text{s}^2}\frac{\text{s}}{\text{m}^2} = \frac{\text{kg}}{\text{m-s}} \tag{11.23}$$

Both of these representations are correct, the form of Equation 11.23 is the one most often seen.

Another term seen occasionally is the **poise** defined as 1 g/cm-s. Units for the kinematic viscosity are seen to be m^2/s and an additional unit is the **stoke** defined as 1 cm^2/s.

The use of Equation 11.18 is demonstrated in Examples 11.2 and 11.3.

Example 11.2 The hydraulic ram or hoist in an automobile lift consists of a circular shaft with an outer diameter of 37.88 cm, which moves inside a stationary cylindrical housing having an inside diameter of 37.91 cm. The space between the stationary and moving surfaces is filled with hydraulic fluid having a viscosity of 0.304 kg/m-s. The desired rate of travel of the hoist is 0.18 m/s. (a) Determine the frictional force (resistance) for a 2.5-m long hoist. (b) Assume that a length of 2.5 m of the ram (the moving cylinder) is engaged and determine the rate of sink of a ram loaded with 700 kg when the downward gravitational force is balanced by the frictional force due to the viscosity of the hydraulic fluid (Figure 11.13).

Solution

Assumptions and Specifications

(1) Steady-state conditions exist.
(2) The fluid is incompressible.

(a) An expanded view of the space between the two surfaces is shown in Figure 11.13b. Because the gap thickness, $\Delta = 0.015$ cm, is so small relative to the ram diameter,

$$\frac{\Delta}{D} = \frac{0.015\,\text{cm}}{37.88\,\text{cm}} = 3.960 \times 10^{-4}$$

we will neglect any influence of the surface curvature.

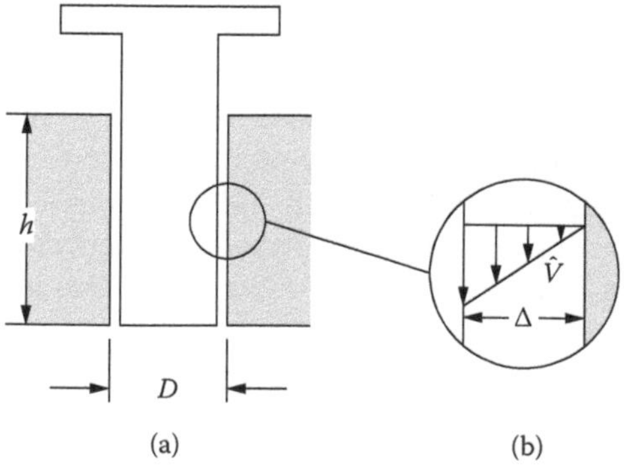

FIGURE 11.13
Hydraulic ram for Example 11.2: (a) the configuration and (b) a detail of the space between the shaft and the housing.

The resisting force is evaluated as

$$F = \int_S \tau \, dS = \int \tau \pi D \, dy = \tau \pi D h$$

and with the curvature neglected

$$\tau = \mu \frac{d\hat{V}}{dx} = \mu \frac{\hat{V}}{\Delta}$$

Therefore,

$$\begin{aligned} F &= \mu \frac{\hat{V}}{\Delta} \pi D h \\ &= \frac{(0.304\,\text{kg/m-s})(0.18\,\text{m/s})(\pi)(0.3788\,\text{m})(2.5\,\text{m})}{1.5 \times 10^{-4}\,\text{m}} \\ &= 1085\,\text{N} \Longleftarrow \end{aligned}$$

(b) With the frictional force set equal to the load

$$F = \mu \frac{\hat{V}}{\Delta} \pi D h = mg$$

we have

$$\begin{aligned} \hat{V} &= \frac{mg\Delta}{\mu \pi D h} \\ &= \frac{(700\,\text{kg})(9.81\,\text{m/s}^2)(1.5 \times 10^{-4}\,\text{m})}{(0.304\,\text{kg/m-s})(\pi)(0.3788\,\text{m})(2.5\,\text{m})} \\ &= 1.14\,\text{m/s} \Longleftarrow \end{aligned}$$

Example 11.3 A circular disk has its plane surface parallel to a stationary plate (Figure 11.14). The space between the two surfaces, a distance Δ apart, contains a fluid with a viscosity, μ. Develop a relationship for the torque, T, required to rotate the disk at a uniform angular velocity, ω.

Solution

Assumptions and Specifications

(1) Steady-state conditions exist.
(2) The fluid in the space between the surfaces is incompressible.
(3) There are no edge effects.

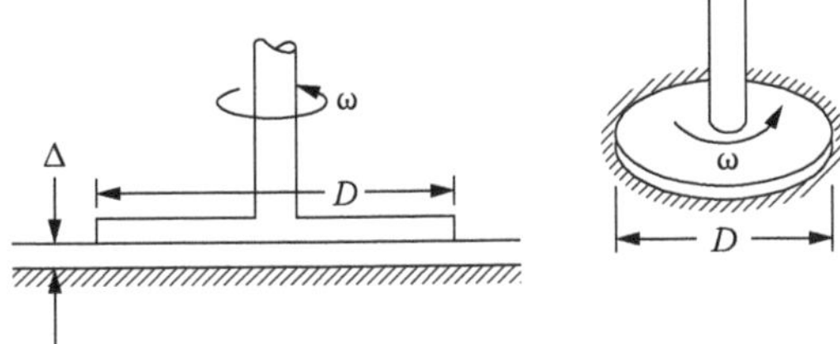

FIGURE 11.14
Rotating disk for Example 11.3.

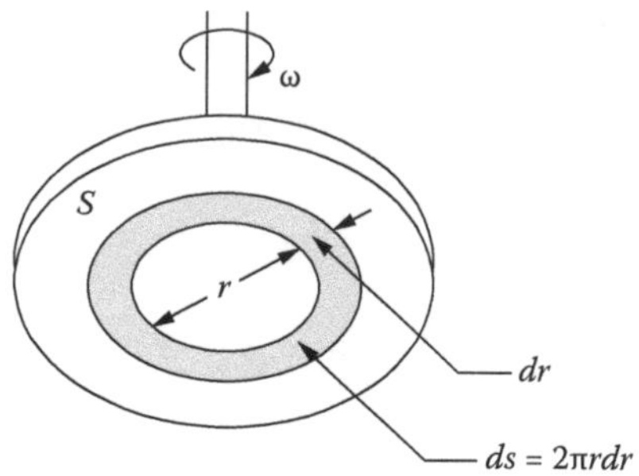

FIGURE 11.15
Detail of rotating disk.

This is another application of Newton's viscosity relationship, but in this case, because the moving surface is rotating, the velocity is a variable and is a function of the distance from the axis of rotation.

For this case, the viscous force, dF, associated with the area dS, as shown in Figure 11.15, will be

$$dF = \tau dS = \tau(2\pi r dr)$$

The area, dS, is chosen in this manner because it lies at a distance, r, from the axis of rotation. The velocity gradient is therefore,

$$\frac{\hat{V}_r}{\Delta} = \frac{r\omega}{\Delta}$$

where Δ is the separation between the disk and the plate.

The problem is now solved in a straightforward manner

$$T = \int r\,dF = \int r\tau dS$$

$$= \int_0^{D/2} r\mu\frac{r\omega}{\Delta}(2\pi r dr)$$

$$= \mu\frac{2\pi\omega}{\Delta}\int_0^{D/2} r^3 dr$$

or

$$T = \mu\frac{2\pi\omega}{\Delta}\frac{r^4}{4}\Big|_0^{D/2} = \frac{\mu\pi\omega D^4}{32\Delta} \Longleftarrow$$

A device that is used to measure the viscosity of a gas or a liquid is called a **viscometer.** Such a device may take on a variety of forms and Design Example 4, which now follows, illustrates the use of a rotating cylinder as a viscometer.

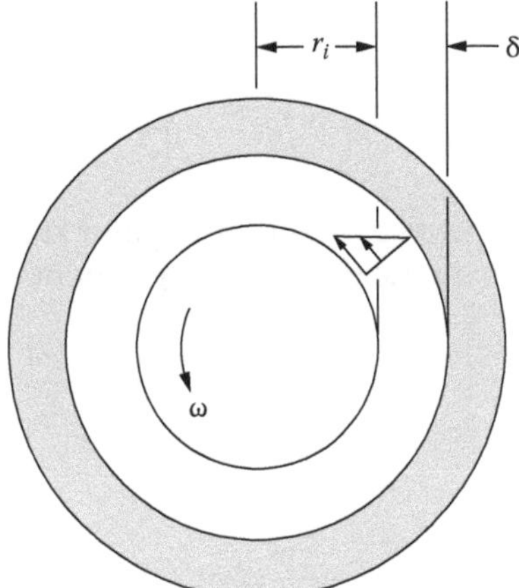

FIGURE 11.16
Rotating drum viscometer.

11.8 Design Example 4

One type of viscometer is constructed of two concentric cylinders separated by a very narrow gap. Such a device is shown in Figure 11.16 where the diameter of the inner (rotating) cylinder is $d_i = 16\,\text{cm}$, the diameter of the outer (stationary) cylinder is $d_o = 16.30\,\text{cm}$ and the length of both cylinders is $L = 50\,\text{cm}$. The gap is filled with an oil of unknown viscosity and the inner cylinder is rotated at 300 rpm. If the measured torque is 1.84 N-m, determine the viscosity of the oil.

Solution

Assumptions and Specifications

(1) Steady-state conditions exist.
(2) There is no flow of fluid.
(3) The fluid is incompressible.
(4) There are no end effects.
(5) The shear stress in the gap is directly proportional to the velocity gradient.

The torque necessary to produce a steady-state rotation of the inner cylinder is equal to the product of the viscous force imparted to the inner surface by the oil and the distance, r_i, from the axis of rotation.

The viscous shear stress is given by

$$\tau = \mu \left. \frac{d\hat{V}}{dr} \right|_{r=r_i}$$

The resulting viscous force is

$$F = \tau A = \tau(2\pi r_i L) = \left[\mu \left. \frac{d\hat{V}}{dr} \right|_{r=r_i} \right] (2\pi r_i L)$$

and the required torque is

$$T = Fr_i = \left[\mu \frac{d\hat{V}}{dr}\bigg|_{r=r_i} \right] (2\pi r_i^2 L)$$

When the size of the gap, δ, between the rotating and stationary cylinders is relatively small ($\delta/r_i \ll 1$), we may approximate the velocity gradient by the linear relationship

$$\frac{d\hat{V}}{dr}\bigg|_{r=r_i} \simeq \frac{\Delta\hat{V}}{\Delta r} = \frac{\hat{V}_i - 0}{\delta} = \frac{r_i\omega}{\delta}$$

Our torque expression is now

$$T = \mu\frac{r_i\omega}{\delta}(2\pi r_i^2 L)$$

and the viscosity can be evaluated as

$$\mu = \frac{T\delta}{2\pi r_i^3 L\omega}$$

and with

$$T = 1.84\,\text{N-m} \qquad \text{(specified)}$$

$$\delta = \frac{d_o - d_i}{2} = \frac{16.30\,\text{cm} - 16.00\,\text{cm}}{2} = 0.15\,\text{cm}$$

$$r_i = \frac{16.00\,\text{cm}}{2} = 8.00\,\text{cm}$$

$$L = 50\,\text{cm}$$

and

$$\omega = \frac{300\,\text{rev/min}}{60\,\text{s/min}}(2\pi\,\text{rad/rev}) = 10\pi\,\text{rad/s}$$

we have

$$\mu = \frac{(1.84\ \text{N-m})(0.0015\,\text{m})}{2\pi(0.080\,\text{m})^3(0.50\,\text{m})(10\pi\,\text{rad/s})} = 0.0546\,\text{kg/m-s} \Longleftarrow$$

11.9 Other Fluid Properties

11.9.1 Specific Gravity and Specific Weight

Density, ρ, specific volume, v, specific gravity, SG, and specific weight, γ, were considered in detail in Section 2.4.

11.9.2 Compressibility

Fluids are designated as **compressible** or **incompressible** depending on whether their density is variable or constant. We generally consider liquids to be incompressible and gases to be compressible and we will discuss certain exceptions to these ideas in subsequent chapters.

The **bulk modulus of elasticity**, frequently referred to as just the **bulk modulus**, is a fluid property that characterizes the compressibility. It can be written as

$$\beta = \frac{dP}{dV/V} \tag{11.24a}$$

or as

$$\beta = -\frac{dP}{d\rho/\rho} \tag{11.24b}$$

and has the dimensions N/m^2.

11.9.3 The Speed of Sound

Disturbances that are introduced at some point in a fluid propagate at a finite velocity called **acoustic velocity** or **speed of sound**, c. It can be shown that the speed of sound is related to changes in pressure and density by

$$c = \left(\frac{dP}{d\rho}\right)^{1/2} \tag{11.25a}$$

and use of Equation 11.24b gives us the alternate form

$$c = \left(-\frac{B}{\rho}\right)^{1/2} \tag{11.25b}$$

Moreover, for gases undergoing an isentropic process where $PV^k = C$, a constant, we have

$$c = \left(\frac{kP}{\rho}\right)^{1/2}$$

or

$$c = (kRT)^{1/2} \tag{11.25c}$$

where R in the SI System has units of J/kg-K.

Often, in gas flows, even though the fluids are obviously compressible, density variations are negligibly small so that we may treat the flow as incompressible. A common criterion for such a consideration is the **Mach number**, which is defined as the ratio of the fluid velocity of interest, $\hat{V}$, to the speed of sound, c, in the fluid

$$\mathrm{M} = \frac{\hat{V}}{c} \tag{11.26}$$

If $M < 0.2$, the flow may be treated as incompressible with negligible error.

Example 11.4 A jet aircraft flies at an altitude of 9500 m where the air temperature is 226.41 K. Determine whether or not the compressible effects are significant when the aircraft speed is (a) 900 km/h and (b) 200 km/h.

Solution

Assumptions and Specifications

(1) The flow is steady.

(2) The air may be taken as an ideal gas.

To determine the compressibility effects, we calculate the Mach number for the air at the two speeds. With $k = 1.4$, $R = 287\,\text{J/kg-K}$ (Table A.1 for air) and $T = 226.41\,\text{K}$, Equation 11.25c gives

$$c = (kRT)^{1/2} = [(1.4)(287\,\text{J/kg-K})(226.41\,\text{K})]^{1/2} = 90{,}972\,\text{m}^2/\text{s}^2 = 301.61\,\text{m/s}$$

For $\hat{V} = 900\,\text{km/h}$ (250 m/s), the Mach number is

$$M = \frac{\hat{V}}{c} = \frac{250\,\text{m/s}}{301.61\,\text{m/s}} = 0.829$$

Because $M > 0.2$, the compressibility effects are significant.

For $\hat{V} = 200\,\text{km/h}$ (55.5 m/s), the Mach number is

$$M = \frac{\hat{V}}{c} = \frac{55.6\,\text{m/s}}{301.61\,\text{m/s}} = 0.184$$

Because $M < 0.2$, the air flow over the aircraft may be regarded as incompressible with negligible error.

11.9.4 Surface Tension and Capillary Action

We are familiar with the condition where a small amount of unconfined liquid forms a spherical drop. The reason for this occurrence is the attraction that exists between the liquid molecules. Within any particular drop, any molecule is surrounded by many others and the average force of attraction is the same in all directions. On the other hand, particles close to the surface will have an imbalance of net force because the number of adjacent particles are not uniform. The extreme condition, of course, is the density discontinuity. Particles right at the surface experience a relatively strong inward attractive force.

With these ideas in mind, it is clear that some work must be done when a particle is moved toward the surface. If fluid is added to a drop, it will expand and additional surface will be created. The work needed to create new surface is the **surface tension**, designated by σ. Quantitatively, the surface tension is the work per unit area in N-m/m^2 or force per unit length of interface in N/m.

Both phases that are separated by an interface have the property of surface tension. The most common materials are water and air but many combinations are obviously possible. Surface tension, for a given interfacial composition, is a function of temperature and pressure although it is a much stronger function of temperature. Selected values for σ for a few fluids in air are given in Table 11.1 and, for water in air at 1 atmosphere, the equation

$$\sigma = 0.1185(1 - 0.001325T) \qquad \text{N/m} \tag{11.27}$$

can be used to provide σ in N/m when T in K.

The name, surface tension, suggests the presence of a skin-like membrane that holds the droplet together. Although this is not a correct physical concept quantitatively, the free surface can be treated as a membrane under uniform tension with magnitude, σ. Consistent

TABLE 11.1

Surface Tensions of Some Fluids in Air at 1 atm at 20°C as a Function of Temperature (Extracted from *Handbook of Chemistry and Physics*)

Fluid	σ, N/m
Ammonia	0.021
Ethyl alcohol	0.028
Gasoline	0.022
Glycerin	0.063
Kerosene	0.028
Mercury	0.440
Soap solution	0.025
SAE 30 oil	0.035

with this concept of surface tension, is the idea that the curved surface will be associated with a pressure difference. Because the sphere represents the minimum surface area for a prescribed volume, we show a free-body diagram of the forces in a hemisphere of radius, r (Figure 11.17). The difference in pressure between the inside and outside of the hemisphere, ΔP, acting over the area, πr^2, is balanced on the hemisphere periphery by the force due to the surface tension, $2\pi r\sigma$. Hence, the required force balance is

$$\pi r^2 \Delta P = 2\pi r\sigma$$

which shows that the difference in pressure is

$$\Delta P = \frac{2\sigma}{r} \tag{11.28}$$

For a soap bubble, which has an extremely thin wall, the pressure difference across the *two* interfaces present will be

$$\Delta P = \frac{4\sigma}{r} \tag{11.29}$$

We note that as the radius of curvature, r, increases, the resulting ΔP will decrease. Thus, in the case of no curvature (a flat surface), there is no net pressure difference due to surface tension.

An important result of the surface-tension-caused pressure difference across a curved surface is the phenomenon of **capillary action**. This effect, which can cause a liquid column to rise in a tube, is a function of how well a liquid **wets** a solid boundary. The **contact angle**

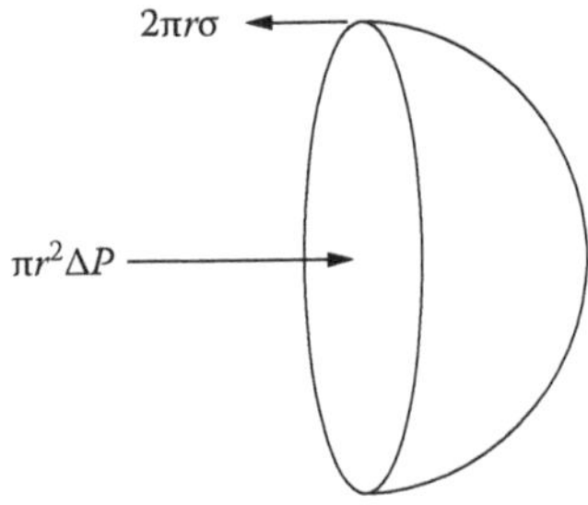

FIGURE 11.17
Free-body diagram of a hemispherical droplet showing the balance between pressure and tension forces.

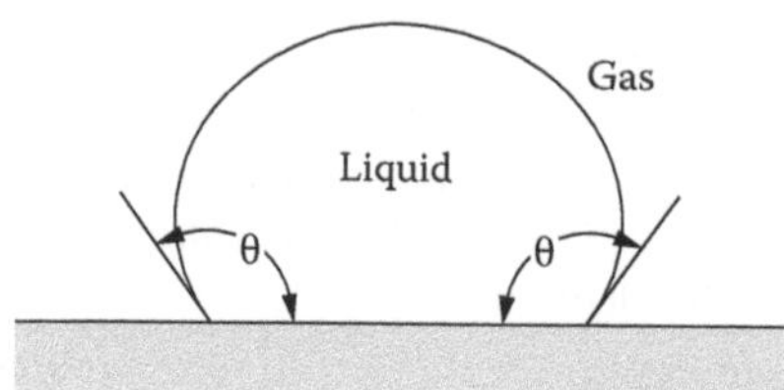

FIGURE 11.18
The contact angle associated with a gas-liquid-vapor interface for the nonwetting case.

associated with a gas-liquid-solid interface is the indicator of a wetting or nonwetting case. As shown in Figure 11.18, the contact angle, θ, is measured in the liquid and the delineation between wetting and nonwetting liquids is in the magnitude of the contact angle. For wetting liquids, $\theta < 90^\circ$ and for nonwetting liquids, $\theta > 90^\circ$. For example, mercury in contact with a clean glass tube has a value of $\theta \approx 130^\circ$. For water in a clean glass tube, θ may be taken as $\theta = 0^\circ$ because the water almost completely wets the surface.

In Figure 11.19a, we see that a small diameter tube has been inserted into a pool of water. The liquid level interface in the tube makes an angle, θ, with the wall of the tube and rises a distance, h, above the level of the water in the pool. In this case, an attraction or adhesion between the wall of the tube and the liquid molecules overcomes the mutual attraction of the molecules (cohesion) at the surface of the liquid. Here, the water appears to be "pulled up" into the tube and is said to **wet** the surface of the tube. Figure 11.19b shows a tube inserted into a pool of mercury where the adhesion of the mercury molecules to the solid surface is weak in comparison with the cohesive forces at the liquid surface. In this case, the mercury is depressed a distance, h, below the level of the pool of mercury.

A free-body diagram of a wetting liquid within a tube is shown in Figure 11.19c. Here the vertical force due to the surface tension

$$2\pi r \sigma \cos \theta$$

must balance the force due to the weight of the fluid volume, $V = \pi r^2 h$

$$\rho \pi r^2 h g$$

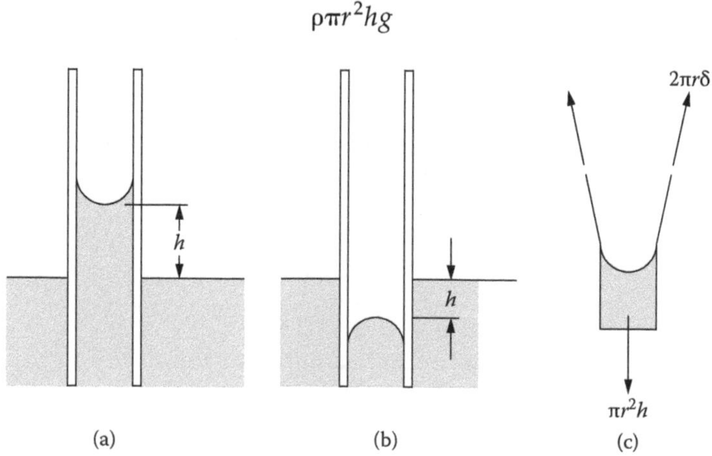

FIGURE 11.19
(a) A tube immersed in water, (b) a tube immersed in mercury, and (c) a free-body diagram of a wetting liquid inside a tube.

Hence,

$$2\pi r\sigma\cos\theta = \rho\pi r^2 hg$$

and the height, h, is given by

$$h = \frac{2\sigma\cos\theta}{\rho g r} \qquad (11.30)$$

Example 11.5 A 5-mm-radius glass tube is inserted into a pool of mercury at 20°C. The surface tension of the mercury and air is 0.47 N/m and the contact angle is 130°. Determine the distance that the mercury is depressed below the pool level (Figure 11.19b).

Solution

Assumptions and Specifications

(1) Steady state exists.
(2) The acceleration of gravity is assumed to be $9.81\,\text{m/s}^2$.
(3) The density of water is $1000\,\text{kg/m}^3$.
(4) Mercury is an incompressible liquid and has a specific gravity of 13.6.

For mercury, $\rho = 13{,}600\,\text{kg/m}^3$ and use of Equation 11.30 gives

$$h = \frac{2\sigma\cos\theta}{\rho g r} = \frac{2(0.47\ \text{N/m})(\cos 130^\circ)}{(13{,}600\,\text{kg/m}^3)(9.81\,\text{m/s}^2)(5\times10^{-3}\,\text{m})}$$

or

$$h = -9.06\times10^{-4}\,\text{m} \qquad (0.906\,\text{mm}) \Longleftarrow$$

11.10 Summary

In this introductory chapter to the subject of fluid mechanics, a number of concepts and definitions have been introduced. Most of them will be encountered quite often in the remainder of this book.

The concept of a fluid being a material that deforms **continuously** when subjected to a shearing stress led to a discussion of stress and how to characterize stress at a point, as well as the manner in how it changes throughout a fluid continuum.

A property that can vary from point to point was examined, and the rate of change was seen to vary from zero along an **isoline** to a maximum in the direction of the **gradient**.

In this book, gases and liquids, both fluids, are considered as being continuously distributed throughout the region so that the fluid may be considered as a continuum. The density, ρ, was the primary property employed to determine if the continuum concept is appropriate.

Laminar and turbulent flow were discussed. Smooth and orderly flow is termed laminar. The flow is layer-like and fluid particles in a given layer retain their identity as long as laminar flow continues. In contrast, turbulent flow is characterized by apparently random motion of large numbers of fluid particles.

The fluid property **viscosity** was defined and its role in producing a force between a moving surface and a viscous fluid was illustrated. Newton's viscosity relationship of the form

$$\tau_{yx} = \mu \frac{d\hat{V}_x}{dy} \tag{11.19}$$

is the defining relationship for viscosity.

11.11 Problems

Evaluation of the Gradient

11.1: A scalar field, ϕ, is represented by

$$\phi = xy^2(z-1)^3$$

Determine grad ϕ and evaluate grad ϕ at $x = 1,\ y = 3$, and $z = -1$.

11.2: A scalar field, ϕ, is represented by

$$\phi = x^3(y-2)$$

Determine grad ϕ and evaluate grad ϕ at $x = 2,\ y = 2$, and $z = 3$

11.3: A scalar field, ϕ, is represented by

$$\phi = (x-2)^2yz^2$$

Determine grad ϕ and evaluate grad ϕ at $x = -1,\ y = -2$, and $z = 3$.

11.4: A scalar field, ϕ, is represented by

$$\phi = x^2y^3z^4$$

Determine grad ϕ and evaluate grad ϕ at $x = 1,\ y = -2$, and $z = -3$.

11.5: A scalar field, ϕ, is represented by

$$\phi = x(x^2-1)yz$$

Determine grad ϕ and evaluate grad ϕ at $x = 1,\ y = 2$, and $z = -3$.

11.6: A scalar field, ϕ, is represented by

$$\phi = xy(y+3)z^2$$

Determine grad ϕ and evaluate grad ϕ at $x = 1,\ y = 3$, and $z = 5$.

11.7: A scalar field, ϕ, is represented by

$$\phi = x^2(z^2-4z)$$

Determine grad ϕ and evaluate grad ϕ at $x = -1,\ y = 3$, and $z = -5$.

11.8: A scalar field, ϕ, is represented by

$$\phi = 3x^2y + 4y^2$$

Determine grad ϕ at the point $(3, 5)$.

11.9: Show that the unit vectors $\boldsymbol{r}$ and $\boldsymbol{\theta}$ are related to the unit vectors $\boldsymbol{x}$ and $\boldsymbol{y}$ by

$$\boldsymbol{r} = \boldsymbol{x}\cos\theta + \boldsymbol{y}\sin\theta$$

and

$$\boldsymbol{\theta} = \boldsymbol{x}\sin\theta + \boldsymbol{y}\cos\theta$$

11.10: Using the results of Problem 11.9, show that

$$\frac{d\boldsymbol{r}}{d\theta} = \boldsymbol{\theta} \quad \text{and} \quad \frac{d\boldsymbol{\theta}}{d\theta} = -\boldsymbol{r}$$

11.11: Using the results of Problem 11.9, and, in addition

$$\frac{\partial}{\partial x} = \cos\theta\frac{\partial}{\partial r} + \frac{\cos\theta}{r}\frac{\partial}{\partial\theta}$$

and

$$\frac{\partial}{\partial y} = \sin\theta\frac{\partial}{\partial r} - \frac{\sin\theta}{r}\frac{\partial}{\partial\theta}$$

transform the operator ∇ to the cylindrical coordinates, r, θ, and z.

11.12: Using the expression for the gradient in polar coordinates, determine the gradient of $\psi(r, \theta)$ and the point of maximum gradient when A and a are constants and

$$\psi = Ar\sin\theta\left(1 - \frac{a^2}{r^2}\right)$$

11.13: A pressure field is given by

$$P = P_o + \frac{1}{2}\rho\hat{V}_\infty^2\left[2\frac{xyx}{L_3} + 3\left(\frac{x}{L}\right)^2 + \frac{\hat{V}_\infty t}{L}\right]$$

where x, y, and z are space coordinates, t is time, and P_o, ρ, $\hat{V}_\infty$, and L are constants. Find the pressure gradient.

Density

11.14: Which of the following conditions are flow properties and which are fluid properties?

pressure	temperature	velocity
density	stress	speed of sound
specific heat	pressure gradient	

11.15: The molecular weight of carbon monoxide is 28.01. Determine its density in kg/m^3 when the absolute pressure is 310 kPa and the temperature is 50°C.

11.16: The molecular weight of chlorine is 70.9. Determine its density in kg/m^3 at a gage pressure of 101 kPa and a temperature of 27°C.

11.17: A tire having a volume of $0.08 m^3$ contains air at a gage pressure of 210 kPa and a temperature of 20°C. Determine (a) the density of the air and (b) the weight of the air continued in the tire.

11.18: A compressed air tank contains 10 kg of air at a temperature of 80°C. If a gage on the tank reads 250 kPa, determine the volume of the tank.

11.19: A large dirigible having a volume of 100,000 m^3 contains helium under standard atmospheric conditions (101 kPa and 15°C). Determine the density and the total weight of the helium.

11.20: A quart of SAE 30 oil weighs about 0.84 kg. Assume that the density of water is 1000 kg/m^3 and determine the mass density, the specific weight, and the specific gravity of the oil.

11.21: Gasoline has an approximate specific weight of 7150 N/m^3. Assume that the density of water is 1000 kg/m^3 and determine its mass density, specific volume, and specific gravity.

11.22: A conical tank has a base radius of 25 cm and a height of 50 cm. Assume that the density of water is 1000 kg/m^3, and determine the water level above the base of the tank if 0.024 m^3 of water are poured into the tank.

11.23: Under standard conditions, a quantity of gas weighs 0.15 N/m^3. Determine its density and specific volume.

11.24: Determine the specific weight of a gas having a specific volume of 0.125 m^3/kg.

11.25: A gas begins to deviate from the continuum when it contains less than about 10^{12} molecules per cubic millimeter. To what absolute pressure in pascals at 300 K does this limit for air correspond?

11.26: For a fluid density, ρ, in which solid particles of density, ρ_s, are uniformly dispersed, show that if x is the mass fraction of solid in the mixture, the density is given by

$$\rho_{\text{mixture}} = \frac{\rho_s \rho}{\rho x + \rho_s(1 - x)}$$

11.27: Assuming that the fluid of density, ρ, in Problem 11.26 obeys the ideal gas law, obtain the equation of state of the mixture, that is, $P = f(\rho_s,\ RT/M,\ \rho_{\text{mix}},\ x)$.

Viscosity

11.28: Crude oil having a viscosity of 0.046 kg/m-s is contained between parallel plates. The bottom plate is fixed and the upper plate moves when a force is applied (Figure P11.28). If the distance between the two plates is 5 mm and the effective area of the upper plate is 0.096 m^2, determine the value of P required to move the upper plate with a velocity of 6 cm/s.

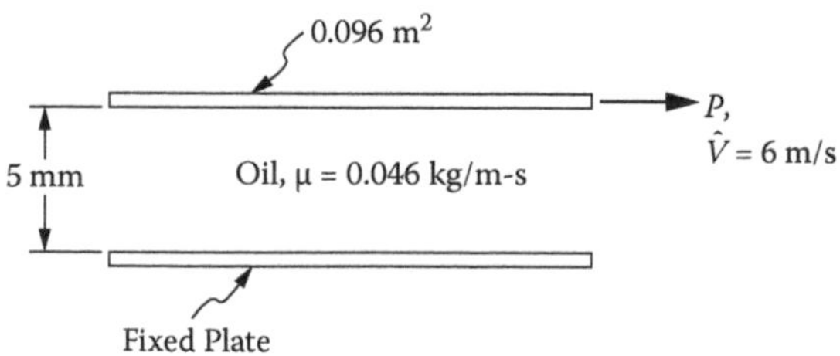

FIGURE P11.28

11.29: A cubical block 0.250 m on a side and having a mass of 0.2548 kg slides down a 22.5° incline on a film of oil that is 0.626-μm thick. Assume a linear velocity profile in the oil having a viscosity of 0.007 kg/m-s and determine the velocity of the block.

11.30: A Newtonian fluid having a density of 920 kg/m^3 and a kinematic viscosity of 5×10^{-4} m/s^2 flows past a fixed surface. The velocity profile near the surface is shown in Figure P11.30. Determine the magnitude and the direction of the shearing stress developed on the plate expressing your result in terms of $\hat{V}$ and Δ in units of m/s and m, respectively.

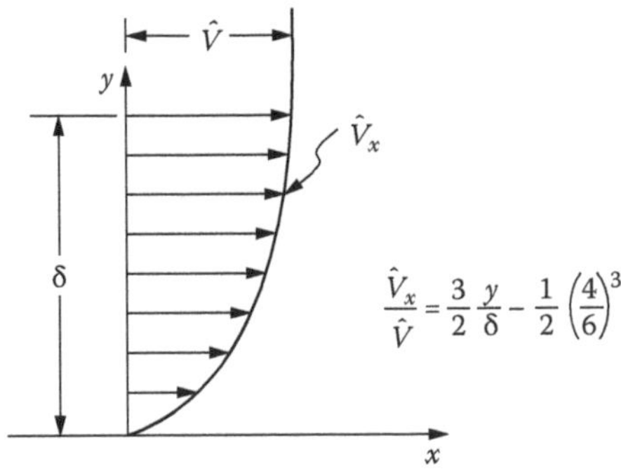

FIGURE P11.30

11.31: A steel (7850 kg/m^3) shaft that is 6 cm in diameter and 30-cm long falls of its own weight inside a vertical open tube that is 4.03 cm in diameter. The clearance, assumed to be uniform, contains a film of glycerin at 20°C. Determine the velocity of the shaft at terminal conditions.

11.32: In Figure P11.32, a solid cone of angle 2θ, base radius, r_o, and density ρ_c is rotating with angular velocity ω_o inside a conical seat. The clear space, designated by h, is filled with an oil of viscosity, μ. Neglecting air drag, derive an expression for the time required to reduce the angular velocity of the cone from ω_o to 0.10 ω_o.

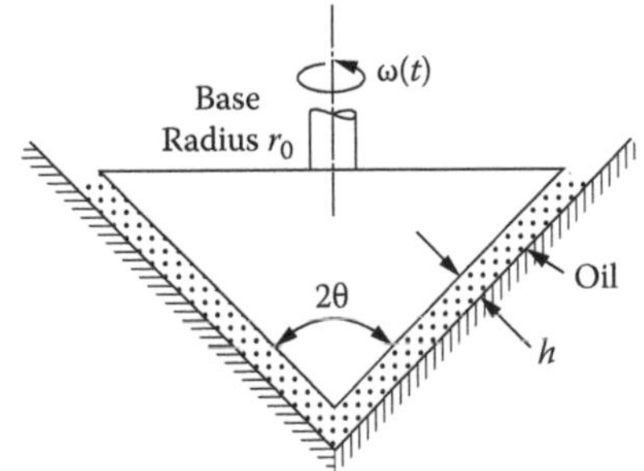

FIGURE P11.32

11.33: A disk of radius R rotates at an angular velocity, ω, inside an oil bath of viscosity μ as indicated in Figure P11.33. Assuming a linear velocity profile and neglecting shear on the outer disk edges, derive an expression for the viscous torque on the disk.

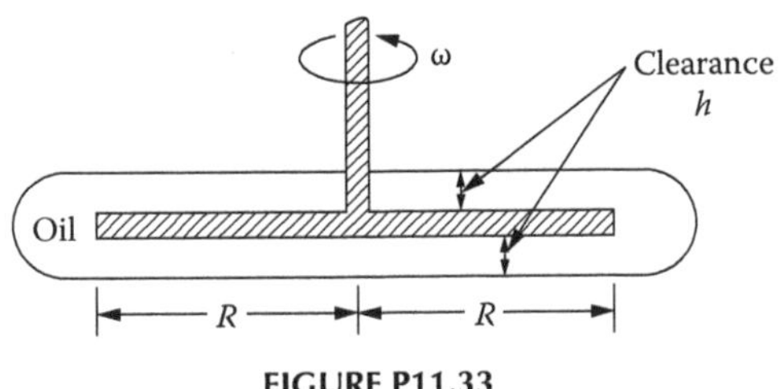

FIGURE P11.33

11.34: A 35-mm-diameter shaft is pulled through a cylindrical bearing is shown in Figure P11.34. The lubricant that fills the 0.3-mm gap between the shaft and the bearing is an oil having a kinematic viscosity of $8 \times 10^{-4}\,\text{m/s}^2$ and a density of $910\,\text{kg/m}^3$. Assume the velocity distribution in the gap to be linear and determine the force, P, required to pull the shaft at a velocity of 4 m/s.

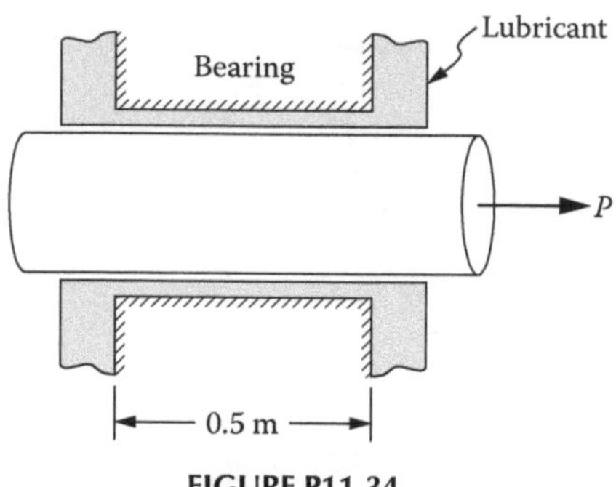

FIGURE P11.34

11.35: Air at 20°C forms a boundary near a solid wall (Figure P11.35) with a sine-wave-shaped velocity profile. The boundary layer thickness is 6 mm and the peak velocity is 10 m/s. Compute the shear stress in the boundary layer in pascals at y equal to (a) 0 mm, (b) 3 mm, and (c) 6 mm.

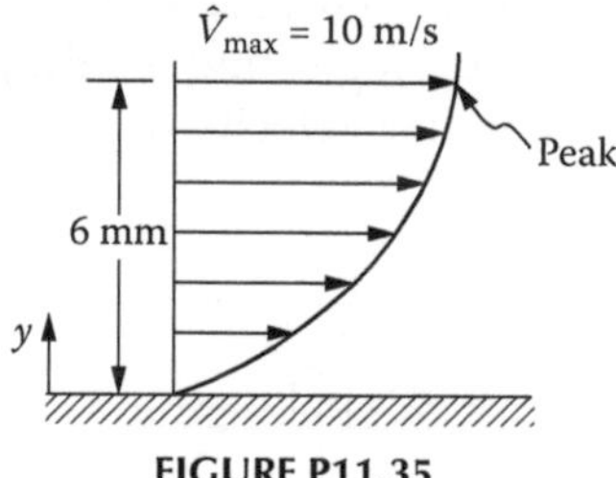

FIGURE P11.35

11.36: A layer of water flows down an inclined fixed surface with the velocity profile shown in Figure P11.36. Determine the magnitude and direction of the shearing stress that the water exerts on the fixed surface for $\hat{V} = 2\,\text{m/s}$ and $h = 0.10\,\text{m}$.

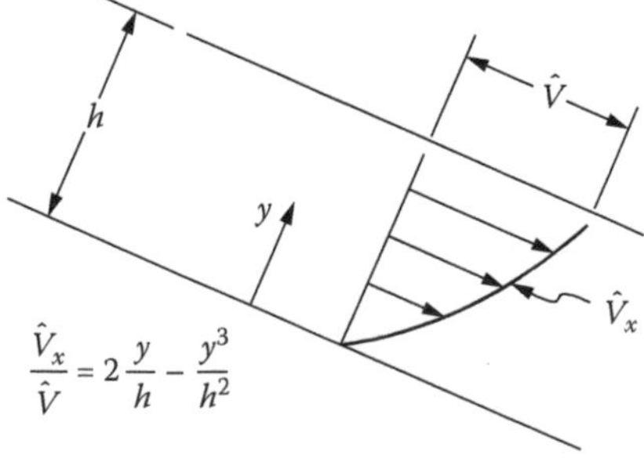

FIGURE P11.36

11.37: A piston that is 75-cm long moves through a 12.5-cm-diameter cylinder. The film separating the piston and the cylinder is 0.025-cm thick and contains oil with a viscosity of 0.975 kg/m-s. Determine the force required to maintain a velocity of 6 m/s.

11.38: A piston having a mass of 0.75 kg, a length of 15 cm and a diameter of 12.5 cm slides through a 12.005-cm inner-diameter vertical pipe. The piston is decelerating at a rate of 0.725 m/s^2 when the velocity is 7.25 m/s. Determine the viscosity of the oil that occupies the space between the piston and the cylinder.

11.39: Two 15-cm-long cylinders, of diameters 7.5 cm and 8.0 cm, fit inside each other. The gap is filled with an oil having a viscosity of 0.425 kg/m-s. Assuming the outer cylinder to be fixed and that the velocity profile in the oil is linear, determine (a) the torque and (b) the power required to maintain a rotation of the inner cylinder at 6.25 rad/s.

11.40: A 25-cm-diameter shaft 75 cm in length is pulled through a cylindrical bearing. Oil having a kinematic viscosity of 8.75×10^{-4} m/s^2 and a specific gravity of 0.92 fills the 0.325-mm gap between the stationary and moving surfaces. Assuming that the velocity distribution in the gap is linear, determine the force required to pull the shaft through the bearing at a velocity of 4 m/s.

11.41: A disk of radius, r, rotates with an angular velocity of ω (Figure P11.41). The clearance, as indicated in the figure is $h = 0.55$ mm, the disk radius is $r = 0.80$ mm, the angular velocity is $\omega = 0.375$ rad/s and the viscosity of the oil is $\mu = 0.008$ kg/m-s. Neglecting end effects and assuming the velocity profile in the fluid to be linear, determine the damping torque.

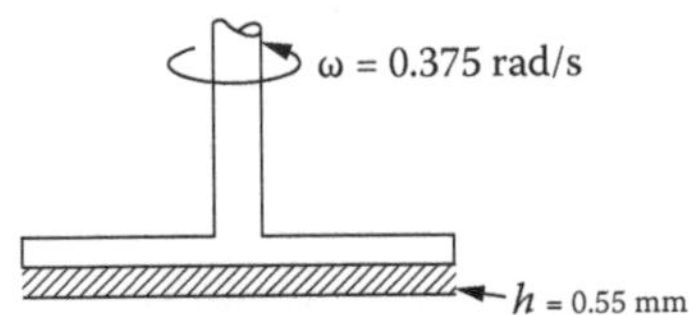

FIGURE P11.41

11.42: A 25-cm-diameter plunger slides inside a 25.02-cm-diameter cylinder. The annular space is filled with oil having a kinematic viscosity of 3.75×10^{-4} m/s^2 and a specific gravity of 0.865. Assume a linear velocity distribution in the oil and, when the plunger moves at 0.18 m/s, determine the frictional resistance when 2.75 cm of the plunger is in the cylinder.

11.43: A 15-cm-diameter shaft is turning in a 15.05-cm-diameter sleeve that is 20-cm long. The gap is filled with an oil having a viscosity of 0.0875 kg/m-s. If the velocity distribution in the oil is linear and the shaft turns at 90 rpm, determine the rate of heat generation.

11.44: A 40-cm square block slides on oil with viscosity of 0.81 kg/m-s over a large smooth surface. The film between the block and the surface is 5-mm thick and the velocity distribution in the oil is linear. Determine the force required to move the block at a velocity of 4 m/s.

11.45: Two plates are separated by an 8-mm gap that contains oil with a viscosity of 0.450 kg/m-s. The bottom plate is stationary and the upper plate moves at 3.75 m/s; the velocity distribution in the oil is linear determine the shear stress.

11.46: A large plate moves with a velocity of 5 m/s over a stationary plate on a layer of oil ($\mu = 0.42$ kg/m-s) that is 5 mm thick. A parabolic profile is assumed for the velocity distribution. Determine the shear stress on the upper plate.

11.47: A 24-kg cube slides down a 20° inclined plane on a 4-mm-thick oil film. The contact area between the cube and the plane is 0.25 m^2 and the velocity profile may be assumed linear; the velocity is 6 m/s. Determine the viscosity of the oil.

11.48: A shaft that is 8 cm in diameter is rotated at 2400 rpm inside a sleeve that is 8.2 cm in diameter. The length of the sleeve is 20 cm and oil, with a kinematic viscosity of

$48 \times 10^{-4}\,m/s^2$ and a specific gravity of 0.925, fills the gap. If the velocity distribution in the oil is linear. Determine the power required to rotate the shaft.

11.49: Two large plane surfaces are separated by a gap of 1.2 cm, which is filled with oil having a viscosity of 0.80 kg/m-s. A thin plate having an area of $0.50\,m^2$ and a thickness of 2 mm is placed midway between the two large plates and dragged through the gap at 0.175 m/s. If the velocity distribution in the oil is linear, determine the force required.

Additional Fluid Properties

11.50: A vessel that contains $1.25\,m^3$ of water at atmospheric pressure is heated from 60 to 95°C. How much water must be removed to keep the volume unchanged?

11.51: A vertical cylindrical tank having a base diameter of 10 m and a height of 6 m is filled to the top with water at 20°C. How much water will overflow if the water is heated to 60°C?

11.52: A liquid in a cylinder has a volume of $1200\,cm^3$ at 1.25 MPa and a volume of $1188\,m^3$ at 2.50 MPa. Determine its bulk modulus of elasticity.

11.53: A differential pressure of 10 MPa applied to $0.25\,m^3$ of a liquid causes a volume reduction of $0.005\,m^3$. Determine the bulk modulus of elasticity.

11.54: The bulk modulus of elasticity for water is 2.205 GPa. Determine the change in pressure required to reduce a given volume by 0.75%.

11.55: Given the bulk modulus of elasticity for water is 2.205 GPa, determine the change in volume of $1\,m^3$ of water when subjected to a change in pressure of 20 MPa.

11.56: Water in a container is originally at 100 kPa. The bulk modulus of elasticity for water is 2.205 GPa and the water is subjected to a pressure of 120 MPa. Determine the percentage decrease of its volume.

11.57: Given the bulk modulus of elasticity for water of 2.205 GPa, determine the change in pressure required to reduce a given volume by 1.25%.

11.58: The viscosity of water is 0.0100 poise at 20°C. Determine its kinematic viscosity.

Surface Tension and Capillary Action

11.59: A small droplet of water at 25°C is in contact with air and has a diameter of 0.05 cm. The difference in pressure between the inside and outside of the droplet is 565 Pa. Determine the value of the surface tension.

11.60: Determine the height to which water at 68°C will rise in a clean capillary tube having a diameter of 0.2875 cm.

11.61: Two clean and parallel glass plates, separated by a gap of 1.625 mm, are dipped in water. If $\sigma = 0.0735\,N/m$, determine how high the water will rise.

11.62: A glass tube having an inside diameter of 0.25 mm and an outside diameter of 0.35 mm is inserted into a pool of mercury at 20°C such that the contact angle is 130°. Determine the upward force on the glass.

11.63: In Problem 11.62, determine the depth of the depression.

11.64: Determine the capillary rise for a water-air-glass interface at 40°C in a clean glass tube of radius, $r = 1\,mm$.

11.65: Determine the capillary rise if the tube in Problem 11.64 is immersed in mercury at 20°C such that the contact angle is 126°.

11.66: Determine the difference in pressure between the inside and outside of a soap film bubble at 20°C if the diameter of the bubble is 4 mm.

11.67: At 25°C, determine the maximum diameter clean glass tube necessary to keep the capillary height at a maximum of 2.5 mm.

11.68: Assuming that the difference in the inside and outside radii of a soap bubble is negligible, determine the surface tension for a 2.25-cm-diameter bubble subjected to a difference in pressure between the inside and outside of $\Delta P = 31$ Pa.

11.69: Two vertical, parallel, clean, glass plates are spaced a distance of 3 mm apart and they are placed in water. Determine how high the water will rise between the plates due to capillary action.

11.70: An open, clean glass tube, having a diameter of 3 mm, is inserted vertically into a dish of mercury at 20°C. Determine how far the column of mercury in the tube will be depressed.

11.71: At 60°C, the surface tensions of water and mercury are water: $\sigma = 0.0662$ N/m and mercury: $\sigma = 0.47$ N/m. Determine the capillary height changes in these two fluids when they are in contact with air in a glass tube of diameter 0.55 mm. Contact angles are 0° for water and 130° for mercury.

11.72: Determine the diameter of the glass tube necessary to keep the capillary-height change of water at 30°C less than 1 mm.

12

Fluid Statics

Chapter Objectives

- To develop the basic equation of fluid statics.
- To describe the use of manometers to measure pressure.
- To evaluate hydrostatic forces on planar and nonplanar surfaces immersed in a fluid.
- To develop the concept of buoyancy.
- To examine the stability of floating objects.
- To describe hydrostatic forces under conditions of uniform rectilinear acceleration.

12.1 Introduction

In Chapter 11, we defined a fluid according to its response to a shearing stress and we commented that a fluid at rest would therefore not be subjected to shear stresses. In this chapter we will consider the behavior of fluids at rest, that is, in a **static** condition.

We should distinguish, at the outset, between **inertial** and **noninertial** reference frames. In general, we will consider a static situation to be one that is stationary relative to the earth's surface; this is what is meant by an inertial reference frame. If a fluid is stationary relative to a coordinate system that has some acceleration of its own, such a reference frame is termed noninertial. An example of a noninertial reference frame would be a fluid inside an aircraft as it executes a maneuver.

Our considerations will be, in general, for inertial reference frames and we will note, as exceptions, those occasions where noninertial reference frames are employed.

12.2 Pressure Variation in a Static Field

In the absence of shear stresses a fluid will experience only gravitational and pressure forces. A representative fluid element of differential size in a static fluid is shown in Figure 12.1. Its dimensions, as indicated, are dx, dy, and dz.

Fluid pressure, $P(x, y, z)$, which acts on all six faces of the element, will produce force components in each of the coordinate directions. Because pressure always acts inward, the x-component pressure forces, which result from the pressure exerted on the x-faces are

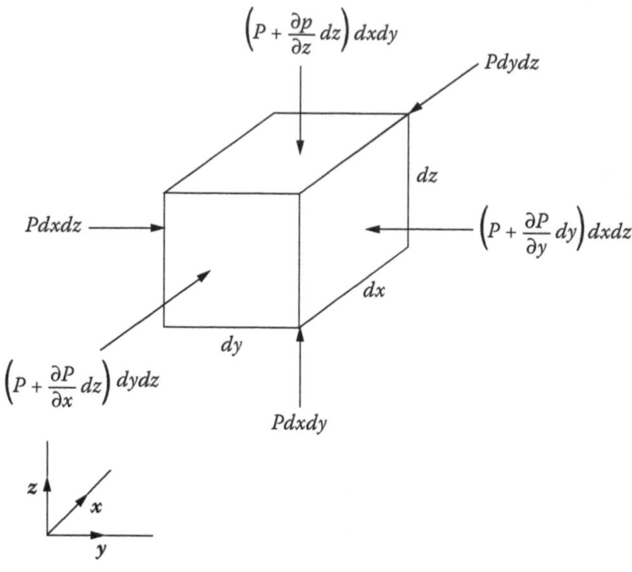

FIGURE 12.1
Pressure forces exerted on a static fluid element.

expressed as

$$\left[Pdydx - \left(P + \frac{\partial P}{\partial x}dx\right)dydx\right]\mathbf{x}$$

After performing the cancellation, we write the net x-component pressure force:

$$P_x = -\frac{\partial P}{\partial x}dxdydz$$

Doing the same operation in the y-coordinate direction, we have

$$\left[Pdzdx - \left(P + \frac{\partial P}{\partial y}dy\right)dzdx\right]\mathbf{y}$$

which reduces to

$$P_y = -\frac{\partial P}{\partial y}dxdydz$$

and in the z-coordinate direction

$$\left[Pdxdy - \left(P + \frac{\partial P}{\partial z}dz\right)dxdy\right]\mathbf{z}$$

which becomes the z-component

$$P_z = -\frac{\partial P}{\partial z}dxdydz$$

The gravitational force is expressed, in general form, as

$$\rho \mathbf{g}\, dxdydz$$

where the direction of $\mathbf{g}$ is left in a completely general form.

Now, because $\sum \boldsymbol{F} = 0$, we have

$$\sum \boldsymbol{F} = P_x \boldsymbol{x} + P_y \boldsymbol{y} + P_z \boldsymbol{z} + \rho \boldsymbol{g}\, dxdydx = 0$$

or

$$-\left[\frac{\partial P}{\partial x}\boldsymbol{x} + \frac{\partial P}{\partial y}\boldsymbol{y} + \frac{\partial P}{\partial z}\boldsymbol{z}\right] dxdydz + \rho \boldsymbol{g}\, dxdydx = 0$$

This expression can be simplified further to the form

$$\frac{\partial P}{\partial z}\boldsymbol{x} + \frac{\partial P}{\partial y}\boldsymbol{y} + \frac{\partial P}{\partial z}\boldsymbol{z} = \rho \boldsymbol{g} \tag{12.1}$$

Recalling the definition of the gradient from Chapter 11, Equation 12.1 may be written in the more compact form

$$\nabla P = \rho \boldsymbol{g} \tag{12.2}$$

Equation 12.2 is generally referred to as the **basic equation of fluid statics**. We will devote the remaining sections of this chapter to applications of this fundamental relationship.

As a final comment before proceeding with applications, we observe that Equation 12.2 is a general vector relationship that is valid for *any* orthogonal coordinate system even though it was derived from consideration of an element in Cartesian coordinates. For any element, which is essentially a point, an orthogonal coordinate system would consist of three unit vectors that are mutually perpendicular. We could designate these directions as x, y, and z or r, θ, and z or in any way we choose. The choice would depend on how we wish to build up (or integrate) from the point to conform to boundaries or coordinates conveniently.

12.3 Hydrostatic Pressure

We have already discussed the consequence of Equation 12.2 for the directions of maximum pressure variation and of zero (constant pressure) variation. A direct application of this idea involves a pressure measuring device called a **manometer**. We can use a U-tube manometer (Figure 12.2) to evaluate the pressure at point A in the chamber by measuring the levels of the manometer fluid on the two sides of the U-tube.

If we start with Equation 12.2, we may write for Cartesian coordinates

$$\frac{\partial P}{\partial x}\boldsymbol{x} + \frac{\partial P}{\partial y}\boldsymbol{y} + \frac{\partial P}{\partial z}\boldsymbol{z} = -\rho g \boldsymbol{z}$$

The next step is to equate the coefficients of the unit vectors and, in doing so, we obtain two scalar equations

$$\frac{\partial P}{\partial x} = 0 \tag{12.3a}$$

$$\frac{\partial P}{\partial y} = 0 \tag{12.3b}$$

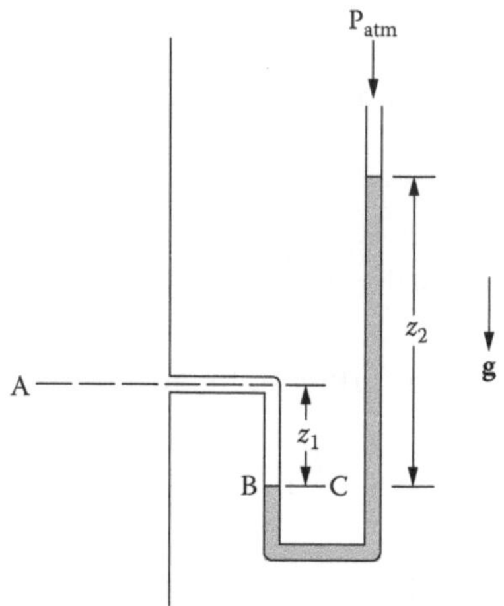

FIGURE 12.2
A U-tube manometer. Note that g is in the negative z direction.

showing that the pressure does not vary in the x and y directions, and a third:

$$\frac{\partial P}{\partial z} = -\rho g = -\gamma \tag{12.3c}$$

Each of these scalar relationships provides us with some information and the useful one is Equation 12.3c. If we apply this equation to our manometer between points C and D, we may separate the variables

$$dP = -\rho g\, dz$$

and then integrate between the appropriate limits

$$\int_{P_C}^{P_D} dP = -\rho g \int_0^{z_2} dz$$

to obtain

$$P_C - P_D = \rho g z_2 = \gamma z_2 \tag{12.4}$$

or, because P_D is atmospheric pressure

$$P_C = P_{atm} + \rho g z_2 = P_{atm} + \gamma z_2 \tag{12.5}$$

This result indicates that, for a vertical tube open to the atmosphere, the pressure at a point below the free liquid surface is greater than atmospheric by the amount $\rho g z$, which is clearly proportional to, z, the height of fluid. This expression will be used without detailed derivation in future discussion in examples involving manometers. In the development of Equation 12.5, the fluid density was taken as constant. In one of the examples soon to follow, we will illustrate the procedure for variable density.

Because points B and C are at the same level and are connected by a continuous column of manometer fluid, our earlier discussion indicates that they are at the same pressure. The pressure at point A can be evaluated, beginning with the statement of this equality

$$P_B = P_C$$

by using ρ_s for the "system" fluid and ρ_m for the "manometer" fluid and then applying Equation 12.4 twice

$$P_B = P_A + \rho_s g z_1$$

and

$$P_C = P_{atm} + \rho_m g z_2$$

Substitution then yields the result

$$P_A + \rho_s g z_1 = P_{atm} + \rho_m g z_2$$

which may be written as

$$P_A = P_{atm} + \rho_m g z_2 - \rho_s g z_1 \quad (12.6)$$

or

$$P_{A,g} = \rho_m g z_2 - \rho_s g z_1 \quad (12.7)$$

Equation 12.5 indicates that the **absolute** pressure at point C is the sum of the atmospheric pressure and the quantity $\rho g z_2$. The difference between atmospheric pressure and absolute pressure is designated the **gage** pressure if $\rho g z_2$ is positive or the **vacuum** pressure if $\rho g z_2$ is negative. An alternative way of writing Equation 12.4 in terms of the gage pressure, in this case, $P_{C,g}$, is

$$P_{C,g} = \rho g z_2 \quad (12.8)$$

Examples 12.1 and 12.2 that now follow will illustrate the application of Equation 12.5 to configurations that involve manometers.

Example 12.1 A fan draws air from the atmosphere through a round duct as shown in Figure 12.3. A differential manometer connected to an opening in the duct wall shows a vacuum pressure of 2.5 cm of water. Determine the air pressure at this point.

Solution

Assumptions and Specifications

(1) Steady-state conditions exist.

(2) The water in the manometer is static and incompressible.

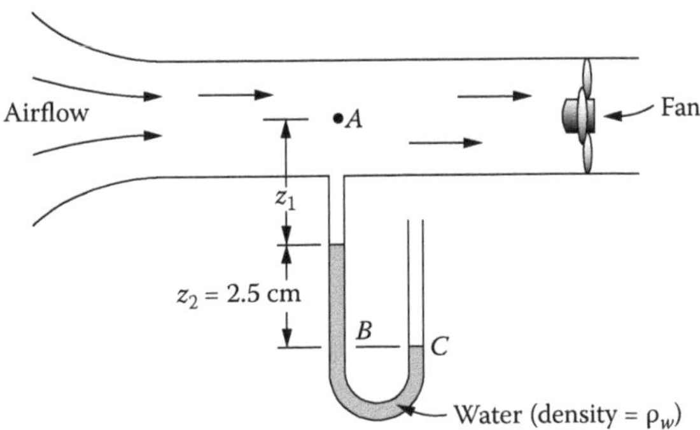

FIGURE 12.3
Airflow through a duct for Example 12.1.

(3) The acceleration of gravity is assumed to be $9.81\,\text{m/s}^2$.
(4) The density of the water is specified.

We assume that the temperature at point A is at 15°C where $\rho_a = 1.234\,\text{kg/m}^3$ and we note at the outset that we already know the pressure at point A, $P_A = 2.5\,\text{cm}$ of water *vacuum*. Thus,

$$P_C = P_{\text{atm}} = P_B$$

and with the densities of air and water designated with subscripts a and w

$$P_B = P_A + \rho_a g z_1 + \rho_w g z_2$$

or

$$P_A = P_{\text{atm}} - \rho_a g z_1 - \rho_w g z_2$$

The gage pressure at point A will be

$$P_{A,g} = -\rho_a g z_1 - \rho_w g z_2$$

and this can be written as

$$P_{A,g} = \rho_w g \left(-\frac{\rho_a}{\rho_w} z_1 - z_2\right)$$

The density ratio at the standard conditions with $T = 15$°C can be evaluated. For $\rho_w = 1000\,\text{kg/m}^3$ as

$$\frac{\rho_a}{\rho_w} = \frac{1.234\,\text{kg/m}^3}{1000\,\text{kg/m}^3} = 1.234 \times 10^{-3}$$

and it is clear that the first term within the parentheses will be negligibly small and may thus be neglected. This leaves

$$P_{A,g} = -\rho_w g z_2$$

which may be converted to Pascals in the usual process of unit conversion

$$P_{A,g} = -\frac{(1000\,\text{kg/m}^3)(9.81\,\text{m/s}^2)(0.025\,\text{m})}{1\,\text{kg-m/N-s}^2} = -245.3\,\text{Pa} \Longleftarrow$$

Example 12.2 Determine the gage pressure in the bulb at point *A* for the conditions shown in Figure 12.4.

Solution

Assumptions and Specifications

(1) Steady-state conditions exist.
(2) The fluids are static and incompressible.

The three material densities may be designated with the subscripts a, oil, and Hg for the air, oil, and mercury, respectively. We next apply the principles that have been developed

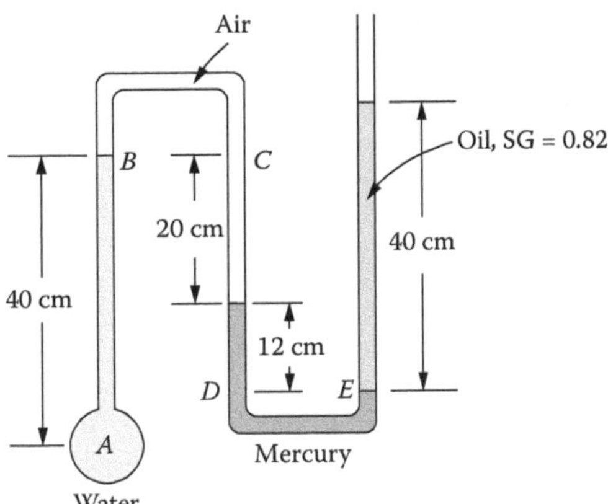

FIGURE 12.4
The compound manometer discussed in Example 12.2.

to express the following:

$$P_D = P_E = P_{atm} + \rho_{oil}g(40\,\text{cm})$$
$$P_D = P_C + \rho_a g(20\,\text{cm}) + \rho_{Hg}g(12\,\text{cm})$$
$$P_B = P_C$$
$$P_A = P_B + \rho_w g(40\,\text{cm})$$

An algebraic exercise that combines these expressions leads to the expression for P_A

$$P_A = P_{atm} + \rho_w g(40\,\text{cm}) + \rho_{oil}g(40\,\text{cm}) - \rho_a g(20\,\text{cm}) - \rho_{Hg}g(12\,\text{cm})$$

If all of the densities are expressed in terms of ρ_w we obtain

$$P_A = P_{atm} + \rho_w g(40\,\text{cm}) + 0.82\rho_w(40\,\text{cm}) - \frac{\rho_a}{\rho_w}\rho_w g(20\,\text{cm}) - 13.6\rho_w g(12\,\text{cm})$$

which can be simplified, with $\rho_w = 1000\,\text{kg/m}^3$, to

$$P_A = P_{atm} + \rho_w g\left[40\,\text{cm} + 32.8\,\text{cm} - \frac{\rho_a}{\rho_w}(20\,\text{cm}) - 163.2\,\text{cm}\right]$$

As before, because $\rho_a/\rho_w \ll 1$, we can neglect the air pressure term. Combination of the rest of the terms yields the gage pressure

$$P_{A,g} = P_A - P_{atm} = \rho_w g(-90.4\,\text{cm})$$

or

$$P_{A,g} = 8.87\,\text{kPa} \quad (\text{vacuum}) \Longleftarrow$$

The next example considers the density of a gas that varies with altitude.

Example 12.3 A temperature of 5°C and a pressure of 0.612 bar are noted on the instrument panel of an aircraft in level flight at an altitude of h m. Surface conditions are 760 mm of mercury and 25°C. Assuming that the atmosphere behaves as a perfect gas and that the air temperature variation between the surface of the earth and the aircraft is linear, estimate the altitude of the aircraft.

Solution

Assumptions and Specifications

(1) Steady-state conditions exist.
(2) The standard barometer is 760 mm of mercury.
(3) Atmospheric air may be considered as a perfect gas.
(4) The air temperature variation is linear with altitude.
(5) The acceleration of gravity has a value of $g = 9.81\,\text{m/s}^2$.

At the surface of the earth, standard atmospheric pressure of 760 mm of mercury is specified and, in accordance with the problem statement, the specified linear temperature variation would be described by

$$T = T_o - \frac{\Delta T}{h} z$$

where

$$T_o = 25°\text{C} \qquad \text{and} \qquad \Delta T = 25°\text{C} - 5°\text{C} = 20°\text{C} \quad (20\,\text{K})$$

Moreover, we have the equation of state for the ideal gas

$$\rho = \frac{P}{RT}$$

where, by Table A.1 in Appendix A, $R = 287\,\text{N-m/kg-K}$.

The basic equation of fluid statics gives

$$\frac{dP}{dz} = -\rho g = -\frac{P}{RT} g$$

We next relate the temperature variation with altitude and separate the variables to obtain

$$\frac{dP}{P} = -\left(\frac{g/R}{T_o - \dfrac{\Delta T}{h} z}\right) dz$$

Now, integration gives

$$\int_{P_o}^{P} \frac{dP}{P} = -\frac{g}{R} \int_0^h \frac{dz}{T_o - \dfrac{\Delta T}{h} z}$$

$$\ln \frac{P}{P_o} = \frac{g}{R} \frac{h}{\Delta T} \ln \left(1 - \frac{\Delta T}{h}\right)$$

or

$$\frac{P}{P_o} = \left(1 - \frac{\Delta T}{T_o}\right)^{gh/R\Delta T}$$

Here we note that T_o need not be converted to K as it plays no thermodynamic role in the foregoing relationship. Then we evaluate P/P_o with $P_o = 1.013 \times 10^5$ Pa (760-mm Hg) as

$$\frac{P}{P_o} = \frac{\text{460-mm Hg}}{\text{760-mm Hg}} = 0.6053$$

and

$$\frac{gh}{R\Delta T} = \frac{(9.81\,\text{m/s}^2)h}{(287\,\text{N-m/kg-K})(20\,\text{K})} = (1.709 \times 10^{-3}\,\text{m}^{-1})h$$

so that

$$\frac{P}{P_o} = \left(1 - \frac{\Delta T}{T_o}\right)^{gh/R\Delta T}$$

$$0.6053 = \left(1 - \frac{20°\text{C}}{25°\text{C}}\right)^{(1.709\times 10^{-3}\,\text{m}^{-1})h}$$

$$= (0.20)^{(1.709\times 10^{-3}\text{m}^{-1})h}$$

$$(1.709 \times 10^{-3}\text{m}^{-1})h = 0.312$$

$$h = 182.5\,\text{m} \Longleftarrow$$

12.4 Hydrostatic Forces on Plane Surfaces

When a solid object is immersed in a fluid, the force exerted on each of its surfaces is due to fluid pressure. Accordingly, we may use our fluid statics relationships to evaluate these forces. A plane surface immersed in a liquid at an angle θ relative to the liquid surface is shown in Figure 12.5. The total surface area is, S, the fluid has a uniform density, ρ, and we are using h as the depth and y as the length along the surface as shown.

For each point on the surface, S, at the same depth, h, the gage pressure will be $\rho g h$. This will be the case for the differential area, dS, which is at a uniform depth, $h = y \sin\theta$, below the fluid surface. The pressure force on dS is

$$dF = PdS = \rho g y \sin\theta dS \tag{12.9}$$

and the total force can be obtained by integration

$$F = \int dF = \rho g \sin\theta \int_S y\,dS \tag{12.10}$$

If we recall the definition of the centroid of an area

$$\bar{y} = \frac{1}{S}\int y\,dS \tag{12.11}$$

we may write Equation 12.10 in an alternate form as

$$F = \rho g \sin\theta S\bar{y} \tag{12.12}$$

Equation 12.12 indicates that the force on a submerged surface is equal to the pressure evaluated at its centroid multiplied by the submerged surface area, S.

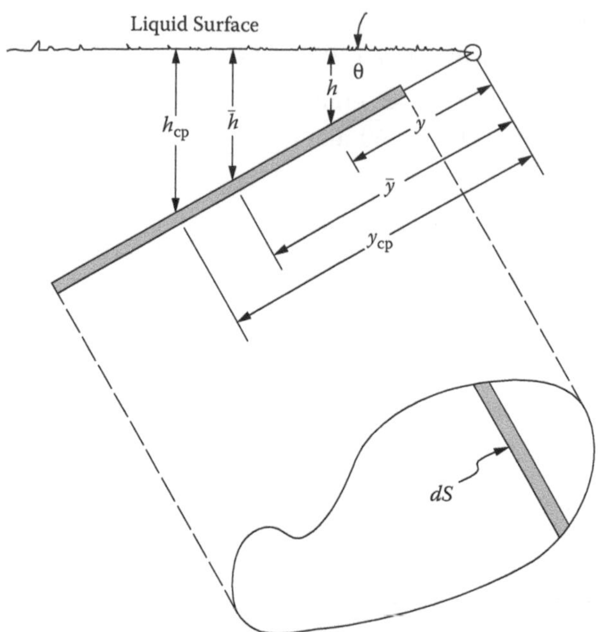

FIGURE 12.5
A submerged plane surface.

If we were to consider a pressure force on a given surface in a static analysis, we would need both its magnitude, which has just been evaluated, and its location, which has yet to be established. The location of interest is designated as the **center of pressure** and its position will be evaluated by referring, once again, to Figure 12.5. We seek the value of $y = y_{cp}$, where the total pressure force will act so as to produce the same moment as the distributed pressure. Using point 0 as reference, we have

$$F y_{cp} = \int_S y\, dF = \int_S yP\, dS \tag{12.13}$$

and substituting for the pressure, we have

$$F y_{cp} = \int_S \rho g \sin\theta y^2 dS \tag{12.14}$$

Using Equation 12.12 for the force provides

$$y_{cp} = \frac{1}{S\bar{y}} \int_S y^2\, dS = \frac{I_{cp}}{S\bar{y}} \tag{12.15}$$

where I_{cp} is the moment of inertia of the surface area about its center of pressure. An alternative expression can be obtained if the moment of inertia is evaluated about an axis through its centroid rather than the center of pressure. The transfer theorem relating I_o (the moment of inertia about the centroidal axis of S) and I_{cp} is

$$I_{cp} = I_o + S\bar{y}^2$$

Substituting this expression into Equation 12.15, we obtain

$$y_{cp} = \frac{I_o + S\bar{y}^2}{S\bar{y}} = \frac{I_o}{S\bar{y}} + \bar{y} \tag{12.16}$$

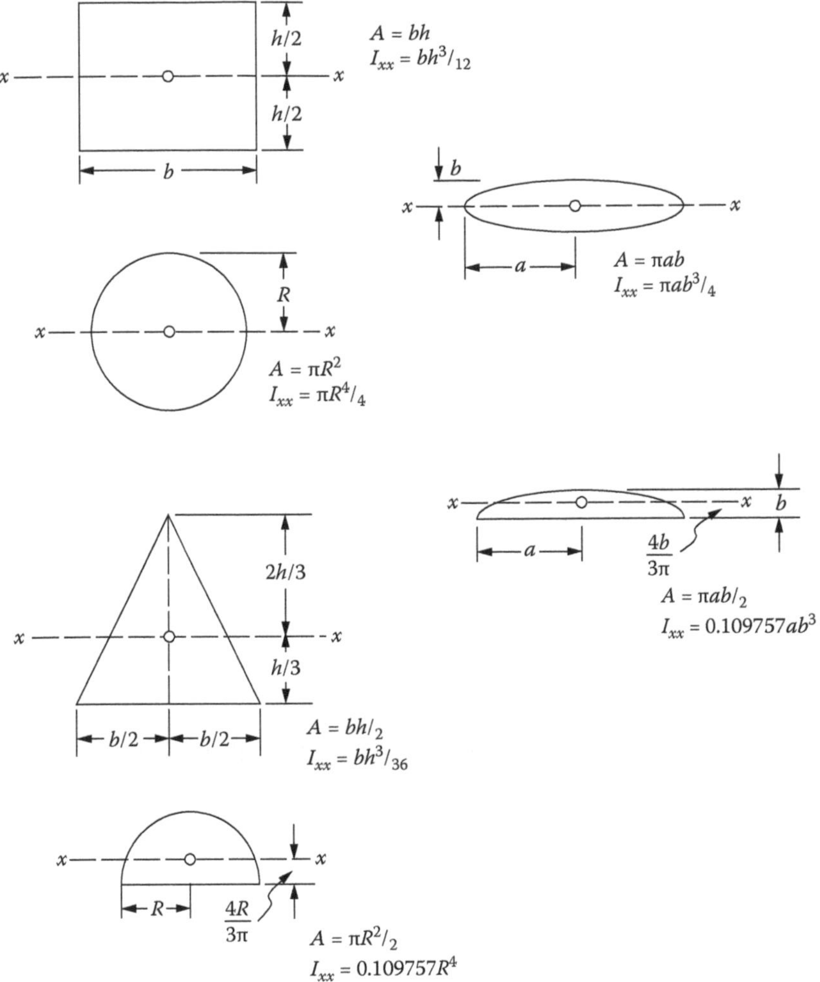

FIGURE 12.6
Moments of inertia and centroidal axes for six plane areas.

which indicates that the center of pressure will always be *below* the centroidal axis of the area by the amount $I_o/S\bar{y}$.

Figure 12.6 provides a guide to the location of the centroidal axes and the equations for the moments of inertia about the centroidal axes for several plane areas.

Example 12.4 Determine the minimum fluid level h for which the rectangular gate shown in Figure 12.7a will rotate in a counterclockwise direction. The cross section of the gate (Figure 12.7b) is 2 m ×2 m and friction in the hinge may be neglected.

Solution

Assumptions and Specifications

(1) Steady-state conditions exist.
(2) There is no flow of fluid.
(3) The fluid is incompressible.
(4) The friction in the hinge may be neglected.

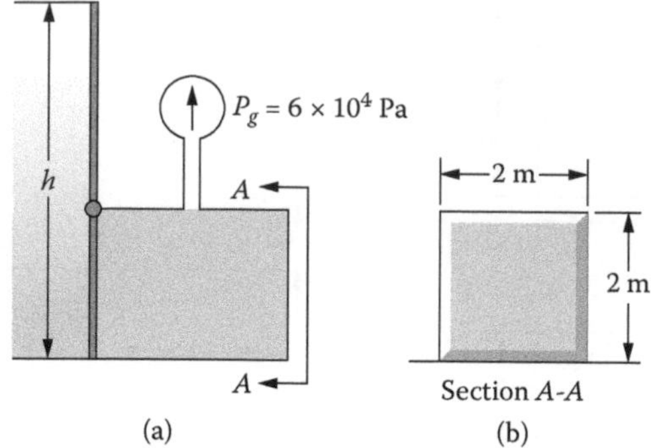

FIGURE 12.7
A rectangular gate separating the fluid column and air chamber in Example 12.4.

(5) The acceleration of gravity may be taken as $g = 9.81\,\text{m/s}^2$.

(6) The density of water may be taken as $1000\,\text{kg/m}^3$.

Figure 12.8 is a free-body diagram of the gate showing the water force F_w. With $y = (h-1)\,\text{m}$, $\rho_w = 1000\,\text{kg/m}^3$ and $\theta = 90^\circ$

$$\begin{aligned} F_w &= \rho_w g S\bar{y} \\ &= (1000\,\text{kg/m}^3)(9.81\,\text{m/s}^2)(2\,\text{m})(2\,\text{m})[(h-1)\,\text{m}] \\ &= 39,240(h-1)\,\text{N} \end{aligned}$$

The location of this force is at y_{cp} from the hinge

$$y_{cp} = 1\,\text{m} + \frac{I_o}{S\bar{y}}$$

and, evaluating I_o as

$$I_o = \frac{bh^3}{12} = \frac{(2\,\text{m})(2\,\text{m})^3}{12} = \frac{4}{3}\,\text{m}^4$$

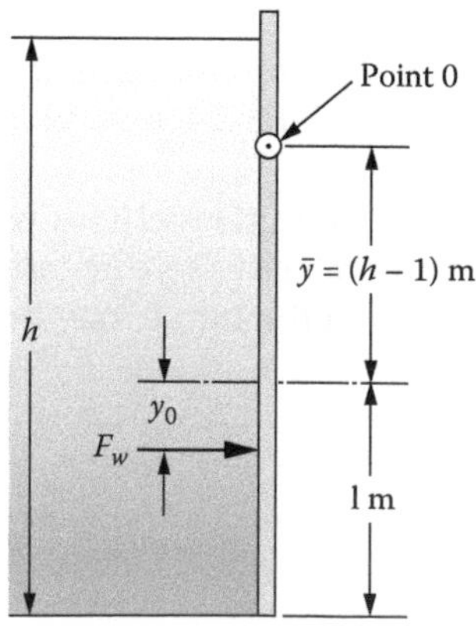

FIGURE 12.8
A detail of the gate in Example 12.4 showing the water force.

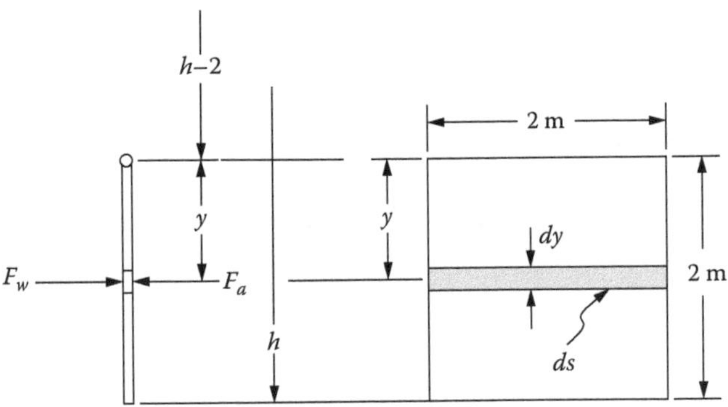

FIGURE 12.9
Detail of gate in Example 12.4 showing both the water force and the air force.

we have

$$y_{cp} = 1\,\text{m} + \frac{4/3\,\text{m}^4}{(2\,\text{m})(2\,\text{m})[(h-1)\,\text{m}]}$$

$$= 1\,\text{m} + \frac{1\,\text{m}}{3(h-1)} = \frac{(3h-2)\,\text{m}}{3(h-1)}$$

We may proceed to solve for h by taking moments about the hinge.

$$F_w y_{cp} = P_a S \bar{y}_{cp}$$

$$(39{,}240\,\text{N})(h-1)\frac{3h-2}{3(h-1)} = (6\times 10^4\,\text{N/m}^2)(4\,\text{m}^2)(1\,\text{m})$$

$$3h - 2\,\text{m} = \frac{24\times 10^4\,\text{N-m}}{13{,}080\text{N}}$$

$$3h - 2\,\text{m} = 18.35\,\text{m}$$

$$3h = 20.35\,\text{m}$$

$$h = 6.78\,\text{m} \Longleftarrow$$

We can check this result via an alternative and equivalent procedure. The net force on dS in Figure 12.9 is

$$dF = (P_w - P_a)dS = (P_w - P_a)2\,dy$$

With P_w and P_a given by

$$P_w = P_{\text{atm}} + \rho_w g(h - 2 + y)$$

and

$$P_a = P_{\text{atm}} + P_g$$

we have

$$dF = [P_{\text{atm}} + \rho_w g(h-2+y) - (P_{\text{atm}} + P_g)]2\,dy$$

or

$$dF = [\rho_w g(h - 2 + y) - P_g]2\,dy$$

The problem statement requires that the following relation holds:

$$\int_A dm = \int_0^{2\,\text{m}} y\,dF = 0$$

Thus,

$$\int_0^{2\,\text{m}} y[\rho_w g(h - 2 + y) - P_g]2\,dy = 0$$

$$\int_0^{2\,\text{m}} [\rho_w g(hy - 2y + y^2) - yP_g]dy = 0$$

$$\rho_w g\left(\frac{hy^2}{2} - y^2 + \frac{y^3}{3}\right)\Big|_0^{2\,\text{m}} - P_g\frac{y^2}{2}\Big|_0^{2\,\text{m}} = 0$$

$$\rho_w g\left(2h - 4\,\text{m} + \frac{8}{3}\,\text{m}\right) = 2P_g$$

Then

$$2h - \frac{4}{3}\,\text{m} = \frac{2P_g}{\rho_w g}$$

$$h = \frac{2}{3}\,\text{m} + \frac{P_g}{\rho_w g}$$

$$= \frac{2}{3}\,\text{m} + \frac{6 \times 10^4\,\text{N/m}^2}{(1000\,\text{kg/m}^3)(9.81\,\text{m/s}^2)}$$

$$= \frac{2}{3}\,\text{m} + 6.11\,\text{m} = 6.78\,\text{m} \Longleftarrow$$

We now turn to Design Example 5, which deals with the significant dimension in a retaining wall.

12.5 Design Example 5

An earth-filled retaining wall having a density of $1700\,\text{kg/m}^3$ and with the profile shown in Figure 12.10a is used to hold back seawater with a density of $1025\,\text{kg/m}^3$. Determine the dimension, L, which will produce a zero moment about point O under the foregoing conditions.

Solution

Assumptions and Specifications

(1) Steady-state conditions exist.
(2) The densities of the retaining wall and seawater are specified.
(3) All forces and moments are evaluated per meter of retaining wall depth.

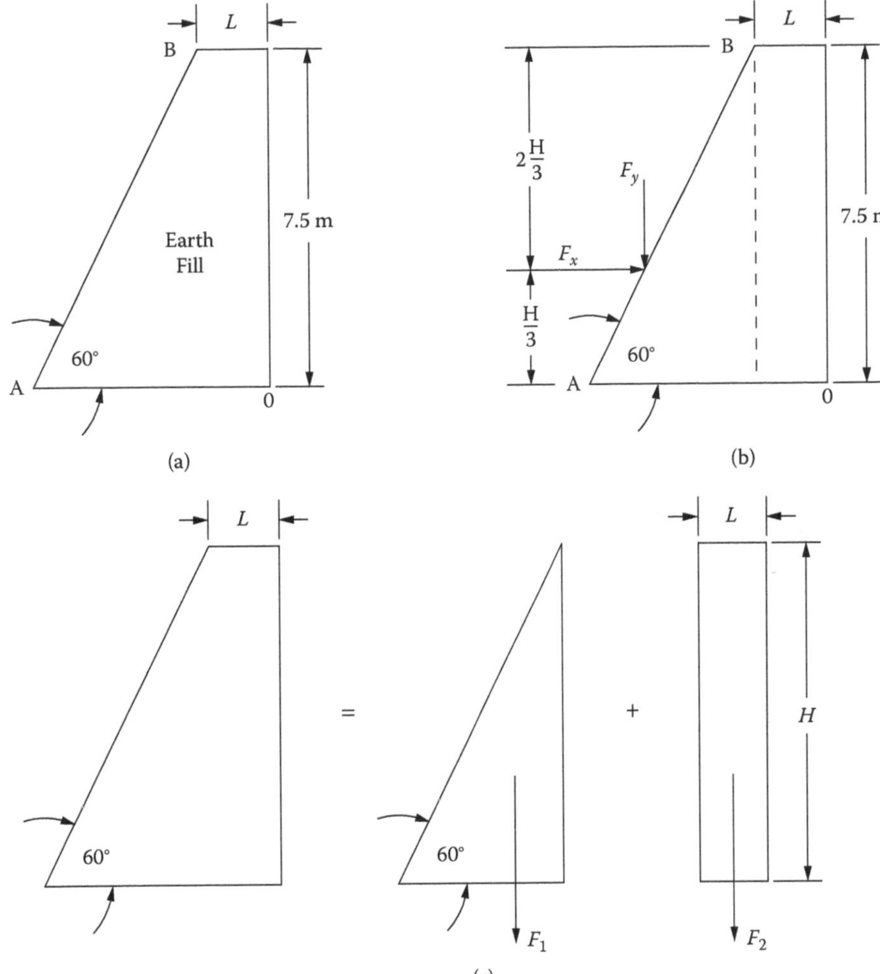

FIGURE 12.10
(a) Retaining wall in Design Example 12.5, (b) wall showing horizontal and vertical seawater forces, and (c) the wall broken into triangular and rectangular segments.

(4) The acceleration of gravity may be taken as $g = 9.81\,\text{m/s}^2$.

(5) The density of water may be taken as $1000\,\text{kg/m}^3$.

We designate

$$\rho_{\text{w}} = 1000\,\text{kg/m}^3, \quad \rho_{\text{sw}} = 1.025\rho_{\text{w}}, \quad \text{and} \quad \rho_{\text{wall}} = 1.700\rho_{\text{w}}$$

as the densities of water, seawater, and earth-filled wall, respectively. We then determine the horizontal and vertical forces due to the seawater on the slanted surface of the retaining wall. These forces are shown in Figure 12.10b.

The horizontal force will be

$$F_x = \int_0^H \rho_{\text{sw}} g y \, dy = 1.025\rho_{\text{w}} g \frac{y^2}{2}\bigg|_0^H = 1.025\rho_{\text{w}} g \frac{H^2}{2} \quad (\text{N})$$

and with $H = 7.5\,\text{m}$

$$F_x = 1.025\rho_w g \frac{(7.5\,\text{m})^2}{2} = 28.83\rho_w \quad \text{(N)}$$

The vertical (downward) seawater force is due to the weight of the seawater above the slanted surface between A and B. With a volume of water of

$$V = \left(\frac{H\cos 60^\circ}{2}\right) H = \frac{H^2}{4} = \frac{(7.5\,\text{m})^2}{4} = 14.06\,\text{m}^3$$

we have

$$F_y = \rho_{sw} g V = 1.025\rho_w g(14.06\,\text{m}^3) = 14.41\rho_w g \qquad \text{(N)}$$

The next step is to determine the moments due to the seawater. With clockwise moments taken as positive, the moment about point O produced by F_x will be

$$M_{zx} = F_x \frac{H}{3} = (28.83\rho_w g)\left(\frac{7.5\,\text{m}}{3}\right) = 72.07\rho_w g \quad \text{(N-m)}$$

A counterclockwise moment is produced by the weight of the water

$$M_{zy} = F_y\left[L + \left(\frac{2H}{3}\right)\cos 60^\circ\right] = F_y(L + 2.5\,\text{m})$$

or

$$M_{zy} = (14.41\rho_w g)(L + 2.5\,\text{m}) \quad \text{(N-m)}$$

The earth-filled retaining wall may be broken up into two component sections as indicated in Figure 12.10c. For F_1, we have the volume (for a wall depth of 1 m)

$$V = \frac{1}{2}(H\cos 60^\circ)H = \frac{1}{2}\left(\frac{H^2}{2}\right) = \frac{1}{2}\frac{(7.5\,\text{m})^2}{2} = 14.06\text{m}^3$$

and

$$F_1 = 1.70\rho_w g V = 1.70\rho_w g(14.06\text{m}^3) = 23.91\rho_w g \qquad \text{(N)}$$

The counterclockwise moment is

$$M_{z1} = F_1\left[L + \frac{H\cos 60^\circ}{3}\right] = 23.91\rho_w g(L + 1.25\,\text{m}) \qquad \text{(N-m)}$$

Finally, for F_2 and a wall depth of 1 m

$$V = HL = (1\,\text{m})(7.5\,\text{m})(1\,\text{m}) = 7.5\,\text{m}^3$$

$$F_2 = 1.70\rho_w g V = 1.70\rho_w g(7.5L) = 12.75\rho_w g \qquad \text{(N)}$$

and the counterclockwise moment is

$$M_{z2} = F_2\left(\frac{L}{2}\right) = 12.75\rho_w g\left(\frac{L}{2}\right) = 6.375\rho_w g L^2 \qquad \text{(N-m)}$$

We now take moments about point 0 ($\sum M = 0$)

$$M_{zx} - M_{zy} - M_{z1} - M_{z2} = 0$$

and with the substitution of the calculated moments

$$72.07\rho_w g - (14.41\rho_w g)(L + 2.5\,\text{m}) \\ -(23.91\rho_w g)(L + 1.25\,\text{m}) - 6.375\rho_w g L^2 = 0$$

Cancelation of the common $\rho_w g$ terms and simplification gives

$$L^2 + 2.26(L + 2.5\,\text{m}) + 3.75(L + 1.25\,\text{m}) - 11.30 = 0$$

or

$$L^2 + 6.01L - 0.96 = 0$$

and the solution to this quadratic gives the result sought

$$L = 0.155\,\text{m} \qquad (15.5\,\text{cm}) \Longleftarrow$$

12.6 Hydrostatic Forces on Curved Surfaces

We next examine the forces exerted on a submerged plane surface in a more general sense than was considered in Section 12.4. The simplest approach to the evaluation of the force exerted on the top of the curved surface, AB, shown in profile in Figure 12.11a, is to consider a free-body diagram of the fluid in the column ABCD above the surface shown in Figure 12.11b. It is clear that the net horizontal force exerted on the surface, (F_2 in Figure 12.12b), is the pressure force exerted by the fluid on a vertical projection of surface AB. Moreover, it is apparent that the vertical force on surface AB is merely the weight of a column of fluid extending from the surface AB to the free-liquid surface. The vector sum of these two components will be the total hydrostatic force exerted on the curved surface AB.

Evaluation of these two components is straightforward. There may be some mathematical complexity if the surface has an odd configuration.

Example 12.5 Suppose that the curved surface, AB, in Figure 12.11 is a quarter circle of radius 50 cm while BD, the vertical surface, is 40-cm high. Both surfaces are 20-cm deep (into the plane of the paper). Determine the horizontal and vertical components of the hydrostatic force acting on surfaces AB and BD per unit width if the surfaces are in contact with water having a density of $1000\,\text{kg/m}^3$.

Solution

Assumptions and Specifications

(1) Steady-state conditions exist.
(2) The fluid is incompressible.
(3) The acceleration of gravity is assumed to be $9.81\,\text{m/s}^2$.
(4) The density of the water is specified.

With reference to Figure 12.11b, the horizontal component, F_1, on the surface BD is

$$F_1 = P_{cg}S = \rho g \bar{y} S \\ = (1000\,\text{kg/m}^3)(9.81\,\text{m/s}^2)(0.20\,\text{m})(0.20\,\text{m})(1\,\text{m}) = 392.4\,\text{N}$$

Because the surface, BD, is vertical, the vertical component of the hydrostatic force is zero.

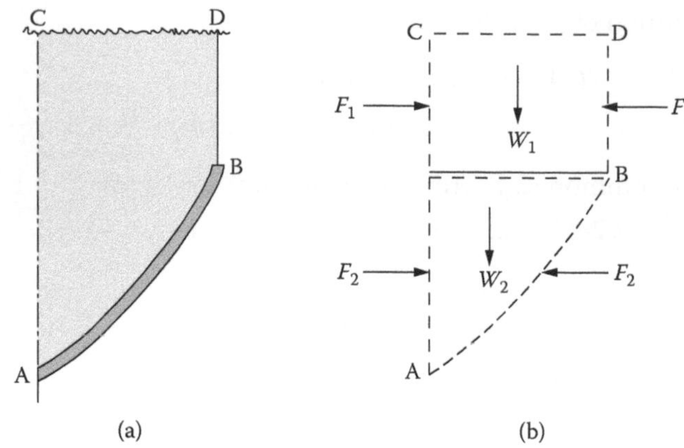

FIGURE 12.11
(a) A submerged curved surface shown in profile and (b) a free-body diagram of the fluid in the column ABCD above the surface.

The horizontal component, F_2, on the surface, AB, is based on the projected surface area, S_{proj}

$$
\begin{aligned}
F_2 &= \rho g \bar{y} S_{proj} \\
&= (1000\,\text{kg/m}^3)(9.81\,\text{m/s}^2)(0.40\,\text{m} + 0.25\,\text{m})(0.50\,\text{m})(1\,\text{m}) \\
&= 3188.3\,\text{N} \Longleftarrow
\end{aligned}
$$

The vertical component, $W_1 + W_2$, is the weight of the column of water standing above the surface, AB

$$
\begin{aligned}
W_1 + W_2 &= (1000\,\text{kg/m}^3)(9.81\,\text{m/s}^2) \\
&\quad \times \left[\left(\frac{\pi}{4}\right)(0.5\,\text{m})^2 + (0.50\,\text{m})(0.40\,\text{m})\right](1\,\text{m}) = 3888.2\,\text{N} \Longleftarrow
\end{aligned}
$$

12.7 Buoyancy

The well-known principle of Archimedes, dating from the third century BC, applies to a body that is totally immersed in a fluid:

> An immersed body is buoyed up by a **buoyant force** that is equivalent to the weight of the fluid displaced by the body.

If the body is less dense than the fluid in question, the buoyant force will be equal to the product of the fluid density and the volume of the displaced fluid. In this case, the body will float because the weight of the volume of fluid required to equal the weight of the body is less than the weight of the total volume of the body.

The buoyant force is clearly a manifestation of the hydrostatic forces exerted by the fluid on the *surfaces* of an immersed object. This can be visualized quite readily by an example.

Example 12.6 A concrete block lies, half-buried, at the bottom of a lake that is 7 m deep. The density of the concrete is 2310 kg/m^3 and the block is rectangular measuring 1 m by 1 m

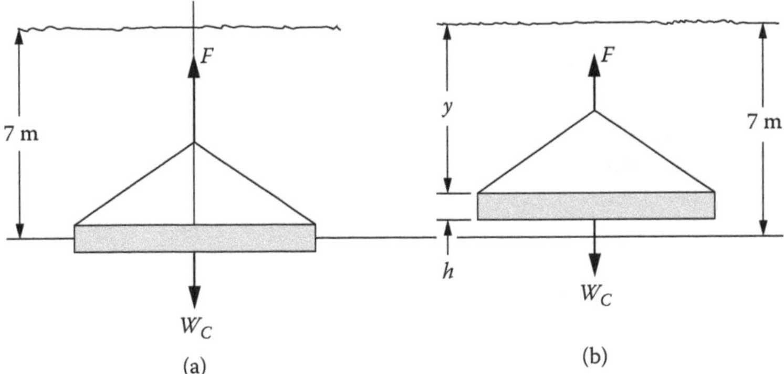

FIGURE 12.12
The concrete block for Example 12.6 (a) originally half-buried and (b) just lifted free.

by 0.15 m. The 0.15-m side is the one that is half-buried. Determine (a) the force required to lift the block free from the bottom of the lake and (b) once pulled free, the force necessary to maintain its position (Figure 12.12).

Solution

Assumptions and Specifications

(1) Steady-state conditions exist.
(2) There is no flow of fluid.
(3) The fluid is incompressible.
(4) The acceleration of gravity is assumed to be $9.81\,\text{m/s}^2$.
(5) The density of the water is $1000\,\text{kg/m}^3$.

(a) To obtain the force required to lift the block free, we write a force balance in the vertical direction. The required freeing force, F, is the weight of a column of water, W_w, that is 6.925 m deep above the top surface of the block plus W_c, the weight of the block itself

$$F = W_w + W_c$$

Here, with $\rho_w = 1000\,\text{kg/m}^3$ and the top surface area equal to $1\,\text{m}^2$, we can write

$$\begin{aligned} W_w &= \rho_w g S y \\ &= \frac{(1000\,\text{kg/m}^3)(9.81\,\text{m/s}^2)(1\,\text{m}^2)(6.925\,\text{m})}{1\,\text{kg-m/s}^2\text{-N}} \\ &= 67.93\,\text{kN} \end{aligned}$$

and

$$\begin{aligned} W_c &= \rho_c g S h \\ &= \frac{(2310\,\text{kg/m}^3)(9.81\,\text{m/s}^2)(1\,\text{m}^2)(0.15\,\text{m})}{1\,\text{kg-m/s}^2\text{-N}} \\ &= 3.40\,\text{kN} \end{aligned}$$

Hence,

$$F = W_w + W_c = 67.93\,\text{kN} + 3.40\,\text{kN} = 71.33\,\text{kN} \Longleftarrow$$

(b) If we consider the block to be at some arbitrary distance, y, below the liquid surface, the hydrostatic forces on the top and bottom surfaces with $S = 1\,\text{m}^2$ will be

$$F_{top} = P_{top}S = \rho_w g y S$$

and

$$F_{bot} = P_{bot}S = \rho_w g(y + h)S$$

The net force, F, required to maintain this position will be

$$\begin{aligned} F &= F_{top} - F_{bot} + W_c \\ &= \rho_w g y S - \rho_w g(y + h)S + \rho_c g Sh \\ &= -\rho_w g h S + \rho_c g Sh \end{aligned}$$

The first of these two terms is seen to be equal to the weight of water contained in the volume Sh and is the upward or buoyant force described by Archimedes. Thus,

$$\begin{aligned} F &= -\frac{(1000\,\text{kg/m}^3)(9.81\,\text{m/s}^2)(1\,\text{m}^2)(0.150\,\text{m})}{1\,\text{kg-m/s}^2\text{-N}} + 3.40\,\text{kN} \\ &= -1.47\,\text{kN} + 3.40\,\text{kN} \\ &= 1.93\,\text{kN} \Longleftarrow \end{aligned}$$

We note that this result is independent of the depth, y, and is a striking example of the effect of buoyancy that is really an imbalance in hydrostatic pressure around an immersed body.

12.8 Stability

We are not only concerned with the force balance on a body that is floating or immersed in a fluid but also with its **stability**. We are able to evaluate the **static stability** of a body by perturbing or disturbing its position slightly and then seeing whether or not the forces on it tend to return it to its original position. If the body will return it is said to be **stable.** If, when perturbed, it continues to move toward another equilibrium position it is **statically unstable.** We are familiar with unstable situations such as a pencil standing on its point, a clown standing on his head, and a thin coin standing on edge.

For a body totally immersed in a fluid, such as the balloon and its passenger shown in Figure 12.13, the situation is stable as long as the center of buoyancy (point B) is above the center of gravity (point G). If the balloon system is displaced from its stable configuration in Figure 12.13a by, say, a gust of wind, it will assume the position illustrated in Figure 12.13b. The moment produced when the weight and buoyant forces are not in line will tend to bring the balloon back to the position in Figure 12.13a, which is the stable configuration.

When a body floats at the interface of two fluids, such as at the surface of a body of water, the circumstances are less obvious. Consider, for example, a rectangular solid block

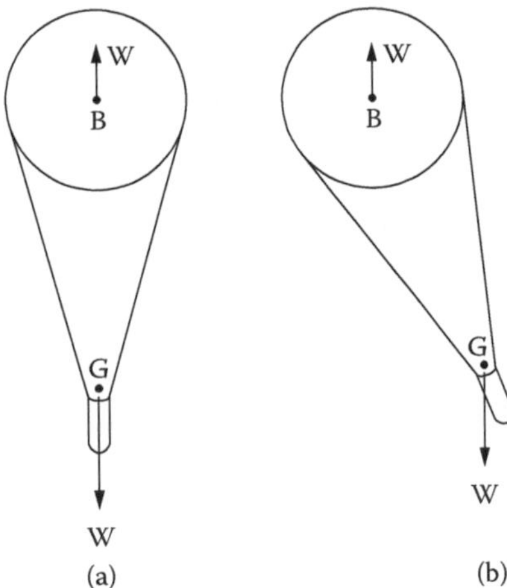

FIGURE 12.13
The stability of a balloon system (a) the stable configuration and (b) the perturbed system.

floating in water, as indicated in Figure 12.14a, with its weight and its buoyant force in line. We note that the weight acts through point G, the center of gravity of the block, and the buoyant force acts through point B, the center of buoyancy, which is at the center of gravity of the *displaced* volume. Clearly, point G is above point B but the block may still be stable.

Our analysis now involves the perturbed state shown in Figure 12.14b, where the two forces are not in line due to a displacement of the block. The weight of the block can still be considered a concentrated force through point G. If the buoyant force continued to act through point B, the block would continue to tumble and would represent an unstable condition.

The tricky part now is to observe that, due to the tipping, volume *Oab* in Figure 12.14b, which was previously submerged, is now above the water and volume *Ocd*, which was formerly above the water, is now submerged. The result of both of these changes is that, compared to the original upright case, there is a greater buoyant force to the left of the centerline and a lesser one to the right. The result of this is that the new center of buoyancy, B', is to the left of the previous position at B. If it is far enough to the left, then the condition is a stable one and the block will right itself. However, if the new center of buoyancy is less than some required amount, the condition will remain unstable.

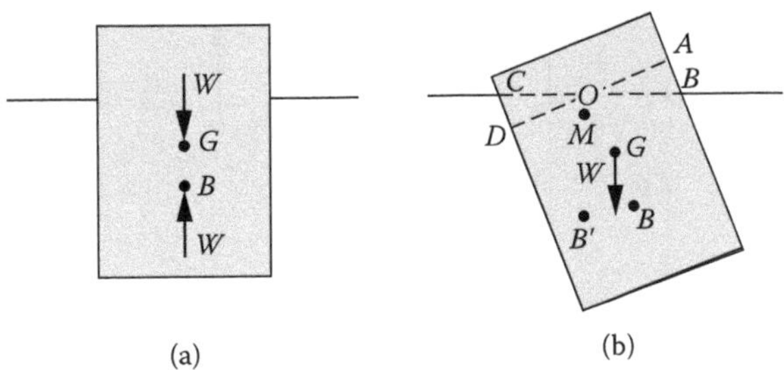

FIGURE 12.14
The stability of a floating block: (a) the stable configuration and (b) the perturbed system.

To determine the stability of the new arrangement, we locate the new center of buoyancy, point B′, and draw a vertical line through this point intersecting the symmetry axis of the block at point M. It is clear that if point M is above point G the block will be stable. Point M is designated as the **metacenter** and the distance $\overline{MG}$ is the **metacentric height**. The condition for stability is that point M be above point G or that the metacentric height be positive.

Example 12.7, which now follows, should aid in illustrating this concept.

Example 12.7 A freighter is idealized as having a rectangular cross section as shown in Figure 12.15. It has a draft height of 2 m and a width of $2L$. We wish to find the limiting ratio of L/h for which this vessel will be stable. The center of gravity may be assumed to be at the waterline as shown.

Solution

Assumptions and Specifications

(1) Steady-state conditions exist.
(2) There is no flow of fluid.
(3) The fluid is incompressible.

The vessel is shown in the tipped configuration in Figure 12.15b. Because the cross section is symmetrical about the original vertical axis (Figure 12.15a) and rectangular, we see that the areas Gab and Gcd are equal. If we take moments about point G, the limiting condition will be for the case where $\sum M = 0$. Proceeding, we may write this balance of moments for small displacement angles $\Delta\theta$. We note that there are two new forces for the two new triangular volumes above and below the waterline. These two forces are located at a distance $2L/3$ from point G which, as Figure 12.15 shows, is the distance to the centroid of each of the triangular volumes being considered.

Taking moments about point G gives

$$\left(\frac{1}{2}L\overline{cd}\right)\left(\frac{2}{3}L\right) + \left(\frac{1}{2}L\overline{ab}\right)\left(\frac{2}{3}L\right) - 2Lh(\overline{GB}) = 0$$

The distances $\overline{cd}$, $\overline{ab}$, and $\overline{GB}$ may be approximated for small displacements as

$$\overline{cd} = \overline{ab} = L\Delta\theta$$

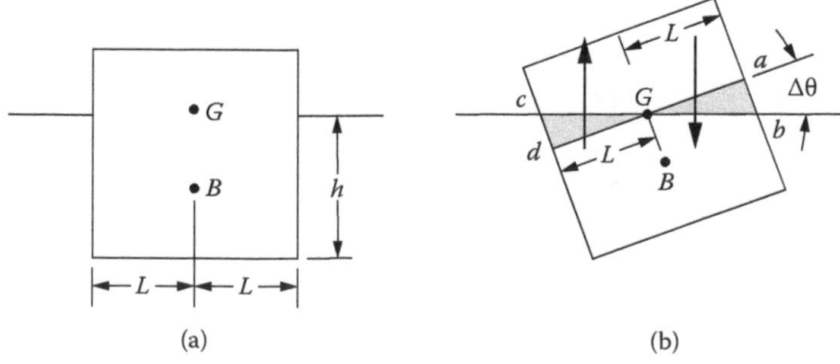

FIGURE 12.15
The cross section of the freighter of Example 12.7: (a) the stable configuration and (b) the perturbed system.

and

$$\overline{GB} = \frac{h}{2}\Delta\theta$$

After substitution of these into the moment equation, we obtain

$$\left(\frac{1}{2}L^2\right)\left(\frac{2}{3}L\right) + \left(\frac{1}{2}L^2\right)\left(\frac{2}{3}L\right) - 2Lh\frac{h}{2} = 0$$

This yields

$$\frac{2}{3}L^2 - h^2 = 0$$

or

$$\frac{L}{h} = \left(\frac{3}{2}\right)^{1/2} = 1.225 \Longleftarrow$$

This is the lower limit of the ratio, L/h for the ship to be stable. If the ship is made wider, that is, if $L > 1.225h$, the stability will increase accordingly.

12.9 Uniform Rectilinear Acceleration

Recall that in Section 12.1, we made the distinction between inertial and noninertial reference frames. We will now consider the special noninertial case of a coordinate system that is accelerating uniformly. Newton's law for such a system equates the sum of all of the forces to the mass acceleration product

$$\sum \boldsymbol{F} = m\boldsymbol{a} \tag{12.17}$$

A development that precisely follows the steps in Section 12.1 would lead, with the fluid at rest relative to a system with constant acceleration, to the expression

$$-\nabla P + \rho\boldsymbol{g} = \rho\boldsymbol{a} \tag{12.18}$$

An alternate form that is consistent with our fundamental fluid statics relation of Equation 12.2 is

$$\nabla P = \rho(\boldsymbol{g} - \boldsymbol{a}) \tag{12.19}$$

Equation 12.19 states that the maximum rate of change of pressure is in the $\boldsymbol{g} - \boldsymbol{a}$ direction. If the magnitude of the acceleration, $a = 0$, Equation 12.19 reverts to Equation 12.2.

Equation 12.19 indicates that the fluid surface in a tank car, initially at rest, as shown in Figure 12.16a will assume the configuration shown in Figure 12.16b when the tank and its contents are accelerated to the right with magnitude, a. As indicated by the vector triangle in Figure 12.16b, the fluid surface, which will be at constant pressure will be perpendicular to the direction of $\boldsymbol{g} - \boldsymbol{a}$. At point b, for example, an analysis of the case shown will yield

$$\frac{dP}{dz}y = -\rho|\boldsymbol{g} - \boldsymbol{a}|y = -\rho(g^2 + a^2)^{1/2}y$$

where $|\boldsymbol{g} - \boldsymbol{a}|$ indicates the magnitude of $\boldsymbol{g} - \boldsymbol{a}$.

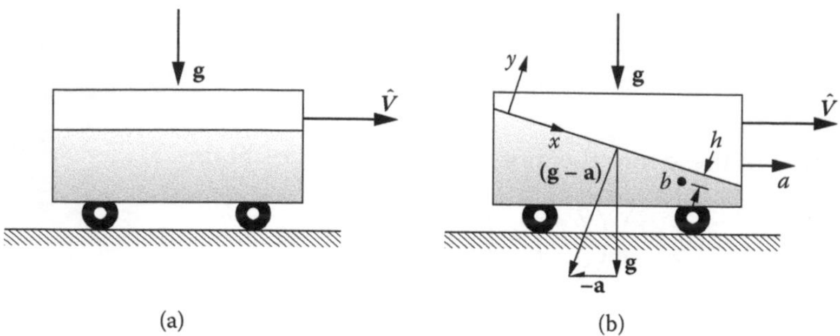

FIGURE 12.16
A uniformly accelerated fuel tank (a) at constant velocity and (b) under uniform acceleration.

Separation of the variables and integration between $y = 0$ and $y = -h$ will give the result

$$\int_{P_{\text{atm}}}^{P_b} dP = -\rho(g^2 + a^2)^{1/2} \int_0^{-h} dy$$

or

$$P_b - P_{\text{atm}} = \rho(g^2 + a^2)^{1/2} h \tag{12.20}$$

A common example of uniform acceleration is the case of a spinning container that contains a liquid at atmospheric pressure. Example 12.8, which now follows, should aid in illustrating this concept.

Example 12.8 Determine the shape of the free surface of the liquid in the rotating cylindrical container shown in Figure 12.17. The container, with inside radius, r, spins with a uniform angular velocity, ω.

Solution

Assumptions and Specifications

(1) Steady-state conditions exist.
(2) The fluid is incompressible.

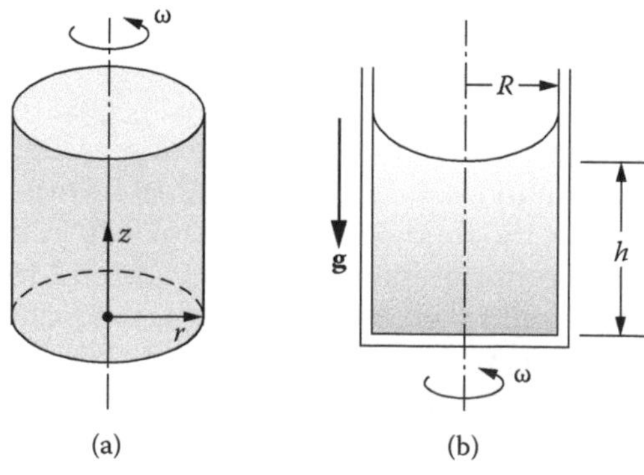

FIGURE 12.17
The spinning cylinder containing a liquid of Example 12.8.

(3) The angular velocity, ω, is specified.

(4) The acceleration of gravity is assumed to be 9.81 m/s^2.

We will employ a cylindrical coordinate system and, for a constant ω, we may presume that the free-liquid surface will be symmetric about the centerline. Thus, the pressure will be a function of the coordinates r and z and we can write

$$dP = \frac{\partial P}{\partial r}dr + \frac{\partial P}{\partial z}dz$$

As we have noted, the pressure is constant along the free surface so that $dP = 0$ and the differential equation for such a constant pressure case is

$$\frac{dz}{dr} = -\frac{\partial P/\partial r}{\partial P/\partial z}$$

The necessary values for these partial derivatives may be obtained from the fluid statics relationship of Equation 12.19,

$$\frac{\partial P}{\partial r}\mathbf{r} + \frac{\partial P}{\partial z}\mathbf{z} = \rho(\mathbf{g} - \mathbf{a})$$

The gravitational acceleration is constant, that is

$$\mathbf{g} = -g\mathbf{z}$$

and the radial acceleration is a function of r and may be expressed as

$$\mathbf{a} = -r\omega^2\mathbf{r}$$

With these in hand, we may write Equation 12.18 as

$$\frac{\partial P}{\partial r}\mathbf{r} + \frac{\partial P}{\partial z}\mathbf{z} = -\rho g\mathbf{z} + \rho r\omega^2\mathbf{r}$$

We now equate the coefficients of the unit vectors to obtain

$$\frac{\partial P}{\partial r} = \rho r\omega^2$$

and

$$\frac{\partial P}{\partial z}\mathbf{z} = -\rho g$$

These are the terms required to solve for the shape of the free-liquid surface, $y(r)$. We may make the appropriate substitutions and obtain

$$\frac{dz}{dr} = \frac{\rho r\omega^2}{\rho g}$$

and the normal separation of the variables and integration yields

$$\int_h^y dz = \frac{\omega^2}{g}\int_0^r r\,dr$$

$$y - h = \frac{\omega^2 r^2}{2g}$$

The equation of the free surface is therefore,

$$y = h + \frac{\omega^2 r^2}{2g} \Longleftarrow$$

which is a parabola with its vertex at the point h on the cylinder centerline.

12.10 Summary

In this chapter we have examined the behavior of static fluids, which are motionless relative to a coordinate system that is either motionless or accelerating uniformly. The basic equation known as the fundamental equation of fluid statics is

$$\nabla P = \rho(\boldsymbol{g} - \boldsymbol{a}) \tag{12.19}$$

where the acceleration of the coordinate system is either zero or a constant. The usual situation is where $\boldsymbol{a} = 0$ in which case

$$\nabla P = \rho \boldsymbol{g} \tag{12.2}$$

Thus, fluid statics deals with problems that are associated with fluids at rest.

A common pressure measuring device, the **manometer**, is based on principles of fluid statics. This chapter also discussed static situation forces on submerged surfaces, buoyancy, and stability.

The stability of floating and immersed objects has been examined. A totally immersed system is stable as long as the center of mass is below the center of buoyancy. On the other hand, a floating object can be stable, unstable, or neutrally stable depending on the **metacentric height**, defined as the distance between the center of mass and the center of buoyancy. The floating body, such as the hull of a vessel, will be stable as long as the metacenter is above the center of buoyancy (as long as the metacentric height is positive).

12.11 Problems

General Considerations

12.1: Using standard conditions to determine the density of air, determine the height of the atmosphere if it were incompressible.

12.2: The isothermal compressibility of a substance is given by $\beta = \rho(\partial P/\partial \rho)_T = -V(\partial P/\partial V)_T$. Determine β for a perfect gas.

12.3: In water, the modulus, β, defined in Problem 12.2, is nearly constant and has a value of 2.068×10^6 kPa. Determine the percentage volume change in water due to a pressure of 20,000 kPa.

12.4: On a certain day, the barometric pressure at sea level is 30 in of mercury and the temperature is 20°C. The pressure gage in an aircraft in flight indicates a pressure of 73 kPa and a temperature gage shows the air to be at 8°C. Estimate the altitude of the aircraft above sea level.

12.5: If the density of seawater is approximated by the equation

$$\rho = \rho_o e^{(P - P_{atm})/\beta}$$

where ρ_o is the density at atmospheric pressure and β is the compressibility, take $\rho_o = 1025\,\text{kg/m}^3$ and $\beta = 2.068 \times 10^6$ kPa and determine the pressure and density at a point 9750 m below the surface.

12.6: The practical depth limit for a suited diver is about 18 m. If the specific gravity of seawater is 1.025, determine the gage pressure in the seawater at that depth.

12.7: The deepest known point in the ocean is 11,034 m in the Mariana Trench in the Pacific. Assuming seawater to have a constant density of $1050\,\text{kg/m}^3$, determine the pressure at this point in atmospheres.

12.8: Determine the depth change to cause a pressure increase of 1 atm for (a) water, (b) seawater (SG = 1.0250), and (c) mercury (SG = 13.6).

Hydrostatic Forces

12.9: A tank is filled with glycerin (SG = 1.258) to a level of 1.625 m. If the air space above the glycerin is pressurized to 40 kPa, determine the pressure at the bottom of the tank.

12.10: A tank is filled with mercury (SG = 13.6) such that the pressure on the bottom of the tank is 112.04 kPa. Determine the height of the mercury.

12.11: An open tank contains 0.50 m of oil with a specific gravity of 0.825 on top of 2.65 m of water. The barometer reads 29.72-in Hg. Determine (a) the pressure at the oil-mercury interface and (b) the pressure at the bottom of the tank.

12.12: A column of 1.25 m of lubricating oil (SG = 0.885), 2.875 m of water, and 6.48 m of a heavy oil yields a pressure of 228.47 kPa at the bottom of the column. If the atmospheric pressure is 100.23 kPa, determine the specific gravity of the heavy oil.

12.13: Mercury (SG = 13.6) is poured in the column of fluid considered in Problem 12.4 so that the absolute pressure at base of the column becomes 473.59 kPa, determine the height of the mercury in the column.

12.14: In Figure P12.14, the fluid is water and the left-hand side is open to an atmospheric pressure of 100 kPa. Determine the pressure P_A.

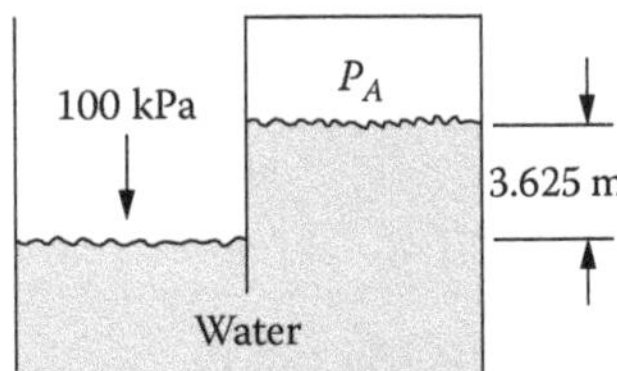

FIGURE P12.14

12.15: In Figure P12.15, the left-hand side is open to the atmosphere at 1.013 bar. Determine the pressure in the air space (a) above column B and (b) above column C.

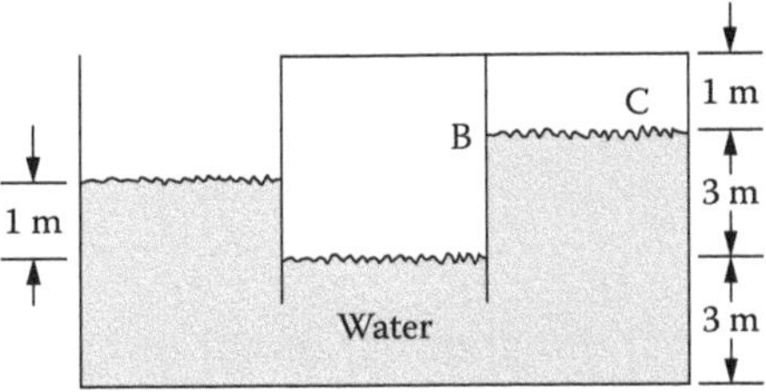

FIGURE P12.15

12.16: In Figure P12.16, $P_A = 200\,\text{kPa}$. Determine the value of P_B.

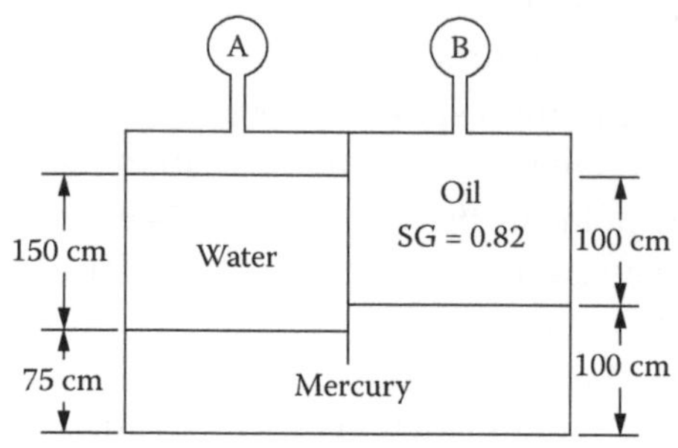

FIGURE P12.16

12.17: In the closed tank of Figure P12.17, gage A reads 305 kPa. Determine (a) the value of h and (b) the reading on gage B.

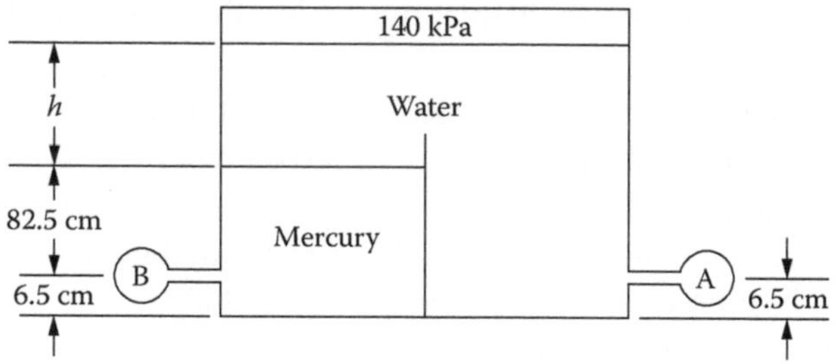

FIGURE P12.17

12.18: A hydraulic jack is sketched in Figure P12.18. Determine the force, F, required to hold the 225-kg mass in place.

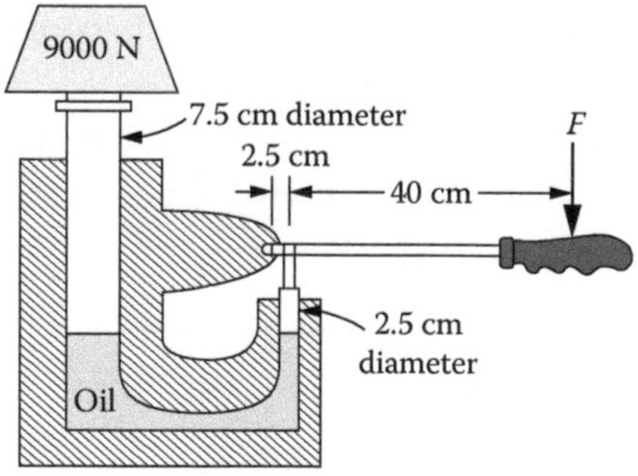

FIGURE P12.18

Barometers and Manometers

12.19: In Figure P12.19, determine the absolute pressure at point A if the liquid in the manometer has a specific gravity of 1.55 and the barometric pressure is 100 kPa.

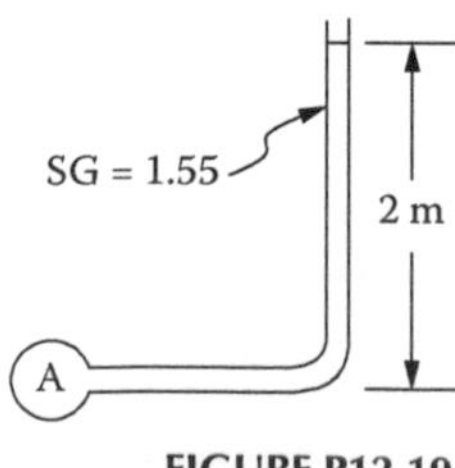

FIGURE P12.19

12.20: The tank shown in Figure P12.20 is closed and the air pressure above the oil is 128.23 kPa. The manometer is open to the atmosphere and the barometric pressure is 29.92 in of mercury. Determine the height, h, of the oil in the manometer.

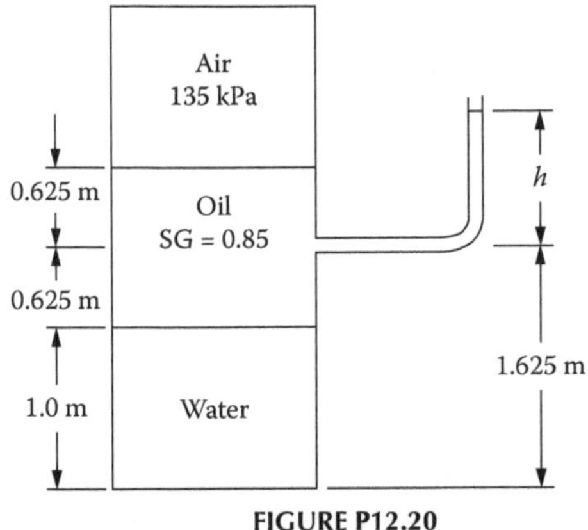

FIGURE P12.20

12.21: The manometer shown in Figure P12.21 is open to the atmosphere. Determine the gage pressure at point A.

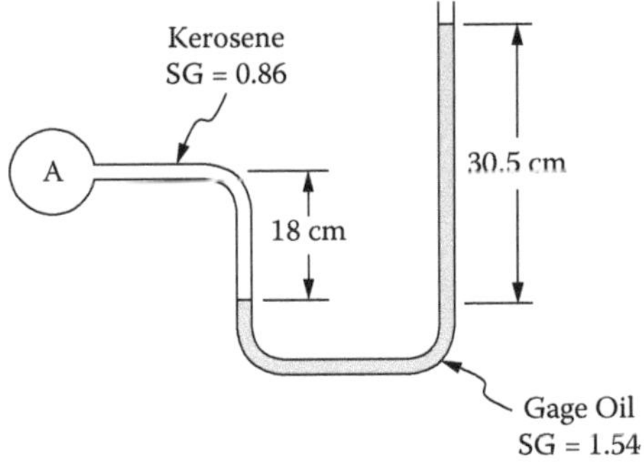

FIGURE P12.21

12.22: A pipe carrying water is tapped by a manometer that is open to the atmosphere as shown in Figure P12.22. Determine the gage pressure at point A.

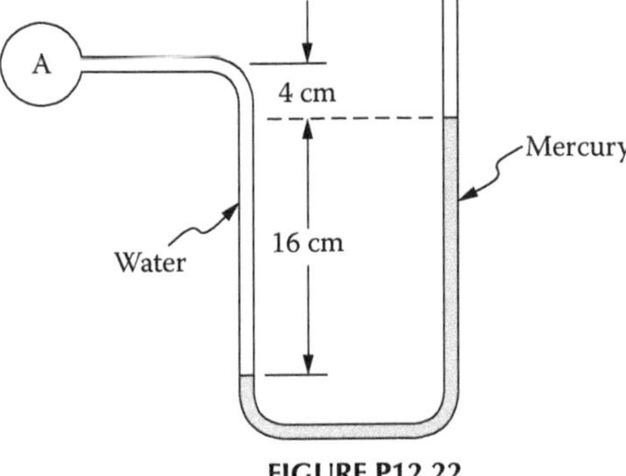

FIGURE P12.22

12.23: The tube shown in Figure P12.23 is open to the atmosphere at both ends. Determine the specific gravity of the oil.

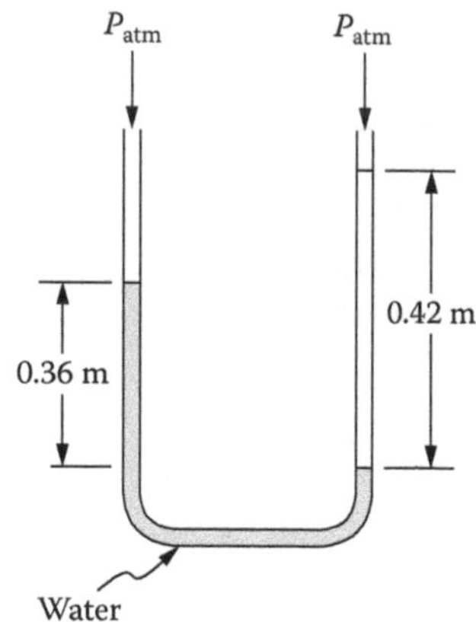

FIGURE P12.23

12.24: A differential manometer is connected at points A and B to a vertical pipe as shown in Figure P12.24. Determine the pressure difference between points A and B.

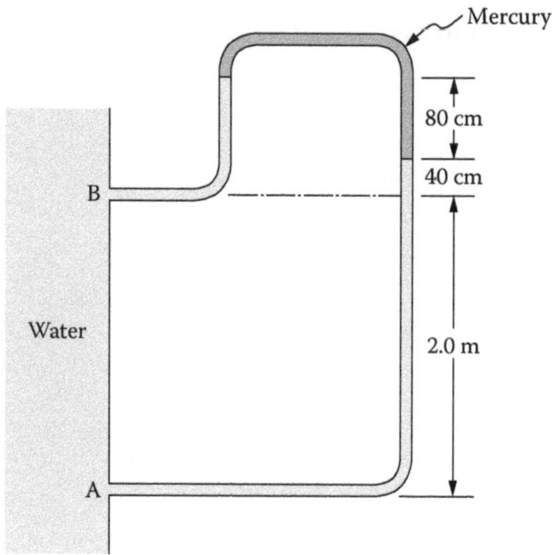

FIGURE P12.24

12.25: In Figure P12.25, determine the mass of the piston if the pressure gage reads 124.85 kPa.

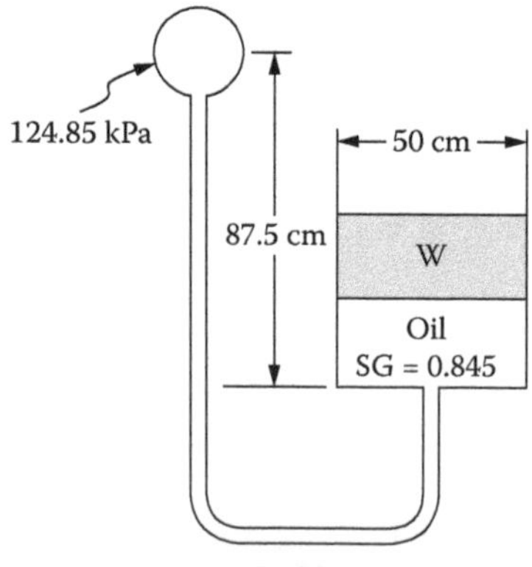

FIGURE P12.25

12.26: In Figure P12.26, tank 1 contains water and tank 2 contains a more dense liquid. Both tanks are open to the atmosphere. For the configuration shown, what is the value for the specific gravity of the oil?

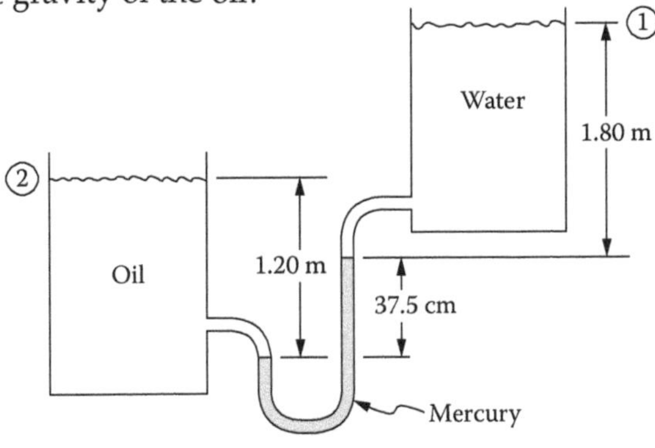

FIGURE P12.26

12.27: In Problem 12.26, determine the level of the water that will cause the pressure difference to be zero.

12.28: A mercury manometer, open to the atmosphere, is connected to a pipe containing oil as indicated in Figure P12.28. Determine the gage reading.

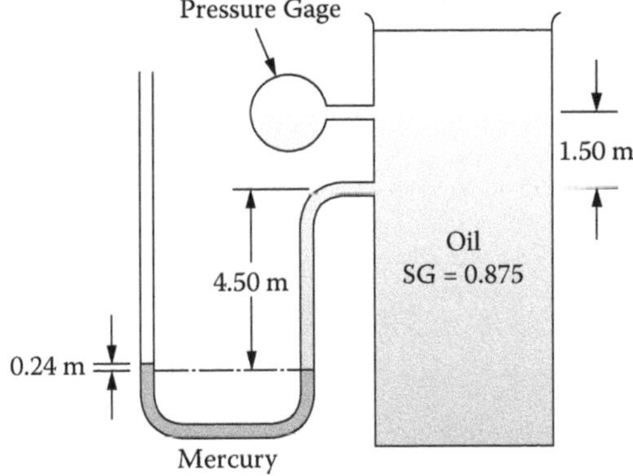

FIGURE P12.28

12.29: In Figure P12.29, point B is at 200 Pa. Liquids 1 and 2 have densities of 857 kg/m^3 and 1261 kg/m^3, respectively. Determine the pressure at point A.

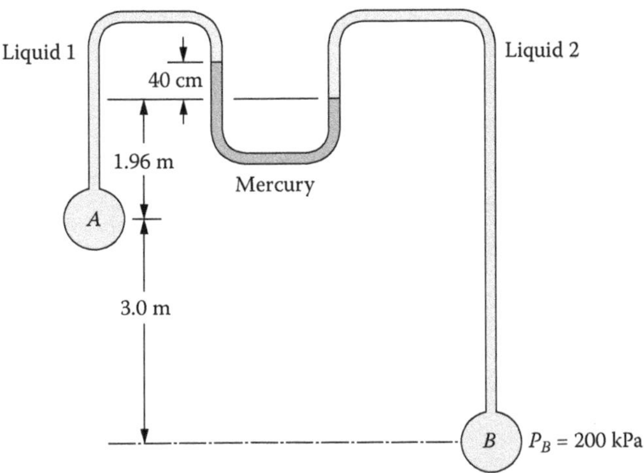

FIGURE P12.29

12.30: Determine the difference in pressure between points A and B in Figure P12.30.

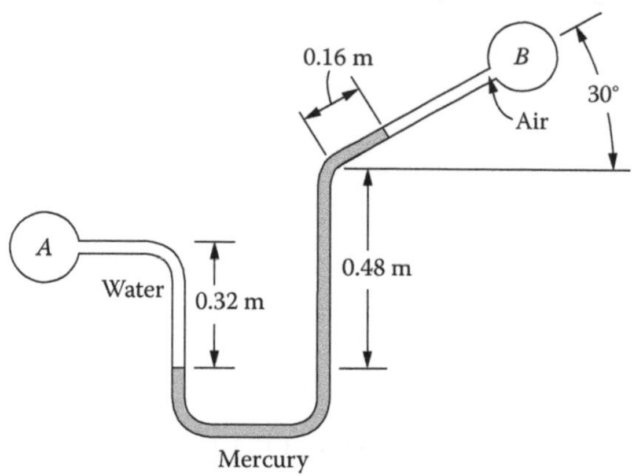

FIGURE P12.30

12.31: The specific gravity of the oil in Figure P12.31 is 0.80 and the manometer is open to the atmosphere. Determine the pressure at point A.

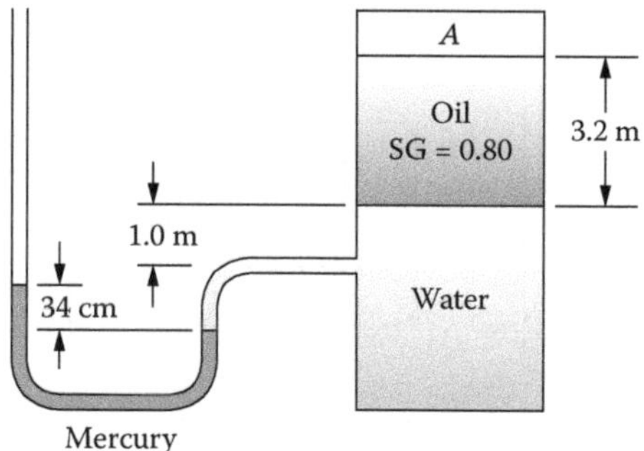

FIGURE P12.31

12.32: In Figure P12.32, determine the height of the oil indicated by h if both manometers are open to the atmosphere.

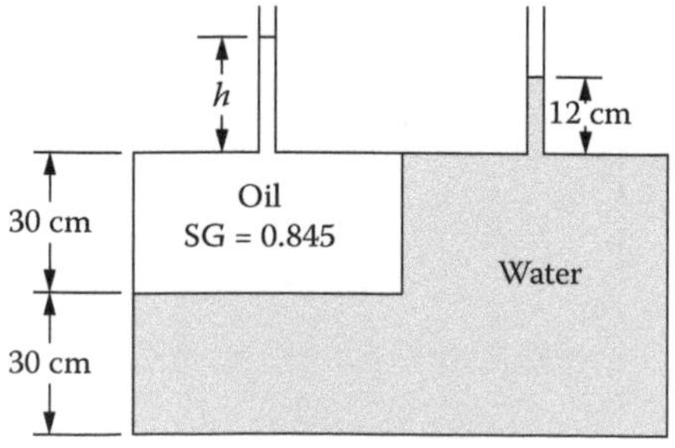

FIGURE P12.32

12.33: A differential manometer is used to measure the pressure change caused by a flow constriction in a piping system as shown in Figure P12.33. Determine (a) the pressure difference between points A and B (Pa) and (b) which section has the higher pressure.

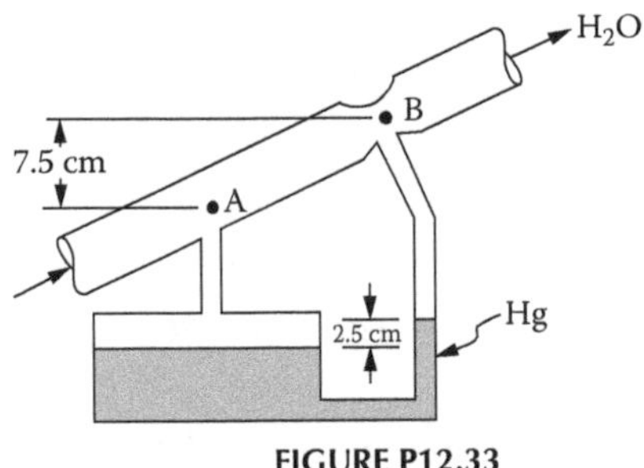

FIGURE P12.33

12.34: Determine the pressure at point A in Figure P12.34.

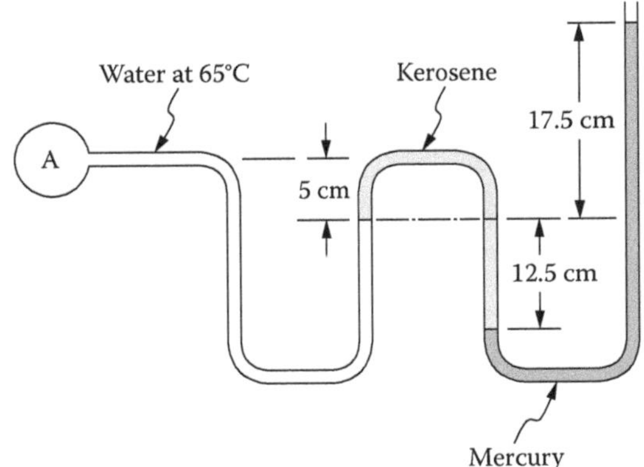

FIGURE P12.34

12.35: In the system shown in Figure P12.35, all fluids are at 20°C. Determine the pressure difference between points A and B.

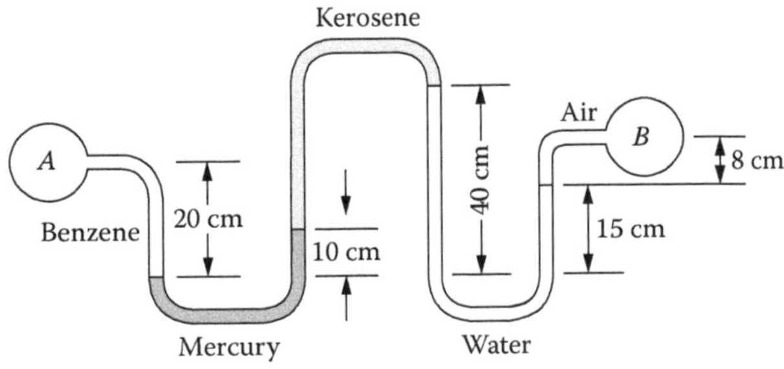

FIGURE P12.35

12.36: In the system shown in Figure P12.36, fluid 1 is glycerin and fluid 2 is carbon tetrachloride and $P_a = 101$ kPa. Determine the absolute pressure at point A.

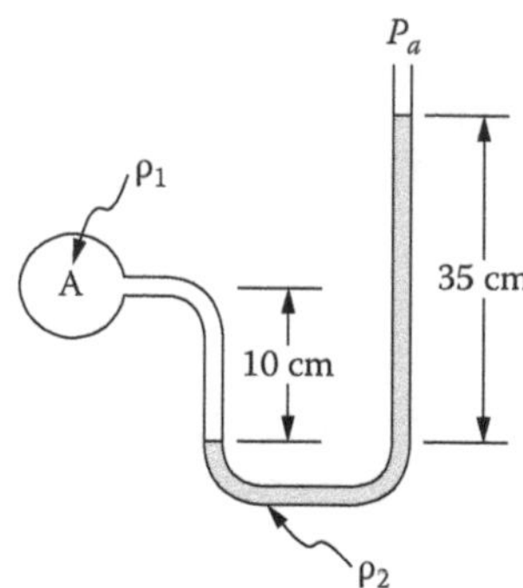

FIGURE P12.36

12.37: The U-tube manometer shown in Figure P12.37 has a 1 cm-ID and contains mercury. Determine the free-surface height in each leg at equilibrium if $10\,\text{cm}^3$ of water is poured into the right-hand leg.

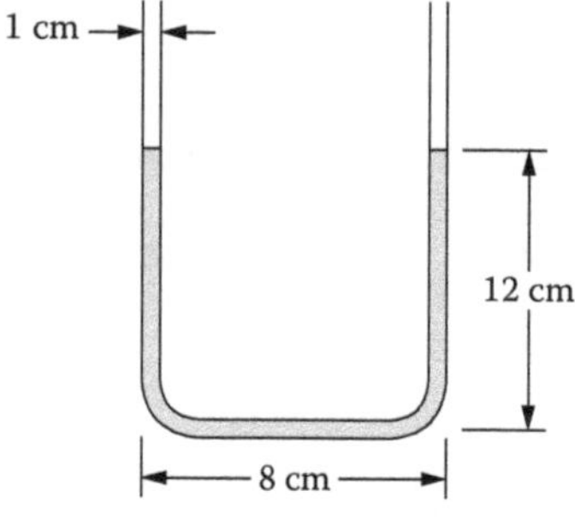

FIGURE P12.37

12.38: The hydraulic jack shown in Figure P12.38 is filled with a hydraulic fluid having a density of $865\,\text{kg/m}^3$. Neglect the weight of the two pistons and determine what force on the handle is required to support the 100-N weight.

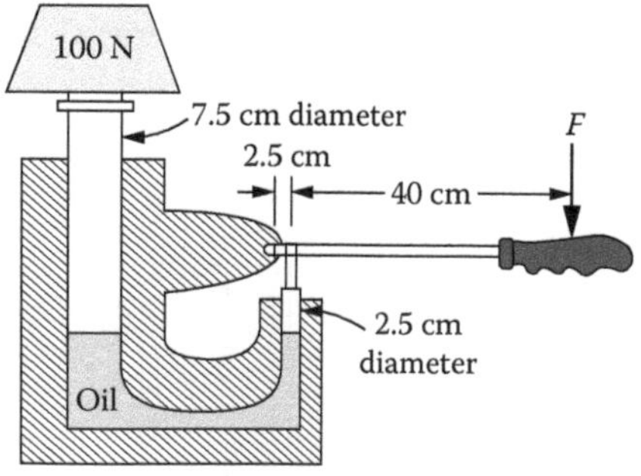

FIGURE P12.38

Forces on Plane Areas

12.39: Freshly poured concrete approximates a fluid with a specific gravity of 2.40. Figure P12.39 shows a 2.5 m × 3 m × 0.200 m slab that is poured between two wooden forms, which are connected by four corner bolts. Neglect end effects and compute the forces in all four bolts.

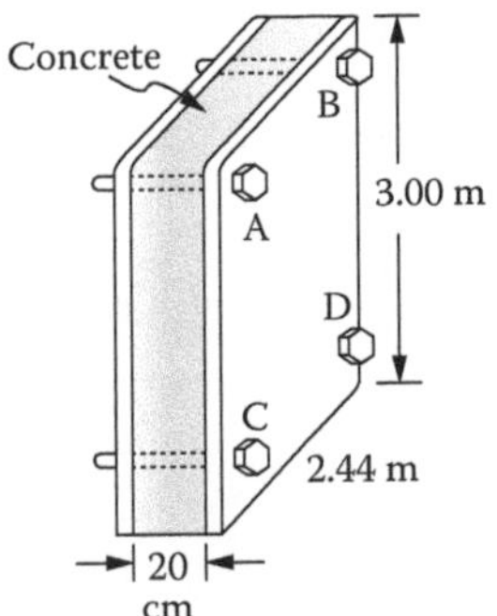

FIGURE P12.39

12.40: Glass viewing windows are to be installed in an aquarium. Each window is to be 0.6 m in diameter and centered 2 m below the surface level. Determine the magnitude and location of each force acting on the window.

12.41: Determine the minimum value of h for which the gate shown in Figure P12.41 will rotate counterclockwise if the gate cross section is triangular, 1.2 m at the base, and 1.2-m high. Neglect bearing friction.

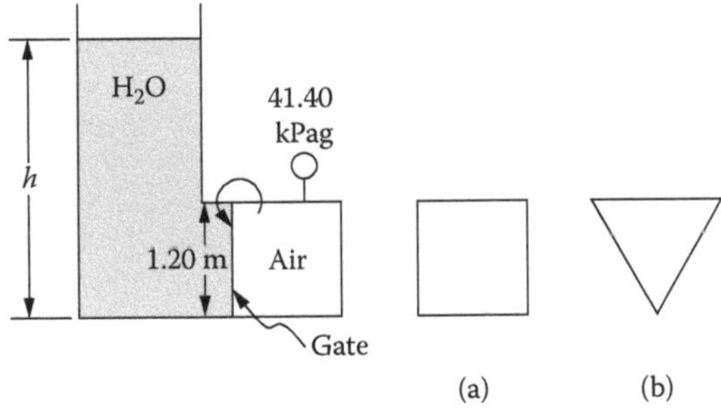

FIGURE P12.41

12.42: A dam spillway gate holds back water of depth, h, as shown in Figure P12.42. The gate weighs 6000 N and it is hinged at A. Determine the depth of water for which the gate will raise up and permit water to flow under it.

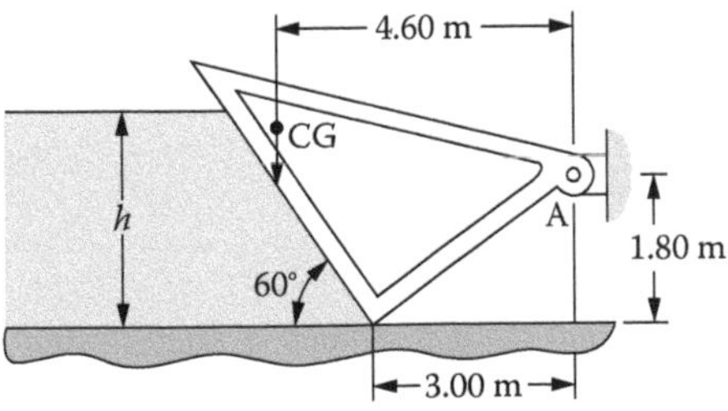

FIGURE P12.42

12.43: A watertight bulkhead, 6.75-m high, forms a temporary dam for some construction work. As indicated in Figure P12.43, the top 3.75 m contains seawater with $\rho = 1030\,kg/m^3$ behind the bulkhead, and the bottom 3 m contains a mixture of mud and water that can be considered as a fluid having a density of $2100\,kg/m^3$. Determine the total horizontal load per unit width and the location of the center of pressure measured from the bottom.

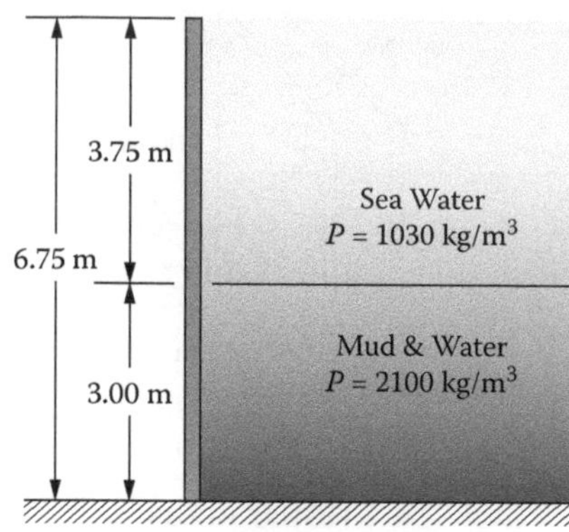

FIGURE P12.43

12.44: The circular gate, ABC, shown in Figure P12.44, has a 1-m radius and is hinged at B. Neglect atmospheric pressure and determine the force, P, just sufficient to keep the gate from opening when $h = 12\,\text{m}$.

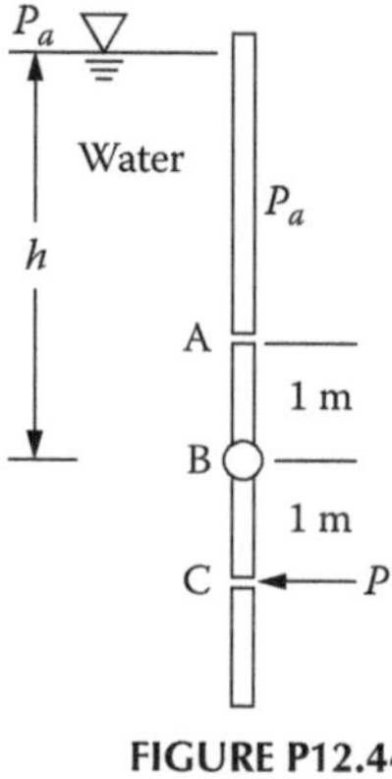

FIGURE P12.44

12.45: As indicated in Figure P12.45, gate AB is semicircular, hinged at B, and held by a horizontal force, P at A. Determine the force, P, required for equilibrium.

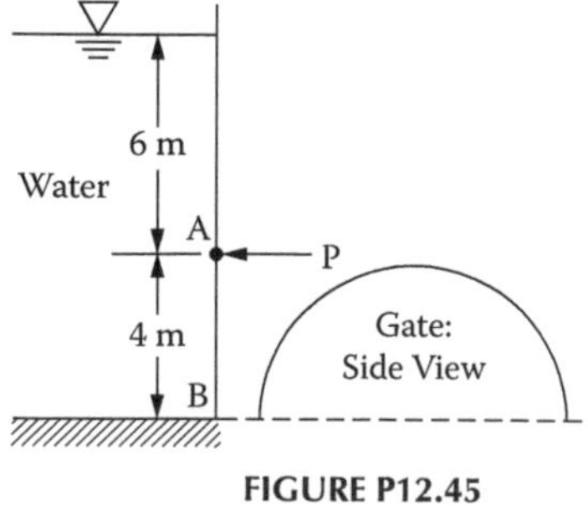

FIGURE P12.45

12.46: Gate AB in Figure P12.46 is hinged at point A located 3.2 m above the surface of the water on its left. The gate is 3.6-m wide and the height of the water is 10.8 m. A volume of seawater (SG = 1.023), to a depth of *h* m, is on the right-hand side of the gate. Determine *h*, in meters, to just permit the gate to open.

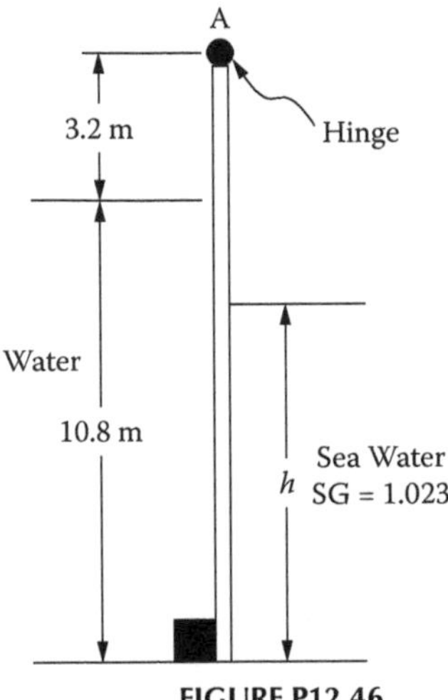

FIGURE P12.46

12.47: The gate, AB, in Figure P12.47 is hinged at B and has a weight of 9000 N. It rests against a smooth wall at A. Water is present on both sides; on the left to an unknown height, *h*, and on the right to a depth of 14 m. The gate is 6-m wide. Determine the height, *h*, which will just cause the gate to open.

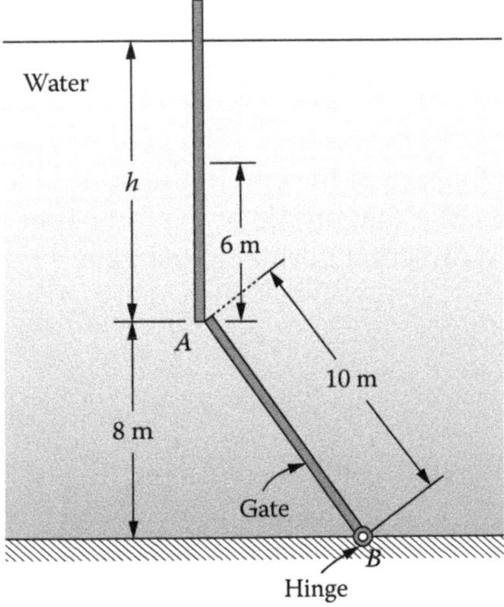

FIGURE P12.47

12.48: The vertical plate in Figure P12.48 is submerged vertically in water. Determine the magnitude of the hydrostatic force on one side and the depth to the center of pressure.

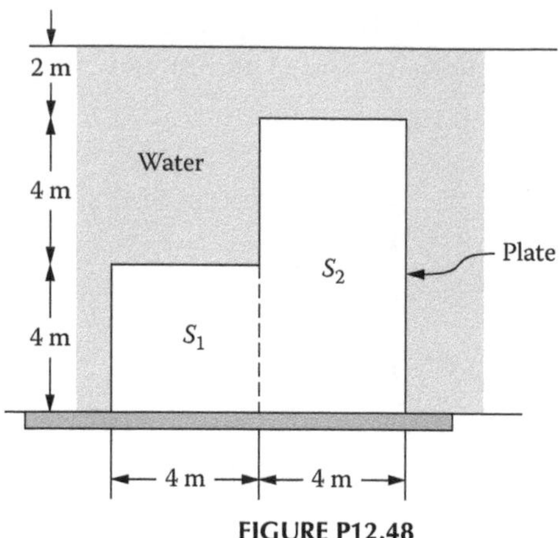

FIGURE P12.48

12.49: Gate AB in Figure P12.49 is hinged at B and, when opened, rotates in the clockwise direction. The width of the gate is 2 m. Determine the counterclockwise torque required to hold the gate shut.

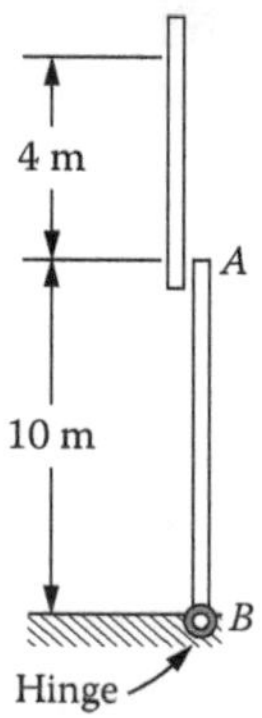

FIGURE P12.49

12.50: To control the water surface in a reservoir, a hinged leaf gate having the dimensions indicated in Figure P12.50 is provided. The gate is shown in its upper limiting position and presses on the seat at the end of the bottom leaf. If the weight of the gate is neglected, determine the magnitude of the force (per linear meter of crest) that is exerted at the seat in a direction normal to the gate.

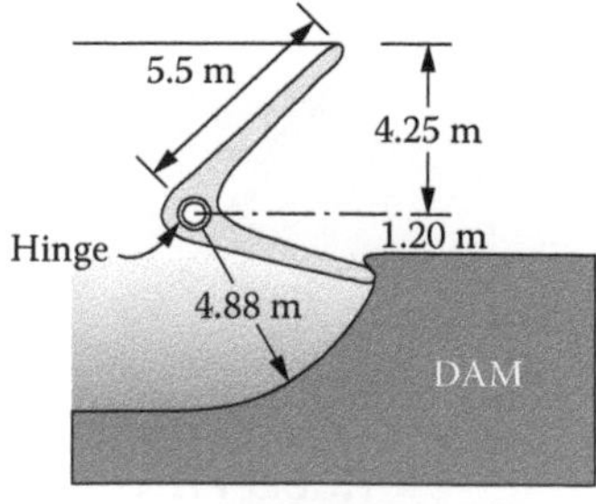

FIGURE P12.50

12.51: A tank that is 2.5-m wide contains 4 m of oil (SG = 0.780) and 2.4 m of water as indicated in Figure P12.51. Determine the magnitude and depth of the total force on side ABC.

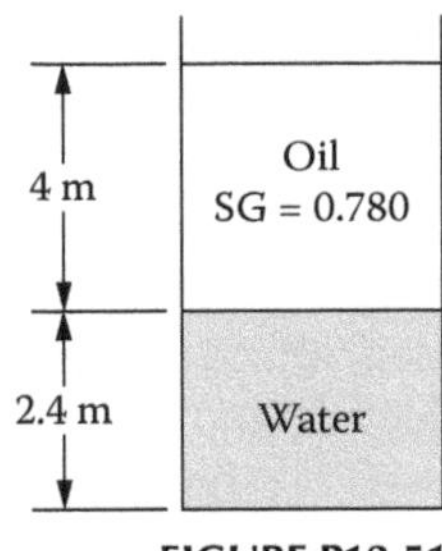

FIGURE P12.51

12.52: Figure P12.52 shows an open triangular channel in which the two sides, AB and AC, are held together by cables, spaced 1 m apart, between B and C. Determine the cable tension.

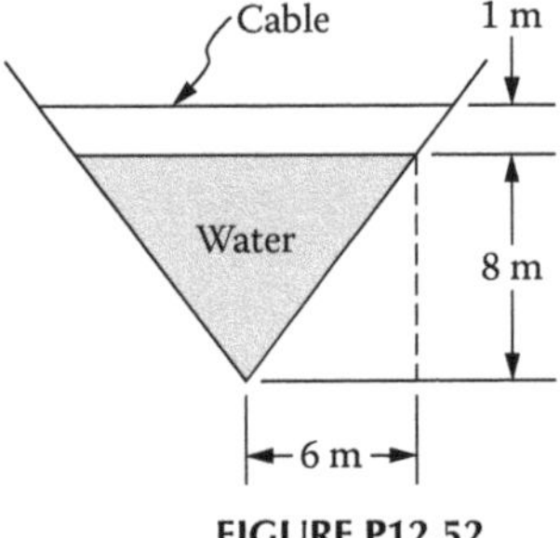

FIGURE P12.52

12.53: The dam shown in Figure P12.53 is 100-m wide. Determine the magnitude and location of the force on the inclined surface.

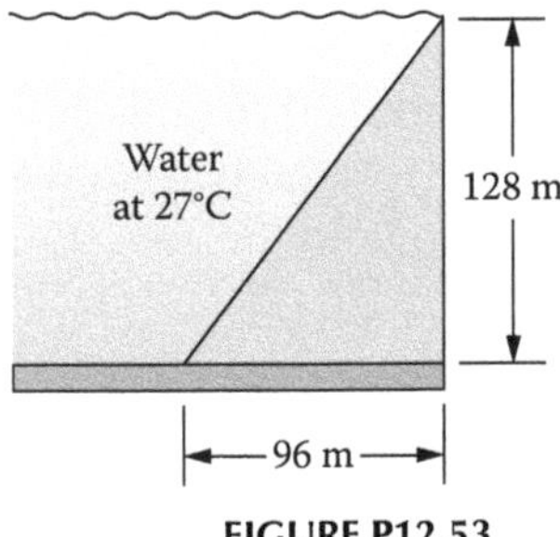

FIGURE P12.53

12.54: A vertical rectangular gate, having a width of 3 m, is located in a wall with water on the right-hand side as shown in Figure P12.54. Determine the force on the gate and its location.

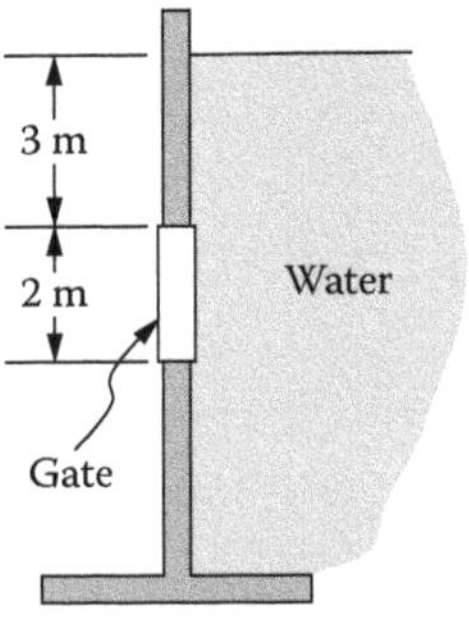

FIGURE P12.54

12.55: Repeat Problem 12.54 using an isosceles triangular gate with a base length of 2 m and with the vertex facing downward.

Forces on Curved Areas

12.56: A dam structure has a rounded shape at its bottom as shown in Figure P12.56. The dam width is 50 m. Neglecting the effect of atmospheric pressure, determine the horizontal and vertical forces on the dam structure and determine their line of action.

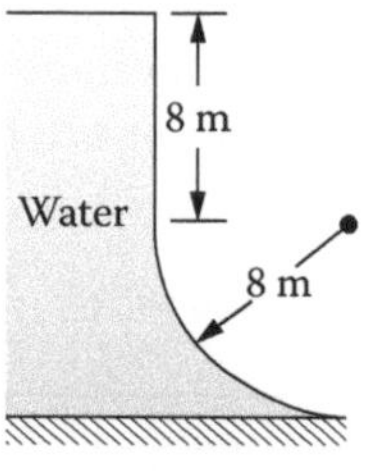

FIGURE P12.56

12.57: Neglecting atmospheric pressure and assuming the width of the curved surface AB in Figure P12.57 to be 4 m, determine the total hydrostatic force exerted on the surface, AB.

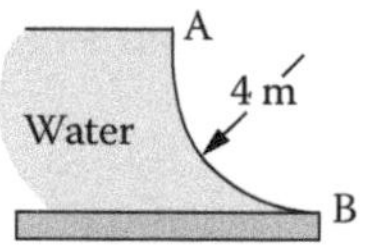

FIGURE P12.57

12.58: A quarter circle gate, AB, of radius 0.70 m, is hinged at B as indicated in Figure P12.58. The concave surface of the gate is wetted by a fluid of density, 920 kg/m^3. Determine the force necessary to hold the gate stationary.

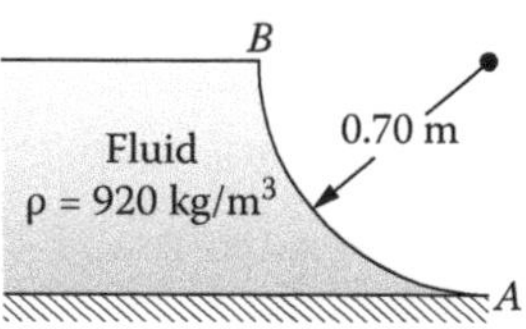

FIGURE P12.58

12.59: A 4-m-wide gate in the form of a quarter circle radius 3 m holds back water to a level of 1.8 m as shown in Figure P12.59. Determine the magnitude and direction of the resultant hydrostatic force.

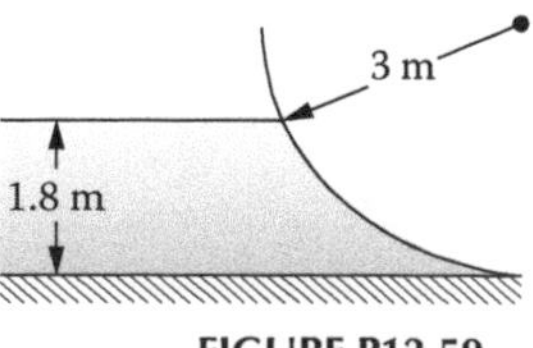

FIGURE P12.59

12.60: The flow from a water reservoir is controlled by a sluice gate 3-m wide. The gate is in the form of a sector of a circle with a radius of 1.2 m and an angle of 18° as shown in Figure P12.60. Determine the magnitude and direction of the resultant force on the gate.

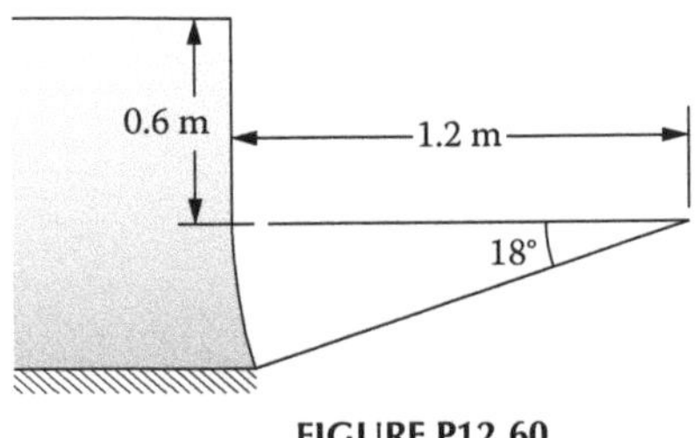

FIGURE P12.60

12.61: As shown in Figure P12.61, a cylinder having a radius of 1.4 m holds back, on its left half, a fluid with a specific gravity of 0.84. The cylinder has a length of 2.5 m and is made of a material with a density of 145 kg/m^3. Determine the forces, F_1 and F_2, that must be applied to hold the cylinder stationary.

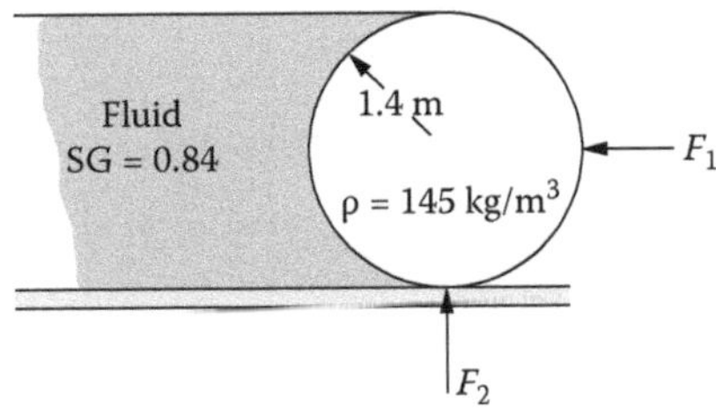

FIGURE P12.61

12.62: As indicated in Figure P12.62, in the flat side of a tank holding water, there is a hemispherical surface of diameter, d, whose centerline is located h below the water surface. Derive expressions for the horizontal force, F_h, and the upward force, F_u, exerted on the hemispherical surface by the surrounding water.

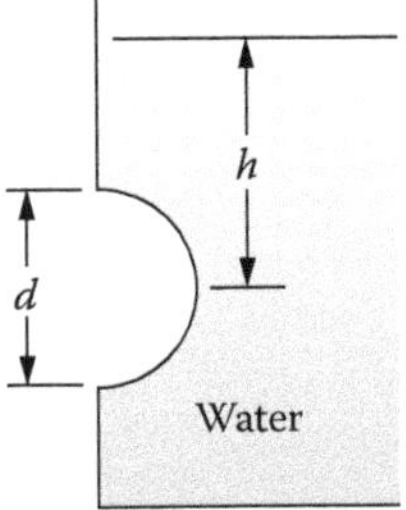

FIGURE P12.62

12.63: In Figure P12.63, compute the magnitude of the horizontal and vertical components of hydrostatic force acting on the quarter circle of radius 0.75 m and width 1.5 m. The density of the oil is 850 kg/m^3.

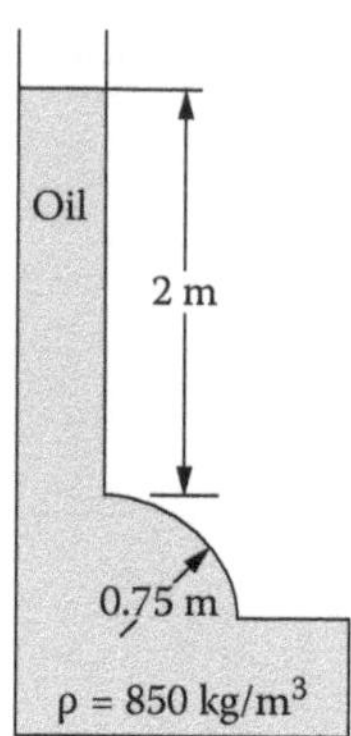

FIGURE P12.63

Buoyancy

12.64: The open end of a cylindrical tank that is 1 m in diameter and 1.5-m high is submerged in water as shown in Figure P12.64. The local barometric pressure is 101.3 kPa. The tank weighs 750 N and the thickness of the tank walls may be neglected. To what depth, h, will the tank submerge?

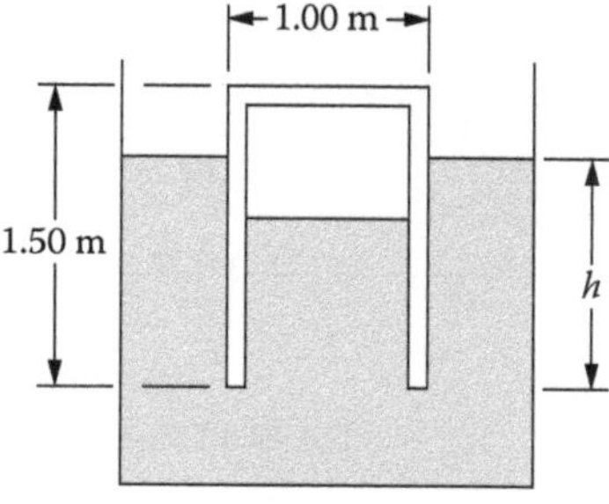

FIGURE P12.64

12.65: In Problem 12.64, determine (a) the depth at which the net force on the tank is zero and (b) the additional force required to bring the top of the tank flush with the water surface.

12.66: A 12 m diameter balloon (Figure P12.66) is filled with helium (molecular weight 4.00) and is pressurized to 105 kPa. The surrounding air is at 100 kPa and 20°C. Determine the tension in the mooring line.

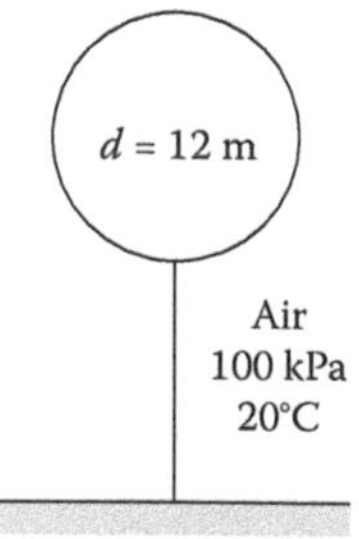

FIGURE P12.66

12.67: The **hydrometer** shown in Figure P12.67 floats at a level that is a measure of the specific gravity of the liquid. The stem is of constant diameter, d, and a weight in the bottom bulb stabilizes the body. The total hydrometer weight is 0.375 N and the stem diameter is 0.08 m. Determine the height at which the device floats when the liquid has a specific gravity of 1.20.

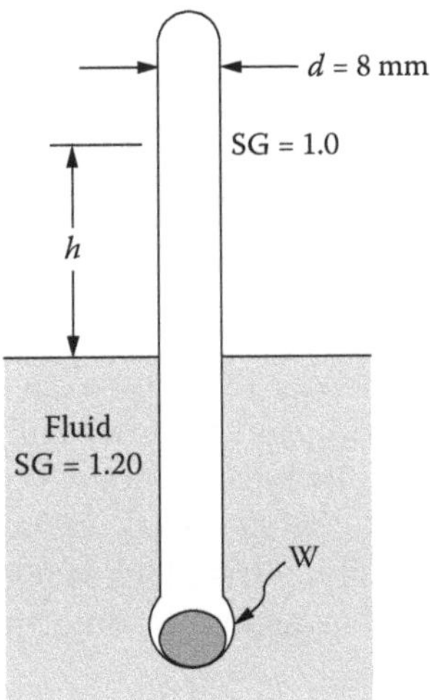

FIGURE P12.67

12.68: A barge weighs 360,000 N empty and is 8-m wide × 16-m long × 2.25-m high. Determine the depth below the waterline when loaded with 1.335×10^6 N of gravel floating in seawater (SG = 1.025).

12.69: A cube of side, L, and density, ρ, floats in water with density, ρ_w, as shown in Figure P12.69. Show that the depth of submergence, h, is given by

$$h = \frac{L\rho}{\rho_w}$$

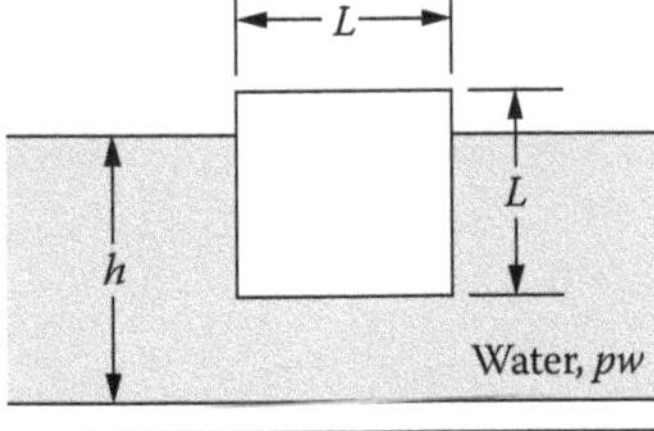

FIGURE P12.69

12.70: The float in a toilet tank is a sphere of radius R and is made of a material with density, ρ. An upward buoyant force, F, is required to shut the ball cock valve. The density of water is designated ρ_w. Develop an expression for the fraction of the float submerged, x, in terms of R, ρ, F, g, and ρ_w.

12.71: An elliptical cylinder has a major axis of 10 cm, a minor axis of 4 cm, and a length of 50 cm. Its density of $\rho = 545\,\text{kg/m}^3$. The cylinder is kept submerged in a water container with the help of a string attached to the bottom of the container as shown in Figure P12.71. Assuming the density of water to be $1000\,\text{kg/m}^3$, determine the tension, T, in the string.

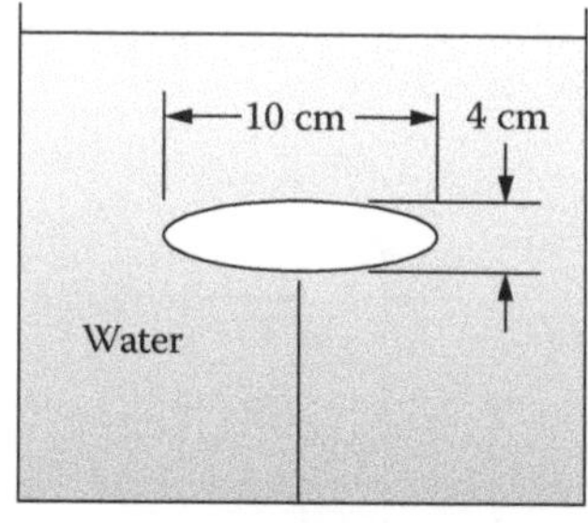

FIGURE P12.71

12.72: A rectangular block of dimensions 12 cm by 8 cm is 80-cm long. The block is held submerged in water with a string and a horizontal pivoted rod 4 cm long, as shown in Figure P12.72. The mass in the block is not evenly distributed and consequently its center of mass is 2 cm from its geometric center. Assuming that the density of water is 1000 kg/m^3, the density of the block is 500 kg/m^3, $g = 9.81$ m/s^2 and that the mass of the horizontal rod is negligible, determine the tension, T, in the string.

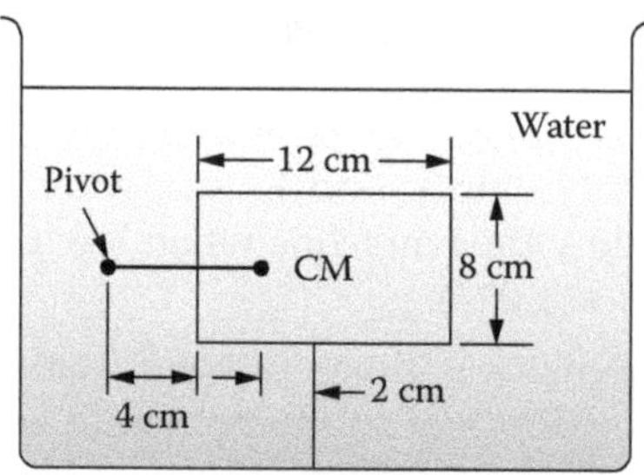

FIGURE P12.72

12.73: As shown in Figure P12.73, a rectangular block with sides, a and b, length, L, and density, ρ_3, floats at the interface between fluids of density, ρ_1 and ρ_2. Show that the ratio, b_1/b_2 is given by

$$\frac{b_1}{b_2} = \frac{\rho_3 - \rho_2}{\rho_1 - \rho_3}$$

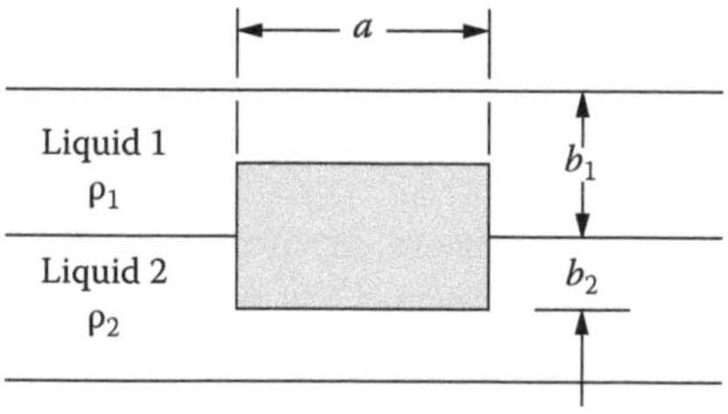

FIGURE P12.73

12.74: Figure P12.74 shows a cylinder of 25-cm diameter and 75-cm height, resting on the floor of a container filled with a liquid of density, $\rho = 800$ kg/m^3. The density of the cylinder is 250 m^3. Determine the magnitude of the force, F.

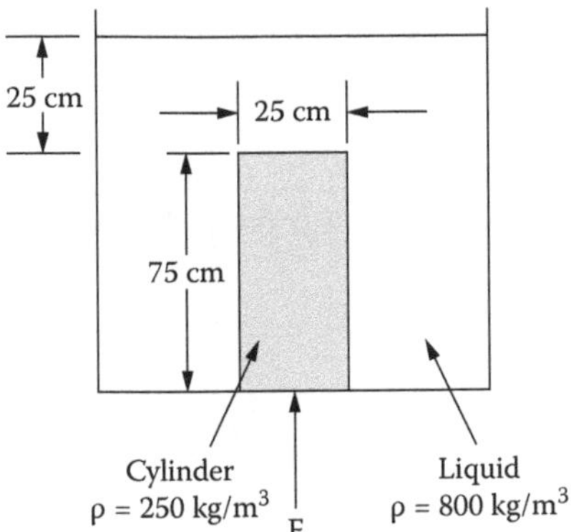

FIGURE P12.74

12.75: Rework Problem 12.74 with the cylinder replaced by a right circular cone with a base diameter of 25 cm and a height of 75 cm.

12.76: A spherical balloon that is 8 m in diameter is filled with helium ($\rho = 0.15\,\text{kg/m}^3$). A cage having a mass of 15 kg capable of carrying a load of 160 kg is attached to the bottom of the balloon. The balloon is released in the air ($\rho = 1.2\,\text{kg/m}^3$) with its full load. Determine the net upward force exerted on the balloon.

12.77: The truncated cone shown in Figure P12.77 is completely immersed in a fluid of density, $\rho = 760\,\text{kg/m}^3$. Determine the density of the truncated cone if the cone is balanced by a pulley-string arrangement that carries a mass of 5 kg.

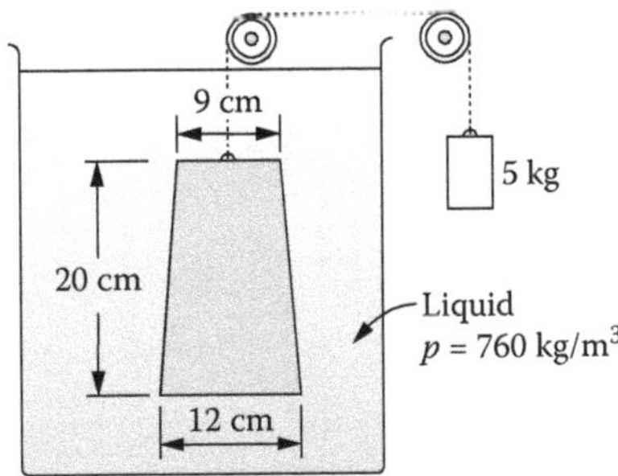

FIGURE P12.77

Stability

12.78: A cubical piece of wood with specific gravity, 0.90, and edge length, L, floats in water with the right-hand edge of the block flush with the water as shown in Figure P12.78. What moment, M, is required to hold the cube in the position shown?

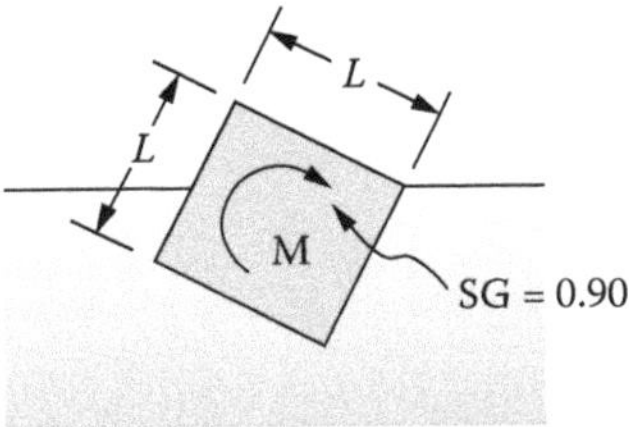

FIGURE P12.78

12.79: A cube measuring 50 cm on a side and made of a material with a density of 400 kg/m^3 floats partially submerged in saltwater ($\rho = 1050$ kg/m^3). Determine the metacentric height.

12.80: A cylinder of radius, R, and length, L, floats in water as shown in Figure P12.80. Show that when the cylinder is in neutral equilibrium, the metacentric height is zero.

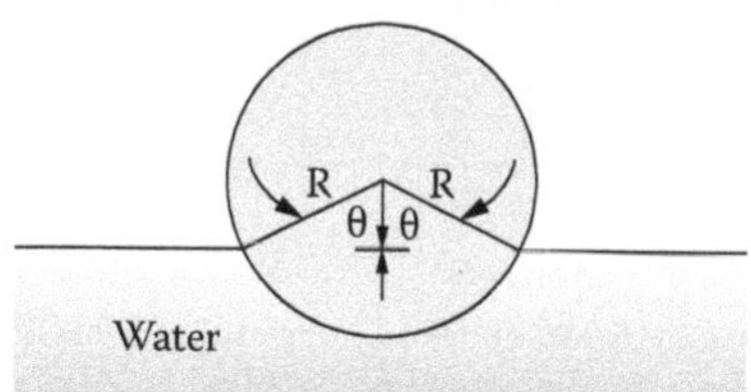

FIGURE P12.80

12.81: Figure P12.81 shows a cylinder of radius, R, and height, H, floating in water. The specific gravity of the cylinder is S. Show that the metacentric height, MG, is given by

$$MG = \frac{1}{2}\left[\frac{R^2}{2HS} - H(1 - S)\right]$$

and that for stable equilibrium

$$\frac{R}{H} > [2S(1 - S)]^{1/2}$$

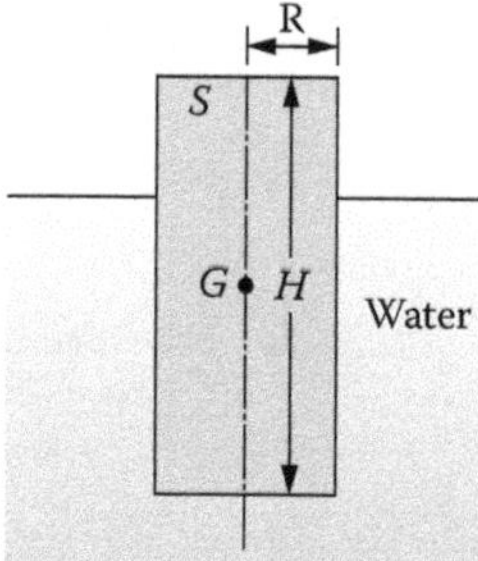

FIGURE P12.81

12.82: A rectangular barge is 16-m long, 6-m wide, and 2-m high. The barge has a mass of 80,640 kg when empty. Given that the density of seawater is 1050 kg/m^3, determine H, from the waterline to the bottom of the barge when it is floating.

13

Control Volume Analysis—Mass and Energy Conservation

Chapter Objectives

- To apply the law of mass conservation to fluid mechanics.
- To apply the first law of thermodynamics to fluid mechanics.
- To provide applications of fluid mechanics to the control volume expression for the first law of thermodynamics.
- To introduce the Bernoulli equation.

13.1 Introduction

Fluid motion is ubiquitous in the world around us. There are countless situations in which the understanding and harnessing of fluid motion provide the means for improving our lives. The heating and cooling of our homes and workplaces, the operation of engines for power and transportation and the transport of working fluids through pipelines and channels are a few examples. Our task in this and the next chapter is to develop a set of concepts that will allow a systematic examination of fluids in motion and their related effects. This is a logical progression beyond the fluid statics considerations examined in the previous chapter.

13.2 Fundamental Laws

All analyses of fluids in motion involve applications of three fundamental physical laws that are

- The law of mass conservation (**continuity**)
- The first law of thermodynamics (**energy**)
- Newtons's second law of motion (**momentum**)

Continuity, energy, and **momentum** are terms that are often used when referring to a specific law. For example, the first law of thermodynamics is often referred to as the energy equation. We will use these designations interchangeably throughout this text; such usage is commonplace in the fluid mechanics community.

We will deal exclusively with incompressible flows in this text. The analysis of compressible flows, which is beyond our scope, requires the application of the second law of thermodynamics in addition to the three laws just cited.

There are also a number of other useful relationships that are often referred to as "laws." Such relationships are more properly designated as **constitutive relationships** because they generally define a fluid property. One of these was introduced in Chapter 11: Newton's viscosity relationship, which some treatises call Newton's "law" of viscosity. In this text, we will reserve the term "law" for one of the three fundamental laws.

13.3 Conservation of Mass

The mass balance for a control volume was developed in Chapter 5 and expressed in operational form as Equation 5.1, which is restated here for reference

$$\frac{m_{cv}}{dt} = \sum_i \dot{m}_i - \sum_e \dot{m}_e \tag{13.1}$$

In words, Equation 13.1 states that the rate of mass accumulation in a control volume is equal to the total rate of mass inflow entering, (subscript i) minus the total rate of mass outflow exiting (subscript e).

13.4 Mass Conservation Applications

Equation 13.1 is completely general. It will be instructive to now consider some specific situations in which this expression may be applied. Two important cases will be considered initially: **incompressible flow** and **steady flow**.

Incompressible flow, as explained earlier, is flow in which the changes in density are so negligibly small that the density may be treated as constant. Keep in mind that we may often treat the flow of a compressible fluid, such as air, as incompressible.

For the case of incompressible flow, Equation 13.1 takes the form

$$\rho \frac{dV}{dt} = \sum_i \rho_i A_i \hat{V}_i - \sum_e \rho_e A_e \hat{V}_e$$

or

$$\frac{dV}{dt} = \sum_i A_i \hat{V}_i - \sum_e A_e \hat{V}_e \tag{13.2}$$

where we have treated the density as constant.

In steady flow, there is no variation of conditions with time. Operationally, this means that time derivatives vanish. Thus, Equation 13.1 reduces to

$$\sum_i \dot{m}_i - \sum_e \dot{m}_e = 0 \tag{13.3}$$

and for the situation where the flow is both steady and incompressible

$$\sum_i A_i \hat{V}_i - \sum_e A_e \hat{V}_e = 0 \tag{13.4}$$

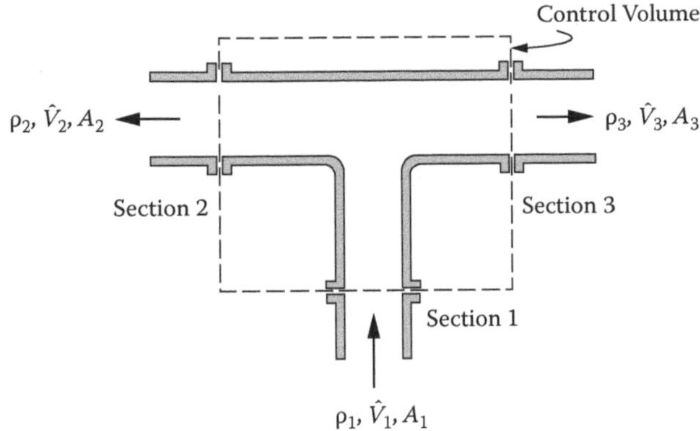

FIGURE 13.1
Flow through a T section in Example 13.1.

We must also keep in mind that Equations 13.3 and 13.4 do not mean that there is no flow but merely that the rates of inflow and outflow are the same: that is, there is zero accumulation.

We now consider three illustrative examples.

Example 13.1 Consider the T section shown in Figure 13.1. Flow enters at section 1 and is then split into two streams, which leave at sections 2 and 3. Express the mass balance for this T section under steady flow conditions.

Solution

Assumptions and Specifications

(1) The control volume is enclosed by dashed lines in Figure 13.1.
(2) Steady-state conditions exist.
(3) The flow is incompressible.
(4) Uniform properties exist at each cross section where fluid enters or leaves the control volume.

Because this is a condition of steady flow, the applicable mass conservation expression for a control volume is Equation 13.3

$$\sum_i \dot{m}_i - \sum_e \dot{m}_e = 0 \tag{13.3}$$

We will apply this equation to the control volume shown in Figure 13.1 (dashed lines). Note that mass crosses the control surface at only three places, at sections 1, 2, and 3. Evaluation of the flow terms provides

$$\sum_i \dot{m}_i = \rho_1 \hat{V}_1 A_1$$

$$\sum_e \dot{m}_e = \rho_2 \hat{V}_2 A_2 + \rho_3 \hat{V}_3 A_3$$

and, in accordance with Equation 13.3, we have

$$\rho_1 \hat{V}_1 A_1 = \rho_2 \hat{V}_2 A_2 + \rho_3 \hat{V}_3 A_3 \tag{13.5}$$

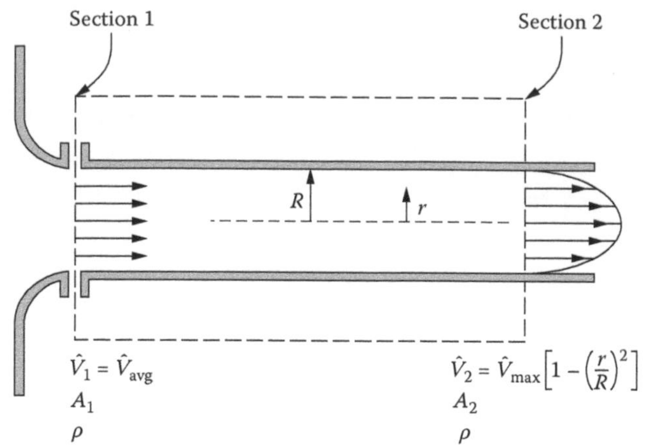

FIGURE 13.2
Steady incompressible flow in a circular pipe in Example 13.2.

This result states that the mass flow in at section 1 in the control volume is equal to the sum of mass flows out at sections 2 and 3.

If, in addition, the flow were incompressible, we would have $\rho_1 = \rho_2 = \rho_3$ and our result would have the form

$$\hat{V}_1 A_1 = \hat{V}_2 A_2 + \hat{V}_3 A_3 \tag{13.6}$$

which is in terms of the volume flow rates in and out.

In Example 13.1, we have used average values for the densities at the three sections. This is certainly allowable if these values are known. Example 13.2 will relate local and average values for a specific case.

Example 13.2 Water enters a straight section of pipe from a reservoir as shown in Figure 13.2. At some distance downstream from the entrance, the velocity profile is known to have a parabolic form expressed as

$$\hat{V}_{av} = \hat{V}_{max}\left[1 - \left(\frac{r}{R}\right)^2\right] \tag{13.7}$$

where $\hat{V}_{max}$ is the value of velocity at the centerline of the pipe (at $r = 0$). From the no-slip condition, the velocity is zero at the pipe wall where $r = R$. For the case of incompressible steady flow, we are to evaluate the value for $\hat{V}_{av}$ or the average velocity in terms of $\hat{V}_{max}$.

Solution

Assumptions and Specifications

(1) Steady-state conditions exist.

(2) The water is incompressible.

(3) Uniform properties exist where the fluid enters the pipe.

For steady incompressible flow, the applicable form of the law of conservation of mass is Equation 13.4. Our control volume is enclosed by the dashed line in Figure 13.2. For this case, we can evaluate Equation 13.4 in terms of the volumetric flow rate, $\dot{V} = A\hat{V}$

$$\sum_i \dot{V}_i - \sum_e \dot{V}_e = 0$$

At section 1, with $\hat{V}_1 = \hat{V}_{avg}$, we have

$$\sum_i \dot{V}_i = \dot{V}_{avg} = A_1 \hat{V}_{av} = \pi R^2 \hat{V}_{av}$$

and at section 2 where we have the parabolic velocity distribution

$$\sum_e \dot{V}_e = \dot{V}_2 = \int_0^R \hat{V}_{max}\left[1 - \left(\frac{r}{R}\right)^2\right] 2\pi r\, dr$$

or

$$\begin{aligned} \dot{V}_2 &= 2\pi \hat{V}_{max} \int_0^R \left(r - \frac{r^3}{R^2}\right) dr \\ &= 2\pi \hat{V}_{max}\left[\frac{r^2}{2} - \frac{r^4}{4R^2}\right]_0^R \\ &= \pi \hat{V}_{max}\frac{R^2}{2} \end{aligned}$$

When we put these expressions for the volumetric flow rate at sections 1 and 2 together, we achieve the result

$$\hat{V}_1 = \hat{V}_{avg} = \frac{\hat{V}_{max}}{2} \tag{13.8}$$

Flow measurements are made by a variety of devices. The basic ideas of these devices will be presented in a later chapter. One of these, the **pitot tube**, measures the velocity at a point. An approximation for the velocity profile across a pipe or duct can then be made and the total flow evaluated as illustrated in Example 13.3.

Example 13.3 A pitot tube that traverses a circular duct having an inside diameter of 50 cm yields the following velocity values:

Distance from Center, cm	Velocity, m/s	Distance from Center, cm	Velocity, m/s
0	2.29	19.7	1.67
8.3	2.16	21.3	1.55
11.3	2.06	22.7	1.37
13.9	1.96	24.1	1.16
16.1	1.87	25.0	0.73
18.0	1.77		

From these data find (a) the flow rate in m^3/s and (b) the average velocity.

Solution

Assumptions and Specifications

(1) Steady-state conditions exist.
(2) The flow is incompressible.
(3) Uniform properties exist at the cross section of interest.

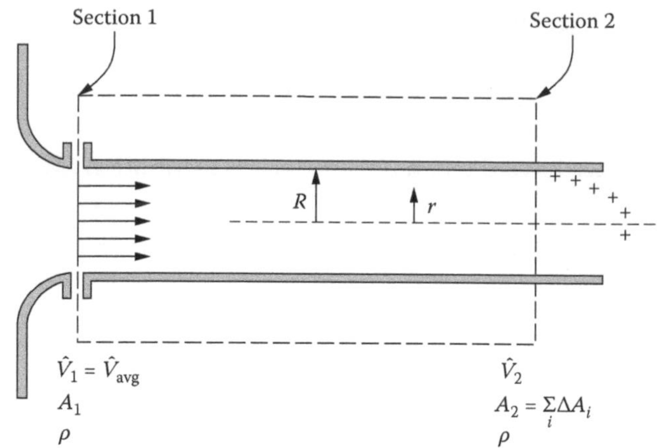

FIGURE 13.3
Steady incompressible flow in a circular duct in Example 13.2.

The information provided is in discrete form rather than in continuous form as in an equation. Nevertheless, the governing relationship remains the conservation of mass, Equation 13.1. For the case of steady, incompressible flow, Equation 13.1 reduces to

$$\sum_i \dot{m}_i - \sum_e \dot{m}_e = 0 \tag{13.3}$$

This situation is indicated in Figure 13.3 with the velocity at section 2 represented as discrete points rather than a smooth curve as shown in Figure 13.2. The solution to our form of the mass conservation expression can be written as

$$\sum_i \dot{V}_i - \sum_e \dot{V}_e = 0$$

which reduces further to

$$\sum_i \hat{V}_i \Delta A_i - \hat{V}_{avg} A = 0$$

or

$$\hat{V}_{avg} A = \sum_i \hat{V}_i \Delta A_i \tag{13.9}$$

and finally

$$\hat{V}_{avg} = \sum_i \hat{V}_i \frac{\Delta A_i}{A} \tag{13.10}$$

In Equation 13.9, the product $\hat{V}_i \Delta A_i$ is a volumetric flow rate through a portion of the cross section, designated as A_i, where $\hat{V}_i$ is the average velocity over the section, ΔA_i.

Referring now to the data provided in the problem statement, we may consider each of the velocity values to be $\hat{V}_i$, applying over area ΔA_i as indicated in Figure 13.4.

For example, the velocity, $\hat{V}_i = 1.87\,\text{m/s}$, is associated with a segment of the total cross section lying between

$$r_1 = \frac{13.9\,\text{cm} + 16.1\,\text{cm}}{2} = 15.00\,\text{cm} \quad \text{and} \quad r_2 = \frac{16.1\,\text{cm} + 18.0\,\text{cm}}{2} = 17.05\,\text{cm}$$

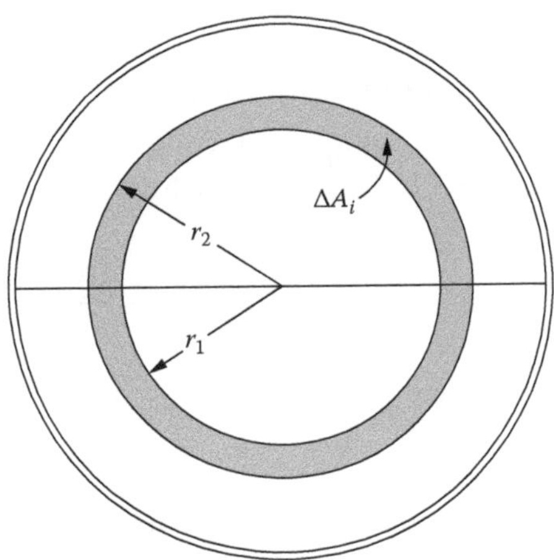

FIGURE 13.4
Cross section of duct in Example 13.3.

The corresponding value for ΔA_i is therefore,

$$\Delta A_i = \pi[(17.05\,\text{cm})^2 - (15.00\,\text{cm})^2]$$
$$= 206.4\,\text{cm}^2 \quad \text{or} \quad 0.0206\,\text{m}^2$$

In a similar fashion, each $\hat{V}_i$ can be related to ΔA_i. A summary of this procedure is shown in the following table.

$\hat{V}_i$, m/s	ΔA_i, m²	$\hat{V}_i \Delta A_i$, m³/s
2.29	0.0050	0.0115
2.16	0.0242	0.0523
2.06	0.0206	0.0424
1.96	0.0208	0.0408
1.87	0.0206	0.0385
1.77	0.0203	0.0359
1.67	0.0204	0.0341
1.55	0.0200	0.0310
1.37	0.0200	0.0274
1.16	0.0173	0.0201
0.73	0.0070	0.0051
	$\sum \Delta A_i = 0.1962\,\text{m}^2$	$\sum \hat{V}_i \Delta A_i = 0.3391\,\text{m}^3/\text{s}$

The total flow rate is seen to be

$$\dot{V} = 0.3391,\text{m}^3/\text{s} \Longleftarrow$$

and the average velocity is given by

$$\hat{V}_{\text{avg}} = \frac{\sum_i \hat{V}_i A_i}{\sum_i \Delta A_i} = \frac{0.3391\,\text{m}^3/\text{s}}{0.0192\,\text{m}^2} = 1.73\,\text{m/s} \Longleftarrow$$

13.5 The First Law of Thermodynamics for a Control Volume

The governing control volume relationship for the first law of thermodynamics was developed in Section 5.4.4 as Equation 5.11. This relationship may be rewritten here as

$$\frac{dE_{cv}}{dt} = \dot{Q}_{cv} - \dot{W}_{cv} + \sum_i \dot{m}_i \left(h_i + \frac{\hat{V}_i^2}{2} + gz_i \right) - \sum_e \dot{m}_e \left(h_e + \frac{\hat{V}_e^2}{2} + gz_e \right) \tag{13.11}$$

Equation 13.11 will be the starting point for all subsequent analyses in this chapter.

The reader is reminded that

- dE_{cv}/dt is the rate of increase of the total energy within the control volume,
- $\dot{Q}_{cv}$ is the net rate of heat transfer into the control volume from the surroundings,
- $\dot{W}_{cv}$ is the net rate of work done by the control volume on its surroundings,
- $\dot{m}_i$ is the mass flow rate into the control volume by means of inlet stream, i,
- $h_i = u_i + P_i/\rho_i$ is the enthalpy of the inlet stream, i,
- $\hat{V}_i^2/2$ is the kinetic energy of the inlet stream, i, and
- gz_i is the potential energy of the inlet stream, i.

Each of the terms written with the subscript, i, has a counterpart with the same meaning for the exiting streams bearing the subscript, e.

To review the specific energy terms, observe that in Equation 13.11, the energy per unit mass of a quantity of fluid flowing into or out of a control volume may be written in terms of its component parts as

$$e + \frac{P}{\rho} = gz + \frac{\hat{V}^2}{2} + u + \frac{P}{\rho} \tag{13.12}$$

or

$$e + \frac{P}{\rho} = gz + \frac{\hat{V}^2}{2} + h \tag{13.13}$$

where gz is the specific potential energy, $\hat{V}^2/2$ is the specific kinetic energy, $h = u + P/\rho$ is the specific enthalpy, and u is the specific internal energy. The reader is cautioned at this point to observe that P/ρ and the enthalpy, h, represent valid energy terms only at a flow boundary and that the accumulation term does not include these components.

Equation 13.11 now assumes its place as one of the principle control volume equations that will form a basis for fluid flow analysis. All quantitative descriptions of fluid flow to be considered in this text are based on this expression, along with Equation 13.1 and the yet-to-be-developed expressions for linear momentum and moment of momentum.

The use of Equation 13.11 and the interpretation of the various terms that it contains will be clarified in the next section where a number of practical examples will be considered in detail.

13.6 Applications of the Control Volume Expression for the First Law

A series of examples is presented in this section to provide the reader with some insights concerning the energy equation.

Example 13.4 For the nozzle shown in Figure 13.5, we wish to know the magnitude of the change in internal energy of the flow. This can be thought of as the energy loss to friction that is rendered unavailable to do mechanical work. The upstream pressure in the pipe is 530 kPa (absolute).

Solution

Assumptions and Specifications

(1) Steady-state conditions exist.
(2) The flow is incompressible.
(3) Uniform properties exist at the cross section at the inlet and outlet of the nozzle.
(4) Atmospheric pressure may be taken as 101 kPa.

Applying Equation 13.11 to the control volume indicated in Figure 13.5, we first note that $dE_{cv}/dt = 0$ so that

$$\dot{Q}_{cv} = \dot{W}_{cv} = E_e - E_i$$

and with both $\dot{Q}_{cv} = \dot{W}_{cv} = 0$,

$$\sum_i \dot{m}_i \left(h_i + \frac{\hat{V}_i^2}{2} + gz_i\right) - \sum_e \dot{m}_e \left(h_e + \frac{\hat{V}_e^2}{2} + gz_e\right)$$

$$= \dot{m}_1 \left(u_1 + \frac{P_1}{\rho} + \frac{\hat{V}_1^2}{2} + gz_1\right) - \dot{m}_2 \left(u_2 + \frac{P_2}{\rho} + \frac{\hat{V}_2^2}{2} + gz_2\right) = 0$$

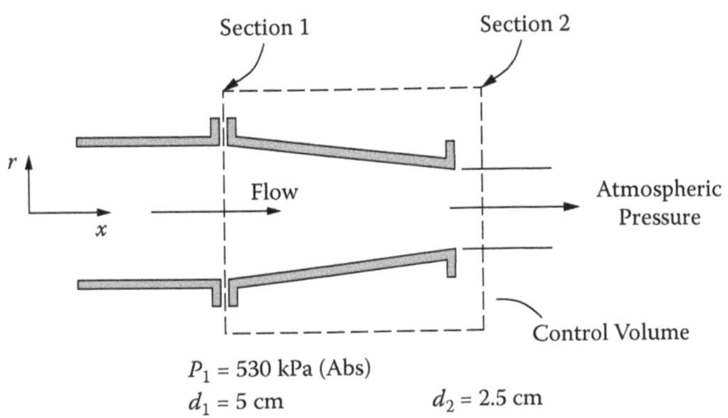

FIGURE 13.5
Water flowing through a horizontal nozzle in Example 13.4.

Because mass conservation dictates that $\dot{m}_1 = \dot{m}_2$, our expression becomes

$$u_1 - u_2 + \frac{P_1 - P_2}{\rho} + \frac{\hat{V}_1^2 - \hat{V}_2^2}{2} + g(z_1 - z_2) = 0 \tag{13.14}$$

The quantity sought is $u_2 - u_1$, which can be expressed as

$$u_2 - u_1 = \frac{P_1 - P_2}{\rho} + \frac{\hat{V}_1^2 - \hat{V}_2^2}{2} + g(z_1 - z_2)$$

The terms on the right-hand side of this expression are evaluated as follows:

$$\frac{P_1 - P_2}{\rho} = \frac{530\,\text{kPa} - 101\,\text{kPa}}{1000\,\text{kg/m}^3} = 429\,\text{m}^2/\text{s}^2$$

$$\begin{aligned}\frac{\hat{V}_1^2 - \hat{V}_2^2}{2} &= \frac{\dot{V}^2}{2}\left(\frac{1}{A_1^2} - \frac{1}{A_2^2}\right)\\ &= \frac{(53\,\text{m}^3/\text{h})^2}{2(3600\,\text{s/h})^2}\left\{\left[\frac{4}{\pi(0.050\,\text{m})^2}\right]^2 - \left[\frac{4}{\pi(0.025\,\text{m})^2}\right]^2\right\}\\ &= -422\,\text{m}^2/\text{s}^2\end{aligned}$$

and

$$g(z_1 - z_2) = 0$$

The resulting value of $u_2 - u_1$ thus becomes

$$\begin{aligned}u_2 - u_1 &= \frac{P_1 - P_2}{\rho} + \frac{\hat{V}_1^2 - \hat{V}_2^2}{2} + g(z_1 - z_2)\\ &= 429\,\text{m}^2/\text{s}^2 - 422\,\text{m}^2/\text{s}^2\\ &= 7\,\text{m}^2/\text{s}^2 \Longleftarrow\end{aligned}$$

This result indicates that $7\,\text{m}^2/\text{s}^2$ of energy is lost due to frictional effects and other dissipative processes in the nozzle. It is reassuring that the answer is positive. A negative result would have suggested that dissipative processes are **increasing** the useful energy potential of the leaving stream that is clearly not physically possible.

The units of this result may be confusing at first glance. The terms in Equation 13.14 are all appropriate forms of energy per unit mass although they may not seem so. The units for $g\Delta z$, the potential energy change and for $\Delta\hat{V}^2/2$, the kinetic energy change, are in m^2/s^2. The internal energy terms and the flow energy change are in N-m/kg. The conversion is straightforward in this example

$$7\,\text{m}^2/\text{s}^2\left(\frac{\text{s}^2\text{-N}}{1\,\text{kg-m}}\right) = 7\,\text{N-m/kg}$$

This conversion is relatively simple in SI units where the numerical value of g_c is unity, The reader is hereby alerted to the situation where unit conversions are likely to be a major source of effort and potential error in using the energy equation. One should be careful in the process of obtaining a numerical result in a specific set of units.

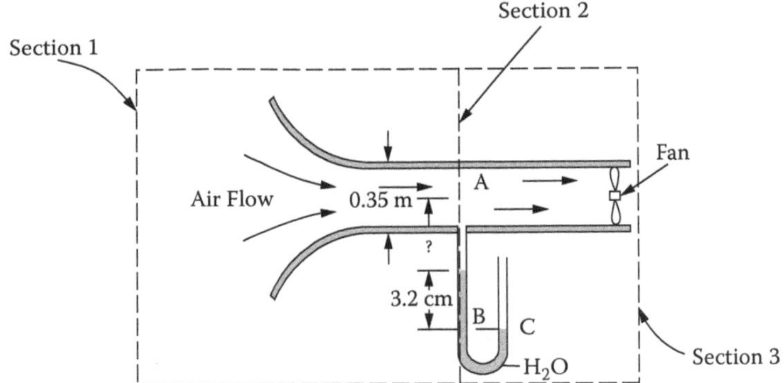

FIGURE 13.6
Induced-draft fan and duct assembly in Example 13.5.

Example 13.5 provides another illustration of the use of the control volume expression of the first law of thermodynamics.

Example 13.5 Consider the fan and duct system shown in Figure 13.6. The duct has a diameter of 0.35 m and its entrance is smoothly rounded to minimize losses. Air is drawn through the duct by the induced-draft fan and the indicated differential manometer shows a vacuum pressure of 3.2 cm of water. The density of air may be taken as $1.22\,\text{kg/m}^3$. Determine (a) the volumetric flow rate of the air (m^3/s) and (b) the power required to drive the fan (kW).

Solution

Assumptions and Specifications

(1) Steady-state conditions exist.
(2) The liquid in the manometer is incompressible.
(3) Uniform properties exist at each cross section of interest.
(4) No change in internal energy will be assumed.

(a) We choose a control volume that will allow the direct determination of the volumetric flow rate. Because the power required to drive the fan is not known, we should *not* include the fan in our control volume. The dashed lines indicate an appropriate choice for the control volume. We know the conditions at both sections 1 and 2 and, because nothing has been given concerning Δu, the frictional losses, we will neglect frictional effects. The steady flow form of the energy equation applies and we have

$$g(z_2 - z_1) + \frac{\hat{V}_2^2 - \hat{V}_1^2}{2} + u_2 - u_1 + \frac{P_2 - P_1}{\rho} = 0$$

At section 1

$$P_1 = P_{\text{atm}} \quad \text{and} \quad \hat{V}_1 = 0$$

and at section 2

$$P_2 = P_{\text{atm}} - 3.2\,\text{cm}\,\text{H}_2\text{O} \quad \text{and} \quad \hat{V}_2 = \frac{\dot{V}}{A_2}$$

Then Equation 13.14

$$u_1 - u_2 + \frac{P_1 - P_2}{\rho} + \frac{\hat{V}_1^2 - \hat{V}_2^2}{2} + g(z_1 - z_2) = 0 \tag{13.14}$$

gives us with, $\Delta u = \Delta z$ and $\hat{V}_1 = 0$

$$\frac{\hat{V}^2}{2} = \frac{1}{2}\left(\frac{\dot{V}}{A_2}\right)^2 = \frac{P_1 - P_2}{\rho}$$

Then with 1 cm of water = 98.1 N/m^2 and with

$$A_2 = \frac{\pi}{4}(0.35\,\text{m})^2 = 0.0962\,\text{m}^2$$

we write

$$\begin{aligned}
\dot{V} &= A_2\left[\frac{2(P_1 - P_2)}{\rho}\right]^{1/2} \\
&= A_2\left[\frac{2[P_{\text{atm}} - (P_{\text{atm}} - 3.2\,\text{cm water})]}{1.22\ \text{kg/m}^3}\left(\frac{98.1\,\text{N/m}^2}{\text{cm water}}\right)\left(1\,\frac{1\,\text{kg-m}}{\text{N-s}^2}\right)\right]^{1/2} \\
&= A_2\left[\frac{2(3.2\,\text{cm water})}{1.22\ \text{kg/m}^3}\left(\frac{98.1\,\text{N/m}^2}{\text{cm water}}\right)\left(1\,\frac{1\,\text{kg-m}}{\text{N-s}^2}\right)\right]^{1/2} \\
&= 22.69\,A_2\,\text{m/s}
\end{aligned}$$

Thus,

$$\dot{V} = (22.69\,\text{m/s})A_2 = (22.69\,\text{m/s})(0.0962\,\text{m}^2) = 2.18\,\text{m}^3/\text{s} \Longleftarrow$$

(b) To establish the power to drive the fan, we need to define a control volume that includes it. The choice made, as indicated in Figure 13.6, is between sections 1 and 3. For the case of steady flow, we employ Equation 13.11 with $dE_{\text{cv}}/dt = 0$

$$\dot{Q}_{\text{cv}} - \dot{W}_{\text{cv}} + \sum_i \dot{m}_i\left(h_i + \frac{\hat{V}_i^2}{2} + gz_i\right) - \sum_e \dot{m}_e\left(h_e + \frac{\hat{V}_e^2}{2} + gz_e\right) = 0$$

The summation terms may be rearranged to the form

$$\begin{aligned}
&\sum_i \dot{m}_i\left(h_i + \frac{\hat{V}_i^2}{2} + gz_i\right) - \sum_e \dot{m}_e\left(h_e + \frac{\hat{V}_e^2}{2} + gz_e\right) \\
&\quad = \dot{m}\left[(u_1 - u_3) + \frac{P_1 - P_3}{\rho} + \frac{V_1^2 - V_3^2}{2} + g(z_1 - z_3)\right]
\end{aligned}$$

where $\dot{m} = \dot{m}_1 = \dot{m}_3$. Now, with

$$\dot{Q}_{cv} = 0$$

$$\hat{V}_1 = 0$$

$$u_1 - u_3 = 0$$

$$P_1 = P_3 = P_{atm}$$

$$z_1 - z_3 = 0$$

and $\dot{W}_{cv} = -\dot{W}_{fan}$, the expression to be solved is

$$\dot{W}_{fan} = -\dot{m}\left[\frac{1}{2}\left(\frac{\dot{V}}{A_3}\right)^2\right]$$

and with $A_3 = A_2$, this yields

$$\dot{W}_{fan} = -\frac{(1.22\,\text{kg/m}^3)(2.18\,\text{m}^3/\text{s})}{2}\left[\frac{2.18\,\text{m}^3/\text{s}}{(\pi/4)(0.35\,\text{m})^2}\right]^2$$

$$= -683\,\text{W} = -0.683\text{kW} \Longleftarrow$$

The negative sign indicates that the power is transmitted from the surroundings to the control volume. Because the fan is a part of the control volume, the power to drive the fan is the unknown quantity sought, and thus the negative sign makes sense.

13.7 The Bernoulli Equation

The energy equation, under certain conditions, reduces to a form that is referred to as the **Bernoulli equation**. This equation is sufficiently important and useful that we devote an entire section to its discussion.

The following conditions apply:

- Adiabatic flow (no heat transferred), $\dot{Q} = 0$
- No work done, $\dot{W} = 0$
- Steady flow
- Incompressible flow
- Negligible viscous effects (**inviscid** flow)

Application of the first four of the foregoing conditions yields the steady-flow form of the energy equation, Equation 13.14, which was derived in a Example 13.5. An added condition for this case is that of inviscid flow, which gives $\Delta u = 0$. The resulting form of the control volume expression for energy is the **Bernoulli equation**

$$gz + \frac{\hat{V}^2}{2} + \frac{P}{\rho} = C \quad \text{(a constant)} \tag{13.15}$$

The Bernoulli equation establishes that, for the five conditions specified, the total energy in a fluid is manifested as the sum of potential, kinetic, and flow energies. Any change in one of these forms requires appropriate changes in the others.

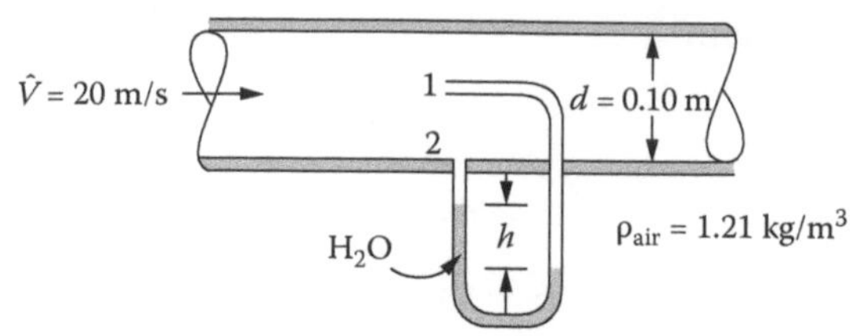

FIGURE 13.7
A pitot-static tube in an air duct.

As expressed by Equation 13.15, the units of each term are in m^2/s^2. If each term is divided by g, the result, given in meters is

$$z + \frac{\hat{V}^2}{2g} + \frac{P}{\rho g} = C \quad \text{(a constant)} \tag{13.16}$$

In this form, each energy term is designated as a "**head**" due to elevation (potential energy), velocity (kinetic energy), and pressure (flow energy).

Two examples involving the Bernoulli equation will now be considered.

A **pitot static** or pitot tube is employed in Example 13.6 to measure velocity. In Figure 13.7, the two pressure lines are situated differently relative to the flow. At point 1, because of the orientation of the tube opening with respect to the flow, the fluid is brought to rest. The pressure at point 1 is designated the **stagnation** pressure or **impact** pressure. At point 2, the orientation of the tube opening is such that it does not obstruct the flow, and the pressure sensed at point 2 is known as the **static** pressure.

Example 13.6 Air flows in a duct with pressure probes is attached to a manometer as shown in Figure 13.7. If the density of air is 1.21 kg/m^3 and that of water as 1000 kg/m^3 determine the elevation difference, h, between the two legs of the manometer.

Solution

Assumptions and Specifications

(1) Steady-state conditions exist.
(2) The flow is incompressible.
(3) Uniform properties exist at each cross section.

We select the centerline of the duct as the reference elevation and note in Figure 13.7 that the difference in elevation between points 1 and 2 is half of the duct diameter, or 0.05 m. An application of the Bernoulli equation at both points yields the relation

$$z_1 + \frac{\hat{V}_1^2}{2g} + \frac{P_1}{\rho g} = z_2 + \frac{\hat{V}_2^2}{2g} + \frac{P_2}{\rho g}$$

Noting that $\hat{V}_1 = 0$ (at the stagnation point) and that $z_2 - z_1 = -0.05$ m, we obtain

$$\begin{aligned}\frac{P_1 - P_2}{\rho g} &= z_2 - z_1 + \frac{\hat{V}_2^2}{2g}\\ &= -0.05\,\text{m} + \frac{(20\,\text{m/s})^2}{2(9.81\,\text{m/s}^2)}\\ &= 20.3\,\text{m}\end{aligned}$$

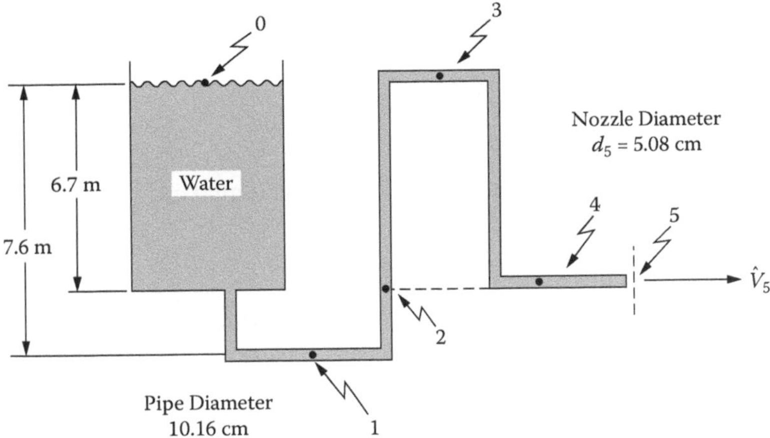

FIGURE 13.8
A water tank and discharge system in Example 13.7.

This is the pressure head that is required to bring the velocity to zero but is not the manometer displacement. The manometer will register the same pressure difference in units of meters of water. Hence,

$$\frac{\Delta P}{\rho} = 20.3\,\text{m of air}\left(\frac{1.21\,\text{kg/m}^3\text{ of air}}{1000\,\text{kg/m}^3\text{ of water}}\right)$$

$$= 0.0246\,\text{m of water} \quad \text{or} \quad 2.46\,\text{cm of water} \Longleftarrow$$

The use of pitot tubes to evaluate velocity is common and we will see them again. Usually, the velocity is evaluated by reading the displacement of fluid in the manometer. By positioning the impact pressure probe at a number of positions across the flow passage, the series of local velocities measured will provide a velocity profile of the flow. This was the case considered in Example 13.3.

Example 13.7 An open water tank and piping system is shown in Figure 13.8. The water discharges downstream through a 5.08-cm-diameter nozzle. Assuming that the liquid level in the tank remains constant and neglecting frictional losses, determine (a) the rate of water discharge through the nozzle and (b) the pressure and velocity at points 1, 2, 3, 4, and 5.

Solution

Assumptions and Specifications

(1) Steady-state conditions exist.
(2) The flow is incompressible.
(3) Uniform properties exist at each cross section considered.

For the conditions specified in the problem statement, the Bernoulli equation applies everywhere and can be employed to relate conditions between the various points.

(a) At the outset, consider the surface of the fluid in the tank at point 0, where $\hat{V}_0 = 0$ and the exit from the nozzle at point 5. Both locations are at atmospheric pressure.

Equation 13.16 gives

$$z_0 + \frac{P_{atm}}{\rho g} = z_5 + \frac{\hat{V}_5^2}{2g} + \frac{P_{atm}}{\rho g}$$

from which the exit velocity from the nozzle can be calculated as

$$\begin{aligned}\hat{V}_5 &= [2g(z_0 - z_5)]^{1/2} \\ &= [2(9.81\,\text{m/s}^2)(6.7\,\text{m})]^{1/2} \\ &= (131.45\,\text{m}^2/\text{s}^2)^{1/2} \\ &= 11.47\,\text{m/s} \Longleftarrow\end{aligned}$$

(b) Because $\hat{V}_5 = 11.47\,\text{m/s}$ is the velocity through the 5.08-cm nozzle, the velocity through the 10.16-cm-diameter pipe will be

$$\hat{V} = 11.47\,\text{m/s}\left(\frac{5.08\,\text{cm}}{10.16\,\text{m/s}}\right)^2 = 2.87\,\text{m/s}$$

This is the velocity at all other points of interest

$$\hat{V}_1 = \hat{V}_2 = \hat{V}_3 = \hat{V}_4 = 2.87\,\text{m/s}$$

Between points 0 and 1, we have

$$z_0 + \frac{\hat{V}_0^2}{2g} + \frac{P_{atm}}{\rho} = z_1 + \frac{\hat{V}_1^2}{2g} + \frac{P_1}{\rho}$$

which, with $\hat{V}_0 = 0$, gives

$$\begin{aligned}\frac{P_1 - P_{atm}}{\rho} &= z_0 - z_1 - \frac{\hat{V}_1^2}{2g} \\ &= 7.6\,\text{m} - \frac{(2.87\,\text{m/s})^2}{2(9.81\,\text{m/s})} \\ &= 7.18\,\text{m of water}\end{aligned}$$

This is the gage pressure at point 1.

We note, by inspection, that conditions at points 2 and 4 are identical because the velocities and elevations are the same at these points. The pressure at point 2 may be determined by applying the Bernoulli equation between points 0 and 2.

$$z_0 + \frac{\hat{V}_0^2}{2g} + \frac{P_{atm}}{\rho} = z_2 + \frac{\hat{V}_2^2}{2g} + \frac{P_2}{\rho}$$

which with $\hat{V}_0 = 0$ gives

$$\frac{P_2 - P_{atm}}{\rho} = z_0 - z_2 - \frac{\hat{V}_2^2}{2g}$$

$$= 6.70\,\text{m} - \frac{(2.87\,\text{m/s})^2}{2(9.81\,\text{m/s})}$$

$$= 6.28\,\text{m of water}$$

This is the gage pressure at points 2 and 4.

The final unknown is the pressure at point 3. We again apply the Bernoulli equation, this time between points 0 and 3, and obtain

$$z_0 + \frac{\hat{V}_0^2}{2g} + \frac{P_{atm}}{\rho} = z_3 + \frac{\hat{V}_3^2}{2g} + \frac{P_3}{\rho}$$

which, with $\hat{V}_0 = 0$, gives

$$\frac{P_3 - P_{atm}}{\rho} = z_0 - z_3 - \frac{\hat{V}_3^2}{2g}$$

$$= 0.00\,\text{m} - \frac{(2.87\,\text{m/s})^2}{2(9.81\,\text{m/s})}$$

$$= 0.00\,\text{m} - 0.42\,\text{m}$$

$$= -0.42\,\text{m of water}$$

This is the gage pressure at point 3.

At point 3, the pressure is 0.42 m of water **vacuum**, that is, the pressure is below atmospheric pressure. It is clear that, as the part of the piping system containing point 3 is elevated, the pressure at the point will continue to decrease. It is of interest to think about the maximum height that might be reasonable. Is it possible for the absolute pressure at point 3 to be negative?

An application of the Bernoulli equation to the design of a drinking fountain now follows.

13.8 Design Example 6

A public drinking fountain is to be designed. The fountain employs a nozzle to discharge drinking water vertically upward to a level that is 4 cm above the nozzle discharge. The fountain configuration is shown in Figure 13.9. In the absence of viscous effects, determine (a) the pressure at the gage indicated to produce this condition and (b) the flow rate of the water achieved.

Solution

Assumptions and Specifications

(1) Steady-state conditions exist.

(2) The fluid is incompressible.

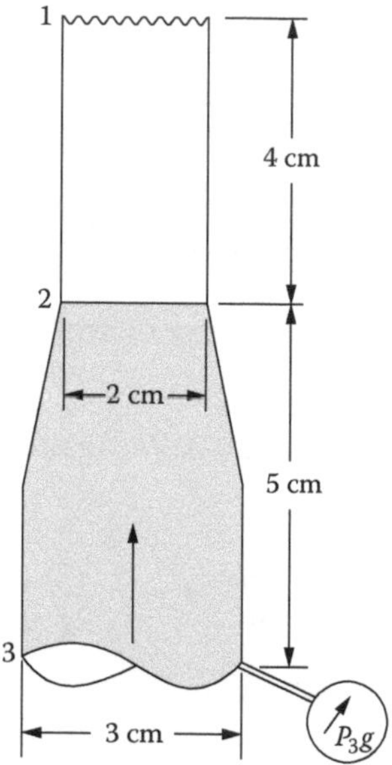

FIGURE 13.9
Artist's conception of a bubble drinking fountain.

(3) Uniform properties exist at each cross section.

(4) The acceleration of gravity may be taken as $9.81\,\text{m/s}^2$.

(5) The density of water may be taken as $1000\,\text{kg/m}^3$.

Because the flow is considered incompressible, the Bernoulli equation may be applied between locations of interest. We will break the configuration into two control volumes as indicated in Figure 13.10.

(a) In Figure 13.10, control volume A extends from the fountain discharge upward to the level where the vertical water jet velocity reaches zero. The Bernoulli equation between levels 1 and 2 is

$$\frac{P_1}{\rho_1} + gy_1 + \frac{\hat{V}_1^2}{2} = \frac{P_2}{\rho_2} + gy_2 + \frac{\hat{V}_2^2}{2}$$

Noting that $\hat{V}_1 = 0$ and that $P_1 = P_2 = P_{\text{atm}}$, this expression reduces to

$$g(y_1 - y_2) = \frac{\hat{V}_2^2}{2}$$

or

$$\begin{aligned} \hat{V}_2^2 &= 2g(y_1 - y_2) \\ &= 2(9.81\,\text{m/s}^2)(0.04\,\text{m}) = 0.785\,\text{m}^2/\text{s}^2 \\ \hat{V}_2 &= 0.886\,\text{m/s} \end{aligned}$$

For control volume B, extending between levels 2 and 3, we may write

$$\frac{P_2 - P_3}{\rho} + g(y_2 - y_3) + \frac{\hat{V}_2^2 - \hat{V}_3^2}{2} = 0$$

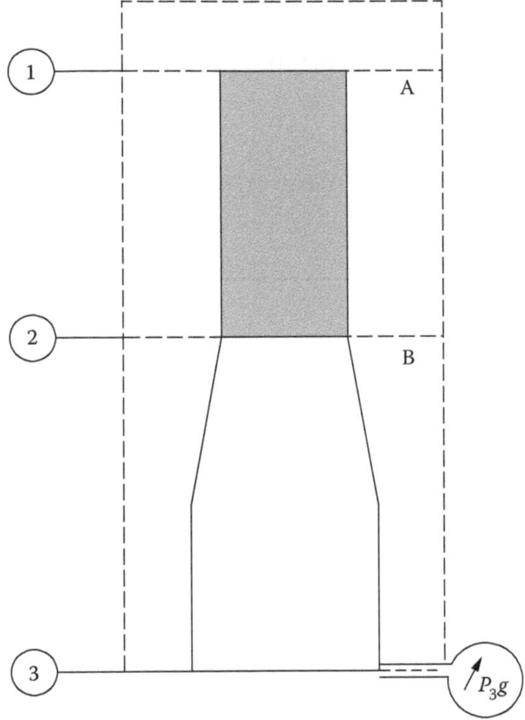

FIGURE 13.10
Control volumes used in the design of a bubble drinking fountain.

The quantity sought, P_3, may now be expressed as

$$\frac{P_3 - P_2}{\rho} = g(y_2 - y_3) + \frac{\hat{V}_2^2 - \hat{V}_3^2}{2}$$

We note that

$$P_3 - P_2 = P_3 - P_{atm} = P_{3g}$$

and that, by continuity,

$$\hat{V}_3 = \hat{V}_2\left(\frac{A_2}{A_3}\right) = \hat{V}_2\left(\frac{d_2}{d_3}\right)^2$$

Thus,

$$\begin{aligned}\frac{P_{3g}}{\rho} &= g(y_2 - y_3) + \frac{\hat{V}_2^2}{2}\left[1 - \left(\frac{d_2}{d_3}\right)^4\right] \\ &= (9.81\,\text{m/s}^2)(0.05\,\text{m}) + \frac{(0.886\,\text{m})^2}{2}\left[1 - \left(\frac{0.02\,\text{m}}{0.03\,\text{m}}\right)^4\right] \\ &= 0.491\,\text{m}^2/\text{s}^2 + 0.315\,\text{m}^2/\text{s}^2 = 0.806\,\text{m}^2/\text{s}^2\end{aligned}$$

giving

$$P_{3g} = \rho(0.806\,\text{m}^2/\text{s}^2) = (1000\,\text{kg/m}^3)(0.806\,\text{m}^2/\text{s}^2) = 806\,\text{N/m}^2 = 806\,\text{Pa} \Longleftarrow$$

(b) The water flow rate is

$$\dot{V} = A_2\hat{V}_2 = \frac{\pi}{4}(0.02\,\text{m})^2(0.886\,\text{m/s}) = 2.78 \times 10^{-4}\,\text{m}^3/\text{s} \Longleftarrow$$

13.9 Summary

The law of mass conservation and the first law of thermodynamics, in forms that are applicable to a control volume, have been considered in this chapter. The operational equations for these laws are the conservation of mass equation

$$\frac{m_{\text{cv}}}{dt} = \sum_i \dot{m}_i - \sum_e \dot{m}_e \tag{13.1}$$

and the first law of thermodynamics

$$\frac{dE_{\text{cv}}}{dt} = \dot{Q}_{\text{cv}} - \dot{W}_{\text{cv}} + \sum_i \dot{m}_i\left(h_i + \frac{\hat{V}_i^2}{2} + gz_i\right) - \sum_e \dot{m}_e\left(h_e + \frac{\hat{V}_e^2}{2} + gz_e\right) \tag{13.11}$$

We considered several examples that involved applications of these expressions to some situations of practical importance.

Under certain conditions, the energy equation reduces to the Bernoulli equation, which is broadly applicable to a number of physical situations

$$gz + \frac{\hat{V}^2}{2} + \frac{P}{\rho} = C \quad \text{(a constant)} \tag{13.15}$$

The conditions for which the Bernoulli equation applies are as follows:

- Adiabatic flow (no heat transferred), $\dot{Q} = 0$
- No work done, $\dot{W} = 0$
- Steady flow
- Incompressible flow
- Negligible viscous effects (**inviscid** flow)

13.10 Problems

Conservation of Mass

13.1: Oil with a specific gravity of 0.85 flows through a 7.62-cm-diameter pipe at a velocity of 3 m/s. Determine (a) the volumetric flow rate and (b) the mass flow rate.

13.2: A tapering pipe has an inside diameter of 30 cm at section 1 and 45 cm at section 2. Water flows through the pipe at a velocity of 5 m/s at section 2. Determine (a) the velocity at section 1, (b) the volumetric flow rate at section 1, (c) the volumetric flow rate at section 2, and (d) the mass flow rate.

13.3: Air flows through a 40-cm × 80-cm rectangular duct. The air is at 27°C and 125 kPa and it flows at a rate of 1.25 kg/s. Determine (a) the average velocity and (b) the volumetric flow rate.

13.4: A chamber has one inlet and two outlets. The inlet (port 1) has a diameter of 10 cm and the volumetric flow rate is 0.0525 m^3/s. One outlet (port 2) has a diameter of 7.5 cm and an average velocity of 10 m/s. At the other outlet (port 3) the diameter is 2.5 cm. Determine (a) the volumetric flow rate at port 3 and (b) the average velocity at port 3.

13.5: Water flows through a converging nozzle at a mass flow rate of 80 kg/s. The nozzle inlet (section 1) has a diameter of 24 cm, and the outlet (section 2) has a diameter of 8 cm. Determine the average velocity at (a) the inlet and (b) the outlet.

13.6: A 12.5-cm plunger is pushed at a rate of 8 cm/s into a tank filled with oil having a specific gravity of 0.775. The tank outlet has a diameter of 1.5 cm. Determine the mass flow of fluid at the outlet.

13.7: A tank has two inlets and one outlet. Water enters at port 1, which has a diameter of 3.75 cm, at a velocity of 15.5 m/s; at port 2 water enters with a volumetric flow rate of 0.0192 m^3/s. The water leaves port 3, which has a diameter of 5 cm. The level of water in the tanks remains constant. Determine the exit velocity.

13.8: Oil with a specific gravity of 0.862 flows through a 75-cm-diameter pipeline at a flow rate of 100 L/s. Determine (a) the velocity and (b) the mass flow rate.

13.9: Air enters a 20-cm-diameter pipe at a temperature of 22°C at a velocity of 70 m/s and a pressure 105 kPa. The pipe tapers to a diameter of 8 cm, and the air leaves at 667°C at a pressure of 1.4 MPa. Determine the exit velocity.

13.10: For the pipe arrangement shown in Figure P13.10, the volumetric flow rate in pipe 2 is 40% of the volumetric flow rate in pipe 1. Assume that the fluid is incompressible and determine the mean velocities and the volumetric flow rates in pipes 2 and 3.

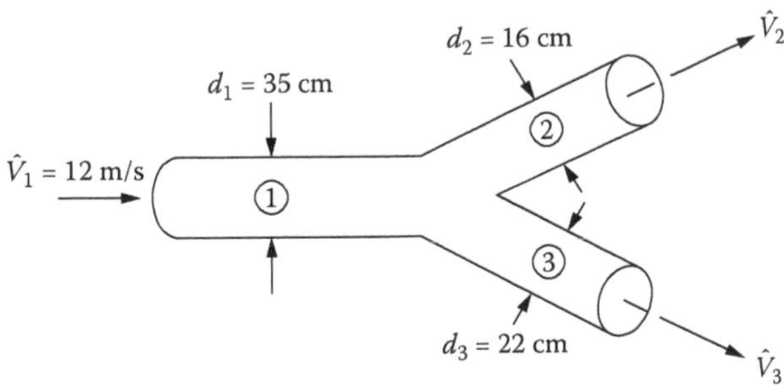

FIGURE P13.10

13.11: As indicated in Figure P13.11, water flows in a 0.6-m-diameter pipe with a parabolic profile given by

$$\hat{V} = 16\left(1 - \frac{r^2}{0.09}\right) \qquad (0 \le r \le 0.3)$$

where r is the radial coordinate in meters and $\hat{V}$ is the velocity in m/s. The water is discharged through a sudden contraction into a 0.20-m-diameter pipe. Determine the average velocity in this pipe.

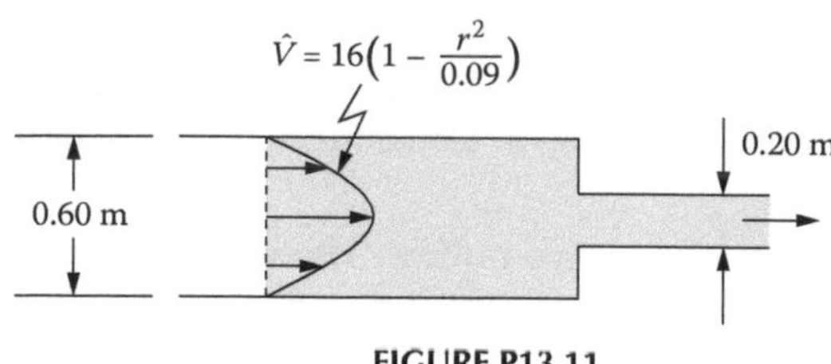

FIGURE P13.11

13.12: Oil with a specific gravity of 0.82 flows through a 10-cm-diameter pipe at $0.15\,\text{m}^3/\text{s}$. Assuming the density of water to be $1000\,\text{kg/m}^3$, determine (a) the average velocity in m/s and (b) the mass flow rate in kg/s.

13.13: Air flows through a conical diffuser. At the inlet $\hat{V}_i = 5\,\text{m/s}$, $A_i = 20\,\text{cm}^2$ and $\rho_i = 1.02\,\text{kg/m}^3$. The desired velocity at the exit, $\hat{V}_e$, is 1 m/s, and $\rho_e = 0.98\,\text{kg/m}^3$. Determine (a) the exit area needed to provide this exit density and (b) the mass flow rate through the diffuser.

13.14: A fluid of constant density flows in the passage formed by two parallel plates. Letting L be the separation distance between the plates and y the distance from the lower plate, derive an expression for the average velocity if the velocity distribution is given by (a) $\hat{V} = C_1 y$, (b) $\hat{V} = C_2 y^{1/2}$, and (c) $\hat{V} = C_3(Ly - y^2)$.

13.15: Assuming ideal gas behavior and that air is compressible, consider Figure P13.15, which shows a convergent-divergent duct. The flow is one dimensional and the diameter, pressure, and temperature at stations 1, 2, and 3 are

$$
\begin{aligned}
d_1 &= 40\,\text{cm} \quad & P_1 &= 200\,\text{kPa} \quad & T_1 &= 800\,\text{K} \\
d_2 &= 8\,\text{cm} \quad & P_2 &= 100\,\text{kPa} \quad & T_2 &= 450\,\text{K} \\
d_3 &= 25\,\text{cm} \quad & P_3 &= 60\,\text{kPa} \quad & T_3 &= 320\,\text{K}
\end{aligned}
$$

The airflow in the duct is 0.4 kg/s. Determine the average velocities, $\hat{V}_1$, $\hat{V}_2$, and $\hat{V}_3$.

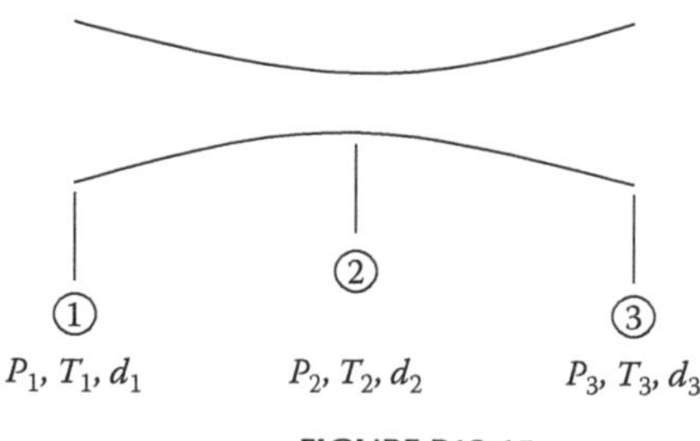

FIGURE P13.15

13.16: As indicated in Figure P13.16, water enters one end of a perforated pipe (20 cm in diameter) with a velocity of 6 m/s. The discharge through the pipe wall is approximated by a linear profile and the flow is steady. Determine the discharge velocity $\hat{V}$.

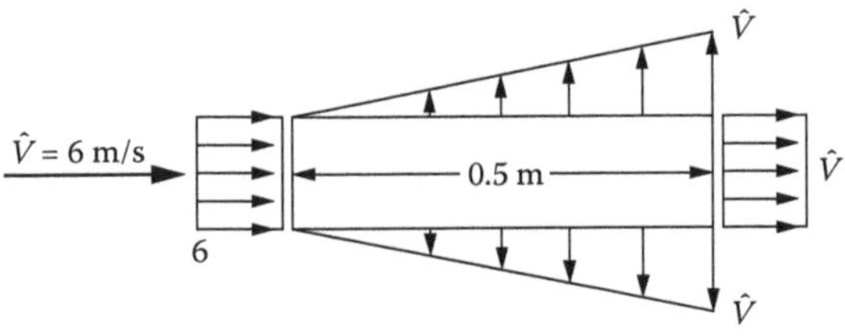

FIGURE P13.16

13.17: A water jet pump possesses coaxial streams as shown in Figure P13.17. The jet emerging from the inner tube of 2-cm diameter has a velocity of 25 m/s. The jet emerging from the coaxial tube with a 15-cm diameter has a velocity of 4 m/s. The two streams mix thoroughly and flow through the exit. Determine the average exit velocity $\hat{V}_e$.

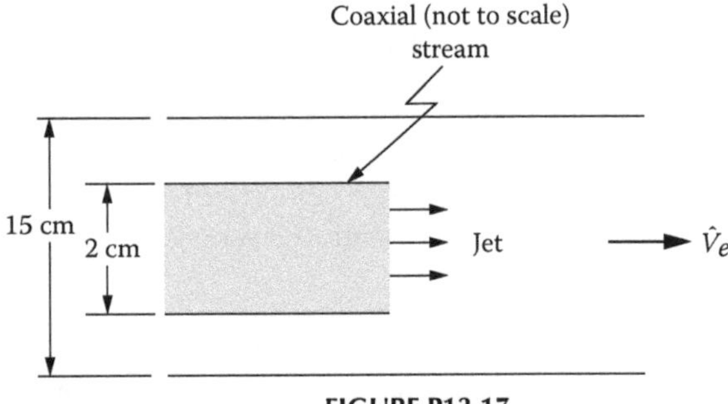

FIGURE P13.17

13.18: Air, which may be taken as an ideal gas with a constant density of 1.05 kg/m^3, flows through a 5-cm-diameter pipe with a velocity profile of

$$\hat{V} = 12\left[1 - \sin\left(\frac{\pi r}{5}\right)\right] \qquad (0 < r < 2.5\,\text{cm})$$

where $\hat{V}$ is the velocity in m/s and r is the radial coordinate in centimeters. Determine (a) the average velocity and (b) the mass flow rate.

13.19: The water tank shown in Figure P13.19 is being filled through section 1 at $\hat{V}_1 = 5$ m/s and through section 3 at $\dot{V}_3 = 0.01$ m^3/s; the water level, h, is constant. Determine $\hat{V}_2$.

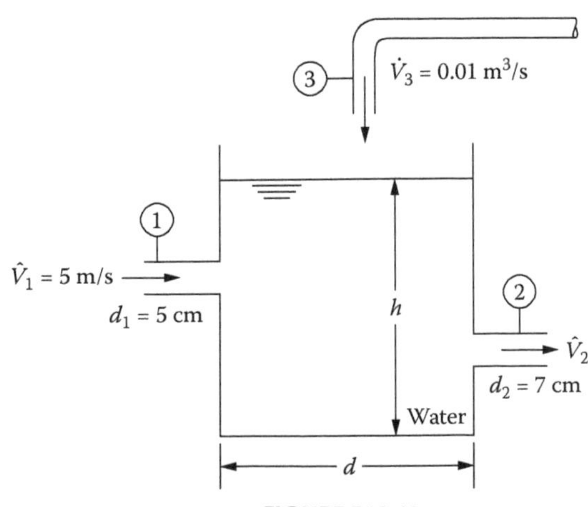

FIGURE P13.19

13.20: In Figure P13.20, oil with a density of 850 kg/m^3 flows at a velocity of 2 m/s in a pipe with diameter of 38 cm. A sudden contraction reduces the pipe diameter to 25 cm. The pipe then splits into branches of diameters 15 and 10 cm. The velocity in the 15-cm-diameter pipe is 3.2 m/s. Determine (a) the mass flow rates and (b) the average velocity in the 10-cm branch.

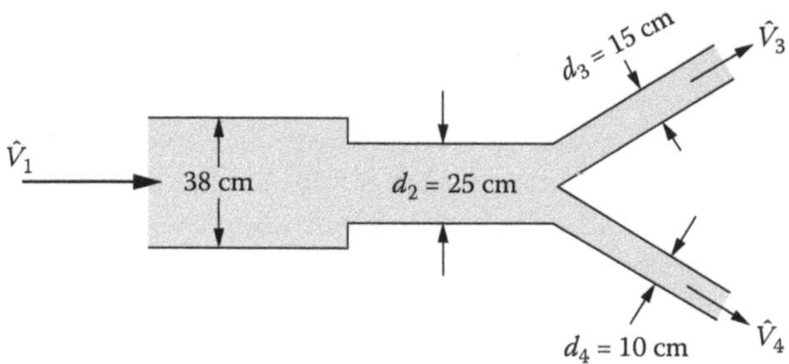

FIGURE P13.20

13.21: Water flows steadily in the control volume shown in Figure P13.21 in three circular sections. Section 1 has a diameter of 1.60 cm and the flow is 0.028 m^3/s. Section 2 has a diameter of 0.60 cm at an average velocity of 1.4 m/s. Determine the average velocity and volumetric flow rate at section 3 where, the diameter is 8 mm.

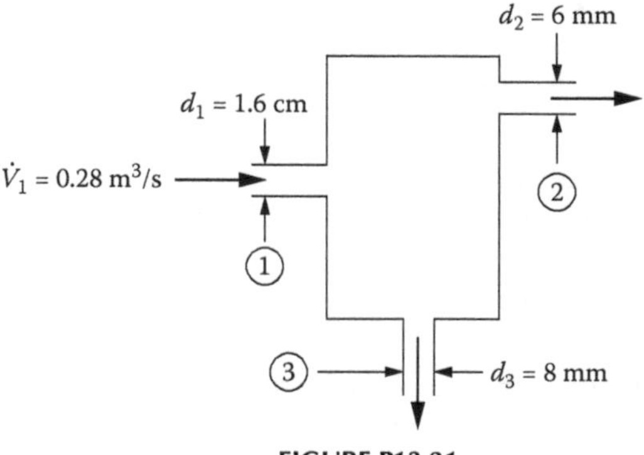

FIGURE P13.21

13.22: As indicated in Figure P13.22, a fluid fills the space between two very long plates of length, $2L$, which are separated by a distance b. The upper plate moves downward with at a rate $\hat{V}$. Determine the mass flow rate and maximum velocity for (a) uniform exit velocity and (b) parabolic exit velocity.

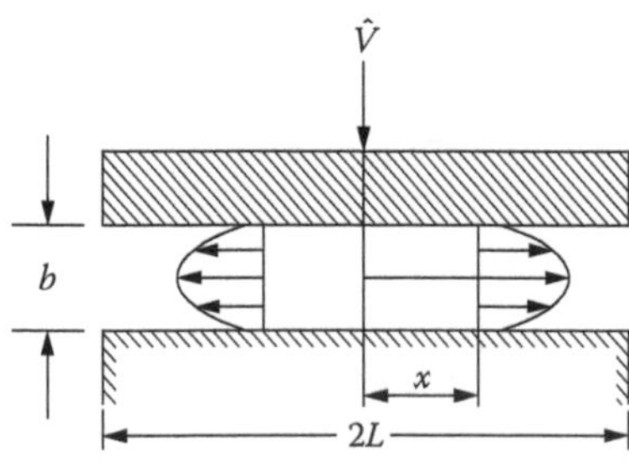

FIGURE P13.22

13.23: In some wind tunnels, the test section wall is porous or perforated; fluid is sucked out to provide a thin viscous boundary layer. The wall in Figure P13.23 is 4-m long and contains 800 holes of 6-mm diameter per square meter of area. The suction velocity out of each hole is $\hat{V}_s = 12\,\text{m/s}$, and the test section entrance velocity is $\hat{V}_1 = 45\,\text{m/s}$. Assuming incompressible flow of air at 20°C and a pressure of 1 atm, determine (a) $\hat{V}_0$, (b) the total wall suction volume flow, (c) $\hat{V}_2$, and (d) $\hat{V}_f$.

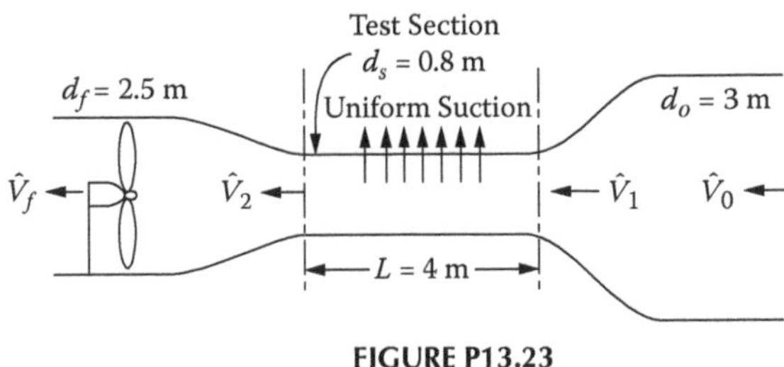

FIGURE P13.23

13.24: The velocities in a circular duct of 1-m diameter were measured as follows:

Distance from Center, cm	Velocity, m/s	Distance from Center, cm	Velocity, m/s
0.0	2.52	30.0	1.50
5.0	2.30	35.0	1.31
10.0	2.26	40.0	1.02
15.0	2.01	45.0	0.84
20.0	1.92	50.0	0.70
25.0	1.80		

Determine (a) the average velocity and (b) the volumetric flow rate in m^3/s.

13.25: Water is flowing through the circular conduit shown in Figure P13.25 with a velocity profile given by

$$\hat{V} = \hat{V}_{\text{max}}\left(1 - \frac{r^2}{R^2}\right)$$

where $\hat{V}_{\text{max}} = 3\,\text{m/s}$ and $R = 1.25\,\text{m}$.

Determine the average water velocity in the 80-cm-diameter pipe.

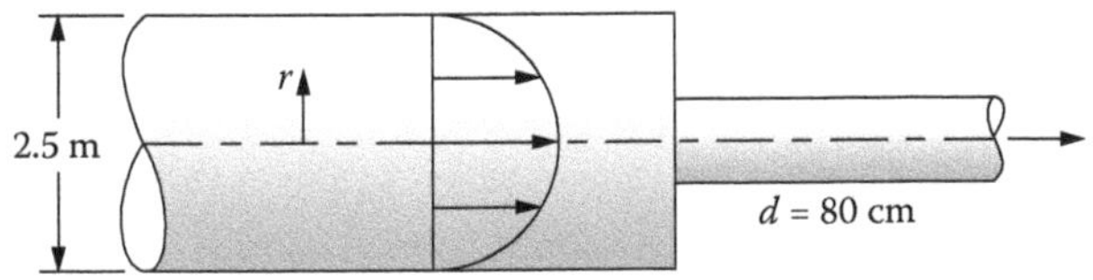

FIGURE P13.25

13.26: Water enters the conical diffusing passage shown in Figure P13.26 with an average velocity of $4\,\text{m/s}$. The entrance cross-sectional area is $0.0925\,\text{m}^2$. Determine the exit area needed to reduce the velocity to $0.40\,\text{m/s}$.

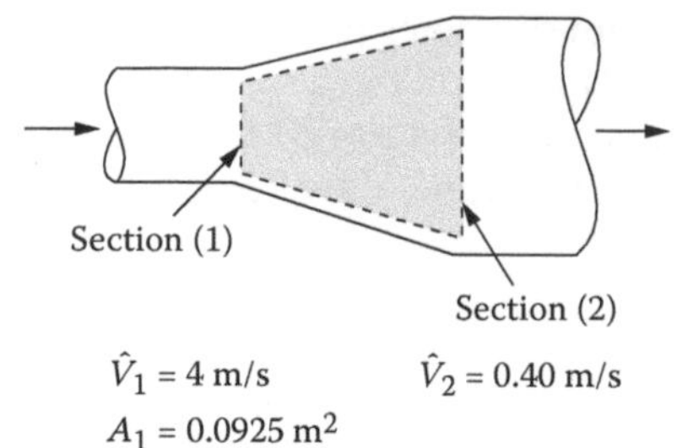

FIGURE P13.26

13.27: Water enters the 10-cm-square channel shown in Figure P13.27 at a velocity of 3.2 m/s. The channel converges to a 5-cm square at the outlet section, which is cut at a 30° angle to the vertical. Determine the outlet water velocity and the total rate of flow if the mean velocity of the discharging water remains horizontal.

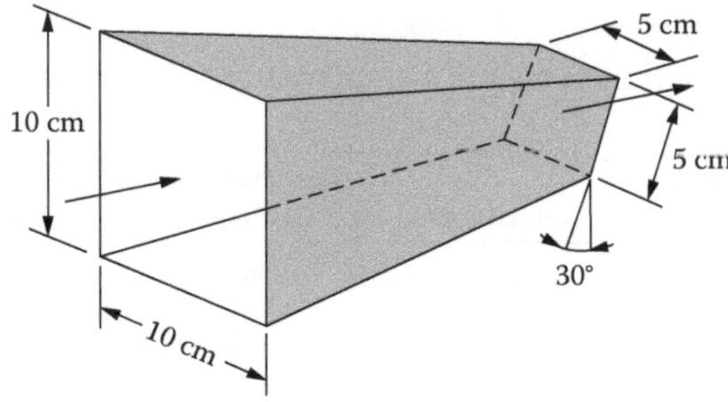

FIGURE P13.27

13.28: Water flows steadily through a nozzle (Figure P13.28) at 60 kg/s. The diameters are 25 cm at the left (d_1) and 8 cm at the right (d_2). Determine the average velocities at sections 1 and 2 in m/s.

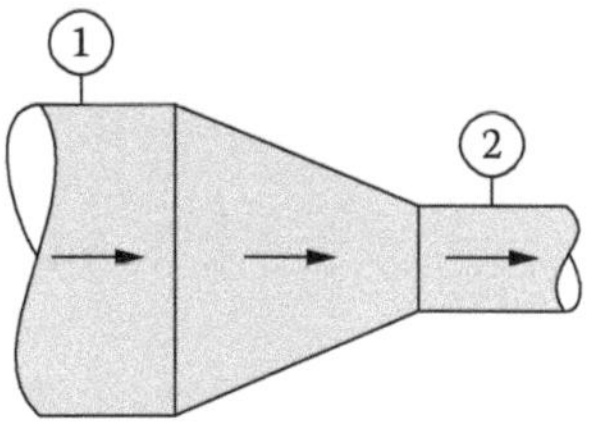

FIGURE P13.28

13.29: The hypodermic syringe shown in Figure P13.29 contains serum (SG = 1.02). Assume that there is no leakage past the plunger, which is pushed in steadily at 1.5 cm/s. Determine the exit velocity, $\hat{V}_2$ in m/s.

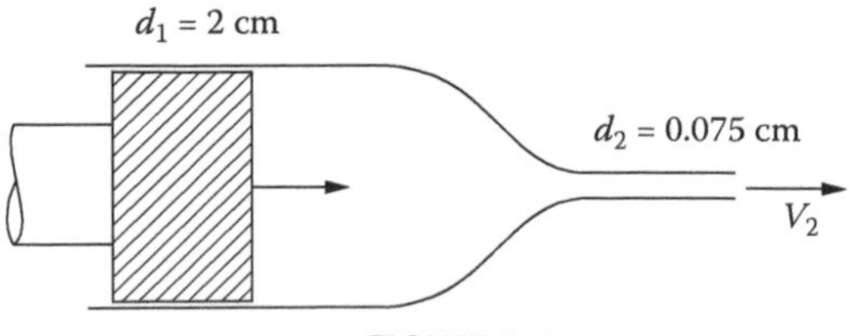

FIGURE P13.29

13.30: Given a garden hose with a flow rate of of 1 L/s, how long would it take to fill to a depth of 2 m a cylindrical swimming pool having a diameter of 8 m?

The First Law of Thermodynamics

13.31: A jet of water is projected vertically with a velocity of 22.5 m/s. Neglect air resistance and determine how high the jet will rise.

13.32: A water jet is inclined upward at an angle of 36.87° from the horizontal. Neglecting air resistance, determine the velocity required to allow the jet to clear a 3.5-m wall at a distance of 20 m.

13.33: Water flows in an inclined pipe from point 1 where the diameter is 30 cm, the elevation is 12 m, and the pressure head is 8 m, to point 2 where the diameter is 60 m and the elevation is 8 m. The volumetric flow rate is 0.375 m^3/s. Assume that there are no losses and determine the pressure head at point 2.

13.34: A water jet is inclined upward at an angle of 36.87° from the horizontal. Neglecting air resistance, determine the velocity required to allow the jet to clear a 3.5-m wall at a distance of 20 m.

13.35: Water flows in an inclined pipe from point 1 where the diameter is 30 cm, the elevation is 12 m, and the pressure head is 8 m, to point 2 where the diameter is 60 cm and the elevation is 8 m. The volumetric flow rate is 0.375 m^3/s. Assume that there are no losses and determine the pressure head at point 2.

13.36: Seawater ($\rho = 1025\ kg/m^3$) flows through a pump at a flow rate of 0.14 m^3/s. The pump inlet is 0.25 m in diameter. At the inlet, the pressure is −15 cm of water. The pump outlet, which is 0.152 m in diameter, is located 1.8 m above the inlet, and the outlet pressure is 1.75 kPa. The inlet and exit temperatures are equal. Determine the power that the pump adds to the fluid.

13.37: A liquid is heated in a 15-m-long vertical tube with a constant diameter of 5 cm. The flow is upward and at the entrance the average velocity is 1 m/s, the pressure is 340 kPa, and the density is 1000 kg/m^3. The increase in internal energy is 200 kJ/kg. Determine the amount of heat added to the fluid.

13.38: For the convergent nozzle shown in Figure P13.38, assume that the work done is zero and that the nozzle is adiabatic and determine the change in the internal energy of the water.

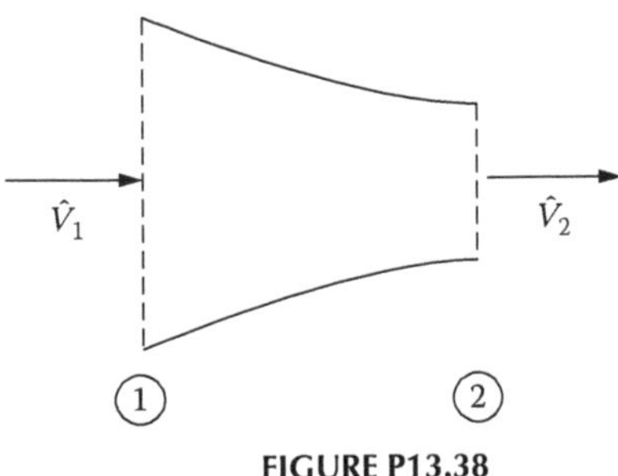

FIGURE P13.38

13.39: As indicated in Figure P13.39, a fan draws air from the atmosphere through a 0.30-m-diameter round duct that has a smoothly rounded entrance. A differential manometer connected to an opening in the wall of the duct shows a vacuum pressure of 2.5 cm of water. Given the density of air (1.22 kg/m^3), determine (a) the volumetric rate of airflow and (b) the work done by the fan.

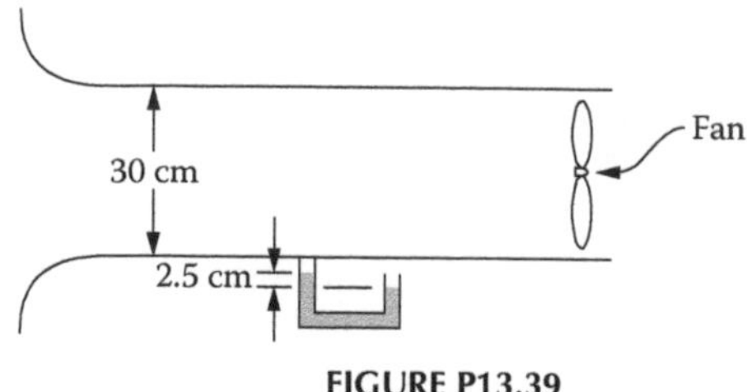

FIGURE P13.39

13.40: A liquid of density 800 kg/m^3 flows through the diffuser shown in Figure P13.40. The mass flow rate of the liquid through the diffuser is 10 kg/s and the change of internal energy between the inlet and outlet is 45 m^2/s^2. Assuming steady flow with no external heat or work interactions, determine the outlet pressure P_0.

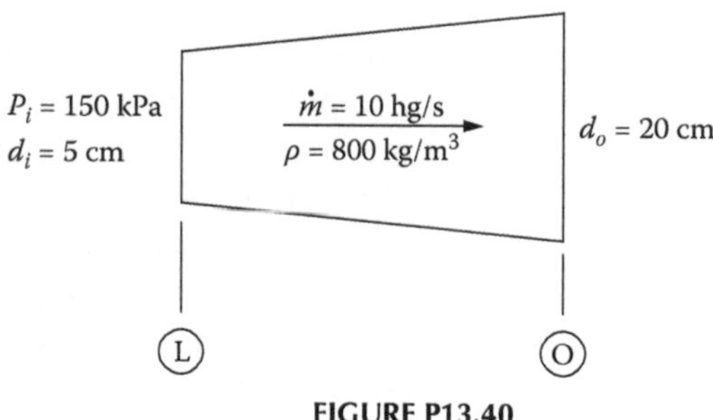

FIGURE P13.40

13.41: Water flows through the reducer shown in Figure P13.41 at the rate of 0.05 m^3/s. The diameters of the reducer are 12 cm and 7 cm. The pressure drop across the reducer is 95 kPa and the difference in elevation is 0.4 m. Determine the change in internal energy of the water as it flows through the reducer.

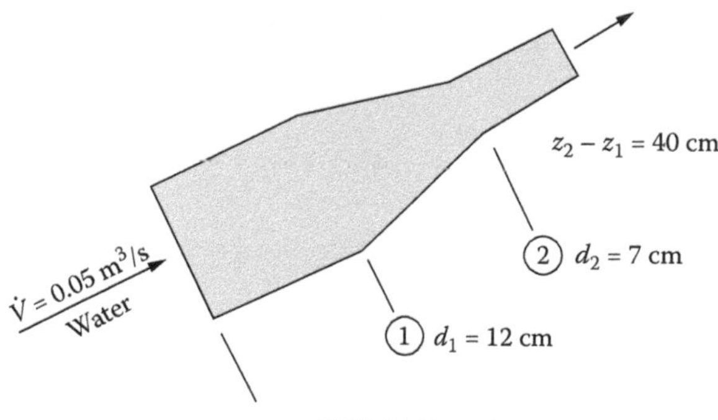

FIGURE P13.41

13.42: Water flows through a valve in a horizontal constant area pipe such that the change in pressure is 5 kPa. Determine the change in internal energy of the water.

13.43: Air at 20°C flows into a 0.285 m^3 reservoir at a velocity of 20 m/s. The reservoir temperature and pressure are 20°C and 96 kPa, respectively. Assume that the incoming air is at reservoir pressure and flows through a 20-cm-diameter pipe. Determine the rate of temperature increase in the reservoir.

13.44: Water flows through a 5-cm-diameter horizontal pipe at 2.25×10^{-3} m^3/s. Heat transfer can be neglected and frictional forces cause a pressure drop of 480 Pa. Determine the rate at which heat is added to the water.

The Bernoulli Equation

13.45: The pressurized tank shown in Figure P13.45 has a circular cross section of 1.8-m diameter. Oil is drained through a nozzle 5 cm in diameter in the side of the tank. The specific gravities of the oil and mercury are 0.85 and 13.6, respectively. Assuming that the air pressure is kept constant, determine how long it will take to lower the surface of the oil in the tank by 1.50 m.

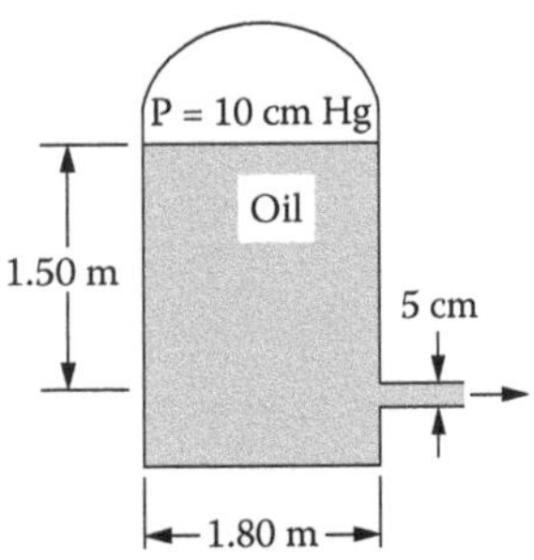

FIGURE P13.45

13.46: Water in an open cylindrical tank, 3 m in diameter, discharges to the atmosphere through a nozzle, 5 cm in diameter. Neglecting friction and any unsteadiness in the flow, determine the time required for the water in the tank to drop from a level of 8.52 m above the nozzle to the 1.20-m level.

13.47: Water flows in an inclined pipe from point 1 where the diameter is 30 cm, the elevation is 12 m, and the pressure head is 8 m. At point 2, the diameter is 60 m and the elevation is 8 m. The volumetric flow rate is 0.375 m^3/s. Assume that there are no losses and determine the pressure head at point 2.

13.48: A pipe carries water from point 1 where the diameter is 8 cm and the elevation above the datum is 4 m, to point 2 where the diameter is 4 cm and the elevation above the datum is 10 m. The volumetric flow rate is 0.0375 m^3/s, and the pressure at point 1 is 988 kPa. Assuming no frictional losses, determine P_2.

13.49: An inclined pipe, carrying oil with a specific gravity of 0.885, changes in diameter from 15 cm at section 1 to 46 cm at section 2. Section 1 is 4 m lower than section 2 and the volumetric flow rate is 0.142 m^3/s. The pressures at sections 1 and 2 are 90 and 60 kPa, respectively, and losses are considered to be negligible. Determine the direction of flow.

13.50: Water flows from an opening in the bottom of a reservoir through a 5-cm-diameter pipe to a discharge point below the reservoir surface as shown in Figure P13.50. The volumetric flow rate is 0.01875 m^3/s, and no head is lost in the system. Determine the difference in height, h, between the reservoir surface and the discharge.

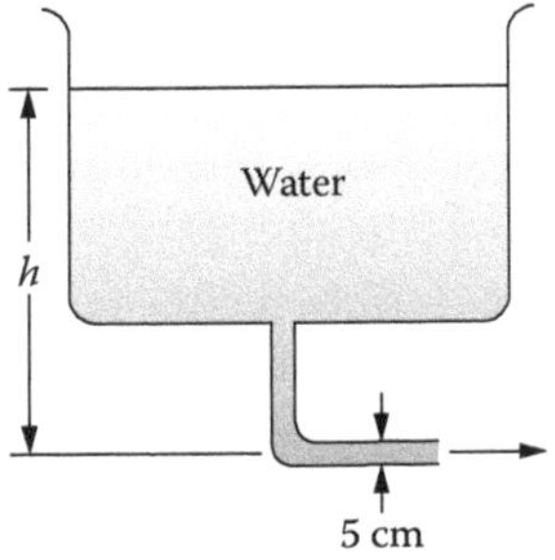

FIGURE P13.50

13.51: A nozzle with a diameter of 5 cm is attached to a horizontal pipe with a diameter of 10 cm. Assuming that head losses are negligible and the pressure upstream of the nozzle is 480 kPa, determine the water jet velocity.

13.52: Consider Figure P13.52 and show that when the siphon is running continuously that (a) the lowest pressure depends on $y_1 + y_2$ and is located at point 3 and (b) that the exit velocity, $\hat{V}_2$ depends upon both y_1 and g.

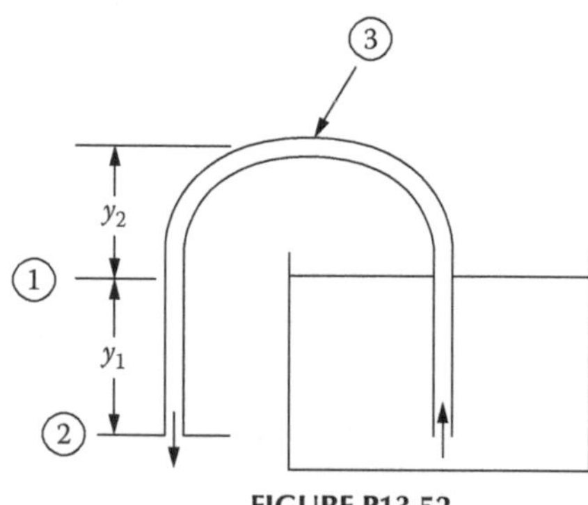

FIGURE P13.52

13.53: Water flows from station 1 to station 2 in a pipe. Station 2 is 2.25-m lower than station 1. Assuming $\hat{V}_1 = 2.4\,\text{m/s}$. $P_1 = 325\,\text{kPa}$, $d_1 = 10\,\text{cm}$, $d_2 = 5\,\text{cm}$, and no losses, determine P_2.

13.54: A large tank possesses a small, well-rounded opening, 25 m below the surface of the water in the tank. Determine the velocity of the water passing through the hole.

13.55: Oil with a specific gravity of 0.875 flows from a tank through a 15-cm-diameter pipe. Point 1 at an elevation of 36 m is under an unknown air pressure. Point 2 at an elevation of 4.5 m is at the discharge. If head losses are negligible, determine the air pressure P_1 necessary to discharge $0.625\,\text{m}^3/\text{s}$ of oil.

13.56: A pump delivers $0.0375\,\text{m}^3/\text{s}$ of oil, having a specific gravity of 0.875, through a 10-cm suction pipe. The pressure at point 1, which is 1.25 m below the datum that runs through the centerline of the pump, is a vacuum of 200 mm of mercury. Determine the total energy head at point 1.

13.57: A horizontal air duct has an area reduction from $0.0375\,\text{m}^2$ at station 1 to $0.0100\,\text{m}^2$ at station 2. For a mass flow of air at $\dot{m} = 0.075\,\text{kg/s}$, with a density ρ of $0.0995\,\text{kg/m}^3$, determine the pressure change between the stations when frictional losses are neglected.

13.58: Oil with a specific gravity of 0.825 is flowing through a pipe of 3-cm inner diameter at a pressure of 100 kPa. The total energy, relative to a reference level 24 m below the centerline of the pipe is 225 m. Determine the flow rate of the oil.

13.59: Water in an open cylindrical tank, 3 m in diameter, discharges through a nozzle that is 5 cm in diameter. Neglecting losses, determine the time required for the water in the tank to drop from a level of 6 m above the nozzle inlet to a level of 1.5 m.

13.60: A venturi meter is a carefully designed constriction whose pressure difference is a measure of the flow rate in a pipe. A venturi like the one shown in Figure P13.60, is connected to a horizontal pipe of 20-cm diameter. The water that flows through one pipe may have a maximum flow rate of $8.07\,\text{m}^3/\text{s}$. The pressure head at the inlet for this flow is 18 m above atmospheric, and the pressure head at the throat must not be lower than 7 m below atmospheric. Between the inlet and the throat, there is an estimated frictional loss of 10% of the difference in pressure head between these points. Determine the minimum allowable diameter of the throat.

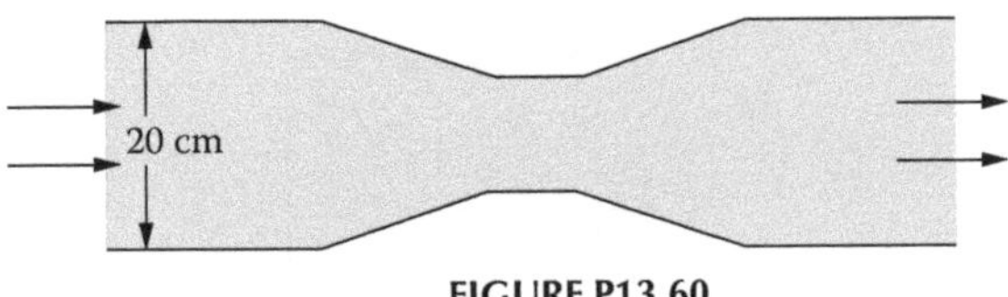

FIGURE P13.60

13.61: A horizontal venturi meter has a diameter of 8 cm at its inlet and 4 cm at its throat. The pressure difference between the inlet and the throat is 25 kPa. Determine the volumetric rate of water flow through the venturi when no losses need be accounted for.

13.62: In Figure P13.62, the flowing fluid is air at a density ρ of 1.04 kg/m^3 and the manometer fluid is oil with a specific gravity of 0.827. Assuming no losses, compute the flow rate in m^3/s.

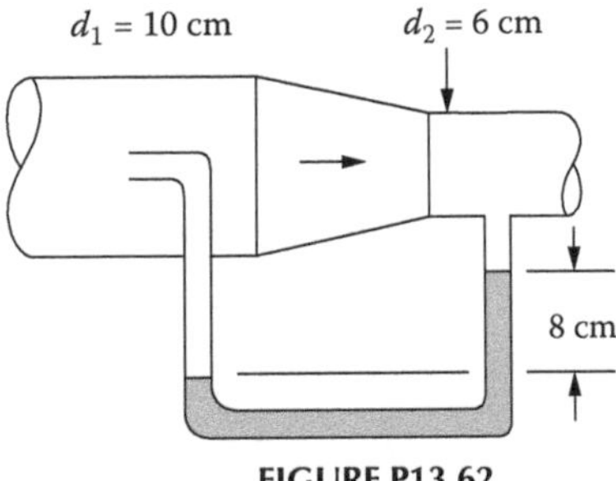

FIGURE P13.62

13.63: Using Bernoulli's equation for steady incompressible flow with no losses, show that the volumetric flow rate through a venturi meter is related to the manometer reading, h by

$$\dot{V} = \frac{A_2}{\sqrt{1 - \left(\dfrac{d_2}{d_1}\right)^4}} \sqrt{\frac{2gh(\rho_m - \rho)}{\rho}}$$

where ρ_m is the density of the manometer fluid.

13.64: In Figure P13.64, the manometer fluid is mercury. Neglecting losses, determine the flow rate in the tube in m^3/s if the flowing fluid is (a) water with $\rho = 1000\,\text{kg/m}^3$ and (b) air with $\rho = 1.02\,\text{kg/m}^3$.

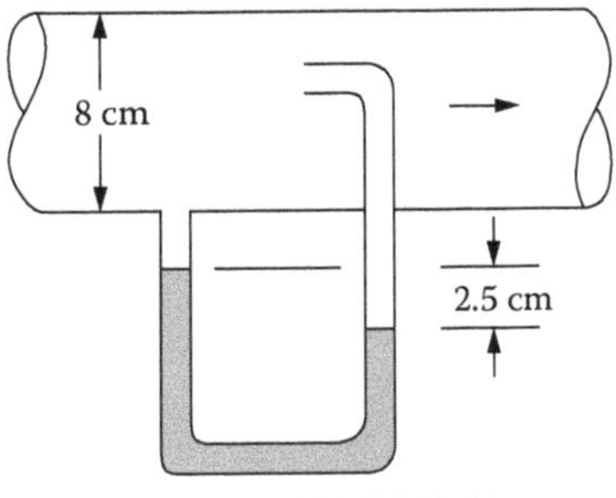

FIGURE P13.64

13.65: Oil with a specific gravity of 0.900 flows downward through a vertical contraction as shown in Figure P13.65. The mercury manometer reading h is 12 cm. Determine the volumetric flow rate for frictionless flow and state whether the actual flow rate is more or less than the value for frictionless flow.

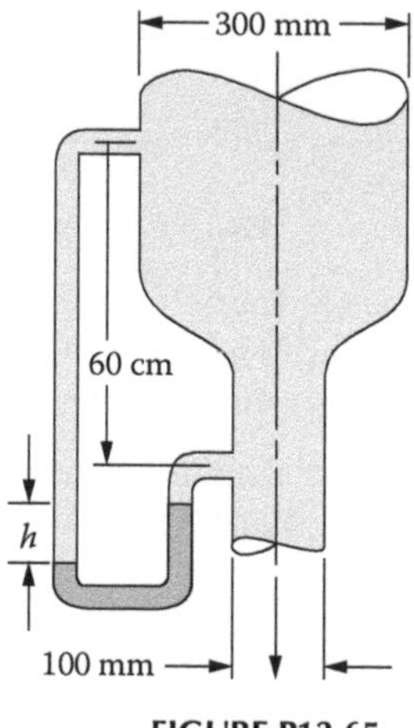

FIGURE P13.65

13.66: Air of density 1.21 kg/m^3 is flowing as shown in both parts of Figure P13.66, and $\hat{V} = 15\,\text{m/s}$. Determine the readings on the manometers.

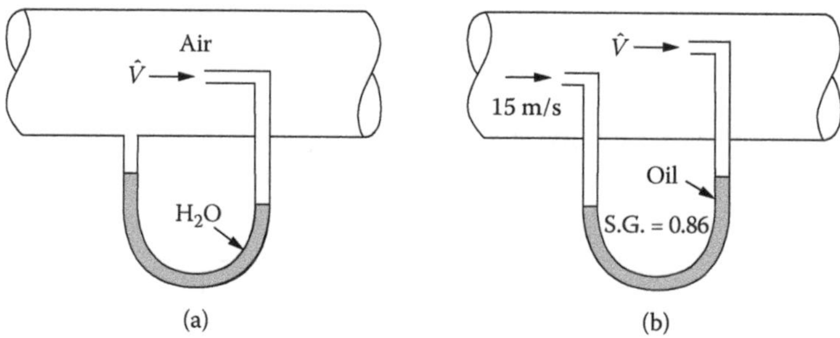

FIGURE P13.66

13.67: In Figure P13.67, the flowing fluid is air with a density ρ of 1.2 kg/m^3, and the manometer fluid is oil with a specific gravity, SG = 0.827. Assuming no losses, compute the flow rate in m^3/s.

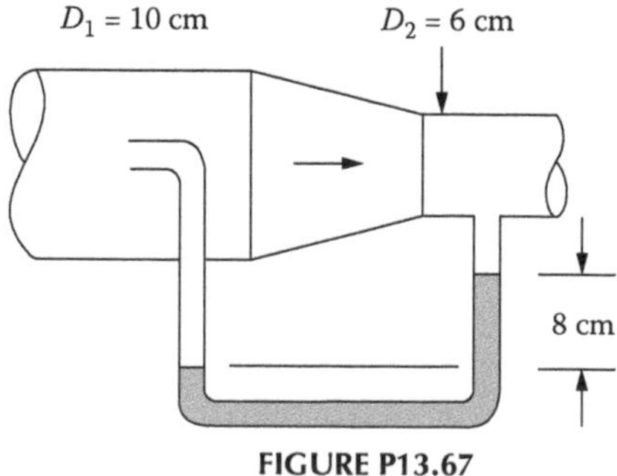

FIGURE P13.67

13.68: Water flows through the vertical pipe in Figure P13.68. Is the flow in the pipe up or down?

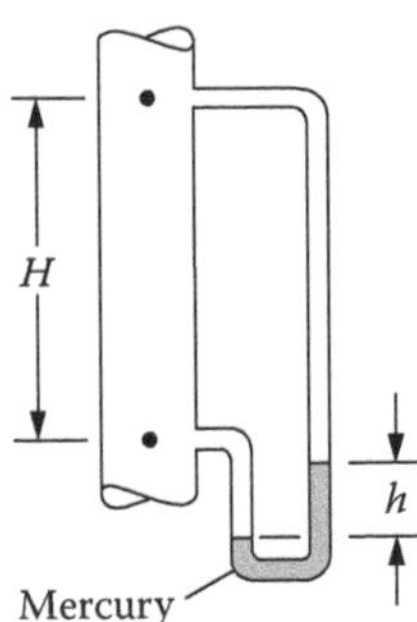

FIGURE P13.68

13.69: A liquid flows in the horizontal pipeline shown in Figure P13.69. The velocity is 0.65 m/s with a friction loss of 0.1375 m of flowing fluid. For a pressure head at B of 0.60 m, determine the pressure head at A.

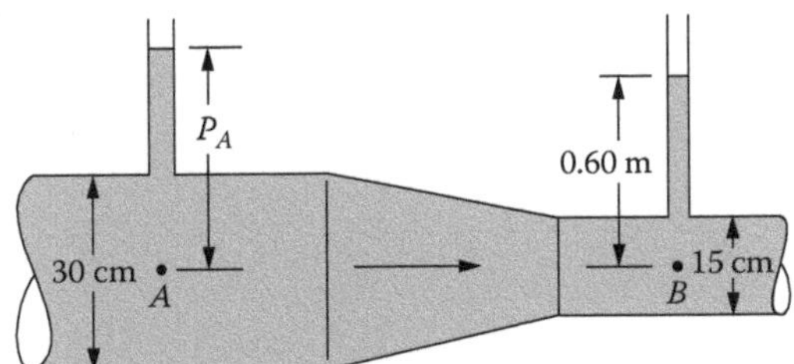

FIGURE P13.69

13.70: Water flows steadily upward in the vertical pipe shown in Figure P13.70. It is then deflected to flow outward with uniform radial velocity. Neglecting friction and assuming that the pressure at A is 70 kPa, determine the volumetric flow rate of the water through the pipe.

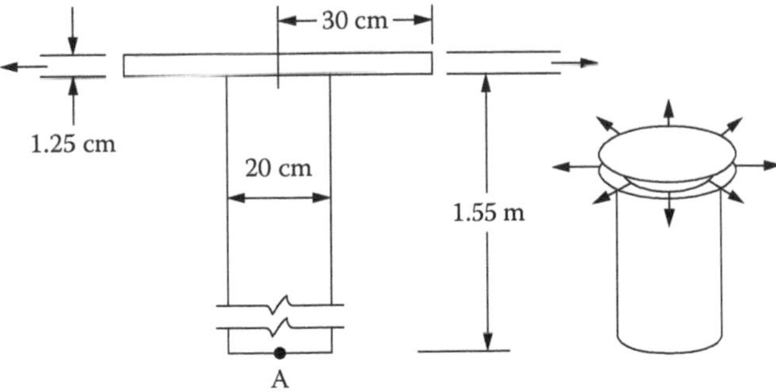

FIGURE P13.70

14

Newton's Second Law of Motion

Chapter Objectives

- To introduce the control volume relationship for linear momentum.
- To provide applications of the momentum theorem.
- To consider the control volume relationship for the moment of momentum.
- To use applications taken from a number of physical situations to show applications of the moment of momentum relationship.

14.1 Introduction

The third fundamental law to be considered is **Newton's second law**. As in Chapter 13, we will develop control volume expressions, which, in this case, will be related to both linear and angular motion. The basic expressions will then be applied to a number of physical situations.

14.2 Linear Momentum

Newton's second law of motion may be stated as

> The time rate of change of momentum of a system is equal to the net force on the system and takes place in the direction of the net force.

This statement is notable in two ways. First, it is cast in a form that includes both magnitude and direction and is, therefore, a vector expression. Second, it refers to a system rather than a control volume. As we know by now, a system is a fixed collection of mass, whereas a control volume is a fixed region in space that encloses a different mass of fluid (or system) at different times.

The transformation of the second law statement from a system to a control volume point of view is dealt with in numerous texts. We will presume the correctness of the following word equation:

$$\left\{\begin{array}{c}\text{Net force}\\ \text{acting on the}\\ \text{control volume}\end{array}\right\} = \left\{\begin{array}{c}\text{Rate of momentum}\\ \text{out of the control}\\ \text{volume by mass flow}\end{array}\right\} - \left\{\begin{array}{c}\text{Rate of momentum}\\ \text{into the control}\\ \text{volume by mass flow}\end{array}\right\} + \left\{\begin{array}{c}\text{Rate of accumulation}\\ \text{of momentum within}\\ \text{the control volume}\end{array}\right\} \qquad (14.1)$$

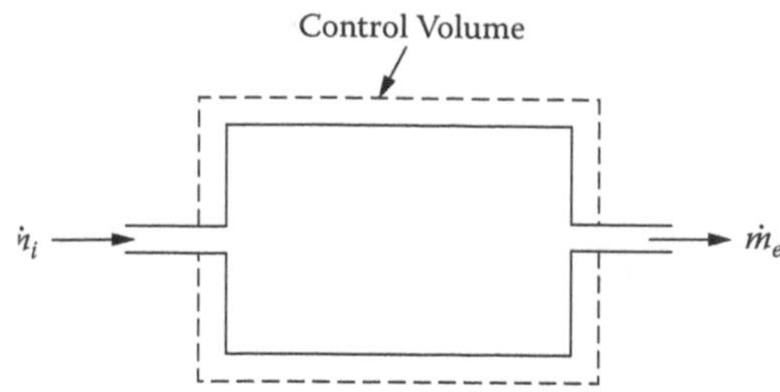

FIGURE 14.1
A general control volume and flow field.

The control volume shown in Figure 14.1 has the same general features that are shown in Figure 13.1. The total force acting on this control volume, due to both surface and body forces, will be treated as a single vector term $\sum \boldsymbol{F}$.

As with our earlier control volume considerations, the rate of mass flow through any single "passageway" across the control surface is written as $\dot{m}_i$ or $\dot{m}_e$ with the subscripts indicating whether the mass flow is entering (subscript i) or exiting (subscript e).

The **momentum fluxes** associated with these mass flows are expressed as $\hat{\boldsymbol{V}}_i\dot{m}_i$ and $\hat{\boldsymbol{V}}_e\dot{m}_e$. The velocity vectors $\hat{\boldsymbol{V}}_i$ and $\hat{\boldsymbol{V}}_e$ are the velocities of the corresponding mass flows relative to the defined coordinate system. This coordinate system may be defined in whatever manner is most convenient in a given application and it may either be stationary or moving. This choice will become clearer when examples are considered in the next section.

The accumulation term in Equation 14.1 may be written as

$$\left\{\begin{matrix}\text{Rate of accumulation}\\ \text{of momentum within}\\ \text{the control volume}\end{matrix}\right\} = \frac{d}{dt}\hat{\boldsymbol{V}}_{\text{cv}}m_{\text{cv}}$$

where the velocity vector, $\boldsymbol{V}_{\text{cv}}$, is the velocity at which the control volume is moving relative to the defined coordinate system.

The resulting control volume expression for linear momentum can now be written as

$$\sum \boldsymbol{F} = \sum_e \hat{\boldsymbol{V}}_e\dot{m}_e - \sum_i \hat{\boldsymbol{V}}_i\dot{m}_i + \frac{d}{dt}\hat{\boldsymbol{V}}_{\text{cv}}m_{\text{cv}} \tag{14.2}$$

The momentum flux terms are now shown as the sums of the contributions of all entering and exiting streams.

Equation 14.2 is often referred to as the **momentum theorem**. It is the third of our fundamental control volume relationships. The similarity between Equation 14.2 and its counterparts, Equations 13.4 and 13.11, is clear. There are, likewise, some differences.

The first difference of note is that the sum of all momentum terms is not zero because there is a force term involved. The second difference, as mentioned earlier, is the vector character of Equation 14.2 in contrast to both Equations 13.4 and 13.11, which are scalar. For a rectangular coordinate system, Equation 14.2 can be written in component form as Equations 13.4 and 13.11

$$\sum F_x = \sum_e \hat{V}_{xe}\dot{m}_e - \sum_i \hat{V}_{xi}\dot{m}_i + \frac{d}{dt}\hat{V}_{x,\text{cv}}m_{\text{cv}} \tag{14.3a}$$

$$\sum F_y = \sum_e \hat{V}_{ye}\dot{m}_e - \sum_i \hat{V}_{yi}\dot{m}_i + \frac{d}{dt}\hat{V}_{y,\text{cv}}m_{\text{cv}} \tag{14.3b}$$

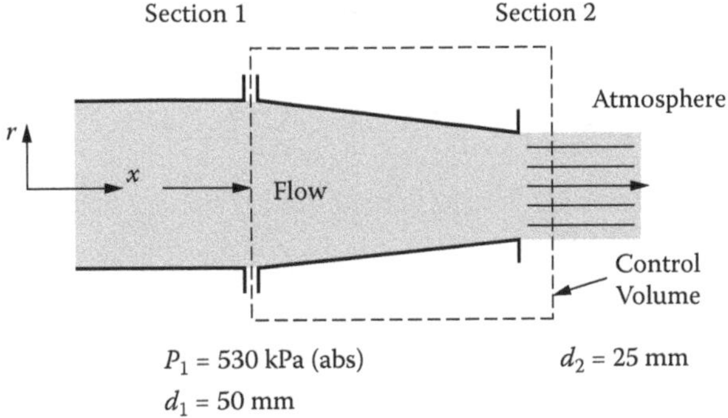

FIGURE 14.2
Water flowing through a horizontal nozzle in Example 14.1.

and

$$\sum F_z = \sum_e \hat{V}_{ze}\dot{m}_e - \sum_i \hat{V}_{zi}\dot{m}_i + \frac{d}{dt}\hat{V}_{z,\text{cv}}m_{\text{cv}} \tag{14.3c}$$

The utility of these expressions will become more obvious as we consider several examples.

14.3 Applications of the Control Volume Expression

Two critical choices must be made when applying the momentum theorem to a specific case. The choices pertain to the control volume and to the coordinate system. Various combinations will be illustrated in the examples to follow.

Example 14.1 Water flows through the nozzle shown in Figure 14.2, discharging to the atmosphere. The flow rate is $53\,\text{m}^3/\text{h}$ with the other dimensions and conditions shown. Determine the force, including its direction, exerted on the nozzle by the coupling used to hold it in place.

Solution

Assumptions and Specifications

(1) Steady-state conditions exist.
(2) The flow is incompressible.
(3) Uniform properties exist at each cross section considered.

In this case, the choices of coordinate system and control volume are straightforward. Because the nozzle is stationary, we choose a coordinate system that is fixed with the x-axis oriented along the direction of flow. Moreover, because we are interested in the interaction between the nozzle and the coupling, we will choose the control volume that cuts through the point of contact, section 1, shown as a dashed line in Figure 14.2.

The x-directional scalar form of the control volume applies so that the basic equation to be solved is Equation 14.3a

$$\sum F_x = \sum_e \hat{V}_{xe}\dot{m}_e - \sum_i \hat{V}_{xi}\dot{m}_i + \frac{d}{dt}\hat{V}_{x,\text{cv}}m_{\text{cv}} \tag{14.3a}$$

Moreover, because we are dealing with a steady-flow process, the accumulation term may be suppressed, leaving

$$\sum F_x = \sum_e \hat{V}_{xe}\dot{m}_e - \sum_i \hat{V}_{xi}\dot{m}_i$$

The force may be expressed as

$$\sum F_x = F_x + P_{1g}A_1$$

where the term F_x is the force exerted on the control volume at the location where the control surface interacts with its surroundings. In this case, F_x is exerted *on* the control volume *by* the coupling and is the unknown that we are seeking. Had the problem asked for the force that the nozzle exerts on the coupling, we would be looking for the reaction to F_x, namely $R_x = -F_x$. The unknown, F_x, is assumed to have its positive sense in the positive x direction. The solution will then indicate whether it is oriented in this manner; a positive result for F_x will mean that the resulting force on the control volume is to the right and a negative result will mean that it is to the left.

The lone pressure-force term, $P_{1g}A_1$, is the resulting x-directed pressure force because atmospheric pressure, exerted over the entire control surface, cancels. P_{1g} is, thus, the net pressure and it is exerted *on* the control surface at section 1. This is in the positive x direction and this is the reason for its positive sign.

We now examine the summation terms on the right-hand side of our equation. For this case, mass crosses the control surface only at sections 1 and 2. Thus, we can write

$$\sum_e \hat{V}_{xe}\dot{m}_e - \sum_i \hat{V}_{xi}\dot{m}_i = -\dot{m}_1\hat{V}_{x1} + \dot{m}_2\hat{V}_{x2}$$

Because the flow is *in* at section 1, the sign is negative and the positive sign indicates that the flow is *out* at section 2. Both $\hat{V}_{x1}$ and $\hat{V}_{x2}$ are to the right and have positive signs.

A further simplification is possible when, from a mass balance, we note that $\dot{m}_1 = \dot{m}_2$. The subscripts can then be dropped and the momentum flux terms become

$$\sum_e \hat{V}_{xe}\dot{m}_e - \sum_i \hat{V}_{xi}\dot{m}_i = \dot{m}(\hat{V}_{x2} - \hat{V}_{x1})$$

Thus, the complete momentum equation reduces to

$$F_x + P_{1g}A_1 = \dot{m}(\hat{V}_{x2} - \hat{V}_{x1})$$

and the unknown, F_x, may be expressed as

$$F_x = \dot{m}(\hat{V}_{x2} - \hat{V}_{x1}) - P_{1g}A_1$$

The velocities $\hat{V}_{x1}$ and $\hat{V}_{x2}$ can be expressed as

$$\hat{V}_{x1} = \frac{\dot{V}}{A_1} = \frac{(53\,\text{m}^3/\text{h})(1\,\text{h}/3600\,\text{s})}{(\pi/4)(0.050\,\text{m})^2} = 7.5\,\text{m/s}$$

and

$$\hat{V}_{x2} = 4\hat{V}_{x1} = 4(7.5\,\text{m/s}) = 30\,\text{m/s}$$

We are now able to solve for F_x. With the density of water taken as $1000\,\text{kg/m}^3$ and atmospheric pressure taken as $P_{atm} = 101\,\text{kPa}$, we have

$$\begin{aligned} F_x &= \dot{m}(\hat{V}_{x2} - \hat{V}_{x1}) - P_{1g}A_1 \\ &= \frac{(53\,\text{m}^3/\text{h})(1000\,\text{kg/m}^3)(30\,\text{m/s} - 7.5\,\text{m/s})}{(3600\,\text{s/h})(1\,\text{kg-m/N-s}^2)} \\ &\quad -[(530 - 101) \times 10^3\,\text{N/m}^2]\left(\frac{\pi}{4}\right)(0.050\,\text{m})^2 \\ &= 331\,\text{N} - 842\,\text{N} = -511\,\text{N} \Longleftarrow \end{aligned}$$

The force is negative, that is, it is acting toward the left.

Example 14.2 Consider the nozzle of Example 14.1 with all conditions the same except that the nozzle discharge is inclined at an angle, θ, relative to the x direction. Determine (a) the force exerted on the nozzle by the coupling for the general case shown and (b) the values of F_x for $\theta = 90°$ and $\theta = 180°$.

Solution

Assumptions and Specifications

(1) Steady-state conditions exist.
(2) The flow is incompressible.
(3) Uniform properties exist at each cross section considered.

(a) The control volume and the coordinate system are indicated in Figure 14.3. For this case, the total force exerted by the coupling, $\boldsymbol{F}$, will have components in both the x- and y-coordinate directions. We must consider the two scalar component expressions

$$\sum F_x = \sum_e \hat{V}_{xe}\dot{m}_e - \sum_i \hat{V}_{xi}\dot{m}_i$$

and

$$\sum F_y = \sum_e \hat{V}_{ye}\dot{m}_e - \sum_i \hat{V}_{yi}\dot{m}_i$$

where, because this is a steady-flow case, the time-dependent terms have been suppressed.

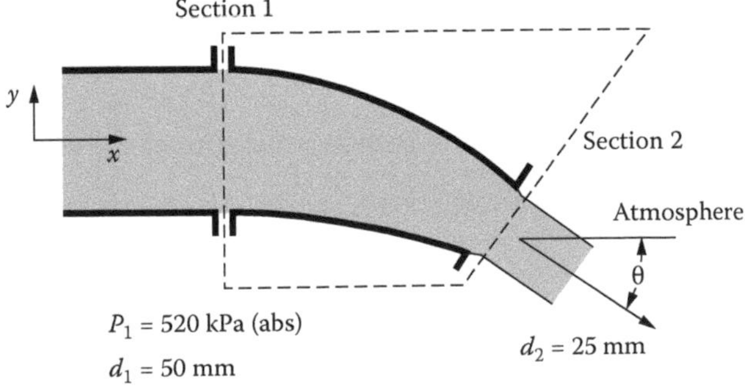

FIGURE 14.3
Water flowing through a reducing pipe bend in Example 14.2.

Considering the first of these expressions, in the x direction, we have

$$F_x + P_{1g}A_1 = \dot{m}(\hat{V}_{x2} - \hat{V}_{x1})$$

and for the y direction, we may write

$$F_y = \dot{m}\hat{V}_{y2}$$

Note that the atmospheric pressure exerted over the entire surface cancels leaving $P_{1g}A_1$ at section 1 as the only x directed net pressure force. Thus, it appears only in the x-component expression. One should also observe that the momentum influx, due to mass flow, occurs only at section 1 and, because $\hat{V}_1$ is entirely x directed, this term also appears only in the x-component direction.

We may now complete the solution for the total force. The operational equations for our two unknown force components are

$$F_x = \dot{m}(\hat{V}_{x2}\cos\theta - \hat{V}_{x1}) - P_{1g}A_1$$

and

$$F_y = \dot{m}\hat{V}_2\sin\theta$$

Using values already computed in Example 14.1, we obtain for F_x

$$\begin{aligned} F_x &= \dot{m}(\hat{V}_{x2}\cos\theta - \hat{V}_{x1}) - P_{1g}A_1 \\ &= \frac{(53\,\mathrm{m^3/h})(1000\ \mathrm{kg/m^3})(30\cos\theta\,\mathrm{m/s} - 7.5\,\mathrm{m/s})}{(3600\,\mathrm{s/h})(1\,\mathrm{kg\text{-}m/N\text{-}s^2})} \\ &\quad -[(530-101)\times 10^3\,\mathrm{N/m^2}]\left(\frac{\pi}{4}\right)(0.050\,\mathrm{m})^2 \\ &= (110.4\,\mathrm{N})(4\cos\theta - 1)\,\mathrm{N} - 842.4\,\mathrm{N} \\ &= 441.6\cos\theta\,\mathrm{N} - 952.8\,\mathrm{N} \end{aligned}$$

and for F_y

$$\begin{aligned} F_y &= \dot{m}\hat{V}_{y2}\sin\theta \\ &= \frac{(53\,\mathrm{m^3/h})(1000\,\mathrm{kg/m^3})(30\ \sin\theta\,\mathrm{m/s})}{(3600\,\mathrm{s/h})(1\,\mathrm{kg\text{-}m/N\text{-}s^2})} \\ &= 441.6\sin\theta\,\mathrm{N} \end{aligned}$$

The force on the nozzle by the coupling is the vector sum of these components,

$$\boldsymbol{F} = (441.6\cos\theta - 952.8)\boldsymbol{x}\,\mathrm{N} + 441.6\sin\theta\boldsymbol{y}\,\mathrm{N} \Longleftarrow$$

where $\boldsymbol{x}$ and $\boldsymbol{y}$ are unit vectors in the x and y directions.

We may note that when the nozzle is horizontal, $\theta = 0°$, this expression yields a force in the x direction of

$$\boldsymbol{F} = -511.2\boldsymbol{x}\,\mathrm{N}$$

which checks with the solution for Example 14.1

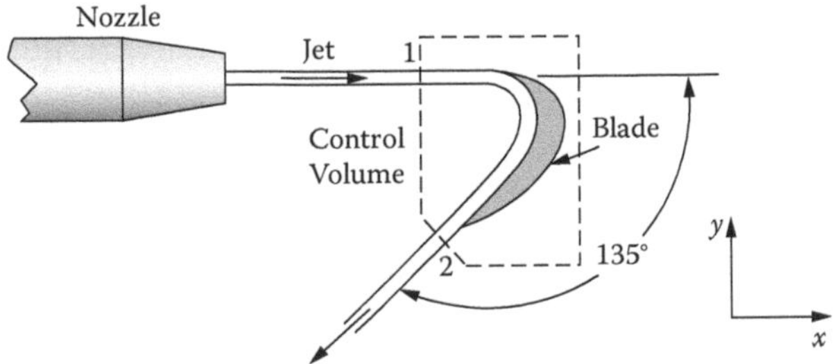

FIGURE 14.4
Free-water jet striking a turning vane for Example 14.3.

(b) We can use our solution for part (a)

$$\boldsymbol{F} = (441.6\cos\theta - 952.8)\boldsymbol{x}\,\text{N} + 441.6\sin\theta\boldsymbol{y}\,\text{N}$$

to determine the force for any angle. If $\theta = 90,^\circ \cos\theta = 0$, and $\sin\theta = 1.0$ so that

$$\boldsymbol{F} = -952.8\boldsymbol{x}\,\text{N} + 441.6\boldsymbol{y}\,\text{N} \Longleftarrow$$

with a magnitude of

$$F = 1050.2\,\text{N}$$

and if $\theta = 180^\circ$, $\cos\theta = -1$, and $\sin\theta = 0$

$$\boldsymbol{F} = -1394\,\text{N} \Longleftarrow$$

Most of us have observed, firsthand, the behavior of a hose with water being discharged through a nozzle. Some thought about our findings in Examples 14.1 and 14.2 should reconcile our experiences relative to the results obtained.

Our next example examines the effect of a free jet, that is one that issues from a nozzle and is not constrained by a tube wall, as it strikes an object that deflects the stream.

Example 14.3 A free jet of water flowing at a rate of $0.06\,\text{m}^3/\text{s}$ and a velocity of $7.6\,\text{m/s}$ strikes a fixed blade that turns the flow as indicated in Figure 14.4. Neglecting friction between the blade and the water jet (a) evaluate the force exerted by the blade on the water jet and (b) determine the force exerted on the jet if the blade moves to the right at a steady velocity of $3\,\text{m/s}$.

Solution

Assumptions and Specifications

(1) Steady-state conditions exist.
(2) The flow is incompressible.
(3) Uniform properties exist at each cross section considered.
(4) Negligible friction is specified.

(a) We first specify the coordinate system to be fixed, with the x and y directions selected as indicated in Figure 14.4. The control volume, enclosed in the dashed line, is conveniently chosen so that the entering and leaving liquid streams are normal to the boundary. It is important to realize that the control volume will also cut through the blade support and that the net force exerted *on* the blade *by* the surroundings is the force exerted by the blade support.

The x- and y-scalar forms of the momentum theorem, for steady flow, may be written as

$$F_x = \sum_e \hat{V}_{xe}\dot{m}_e - \sum_i \hat{V}_{xi}\dot{m}_i = \dot{m}(\hat{V}_{x2} - \hat{V}_{x1})$$

$$= \dot{m}(-\hat{V}\cos 45^\circ - \hat{V}) = -\dot{m}\hat{V}(\cos 45^\circ + 1) = -1.707\dot{m}\hat{V}$$

and

$$F_y = \sum_e \hat{V}_{ye}\dot{m}_e - \sum_i \hat{V}_{yi}\dot{m}_i$$

$$= \dot{m}(\hat{V}_{y2} - \hat{V}_{y1}) = -\dot{m}\hat{V}\sin 45^\circ = -0.707\hat{V}$$

With the density of water taken as $1000\,\text{kg/m}^3$, insertion of the proper numerical values allows the solution to be completed.

$$F_x = -\frac{(0.06\,\text{m}^3/\text{s})(1000\,\text{kg/m}^3)(7.6\,\text{m/s})(1.707)}{1\,\text{kg-m/N-s}^2} = -778.4\,\text{N}$$

and

$$F_y = -\frac{(0.06\,\text{m}^3/\text{s})(1000\,\text{kg/m}^3)(7.6\,\text{m/s})(0.707)}{1\,\text{kg-m/N-s}^2} = -322.4\,\text{N}$$

These are the components of the force exerted by the surroundings to hold the blade in place under the influence of the water jet. Because the problem statement specifies that we find the force exerted by the water on the blade, the correct answer to our problem is the negative of the force, $\boldsymbol{F}$. Hence, the result is

$$\boldsymbol{F} = 778.4\,\boldsymbol{x}\,\text{N} + 322.4\boldsymbol{y}\,\text{N} \Longleftarrow$$

(b) When the blade is moving to the right, the choice of the coordinate system and control volume is less obvious. To illustrate the effect of these choices, we will work the problem twice.

First, we take the control volume as shown in Figure 14.4, that is, it moves *with the blade* and always has the same orientation relative to the blade. We will also choose a coordinate system that is fixed to the blade and is thus moving to the right at a fixed velocity of 3 m/s.

Having made these choices, we can write the expressions for F_x and F_y as

$$F_x = \sum_e \hat{V}_{xe}\dot{m}_e - \sum_i \hat{V}_{xi}\dot{m}_i = \dot{m}(\hat{V}_{x2} - \hat{V}_{x1})$$

and

$$F_y = \sum_e \hat{V}_{ye}\dot{m}_e - \sum_i \hat{V}_{yi}\dot{m}_i = \dot{m}(\hat{V}_{y2} - \hat{V}_{y1})$$

which are the same expressions as in part (a). The quantities $\dot{m}$ and $\hat{V}$, however, are different in this case. The mass flow rate, $\dot{m}$, across the boundary is equal to $\rho A\hat{V}_r$ where $\hat{V}_r$ is the velocity *relative the boundary*. With the density of water taken as $\rho = 1000\,\text{kg/m}^3$

$$\dot{m} = \rho A \hat{V}_r$$

and with $A = \dot{V}/\hat{V}_r$

$$\dot{m} = (1000\,\text{kg/m}^3)\left(\frac{0.06\,\text{m}^3/\text{s}}{7.6\,\text{m/s}}\right)(7.6\,\text{m/s} - 3.0\,\text{m/s}) = 36.3\,\text{kg/s}$$

The velocities, $\hat{V}_{x1}$, $\hat{V}_{x2}$, $\hat{V}_{y1}$, and $\hat{V}_{y2}$ are all for the fluid stream crossing the boundary *relative to the selected coordinate system*. In our case, we have

$$\hat{V}_{x1} = 7.6\,\text{m/s} - 3.0\,\text{m/s} = 4.6\,\text{m/s}$$

$$\hat{V}_{x2} = -4.6\cos 45^\circ\,\text{m/s}$$

$$\hat{V}_{y1} = 0$$

$$\hat{V}_{y2} = -4.6\sin 45^\circ\,\text{m/s}$$

We may now solve for F_x and F_y

$$\begin{aligned} F_x &= \dot{m}(\hat{V}_{x2} - \hat{V}_{x1}) \\ &= \frac{(36.3\,\text{kg/s})(-4.6\,\text{m/s})(\cos 45^\circ + 1)}{1\,\text{kg-m/N-s}^2} \\ &= \frac{(36.3\,\text{kg/s})(-4.6\,\text{m/s})(1.707)}{1\,\text{kg-m/N-s}^2} \\ &= -285.0\,\text{N} \end{aligned}$$

and

$$\begin{aligned} F_y &= \dot{m}(\hat{V}_{y2} - \hat{V}_{y1}) \\ &= \frac{(36.3\,\text{kg/s})(-4.6\,\text{m/s})(\sin 45^\circ)}{1\,\text{kg-m/N-s}^2} \\ &= \frac{(36.3\,\text{kg/s})(-4.6\,\text{m/s})(0.707)}{1\,\text{kg-m/N-s}^2} \\ &= -118.1\,\text{N} \end{aligned}$$

These are the components of the force exerted by the surroundings as discussed earlier. The force exerted by the water on the blade will be

$$\boldsymbol{F} = 285.0\boldsymbol{x}\,\text{N} + 118.1\,\boldsymbol{y}\,\text{N} \Longleftarrow$$

As a second approach to this problem, we will use the same moving control volume but choose a coordinate system fixed in space. This would be the case of a distant stationary observer watching as the blade passes by. The same equations for F_x and F_y apply, that is

$$F_x = \dot{m}(\hat{V}_{x2} - \hat{V}_{x1})$$

and

$$F_y = \dot{m}(\hat{V}_{y2} - \hat{V}_{y1})$$

Because the control volume is still attached to the moving blade, the mass flow rate, $\dot{m}$, will still be 36.3 kg/s. However, the velocity components will differ and will now be

$$\hat{V}_{x1} = 7.6\,\text{m/s}$$

$$\hat{V}_{x2} = [(7.6 - 3)(-\cos 45^\circ)\,\text{m/s})] + 3\,\text{m/s}$$

$$\hat{V}_{y1} = 0$$

$$\hat{V}_{y2} = (7.6 - 3)(-\sin 45^\circ)\,\text{m/s}$$

Here, $\hat{V}_{x1}$ and $\hat{V}_{y1}$, the velocity components at section 1 are straightforward; the velocities at this point are those of the jet. At section 2, the relationships are a bit more complex. The foregoing values of $\hat{V}_{x1}$ and $\hat{V}_{y1}$ are the vector sums of the velocity of the water relative to the blade *plus* the velocity of the blade relative to the fixed coordinate system.

When we substitute into the expressions for F_x and F_y, we obtain

$$F_x = \frac{(36.3\,\text{kg/s})\{[(7.6\,\text{m/s} - 3\,\text{m/s})(-\cos 45^\circ)] + 3\,\text{m/s} - 7.6\,\text{m/s}\}}{1\,\text{kg-m/N-s}^2}$$

$$= \frac{(36.3\,\text{kg/s})(7.6\,\text{m/s} - 3\,\text{m/s})(-\cos 45^\circ - 1)}{1\,\text{kg-m/N-s}^2} = -285.0\,\text{N}$$

and

$$F_y = \frac{(36.3\,\text{kg/s})[(7.6\,\text{m/s} - 3\,\text{m/s})(-\sin 45^\circ) - 0\,\text{m/s}]}{1\,\text{kg-m/N-s}^2} = -118.1\,\text{N}$$

Thus, the force exerted by the water on the blade will be

$$\boldsymbol{F} = 285\boldsymbol{x}\,\text{N} + 118.1\boldsymbol{y}\,\text{N} \Longleftarrow$$

It is reassuring to note that the same result was achieved even though the problem starting conditions were posed differently. The reader will likely prefer one approach over the other but proper analysis of any choice will yield a correct solution as long as all of the terms are evaluated correctly.

In our next example, we will consider a case where the accumulation term is nonzero.

Example 14.4 A tank car with an open top is shown in Figure 14.5 at a time when it is in the path of a jet of water. The jet issues from a 1-cm-diameter nozzle and has a velocity of 15 m/s. The tank car moves to the right at 5 m/s. Determine the force exerted on the tank car by the water jet.

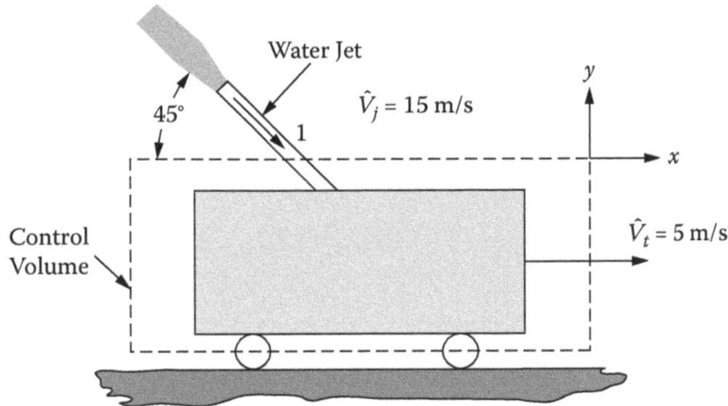

FIGURE 14.5
Tank car receiving water from a jet in Example 14.4.

Solution

Assumptions and Specifications

(1) Steady-state conditions exist.
(2) The flow is incompressible.
(3) Uniform properties exist at each cross section considered.

The control volume, which is moving to the right with the tank car, is shown in Figure 14.5 by the dashed lines. We will first choose a coordinate system that also moves with the tank car.

The x- and y-component forms of the momentum theorem are given by Equations 14.3a and 14.3b

$$\sum F_x = \sum_e \hat{V}_{xe}\dot{m}_e - \sum_i \hat{V}_{xi}\dot{m}_i + \frac{d}{dt}\hat{V}_{x,\text{cv}}m_{\text{cv}} \tag{14.3a}$$

and

$$\sum F_y = \sum_e \hat{V}_{ye}\dot{m}_e - \sum_i \hat{V}_{yi}\dot{m}_i + \frac{d}{dt}\hat{V}_{y,\text{cv}}m_{\text{cv}} \tag{14.3b}$$

We first consider the x direction where Equation 14.3a reduces to

$$F_x = -\hat{V}_{x1}(\rho A_1 V_1) + \frac{d}{dt}\hat{V}_{x,t}\rho\dot{m}_{\text{cv}} = -\dot{m}(\hat{V}_j \cos 45^\circ - \hat{V}_t) - \dot{m}(0)$$

In this expression, we have noted that, by mass conservation, the rate of mass accumulation in the tank is equal to the rate at which mass enters across the top boundary. The velocity of this fluid in the tank is zero relative to the coordinate system, which also moves with velocity $\hat{V}_t$.

The x-component velocity relative to the moving coordinate system is seen to be $\hat{V}_j$ cos 45° $-$ $\hat{V}_t$. Solving for F_x, with the density of water taken as 1000 kg/m^3 yields

$$F_x = -\frac{(1000\,\text{kg/m}^3)\left(\frac{\pi}{4}\right)(0.15\,\text{m})^2(15\,\text{m/s})(15\cos 45^\circ\,\text{m/s} - 5\,\text{m/s})}{1\,\text{kg-m/N-s}^2}$$

$$= -1.486\,\text{kN}$$

For F_y, which is less complicated, we obtain

$$F_y = -\hat{V}_{y1}\dot{m}_1 + \frac{d}{dt}\hat{V}_{y,t}\dot{m}_{cv}$$

$$= -\dot{m}(-\hat{V}_j \sin 45^\circ) + \dot{m}(0)$$

$$= -\frac{1000\,\text{kg/m}^3)\left(\frac{\pi}{4}\right)(0.15\,\text{m})^2(15\,\text{m/s})(-15\sin 45^\circ\,\text{m/s})}{1\,\text{kg-m/N-s}^2}$$

$$= 2.811\,\text{kN}$$

The force exerted by the water on the tank car has components that are of the opposite sign. Hence, the desired solution for the water force is

$$\boldsymbol{F}_{\text{w}} = 1.486\boldsymbol{x}\,\text{kN} + 2.811\boldsymbol{y}\,\text{kN} \Longleftarrow$$

A stationary coordinate system will be chosen for an alternate solution. In this case, we may write for F_x and F_y

$$F_x = -\hat{V}_{x1}\dot{m}_1 + \frac{d}{dt}\left(\hat{V}_{x,\text{cv}}\dot{m}_{\text{cv}}\right)$$

$$= -\dot{m}(\hat{V}_j \cos 45^\circ) + \dot{m}\hat{V}_t = -\dot{m}(\hat{V}_j \cos 45^\circ - \hat{V}_t)$$

and

$$F_y = -\hat{V}_{y1}\dot{m}_1 + \frac{d}{dt}\left(\hat{V}_{y,\text{cv}}\dot{m}_{\text{cv}}\right) = -\dot{m}(-\hat{V}_j \sin 45^\circ) + \dot{m}(0)$$

These two solutions, although developed differently, are identical to those obtained in the previous approach for a moving coordinate system.

We further illustrate the momentum theorem in Design Example 7 that now follows.

14.4 Design Example 7

A concrete slurry mixture with a density of 2500 kg/m^3 is dumped from a hopper onto a conveyer belt as shown in Figure 14.6. The slurry is delivered to the conveyer at a rate of 600 m^3/h. For conveyer drive wheels of 50-cm diameter rotating at 300 rpm, estimate the power required to drive the conveyer.

Solution

Assumptions and Specifications

(1) Steady-state conditions exist.
(2) The concrete slurry is incompressible.
(3) Uniform properties exist at each cross section considered.

A stationary control volume is chosen as shown in Figure 14.7 and we may apply the directional component form of the momentum equation to this stationary control volume.

$$F_x = \sum \hat{V}_{xe}\dot{m}_e - \sum \hat{V}_{xi}\dot{m}_i$$

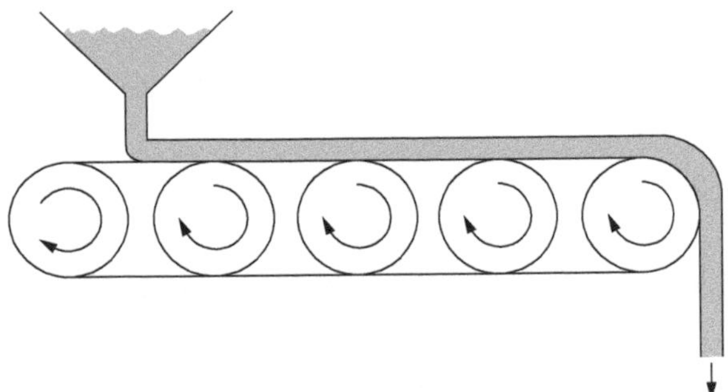

FIGURE 14.6
Conveyor belt for Design Example 7.

and for this single input-single output situation, we have $\dot{m}_e = \dot{m}_i = \dot{m}$ and

$$F_x = \dot{m}(\hat{V}_{xe} - \hat{V}_{xi})$$

We note that the slurry entering the control volume at point 1 has no horizontal velocity component. Thus, $\hat{V}_{xe} = \hat{V}_x$ and

$$F_x = \dot{m}\hat{V}_x$$

Because the required power is $\dot{W} = F_x\hat{V}_x$, we have

$$\dot{W} = F_x\hat{V}_x = (\dot{m}\hat{V}_x)\hat{V}_x = \dot{m}(\hat{V}_x)^2$$

The velocity $\hat{V}_x$ is related to the velocity of the drive wheels

$$\hat{V}_x = r\omega = (0.25\,\text{m})(2\pi\,\text{rad/rev})(300\,\text{rev/min})\left(\frac{1}{60}\,\text{min/s}\right) = 7.85\,\text{m/s}$$

and with

$$\dot{m} = (600\,\text{m}^3/\text{h})(2500\,\text{kg/m}^3)(1\,\text{h}/3600\,\text{s}) = 416.7\,\text{kg/s}$$

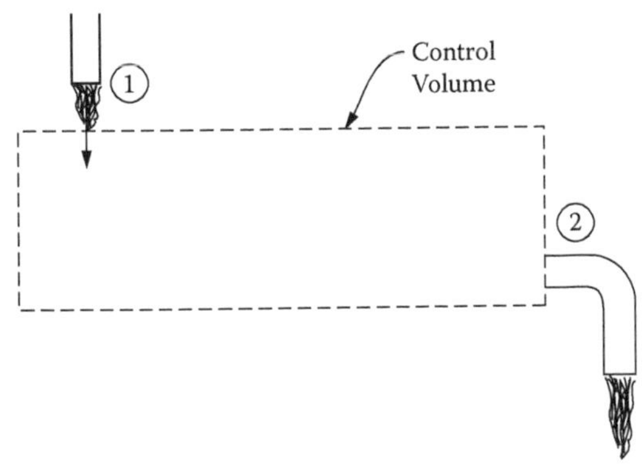

FIGURE 14.7
Stationery control volume for Design Example 7.

we have

$$\dot{W} = \dot{m}(\hat{V}_x)^2 = (416.7\,\text{kg/s})(7.85\,\text{m/s})^2 = 25{,}700\,\text{W} \quad (25.7\,\text{kW}) \Longleftarrow$$

14.5 The Control Volume Relation for the Moment of Momentum

The treatment of angular momentum or moment of momentum is a direct extension of that just considered for linear momentum. Operationally, we will produce a moment by forming the **vector** or **cross product** of a displacement vector, $\boldsymbol{r}$, with each term in our fundamental relationship, Equation 14.2,

$$\boldsymbol{r} \times \sum \boldsymbol{F} = \boldsymbol{r} \times \sum_e \hat{\boldsymbol{V}}_e \dot{m}_e - \boldsymbol{r} \times \sum_i \hat{\boldsymbol{V}}_i \dot{m}_i + \boldsymbol{r} \times \frac{d}{dt} \hat{\boldsymbol{V}}_{\text{cv}} m_{\text{cv}} \tag{14.4}$$

An evaluation of the left-hand side produces

$$\boldsymbol{r} \times \sum \boldsymbol{F} = \sum \boldsymbol{r} \times \boldsymbol{F} = \sum \boldsymbol{M}$$

where $\boldsymbol{M}$ is the net external moment exerted on the control volume by its interaction with the surroundings.

The terms on the right-hand side become

$$\begin{aligned} \boldsymbol{r} \times \sum_e \hat{\boldsymbol{V}}_e \dot{m}_e - \boldsymbol{r} \times \sum_i \hat{\boldsymbol{V}}_i \dot{m}_i + \boldsymbol{r} \times \frac{d}{dt} \hat{\boldsymbol{V}}_{\text{cv}} m_{\text{cv}} \\ = \sum_e (\boldsymbol{r} \times \hat{\boldsymbol{V}}_e) \dot{m}_e - \sum_i (\boldsymbol{r} \times \hat{\boldsymbol{V}}_i) \dot{m}_i + \frac{d}{dt} (\boldsymbol{r} \times \hat{\boldsymbol{V}}_{\text{cv}}) m_{\text{cv}} \end{aligned}$$

With these modifications, Equation 14.4 may be written in operational form as

$$\sum \boldsymbol{M} = \sum_e (\boldsymbol{r} \times \hat{\boldsymbol{V}}_e) \dot{m}_i - \sum_i (\boldsymbol{r} \times \hat{\boldsymbol{V}}_i) \dot{m}_i + \frac{d}{dt} (\boldsymbol{r} \times \hat{\boldsymbol{V}}_{\text{cv}}) m_{\text{cv}} \tag{14.5}$$

This expression is fundamental for the application of Newton's second law to a rotating system. In Cartesian coordinates, the component forms of Equation 14.5 are

$$\sum M_x = \sum_e (\boldsymbol{r} \times \hat{\boldsymbol{V}}_e)\Big|_x \dot{m}_e - \sum_i (\boldsymbol{r} \times \hat{\boldsymbol{V}}_i)\Big|_x \dot{m}_i + \frac{d}{dt} (\boldsymbol{r} \times \hat{\boldsymbol{V}}_{\text{cv}})\Big|_x m_{\text{cv}} \tag{14.6a}$$

$$\sum M_y = \sum_e (\boldsymbol{r} \times \hat{\boldsymbol{V}}_e)\Big|_y \dot{m}_e - \sum_i (\boldsymbol{r} \times \hat{\boldsymbol{V}}_i)\Big|_y \dot{m}_i + \frac{d}{dt} (\boldsymbol{r} \times \hat{\boldsymbol{V}}_{\text{cv}})\Big|_y m_{\text{cv}} \tag{14.6b}$$

$$\sum M_z = \sum_e (\boldsymbol{r} \times \hat{\boldsymbol{V}}_e)\Big|_z \dot{m}_e - \sum_i (\boldsymbol{r} \times \hat{\boldsymbol{V}}_i)\Big|_z \dot{m}_i + \frac{d}{dt} (\boldsymbol{r} \times \hat{\boldsymbol{V}}_{\text{cv}})\Big|_z m_{\text{cv}} \tag{14.6c}$$

It is worth noting some characteristics of the vector operations that are indicated in Equations 14.5 and 14.6. First, one must be careful to observe that the vector cross product is not commutative, that, is $\boldsymbol{r} \times \hat{\boldsymbol{V}} \neq \hat{\boldsymbol{V}} \times \boldsymbol{r}$. For these equations to be valid, care should be taken to always form the cross product as $\boldsymbol{r} \times \hat{\boldsymbol{V}}$.

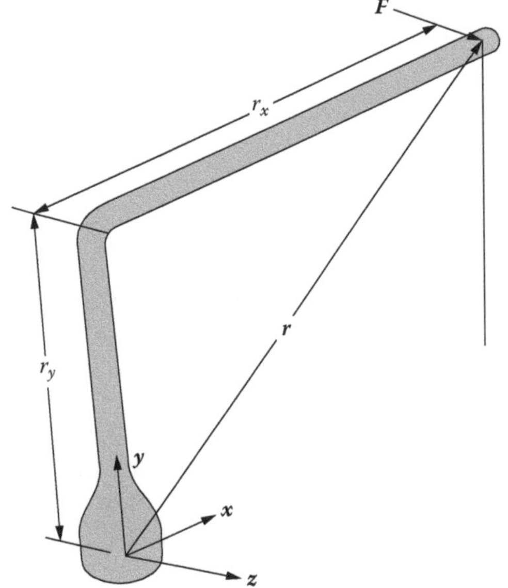

FIGURE 14.8
Vector representation for a moment.

We may also recall that a convenient means for evaluating a cross product is by expanding a 3 × 3 determinant as

$$\boldsymbol{r} \times \boldsymbol{V} = \begin{vmatrix} \boldsymbol{x} & \boldsymbol{y} & \boldsymbol{z} \\ r_x & r_y & r_z \\ \hat{V}_x & \hat{V}_y & \hat{V}_z \end{vmatrix}$$

$$= (r_y\hat{V}_z - r_z\hat{V}_y)\boldsymbol{x} + (r_z\hat{V}_x - r_x\hat{V}_z)\boldsymbol{y} + (r_x\hat{V}_y - r_y\hat{V}_x)\boldsymbol{z}$$

and we take note of the components of $\boldsymbol{r} \times \boldsymbol{V}$

$$(\boldsymbol{r} \times \boldsymbol{V})\Big|_x = r_y\hat{V}_z - r_z\hat{V}_y$$

$$(\boldsymbol{r} \times \boldsymbol{V})\Big|_y = r_z\hat{V}_x - r_x\hat{V}_z$$

$$(\boldsymbol{r} \times \boldsymbol{V})\Big|_z = r_x\hat{V}_y - r_y\hat{V}_x$$

The result of the vector product, $\boldsymbol{r} \times \boldsymbol{F}$, is a vector whose direction is perpendicular to the plane of the two vectors, $\boldsymbol{r}$ and $\boldsymbol{F}$. We may recall, from solid mechanics, that a force, $\boldsymbol{F}$, applied to a socket wrench, as shown in Figure 14.8, will have a vector character that identifies the axis about which the turning tendency is induced. If, in Figure 14.8, the force, $\boldsymbol{F}$, is in the z direction, then the resulting moment is evaluated as

$$\boldsymbol{M} = \boldsymbol{r} \times \boldsymbol{F} = \begin{vmatrix} \boldsymbol{x} & \boldsymbol{y} & \boldsymbol{z} \\ r_x & r_y & 0 \\ 0 & 0 & F \end{vmatrix} = r_y F\boldsymbol{x} - r_x F\boldsymbol{y}$$

In this case, we see that the moment, $\boldsymbol{M}$, is normal to the plane of the two vectors, $\boldsymbol{r}$ and $\boldsymbol{F}$. It is also clear that if $r_y = 0$, $\boldsymbol{r} \times \boldsymbol{F}$ would equal $-(\boldsymbol{F} \times \boldsymbol{r})$, that is, the vector $\boldsymbol{M}$ will point in the $-y$ direction.

These ideas and conventions will become clearer as we consider some examples of moment of momentum applications.

14.6 Applications of the Moment of Momentum Relationship

Example 14.5 Consider Figure 14.9 in which the free-water jet is issuing from a nozzle and striking a fixed blade as was analyzed in Example 14.3. In this example, we want to consider that the blade is connected to a shaft at some distance, r, which is normal to the plane of Figure 14.9. The hub of this blade is oriented such that all actual or induced rotary motion will be about the y axis. The displacement vector, $\boldsymbol{r}$, would then be equal to $r\boldsymbol{z}$. For this case, evaluate the moment about the y axis produced by the water jet as it strikes the blade and has its direction changed. Evaluate the following cases: (a) the blade is stationary and (b) the blade possesses a tangential velocity of 3 m/s at the point of application of the water jet. The radial distance from the axis of rotation is 12.5 cm. As in Example 14.3, the jet exits the nozzle with $\dot{m} = 0.06$ kg/s and $\hat{V} = 7.6$ m/s.

Solution

Assumptions and Specifications

(1) Steady-state conditions exist.
(2) The flow is incompressible.
(3) Uniform properties exist at each cross section considered.

The control volume choice is indicated by the dashed lines in Figure 14.9. The control volume extends in the z direction far enough to include the axis of rotation and cuts through the shaft that is oriented with its axis in the y direction. We will consider a coordinate system that is fixed on this axis, with $\boldsymbol{x}$ and $\boldsymbol{y}$ pointing radially outward. The starting point for the solution of this problem is Equation 14.6b.

(a) Because steady flow exists, Equation 14.6b with the accumulation term suppressed is

$$\sum M_y = \sum_e (\boldsymbol{r} \times \boldsymbol{V}_e)\Big|_y \dot{m}_e - \sum_i (\boldsymbol{r} \times \boldsymbol{V}_i)\Big|_y \dot{m}_i \tag{14.6b}$$

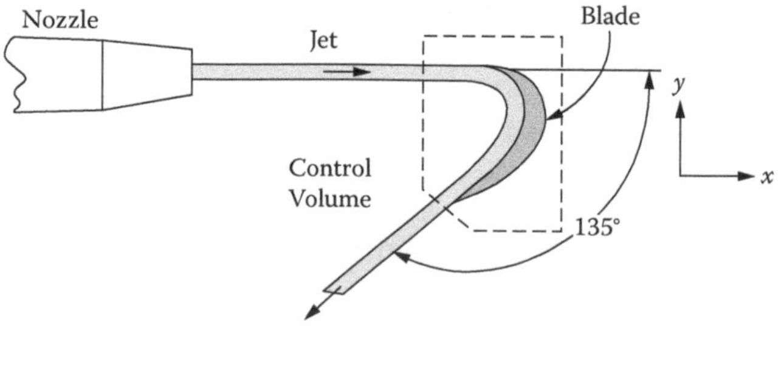

FIGURE 14.9
Free-water jet striking a turning vane for Example 14.5.

Then, observing that mass crosses the boundary at sections 1 and 2, we have

$$M_y = (-r\hat{V}_j \cos 45^\circ \dot{m}) - (r\hat{V}_j\dot{m}) = -r\hat{V}_j\dot{m}(1+\cos 45^\circ)$$

With appropriate numerical values substituted, we obtain

$$M_y = -\frac{(0.125\,\text{m})(7.6\,\text{m/s})(1000\,\text{kg/m}^3)(0.06\,\text{m}^3/\text{s})(1.707)}{1\,\text{kg-m/n-s}^2} = -97.3\,\text{N-m}$$

This is the moment exerted by the surroundings (the shaft in this case) to maintain the blade stationary under the influence of the jet. The quantity that we seek is the moment produced by the water jet, which is the negative of M_y. Thus,

$$M_{y,\text{water}} = 97.3\,\text{N-m} \Longleftarrow$$

(b) The same coordinate system and control volume employed in (a) will be used here. The expression that applies, obtained from Equation 14.6b, is

$$M_y = r\hat{V}_x\bigg|_2 \dot{m} - r\hat{V}_x\bigg|_1 \dot{m}$$

where, because of the motion of the control volume, $\dot{m}$, $\hat{V}_{x1}$ and $\hat{V}_{x2}$ will have values different than those used in (a). The proper values for these quantities are

$$\dot{m} = (\rho A\hat{V})_{\text{in}} = (\rho A\hat{V})_{\text{out}}$$

$$\hat{V}_{x1} = 7.6\,\text{m/s}$$

and

$$\begin{aligned}\hat{V}_{x2} &= (7.6\,\text{m/s} - 3.0\,\text{m/s})(-\cos 45^\circ) + 3.0\,\text{m/s}\\ &= -3.25\,\text{m/s} + 3.0\,\text{m/s}\\ &= -0.25\,\text{m/s}\end{aligned}$$

Having these values, we may complete the solution to obtain

$$\begin{aligned}M_y &= \dot{m}r(\hat{V}_{x2} - \hat{V}_{x1})\\ &= \frac{(0.125\,\text{m})(1000\,\text{kg/m}^3)(0.06\,\text{m}^3/\text{s})(-0.25\,\text{m/s} - 7.6\,\text{m/s})}{1\,\text{kg-m/n-s}^2}\\ &= -58.87\,\text{N-m}\end{aligned}$$

As before, the moment produced on the shaft by the water is

$$M_{y,\text{water}} = 58.87\,\text{N-m} \Longleftarrow$$

In the foregoing example, we have performed the basic analysis of a **water turbine**. In the case of a real turbine, there would, naturally, be many blades attached to a rotating hub, where as one blade rotates out of the path of the water jet, another rotates into position, so that the process is continued.

We might also note that **maximum moment** is produced when there is no rotation. However, the desired output from a turbine is **power**, which is the product of angular velocity and the moment (or torque) produced. Thus, we are interested in the combination of angular velocity and moment that will produce maximum power.

A turbine is one type of rotating machine that is of interest to engineers. Chapter 18 discusses, in some detail, the concepts relating to rotating machinery and their operational features.

Example 14.6 A schematic of a lawn sprinkler is shown in Figure 14.10. Water flows through the sprinkler at a rate of 0.001 m^3/min and the sprinkler rotates at 40 rpm. Determine the torque due to friction at the sprinkler hub.

Solution

Assumptions and Specifications

(1) Steady-state conditions exist.
(2) The flow is incompressible.
(3) Uniform properties exist at each cross section considered.

The control volume and coordinate system are shown in Figure 14.10. The sought-after moment will be z directed so the moment of momentum expression to be solved is Equation 14.6c

$$\sum M_z = \sum_e (\boldsymbol{r} \times \boldsymbol{V}_e)\bigg|_z \dot{m}_e - \sum_i (\boldsymbol{r} \times \boldsymbol{V}_i)\bigg|_z \dot{m}_i + \frac{d}{dt}(\boldsymbol{r} \times \boldsymbol{V}_{\text{cv}})\bigg|_z m_{\text{cv}} \tag{14.6c}$$

where, for steady-state operation, the accumulation term will be zero, leaving

$$M_z = \sum_e (\boldsymbol{r} \times \boldsymbol{V}_e)\bigg|_z \dot{m}_e - \sum_i (\boldsymbol{r} \times \boldsymbol{V}_i)\bigg|_z \dot{m}_i$$

In evaluating the remaining terms, we note that mass crosses the control surface at three locations; it enters vertically through the hub and exits from the two sprinkler heads. Accordingly, the expression for M_z reduces to

$$M_z = \sum_2 (\boldsymbol{r} \times \boldsymbol{V}_2)\dot{m}_2 - \sum_1 (\boldsymbol{r} \times \boldsymbol{V}_1)\dot{m}_1$$

where section 1 is the entering section and section 2 represents the surface across which the flow leaves.

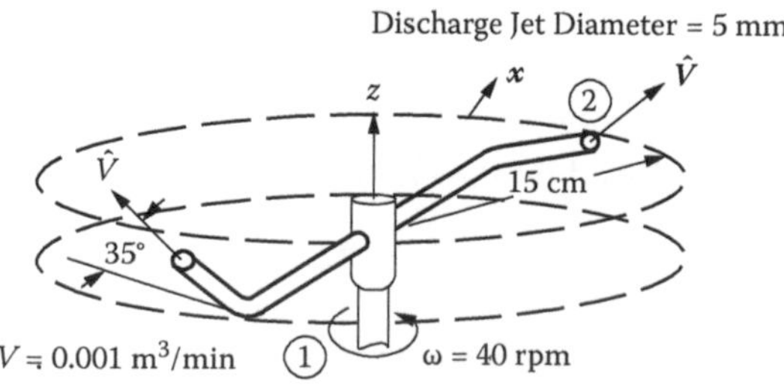

FIGURE 14.10
Lawn sprinkler for Example 14.6.

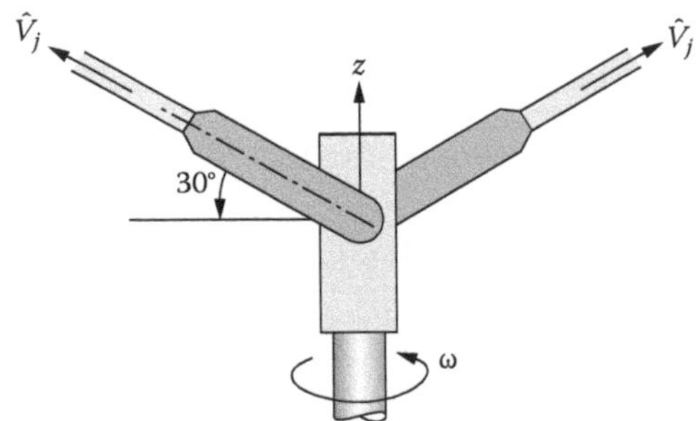

FIGURE 14.11
View of the lawn sprinkler of Example 14.6 looking down at the radial part of the sprinkler arm.

Because the coordinate reference frame is stationary, we must think through the representation of the exiting velocity. Figure 14.11 examines the exit flow from a perspective looking down the radial part of the sprinkler arm. With the cross-sectional area of the jet openings expressed as

$$A_j = 2\left(\frac{\pi}{4}\right)(0.005\,\mathrm{m})^2 = 3.927 \times 10^{-5}\,\mathrm{m}^2$$

the magnitude of the jet velocity relative to the sprinkler arm is seen to be

$$\hat{V}_j = \frac{\dot{V}}{A_j} = \frac{(0.001\,\mathrm{m}^3/\mathrm{min})(1\,\mathrm{min}/60\,\mathrm{s})}{3.927 \times 10^{-5}\mathrm{m}^2} = 0.424\,\mathrm{m/s}$$

The tangential velocity of the sprinkler arm is

$$\hat{V}_t = r\omega = (0.15\,\mathrm{m})(40\,\mathrm{rev/min})(2\pi\,\mathrm{rad/rev})(1\,\mathrm{min}/60\,\mathrm{s}) = 0.628\,\mathrm{m/s}$$

The velocity of one exit stream, relative to a fixed coordinate frame, is the vector sum of the jet and tangential velocities. The absolute velocity can thus be expressed as

$$\begin{aligned} \boldsymbol{V}_2 &= (-\hat{V}_j \cos 30^\circ \boldsymbol{\theta} + \hat{V}_j \sin 30^\circ \boldsymbol{z} + \hat{V}_t \boldsymbol{\theta})\,\mathrm{m/s} \\ &= [(-0.424\cos 30^\circ + 0.628)\boldsymbol{\theta} + 0.424 \sin 30^\circ \boldsymbol{z}]\,\mathrm{m/s} \\ &= 0.261\boldsymbol{\theta}\,\mathrm{m/s} + 0.212\boldsymbol{z}\,\mathrm{m/s} \end{aligned}$$

We may also represent the inlet velocity as

$$\boldsymbol{V}_1 = \hat{V}_1 \boldsymbol{z}$$

and the solution for $\boldsymbol{M}$, a vector, is now a matter of substituting the foregoing quantities into

$$\boldsymbol{M} = 2(\boldsymbol{r} \times \boldsymbol{V}_2)\dot{m} - (\boldsymbol{r} \times \boldsymbol{V}_1)\dot{m}$$

or

$$\boldsymbol{M} = 2\begin{vmatrix} \boldsymbol{r} & \boldsymbol{\theta} & \boldsymbol{z} \\ -0.15 & 0 & 0 \\ 0 & 0.261 & 0.212 \end{vmatrix}\dot{m} - \begin{vmatrix} \boldsymbol{r} & \boldsymbol{\theta} & \boldsymbol{z} \\ 0 & 0 & 0 \\ 0 & 0 & \hat{V}_1 \end{vmatrix}(-\dot{m})$$

or

$$\boldsymbol{M} = (-0.15)(0.261)\boldsymbol{z}\,\text{N-m} - (-0.15)0.212)\boldsymbol{\theta}\,\text{N-m}$$

We may now obtain a numerical value for the component of $\boldsymbol{M}$ in the z-coordinate direction as

$$M_z = \frac{2(-0.15\,\text{m})(0.001\,\text{m}^3/60\,\text{s})(0.261\,\text{m/s})(1000\,\text{kg/m}^3)}{1\,\text{kg-m/N-s}^2}$$
$$= -0.00131\,\text{N-m} \Longleftarrow$$

This is the result sought, the torque imposed on the sprinkler shaft by friction. This torque balances the moment of momentum produced by the water jets. Notice that the external moment exerted on the shaft is negative.

14.7 Summary

In this chapter we have developed and applied the control volume expression for Newton's second law of motion. The two fundamental integral expressions are Equation 14.2 for linear momentum

$$\sum \boldsymbol{F} = \sum_e \boldsymbol{V}_e \dot{m}_e - \sum_i \boldsymbol{V}_i \dot{m}_i + \frac{d}{dt}\boldsymbol{V}_{\text{cv}} m_{\text{cv}} \qquad (14.2)$$

and Equation 14.5 for the moment of momentum

$$\sum \boldsymbol{M} = \sum_e (\boldsymbol{r} \times \boldsymbol{V}_e)\dot{m}_e - \sum_i (\boldsymbol{r} \times \boldsymbol{V}_i)\dot{m}_i + \frac{d}{dt}(\boldsymbol{r} \times \boldsymbol{V}_{\text{cv}}) m_{\text{cv}} \qquad (14.5)$$

These expressions are extremely powerful and form the basis for a significant portion of fluid flow analysis.

14.8 Problems

The Cross Product

14.1: Form the cross product $\boldsymbol{A} \times \boldsymbol{B}$ if

$$\boldsymbol{A} = 2\boldsymbol{x} + 3\boldsymbol{y} + 4\boldsymbol{z} \quad \text{and} \quad \boldsymbol{B} = -\boldsymbol{x} + 2\boldsymbol{z}$$

14.2: Form the cross product $\boldsymbol{A} \times \boldsymbol{B}$ if

$$\boldsymbol{A} = \boldsymbol{x} + 2\boldsymbol{y} + 4\boldsymbol{z} \quad \text{and} \quad \boldsymbol{B} = \boldsymbol{x} + 4\boldsymbol{z}$$

14.3: Form the cross product $\boldsymbol{A} \times \boldsymbol{B}$ if

$$\boldsymbol{A} = 2\boldsymbol{y} \quad \text{and} \quad \boldsymbol{B} = -8\boldsymbol{z}$$

14.4: Form the cross product $\boldsymbol{A} \times \boldsymbol{B}$ if

$$\boldsymbol{A} = \boldsymbol{x} + \boldsymbol{y} + \boldsymbol{z} \quad \text{and} \quad \boldsymbol{B} = 2(\boldsymbol{x} - \boldsymbol{y} - \boldsymbol{z})$$

14.5: Form the cross product $\boldsymbol{A} \times \boldsymbol{B}$ if

$$\boldsymbol{A} = \boldsymbol{x} + 2\boldsymbol{y} - 3\boldsymbol{z} \quad \text{and} \quad \boldsymbol{B} = -\boldsymbol{y} + 2\boldsymbol{z}$$

14.6: Form the cross product $A \times B$ if

$$A = x - 2y - 3z \quad \text{and} \quad B = x + y + 2z$$

14.7: Form the cross product $A \times B$ if

$$A = 2x + y \quad \text{and} \quad B = x - 2z$$

14.8: Form the cross product $A \times B$ if

$$A = 2z \quad \text{and} \quad B = -8x$$

14.9: Form the cross product $A \times B$ if

$$A = 24x - 36y + 48y \quad \text{and} \quad B = 12x - 24y$$

14.10: Form the cross product $A \times B$ if

$$A = 6x - 8y + 10z \quad \text{and} \quad B = 2(3x + 4y - 5z)$$

Forces Developed by Fluids in Motion

14.11: Mud, at a rate of 3200 kg/s, is discharged to a barge at a velocity of 1.25 m/s as shown in Figure P14.11. Determine the tension on the mooring line.

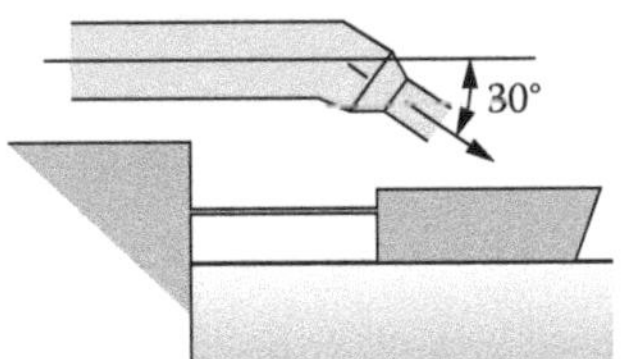

FIGURE P14.11

14.12: Water is pumped through a 30-cm-diameter pipe entering the bow of a boat at $\hat{V} = 2.05$ m/s. The pipe tapers to a diameter of 20 cm at the stern of the boat where it is discharged. Determine the thrust developed.

14.13: In Figure P14.13, the vane manages to turn the jet completely around. If the maximum support force is F_o, determine an expression for the maximum water jet velocity, $\hat{V}_o$.

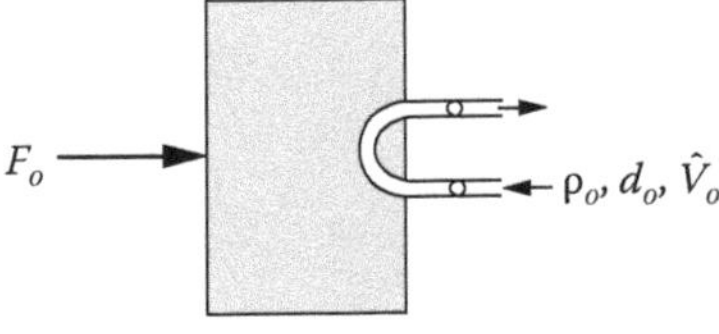

FIGURE P14.13

14.14: In the sluice gate arrangement shown in Figure P14.14, uniform flow and hydrostatic pressure may be assumed at section 1 upstream of the gate and section 2 downstream of the gate. Derive an expression for the force, F, required to hold the gate in place as a function of ρ, $\hat{V}$, g, h_1, and h_2.

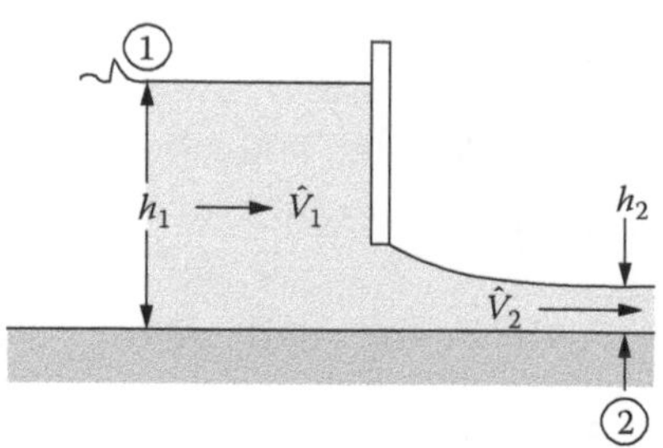

FIGURE P14.14

14.15: A small box rests on a platform which, in turn, rests on a steady water jet with a flow area of 0.03 m^2. If the total weight is 750 N, determine the jet velocity.

14.16: The tank in Figure P14.16 has an empty weight of 925 N and holds 1.125m^3 of water at 20°C. The entrance and exit pipes both have a 6-cm diameter and both carry 82.5 L/min. Determine the value of the scale reading, *W*.

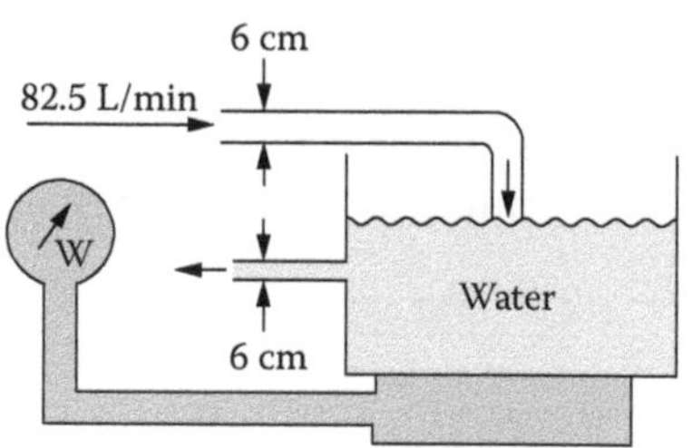

FIGURE P14.16

14.17: A 2.5-cm-diameter jet of water hits a flat plate held normal to the axis of the jet. The force exerted by the jet on the plate is 650 N. Determine the volumetric flow rate of the water.

14.18: A jet of water having a velocity of 12.5 m/s and a volumetric flow rate of 6×10^{-4} m^3/s is deflected through a right angle as shown in Figure P14.18. Neglect friction and determine the magnitude and direction of the resultant force.

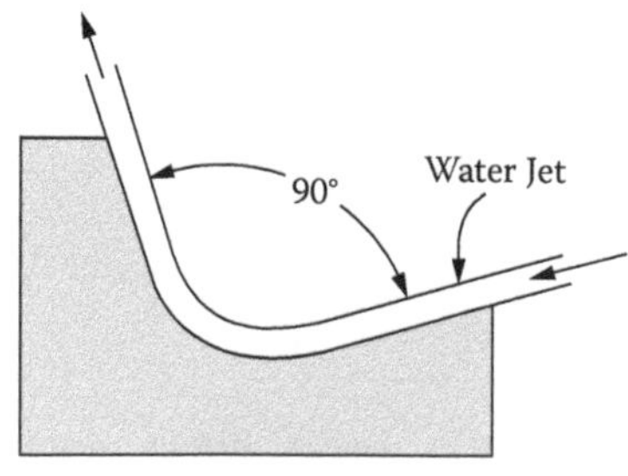

FIGURE P14.18

14.19: In Figure P14.19, the spring has a stiffness of 1.625 kN/m. Determine the force on the wheels caused by the deflection of the water jet and the compression of the spring.

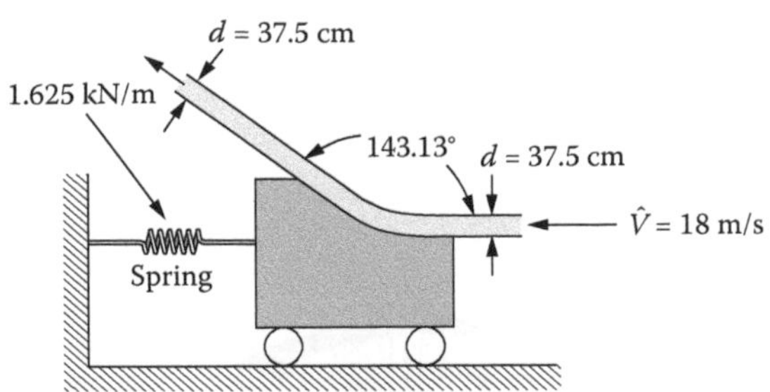

FIGURE P14.19

14.20: The water tank in Figure P14.20 sits on a frictionless cart and feeds a jet of of 7.5 cm in diameter at a velocity 12 m/s. Determine the tension in the cable.

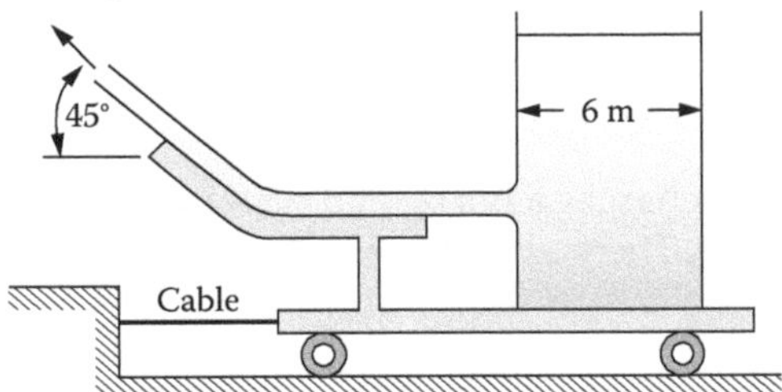

FIGURE P14.20

14.21: A free jet of water is deflected by a 90° curved vane as indicated in Figure P14.21. The water jet has a diameter of 4 cm and a velocity of 8 m/s. Determine the force required to hold the vane in place.

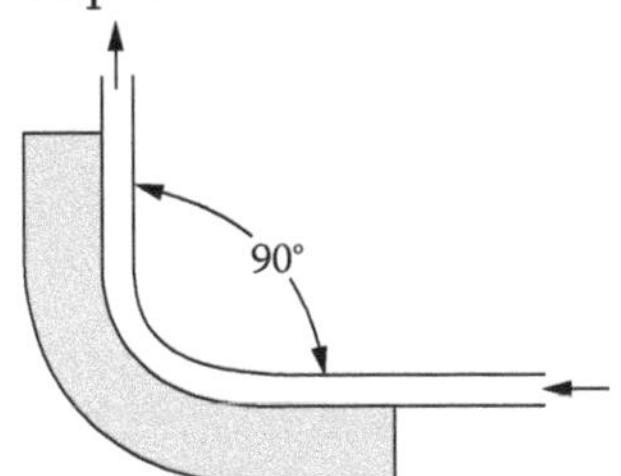

FIGURE P14.21

14.22: Refer once again to Problem 14.14 and the sluice gate in Figure P14.14. If $\hat{V}_1 = 1.375\,\text{m/s}$, $h_1 = 4.125\,\text{m}$, and $h_2 = 0.75\,\text{m}$, determine the force per unit width required to hold the gate in place.

14.23: The water jet in Figure P14.23 moving at 15 m/s strikes a stationary splitting vane such that 40% of the water moves downward. Assume ideal flow in the horizontal plane and determine the magnitude and direction of the force on the splitter.

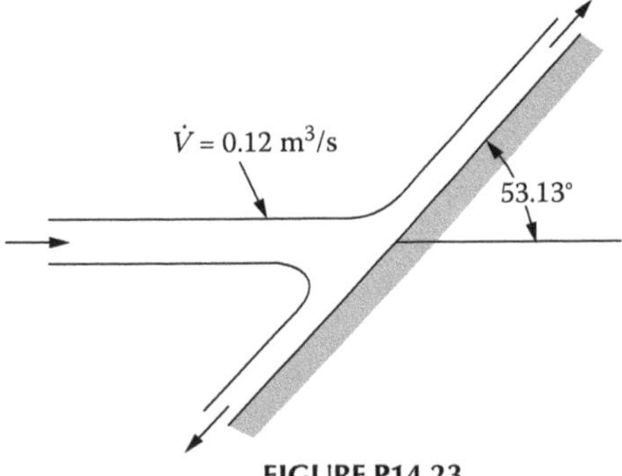

FIGURE P14.23

14.24: The horizontal nozzle in Figure P14.24 has diameters of $d_1 = 25\,\text{cm}$ and $d_2 = 15\,\text{cm}$. The inlet pressure is 400 kPa and the exit velocity is 27.5 m/s. Assuming incompressible steady flow, determine the total tensile force required to hold the nozzle stationary.

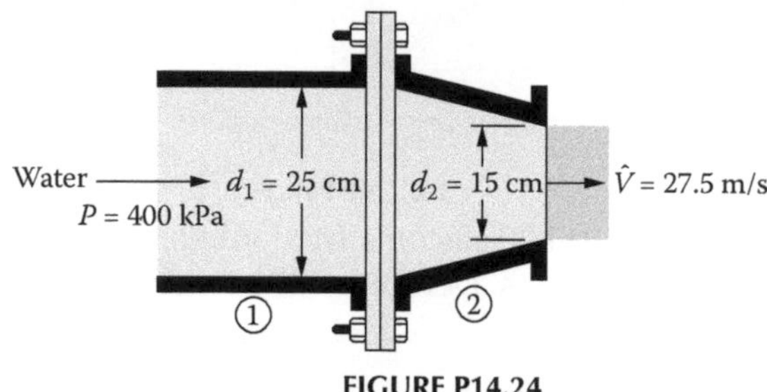

FIGURE P14.24

14.25: A hose and nozzle discharges a horizontal water jet against a vertical plate as indicated in Figure P14.25. The volumetric flow rate of the water is $0.045\,\text{m}^3/\text{s}$ and the diameter of the nozzle is 5 cm. Determine the force necessary to hold the plate in place.

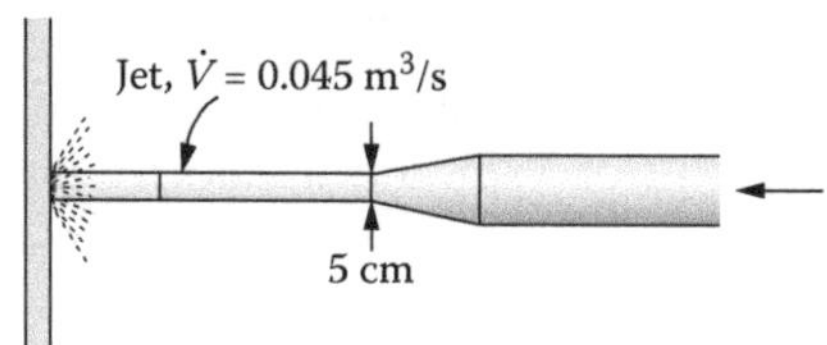

FIGURE P14.25

Linear Momentum

14.26: A two-dimensional object is placed in a 1.6-m-wide water tunnel as shown in Figure P14.26. The upstream velocity, $\hat{V}_1$, is uniform across the cross section. For the downstream velocity profile shown, determine the value of $\hat{V}_2$.

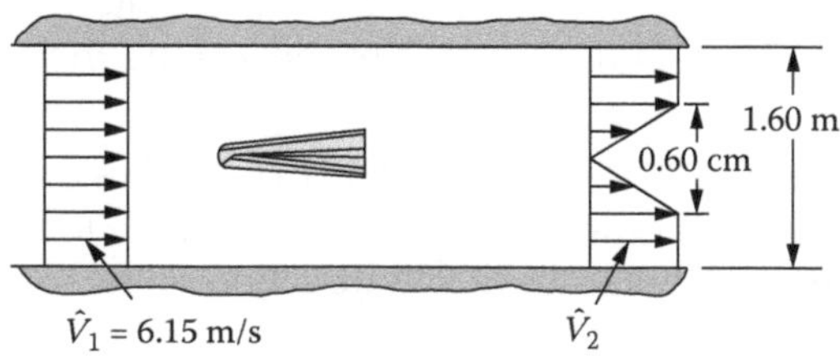

FIGURE P14.26

14.27: A water jet pump has an area, $A_j = 7.50 \times 10^{-3}\,\text{m}^2$ and a jet velocity of $\hat{V}_j = 17.5\,\text{m/s}$, which entrains a second stream of water having a velocity of $\hat{V}_s = 3.2\,\text{m/s}$ in a constant area pipe of total area $A = 0.0560\,\text{m}^2$. The configuration is shown in Figure P14.27 and, at section 2, the water is throughly mixed. Assuming one-dimensional flow and neglecting wall shear, determine (a) the average velocity of mixed flow at section 2 and (b) the pressure rise, $P_2 - P_1$, assuming the pressure of the jet and secondary stream to be the same at section 1.

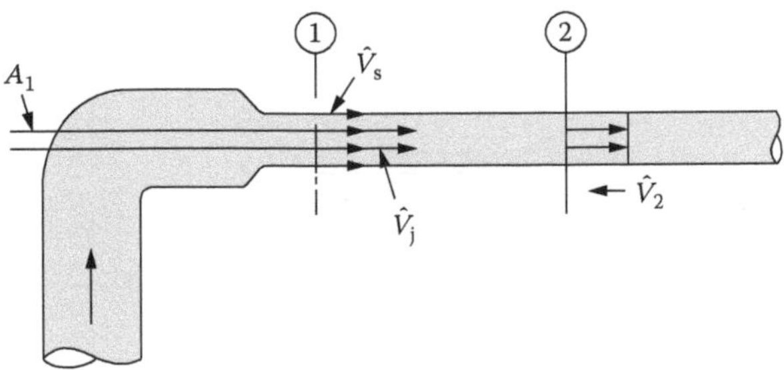

FIGURE P14.27

14.28: The pressure on the control volume and the x components of velocity are as illustrated in Figure P14.28. Assuming incompressible flow, determine the forces exerted on the cylinder by the field.

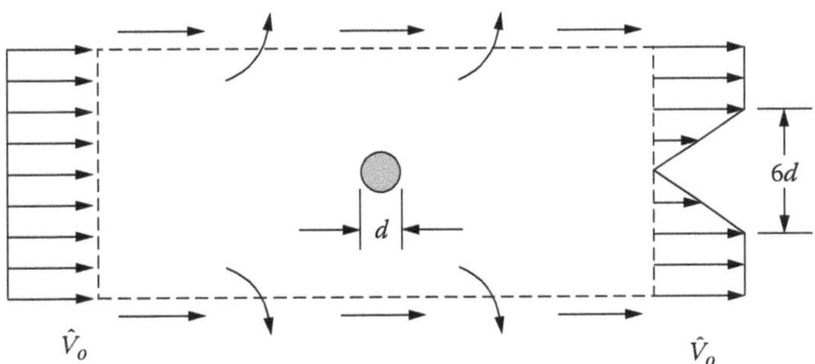

FIGURE P14.28

14.29: The volumetric flow rate of a jet of water is $0.056\,\text{m}^3/\text{s}$ and its velocity is $8\,\text{m/s}$. Determine (a) the magnitude of the x and y components of the force exerted on the fixed blade in Figure P14.29 and (b) the magnitude and velocity of the water jet leaving the blade if the blade is moving to the left at $4.8\,\text{m/s}$.

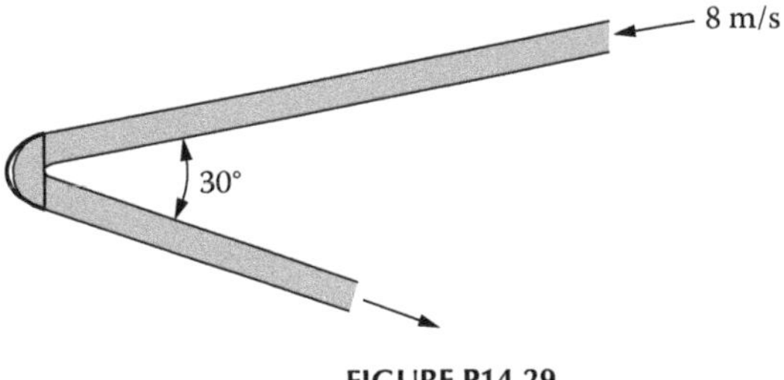

FIGURE P14.29

14.30: A plate moves perpendicularly toward a discharging jet at $1.5\,\text{m/s}$. The jet discharges water at the rate of $0.056\,\text{m}^3/\text{s}$ and a velocity of $9.6\,\text{m/s}$; the flow is frictionless. Determine (a) the force of the fluid on the plate and (b) the force of the fluid on the plate if the plate were stationary.

14.31: Air with a density of $\rho = 1.2\,\text{kg/m}^3$ flows in the 25-cm-diameter duct at $10\,\text{m/s}$. The duct is choked at its exit by the presence of a 90° cone as shown in Figure P14.31. Neglect air friction and estimate the force of the airflow on the cone.

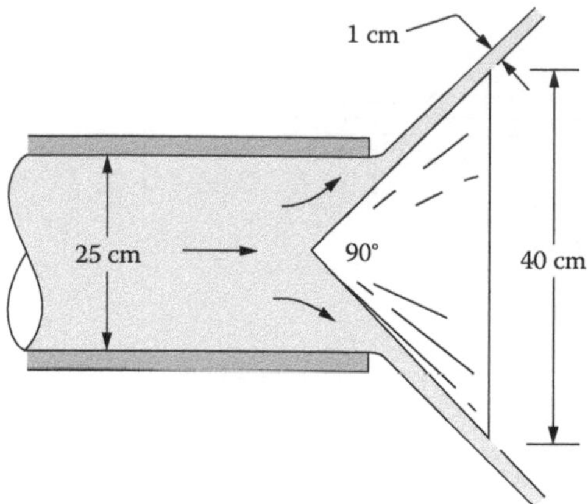

FIGURE P14.31

14.32: The water jet in Figure P14.32, is 12 cm in diameter with a volumetric flow rate of $0.10\,m^3/s$. It flows over a fixed cone with a base diameter of 40 cm. The water leaves as a conical sheet with the same velocity as the jet. Determine the force in newton required to hold the cone fixed.

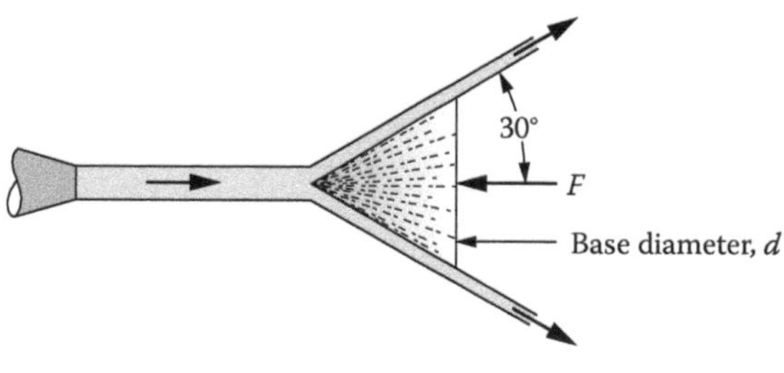

FIGURE P14.32

14.33: Oil with a specific gravity of 0.80 flows smoothly through the circular reducing section shown in Figure P14.33 at a volumetric flow rate of $0.085\,m^3/s$. The entering and leaving velocity profiles are uniform. Estimate the force that must be applied to the reducer to hold it in place.

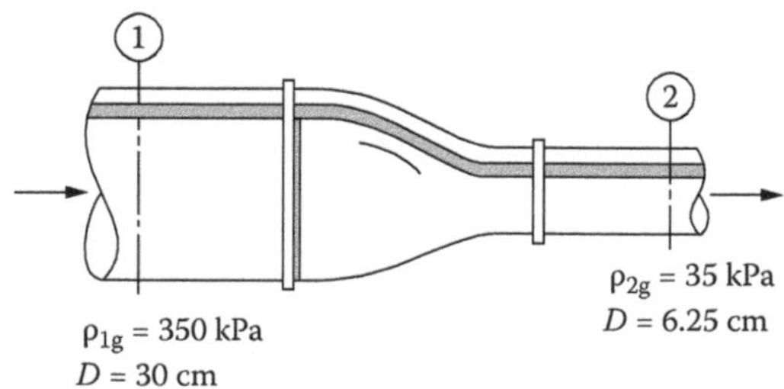

FIGURE P14.33

14.34: A nozzle that discharges a jet of water having a diameter of 3.60 cm into the atmosphere is at the end of a 7.5-cm-diameter pipe. The pressure in the pipe is 400 kPa gage and the rate of discharge is $1.50\,m^3/min$. Determine the magnitude and direction of the force required to hold the nozzle to the pipe.

14.35: Water flows steadily through the horizontal 30° pipe bend shown in Figure P14.35. At station 1, the diameter is 30 cm, the velocity is 12 m/s and the pressure is 128 kPa gage. At station 2 the diameter is 38 cm and the pressure is 145 kPa gage. Determine the forces, F_x and F_y, necessary to hold the pipe bend stationary.

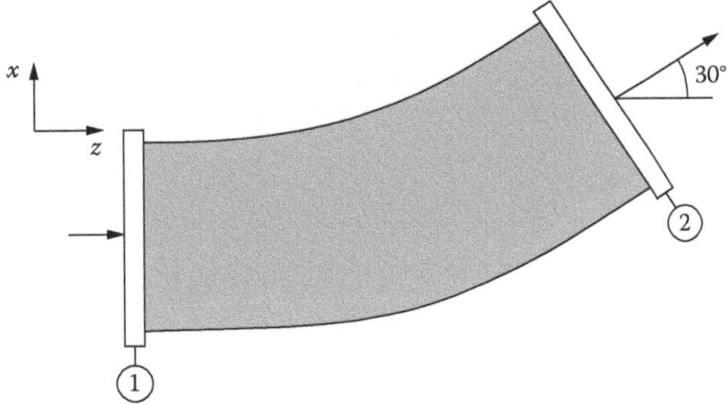

FIGURE P14.35

14.36: The rocket nozzle shown in Figure P14.36 consists of three welded sections. Determine the axial stress at junctions 1 and 2 when the rocket is operating at sea level and the mass flow rate is 350 kg/s.

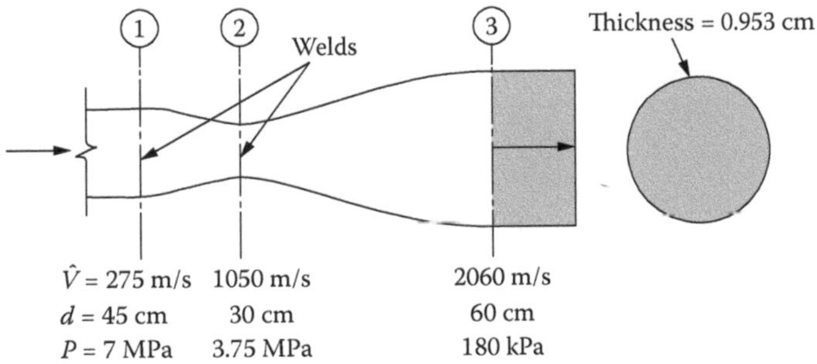

FIGURE P14.36

14.37: Figure P14.37 represents an open tank car that travels to the right at a uniform velocity of 4.5 m/s. At the instant shown, the car passes under a jet of water issuing from a 10-cm-diameter pipe at a velocity of 10 m/s. Determine the force exerted on the tank car by the water jet.

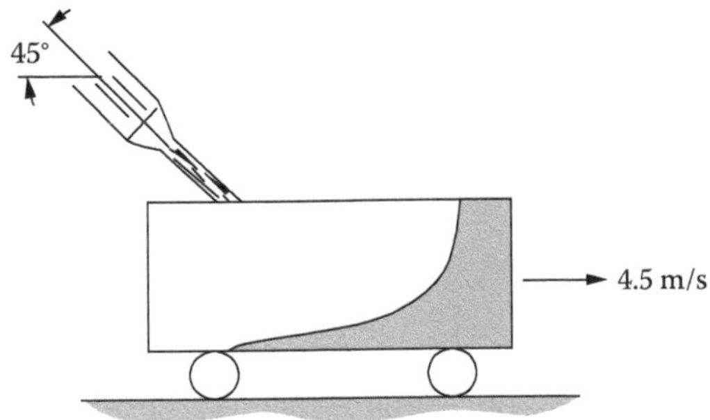

FIGURE P14.37

14.38: The open tank car, shown in Figure P14.38, travels to the right at a velocity of $\hat{V}_c$ m/s. A jet of area A_j exhausts the fluid of density ρ at a velocity of $\hat{V}_j$ m/s relative to the tank car. The tank car, at the same time, collects from an overhead sprinkler fluid that directs fluid downward with velocity $\hat{V}_g$. Assume that the sprinkler flow is uniform over the car area. In terms of the quantities indicated and determine the net force of the fluid on the tank car.

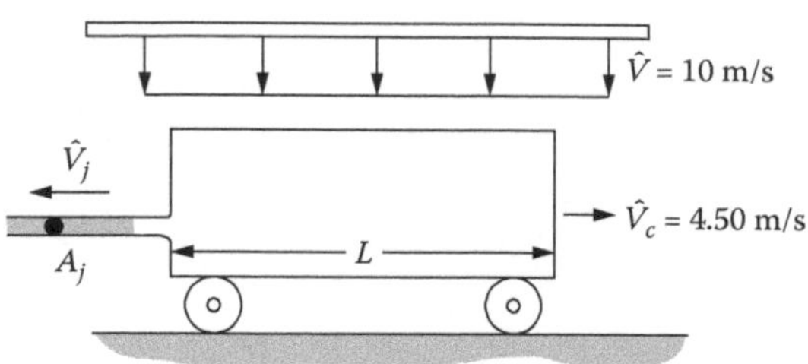

FIGURE P14.38

14.39: The cart shown in Figure P14.39 is supported by wheels and a linear spring. The water jet is deflected 45° by the cart. Determine (a) the force on the wheels caused by the jet and (b) the spring deflection compared with its unstressed position.

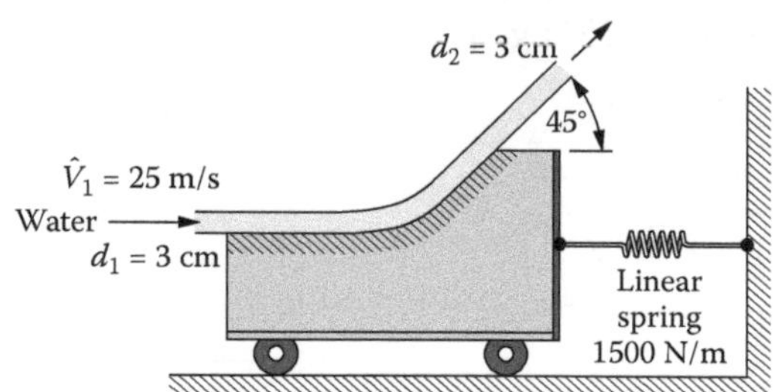

FIGURE P14.39

14.40: The test section of a stationary jet engine is shown in Figure P14.40. Air with density 1.275 kg/m^3 enters as shown. The inlet and outlet cross-sectional areas are both 1 m^2 and the mass of fuel consumed is 1% of the mass of the air entering the test section. For these conditions, determine the thrust developed by the engine.

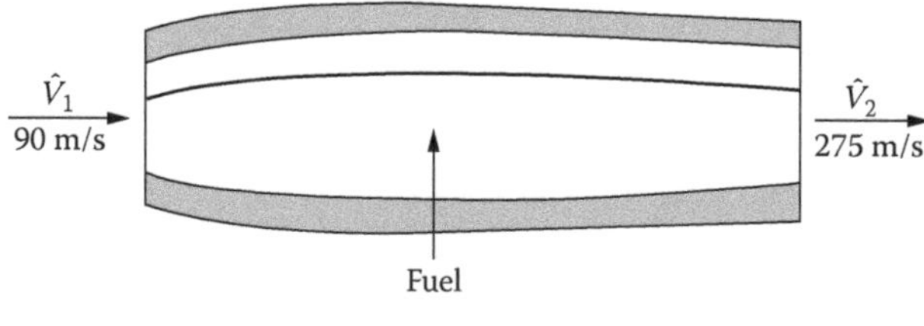

FIGURE P14.40

14.41: For the system of Problem 14.26, the total drag on the object is measured to be 600 N/m of length normal to the direction of flow. Neglecting frictional forces, determine the pressure difference between the inlet and outlet conditions.

14.42: The pump in the boat shown in Figure P14.42 pumps 0.25 m^3/s through the submerged water passage that has an area of 0.030 m^2 at the bow and 0.180 m^2 at the stern. Assuming that the inlet and exit pressures are equal, determine the tension in the retraining rope.

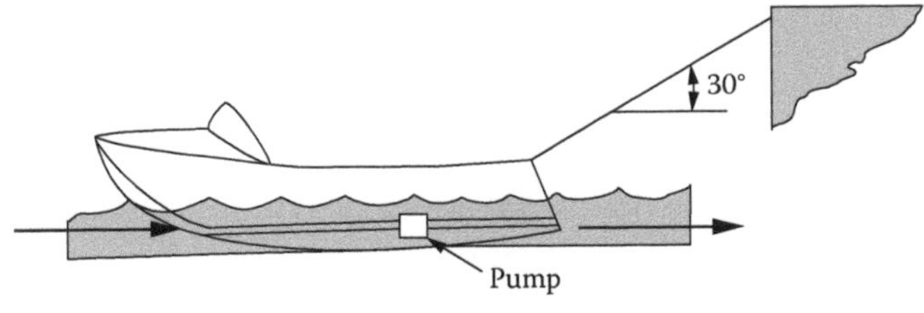

FIGURE P14.42

14.43: The boat shown in Figure P14.43 is being driven at a steady speed, $\hat{V}_o$, by a jet of compressed air issuing from a 3-cm-diameter hole at $\hat{V}_e = 350\,\text{m/s}$. Jet exit conditions are $P_e = 100\,\text{kPa}$ and $T_e = 20°\text{C}$. The drag force on the hull as it moves through the water is given by $K\hat{V}_o^2$, where $K = 20\,\text{N/(m/s)}^2$. The air drag on the upper surfaces of the boat is negligible. Determine the steady boat speed, $\hat{V}_o$.

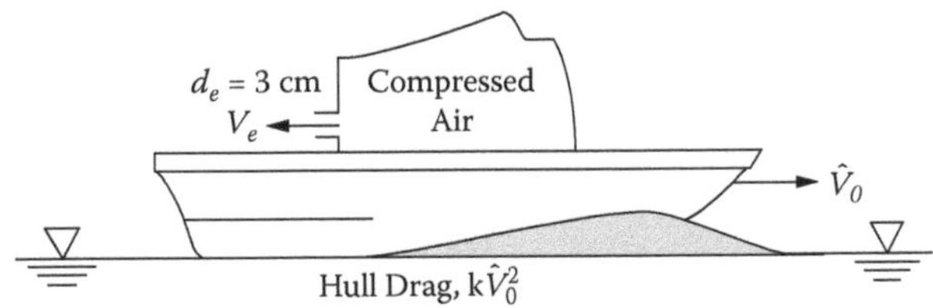

FIGURE P14.43

14.44: Figure P14.44 considers a shock wave moving to the right at $\hat{V}_o$ m/s. The properties ahead and behind the shock wave are not a function of time. By using the illustrated control volume, show that the pressure difference across the shock is

$$P_2 - P_1 = \rho_1 \hat{V}_w \hat{V}_2$$

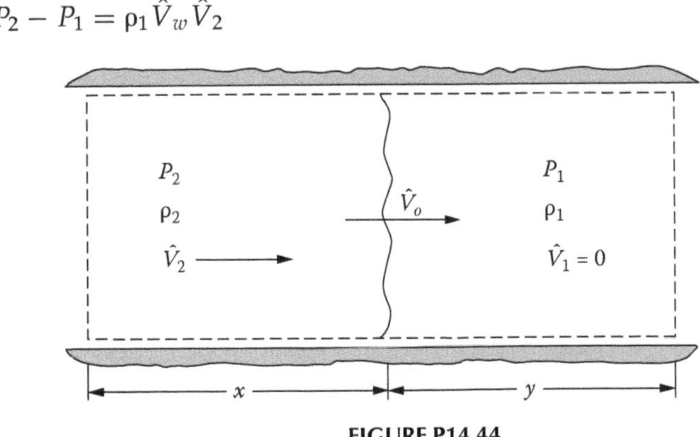

FIGURE P14.44

14.45: If the shock-wave velocity in Problem 14.44 is approximately the speed of sound, determine the pressure change required to cause a velocity change of 3 m/s in (a) air at standard conditions ($\hat{V}_c = 344\,\text{m/s}$) and (b) water ($\hat{V}_c = 1430\,\text{m/s}$).

14.46: Water enters section 1 in the pipe bend shown in Figure P14.46 at a volumetric flow rate of 0.20 m^3/s and exits at a 30° angle at section 2. With r as a radial coordinate and R as the pipe radius, section 1 has the laminar profile

$$\hat{V} = \hat{V}_{m1}\left(1 - \frac{r^2}{R^2}\right)$$

while section 2 has changed to a turbulent profile

$$\hat{V} = \hat{V}_{m2}\left(1 - \frac{r}{R}\right)^{1/7}$$

For steady incompressible flow, what are the maximum velocities, $\hat{V}_{m1}$ and $\hat{V}_{m2}$, in meters per second?

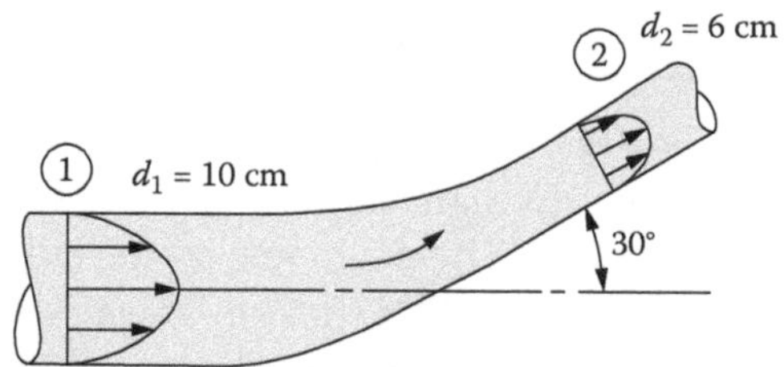

FIGURE P14.46

14.47: The plate shown in Figure P14.47 is inclined at an angle of 53.13° and the flow is frictionless. Determine the forces F_x and F_y necessary to maintain its position.

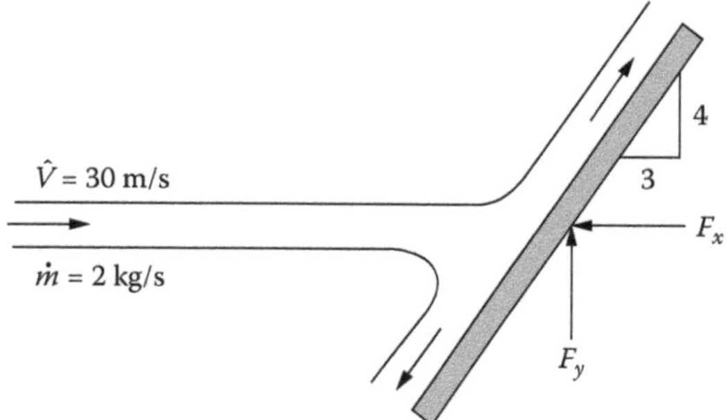

FIGURE P14.47

14.48: A steady, incompressible, frictionless, two-dimensional jet of fluid with breadth, h, velocity, $\hat{V}$, and unit width impinges on a flat plate held at an angle, α, to its axis as indicated in Figure P14.48. Neglecting gravity forces, determine (a) the total force on the plate and the breadths a and b of the two branches and (b) the distance, ℓ, to the center of pressure along the pipe from point 0.

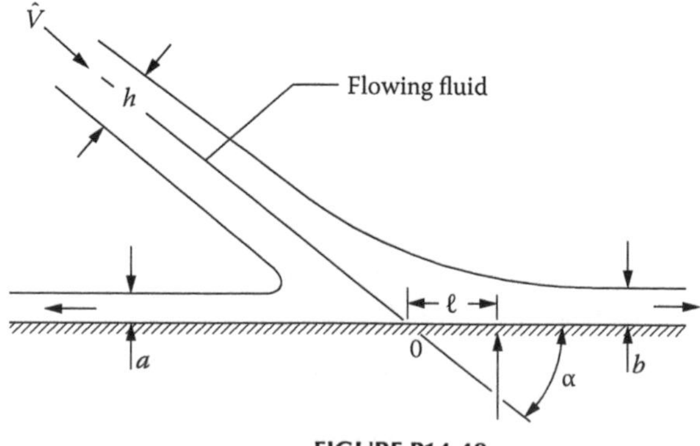

FIGURE P14.48

14.49: Water flows in a pipe at 4 m/s. A valve at the downstream end of the pipe is suddenly closed. Determine the pressure rise in the pipe.

14.50: A liquid of density, ρ, flows through a sluice gate as shown in Figure P14.50. The upstream and downstream flows are uniform and parallel so that the pressure variations at stations 1 and 2 may be considered to be hydrostatic. Determine (a) the velocity at station 2 and (b) the force per unit width R necessary to hold the sluice gate in place.

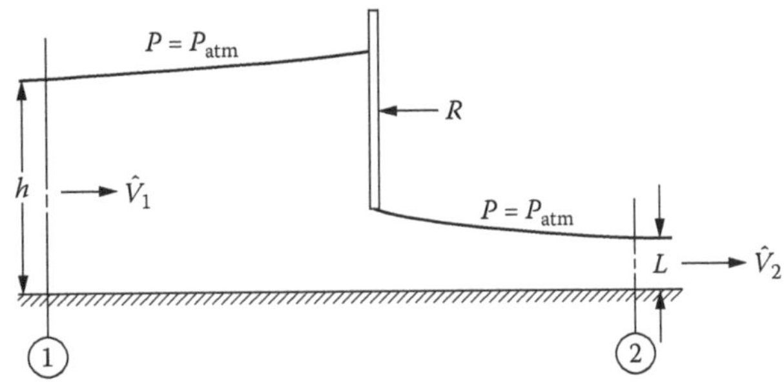

FIGURE P14.50

14.51: For the pipe flow reducing section shown in Figure P14.51, $d_1 = 8\,\text{cm}$, $d_2 = 5\,\text{cm}$, and P_2 is approximately atmospheric. The entrance velocity $\hat{V}_1$ is 6 m/s and the manometer reading is $h = 28\,\text{cm}$. Estimate the force resisted by the flange bolts.

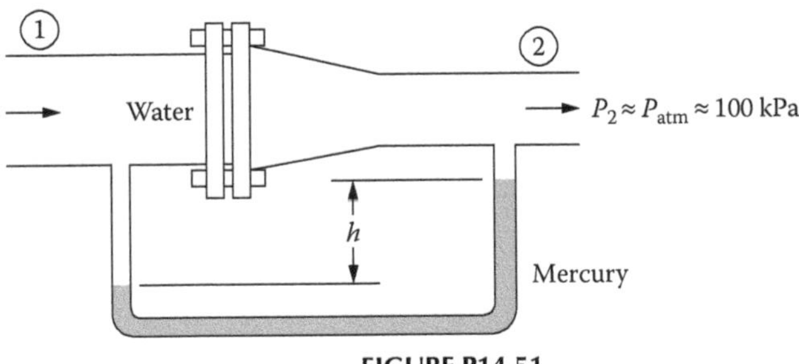

FIGURE P14.51

14.52: A jet engine on a test stand (Figure P14.52) takes in atmospheric air at 20°C at section 1, where $\hat{V}_1 = 200\,\text{m/s}$ and $A_1 = 0.30\,\text{m}^2$. The fuel to air ratio is 1:25 and the air leaves at section 2 where the area is $A_2 = 25\,\text{m}^2$ at atmospheric pressure with a velocity of $\hat{V}_2 = 1000\,\text{m/s}$. Determine the test stand reaction that balances the thrust of the engine in steady flow, R_x.

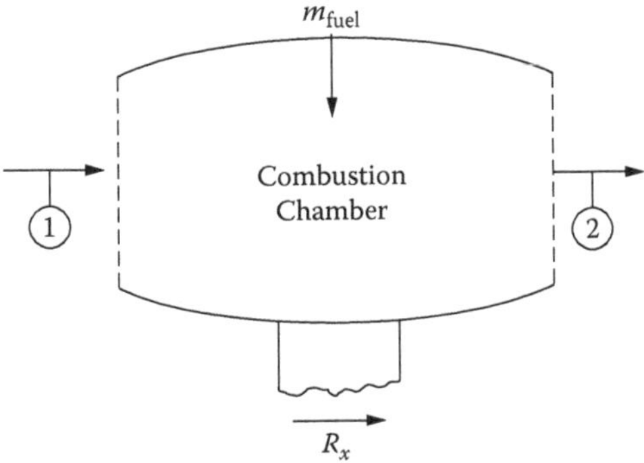

FIGURE P14.52

14.53: In Figure P14.53, the water jet strikes normal to the plate. Neglect gravity and friction and determine the force, F, in newton required to hold the plate fixed.

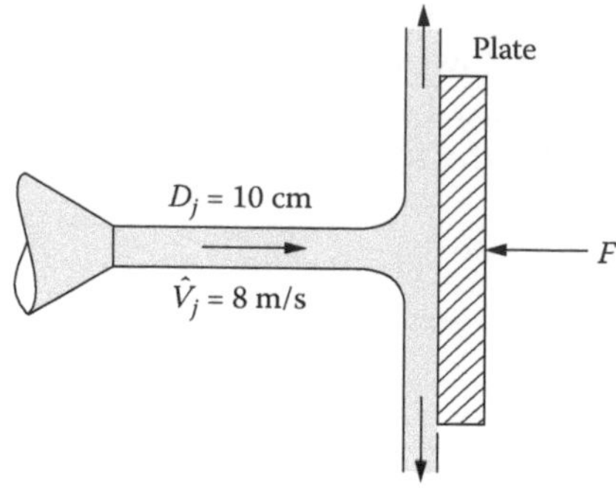

FIGURE P14.53

14.54: The steady-flow entrance case, shown in Figure P14.54, develops from uniform flow with velocity, $\hat{V}_o$, at section 1 to the laminar profile

$$\hat{V} = \hat{V}_{\max}\left(1 - \frac{r^2}{R^2}\right)$$

at section 2. Determine the wall drag as a function of P_1, P_2, ρ, $\hat{V}_o$, and R.

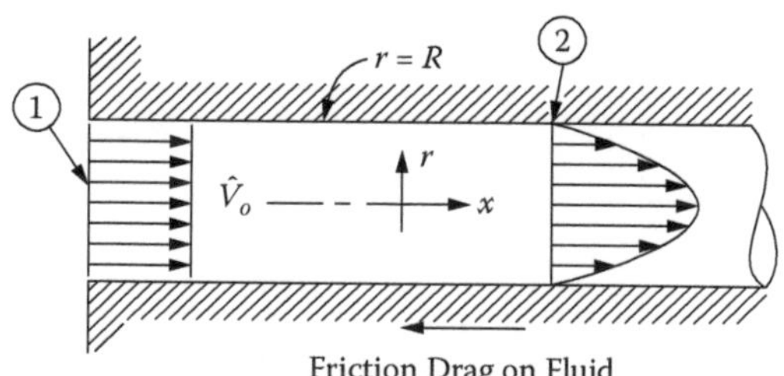

FIGURE P14.54

14.55: A vane of turning angle, $\theta = 120°$, is mounted on a water tank as shown in Figure P14.55. It is struck by a 5-cm-diameter water jet flowing with a velocity of 16 m/s, which turns and falls into the water without spilling. Determine the force required to hold the tank stationary if $\theta = 45°$, $60°$, and $90°$.

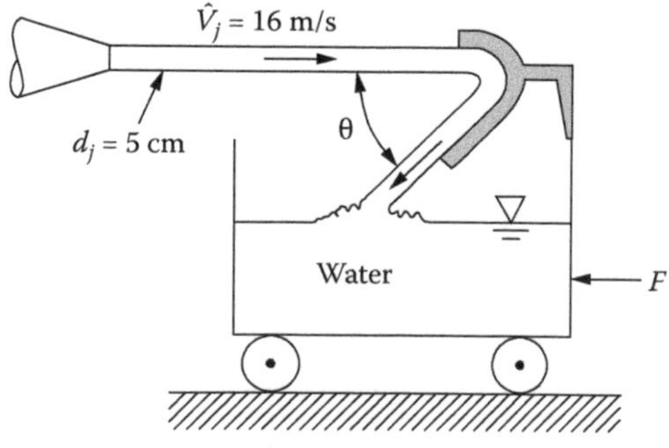

FIGURE P14.55

14.56: A dam discharges into a channel of constant width as shown in Figure P14.56. It is observed that water backs up behind the jet to a height, H. The velocity and height of the flow in the channel are given as $\hat{V}$ and h, respectively, and the density of the water is ρ. Neglect the horizontal momentum of the flow that is entering the control volume from above and assume friction to be negligible. Taking the air pressure in the cavity below the crest of falling water as atmospheric, use the momentum theorem and the control surface indicated to determine H.

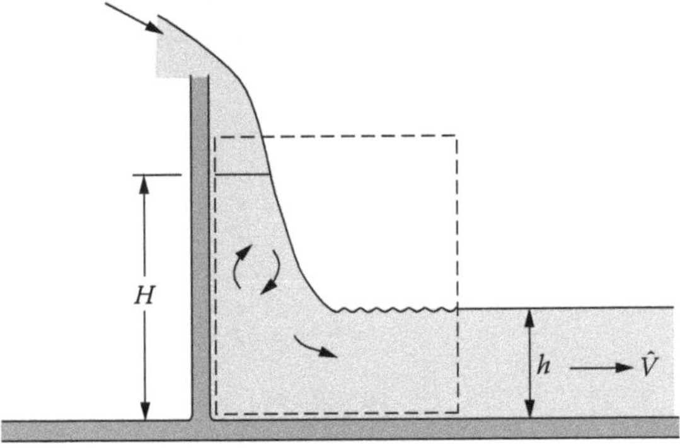

FIGURE P14.56

Moment of Momentum

14.57: The pipe shown in Figure P14.57 has a slit of thickness 0.0984 cm, so shaped that a sheet of water of uniform thickness, 0.0984 cm, issues out radially from the pipe. The velocity is constant along the pipe as shown and a volumetric flow rate of $0.90\ \mathrm{m^3/s}$ enters at the top. Determine the moment on the tube about the axis-BB.

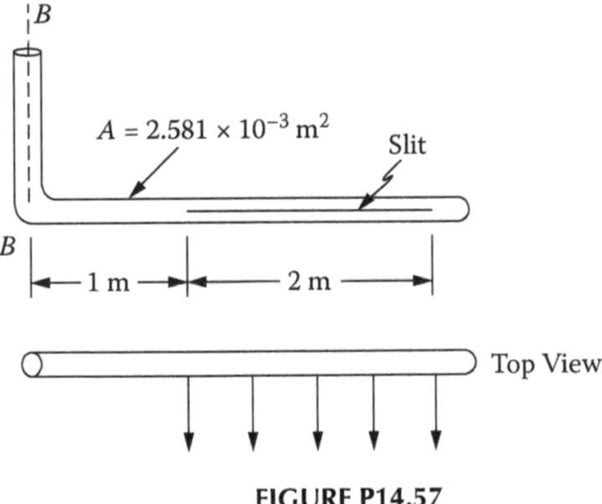

FIGURE P14.57

14.58: Figure P14.58 shows a vane with a turning angle, θ, which moves with a steady speed of $\hat{V}_c$. The vane receives a jet that leaves a fixed nozzle with speed, $\hat{V}$. (a) Assuming that the vane is mounted on rails as indicated in Figure P14.58, show that the power transmitted to the cart is $\hat{V}_c/\hat{V} = 1/3$ and (b) assuming that there are a large number of such vanes mounted to a rotating wheel with a peripheral speed of $\hat{V}_c$, show that the power transmitted is maximum when $\hat{V}_c/\hat{V} = 1/2$.

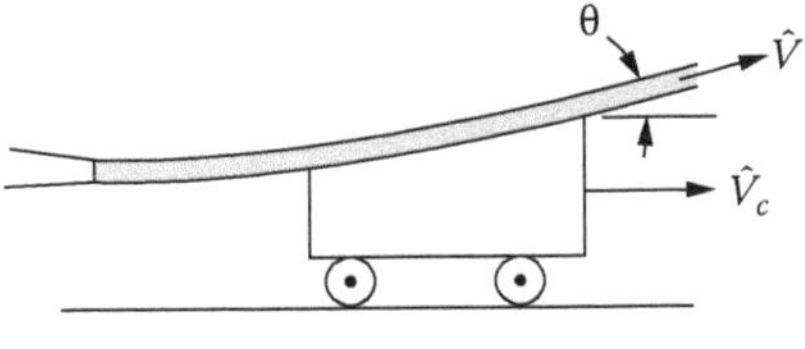

FIGURE P14.58

14.59: A lawn sprinkler consists of two sections of curved pipe rotating about a vertical axis as shown in Figure P14.59. The sprinkler rotates with an angular velocity, ω, and the effective discharge area is A. Thus, the water is discharged at the rate of $\dot{V} = 2\hat{V}_r A$, where $\hat{V}_r$ is the velocity of the water relative to the rotating pipe. A constant friction torque resists the motion of the sprinkler. Find an expression for the speed of the sprinkler in terms of the significant variables.

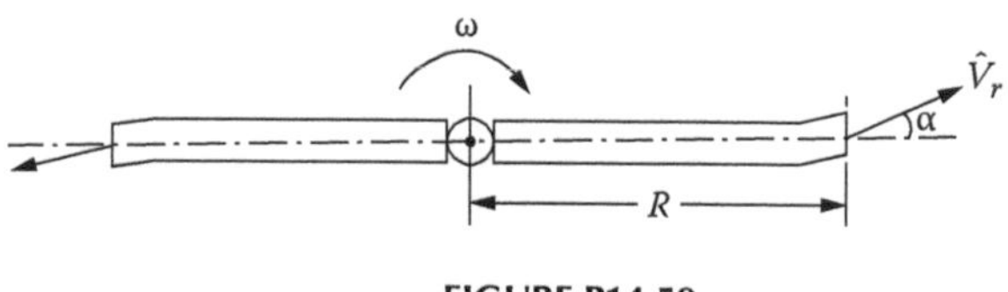

FIGURE P14.59

14.60: An air jet of diameter, d_1 enters a series of moving blades as indicated in Figure P14.60 at absolute velocity, $\hat{V}_1$ and angle β_1 and leaves at absolute velocity, $\hat{V}_2$, and angle, β_2. The blades move at speed, $\hat{V}$. Assuming incompressible flow, if $\hat{V}_1 = 60\,\text{m/s}$, $\hat{V}_2 = 40\,\text{m/s}$, $\beta_1 = 30°$, and $\beta_2 = 60°$, (a) find the velocity of the blades, $\hat{V}$ and (b) if $\rho = 1.2\,\text{kg/m}^3$ and $d_1 = 5\,\text{cm}$, and (c) find the power in watts applied to the blade for $\rho = 1.2\,\text{kg/m}^3$ and $d_1 = 5\,\text{cm}$.

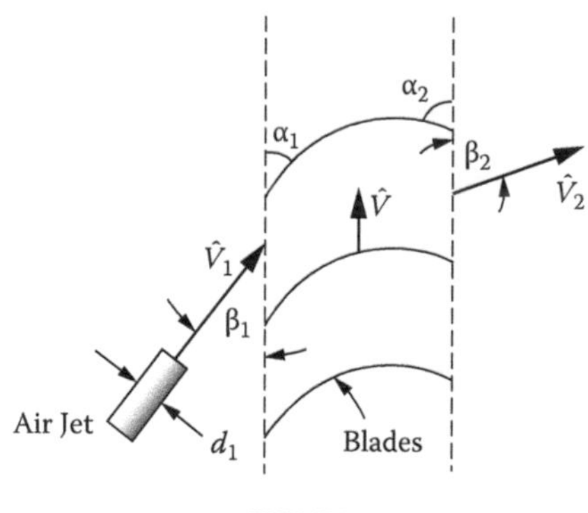

FIGURE P14.60

14.61: Seawater, $\rho = 1025\,\text{kg/m}^3$, flows through the impeller of the centrifugal pump shown in Figure P14.61 at a volumetric flow rate of $0.0520\,\text{m}^3/\text{s}$. Data and dimensions are

$n = 1180\,\text{rpm}$	$\delta_1 = 5\,\text{cm}$
$r_1 = 5.00\,\text{cm}$	$\theta_2 = 135°$
$r_2 = 20\,\text{cm}$	$\delta_2 = 1.5\,\text{cm}$

Assuming that the absolute velocity of the water entering the impeller is radial, determine the torque exerted on the impeller by the fluid and the power required to drive the pump.

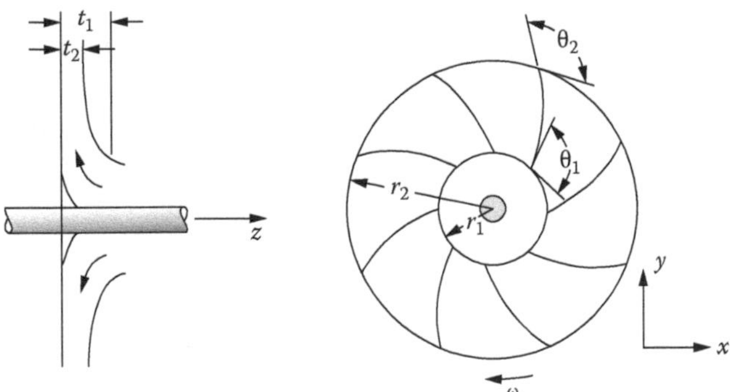

FIGURE P14.61

14.62: In Problem 14.61, determine (a) the angle θ_1 such that the entering flow is parallel to the vanes and (b) the axial load on the shaft for a shaft diameter of 2.5 cm and atmospheric pressure at the pump inlet.

14.63: A water sprinkler consists of two 1-cm-diameter jets at the ends of a rotating hollow rod as shown in Figure P14.63. The water leaves at 6 m/s. Determine the torque necessary to hold the sprinkler in place.

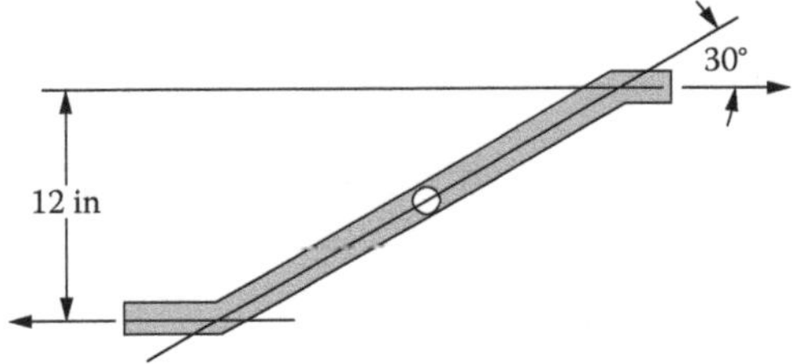

FIGURE P14.63

14.64: Water enters the rotor shown in Figure P14.64 with a volumetric flow rate of $0.015\,m^3/s$, along the axis of rotation. The cross-sectional area of each of the three nozzles is $2\,cm^2$. Determine (a) the magnitude of the resisting torque necessary to hold the rotor stationary for $\theta = 30°$ and (b) how fast the rotor will spin steadily if the resisting torque is reduced to zero and $\theta = 0°$, $30°$, and $60.°$

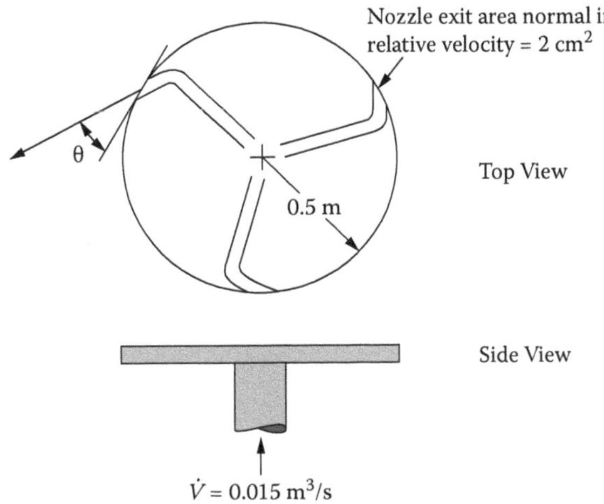

FIGURE P14.64

14.65: Front and side views of a centrifugal pump rotor or impeller are shown in Figure P14.65. The pump delivers 220 L/s of water and the blade exit angle is 35° from the tangential direction. The flow entering the rotor blade row is essentially radial as viewed from a stationary frame. Determine the power requirement.

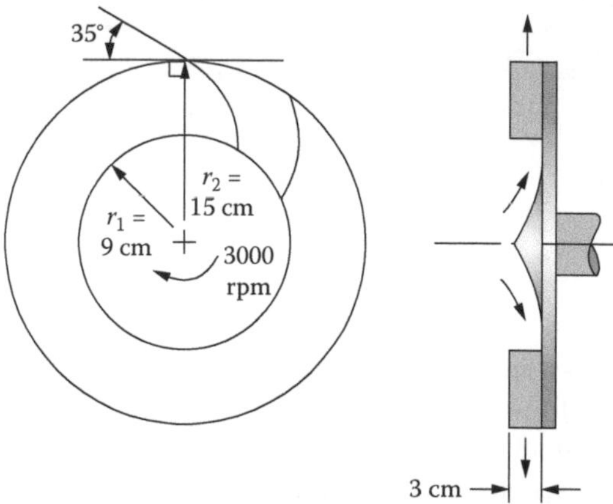

FIGURE P14.65

Jets and Rockets

14.66: A rocket engine burns 5.5 kg of fuel and oxidizer per second and the combustion products leave the rocket at a velocity of 565 m/s relative to the rocket at the ambient air pressure. Determine the total thrust produced by the rocket engine.

14.67: A jet aircraft traveling at a rate of 200 m/s takes in air at the rate of 22.5 kg/s. The air fuel ratio is 25:1 and the exhaust velocity relative to the jet is 600 m/s. Determine the thrust produced by the engine.

14.68: A jet engine is being tested in the laboratory. The engine consumes 22.75 kg/s of air and 0.225 kg/s of fuel. The exit velocity of the gases is 475 m/s. Determine the thrust.

14.69: A jet engine under test uses 22.65 kg/s of air and 0.2265 kg/s of fuel. The exit velocity of the products of combustion is 460 m/s. Determine the thrust being developed by the engine.

14.70: A jet engine in an aircraft flying at 182.5 m/s uses 22.5 kg/s of air. Determine the velocity of the combustion products in order to develop a thrust of 6720 N.

14.71: The mass flow rate of the products of combustion for a rocket engine is 5.25 kg/s and these products exit from the engine at 560 m/s. If the pressure at the exhaust is approximately equal to the ambient air pressure, determine the thrust produced by the engine.

14.72: A rocket burns fuel at rate of 6.875 kg/s. The exhaust gases leave the rocket at a relative velocity of 1100 m/s and at atmospheric pressure. The gross weight of the rocket is 2200 N, the exhaust nozzle area is 325 cm^2 and 2500 hp are developed by the rocket. Determine its velocity.

15

Dimensional Analysis and Similarity

Chapter Objectives

- To develop the rationale for dimensional analysis.
- To introduce the Buckingham Pi theorem.
- To apply the Buckingham Pi theorem to the case of rotating machines.
- To establish criteria for geometric, kinematic, and dynamic similarity.
- To employ similarity concepts in establishing scaling relationships for fluid mechanical systems.

15.1 Introduction

Dimensional homogeneity has been an important consideration in all of the material that has been presented thus far. Clearly, it is necessary that all quantitative relationships be dimensionally consistent and certain conversion factors are often used to achieve this consistency. In this chapter, we will use the principle of dimensional homogeneity in another way. By invoking a formalism to this concept, using **dimensional analysis**, we will combine important variables that describe a physical situation into a smaller number of dimensional groups.

Such a process has a number of benefits. One benefit is that, in an experimental study involving a number of variables, the combination of variables into a smaller number of dimensionless parameters can greatly simplify the analysis and the presentation of experimental results.

A second attribute of dimensionless groups is **scaling**. Scaling is the mathematical process of utilizing experimental results from a scale model to predict the performance of a full-sized device or system. A requirement of scaling is that the model and prototype be similar in all respects. The meaning and implications of scaling will be considered in detail in this chapter.

15.2 Fundamental Dimensions

In expressing the units of certain quantities, some have dimensions that are fundamental, and others are simply combinations of the fundamental dimensions. For example, if we consider mass, M, length, L, and time, T, as fundamental, we can express the dimensions

TABLE 15.1

Some Important Variables and Their Representations in the M, L, T System

Variable	Symbol	Dimensions
Mass	m	M
Length	L	L
Time	t	T
Area	A	L^2
Volume	V	L^3
Velocity	$\hat{V}$	L/T
Acceleration	a	L/T^2
Density	ρ	M/L^3
Viscosity	μ	M/LT
Force	F	ML/T^2
Torque	τ	ML^2/T^2
Pressure	P	M/LT^2
Surface tension	σ	M/T^2

of a large number of quantities as combinations of these. Table 15.1 presents a list of such quantities and their representation in the M, L, T system.

We should note that the quantity, F, or force, appears on this list. Force and mass are related by Newton's second law, $F = ma$, and conversion from force to mass involves the correction factor, g_c. In some books dealing with the SI system, there is no reference to the factor, g_c. Dimensional consistency in SI units is assured when proper units are employed for mass, force, displacement, and time. This is a consequence of the numerical value for g_c being 1 (or unity). The reader is reminded that the conversion between mass and force does possess units even though the numerical value may be unity. In the SI system, g_c has the numerical value of 1 kg-m/N-s^2 and in the English system the numerical value is 32.17 lb$_m$-ft/s^2-lb$_f$.

We could have made another choice and selected F, L, and T for the fundamental dimensions. In this case, mass would be represented in a list such as Table 15.1 with dimensions FT^2/L. The choice of M, L, and T is arbitrary and we will use these as fundamental dimensions exclusively in this book.

In a more extensive treatment of transport phenomena, we would need at least one more fundamental dimension: temperature, θ. In the fluid mechanics portion of this text (Chapters 11 through 18), we will be considering isothermal situations and, thus, M, L, and T will be sufficient for our purposes.

15.3 The Buckingham Pi Theorem

One means of generating dimensionless groups in a fluid mechanics application is to go through a formal procedure to nondimensionalize the governing differential equation for the process or phenomenon of interest. This approach will be discussed in Section 15.4.

We often find that a system or device of interest is not easily described by a differential equation or, if it is, we don't know its form ahead of time. An example of a complex system that is not amenable to description by differential equations would be a rotating machine such as a pump or turbine. The only possible way to evaluate such devices is by experimentation. Experimental results are then communicated in the form of dimensionless parameters. We will discuss rotating machinery in a more complete fashion in Chapter 18.

For now we are interested in two questions regarding the appropriate dimensionless parameters:

- How many are there?
- What are they?

Several methods exist for obtaining the answers to these questions and all of them have some common features. We will consider only one such procedure known as the Buckingham[1] method. The method consists of two parts that relate, in turn, to answering the questions of "How many are there?" and "What are they?"

The procedure for answering the question "How many are there?" involves the use of a straightforward rule known as the **Buckingham Pi theorem**, proposed in 1914. The designation, Pi, refers to the notation used for the dimensionless parameters, π, which means "product of variables." The idea is to first determine how many π groups are to be formed from the primitive variables that pertain to a given physical situation.

The Buckingham Pi theorem is stated as

$$i = n - r \tag{15.1}$$

where

i = number of independent dimensionless π groups

n = number of variables involved

r = rank of the dimensional matrix

The dimensional matrix is formed by tabulating the exponents on the fundamental dimensions, M, L, and T, which represent each variable involved. Example 15.1 provides an example of such a matrix and the evaluation of i, n, and r.

Example 15.1 Determine the number of dimensionless groups needed to represent the variables involved in describing the effects of flow normal to a cylinder. The variables involved are the force, F, the diameter, d, the velocity, $\hat{V}$, the density, ρ, and the viscosity, μ.

Solution

The usual first step is to list all of the variables and their dimensional representation in the following form.

Variable	Symbol	Dimensions
Force	F	ML/T^2
Diameter	d	L
Velocity	$\hat{V}$	L/T
Density	ρ	M/L^3
Viscosity	μ	M/LT

The second step is the formation of an array of exponents in the form

$$\begin{array}{cccccc} & F & d & \hat{V} & \rho & \mu \\ M: & 1 & 0 & 0 & 1 & 1 \\ L: & 1 & 1 & 1 & -3 & -1 \\ T: & -2 & 0 & -1 & 0 & -1 \end{array}$$

[1] Edgar Buckingham (1867–1940) was an American physicist who worked in thermodynamics and dimensional analysis and who developed the Buckingham π theorem.

where the numbers coincide with the powers of M, L, and T in the dimensions of each variable. As an example, the dimensions of μ are M/LT or $M^1L^{-1}T^{-1}$. Thus, the values of 1, −1, and −1 appear in the foregoing array. Doing this for all of the variables establishes the dimensional matrix which, for this problem, is

$$\begin{bmatrix} 1 & 0 & 0 & 1 & 1 \\ 1 & 1 & 1 & -3 & -1 \\ -2 & 0 & -1 & 0 & -1 \end{bmatrix}$$

This is a 3 × 5 matrix (one that contains 3 rows and 5 columns).

The pi theorem is expressed in terms of the **rank** of this matrix. The rank of a matrix is the number of rows and columns in the largest nonzero determinant that can be formed from the matrix. In this case, the maximum value of r is three ($r = 3$), which can be verified by taking the first three columns of our matrix and obtaining the determinant

$$\det\begin{bmatrix} 1 & 0 & 0 \\ 1 & 1 & 1 \\ -2 & 0 & -1 \end{bmatrix} = -1$$

which is nonzero. This shows that the rank, in this case, is 3. We note that r is also the number of fundamental dimensions. Some treatises express the pi theorem with r defined as the number of fundamental dimensions and, in a great majority of cases, this will be true. Exceptions, however, may occur and one will never experience difficulty if the pi theorem is interpreted as we have specified.

Equation 15.1 now provides

$$i = n - r = 5 - 3 = 2$$

which shows that there are two dimensionless groups. Our task is to determine the composition of the two dimensionless groups and Example 15.2 will demonstrate this process.

Example 15.2 From the results of Example 15.1, determine the combinations of variables that make up the two dimensionless parameters that are relevant to the case of flow normal to a cylinder.

Solution

We designate the π groups to be formed as π_1 and π_2. Our procedure will be to first determine a **core group** of variables. The number of variables in the core will be r which, in our case, is 3. The process of evaluating the two π groups will consist of the following steps:

1. A core group of r variables (three for this example) will be chosen.
2. The core group will be combined with the remaining variables, one at a time.
3. The combination formed in step 2 will be made dimensionless by suitable choices of the exponents on the core variables.

The variables in the core group must, among them, contain all of the fundamental dimensions. A feature of the process outlined in the foregoing list is that the core variables may appear in each dimensionless group. A guiding principle in choosing the core variables is to then exclude from the core those variables whose effect one wishes to isolate. In this example we are principally interested in the variation of the drag force, F, so we will not choose it to be in the core. The dynamic viscosity, μ, will be arbitrarily chosen as the other variable to be excluded. The remaining variables, d, $\hat{V}$, and ρ will then constitute the core group. The two π groups will thus be

$$\pi_1 = d^a \hat{V}^b \rho^c F$$

and

$$\pi_2 = d^a \hat{V}^b \rho^c \mu$$

Our remaining task is to evaluate the exponents so that π_1 and π_2 are dimensionless.

First, for π_1, we write

$$\pi_1 = d^a \hat{V}^b \rho^c F$$

Dimensionally, this becomes

$$M^0 L^0 T^0 = L^a \left(\frac{L}{T}\right)^b \left(\frac{M}{L^3}\right)^c \frac{ML}{T^2}$$

We now equate the exponents on M, L, and T to form three simple linear algebraic equations in three unknowns

$$\begin{aligned} M &: 0 = \quad\quad\quad\quad c + 1 \\ L &: 0 = a + b - 3c + 1 \\ T &: 0 = \quad -b \quad\quad\; - 2 \end{aligned}$$

and the reader may verify that the solution of this set is

$$a = -2, \quad b = -2, \quad \text{and} \quad c = -1$$

These make π_1

$$\pi_1 = \frac{F}{d^2 \hat{V}^2 \rho}$$

The second π group, π_2, can be formed in the same manner

$$\pi_2 = d^a \hat{V}^b \rho^c \mu$$

Dimensionally, this becomes

$$M^0 L^0 T^0 = L^a \left(\frac{L}{T}\right)^b \left(\frac{M}{L^3}\right)^c \frac{M}{LT}$$

and when we equate the exponents on M, L, and T, we obtain another set of three simple linear algebraic equations in three unknowns

$$\begin{aligned} M &: 0 = \quad\quad\quad\quad c + 1 \\ L &: 0 = a + b - 3c - 1 \\ T &: 0 = \quad -b \quad\quad\; - 1 \end{aligned}$$

The solution of this set yields the values

$$a = -1, \quad b = -1, \quad \text{and} \quad c = -1$$

These make our parameter

$$\pi_2 = \frac{\mu}{d\hat{V}\rho}$$

We see that, from dimensional analysis, in the case of a cylinder in cross flow, there are two applicable dimensionless parameters related functionally as

$$\pi_1 = f(\pi_2)$$

Experimental data must be obtained for a determination of this function. We shall encounter this relationship in Chapter 16 where, the function, f will be determined.

The two dimensionless groups that have been generated in this exercise are possibly new to the reader. In the fluid mechanics community, they, or similar forms, are often seen. The first of these groups,

$$\pi_1 = \frac{F}{d^2\hat{V}^2\rho}$$

can also be written in the form

$$\pi_1 = \frac{P}{\hat{V}^2\rho}$$

where the numerator is the pressure, $P = F/d^2$, with units of $\mathrm{N/m^2}$. This parameter is considered, physically, to be the ratio of pressure forces to inertial forces and is given different symbols and names, depending on the branch of fluid mechanics involved. Often, it is designated as the **Euler number** symbolized as Eu. In future chapters, we will also see it designated as the **drag coefficient**, C_D, and as the **coefficient of skin friction**, designated as C_f.

The second π group formed

$$\pi_2 = \frac{\mu}{d\hat{V}\rho}$$

is most often seen in the form

$$\pi_2 = \frac{\rho\hat{V}d}{\mu}$$

and is designated as the **Reynolds number** symbolized as Re. The Reynolds number is, physically, the ratio of the inertial forces to viscous forces. The Reynolds number, named for Osborne Reynolds,[2] a pioneer in the fluid mechanics field, is generally considered as the most important parameter in fluid mechanics. We will see these parameters, particularly the Reynolds number, again.

---

[2] Osborne Reynolds (1842–1912) was a British engineer and physicist who demonstrated streamlines and turbulent flow in pipes and showed the transition between them occurs at a critical velocity determined by the Reynolds number.

15.4 Reduction of Differential Equations to a Dimensionless Form

An in-depth analysis of fluid mechanics generally involves the derivation and solution, where possible, of the governing differential equations. These equations also provide the basis for a considerable number of sophisticated computational codes. Such applications are beyond the scope of this text.

When the differential equations that describes a given flow situation are known, they may be made nondimensional by using a relatively simple process that requires only a modest level of insight. A by-product of such a procedure is the formulation of dimensionless parameters associated with the flow.

This process will now be illustrated. The starting equations for this exercise will not be derived. Their validity can be validated examining any of a large number of texts such as Welty et al. (2001) and Munson et al. (1998).

A representative case to examine is two-dimensional compressible flow involving a velocity vector expressed in terms of its components

$$\mathbf{V} = \hat{V}_x\mathbf{x} + \hat{V}_y\mathbf{y}$$

in a situation described by the equations

$$\frac{\partial \hat{V}_x}{\partial x} + \frac{\partial \hat{V}_y}{\partial y} \tag{15.2}$$

and

$$\rho\left(\frac{\partial \mathbf{V}}{\partial t} + \hat{V}_x\frac{\partial \mathbf{V}}{\partial x} + \hat{V}_y\frac{\partial \mathbf{V}}{\partial y}\right) = \rho\mathbf{g} - \nabla P + \mu\left(\frac{\partial^2 \mathbf{V}}{\partial x^2} + \frac{\partial^2 \mathbf{V}}{\partial y^2}\right) \tag{15.3}$$

Equation 15.2 is the continuity equation for two-dimensional incompressible flow. Its derivative is based on the principle of mass conservation, Equation 15.3 is a differential expression of Newton's second law of motion. This equation is the two-dimensional form of the Navier[3]-Stokes[4] equations.

Some physical insight is required at this point. We need to establish reference conditions for the flow and the following choices are made:

- Reference length $= L$
- Reference velocity $= \hat{V}_\infty$

Nondimensional quantities (those with asterisks) in our differential equations will now be defined as follows:

$$x^* = \frac{x}{L} \qquad y^* = \frac{y}{L} \qquad t^* = \frac{t\hat{V}_\infty}{L}$$

$$\hat{V}_x^* = \frac{\hat{V}_x}{\hat{V}_\infty} \qquad \hat{V}_y^* = \frac{\hat{V}_y}{\hat{V}_\infty} \qquad \hat{V}^* = \frac{\hat{V}}{\hat{V}_\infty}$$

$$\nabla^* = L\nabla$$

[3] C. L. M. H. Navier (1785–1836) was a French physicist and engineer who studied analytical mechanics and its application to the strength of materials, machines, and the motions of solid and liquid bodies.

[4] G. C. Stokes (1819–1903) was a British mathematician and physicist who originated the idea of determining the chemical composition of the sun and stars from their spectra and also studied double refraction and electromagnetic waves.

The last category in the foregoing tabulation, the gradient, is composed of first derivatives with respect to space coordinates. Hence, the length scale, L, is employed to make the gradient dimensionless.

Our two differential equations may now be put into a dimensionless form. This process, in general, uses the chain rule of differentiation. For example, the continuity equation involves the terms

$$\frac{\partial \hat{V}_x}{\partial x} = \frac{\partial \hat{V}_x^*}{\partial x^*}\frac{\partial \hat{V}_x}{\partial \hat{V}_x^*}\frac{\partial x^*}{\partial x} = \frac{\partial \hat{V}_x^*}{\partial x^*}(\hat{V}_\infty)\left(\frac{1}{L}\right) = \frac{\hat{V}_\infty}{L}\frac{\partial \hat{V}_x^*}{\partial x^*}$$

and

$$\frac{\partial \hat{V}_y}{\partial y} = \frac{\partial \hat{V}_y^*}{\partial y^*}\frac{\partial \hat{V}_y}{\partial \hat{V}_y^*}\frac{\partial y^*}{\partial y} = \frac{\hat{V}_\infty}{L}\frac{\partial \hat{V}_y^*}{\partial y^*}$$

which can be substituted into Equation 15.2 to give

$$\frac{\partial \hat{V}_x^*}{\partial x^*} + \frac{\partial \hat{V}_y^*}{\partial y^*} = 0 \tag{15.4}$$

This dimensional equation has the same form in terms of nondimensional quantities as it had before.

Utilizing the chain rule, as illustrated in the foregoing, the equation of motion is transformed to

$$\frac{\rho \hat{V}_\infty^2}{L}\left(\frac{\partial \mathbf{V}^*}{\partial t^*} + \hat{V}_x\frac{\partial \mathbf{V}^*}{\partial x^*} + \hat{V}_y\frac{\partial \mathbf{V}^*}{\partial y^*}\right)$$

$$= \rho \mathbf{g} - \frac{1}{L}\nabla^* P + \frac{\mu \hat{V}_\infty}{L^2}\left(\frac{\partial^2 \mathbf{V}^*}{\partial x^{*2}} + \frac{\partial^2 \mathbf{V}^*}{\partial y^{*2}}\right) \tag{15.5}$$

Note that every term in this equation has the units, M/L^2T^2 or F/L^3. Each term represents a certain kind of force, that is $\rho\hat{V}_\infty^2/L$ is an inertial force, $\mu\hat{V}_\infty/L^2$ is a viscous force, ρg is the gravitational force, and P/L is the pressure force.

If we now divide by the quantity $\rho\hat{V}_\infty^2/L$, our dimensionless equation becomes

$$\frac{\partial \mathbf{V}^*}{\partial t^*} + \hat{V}_x\frac{\partial \mathbf{V}^*}{\partial x^*} + \hat{V}_y\frac{\partial \mathbf{V}^*}{\partial y^*}$$

$$= \frac{\mathbf{g}L}{\hat{V}_\infty^2} - \frac{\nabla^* P}{\rho\hat{V}_\infty^2} + \frac{\mu}{\rho L\hat{V}_\infty}\left(\frac{\partial^2 \mathbf{V}^*}{\partial x^{*2}} + \frac{\partial^2 \mathbf{V}^*}{\partial y^{*2}}\right) \tag{15.6}$$

This equation looks very much like the original expression except that in the transformation to a dimensionless form, the gravitational and pressure terms have been modified and the coefficient of the viscous term is a group of variables. Moreover, an examination of these terms reveals that each is dimensionless. It is also possible to interpret these parameters, in physical terms, in light of the way they were developed.

An initial consideration of the gravitational term $gL/\hat{V}_\infty^2$ establishes, first, that it is indeed a dimensionless ratio. The choice of $gL/\hat{V}_\infty^2$ or $\hat{V}_\infty^2/gL$ is arbitrary—clearly both forms are dimensionless. The conventional choice is the latter; the **Froude**[5] **number** (pronounced

[5] William Froude (1810–1879) was a British engineer who promulgated the Froude law of comparison when an object is towed in a liquid.

Frōōd), defined by

$$\text{Fr} \equiv \frac{\hat{V}_{\infty}^2}{gL} \tag{15.7}$$

From its form, we may infer that the Froude number is a ratio of the inertial forces to the gravitational forces. The Froude number, an important scaling parameter in flows that involve a free surface, is encountered in the analysis of open channel flows.

Another parameter associated with viscous terms, relating inertial forces and viscous forces, is the **Reynolds number**

$$\text{Re} \equiv \frac{L\hat{V}_{\infty}^2\rho}{\mu} \tag{15.8}$$

which we encountered in Example 15.2.

A third dimensionless group is present in Equation 15.6. The pressure term now involves the group $P/\rho\hat{V}_{\infty}^2$, which is dimensionless and can be interpreted as the ratio of the pressure forces to the inertial forces. Formally, this parameter is designated as the **Euler[6] number**, which is given by

$$\text{Eu} \equiv \frac{P}{\rho\hat{V}_{\infty}^2} \tag{15.9}$$

The Euler number was derived earlier using the Buckingham Pi method in Section 15.3.

15.5 Dimensional Analysis of Rotating Machines

One of the most important areas for applying the Buckingham Pi theorem is that of rotating machines such as compressors, fans, and turbines. These machines are sufficiently complex to preclude the possibility of a precise theoretical analysis.

For a centrifugal pump, with the configuration shown in Figure 15.1, the important performance variables are listed in Table 15.2.

We should note, before continuing, that some physical insight was necessary in developing the list of variables in Table 15.2. The variables ρ and μ represent fluid properties that would be important in evaluating pump operation. The diameter, d, is a logical geometric parameter and the shaft speed (or angular velocity), n, which is normally controllable is also a straightforward choice. The other three variables, gh, P, and $\dot{Q}$, are all performance quantities of interest. It is conventional practice in the pump industry to represent total head as gh in units of m^2/s^2 rather than simply h in units of m.

For a pump, operating on a system, the **flow rate** of fluid at a given **pressure** would naturally be of primary interest. It is important to know the **power** required to achieve this performance. Actual laboratory testing of a pump, with performance variables grouped according to dimensional analysis, is an important part of performance analysis that may lead to design modifications.

[6] Leonhard Euler (1707–1783) was a Swiss mathematician who contributed to algebraic series and differential and integral calculus and realized the significance of the Euler coefficients in certain trigonometric expansions.

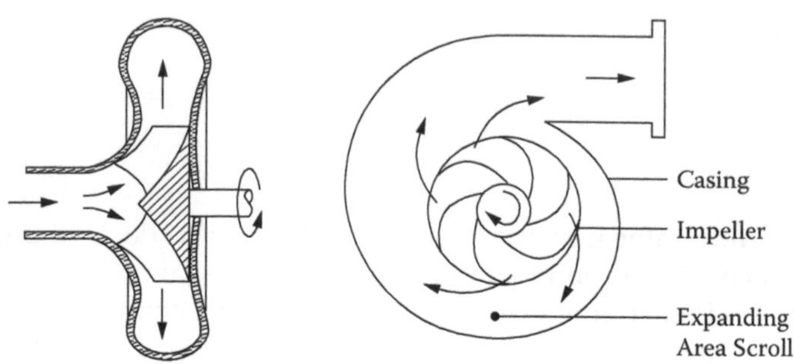

FIGURE 15.1
Cutaway view of a centrifugal pump.

Now, to continue the dimensional grouping of the seven variables in Table 15.2, we first form the array

$$\begin{array}{rrrrrrrr} & gh & P & \dot{V} & d & n & \rho & \mu \\ M: & 0 & 1 & 0 & 0 & 0 & 1 & 1 \\ L: & 2 & 2 & 3 & 1 & 0 & -3 & -1 \\ T: & -2 & -3 & -1 & 0 & -1 & 0 & -1 \end{array}$$

and this leads to the dimensional matrix

$$\begin{bmatrix} 0 & 1 & 0 & 0 & 0 & 1 & 1 \\ 2 & 2 & 3 & 1 & 0 & -3 & -1 \\ -2 & -3 & -1 & 0 & -1 & 0 & -1 \end{bmatrix}$$

The rank of this matrix is 3 and the pi theorem tells us that the number of dimensionless groups to be formed is

$$i = n - r = 7 - 3 = 4$$

The next step is to choose the core group, which will consist of $r = 3$ variables. Clearly, we do not want gh, P, or $\dot{V}$ to be in the core. One additional variable will be excluded from the core and we choose the fluid viscosity, μ. With these choices, the four pi groups consist

TABLE 15.2
Pump Performance Variables: Power Is Designated as P Rather Than $\dot{W}$

Variable	Symbol	Dimensions
Total head	gh	L^2/T^2
Power	P	ML^2/T^3
Flow rate	$\dot{V}$	L^3/T
Impeller diameter	d	L
Shaft speed	n	$1/T$
Fluid density	ρ	M/L^3
Fluid viscosity	μ	M/LT

of the following:

$$\pi_1 = d^a n^b \rho^c (gh)$$

$$\pi_2 = d^a n^b \rho^c P$$

$$\pi_3 = d^a n^b \rho^c \dot{V}$$

$$\pi_4 = d^a n^b \rho^c \mu$$

Here we evaluate π_1 and ask the reader to verify the details for the evaluation of π_2 through π_4. For π_1, we have

$$\pi_1 = d^a n^b \rho^c (gh)$$

Dimensionally, this becomes

$$M^0 L^0 T^0 = L^a (T^{-b})(M^c L^{-3c})(L^2 T^{-2})$$

When we equate the exponents on M, L, and T, we obtain

$$M: \; 0 = c$$

$$L: \; 0 = a - 3c + 2$$

$$T: \; 0 = -b + 2$$

The solution of this set yields the values

$$a = -2, \quad b = -2, \quad \text{and} \quad c = 0$$

These make our parameter

$$\pi_1 = \frac{gh}{d^2 n^2}$$

Now, the reader may check the values of π_2 through π_4. The resulting forms are

$$\pi_1 = \frac{gh}{d^2 n^2} \quad \text{(head coefficient)} \tag{15.10}$$

$$\pi_2 = \frac{P}{\rho n^3 d^5} \quad \text{(power coefficient)} \tag{15.11}$$

$$\pi_3 = \frac{\dot{V}}{n d^3} \quad \text{(flow coefficient)} \tag{15.12}$$

and

$$\pi_4 = \frac{\mu}{d^2 n \rho} \quad \text{(a form of Reynolds number)} \tag{15.13}$$

These four parameters are referred to as indicated.

Typical pump performance curves are shown on coordinates with π_1 and π_2 plotted as a function of π_3. An example of such a performance plot is shown in Figure 15.2.

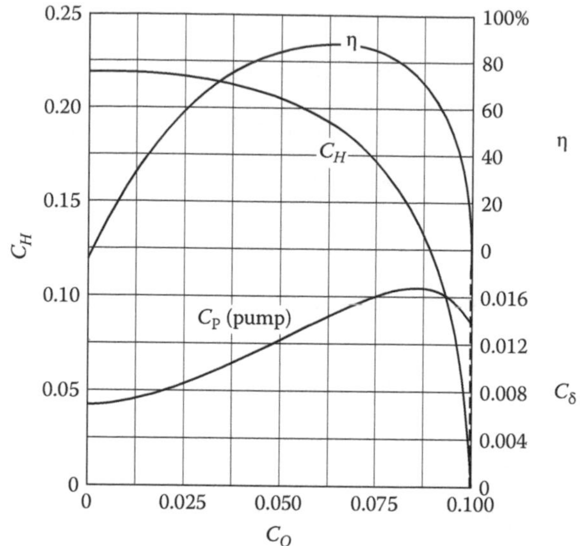

FIGURE 15.2
Dimensionless pump performance curves.

15.6 Similarity

We have, on several occasions, alluded to the process whereby data acquired for a model of a device or system can be used to predict the performance of a full-scale operating entity. The idea of testing a model of an aircraft in a wind tunnel is obviously less expensive and faster than actually fabricating an aircraft and flight testing it. The question of **scaling** has to do with how we must interpret model behavior in order to predict the performance of the full-scale prototype. The concept of **similarity** is critical in this process. There are three types of similarity:

- Geometric similarity
- Kinematic similarity
- Dynamic similarity

Geometric similarity is an obvious requirement between a model and a prototype. A simple statement of geometric similarity involves the dimensions of the two systems being considered:

> Two systems are geometrically similar when the ratios of significant dimensions are equal for both.

An example of the concept of two geometrically similar systems is illustrated in Figure 15.3.

In Figure 15.3a, the two cylinders will be geometrically similar when

$$\frac{L_1}{d_1} = \frac{L_2}{d_2} \tag{15.14}$$

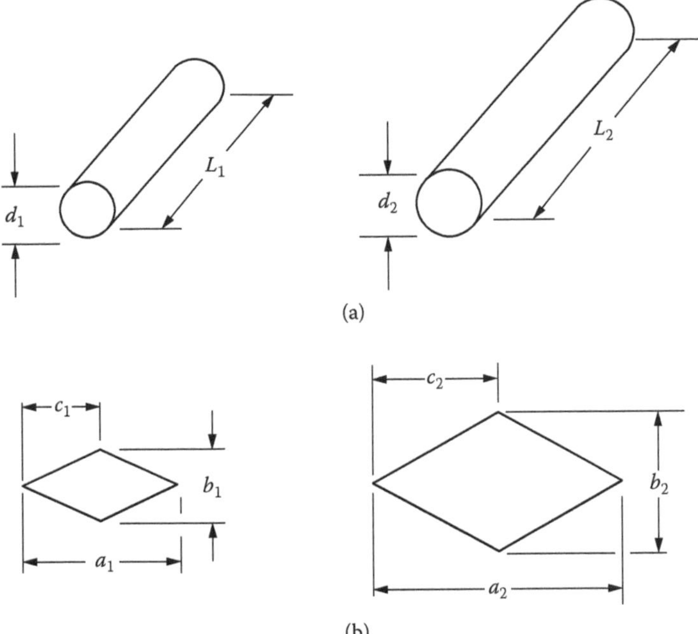

FIGURE 15.3
Examples of geometrically similar systems (a) cylinders and (b) diamond shapes.

and, for the diamond-shape objects (Figure 15.3b), the requirement is

$$\frac{a_1}{b_1} = \frac{a_2}{b_2} \quad \text{and} \quad \frac{a_1}{c_1} = \frac{a_2}{c_2}$$

or

$$\frac{a_1}{a_2} = \frac{b_1}{b_2} = \frac{c_1}{c_2} \tag{15.15}$$

With regard to kinematic similarity:

> Two systems exhibit kinematic similarity when they are geometrically similar and when they are exposed to the same flows.

Quantitatively, kinematic similarity corresponds to a case when, between systems 1 and 2,

$$\frac{\hat{V}_{x1}}{\hat{V}_{x2}} = \frac{\hat{V}_{y1}}{\hat{V}_{y2}} = \frac{\hat{V}_{z1}}{\hat{V}_{z2}} \tag{15.16}$$

Here, it is presumed that the type of mean flow, either laminar or turbulent, applies to both systems.

Finally, for dynamic similarity:

> Two systems exhibit dynamic similarity when they are geometrically and kinematically similar and when the ratios of significant forces between the systems are equal.

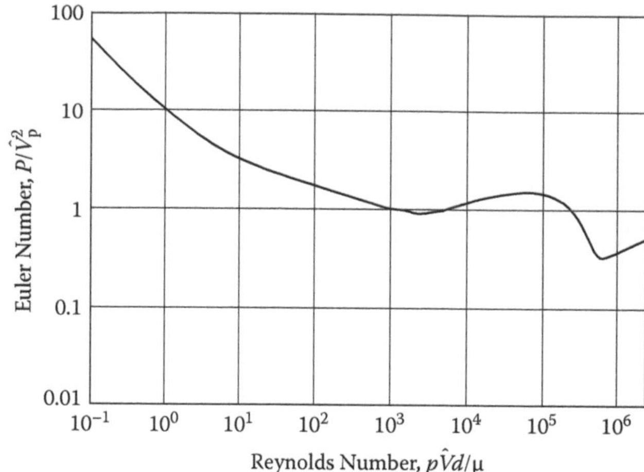

FIGURE 15.4
Euler number variation as a function of Reynolds number for a cylinder in cross flow.

Because the dimensionless parameters generated earlier (such as the Reynolds and Euler numbers) physically represent such force ratios, dynamic similarity is exhibited when the parameters, which relate **significant** forces, are the same between the systems.

In Example 15.2, it was determined that, in the case of flow across a cylinder, the dimensionless force ratios constituting the Reynolds and Euler numbers apply. Two such systems would thus be dynamically similar when

$$\text{Re}_1 = \text{Re}_2 \qquad \text{and} \qquad \text{Eu}_1 = \text{Eu}_2$$

provided they are also geometrically and kinematically similar. This simple concept suggests that a plot of Eu as a function of Re, as shown in Figure 15.4, generated in a well-controlled experiment in a laboratory wind tunnel, can be used to predict the pressure drag force for **any** cylinder influenced by the flow of **any** fluid. The importance of such scaling is enormous.

Example 15.3 will illustrate the use of scaling in predicting equipment performance.

Example 15.3 A common method for scaling cylindrical mixing tanks and impellers is to maintain a constant ratio of power to unit volume. If it is desired to increase the volume of a properly baffled liquid mixer by a factor of 5, by what ratio must the tank diameter and impeller speed be changed? We may presume that the tanks are geometrically similar and that both will operate in the fully turbulent regime. The power, P, supplied to the impeller may be assumed to be a function of the impeller diameter, d, its rotational velocity, n, and the liquid density, ρ.

Solution

The tank diameter ratio will be scaled by the requirement of geometric similarity. For cylindrical geometry, the two tanks, shown in Figure 15.5 have volumes

$$V_a = \frac{\pi}{4} d_a^2 L_a \quad \text{and} \quad V_b = \frac{\pi}{4} d_b^2 L_b$$

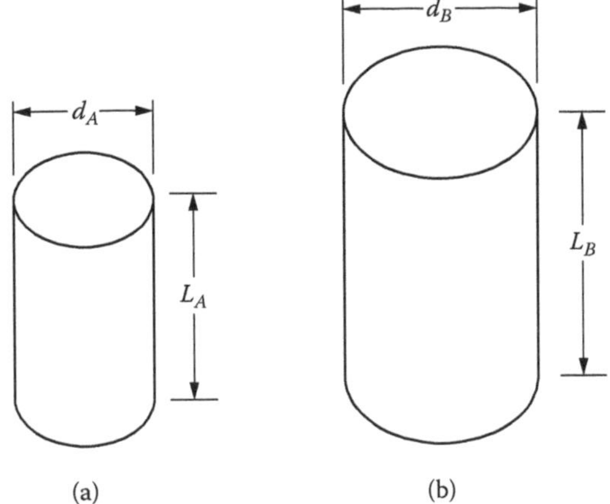

FIGURE 15.5
Cylindrical mixing tanks for Example 15.3.

We may combine these expressions for the individual volumes in accordance with the stated relationship that $V_b/V_a = 5$. Hence,

$$\frac{V_b}{V_a} = \frac{\frac{\pi}{4}d_b^2 L_b}{\frac{\pi}{4}d_a^2 L_a} = 5$$

or

$$\left(\frac{d_b}{d_a}\right)^2 \frac{L_b}{L_a} = 5$$

The condition of geometric similarity requires that

$$\frac{d_a}{L_a} = \frac{d_b}{L_b}$$

or

$$\frac{L_b}{L_a} = \frac{d_b}{d_a}$$

Making this substitution yields

$$\left(\frac{d_b}{d_a}\right)^3 = 5$$

which gives us the required ratio of the diameters.

$$\frac{d_b}{d_a} = 5^{1/3} = 1.710 \Longleftarrow$$

The ratio of impeller speeds is dictated by dynamic similarity requirements. Because we know the dependence of power on the characteristics of the impeller, $P = f(d, n, \rho)$, we may combine these quantities into the dimensionless ratio

$$\frac{P}{\rho n^3 d^5}$$

which was earlier designated as the power coefficient in Equation 15.11. For this parameter to be a constant between tanks a and b, we have

$$\frac{P_a}{\rho_a n_a^3 d_a^5} = \frac{P_b}{\rho_b n_b^3 d_b^5}$$

and for a constant power ratio, the speed ratio is

$$\frac{n_b}{n_a} = \left[\frac{P_b}{P_a}\left(\frac{d_a}{d_b}\right)^5 \frac{\rho_a}{\rho_b}\right]^{1/3}$$

$$= [(5)(5)^{-5/3}(1)]^{1/3} = 5^{-2/9} = 0.699 \Longleftarrow$$

The larger tank must have a diameter that is 1.71 times the diameter of the smaller one and an impeller speed that is approximately 0.70 times that of the smaller system. These results are not at all intuitive.

15.7 Summary

This chapter has introduced the formalism associated with the concepts of dimensional analysis, which is basically an application of dimensional homogeneity to a given situation.

The Buckingham Pi theorem is a means to determine how many dimensionless groups may be formed from the appropriate listing of primitive variables that are significant in the analysis of a given flow situation. Once the number of π groups has been determined, a straightforward method for identifying the π groups may be implemented.

The requirements of geometric, kinematic, and dynamic similarity enable one to use model data to predict the behavior of a prototype or full-size piece of equipment. **Similitude** or **model theory** is a very important application of the parameters obtained from dimensional analysis.

Several applications of dimensional analysis will be considered in subsequent chapters.

15.8 Problems

Dimensionless Parameters and Dimensionless Forms

15.1: A car with a length of 5.8 m and a radio antenna of 6.4 mm is traveling along a road at 22.2 m/s. Determine the Reynolds number (a) based on the length of the car and (b) based on the diameter of the radio antenna.

15.2: Some common variables employed in the study of fluid mechanics include the volumetric flow rate, $\dot{V}$, the acceleration of gravity, g, the viscosity, μ, the density, ρ, and a length, L. Determine which of the following variables are dimensionless: (a) $\dot{V}^2/gL^2$, (b) $\rho\dot{V}/\mu L$, (c) $gL^2/\dot{V}$, and (d) $\rho\dot{V}L/\mu$.

15.3: In a gas, the velocity of sound, c, is known to depend on the pressure, P, and the density, ρ. Use dimensional reasoning to establish (to within a constant) the relationship between c, P, and ρ.

15.4: A cylindrical container is rotated around its vertical axix. The pressure, P, in the radial direction, depends on the rotational speed, ω, the location in the radial direction, r, and the density of the fluid, ρ. Determine the form of the equation for P.

15.5: For the flow of a thin film of liquid with a depth, h, and a free surface, two important dimensionless parameters are the Froude number, $\text{Fr} = \hat{V}/\sqrt{gh}$ and the Weber number, $\text{We} = \rho\hat{V}^2h/\sigma$. Determine the value of both of these parameters for water at 20°C flowing with a velocity of 0.50 m/s at a depth of 3 mm. At 20°C, the surface tension of water, $\sigma = 0.0705\,\text{N/m}$.

15.6: Standard air with a velocity, $\hat{V}$, flows past an airfoil having a chord length, b, of 2.5 m. Determine (a) the Reynolds number, $\rho\hat{V}b/\mu$ for $\hat{V} = 90\,\text{m/s}$ and (b) if this airfoil were attached to an aircraft flying at the same speed in a standard atmosphere at an altitude of 3250 m.

15.7: The ratio of the viscous dissipation to the heat conduction in a fluid is given by the **Brinkman number**, Br. The Brinkman number is a combination of flow velocity, $\hat{V}$, viscosity, μ, thermal conductivity, k and temperature, T. If the Brinkman number is proportional to viscosity, derive the Brinkman number. The thermal conductivity can be expressed dimensionally as $k \sim ML/T^3\theta$, where θ is temperature.

15.8: The velocity of propagation, C, of a capillary wave in deep water is a function of the wavelength, λ, the surface tension, σ, and the density, ρ. Using a dimensionless constant, K, (a) determine the proper functional relationship for this phenomenon and (b) how the propagation velocity changes when the surface tension increases by a factor of 3.

15.9: Inside a soap bubble, the excess pressure, ΔP, varies with surface tension, σ, and radius, r. Using dimensional analysis, determine how this pressure difference varies if (a) the radius increases by a factor of 1.5 and (b) the surface tension is doubled.

15.10: The volumetric flow rate, $\dot{V}$, of an ideal liquid through an orifice is a function of the density of the liquid, ρ, the diameter of the orifice, d, and the pressure difference across the orifice, ΔP. Use dimensional analysis to determine the parameters involved in relating these variables.

15.11: Assuming the pressure is a function of density and velocity, determine the dynamic pressure exerted on an immersed object by a flowing incompressible fluid.

Reduction of Differential Equations to a Dimensionless Form

15.12: The flow between two concentric cylinders (Figure P15.12) is governed by the differential equation

$$\frac{d^2\hat{V}_\theta}{dr^2} + \frac{d}{dr}\left(\frac{\hat{V}_\theta}{r}\right) = 0$$

where $\hat{V}_\theta$ is the tangential velocity at any local radial location, r. The inner cylinder is fixed and the outer cylinder rotates with an angular velocity, ω. Express the equation in dimensionless form using R_o and ω as reference parameters.

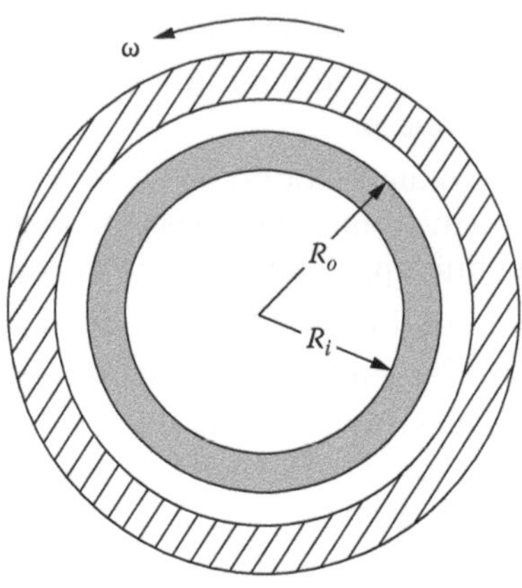

FIGURE P15.12

15.13: An incompressible fluid is contained between two large parallel plates as shown in Figure P15.13. The upper plate is fixed, the fluid is initially at rest and if the bottom plate suddenly starts to move with a constant velocity, $\hat{V}$, the differential equation governing the motion is

$$\rho \frac{\partial \hat{V}_x}{\partial t} = \mu \frac{\partial^2 \hat{V}_x}{\partial y^2}$$

where $\hat{V}_x$ is the velocity in the x direction and ρ and μ are the fluid density and viscosity, respectively. Rewrite the differential equation and the initial and boundary conditions in dimensionless form using h and $\hat{V}$ as reference parameters for length and velocity and $h^2\rho/\mu$ as a reference parameter for time.

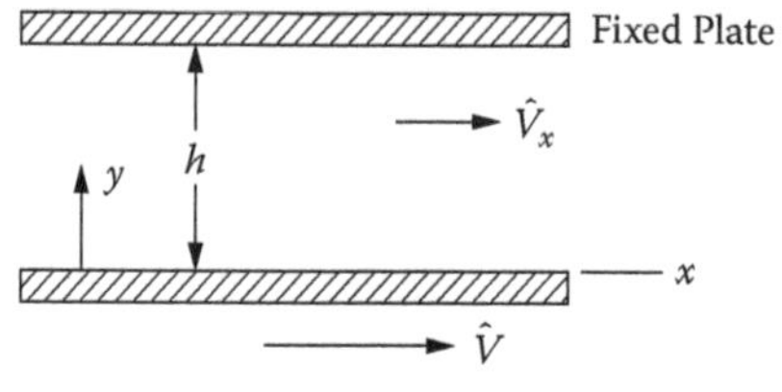

FIGURE P15.13

15.14: The deflection of a cantilever beam shown in Figure P15.14 is governed by the differential equation

$$EI\frac{d^2y}{dx^2} = P(x - L)$$

where E is the modulus of elasticity and I is the moment of inertia of the beam cross section. The boundary conditions are $y = 0$ at $x = 0$ and $dy/dx = 0$ at $x = 0$. (a) Rewrite the equation and the boundary conditions in dimensionless form using the beam length, L, as the reference length and (b) based on the results of part (a), determine the equation to predict deflection.

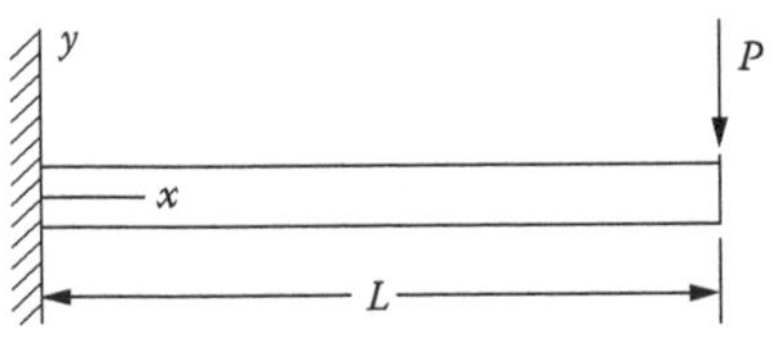

FIGURE P15.14

15.15: A liquid is contained in a pipe that is closed at one end as shown in Figure P15.15. Initially, the liquid is at rest, but if the end is suddenly opened, the liquid starts to move. Assuming that the pressure, P_1, remains constant, the differential equation that describes the resulting motion of the liquid is

$$\rho\frac{\partial \hat{V}_z}{\partial t} = \frac{P_1}{L} + \mu\left(\frac{\partial^2 \hat{V}_z}{\partial r^2} + \frac{1}{r}\frac{\partial \hat{V}_z}{\partial r}\right)$$

where $\hat{V}_z$ is the velocity at any radial location, r, and t is time. Rewrite this equation in dimensionless form using the pipe radius, R, the mean velocity, $\hat{V}_{avg}$, as the reference dimensions. Note that the reference time becomes $R/\hat{V}_{avg}$.

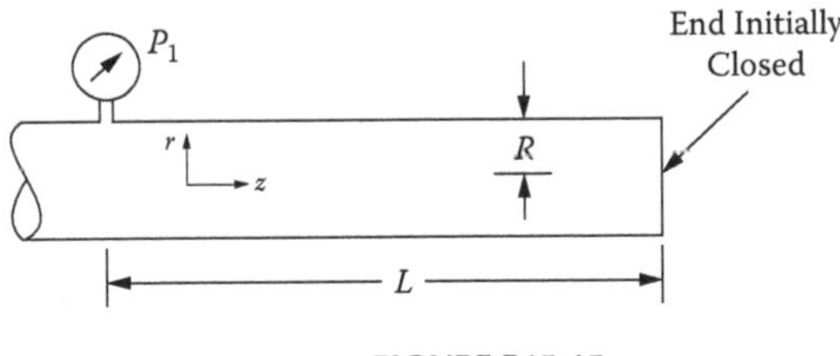

FIGURE P15.15

15.16: A plane wall of infinite extent in the y and z directions has the origin of its x-coordinate system at the centerline of the wall. The cooling of the wall of thickness $2L$ is governed by the Fourier equation

$$\frac{\partial T}{\partial x^2} = \frac{1}{\alpha}\frac{\partial T}{\partial t}$$

where α is the thermal diffusivity of the wall material, T is the temperature, and t is the time. Express the differential equation in a nondimensional form using x_o as the reference, $\Delta T = T_i - T_\infty$ as the reference temperature and x_o/α as the reference time.

Use of the Buckingham π Theorem

15.17: Using the Buckingham Pi theorem, generate the dimensionless parameters that will correlate the variables in Problem 15.13. The velocity and time should be in separate parameters.

15.18: The power output of a hydraulic turbine depends on the diameter, d, of the turbine, the density, ρ, of the water, the height, h, of the water surface above the turbine, the gravitational acceleration, g, the angular velocity, ω, of the turbine wheel, the discharge, $\dot{V}$, of water through the turbine, and the efficiency, η, of the turbine. By dimensional analysis, generate a set of appropriate dimensional groups.

15.19: The pressure rise across a pump, ΔP (this term is proportional to the head developed by the pump) may be considered to be affected by the fluid density, ρ, the angular velocity, ω, the impeller diameter, d, the volumetric rate of flow, $\dot{V}$, and the fluid viscosity, μ. Find the pertinent dimensionless groups, choosing them so that P, $\dot{V}$, and μ each appear in only one group.

15.20: The power, P, required to run a compressor varies with the angular velocity, ω, the impeller diameter, d, the volumetric rate of flow, $\dot{V}$, the fluid density, ρ, and the fluid viscosity, μ. Develop a relationship between these variables using dimensional analysis where fluid viscosity and angular velocity appear in only one dimensionless parameter.

15.21: The performance of a journal bearing around a rotating shaft is a function of the following variables: the volumetric rate of flow of the lubricating oil, $\dot{V}$, the lubricant viscosity, μ, the fluid density, ρ, the bearing diameter, d, the shaft speed in revolutions per minute, N, and the surface tension of the lubricating oil, σ. Suggest appropriate parameters that may be employed in the correlation of experimental data for such a system.

15.22: The mass, m, of drops formed by a liquid discharging by gravity from a vertical tube is a function of the liquid density, ρ, the tube diameter, d, the surface tension, σ, and the acceleration of gravity, g. Neglecting any effects of viscosity, determine the independent dimensionless groups that would allow the surface tension effect to be analyzed.

15.23: The rate at which metallic ions are electroplated from a dilute electrolytic solution onto a rotating disk electrode is usually governed by the mass diffusion rate of ions to the disk. This process is believed to be controlled by the following variables:

k = mass transfer coefficient, m/s

D = diffusion coefficient, m^2/s

d = disk diameter, m

ω = angular velocity, 1/s

ρ = density, kg/m^3

μ = viscosity, kg/m-s

Obtain a set of dimensionless groups for these variables keeping k, μ, and D in separate groups.

15.24: The fundamental frequency, n, of a stretched string is a function of the string length, L, its diameter, d, the mass density, ρ, and the applied tensile force, T. Suggest a set of dimensionless parameters relating these variables.

15.25: A thin elastic wire is placed between supports. A fluid flows past the wire and it is desired to study the static deflection, δ, at the center of the wire due to the fluid drag. Assume that

$$\delta = f(L, d, \rho, \mu, \hat{V}, E)$$

where L is the length of the wire, d is the wire diameter, ρ is the fluid density, μ is the fluid viscosity, $\hat{V}$ is the fluid velocity, and E is the modulus of elasticity of the wire material. Develop a suitable set of pi terms for this problem.

15.26: Through a series of tests on pipe flow, Darcy[7] derived an equation for the friction loss in pipe flow as

$$h_L = f\frac{L}{d}\frac{\hat{V}^2}{2g}$$

in which f is a dimensionless coefficient that depends on the average velocity of the pipe flow, $\hat{V}$, the pipe diameter, d, the pipe length, L, the fluid density, ρ, the fluid viscosity, μ, and the average pipe wall roughness or unevenness, e, which is a length term. Using the Buckingham Pi theorem, find a dimensionless function for the coefficient, f.

15.27: The maximum pitching moment that is developed by the water on a flying boat as it lands is denoted as C_{max}. The following are the variables involved in this action:

α = angle made by the flight path of the aircraft with the horizontal

β = angle defining the attitude of the aircraft

m = mass of the aircraft

L = length of the hull

ρ = density of water

g = acceleration of gravity

R = radius of gyration of the aircraft about the pitching axis

Use the Buckingham Pi theorem to determine (a) the number of independent dimensionless groups needed to characterize this problem, (b) the dimensional matrix of the problem, and (c) the rank of the dimensional matrix.

15.28: Assume that the volumetric flow rate, $\dot{V}$, of a gas from a smokestack is a function of the density of ambient air, ρ_a, the density of the gas within the stack, ρ_g, the acceleration of gravity, g, the height of the stack, h, and the diameter of the stack, d. Use ρ_a, d, and g as repeating variables to develop a set of pi terms that could be employed to describe this problem.

15.29: In a fuel injection system, small droplets are formed due to the breakup of the liquid jet. Assume that the droplet diameter, d, is a function of the liquid density, ρ, the fluid viscosity, μ, the surface tension, σ, the jet velocity, $\hat{V}$ and the diameter, d. Form an appropriate set of dimensionless parameters using μ, $\hat{V}$, and d as repeating variables.

15.30: A liquid flows with a velocity, $\hat{V}$, through a hole in the side of a large tank. Assume that

$$\hat{V} = f(h, g, \rho, \sigma)$$

where h is the depth of the fluid above the hole, g is the acceleration of gravity, ρ is the fluid density and σ is the surface tension. The following data were obtained by changing h and measuring $\hat{V}$ with a fluid having a density of 1000 kg/m^3 and a surface tension of 0.74 N/m:

$\hat{V}$, m/s	3.13	4.43	5.42	6.25	7.00
h, m	0.50	1.00	1.50	2.00	2.50

[7] H. P. G. Darcy (1803–1858) performed extensive tests on filtration and pipe resistance and initiated open-channel studies.

Plot these data by using appropriate dimensionless variables and comment as to whether any of the original variables could have been omitted.

15.31: As indicated in Figure P15.31, the pressure drop across a short hollowed plug placed in a circular tube through which a liquid flows can be expressed as

$$\Delta P = f(\rho, \hat{V}, D, d)$$

where ρ is the fluid density and $\hat{V}$ is the mean velocity in the tube. Some experimental data obtained with $D = 61$ cm, $\rho = 1031\ \text{kg/m}^3$, and $\hat{V} = 0.61$ m/s are given in the following table:

d, cm	1.829	2.438	3.048	4.572
ΔP, N/m^2	23,643	7479	3064	603

Plot these data by using appropriate dimensionless variables.

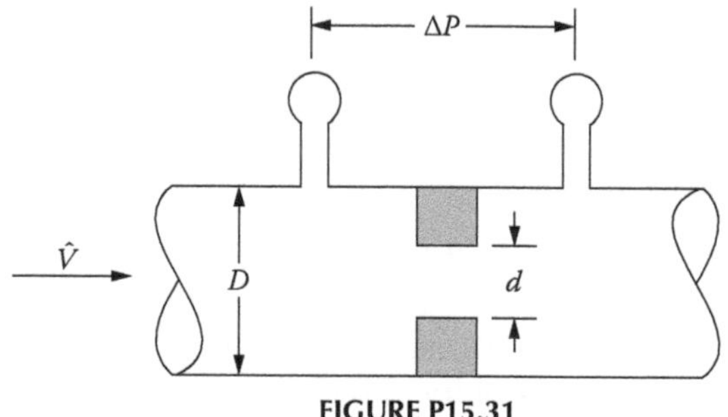

FIGURE P15.31

15.32: One type of viscosimeter consists of an open reservoir with a small diameter tube at the bottom as illustrated in Figure P15.32. To measure the viscosity, the system is filled with the liquid of interest and the time required for the liquid to fall from h_i to h_f is determined. Variables involved are the initial head, h_i, and the final head, h_f ($\Delta h = h_i - h_f$), the density, ρ, the viscosity, μ, the time, t, the acceleration of gravity, g, and the diameter, d. Determine the dimensionless parameters that would be appropriate in presenting test measurements for this system.

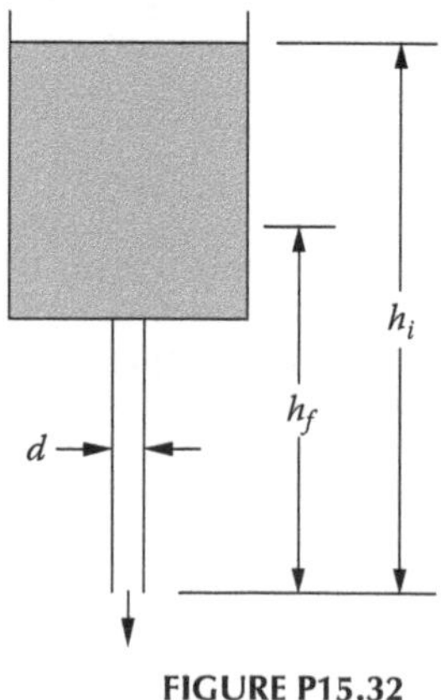

FIGURE P15.32

15.33: The concentric cylinder device of the type shown in Figure P15.33 is commonly used to measure the viscosity, μ, of liquids by relating the angle of twist, θ, to the angular velocity, ω, of the outer cylinder. Assume that

$$\theta = f(\omega, \mu, K, d_1, d_2, L)$$

where K depends on the suspending wire properties and has the dimensions, FL. The following data were obtained in a series of tests for which $\mu = 0.01488\,\text{kg/m-s}$, $K = 13.588\,\text{J}$, $L = 0.3048\,\text{m}$, and d_1 and d_2 were constant:

θ, rad	ω, rad/s
0.89	0.30
1.50	0.50
2.51	0.82
3.05	1.05
4.28	1.43
5.52	1.86
6.40	2.14

From these data, determine, with the aid of dimensional analysis, the relationship between θ, ω, and μ for this particular apparatus. *Hint:* Plot the data, using appropriate dimensionless parameters, and determine the equation of the result using a standard curve-fitting technique. The equation so obtained should satisfy the condition that $\theta = 0$ for $\omega = 0$.

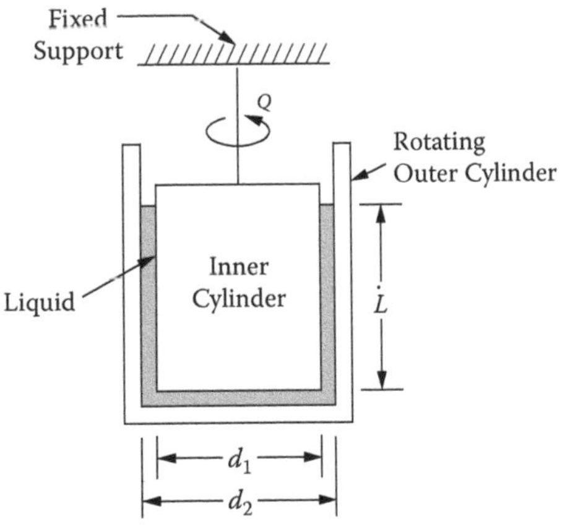

FIGURE P15.33

15.34: An estimate is needed of the lift provided by a hydrofoil wing section when it moves through water at 27.5 m/s. Test data are available for this purpose from experiments in a pressurized wind tunnel with an airfoil section model geometrically similar to but twice the size of the hydrofoil. The lift, F, is a function of the density of the fluid, ρ, the velocity of the flow, $\hat{V}$, the angle of attack, θ, the chord length, d, and the viscosity, μ. Take the density of water as $1000\,\text{kg/m}^3$ and assume that the angle of attack is the same in both cases, the density of the air in the pressurized wind tunnel is $2.58\,\text{kg/m}^3$, the kinematic viscosity of the air is $7.43 \times 10^{-6}\,\text{m/s}^2$, and the kinematic viscosity of the water is $0.929 \times 10^{-6}\,\text{m/s}^2$. Determine the velocity of flow in the wind tunnel that would correspond to the hydrofoil velocity for which the estimate is desired.

Similarity

15.35: The pressure drop in a Venturi meter is known to vary only with the fluid density, ρ the velocity of approach, $\hat{V}$, and the diameter ratio of the meter. Meter 1 in water shows a 6.25-kPa drop when the velocity of approach is 4.875 m/s. Meter 2, which is geometrically similar to meter 1 is used with benzene ($\rho = 682\ \text{kg/m}^3$) in an application where $\dot{V} = 0.1688\ \text{m}^3/\text{s}$. Determine the pipe diameter of meter 2 that will yield a 16.25-kPa pressure drop in meter 2.

15.36: In order to find the drag on a 0.10-cm-diameter sphere in slowly flowing water at 20°C a 0.925-cm-diameter sphere is tested in glycerin at 320 K and $\hat{V} = 0.2875\ \text{m/s}$ with a measured drag on the model of 1.325 N. Determine the water velocity and the drag on the smaller sphere in water.

15.37: It is proposed that a subsonic wind tunnel be used to determine the drag on an aircraft whose velocity is 125 m/s using a 1:24 scale model at the same pressure and temperature. Can this be done?

15.38: A ship that is 195-m long is to operate at 10 m/s in seawater ($\nu = 12.61 \times 10^{-6}\ \text{m/s}^2$). Determine the kinematic viscosity of a fluid employed with a 3.35-m model that preserves the equality of both the Reynolds and the Froude numbers.

15.39: A model of a ship propeller is to be tested in water at the same temperature that would be encountered by a full-scale propeller. Over the speed range considered, it is assumed that there is no dependence on either the Reynolds or the Euler numbers but only on the Froude number (based on the forward velocity, $\hat{V}$, and propeller diameter, d). In addition, the ratio of the forward to rotational velocity of the propeller ($\hat{V}/nd$) where n is the propeller rpm, must be constant. With a model 0.41 m in diameter, a forward speed of 2.58 m/s and a rotational speed of 450 rpm, determine the forward and rotational speeds of a 2.45-m prototype.

15.40: A 40% scale model of an aircraft is to be tested in a flow regime where unsteady flow effects are important. The full-scale vehicle experiences the unsteady effects at a Mach number of 1.00 at an altitude of 12,000 m. The model is to be tested at 22°C and noting that the speed of sound is proportional to the square root of the temperature, determine the temperature that the model must be tested at to produce an equal Reynolds number.

15.41: During the development of a ship having a length of 125 m, it is desired to test a 10% model in a towing tank to determine the drag characteristics of the hull. Determine how the model is to be tested if the Froude number is to be duplicated.

15.42: A 25% scale model of an undersea vehicle that has a maximum speed of 16 m/s is to be tested in a wind tunnel with a pressure of 6 atmospheres to determine the drag characteristics of the full-scale vehicle. The model is 3-m long. Determine (a) the air speed required to test the model and (b) the ratio of the model drag to the full-scale drag.

15.43: A one-sixth scale model of a torpedo is tested in a water tunnel to determine drag characteristics. Determine (a) the prototype resistance if the model resistance is 45 N and (b) the velocity that corresponds to a torpedo velocity of 20 knots.

15.44: For a certain fluid flow problem, it is known that both the Froude and Weber numbers are important dimensionless parameters. The problem is to be studied using a 1/10 scale model and the model and the prototype operate in the same gravitational field. Determine the required surface tension scale if the density scale is equal to 1.00. The Weber number, We, relates inertial forces to surface tension forces and has the form $\text{We} = \rho\hat{V}^2 L/\sigma$.

15.45: A large, rigid, rectangular billboard is supported by an elastic column as shown in Figure P15.45. There is concern about the deflection, δ, of the top of the structure during a high wind of velocity, $\hat{V}$. A wind-tunnel test is to be conducted with a 1/15 scale model. Assume that the pertinent column variables are its length, cross-sectional dimensions, and the modulus of elasticity of the material used for the column. The only important "wind" variables are the air density and the velocity. Determine (a) the model design conditions and the prediction equation for the deflection and (b) if the same structural materials are used for both the model and the prototype and the wind tunnel operates under standard atmospheric conditions, find the required wind-tunnel velocity to match an 80-km/h wind.

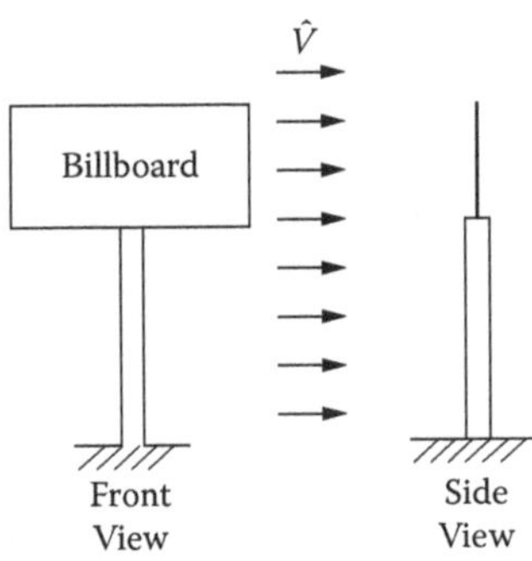

FIGURE P15.45

15.46: The pressure drop between the entrance and exit of a 150-mm-diameter 90° elbow through which ethyl alcohol at 300 K is flowing is to be determined with a geometrically similar model. The velocity of the alcohol is 5 m/s and the model fluid is to be water at 20°C with a velocity limited to 10 m/s. Determine (a) the required diameter of the model elbow to maintain dynamic similarity and (b) the prototype pressure drop if the measured pressure drop in the model is 20 kPa.

15.47: A solid sphere having a diameter, d, and a density, ρ_s, is immersed in a liquid having a density, ρ_f ($\rho_f > \rho_s$) and then released. It is desired to use a model system to determine the maximum height, h, above the liquid surface to which the sphere will rise, upon release, from a depth, H. It can be assumed that the important liquid properties are the density, ρ_f and the viscosity, μ_f. Establish the model design conditions and the prediction equation for the system.

15.48: The drag characteristics for a newly designed automobile having a maximum characteristic length of 6.10 m are to be determined through a model study. The characteristics at both low speed (approximately 8.94 m/s) and high speed (44.70 m/s) are of interest. For a series of projected model tests, an unpressurized wind tunnel will be used to accommodate a model with a maximum characteristic length of 1.21 m. Determine the range of air velocities that would be required for the wind tunnel if Reynolds number similarity is desired and comment on the suitability of the velocities.

15.49: Determine the drag due to an 80-km/h wind on a 2-m-diameter satellite dish. A wind-tunnel test is to be performed using a geometrically similar 0.5-m-diameter model dish. Assume standard air for both model and prototype and determine (a) the air speed at which the model should be run and (b) the predicted drag on the prototype dish with all similarity conditions satisfied and a measured drag on the model of 200 N. What is the predicted drag on the satellite dish?

15.50: An 8-cm sphere is tested in 30°C water flowing at 5 m/s, yields a measured drag of 12 N. For a 2.875-m weather balloon moving in air at 7°C and one atmosphere, determine the velocity and drag force.

15.51: Is it reasonable to use air in a subsonic wind tunnel at the same temperature and pressure for a test to find the drag on a 1:24 scale model of an aircraft whose velocity is 4.5 km/h.

15.52: A 1:24 scale model of a submarine is tested at 60 m/s in a wind tunnel using standard air at sea level. (a) Determine the prototype speed in seawater at 20°C and (b) for a model drag is 7.20 N, determine the prototype drag.

15.53: A model airplane represents a 1:24 scale prototype. The prototype is required to fly at 765 km/h, the Reynolds numbers of both model and prototype are required to be identical and the air temperature and pressure are to remain the same. Determine the air velocity in the wind tunnel.

15.54: A freighter that is 200-m long must operate at 0.975 m/s in seawater whose kinematic viscosity is 0.9982×10^{-6} m/s^2. Determine the kinematic viscosity of the fluid used in a test of a 3.50-m model so that both the Reynolds and Froude number are equivalent.

15.55: To assess overall head losses, a model of a venturi meter has linear dimensions that are 15% of those of the prototype. Both the model and the prototype operate in water, the model at 10°C and the prototype at 100°C. For a throat velocity of 6.75 m/s and a throat diameter of 67.5 cm, determine the flow through the meter.

15.56: Air at 20°C at a velocity of 2.15 m/s flows through a 6-cm pipe. For dynamic similarity, determine the size of pipe to carry water at 15°C and a velocity of 1.15 m/s.

15.57: A blimp is designed to move in air at 27°C at 6.75 m/s. A 1:20 scale model is used in water at 20°C with a measured drag of 3.00 kN. Determine (a) the water velocity of the model and (b) the drag on the prototype, and (c) the power required for its propulsion.

15.58: A 1:8 scale model of an automobile is tested in a wind tunnel having the same properties as the prototype. The velocity of the prototype is 45 km/h and, for dynamically similar airflow conditions, the model drag is 350 N. Determine the drag on the prototype and the power required for model propulsion.

15.59: A one-sixth scale model of a torpedo is tested in a wind tunnel. The prototype is expected to attain a velocity of 6.25 m/s in water at 60°C. Determine the model speed for the wind-tunnel pressure of 2 MPa and its temperature 80°C.

15.60: Water at 20°C flows at 4 m/s in a 15-cm-diameter pipe. For dynamic similarity, determine the velocity of a fuel oil having a kinematic viscosity of 8.012×10^{-6} m/s^2 in a 7.5-cm-diameter pipe.

15.61: Air at 27°C is to flow through a 60-cm-diameter pipe at an average velocity of 2.25 m/s for dynamic similarity to be maintained. Determine the size of pipe required to carry 20°C water at 1.25 m/s.

16

Viscous Flow

Chapter Objectives

- To review Reynolds' classic experiment.
- To consider the boundary layer concept and it implications.
- To develop concepts relative to viscous flow over bluff bodies.
- To establish concepts relative to viscous flow along plane surfaces.

16.1 Introduction

In this chapter, we will combine some of ideas that have been developed earlier to describe the total interaction between a solid body and a fluid flowing through it or around it. The chapter title, "Viscous Flow," correctly implies that we will be considering viscous effects in contrast to the analysis of **inviscid flow**, which is a subject of particular interest to aerospace engineers. The subject of inviscid flow tends to be quite sophisticated mathematically and is beyond the scope of this book.

One of the major success stories in the annals of fluid mechanics is the concept of the **boundary layer** proposed by Ludwig Prandtl[1] around the beginning of the twentieth century. The idea of a thin region, near a solid boundary, where viscous flow effects are manifested, has led to much of the useful modeling on which we now rely to describe fluid flow effects. Outside the boundary layer, viscous flow effects are negligibly small and the flow can be idealized to follow the Bernoulli equation.

We will now discuss some practical aspects of viscous flow.

16.2 Reynolds' Experiment

The existence of two kinds of viscous flows, **laminar** and **turbulent**, is a universally accepted concept. Water flowing slowly from a faucet proceeds smoothly and uniformly for a short distance and, then, abruptly changes into an unstable, irregular behavior. Similar behavior can be observed for smoke emanating from a slowly burning object.

[1] L. Prandtl (1875–1953) was a German physicist who contributed to fluid mechanics, particularly aerodynamics, and introduced the concept of the boundary layer.

The well-ordered type of flow has already been designated as **laminar** to indicate the layer-like or laminar behavior as adjacent fluid layers slide smoothly over one another without significant mixing between them.

The other flow behavior, in which packets of fluid particles move between layers giving it a fluctuating character, is designated as **turbulence** and the result is turbulent flow.

In 1883, Osborne Reynolds first described an experiment that, quantitatively, established the transition between laminar and turbulent flow. His experiment, which has been duplicated countless times, was conducted in an apparatus with general features as shown in Figure 16.1a.

Water was allowed to flow through a transparent tube at rate controlled by the valve. A dye with a specific gravity of unity, to avoid buoyancy effects, was introduced into the flow at a location near the flow entrance, and the appearance of the dye streak was observed as the valve opening was increased to provide greater flow rates.

As the valve was opened slowly, the dye streak retained its regular flow character, forming a single line of color as illustrated in Figure 16.1b (top). As the valve opening was increased, producing ever larger flow rates, it was observed that the dye suddenly became dispersed through the cross section (Figure 16.1b bottom). This behavior was clearly due to the orderly character of laminar flow in the first case and the chaotic nature of turbulent flow in the latter case.

It was clear, in Reynolds' experiment, that the transition from laminar to turbulent flow was velocity dependent. In investigating other variables affecting this transition, he found that pipe diameter, fluid density, and fluid viscosity were also important. The combination of these four variables into a dimensionless parameter produced the **Reynolds number**

$$\mathrm{Re} = \frac{\rho \hat{V} d}{\mu} \tag{16.1}$$

named in honor of the British engineer, Sir Osborne Reynolds, whose experiment first demonstrated its importance. The Reynolds number is often specified on the basis of the length dimension employed. For example, when the Reynolds number is based on the diameter, Re_d is often used; when a length or a length coordinate is employed, we frequently see the Reynolds number indicated by Re_L or Re_x.

The transition from laminar to turbulent flow in a pipe is generally agreed to occur at a value of Re equal to 2300. Below this value, the flow is definitely laminar. For carefully controlled experiments, laminar flow has been reported for values of Re as high as 40,000. However, at high values of Re, a slight disturbance will cause the flow to become turbulent. Laminar flows at a Reynolds number above 2300 are therefore **unstable** and this value, 2300, is the generally accepted value for the **critical** Reynolds number for pipe flow.

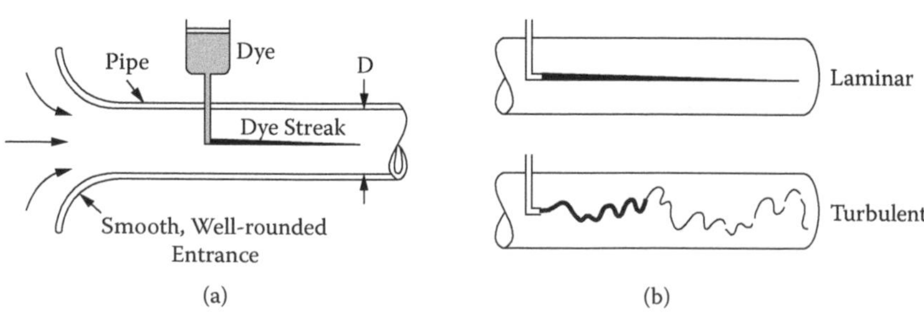

FIGURE 16.1
(a) Features of the Reynolds' experiment and (b) flow behaviors.

16.3 Fluid Drag

In Example 15.1, in the preceding chapter, we used dimensional analysis to determine that the effects of flow normal to a cylinder are related by two parameters

$$\pi_1 = \frac{F/d^2}{\rho \hat{V}^2} \qquad \text{and} \qquad \pi_2 = \frac{\mu}{d\hat{V}\rho}$$

The second of these dimensionless quantities, π_2, is observed to be 1/Re or we could simply use Re, which is the better-accepted dimensionless form.

The other dimensionless parameter, π_1, relates the resisting or **drag force** per length squared (we might think of this as an area) to the kinetic energy of flow. An equivalent dimensionless form is

$$\frac{F}{A} = \pi_1 \frac{\rho \hat{V}_\infty^2}{2}$$

where F = the drag force
A = the area
ρ = the fluid density
$\hat{V}_\infty$ = the free-stream velocity

Note that the kinetic energy has been written with a 2 in the denominator, which is the generally accepted form.

The coefficient, π_2, which can be thought of as a proportionality factor, is generally given the symbol, C.

These coefficients are commonly encountered in two forms. One, designated C_f, is the **coefficient of skin friction**. It relates the viscous or frictional drag force to the kinetic energy of flow according to the expression

$$\frac{F}{A} = C_f \frac{\rho \hat{V}_\infty^2}{2} \tag{16.2}$$

where F = the frictional drag force
A = the surface area of the solid body contacting the flowing fluid
C_f = the coefficient of skin friction

As we will see directly, Equation 16.2 applies for flow over streamlined bodies where flow generally conforms to the shape of the solid boundary.

The other coefficient, designated as C_D, is termed the **drag coefficient** and relates the drag force due to pressure differences fore-and-aft to the kinetic energy of the flow. The relationship that applies in this case is

$$\frac{F}{A} = C_D \frac{\rho \hat{V}_\infty^2}{2} \tag{16.3}$$

where F = the total drag force
A = the **projected** area of the body normal to the direction of flow
C_D = the drag coefficient

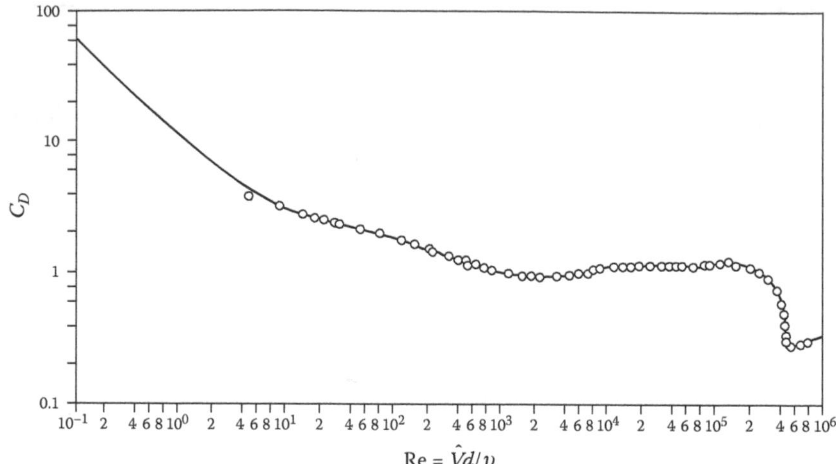

FIGURE 16.2
Drag coefficient variation with the Reynolds number for flow across a circular cylinder.

The drag force given by Equation 16.3 is generally associated with bluff (nonstreamlined) bodies where the streamlines actually separate from the surface giving rise to significant pressure differences and, consequently, significant drag forces.

It should also be noted that the measured drag force for flow over bluff bodies includes both those due to friction and pressure. The magnitude of viscous drag is usually much the smaller of the two. The drag coefficient, as defined by Equation 16.3, may be thought of as that due to pressure effects but does, in fact, include the relatively smaller viscous effects as well.

Recall that dimensional analysis indicates these coefficients are both functions of the Reynolds number. Figure 16.2 shows the variation in C_D with Re_d for flow normal to a circular cylinder. The characteristics shown are the results of many experimental studies.

It is important that we understand the mechanical phenomena responsible for the behavior illustrated in Figure 16.2. We will discuss this behavior, proceeding as the flow velocity (thus Re_d) increases through four regimes.

- **Regime I:** At very small Reynolds numbers, generally in the regime $\text{Re}_d < 1$, the condition is often designated as one of **creeping flow**. Here, the flow is completely laminar and is slow enough to completely follow the shape of the body. Under these conditions, drag is totally due to viscous effects that extend throughout the flow field. Flow is stable and the downstream flow known as the **wake** does not oscillate. Typical flow patterns for a cylinder in cross flow are shown in Figure 16.3. The creeping flow case is represented in regime I.
- **Regime II:** In Regime II, which extends in the Re_d range from 1 to approximately 10^3, small eddies begin to form near the aft **stagnation point**. As the velocity increases, the eddies also grow in extent and are shed, alternately, from the upper and lower portions of the separated region. As they are swept downstream, they impart an oscillating character to the wake. At values of Re_d where this oscillatory behavior is distinct, the pattern is referred to as a **Von Karman**[2] **vortex** "street"

[2] T. Von Karman (1881–1963) was an American aerodynamicist who made theoretical contributions to aerodynamics and who formulated von Karman's theory of vortex sheets, an early step in the mathematical treatment of turbulent motion.

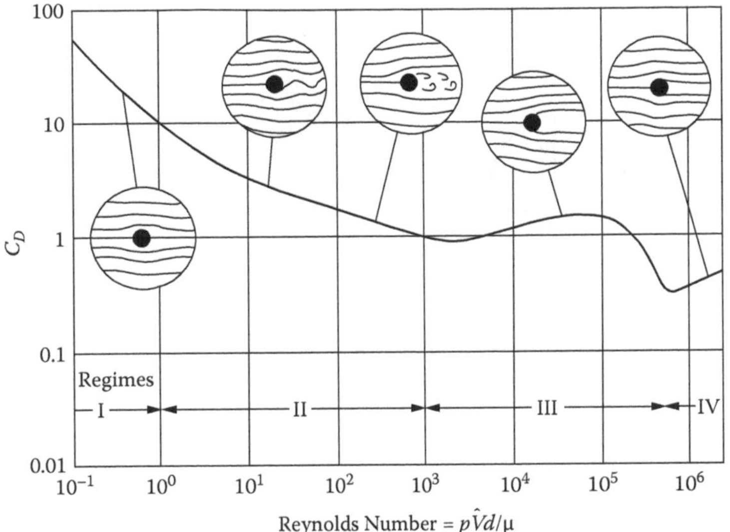

FIGURE 16.3
Drag coefficient variation with the Reynolds number for flow across a circular cylinder.

or trail. In this region, the wake is unsteady and the streamlines actually separate from the rear of the cylinder; θ, the angle between the forward stagnation point and the separation point, becomes smaller for increasing values of Re_d. In this regime, the drag coefficient decreases in a regular manner as Re_d increases.

- **Regime III:** As the velocity increases, the separation point continues to move forward, stabilizing at a position approximately 80° from the forward stagnation point. In this region, extending roughly between Re_d values from 10^3 to 3×10^5, the value of C_D remains near a value of 1. Over the portion of the surface where the streamlines are attached, the flow is laminar. The wake region is no longer characterized by large eddies but remains stable.
- **Regime IV:** An interesting and important phenomenon occurs for a value of Re_d near 5×10^5, where the drag coefficient suddenly decreases from a value near 1 to approximately 0.3. The explanation for this behavior is that boundary layer flow in the attached region becomes turbulent and, in turn, the separation point moves rearward from $\theta = 80°$ to approximately 145°. As a result of this observation, we may conclude that a turbulent boundary layer tends to remain attached longer or, in other words, resists separation more, than does a laminar boundary layer. The direct result of a greater region of boundary layer attachment is the sudden decrease in C_D that is evident in Figure 16.2. As Re_d increases beyond the transition value, C_D will, again, increase slowly.

Figure 16.4 illustrates the pressure distribution about the cylindrical surface for values of Re_d typical of the flow regimes discussed.

For the case of attached streamlines, the pressure distribution over the front and back portions of the cylinder is seen to be the same. Thus, there is no net pressure force.

However, when boundary layer separation occurs, the effective pressure force over the forward portion of the cylinder is greater than over the rear and a net pressure force will exist. There will still be some viscous drag contributed by the region where the boundary

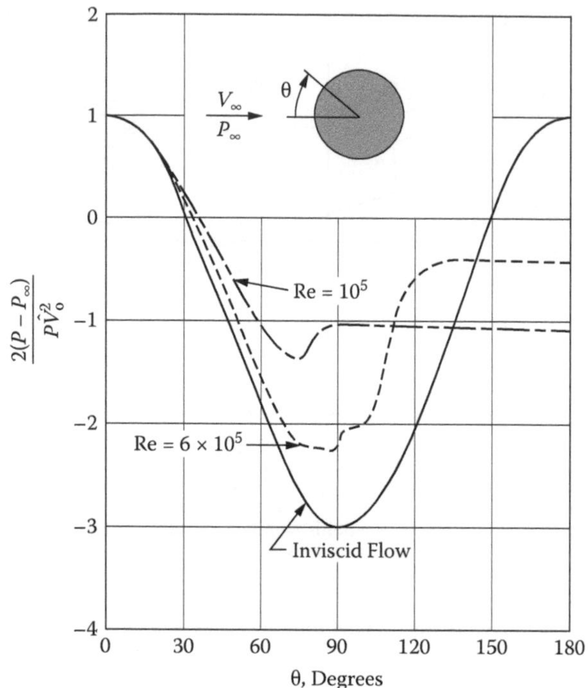

FIGURE 16.4
Pressure distribution over a cylinder in cross flow for different values of Re_d.

layer is attached but this will be small compared with the drag due to the pressure imbalance. When separation exists the coefficient used to evaluate drag is, thus, C_D.

The pressure distributions for laminar and turbulent boundary layer flows show why the value of C_D for the turbulent case is lower than for the laminar case. This is a direct consequence of boundary layer separation.

The situation for flow past a sphere is very much like that just discussed for the case of the cylinder except that, at the same values of Re_d, C_D will differ between the two shapes. Figure 16.5 shows C_D as a function of Re_d for three other plane surface cases oriented perpendicular to the free-stream velocity.

We note that for $Re_d < 1$, the asymptotic limit for C_D in the case of spheres is described by the expression

$$C_D = \frac{24}{Re_d}$$

This is the **Stokes' limit** or **Stokes' law**, achieved analytically by Stokes.

Table 16.1 gives values of C_D for a number of bluff objects.

Example 16.1 Evaluate the steady-state terminal velocity of a glass sphere measuring 1 cm in diameter falling freely through (a) glycerin at 300 K, (b) water at 300 K, and (c) air at 300 K. The density of glass may be taken as 2250 kg/m^3.

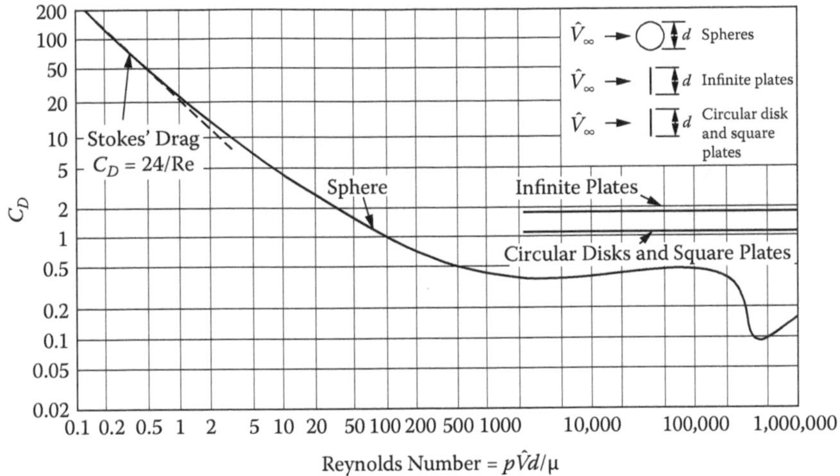

FIGURE 16.5
Drag coefficients as a function of Re_d for various bluff bodies.

Solution

Assumptions and Specifications

(1) Steady-state conditions exist.
(2) The glycerin is incompressible.

A freely falling object will reach a steady, terminal velocity when the forces due to its weight and to fluid drag are in balance. The expression to be used equates the weight

$$\rho_s \frac{\pi d^3}{6} g$$

with the drag

$$C_D \rho_f \frac{\hat{V}_\infty^2}{2} \frac{\pi d^2}{4}$$

TABLE 16.1
Drag Coefficient Values for Various Blunt Shapes: The Value for the Ring is Based on the Ring Area

Shape	C_D
Disk	1.17
Ring	1.20
Finite Cylinder	
$w/h = 1$	1.12
$w/h = 10$	1.31
$w/h = 20$	1.44
Hemisphere	
Open end facing flow	1.42
Open end facing downstream	0.38
Half-Cylinder	
Open end facing flow	2.30
Open end facing downstream	1.20

to yield

$$C_D \hat{V}_\infty^2 = \frac{4}{3} \frac{\rho_s}{\rho_f} dg$$

We may recall from Figure 16.5, that the drag coefficient for a sphere is a function of the Reynolds number, Re_d. Because Re_d is a function of $\hat{V}_\infty$, it is not possible to solve this expression explicitly for $\hat{V}_\infty$ unless $\text{Re}_d < 1$, which would permit the use of Stokes' law in expressing C_D.

In this case, we must conclude that a trial-and-error solution is required. The conditions to be satisfied are the force balance and the figure that gives C_D as a function of Re_d.

The results for the three fluids are

$$\begin{aligned}
\text{(a) For glycerin}\quad & C_D \cong 5 \\
& \text{Re}_d \cong 8.6 \\
& \hat{V}_\infty \cong 0.305\,\text{m/s} \\
\text{(b) For water}\quad & C_D \cong 0.42 \\
& \text{Re}_d \cong 3.1 \times 10^4 \\
& \hat{V}_\infty \cong 1.34\,\text{m/s} \\
\text{(c) For air}\quad & C_D \cong 0.46 \\
& \text{Re}_d \cong 4.19 \times 10^6 \\
& \hat{V}_\infty \cong 32.9\,\text{m/s}
\end{aligned}$$

An interesting example of the application of these ideas is the golf ball, with its dimpled surface. The purpose of this deliberate roughing of the surface is to induce turbulence in the boundary layer so that boundary layer separation will be reduced and the drag force will, in turn, be decreased. Because its drag is reduced, a dimpled golf ball will travel significantly further than would a smooth one.

Design Example 8, which now follows, shows how the power required to overcome vehicle drag may be determined.

16.4 Design Example 8

The drag coefficient for a late model automobile having a frontal area of 2.354 m^2 is specified as $C_D = 0.28$. (a) Plot the power necessary to overcome aerodynamic drag as a function of vehicle speed from 40 mph to 80 mph in the air at 16°C and (b) determine the vehicle speed for an applied 20 hp.

Solution

Assumptions and Specifications

(1) Steady-state conditions exist.

(2) The surrounding air is incompressible.

TABLE 16.3

Summary of Computations to Determine the Power Needed to Overcome Vehicle Drag

Assume $\hat{V}$	mph	40	50	60	70	80
$\hat{V}$	m/s	17.88	22.35	26.82	31.29	35.76
$\hat{V}^3$	m^3/s^3	5717	11,166	19,292	30,635	45,729
$\dot{W}$	kW	2.32	4.52	7.81	12.41	18.52
$\dot{W}$	hp	3.11	6.06	10.48	16.64	24.84

(a) For air at 16°C (289 K), Table A.13 in Appendix A gives (with interpolation)

$$\rho = 1.229\,\text{kg/m}^3$$

The drag force on a bluff body is related to the velocity according to

$$F_d = AC_D\rho\frac{\hat{V}^2}{2}$$

and the power required is

$$\dot{W} = F_d\hat{V} = \left(AC_D\rho\frac{\hat{V}^2}{2}\right)\hat{V} = AC_D\rho\frac{\hat{V}^3}{2}$$

We observe that the power varies as the third power of the velocity.

Table 16.3 provides a summary of the data needed to be generated in order to plot $\dot{W}$ in hp as a function of velocity. These data are plotted in Figure 16.6.

(b) For a 20-hp application, the speed of the vehicle is approximately 74 mph.

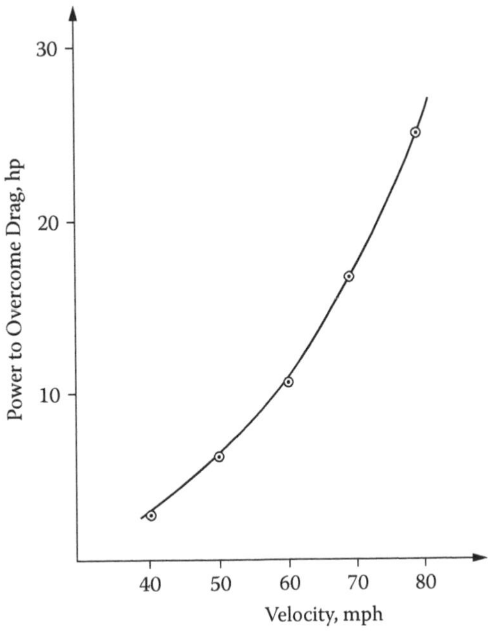

FIGURE 16.6
Power required to overcome drag as a function of vehicle velocity for a late-model automobile.

16.5 Boundary Layer Flow over a Flat Plate

A classic problem in fluid mechanics is the case of flow associated with a plane surface oriented parallel to the direction of flow. The boundary layer concept of Prandtl, dating to 1904, is that all viscous effects are confined to a thin region close to the solid surface. In this boundary layer region, the velocity changes from zero at the wall (owing to the no-slip condition) to the free-stream value at its outer edge.

Figure 16.7, illustrates the velocity profiles at several positions along a flat plate. This figure is not to scale. In reality, the boundary layer region is much thinner than as shown in the figure where the vertical scale has been exaggerated for clarification.

In examining these velocity profiles, certain qualitative conclusions may be reached:

(a) The boundary layer region grows in a regular manner with increasing distance, x, from the leading edge.
(b) The velocity profiles near the leading edge are similar in shape, the major difference between them being the distance from the surface where the velocity reaches the free-stream value, $\hat{V}_\infty$. The boundary layer flow in this region is laminar.
(c) At large values of x, there is a distinct change in the shape of the velocity profiles. Near the wall, the velocity changes rapidly; then the slope of the profile becomes quite small for a considerable distance until, finally, $\hat{V}_x = \hat{V}_\infty$. This is the turbulent boundary layer region with laminar flow hypothesized to occur in the very thin region near the surface where the velocity gradient (the rate at which the velocity changes with the y-direction near the no-slip boundary) is large. This thin region is called the **laminar sublayer**.

If we connect the loci of the points at which $\hat{V}_x$ reaches $\hat{V}_\infty$, a representative profile showing the boundary layer growth is obtained.

For the case of the boundary layer for flow parallel to a plane surface, the criterion for the type of flow is the **local Reynolds number**, Re_x, which is defined as

$$\text{Re}_x = \frac{x\hat{V}\rho}{\mu} = \frac{x\hat{V}}{\nu} \tag{16.4}$$

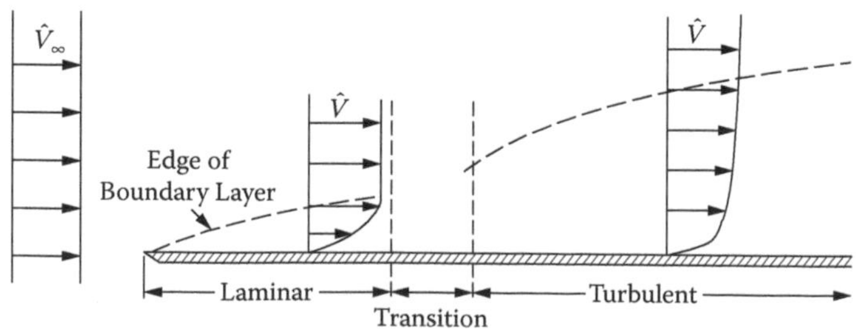

FIGURE 16.7
Velocity profiles for flow along a flat surface.

where the length scale, x, is the distance along the surface measured from the leading edge. The values of Re_x which relate to the type of flow are given for three conditions:

(a) For $0 < \mathrm{Re}_x < 2 \times 10^5$, the boundary layer flow is laminar.
(b) For $2 \times 10^5 < \mathrm{Re}_x < 3 \times 10,^6$ the boundary layer flow may be either laminar or turbulent.
(c) For $3 \times 10^6 < \mathrm{Re}_x$, the boundary layer flow is turbulent.

Because there is no precise location where viscous forces suddenly become negligible, the thickness of the boundary layer, normally designated, δ, is fictitious. The criterion that is normally used is that the boundary layer is considered to extend away from the surface a distance such that

$$\frac{\hat{V}_x}{\hat{V}_\infty} = 0.99$$

or when the velocity is within 1% of the free stream value.

With this concept, Prandtl divided the flow regime between the boundary layer, where viscous effects are significant, and the free stream, where the flow may be treated as inviscid.

Analytical solutions that give quantitative information for viscous flow effects in the boundary layer have been achieved for certain special cases. Flow parallel to a plane surface is the case usually examined in the first course in fluid mechanics. A detailed study of such boundary layer flows involves mathematical techniques that are beyond the scope of this work. The interested reader may refer to the original work of Blasius (1908) or to Welty et al. (2001) for a discussion of this analysis. For the present, we will accept the results of such analyses.

For laminar boundary layer flow along a plane surface, the following results have been achieved:

- The thickness of the boundary layer, δ, can be evaluated according to

$$\delta = 5\frac{x}{\mathrm{Re}_x^{1/2}} \tag{16.5}$$

- The local coefficient of skin friction, C_{fx}, can be determined from

$$C_{fx} = 0.664\mathrm{Re}_x^{-1/2} \tag{16.6}$$

The use of these expressions will be demonstrated in Example 16.2 that follows.

Example 16.2 For the case of air at 300 K (27°C) flowing at 15 m/s along a plane surface, determine

(a) The boundary layer thickness at a location 20 cm from the leading edge
(b) The value of C_{fx} at this location
(c) The total viscous drag exerted by the air on the surface whose dimension is 0.20 m in the flow direction
(d) The distance, x, from the leading edge where the transition region would begin
(e) The distance from the leading edge where the turbulent boundary layer region would begin

Solution

Assumptions and Specifications

(1) Steady-state conditions exist.
(2) The flow is incompressible.
(3) The surface is infinitely wide.
(4) Uniform properties exist at each point considered.

(a) We will need the value of the Reynolds number. Table A.13 gives at 300 K, $\nu = 15.69 \times 10^{-6}\,\text{m/s}^2$ so that

$$\text{Re}_x = \frac{x\hat{V}_x}{\nu} = \frac{(0.20\text{m})(15\text{ m/s})}{15.69 \times 10^{-6}\,\text{m/s}^2} = 1.912 \times 10^5$$

Because $\text{Re}_x < 2 \times 10^5$, the boundary layer is laminar. Using Equation 16.5, we determine δ as

$$\delta = 5(20\,\text{cm})(1.912 \times 10^5)^{-1/2} = 0.229\,\text{cm} \Longleftarrow$$

(b) Now that we know Re_x, we can use Equation 16.6 to evaluate C_{fx} as

$$C_{fx} = (0.664)(1.912 \times 10^5)^{-1/2} = 0.00152 \Longleftarrow$$

With this quantity we can evaluate the **local** drag force, that is, the drag force associated with a differential element that is located 20 cm from the leading edge.

(c) A much more interesting quantity is the *total* drag force over all of the differential areas between $x = 0$ and $x = 20\,\text{cm}$. This is the information sought in part (c) which is clearly obtained by integration of Equation 16.2.

$$F_d = \int_A dF_d = \int_A C_{fx}\frac{\rho\hat{V}_\infty^2}{2}\,dA$$

or

$$F_d = \frac{\rho\hat{V}_\infty^2}{2}\int_0^L 0.664\text{Re}_x^{-1/2}W dx$$

where W is the plate width. We may continue the calculation to obtain

$$\frac{F_d}{W} = \frac{\rho\hat{V}_\infty^2}{2}(0.664)\int_0^L \left(\frac{\nu}{\hat{V}_\infty}\right)^{1/2} x^{-1/2}dx$$

$$= \frac{\rho\hat{V}_\infty^2}{2}(0.664)\left(\frac{\nu}{\hat{V}_\infty}\right)^{1/2} 2L^{1/2}$$

$$= \frac{\rho\hat{V}_\infty^2}{2}(1.328)L\text{Re}_L^{-1/2}$$

Thus, with $\rho = 1.1774\,\text{kg/m}^3$ at 300 K (Table A.13)

$$\frac{F_d}{W} = \frac{(1.1774\,\text{kg/m}^3)(15\,\text{m/s})^2}{2}\frac{(1.328)(0.20\,\text{m})}{(1.912 \times 10^5)^{1/2}}$$

or

$$\frac{F_d}{W} = 0.0805\,\text{N/m of width} \Longleftarrow$$

(d) The distance from the leading edge where the transition region begins can be found from Re_x

$$x = \frac{\nu}{\hat{V}_\infty}\text{Re}_x$$

and for $\text{Re}_x = 2 \times 10^5$, the result is

$$x = \frac{15.69 \times 10^{-6}\,\text{m}^2/\text{s}}{15\,\text{m/s}}(2 \times 10^5) = 0.209\,\text{m} \Longleftarrow$$

(e) In a similar fashion, the distance from the leading edge where the turbulent boundary layer region begins will be at $\text{Re}_x = 3 \times 10^6$, and we have

$$x = \frac{15.69 \times 10^{-6}\,\text{m}^2/\text{s}}{15\,\text{m/s}}(3 \times 10^6) = 3.14\,\text{m} \Longleftarrow$$

The results of the foregoing example provide some interesting insights as to the magnitude of some of the quantities that are being discussed. For example, at a point 20 cm (approximately 7 3/4 in) from the leading edge, the boundary layer is 2.3-mm (0.091 in) thick. Over the first 20 cm of a 1 m wide plate, the drag force exerted by the air is 0.0805 N (0.358 lb_f).

It was also demonstrated that the total drag force could be obtained by evaluating the sum of all local contributions up to a specific position. A more useful way to evaluate the total drag is to use the **mean coefficient of skin friction** defined as

$$\bar{C}_f = \frac{F/A}{\rho\hat{V}_\infty^2/2} \tag{16.7}$$

Analysis has given us an expression for C_{fx}, the local coefficient. The mean coefficient is related to C_{fx} as

$$F = \bar{C}_f\frac{\rho\hat{V}_\infty^2}{2}A = \int_A C_{fx}\frac{\rho\hat{V}_\infty^2}{2}dA$$

In general then

$$\bar{C}_f = \frac{1}{A}\int_A C_{fx}dA \tag{16.8}$$

and for a fixed-plate width, W

$$\bar{C}_f = \frac{1}{L}\int_0^L C_{fx}dx \tag{16.9}$$

Now, because we have an expression for C_{fx} for laminar boundary layer flow, its substitution into Equation 16.9 yields

$$\bar{C}_f = \frac{1}{L}\int_0^L 0.664\,\mathrm{Re}_x^{-1/2}\,dx$$

$$= \frac{1}{L}\int_0^L 0.664\left(\frac{\nu}{\hat{V}_\infty}\right)^{1/2} x^{-1/2}\,dx = 1.328\left(\frac{\nu}{L\hat{V}_\infty}\right)^{1/2}$$

or

$$\hat{C}_f = 1.328\mathrm{Re}_L^{-1/2} \tag{16.10}$$

The parameter, Re_L, is a *total* Reynolds number with length scale, L, being the dimension of the area of interest along the direction of flow. The local Reynolds number, Re_x, contains a length scale, x, which applies at a particular point in the flow direction. For a given length of plate, there is only one value of Re_L but an infinite number of values for Re_x, one being at the **point** where $x = L$ or, in magnitude, $\mathrm{Re}_x = \mathrm{Re}_L$. This distinction between the local and mean parameters will be made again as we continue.

For turbulent boundary layer flow along a plane surface, these same quantities, δ, C_{fx}, and C_{fL} have been determined. However, the techniques involved are less precise than for laminar flow. The reader is again referred to Welty et al. (2001) for the relevant details of this evaluation.

The following expressions apply for turbulent boundary layer flow along a plane surface. Recall that boundary layer flow will be turbulent for $\mathrm{Re}_x > 3 \times 10^6$.

$$\delta = \frac{0.376x}{\mathrm{Re}_x^{1/5}} \tag{16.11}$$

$$C_{fx} = \frac{0.0576}{\mathrm{Re}_x^{1/5}} \tag{16.12}$$

and

$$C_{fL} = \frac{0.072}{\mathrm{Re}_x^{1/5}} \tag{16.13}$$

The remaining question to be considered in this discussion of boundary layer flow along a plane wall is "How do we characterize flow along that portion of the surface where $2 \times 10^5 < \mathrm{Re}_x < 3 \times 10^6$?"

This is the so-called "transition region" between fully laminar and fully turbulent boundary layer flow. At a location within this region, flow will fluctuate between laminar and turbulent. Near the lower limit ($\mathrm{Re}_x \approx 2 \times 10^5$), the flow will be laminar most of the time and occasionally turbulent. Near the upper limit ($\mathrm{Re}_x \approx 3 \times 10^6$), the flow will be turbulent most of the time and occasionally laminar.

16.6 Summary

Viscous flow effects have drawn our attention throughout this chapter. We have been particularly interested in the force exerted by a viscous fluid as it flows around or past or solid surface.

Initially, the Reynolds number criterion for the transition between laminar and turbulent flows was reviewed. The Reynolds number, earlier claimed to be "the most important dimensionless parameter in fluid mechanics," was encountered throughout the treatment in this chapter.

Viscous flow about bluff bodies was discussed with the drag coefficient, C_D, and the coefficient of skin friction, C_f, defined. Particular attention was paid to a circular cylinder in cross flow. Boundary layer separation effects were discussed and quantified.

Approximately half of this chapter has been devoted to two-dimensional, steady, incompressible flow parallel to a plane surface. The classic solution, due to Blasius, was considered in some detail. A less precise but more versatile approach was next discussed and the results of this method were found to agree well with the Blasius results. Information obtained from boundary layer analysis included the boundary layer thickness, $\delta(x)$, and the local and mean skin friction coefficients, C_{fx} and C_f, respectively.

16.7 Problems

Boundary Layers

16.1: If a Reynolds' experiment were to be performed with a 30-mm pipe, determine the flow velocity that occurs at transition.

16.2: An aircraft flies at a speed of 150 km/h at an altitude of 3000 m. Assuming a transitional Reynolds number of $\mathrm{Re}_x = 5 \times 10^5$ and behavior of the boundary layer on the wing surfaces like that on a flat plate, estimate the extent of the laminar boundary layer flow along the wing.

16.3: A viscous fluid flows past a flat plate such that the boundary layer thickness at a distance of 1.40 m from the leading edge is 12 mm. Assuming laminar flow throughout, determine the boundary layer thickness at distances of 0.20, 2.0, and 20 m from the leading edge.

16.4: If $\hat{V}_\infty$, the upstream velocity of the flow in Problem 16.3 is 1.5 m/s, determine the kinematic viscosity of the fluid.

16.5: Consider the flow of a viscous fluid across a flat plate. The boundary layer thickness at a distance of 1.25 m from the leading edge is 1.25 cm. Assuming laminar flow, determine the boundary layer thicknesses at distances from the leading edge of (a) 0.25 m, (b) 2.0 m, and (c) 12.5 m.

16.6: In Problem 16.5, if the upstream velocity is $\hat{V} = 1.625$ m/s, determine the kinematic viscosity of the fluid.

16.7: Water flows past a plate with an upstream velocity of $\hat{V} = 0.25$ m/s. Determine the boundary layer thickness at distances of (a) 1.75 cm and (b) 8.75 cm.

16.8: Consider laminar flow over a smooth flat plate with a length of 5 m and a breadth of 3 m. The plate is placed in the air with an upstream velocity of 0.625 m/s and 300 K. Determine the boundary layer thickness and wall shear stress at (a) the center of the plate and (b) the trailing edge of the plate.

16.9: An aircraft flies at a velocity of 640 km/h and an altitude of 3500 m. Suppose that the boundary layers on the wing surfaces approximate those on a flat plate and for a transitional Reynolds number of 5×10^5, determine the extent of the laminar boundary layer flow along the wings.

16.10: A fluid flows across a flat plate at 3 m/s. Determine the thickness and location where the boundary layer becomes turbulent for (a) water at 20°C, (b) air at 300 K, and atmospheric pressure, and (c) ethylene glycol at 27°C.

Fluid Drag

16.11: A two-door hardtop has a drag coefficient of 0.45 at road speeds using a reference area of 2.29 m^2. Determine the horsepower required to overcome the drag at a velocity of 30 m/s. Compare this figure with the case of head- and tail-winds of 6 m/s.

16.12: What diameter plate would have the same drag as the automobile in Problem 16.11?

16.13: The drag coefficient for a smooth sphere is shown in Figure P16.13. Determine the speed at the critical Reynolds number for a 140-mm-diameter sphere in the air.

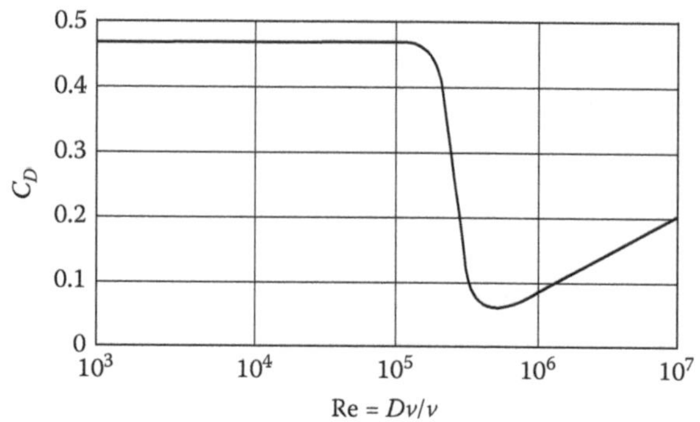

FIGURE P16.13

16.14: Using the information provided in Figure P16.13, plot a curve of drag as a function of velocity for a 4.166-cm sphere between velocities of 15 m/s and 110 m/s.

16.15: The lift coefficient is defined as

$$C_L = \frac{F_L}{\rho \hat{V}_x^2 A/2}$$

where F_L is the lifting force. If the lift coefficient for the automobile in Problem 16.11 is 0.40, determine the lift force at a road speed of 62.5 km/h.

16.16: A baseball having a diameter of 7.163 cm is thrown through the air at sea level and 15°C. Determine the drag force on the ball if its velocity is 43 m/s.

16.17: A slender metal rod is in the shape of a cylinder, 16 m high and 12 cm in diameter. When placed in a vertical position, wind at 16 m/s and 20°C blows across it. Neglect end effects and determine the bending moment about the base of the cylinder.

16.18: Modern subsonic aircraft have been refined to such an extent that 75% of the parasitic drag (that portion of the total aircraft drag not directly associated with producing lift) can be attributed to friction along the external surfaces. For a typical subsonic jet, the parasitic drag coefficient based on the wing area is 0.011. If the wing area is 225 m^2, determine the friction drag on such an aircraft flying at 312.5 km/h at an altitude of 10,000 m.

16.19: A silo in western Iowa is 3.75 m in diameter and 10-m tall. Wind at 25°C is blowing across the silo at 8.75 km/h. Determine the bending moment about the base of the silo.

16.20: Lubricating oil (≈ SAE 50) is flowing at 20°C and 0.75 m/s across a smooth flat plate that is 12.5-cm wide and 42-cm long. Determine the friction drag on one side of the plate.

16.21: A 20.32-cm-diameter spherical balloon travels at 8 m/s at sea level. Determine the speed at an altitude of 3 km to yield the same drag force.

16.22: The frequency of shedding vortices for a cylinder is predicted by the equation

$$f = 0.198\frac{\hat{V}_x}{d}\left(1 - \frac{19.7}{\text{Re}}\right)$$

where f is the frequency of the vortices shed from one side of the cylinder and Re is the free-stream Reynolds number. Determine (a) the frequency of vortex shedding of a 0.50-cm-diameter wire in a 12.5-km/h wind and (b) the wind velocity required to make the vortex shedding equal to zero.

16.23: A sphere of diameter, d, and density, ρ_s, falls at a steady rate through a liquid of density, ρ, and viscosity, μ. If the Reynolds number, $\text{Re} = \rho d\hat{V}/\mu$, is less than unity, show that the speed of fall is

$$\hat{V} = \frac{gd^2(\rho_s - \rho)}{18\mu}$$

16.24: For a given vehicle, being driven at 35 km/h, 30 HP is needed to overcome aerodynamic drag. Estimate the horsepower at 45 km/h.

16.25: Suppose that a parachute is to be designed to safely land a 120-kg paratrooper as if he had jumped without a parachute from a 3.25-m wall. Assume properties of standard air at sea level and determine the diameter of the parachute.

16.26: A sharp, flat plate that is 3-m wide and 1-m long is parallel to an airstream at 27°C flowing at 2.15 m/s. Determine (a) the drag on one side of the plate and (b) the thickness of the boundary layer at the trailing edge of the plate.

16.27: Solve Problem 16.26 for water at the same temperature and velocity.

16.28: A 32-km/h wind blows against an outdoor movie screen that is 18-m wide and 6-m tall. Estimate the wind force on the screen.

16.29: For a frontal area of 0.335 m^2 and a drag coefficient of $C_D = 0.85$, determine how much more power is required to pedal a bicycle at 9.50 km/h into a 12.5-km/h headwind than at 9.50 km/h through still air.

16.30: Consider the flow of air at 30 m/s along a flat plate. Determine at what distance from the leading edge transition occurs.

16.31: Estimate the drag force on a 1-m-long radio antenna with an average diameter of 0.50 cm moving at a speed of 37.5 km/h.

16.32: A very wide flat plate 1-m long is immersed in a stream of sea level standard air moving at 1.2 m/s. Determine the drag per unit width of the plate.

16.33: A ship is towing a sonar array that approximates a submerged cylinder 0.325 m in diameter and 9.15-m long with its axis normal to the direction of tow. The tow speed is 15 knots (1 knot = 0.515 m/s). Estimate the horsepower required to tow this cylinder.

16.34: A helium-filled balloon is required to have a sea level terminal ascent velocity of 6 m/s. The helium pressure is 125 kPa and the balloon payload weight (not including the helium) is 280 N. Neglecting payload drag, estimate the proper balloon diameter.

16.35: A hot-air balloon that is roughly spherical in shape has a volume of 0.5296 m^3 and a total weight of 0.0508 N. The outside air temperature is 26°C, the temperature within the balloon is 81°C, and the pressure is atmospheric. Estimate the rate at which the balloon will rise under steady-state conditions.

16.36: A flagpole standing in sea level standard air is 12 cm in diameter and 15-m high. Estimate the wind-induced bending moment about the base of the flagpole when the wind velocity is 20 knots (1 knot = 0.515 m/s).

16.37: A circular chimney that has a diameter of 2.15 m and a height of 42.5 m is exposed to a sea level storm with winds at 27.5 m/s. Estimate (a) the drag force and (b) the bending moment about the bottom of the chimney.

16.38: A vertical 25 cm diameter piling in seawater is 5-m deep and the current velocity is 2.2 m/s. Estimate the drag on the piling.

16.39: A hydrofoil that is 0.375-m wide and 1.875-m long is placed in water at 30°C with a flow of 12.5 m/s. Estimate the boundary layer thickness at the trailing edge of the plate.

16.40: For the data of Problem 16.39, estimate the friction drag for the case of turbulent boundary layer flow over the total surface.

16.41: Solve Problem 16.40 for laminar flow at the leading edge and a transition Reynolds number of 5×10^5.

16.42: Consider a heavy solid sphere of diameter, d, and density, ρ_s suddenly dropped from rest into a fluid of density, ρ, and viscosity, μ. Assuming that C_D is constant, set up a differential equation and solve for (a) the time history, $\hat{V}(t)$, (b) the final or terminal velocity, $\hat{V}_f$, and (c) the time required for $\hat{V}$ to reach 99% of $\hat{V}_f$.

16.43: A 3.835-cm-diameter table tennis ball, weighing 0.0247 N, is released from the bottom of a swimming pool. Determine its velocity when it breaks the surface if the pool is 2.5-m deep.

16.44: The ping-pong ball with specifications given in Problem 16.43 can be supported by an air jet from a vacuum cleaner exhaust. For sea level standard air, determine the jet velocity required (Figure P16.44).

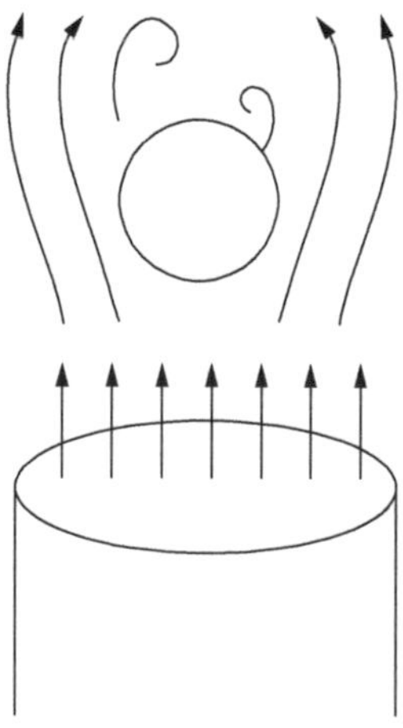

FIGURE P16.44

16.45: Assume that a radioactive dust particle approximates a sphere with a specific weight of 25 kN/m^3. Determine how long it will take such a particle to settle to earth from an altitude of 10 km if the particle diameter is (a) 1 μm and (b) 10 μm.

16.46: Assuming each to behave as a solid sphere, compare the rise velocity of a 0.3175-cm-diameter air bubble in water to the fall velocity of a 0.3175-cm-diameter water drop in the air.

16.47: Estimate the normal force on a circular sign 2.5 m in diameter during a hurricane wind of 75 km/h.

16.48: A square 15-cm piling is subject to a water flow of 1.75 m/s at a depth of 6.75 m. Determine the bending moment at the bottom of the piling.

16.49: A thin flat plate that is 0.625 m by 1.25 m is immersed in a stream of ethylene glycol at 17°C flowing at 6 m/s. The stream is parallel to the short side. Determine the viscous drag.

16.50: Rework Problem 16.49 if the flow of the stream is parallel to the long side.

16.51: As indicated in Figure P16.51, a heavy sphere attached to a string should hang at angle θ when immersed in a stream of velocity, $\hat{V}$. Neglecting the string drag and assuming that the fluid is sea level standard air at $\hat{V} = 40$ m/s (a) derive an expression for θ as a function of sphere and flow properties and (b) determine θ for a steel sphere with a specific gravity of 7.86 and a diameter of 4 cm.

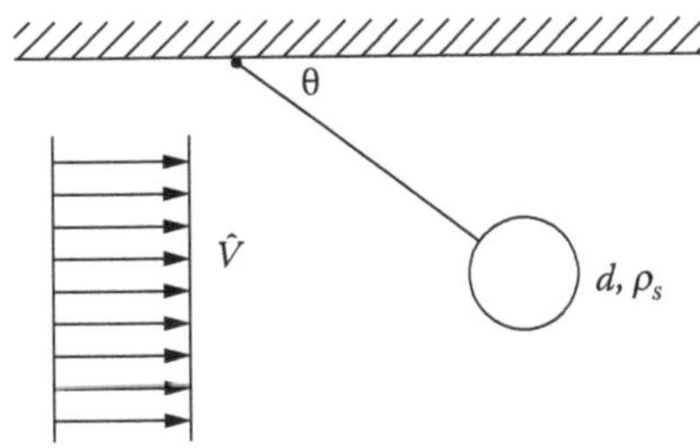

FIGURE P16.51

16.52: A small spherical water drop of diameter 0.005 cm exists in the atmosphere at an elevation of 1600 m. Determine whether the drop will rise or fall if it is in an upwardly flowing thermal of (a) 1.25 m/s, (b) 0.300 m/s, and (c) 0.0300 m/s.

16.53: A pickup truck has a projected area of 10 m^2 and a measured drag coefficient, $C_D =$ 0.35. Estimate the horsepower required to drive the truck at 35 km/h (a) clean and (b) with a 1 m × 2 m sign installed.

16.54: Determine the terminal velocity of a roughly spherical 180-N rock of specific gravity, SG = 1.93, if the rock falls through (a) air and (b) water.

16.55: A parachutist is to land at sea level at a vertical velocity of 15 km/h. The parachutist and pack weigh 935 N. Determine the diameter of the chute.

16.56: A smooth steel (SG = 7.86) sphere is immersed in a stream of ethanol at 17°C moving at 1.5 m/s. Take the diameter as 3 cm and determine (a) the drag in Newtons and (b) the stream velocity that would quadruple the drag.

16.57: Determine the drag on a small circular disk of 0.665-cm diameter moving at 0.0305 m/s through oil with a specific gravity of 0.87. The disk is oriented normal to the upstream velocity.

16.58: An aircraft travels at 225 m/s at an altitude of 8.75 km. It has a smooth wing that is 6.75 m × 48 m. Determine the power required to overcome friction drag.

17

Flow in Pipes and Pipe Networks

Chapter Objectives

- To introduce the concept of frictional head loss in pipes.
- To provide a framework for head loss as a function of Reynolds number, relative roughness, and the Fanning friction factor, f_f.
- To develop the Hagen-Poiseuille equation for f_f in laminar pipe flow.
- To introduce the Moody diagram for evaluating f_f in transition flow and in fully turbulent flow.
- To employ f_f in evaluating flow in single-path applications.
- To use f_f in evaluating flow in multiple-path applications.

17.1 Introduction

Among the things that everyone takes for granted is the manner in which fluids are transported from one place to another so that certain desirable things can occur. Examples of fluids delivered through pipes, ducts, and other confining conduits include building heating and ventilating systems, automobile cooling systems, water systems in municipalities and homes, gas and oil furnaces, hydraulic systems, petroleum refineries, and cooling channels in integrated circuits.

Our purpose, in this chapter, is to bring the accumulated knowledge generated in earlier chapters to bear on the questions related to the delivery of fluids from one place to another through **conduits**. The term "conduit" applies to any closed-flow passage and the shape can be quite irregular in cross section or quite regular (e.g., square or circular). For purposes of this discussion, the generic term, "pipes," implying a circular cross section, will be used. Later in the chapter, it will be shown that relationships for pipe flow can be altered in relatively simple fashion to apply to ducts of any other shape, no matter how irregular that shape may be.

The examination of pipe flow is generally undertaken for two reasons. First, we wish to determine what operating conditions are necessary to achieve a desired rate of flow through an existing flow system (the **analysis** problem). The other reason is to specify the configuration of a piping system that will deliver the fluid at a desired rate between two points, a source and a sink, where conditions are known (the **design** or **synthesis** problem). Involved in such problems are devices that produce the flow such as pumps and fans and devices that control the flow such as valves and fittings.

Pipes may be as large as a crude oil pipeline or as small as the cooling channel in an integrated circuit and flows may be laminar or turbulent. A system may be simple, as with a pipe of a constant cross section connecting a source and a sink, or complex, as with a municipal water system consisting of myriad pipe sizes and lengths along with multiple valves and fittings, where flow may be through numerous pipes in a series, or in parallel, or in any combination.

After beginning with the simple case of a single pipe with a constant cross section connecting two locations, we will develop the ideas needed in the analysis of more complex systems.

17.2 Frictional Loss in Pipes

The simplest pipe flow case is one in which the source and the sink are at the same elevation, the pipe connecting them has a constant cross section and the flow is steady. Such a case is shown in Figure 17.1

The first law of thermodynamics for this case, written between the upstream (point 1) and downstream (point 2) ends of the pipe reduces to the form

$$gz_1 + \frac{\hat{V}_1^2}{2} + \frac{P_1}{\rho} + u_1 = gz_2 + \frac{\hat{V}_2^2}{2} + \frac{P_2}{\rho} + u_2$$

or, equivalently

$$g(z_1 - z_2) + \frac{\hat{V}_1^2 - \hat{V}_2^2}{2} + \frac{P_1 - P_2}{\rho} = u_2 - u_1 \tag{17.1}$$

Note at this point that all quantities with subscripts (e.g., $\hat{V}_1$) are the average values at the designated location.

For this specific application with points 1 and 2 at the same elevation, $z_1 - z_2 = 0$, and for steady flow through a constant diameter pipe, $\hat{V}_1^2 - \hat{V}_2^2 = 0$. Equation 17.1 thus reduces to the very simple form that relates the pressure drop as a function of the internal energy increase.

$$\frac{P_1 - P_2}{\rho} = u_2 - u_1 \tag{17.2}$$

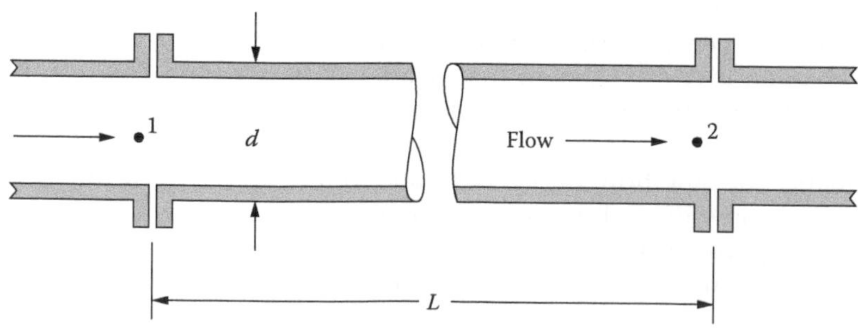

FIGURE 17.1
Single-path pipe flow.

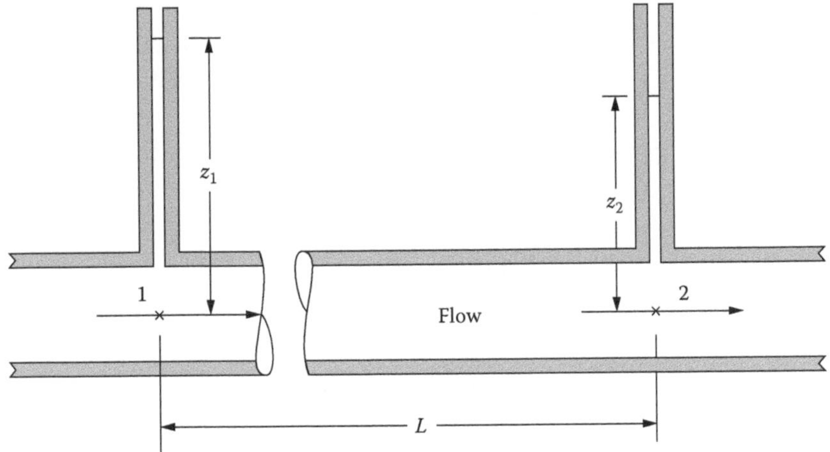

FIGURE 17.2
Single-path pipe flow with manometers to indicate head at stations 1 and 2.

The internal energy will increase because of frictional effects between the flowing fluid and the pipe wall. Our immediate task is that of evaluating this frictional energy loss that can be related to a pressure drop.

If the system shown in Figure 17.1 is modified slightly so that the pressures at points 1 and 2 are indicated by the heights of fluid supported in the vertical tubes at those locations as indicated in Figure 17.2 where the tops of the tubes are open to the atmosphere, the equivalent form of Equation 17.2 is

$$g(z_1 - z_2) = u_2 - u_1 \tag{17.3}$$

where the internal energy increase is equivalent to an elevation change or **head loss** due to friction. The term "head loss," often used in this context is symbolized by h_L.

Head loss represents a decrease in the capacity of a flow system to do useful mechanical work, hence the term **loss**. It can be caused by a number of factors that we will consider in this chapter. At this time, for the case described by Equation 17.3, the head loss is related solely to pipe friction.

17.3 Dimensional Analysis of Pipe Flow

As an initial approach to expressing head loss in the pipe flow condition shown in Figure 17.2, we will use dimensionless analysis. Table 17.1 lists the important variables for this case.

With the exception of pipe roughness, the quantities listed are familiar by virtue of earlier use. Pipe roughness, designated e, can be thought of as the characteristic height of projections from the pipe wall; hence, its units of length. The term **roughness** represents the pipe surface condition.

A dimensional analysis performed with this set of variables, with the core group composed of $\hat{V}$, d, and ρ, proceeds as follows:

The Buckingham Pi theorem gives

$$i = n - r = 7 - 3 = 4$$

TABLE 17.1

Significant Variables for Dimensional Analysis of Pipe Fow

Variable	Symbol	Dimensions
Pressure loss	ΔP	M/LT^2
Velocity	$\hat{V}$	L/T
Pipe diameter	d	L
Pipe length	L	L
Pipe roughness	e	L
Fluid density	ρ	M/L^3
Fluid viscosity	μ	M/LT

which shows that there will be four pi groups. They are formed as follows:

$$\pi_1 = \hat{V}^a d^b \rho^c \Delta P$$

$$\pi_2 = \hat{V}^a d^b \rho^c L$$

$$\pi_3 = \hat{V}^a d^b \rho^c e$$

$$\pi_4 = \hat{V}^a d^b \rho^c \mu$$

Proceeding as discussed in Chapter 15, we achieve the following forms for the pi groups:

$$\pi_1 = \frac{\Delta P}{\rho \hat{V}^2}$$

$$\pi_2 = \frac{L}{d}$$

$$\pi_3 = \frac{e}{d}$$

and

$$\pi_4 = \frac{\mu}{d\hat{V}\rho}$$

The first pi group is the Euler number. It contains the frictional pressure drop in the form represented by the left side of Equation 17.2. This term can be expressed in terms of head loss by replacing $\Delta P/\rho$ by its equivalent, gh_L, from Equation 17.3. Thus, an equivalent form for π_1 is

$$\pi_1 = \frac{gh_L}{\hat{V}^2}$$

The parameters π_2 and π_3 are clearly dimensionless because they are ratios of quantities having units of length. The ratio, $\pi_3 = e/d$ is commonly referred to as the **relative roughness**. The fourth pi group is the reciprocal of the Reynolds number, Re_d.

In functional form, the pi groups can be written as

$$\frac{h_L}{\hat{V}^2/g} = f\left(\frac{L}{d}, \frac{e}{d}, \mathrm{Re_d}\right) \tag{17.4}$$

Experience has shown that the head loss in fully developed pipe flow is directly proportional to the ratio, L/d. This dimensionless ratio is, thus, not included as a part of the function, f, and Equation 17.4 becomes modified as

$$\frac{h_L}{\hat{V}^2/g} = \frac{L}{d} f\left(\frac{e}{d}, \text{Re}_\text{d}\right) \tag{17.5}$$

The function

$$f\left(\frac{e}{d}, \text{Re}_\text{d}\right)$$

which varies with relative roughness and Reynolds number is yet to be evaluated. The head loss expression, in its general operational form, is a modified version of Equation 17.5 and this form is

$$h_L = 2 f_f \frac{L}{d} \frac{\hat{V}^2}{g} \tag{17.6}$$

where the factor 2 is associated with f_f, the **Fanning friction factor**.

In a variation of Equation 17.5, often seen, we write

$$h_L = f_d \frac{L}{d} \frac{\hat{V}^2}{2g} \tag{17.7}$$

where f_d is the **Darcy friction factor**. It is obvious that $f_d = 4 f_f$.

The constant, 2, in Equation 17.6 is conventionally associated with the head loss expression incorporating the Fanning friction factor. In this form $f_f = C_f$, the coefficient of skin friction, which was introduced in Chapter 16. Equation 16.6, by convention, is written with the 2 in the denominator so that the right-hand side has the form of kinetic energy. In this book, the Fanning friction factor will be used exclusively.

Our immediate task is to determine suitable relations for f_f from both theory and experimental data.

17.4 Fully Developed Flow

When a fluid such as water enters a pipe, its velocity profile will be altered as viscous effects propagate from the pipe wall into the fluid. The entering profile may be square (this is generally referred to as **slug flow**) and will be altered owing to the effect of viscosity as the fluid proceeds down its flow path. Figure 17.3 shows the propagation of the velocity profiles between the entrance at $x = 0$ to the distance, L_e, where viscous flow effects have proceeded to the center of the pipe.

If the flow remains laminar (recall that this will be true for $\text{Re}_d < 2300$), the fully developed velocity profile will be parabolic. The fully developed condition will occur at some distance, L_e, downstream from the entrance. The distance, L_e, is designated as the **entrance length**. For $x < L_e$ the velocity profile will change with the distance, x. In this **entry** region there will be two regimes of flow. The region affected by the presence of the pipe wall is called the **viscous region**. This is where velocity gradients exist, that is, $\partial \hat{V}_x/\partial r \neq 0$. Near the center of the pipe, the velocity profile remains flat, that is $\partial \hat{V}_x/\partial r = 0$. This is called the **core flow** region. For $x > L_e$, the viscous region completely fills the pipe and the velocity profiles will no longer vary with x. This condition is referred to as **fully developed flow**. Mathematically, fully developed flow requires that $\partial \hat{V}_x/\partial x$ be zero.

We should recall from the discussion in Chapter 13 that conservation of mass requires that the flow rate be constant for all values of x. Because the velocity near the wall is decreasing,

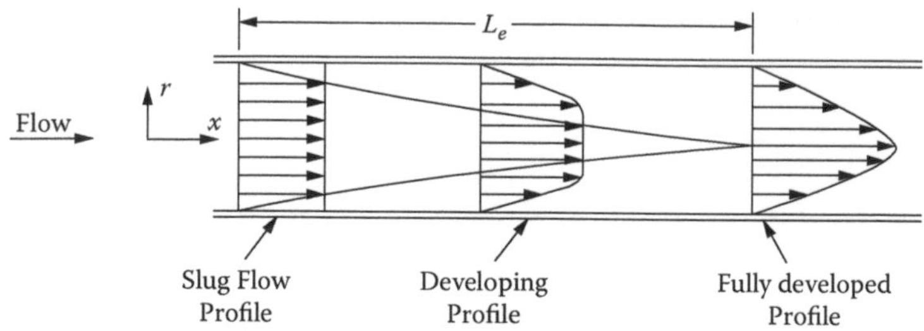

FIGURE 17.3
Velocity profiles for flow in pipes or tubes.

the rate of flow in the core must increase, eventually reaching a value $\hat{V}_{max} = 2\hat{V}_{av}$ at the centerline for fully developed laminar flow.

The concepts of entrance length and fully developed flow are also valid when the flow is turbulent; the velocity profiles will be different, however, and the fully developed profile will not be parabolic.

17.5 Friction Factors for Fully Developed Flow

17.5.1 Laminar Flow

The case of fully developed laminar flow in a circular flow passage is amenable to a straightforward analytical description. The interested reader may refer to a variety of fluid mechanics texts for a detailed discussion [Fox and McDonald (1985), Welty et al. (2008), and White (1999)]. The result of such analyses is the classic **Hagen[1]-Poiseuille[2]** equation, which is stated here without proof

$$-\frac{dP}{dx} = 32\mu\frac{\hat{V}}{d^2} \tag{17.8}$$

This equation is easily solved by separation of the variables and integration

$$-\int_{P_0}^{P} dP = 32\mu\frac{\hat{V}}{d^2}\int_0^L dx$$

or

$$\Delta P = P_0 - P = 32\mu\hat{V}\frac{L}{d} \tag{17.9}$$

The ratio $\Delta P/\rho$ is the internal energy change as expressed by Equation 17.2. We can thus write

$$\frac{\Delta P}{\rho} = gh_L = 32\frac{\mu}{\rho}\hat{V}\frac{L}{d} \tag{17.10}$$

[1] G. H. Hagen (1797–1884) was a German hydraulic engineer who developed the Hagen-Poiseuille law independent of J. L. M. Poiseuille and directed construction of dikes, harbor installations, and dune fortifications

[2] J. L. M. Poiseuille (1797–1869) was a French physiologist and physicist who studied the physiology of arterial circulation, invented improved methods for measuring blood pressure, and discovered the Hagen-Poiseuille law independently of G. H. Hagen.

which can be related to Equation 17.6 to introduce f_f. Doing so, we have

$$h_L = 32\frac{\mu}{\rho g}\hat{V}\frac{L}{d} = 2f_f\frac{\hat{V}^2}{\rho g}\frac{L}{d}$$

The resulting expression for f_f in laminar flow is

$$f_f = \frac{16}{\rho\hat{V}d/\mu} = \frac{16}{\mathrm{Re}_d} \qquad (17.11)$$

Equation 17.11 verifies the dimensional analysis result that the friction head loss is functionally related to the Reynolds number. This result has also been verified experimentally. The absence of any dependency on pipe roughness in laminar flow has also been experimentally verified. While this may seem counterintuitive, the physical explanation is that any flow disruption due to wall roughness is dampened out by viscous effects. Recall that for values of Re_d in pipe flow less than 2300, viscous effects are predominant over inertial effects.

This discussion suggests that roughness may be a significant factor with turbulent flow. This case will be examined next.

17.5.2 Turbulent Flow

In the case of turbulent flow, there is no neat and simple relationship such as the Hagen-Poiseuille equation to evaluate the frictional pressure drop.

For values of $\mathrm{Re}_d > 2300$, a range that encompasses both the transition and fully turbulent regimes (we will identify these regimes shortly), the relationships for f_f have been obtained by incorporating both analytical approaches, using turbulent flow analysis and empiricism. For details, the interested reader is, again, referred to the available fluid mechanics texts such as Welty et al. (2001).

The following relationships and range of application for f_f in pipe flow are now summarized:

- Laminar flow, $\mathrm{Re}_d < 2300$ (repeated here in order to show a complete catalog)

$$f_f = \frac{16}{\mathrm{Re}_d} \qquad (17.11)$$

- Turbulent flow, $\mathrm{Re}_d > 2300$, hydraulically smooth pipe

$$\frac{1}{\sqrt{f_f}} = 4.0\log_{10}[\mathrm{Re}_d\sqrt{f_f})] - 0.40 \qquad (17.12)$$

- Turbulent flow, $\mathrm{Re}_d > 2300$, rough pipe, $(d/e)/\mathrm{Re}_d\sqrt{f_f} < 0.01$

$$\frac{1}{\sqrt{f_f}} = 4.0\log_{10}\frac{d}{e} + 2.28 \qquad (17.13)$$

- Transition flow, $\mathrm{Re}_d > 2300$, $(d/e)/\mathrm{Re}_d\sqrt{f_f} > 0.01$

$$\frac{1}{\sqrt{f_f}} = 4.0\log_{10}\frac{d}{e} + 2.28 - 4.0\log_{10}\left[4.67\frac{d/e}{\mathrm{Re}_d\sqrt{f_f}} + 1\right] \qquad (17.14)$$

These four equations provide the basis for a graphical representation of f_f as a function of Re_d, which is a standard tool for pipe flow analysis.

The range of applications for Equations 17.12 through 17.14 includes the terms "smooth" and "rough." The accepted criterion for the term "hydraulically smooth" is when the parameter

$$\frac{d/e}{\mathrm{Re}_d\sqrt{f_f}} < 0.01 \tag{17.15}$$

One may observe that for fully turbulent flow in smooth pipes, according to Equation 17.12, the Fanning friction factor is not functionally dependent on the relative roughness and is a function only of Re_d. Similarly, for rough pipes in fully turbulent flow, f_f is solely a function of relative roughness.

17.6 Friction Factor and Head Loss Determination for Pipe Flow

17.6.1 Pipe Friction Factor

Figure 17.4 plots f_f as a function of Re_d, with separate lines for a range of e/d values in the turbulent regime. This Figure is similar to a plot presented by Moody (1944) and such plots are frequently referred to as **Moody diagrams** or **Moody charts**.

One may note the three regimes shown in Figure 17.4. They are

- The laminar flow regime where

$$f_f = f(\mathrm{Re}_d)$$

- The fully turbulent regime where

$$f_f = f\left(\frac{e}{d}\right)$$

- The transition regime where

$$f_f = f\left(\mathrm{Re}_d, \frac{e}{d}\right)$$

The remaining piece of information necessary for the use of Figure 17.4 is the value for the relative roughness, e. Some typical values of the pipe roughness for common construction materials are listed in Table 17.2.

One should be aware that these values for pipe roughness are reasonable approximations when pipe surfaces are free from corrosion or foreign material. The values listed in Table 17.2 are used, in general, along with the friction factor plot, to solve pipe flow analysis or design problems.

17.6.2 Head Loss Due to Fittings and Valves

Additional head losses will occur in a piping system due to the presence of valves, elbows, and a variety of other fittings that involve changes in flow direction, changes in the size of the flow passages or control of the flow rate and/or the pressure. Head losses associated with such elements are functions of geometry, roughness, and the Reynolds number. The Reynolds number dependence being relatively small, the head loss associated with valves and fittings can be reasonably approximated as

$$h_L = \frac{\Delta P}{\rho g} = K\frac{\hat{V}^2}{2g} \tag{17.16}$$

where the factor, K, has a specific value for a given fitting or valve.

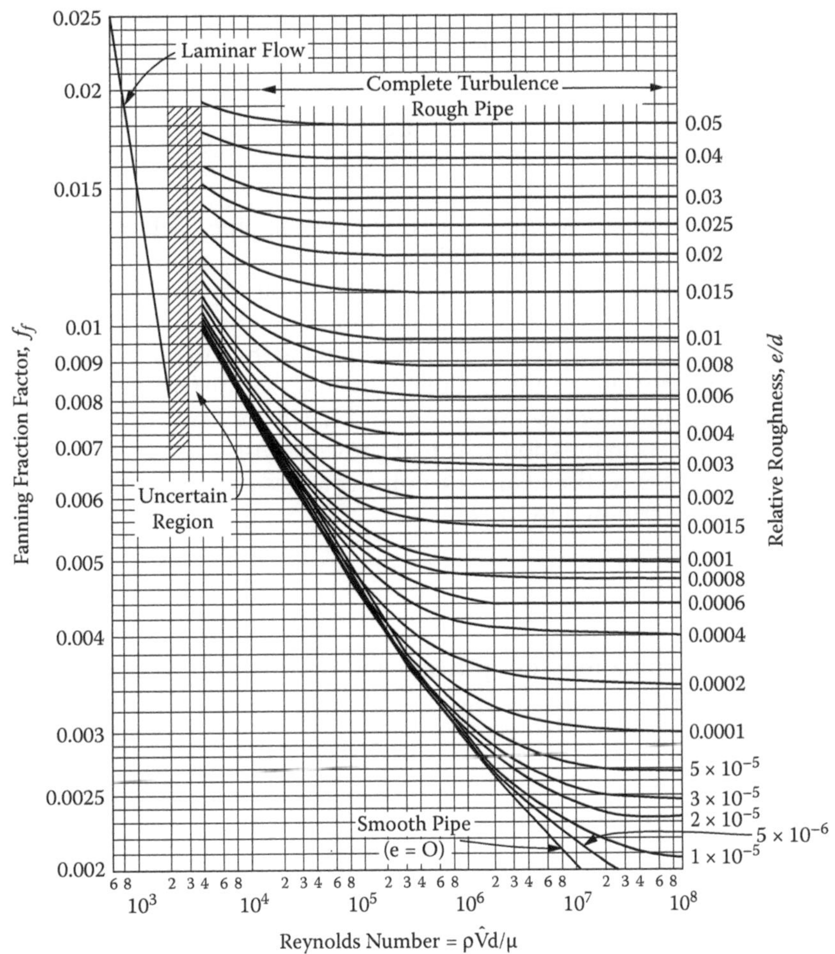

FIGURE 17.4
The Fanning friction factor as a function of the Reynolds number and relative pipe roughness.

TABLE 17.2
Values of Pipe Roughness, *e*, for Several Materials

Material	*e*, m $\times 10^4$
Drawn tubing	0.0152
Commercial steel	0.457
Wrought iron	0.457
Asphalted cast iron	1.22
Galvanized iron	1.52
Cast iron	2.59
Wood stave	3.0–9.1
Concrete	30.5
Riveted steel	9.1–91.4

Another approach to accounting for the head loss due to minor losses is the concept of an **equivalent length**. The idea here is to determine the length of straight pipe that would involve the same head loss as a particular valve or fitting. Thus, a single expression of the form

$$h_L = 2f_f \frac{L_{eq}}{d} \frac{\hat{V}^2}{2g} \tag{17.17}$$

can be used for a given system where L_{eq} accounts for the pipe itself plus all of the minor losses present in the form of valves and fittings.

In comparing Equations 17.16 and 17.17 it is clear that the factor K is equivalent to $4f_f L_{eq}/d$. It appears that Equation 17.17 is Reynolds number dependent because it includes f_f, but this is not the case. The assumption is made that, in both cases, the Reynolds number is so large that the flow is fully turbulent. In this case, the friction factor will depend only on the pipe roughness. Some values for K and L_{eq}/d are listed in Table 17.3.

An additional head loss expression that will prove useful is associated with a sudden expansion. When the cross-sectional area of a pipe changes abruptly, there is an associated head loss that can be determined from

$$h_L = \frac{\hat{V}_1^2}{2g}\left(1 - \frac{A_1}{A_2}\right)^2 \tag{17.18}$$

where the subscripts 1 and 2 refer to upstream and downstream conditions, respectively. The term

$$\left(1 - \frac{A_1}{A_2}\right)^2$$

TABLE 17.3

Friction Coefficients and Equivalent Lengths for Pipe Fittings

	Fitting	K	L_{eq}, m
Valves	Globe, fully open	10	470
	Angle, fully open	2	90
	Gate, fully open	0.15	7
	Gate, 3/4 open	0.26	7
	Gate, 1/2 open	2.1	12
	Gate, 1/4 open	17	800
	Ball, fully open	0.05	2
	Ball, 2/3 open	5.5	260
	Ball, 1/3 open	210	10,000
Elbows	Regular, 90°, flanged	0.3	14
	Regular, 90°, threaded	1.5	70
	Long radius, 90°, flanged	0.2	9
	Long radius, 90°, threaded	0.7	30
	Regular, 45°, threaded	0.4	19
	Long radius, 45°, flanged	0.2	9
Bends	180°, threaded	1.5	70
	180°, flanged	0.2	9
Tees	Straight through flanged	0.2	9
	Straight through threaded	0.9	40
	Side outlet flanged	1	50
	Side outlet threaded	2	90
Union	Threaded	0.08	4

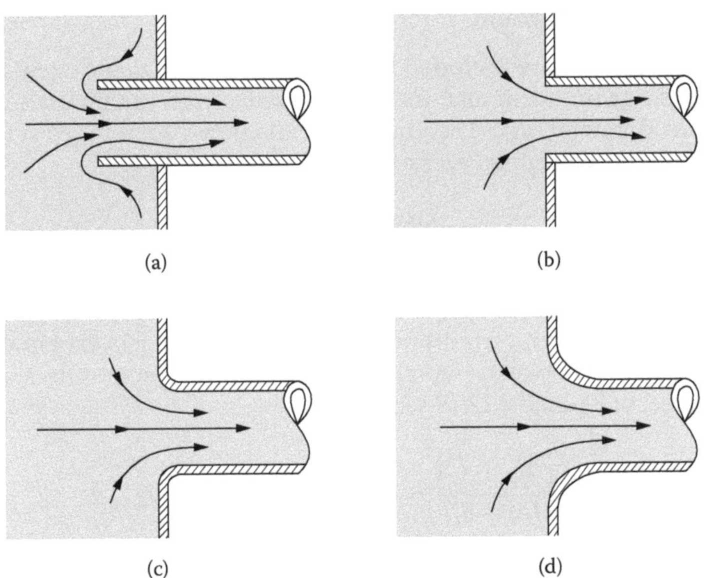

FIGURE 17.5
Entrance shapes for four typical inlet geometries.

represents the loss coefficient, K, in this case. The value of K for a sudden expansion is seen to vary when A_2 is very large, so that as $A_1/A_2 \longrightarrow 0$, $K = 1$, and as $A_1/A_2 \longrightarrow 1$, $K = 0$.

In the reverse case, that of a sudden contraction in the flow passage, the loss coefficient will have a range in values that will vary with the shape of the inlet. Figure 17.5 illustrates four representative inlet shapes for the case of flow from a large tank or reservoir into a circular pipe ($A_2/A_1 \approx 0$).

In the case of flow from a larger pipe to a smaller one, with $0 < A_2/A_1 < 1$, the corresponding value of K will vary from 0.5 when $A_2/A_1 \approx 0$, as in Figure 17.5b, to 0 when $A_2/A_1 \approx 1$. This variation of K as a function of A_2/A_1 is shown in Figure 17.6.

The information presented in Table 17.3 and in Figures 17.5 and 17.6 is representative, but by no means exhaustive, of the various cases for which loss coefficients in piping systems may be determined. More detailed and complete information of this kind is available in numerous references such as Munson et al. (1994).

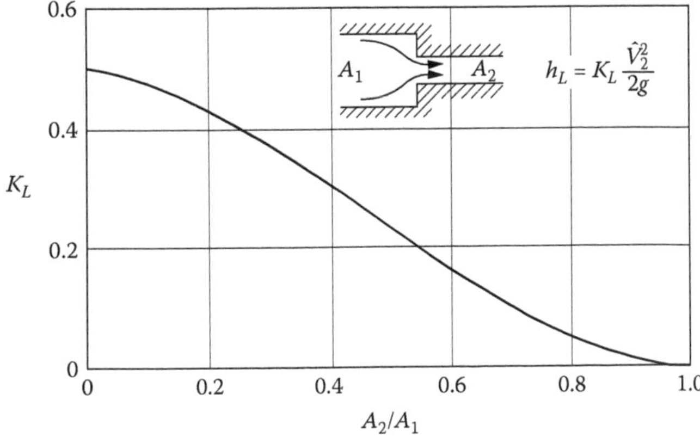

FIGURE 17.6
Loss coefficient for a sudden contraction.

17.6.3 Noncircular Flow Passages

All equations used this far have included the diameter, d, implying a circular cross section. When the flow passage is noncircular, these same expressions can be used with d replaced by the **equivalent diameter,** d_{eq}. The equivalent diameter, sometimes referred to as the **hydraulic diameter,** is determined according to

$$d_{eq} = 4\frac{\text{cross-sectional area}}{\text{wetted perimeter}} \tag{17.19}$$

It is a simple matter to verify that $d_{eq} = d$ for a circular flow passage. For other shapes, d_{eq} may have values which one would not predict intuitively. For example, in the case of an annular flow passage between two circular pipes (inner pipe with d_i and outer pipe with d_o), application of Equation 17.19 yields

$$d_{eq} = 4\frac{\frac{\pi}{4}(d_o^2 - d_i^2)}{\pi(d_o + d_i)} = \frac{(d_o + d_i)(d_o - d_i)}{d_o + d_i} = d_o - d_i$$

17.6.4 Single-Path Pipe Systems

Examples 17.1 through 17.3 that follow, illustrate the application of the material presented in earlier sections to the analysis and design of single-path piping systems.

Example 17.1 In one section of the Alaska pipeline, (Figure 17.7), crude oil at 55°C ($\rho = 864\,\text{kg/m}^3$ and $\mu = 4.32 \times 10^{-3}\,\text{kg/m-s}$) is pumped at a rate of $3.4\,\text{m}^3/\text{s}$ through a 1.22-m-diameter steel pipe. Determine the pumping power required to produce this flow over a length of 850 miles (1370 km) between two pumping stations.

Solution

Assumptions and Specifications

(1) Steady-state conditions exist.
(2) The flow is incompressible.
(3) There are no entrance and exit losses to the pipe.
(4) There are no losses due to bends, pipes, and fittings.
(5) Heat transfer and potential energy changes are negligible.
(6) The pressures at the upstream and downstream end are equal.
(7) The pump is considered as exterior to the pipe section.
(8) Uniform properties exist at each cross section considered.

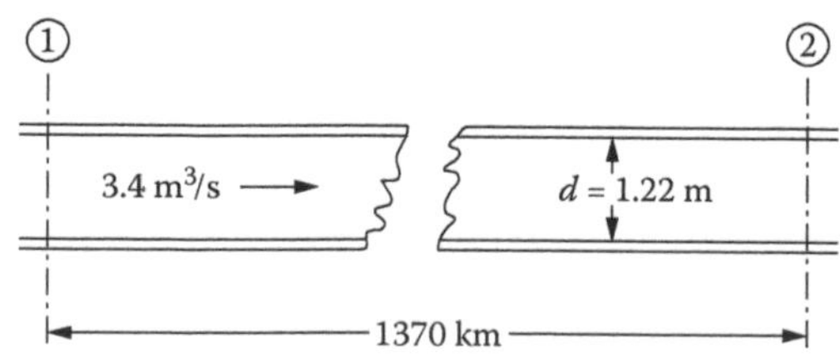

FIGURE 17.7
Piping configuration for Example 17.1.

This is an analysis problem because the size, length, and configuration of the pipe are established. The pumping power is related to the other quantities by the first law of thermodynamics. For the 1370-km-long control volume between the upstream location (point 1) and the downstream location (point 2), the steady-flow statement of the first law is

$$\dot{Q}_{cv} - \dot{W}_{cv} + \dot{m}\left[g(z_1 - z_2) + \frac{\hat{V}_1^2 - \hat{V}_2^2}{2} + \frac{P_1 - P_2}{\rho} + (u_1 - u_2)\right] = 0$$

For this case, the terms are evaluated as:

The heat transfer and potential energy changes are assumed negligible

$$\dot{Q}_{cv} = 0 \qquad \text{and} \qquad g(z_1 - z_2) = 0$$

- For steady flow through a constant diameter pipe, $\hat{V}_1 = \hat{V}_2$

$$\frac{\hat{V}_1^2 - \hat{V}_2^2}{2} = 0$$

- For upstream and downstream conditions at equal pressures

$$\frac{P_1 - P_2}{\rho} = 0$$

- There is a relationship between the head loss and the change in specific internal energy

$$u_2 - u_1 = gh_L = g\left(2f_f \frac{L}{d}\frac{\hat{V}^2}{g}\right)$$

and the first law relationship reduces to

$$-\dot{W}_{cv} = \dot{m}\left[2f_f \frac{L}{d}\hat{V}^2\right]$$

For crude oil at 55°C, the following properties apply:

$$\rho = 864\,\text{kg/m}^3 \quad \text{and} \quad \mu = 4.32 \times 10^{-3}\,\text{N-s/m}^2$$

The Reynolds number is evaluated as

$$\text{Re}_d = \frac{d\hat{V}\rho}{\mu} = \frac{d\dot{V}\rho}{A\mu} = 4\frac{\dot{V}\rho}{\pi d\mu}$$

$$= \frac{4(3.4\,\text{m}^3/\text{s})(864\,\text{kg/m}^3)}{\pi(1.22\,\text{m})(4.32 \times 10^{-3}\,\text{N-s/m}^2)}\left(\frac{1\,\text{N-s}^2}{\text{kg-m}}\right) = 709{,}000$$

and the flow is turbulent.

With the value of the pipe roughness, $e = 0.457 \times 10^{-4}$ m taken from Table 17.2 for commercial steel, we have the relative roughness

$$\frac{e}{d} = \frac{0.457 \times 10^{-4}\,\text{m}}{1.22\,\text{m}} = 3.75 \times 10^{-5}$$

and the value of f_f can be read from Figure 17.4

$$f_f = 0.0033$$

Now with

$$\hat{V} = \frac{\dot{V}}{A} = \frac{4\dot{V}}{\pi d^2} = \frac{4(3.4\,\text{m}^3/\text{s})}{\pi(1.22\,\text{m})^2} = 2.909\,\text{m/s}$$

we can now complete the solution

$$-\dot{W}_{cv} = \dot{m}\left(2f_f\frac{L}{d}\hat{V}^2\right)$$

$$= (3.4\,\text{m}^3/\text{s})(864\,\text{kg/m}^3)\left[2(0.0033)\left(\frac{1.37\times 10^6\,\text{m}}{1.22\,\text{m}}\right)(2.909\,\text{m/s})^2\right]$$

or

$$\dot{W}_{cv} = \left(-18.42\times 10^6\frac{\text{kg-m}^2}{\text{s}^2}\right)\left(\frac{1\,\text{N-s}^2}{\text{kg-m}}\right)\text{W} = -184.2\,\text{MW}$$

The negative value indicates that work is being done *on* the fluid.

Example 17.2 A pump delivers $0.03\,\text{m}^3/\text{s}$ of water through a 15.5-cm-diameter commercial steel pipeline as displayed in Figure 17.8, which shows two long radius 45° elbows. For a pump discharge pressure of 810 kPa, determine the pressure at the pipe exit.

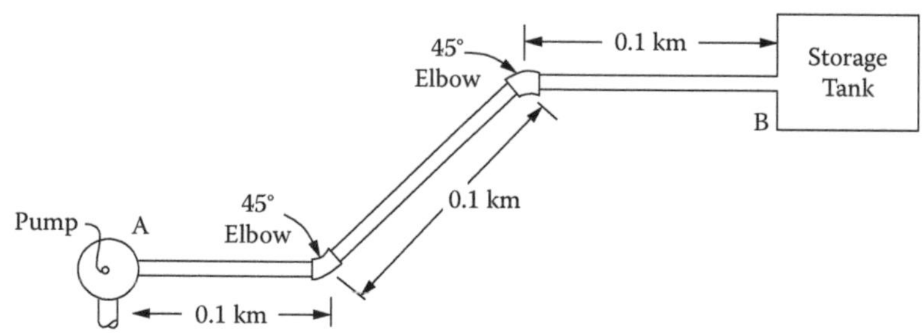

FIGURE 17.8
Pumping configuration for Example 17.2.

Solution

Assumptions and Specifications

(1) Steady-state conditions exist.
(2) The flow is incompressible.
(3) There are no entrance and exit losses to the pipe.
(4) There are no losses due bends, pipes, and fittings.
(5) Heat transfer is negligible.
(6) The pump is treated as exterior to the pipe section.
(7) Uniform properties exist at each cross section considered.

The analysis will be based on a control volume that encloses the pipe and extends from the pump discharge (point A) to the entrance of the storage tank (point B). For this control volume, the energy equation takes the form

$$g(z_B - z_A) + \frac{P_B - P_A}{\rho} + u_B - u_A = 0$$

which may be written as

$$\frac{P_A - P_B}{\rho} = g(z_B - Z_A) + gh_L$$

where P_B is the unknown being sought.

The potential energy term is evaluated as

$$g(z_B - z_A) = (9.81\,\text{m/s}^2)(100 \sin 45^\circ\,\text{m}) = 694\,\text{m}^2/\text{s}^2$$

The head-loss term includes the frictional effect of flow through 300 m of pipe plus the loss due to two 45° elbows.

To evaluate the pipe friction, we must determine the Reynolds number. Using the properties of water at 20°C from Table A.14 in Appendix A, we find

$$\rho = 998.0\,\text{kg/m}^3 \qquad \text{and} \qquad \nu = 1.012 \times 10^{-6}\,\text{m}^2/\text{s}$$

Then we have

$$\text{Re}_d = \frac{d\hat{V}}{\nu} = \frac{4\dot{V}}{\pi \nu d} = \frac{4(0.03\,\text{m}^3/\text{s})}{\pi(1.012 \times 10^{-6}\,\text{m/s}^2)(0.155\,\text{m})} = 243{,}500$$

which is well into the turbulent range.

The relative roughness is evaluated as

$$\frac{e}{d} = \frac{0.457 \times 10^{-4}\,\text{m}}{0.155\,\text{m}} = 2.95 \times 10^{-4}$$

With this information, the Fanning friction factor is determined from Figure 17.4 as

$$f_f = 0.0044$$

The determination of the minor losses involves Equation 17.19. For a 45° elbow (flanged) the value of K from Table 17.3 is 0.2. Then, the head loss for two elbows and the pipe friction will be

$$\begin{aligned} gh_L &= 2f_f \frac{L}{d}\hat{V}^2 + 2K\hat{V}^2 \\ &= 2\hat{V}^2\left(f_f\frac{L}{d} + K\right) \\ &= 2\left[\frac{4(0.30\,\text{m}^3/\text{s})}{\pi(0.155\,\text{m})^2}\right]^2\left[(0.0044)\left(\frac{300\,\text{m}}{0.155\,\text{m}}\right) + 0.2\right] \\ &= 44.1\,\text{m}^2/\text{s}^2 \end{aligned}$$

Returning to our energy equation expression, we now have

$$\frac{P_A - P_B}{\rho} = 694\,\text{m}^2/\text{s}^2 + 44.1\,\text{m}^2/\text{s}^2 = 738.1\,\text{m}^2/\text{s}^2$$

Finally, the pipe exit pressure is evaluated as

$$\begin{aligned} P_B &= P_A - \rho(738.1\,\text{m}^2/\text{s}^2) \\ &= 810{,}000\,\text{N/m}^2 - \frac{(998.1\,\text{kg/m}^3)(738.1\,\text{m}^2/\text{s}^2)}{1\,\text{kg-m/N-s}^2} \\ &= 810{,}000\,\text{N/m}^2 - 736{,}700\,\text{N/m}^2 \\ &= 73{,}300\,\text{N/m}^2 = 73.3\,\text{kPa} \Longleftarrow \end{aligned}$$

Example 17.3 Water at a temperature of 20°C is pumped through the system shown in Figure 17.9 by a pump that develops 164 kW. The discharge pressure is 279 kPa (gage). Determine the volumetric flow rate of the water.

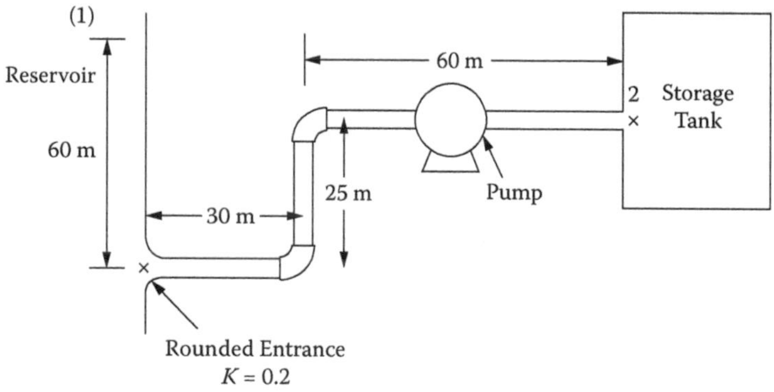

FIGURE 17.9
Piping configuration for Example 17.3.

Solution

Assumptions and Specifications

(1) Steady-state conditions exist.
(2) The flow is incompressible.
(3) The pump is considered as exterior to the pipe section.
(4) Uniform properties exist at each cross section considered.

This is a variation of the analysis problem where the unknown quantity is the flow rate. A convenient choice for the control volume is between point 1 at the surface of the reservoir and point 2 where the storage tank pressure is known. The first law of thermodynamics between these points is

$$-\dot{W}_{cv} - \dot{m}\left[g(z_2 - z_1) + \frac{\hat{V}_2^2 - \hat{V}_1^2}{2} + \frac{P_2 - P_1}{\rho} + u_2 - u_1\right] = 0$$

A term-by-term evaluation with $\rho = 998.1\,\text{kg/m}^3$ and $\nu = 1.012 \times 10^{-6}\,\text{m/s}^2$ (Table A.14) yields

$$-\dot{W}_{cv} = -(-164\,\text{kW}) = 164\,\text{kW}$$

$$\dot{m} = \rho\dot{V} = 998.1\dot{V}\,\text{kg/s}$$

$$g(z_2 - z_1) = (9.81\,\text{m/s}^2)(25\,\text{m} - 60\,\text{m}) = -343.4\,\text{m}^2/\text{s}^2$$

$$\frac{\hat{V}_2^2 - \hat{V}_1^2}{2} = \frac{1}{2}\left\{\left[\frac{4\dot{V}}{\pi(0.20\,\text{m})^2}\right]^2 - 0\right\} = 507\dot{V}^2\,\text{m}^2/\text{s}^2$$

$$\frac{P_2 - P_1}{\rho} = \frac{279\,\text{kPa} + P_{\text{atm}} - P_{\text{atm}}}{998.1\,\text{kg/m}^3} = 280\,\text{m}^2/\text{s}^2$$

and with two 90°-flanged elbows where $K = 0.30$

$$\begin{aligned}
u_2 - u_1 &= gh_L \\
&= 2f_f\frac{L}{d}\hat{V}^2 + \sum K\frac{\hat{V}^2}{2} \\
&= \left(2f_f\frac{L}{d} + \frac{\sum K}{2}\right)\left(\frac{4\dot{V}}{\pi d^2}\right)^2 \\
&= \left[2f_f\frac{115\,\text{m}}{0.2\,\text{m}} + \frac{1}{2}(0.2 + 0.6)\right]\left[\frac{4\dot{V}}{\pi(0.2\,\text{m})^2}\right]^2 \\
&= (1150f_f + 0.40)(1013\dot{V}^2)\,\text{m}^2/\text{s}^2 \\
&= (1.165 \times 10^6 f_f + 405)\dot{V}^2\,\text{m}^2/\text{s}^2
\end{aligned}$$

The $u_2 - u_1$ term cannot be evaluated further at this time because f_f is a function of Re_d which, in turn, is a function of the flow rate.

Dividing both sides of the energy equation by $\dot{m}$ and substituting terms that have been evaluated in the foregoing itemized term-by-term evaluation we obtain

$$\begin{aligned}
\frac{164 \times 10^3\,\text{W}}{998.1\dot{V}} &= -343.4\,\text{m}^2/\text{s}^2 + 507\,\dot{V}^2\,\text{m}^2/\text{s}^2 + 280\,\text{m}^2/\text{s}^2 \\
&\quad + (1.165 \times 10^6 f_f + 405)\dot{V}^2\,\text{m}^2/\text{s}^2
\end{aligned}$$

and we see that all terms have the same units (m^2/s^2). When we perform some algebra, this expression simplifies to

$$164 = -63.4\dot{V} + (1.165 \times 10^6 f_f + 912)\dot{V}^3 \tag{a}$$

where $\dot{V}$ is in m^3/s and f_f is a function of the Reynolds number

$$\text{Re}_d = \frac{d\hat{V}}{\nu} = \frac{4\dot{V}}{\pi d \nu}$$

Using properties of water at 20°C, the Reynolds number becomes

$$\text{Re}_d = \frac{4\dot{V}}{\pi(0.20\,\text{m})(1.012 \times 10^{-6}\,\text{m/s}^2)} = 6.292 \times 10^6 \dot{V} \qquad \text{(b)}$$

To complete the solution, an additional piece of independent information relating f_f and $\dot{V}$ is needed. Such information is provided by the friction factor plot given in Figure 17.4. Because this second piece of information is given in graphical form, an iterative, or trial-and-error solution is required.

The solution proceeds as follows:

1. Assume a value for f_f.
2. With this value for f_f, use Figure 17.4 to find the corresponding value for Re_d.
3. Having Re_d, solve Equation (b) for $\dot{V}$.
4. Use the assumed value for f_f, solve Equation (a) for $\dot{V}$.
5. If the values for $\dot{V}$ as calculated in steps 3 and 4 agree, the desired result has been found and you may stop.
6. If the two values of $\dot{V}$ do not agree, adjust the value of f_f and go back to step 2.
7. Repeat the procedure until the values of $\dot{V}$ agree.

To illustrate this procedure, we will first assume the flow to be fully turbulent. For such a case, f_f is a function only of e/d, which for this problem is

$$\frac{e}{d} = \frac{0.457 \times 10^{-4}\,\text{m}}{0.20\,\text{m}} = 0.000229$$

giving $f_f = 0.0037$.

To start the iterative process, we go to step 4 and evaluate $\dot{V}$ from Equation (a) to obtain

$$\dot{V} = 0.328\,\text{m}^3/\text{s}$$

Then using Equation (b) we find, for this value of $\dot{V}$, $\text{Re}_d = 2.013 \times 10^6$ and, using Figure 17.4, we find that for $e/d = 0.000229$ and $\text{Re}_d = 2.013 \times 10^6$, the value of $f_f \cong 0.0040$, a slightly different value than we assumed.

Then, using $f_f = 0.0037$ and calculating $\dot{V}$ from Equation (a), we get

$$\dot{V} = 0.331\,\text{m}^3/\text{s}$$

for which, from Equation (b)

$$\text{Re}_d = 2.082 \times 10^6$$

A check of Figure 17.4 indicates that these values are in agreement so that the solution may be considered as completed. The flow rate of water delivered by the pump is 0.331 m^3/s.

We now turn to three design examples.

- Design Example 9 (Section 17.7) involves the sizing of a piping system that employs commercial steel pipe, contains 90° elbows, and a hydraulic turbine.

- Design Example 10 (Section 17.8) considers an irrigation system in which the size of PVC pipe is to be obtained. The system also contains a pump.
- Design Example 11 (Section 17.9) pertains to the transport of crude oil from an unloading area to a refinery located several kilometers away. A heater is placed at a strategic location to modify the flow properties of the oil and the heater output for a prescribed pumping head is sought.

17.7 Design Example 9

In the system shown in Figure 17.10, water at 16°C flows from a reservoir through a piping system that includes a hydraulic turbine. The discharge pressure is atmospheric, the water flow rate is 1.6 m^3/s, and 75 kW is generated by the turbine. Determine the size of commercial steel pipe required.

Solution

Assumptions and Specifications

(1) Steady-state conditions exist.
(2) The flow is incompressible.
(3) Minor losses for two 90° bends are to be included.
(4) The entrance to the pipe from the reservoir is assumed sharp-edged.
(5) Heat transfer is negligible.
(6) The pressures at the upstream and downstream ends are equal.
(7) The turbine is considered as exterior to the pipe section.
(8) Uniform properties exist at each cross section considered.
(9) Each of the 90°elbows is flanged.

We apply the energy equation between the reservoir surface (point 1) and the pipe discharge (point 2) and obtain

$$-\dot{W}_{cv} = \dot{m}\left[g(z_2 - z_1) + \frac{\hat{V}_2^2 - \hat{V}_1^2}{2} + \frac{P_2 - P_1}{\rho} + u_2 - u_1\right] = 0$$

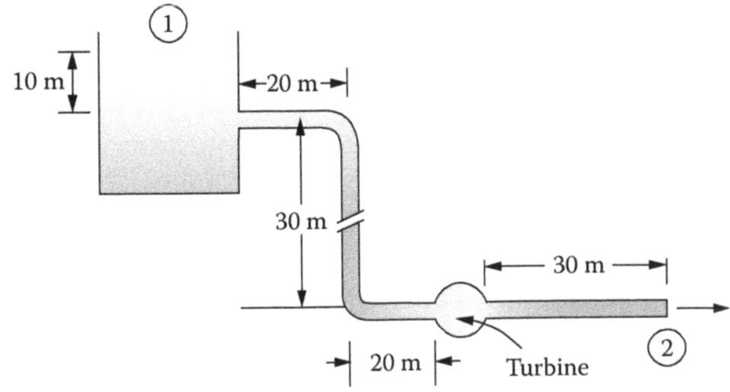

FIGURE 17.10
Piping configuration for Design Example 9.

A term-by-term evaluation yields the following:

$$-\dot{W}_{cv} = -75 \text{ W}$$

$$\dot{m} = \rho \dot{V} = (1000\,\text{kg/m}^3)(1.6\,\text{m}^3/\text{s}) = 1600\,\text{kg/s}$$

$$g(z_2 - z_1) = (9.81\,\text{m/s}^2)(0\,\text{m} - 40\,\text{m}) = -392.4\,\text{m}^2/\text{s}^2$$

$$\frac{\hat{V}_2^2 - \hat{V}_1^2}{2} = \frac{\hat{V}_2^2 - 0}{2} = \frac{1}{2}\left(\frac{4\dot{V}}{\pi d^2}\right)^2 = \frac{1}{2}\left\{\frac{4(1.6\,\text{m}^3/\text{s})}{\pi d^2}\right\}^2 = \frac{2.08}{d^4}\,\text{m}^2/\text{s}^2$$

where d is in meters. Because $P_1 = P_2 = P_{atm}$

$$\frac{P_2 - P_1}{\rho} = 0$$

and

$$u_2 - u_1 = gh_L = 2f_f\frac{L}{d}\hat{V}^2 + \sum K\frac{\hat{V}^2}{2} = \left(2f_f\frac{L}{d} + \frac{\sum K}{2}\right)\hat{V}^2$$

Minor losses will include the effects of two 90° elbows and the entrance from the reservoir into the pipe section that will be assumed to be sharp edged. With these coefficients chosen from Table 17.3 and Figure 17.6, the head loss term becomes

$$\begin{aligned} u_2 - u_1 &= gh_L \\ &= \left[2f_f\frac{100\,\text{m}}{d} + \frac{1}{2}(0.3 + 0.3 + 0.5)\right]\hat{V}^2 \end{aligned}$$

which results in

$$gh_L = \left(200\frac{f_f}{d} + 0.55\right)\hat{V}^2$$

With these terms substituted into the first law expression, we have

$$\begin{aligned} -75\,\text{kW} = (1600\,\text{kg/s})\Bigg[&-392.4\,\text{m}^2/\text{s}^2 + \frac{2.08}{d^4}\,\text{m}^2/\text{s}^2 \\ &+ \left(200\frac{f_f}{d} + 0.55\right)\hat{V}^2\,\text{m}^2/\text{s}^2\Bigg] \end{aligned}$$

Dividing by $\dot{m}$ and further simplifying, we have

$$-46.9 = -392.4 + \frac{2.08}{d^4} + \left[200\frac{f_f}{d} + 0.55\right]\left(\frac{4\dot{V}}{\pi d^2}\right)^2$$

where all of the terms have units of m^2/s^2. In its simplest form, the energy equation is

$$345.5 = \frac{4.36}{d^4} + 830\frac{f_f}{d^5} \tag{a}$$

The Fanning friction factor and pipe diameter are related by the friction factor plot in Figure 17.4 where the Reynolds number has the form

$$\text{Re}_d = \frac{d\hat{V}}{\nu} = \frac{4\dot{V}}{\pi d\nu}$$

Using properties of water at 16°C,

$$\rho = 1000\,\text{kg/m}^3 \qquad \text{and} \qquad \nu = 1.310 \times 10^{-6}\,\text{m/s}^2$$

the Reynolds number becomes

$$\text{Re}_d = \frac{4(1.6\,\text{m}^3/\text{s})}{\pi d(1.310 \times 10^{-6}\,\text{m/s}^2)}$$

or

$$\text{Re}_d = \frac{1.555 \times 10^6}{d} \tag{b}$$

As was the case in Example 7.2, the solution to this problem will require trial and error because the two conditions to be satisfied, Equation (a) and Figure 17.4, are not in forms that allow d to be determined explicity. The procedure to evaluate d will be as follows:

1. Assume a value for $f_{f,\text{old}}$.
2. With this value, solve Equation (a) for d.
3. Use Equation (b) to evaluate Re_d.
4. Evaluate e/d (for commercial steel, $e = 0.457 \times 10^{-4}$ m).
5. Using Re_d and e/d, read $f_{f,\text{new}}$ from Figure 17.4.
6. If $f_{f,\text{new}} = f_{f,\text{old}}$, you may stop because d has the value calculated in step 2.
7. If $f_{f,\text{new}} \neq f_{f,\text{old}}$, adjust the value of $f_{f,\text{old}}$ ($f_{f,\text{old}} = f_{f,\text{new}}$ is a possible choice) and go back to step 2.
8. Repeat the procedure until the values of $f_{f,\text{new}}$ and $f_{f,\text{old}}$ agree.

Employing this procedure, we obtain the following results:

1. Assume $f_{f,\text{old}} = 0.003$.
2. Equation (a) becomes

$$345.5 = \frac{1}{d^4}\left(4.36 + \frac{2.49}{d}\right)$$

   from which, by trial and error, we find

$$d = 0.416\,\text{m}$$

3. Using Equation (b) we obtain $\text{Re}_d = 3.738 \times 10^6$.
4. The relative roughness is

$$\frac{e}{d} = \frac{0.457 \times 10^{-4}\,\text{m}}{0.416\,\text{m}} = 1.099 \times 10^{-4}$$

5. From Figure 17.4, we obtain $f_{f,\text{new}} = 0.00315$.
6. Because $f_{f,\text{new}} \neq f_{f,\text{old}}$, we let $f_{f,\text{old}} = f_{f,\text{new}} = 0.00315$ and return to step 2.

Repeating this procedure with $f_f = 0.00315$, the value of d is 0.418 m, which is the solution to our problem. The required pipe diameter is 0.418 m.

Because pipes are made in standard sizes, in an actual design one would select a pipe whose inside diameter is as close as possible to this value. Generally, the deviation from operational specifications will be quite small when such a choice is made.

17.8 Design Example 10

A wheat farmer in eastern Oregon plans to use a spray irrigation system for one of his fields. His system will utilize water from a large holding pond that is 38 m below the intended sprayer location. The irrigation system will require that 90 gal/min (0.00568 m^3/s) be delivered to the sprayer at 34 psig (234.4 kPa)$_g$. The pump that is available is capable of providing the needed flow rate at a discharge pressure of 90 psig (620.6 kPa gage). The pump will be sited at a location that is 67 m from the sprayer.

Neglecting any minor losses, determine the minimum diameter of PVC pipe that will be suitable for this application.

Solution

Assumptions and Specifications

(1) Steady-state conditions exist.
(2) The PVC pipe may be considered hydraulically smooth.
(3) Minor losses are negligible.
(4) Water properties are to be taken at 50°C.

Strategy

1. The system to be evaluated extends from the pump discharge to the sprayer intake and is composed of PVC pipe acting as a conduit for the water flow.
2. The energy equation (First Law of Thermodynamics), Equation 17.1

$$g(z_1 - z_2) + \frac{\hat{V}_1^2 - \hat{V}_2^2}{2} + \frac{P_1 - P_2}{\rho} = u_2 - u_1 \tag{17.1}$$

must be satisfied. It is clear that

- The required flow rate is specified as 0.00568 m^3/s.
- The velocity will be unknown until the pipe diameter is determined.

3. The frictional head loss between the pipe entrance and discharge must satisfy the Fanning head loss expression of Equation 17.6

$$h_L = 2f_f \frac{L}{d}\frac{\hat{V}^2}{g} \tag{17.6}$$

where
d = diameter of the pipe which is the parameter sought
$\hat{V}$ = velocity that is directly proportional to the diameter

f_f = Fanning friction factor that is determined using Figure 17.4. Note that the independent variable is the Reynolds number which includes both diameter and velocity

4. With the energy equation involving the unknowns, $\hat{V}$ and d, in complex ways (f_f, d, and $\hat{V}$ are all related graphically in Figure 17.4), it is clear that a trial-and-error procedure must be employed.

Now that the strategy has been established, we proceed to the solution of the problem. The energy equation applied to a control volume, which includes the pipe with inflow at the pump discharge (section 1) and outflow at the sprayer inlet (section 2), takes the form

$$h_L = (z_1 - z_2) + \frac{\hat{V}_1^2 - \hat{V}_2^2}{2g} + \frac{P_1 - P_2}{\rho g} = \frac{u_2 - u_1}{g}$$

Note that there is no heat transfer and no shaft work done on or by the control volume and that for water at 50°C, Table A.14 provides

$$\rho = 988.1\,\text{kg/m}^3 \qquad \text{and} \qquad \nu = 0.5435 \times 10^{-6}\,\text{m/s}^2$$

Evaluating each term in this relationship, we have

$$z_1 - z_2 = -38\,\text{m}$$

$$\frac{\hat{V}_1^2 - \hat{V}_2^2}{2g} = 0$$

$$\frac{P_1 - P_2}{\rho g} = \frac{620.6\,\text{kPa} - 234.4\,\text{kPa}}{(988.1\,\text{kg/m}^3)(9.81\,\text{m/s}^2)} = 39.84\,\text{m}$$

Moreover, with

$$\hat{V} = \frac{4\dot{V}}{\pi d^2} = \frac{4}{\pi}\left(\frac{0.00568\,\text{m}^3/\text{s}}{d^2}\right) = \frac{7.232 \times 10^{-3}}{d^2}\,\text{m/s}$$

the head loss term may be evaluated as

$$h_L = \frac{2f_f}{g}\frac{L}{d}\hat{V}^2$$

$$= \frac{2f_f}{9.81\,\text{m/s}^2}\frac{67\,\text{m}}{d}\left(\frac{7.232 \times 10^{-3}}{d^2}\,\text{m/s}\right)^2$$

$$= \frac{7.144 \times 10^{-4}}{d^5} f_f\,\text{m}$$

The energy equation may be evaluated as

$$-38\,\text{m} + 39.84\,\text{m} = \frac{7.144 \times 10^{-4}}{d^5} f_f$$

or

$$1.84\,\text{m} = \frac{7.144 \times 10^{-4}}{d^5} f_f\,\text{m}$$

As discussed earlier, f_f is a function of the Reynolds number that is

$$\mathrm{Re}_d = \frac{d\hat{V}}{\nu} = \frac{d}{\nu}\left(\frac{7.232 \times 10^{-3}}{d^2}\,\mathrm{m/s}\right)$$

$$= \frac{(7.232 \times 10^{-3}/d^2)\,\mathrm{m/s}}{0.5435 \times 10^{-6}\,\mathrm{m/s^2}} = \frac{13{,}300}{d}$$

We are now ready to proceed with our trial-and-error solution. The computations are summarized in Table 17.4.

TABLE 17.4

Computations to Establish the PVC Pipe Diameter

Assume d, m	0.08	0.065	0.068	0.0689
Re_d	169,290	208,350	199,160	193,000
f_f	0.0042	0.0039	0.0040	0.0040
$7.1441 \times 10^{-4} f_f/d^5$, m^{-5}	0.017	2.403	1.967	1.840

The required conditions are met for $d = 0.0689$ m. Thus, our required pipe diameter is

$$0.0689\,\mathrm{m} \qquad \text{or} \qquad 2.71\,\mathrm{in} \Longleftarrow$$

It is frequently necessary to transport crude oil from an unloading area to a refinery. In such a case, a system is selected and the system output is calculated. This design problem considers the capacity of an available pipeline as well as the economic advantage of heating the oil to increase the capacity.

17.9 Design Example 11

An oil refinery is located 11 km from a marine terminal where oil tankers are unloaded. Two sections of pipe are available for use and each section is to be equipped with a pump located at the beginning of the section. Pertinent data for the piping system containing smooth pipe is as follows: Pipe 1 is 3-km long, has a diameter of 80 cm and its pump has an output of 1000 kPa. Pipe 2 is 11-km long, has a diameter of 100 cm and its pump has an output of 860 kPa (Figure 17.11).

Section 2 is equipped with a heater that will heat the oil as needed to modify its flow properties. The oil leaves the tanker and enters the system at 20°C. The cargo space in the tanker is 160,000m^3.

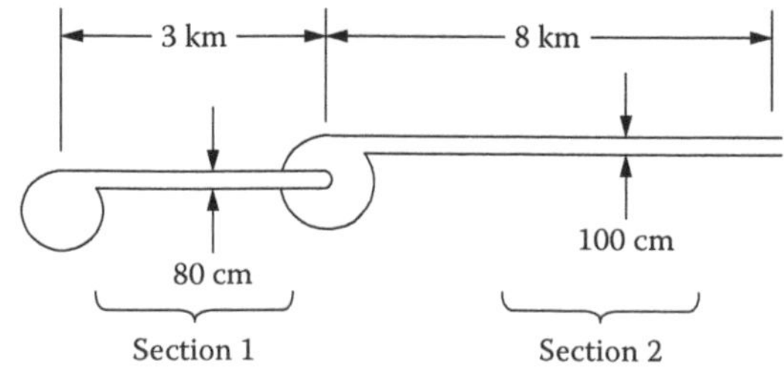

FIGURE 17.11
Piping configuration for Design Example 11.

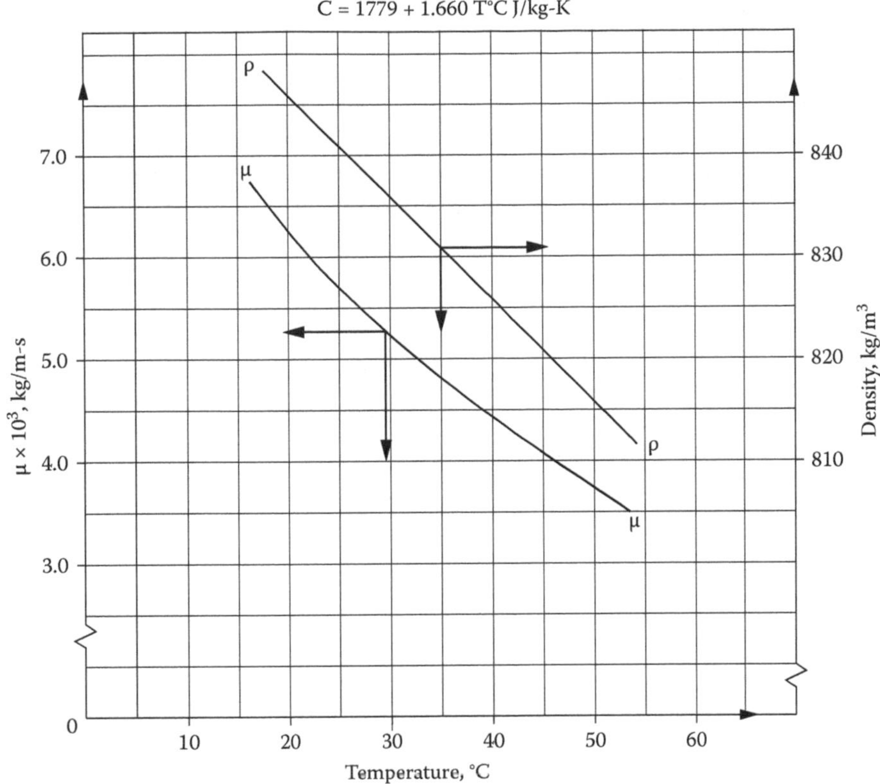

FIGURE 17.12
Thermal properties of 34° mid-continent crude oil. (Taken from Kern, D. Q., *Process Heat Transfer*, McGraw-Hill Book Company, New York, NY, 1951.)

The oil is a mid-continent crude (34° API) and thermal properties are provided in Figure 17.12. Determine (a) the capacity of the piping system, (b) the required power, if any, for heating the oil, and (c) the time required to completely empty the tanker using the pump at the entrance to Section 1.

Solution

Assumptions and Specifications

(1) Steady-state conditions exist.
(2) Data for the piping system is specified.
(3) Pertinent properties of mid-continent crude are given in Figure 17.12.
(4) Physical properties of the oil are uniform at any cross section.
(5) Both pipes may be considered as smooth.

Strategy

Each pump is to be operated at its maximum capacity. We will use section 1 and section 2 in that order because the pump 1 operates at a higher pressure and section 1 is not equipped with a heater. One procedure is

1. Assume values of the oil velocity in section 1 and find, by trial and error, the velocity that will yield a head loss in the system of 1000 kPa.

2. Once the velocity is established, we will be able to determine the capacity of the system in both kilogram per second and gallons per minute.
3. We will see at 20°C whether the flow established in section 1 can be accommodated in section 2. If so, we will determine whether the head loss is section 2 is equal to the pump output of 875 kPa. If not, we will alter the temperature of the oil in section 2. This will adjust the flow so as to make the system optimum. In this case, "optimum," means that both pumps are operating at their rated output pressures.

Section 1 using properties from Figure 17.12

$$L = 3.0\,\text{km} \qquad d = 0.80\,\text{m}$$

$$T = 20^\circ\text{C} \qquad L/d = 3000\,\text{m}/0.80\,\text{m} = 3750$$

$$\rho = 846.3\,\text{kg/m}^3 \qquad \mu = 0.0062\,\text{kg/m-s}$$

With the head lost given by Equation 17.6

$$h_L = 2f_f \frac{L}{d}\frac{\hat{V}^2}{g}$$

values of $\hat{V}$ can be assumed so that

$$h_L = 2f_f \frac{L}{d}\frac{\hat{V}^2}{g} = \frac{2(3000\,\text{m})}{(0.80\,\text{m})(9.81\,\text{m/s}^2)} f_f \hat{V}^2 = 764.53\, f_f \hat{V}^2$$

Then with

$$\text{Re}_d = \frac{\rho d}{\mu}\hat{V} = \frac{(846.3\,\text{kg/m}^3)(0.80\,\text{m})}{0.0062\,\text{kg/m-s}}\hat{V} = 109,200\,\hat{V}$$

where $\hat{V}$ is in m/s. Then, by trial-and error, thus with

Assume $\hat{V}$, m/s	8.0	7.25	7.00	7.05
Find $\text{Re}_d \times 10^3$	130	87.3	79.2	77.0
Obtain, f_f	0.00285	0.00300	0.00350	0.00318
h_L, m of oil	313.76	146.8	126.6	120.8
h_l, kPa	2605	1218.7	1050.9	1003

$$A = \frac{\pi}{4}d^2 = \frac{\pi}{4}(0.80\,\text{m})^2 = 0.5027\,\text{m}^2$$

we have with $\hat{V} = 7.05$ from the foregoing table

$$\dot{m} = \rho A\hat{V} = (846.3\,\text{kg/m}^3)(0.5027\,\text{m}^2)(7.05\,\text{m/s}) = 3000\,\text{kg/s}$$

(a) The capacity of the system is

$$\dot{V} = \hat{V}A = (7.05\,\text{m/s})(0.5027\,\text{m}^2) = 3.584\,\text{m}^3/\text{s} \Longleftarrow$$

Section 2, again with properties taken from Figure 7.12

$$L = 8.0\,\text{km} \qquad d = 1.00\,\text{m}$$
$$\dot{m} = 3000\,\text{kg/s} \qquad L/d = 8000\,\text{m}/1.00\,\text{m} = 8000$$

With

$$A = \frac{\pi}{4}d^2 = \frac{\pi}{4}(1.00\,\text{m})^2 = 0.7854\,\text{m}^2$$

we have

Assume T	°C	20.0	30.0	40.0	50
ρ	kg/m^3	870	865	860	855
μ	kg/m-s	0.0068	0.00525	0.0044	0.0038
$\hat{V} = \dot{m}/\rho A$	m/s	4.577	4.603	4.630	4.657
$\text{Re}_d \times 10^6$		0.5856	0.7584	0.9050	1.048
f_f		0.00303	0.00292	0.00286	0.00280
h_L	N/m^2	893.5	856.3	843.6	830.5

These data are plotted in Figure 17.13 where it may be observed that to meet the pressure at the outlet of pump 2, the temperature should be

$$T = 37.5°\text{C}$$

(b) With the specific heat taken from Figure 17.11 at 35°C is 1745 kJ/kg-K so that the heat supplied to the oil will be

$$\dot{Q} = \dot{m}c\Delta T = (3172\,\text{kg/s})(1745\,\text{kJ/kg-K})(17.5°\text{C}) = 49.54\,\text{kW} \Longleftarrow$$

(c) The time to empty the tanker at the volumetric flow rate established in (a) will be

$$t = \frac{160,000\text{m}^3}{(3.584\,\text{m}^3/\text{s})(3600\,\text{s/h})} = 12.37\,\text{h} \Longleftarrow$$

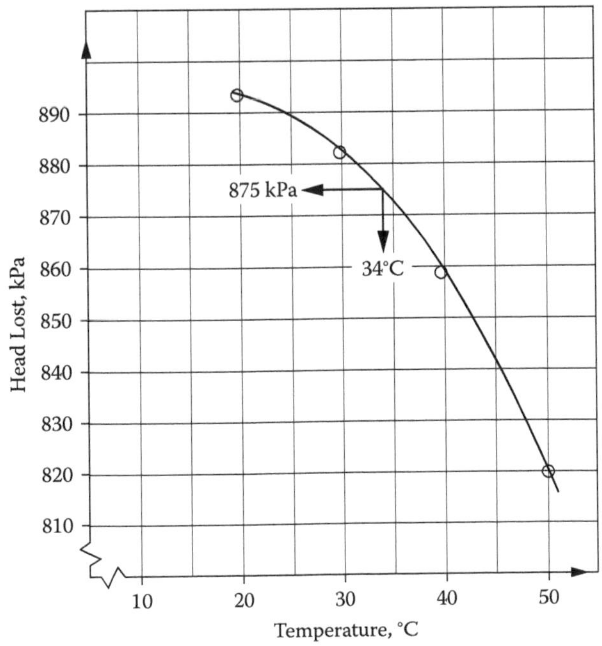

FIGURE 17.13
Head loss as a function of temperature for pipe 2 in Design Example 11.

17.10 Multiple-Path Pipe Systems

Many industrial piping systems involve multiple paths for fluid flow between source and destination. Such systems may be quite involved and any solution will likely be very laborious and time consuming. The design of new multiple-path systems or the analysis of the performance of an existing piping network will likely entail the use of a computer code. In this section, we will illustrate the process of analysis by using a relatively simple example. The steps in such an analysis are those that are used in more complex systems.

A flow system, such as the one shown in Figure 17.14, is directly analogous to a parallel electrical circuit. There are two branches (numbered 1 and 2) between two nodes (A and B) for the multiple-path portion of the flow system shown. For this two-node system, it is clear that, for each branch, there is a common pressure difference, $P_A - P_B$, and a common elevation change, $z_B - z_A$. Furthermore, continuity at each node will require that

$$\dot{V}_A = \dot{V}_B = \dot{V}_1 + \dot{V}_2$$

For each of the two branches, an equation of the form

$$g(z_B - z_A) + \frac{\hat{V}_B^2 - \hat{V}_A^2}{2} + \frac{P_B - P_A}{\rho} + u_B - u_A = 0$$

will apply and, as usual, the internal energy increase can be written as

$$u_B - u_A = gh_L = 2f_f\frac{L}{d}\hat{V}^2 + \sum K\frac{\hat{V}^2}{2}$$

Analysis of a two-node piping system will be of two general types:

1. Given the flow conditions at A, along with P_A, the pipe size, geometry, and roughness, find P_B.
2. Given P_A and P_B, along with the pipe size, geometry, and roughness, evaluate the total flow, $\dot{V}_A + \dot{V}_B$ as well as $\dot{V}$ in each branch.

Example 17.4 presents the steps in analyzing a two-branch, two-node system.

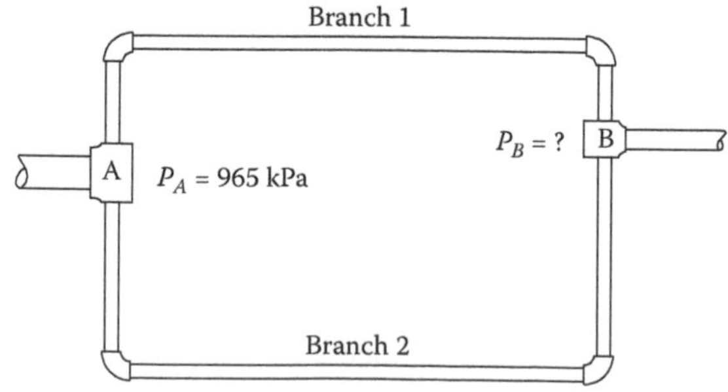

FIGURE 17.14
A multiple-path piping system with two branches and two nodes.

Example 17.4 In the piping system shown in Figure 17.14, the pressure at point A is 965 kPa (gage) and the total flow rate of water at 16°C is 0.70 m^3/s. Pipe lengths and diameters are as follows: for pipe 1, $L_1 = 550$ m and $d_1 = 0.36$ m and for pipe 2, $L_2 = 850$ m and $d_2 = 0.50$ m. Neglect minor losses and determine the pressure at node B.

Solution

Assumptions and Specifications

(1) Steady-state conditions exist.
(2) The flow is incompressible.
(3) Minor losses are neglected.
(4) Heat transfer is negligible.
(5) Uniform properties exist at each cross section considered.

The constraints on this system include continuity (i.e., the total flow in branches 1 and 2 must equal 0.70 m^3/s); the energy equation, which relates potential energy, kinetic energy, pressure change, and head loss along each of the branches connecting nodes A and B; and the frictional head loss, which must be compatible with Figure 17.4 for both branches.

As a first step we can write the first law expression for each branch. For branch 1

$$g(z_B - z_A) + \frac{\hat{V}_{1B}^2 - \hat{V}_{1A}^2}{2} + \frac{P_B - P_A}{\rho} + u_{1B} - u_{1A} = 0 \tag{a}$$

and for branch 2

$$g(z_B - z_A) + \frac{\hat{V}_{2B}^2 - \hat{V}_{2A}^2}{2} + \frac{P_B - P_A}{\rho} + u_{2B} - u_{2A} = 0 \tag{b}$$

Noting that

$$\frac{\hat{V}_{1B}^2 - \hat{V}_{1A}^2}{2} = \frac{\hat{V}_{2B}^2 - \hat{V}_{2A}^2}{2} = 0$$

and that the potential energy and flow energy change are the same for both branches, it is clear that $\Delta u_1 = \Delta u_2$ or

$$h_{L1} = h_{L2} \tag{c}$$

With the system constraints in mind, we can proceed with a trial-and-error solution to our problem. The procedure to be used is the following:

1. Assume a flow rate, $\dot{V}_1$, through one branch and determine the associated head loss.
2. Since, according to Equation (c), $h_{L1} = h_{L2}$, solve for the flow rate through the other branch, $\dot{V}_2$.
3. Assuming that the actual flow divides in the same ratio as $\dot{V}_1/\dot{V}_2$, adjust the values for $\dot{V}_1$ and $\dot{V}_2$ so that $\dot{V}_1 + \dot{V}_2 = \dot{V}_A$.
4. With these values for $\dot{V}_1$ and $\dot{V}_2$, solve for h_{L1} and h_{L2}.
5. If $h_{L1} = h_{L2}$, the solution is complete.
6. If $h_{L1} \neq h_{L2}$, adjust the assumed value for $\dot{V}_1$ and return to step 2.

We now follow this procedure for the velocities specified in the problem statement.

1. Assume $\dot{V}_1 = 0.20\,\text{m}^3/\text{s}$ and solve for h_{L1}. With

$$\hat{V} = \frac{4\dot{V}}{\pi d^2}$$

we have

$$gh_{L1} = 2f_{f1}\frac{L_1}{d_1}\left(\frac{4\dot{V}_1}{\pi d_1^2}\right)^2 = \frac{32}{\pi^2}f_{f1}\frac{L_1}{d_1^5}\dot{V}_1^2$$

Because the pipe is hydraulically smooth, we must determine f_{f1} from Figure 17.4. The Reynolds number is evaluated as

$$\text{Re}_1 = \frac{d_1\hat{V}_1}{\nu} = \frac{d_1}{\nu}\left(\frac{4\dot{V}_1}{\pi d_1^2}\right) = \frac{4\dot{V}_1}{\pi d_1 \nu}$$

For water at 16°C, $\nu = 1.133 \times 10^{-6}\,\text{m/s}^2$, we have

$$\text{Re}_1 = \frac{4(0.20\,\text{m}^3/\text{s})}{\pi(0.36\,\text{m})(1.133 \times 10^{-6}\,\text{m}^2\text{s})} = 6.25 \times 10^5$$

The corresponding value of f_{f1} from Figure 17.4 is

$$f_{f1} = 0.0031$$

The value of gh_{L1} can now be evaluated:

$$gh_{L1} = \frac{32}{\pi^2}(0.0031)\frac{(550\,\text{m})}{(0.36\,\text{m})^5}(0.020\text{m}^3/\text{s})^2 = 36.57\,\text{m}^2/\text{s}^2$$

2. With $gh_{L1} = gh_{L2} = 36.57\,\text{m}^2/\text{s}^2$, we can calculate $\dot{V}_2$ from the expression

$$gh_{L2} = \frac{32}{\pi^2}f_{f2}\frac{L_2}{d_2^5}\dot{V}_2^2$$

Solving for $\dot{V}_2$, we have

$$\dot{V}_2^2 = \frac{(36.57\,\text{m}^2/\text{s}^2)\pi^2(0.50\text{ m})^5}{32(850\,\text{m})f_{f2}} = \frac{4.147 \times 10^{-4}}{f_{f2}}$$

By trial and error, this can be solved to provide

$$\dot{V}_2 = 0.372\,\text{m}^3/\text{s}$$

We note that

$$\dot{V}_1 + \dot{V}_2 = 0.20\,\text{m/s} + 0.372\,\text{m/s} = 0.572\,\text{m}^3/\text{s}$$

3. Adjusted values are determined as

$$\dot{V}_1 = \left(\frac{0.20}{0.572}\right)(0.70\,\text{m}^3/\text{s}) = 0.245\,\text{m}^3/\text{s}$$

and

$$\dot{V}_2 = \left(\frac{0.372}{0.572}\right)(0.70\,\text{m}^3/\text{s}) = 0.455\,\text{m}^3/\text{s}$$

4. The corresponding head loss values are determined once more

$$\mathrm{Re}_1 = \frac{4(0.245\,\mathrm{m}^3/\mathrm{s})}{\pi(0.36\,\mathrm{m})(1.133\times 10^{-6}\,\mathrm{m}^2/\mathrm{s})} = 7.644\times 10^5$$

with the corresponding value of f_{f1} from Figure 17.4 of

$$f_{f1} = 0.00305$$

The value of gh_{L1} is now evaluated as

$$gh_{L1} = \frac{32}{\pi^2}(0.00305)\frac{(550\,\mathrm{m})}{(0.36\,\mathrm{m})^5}(0.245\,\mathrm{m}^3/\mathrm{s})^2 = 54.0\,\mathrm{m}^2/\mathrm{s}^2$$

These same steps carried through for pipe 2 yield

$$\mathrm{Re}_2 = 8.356\times 10^5$$

and

$$f_{f2} = 0.0030$$

so that

$$gh_{L2} = \frac{32}{\pi^2}(0.0030)\frac{(850\,\mathrm{m})}{(0.50\,\mathrm{m})^5}(0.455\,\mathrm{m}^3/\mathrm{s})^2 = 54.7\,\mathrm{m}^2/\mathrm{s}^2$$

5. These two values of head loss compare within 1.3%, which is an acceptable accuracy. We thus conclude that the values for $\dot{V}_1$ and $\dot{V}_2$ obtained in step 3 are satisfactory.
6. With the average value of gh_L taken as $54.35\,\mathrm{m}^2/\mathrm{s}^2$, we may use either Equation (a) or (b) to evaluate P_B. The solution is

$$\begin{aligned}\frac{P_A - P_B}{\rho} &= g(z_B - z_A) + u_B - u_A \\ &= (9.81\,\mathrm{m}/\mathrm{s}^2)(12.2\,\mathrm{m}) + 54.35\,\mathrm{m}^2\mathrm{s}^2 \\ &= 174.0\,\mathrm{m}^2/\mathrm{s}^2\end{aligned}$$

The final result is now obtained as

$$\begin{aligned}P_B &= P_A - \rho(174\,\mathrm{m}^2/\mathrm{s}^2) \\ &= 965\,\mathrm{kPa} - (1000\,\mathrm{kg}/\mathrm{m}^3)(174\,\mathrm{m}^2/\mathrm{s}^2) = 791\,\mathrm{kPa} \Longleftarrow\end{aligned}$$

The foregoing example has illustrated a solution procedure for a representative two-branch pipe flow problem of the type where the flow conditions at A, along with P_A, the pipe size, geometry, and roughness are given and P_B is to be found. The procedure for solving a problem where P_A and P_B, along with pipe size, geometry, and roughness are given and the total flow as well as $\dot{V}$ in each branch is to be found, would follow a similar thought process.

As stated earlier, the two-node, two-branch problem is as simple as multiple-path analysis gets. Myriad paths and nodes may exist in a real system. The solution for such a flow network is applied to analyze the individual relationships that compound to become a complex network.

17.11 Summary

This chapter has dealt with the common and very important subject of pipe flow.

Initially, the relationships for frictional pressure drop associated with both laminar and turbulent flow were examined. It was determined that the Fanning friction factor is functionally related to the Reynolds number, Re_d, and the relative roughness, e/d, as represented by the friction factor plot of Figure 17.4.

A means for accounting for minor losses associated with valves, fittings, and entrance and exit effects was also considered. A suitable relationship, Equation 17.19, along with coefficients listed in chapter tables and plots, is available to compute these effects.

The chapter concluded with a series of problems dealing with single- and multiple-path flows. Most problems of these type involve trial and error to match the constraints of friction factor along with the satisfaction of mass conservation and the first law of thermodynamics.

17.12 Problems

Mass Conservation Revisited

17.1: A circular pipe increases from an inlet (section 1) with a diameter of 30 cm to an outlet (section 2) with a diameter of 45 cm. Water flows through the pipe at a velocity of 5 m/s at section 2. Determine (a) the velocity at section 1, (b) the volumetric flow rate at section 1, (c) the volumetric flow rate at section 2, and (d) the mass flow rate.

17.2: A gas flows through a tapering square conduit. At section 1, the conduit has sides that are 0.08 m in length and at section 2, the sides are 0.24-m long. The gas velocity is 8.25 m/s at section 1 and its density is 1.092 kg/m^3. At section 2, the velocity is 1.98 m/s. Determine the mass flow rate of the gas and its density at section 2.

17.3: Water enters a tank at sections 1 and 2 and leaves at section 3. The velocity of the water entering the tank at section 1 is 4.875 m/s and the volumetric flow rate at section 2 is $\dot{V} = 0.0125\,\text{m}^3/\text{s}$. The diameters of the conduits at sections 1 and 3 are 4 cm and 6 cm, respectively. The water level in the tank remains constant. Determine the exit velocity, $\hat{V}_3$.

17.4: Air at 27°C and 108 kPa flows at 1.5625 kg/s through a circular annular passage with an inner diameter of 16 cm and an outer diameter of 32 cm. Determine (a) the average velocity and (b) the volumetric flow rate.

17.5: Oil with a specific gravity of 0.858 flows through a 75-cm-diameter pipe at a flow rate of 8760 gal/min. Determine (a) the volumetric flow rate in m^3/s, (b) the velocity, and (c) the mass flow rate.

17.6: A mixing chamber has an inlet (section 1) and an outlet (section 2). A third conduit, which can be used as either an inlet or an outlet (section 3), is connected to the chamber. At section 1, the diameter is 5 cm and the volumetric flow rate is 0.60 m^3/s. At section 2, the diameter is 7.5 cm and the velocity of egress is 12 m/s. If the diameter at section 3 is 2.5 cm, determine whether the flow at section 3 is into or out of the mixing chamber.

17.7: Consider a plunger-exit pipe arrangement in which the plunger at section 1 is being pushed into a tank with a fluid having a specific gravity of 0.72. At section 1 the velocity of the plunger is 6.25 cm/s. A pipe with a diameter of 2 cm leads from the tank at section 2. Determine the mass rate of flow at section 2.

17.8: A gasoline pump is used to fill a 100-L tank in 90 s. The pump has an exit diameter of 3.75 cm. Determine the exit velocity from the pump.

Flow Regimes

17.9: Unused engine oil at 27°C flows in a 24-cm pipe. Determine the maximum velocity to assure laminar flow.

17.10: For a Reynolds number of 300,000, determine the velocity of an airstream at 27°C flowing past a 12-cm-diameter sphere.

17.11: Air flows in a circular pipe at a rate of 0.04 kg/s; the air temperature is 27°C. Determine the minimum allowable pipe diameter for the flow to remain laminar.

17.12: Water, at 15°C, flows in an annular space between two concentric pipes. The inner and outer diameters of the annular region are 7.5 cm and 11.5 cm, respectively. (a) Determine the volumetric flow rate, in m^3/s, that will be achieved for a water velocity of 3 m/s and (b) specify whether this flow is laminar or turbulent.

17.13: For 15°C water flowing in the annular region specified in Problem 17.12, determine what maximum water velocity can be obtained if the flow is to be laminar.

17.14: Solve Problem 17.13 if the fluid is air at 25°C.

17.15: Rework Problem 17.10 for flowing fluids of (a) water at 20°C and (b) nitrogen at 27°C.

17.16: For fluid flow at 900 cm^3/s through a 6.5-cm-diameter duct, determine whether the flow is laminar or turbulent for each of the following fluids at 20°C: (a) air, (b) carbon dioxide, (c) ethylene glycol, (d) water, (e) unused engine oil, and (f) glycerin.

17.17: SAE 50 oil flows through a 10-cm-diameter pipe. Determine the flow rate, in m^3/s, for transition to turbulent flow for (a) 27°C and (b) 100°C.

17.18: For glycerin flowing through a 5-cm-diameter pipe, determine the volumetric flow rate, in m^3/s, for transition to turbulence at (a) 27°C and (b) 47°C.

17.19: The Reynolds number for a fluid in a 25-cm-diameter pipe is 2000. A pipe that is 15-cm diameter forms an extension to the 25-cm-diameter pipe. The flow is incompressible. Determine the Reynolds number in the 15-cm-diameter pipe.

17.20: Water flows through two pipes 1 and 2 that intersect to form a single conduit (pipe 3). Pipe 1 has a diameter of 5 mm and the volumetric flow rate in the two pipes are equal. The diameters of pipes 2 and 3 are 4 mm and 6 mm, respectively. The water temperature is 27°C. Determine the maximum volumetric flow rate, $\dot{V}_2$, in pipe 2 in order to keep the flow in pipe 3 laminar.

17.21: Compare the maximum velocity to yield laminar flow in a 15-cm pipe for (a) a fuel oil at 15°C with a kinematic viscosity of $\nu = 4.42 \times 10^{-6}\,m/s^2$ and (b) water at 15°C.

17.22: For laminar flow conditions, determine the diameter of the pipe needed to deliver 10 gal/min of a fuel oil having a viscosity of $\nu = 5.92 \times 10^{-7}\,m/s^2$.

17.23: A fluid with a kinematic viscosity of $\nu = 1.35 \times 10^{-5}\,m/s^2$ flows through a 24-cm-diameter pipe. Determine the maximum velocity to keep the flow laminar.

17.24: Determine the Reynolds number of oil flowing through a 42-cm-diameter pipe if the oil has a specific gravity of 0.852, a dynamic viscosity of 0.0258 kg/m-s and a volumetric flow rate of 0.385 m^3/s.

17.25: Determine the type of flow occurring in a 30-cm-diameter pipe when (a) water at 15°C flows at a velocity of 1.25 m/s and (b) a fuel oil with $\nu = 2 \times 10^{-5}\,\text{m/s}^2$ flowing at the same velocity.

17.26: An oil with a specific gravity of 0.862 and a kinematic viscosity of $\nu = 2 \times 10^{-5}\,\text{m/s}^2$ flows through a pipe that has a diameter of 25 cm. The volumetric flow rate is 0.625 L/s. Determine whether the flow is laminar or turbulent.

Head Loss—Laminar Flow

17.27: An oil with a density of 910 kg/m^3 and a kinematic viscosity of $7.4 \times 10^{-6}\,\text{m/s}^2$ flows through a horizontal tube 6 mm in diameter at a volumetric flow rate of 0.0838 m^3/h. Determine the pressure drop that will be achieved if the tube is 10-m long.

17.28: For the 10-m-long tube described in Problem 17.27, determine the maximum volumetric flow rate in m^3/s to keep the flow laminar.

17.29: A tube used in lubrication is 75-cm long and has a diameter of 2.5 mm; the oil properties are as given in Problem 17.27. Determine the flow rate of the oil when the pressure drop is 100 kPa.

17.30: A viscometer for measuring the viscosity of liquids commonly consists of a relatively large reservoir connected to a relatively small tube as shown in Figure P17.30. An oil flows out of the viscometer at a rate of 29 cm^3/s into a 1.8-mm-diameter tube. Determine the viscosity of the oil.

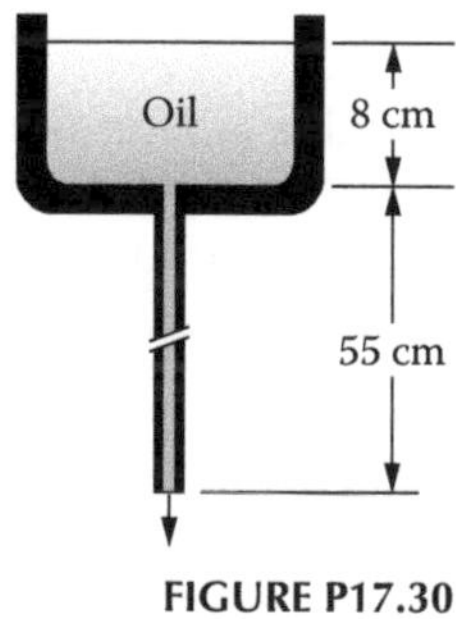

FIGURE P17.30

17.31: For the viscometer specified in Problem 17.30 and sketched in Figure P17.30, water at 15°C is used as the viscous fluid. Determine the expected flow rate of the water in cm^3/s.

17.32: Oil with a kinematic viscosity of $4.5 \times 10^{-6}\,\text{m/s}^2$ and a density of 800 kg/m^3 is to be pumped at a rate of 3.5 m^3/s through a 260-km-long pipeline, 62 cm in diameter, made of commercial steel. Determine the pumping power required to achieve these conditions.

17.33: Rework Problem 17.32 when the flowing fluid is water at 40°C.

17.34: A 65-km-long pipe line delivers petroleum at a rate of 5000 barrels per day (795 m^3/day) with a pressure drop, between origin and destination, of 4.07 MPa. Assuming that the flow in the pipe to be at its maximum value for laminar flow, determine the diameter of the pipe.

17.35: For the case in Problem 17.34, consider that a parallel section of the same diameter pipe is added over the last 40 km of the system and, still assuming the flow in the system to be laminar, and the pressure drop for the total system to remain at 4.07 MPa, determine the new capacity of the piping system.

17.36: Solve Problem 17.34 for the case with two additional sections of the same diameter pipe added over the last 40 km of the pipeline. All other conditions remain as given in Problem 17.34. Determine the new capacity of the piping system.

17.37: A liquid flowing under the influence of gravity down an inclined plane surface, with its surface exposed to the atmosphere, is referred to as a **falling film.** For such a film in steady, laminar, and incompressible flow, the velocity profile between the wall and the free surface varies with the distance, y, as shown in Figure P17.37, according to the expression

$$\hat{V}_x = \frac{\rho g L^3 \sin\theta}{\mu}\left(\frac{y}{L} - \frac{y^2}{2L^2}\right)$$

Determine the values of (a) $\hat{V}_{x,\max}$ and (b) $\hat{V}_{x,\text{avg}}$.

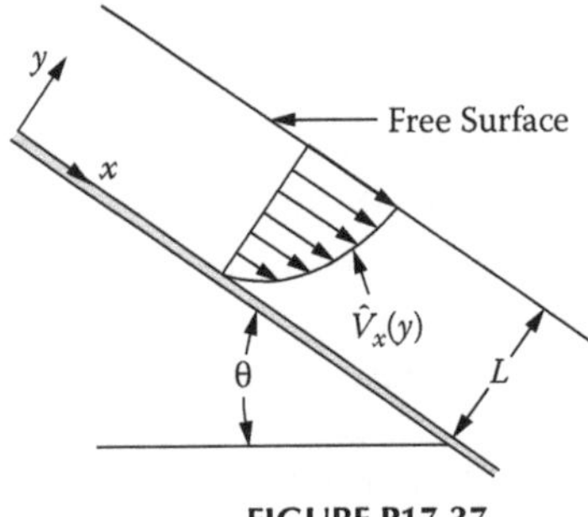

FIGURE P17.37

17.38: A horizontal capillary tube with an inside diameter of 5 mm is used to measure the viscosity of thick fluids such as oils. In such a system, the pressure gradient is measured to be 380 kPa/m of tube length when the flow rate of the liquid is 0.073 m^3/h. Determine the fluid viscosity.

17.39: The capillary tube viscometer of Problem 17.38 is to be used with glycerin at 20°C, determine the flow rate, in m^3/h, that is achieved for a Reynolds number of 2000.

17.40: Water is being siphoned from a reservoir as shown in Figure P17.40. For $h = 100$ cm, will the flow be laminar or turbulent?

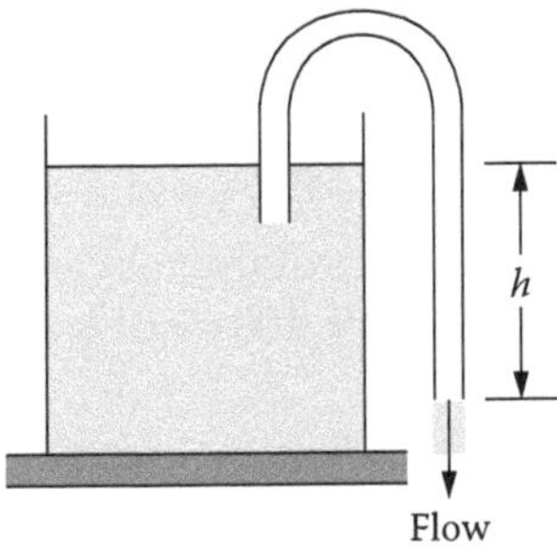

FIGURE P17.40

17.41: An oil, having a density of $\rho = 870\,\text{kg/m}^3$ and a dynamic viscosity of $\mu = 0.065$ kg/m-s, is pumped through a horizontal pipe, 1.2-km long, with an input power of 1 kW. If laminar flow is to be maintained, determine (a) the maximum flow that can be achieved and (b) the corresponding pipe diameter.

17.42: The piston shown in Figure P17.42 provides constant flow through the hypodermic needle. Determine the force necessary to achieve a flow rate of $0.15\,cm^3/s$, with a fluid having a density ρ of $870\,kg/m^3$ and a viscosity of $\mu = 0.002\,kg/m\text{-}s$.

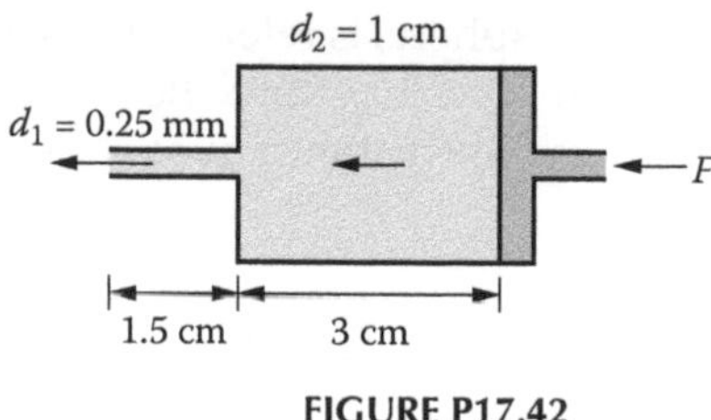

FIGURE P17.42

Head Loss—Other Types of Flow

17.43: Water at 20°C flows through a cast-iron pipe at a velocity of 3.2 m/s. The pipe is 400 m long and has a diameter of 15 cm. Determine the head loss due to friction.

17.44: A 2.44-m-diameter pipe carries water at 15°C. The head loss due to friction is 0.500 m per 300 m of pipe. Determine the volumetric flow rate of the water leaving the pipe.

17.45: Water at 20°C is being drained from an open tank through a cast-iron pipe 60 cm in diameter and 40-m long. The surface of the water in the pipe is at atmospheric pressure and at an elevation of 45.9 m, and the pipe discharges to the atmosphere at an elevation of 30 m. Neglecting minor losses due to configuration, bends, valves, and fittings, determine the volumetric flow rate of the water leaving the pipe.

17.46: Unused engine oil is being pumped at a volumetric flow rate of $0.20\,m^3/s$ through a level, 15-cm-diameter wrought-iron pipe. Determine the pressure loss in Pa per diameter of pipe.

17.47: SAE 50 oil at 20°C is to be pumped at a volumetric flow rate of $0.275\,m^3/s$ through a level cast-iron pipe. The allowable pipe friction loss is 70,000 Pa per 300 m of pipe. Determine the size of commercial pipe to be used.

17.48: Water at 20°C is to be pumped through a welded-steel pipe ($e = 0,457 \times 10^{-4}$ m) at 4.25 m/s. The pipe is 400-m long and has a diameter of 17.5 cm. Determine the head loss due to friction.

17.49: Gasoline with a specific gravity of 0.714 and a dynamic viscosity of $2.96 \times 10^{-4}\,kg/m\text{-}s$ is being discharged from an inclined pipe in which section 1, at the inlet, is 16-m higher than the outlet. The pressure at section 1 is 2400 kPa and the discharge is at atmospheric pressure. The pipe roughness, e, is 0.480 mm and the pipe is 950-km long. Determine the diameter needed to yield a volumetric flow rate of $0.125\,m^3/s$.

17.50: Water at 20°C flows through a 10-cm-diameter cast-iron pipe at a velocity of 4.8 m/s. Determine (a) the pressure loss per 100 m of pipe and (b) the power lost due to friction.

17.51: A 15-cm wrought-iron pipe is to carry water at 20°C. Assuming level pipe, determine the volumetric flow rate at the discharge if the pressure loss is not to exceed 32.5 kPa per 100 m of pipe length.

17.52: An oil having a kinematic viscosity of $8.30 \times 10^{-5}\,m/s^2$ flows through a 600-m horizontal pipe with a diameter of 7.5 cm. For an upstream pressure of 33.5 kPa and a downstream pressure of 24.0 kPa, determine the mass flow rate in kg/s.

17.53: Alcohol with a specific gravity of 0.605 and a kinematic viscosity of $5 \times 10^{-7}\,m/s^2$ is drawn from a tank through a hose having an inside diameter of 2.5 cm as shown in Figure P17.53. The hose has a relative roughness of $e = 0.0004$ m. The total length of the hose is 8.75 m and we are to neglect minor losses due to configuration, bends, valves, and fittings. Determine (a) the volumetric flow rate of the alcohol leaving the hose and (b) the minimum pressure in the hose.

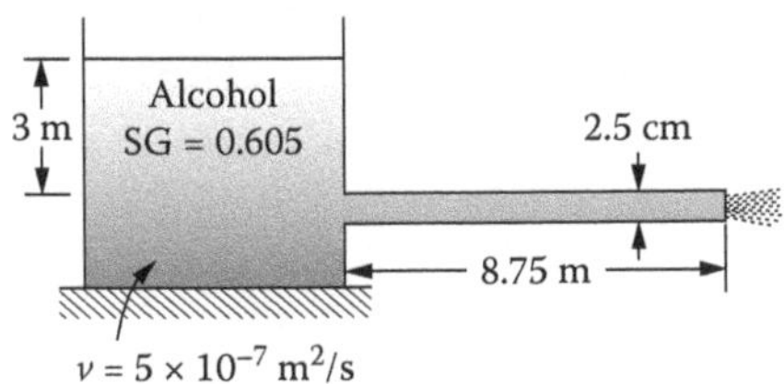

FIGURE P17.53

17.54: A level 100-m-long water pipe has a manometer at both the inlet and the outlet. The manometers indicate pressure heads of 1.5 and 0.2 m, respectively. The pipe diameter is 15 cm and the pipe roughness is 0.0004 m. Determine the mass flow rate in the pipe in kg/s.

17.55: Level wrought-iron pipe having a diameter of 4.5 cm and a length of 2 km carries water at 20°C at a velocity of 7 m/s. Determine (a) the head loss and (b) the pressure drop.

17.56: Mercury (SG = 13.6, $\mu = 1.56 \times 10^{-6}$ kg/m-s) flows through 2.75 m of 6-mm-diameter glass tubing with a velocity of 2 m/s. Determine (a) the head loss and (b) the pressure drop.

17.57: Acetic acid (SG = 0.86, $\nu = 2.60 \times 10^{-6}\ \text{m/s}^2$) flows at 0.325 m/s through cast-iron pipe having a diameter of 15 cm. The pipe is 0.626-m long and slopes upward from inlet to outlet at an angle of 12° between inlet and outlet. Determine (a) the head loss and (b) the pressure change between inlet and outlet.

17.58: Determine the depth of water behind the dam in Figure P17.58 that will provide a flow rate of $5.675 \times 10^{-4}\ \text{m}^3/\text{s}$ through an 18-m long, 1.25-cm-diameter commercial steel pipe.

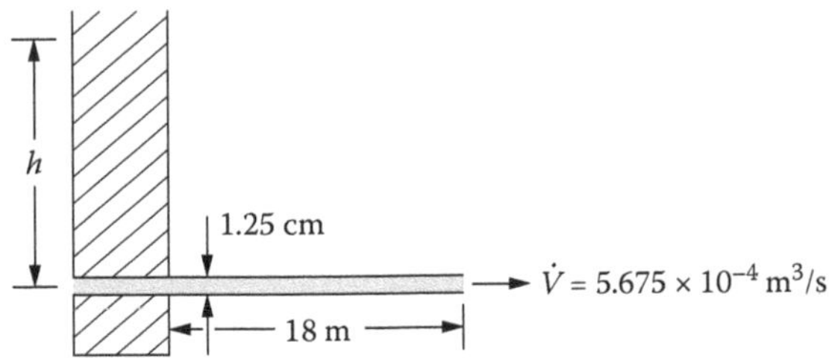

FIGURE 17P.58

17.59: Using minor loss coefficients for the elbows, globe valve and entrance of 0.90 (each), 10.0, and 0.50, respectively, determine the volumetric flow rate in the 1.5-cm pipeline in Figure P17.59 for $h = 12$ m as shown, the commercial steel pipe is 140-m long.

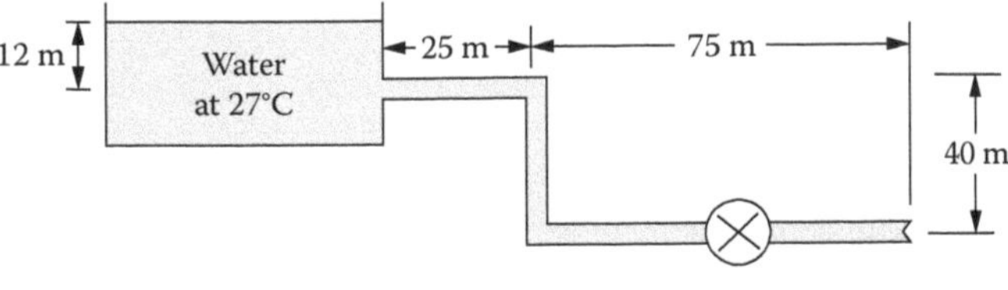

FIGURE P17.59

17.60: A pipeline consists of 75 m of 10-cm-diameter steel pipe. The pipeline includes a 90° bend, a fully open gate valve, an additional 35 m of 10-cm pipe with a 20° taper, an abrupt contraction to 7.5-cm, and an additional 20 m of 7.5-cm steel pipe. The discharge rate through this piping system is 0.0425 m^3/s. Determine the head lost.

17.61: Water flows at 20°C from reservoir 1 to reservoir 2 through 150 m of 10-cm diameter of welded-steel pipe as indicated in Figure P17.61. Determine the volumetric flow rate in the pipe.

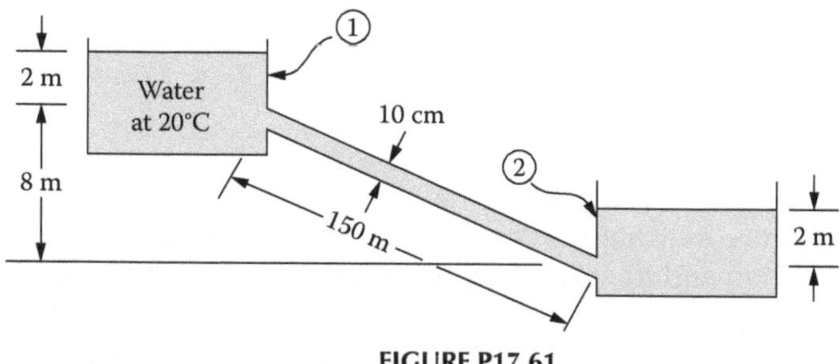

FIGURE P17.61

17.62: Figure P17.62 shows two reservoirs holding water at 40°C. They are connected by 35 m of commercial steel pipe of diameter, 10 cm, containing a sharp-edged entrance, two 90° elbows and a gate valve that is 50% closed. Determine the volumetric flow rate in the piping system between the reservoirs when the water surface elevation in the reservoirs are as shown.

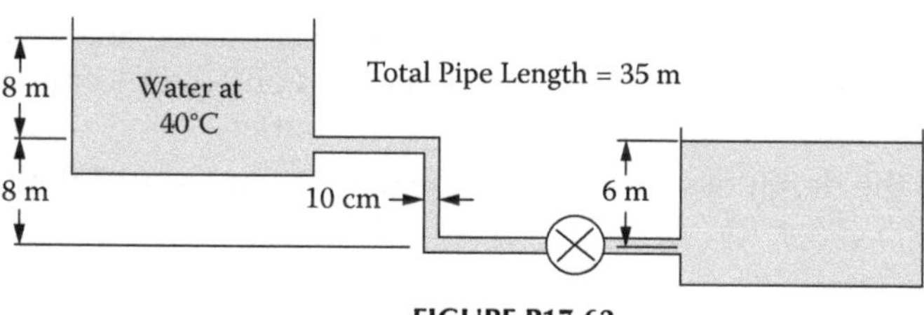

FIGURE P17.62

17.63: A piping system carries water from a reservoir and discharges it as a free jet as indicated in Figure P17.63. Determine the flow through the 20-cm-commercial steel pipe with the fittings shown.

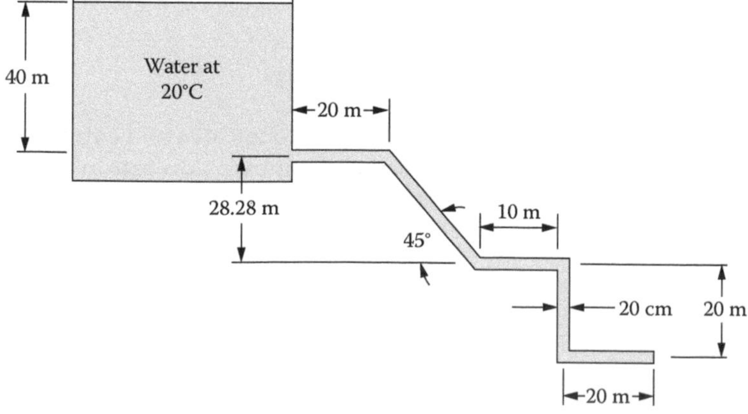

FIGURE P17.63

17.64: The piping system shown in Figure P17.64 is to be used to transport oil from tank 1 to tank 2 at a maximum volumetric flow rate of 0.0285 m^3/s. Determine the diameter of the steel pipe required.

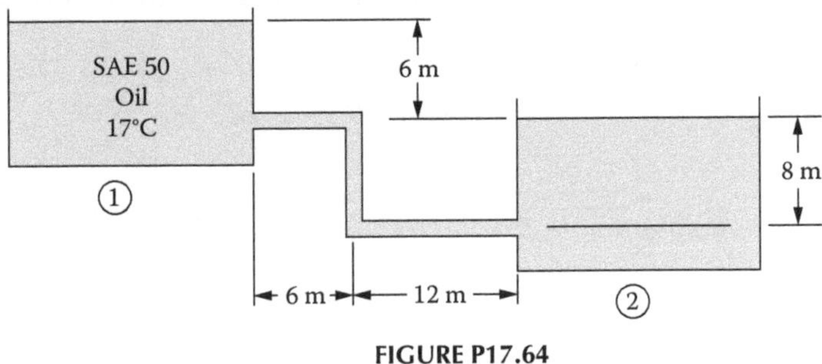

FIGURE P17.64

Multiple Pipe Systems

17.65: Water flows at a volumetric flow rate of 0.225 m^3/s and 20°C from reservoir 1 to reservoir 2 through three concrete pipes connected in a series. Pipe 1 is 1000-m long and has a diameter of 16 cm. Pipe 2 has a length of 1500 m and a diameter of 20 cm. Pipe 3 is 800-m long and the diameter is 18 cm. Neglecting minor losses, determine the difference in surface elevations.

17.66: A piping system consists of three pipes in a series. The total pressure drop is 160,000 Pa and the difference in elevation is 5 m. Data for the three pipes are

Pipe	Length, m	Diameter, cm	Roughness, mm
1	125	8	0.240
2	150	6	0.120
3	100	4	0.200

For water at 20°C, neglect minor losses and determine the volumetric flow rate in the system.

17.67: Two concrete pipes are connected in series. The flow rate of water at 20°C through the pipes is 0.15 m^3/s, with a total head loss of 15 m for both pipes. Each pipe has a length of 312.5 m and a relative roughness of 0.0035 m. Neglecting minor losses, if one pipe has a diameter of 30 cm, determine the diameter of the other.

17.68: A piping system containing a loop is shown in Figure P17.68. The total volumetric flow of water at 20°C is 0.20 m^3/s. Determine the volumetric flow rate in pipes C and D and the head lost from point 1 to point 2.

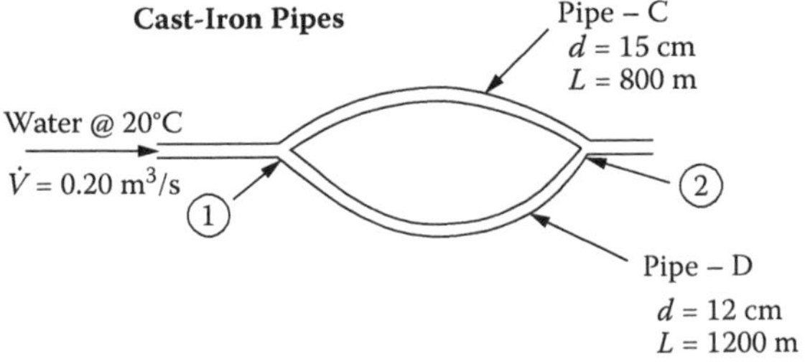

FIGURE P17.68

17.69: A commercial pipe network contains two branches in parallel as shown in Figure P17.69. A volumetric flow of $0.575\,\text{m}^3/\text{s}$ of a liquid with $\rho = 1000\,\text{kg/m}^3$ and $\nu = 0.1025 \times 10^{-4}\,\text{m/s}^2$ passes through the parallel combination. The pressure is 700 kPa at node 1. Neglecting minor losses, determine the pressure at node 2.

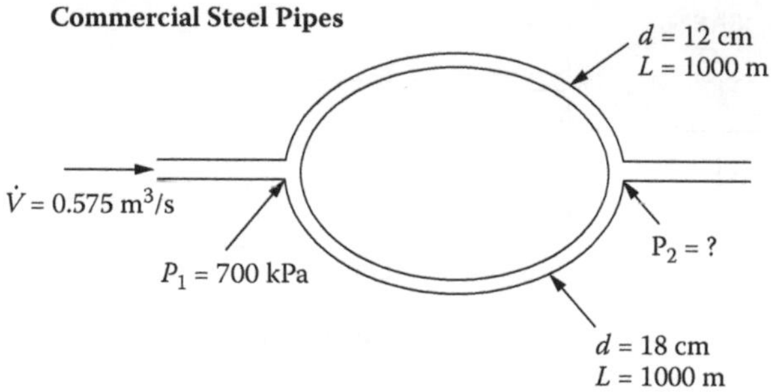

FIGURE P17.69

17.70: A piping system consists of three pipes in parallel with a total loss of head of 21 m. Data for the three pipes are as follows:

Pipe	Length, m	Diameter, cm	Roughness, mm
1	100	8	0.240
2	150	6	0.120
3	80	4	0.200

For water at 20°C, neglect minor losses and determine the volumetric flow rate in the system.

18

Fluid Machinery

Chapter Objectives

- To describe the various types of fluid machines.
- To provide the theoretical considerations for centrifugal pumps.
- To discuss the problem of cavitation and the use of the net positive suction head.
- To show how the performance of a system is matched to the performance of a centrifugal pump the system contains.
- To derive the scaling laws for pumps and fans.
- To discuss axial and mixed flow pumps and turbines.

18.1 Introduction

The term **fluid machinery** is commonly used to categorize mechanical devices that exchange fluid energy and mechanical work. When mechanical energy is applied to a fluid, producing flow or higher pressure—or both, the machine is a **pump**. When the reverse is true, and fluid energy drives the machine to produce mechanical work, we call the machine a **turbine**.

There are two principal types of fluid machines, **positive displacement machines** and **turbo machines.** In a positive displacement machine, a fluid is confined in a chamber whose volume is varied. The human heart and a bicycle tire pump are both positive displacement pumps. Examples of such devices are shown in Figure 18.1.

Turbo machines involve rotary motion as the name implies. Window fans and aircraft propellers are examples of **unshrouded** turbo machines. Pumps used with liquids generally have a **shroud** that surrounds the impeller and thus contains and guides the flow. Two categories of pumps are shown in Figure 18.2, the **radial flow** pump and the **axial flow** pump. The designations **radial** and **axial** refer to the direction of flow relative to the axis of rotation of the blades.

The term pump is generally used with flows of liquids. When a gas or vapor is the fluid of interest, the following terms apply:

- **Fans**, which are generally associated with relatively low pressure changes with ΔP in the neighborhood of 35-cm H_2O (1/2 psi).
- **Blowers**, which are in positive and variable displacement configurations with ΔP up to 2.8-m H_2O (40 psi).

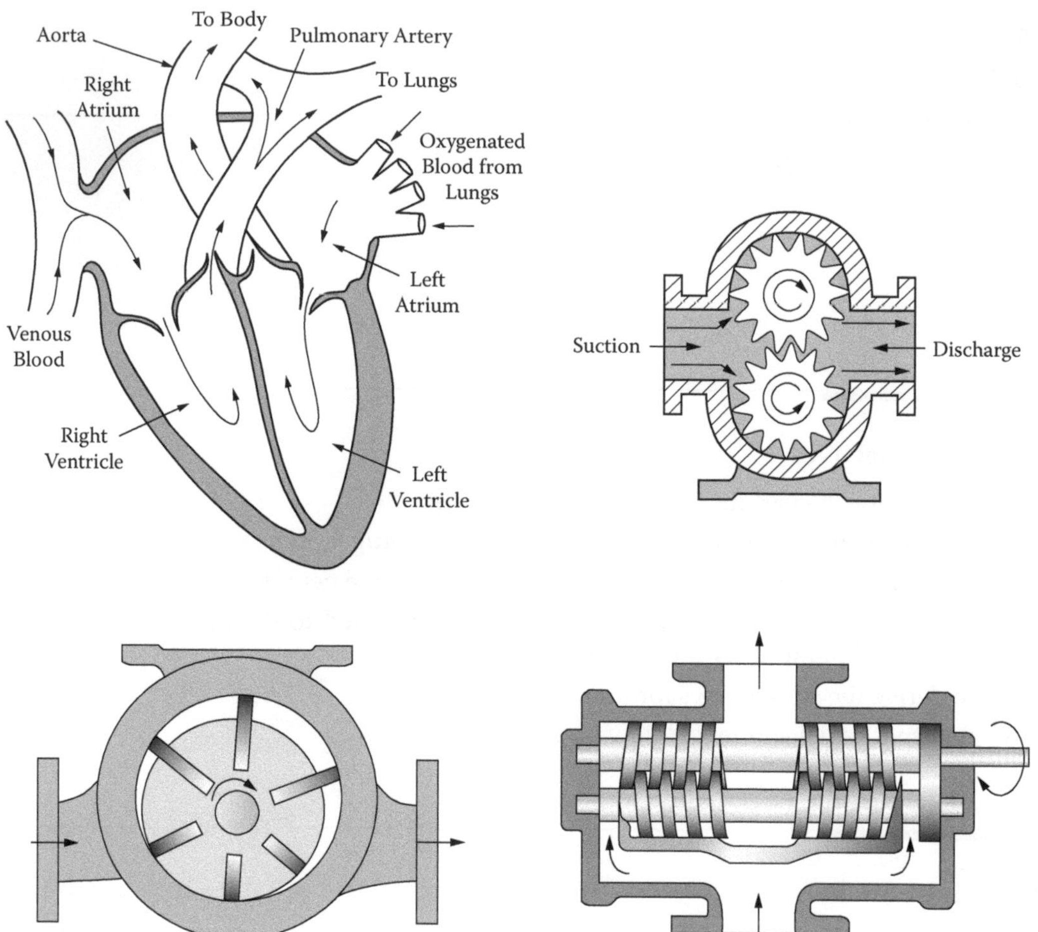

FIGURE 18.1
Examples of positive displacement pump configurations.

- **Compressors**, which are in positive and variable displacement configurations with delivery pressures as high as 69 MPa (10^4 psi).

18.2 The Centrifugal Pump

18.2.1 Introduction

We will first discuss the operation of centrifugal pumps that are commonly used in domestic and industrial applications.

A **performance curve** for a centrifugal pump is basically a plot of pressure rise (or head increase) between pump inlet and discharge as a function of flow rate. Typical performance for a centrifugal pump is shown in Figure 18.3. For comparison purposes, the performance of a positive displacement pump is also shown.

In the United States, conventional units for head are feet of water and for flow rate, gallons per minute. In keeping with our general use of SI units in this text, we will use meters of water (m H_2O) and m^3/s, respectively, in all subsequent discussion.

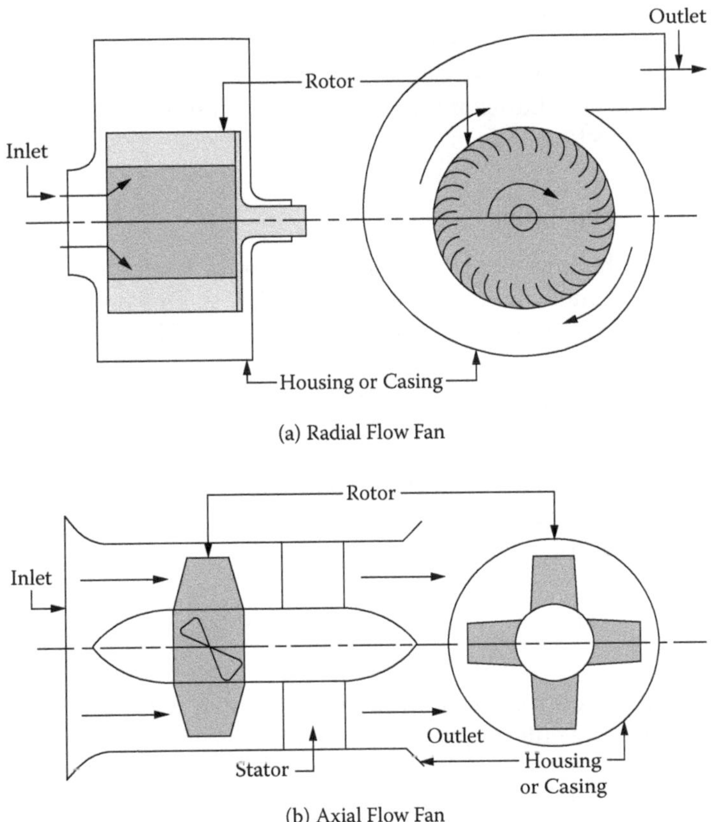

FIGURE 18.2
(a) Radial flow and (b) axial flow configurations.

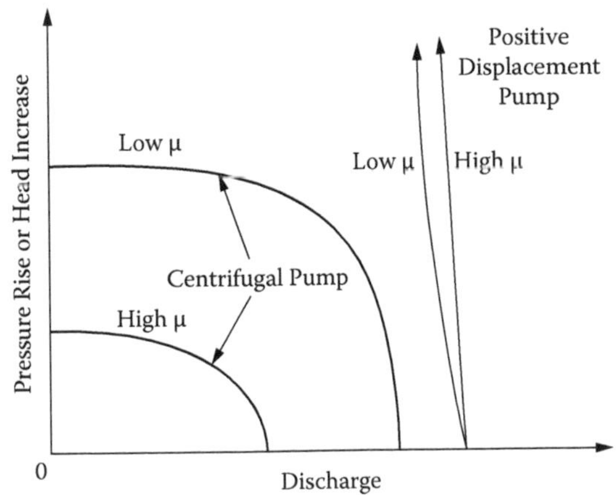

FIGURE 18.3
Performance of a typical centrifugal pump. The symbol, μ, indicates viscosity.

The shape of a performance curve, and values for pressure head and flow rate, are unique for a given pump configuration at a particular speed of rotation. Considerable art and empirical judgment are used in designing pumps. Once a pump has been fabricated and a baseline performance curve has been determined, rules exist for predicting performance at other rotational speeds and characteristic sizes and for other fluids. We will address these **pump scaling laws** in Section 18.5.

18.2.2 Theoretical Considerations

Figure 18.4 represents the impeller blade configuration for **backward curved blades**. The impeller rotates, as indicated, with an angular velocity, ω. Our analysis will involve the moment-of-momentum expression presented in Chapter 14. The governing equation is

$$\sum \boldsymbol{M} = \sum_e (\boldsymbol{r} \times \boldsymbol{V}_e)\dot{m}_e - \sum_i (\boldsymbol{r} \times \boldsymbol{V}_i)\dot{m}_i + \frac{d}{dt}(\boldsymbol{r} \times \boldsymbol{V}_{\text{cv}})m_{\text{cv}} \tag{15.5}$$

In the present analysis we will consider the coordinate origin to be at the axis of rotation so that rotation is about the z axis. For one-dimensional steady flow, the scalar form of Equation 15.5, written for the z-direction, is

$$M_z = \sum_e (\vec{r} \times \boldsymbol{V}_e)\Big|_z \dot{m}_e - \sum_i (\vec{r} \times \vec{V}_i)\Big|_z \dot{m}_i \tag{18.1}$$

Equation 18.1 will be solved for the control volume shown with the dashed lines in Figure 18.4 (left). Note that the entering flow is along the axis of rotation (the z direction) and that the exiting flow occurs along the outer circumference of the impeller. An expanded view of the velocity components entering and exiting the impeller is shown in Figure 18.5.

Equation 18.1 can now be evaluated: by definition, $\hat{V}_{\theta 1} = \hat{V}_{z1} = 0$

$$M_z = \dot{m}\left[\begin{vmatrix} \boldsymbol{r} & \boldsymbol{\theta} & z \\ r & 0 & 0 \\ \hat{V}_r & \hat{V}_\theta & \hat{V}_z \end{vmatrix}_2 - \begin{vmatrix} \boldsymbol{r} & \boldsymbol{\theta} & z \\ r & 0 & 0 \\ \hat{V}_r & \hat{V}_\theta & \hat{V}_z \end{vmatrix}_1\right]_z$$

or

$$M_z = \dot{m}[(r_2\hat{V}_{\theta 2}) - 0] = \rho\dot{V}r_2\hat{V}_{\theta 2} \tag{18.2}$$

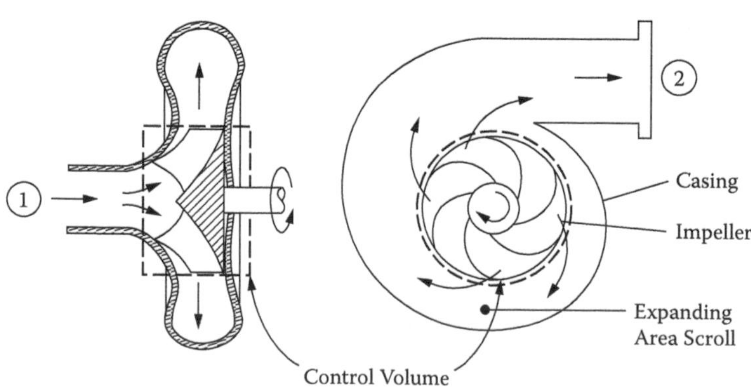

FIGURE 18.4
Flow through the impeller of a centrifugal pump.

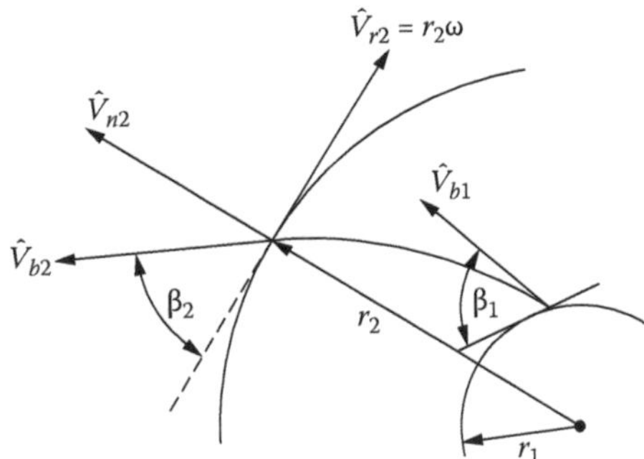

FIGURE 18.5
Velocity diagrams for conditions at the exit of a centrifugal pump impeller.

The challenge here is in the evaluation of $\hat{V}_{\theta 2}$. By definition, $\hat{V}_{\theta 2}$ is the tangential component of the exiting fluid stream *relative* to the fixed coordinate system. The velocity diagram in Figure 18.6 is useful in evaluating this quantity.

The absolute velocity of the exiting flow, $\mathbf{V}_2$, is the vector sum of the velocity relative to the impeller blade and the blade tip velocity relative to our fixed coordinate frame. In terms of the dimensions shown in Figure 18.6, we express the following:

The normal velocity of flow for blade length, L, at r_2

$$\hat{V}_{n2} = \frac{\dot{V}}{2\pi r_2 L}$$

The velocity of flow along the blade

$$\hat{V}_{b2} = \frac{\hat{V}_{n2}}{\sin \beta_2}$$

The blade tip velocity

$$\hat{V}_{t2} = r_2 \omega$$

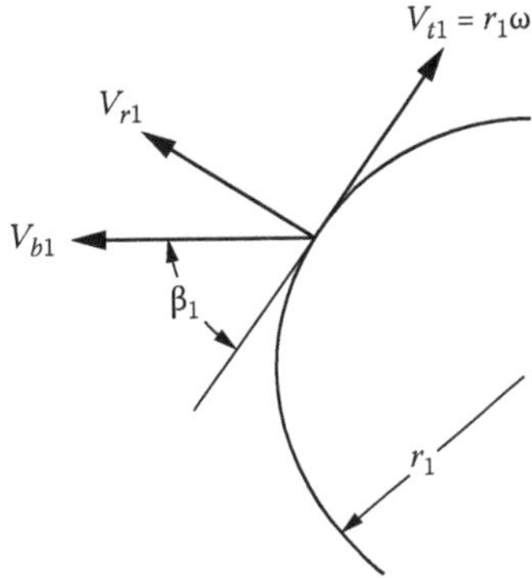

FIGURE 18.6
Velocity relationships at blade inlet.

We can now evaluate the quantity, $\hat{V}_{\theta 2}$, which is normal to r_2 as

$$\hat{V}_{\theta 2} = r_2\omega - \hat{V}_{b2}\cos\beta_2 = r_2\omega - \frac{\hat{V}_{n2}}{\sin\beta_2}\cos\beta_2$$

or

$$\hat{V}_{\theta 2} = r_2\omega - \frac{\dot{V}}{2\pi r_2 L}\cot\beta_2 \tag{18.3}$$

The moment, M_z, is thus,

$$M_z = \rho\dot{V}r_2\left[r_2\omega - \frac{\dot{V}}{2\pi r_2 L}\cot\beta_2\right] \tag{18.4}$$

and the power delivered to the fluid is by definition, $M_z\omega$ or

$$\dot{W} = \rho\dot{V}r_2\omega\left[r_2\omega - \frac{\dot{V}}{2\pi r_2 L}\cot\beta_2\right] \tag{18.5}$$

We will keep this result in mind as we next examine our pump from an energy point of view.

The first law of thermodynamics applied to the same control volume as before will yield the following. Beginning with Equation 13.11

$$\frac{d}{dt}E_{\text{cv}} = \dot{Q}_{\text{cv}} - \dot{W}_{\text{cv}}$$

$$+\sum_i \dot{m}_i\left(h_i + \frac{\hat{V}_i^2}{2} + gz_i\right) - \sum_e \dot{m}_e\left(h_e + \frac{\hat{V}_e^2}{2} + gz_e\right) \tag{13.11}$$

we have, for steady flow without heat transfer,

$$-\dot{W}_{\text{cv}} = \dot{m}\left[h_2 - h_1 + \frac{\hat{V}_2^2 - \hat{V}_1^2}{2} + g(z_2 - z_1)\right]$$

It is customary to neglect the relatively small differences in velocity and elevation between points 1 and 2, that is

$$\hat{V}_2^2 - \hat{V}_1^2 \approx 0 \qquad \text{and} \qquad z_2 - z_1 \approx 0$$

The expression that remains is

$$-\dot{W}_{\text{cv}} = \dot{m}\left(u_2 - u_1 + \frac{P_2 - P_1}{\rho}\right) \tag{18.6}$$

We recall that the term $u_2 - u_1$ represents the loss in head due to friction and other irreversible effects. Thus, Equation 18.6 provides an expression for the net pressure head produced in the pump

$$\frac{P_2 - P_1}{\rho} = -\frac{\dot{W}_{\text{cv}}}{\dot{m}} - h_L \tag{18.7}$$

Equations 18.5 and 18.7 can be combined to express the net pressure head as a function blade angle, flow rate, angular velocity, and impeller radius. Head losses must be determined experimentally.

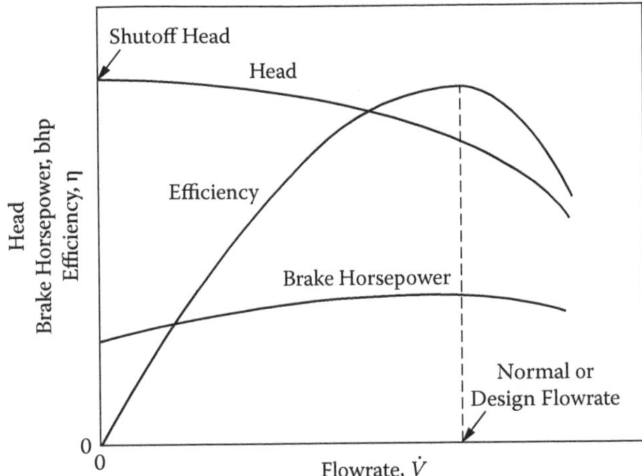

FIGURE 18.7
Performance curves for a centrifugal pump impeller.

The foregoing results can now be used to evaluate another important performance parameter, the **efficiency**. In broad terms, efficiency is expressed as the ratio of actual output to required input. For the centrifugal pump, the efficiency, designated as η, will be

$$\eta = \frac{\text{power added to the fluid}}{\text{shaft power to the impeller}}$$

The power required by the fluid has been expressed as

$$\dot{W}_{\text{fluid}} = \dot{m}\left(\frac{P_2 - P_1}{\rho}\right) \tag{18.8}$$

so that the efficiency can be expressed as

$$\eta = \frac{\dot{m}(P_2 - P_1)}{\rho \dot{W}_{\text{cv}}} \tag{18.9}$$

The performance curves presented in Figure 18.4 can now be enhanced to include the efficiency and the input shaft power that is generally referred to as brake power. Figure 18.7 includes these quantities in addition to the developed head.

For minimal friction loss at the radial location, r_1, where the inlet flow enters the impeller, it is desirable that flow relative to the blade be along the surface, oriented at the angle β_1. An expanded view of this section is shown in Figure 18.8. The design point for minimum losses is achieved when

$$\hat{V}_{b1} \cos \beta_1 = r_1 \omega$$

or, equivalently, when

$$\hat{V}_{r1} = \hat{V}_{b1} \sin \beta_1 = r_1 \omega \frac{\sin \beta_1}{\cos \beta_1}$$

and, finally, when

$$\hat{V}_{r1} = r_1 \omega \tan \beta_1 \tag{18.10}$$

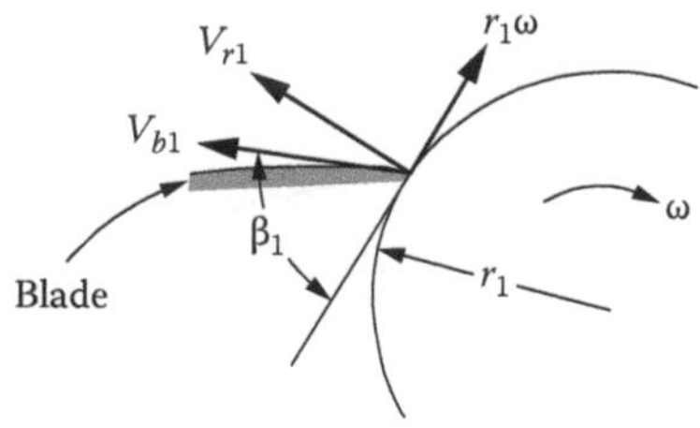

FIGURE 18.8
Expanded velocity diagram of conditions in Figure 18.6.

The following example demonstrates how the foregoing analysis relates to centrifugal pump performance.

Example 18.1 A centrifugal water pump with $r_2 = 9\,\text{cm}$, $r_1 = 5.125\,\text{cm}$, and blade length, $L = 4.5\,\text{cm}$, is designed with backward curved blades having $\beta_1 = 30°$ and $\beta_2 = 20°$. For a rotational speed of 1400 rpm, and given $\rho = 1000\,\text{kg/m}^3$ for water, estimate (a) the design-point volume flow rate, (b) the power gained by the water, and (c) the discharge pressure head.

Solution

Assumptions and Specifications

(1) Steady flow exists.
(2) The flow is incompressible.
(3) There is uniform flow at the inlet and outlet sections.
(4) Torques due to body and surface torques may be neglected.

(a) To establish the design-point volume flow rate, we note that for operation at the design condition, Equation 18.10 must be satisfied. Thus, we have

$$\omega = (1400\,\text{rev/min})(1\,\text{m}/60\ \text{s})(2\pi\,\text{rad/rev}) = 146.6\,\text{rad/s}$$

$$\begin{aligned}\hat{V}_{r1} &= r_1\omega\tan\beta_1\\ &= (0.05125\,\text{m})(146.6\,\text{rad/s})\tan 30°\\ &= 4.338\,\text{m/s}\end{aligned}$$

The corresponding flow rate is

$$\begin{aligned}\dot{V} &= 2\pi r_1 L\hat{V}_{r1}\\ &= 2\pi(0.05125\,\text{m})(0.045\,\text{m})(4.338\,\text{m/s})\\ &= 0.0629\,\text{m}^3/\text{s} \qquad (996\,\text{gpm}) \Longleftarrow\end{aligned}$$

(b) The power added to the fluid can be determined using Equation 18.5

$$\dot{W} = \rho\dot{V}r_2\omega\left[r_2\omega - \frac{\dot{V}}{2\pi r_2 L}\cot\beta_2\right] \tag{18.5}$$

With

$$\rho\dot{V}\omega r_2 = (1000\,\text{kg/m}^3)(0.0629\,\text{m}^3/\text{s})(146.6\,\text{rad/s})(0.09\,\text{m})$$
$$= 830.29\,\text{kg-m/s}^2$$

and

$$\frac{\dot{V}}{2\pi r_2 L} = \frac{0.0629\,\text{m}^3/\text{s}}{2\pi(0.09\,\text{m})(0.045\,\text{m})} = 2.47\,\text{m/s}$$

we obtain

$$\dot{W} = (830.29\,\text{kg-m/s}^2)[(0.09\,\text{m})(146.7\,\text{rad/s}) - (2.47\,\text{m/s})\cot 20°]$$
$$= 5314\,\text{W} \Longleftarrow$$

This is equivalent to 7.13 hp

(c) The ideal discharge pressure (when losses are neglected) can be determined using Equation 18.7. The discharge pressure head is expressed as

$$\frac{P_2 - P_1}{\rho g} \simeq \frac{\dot{W}_{cv}}{\dot{m} g}$$
$$\simeq \frac{(5314\,\text{W})(1\,\text{kg-m}^2/\text{W-s}^3)}{(1000\,\text{kg/m}^3)(0.0629\,\text{m}^3/\text{s})(9.81\,\text{m/s}^2)}$$
$$\simeq 8.61\,\text{mH}_2\text{O} \Longleftarrow$$

This result is the maximum that can be achieved. The actual value will be less than this due to friction and other irreversible losses.

18.3 The Net Positive Suction Head

A major cause of degradation of pump performance is the presence of **cavitation**. Cavitation occurs when the liquid being pumped vaporizes or boils whereupon the vapor bubbles that form cause a decrease in efficiency and structural damage to the pump, which sometimes results in catastrophic failure. The parameter that characterizes the potential for cavitation to occur is the **net positive suction head (NPSH)**.

At the impeller, the total head on the suction side, which is the low pressure location at which cavitation will first occur, is related to the NPSH according to

$$\text{NPSH} + \frac{P_v}{\rho g} = \frac{\hat{V}_i^2}{2g} + \frac{P_i}{\rho g} \tag{18.11}$$

where $\hat{V}_i$ and P_i are the velocity and pressure at the pump inlet and P_v is the liquid vapor pressure. Values of NPSH are generally determined experimentally for a given pump over a range of flow rates. Typically, the variation of NPSH appears as shown in Figure 18.9.

In the representative installation indicated in Figure 18.10, the liquid being pumped is drawn from a reservoir whose level is the distance, z, below the pump inlet. An energy

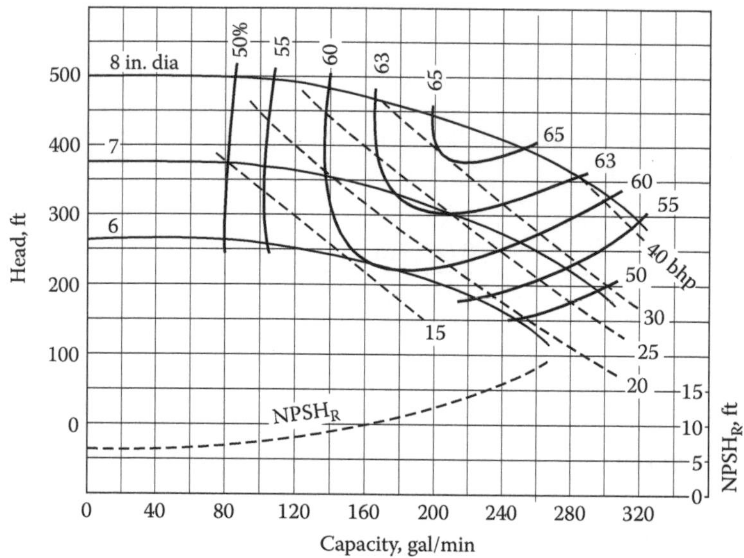

FIGURE 18.9
Pump performance including NPSH behavior.

balance between the reservoir surface and the pump inlet yields

$$\frac{P_{atm}}{\rho g} = z_2 + \frac{P_2}{\rho g} + \frac{\hat{V}_2^2}{2g} + \sum h_L \tag{18.12}$$

where $\sum h_L$ represents the head losses between points 1 and 2.

Equation 18.12 can be rewritten as

$$\frac{P_2}{\rho g} + \frac{\hat{V}_2^2}{2g} = \frac{P_{atm}}{\rho g} - z_2 - \sum h_L$$

and, with the incorporation of Equation 18.11 for the NPSH, we have

$$\text{NPSH} = \frac{P_{atm}}{\rho g} - z_2 - \frac{P_v}{\rho g} - \sum h_L \tag{18.13}$$

From these results, and a performance plot such as Figure 18.9, a pump system designer can establish operating conditions to avoid the possibility of cavitation. The value of NPSH

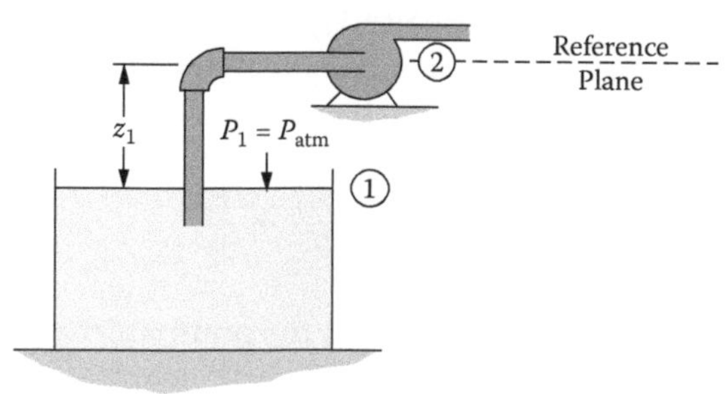

FIGURE 18.10
Pump installation drawing water from a reservoir.

determined from Equation 18.13 should be greater than the value given on a performance plot at the same flow rate.

Example 18.2 A centrifugal pump is needed to pump water from an open tank with a system configuration like that shown in Figure 18.10. The desired flow rate of water through a 10-cm-diameter pipe is $0.02\,\text{m}^3/\text{s}$, the manufacturer's specifications show an NPSH value of 4.5 m at this flow rate. Determine the maximum height, z, above the water surface at which the pump should be placed to avoid cavitation.

For this problem, the minor loss coefficient has a value of $K_L = 12$ and all other losses may be neglected. Properties of water may be valuated at 27°C (300 K).

Solution

Assumptions and Specifications

(1) Steady flow exists.
(2) The flow is incompressible.
(3) There is uniform flow at the inlet and outlet sections.
(4) Torques due to body and surface torques may be neglected.
(5) Atmospheric pressure is taken as 101,360 Pa.

From Equation 18.13, the available NPSH is

$$\text{NPSH} = \frac{P_{\text{atm}}}{\rho g} - z_2 - \sum h_L - \frac{P_v}{\rho g}$$

which must be solved for z_2

$$z_2 \leq \frac{P_{\text{atm}} - P_v}{\rho g} - \sum h_L - \text{NPSH}$$

At 27°C (300 K), water properties of interest include

$$P_v = 3495\,\text{Pa} \qquad \text{and} \qquad \rho = 997\,\text{kg/m}^3$$

We are now able to evaluate $\hat{V}$ and $\sum h_L$

$$\hat{V} = \frac{\dot{V}}{A} = \frac{4(0.02\,\text{m}^3/\text{s})}{\pi(0.10\,\text{m})^2} = 2.55\,\text{m/s}$$

$$\sum h_L = K_L \frac{\hat{V}^2}{2g} = \frac{12(2.55\,\text{m/s})^2}{2(9.81\,\text{m/s}^2)} = 3.98\,\text{m}$$

so that the value of z_2 can be determined as

$$z_2 \leq \frac{101{,}360\,\text{Pa} - 3495\,\text{Pa}}{(997\,\text{kg/m}^3)(9.81\,\text{m/s}^2)} - 3.98\,\text{m} - 4.5\,\text{m} \simeq 1.526\,\text{m} \Longleftarrow$$

The pump should be located no more than 1.526 m above the water level in the tank.

18.4 Combining Pump and System Performance

In the preceeding discussion, we combined system characteristics with the realistic concern of avoiding cavitation, which will degrade pump performance and, eventually, lead to failure.

We now examine the interaction between pump performance as described by operating curves of head, efficiency, and NPSH as a function of volumetric flow rate and the characteristics of the system in which the pump produces flow.

Figure 18.11 illustrates a simple system configuration for a pump that produces flow between two reservoirs that are at different elevations. An energy balance between the surface of the lower reservoir, point 1, and the upper reservoir surface, point 2, gives

$$-\frac{\dot{W}}{\dot{m}} = g(z_2 - z_1) + \frac{P_2 - P_1}{\rho} + (u_2 - u_1) \tag{18.14}$$

Noting that $P_1 = P_2 = P_{atm}$ and writing $u_2 - u_1 = g\sum h_L$, we arrive at the head loss resulting from pipe friction and minor losses:

$$-\dot{W} = \dot{m}g\left[z_2 - z_1 + \sum h_L\right]$$

or

$$-\frac{\dot{W}}{\dot{m}g} = z_2 - z_1 + \sum h_L \tag{18.15}$$

We noted in Chapter 17 that h_L is proportional to $\hat{V}^2$. Thus,

$$h_L = K\hat{V}^2$$

and Equation 18.15 can be written as

$$-\frac{\dot{W}}{\dot{m}g} = z_2 - z_1 + K\hat{V}^2 \tag{18.16}$$

where K includes contributions from pipe friction and other minor losses due to items such as valves, elbows, and fittings.

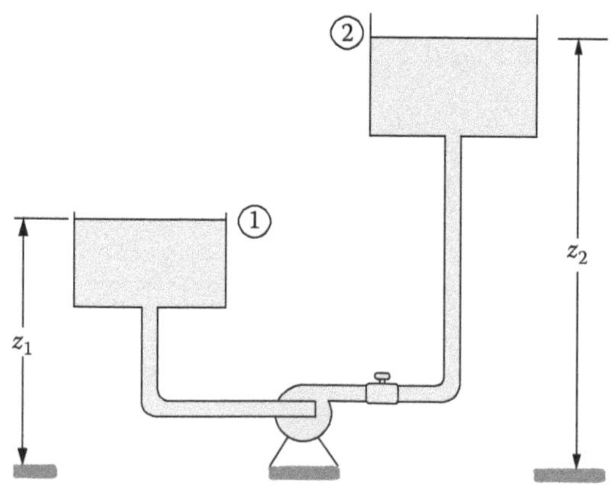

FIGURE 18.11
Pump operating system.

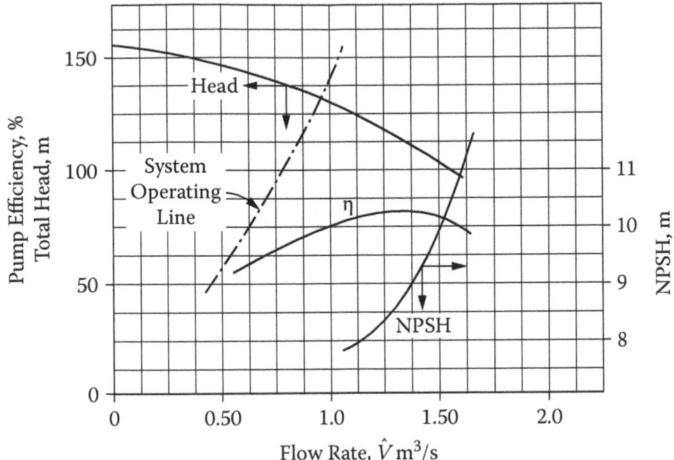

FIGURE 18.12
System operation and pump performance.

When the system operating line is plotted on the characteristic performance plot of the pump, the combined plot appears as shown in Figure 18.12. We note that the operating lines for the system and for the pump will intersect at a flow rate where the pumping head is equal to the required head needed for that flow rate to be achieved within the system. This **operating point** is associated with a corresponding pump efficiency as indicated in the figure. Naturally, the designer wants a given system operation to occur near the flow rate at which the pump efficiency is at a maximum. If such a match is not achieved, changes must be made either to the system (which is not likely to be feasible) or in the pump operating conditions.

The **scaling rules** for pump performance will be considered in the next section. First, we work the following example where pump performance and system characteristics define a specific operating condition.

Example 18.3 The performance of a centrifugal pump is presented in Figure 8.12 for operation at 1200 rpm. Water at 27°C (300 K) is to be pumped in the system shown in Figure 18.11. The two reservoirs differ in elevation by 40 m and the total length of 40-cm-diameter pipe is 480 m with a friction factor of $f_f = 0.0075$. Determine the rate of flow for this system in m^3/s and evaluate the corresponding pump efficiency.

Solution

Assumptions and Specifications

(1) Steady flow exists.
(2) The flow is incompressible.
(3) There is uniform flow at the inlet and outlet sections.
(4) Torques due to body and surface torques may be neglected.

The energy equation for this system is

$$-\frac{\dot{W}_S}{\dot{m}g} = z_2 - z_1 + 2f_f\frac{L}{d}\frac{\hat{V}^2}{g}$$

and when we insert the given numerical values, we obtain

$$-\frac{\dot{W}_S}{\dot{m}g} = 40\,\text{m} + 2(0.0075)\left(\frac{480\,\text{m}}{0.4\,\text{m}}\right)\left(\frac{\hat{V}^2}{9.81\,\text{m/s}^2}\right)$$

or

$$-\frac{\dot{W}_S}{\dot{m}g} = 40\,\text{m} + 1.835\hat{V}^2\,\text{m}$$

For use with Figure 18.12, this expression must be in terms of $\dot{V}$ and we note that the relationship between $\hat{V}$ and $\dot{V}$ is

$$\dot{V} = \hat{V}\left(\frac{\pi d^2}{4}\right) = \hat{V}\left[\frac{\pi(0.4\,\text{m})^2}{4}\right] = 0.1257\hat{V} \quad (\text{m}^3/\text{s})$$

Substitution into the system energy balance expression yields

$$-\frac{\dot{W}_S}{\dot{m}g} = 40\,\text{m} + 1.835\left(\frac{\dot{V}}{0.1257}\right)^2\,\text{m}$$

or

$$-\frac{\dot{W}_S}{\dot{m}g} = 40\,\text{m} + 116\dot{V}^2\,\text{m}$$

The operating line corresponding to this expression is shown in Figure 18.12. The intersection of the pump and system operating lines is at a flow rate of

$$\dot{V} = 0.90\,\text{m}^3\text{s} \Longleftarrow$$

where the efficiency is approximately

$$\eta = 0.74 \Longleftarrow$$

18.5 Scaling Laws for Pumps and Fans

Concepts of similarity were introduced and demonstrated in Chapter 15. As discussed at that point, the requirements of geometric, kinematic, and dynamic similarity provide the basis for scaling, which is an extremely important process in predicting the performance of fluid machinery from laboratory- or bench-scale test results. The reader is referred to Chapter 15 for a review of the different types of similarity and how these concepts lead to predicting prototype performance.

A dimensional analysis was performed for a centrifugal pump in Section 15.4. The relevant operating variables that were listed in Table 15.2 are repeated here for reference as Table 18.1.

Then, by employing the formalism of the Buckingham method, the following dimensionless parameters were obtained

$$\pi_1 = \frac{gh}{n^2d^2} \qquad \pi_3 = \frac{\dot{V}}{nd^3}$$

$$\pi_2 = \frac{P}{\rho n^3 d^5} \qquad \pi_4 = \frac{\mu}{d^2 n \rho}$$

It is easily demonstrated that these parameters are all dimensionless.

TABLE 18.1

Pump Performance Variables: Power Is Designated as P Rather Than $\dot{W}$

Variable	Symbol	Dimensions
Total head	gh	L^2/T^2
Power	P	ML^2/T^3
Flow rate	$\dot{V}$	L^3/T
Impeller diameter	d	L
Shaft speed	n	$1/T$
Fluid density	ρ	M/L^3
Fluid viscosity	μ	M/LT

The resulting parameters used in scaling pumps and fans are

$$C_H = \text{the head coefficient, } \frac{gh}{n^2d^2} \tag{18.17}$$

$$C_Q = \text{the capacity coefficient, } \frac{\dot{V}}{nd^3} \tag{18.18}$$

and

$$C_P = \text{the power coefficient, } \frac{P}{\rho n^3 d^5} \tag{18.19}$$

The relationships between these parameters provide the **scaling laws** we seek.

For geometrically and kinematically similar pump families, dynamic similarity requires that

$$C_H = f_1(C_Q) \tag{18.20}$$

and

$$C_P = f_2(C_Q) \tag{18.21}$$

We already know the other performance parameter as the efficiency, η, which, by definition, is dimensionless. Using the coefficients just defined, we can write

$$\eta = \frac{C_H C_Q}{C_P} = f_3(C_Q) \tag{18.22}$$

and η is also seen to be a function of C_Q.

The relationships expressed in Equations 18.20 through 18.22 indicate that pump performance curves as introduced earlier can also be plotted using dimensionless quantities.

Last, our three coefficients, C_H, C_Q, and C_P, will provide the basis for scaling. Similarity requirements dictate that for two pumps that possess geometric, kinematic, and dynamic similarity, designated as pumps 1 and 2, the following relationships apply:

$$C_{H1} = C_{H2}$$

or

$$\frac{gh_1}{n_1^2 d_1^2} = \frac{gh_2}{n_2^2 d_2^2}$$

so that

$$\frac{h_2}{h_1} = \left(\frac{n_2}{n_1}\right)^2 \left(\frac{d_2}{d_1}\right)^2 \tag{18.23}$$

Performing the same operations on C_Q and C_P, we obtain

$$\frac{\dot{V}_2}{\dot{V}_1} = \frac{n_2}{n_1}\left(\frac{d_2}{d_1}\right)^3 \tag{18.24}$$

and

$$\frac{P_2}{P_1} = \frac{\rho_2}{\rho_1}\left(\frac{n_2}{n_1}\right)^3 \left(\frac{d_2}{d_1}\right)^5 \tag{18.25}$$

These three expressions comprise the **pump laws** or **fan laws**, which are universally used in scaling rotating machines and their performance.

The following example demonstrates the utility of these scaling relationships.

Example 18.4 The pump described in Example 18.3 is operated at 1200 rpm against a head of 132 m H_2O. For a geometrically similar pump with a 20% larger impeller diameter, operating at the same rotational speed, what flow rate can be expected?

In addition to the increase in the size of the larger pump, its speed is increased to 1400 rpm, determine the new values of flow rate, total head, and power required.

Solution

Assumptions and Specifications

(1) Steady flow exists.
(2) The flow is incompressible.
(3) There is uniform flow at the inlet and outlet sections.
(4) Torques due to body and surface torques may be neglected.

Designating pump 1 as the one specified in Example 18.3 and pump 2 with $d_2 = 1.2d_1$, we have for $n_1 = n_2$,

$$\begin{aligned}\dot{V}_2 &= \dot{V}_1 \frac{n_2}{n_1}\left(\frac{d_2}{d_1}\right)^3 \\ &= (0.90\,\text{m}^3/\text{s})(1)(1.2)^3 \\ &= 1.555\,\text{m}^3/\text{s} \Longleftarrow\end{aligned}$$

Then, for $n_2 = 1400$ rpm and $d_2 = 1.2d_1$, another application of Equation 18.24 provides

$$\begin{aligned}\dot{V}_2 &= \dot{V}_1 \frac{n_2}{n_1}\left(\frac{d_2}{d_1}\right)^3 \\ &= (0.90\,\text{m}^3/\text{s})\left(\frac{1400\,\text{rpm}}{1200\,\text{rpm}}\right)(1.2)^3 \\ &= (0.90\,\text{m}^3/\text{s})(1.167)(1.2)^3 \\ &= 1.814\,\text{m}^3/\text{s} \Longleftarrow\end{aligned}$$

The total head is determined from Equation 18.23. With $h_1 = 132\,\text{mH}_2\text{O}$,

$$\begin{aligned}h_2 &= h_1\left(\frac{n_2}{n_1}\right)^2\left(\frac{d_2}{d_1}\right)^2 \\ &= (132\,\text{mH}_2\text{O})\left(\frac{1400\,\text{rpm}}{1200\,\text{rpm}}\right)^2(1.2)^2 \\ &= (132\,\text{mH}_2\text{O})(1.167)^2(1.2)^2 \\ &= 258.7\,\text{mH}_2\text{O} \Longleftarrow\end{aligned}$$

The power ratio is given by Equation 18.25. With $\rho_2 = \rho_1$

$$\begin{aligned}\frac{P_2}{P_1} &= \frac{\rho_2}{\rho_1}\left(\frac{n_2}{n_1}\right)^3\left(\frac{d_2}{d_1}\right)^5 \\ &= (1)\left(\frac{1400\,\text{rpm}}{1200\,\text{rpm}}\right)^3(1.2)^5 \\ &= 3.95\end{aligned}$$

The solution for P_2 requires that we calculate P_1. The operating conditions for the original pump included

$$\begin{aligned}\dot{V}_1 &= 0.90\,\text{m}^3/\text{s} \\ h_1 &= 132\,\text{mH}_2\text{O} \\ \eta &= 0.74\end{aligned}$$

Neglecting minor losses for the time being, the power to the fluid is

$$-\dot{W} = P_{\text{H}_2\text{O}} = mgh_1 = \rho\dot{V}gh_1$$

and with $\rho = 997\,\text{kg/m}^3$

$$P_{\text{H}_2\text{O}} = (997\,\text{kg/m}^3)(0.90\,\text{m}^3/\text{s})(9.81\,\text{m/s}^2)(132\,\text{m}) = 1162\,\text{kW}$$

Then

$$P_1 = \frac{P_{\text{H}_2\text{O}}}{\eta} = \frac{1162\,\text{kW}}{0.74} = 1570\,\text{kW}$$

and our final result for P_2 is

$$P_2 = 3.95P_1 = 3.95(1570\,\text{kW}) = 6201\,\text{kW} \Longleftarrow$$

18.6 Axial and Mixed Flow Pumps

Our examination of pumps has, thus far, considered only centrifugal pumps. It was mentioned earlier that axial flow pumps are encountered in numerous applications.

The distinction between axial and centrifugal flow is in the direction of fluid flow through the pump. In axial flow, the fluid moves parallel to the axis of rotation and in the centrifugal case, the flow is turned 90° normal to the rotational axis. It seems reasonable that there is an intermediate case in which the flow has both axial and normal components. This intermediate configuration is called **mixed flow**.

The choice of a centrifugal, mixed or axial flow pump is determined by choosing an appropriate combination of head and volumetric flow rate that will achieve the peak efficiency for a given pump configuration.

A useful parameter, in this regard, is obtained by eliminating the diameter between Q_Q and C_H. This operation yields the **specific speed**, N_S, defined as

$$N_S = \frac{C_Q^{1/2}}{C_H^{3/4}} \tag{18.26}$$

Values of N_S, at peak efficiency, for pump configurations ranging from centrifugal to axial are shown in the Figure 18.13 (lower axis). Also shown in the figure (upper axis) are values of N_S expressed in what we call **U. S. Customary Units**. This parameter is dimensional and has numerical values determined using English-system dimensions as

$$N_{sd} = \frac{(\omega, \text{rpm})[\dot{V}, \text{gpm}]^{1/2}}{[h, \text{ft}]^{3/4}} \tag{18.27}$$

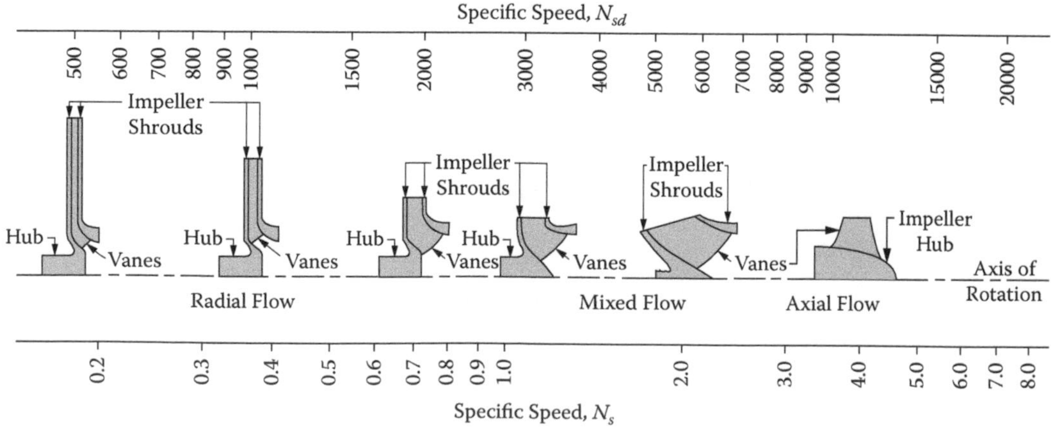

FIGURE 18.13
Specific speed variation for various impeller configurations. (Adapted from *Hydraulic Institute Standards*, 14th ed., Hydraulic Institute, Cleveland, OH, 1983.)

The basic message of this discussion, which is quantified in Figure 8.13, is that relatively large flow rates require the use of mixed flow and axial flow configurations.

18.7 Turbines

As stated at the beginning of this chapter, turbo machines either *require* power to produce flow and/or higher pressure or they *produce* power from a high-energy fluid. Machines that generate power are called **turbines.**

Turbine operation involves a fluid that emanates from a nozzle, which interacts with blades attached to the periphery of a rotating unit, the **rotor**. The change in direction of the fluid flow, caused by the blades, results in a momentum exchange that produces power at the rotor shaft. The analysis of this momentum exchange process was examined in Chapter 14.

Turbines are classified into two categories. In **impulse turbines,** there is no pressure drop across the rotor. In a **reaction turbine**, there is both a pressure drop and a change in the velocity of fluid flow through the rotor. Impulse turbines characteristically involve high head and low-flow rates and reaction turbines are low head, high-flow rate devices.

A detailed discussion of turbines is beyond the scope of this book. The interested reader may consult any of a large number of treatises on this subject, some of which are cited in Fox and McDonald (1999), Munson et al. (1998), Potter and Wiggert (1997), Roberson and Grove (1997), and White (1999).

18.8 Summary

Turbo machines have been examined in this chapter. Included in our discussion and analysis have been those machines that require power to produce higher pressure, higher flow, or both. These machines are designated as pumps or fans. A machine that does the reverse, that is, produces power from a fluid, is designated a turbine.

Pumps and fans are described in terms of the flow direction through the rotor. In centrifugal pumps (or fans), flow is turned 90° and in axial pumps (or fans), flow is along the rotational axis. A machine in which flow has both axial and radial components is designated as a mixed flow device.

Each pump design is unique and is described by performance curves that are different in shape and operating values from other configurations.

Operating variables of interest are pump head, flow rate, speed, characteristic size, power, and efficiency. A standard performance plot will show head, power, and efficiency as functions of the flow rate for a pump of characteristic impeller diameter at a particular speed of rotation.

Scaling laws were obtained using dimensional analysis. For a family of geometrically similar pumps which also exhibit kinematic similarity, the scaling laws relate operating variables in the following ways:

$$\frac{h_2}{h_1} = \left(\frac{n_2}{n_1}\right)^2 \left(\frac{d_2}{d_1}\right)^2 \tag{18.23}$$

$$\frac{\dot{V}_2}{\dot{V}_1} = \frac{n_2}{n_1} \left(\frac{d_2}{d_1}\right)^3 \tag{18.24}$$

and

$$\frac{P_2}{P_1} = \frac{\rho_2}{\rho_1}\left(\frac{n_2}{n_1}\right)^3\left(\frac{d_2}{d_1}\right)^5 \tag{18.25}$$

18.9 Problems

Centrifugal Pumps

18.1: A centrifugal pump delivers 0.2 m^3/s of water when operating at 850 rpm. Pertinent impeller dimensions are as follows: outer diameter = 0.5 m, blade length = 50 cm, and blade exit angle = 25°. Determine (a) the torque and power required to drive the pump and (b) the maximum pressure increase across the pump.

18.2: A centrifugal pump is used with gasoline (ρ = 680 kg/m^3). Pertinent dimensions are as follows: d_1 = 15 cm, d_2 = 28 cm, L = 9 cm, β_1 = 25°, and β_2 = 40°. The gasoline enters the pump parallel to the pump shaft when the pump operates at 1160 rpm. Determine (a) the flow rate in m^3/s, (b) the power delivered to the gasoline, and (c) the head in meters.

18.3: A centrifugal pump has the following dimensions: d_2 = 45 cm, L = 5 cm, and β_2 = 35°. It rotates at 1100 rpm. The head generated is 50 m of water. Assuming radial entry flow, determine the theoretical values for (a) the flow rate and (b) the power.

18.4: A centrifugal pump has the configuration and dimensions shown in Figure P18.4. For water flowing at a rate of 0.0071 m^3/s and an impeller speed of 980 rpm, determine the power required to drive the pump. The inlet flow is directed radially outward and the exiting velocity may be assumed to be tangent to the vane at its trailing edge.

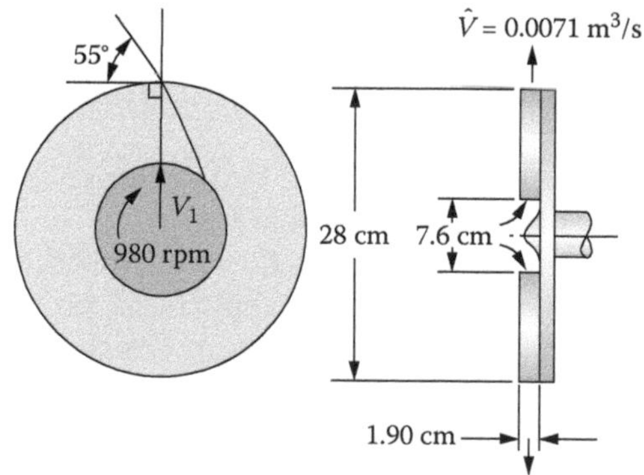

FIGURE P18.4

18.5: A centrifugal pump is being used to pump water at a flow rate of 0.016 m^3/s and the required power is measured to be 4.4 kW. If the pump efficiency is 65%, determine the head generated by the pump.

18.6: A centrifugal pump having the dimensions shown in Figure P18.6, develops a flow rate of 0.028 m^3/s when pumping gasoline (ρ = 680 kg/m^3). The inlet flow may be assumed to be radial. Estimate (a) the theoretical horsepower, (b) the head increase, and (c) the proper blade angle at the impeller inlet.

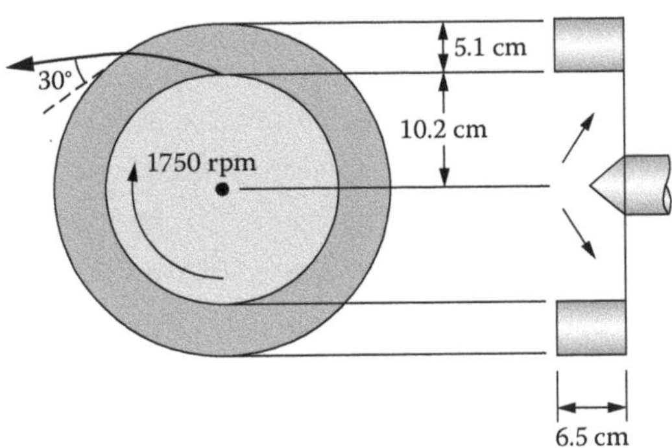

FIGURE P18.6

18.7: A centrifugal water pump operates at 1500 rpm. The dimensions are the following:

$$r_1 = 10\,\text{cm} \qquad \beta_1 = 30^\circ$$

$$r_2 = 17.5\,\text{cm} \qquad \beta_2 = 20^\circ$$

$$L = 4.5\,\text{cm}$$

Determine (a) the design point discharge rate, (b) the water horsepower, and (c) the discharge head.

18.8: Figure P18.8 represents performance, in a nondimensional form, for a family of centrifugal pumps. For a pump from this family with a characteristic diameter of 55 cm operating at 1500 rpm pumping water at 15°C, operating at maximum efficiency, estimate (a) the head, (b) the discharge rate, (c) the pressure rise, and (d) the brake horsepower.

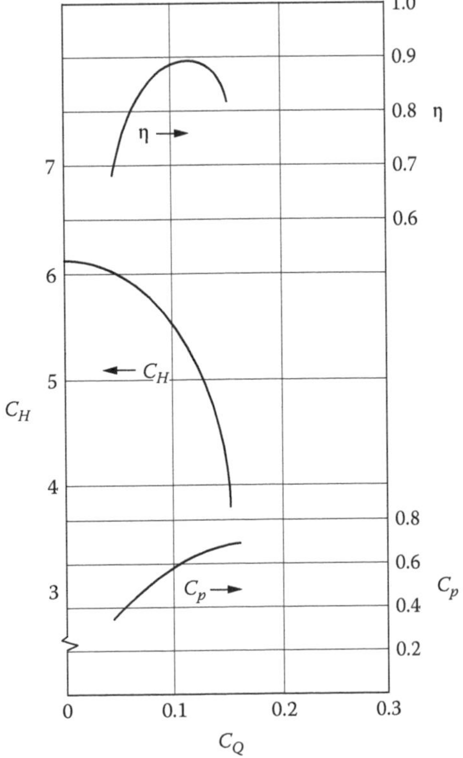

FIGURE P18.8

18.9: We wish to build a pump having the characteristics of Figure P18.8 that will deliver water at a rate of $0.2\,\text{m}^3/\text{s}$ when operating at best efficiency and a rotational speed of 1200 rpm. Estimate (a) the impeller diameter and (b) the maximum pressure rise.

18.10: Rework Problem 18.8 for a pump diameter of 35 cm operating at 2200 rpm.

18.11: Rework Problem 18.8 for a pump diameter of 30 cm operating at 2400 rpm.

18.12: Rework Problem 18.9 for desired flow rate of $0.25\,\text{m}^3/\text{s}$ at 1800 rpm.

18.13: Rework Problem 18.9 for desired flow rate of $0.20\,\text{m}^3/\text{s}$ at 1800 rpm.

18.14: A centrifugal pump with a 15-cm-diameter impeller has performance characteristics as indicated in Figure P18.14. The pump is to be used to pump gasoline ($\rho = 680\,\text{kg/m}^3$ and $\nu = 0.78 \times 10^{-6}\,\text{m/s}^2$) through a 7.5-cm-diameter commercial steel pipe that has a length of 1200 m between two reservoirs at the same elevation. Determine the expected flow rate.

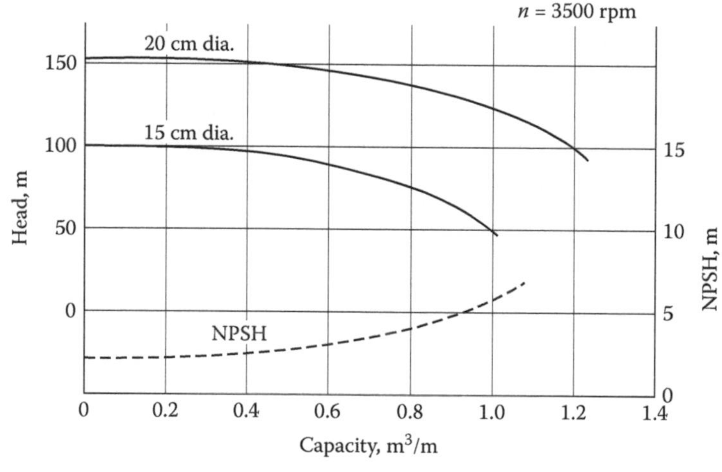

FIGURE P18.14

18.15: Determine the flow rate for the system described by Problem 18.14 if the pipe diameter is increased to 10 cm.

18.16: Rework Problem 18.14 for an impeller that has a diameter of 16.5 cm.

18.17: Rework Problem 18.14 if the impeller diameter is increased to 18 cm.

Scaling Laws

18.18: Performance curves for an operating centrifugal pump are shown in Figure P18.18a and in a dimensionless form in Figure P18.18b. This pump is used to pump water at maximum efficiency at a head of 80 m. Determine, at these new conditions, (a) the pump speed required and (b) the rate of discharge.

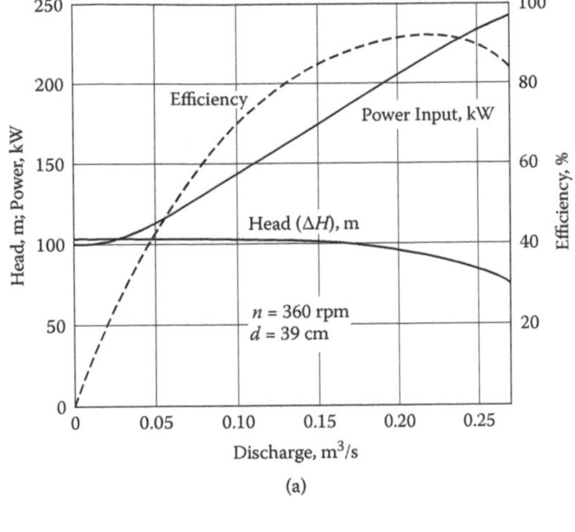

(a)

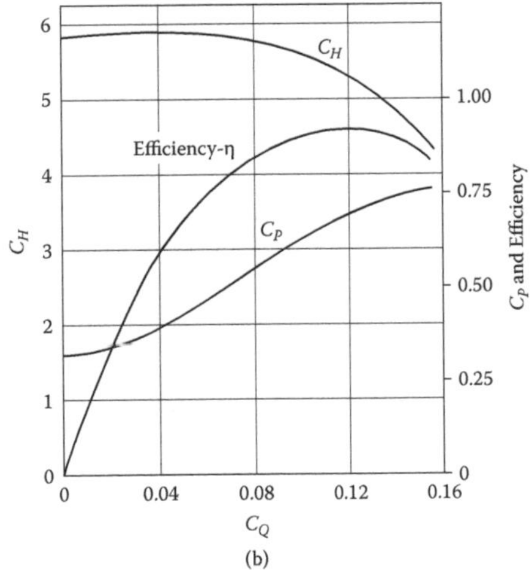

(b)

FIGURE P18.18

18.19: The pump having the characteristics shown in Figure P18.18 was used as a model for a prototype that is to be 6 times larger. If this prototype operates at 400 rpm, what (a) power, (b) head, and (c) discharge flow rate should be expected at maximum efficiency?

18.20: For the pump having the characteristics shown in Figure P18.18 operating at the conditions specified with the speed increased to 1000 rpm, what will be the (a) new discharge flow rate and (b) the power required at this new speed?

18.21: The pump having the characteristics shown in Figure P18.18 is to be operated at 750 rpm. What discharge rate is to be expected if the head developed is 424 m?

18.22: If the pump having the characteristics shown in Figure P18.18 is tripled in size but halved in rotational speed, what will be the discharge rate and head when operating at maximum efficiency?

18.23: Plot the head-discharge curve for a centrifugal pump geometrically similar to the one having the characteristics shown in Figure P18.18 but with a diameter of 1.6 m operating at 650 rpm.

18.24: A 12.5-cm-diameter centrifugal pump operating at 1800 rpm displays the following characteristics when pumping water:

$\dot{V}$, m^3/s	0.03	0.09	0.15	0.21	0.27
h, m	10.6	11.9	12.0	10.7	7.3
η, percent	18	66	84	73	25

Determine the power required when this pump runs at 2400 rpm and delivers a flow of 0.170 m^3/s.

18.25: A pump that is geometrically similar to the one specified in Problem 18.24 is to deliver 0.027 m^3/s when operating at 3000 rpm, determine the power required.

18.26: Performance curves for an axial flow pump are shown in Figure P18.26a and, in a dimensionless form, in Figure P18.26b. When this pump operates at 690 rpm in a system requiring 3 m of head, what flow rate will be produced and what is the required power?

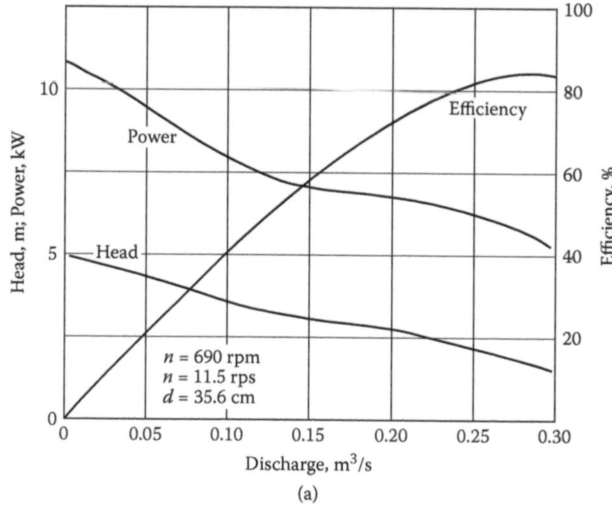

(a)

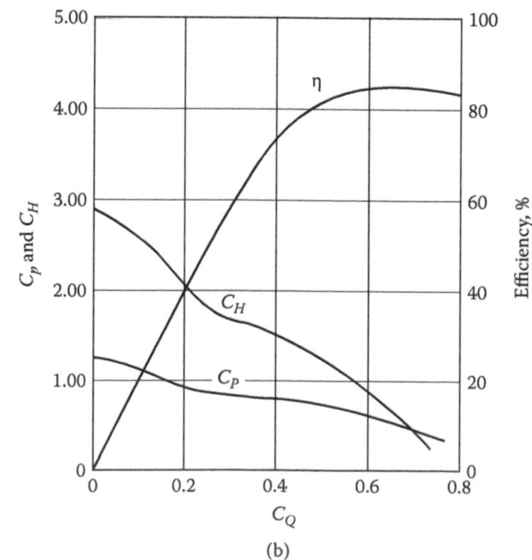

(b)

FIGURE P18.26

18.27: A prototype axial flow pump is geometrically similar to the one whose operating characteristics are shown in Figure P18.26a and P18.26b. The pump has a characteristic diameter of 36 cm and is to operate at 900 rpm to produce a volumetric flow rate of 1.5 m^3/s. Determine (a) the head generated and (b) the level of power required.

18.28: A pump, which is geometrically similar to the one whose operating characteristics are shown in Figure P18.26a, is operated at 720 rpm and has a characteristic diameter of 80 cm. Determine (a) the discharge rate and (b) power required when the head is 150 m.

18.29: A pump that is geometrically similar to the one whose operating characteristics are shown in Figure P18.26 is to have a characteristic diameter of 65 cm. Determine (a) the discharge rate, (b) the head produced, and (c) the required power when the pump is operating at maximum efficiency at 900 rpm.

18.30: Rework Problem 18.7 for a centrifugal pump that is geometrically similar to the pump specified but 50% larger in its physical dimensions. The larger pump will operate at 1500 rpm.

18.31: Rework Problem 18.7 with all specifications remaining the same except for a new rotational speed of 1220 rpm.

18.32: Rework Problem 18.7 with all specifications remaining the same except for a new rotational speed of 1850 rpm.

18.33: A centrifugal pump with an impeller diameter of 1.2 m is being constructed to provide a flow rate of 4.5 m^3/s of water with a head rise of 180 m when it operates at 1400 rpm. A geometrically similar model at 1/8 scale is to be tested in the laboratory. Assuming both model and prototype to operate at the same rotational speed, determine (a) the model discharge flow rate and (b) the model head rise.

18.34: Determine and plot values for the dimensionless coefficients, C_H, C_P, C_Q, and η for the pump data given in Problem 18.24.

18.35: A centrifugal pump with a 30-cm-diameter impeller requires input power of 45 kW when producing 0.20 m^3/s of flow against a head of 61 m. Determine (a) the expected flow rate, (b) head, and (c) the power if the pump operates at the same speed as an impeller having a diameter of 25 cm.

18.36: A pump, which is geometrically similar to the one whose operating characteristics are shown in Figure P18.26, has a characteristic diameter of 70 cm. It is desired to operate this pump at 900 rpm and at maximum efficiency. Determine (a) the discharge rate, (b) the head, and (c) the power required.

Pump/System Compatability

18.37: The pump having the characteristics shown in Figure P18.18 is used to pump water from one reservoir to another that is 90 m higher in elevation. The water will flow through a commercial steel pipe, 32 cm in diameter and 520 m in length. Determine the discharge rate.

18.38: A pump whose operating characteristics are shown in Figure P18.26 is to be used in the system in Figure P18.38. Determine (a) the discharge rate and (b) power required.

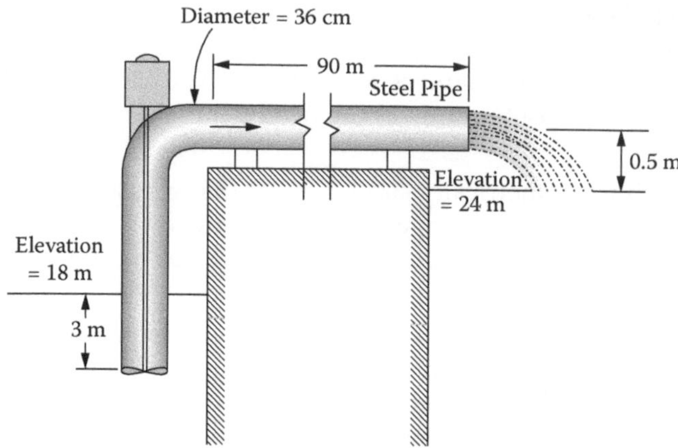

FIGURE P18.38

18.39: For the same pump and system operation described in Problem 18.38, determine (a) the discharge rate and (b) power required when the pump operates at 900 rpm.

18.40: Water at 20°C is to be pumped through the system shown in Figure P18.40. The operating data for a motor-driven pump are as follows:

Capacity, m^3/s	Developed Head, m	Efficiency, %
0	36.6	0
10	35.9	19.1
20	34.1	32.9
30	31.2	41.6
40	27.5	42.2
50	23.3	39.7

The inlet pipe to the pump is commercial steel, 6.5 cm in diameter and 7.6 m in length. The discharge line consists of 64 m of 16.5-cm-diameter steel pipe and all valves are fully open globe valves. Determine the flow rate through the system and the electrical power necessary to achieve this flow.

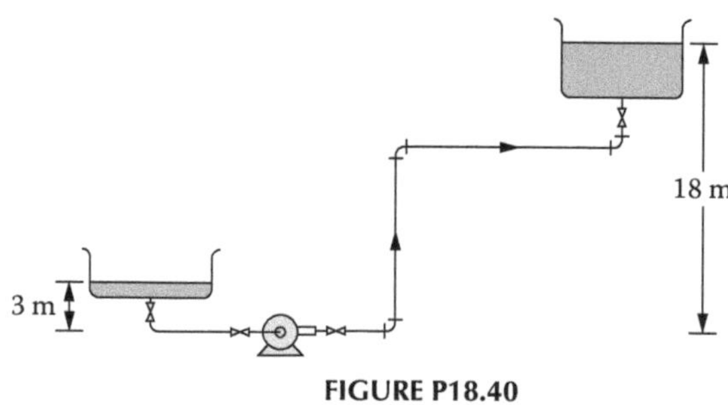

FIGURE P18.40

Net Positive Suction Head

18.41: A 20-cm pump delivers 20°C water ($P_v = 2.34$ kPa) at 0.0631 m^3/s and 2200 rpm. The pump begins to cavitate when the inlet pressure is 82.7 kPa and the inlet velocity is 6.1 m/s. Determine the corresponding NPSH.

18.42: For the pumping system described in Problem 18.41, how will the maximum elevation above the surface of the reservoir change if the water temperature is 80°C ($P_v = 47.35$ kPa)?

18.43: The performance curves for a centrifugal pump with an impeller diameter of 20 cm are shown in Figure P18.43. This pump is to be used to pump water ($\rho = 1000\,\text{kg/m}^3$) with the pump inlet located 3.5 m above the surface of the supply reservoir. At a flow rate of $0.760\,\text{m}^3/\text{s}$, the head loss between the reservoir surface and the pump inlet is 1.83 m of water. Would you expect cavitation to occur?

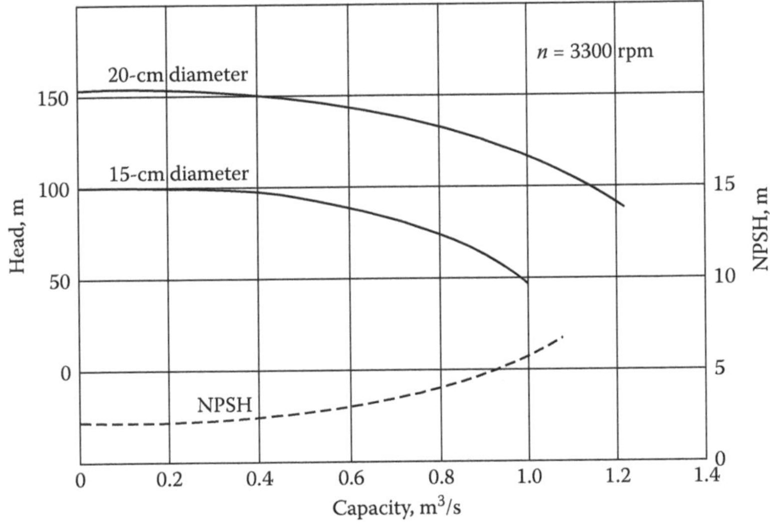

FIGURE P18.43

18.44: The pump whose performance is given in Figure P18.43 is to be placed above a large open tank and is to pump water at a rate of $0.2\,\text{m}^3/\text{s}$. The system head loss can be assumed to have a coefficient of $K = 0.3$ and the pump inlet has a diameter of 12 cm. Determine the maximum height at which the pump can be located without experiencing cavitation.

18.45: A prototype pump, geometrically similar to the one specified in Problem 18.41, is 5 times larger and runs at 1200 rpm. What is the required NPSH for the prototype?

18.46: A centrifugal pump is installed *z* meters above a reservoir. The manufacturer's specifications indicate the value of NPSH to be 4.8 m when the system is pumping water at 20°C ($P_v = 2.34\,\text{kPa}$) at a rate of $0.014\,\text{m}^3/\text{s}$. All losses in the installation combine to yield a loss coefficient of $K_L = 20$ and the inlet pipe has a diameter of 10 cm. Determine the maximum elevation, *z*, above the reservoir surface at which the pump can be positioned without experiencing cavitation.

Specific Speed

18.47: Pumps used in an aqueduct operate at 400 rpm and deliver a flow of $200\,\text{m}^3/\text{s}$ against a head of 450 m. What types of pumps are they?

18.48: A pump is required to deliver 60,000 gpm against a head of 290 m when operating at 2000 rpm. What type of pump should be specified?

18.49: An axial flow pump has a specified specific speed of 5.0. The pump must deliver 2500 gpm against a head of 16 m. Determine the required operating rpm of the pump.

18.50: A pump operating at 400 rpm has the capability of producing $3\,\text{m}^3/\text{s}$ of water flow against a head of 18 m. What type of pump is this?

18.51: A pump operating at 1800 rpm produces a head rise of 27 m and a discharge flow rate of $0.045\,\text{m}^3/\text{s}$ when the power required is 18.5 kW. Determine (a) the specific speed for this pump and the new values of (b) head rise and (c) shaft horsepower if the pump speed is decreased to 1400 rpm.

18.52: A centrifugal pump achieves a discharge flow rate of $0.032\,m^3/s$ when operating at 1800 rpm against a head of 60 m. Determine the new (a) flow rate and (b) developed head if the rotational speed is increased to 3600 rpm.

18.53: It has been determined that, in a specific application, a pump must deliver $0.32\,m^3/s$ of flow against a head of 90 m when operating at 1200 rpm. What type of pump would you recommend?

18.54: A pump operating at 2160 rpm delivers $3.5\,m^3/s$ of water against a head of 20 m. Is this pump an axial flow, mixed flow, or radial flow machine?

18.55: An axial flow pump with a specific speed of 6.0 is required to deliver $0.20\,m^3/s$ of water against a head of 4.5 m. At what speed should this pump be operated?

19

Introduction to Heat Transfer

Chapter Objectives

- To introduce and describe the three modes of heat transfer: conduction, convection, and radiation.
- To define thermal conductivity, a heat transfer property.
- To develop the concept of thermal resistance.
- To consider the overall heat transfer coefficient.

19.1 Introduction

The remaining eight chapters of our study deal with heat transfer. The quantity, heat, was defined in earlier chapters and it was quantitatively related to work and system behavior through the first law of thermodynamics in Chapter 5. The first law remains a fundamental precept as we now devote our attention to the *rate* of heat exchange.

As discussed earlier, heat will be exchanged between systems when they are at different temperatures. The second law of thermodynamics stipulates that this exchange will be from the higher-temperature body or system, toward the bodies or systems at lower temperatures.

Practical considerations regarding heat transfer processes involve the rate at which heat transfer occurs. All decisions that involve equipment and material specifications require that heat transfer rates and process temperatures be determined.

Our goal, in this chapter, is to examine the mechanisms of heat transfer and to introduce the equations that are fundamental to the evaluation of rates at which energy transfer, due to temperature difference, occurs.

19.2 Conduction

Energy transfer by conduction is accomplished via two mechanisms. The first mechanism is by molecular interaction, in which the greater motion of a molecule at a higher-energy (temperature) level imparts energy to adjacent molecules at lower-energy levels. This type of energy transfer is present, to some degree, in all systems in which a temperature gradient exists and in which molecules of a solid, liquid, or gas are present.

The second mechanism of conduction heat transfer is by "free" electrons. The free-electron mechanism is significantly greater in pure metallic solids; the concentration of free electrons varies considerably for alloys and becomes very low for nonmetallic solids. The ability of solids to conduct heat varies directly with the concentration of free electrons. Thus, it is not surprising that pure metals are the best heat conductors, as our experience has indicated.

Because heat conduction is primarily a molecular phenomenon, we might expect that the basic equation used to describe this process is similar to the expression used in the molecular transfer of momentum, Equation 11.19. Such an equation was first stated in 1822 by Fourier in the form

$$\dot{q}_x = \frac{\dot{Q}_x}{A} = -k\frac{dT}{dx} \tag{19.1}$$

where $\dot{q}_x$ is the **heat flux**, defined as the **rate of heat transfer**, $\dot{Q}_x$, per unit of area normal to the direction of heat flow, dT/dx is the **temperature gradient**, and k is the **thermal conductivity** in W/m-K. A more general relation for the heat flow is

$$\dot{q} = \frac{\dot{Q}}{A} = -k\nabla T \tag{19.2}$$

which expresses the heat flux (and rate of heat transfer) as proportional to the temperature gradient. The proportionality constant is the thermal conductivity that plays a role similar to that of viscosity in momentum transfer. The negative signs in both Equations 19.1 and 19.2 indicate that the heat flow is in the direction of a negative temperature gradient. Equation 19.2 is the vector form of the **Fourier rate equation** and is often referred to as **Fourier's law of heat conduction.**

The thermal conductivity, k, which is defined by Equation 19.1 is assumed independent of direction in Equation 19.2. Thus, Equation 19.2 applies only to an **isotropic medium.** The thermal conductivity is a property of a conducting medium and, like the viscosity, is primarily a function of temperature, varying significantly with pressure only in the case of gases subjected to high pressures.

Under steady-state conditions, Equation 19.1 can be solved to evaluate the heat transfer by conduction between two locations.

Its solution is achieved by the method of separation of the variables and invoking the boundary conditions where $T(x = x_1) = T_1$, $T(x = x_2) = T_2$, to yield

$$\dot{Q} = \frac{kA}{L}(T_1 - T_2) \tag{19.3}$$

where $L = x_2 - x_1$.

For one-dimensional, steady, heat flow in cylindrical coordinates where the heat flow is in the radial direction, Equation 19.1 becomes

$$\dot{Q} = -kA\frac{dT}{dr} = -k(2\pi r L)\frac{dT}{dr} \tag{19.4}$$

where $A = 2\pi r L$. For one-dimensional, steady, heat flow in spherical coordinates where the heat flow is in the radial direction, Equation 19.1 becomes

$$\dot{Q} = -kA\frac{dT}{dr} == -k(4\pi r^2)\frac{dT}{dr} \tag{19.5}$$

where $A = 4\pi r^2$.

Equations 19.4 and 19.5 may be solved using separation of the variables subject to the boundary conditions

$$T(r = r_1) = T_1 \quad \text{and} \quad T(r = r_2) = T_2$$

to yield for the cylinder

$$\dot{Q} = \frac{2\pi k L(T_1 - T_2)}{\ln r_2/r_1} \tag{19.6}$$

and for the sphere

$$\dot{Q} = \frac{4\pi k(T_1 - T_2)}{\dfrac{1}{r_1} - \dfrac{1}{r_2}}$$

With

$$\frac{1}{r_1} - \frac{1}{r_2} = \frac{r_2 - r_1}{r_1 r_2}$$

we have

$$\dot{Q} = \frac{4\pi k r_1 r_2(T_1 - T_2)}{r_2 - r_1} \tag{19.7}$$

19.3 Thermal Conductivity

Equation 19.1 is the defining equation for the thermal conductivity and experimental measurements made on the basis of this definition may be employed to evaluate the thermal conductivities of different materials.

In the solid phase, thermal conductivity is attributed to both molecular interaction and free-electron drift that is present, primarily, in pure metals. The solid phase is amenable to quite precise measurements of thermal conductivity because there is no effect of convection currents. The thermal properties of most solids of engineering interest have been evaluated. Table A.17 (for metals) and Table A.18 (for nonmetallic solids) listing thermal conductivities and other properties are included in Appendix A.

The thermal conductivity of a liquid is not amenable to any simplified kinetic-theory development because the molecular behavior of the liquid phase is not clearly understood and no universally accurate mathematical model presently exists. Some empirical correlations have met with reasonable success, but they are so specialized they will not be included in this book. A general observation about liquid thermal conductivities is that they vary only slightly with temperature and are relatively independent of pressure. One problem in experimentally determining values of the thermal conductivity in a liquid is making sure that the liquid is free of convection currents.

Table A.14 gives the thermal conductivity of water as well as other properties and Table A.16 provides similar properties for several other liquids.

The **kinetic theory of gases** can be used to predict the thermal conductivity of gases and experiments confirm that the thermal conductivity is proportional to the absolute temperature, Table A.13 gives the thermal conductivity of air as well as other properties and Table A.15 provides similar properties for several other gases.

Figure 19.1 illustrates the thermal conductivity variation with temperature of several important materials in solid, liquid, and gas phases.

Example 19.1 As indicated in Figure 19.2, a pure aluminum cylinder having an inside diameter of 1.880 cm and a wall thickness of 0.391 cm is exposed to inside and outside surface temperatures of 94°C and 71°C. Determine, per meter of pipe length (a) the heat flow rate and (b) the heat fluxes based on both the inside and outside surface area.

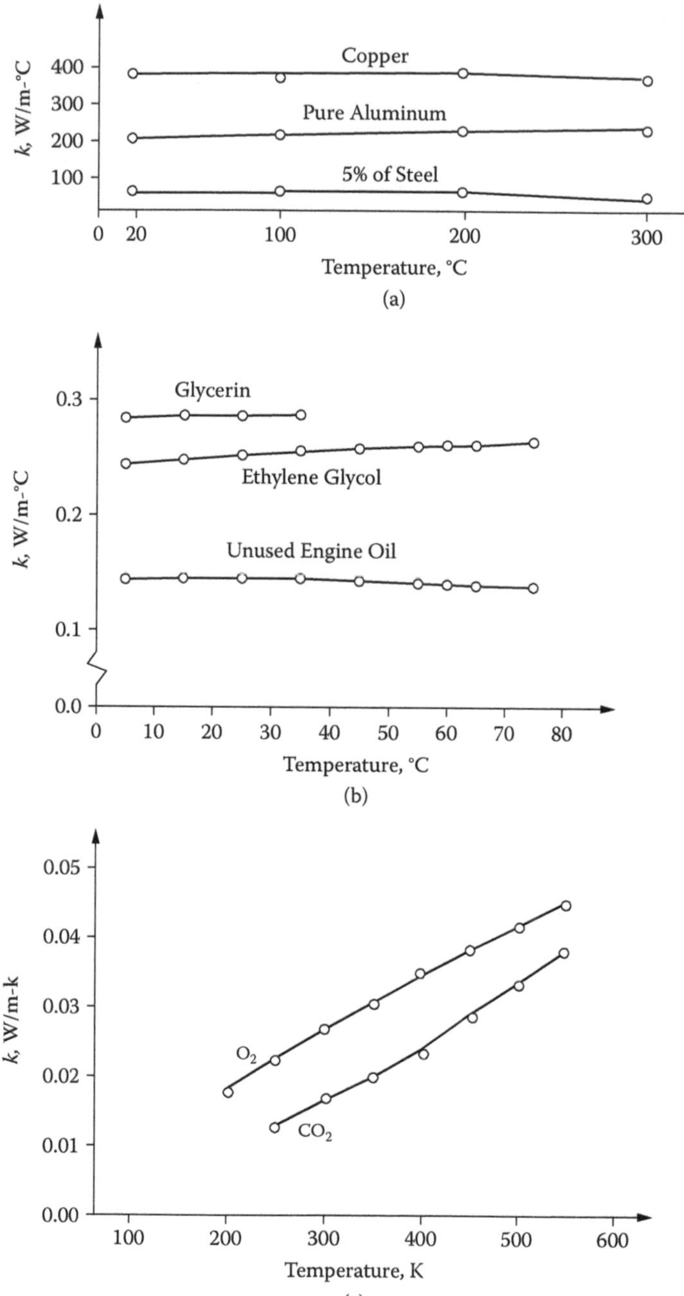

FIGURE 19.1
Representative thermal conductivities of (a) metals, (b) liquids, and (c) gases as functions of temperature.

Solution

Assumptions and Specifications

1. The heat flow is steady.
2. The heat flow is one dimensional.

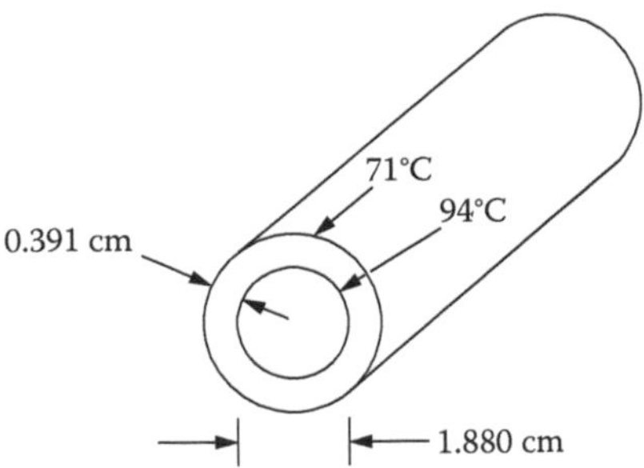

FIGURE 19.2
Configuration for Example 9.1.

Here, r_1 and r_2 designate the inner and outer radii, respectively, and because the heat flow is in the radial direction, Equation 19.4 applies

$$\dot{Q} = -kA\frac{dT}{dr} = -k(2\pi r L)\frac{dT}{dr}$$

With the variables separated this becomes

$$\dot{Q}\frac{dr}{r} = -2k\pi L dT$$

and the integration gives

$$\dot{Q}\frac{dr}{r}\Bigg|_{r_1}^{r_2} = \dot{Q}\ln r\Bigg|_{r_1}^{r_2} = \dot{Q}\ln r_2/r_1 = -k(2\pi L)T\Bigg|_{T_1}^{T_2} = 2\pi k L(T_1 - T_2)$$

or finally

$$\dot{Q} = \frac{2\pi k L(T_1 - T_2)}{\ln r_2/r_1} = \frac{2\pi k L(T_1 - T_2)}{\ln d_2/d_1} \tag{19.6}$$

Notice that the radius ratio r_2/r_1 can be replaced by the diameter ratio, d_2/d_1. Here, the radius ratio will be used and with

$$d_i = d_1 = 1.880\,\text{cm} \quad \text{and} \quad d_o = d_2 = 1.880\,\text{cm} + 2(0.391\,\text{cm}) = 2.662\,\text{cm}$$

These make

$$r_1 = 0.94\,\text{cm} \quad \text{and} \quad r_2 = 1.331\,\text{cm}$$

The arithmetic average temperature in the pipe wall is

$$T_{\text{av}} = \frac{94°\text{C} + 71°\text{C}}{2} = \frac{165°\text{C}}{2} = 82.5°\text{C}$$

Table A.17 reveals (with interpolation) that the thermal conductivity of pure aluminum at $T_{\text{av}} = 82.5°\text{C}$ is $k = 205.6\,\text{W/m-K}$.

(a) The heat flow will be

$$\dot{Q} = \frac{2\pi(205.6\,\text{W/m-K})(1.00\,\text{m})(94°\text{C} - 71°\text{C})}{\ln(0.01331\,\text{m}/0.0094\,\text{m})} = \frac{26,712\,\text{W}}{0.3478} = 85.43\,\text{kW} \Longleftarrow$$

(b) The heat flux varies as a function of location. At the inner surface

$$\dot{q} = \frac{85.43\,\text{kW}}{2\pi(0.0094\,\text{m})(1.00\,\text{m})} = 1446\,\text{kW/m}^2 \Longleftarrow$$

and at the outer surface

$$\dot{q} = \frac{85.43\,\text{kW}}{2\pi(0.01331\,\text{m})(1.00\,\text{m})} = 1021.5\,\text{kW/m}^2 \Longleftarrow$$

Notice that for the same amount of heat flow, the heat fluxes based on the inside and outside areas of the pipe are vastly different, differing by about 42%.

19.4 Convection

Heat transfer by convection occurs between a surface and an adjacent fluid. When the fluid is induced to flow past the surface by an external agency such as a pump or fan, the condition is that of **forced convection.** When fluid motion is induced by density differences in the fluid caused by the heating or cooling of the fluid through the heat exchange process, the process is called **natural convection** or **free convection.**

The rate equation for convective heat transfer is attributed to Isaac Newton (1701). The **Newton rate equation,** sometimes referred to as "Newton's law of cooling," is

$$\dot{q} = \frac{\dot{Q}}{S} = h\Delta T \tag{19.8}$$

where $\dot{Q}$ is the rate of heat transfer in W, S is the surface area,[1] and h is the **coefficient of heat transfer** in W/m^2-K. Equation 19.8 is, fundamentally, the defining relationship of h. One of the principal challenges in quantifying convective heat transfer is that of determining the appropriate value for this quantity. The coefficient, h, is a function of the system geometry, fluid and flow properties, and the magnitude of the temperature difference, ΔT.

Because fluid behavior is involved in convective heat transfer, it will be necessary to use our knowledge of fluid mechanics considered in earlier chapters of this text. We will need to characterize the flow as laminar or turbulent, and will have occasion to use many of the concepts and methods of analysis developed previously.

Heat transfer involving a change of phase, while not technically convective processes, are still treated quantitatively using Equation 19.8. When either **boiling** or **condensation** occur, relatively high rates of heat transfer are involved even though the magnitude of ΔT may be relatively small. Table 19.1 lists values of h typical of five different convective mechanisms.

19.5 Radiation

Radiant heat transfer differs from conduction and convection in that no medium is required for its propagation. Indeed, energy transfer by radiation is at maximum when the two surfaces that are exchanging energy are separated by a perfect vacuum.

[1] In this book, a distinction is made between cross-sectional area, A, and surface area, S.

TABLE 19.1

Approximate Values of the Convection Heat Transfer Coefficient

Mechanism	h, **W/m²-K**
Free convection, air	5–50
Forced convection, air	25–250
Forced convection, water	250–15,000
Boiling water	2500–25,000
Condensing water vapor	5000–100.000

The rate of energy emission from a perfect radiator or **blackbody** is given by

$$\dot{Q} = \sigma S T^4 \tag{19.9}$$

where σ is the **Stefan-Boltzmann constant**

$$\sigma = 5.67 \times 10^{-8}\ \text{W/m}^2\text{-K}^4$$

In Equation 19.9, $\dot{Q}$ is the rate of radiant energy transfer in W, S is the surface area of the emitting surface, and T is the absolute temperature of the emitting surface in K. The proportionality constant, σ, relating the radiant energy flux, $\dot{q} = \dot{Q}/S$, to the fourth power of the absolute temperature is named after Stefan who, from experimental observations, proposed Equation 19.9 in 1879, and Boltzmann, who derived this equation theoretically in 1884. Equation 19.9 is most often referred to as the **Stefan-Boltzmann law** of thermal radiation.

Certain modifications will be made to Equation 19.9 to account for the net energy transfer between two surfaces, the degree of departure of the emitting surface and the receiving surface from ideal emitter and receiver behavior, and geometric factors associated with radiant interchange between a surface and its surroundings.

If the surface and geometric effects are lumped into a factor $\mathcal{F}$, then, we may write for the heat transfer by radiation between surface 1 at T_1 to surface 2 at T_2

$$\dot{Q} = \sigma S \mathcal{F}(T_1^4 - T_2^4) \tag{19.10}$$

19.6 Thermal Resistance

The steady-state heat flow in one dimension through a plane wall is represented by Equation 19.4

$$\dot{Q} = \frac{kA}{L}(T_1 - T_2) \tag{19.4}$$

where L is the wall thickness and T_1 and T_2 are the surface temperatures. This equation may be put into the form of a thermal resistance

$$R_{\text{cond}} \equiv \frac{T_1 - T_2}{\dot{Q}} = \frac{\Delta T}{\dot{Q}} = \frac{L}{kA} \tag{19.11a}$$

which, not only defines a **thermal resistance in conduction**, but also suggests an **electrothermal analogy** with Ohm's law

$$R = \frac{\Delta V}{I}$$

in which

$$\Delta V \Longleftrightarrow \Delta T$$

$$I \Longleftrightarrow \dot{Q}$$

$$R \Longleftrightarrow R_{\text{thermal}}$$

The thermal resistances of the cylinder and the sphere in conduction are respectively,

$$R_{\text{cond}} = \frac{\ln r_2/r_1}{2\pi k L} \tag{19.11b}$$

and

$$R_{\text{cond}} = \frac{r_2 - r_1}{4\pi k r_1 r_2} \tag{19.11c}$$

Using this same reasoning, Equation 19.8, written in the form,

$$\dot{Q} = hS\Delta T$$

becomes, upon rearrangement,

$$R_{\text{conv}} = \frac{\Delta T}{\dot{Q}} = \frac{1}{hS} \tag{19.12}$$

For the case of radiation, Equation 19.8

$$\dot{Q} = \sigma \mathcal{F} S\left(T_1^4 - T_2^4\right) \tag{19.8}$$

the thermal resistance is

$$R_{\text{rad}} = \frac{1}{h_{\text{rad}} S} \tag{19.13}$$

in which

$$h_{\text{rad}} = \sigma \mathcal{F}\left(T_1^2 + T_2^2\right)(T_1 + T_2) \tag{19.14}$$

19.7 Combined Mechanisms of Heat Transfer

The three heat transfer modes have been introduced in Sections 19.2, 19.4, and 19.5. Most often more than one of these mechanisms occur in combination. In this section, we examine some typical situations where more than one heat transfer mode is involved.

We first consider the plane wall shown in Figure 19.3, with steady-state conduction occurring through the wall whose surfaces are at temperatures T_1 and T_4, respectively. The wall consists of three different materials, having thermal conductivities, k_2 through k_4 and

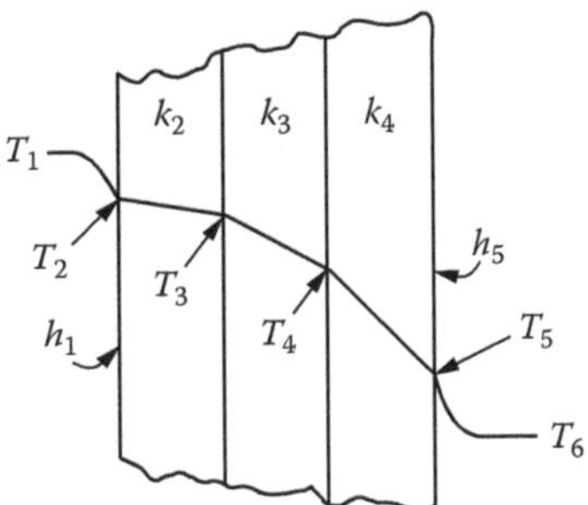

FIGURE 19.3
Composite plane wall with convection at its faces.

each external face of the wall is exposed to a convective environment. At the left face, the environmental temperature is T_1 and the heat transfer coefficient is h_1 and at the right face, the environmental temperature is T_6 and the heat transfer coefficient is h_6.

Observe that there are five temperature differences of interest and that each of them is related to the common heat flow through them via an appropriate thermal resistance:

$$T_1 - T_2 = \dot{Q}R_1 = \dot{Q}\left(\frac{1}{h_1 S}\right)$$

$$T_2 - T_3 = \dot{Q}R_2 = \dot{Q}\left(\frac{L_2}{k_2 A}\right)$$

$$T_3 - T_4 = \dot{Q}R_3 = \dot{Q}\left(\frac{L_3}{k_3 A}\right)$$

$$T_4 - T_5 = \dot{Q}R_4 = \dot{Q}\left(\frac{L_4}{k_4 A}\right)$$

$$T_5 - T_6 = \dot{Q}R_5 = \dot{Q}\left(\frac{1}{h_5 S}\right)$$

Adding these yields

$$T_1 - T_6 = \dot{Q}\left(\frac{1}{h_1 S} + \frac{L_2}{k_2 A} + \frac{L_3}{k_3 A} + \frac{L_4}{k_4 A} + \frac{1}{h_5 S}\right) \tag{19.15}$$

and because $A = S$ for the plane wall, this is equivalent to

$$T_1 - T_6 = \frac{\dot{Q}}{S}\left(\frac{1}{h_1} + \frac{L_2}{k_2} + \frac{L_3}{k_3} + \frac{L_4}{k_4} + \frac{1}{h_5}\right)$$

The electrothermal analog is shown in Figure 19.4.

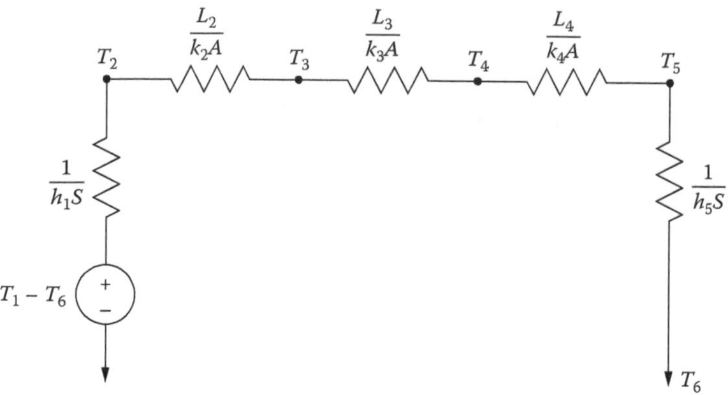

FIGURE 19.4
Electrothermal analog network for the plane wall of Figure 19.3.

Example 19.2 Saturated steam at 0.276 MPa flows inside of a steel pipe having a thermal conductivity of $k = 42.9\,\text{W/m-K}$, an inside diameter of 2.09 cm, and an outside diameter of 2.67 cm. The convective coefficients on the inner and outer pipe surfaces may be taken as $5680\,\text{W/m}^2\text{-K}$ and $22.7\,\text{W/m}^2\text{-K}$, respectively. The surrounding air is at 21°C. Find the heat loss per meter of bare pipe.

Solution

Assumptions and Specifications

1. The heat flow is steady.
2. The heat flow is one dimensional.

Here, the convective resistances are designated as R_1 at the inside and R_3 at the outside of the pipe. The conduction resistance for the pipe wall is designated as R_2. These three thermal resistances are evaluated first. With r_i and r_o taken as the inside and outside radii, respectively

$$R_1 = \frac{1}{h_i S_i} = \frac{1}{2\pi h_i r_i L}$$

$$= \frac{1}{2\pi(5680\,\text{W/m}^2\text{-K})(0.01045\,\text{m})(1.00\,\text{m})} = 0.00268\,\text{K/W}$$

$$R_2 = \frac{\ln r_2/r_1}{2\pi k L}$$

$$= \frac{\ln(1.335\,\text{cm}/1.045\,\text{cm})}{2\pi(42.9\,\text{W/m-K})(1.00\,\text{m})} = 0.00091\,\text{K/W}$$

$$R_3 = \frac{1}{h_o S_o} = \frac{1}{2\pi h_o r_o L}$$

$$= \frac{1}{2\pi(22.7\,\text{W/m}^2\text{-K})(0.01335\,\text{m})(1.00\,\text{m})} = 0.5252\,\text{K/W}$$

The total thermal resistance is

$$\begin{aligned} R_T &= R_1 + R_2 + R_3 \\ &= 0.00268\,\text{K/W} + 0.00091\,\text{K/W} + 0.5252\,\text{K/W} \\ &= 0.5288\,\text{K/W} \end{aligned}$$

Table A.4 (with interpolation) gives the saturation temperature of steam at 276 kPa as 131°C so that the heat loss per meter of pipe length is

$$\dot{Q} = \frac{\Delta T}{R_T} = \frac{131°\text{C} - 21°\text{C}}{0.5288\,\text{K/W}} = 208\,\text{W} \Longleftarrow$$

19.8 The Overall Heat Transfer Coefficient

An alternative means for treating a heat transfer situation involving combined modes is through the use of **the overall heat transfer coefficient,** U. This quantity is defined according to the relationship

$$\dot{Q} = US\Delta T \tag{19.16}$$

where U accounts for all of the thermal resistances between the two extreme temperatures.

Equation 19.16 is straightforward when a plane composite wall is involved but when the system is cylindrical or spherical, the value of U will vary depending on the choice of the reference area, S. For the case with cylindrical geometry, as in Example 19.2, we could use either of the forms

$$\dot{Q} = U_o S_o \Delta T \qquad \text{or} \qquad \dot{Q} = U_i S_i \Delta T \tag{19.17}$$

In the two equivalent forms given by Equation 19.17, the coefficients , U_o and U_i, are associated with outside (outer) and inside (inner) surface areas, respectively.

In Example 19.2, the total resistance for the three resistance configuration could have been written as

$$R_T = \frac{1}{h_i S_i} + \frac{\ln r_2/r_1}{2\pi k L} + \frac{1}{h_o S_o}$$

Suppose that the outer surface is taken as the reference. Then

$$U_o = \frac{\dot{Q}}{S_o \Delta T} = \frac{\Delta T/R_T}{S_o \Delta T} = \frac{1}{S_o R_T}$$

and

$$U_o = \frac{1}{\left(\dfrac{S_o}{h_i S_i} + \dfrac{S_o \ln r_2/r_1}{2\pi k L} + \dfrac{1}{h_o}\right)} \tag{19.18}$$

and if the inner surface is taken as the reference

$$U_i = \frac{\dot{Q}}{S_i \Delta T} = \frac{\Delta T / R_T}{S_i \Delta T} = \frac{1}{S_i R_T}$$

and

$$U_i = \frac{1}{\left(\dfrac{1}{h_i} + \dfrac{S_i \ln r_2/r_1}{2\pi k L} + \dfrac{S_i}{h_o S_o}\right)} \tag{19.19}$$

Example 19.3 Rework Example 19.2 using the overall heat transfer coefficient based on the outside surface.

Solution

Assumptions and Specifications

1. The heat flow is steady.
2. The heat flow is one dimensional.

Here,

$$S_o = 2\pi r_o L = 2\pi(0.01335\,\text{m})(1.00\,\text{m}) = 0.0839\,\text{m}^2$$

and

$$S_i = 2\pi r_i L = 2\pi(0.01045\,\text{m})(1.00\,\text{m}) = 0.0657\,\text{m}^2$$

With

$$U_o = \frac{1}{S_o R_T} \quad \text{and} \quad U_i = \frac{1}{S_i R_T}$$

we can take R_T from Example 9.2

$$R_T = 0.5288\,\text{K/W}$$

and obtain

$$U_o = \frac{1}{S_o R_T} = \frac{1}{(0.0839\,\text{m}^2)(0.5288\,\text{K/W})} = 22.54\,\text{W/m}^2\text{-K}$$

and

$$U_i = \frac{1}{S_i R_T} = \frac{1}{(0.0657\,\text{m}^2)(0.5288\,\text{K/W})} = 28.80\,\text{W/m}^2\text{-K}$$

Thus,

$$\dot{Q} = U_o S_o \Delta T = (22.54\,\text{W/m}^2\text{-K})((0.0839\,\text{m}^2)(110°\text{C}) = 208\,\text{W} \Longleftarrow$$

and

$$\dot{Q} = U_i S_i \Delta T = (28.78\,\text{W/m}^2\text{-K})((0.0657\,\text{m}^2)(110°\text{C}) = 208\,\text{W} \Longleftarrow$$

19.9 Summary

The rate equations for heat transfer in the steady state and in one dimension are as follows:
For conduction in rectangular coordinates

$$\dot{Q}_x = -kA\frac{dT}{dx} \tag{19.1}$$

and for conduction in cylindrical and spherical coordinates

$$\dot{Q}_r = -kA\frac{dT}{dr} \tag{19.4, 19.5}$$

For convection

$$\dot{Q} = hS\Delta T \tag{19.8}$$

and for radiation

$$\dot{Q} = \sigma S T^4 \tag{19.10}$$

Each mode may be represented by a thermal resistance:

$$R_{\text{cond}} = \frac{L}{kA} \tag{19.11a}$$

for plane surfaces

$$R_{\text{cond}} = \frac{\ln r_2/r_1}{2\pi k L} \tag{19.11b}$$

for cylinders and

$$R_{\text{cond}} = \frac{r_2 - r_1}{4\pi k r_1 r_2} \tag{19.11c}$$

for spheres. In addition, thermal resistances exist for convection and radiation

$$R_{\text{conv}} = \frac{1}{hS} \tag{19.12}$$

and

$$R_{\text{rad}} = \frac{1}{h_{\text{rad}} S} \tag{19.13}$$

in which

$$h_{\text{rad}} = \sigma \mathcal{F}(T_1^2 + T_2^2)(T_1 + T_2) \tag{19.14}$$

The overall heat transfer coefficient is designated as U and must be linked to a particular surface area. When this is done,

$$\dot{Q} = U_o S_o \Delta T = U_i S_i \Delta T \tag{19.16}$$

19.10 Problems

Conduction

19.1: Heat flows at a rate of 4 kW through an insulating material having a cross-sectional area of 12 m^2 and a thickness of 3.6 cm. The hot surface is at 112°C and the thermal conductivity of the material is 0.25 W/m-K. Determine the cold surface temperature.

19.2: A temperature difference of 80°C exists across a 12-cm layer of insulating material having a thermal conductivity of 0.032 W/m-K. Determine the heat transferred per unit area.

19.3: An igloo in the shape of a perfect hemisphere, has an outside diameter of 4 m and the ice walls are 18-cm thick. If the inner face of the ice is at 0°C and the outer face is at −12°C, determine the heat transferred through the wall of the igloo.

19.4: A pyrex glass beaker with a thermal conductivity of 0.56 W/m-K has an outside diameter of 12.70 cm, a height of 15.24 cm, and walls that are 3.5-mm thick. The beaker contains a hot liquid such that the inside surface is maintained at 92°C. If the outside surface is held at 17°C, determine the heat transferred through the walls of the beaker.

19.5: A beverage carrier has inside dimensions of 50 cm × 25 cm and is 20-cm high. It is fabricated of a material having a thermal conductivity of 0.023 W/m-K. The walls of the carrier are 2.25-cm thick and the outside temperature of the carrier is at 28°C. Suppose that the latent heat of fusion of ice is 333.7 kJ/kg and that the entire inside volume of the carrier is filled with ice. Neglect the heat flow through the base of the carrier and determine the amount of ice that will melt in 1 hour.

19.6: The inside surface of an 8 in, schedule 80, 0.5% carbon steel pipe is maintained at 39°C and the outside surface is held at 21°C. Determine the heat loss per meter of length.

19.7: It is proposed to use Teflon to coat a 2-in-OD tube. If the surface of the tube is held at 82°C and the surface of the Teflon is maintained at 18°C, determine the thickness of the Teflon needed to limit the heat loss to 40 W/m of tube length.

19.8: A basement has a dry concrete floor that is 10-m long, 8-m wide, and 20.32-cm thick. The top and bottom temperatures of the floor are 17°C and 8°C, respectively. The basement is heated by a gas furnace and the gas costs $0.08/MJ. Consider the heat loss through the floor and determine the hourly cost of the gas.

Convection

19.9: A wall that is 8-m high and 6-m wide is held at 37°C. If the heat transfer coefficient between the wall and the surroundings is 7.2 W/m^2-K and the surroundings are at 17°C, determine the heat loss by convection.

19.10: A small 4-cm-diameter sphere is maintained at 62°C in an enclosure where the air is at 44°C. If the heat transfer coefficient is 14 W/m^2-K, determine the heat loss by convection.

19.11: How much heat will be transferred by convection from an 8-m long 3-in schedule 80 pipe held at 77°C in a room held at 17°C where the heat transfer coefficient is 10.25 W/m^2-K?

19.12: An electric heater is embedded in a 3-in tube. When the outside surface of the tube is held at 80°C and the air surrounding the tube is at 16°C, it is noted that 640 W is provided to the heater. If the tube is 2-m long, determine the heat transfer coefficient.

19.13: A power transistor may be treated as a capped cylinder having an outside diameter of 12.5 mm and a height of 15 mm. The transistor is cooled by air at 37°C via a heat transfer coefficient of 75 W/m^2-K and its case is to be held at 72°C to satisfy reliability constraints. Determine the power rating of the transistor for these conditions.

19.14: Consider a square microcircuit chip that is 4 mm on a side and mounted in an insulated substrate. If the surface of the chip is to be held at 77°C because of reliability constraints and air at 17°C under a heat transfer coefficient of 80 W/m^2-K is provided across the chip, determine the power dissipation.

Radiation

19.15: A small sphere, whose surface is an ideal emitter, has a diameter of 4 cm. It is held at 127°C and it radiates to a large enclosure at 27°C. Determine the heat transferred by radiation.

19.16: If the pipe in Problem 19.11 is an ideal emitter and it exists in a vacuum, determine the heat flow by radiation.

19.17: If the surface of the power transistor in Problem 19.13 is an ideal emitter and the enclosure surrounding the transistor is maintained at 37°C, determine the heat loss to the enclosure by radiation.

19.18: If the surface of the pipe in Problem 19.12 is an ideal emitter and 640 W are to be dissipated, determine the surface temperature of the pipe if radiation is the only mode of heat transfer.

19.19: If the surface of the wall in Problem 19.9 is an ideal emitter, determine the heat flow from the wall by radiation.

19.20: A steel plate that is coated so that its surface may be considered as an ideal emitter is suspended by nonconducting cords in an evacuated enclosure with walls at 27°C. The plate is 50-cm square. Determine the surface temperature required to make the dissipation from the plate 68.5 W.

Thermal Resistance

19.21: Determine the thermal resistance of the plate in Problem 19.1.

19.22: Determine the thermal resistance of the walls of the igloo in Problem 19.3.

19.23: Determine the thermal resistance of the walls of the beaker in Problem 19.4.

19.24: Determine the thermal resistance of the walls of the beverage carrier in Problem 19.5.

19.25: Determine the thermal resistance of the pipe in Problem 19.6.

19.26: Determine the thermal resistance of the concrete floor in Problem 19.8.

19.27: Determine the convective thermal resistance of the plate in Problem 19.9.

19.28: Determine the convective thermal resistance of the sphere in Problem 19.10.

19.29: Determine the convective thermal resistance of the pipe in Problem 19.11.

19.30: Determine the radiative thermal resistance of the sphere in Problem 19.15.

19.31: The outside surface of the pipe in Problem 19.11 is an ideal emitter and it is held at 77°C in an enclosure at 17°C. Determine the radiative thermal resistance.

19.32: Determine the radiative thermal resistance of the transistor in Problem 19.17.

19.33: Determine the radiative thermal resistance of the pipe in Problem 19.18 if the pipe is held at 107°C and is put into an enclosure at 7°C.

19.34: Determine the radiative thermal resistance of the wall in Problem 19.19.

Combined Mechanisms of Heat Transfer

19.35: A composite wall has an area of 48 m^2 and consists of a 6.25-cm layer of fiberglass insulation ($k = 0.046$ W/m-K) sandwiched between a pair of 2.5-cm-thick white pine boards ($k = 0.24$ W/m-K). The temperatures at the face of the boards are 30°C and 12°C. Determine (a) the thermal resistance of the fiberglass, (b) the thermal resistance of each board, and (c) the heat flow.

19.36: A wall having an area of 8 m^2 is fabricated of corkboard having a thickness of 1.27 cm and oak ($k = 0.35$ W/m-K) having a thickness of 1.905 cm. The corkboard side is in a room that is maintained at 77°C with a heat transfer coefficient of 5 W/m^2-K. The oak side is exposed to a winter environment at 0°C with a wind blowing such that the heat transfer coefficient is 12 W/m^2-K. Determine the heat transferred.

19.37: A plane wall has a surface area of 4 m^2 and a surface temperature of 327°C. One surface of the wall is insulated and the other surface is exposed to enclosure walls and an air flow at 27°C with a heat transfer coefficient of 14.5 W/m^2-K. The surface may be treated as an ideal emitter. Determine the heat transferred from the wall to the surroundings.

19.38: A 2-in × 10-BWG zinc tube is 2-m long and its inside surface is held at 102°C. The tube is covered with a 1.75-cm layer of asbestos insulation. The asbestos is exposed to an environment at 27°C with a heat transfer coefficient of 40 W/m^2-K. Determine (a) the total thermal resistance, (b) the heat transferred, and (c) the temperature at the zinc-asbestos interface.

19.39: The interior of a 2.5-m long 6-in schedule 80 brass (70% copper and 30% zinc) pipe is held at 227°C. The pipe is covered with 1 cm of 85% magnesia insulation ($k = 0.074$ W/m-K) and is placed in an evacuated space at 27°C. The insulation is presumed to have a surface that is an ideal emitter. Determine the heat flow.

19.40: An insulated 6-in-schedule 80 pipe fabricated of 1% carbon steel and 8-m long carries a liquid at 125°C with a heat transfer coefficient of 242 W/m^2-K. The insulation consists of 1 cm of fiberglass ($k = 0.046$ W/m-K). The pipe-insulation entity is in atmospheric air at 15°C with a natural convection heat transfer coefficient of 14.5 W/m^2-K. Determine (a) the thermal resistance of the pipe, (b) the thermal resistance of the insulation, and (c) the heat transferred.

19.41: The interior surface of an 8-cm-outside-diameter hollow brass sphere is held at 150°C. The wall of the sphere is 0.75-cm thick and it is covered with a 0.625-cm layer of asbestos. The exterior of the asbestos is maintained in a 15°C environment where the heat transfer coefficient is 12.2 W/m^2-K. Determine (a) the heat transferred and (b) the temperature at the brass asbestos interface.

19.42: A brass sphere having a diameter of 10 cm has a surface that is considered an ideal emitter. The surface is held at 227°C. Convection and radiation are present at the exterior of the sphere in a 27°C environment (both airstream and enclosure walls) and the heat transfer coefficient is 40 W/m^2-K. Determine (a) the convection thermal resistance and (b) the heat transferred.

The Overall Heat Transfer Coefficient

19.43: In Problem 19.35, what is the overall heat transfer coefficient?

19.44: In Problem 19.36, what is the overall heat transfer coefficient?

19.45: Use the surface at the exterior of the asbestos in Problem 19.38 to determine the overall heat transfer coefficient.

19.46: Use the surface at the exterior of the fiberglass in Problem 19.40 to determine the overall heat transfer coefficient.

19.47: Use the surface at the interior of the sphere in Problem 19.41 to determine the overall heat transfer coefficient.

19.48: Use the surface at the exterior of the sphere in Problem 19.42 to determine the overall heat transfer coefficient.

20

Steady-State Conduction

Chapter Objectives

- To introduce the Fourier law and the equation that governs steady conduction in one dimension.
- To analyze one-dimensional steady conduction through a plane wall, a hollow cylinder, and a hollow sphere.
- To introduce contact resistance and discuss its role in conduction through composite systems.
- To consider the critical radius of insulation.
- To discuss the effect of uniform heat generation on one-dimensional heat conduction.
- To study the performance of longitudinal and radial fins as heat transfer enhancement devices.
- To provide an introduction to two-dimensional steady conduction and to illustrate the conduction shape factor method.

20.1 Introduction

A wide variety of engineering applications involve heat transfer by **conduction**. Unlike convection that pertains to energy transport due to fluid motion and radiation that can propagate in a perfect vacuum, conduction requires the presence of an intervening medium. The medium can either be a solid or a stationary fluid.

At a microscopic level, conduction in stationary fluids is a consequence of molecules at higher temperature interacting and exchanging energy with molecules at a lower temperature. In solids, the transport of energy by conduction involves phenomena such as lattice waves (phonons) and translational motion of free electrons. This microscopic approach is essential in the study of the thermal behavior of superconducting thin films, microsensors, and other micromechanical devices. For a vast majority of engineering applications, a macroscopic study based on the **Fourier law** suffices.

In the study of heat conduction, a distinction is made among **uniform, nonuniform, steady,** and **nonsteady** or **transient** heat flow. Usually, the words "uniform" and "nonuniform" refer to spacial variations, whereas "steady" and "transient" refer to the time domain. We note that steady-state heat flow exists when the temperature does not vary with time at any point in the heat flow path. Conversely, transient heat flow occurs when the temperatures at any point in the heat flow path varies with time.

20.2 The General Equation of Heat Conduction

It was Fourier,[1] who in 1822, proposed that the rate of heat flow by conduction through a material is proportional to the area normal to the heat flow path and to the temperature gradient along the heat flow path. We assemble these two facts into a proportionality

$$\dot{Q}_x \sim -A\frac{dT}{dx}$$

where the minus sign is employed to allow for a positive heat flow in the presence of a falling temperature gradient along the heat flow path. Insertion of a proportionality constant gives us what is called the **Fourier law**

$$\dot{Q}_x = -kA\frac{dT}{dx} \tag{20.1}$$

which serves to define the proportionality constant, k, as the thermal conductivity, a property of the material

$$k \equiv -\frac{\dot{Q}_x}{A(-dT/dx)}$$

With Equation 20.1 in hand, we may now derive the **general equation of heat conduction** by considering the differential volume shown in Figure 20.1. We write an energy balance equating the rate of heat flow entering the differential volume with the rate of heat flow leaving plus the rate of change of internal energy (actually the heat stored in the differential volume) and obtain

$$\dot{Q}_{\text{in}} + \dot{Q}_{\text{gen}} = \dot{Q}_{\text{out}} + \dot{Q}_{\text{stored}}$$

or

$$\dot{Q}_{\text{in}} - \dot{Q}_{\text{out}} + \dot{Q}_{\text{gen}} = \dot{Q}_{\text{stored}} \tag{20.2}$$

Consider the x direction first. Heat will enter the left face in accordance with Equation 20.1

$$\dot{Q}_x = -kA\frac{\partial T}{\partial x} = -\left(k\frac{\partial T}{\partial x}\right)dydz$$

where the temperature gradient is expressed as a partial derivative because the temperature is a function of x, y, z, and t. Heat will leave the right face at $x + dx$

$$\dot{Q}_{x+dx} = \left\{\left(-k\frac{\partial T}{\partial x}\right) + \frac{\partial}{\partial x}\left(-k\frac{\partial T}{\partial x}\right)dx\right\}dydz$$

[1] J. B. J. Fourier (1768–1830) was a French geometrician and physicist who formalized the law of heat propagation and proposed the Fourier series, which is a trigonometric series for arbitrary functions.

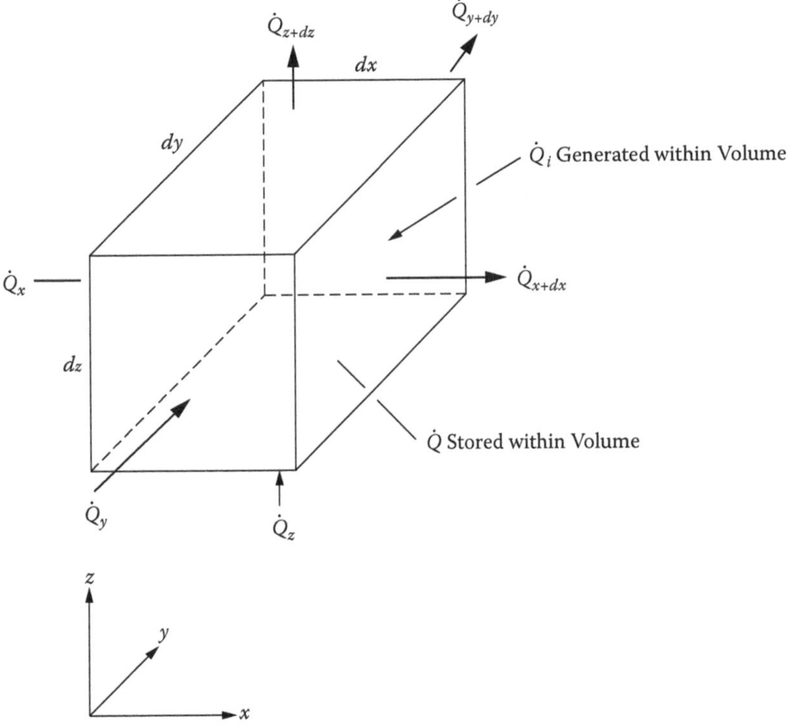

FIGURE 20.1
Differential control volume used to derive the general equation of heat conduction.

so that the difference between the heat entering at x and leaving at $x + dx$ is

$$\dot{Q}_x - \dot{Q}_{x+dx} = \frac{\partial}{\partial x}\left(k\frac{\partial T}{\partial x}\right) dxdydz$$

We may use an identical procedure for the y and z directions

$$\dot{Q}_y - \dot{Q}_{y+dy} = \frac{\partial}{\partial y}\left(k\frac{\partial T}{\partial y}\right) dxdydz$$

and

$$\dot{Q}_z - \dot{Q}_{z+dz} = \frac{\partial}{\partial z}\left(k\frac{\partial T}{\partial z}\right) dxdydz$$

Thus, $\Delta\dot{Q} = \dot{Q}_{\text{in}} - \dot{Q}_{\text{out}}$ in Equation 20.2 is now accounted for and is

$$\Delta\dot{Q} = \left[\frac{\partial}{\partial x}\left(k\frac{\partial T}{\partial x}\right) + \frac{\partial}{\partial y}\left(k\frac{\partial T}{\partial y}\right) + \frac{\partial}{\partial z}\left(k\frac{\partial T}{\partial z}\right)\right] dxdydz \tag{20.3}$$

The heat generated within the differential volume may be an I^2R dissipation, or a chemical or nuclear reaction. With q_i designated as the volumetric heat generation in W/m^3, $\dot{Q}_{\text{gen}}$ in Equation 20.2 is

$$\dot{Q}_{\text{gen}} = q_i dxdydz \tag{20.4}$$

Finally, the heat stored within the differential volume is related to the rate of change of internal energy,

$$\frac{du}{dt} = c(dm)\frac{\partial T}{\partial t}$$

With $dm = \rho dxdydz$, where ρ is the material density, we have

$$\dot{Q}_{\text{stored}} = \frac{du}{dt} = \rho c \frac{\partial T}{\partial t} dxdydz \tag{20.5}$$

When Equations 20.3, 20.4, and 20.5 are inserted into Equation 20.2, the result is

$$\frac{\partial}{\partial x}\left(k\frac{\partial T}{\partial x}\right) + \frac{\partial}{\partial y}\left(k\frac{\partial T}{\partial y}\right) + \frac{\partial}{\partial z}\left(k\frac{\partial T}{\partial z}\right) + q_i = \rho c\frac{\partial T}{\partial t}$$

where the common $dxdydz$ terms have been cancelled. Then, if k, ρ, and c are independent of temperature, position, and time, we obtain the general equation of heat conduction

$$\frac{\partial^2 T}{\partial x^2} + \frac{\partial^2 T}{\partial y^2} + \frac{\partial^2 T}{\partial z^2} + \frac{q_i}{k} = \frac{1}{\alpha}\frac{\partial T}{\partial t} \tag{20.6}$$

where α is called the **thermal diffusivity** of the material

$$\alpha \equiv \frac{k}{c\rho} \qquad (\text{m}^2/\text{s}) \tag{20.7}$$

If the system contains no heat sources, Equation 20.6 becomes the **Fourier equation** (not the Fourier law)

$$\frac{\partial^2 T}{\partial x^2} + \frac{\partial^2 T}{\partial y^2} + \frac{\partial^2 T}{\partial z^2} = \frac{1}{\alpha}\frac{\partial T}{\partial t} \tag{20.8}$$

If steady state is considered, that is, if the temperature distribution does not vary with time, we obtain the **Poisson[2] equation**.

$$\frac{\partial^2 T}{\partial x^2} + \frac{\partial^2 T}{\partial y^2} + \frac{\partial^2 T}{\partial z^2} + \frac{q_i}{k} = 0 \tag{20.9}$$

Finally, in the absence of heat sources and in the steady state, Equation 20.6 reduces to the **Laplace[3] equation**

$$\frac{\partial^2 T}{\partial x^2} + \frac{\partial^2 T}{\partial y^2} + \frac{\partial^2 T}{\partial z^2} = 0 \tag{20.10}$$

which is often written as

$$\nabla^2 T = 0 \tag{20.11}$$

---

[2] S. D. Poisson (1781–1840) was a French mathematician who worked on mathematical physics, contributed to the wave theory of light, and formulated the Poisson ratio governing the elasticity of materials.

[3] P. S. Laplace (1749–1827) was a French astronomer and mathematician who contributed to celestial mechanics, formulated the laws of probability, and discovered the Laplace differential equation.

where ∇^2 is a differential operator known as the **Laplacian**, which in rectangular coordinates, is

$$\nabla^2 \equiv \frac{\partial^2}{\partial x^2} + \frac{\partial^2}{\partial y^2} + \frac{\partial^2}{\partial z^2} = 0$$

20.3 Conduction in Plane Walls

20.3.1 The Single-Material Layer

We first take the simplest possible case, shown in Figure 20.2. In the steady state where there is no heat generation and where the heat flow is in the x direction, Equation 20.11 reduces to the ordinary differential equation

$$\frac{d^2T}{dx^2} = 0 \tag{20.12}$$

which can be integrated twice to yield

$$T = C_1x + C_2 \tag{20.13}$$

The specified boundary conditions are

$$T(x = 0) = T_1 \qquad \text{and} \qquad T(x = L) = T_2$$

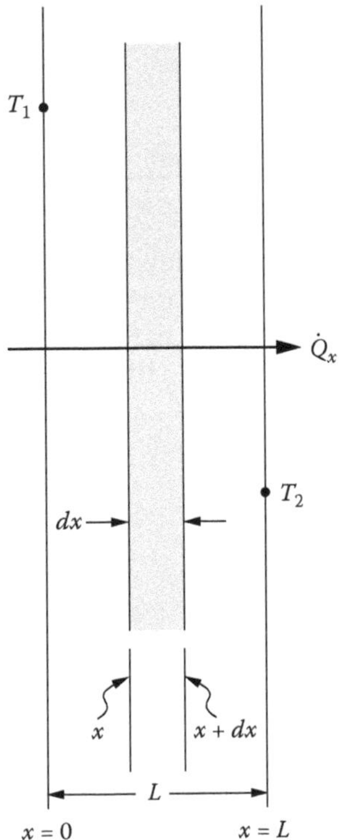

FIGURE 20.2
Single slab used to derive the Fourier law in one dimension.

which may be applied to Equation 20.13 to obtain the pair of simultaneous linear equations

$$T_1 = C_2$$

$$T_2 = C_1 L + C_2$$

The solution to this set is

$$C_2 = T_1 \qquad \text{and} \qquad C_1 = \frac{T_2 - T_1}{L}$$

so that the temperature distribution in the plane wall is

$$T(x) = \frac{T_2 - T_1}{L} x + T_1 \tag{20.14}$$

Two conclusions have emerged from this analysis:

1. The first integration of Equation 20.12 yields dT/dx = constant, which coupled with Equation 20.1, yields $\dot{Q}_x$ = constant. Thus, for steady-state one-dimensional conduction, in a plane wall with no energy generation, the heat flux is constant.
2. The temperature distribution in a plane wall with constant thermal conductivity and no energy generation varies linearly with the thickness coordinate, x.

We can evaluate dT/dx from Equation 20.14

$$\frac{dT}{dx} = \frac{T_2 - T_1}{L}$$

and use this result in Equation 20.1 to obtain

$$\dot{Q}_x = \frac{kA(T_1 - T_2)}{L} \tag{19.3}$$

Example 20.1 Consider the plane wall of Figure 20.3 with the face at $x = 0$ kept at T_1 but with the face at $x = L$ cooled by a fluid at temperature T_∞ under a heat transfer coefficient of h. Determine the temperature distribution in the wall, the surface temperature at $x = L$, and the rate of heat transfer by (a) integrating the heat conduction equation and (b) using the electrothermal analog.

Solution

Assumptions and Specifications

(1) Steady-state conditions exist.
(2) The heat flow is one dimensional.
(3) Thermal properties do not vary with temperature.
(4) Conduction and convection are the modes of heat transfer.

(a) The heat conduction equation

$$\frac{d^2T}{dx^2} = 0 \tag{20.12}$$

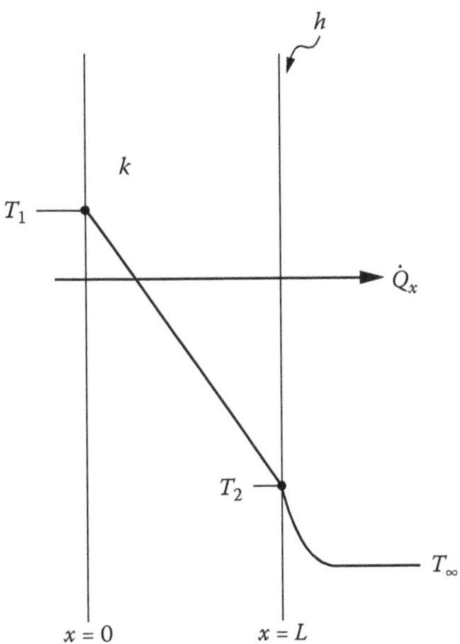

FIGURE 20.3
Plane wall with convection at its right face.

is to be solved subject to the boundary conditions

$$T(x - 0) = T_1 \qquad \text{and} \qquad -k\frac{dT}{dx}\bigg|_{x=L} = h(T_2 - T_\infty)$$

where, for the plane wall, the cross-sectional area, A, is equal to the surface area, S.

Integration of the heat conduction equation and application of the foregoing boundary conditions gives

$$T = T_1 - \frac{h(T_1 - T_\infty)x}{hL + k} \Longleftarrow$$

and the temperature at the right face of the wall at $x = L$ will be

$$T_2 = T_1 - \frac{h(T_1 - T_\infty)L}{hL + k} \Longleftarrow$$

The Fourier law can now be used to determine $\dot{Q}_x$.

$$\dot{Q}_x = -kA\frac{dT}{dx} = \frac{kAh(T_1 - T_\infty)}{hL + k} = \frac{T_1 - T_\infty}{\dfrac{L}{kA} + \dfrac{1}{hA}} \Longleftarrow$$

(b) The thermal network (Figure 20.4) consists of a conduction resistance, L/kA, in a series with a surface convection resistance, $1/hS$. With $S = A$, the value of $\dot{Q}_x$ can be

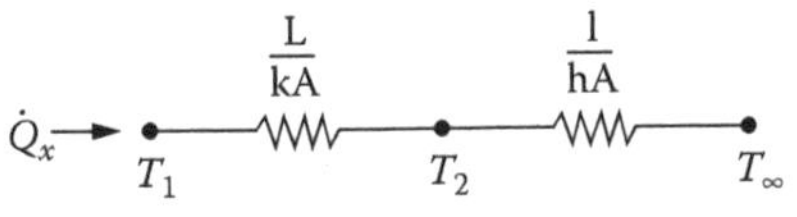

$\dot{Q}_x$ → T_1 —[$\frac{L}{kA}$]— T_2 —[$\frac{1}{hA}$]— T_∞

FIGURE 20.4
Electrothermal analog network for Example 20.1.

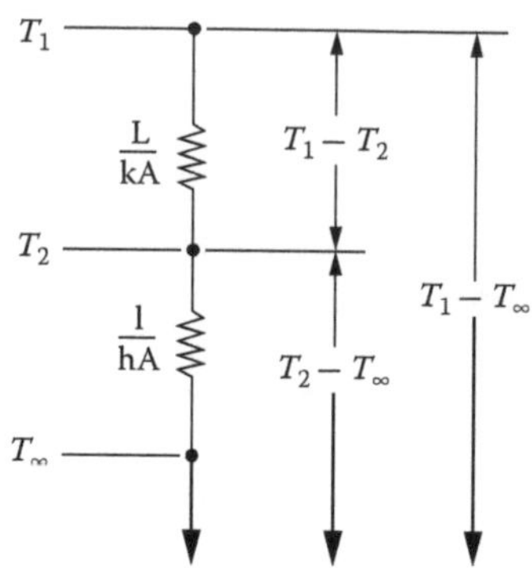

FIGURE 20.5
Electrothermal analog network used to consider the voltage divider concept for Example 20.1.

readily obtained by dividing the overall temperature difference, $T_1 - T_\infty$, by the sum of the resistances, L/kA and $1/hA$:

$$\dot{Q}_x = \frac{T_1 - T_\infty}{\frac{L}{kA} + \frac{1}{hA}} \Longleftarrow$$

This is the rate of heat transfer established in part (a).

Because

$$\dot{Q}_x = \frac{T_1 - T_2}{\frac{L}{kA}}$$

we obtain T_2 as

$$T_2 = T_1 - \frac{\dot{Q}_x L}{kA} = T_1 - \frac{(T_1 - T_\infty)L}{(L/kA + 1/hA)\,kA} = T_1 - \frac{h(T_1 - T_\infty)L}{hL + k} \Longleftarrow$$

Those who are familiar with the voltage divider concept (Figure 20.5) may note that

$$\frac{T_1 - T_2}{T_1 - T_\infty} = \frac{\frac{L}{kA}}{\frac{L}{kA} + \frac{1}{hA}}$$

or

$$T_2 = T_1 - \frac{h(T_1 - T_\infty)L}{hL + k}$$

Finally, the temperature distribution can be expressed in terms of T_1, T_2, x, and L using Equation 20.14

$$T(x) = \frac{T_2 - T_1}{L}x + T_1$$

Thus, when we employ the expression for T_2, we have, after simplification

$$T(x) = T_1 - \frac{h(T_1 - T_\infty)x}{hL + k}$$

Observe that the results of the two approaches are exactly the same.

Example 20.2 A glass window with a thermal conductivity of $k = 1.4\,\text{W-m/K}$ is 90-cm high, 50-cm wide, and 6-mm thick. On a winter day, forced air heating maintains the inside surface of the window at T_1 (Figure 20.6). The outside surface of the window is at $T_2 = 10°\text{C}$ and the window loses heat by convection and radiation to the outside world at $T_\infty = 5°\text{C}$. The convection heat transfer coefficient is $10\,\text{W/m}^2\text{-K}$ and the emissivity of the outside window surface is $\epsilon = 0.92$. Determine (a) the convective and radiative heat losses from the outside surface and (b) the inside surface temperature of the window.

Solution

Assumptions and Specifications

(1) Steady-state conditions exist.
(2) The heat flow is one dimensional.
(3) Thermal properties do not vary with temperature.
(4) Conduction, convection, and radiation are the modes of heat transfer.

(a) In Figure 20.6, we use Equations 19.8 and 19.11 to compute the convection and radiation heat loss. In both cases, $A = S = (0.90\,\text{m})(0.50\,\text{m}) = 0.45\,\text{m}^2$. From Equation 19.8

$$\begin{aligned}\dot{Q}_{\text{conv}} &= hA(T_1 - T_\infty)\\ &= (10\,\text{W/m}^2)(0.45\,\text{m}^2)(10°\text{C} - 5°\text{C}) = 22.50\,\text{W} \Longleftarrow\end{aligned}$$

and from Equation 20.11 with $T_2 = 283\,\text{K}$ and $T_\infty = 278\,\text{K}$

$$\begin{aligned}\dot{Q}_{\text{rad}} &= \epsilon\sigma A(T_1^4 - T_\infty^4)\\ &= (0.92)(5.67 \times 10^{-8}\,\text{W/m}^2\text{-K}^4)(0.45\,\text{m}^2)[(283\,\text{K})^4 - (278\,\text{K})^4]\\ &= 10.36\,\text{W} \Longleftarrow\end{aligned}$$

(b) An energy balance at the outer surface provides

$$\dot{Q}_{\text{cond}} = \dot{Q}_{\text{conv}} + \dot{Q}_{\text{rad}}$$

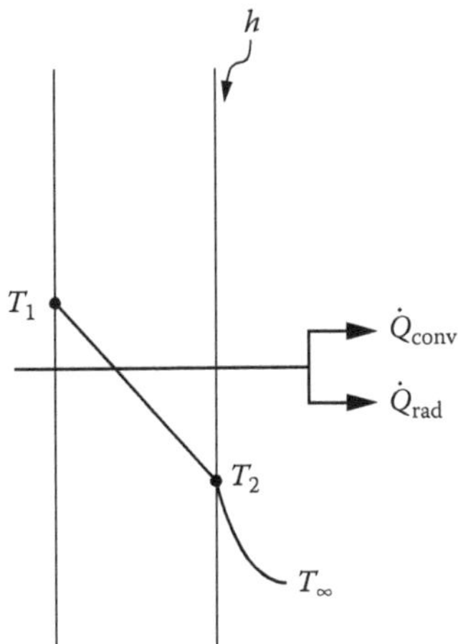

FIGURE 20.6
Configuration for Example 20.2.

or

$$\frac{kA(T_1 - T_2)}{L} = 22.50\,\text{W} + 10.36\,\text{W} = 32.86\,\text{W}$$

We may then solve for T_1

$$T_1 = (32.86\,\text{W})\frac{L}{kA} + T_2$$

$$= \frac{(32.86\,\text{W})(0.006\,\text{m})}{(1.4\text{W/m-})(0.45\text{m}^2)} + 10°\text{C} = 10.31°\text{C} \Longleftarrow$$

20.3.2 The Composite Plane Wall

Most engineering applications involve heat conduction through composite walls consisting of two or more layers of material. A typical example is the building wall, which is a composite of plaster board, fiberglass insulation and either brick, wood, or aluminum siding.

Figure 20.7 shows a composite multilayered wall consisting of n layers and with heat flow from left to right. The ith layer has thickness, L_i and thermal conductivity, k_i, and the temperatures at its extremes are T_1 and T_{n+1}. We allow for a convective heat input at the left (from $T_{\infty,1}$ with h_1) and a heat loss at the right (to $T_{\infty,n+1}$ with h_{n+1}).

The electrothermal analog for the n-layer wall is shown in Figure 20.8a. The heat flow is clearly the same in all layers (convective and conductive) and the total thermal resistance of the n-conduction layers and two convection layers will be

$$R_T = \frac{1}{h_1 S} + \sum_{i=1}^{n} \frac{L_i}{k_i A} + \frac{1}{h_{n+1} S}$$

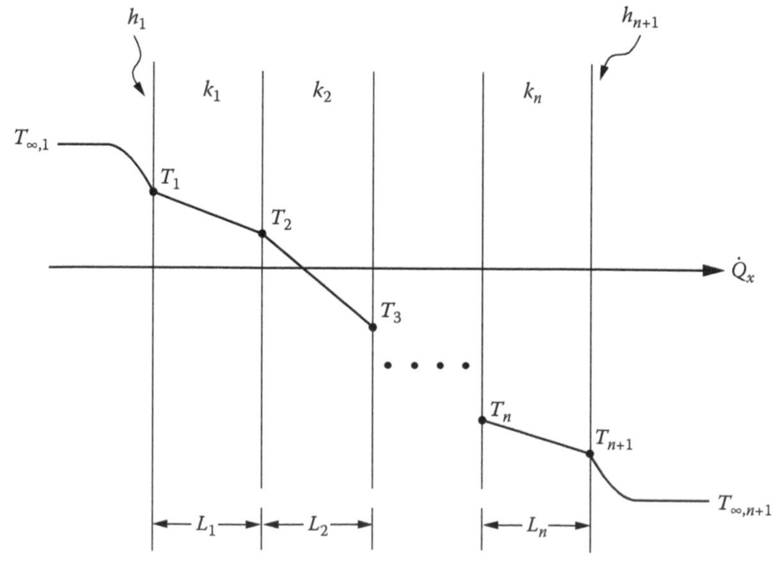

FIGURE 20.7
Composite plane wall with convection on both exterior faces.

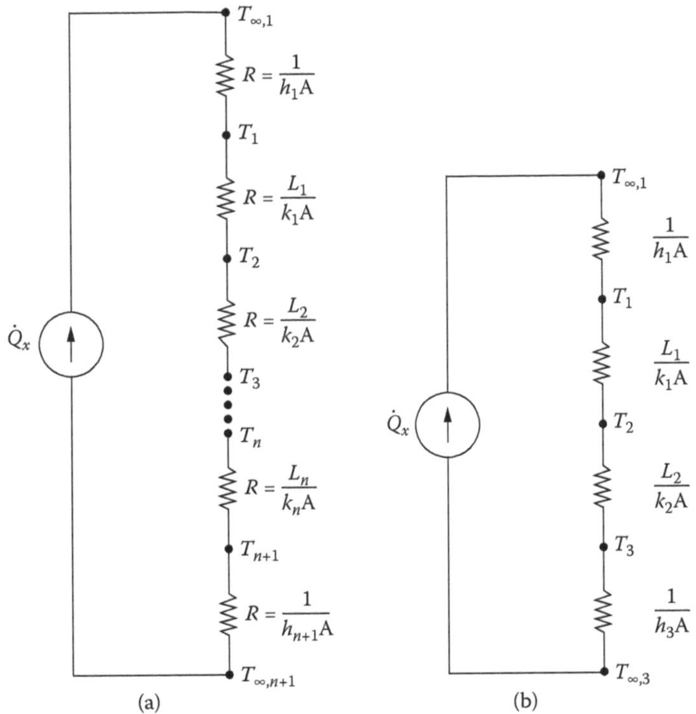

FIGURE 20.8
Electrothermal analog network with convection for (a) the n-layer wall and (b) the two-layer wall.

Then, with $S = A$, we find that the heat flow, $\dot{Q}_x$, will be $\Delta T/R_T$

$$\dot{Q}_x = \frac{T_{\infty,1} - T_{\infty,n+1}}{\dfrac{1}{h_1 S} + \displaystyle\sum_{i=1}^{n} \frac{L_i}{k_i A} + \dfrac{1}{h_{n+1} S}} \tag{20.15}$$

and if we have just a pair of plane layers, again, with convection, (Figure 20.9) with the electrothermal analog in Figure 20.8b,

$$\dot{Q}_x = \frac{T_{\infty,1} - T_{\infty,n+1}}{R_T} = \frac{T_{\infty,1} - T_{\infty,n+3}}{\dfrac{1}{h_1 A} + \dfrac{L_1}{k_1 A} + \dfrac{L_2}{k_2 A} + \dfrac{1}{h_3 A}}$$

Sometimes, the composite wall construction is such that its electrothermal analog network involves resistances in parallel as well as in a series. Such a scenario is illustrated in Figure 20.10. If materials 2 and 3 have comparable thermal conductivities, the heat conduction may be assumed to be one dimensional, that is, the surfaces normal to the direction of heat conduction are isothermal. With the height of each component designated as H and with Z taken as the width normal to the plane of the paper, the rate of heat transfer by conduction can be expressed as

$$\frac{\dot{Q}_x}{W} = \frac{T_1 - T_4}{\dfrac{L_1}{k_1 H_1} + \dfrac{L_2 L_3}{k_2 H_2 L_3 + k_3 H_3 L_2} + \dfrac{L_4}{k_4 H_4}} \tag{20.16a}$$

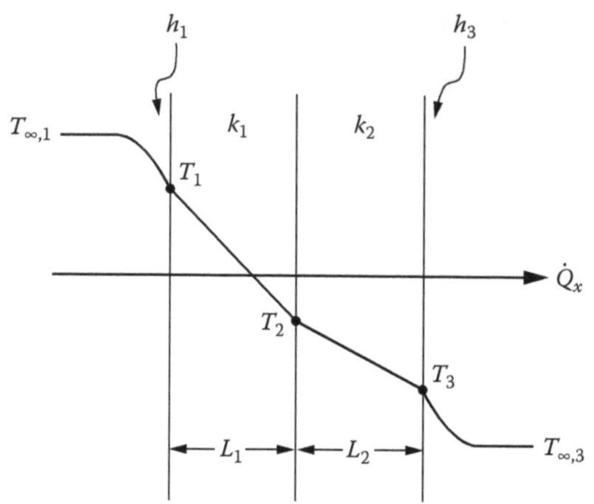

FIGURE 20.9
The composite wall of two layers with convection at both exterior surfaces.

or because $L_2 = L_3$

$$\frac{\dot{Q}_x}{Z} = \frac{T_1 - T_4}{\dfrac{L_1}{k_1 H_1} + \dfrac{L_2}{k_2 H_2 + k_3 H_3} + \dfrac{L_4}{k_4 H_4}} \tag{20.16b}$$

where $H_1 = H_2 + H_3 = H_4$ and the rule for combining resistances in parallel

$$R_{\|} = \frac{R_1 R_2}{R_1 + R_2}$$

has been employed.

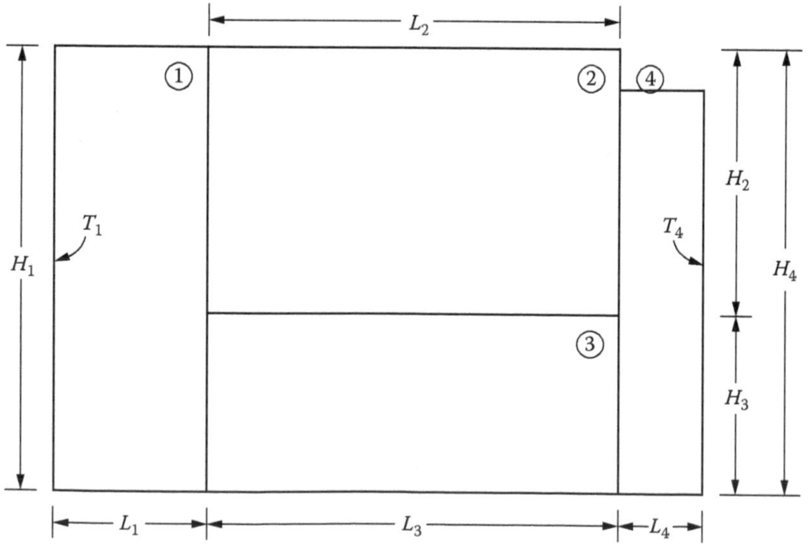

FIGURE 20.10
Multicomponent plane wall with heat flow paths in parallel.

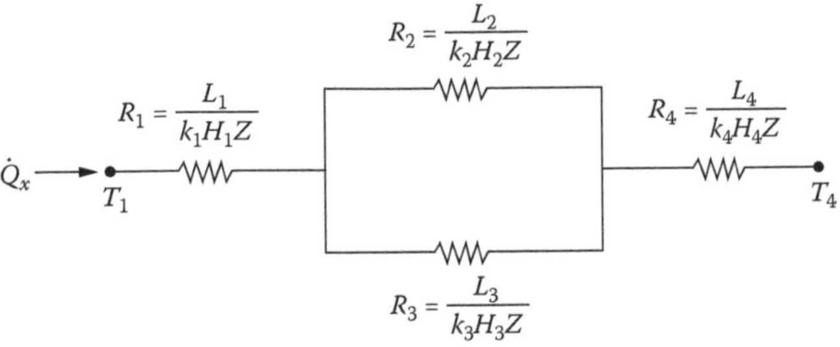

FIGURE 20.11
Electrothermal analog for composite wall in Example 20.3.

Once $\dot{Q}_x$ has been determined, the interface temperatures, T_1 and T_2, may be established

$$T_2 = T_1 - \left(\frac{\dot{Q}_x}{W}\right)\left(\frac{L_1}{k_1 H_1}\right) \tag{20.17}$$

and

$$T_3 = T_4 + \left(\frac{\dot{Q}_x}{W}\right)\left(\frac{L_4}{k_4 H_4}\right) \tag{20.18}$$

The foregoing is illustrated in Example 20.3 that now follows.

Example 20.3 Determine the rate of conduction heat flow through the composite wall structure shown in Figure 20.10. The wall dimensions are $L_1 = 5\,\text{cm}$, $H_1 = 12\,\text{cm}$, $L_2 = L_3 = 10\,\text{cm}$, $H_2 = 5\,\text{cm}$, $H_3 = 7\,\text{cm}$, $L_4 = 4\,\text{cm}$, and $W = 1\,\text{m}$. The thermal conductivities may be taken as $k_1 = 2\,\text{W/m-K}$, $k_2 = 0.4\,\text{W/m-K}$, $k_3 = 0.5\,\text{W/m-K}$, and $k_4 = 12\,\text{W/m-K}$ and the surface temperatures are $T_1 = 80°\text{C}$ and $T_4 = 30°\text{C}$ (Figure 20.10).

Solution

Assumptions and Specifications

(1) Steady-state conditions exist.
(2) The heat flow is one dimensional.
(3) Thermal properties do not vary with temperature.
(4) Conduction is the only mode of heat transfer.
(5) There is no contact resistance between any of the components.

The thermal resistances are calculated first:

$$R_1 = \frac{L_1}{k_1 H_1 W} = \frac{0.05\,\text{m}}{(2\,\text{W/m-K})(0.12\,\text{m})(1\,\text{m})} = 0.2083\,\text{K/W}$$

$$R_2 = \frac{L_2}{k_2 H_2 W} = \frac{0.10\,\text{m}}{(0.4\,\text{W/m-K})(0.05\,\text{m})(1\,\text{m})} = 5.0000\,\text{K/W}$$

$$R_3 = \frac{L_3}{k_3 H_3 W} = \frac{0.10\,\text{m}}{(0.5\,\text{W/m-K})(0.07\,\text{m})(1\,\text{m})} = 2.8571\,\text{K/W}$$

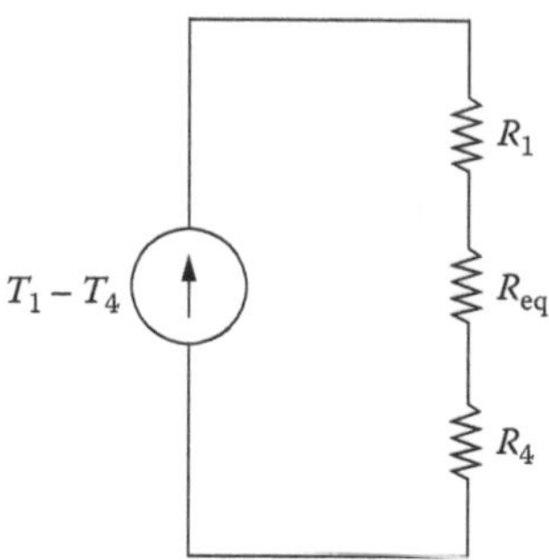

FIGURE 20.12
Simplification of electrothermal analog network for composite wall in Example 20.3.

and

$$R_4 = \frac{L_4}{k_4 H_4 Z} = \frac{0.04\,\text{m}}{(12\,\text{W/m-K})(0.12\,\text{m})(1\,\text{m})} = 0.0278\,\text{K/W}$$

The equivalent of R_2 and R_3 in parallel will be R_{eq} (Figures 20.11 and 20.12)

$$R_{eq} = \frac{R_2 R_3}{R_2 + R_3} = \frac{(5.0000\,\text{K/W})(2.8571\,\text{K/W})}{5.0000\,\text{K/W} + 2.8571\,\text{K/W}} = 1.8182\,\text{K/W}$$

Then, use of the electrothermal analog network in Figure 20.12 provides the rate of heat transfer

$$\dot{Q}_x = \frac{T_1 - T_4}{R_1 + R_e + R_4} = \frac{80°\text{C} - 30°\text{C}}{(0.2083 + 1.8182 + 0.0278)\,\text{K/W}} = 24.3\,\text{W} \Longleftarrow$$

20.3.3 Contact Resistance

The conduction analysis for a composite wall presented in Section 20.3.2 assumed a perfect contact at the interface between the two materials. In reality, the mating surfaces are rough and the actual contact occurs at discrete points (**asperities** or **peaks**) as illustrated in Figure 20.13. The voids between the two surfaces are filled with air or some other material. As indicated in Figure 20.14, the departure from perfect contact is manifested in the form of a temperature drop, ΔT_c in K, across the contact and the ratio of this temperature drop

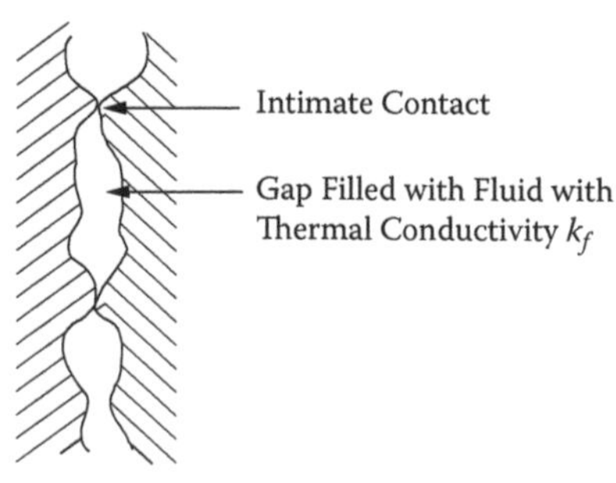

FIGURE 20.13
Artist's conception of two metals in contact.

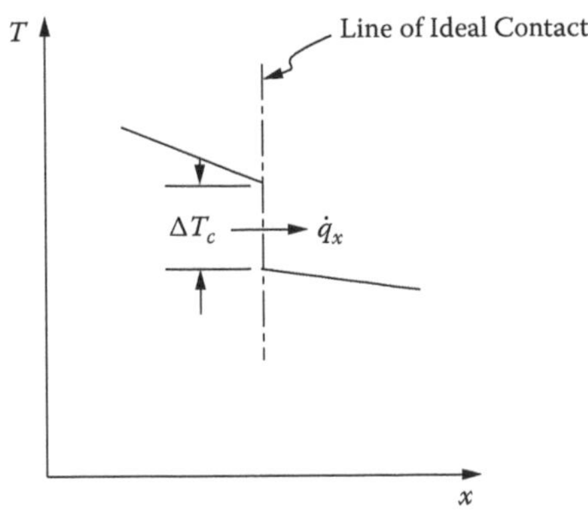

FIGURE 20.14
Temperature drop across contacting surfaces.

to the heat flux, $\dot{q}_x$ in W/m^2, is defined as the contact resistance,

$$R_c = \frac{\Delta T_c}{\dot{q}_x} \qquad (\text{K-m}^2/\text{W}) \tag{20.19}$$

Because R_c is defined on the basis of a heat flux and not the rate of heat transfer, it should be divided by the contact area, A, and then incorporated into the construction of the electrothermal network.

The parameter R_c is a complicated function of the contact pressure, surface roughnesses, thermal conductivities, the microhardness of the mating surfaces and the material trapped in the voids. Additional complications arise when bolts, screws, or rivets are used to fasten the surfaces. An excellent review is provided by Yovanovich and Marotta in Bejan and Kraus (2001).

In the thermal control of microelectronic equipment, it is crucial to minimize the contact resistance between mating surfaces. This is achieved by treating the surfaces with silicon-based thermal grease or a soft metallic foil prior to the fastening of the two materials. Table 20.1 provides some typical contact resistances of metal joints.

Example 20.4 A composite wall is made of two materials with thermal conductivities $k_1 = 5\,\text{W/m-K}$ and $k_2 = 7\,\text{W/m-K}$ and with thicknesses $L_1 = 15\,\text{mm}$ and $L_2 = 25\,\text{mm}$

TABLE 20.1
Contact Resistance of Metal Joints in the Air at Moderate Temperatures

Material	Roughness, μm	Contact Pressure, MPa	R_c, K-m^2/W
Ground 416 stainless steel	2.54	0.3–2.5	2.63×10^{-4}
Ground 304 stainless steel	1.14	4–7	5.26×10^{-4}
Ground aluminum	2.54	1.2–2.5	8.77×10^{-5}
Ground copper	1.27	1.2–20	6.99×10^{-6}
Ground 416 stainless steel	2.54	0.30–2.50	2.63×10^{-4}
Stainless steel-aluminum	20–30	10	3.45×10^{-4}
		20	2.78×10^{-4}
Stainless steel-aluminum	1.3–2.0	10	6.10×10^{-5}
		20	4.81×10^{-5}
Ground aluminum-copper	1.3–1.4	5	2.38×10^{-5}
		15	1.79×10^{-5}
Milled aluminum-copper	4.4–4.5	10	8.33×10^{-5}
	4.4–4.5	20–35	4.45×10^{-4}

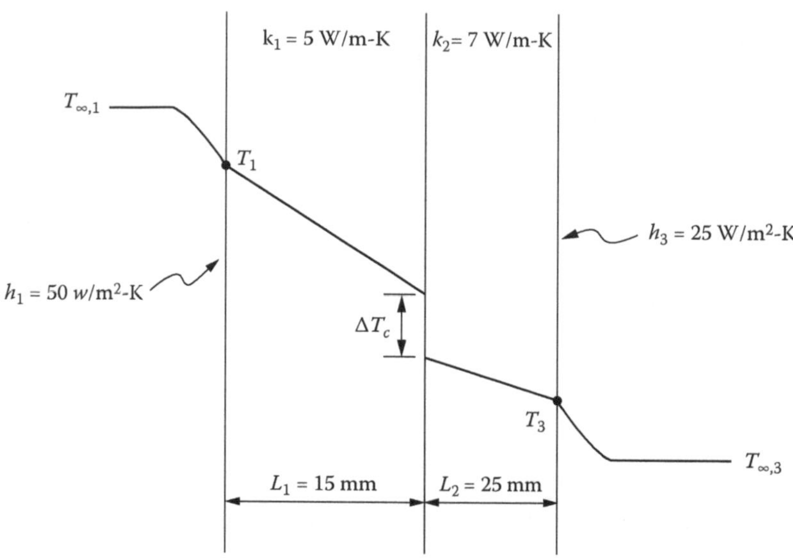

FIGURE 20.15
Configuration for Example 20.4.

(Figure 20.15). The left face of the wall is exposed to a hot fluid at $T_{\infty,1} = 80°\text{C}$ with a heat transfer coefficient of $h_1 = 50\,\text{W/m}^2\text{-K}$. The right face is in contact with cold air at $T_{\infty,3} = 25°\text{C}$ with a heat transfer coefficient of $h_3 = 25\,\text{W/m}^2\text{-K}$. The wall area normal to the heat flow direction is $A = 0.80\,\text{m}^2$. Determine the rate of heat transfer, the surface temperatures, T_1 and T_3, and the temperature drop, ΔT_c, across the contact assuming (a) perfect contact and (b) allowing for a contact resistance of $R_c = 0.02\,\text{m}^2\text{-K/W}$.

Solution

Assumptions and Specifications

(1) Steady-state conditions exist.
(2) The heat flow is one dimensional.
(3) Thermal properties do not vary with temperature.
(4) Conduction and convection are the modes of heat transfer.

(a) Equation 20.15 is applicable, and for the case of no contact resistance may be used to compute $\dot{Q}_x$.

$$\dot{Q}_x = \frac{T_{\infty,1} - T_{\infty,3}}{\dfrac{1}{h_1 A} + \dfrac{L_1}{k_1 A} + \dfrac{L_2}{k_2 A} + \dfrac{1}{h_3 A}}$$

The individual resistances in the denominator are

$$\frac{1}{h_1 A} = \frac{1}{(50\,\text{W/m}^2\text{-K})(0.80\,\text{m}^2)} = 0.0250\,\text{K/W}$$

$$\frac{L_1}{k_1 A} = \frac{0.015\,\text{m}}{(5\,\text{W/m-K})(0.80\,\text{m}^2)} = 3.750 \times 10^{-3}\,\text{K/W}$$

$$\frac{L_2}{k_2 A} = \frac{0.025\,\text{m}}{(7\,\text{W/m-K})(0.80\,\text{m}^2)} = 4.464 \times 10^{-3}\,\text{K/W}$$

and

$$\frac{1}{h_3 A} = \frac{1}{(25\,\text{W/m}^2\text{-K})(0.80\,\text{m}^2)} = 0.0500\,\text{K/W}$$

With these values put into Equation 20.15, we obtain

$$\dot{Q}_x = \frac{80^\circ\text{C} - 25^\circ\text{C}}{(25.000 + 3.750 + 4.464 + 50.000) \times 10^{-3}\,\text{K/W}}$$

$$= \frac{55^\circ\text{C}}{0.0832\,\text{K/W}} = 661\,\text{W} \Longleftarrow$$

The temperatures, T_1 and T_2, derive from Ohm's law in a thermal sense

$$T_1 = T_{\infty,1} - \frac{\dot{Q}_x}{h_1 A} = 80^\circ\text{C} - \frac{661\text{W}}{(50\,\text{W/m}^2\text{-K})(0.80\,\text{m}^2)} = 63.5^\circ\text{C} \Longleftarrow$$

and

$$T_3 = T_{\infty,3} + \frac{\dot{Q}_x}{h_2 A} = 25^\circ\text{C} + \frac{661\,\text{W}}{(25\,\text{W/m}^2\text{-K})(0.80\,\text{m}^2)} = 58.1^\circ\text{C} \Longleftarrow$$

and for a perfect contact

$$\Delta T_c = 0 \Longleftarrow$$

(b) With the inclusion of contact resistance, Equation 20.15 must be modified to

$$\dot{Q}_x = \frac{T_{\infty,1} - T_{\infty,2}}{\dfrac{1}{h_1 A} + \dfrac{L_1}{k_1 A} + \dfrac{R_c}{A} + \dfrac{L_2}{k_2 A} + \dfrac{1}{h_3 A}}$$

With

$$\frac{R_c}{A} = \frac{0.02\,\text{m}^2\text{-K/W}}{0.80\,\text{m}^2} = 0.0250\,\text{K/W}$$

we have

$$\dot{Q}_x = \frac{80^\circ\text{C} - 25^\circ\text{C}}{(0.0832\,\text{K/W} + 0.0250\,\text{K/W})} = \frac{55^\circ\text{C}}{0.1082\,\text{K/W}} = 508\,\text{W} \Longleftarrow$$

and we note that the presence of the contact resistance reduces the rate of heat transfer by about 23%.

We again find T_1 and T_2 from Equation 20.21

$$T_1 = T_{\infty,1} - \frac{\dot{Q}_x}{h_1 A} = 80^\circ\text{C} - \frac{508\,\text{W}}{(50\,\text{W/m}^2\text{-K})(0.80\,\text{m}^2)} = 67.3^\circ\text{C} \Longleftarrow$$

and

$$T_3 = T_{\infty,3} + \frac{\dot{Q}_x}{h_3 A} = 25^\circ\text{C} + \frac{508\,\text{W}}{(25\,\text{W/m}^2\text{-K})(0.80\,\text{m}^2)} = 50.4^\circ\text{C} \Longleftarrow$$

Finally, Equation 20.19 permits us to find ΔT_c

$$\Delta T_c = \frac{R_c \dot{Q}_x}{A} = \frac{(0.02\text{m}^2\text{-K/W})(508\,\text{W})}{0.80\,\text{m}^2} = 12.7^\circ\text{C} \Longleftarrow$$

20.4 Radial Heat Flow

20.4.1 Cylindrical Coordinates

The Laplace equation in cylindrical coordinates can be developed in a manner similar to the development in Section 20.2. In the single, radial dimension, this equation becomes the ordinary differential equation

$$\frac{d}{dr}\left(r\frac{dT}{dr}\right) = 0 \tag{20.20}$$

and we use this equation to develop the heat flow in the radial direction and the temperature distribution in a variety of cylindrical configurations.

20.4.1.1 The Hollow Cylinder

Figure 20.16 shows the end view of a hollow cylinder of inside radius, r_1, outside radius, r_2, length, L, and thermal conductivity, k. The inside and outside surfaces are maintained at the constant temperatures T_1 and T_2 and the heat flows in the radial direction. As indicated in Section 20.4.1, the differential equation that applies in this case is Equation 20.20.

The boundary conditions are

$$T(r = r_1) = T_1 \qquad \text{and} \qquad T(r = r_2) = T_2$$

and solution of this equation with the foregoing boundary conditions yields

$$\dot{Q}_r = \frac{2\pi k L(T_1 - T_2)}{\ln(r_2/r_1)} \tag{20.21}$$

Here it is worth noting that the temperature distribution is logarithmic and that the heat flux, $\dot{q}_r$, varies inversely with the radius but the rate of heat transfer is independent of the radius.

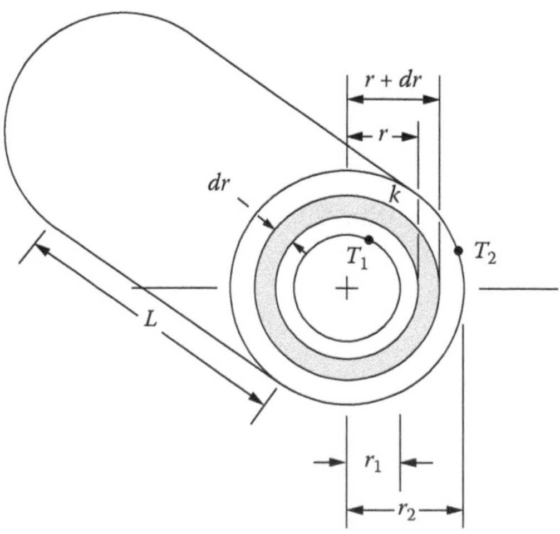

FIGURE 20.16
Hollow cylinder used to establish the Fourier law in cylindrical coordinates.

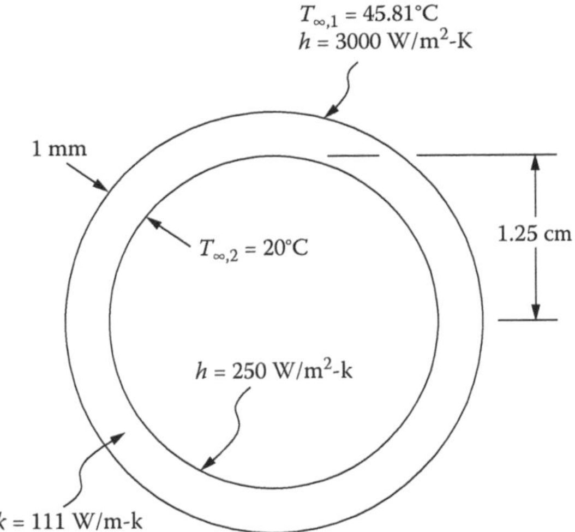

FIGURE 20.17
Hollow cylinder for Example 20.5.

The thermal resistance for the wall of a hollow cylinder is identified by rewriting Equation 20.21 and obtaining Equation 19.12b

$$R_{\text{cond}} = \frac{T_1 - T_2}{\dot{Q}_r} = \frac{\ln(r_2/r_1)}{2\pi k L} \tag{20.22}$$

and if convection occurs on the inside of the cylinder (at $T_{\infty,1}$ with h_1) and on the outside of the cylinder (at $T_{\infty,2}$ with h_2), then the rate of heat transfer may be expressed as

$$\dot{Q}_r = \frac{T_{\infty,1} - T_{\infty,2}}{\dfrac{1}{h_1(2\pi r_1 L)} + \dfrac{\ln(r_2/r_1)}{2\pi k L} + \dfrac{1}{h_2(2\pi r_2 L)}} \tag{20.23}$$

Example 20.5 A typical condenser in a steam power plant consists of hundreds (and sometimes thousands) of metal tubes that are housed in a shell. Steam condenses on the outside of the tubes as the cooling water flows inside. Figure 20.17 shows one such tube, made of brass (k = 111 W/m-K) and having an inside diameter of 2.5 cm and a thickness of 1 mm. The temperature of the condensing steam is 45.81°C and the associated heat transfer coefficient is 3000 W/m^2-K. The cooling water, at an average temperature of 20°C, provides a convection heat transfer coefficient of 250 W/m^2-K. Determine (a) the rate of heat transfer from the steam to the cooling water per meter of length of tube and (b) the rate of condensation of the steam per meter of length of tube if h_{fg} = 2392.8 kJ/kg.

Solution

Assumptions and Specifications

(1) Steady-state conditions exist.
(2) The heat flow is one dimensional.
(3) Thermal properties do not vary with temperature.
(4) Conduction and convection are the modes of heat transfer.

Here

$$r_1 = \frac{d_1}{2} = \frac{2.500\,\text{cm}}{2} = 1.250\,\text{cm} \quad \text{and} \quad r_2 = r_1 + 0.100\,\text{cm} = 1.350\,\text{cm}$$

(a) Equation 20.23 with $T_{\infty,2}$ and $T_{\infty,1}$ reversed because the heat flow is inward

$$\dot{Q}_r = \frac{T_{\infty,2} - T_{\infty,1}}{\dfrac{1}{h_1(2\pi r_1 L)} + \dfrac{\ln(r_2/r_1)}{2\pi k L} + \dfrac{1}{h_2(2\pi r_2 L)}} \tag{20.23}$$

may be used to determine $\dot{Q}_r$. With $L = 1\,\text{m}$, we can compute the individual thermal resistances as

$$\frac{1}{h_1(2\pi r_1 L)} = \frac{1}{(250\,\text{W/m}^2\text{-K})(2\pi)(0.0125\,\text{m})(1\,\text{m})} = 0.0509\,\text{K/W}$$

$$\frac{\ln(r_2/r_1)}{2\pi k_1 L} = \frac{\ln(0.0135/0.0125)}{(2\pi)(111\,\text{W/m-K})(1\,\text{m})} = 1.103 \times 10^{-4}\,\text{K/W}$$

and

$$\frac{1}{h_2(2\pi r_2 L)} = \frac{1}{(3000\,\text{W/m}^2\text{-K})(2\pi)(0.0135\,\text{m})(1\,\text{m})} = 3.930 \times 10^{-3}\,\text{K/W}$$

Then, we have

$$\dot{Q}_r = \frac{45.81^\circ\text{C} - 20^\circ\text{C}}{0.0509\,\text{K/W} + 1.103 \times 10^{-4}\,\text{K/W} + 3.930 \times 10^{-3}\,\text{K/W}}$$

and with a total resistance of 0.0550 K/W, the result is

$$\dot{Q}_r = 469.5\ W \Longleftarrow$$

(b) The rate of condensate per meter of length of tube is

$$\dot{m} = \frac{\dot{Q}_r}{h_{fg}} = \frac{469.5\,\text{W}}{2392.8\,\text{kJ/kg}} = 1.962 \times 10^{-4}\,\text{kg/s} \Longleftarrow$$

20.4.1.2 The Composite Hollow Cylinder

Figure 20.18a shows a composite hollow cylinder composed of n layers. The composite cylinder is heated by convection on the inside surface at r_1 where the heat transfer coefficient is h_1 and the temperature of the fluid on the inside is $T_{\infty,1}$. The outer cylinder is convectively cooled at r_{n+1}, where the heat transfer coefficient is h_{n+1} and the temperature of the fluid on the outside is $T_{\infty,\,n+1}$. Heat flow occurs in the radially outward direction. The electrothermal analog network of Figure 20.18b facilitates the analysis. The rate of heat transfer for n layers can be expressed in the form of Equation 20.23

$$\dot{Q}_r = \frac{T_{\infty,1} - T_{\infty,n+1}}{\dfrac{1}{h_1(2\pi r_1 L)} + \displaystyle\sum_{i=1}^{n} \dfrac{\ln(r_{i+1}/r_i)}{2\pi k_i L} + \dfrac{1}{h_{n+1}(2\pi r_{n+1} L)}} \tag{20.24}$$

Example 20.6 Figure 20.19 shows a steel steam pipe ($k = 15\,\text{W/m-K}$) with an inside radius of 10 cm and an outside radius of 12 cm covered with a 3-cm-thick layer of calcium silicate with a thermal conductivity of $k = 0.2\,\text{W/m-K}$. A 2-cm layer of fiberglass

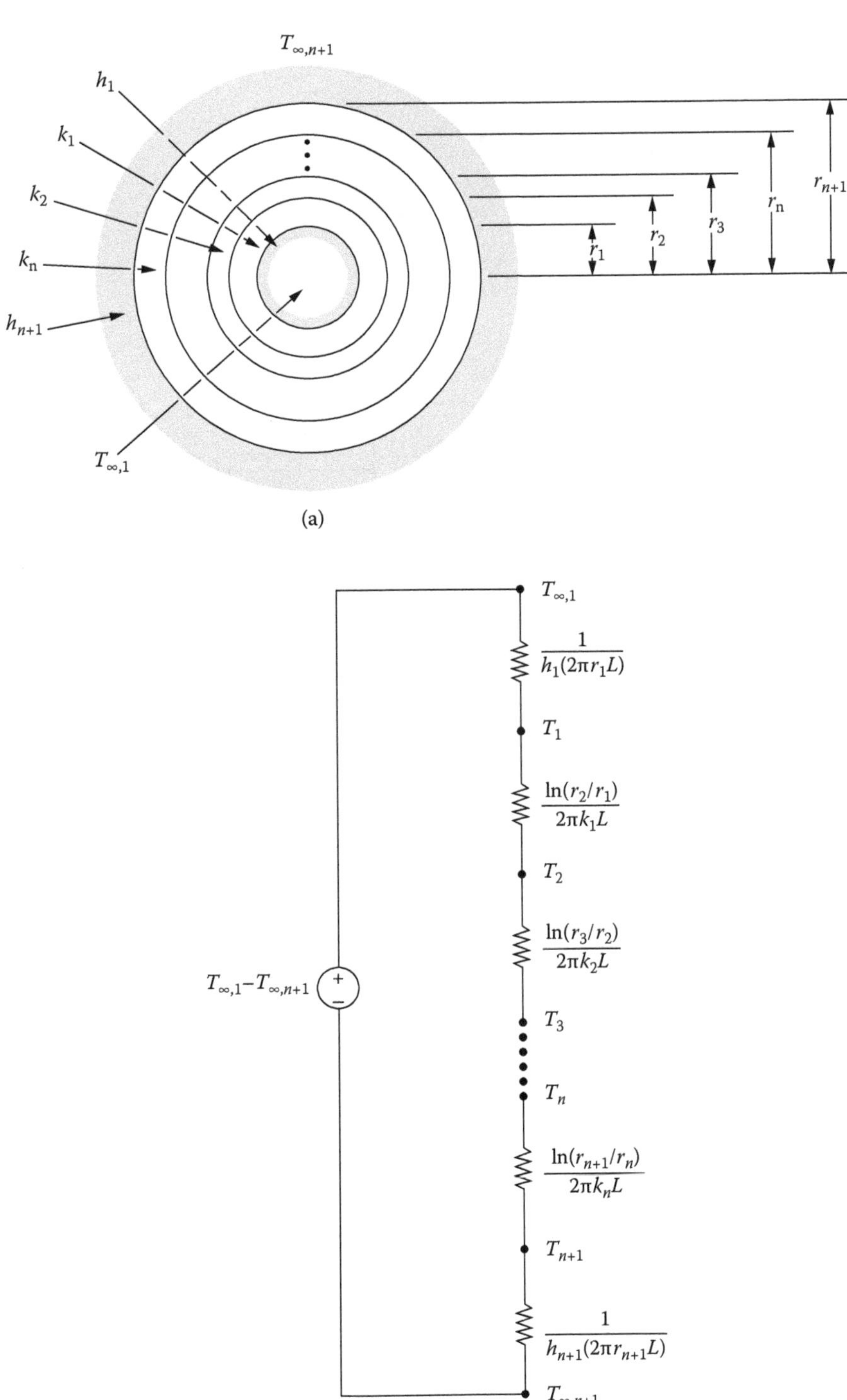

FIGURE 20.18
(a) Composite n-layer hollow cylinder with convection on both the interior and exterior surfaces and (b) electrothermal analog network.

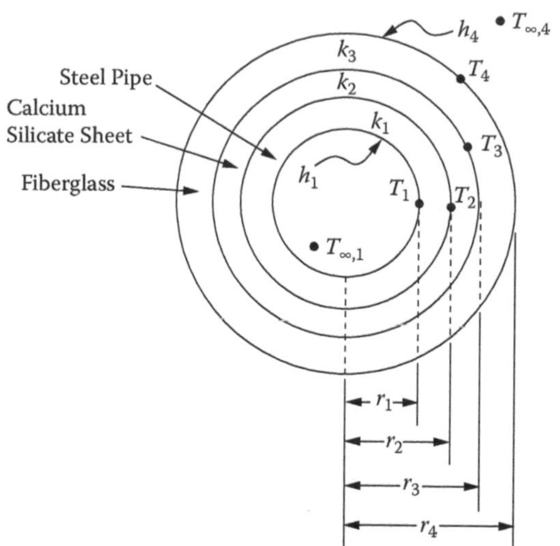

FIGURE 20.19
Three-layer composite hollow cylinder with convection for Example 20.6.

($k = 0.06\,\text{W/cm-K}$) covers the calcium silicate. The steam on the inside, which is at 500°C, provides a heat transfer coefficient of $1500\,\text{W/m}^2\text{-K}$ and the outside surface of the fiberglass is cooled by natural convection with a heat transfer coefficient of $10\,\text{W/m}^2\text{-K}$. The outside surrounding temperature is 20°C. Determine (a) the heat loss per meter of length and (b) the temperature on the outside surface of the fiberglass.

Solution

Assumptions and Specifications

(1) Steady-state conditions exist.
(2) The heat flow is one dimensional.
(3) Thermal properties do not vary with temperature.
(4) Conduction and convection are the modes of heat transfer.
(5) There is no contact resistance between any of the components.

The configuration is shown in Figure 20.19 and we observe that because there are three "solid" layers, we apply Equation 20.23. The five individual resistances can be evaluated as

$$\frac{1}{h_1(2\pi r_1 L)} = \frac{1}{(1500\,\text{W/m}^2\text{-K})(2\pi)(0.100\,\text{m})(1\,\text{m})} = 1.061 \times 10^{-3}\,\text{K/W}$$

$$\frac{\ln(r_2/r_1)}{2\pi k_1 L} = \frac{\ln(0.12/0.10)}{(2\pi)(15\,\text{W/m-K})(1\,\text{m})} = 1.934 \times 10^{-3}\,\text{K/W}$$

$$\frac{\ln(r_3/r_2)}{2\pi k_2 L} = \frac{\ln(0.15/0.12)}{(2\pi)(0.2\,\text{W/m-K})(1\,\text{m})} = 0.1776\,\text{K/W}$$

$$\frac{\ln(r_4/r_3)}{2\pi k_3 L} = \frac{\ln(0.17/0.15)}{(2\pi)(0.06\,\text{W/m-K})(1\,\text{m})} = 0.3320\,\text{K/W}$$

and

$$\frac{1}{h_4(2\pi r_4 L)} = \frac{1}{(10\,\text{W/m}^2\text{-K})(2\pi)(0.17\,\text{m})(1.00\,\text{m})} = 0.0936\,\text{K/W}$$

Thus,

$$R_T = (1.061 \times 10^{-3}\,\text{K/W} + 1.934 \times 10^{-4}\,\text{K/W}$$
$$+ 0.1776\,\text{K/W} + 0.3320\text{K/W} + 0.0936\,\text{K/W}) = 0.6062\,\text{K/W}$$

so that

$$\dot{Q}_r = \frac{500°\text{C} - 20°\text{C}}{R_T} = \frac{480°\text{C}}{0.6062\,\text{K/W}} = 792\,\text{W} \Longleftarrow$$

(b) T_4 will be

$$T_4 = T_{\infty,4} + \frac{\dot{Q}_r}{2\pi h_4 r_4 L} = 20°\text{C} + (792\,\text{W})(0.0936\,\text{K/W}) = 94.1°\text{C} \Longleftarrow$$

20.4.1.3 The Critical Radius of Insulation

We now consider Figure 20.20a, which shows a cylinder with a layer of insulation. Figure 20.20b provides the electrothermal analog network for just the insulation of radius, r_2, which completely surrounds the cylindrical pipe or tube of outer radius, r_1. The thermal conductivity of the insulation is k and the heat transfer coefficient at r_2 is h.

We observe from Figure 20.20b that the total resistance to the flow of heat will be

$$R = \frac{\ln(r_2/r_1)}{2\pi k L} + \frac{1}{2\pi r_2 h L}$$

and because

$$\dot{Q}_r = \frac{T_1 - T_2}{R}$$

we see that the heat flow will be at a maximum when R is at a minimum or when

$$\frac{dR}{dr_2} = \frac{d}{dr_2}\left(\frac{\ln(r_2/r_1)}{2\pi k L} + \frac{1}{2\pi r_2 h L}\right) = 0$$

The differentiation is straightforward and the result is that the minimum resistance, known as the **critical radius of insulation** for the cylinder, $r_{cr,cyl}$, is

$$r_{cr,cyl} = \frac{k}{h} \tag{20.25}$$

and it can be shown that for a sphere

$$r_{cr,sphere} = \frac{2k}{h} \tag{20.26}$$

Thus, for *enhancing* the heat transfer, we make

$$r < r_{cr,cyl} \quad \text{or} \quad r < \frac{k}{h}$$

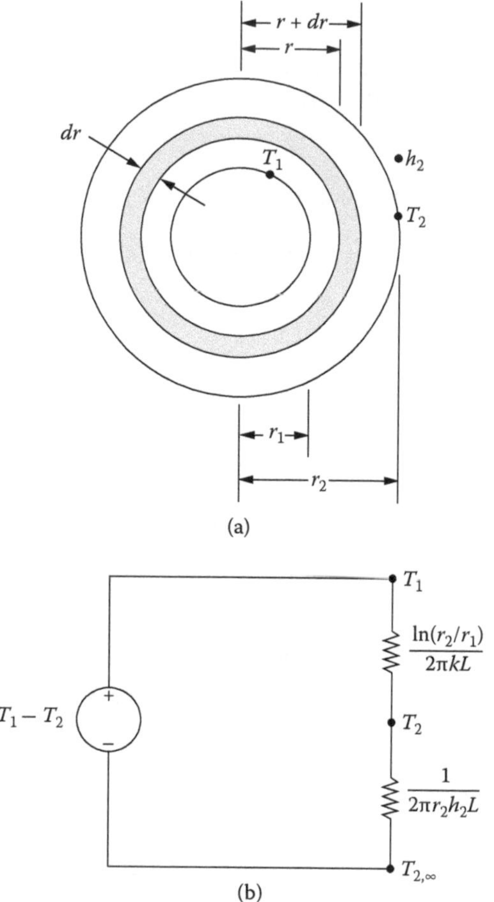

FIGURE 20.20
Configuration for the establishment of the critical radius of insulation: (a) the terminology and (b) the electrothermal analog network.

and for *retarding* the heat transfer, we make

$$r > r_{\text{cr,cyl}} \quad \text{or} \quad r > \frac{k}{h}$$

20.4.2 Spherical Coordinates

The Laplace equation in spherical coordinates can be developed in a manner similar to the development in Section 20.2. In the single, radial dimension, this equation becomes the ordinary differential equation

$$\frac{d}{dr}\left(r^2 \frac{dT}{dr}\right) = 0 \tag{20.27}$$

and we use this equation to develop the heat flow in the radial direction and the temperature distribution in a variety of spherical configurations.

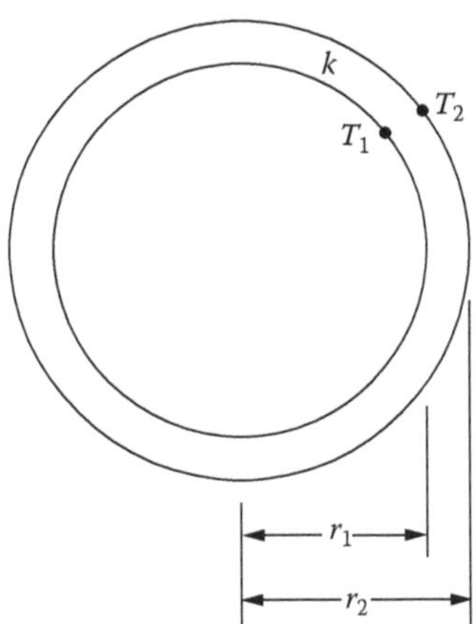

FIGURE 20.21
Hollow sphere.

20.4.2.1 The Hollow Sphere

Figure 20.21 shows a hollow sphere of inside radius, r_1, outside radius, r_2, and thermal conductivity, k. The inside and outside surfaces are maintained at the constant temperatures T_1 and T_2 and the heat flows in the radial direction. The differential equation that applies in this case is Equation 20.27 with the boundary conditions

$$T(r = r_1) = T_1 \qquad \text{and} \qquad T(r = r_2) = T_2$$

Solution of Equation 20.27 with the foregoing boundary conditions yields

$$T = T_1 + \frac{T_1 - T_2}{\dfrac{1}{r_2} - \dfrac{1}{r_1}}\left(\frac{1}{r_1} - \frac{1}{r}\right) \Longleftarrow \tag{20.28}$$

and the heat flow will be

$$\dot{Q}_r = \frac{4\pi k(T_1 - T_2)}{\dfrac{1}{r_1} + \dfrac{1}{r_2}} \tag{20.29}$$

Equation 20.28 shows that the temperature distribution is hyperbolic, that is, it varies inversely with r.

The thermal resistance of a hollow sphere can be identified from Equation 20.29 (or Equation 19.12c) as

$$R_{\text{cond}} = \frac{\dfrac{1}{r_1} - \dfrac{1}{r_2}}{4\pi k} \tag{20.30}$$

and if convection occurs on the inside of the sphere from $T_{\infty,1}$ via h_1 and at the outside of the sphere to $T_{\infty,2}$ via h_2, then the rate of heat transfer may be expressed with

$$\frac{1}{r_1} - \frac{1}{r_2} = \frac{r_2 - r_1}{r_1 r_2}$$

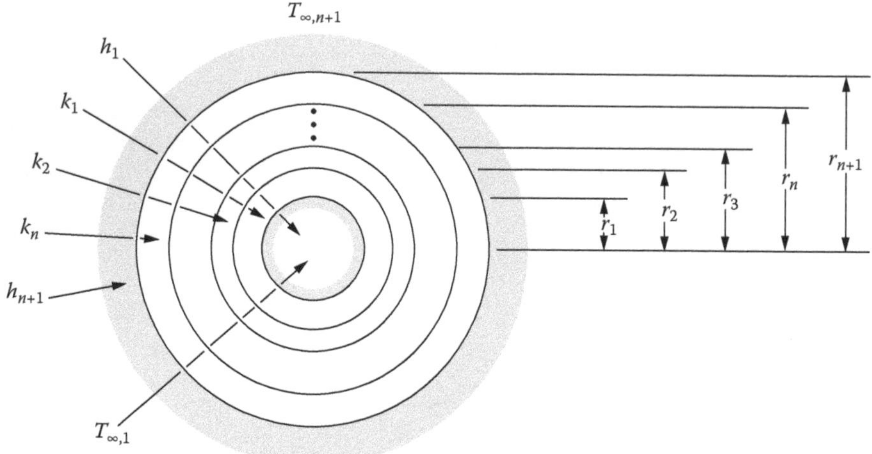

FIGURE 20.22
Composite hollow sphere with n layers and interior and exterior convection.

as

$$\dot{Q}_r = \frac{T_{\infty,1} -, T_{\infty,2}}{\dfrac{1}{4\pi h_1 r_1^2} + \dfrac{r_2 - r_1}{4\pi k r_1 r_2} + \dfrac{1}{4\pi h_2 r_2^2}} \tag{20.31}$$

20.4.2.2 The Composite Hollow Sphere

The composite n-layer hollow sphere with n layers shown in Figure 20.22 (identical to Figure 20.20) is heated by convection on the inside surface and cooled by convection on the outside surface. Heat flow occurs in the radially outward direction. The fluid inside the sphere at r_1 is at $T_{\infty,1}$, the inner surface is at T_1, and the heat transfer coefficient at the inside of the inner shell is h_1. The outside surface is at r_{n+1} where the temperature is T_{n+1} and where the heat transfer coefficient is h_{n+1}. The surrounding temperature on the outside of the composite sphere is at $T_{\infty,n+1}$. To facilitate the analysis, Figure 20.23 provides the electrothermal analog network.

With

$$\frac{1}{r_i} - \frac{1}{r_{i+1}} = \frac{r_{i+1} - r_i}{r_i r_{i+1}}$$

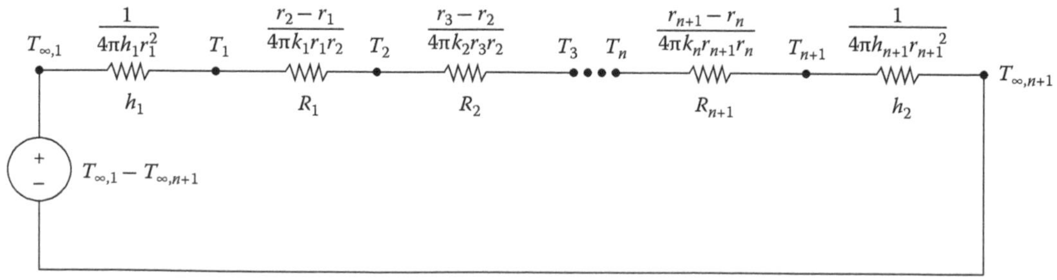

FIGURE 20.23
Electrothermal analog for the n-layer composite hollow sphere with both interior and exterior convection.

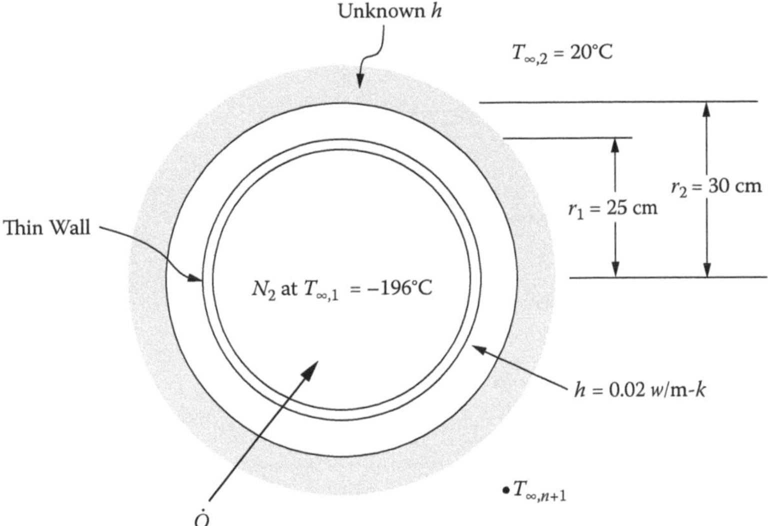

FIGURE 20.24
Configuration for Example 20.7.

the rate of heat transfer can written as

$$\dot{Q}_r = \frac{T_{\infty,1} - T_{\infty,n+1}}{\dfrac{1}{4\pi h_1 r_1^2} + \sum_{i=1}^{n} \dfrac{r_{i+1} - r_i}{4\pi k_i r_i r_{i+1}} + \dfrac{1}{4\pi h_{n+1} r_{n+1}^2}} \tag{20.32}$$

Example 20.7 Liquid nitrogen at −196°C is contained in a thin spherical tank with an outside radius of 25 cm (Figure 20.24). The tank is covered with a 5-cm-thick layer of insulation having a thermal conductivity of $k = 0.02$ W/m-K. The heat gain from the environment causes the liquid nitrogen to vaporize and the nitrogen vapor escapes from the tank through a vent. Determine the maximum natural convection heat transfer coefficient that can be permitted on the outside surface of the insulation if the vapor loss is to be limited to 4×10^{-4} kg/s. Take the surrounding air temperature as 20°C with the boiling point of nitrogen as −196°C and its enthalpy of vaporization as 198 kJ/kg.

Solution

Assumptions and Specifications

(1) Steady-state conditions exist.
(2) The heat flow is one dimensional.
(3) Thermal properties do not vary with temperature.
(4) Conduction and convection are the modes of heat transfer.
(5) There is no contact resistance between any of the components.

The rate of heat transfer into the tank, $\dot{Q}$, can be found from the rate of vaporization of the nitrogen, $\dot{m}$ and the enthalpy of vaporization, h_{fg}. Thus,

$$\dot{Q} = \dot{m} h_{fg} = (4 \times 10^{-4}\,\text{kg/s})(198 \times 10^3\,\text{J/kg}) = 79.2\,\text{W}$$

We assume that the inside surface of the spherical tank is at the temperature of the boiling nitrogen and that the temperature drop through the tank wall is negligible. With the heat flowing radially inward, we may therefore modify Equation 20.32 with $r_1 = 0.25\,\text{m}$ and $r_2 = 0.30\,\text{m}$ to

$$\dot{Q}_r = \frac{T_{\infty,2} - T_{\infty,1}}{\dfrac{r_2 - r_1}{4\pi k_2 r_2 r_1} + \dfrac{1}{4\pi h r_3^2}}$$

The terms in the denominator can be evaluated as

$$\frac{r_2 - r_1}{4\pi k_2 r_2 r_3} = \frac{0.30\,\text{m} - 0.25\,\text{m}}{4\pi(0.02\,\text{W/m-K})(0.30\,\text{m})(0.25\,\text{m})} = 2.6526\,\text{K/W}$$

and

$$\frac{1}{4\pi h r_2^2} = \frac{1}{4\pi h(0.30\,\text{m})^2} = \frac{0.8842}{h}\,\text{K/W}$$

Thus, the heat flow equation can be written as

$$79.2\,\text{W} = \frac{20°\text{C} - (-196°\text{C})}{2.6526\,\text{K/W} + \dfrac{0.8842}{h}\,\text{K/W}}$$

and we find that

$$h = 11.84\,\text{W/m}^2\text{-K} \Longleftarrow$$

Because radiation and convection have comparable effects in a natural convection application, the maximum limit on the natural convection heat transfer coefficient should be conservatively put as 50% of the calculated value of h. Thus, h should not exceed about $5.92\,\text{W/m}^2\text{-K}$ if the boil-off rate is not to exceed 4×10^{-4} kg/s. Coating the surface of the insulation with a low-emissivity material would reduce the radiative heating.

20.5 Simple Shapes with Heat Generation

20.5.1 The Plane Wall

The analysis for the plane wall considered in Section 20.3.1 is now extended to include the effect of internal heat generation. Such heat generation may be due to the passage of electric current or due to chemical or nuclear activity in the wall.

Suppose the plane wall of Figure 20.25 is subjected to a volumetric heat generation at the rate of q_i in W/m^3. In this case, Equation 20.9, which is the Poisson equation, pertains. When this relation is written in one dimension, an ordinary differential equation results

$$\frac{d^2T}{dx^2} + \frac{q_i}{k} = 0 \tag{20.33}$$

and the solution of this equation is subject to the boundary conditions

$$T(x = -L) = T_1 \qquad \text{and} \qquad T(x = L) = T_2$$

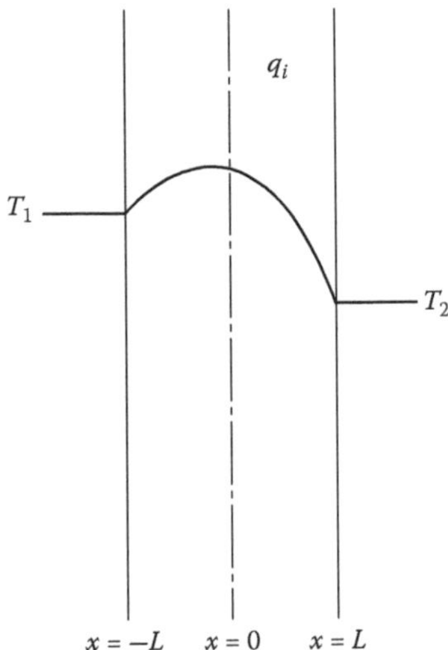

FIGURE 20.25
Plane wall with volumetric heat generation, q_i.

The solution to this equation is obtained via a double integration as

$$T = \frac{q_i}{2k}(L^2 - x^2) + \frac{T_2 - T_1}{2}\frac{x}{L} + \frac{T_1 + T_2}{2} \tag{20.34}$$

Observe that T is a quadratic function of x and that the heat flux at any location will be

$$\dot{q} = -k\frac{dT}{dx} = -k\left(-\frac{q_i x}{k} + \frac{T_2 - T_1}{2L}\right)$$

or

$$\dot{q} = q_i x + k\frac{T_1 - T_2}{2L} \tag{20.35}$$

Equation 20.35 clearly shows that $\dot{q}$ is location dependent.

By setting $dT/dx = 0$ in Equation 20.34, we can obtain the location, x_{max}, at which the maximum temperature occurs

$$x_{\text{max}} = \frac{k(T_2 - T_1)}{2q_i L} \tag{20.36}$$

which when used in Equation 20.34 yields an expression for the maximum temperature in the wall:

$$T_{\text{max}} = \frac{q_i L^2}{2k} + \frac{k(T_2 - T_1)^2}{8q_i L^2} + \frac{T_1 + T_2}{2} \tag{20.37}$$

If the surface temperatures are identical, that is, if $T_1 = T_2 = T_s$, then the temperature distribution is symmetrical about $x = 0$ and it follows from Equation 20.34 that

$$T(x) = \frac{q_i}{2k}(L^2 - x^2) + T_s \tag{20.38}$$

and the maximum temperature at the mid-plane where $x = 0$ will be

$$T_{\max} = \frac{q_i L^2}{2k} + T_s \tag{20.39}$$

Moreover, if identical convection conditions prevail on both faces so that $T_{\infty,1} = T_{\infty,2} = T_\infty$ and $h_1 = h_2 = h$, then it follows that

$$T_s = T_\infty + \frac{q_i L}{h} \tag{20.40}$$

In this case, the temperature distribution provided by Equation 20.38 may be expressed as

$$\frac{T(x) - T_o}{T_s - T_o} = \left(\frac{x}{L}\right)^2$$

where T_o is the temperature at $x = 0$.

A practical situation is one for which the temperature of the adjoining fluid, T_∞, is known. In this case, heat flows between the wall at T_s and the fluid at T_∞ through a convective layer represented by h. In this case, the energy balance reduces to

$$-k\frac{dT}{dx}|_{x=L} = h(T_s - T_\infty)$$

and use of Equation 20.38 to establish the temperature gradient at $x = L$ gives

$$T_s = T_\infty + \frac{q_i L}{h} \tag{20.41}$$

Example 20.8 Rectangular copper bus bars are commonly used in the transmission of electric power. Suppose that such a bar is 10-mm thick, 200-mm high, and 3-m deep as shown in Figure 20.26. The bus bar, which carries a current of 5000 A, is exposed on its 3-m × 200-mm faces to a convective environment at 20°C with a heat transfer coefficient of 25 W/m^2-K. The electrical resistivity and the thermal conductivity of the copper are $\rho_e = 1.68 \times 10^{-8}$ Ω-m and $k = 400$ W/m-K, respectively. Determine (a) the volumetric rate of heat generation, (b) the surface temperature, T_s, and (c) the maximum temperature, $T_{\max}$.

Solution

Assumptions and Specifications

(1) Steady-state conditions exist.
(2) The heat flow is one dimensional.
(3) Thermal properties do not vary with temperature.
(4) Conduction and convection are the modes of heat transfer.

(a) The electrical resistance of the bus bar, R, may be calculated from a knowledge of its resistivity, ρ_e, depth, D, and cross-sectional area, A

$$R = \rho_e \frac{D}{A} = (1.68 \times 10^{-8}\ \Omega\text{-m})\frac{(3\,\text{m})}{(0.20\,\text{m})(0.01\,\text{m})} = 2.52 \times 10^{-5}\ \Omega$$

The volumetric heat generation rate, q_i, will then be

$$q_i = \frac{I^2 R}{\text{volume}} = \frac{(5000\,\text{A})^2(2.52 \times 10^{-5}\ \Omega)}{(3\,\text{m})(0.01\,\text{m})(0.20\,\text{m})} = 105{,}000\ \text{W/m}^3 \Longleftarrow$$

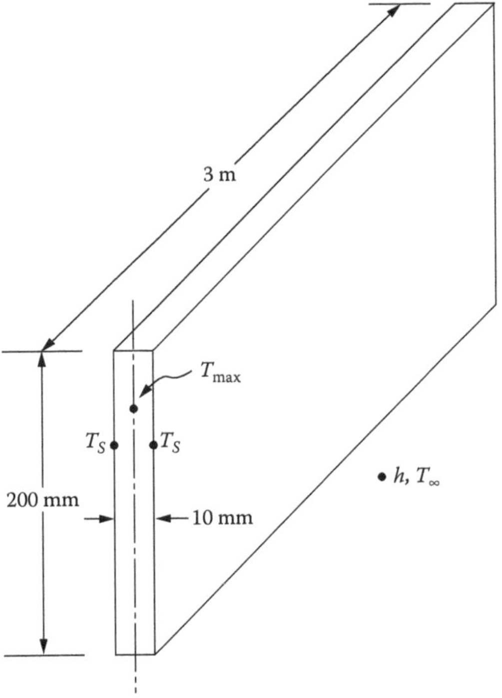

FIGURE 20.26
Configuration for Example 20.8.

(b) The surface temperature can be obtained from Equation 20.41 with $L = 0.01\,\text{m}/2 = 0.005\,\text{m}$.

$$T_s = T_\infty + \frac{q_i L}{k} = 20°\text{C} + \frac{(105{,}000\,\text{W/m}^3)(0.005\,\text{m})}{(25\,\text{W/m}^2\text{-K})} = 41°\text{C} \Longleftarrow$$

(c) The maximum temperature derives from Equation 20.39

$$T_{\text{max}} = \frac{q_i L^2}{2k} + T_s = \frac{(105{,}000\,\text{W/m}^3)(0.005\,\text{m})^2}{2(400\,\text{W/m-K})} + 41°\text{C} \approx 41°\text{C} \Longleftarrow$$

Observe that because of the high-thermal conductivity and small thickness of the bus bar, there is but a negligible difference between the maximum temperature and the surface temperature.

20.5.2 The Cylinder

Suppose that the cylindrical shell of Figure 20.27 is subjected to a volumetric heat generation of q_i (W/m^3). In this case, the Poisson equation in cylindrical coordinates in one dimension gives the governing differential equation

$$\frac{1}{r}\frac{d}{dr}\left(r\frac{dT}{dr}\right) + \frac{q_i}{k} = 0 \tag{20.42}$$

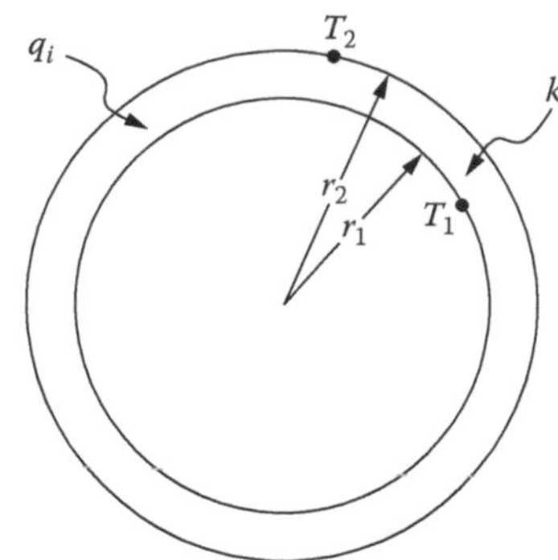

FIGURE 20.27
Hollow cylinder with volumetric heat generation, q_i.

Equation 20.42 may be solved with the boundary conditions

$$T(r = r_1) = T_1 \qquad \text{and} \qquad T(r = r_2) = T_2$$

to give

$$T(r) = T_1 + \frac{q_i}{2k}\left(\frac{r_2^2 - r_1^2}{2} + r_1^2 \ln\frac{r}{r_2}\right) \tag{20.43}$$

Next, consider a long solid cylinder of radius, r_o, generating q_i (W/m^3) and whose surface is maintained at temperature, T_s. In this case, Equation 20.42 must be solved subject to the boundary conditions

$$T(r = r_o) = T_s \qquad \text{and} \qquad \left.\frac{dT}{dr}\right|_{r=0} = 0$$

where $dT/dr|_{r=0}$ is a consequence of thermal symmetry about $r = 0$. The solution for T takes the form

$$T = T_s + \frac{q_i}{4k}(r_o^2 - r^2) \tag{20.44}$$

and the maximum temperature occurs along the centerline (at $r = 0$)

$$T_{\text{max}} = T_s + \frac{q_i r_o^2}{4k} \tag{20.45}$$

If the outside surface of the cylinder is cooled by a coolant at temperature, T_∞, that provides a heat transfer coefficient, h, then, after simplification, the overall energy balance equating the heat generated, $\pi r_o^2 L q_i$, to the heat loss by convection, $(2\pi_o L)h(T_s - T_\infty)$, gives

$$T_s = T_\infty + \frac{q_i r_o}{2h} \tag{20.46}$$

20.5.3 The Sphere

For the spherical shell of Figure 20.28 subjected to a volumetric heat generation of q_i (W/m^3), the Poisson equation in spherical coordinates in one dimension gives the

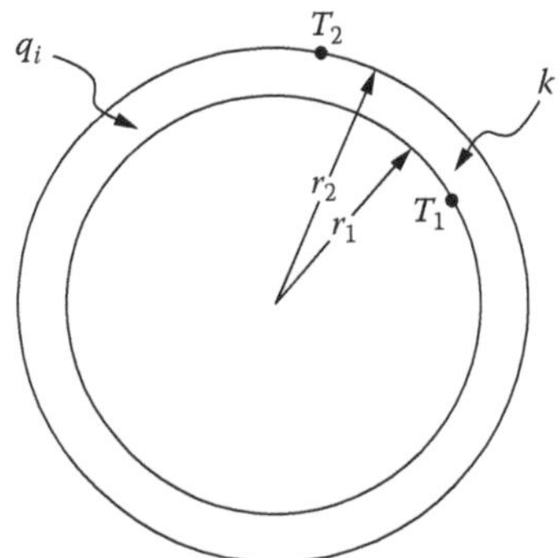

FIGURE 20.28
Hollow sphere with volumetric heat generation, q_i.

governing differential equation

$$\frac{d}{dr}\left(r^2\frac{dT}{dr}\right)+\frac{q_i r^2}{k}=0 \tag{20.47}$$

When this equation is solved with the boundary conditions

$$T(r=r_1)=T_1 \qquad \text{and} \qquad T(r=r_2)=T_2$$

the result is

$$T=T_2+\frac{q_i}{6k}\left[1-\left(\frac{r}{r_2}\right)^2\right]+\frac{q_i r_1^3}{3k}\left(\frac{1}{r_2}-\frac{1}{r}\right) \tag{20.48}$$

Next, consider a solid sphere of radius, r_o, with surface temperature, T_s, and having an internal heat generation of q_i (W/m^3). In this case, Equation 20.47 pertains but the boundary conditions are

$$T(r=r_o)=T_s \qquad \text{and} \qquad \left.\frac{dT}{dr}\right|_{r=r_o}=0 \tag{20.49}$$

The solution for T takes the form

$$T=T_s+\frac{q_i}{6k}(r_o^2-r^2) \tag{20.50}$$

and the maximum temperature occurs at the center where $r=0$

$$T_{\text{max}}=T_s+\frac{q_i r_o^2}{6k} \tag{20.51}$$

If a coolant at temperature T_∞ provides a heat transfer coefficient of h at the outside surface of the sphere, then we can equate the total rate of heat generation to the rate of convection from the outside surface at $r=r_o$

$$q_i\left(\frac{4}{3}\pi r_o^3\right)=h(4\pi r_o^2)(T_s-T_\infty)$$

or

$$T_s=T_\infty+\frac{q_i r_o}{3h} \tag{20.52}$$

20.6 Extended Surfaces

20.6.1 Introduction

When engineers need to increase the heat transfer between a primary surface and its adjoining fluid, it is a common practice to attach an **extended surface** or **fin** to the primary surface. Although the extended surface provides an increased heat transfer surface area, it also decreases the average primary surface temperature. However, if the extended surface is properly designed, the net result is an increase in the rate of heat transfer. Extended surfaces appear in such diverse applications as air-cooled engines, compact heat exchangers, **heat sinks** for electronic equipment cooling, and space radiators.

Decades of research in extended surface technology has resulted in a wide variety of designs. The three basic extended surface elements that appear in most designs are the longitudinal fin of uniform thickness or rectangular profile (Figure 20.29a), the cylindrical spine or pin fin (Figure 20.29b), and the annular or radial fin of uniform thickness (Figure 20.29c). We study these three configurations in this section, both as individual fins and arrays of fins.

We will analyze these fins using the limiting assumptions proposed by Murray (1938) and Gardner (1945) which involve:

- A homogeneous fin material and operation in the steady state.
- Uniform surrounding temperature.
- Uniform heat transfer coefficient and uniform thermal conductivity.
- No temperature gradients along the length or across the thickness of the fin.
- No bond resistance between the primary surface and the fin.
- No heat sources within the fin itself.
- No heat transfer from or to the fin tip.
- No heat transfer from the fin edges.
- Uniform temperature along the fin base.

20.6.2 The Longitudinal Fin of Uniform Thickness

Consider the longitudinal fin of thickness, δ, height, b, and length, L as shown in Figure 20.30. The fin is made of a material having a thermal conductivity, k. The adjoining fluid at temperature, T_∞, provides a convective heat transfer coefficient, h, over the exposed surface of the fin. To derive the differential equation that governs the temperature distribution in the fin, we focus on a differential element of height, dx, and write an energy balance giving

$$\dot{Q}_x - hP(T - T_\infty)dx = \dot{Q}_{x+dx} = \dot{Q}_x + \frac{d\dot{Q}_x}{dx}dx$$

where, in this case, heat transfer from the edges is included, $P = 2(L + \delta)$ is the perimeter of the fin cross section, and $P\,dx$ is the surface area of fin available for convection. Introducing Fourier's law and noting that the area, A, normal to the conduction heat flow path is $A = \delta L$, we have

$$\dot{Q}_x = -kA\frac{dT}{dx} = -k\delta L\frac{dT}{dx}$$

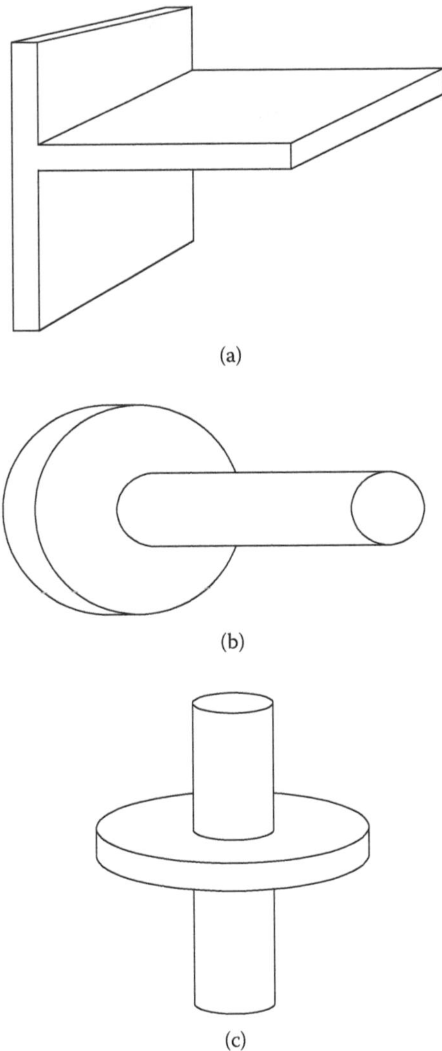

FIGURE 20.29
Extended surfaces: (a) the longitudinal fin of uniform thickness or rectangular profile, (b) the cylindrical spine, and (c) the annular or radial when fin of uniform thickness.

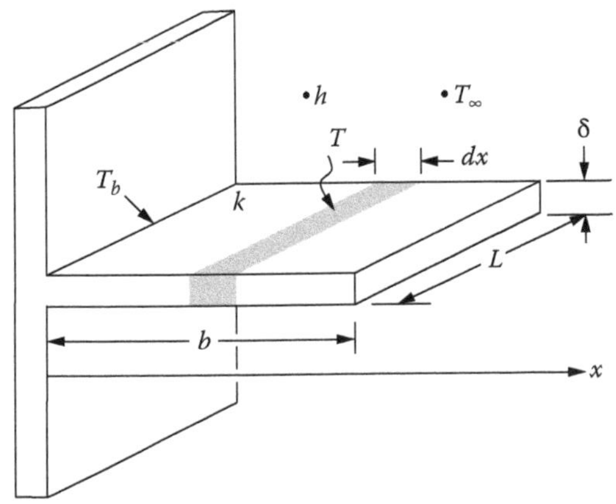

FIGURE 20.30
Coordinate system for the longitudinal fin of uniform thickness.

Then, the energy balance gives rise to the governing differential equation

$$\frac{d^2T}{dx^2} - \frac{hP}{k\delta L}(T - T_\infty) = 0 \tag{20.53}$$

We let the **temperature excess** be

$$\theta = T - T_\infty \tag{20.54a}$$

and with

$$m = \left(\frac{hP}{k\delta L}\right)^{1/2} \approx \left(\frac{2h}{k\delta}\right)^{1/2} \tag{20.54b}$$

we may write Equation 20.53 as

$$\frac{d^2\theta}{dx^2} - m^2\theta = 0 \tag{20.55}$$

which has a general solution that may be written in terms of exponential functions

$$\theta = C_1 e^{mx} + C_2 e^{-mx} \tag{20.56}$$

or hyperbolic functions

$$\theta = C_1 \cosh mx + C_2 \sinh mx \tag{20.57}$$

The evaluation of the constants, C_1 and C_2, requires the specification of two boundary conditions, one at the fin base and one at the fin tip. Depending on the operational conditions, a variety of boundary conditions may be appropriate. For example, the temperature at the base may be assumed constant and equal to that of the primary or base surface to which the fin is attached. This implies zero contact resistance between the primary surface and the base of the fin. It is also implicit that the attachment of the fin does not cause a base temperature depression relative to the unfinned portion of the primary surface.

Likewise, we may envision several types of boundary conditions at the fin tip. The two considered here are a convective heat loss from the tip, which removes one of the Murray-Gardner conditions from consideration, and an insulated tip.

20.6.2.1 Constant Base Temperature with Tip Heat Loss

The boundary conditions are

$$\theta(x = 0) = \theta_b = T_b - T_\infty \qquad \text{and} \qquad -kL\left.\frac{d\theta}{dx}\right|_{x=b} = h_a\theta \tag{20.58}$$

where T_b is the base temperature and h_a is the heat transfer coefficient at the tip. The solution for θ is

$$\theta = \theta_b \frac{\cosh m(b-x) + H \sinh m(b-x)}{\cosh mb + H \sinh mb} \tag{20.59}$$

where $H = h_a/km$.

Because all of the heat dissipated by the fin must, in the steady state, pass the base of the fin, it follows that

$$\dot{Q} = -kA\left.\frac{d\theta}{dx}\right|_{x=0} = km\delta L\theta_b \left(\frac{\sinh mb + H \cosh mb}{\cosh mb + H \sinh mb}\right) \tag{20.60}$$

20.6.2.2 Constant Base Temperature with Insulated Tip

Here, the boundary conditions are

$$\theta(x=0) = \theta_b = T_b - T_\infty \qquad \text{and} \qquad k\delta L \left.\frac{d\theta}{dx}\right|_{x=b} = 0 \tag{20.61}$$

The solutions for θ and $\dot{Q}$ are immediately obtainable by setting $H = 0$.

$$\theta = \theta_b \frac{\cosh m(b-x)}{\cosh mb} \tag{20.62}$$

and

$$\dot{Q} = -kA\left.\frac{d\theta}{dx}\right|_{x=0} = km\delta L\theta_b \tanh mb \tag{20.63}$$

20.6.3 Fin Performance Criteria

Three criteria are commonly employed to assess the performance of a fin. The first one is the **fin effectiveness,** ϵ, which is defined as the ratio of the rate of heat dissipated by the fin to the rate of heat transfer from the prime surface if the fin were not present

$$\epsilon = \frac{\dot{Q}}{hA\theta_b} \tag{20.64}$$

If the use of the fin is to be economically justifiable, ϵ should be greater than unity and as large as possible. For the case of the longitudinal fin of constant thickness with a constant base temperature and an insulated tip,

$$\epsilon = \left(\frac{kP}{Ah}\right)^{1/2} \tanh mb \tag{20.65}$$

The second performance criterion involves the use of the **fin efficiency,** η, which is defined as the ratio of $\dot{Q}$, the heat dissipated by the fin to $\dot{Q}_{\text{id}}$, and the rate of heat transfer from an ideal fin when operating at the base temperature, θ_b, (a fin with infinite thermal conductivity). The fin surface area is $S = Pb$ so that we may express the ideal heat flow from the fin as

$$\dot{Q}_{\text{id}} = hPb\theta_b \tag{20.66}$$

Then, the efficiency of the longitudinal fin with uniform thickness and an insulated tip where the fin surface is $S_f = 2bL$ will be the ratio of Equation 20.63 to Equation 20.66

$$\eta = \frac{\dot{Q}}{\dot{Q}_{\text{id}}} = \frac{k\delta L\theta_b m \tanh mb}{2bL\theta_b} = \frac{\tanh mb}{mb} \tag{20.67}$$

The third criterion of performance may be related to the thermal resistance of the fin, which can be expressed as

$$R_f = \frac{\theta_b}{\dot{Q}} \tag{20.68}$$

For a fin to be effective, its thermal resistance must be less than the thermal resistance of the exposed base, $1/hA = 1/h\delta L$. Thus,

$$R_f = \frac{\theta_b}{\dot{Q}} < \frac{1}{h\delta L} \tag{20.69}$$

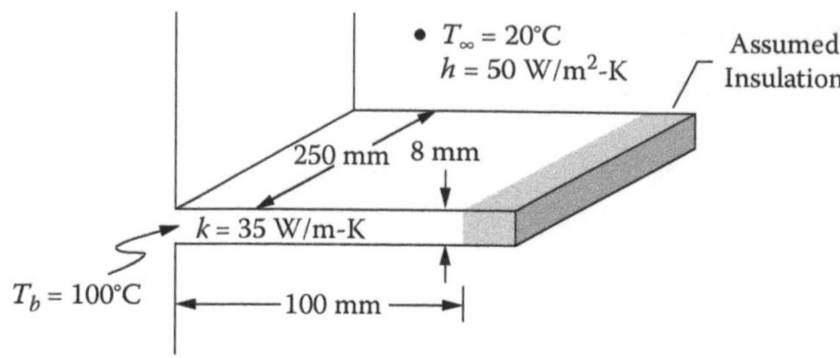

FIGURE 20.31
Longitudinal fin of uniform thickness for Example 20.9.

Example 20.9 A longitudinal fin of uniform thickness (Figure 20.31) operates in a convective environment at a temperature of 20°C, which provides a heat transfer coefficient of 50 W/m²-K. The base temperature of the fin is 100°C, and the fin is fabricated from steel with $k = 35\,\text{W/m-K}$. The fin is 8-mm thick, 100-mm high, and 250-mm long and the fin tip may be assumed to be insulated. Determine (a) the fin tip temperature, (b) the fin heat dissipation, (c) the fin effectiveness, (d) the fin efficiency, and (e) the thermal resistance of the fin.

Solution

Assumptions and Specifications

(1) Steady-state conditions exist.
(2) The heat flow is one dimensional.
(3) Thermal properties do not vary with temperature.
(4) Conduction and convection are the modes of heat transfer.
(5) The Murray-Gardner assumptions apply.
(6) The fin tip is insulated.

We first calculate, θ_b, A, m, and mb.

$$\theta_b = T_b - T_\infty = 100°\text{C} - 20°\text{C} = 80°\text{C}$$

$$A = L\delta = (0.250\,\text{m})(0.008\,\text{m}) = 2.00 \times 10^{-3}\,\text{m}^2$$

$$m = \left(\frac{2h}{k\delta}\right)^{1/2} = \left[\frac{2(50\,\text{W/m}^2\text{-K})}{(35\,\text{W/m-K})(0.008\,\text{m})}\right]^{1/2} = 18.898\,\text{m}^{-1}$$

and

$$mb = (18.898\,\text{m}^{-1})(0.100\,\text{m}) = 1.89$$

(a) We use Equation 20.62 to obtain the tip temperature at $x = b = 0.100\,\text{m}$

$$\theta(x) = \theta_b \frac{\cosh m(b-x)}{\cosh mb}$$

$$\theta(x = b = 0.10\,\text{m}) = \frac{80°\text{C}}{\cosh(1.89)}$$

$$= \frac{80°\text{C}}{3.3852} = 23.6°\text{C}$$

and

$$T(x = b) = 20°\text{C} + 23.6°\text{C} = 43.6°\text{C} \Longleftarrow$$

(b) The fin heat dissipation is found from Equation 20.64

$$\begin{aligned} \dot{Q} &= kA\theta_b m \tanh mb \\ &= (35\,\text{W/m-K})(0.002\,\text{m}^2)(80°\text{C})(18.898\,\text{m}^{-1})\tanh(1.89) \\ &= 101.1\,\text{W} \Longleftarrow \end{aligned}$$

(c) The effectiveness of the fin is given by Equation 20.64

$$\epsilon = \frac{\dot{Q}}{hA\theta_b} = \frac{101.1\,\text{W}}{(50\,\text{W/m}^2\text{-K})(0.002\,\text{m}^2)(80°\text{C})} = 12.64 \Longleftarrow$$

(d) The efficiency may be obtained from Equation 20.67

$$\eta = \frac{\tanh mb}{mb} = \frac{\tanh(1.89)}{1.89} = 0.506 \Longleftarrow$$

(e) The thermal resistance of the fin can be obtained from Equation 20.68

$$R_f = \frac{\theta_b}{\dot{Q}_b} = \frac{80°\text{C}}{101.1\,\text{W}} = 0.792\,\text{K/W} \Longleftarrow$$

20.6.4 The Cylindrical Spine or Pin Fin

Because the longitudinal fin of uniform thickness (Figure 20.29a) and the cylindrical spine (Figure 20.29b) are both fins of a constant cross section, the equations developed for the longitudinal fin of uniform thickness may be employed for the analysis of the cylindrical spine. The only difference is in the surface and cross-sectional area that may be taken as

$$S = \pi d \qquad \text{and} \qquad A = \frac{\pi}{4}d^2 \tag{20.70}$$

and

$$m = \left(\frac{4h}{kd}\right)^{1/2} = \left(\frac{hP}{kA}\right)^{1/2}$$

Example 20.10 A 20-mm-square chip surface is extended by installing 16 aluminum ($k = 238\,\text{W/m-K}$) pin fins forming an aligned array as in Figure 20.32. Each pin fin has a diameter of 2 mm and a height of 14 mm. The assembly is cooled by a fan blowing air at 20°C and providing a heat transfer coefficient of 100 W/m²-K. Assuming that the pin fin tips are adiabatic and the chip surface temperature will not exceed 70°C, determine the maximum dissipation capability of the chip.

Solution

Assumptions and Specifications

(1) Steady-state conditions exist.

(2) The heat flow is one dimensional.

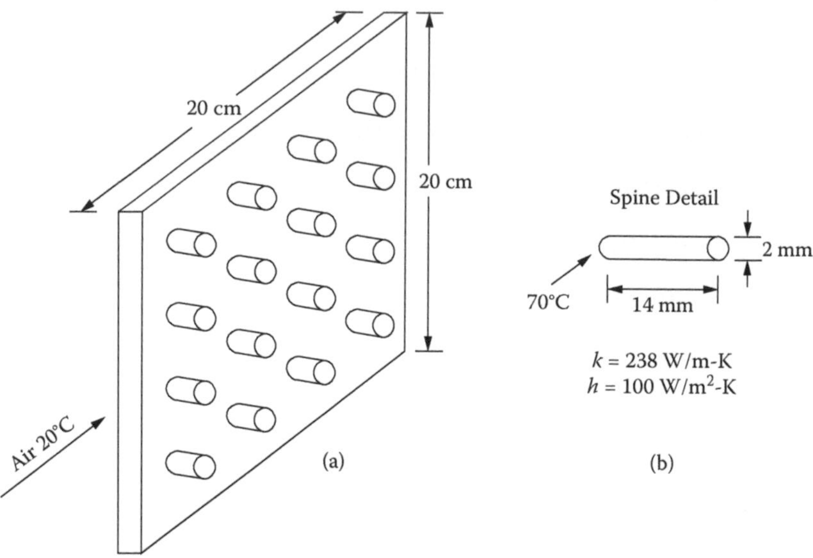

FIGURE 20.32
Array of cylindrical spines for Example 20.10 (a) arrangement and (b) detail.

(3) Thermal properties do not vary with temperature.
(4) Conduction and convection are the modes of heat transfer.
(5) There is no contact resistance between the finned array and the chip surface.
(6) The Murray-Gardner assumptions apply.

We first calculate, θ_b, A, m, and mb for each fin.

$$\theta_b = T_b - T_\infty = 70°\text{C} - 20°\text{C} = 50°\text{C}$$

$$A = \frac{\pi}{4}d^2 = (0.7854)(0.002\,\text{m})^2 = 3.14 \times 10^{-6}\,\text{m}^2$$

$$m = \left(\frac{4h}{kd}\right)^{1/2} = \left[\frac{4(100\,\text{W/m}^2\text{-K})}{(238\,\text{W/m-K})(0.002\,\text{m})}\right]^{1/2} = 28.99\,\text{m}^{-1}$$

and

$$mb = (28.99\,\text{m}^{-1})(0.014\,\text{m}) = 0.406$$

For n pin fins, Equation 20.63 may be adjusted to give with 16 fins

$$\begin{aligned}\dot{Q}_b &= nkA\theta_b m \tanh mb \\ &= (16)(238\,\text{W/m-K})(3.14 \times 10^{-6}\,\text{m}^2)(50°\text{C})(28.99\,\text{m}^{-1}) \tanh(0.406) \\ &= 6.68\,\text{W} \Longleftarrow\end{aligned}$$

The unfinned or prime surface area may be calculated by subtracting the total footprint area of the 16 pin fins from the total chip surface area. Calling this area, A_p, we have

$$A_p = (0.02\,\text{m})^2 - 16(3.14 \times 10^{-6}\,\text{m}^2) = 3.50 \times 10^{-4}\,\text{m}^2$$

and the heat dissipation from the prime or unfinned surface will be

$$\dot{Q}_p = hA_p\theta_b = (100\,\text{W/m}^2\text{-K})(3.50 \times 10^{-4}\,\text{m}^2)(50^\circ\text{C}) = 1.75\,\text{W}$$

The sum of $\dot{Q}_b$ and $\dot{Q}_p$ will provide the heat dissipation capability of the chip

$$\dot{Q} = \dot{Q}_b + \dot{Q}_p = 6.68\,\text{W} + 1.75\,\text{W} = 8.43\,\text{W} \Longleftarrow$$

20.6.5 Annular or Radial Fin of Uniform Thickness

The differential equation that governs the temperature profile in the annular or radial fin of uniform thickness (Figure 20.30c) is Bessel's modified differential equation. The solution to this equation for the temperature profile as well as the fin heat dissipation and its efficiency are in terms of the modified Bessel functions and it is felt that detailing these solutions here will serve no useful purpose. However, these solutions are available in some of the heat transfer textbooks and in Kraus et al. (2001).

Figure 20.33 is a graph of the fin efficiency, η, plotted as a function of $b(2h/k\delta)^{1/2}$ with $b = r_a - r_b$ for selected values of the radius ratio, r_a/r_b. Once η is read from the graph, the heat dissipation from the fin can be determined from

$$\dot{Q}_b = \eta\dot{Q}_{\text{id}} = \eta h[2\pi(r_a^2 - r_b^2) + 2\pi r_a\delta](T_b - T_\infty) \tag{20.71}$$

where $2\pi(r_a^2 - r_b^2)$ represents the surface area of the two faces of the fin and $2\pi r_a\delta$ is the tip surface of the fin.

Example 20.11 The cylinder of an engine made of cast aluminum ($k = 174\,\text{W/m-K}$) has 30 equally spaced radial fins (two of which are shown in Figure 20.34). The cylinder has a diameter of 14 cm and is 20-cm long. Each fin is 2-mm thick and 30-mm high. The outer surface of the cylinder attains a temperature of 120°C and the air flow over the fins produces

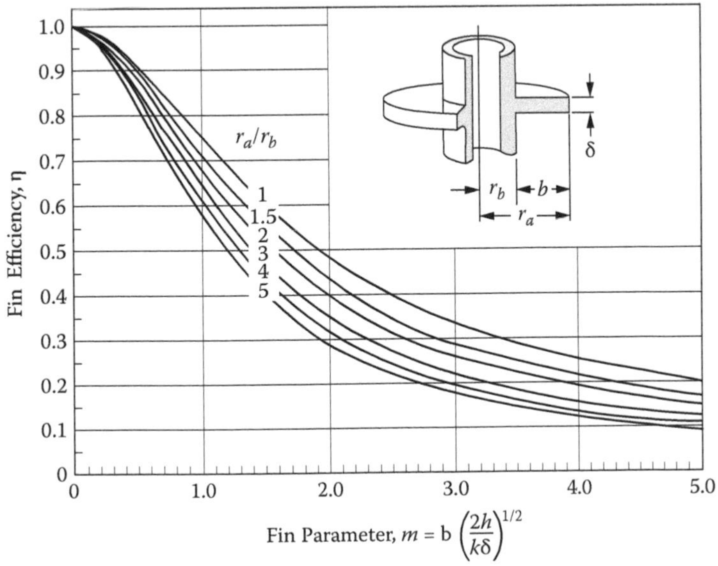

FIGURE 20.33
Efficiency of an annular fin of uniform thickness. (From Gardner, 1945.)

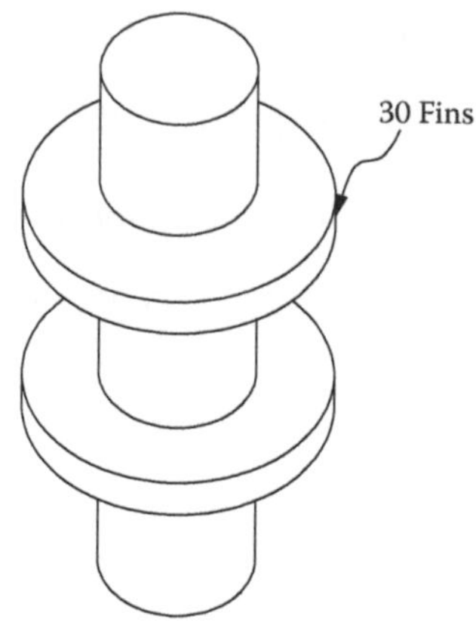

FIGURE 20.34
Array of radial or annular fins for Example 20.11. Observe that the entire array contains 30 fins.

a heat transfer coefficient of 40 W/m^2-K. Determine the heat transfer from the cylinder to the air.

Solution

Assumptions and Specifications

(1) Steady-state conditions exist.
(2) The heat flow is one dimensional.
(3) Thermal properties do not vary with temperature.
(4) Conduction and convection are the modes of heat transfer.
(5) The Murray-Gardner assumptions apply.

The outer radius of the fin is

$$r_a = r_b + 3.00\,\text{cm} = 7.00\,\text{cm} + 3.00\,\text{cm} = 10\,\text{cm}$$

Hence, with $b = r_a - r_b = 3.00\,\text{cm}$

$$b\left(\frac{2h}{k\delta}\right)^{1/2} = (0.030\,\text{m})\left[\frac{2(40\,\text{W/m}^2\text{-K})}{(174\,\text{W/m-K})(0.002\,\text{m})}\right]^{1/2} = 0.455$$

and

$$\frac{r_a}{r_b} = \frac{10\,\text{cm}}{7\,\text{cm}} = 1.429$$

Figure 20.33 may then be employed to show that the fin efficiency, η, will be

$$\eta = 0.93$$

For $n = 30$ fins, Equation 20.71 can be employed to obtain the heat dissipation of the fins with $\theta_b = 120°\text{C} - 20°\text{C} = 100°\text{C}$ and $r_a^2 - r_b^2 = 0.0051\,\text{m}^2$

$$\begin{aligned}\dot{Q}_b &= n\eta h[2\pi(r_a^2 - r_b^2) + 2\pi r_a \delta](T_b - T_\infty)\\ &= (30)(0.93)(40\,\text{W/m}^2\text{-K})\\ &\quad \times 2\pi\{0.0051\,\text{m}^2 + (0.10\,\text{m})(0.002\,\text{m})](120°\text{C} - 20°\text{C})\\ &= 3716\,\text{W} \qquad (3.716\,\text{kW})\end{aligned}$$

Next, we determine the heat transferred from the unfinned portion of the cylinder. The prime surface area is

$$\begin{aligned}A_p &= 2\pi r_b(L - n\delta)\\ &= 2\pi(0.07\,\text{m})[0.20\,\text{m} - 30(0.002\,\text{m})] = 0.0616\,\text{m}^2\end{aligned}$$

and

$$\dot{Q}_p = hA_p(T_b - T_\infty) = (40\,\text{W/m}^2\text{-K})(0.0616\,\text{m}^2)(100°\text{C}) = 246\,\text{W}$$

The total heat transfer is

$$\dot{Q} = \dot{Q}_b + \dot{Q}_p = 3716\,\text{W} + 246\,\text{W} = 3962\,\text{W} \qquad (3.962\,\text{kW}) \Longleftarrow$$

We note that the fins provide about 96% of the total heat dissipation.

20.7 Two-Dimensional Conduction

20.7.1 Introduction

In steady two-dimensional conduction, the temperature is a function of two spatial coordinates. In Cartesian coordinates where there is no heat generation, a form of the Laplace equation given by Equation 20.10 pertains

$$\frac{\partial^2 T}{\partial x^2} + \frac{\partial^2 T}{\partial y^2} = 0 \tag{20.72}$$

The solution to this equation requires four boundary conditions and for the rectangular plate of thickness δ (Figure 20.35) subjected to a sinusoidal temperature variation on the top edge, these boundary conditions are

$$\begin{aligned}T(0, y) &= T_1 \qquad \text{(left edge)}\\ T(L, y) &= T_1 \qquad \text{(right edge)}\\ T(x, 0) &= T_1 \qquad \text{(bottom edge)}\\ T(x, H) &= T_1 + a\sin(\pi x/L) \qquad \text{(top edge)}\end{aligned}$$

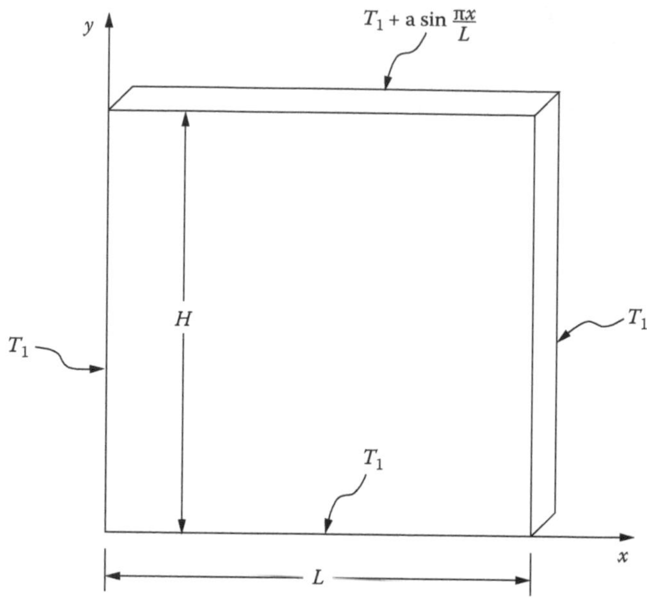

FIGURE 20.35
Rectangular plate subject to a sinusoidal temperature variation at its top edge.

20.7.2 Solution Methods

Equation 20.72 can be solved using its associated boundary conditions either by an exact analytical method such as the **method of separation of the variables** or by an approximate analytical method such as the **integral method**. Numerical methods such as the **finite difference method** or the **finite element method** are also available. A discussion of these methods may be found in many of the heat transfer textbooks such as Bejan (1993), Cengel (2003), Holman (2002), Incropera and DeWitt (2003), and Mills (1999). The method of separation of the variables applied to the rectangular plate of Figure 20.35 subjected to the specified boundary conditions gives the solution

$$T(x, y) = T_1 + \frac{a \sinh(\pi y/L)}{\sinh(\pi H/L)} \sin(\pi x/L)$$

20.7.3 The Conduction Shape Factor Method

The conduction shape factor method provides a simple equation for the two-dimensional rate of heat transfer

$$\dot{Q} = kS(T_1 - T_2) \tag{20.73}$$

where S (which equals A/L for a plane wall) is the conduction shape factor and is a function of the two-dimensional geometry. A thermal resistance for a two-dimensional geometry may be identified by writing Equation 20.73 as

$$R_{\text{cond}} = \frac{T_1 - T_2}{\dot{Q}} = \frac{1}{kS} \tag{20.74}$$

We take note of the fact that the conduction shape factor method does not give the temperature distribution in the configuration.

The conduction shape factor has been worked out for a number of configurations. These are either derived from the exact analytical solutions or approximate analytical or numerical solutions. Table 20.2 provides a brief compilation of conduction shape factors for some useful two-dimensional geometries.

TABLE 20.2

Some Conduction Shape Factors

(1) Isothermal cylinder of length, L, buried in a semi-infinite solid:

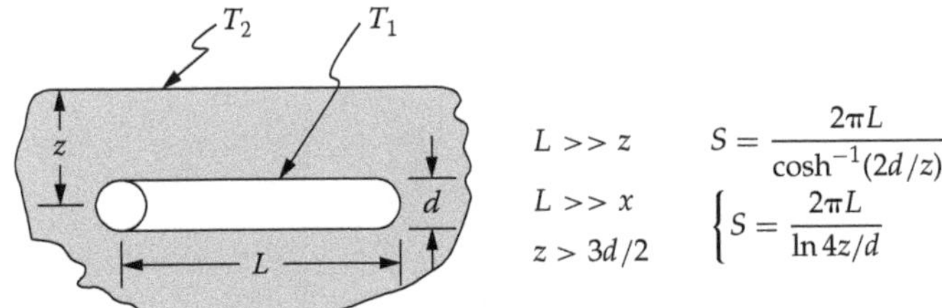

$L >> z$ $\quad S = \dfrac{2\pi L}{\cosh^{-1}(2d/z)}$

$L >> x$ $\quad \left\{ S = \dfrac{2\pi L}{\ln 4z/d} \right.$

$z > 3d/2$

(2) Square of side b and length, L centered in square solid of side, a, and length, L:

$a/b < 1.4$ $\quad S = \dfrac{2\pi L}{0.785 \ln a/b}$

$a/b > 1.4$ $\quad S = \dfrac{2\pi L}{0.970 \ln a/b - 0.50}$

(3) Cylinder of length, L and diameter, d, centered in square solid of side, a, and length, L:

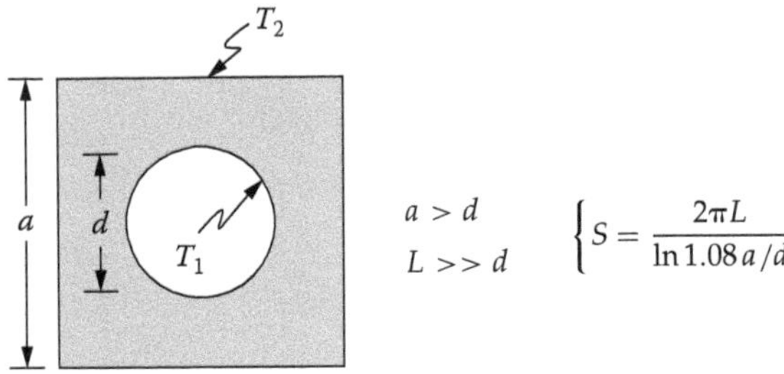

$a > d$ $\quad \left\{ S = \dfrac{2\pi L}{\ln 1.08\, a/d} \right.$

$L >> d$

(4) Isothermal sphere buried in a semi-infinite solid

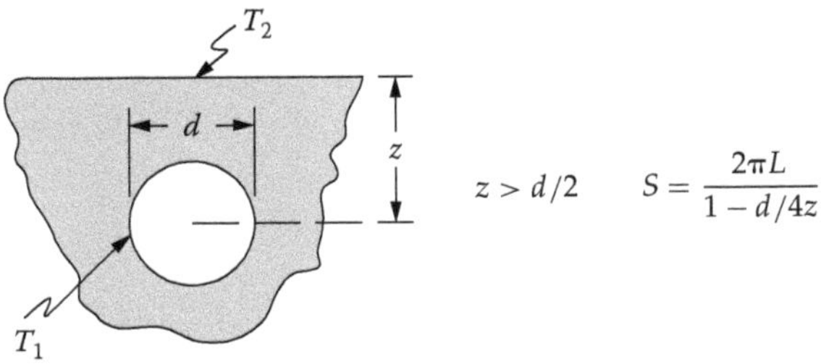

$z > d/2$ $\quad S = \dfrac{2\pi L}{1 - d/4z}$

TABLE 20.2

Some Conduction Shape Factors (*Continued.*)

(5) Two cylinders of diameters, d_1 and d_2 and length, L, placed horizontally in an infinite medium.

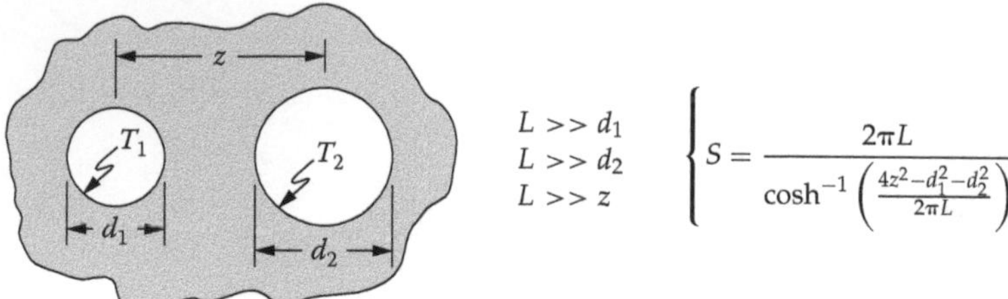

(6) Horizontal cylinder of diameter, d and length, L, placed midway between two plates of equal length and infinite width.

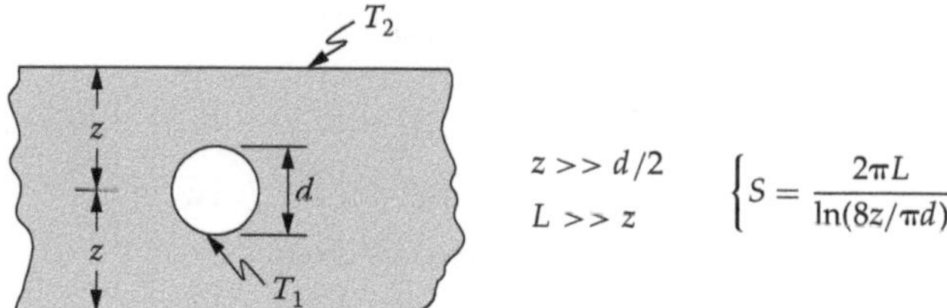

Example 20.12 A long, steel pipe ($k_1 = 35\,\text{W/m-K}$) having an inside radius of $r_1 = 10\,\text{cm}$ and a thickness of 6 mm is laid under the ground ($k_3 = 0.52\,\text{W/m-K}$) such that the centerline of the pipe is 50 cm below the surface of the ground (Figure 20.36). The pipe carries a hot fluid at 120°C that provides a heat transfer coefficient of 80 $\text{W/m}^2\text{-K}$. The pipe is covered with a 10-mm-thick layer of calcium silicate ($k_2 = 0.056\,\text{W/m-K}$). The ground surface is at 20°C; determine the heat loss per meter of pipe.

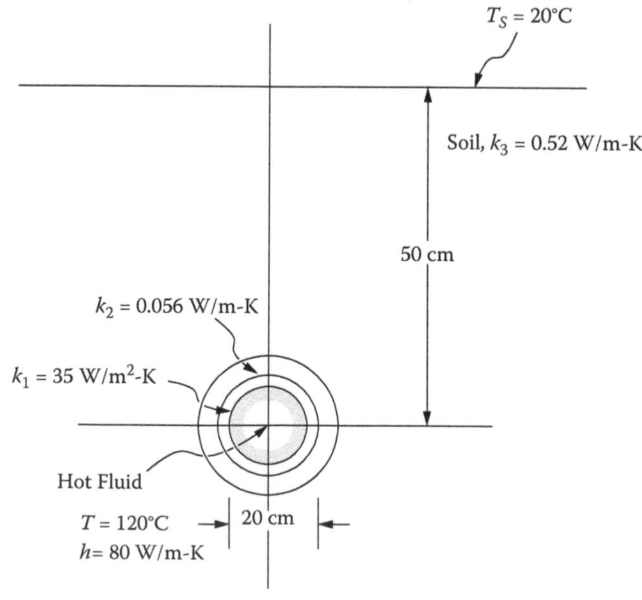

FIGURE 20.36
Buried pipe for Example 20.12.

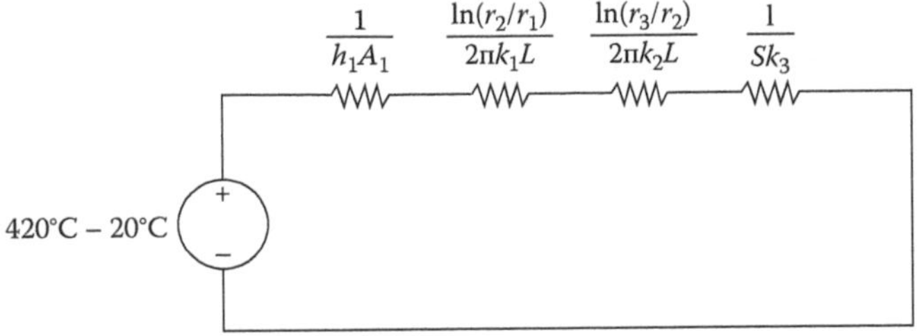

FIGURE 20.37
Electrothermal analog network for Example 20.13.

Solution

Assumptions and Specifications

(1) Steady-state conditions exist.
(2) The heat flow is one dimensional.
(3) Thermal properties do not vary with temperature.
(4) Conduction is the only mode of heat transfer.
(5) There is no contact resistance in this system.

The system may be represented by an electrothermal analog network consisting of the inside convection resistance of the pipe, the conduction resistance of the pipe, the conduction resistance of the calcium silicate layer, and the two-dimensional conduction resistance between the outside surface of the pipe and the ground surface. In this network (Figure 20.37), the outside radius of the pipe, and the inside radius of the calcium silicate), r_2, is 10.6 cm, and the outside radius of the calcium silicate, $r_3 = 11.6$ cm.

We may calculate three of the four thermal resistances. The convection resistance will be

$$R_1 = \frac{1}{hA_1} = \frac{1}{h(2\pi r_1 L)} = \frac{1}{(80\,\text{W/m}^2\text{-K})(2\pi)(0.1\,\text{m})(1\,\text{m})} = 0.0199\,\text{K/W}$$

The resistance of the pipe wall will be

$$R_2 = \frac{\ln(r_2/r_1)}{2\pi k_1 L} = \frac{\ln(0.106\,\text{m}/0.100\,\text{m})}{(2\pi)(35\,\text{W/m-K})(1\,\text{m})}$$

or

$$R_2 = \frac{\ln 1.06}{220\,\text{W/K}} = \frac{0.0583}{220\,\text{W/K}} = 2.650 \times 10^{-4}\,\text{K/W}$$

The resistance of the calcium silicate layer is

$$R_3 = \frac{\ln(r_3/r_2)}{2\pi k_2 L} = \frac{\ln(0.116\,\text{m}/0.106\,\text{m})}{(2\pi)(0.056\,\text{W/m-K})(1\,\text{m})}$$

or

$$R_3 = \frac{\ln 1.0943}{0.352\,\text{W/K}} = \frac{0.0902}{0.352\,\text{W/K}} = 0.256\,\text{K/W}$$

We determine $R_4 = 1/Sk_3$ where we note that configuration 1 (Table 20.2) represents the problem at hand. The condition that $z > 3r_3$ is satisfied and the shape factor, S, is given by

$$S = \frac{2\pi L}{\ln(2z/r_3)}$$

Thus,

$$S = \frac{2\pi(1\,\text{m})}{\ln[4(0.50\,\text{m})/(0.232\,\text{m})]} = \frac{2\pi\,\text{m}}{\ln(8.6206)} = \frac{2\pi\,\text{m}}{2.154} = 2.9170\,\text{m}$$

Then

$$R_4 = \frac{1}{Sk_3} = \frac{1}{(2.9170\,\text{m})(0.52\,\text{W/m-K})} = 0.6593\,\text{K/W}$$

and we can now calculate the heat loss, $\dot{Q}$, as

$$\begin{aligned}\dot{Q} &= \frac{120°\text{C} - 20°\text{C}}{(0.0199 + 2.650 \times 10^{-4} + 0.2561 + 0.6593)} \\ &= \frac{100°\text{C}}{0.9355} \\ &= 106.9\,\text{W}\end{aligned}$$

20.8 Summary

The general equation of heat conduction in rectangular coordinates

$$\frac{\partial^2 T}{\partial x^2} + \frac{\partial^2 T}{\partial y^2} + \frac{\partial^2 T}{\partial z^2} + \frac{q_i}{k} = \frac{1}{\alpha}\frac{\partial T}{\partial t} \tag{20.6}$$

can, in the steady state, and, in the absence of internal volumetric heat generation, be reduced to

$$\frac{d^2 T}{dx^2} = 0 \tag{20.12}$$

in rectangular coordinates and in the x-coordinate direction,

$$\frac{d}{dr}\left(r\frac{dT}{dr}\right) = 0 \tag{20.20}$$

in cylindrical coordinates in the r-coordinate direction and

$$\frac{d}{dr}\left(r^2\frac{dT}{dr}\right) = 0 \tag{20.27}$$

in spherical coordinates in the r-coordinate direction.

The critical radius of insulation in a cylindrical or spherical system pertains to the insulation and yields the minimum resistance to heat flow. It is given by

$$r_{\text{cr,cyl}} = \frac{k}{h} \tag{20.25}$$

for the cylinder and

$$r_{\text{cr,sphere}} = \frac{2k}{h} \tag{20.26}$$

for the sphere.

With a prescribed volumetric heat generation, q_i, Equation 20.6 reduces (in the x direction) to

$$\frac{d^2T}{dx^2} + \frac{q_i}{k} = \frac{1}{\alpha}\frac{\partial T}{\partial t} \tag{20.33}$$

Extended surfaces or fins are used to augment the heat dissipating capability of a surface. For the longitudinal fin of uniform thickness with an insulated tip, the temperature excess, $\theta = T - T_\infty$, at any point is given by

$$\theta = \theta_b \frac{\cosh m(b - x)}{\cosh mb} \tag{20.62}$$

where $m = (2h/k\delta)^{1/2}$. The heat dissipated by the fin is

$$\dot{Q} = km\delta L\theta_b \tanh mb \tag{20.63}$$

The performance of the cylindrical spine is identical to that for the longitudinal fin of uniform thickness except that $m = (4h/kd)^{1/2}$. The fin efficiency in both cases is given by

$$\eta = \frac{\tanh mb}{mb} \tag{20.67}$$

The efficiency of the annular or radial fin of uniform thickness may be found from Figure 20.33 and the heat dissipation is given by

$$\dot{Q}_b = \eta \dot{Q}_{\text{id}} = \eta h[2\pi(r_a^2 - r_b^2) + 2\pi r_a \delta](T_b - T_\infty) \tag{20.71}$$

Conduction shape factors may be employed for two-dimensional heat conduction analysis with the heat flow given by

$$\dot{Q} = kS(T_1 - T_2) \tag{20.73}$$

where S is the shape factor. Several useful shape factors are given in Table 20.2.

20.9 Problems

The Plane Wall

20.1: A metal rod 2.5 cm × 2.5 cm in cross section and 18 cm long is heated at one end and cooled at the other. The heat input at the left end is 2 W and the difference in temperature between the left and right hand ends is 5°C. Determine the thermal conductivity of the rod.

20.2: A horizontal steel plate ($k = 36$ W/m-K) that is 5-cm thick is covered with 1 cm of fiber insulating board ($k = 0.050$ W/m-K). The temperature of the steel and insulating board surfaces are 150°C and 25°C, respectively. The surface area of the steel is 0.625 m^2. Determine (a) the heat transmitted through the composite and (b) the interface temperature.

20.3: A glass window ($k = 1.07\,W/m\text{-}K$) measuring $2\,m \times 3\,m$ is 0.3175-cm thick. The inside air is at 22.2°C and the heat transfer coefficient is $12\,W/m^2\text{-}K$. The outside air is at −23.3°C and the heat transfer coefficient is $20\,W/m^2\text{-}K$. Determine (a) the heat loss and (b) the temperature difference across the glass.

20.4: Glass storm windows are used to decrease heat transfer between the inside of a house and the environment. For a storm window with **two** panes of glass separated by a 7.5-mm air space, use the conditions of Problem 20.3 to determine (a) the heat loss and (b) the temperature drop across the air space.

20.5: A boiler is to be insulated so that its heat loss does not exceed $125\,W/m^2$ of wall surface area. The inside and outside surfaces are to be maintained at 825°C and 200°C. Determine the thickness of glass wool insulation to be used.

20.6: In Problem 20.5, suppose that the glass wool insulation is exposed to the air at 25°C through a heat transfer coefficient of $10\,W/m^2\text{-}K$ and the heat flux of $125\,W/m^2$ is to be maintained. Determine (a) the thickness of the insulation and (b) its surface temperature.

20.7: One side of an aluminum alloy ($k = 182\,W/m\text{-}K$) block is maintained at 180°C. The other side is covered with fiberglass ($k = 0.038\,W/m\text{-}K$) that is 1.25-cm thick and whose outside face is maintained at 60°C. The total heat flow through this composite is 280 W. Determine the cross-sectional area for heat flow.

20.8: The walls of a house are constructed with a 10-cm layer of brick ($k = 0.646\,W/m\text{-}K$), 1.25 cm of celotex ($k = 0.046\,W/m\text{-}K$), an air space 9.25-cm thick, another layer of 1.25 cm celotex, and 0.635 cm of wood paneling ($k = 0.204\,W/m\text{-}K$). The temperature of the outside surface of the brick is 0°C and the inside surface of the wood paneling is 25°C. Assume that the air space heat transfer is solely by conduction and determine (a) the heat transferred and (b) the heat transferred if the air space is filled with glass wool ($k = 0.040\,W/m\text{-}K$).

20.9: A composite slab contains two layers. Layer 1 is brass, which is 2-cm thick and layer 2 is iron, which is 3-cm thick. The iron is exposed to air at 10°C with a heat transfer coefficient of $12\,W/m^2\text{-}K$. Determine the temperature at the brass surface if the brass is exposed to a radiation heat flux of $2000\,W/m^2$.

20.10: The enclosure of a refrigerator is constructed by sandwiching a 4.8-cm layer of fiberglass insulation ($k = 0.038\,W/m\text{-}K$) between two 2.5-mm mild steel plates ($k = 42.9\,W/m\text{-}K$). The outside and inside temperatures are respectively 27°C and 4°C and each convection coefficient on the outside and inside have a value of $5.50\,W/m^2\text{-}K$. Determine the heat gain through $1\,m^2$ of the wall.

20.11: A single-room vacation hut has a floor area of $4\,m \times 6\,m$ and a roof height of 2.25 m. It has a single storm window in the 6-m wall that measures $1.5 \times 1.5\,m$. The effective L/k ratios for the three surfaces are

Item	L/k, $m^2\text{-}K/W$
Window	0.1194
Walls	0.1923
Roof	0.3226

The inside surface of the hut is to be held at 22°C when the outside surface exposed to the environment is at 0°C, and electricity costs \$0.89/kW-h. Determine the cost of heating the hut for a continuous heater operation of 2 weeks.

20.12: Wind blows on the outside of the vacation hut in Problem 20.11 such that the heat transfer coefficient between the outside surface of the hut and the environment at 0°C is 16 W/m^2-K. Determine the cost of heating for the 2-week period.

20.13: An aluminum pot (k = 232 W/m-K) having a diameter of 35 cm holds boiling water (100°C) and is sitting on a heating element that injects heat into the pot at a rate of 750 W. The wall of the pot is 1.125-mm thick and the temperature at the inside surface of the pot is 102°C. Determine (a) the boiling heat transfer coefficient and (b) the temperature at the exterior of the pot just above the heating element.

20.14: The composite wall of a furnace consists of three layers of different materials. Layer 1 (k = 24 W/m-K) is the inner layer and is 30-cm thick. Layer 2 (the middle layer) is 75-cm thick and its thermal conductivity is unknown. Layer 3 (k = 48 W/m-K) is the outer layer and is 15-cm thick. The inside and outside temperatures are 620°C and 20°C, respectively, and the heat flux is measured as 5 kW/m^2. Determine the thermal conductivity of the middle layer.

20.15: The composite wall of a furnace is configured as follows:

Item	Material	k W/m-K	Thickness, cm
1	Fireclay brick	1.07	20
2	Diatomaceous earth brick	0.31	15
3	Common brick	0.69	10

The surface adjacent to the fireclay brick is held at 1075°C and the surface adjacent to the common brick is held at 125°C. Determine (a) the heat loss per m^2 of wall area, (b) the temperatures between the layers, and (c) the temperature at a point 15 cm from the outer surface.

20.16: A slab has a thermal conductivity of 12 W/m-K and a thickness of 10 cm. For unit cross-sectional area and the left-hand and right-hand faces maintained at 130°C and 10°C, respectively, determine the heat flow from left to right.

Contact Resistance

20.17: Reconsider Problem 20.2 with a contact resistance of R_c = 0.0125 m^2-K/W between the steel and the insulating board. Determine (a) the heat transmitted through the composite and (b) the temperature drop due to the contact resistance.

20.18: Suppose that in Problem 20.9, there is a thermal contact resistance of R_c = 0.00875 m^2-K/W but all other conditions remain the same. Determine the temperature of the brass surface.

20.19: Reconsider Problem 20.14 and suppose that, along with the resistances given, there is a contact resistance of R_c = 0.0075 m^2-K/W between layers 1 and 2 and a contact resistance of R_c = 0.0065 m^2-K/W between layers 2 and 3 causing the heat loss to be reduced to 4.479 kW. Determine the thermal conductivity of layer 2.

20.20: Two insulated stainless steel cylindrical rods, 10-cm long with k = 12.8 W/m-K are placed end to end. The interface has a contact resistance of R_c = 0.0048 m^2-K/W. The ends of the rod are maintained at 240°C and 40°C. Determine the diameter of the rods if a heat flow of 50 W is measured.

20.21: An electronic component dissipating 22.5 W is to be conductively cooled via a heat sink having an effective thermal resistance of 0.2431°C/W and two contacts, each with R_c = 10^{-4} m^2-K/W. The area of the contacts is 4.00 cm^2, and the maximum permitted component case temperature is 85°C. Determine if the component must be derated if it sits on a surface at 25°C.

20.22: Two square copper ($k = 385\,\text{W/m-K}$) bars having a length of 22.5 cm and an area of 16 cm^2 are pressed together at their ends so that the contact resistance is $R_c = 2 \times 10^{-4}\,\text{m}^2\text{-K/W}$. The overall temperature difference between the ends is 50°C. Determine the axial heat flow if (a) the contact resistance is neglected and (b) the contact resistance is included.

20.23: Rework Problem 20.22 with one of the bars specified as aluminum ($k = 178\,\text{W/m-K}$) and the other as copper ($k = 360\,\text{W/m-K}$), both having a surface roughness of 50 μm. The contact pressure is 25 MPa.

20.24: Two 3.5-cm-diameter stainless steel bars ($k = 16\,\text{W/m-K}$) having a length of 12.5 cm have a contact between them ($R_c = 5.25 \times 10^{-4}\,\text{m}^2\text{-K/W}$). The overall temperature difference between the ends of the bars is 80°C. Determine (a) the axial heat flow and (b) the temperature drop across the contact.

Cylindrical Pipes and the Critical Radius of Insulation

20.25: A nominal 5-in schedule 40 cast-iron ($k = 46.4\,\text{W/m-K}$) pipe is covered with 4 cm of 85% magnesia ($k = 0.074\,\text{W/m-K}$) insulation. The pipe carries steam such that the inner pipe surface is 435°C and the outer insulation surface is at 40°C. Determine the heat loss for a pipe length of 40 m.

20.26: A nominal 8-in schedule 80 mild steel ($k = 39.6\,\text{W/m-K}$) pipe is covered with a 6 cm of 85% magnesia ($k = 0.074\,\text{W/m-K}$) insulation and then by a 3-cm layer of an insulation with $k = 0.062\,\text{W/m-K}$. The inside pipe surface is at 450°C and the outer insulation surface is at 50°C. For a pipe length of 25 m, determine (a) the heat lost and (b) the temperature between the insulation layers.

20.27: Suppose that the insulation layers in Problem 20.26 are reversed. Determine the heat lost.

20.28: A pipe in a refrigeration system has an outside diameter of 6 cm and a surface temperature of 4°C. It is covered with two layers of insulation, a 2.5-cm layer of 85% magnesia ($k = 0.074\,\text{W/m-K}$) and a 2.5-cm layer of rock wool. Determine which layer should be placed adjacent to the pipe if the surface of the outer layer, whichever one, is to be maintained at 36°C.

20.29: A steel pipe ($k = 25.5\,\text{W/m-K}$) has an outside diameter of 5.50 cm. It is covered with a 0.75-cm layer of insulation ($k = 0.172\,\text{W/m-K}$) and 2.5-cm layer of another insulation ($k = 0.045\,\text{W/m-K}$). The surface temperature of the pipe is 325°C and the outer insulation temperature is 25°C. For a length of 2 m, determine (a) the heat loss and (b) the temperature at the interface between the two insulation layers.

20.30: In Problem 20.29, determine the thickness of the outer layer of insulation to limit the heat loss to 240 W.

20.31: A nominal 1-in schedule 40 wrought iron ($k = 48\,\text{W/m-K}$) is covered with 1.25 cm of insulation ($k = 0.060\,\text{W/m-K}$). The inside surface of the pipe is at 275°C and the temperature at the outer surface of the insulation is 50°C. The pipe length is 1 m. Determine (a) the heat loss and (b) the temperature at the insulation interface.

20.32: In Problem 20.31, determine the required insulation thickness of the outer layer of insulation to limit the heat loss to 120 W.

20.33: Rework Problem 20.25 with convection on both the inside pipe surface ($h = 120\,\text{W/m}^2\text{-K}$) and the outer insulation surface ($h = 12.5\,\text{W/m}^2\text{-K}$). In this case, the temperature of the steam is 450°C and the environmental temperature is 20°C.

20.34: Rework Problem 20.26 with convection on both the inside pipe surface (h = 600 W/m^2-K) and the outer insulation surface (h = 12 W/m^2-K). In this case, the temperature of the steam is 450°C and the environmental temperature is 50°C.

20.35: Rework Problem 20.29 with convection on the outer insulation surface (h = 10 W/m^2-K) with an environmental temperature of 25°C. All other parameters remain the same.

20.36: Steam at 1.4 MPa flows through an 8-in schedule 80 mild steel pipe that is covered with 2.25 cm of insulation (k = 0.058 W/m-K). The heat transfer coefficient at the inner pipe surface is 400 W/m^2-K and the outside surface is exposed to air at 20°C where it loses heat by convection through a heat transfer coefficient of 20 W/m^2-K. The outer surface has an emissivity of 0.78 and the walls of the enclosure are at 20°C. For a pipe length of 1.2 m, determine (a) the heat loss and (b) the temperature at the surface of the insulation.

20.37: A 5-cm-outside-diameter tube with a surface temperature of 240°C passes through an enclosure where the air temperature is 25°C. Insulation (k = 0.205 W/m-K) is to be applied and the natural convection heat transfer coefficient at the surface temperature of the insulation is expected to be h = 4.25 W/m^2-K. Determine (a) the critical radius of the insulation, (b) the heat loss from 1 m of the pipe before application of the insulation, and (c) the heat loss from 1 m of the pipe after application of the insulation.

20.38: A steel tube (k = 44 W/m-K) has an inside diameter of 3.00 cm and a tube wall thickness of 1.5 mm. A fluid at 245°C that produces a heat transfer coefficient of 1600 W/m^2-K flows on the inside of the tube and a second fluid flows across the outside of the tube with a heat transfer coefficient of 200 W/m^2-K at 35°C. The tube is 1-m long. Determine the heat loss.

20.39: To reduce the heat loss in Problem 20.38, the tube is to be equipped with asbestos insulation. All conditions listed in Problem 20.38 are to remain the same. Determine the thickness of insulation required to reduce the heat loss to 500 W.

20.40: A 6-cm-outside-diameter steel pipe with a surface temperature of 385°C is covered with a 1.25-cm layer of an insulation (k = 0.225 W/m-K) followed by a 2-cm layer of another insulation (k = 0.080 W/m-K). The outside of the 2-cm layer is exposed to an environment via a heat transfer coefficient of 25 W/m^2-K, where the air temperature is 25°C. Determine (a) the heat loss and (b) the temperature at the interface between the two layers of insulation.

Hollow Spheres

20.41: A hollow stainless steel sphere (0.1% chrome) has an outer diameter of 80 cm and an inner diameter of 64 cm. Its inside surface is at 40°C and it is placed in a temperature bath at 20°C with a heat transfer coefficient of 125 W/m^2-K. Determine (a) the heat flow from the sphere and (b) the temperature at the surface of the sphere.

20.42: Suppose that a 10-cm layer of insulation (k = 0.165 W/m-K) is added to the outside of the sphere in Problem 20.41 and causes the heat transfer coefficient to drop to 120 W/m^2-K. Determine the heat loss.

20.43: A hollow sphere is fabricated from 22% Si aluminum with inner and outer diameters of 4 cm and 10 cm, respectively. The inside and outside temperatures are 60°C and 20°C. Determine the heat flow from the sphere.

20.44: In an effort to reduce the heat loss, the sphere of Problem 20.43 is to be covered with two layers of insulation, each 2-cm thick. Layer 1 has a thermal conductivity of $k_1 = 0.0625$ W/m-K and layer 2 has a thermal conductivity of $k_2 = 0.125$ W/m-K. The surface temperature of layer 2 is 20°C and the inside of the sphere remains at 60°C. Determine the heat loss.

20.45: Suppose that in Problem 20.44, the two insulation layers are reversed. Determine the heat loss if all other conditions remain the same.

20.46: Reconsider Problem 20.43 with the environment at 20°C and a heat transfer coefficient at the exterior of the sphere at 550 W/m^2-K. Determine the heat loss.

20.47: Suppose that in Problem 20.46, the critical radius of an insulation with $k = 0.938$ is added to the sphere. All other conditions remain the same. Determine the heat loss.

20.48: Rework Problem 20.47 with insulation layers (a) 3-cm thick ($k = 0.15$ W/m-K) and (b) 4-cm ($k = 0.12$ W/m-K) thick.

20.49: A sphere with a diameter of 12.5 cm is electrically heated so that its surface temperature is held at 240°C. The surface is exposed to an environment at 25°C via a heat transfer coefficient of 48 W/m^2-K. Determine the heat loss from the sphere.

20.50: A sphere with a diameter of 12.5 cm is exposed to an environment with a heat transfer coefficient of 12 W/m^2-K and has a surface emissivity of 0.475. Both the enclosure and the environment are at 20°C. Determine the surface temperature if the heat loss is 600 W.

Internal Heat Generation

20.51: A plane wall with a thickness of 16 cm and a thermal conductivity of 9.6 W/m-K has a volumetric heat generation of 160 kW/m.^{3} The left and right sides of the wall are held at 95°C and 75°C. Determine (a) the location of the maximum temperature in the wall and (b) the maximum temperature.

20.52: A plane wall has a thickness of 4 cm and a center plane temperature of 180°C. The wall has a thermal conductivity, k W/m-K, and both faces are maintained at 60°C. Determine the uniform heat generation in the wall.

20.53: Consider a plane slab with thermal conductivity, k, and faces held at T_1 and T_2. The slab is L m thick and the volumetric heat generation is of q_i (W/m^3). With the temperature distribution of Equation 20.34, show that Equation 20.37 results.

20.54: A plane wall with a thickness of 10 cm and a thermal conductivity of 30 W/m-K experiences a volumetric heat generation of 5×10^5 W/m^3. The left face of the wall is in contact with a coolant at 25°C, which provides a heat transfer coefficient of 100 W/m^2-K. The right face of the wall is cooled by a fluid at 30°C with a heat transfer coefficient of 300 W/m^2-K. Determine the temperatures of the faces and the maximum temperature.

20.55: A plane fuel element in a nuclear reactor is 3-cm thick and has a thermal conductivity of 60 W/m-K. The element is exposed on both faces to a coolant at 25°C that creates a heat transfer coefficient of 150 W/m^2-K. Determine the maximum possible volumetric heat generation in the element if the maximum temperature in the fuel element is not to exceed 58°C.

20.56: A 20-cm-thick plane slab has a centerline temperature of 250°C and a surface temperature of 25°C. Determine the distance from the centerline for a temperature of 160°C.

20.57: A current of 180 A is passed through a 2-m-long and 2.5-mm-diameter wire (k = 25 W/m-K and ρ_e = 7.2×10^{-8} *Om*-m). The wire is submerged in a liquid bath at 105°C and experiences a convective dissipation with a constant heat transfer coefficient of 3000 W/m^2-K. Determine the center temperature of the wire.

20.58: A 1-cm-diameter electric wire with a resistance of 0.03 Ω per meter of length carries a current of 80 A under steady-state operation. The conductor dissipates the heat generated to its surroundings at 27°C, which provides a heat transfer coefficient of 950 W/m^2-K. The thermal conductivity of the wire is 25 W/m-K. Determine (a) the volumetric heat generation in the wire, (b) the wire surface temperature, and (c) the maximum temperature in the wire.

20.59: A solid cylinder with a diameter, d_o, of 10 cm and thermal conductivity, k = 18 W/m-K has a surface temperature of 96°C. Determine the volumetric heat generation if the temperature at a point 2.5 cm from its center is 170°C.

20.60: A solid sphere with a radius of 10 cm has a thermal conductivity of 1.2 W/m-K. The center temperature is 240°C and the surface temperature is 40°C. Determine the volumetric heat generation.

20.61: A solid sphere has a radius of 12 cm. The volumetric heat generation is 600,000 W/m^3 and the thermal conductivity of k = 20 W/m-K. The sphere is cooled via a heat transfer coefficient of 200 W/m^2-K in an environment of 20°C. Determine (a) the surface temperature, (b) the center temperature, and (c) the temperature at a point 4 cm from the center.

20.62: A solid sphere 24 cm in diameter has a uniform volumetric heat generation of 800,000 W/m^3 and a thermal conductivity of 16 W/m-K. The sphere has a maximum temperature of 172°C and sits in an environment of 20°C. Determine the heat transfer coefficient.

Extended Surfaces (Fins)

20.63: A cylindrical rod of pyrex glass with a diameter of 1.375 cm and a base temperature of 125°C extends 10 cm into the air at 25°C, where the heat transfer coefficient is 10 W/m^2-K. Determine (a) the temperature at the mid-height, (b) the temperature at the tip, (c) the heat dissipated by the rod, (d) the fin effectiveness, (e) the fin efficiency, and (f) the thermal resistance.

20.64: Reconsider Problem 20.63 with a cast-iron rod (k = 50 W/m-K), all other conditions remaining the same.

20.65: Reconsider Problem 20.63 with an aluminum rod (k = 180 W/m-K), all other conditions remaining the same.

20.66: A longitudinal fin of uniform thickness with a base temperature of 140°C operates in a convective environment with the air at 15°C at a heat transfer coefficient of 50 W/m^2-K. The fin is 4-mm thick, 8-cm high and 40-cm long. The fin is fabricated of a steel having a thermal conductivity k of 27.5 W/m-K. Determine (a) the tip temperature, (b) the heat dissipated by the fin, (c) the fin efficiency, (d) the fin effectiveness, and (e) the thermal resistance.

20.67: Reconsider Problem 20.66 with aluminum (k = 202 W/m-K), all other conditions remaining the same.

20.68: Reconsider Problem 20.66 with copper (k = 385 W/m-K). All other conditions remaining the same.

20.69: Three cylindrical spines, one of glass ($k = 1.12$ W/m-K), one of iron ($k = 55$ W/m-K) and one of aluminum ($k = 205$ W/m-K), have a diameter of 0.75 cm and a height of 6 cm. Their tips are insulated and the heat transfer coefficient in each case is 12 W/m^2-K. The environmental temperature is 25°C and the spine base temperature is 85°C. Determine the tip temperature in each case.

20.70: Rework Problem 20.69 if all three spines dissipate from their tips to the environment via a heat transfer coefficient of 10 W/m^2-K. All other conditions remain the same.

20.71: A 1.75-cm-diameter alloy rod ($k = 50$ W/m-K) with a base temperature of 227°C and a height, b, extends into air at 27°C such that the heat transfer coefficient is 50 W/m^2-K. Neglecting the fin tip heat loss, determine the height, b, necessary to achieve a tip temperature of 172°C.

20.72: A longitudinal fin of uniform thickness has a thermal conductivity of 62.5 W/m^2-K and a thickness of 0.25 cm. It has a base temperature of 80°C and extends into the air at 20°C such that the heat transfer coefficient is 75 W/m^2-K. The fin length is 40 cm and the fin tip is insulated. Determine the heat flow for fin heights of (a) 1.5 cm, (b) 3 cm, and (c) 5 cm.

20.73: Rework Problem 20.72 for the case of a tip heat loss through a heat transfer coefficient having a value of 50 W/m^2-K. All other conditions remain the same.

20.74: An annular fin of uniform thickness has an inner radius of 2.5 cm and an outer radius of 6 cm. The fin thickness is 2.25 cm and the fin possesses a thermal conductivity of 200 W/m-K. The base temperature is 165°C and the fin dissipates to air at 25°C with a heat transfer coefficient of 300 W/m^2-K. Determine (a) the fin efficiency, (b) the heat dissipated by the fin, (c) the fin effectiveness, and (d) the thermal resistance of the fin.

20.75: Rework Problem 20.74 for a fin with a thermal conductivity of 100 W/m-K. All other conditions remain the same.

20.76: One end of an aluminum rod ($k = 208$ W/m-K) is held at 200°C. The other end is held at 100°C and the rod height is 25 cm. The rod has a diameter of 2.5 cm and the heat transfer coefficient between the rod and the surroundings is 20 W/m^2-K. Determine the net rate of heat loss from the rod.

The Shape Factor Method

20.77: A very long pipe with a diameter of 20 cm is buried horizontally in a soil with a thermal conductivity of $k = 1.4$ W/m-K at a depth of 1 m below the surface. The surface temperatures of the pipe and soil are 80°C and 30°C, respectively. Determine the heat loss per unit length.

20.78: Two very long pipes with diameters of $d_1 = 12$ cm and $d_2 = 16$ cm are buried horizontally in a soil with a thermal conductivity of $k = 1.5$ W/m-K at a depth of 6.20 m below the surface. The centerline spacing of the pipes is 1 m and their surfaces are held at $T_1 = 25$°C and $T_2 = 75$°C. Estimate the heat flow between them.

20.79: A tall chimney with the outside cross section of a square has a side dimension of 1 m. A circular flue in the center of the chimney has a diameter of 20 cm and carries a gas such that the flue surface temperature is 180°C. The chimney material has a thermal conductivity of 4 W/m-K and its surface temperature is held at 20°C. Determine the heat loss per meter of chimney height.

20.80: For the chimney in Problem 20.79, consider that the heat transfer coefficient between the outside surface and the environment is 1.25 W/m^2-K. The outside surface is at 20°C. Determine (a) the heat loss per meter of length and (b) the temperature at the surface of the flue.

20.81: Rework Problem 20.79 for a concentric square flue that is 12 cm × 12 cm. All other conditions remain the same.

20.82: An 8-cm-diameter sphere whose surface temperature is 180°C is buried at a depth of 80 cm in soil having a thermal conductivity of 135 W/m-K. The surface of the soil is held at 30°C. Determine the heat loss from the sphere.

20.83: A 1.2-m-long 4-cm-diameter pipe is located at the centerline of a plastic (k = 8 W/m-K) slab that is 40-cm thick. If the pipe surface is held at 100°C and the slab surface is held at 20°C, determine the heat loss from the pipe.

21

Unsteady-State Conduction

Chapter Objectives

- To describe the physics of unsteady or transient conduction.
- To illustrate the use of the lumped capacitance model for the convective cooling of a body of arbitrary shape.
- To develop a criterion for the validity of the lumped capacitance model.
- To describe the semi-infinite model and present solutions for three types of surface thermal boundary conditions.
- To study one-dimensional convective cooling/heating of a plane wall, a solid cylinder and a solid sphere with the help of simple approximate solutions and the Heisler charts.

21.1 Introduction

In **unsteady** or **transient** conduction the temperature of a system changes with time due to a thermal disturbance created within the system and/or the environment. For example, when a power transistor is suddenly energized, the heat generated within the transistor causes its temperature to increase with time. When a hot billet is withdrawn from a furnace and immersed in a coolant pool, it is the change in the environmental condition that causes the temperature of the billet to change with time.

Consider the cooling of a hot solid when it is immersed in a pool of coolant. During the early period, the cooling effect is felt in a thin region of the solid that is closest to the surface. The internal core of the solid is virtually unaffected by the cooling that is occurring at the surface. This early thermal regime that is characterized by steep temperature gradients may be modeled by considering the solid to be semi-infinite in extent. As time progresses, the cooling effect penetrates deeper and deeper into the solid and the temperature gradients begin to moderate. As the late stages of the transient are reached, temperature gradients become so small that the body temperature changes with time but is essentially uniform throughout the solid. It is this behavior that is characterized by the **lumped capacitance model.** This model is the easiest to analyze and is considered first.

21.2 The Lumped Capacitance Model

The lumped capacitance model ignores the spatial temperature variation in the body and assumes that the temperature is a function only of time. Despite its simplicity, the model serves as a useful predictive tool for studying many practical problems involving transient conduction.

21.2.1 Convective Cooling

Consider a solid body of arbitrary shape having volume, V, surface area, S, density, ρ, and specific heat, c. The body is initially at a uniform temperature, T_i, as shown in Figure 21.1. At time, $t \geq 0$, the body is immersed in a coolant, which is at $T_\infty < T_i$. The coolant is assumed to have an infinite thermal capacity so that its temperature is unaffected by any energy that is released from the solid body.

The energy balance dictates that the rate of surface convection must equal the rate of decrease of the internal energy of the solid body, that is

$$-\rho V c \frac{dT}{dt} = hS(T - T_\infty) \tag{21.1}$$

where the minus sign has been inserted because the temperature gradient, dT/dt, is less than zero. The initial condition may be expressed as

$$T(t = 0) = T_i \tag{21.2}$$

To solve Equation 21.1, we separate the variables to yield

$$\frac{dT}{T - T_\infty} = -\frac{hS}{\rho V c} dt \tag{21.3}$$

and then an integration gives

$$\ln(T - T_\infty) = -\frac{hS}{\rho V c} t + C_1 \tag{21.4}$$

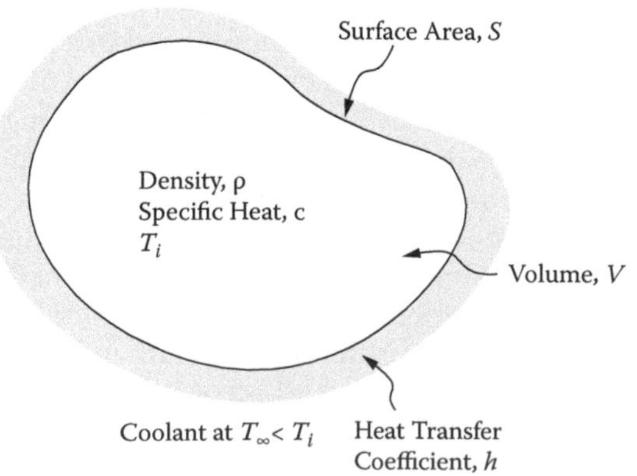

FIGURE 21.1
Solid body of arbitrary shape used to develop the lumped capacitance model for unsteady-state heat conduction.

where C_1 is the constant of integration that is found by applying the initial condition of Equation 21.2

$$C_1 = \ln(T_i - T_\infty)$$

With the substitution of C_1, Equation 21.4 may be written as

$$\frac{T - T_\infty}{T_i - T_\infty} = e^{-(hS/\rho Vc)t} \tag{21.5}$$

The quantity, $\rho Vc/hS$, is the product of the convection resistance, $1/hS$, and, ρVc, a parameter known as the **thermal capacitance** of the body. If we draw the analogy with the **time constant** of a simple first-order electrical network containing a resistor and a capacitor, we see that the quantity, $\rho Vc/hS$, may be called the **thermal time constant**, which we denote as τ.

$$\tau = \rho Vc/hS \tag{21.6}$$

We note that the larger the value of τ resulting from a larger thermal capacitance and/or the smaller the product hS, the slower the cooling of the body and the longer the body will take to reach thermal equilibrium with the coolant.

The cumulative heat loss to the coolant, Q, may be obtained by integrating the instantaneous convective heat loss

$$Q = \int_0^t hS(T - T_\infty)dt \tag{21.7}$$

and with the substitution of $T - T_\infty$ from Equation 21.5, we may perform the integration to obtain

$$Q = \rho Vc(T_i - T_\infty)\left[1 - e^{-(hS/\rho Vc)t}\right] = \rho Vc(T_i - T_\infty)\left(1 - e^{-t/\tau}\right) \tag{21.8}$$

21.2.2 The Validity Criterion

To establish the criterion for the validity of the lumped capacitance model, consider one-dimensional steady conduction in the plane wall of area, A, thickness, L, and thermal conductivity, k. Let the left face of the wall be held at a fixed temperature while the right face loses heat by convection to a fluid with a convection heat transfer coefficient, h. Because the temperature drops in the wall, ΔT_{wall}, and the fluid, ΔT_{fluid}, are respectively proportional to the conduction and convection resistances, L/kA, and $1/hA$, where $A = S$, it follows that

$$\frac{\Delta T_{\text{wall}}}{\Delta T_{\text{fluid}}} = \frac{L/kA}{1/hA} = \frac{hL}{k} \tag{21.9}$$

The ratio of the conduction resistance to the convection resistance is called the **Biot number** or **Biot modulus** and denoted by Bi. Equation 21.9 may be written as

$$\Delta T_{\text{wall}} = (\text{Bi})\Delta T_{\text{fluid}} \tag{21.10}$$

and if Bi is much less than 1, Equation 21.10 indicates that ΔT_{wall} will be much less than ΔT_{fluid} and, therefore, the temperature in the wall may be assumed to be nearly uniform.

Now consider a transient conduction situation where a body, initially at a uniform temperature, is suddenly exposed to a convective environment. If the condition of $\text{Bi} \ll 1$ is satisfied, then the transient conduction in the body can be modeled as a sequence of thermal

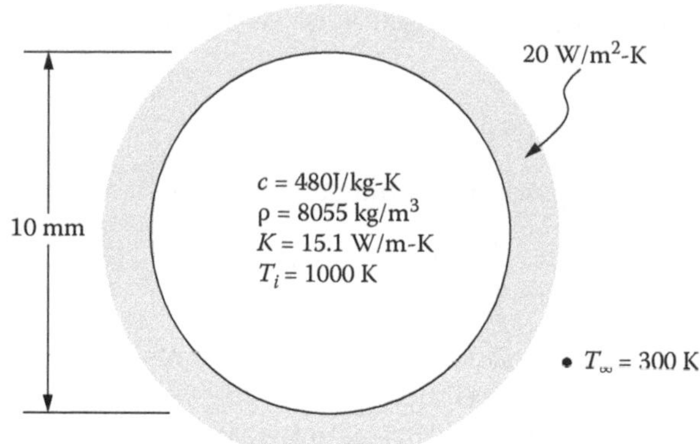

FIGURE 21.2
Configuration for Example 21.1.

states where the temperature of the body changes with time but remains spatially uniform at every instant of time. These are precisely the conditions for the lumped capacitance model and the criterion for its validity may be set as Bi $\ll$ 1. The threshold value of Bi that is commonly used is 0.1 so that the lumped capacitance model is valid if

$$\text{Bi} = \frac{hL_c}{k} < 0.1 \tag{21.11}$$

where $L_c = V/S$ is the characteristic length of the body. For a long cylinder,

$$L_c = \frac{V}{S} = \frac{\pi r_o^2 L}{2\pi r_o L} = \frac{r_o}{2}$$

and for a sphere

$$L_c = \frac{V}{S} = \frac{4\pi r_o^3/3}{4\pi r_o^2} = \frac{r_o}{3}$$

Example 21.1 A 10-mm-diameter AISI 302 stainless steel ball ($c = 480\,\text{J/kg-K}$, $\rho = 8055\,\text{kg/m}^3$, and $k = 15.1\,\text{W/m-K}$) is annealed by heating it to 1000 K in a furnace and then allowed to cool to 400 K in the air at 300 K, which provides a convective heat transfer coefficient of $20\,\text{W/m}^2\text{-K}$. Determine (a) the time required to cool the ball and (b) the cumulative energy loss over the time determined in (a).

Solution

Assumptions and Specifications

(1) Unsteady-state conditions exist.
(2) Thermal properties do not vary with temperature.
(3) There is no radiation exchange with the surroundings.

(a) We first determine the Biot number

$$\text{Bi} = \frac{hr_o}{3k} = \frac{(20\,\text{W/m}^2\text{-K})(0.005\,\text{m})}{3(15.1\,\text{W/m-K})} = 2.21 \times 10^{-3}$$

Because Bi < 0.1, the lumped capacitance model may be employed. Equation 21.5 may be solved to yield

$$t = -\frac{\rho Vc}{hS}\ln\left(\frac{T - T_\infty}{T_i - T_\infty}\right) = -\tau\ln\left(\frac{T - T_\infty}{T_i - T_\infty}\right)$$

With the time constant, τ,

$$\frac{\rho Vc}{hS} = \frac{(8055\,\text{kg/m}^3)(4\pi/3)(0.005\,\text{m})^3(480\text{J/kg-K})}{(20\,\text{W/m}^2\text{-K})(4\pi)(0.005\,\text{m})^2} = 322.2\,\text{s}$$

we have

$$t = -(322.2\,\text{s})\ln\left(\frac{400\,\text{K} - 300\,\text{K}}{1000\,\text{K} - 300\,\text{K}}\right) = -(322.2\,\text{s})\ln(0.143)$$

$$= -(322.2\,\text{s})(-1.946) = 627\,\text{s} \qquad (10.45\,\text{min}) \Longleftarrow$$

(b) Using Equation 21.8, the cumulative heat loss is

$$Q = \rho Vc(T_i - T_\infty)\left(1 - e^{-t/\tau}\right)$$

With

$$\rho Vc = (8055\,\text{kg/m}^3)(4\pi/3)(0.005\,\text{m})^3(480\,\text{J/kg-K}) = 2.024\,\text{J/K}$$

and

$$\frac{t}{\tau} = \frac{627\text{s}}{322.2\text{s}} = 1.946$$

we have

$$Q = (2.024\,\text{J/K})(1000\,\text{K} - 300\,\text{K})(1 - e^{-1.946})$$

$$= (1417.1\,\text{J})(1 - e^{-1.946})$$

$$= (1417.1\,\text{J})(0.857) = 1214.7\,\text{J} \Longleftarrow$$

21.2.3 The Effect of Internal Heat Generation

When an internal heat generation is present, as in the case of an electronic component that is suddenly energized, a heat generation term, $\dot{E}_g$ (in watts), is included in Equation 21.1. With $\dot{E}_g$ included and noting that $dT/dt > 0$ for this case of heating, Equation 21.1 is modified to

$$\rho Vc\frac{dT}{dt} = \dot{E}_g - hS(T - T_\infty) \tag{21.12}$$

The solution to Equation 21.12 is facilitated by introducing a new variable, θ, defined as

$$\theta = \dot{E}_g - hS(T - T_\infty) \tag{21.13}$$

so that

$$\frac{d\theta}{dt} = -hS\frac{dT}{dt}$$

or

$$\frac{dT}{dt} = -\frac{1}{hS}\frac{d\theta}{dt}$$

Then, Equation 21.12 may be written as

$$-\frac{\rho Vc}{hS}\frac{d\theta}{dt} = \theta$$

and after the variables are separated

$$\frac{d\theta}{\theta} = -\frac{hS}{\rho Vc}dt$$

Integration then gives

$$\ln\theta = -\frac{hS}{\rho Vc}t + C_1 = -\frac{t}{\tau} + C_1 \tag{21.14}$$

The initial condition of Equation 21.2 translates to the condition

$$\theta(t=0) = \dot{E}_g - hS(T_i - T_\infty) \tag{21.15}$$

Hence,

$$C_1 = \ln[\dot{E}_g - hS(T_i - T_\infty)]$$

and with the substitution of this into Equation 21.14 we obtain

$$\frac{\dot{E}_g - hS(T - T_\infty)}{\dot{E}_g - hS(T_i - T_\infty)} = e^{-t/\tau} \tag{21.16}$$

The steady-state temperature, T_{ss}, may be found by letting $t \longrightarrow \infty$. Thus, with $t = \infty$ in Equation 21.16

$$T_{ss} = T_\infty + \frac{\dot{E}_g}{hS} \tag{21.17}$$

Example 21.2 An electronic component, which has a heat capacity of $\rho Vc = 60\,\text{J/K}$, generates 8 W when it is suddenly energized. The component, sketched in Figure 21.3, is initially at a temperature of 25°C and a stream of cooling air at 20°C provides a heat transfer coefficient of 100 W/m^2-K as soon as the power is applied. The surface area exposed to the cooling air is 50 cm^2. Assume that the lumped capacitance model is valid and determine

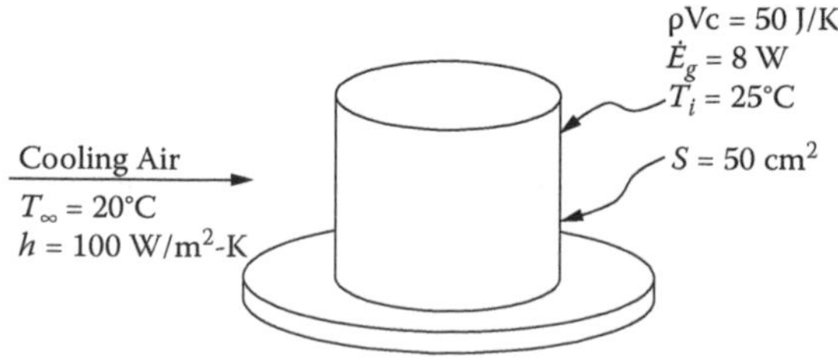

FIGURE 21.3
Configuration for Example 21.2.

(a) the component temperature after one minute of operation and (b) the steady-state temperature of the component.

Solution

Assumptions and Specifications

(1) Unsteady-state conditions exist.
(2) Thermal properties do not vary with temperature.
(3) There is no radiation exchange with the surroundings.
(4) Use of the lumped capacitance model is specified.

(a) Equation 21.16 may be used to determine the temperature after 1 min (60 s). Here, the time constant is

$$\tau = \frac{\rho V c}{hS} = \frac{60\,\mathrm{J/K}}{(100\,\mathrm{W/m^2\text{-}K})(0.005\,\mathrm{m^2})} = 120\,\mathrm{s}$$

so that

$$\frac{t}{\tau} = \frac{60\,\mathrm{s}}{120\,\mathrm{s}} = 0.50$$

Then via Equation 21.16 with $hS = (100\,\mathrm{W/m^2\text{-}K})(0.005\,\mathrm{m^2}) = 0.5\,\mathrm{W/K}$

$$\frac{\dot{E}_g - hS(T - T_\infty)}{\dot{E}_g - hS(T_i - T_\infty)} = e^{-t/\tau}$$

$$\frac{8\,\mathrm{W} - (0.5\,\mathrm{W/K})(T - 20^\circ\mathrm{C})}{8\,\mathrm{W} - 12.5\,\mathrm{W} + 10\,\mathrm{W}} = e^{-0.50} = 0.607$$

$$8\,\mathrm{W} - (0.5\,\mathrm{W/K})(T - 20^\circ\mathrm{C}) = 0.607(5.50\,\mathrm{W})$$

$$-(0.5\,\mathrm{W/K})(T - 20^\circ\mathrm{C}) = -4.662\,\mathrm{W}$$

$$T - 20^\circ\mathrm{C} = 9.32^\circ\mathrm{C}$$

$$T = 29.3^\circ\mathrm{C} \Longleftarrow$$

(b) The steady-state operating temperature is given by Equation 21.17

$$T_{ss} = T_\infty + \frac{\dot{E}_g}{hS} = 20^\circ\mathrm{C} + \frac{8\,\mathrm{W}}{0.50\,\mathrm{W/K}} = 36^\circ\mathrm{C} \Longleftarrow$$

21.3 The Semi-Infinite Solid

As illustrated in Figure 21.4, the semi-infinite model treats the solid as extending to infinity in all dimensions except for one surface where a thermal boundary condition may be imposed. The temperature distribution in the solid is a function of the spatial coordinate, x, and time, t, that is $T(x, t)$, and it is governed by the Fourier equation in one dimension

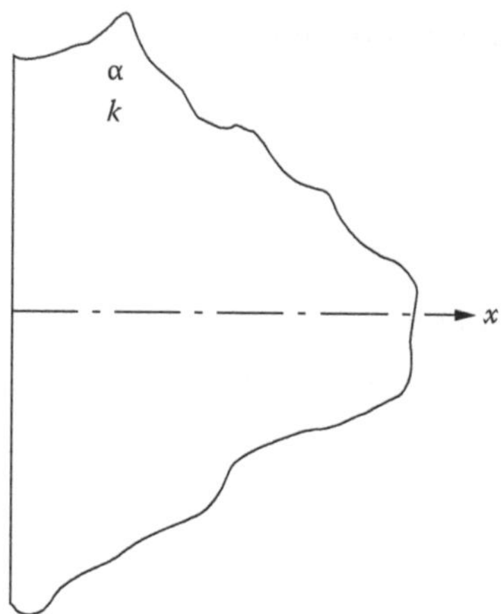

FIGURE 21.4
Semi-infinite solid. All dimensions except x extend to infinity.

given by Equation 20.8

$$\frac{\partial^2 T}{\partial x^2} = \frac{1}{\alpha}\frac{\partial T}{\partial t} \tag{21.18}$$

It will be assumed that the solid is initially at a uniform temperature, T_i, throughout its extent and that the thermal boundary condition at $x = 0$ has no impact at large distances from the surface. These assumptions can be represented as

$$T(x, 0) = T_i \tag{21.19}$$

and

$$T(\infty, t) = T_i \tag{21.20}$$

The three types of boundary conditions may be considered for the surface at $x = 0$ are discussed in Sections 21.3.1 through 21.3.3.

21.3.1 Constant Surface Temperature

The case of constant surface temperature is represented by

$$T(0, t) = T_o \tag{21.21}$$

for which the solution for T will be

$$T(x, t) = T_i + (T_o - T_i)\,\mathrm{erfc}(\phi) \tag{21.22}$$

where

$$\phi = \frac{x}{2(\alpha t)^{1/2}} \tag{21.23}$$

and where erfc denotes the tabulated mathematical relationship known as **complementary error function**

$$\mathrm{erfc}(x) = 1 - \frac{2}{\sqrt{\pi}}\int_0^x e^{-u^2}\,du \tag{21.24}$$

In this case, the surface heat flux

$$\dot{q}_o = -k \frac{\partial T}{\partial x}\bigg|_{x=0}$$

is given by

$$\dot{q}_o = \frac{k(T_o - T_i)}{(\pi \alpha t)^{1/2}} \tag{21.25}$$

21.3.2 Constant Surface Heat Flux

The case of constant surface heat flux is represented by

$$\dot{q}_o = -k \frac{\partial T}{\partial x}(0,\, t) \qquad \text{(specified)} \tag{21.26}$$

and the solution for T will be

$$T(x,\, t) = T_i + \frac{2\dot{q}_o(\alpha t/\pi)^{1/2}}{k} e^{-x^2/4\alpha t} - \frac{\dot{q}_o x}{k} \operatorname{erfc}(\phi) \tag{21.27}$$

Table 21.1 gives values of the complementary error function for arguments between 0.00 and 3.00.

21.3.3 Surface Convection

Here, the surface condition is

$$-k \frac{\partial T}{\partial x}(0,\, t) = h[T_\infty - T(0,\, t)] \tag{21.28}$$

and the solution for T will be

$$T(x,\, t) = T_i + (T_\infty - T_i)$$
$$\times \left\{ \operatorname{erfc}(\phi) \times \left[e^{(hx/k)+(h^2\alpha t/k^2)} \right] \left[\operatorname{erfc}\left(\frac{h}{k}\sqrt{\alpha t} + \phi \right) \right] \right\} \tag{21.29}$$

TABLE 21.1

Values of the Complementary Error Function, erfc(z)

z	erfc(z)	z	erfc(z)	z	erfc(z) $\times 10^3$
0.00	1.0000				
0.10	0.8875	1.10	0.1198	2.10	2.980
0.20	0.7773	1.20	0.0897	2.20	1.863
0.30	0.6714	1.30	0.0660	2.30	1.143
0.40	0.5716	1.40	0.0477	2.40	0.689
0.50	0.4795	1.50	0.0339	2.50	0.407
0.60	0.3961	1.60	0.0237	2.60	0.236
0.70	0.3222	1.70	0.0162	2.70	0.134
0.80	0.2579	1.80	0.0109	2.80	0.075
0.90	0.2031	1.90	0.0072	2.90	0.041
1.00	0.1573	2.00	0.0047	3.00	0.022

Example 21.3 A thick steel slab has a thermal conductivity k of 13.4 W/m-K and a thermal diffusivity α of $3.70 \times 10^{-6}\,\text{m}^2/\text{s}$. The slab has a uniform initial temperature of 27°C. Determine the temperature at a depth of 2 cm and time $t = 5$ min if (a) the surface ($x = 0$) is suddenly changed to and kept at 50°C, and (b) the surface ($x = 0$) is heated with a burner that provides a surface heat flux of 10,000 W/m^2. Then, determine the cumulative energy supplied over the time period.

Solution

Assumptions and Specifications

(1) Unsteady-state conditions exist.
(2) Thermal properties do not vary with temperature.
(3) There is no radiation exchange with the surroundings.
(4) A semi-infinite solid is presumed.

(a) We first determine the quantity, ϕ, given by Equation 21.23

$$\phi = \frac{x}{2(\alpha t)^{1/2}} = \frac{0.02\,\text{m}}{2[(3.70 \times 10^{-6}\,\text{m}^2/\text{s})(300\,\text{s})]^{1/2}} = 0.30$$

and from Table 21.1, we find that erfc(0.30) = 0.6714. Then Equation 21.22 may be used to obtain T at $x = 0.02$ m and $t = 5$ min or 300 s

$$T(x, t) = T_i + (T_o - T_i)\,\text{erfc}(\phi)$$

$$T(0.02\,\text{m}, 300\,\text{s}) = 27^\circ\text{C} + (50^\circ\text{C} - 27^\circ\text{C})\ \text{erfc}(0.30)$$

$$= 27^\circ\text{C} + (23^\circ\text{C})(0.6714)$$

$$= 27^\circ\text{C} + 15.4^\circ\text{C} = 42.4^\circ\text{C} \Longleftarrow$$

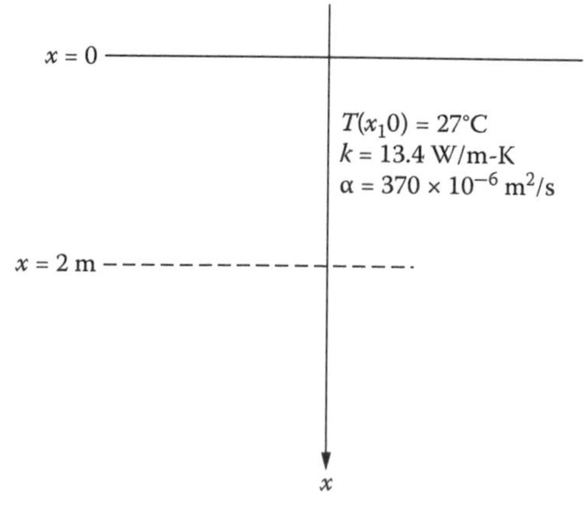

FIGURE 21.5
Configuration for Example 21.3.

The cumulative energy supplied can be determined by integrating Equation 21.25 from $t = 0\,\text{s}$ to $t = 300\,\text{s}$

$$Q_o = \int_0^t \dot{q}_o\,dt = \int_0^t \frac{k(T_o - T_i)}{(\pi\alpha t)^{1/2}}\,dt$$

$$= 2k(T_o - T_i)(t/\pi\alpha)^{1/2}$$

$$= (2)(13.4\,\text{W/m-K})(50^\circ\text{C} - 27^\circ\text{C})[(300\,\text{s}/\pi(370 \times 10^{-6}\,\text{m/s}^2)]^{1/2}$$

$$= 3.131 \times 10^5\,\text{J/m}^2 \qquad (313.1\,\text{kJ/m}^2) \Longleftarrow$$

(b) Here Equation 21.27 is applicable.

$$T(x,\,t) = T_i + \frac{2\dot{q}_o(\alpha t/\pi)^{1/2}}{k}e^{-x^2/4\alpha t} - \frac{\dot{q}_o x}{k}\text{erfc}(\phi)$$

and with

$$\frac{\alpha t}{\pi} = \frac{(3.70 \times 10^{-6}\,\text{m}^2/\text{s})(300\,\text{s})}{\pi} = 3.533 \times 10^{-4}\,\text{m}^2$$

we obtain

$$T(0.02\,\text{m}, 300\,\text{s}) = 27^\circ\text{C}$$

$$+ \frac{2(10^4\,\text{W/m}^2)(3.533 \times 10^{-4}\,\text{m}^2)^{1/2}}{13.4\,\text{W/m-K}}e^{-(0.30)^2}$$

$$- \frac{(10^4\,\text{W/m}^2)(0.02\,\text{m})}{13.4\,\text{W/m-K}}(0.6714)$$

$$= 27^\circ\text{C} + 25.6^\circ\text{C} - 10.0^\circ\text{C} = 42.6^\circ\text{C} \Longleftarrow$$

Because $\dot{q}_o$ is a constant, the cumulative energy supplied over a time, t, is

$$Q_o = \dot{q}_o t = (10^4\,\text{W/m}^2)(300\,\text{s}) = 3.0 \times 10^6\,\text{J/m}^2 \qquad (3000\,\text{kJ/m}^2) \Longleftarrow$$

Water pipes are frequently buried in the ground to prevent their freezing and the minimum depth for a prescribed soil along with the associated temperature conditions is of more than casual interest. A design example that deals with the freezing of buried pipes now follows.

21.4 Design Example 12

Consider a parallel line of water pipes that are buried in a soil having a thermal diffusivity of $\alpha = 0.138 \times 10^{-6}\,\text{m/s}^2$. The surface of the soil is initially at 10°C but suddenly drops to −10°C and remains at this temperature for 80 days. Determine the minimum burial depth to prevent the freezing of the pipes.

Solution

Assumptions and Specifications

(1) Unsteady-state conditions exist.
(2) Thermal properties do not vary with temperature.
(3) There is negligible heat loss to and from the surroundings.
(4) A semi-infinite solid is presumed.
(5) The thermal diffusivity of the soil is specified.

Here, the time of interest is

$$t = (80\,\text{days})(24\,\text{h/day})(3600\,\text{s/h}) = 6.912 \times 10^6\,\text{s}$$

and we solve for x in Equation 21.22. With

$$T(x,t) = 0°\text{C}, \qquad T_o = -10°\text{C} \quad \text{and} \quad T_i = 10°\text{C}$$

a modification of Equation 21.22 yields

$$\frac{T(x,t) - T_i}{T_o - T_i} = \text{erfc}(\phi)$$

where

$$\phi = \frac{x}{2(\alpha t)^{1/2}}$$

Thus,

$$\text{erfc}(\phi) = \frac{0°\text{C} - 10°\text{C}}{-10°\text{C} - 10°\text{C}} = 0.5000$$

and Table 21.1 gives (with interpolation)

$$\phi = 0.4795.$$

The burial depth is

$$\begin{aligned} x &= 2\phi(\alpha t)^{1/2} = 2(0.4795)[(0.138 \times 10^{-6}\,\text{m/s}^2)(6.912 \times 10^6\,\text{s})]^{1/2} \\ &= 2(0.4795)(0.9539\,\text{m}^2)^{1/2} = 0.937\,\text{m} \Longleftarrow \end{aligned}$$

21.5 Finite-Sized Solids

21.5.1 The Long Plane Wall

Figure 21.6 shows a long plane wall of thickness, $2L$. The wall, which has a thermal conductivity, k, a density, ρ, a specific heat, c, and a thermal diffusivity, α, is initially at a uniform temperature, T_i. At time $t = 0$ and for all times thereafter, both surfaces of the wall are suddenly exposed to an environment that is characterized by a temperature, $T_\infty < T_i$ and a

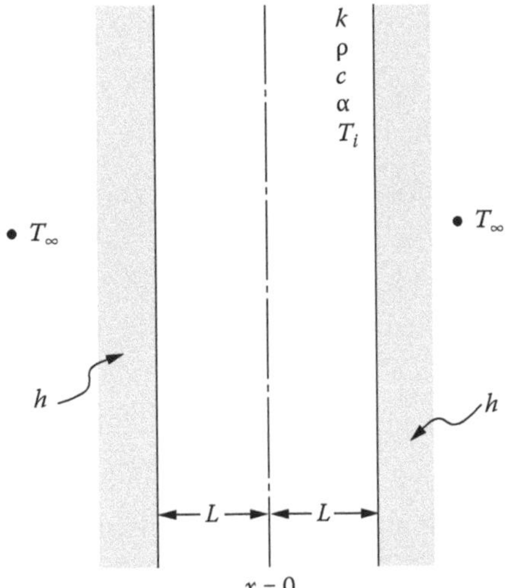

FIGURE 21.6
Long, solid plane wall with thickness $2L$.

heat transfer coefficient, h. Although we are considering convective cooling at the surfaces, the final results are equally applicable to convective surface heating.

The cooling of the wall is governed by the Fourier equation in one dimension.

$$\frac{\partial^2 T}{\partial x^2} = \frac{1}{\alpha}\frac{\partial T}{\partial t} \tag{21.18}$$

The initial and the boundary conditions to be imposed on Equation 21.18 are

$$T(x, 0) = T_i \tag{21.30a}$$

$$\frac{\partial T}{\partial x}(0, t) = 0 \tag{21.30b}$$

and

$$-k\frac{\partial T}{\partial x}(L, t) = h[T(L, t) - T_\infty] \tag{21.30c}$$

where Equation 21.30b is a consequence of the thermal symmetry that exists about $x = 0$ and Equation 21.30c represents the balance between the conductive and convective fluxes at $x = L$.

The solution of Equation 21.18 with the boundary conditions of Equation 21.30 may be obtained using the method of **separation of the variables**. Because this method is somewhat beyond our scope, we list the solution in terms of a dimensionless temperature

$$\theta \equiv \frac{T(x, t) - T_\infty}{T_i - T_\infty}$$

as

$$\theta = 2\sum_{n=1}^{\infty} \frac{\sin\lambda_n L \cos\lambda_n x}{\lambda_n L + \sin\lambda_n L \cos\lambda_n L} e^{-\lambda_n^2 \alpha t} \tag{21.31}$$

where the λ_n's are given by the transcendental equation

$$\lambda_n L \tan \lambda_n L = \text{Bi}$$

with $\text{Bi} = hL/k$.

The instantaneous rate of heat loss from both faces ($x = \pm L$) may be found from

$$\dot{Q} = -2kA \frac{\partial T}{\partial x}(L, t)$$

or

$$\dot{Q} = 4kA(T_i - T_\infty) \sum_{n=1}^{\infty} \frac{\lambda_n \sin^2 \lambda_n L}{\lambda_n L + \sin \lambda_n L \cos \lambda_n L} e^{-\lambda_n^2 \alpha t} \tag{21.32}$$

and the cumulative heat loss over a period of time, t, from

$$Q = \int_0^t \dot{Q} dt$$

or

$$Q = 4\rho c A(T_i - T_\infty) \sum_{n=1}^{\infty} \frac{\sin^2 \lambda_n L}{\lambda_n(\lambda_n L + \sin \lambda_n L \cos \lambda_n L)} (1 - e^{-\lambda_n^2 \alpha t}) \tag{21.33}$$

The evaluation of Equations 21.31, 21.32, and 21.33 involves a laborious computational procedure and we will resort to a graphical method outlined in the paragraphs that follow.

It can be shown using Equation 21.31 that the dimensionless temperature, θ, is a function of $X = x/L$ and the Fourier and Biot moduli

$$\text{Fo} \equiv \frac{\alpha t}{L^2} \quad \text{and} \quad \text{Bi} \equiv \frac{hL}{k}$$

Thus,

$$\theta = f(X, \text{Fo}, \text{Bi})$$

It follows that the dimensionless temperature at $x = 0$ is given by

$$\theta_o = \frac{T(0, t) - T_\infty}{T_i - T_\infty} = \frac{T_o - T_\infty}{T_i - T_\infty} = f(\text{Fo}, \text{Bi})$$

This relationship is plotted in Figure 21.7a.

If the ratio, θ/θ_o, is formed, the ratio becomes a function of x/L and Bi. The ratio, θ/θ_o, is plotted as a function of $1/\text{Bi}$ instead of Bi for several values of x/L in Figure 21.7b.

With

$$Q_i = 2LA\rho c(T_i - T_\infty)$$

Equation 21.33 can be used to create the ratio, Q/Q_i, which is then found to be a function of Fo and Bi. It is noted that Q_i is the initial energy content of the wall above the environment and is the maximum energy that the wall can give to the environment as it cools down to the temperature of the environment. Figure 21.7c provides a graph of Q/Q_i as a function of FoBi^2 for various values of Bi. We note that as $t \longrightarrow \infty$, $\text{Fo} \longrightarrow \infty$, $\text{FoBi} \longrightarrow \infty$ and $Q \longrightarrow Q_i$, or $Q/Q_i = 1$. The graphs in Figure 21.7a and 21.7b were originally developed by Heisler (1947) and are known as **the Heisler charts.**

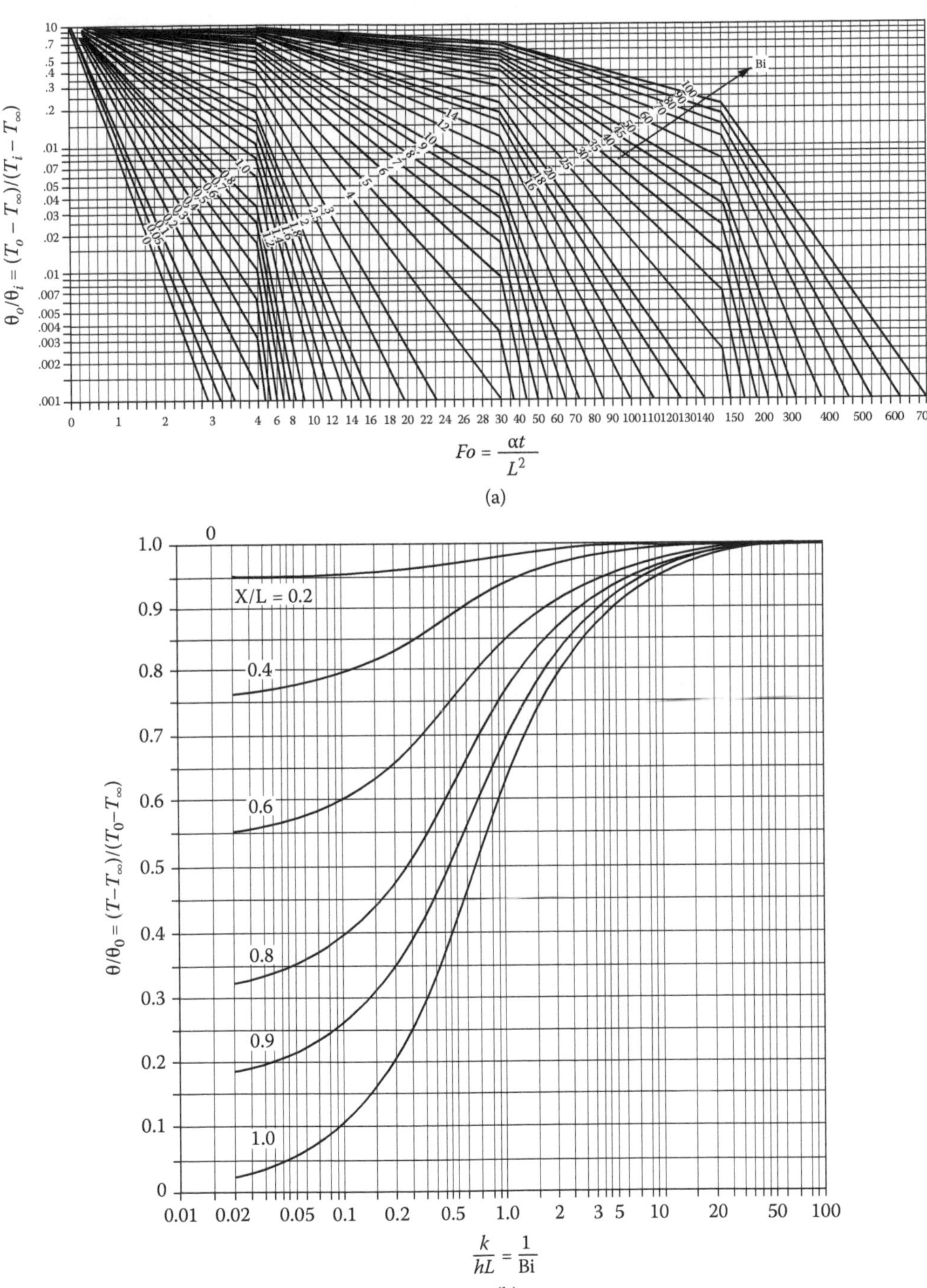

FIGURE 21.7
Charts for unsteady conduction in a long, solid plane wall (a) mid-plane temperature, (b) temperature as a function of the mid-plane temperature, and (c) dimensionless heat loss. Figures 21.7a and 21.7b are reproduced from the original Heisler paper [Heisler (1947), courtesy of ASME]. Figure 21.7c is attributed to Gröeber et al. (1961).

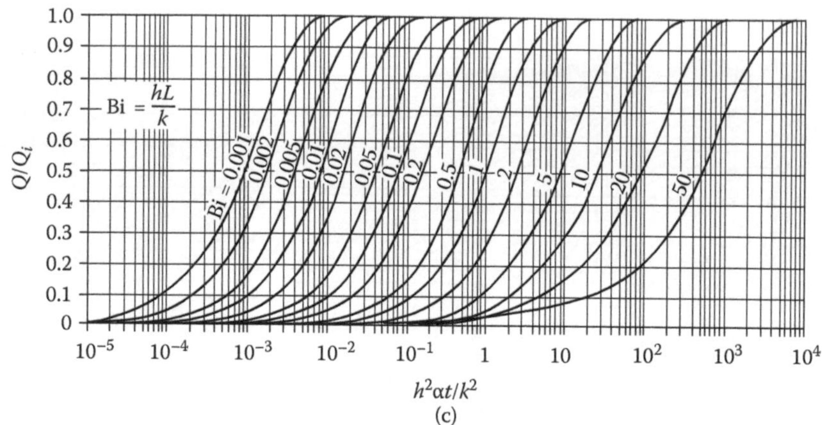

FIGURE 21.7
Continued

Other charts for the rapid solution of unsteady or transient conduction problems were developed by Gurney and Lurie (1925), Gröeber (1925), Boelter et al. (1965), and Schneider (1965).

Example 21.4 illustrates the use of Figure 21.7.

Example 21.4 As sketched in Figure 21.8, an AISI 316 stainless steel plate ($k = 12\,\text{W/m-K}$, $\alpha = 3.48\times10^{-6}\,\text{m}^2/\text{s}$, $\rho = 8238\,\text{kg/m}^4$ and $c = 468\,\text{J/kg-K}$) is 6-cm thick and is originally at a uniform initial temperature of 325°C. The plate is cooled by cold air jets at 25°C that provide a convective heat transfer coefficient of $400\,\text{W/m}^2\text{-K}$. Determine (a) the time required to bring the centerline temperature to 85°C, (b) the surface temperature when the centerline temperature is 85°C, and (c) the cumulative heat loss to the air per unit area of the plate surface.

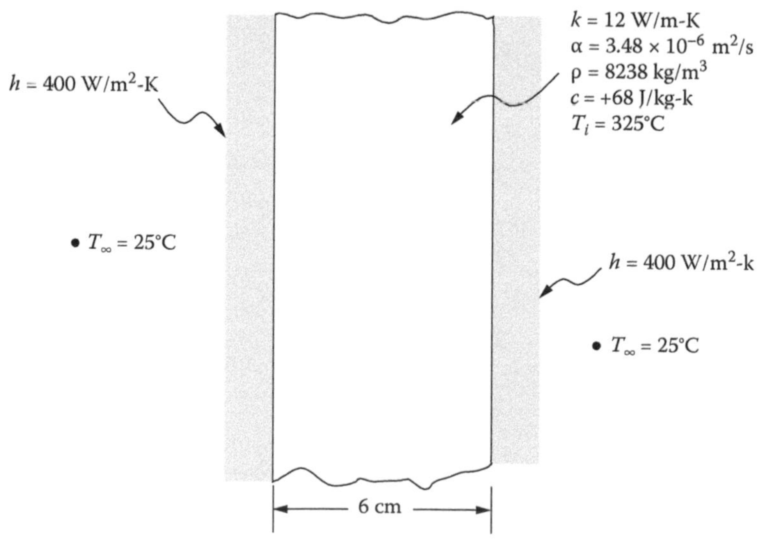

FIGURE 21.8
Configuration for Example 21.4.

Solution

Assumptions and Specifications

(1) Unsteady-state conditions exist.
(2) Thermal properties do not vary with temperature.
(3) There is no radiation exchange with the surroundings.

(a) We first determine the dimensionless temperature, θ_o and the inverse of the Biot number, Bi^{-1}

$$\theta_o = \frac{T(0,\, t) - T_\infty}{T_i - T_\infty} = \frac{85°\text{C} - 25°\text{C}}{325°\text{C} - 25°\text{C}} = 0.20$$

and with $L = 0.06\text{ m}/2 = 0.03\text{ m}$

$$\text{Bi}^{-1} = \frac{k}{hL} = \frac{12\,\text{W/m-K}}{(400\,\text{W/m}^2\text{-K})(0.03\,\text{m})} = 1.0$$

With θ_o and Bi^{-1} known, Figure 21.7a may be read to find

$$\text{Fo} = \frac{\alpha t}{L^2} = 2.40$$

which gives

$$t = \frac{2.40L^2}{\alpha} = \frac{(2.40)(0.03\,\text{m})^2}{3.48 \times 10^{-6}\,\text{m}^2/\text{s}} = 620.7\,\text{s} \qquad (10.35\,\text{min}) \Longleftarrow$$

(b) With $\text{Bi} = 1.0$ and $X = 1.0$ at the surface, we read Figure 21.7b to obtain

$$\frac{T(L,\, t) - T_\infty}{T(0,\, t) - T_\infty} = 0.632$$

Thus, the surface temperature, $T(L,\, t)$ is

$$\begin{aligned} T(L,\, t) &= T_\infty + 0.632[T(0,\, t) - T_\infty] \\ &= 25°\text{C} + 0.632(85°\text{C} - 25°\text{C}) \\ &= 25°\text{C} + 37.9°\text{C} = 62.9°\text{C} \Longleftarrow \end{aligned}$$

(c) With $\text{FoBi}^2 = (2.4)(1.0)^2 = 2.4$ and $\text{Bi} = 1.0$, we may read Figure 21.7c to obtain

$$\frac{Q}{Q_i} = 0.80$$

so that

$$Q = (0.80)\,Q_i = (0.80)[2L\,A\rho c(T_i - T_\infty)]$$

or

$$\begin{aligned} \frac{Q}{A} &= (0.80)(2)(0.03\,\text{m})(8238\,\text{kg/m}^3)(468\,\text{J/kg-K})(325°\text{C} - 25°\text{C}) \\ &= 5.552 \times 10^7\,\text{J/m}^2 \Longleftarrow \end{aligned}$$

TABLE 21.2

Values of C_1 and β_1 Used in the One-Term Approximation to the Series Solution for Transient One-Dimensional Conduction in the Plane Wall

Bi	β_1, rad	C_1	Bi	β_1, rad	C_1
0.01	0.0998	1.0017	0.80	0.7910	1.1016
0.02	0.1410	1.0033	0.90	0.8274	1.1107
0.03	0.1732	1.0049	1.00	0.8603	1.1191
0.04	0.1987	1.0066	2.00	1.0769	1.1725
0.05	0.2217	1.0082	3.00	1.1925	1.2102
0.06	0.2425	1.0098	4.00	1.2646	1.2287
0.07	0.2615	1.0114	5.00	1.3138	1.2402
0.08	0.2791	1.0130	6.00	1.3496	1.2479
0.09	0.2956	1.0145	7.00	1.3766	1.2532
0.10	0.3111	1.0160	8.00	1.3978	1.2570
0.15	0.3779	1.0237	9.00	1.4149	1.2598
0.20	0.4328	1.0311	10.0	1.4289	1.2620
0.25	0.4801	1.0382	21.0	1.4961	1.2699
0.30	0.5218	1.0450	30.0	1.5202	1.2717
0.40	0.5932	1.0580	40.0	1.5325	1.2723
0.50	0.6533	1.0701	50.0	1.5400	1.2727
0.60	0.7051	1.0814	100	1.5552	1.2731
0.70	0.7506	1.0919	∞	1.5707	1.2733

21.5.2 One-Term Approximate Solutions

When $\mathrm{Fo} > 0.20$, the first term of the series in Equations 21.31 and 21.33 provides sufficiently accurate results. These one-term approximate solutions are

$$\theta = \frac{T(x,t) - T_\infty}{T_i - T_\infty} = C_1 e^{-\beta_1^2 \mathrm{Fo}} \cos\left(\beta_1 \frac{x}{L}\right) \tag{21.34}$$

$$\theta_o = C_1 e^{-\beta_1^2 \mathrm{Fo}} \tag{21.35}$$

and

$$\frac{Q}{Q_i} = 1 - \frac{\sin \beta_1}{\beta_1} \theta_o \tag{21.36}$$

The values of C_1 and β_1 for a range of values of Bi are provided in Table 21.2.

Example 21.5 Use the one-term approximate solutions for the long plane wall to develop the solution to Example 21.4.

Solution

Assumptions and Specifications

(1) Unsteady-state conditions exist.
(2) Thermal properties do not vary with temperature.
(3) There is no radiation exchange with the surroundings.
(4) The one-term approximate solution is to be used.

(a) For Bi $= 1$, Table 21.2 gives $\beta_1 = 0.8603$ rad and $C_1 = 1.1191$. Solving Equation 21.35 for Fo with $\theta_o = 0.20$ from Example 21.4 gives

$$\text{Fo} = -\frac{1}{\beta_1^2}\ln\left(\frac{\theta_o}{C_1}\right) = \frac{1}{\beta_1^2}\ln\left(\frac{C_1}{\theta_o}\right) = \frac{1}{(0.8603)^2}\ln\left(\frac{1.1191}{0.2}\right)$$

$$= (1.3511)\ln 5.5955 = (1.3511)(1.7220) = 2.327$$

which then gives

$$t = \frac{2.327L^2}{\alpha} = \frac{2.327(0.03\,\text{m})^2}{3.48\times 10^{-6}\,\text{m}^2/\text{s}} = 601.8\,\text{s} \qquad (10.03\,\text{min}) \Longleftarrow$$

(b) Using Equation 21.34 with $X = x/L = 1$ gives for the surface

$$\frac{T(L,t) - T_\infty}{T_i - T_\infty} = C_1 e^{-\beta_1^2 \text{Fo}} \cos\beta_1$$

$$= 1.1191e^{-(0.8603)^2(2.327)}\cos 0.8603$$

$$= 1.1191e^{-1.7223}(0.6522)$$

$$= 0.130$$

and hence,

$$T(L,t) = 0.130(T_i - T_\infty) + T_\infty$$

$$= 0.130(325°\text{C} - 25°\text{C}) + 25°\text{C} = 64°\text{C} \Longleftarrow$$

(c) Use of Equation 21.36 gives

$$\frac{Q}{Q_i} = 1 - \frac{\sin\beta_1}{\beta_1}\theta_o$$

$$= 1 - \frac{\sin 0.8603}{0.8603}(0.20)$$

$$= 1 - (0.8811)(0.20) = 1 - 0.1762 = 0.8238$$

which then provides

$$Q = (0.8238)\,Q_i = (0.8238)(2LA\rho c(T_i - T_\infty))$$

or

$$\frac{Q}{A} = (0.8238)(2)(0.03\,\text{m})(8238\,\text{kg/m}^3)(468\,\text{J/kg-K})(325°\text{C} - 25°\text{C})$$

$$= 5.72\times 10^7\,\text{J/m}^2 \Longleftarrow$$

These results compare favorably with those obtained in Example 21.4.

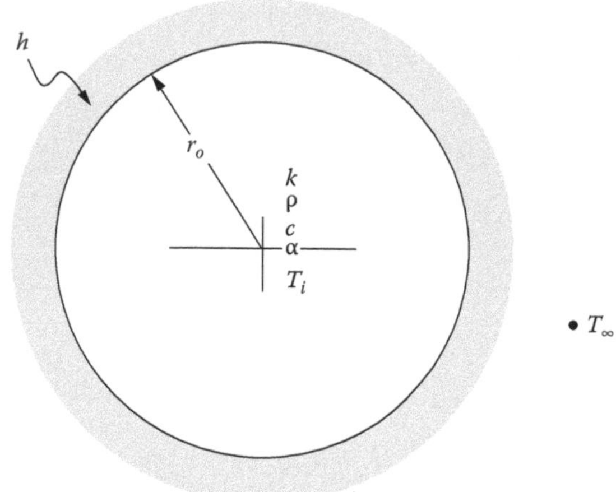

FIGURE 21.9
Long solid cylinder of radius, r_o.

21.5.3 The Long Solid Cylinder

Figure 21.9 shows a long solid cylinder of radius, r_o, initially at a uniform temperature of T_i. The cylinder has thermal conductivity, k, density, ρ, specific heat, c, and thermal diffusivity, α. At time $t = 0$, the cylinder is placed in an environment at temperature, $T_\infty < T_i$, which cools the lateral surface of the cylinder through a heat transfer coefficient, h. Although convective surface cooling is considered, the results obtained are also applicable for the case of convective surface heating.

The partial differential equation describing the cooling of the cylinder is the Fourier equation in one dimension but in cylindrical coordinates

$$\frac{\partial^2 T}{\partial r^2} + \frac{1}{r}\frac{\partial T}{\partial r} = \frac{1}{\alpha}\frac{\partial T}{\partial t} \tag{21.37}$$

The initial and the boundary conditions to be satisfied are:

$$T(r,\, 0) = T_i \tag{21.38a}$$

$$\frac{\partial T}{\partial r}(0,\, t) = 0 \tag{21.38b}$$

and

$$-k\frac{\partial T}{\partial r}(r_o,\, t) = h[T(r_o,\, t) - T_\infty] \tag{21.38c}$$

where Equation 21.38b describes the thermal symmetry about the axis of the cylinder (at $r = 0$) and Equation 21.38c is the result of the surface energy balance at $r = r_o$.

The solution of Equation 21.37, which satisfies the boundary conditions of Equation 21.38, may also be written in terms of a dimensionless temperature

$$\theta \equiv \frac{T(r,\, t) - T_\infty}{T_i - T_\infty}$$

and is

$$\theta = 2\sum_{n=1}^{\infty} \frac{1}{\lambda_n r_o} \frac{J_1(\lambda_n r_o) J_0(\lambda_n r)}{J_0^2(\lambda_n r_o) + J_1^2(\lambda_n r_o)} e^{-\lambda_n^2 \alpha t} \tag{21.39}$$

where the λ_n's are solutions to the transcendental equation

$$\lambda_n r_o \frac{J_1(\lambda_n r_o)}{J_0(\lambda_n r_o)} = \text{Bi} \tag{21.40}$$

where $\text{Bi} = hr_o/k$ and J_0 and J_1 are the Bessel functions of the first kind of order 0 and 1, respectively.

The instantaneous rate of heat loss from the surface of the cylinder is

$$\frac{\dot{Q}}{L} = -2\pi r_o k \frac{\partial T}{\partial r}(r_o, t)$$

or

$$\frac{\dot{Q}}{L} = 4\pi k(T_i - T_\infty) \sum_{n=1}^{\infty} \frac{J_1^2(\lambda_n r_o)}{J_0^2(\lambda_n r_o) + J_1^2(\lambda_n r_o)} e^{-\lambda_n^2 \alpha t} \tag{21.41}$$

and the cumulative heat loss over a period of time, t, will be

$$\frac{Q}{L} = 4\pi \rho c(T_i - T_\infty) \sum_{n=1}^{\infty} \frac{J_1^2(\lambda_n r_o)}{\lambda_n^2 [J_0^2(\lambda_n r_o) + J_1^2(\lambda_n r_o)]} (1 - e^{-\lambda_n^2 \alpha t}) \tag{21.42}$$

In this case, the dimensionless temperature is a function of $R = r/r_o$, $\text{Fo} = \alpha t/r_o^2$ and $\text{Bi} = hr_o/k$, that is

$$\theta = f(R, \text{Fo}, \text{Bi})$$

Similarly,

$$\theta_o = \frac{T(0, t) - T_\infty}{T_i - T_\infty} = \frac{T_o - T_\infty}{T_i - T_\infty} = f(\text{Fo}, \text{Bi})$$

and

$$\frac{Q}{Q_i} = f(\text{Fo}, \text{Bi})$$

where

$$Q_i = \pi r_o^2 L \rho c(T_i - T_\infty)$$

is the initial internal energy of the cylinder above the environment.

Figure 21.10a shows θ_o as a function of Fo and Bi. The ratio, θ/θ_o as a function of Bi for different values of R appears in Figure 21.10b and Figure 21.10c provides a graph of Q/Q_i as a function of Bi^2Fo for various values of Bi. These charts, like those of Figure 21.7, are also Heisler and Gröeber charts.

Example 21.6 illustrates the use of Figure 21.10.

Example 21.6 A cylindrical rod of radius 2 cm is made of a composite material whose properties are $k = 10\,\text{W/m-K}$, $\rho = 1300\,\text{kg/m}^3$, and $c = 1500\text{J/kg-K}$. The rod (Figure 21.11), which is initially at a uniform initial temperature of 25°C, is suddenly exposed to a

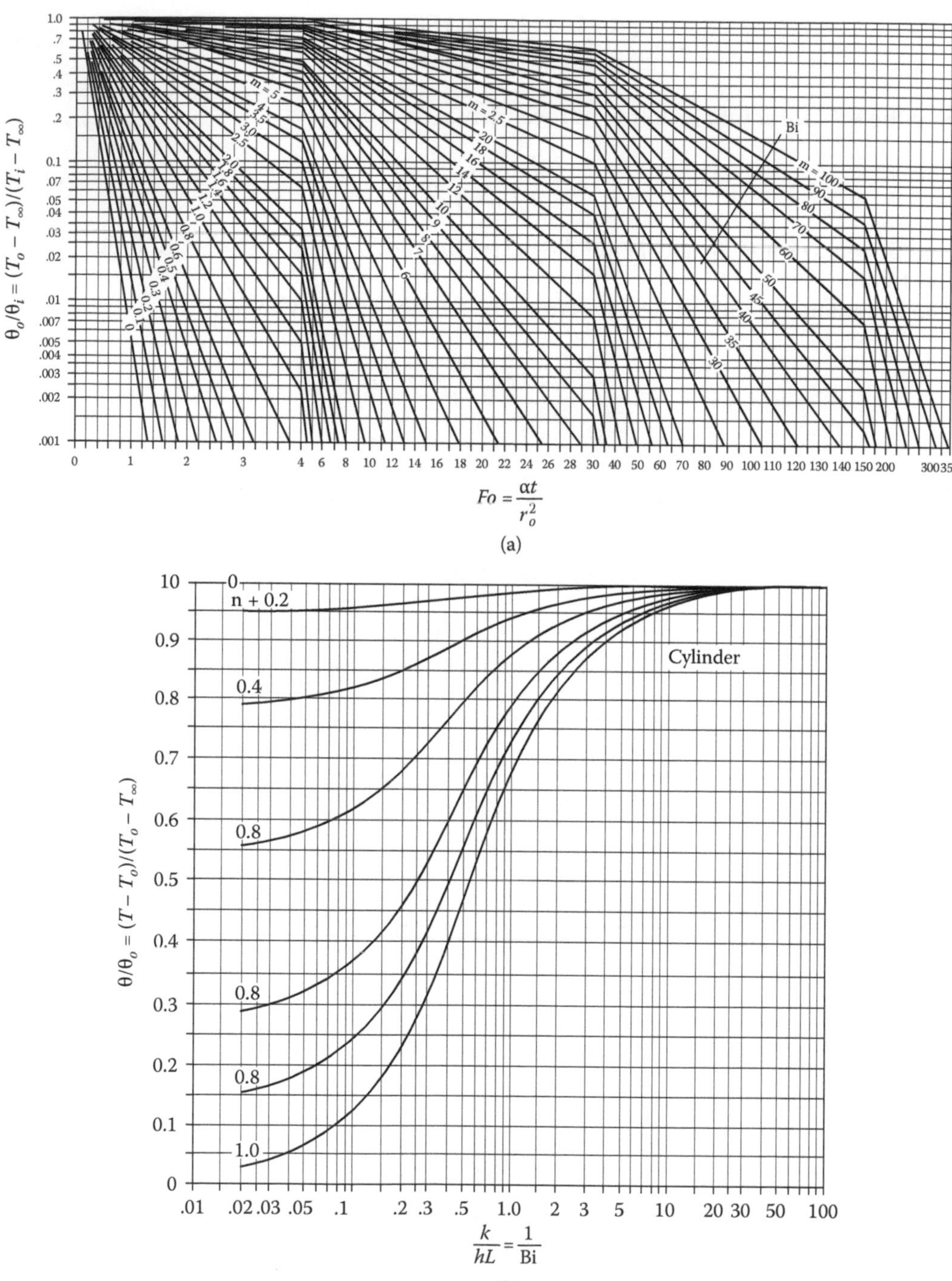

FIGURE 21.10

Charts for unsteady conduction in a long, solid cylinder (a) centerline temperature, (b) temperature as a function of the mid-plane temperature, and (c) dimensionless heat loss. Figures 21.10a and 21.10b are reproduced from the original Heisler paper [Heisler (1947), courtesy of ASME] and Figure 21.10c is attributed to Gröeber et al. (1961).

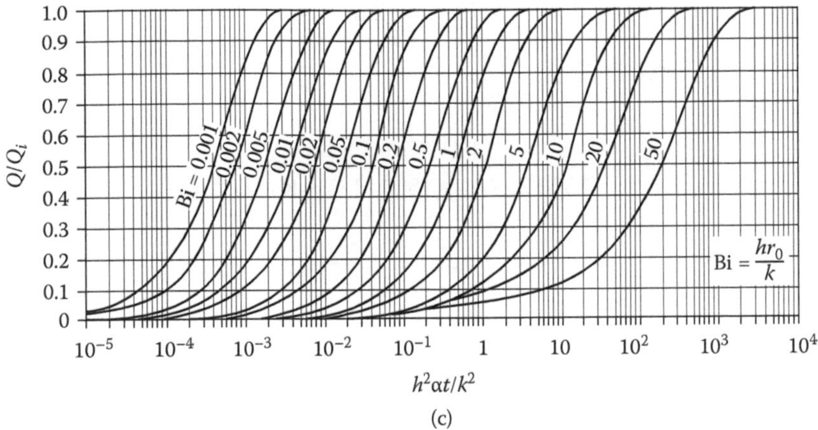

FIGURE 21.10
Continued

stream of hot exhaust gases at 125°C that provides a heat transfer coefficient of 250 W/m^2-K. Determine (a) the temperature at the center of the cylinder 2 min after exposure to the exhaust gases, (b) the temperature at a radial distance of 1.2 cm from the center at 2 min, and (c) the heat gained by the cylinder in 2 min.

Solution

Assumptions and Specifications

(1) Unsteady-state conditions exist.
(2) Thermal properties do not vary with temperature.
(3) There is no radiation exchange with the surroundings.

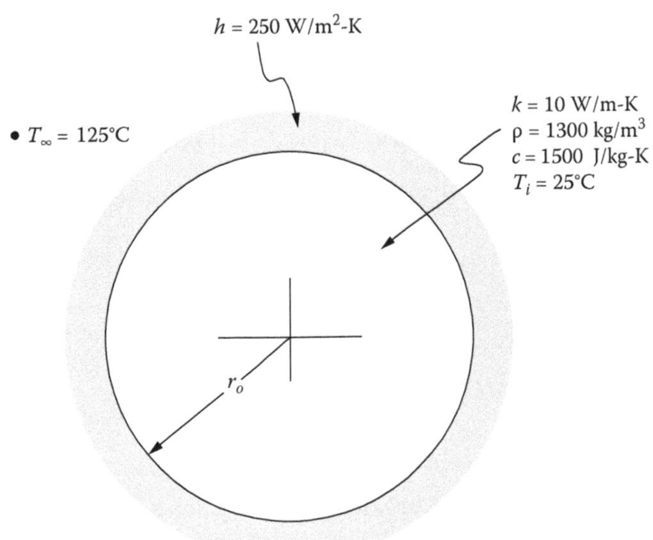

FIGURE 21.11
Configuration for Example 21.6.

(a) We first determine Fo and Bi^{-1}. With $t = (2\,\text{min})(60\,\text{s/min}) = 120\,\text{s}$

$$\text{Fo} = \frac{\alpha t}{r_o^2} = \frac{kt}{\rho c r_o^2}$$

$$= \frac{(10\,\text{W/m-K})(120\,\text{s})}{(1300\,\text{kg/m}^3)(1500\,\text{kJ/kg-K})(0.02\,\text{m})^2}$$

$$= 1.54$$

and

$$\text{Bi}^{-1} = \frac{k}{hr_o} = \frac{10\,\text{W/m-K}}{(250\,\text{W/m}^2\text{-K})(0.02\,\text{m})} = 2.0$$

With Fo and Bi^{-1} known, Figure 21.10a may be read to find

$$\theta_o = \frac{T(0,\,t) - T_\infty}{T_i - T_\infty} = 0.29$$

which then gives

$$T(0,\,t) = (0.29)(T_i - T_\infty) + T_\infty$$

$$= (0.29)(25°\text{C} - 125°\text{C}) + 125°\text{C}$$

$$= -29°\text{C} + 125°\text{C} = 96°\text{C} \Longleftarrow$$

(b) At a radial distance of 1.2 cm

$$R = \frac{r}{r_o} = \frac{1.2\,\text{cm}}{2\,\text{cm}} = 0.60$$

and with $\text{Bi}^{-1} = 2$, Figure 21.10b reveals that

$$\frac{T(r,\,t) - T_\infty}{T(0,\,t) - T_\infty} = 0.92$$

so that

$$T(1.2\,\text{cm},\ 120\,\text{s}) = 0.92(96°\text{C} - 125°\text{C}) + 125°\text{C} = 98.3°\text{C} \Longleftarrow$$

(c) With $\text{FoBi}^2 = (1.54)(0.50)^2 = 0.385$ and $\text{Bi} = 0.5$, we may read Figure 21.10c to obtain

$$\frac{Q}{Q_i} = 0.75$$

so that

$$Q = (0.75)Q_i = (0.75)\pi r_o^2 L\rho c(T_i - T_\infty)$$

or

$$\frac{Q}{L} = (0.75)\pi(0.02\,\text{m})^2(1300\,\text{kg/m}^3)(1500\,\text{J/kg-K}(25°\text{C} - 125°\text{C})$$

$$= -1.837 \times 10^5\,\text{J/m} \Longleftarrow$$

The heat gained by the rod is 183.7 kJ per meter of rod length.

TABLE 21.3

Values of C_1 and β_1 Used in the One-Term Approximation to the Series Solution for Transient One-Dimensional Conduction in the Long Cylinder

Bi	β_1, rad	C_1	Bi	β_1, rad	C_1
0.01	0.1412	1.0025	0.80	1.1490	1.1725
0.02	0.1995	1.0050	0.90	1.2048	1.1902
0.03	0.2439	1.0075	1.00	1.2558	1.2071
0.04	0.2814	1.0099	2.00	1.5995	1.3384
0.05	0.3142	1.0124	3.00	1.7887	1.4191
0.06	0.3438	1.0148	4.00	1.9081	1.4698
0.07	0.3708	1.0173	5.00	1.9898	1.5029
0.08	0.3960	1.0197	6.00	2.0490	1.5253
0.09	0.4195	1.0222	7.00	2.0937	1.5411
0.10	0.4417	1.0246	8.00	2.1286	1.5526
0.15	0.5376	1.0365	9.00	2.1566	1.5611
0.20	0.6170	1.0483	10.0	2.1795	1.5677
0.25	0.6856	1.0598	21.0	2.2881	1.5919
0.30	0.7465	1.0712	30.0	2.3261	1.5973
0.40	0.8516	1.0932	40.0	2.3455	1.5993
0.50	0.9408	1.1143	50.0	2.3572	1.6002
0.60	1.0185	1.1346	100	2.3809	1.6015
0.70	1.0873	1.1539	∞	2.4050	1.6018

21.5.4 One-Term Approximate Solutions

For Fo > 0.20, the one-term approximate solutions are

$$\theta = \frac{T(r,\, t) - T_\infty}{T_i - T_\infty} = C_1 e^{-\beta_1^2 \mathrm{Fo}} J_0(\beta_1 R) \tag{21.43}$$

here $R = r/r_o$.

$$\theta_o = \frac{T(0,\, t) - T_\infty}{T_i - T_\infty} = C_1 e^{-\beta_1^2 \mathrm{Fo}} \tag{21.44}$$

and

$$\frac{Q}{Q_i} = 1 - \frac{2\theta_o}{\beta_1} J_1(\beta_1) \tag{21.45}$$

where $\beta_1 = \lambda_1 r_o$. The values of C_1 and β_1 for a range of values of Bi are provided in Table 21.3 and Table 21.4 provides values of $J_0(x)$ and $J_1(x)$, which are needed when using Equations 21.43 and 21.45.

Example 21.7 Use the one-term approximate solutions for the long cylinder to develop the solution to Example 21.6.

Solution

Assumptions and Specifications

(1) Unsteady-state conditions exist.
(2) Thermal properties do not vary with temperature.
(3) There is no radiation exchange with the surroundings.
(4) The one-term approximate solution is to be used.

TABLE 21.4

Values of the Bessel Functions, $J_0(x)$ and $J_1(x)$

x	$J_0(x)$	$J_1(x)$	x	$J_0(x)$	$J_1(x)$
0.00	1.0000	0.0000			
0.10	0.9975	0.0499	2.10	0.1666	0.5683
0.20	0.9900	0.0995	2.20	0.1104	0.5560
0.30	0.9776	0.1483	2.30	0.0554	0.5399
0.40	0.9604	0.1960	2.40	0.0025	0.5202
0.50	0.9385	0.2423	2.50	−0.0484	0.4971
0.60	0.9120	0.2867	2.60	−0.0968	0.4708
0.70	0.8812	0.3290	2.70	−0.1425	0.4416
0.80	0.8463	0.3688	2.80	−0.1850	0.4097
0.90	0.8075	0.4060	2.90	−0.2243	0.3754
1.00	0.7652	0.4401	3.00	−0.2601	0.3391
1.10	0.7196	0.4709	3.20	−0.3202	0.2613
1.20	0.6711	0.4983	3.40	−0.3643	0.1792
1.30	0.6201	0.5220	3.60	−0.3918	0.0955
1.40	0.5669	0.5420	3.80	−0.4026	0.0128
1.50	0.5118	0.5579	4.00	−0.3972	−0.0661
1.60	0.4554	0.5699	4.20	−0.3766	−0.1387
1.70	0.3980	0.5778	4.40	−0.3423	−0.2028
1.80	0.3400	0.5815	4.60	−0.2961	−0.2566
1.90	0.2818	0.5812	4.80	−0.2404	−0.2985
2.00	0.2239	0.5767	5.00	−0.1776	−0.3276

(a) For Bi = 0.5, Table 21.3 gives for the cylinder, $\beta_1 = 0.9408$ rad and $C_1 = 1.1143$. Solving Equation 21.44 for θ_o with Fo = 1.54 gives

$$\theta_o = C_1 e^{-\beta_1^2 \mathrm{Fo}} = (1.1143)e^{-(0.9408)^2(1.54)}$$

$$= (1.1143)e^{-1.3631} = (1.1143)(0.2559) = 0.2851$$

which then gives

$$T(0,\, t) = (0.2851)(T_i - T_\infty) + T_\infty$$

$$= (0.2851)(25°\mathrm{C} - 125°\mathrm{C}) + 125°\mathrm{C}$$

$$= -28.5°\mathrm{C} + 125°\mathrm{C} = 96.5°\mathrm{C} \Longleftarrow$$

(b) With $R = r/r_o = 0.6$ and $\beta_1 = 0.9408$ rad

$$\beta_1 R = (0.9408)(0.6) = 0.5645$$

and, with interpolation, Table 21.4 provides

$$J_0(0.5645) = 0.9214$$

Then

$$\theta = C_1 e^{-\beta_1^2 \mathrm{Fo}} J_0(\beta_1 R)$$

$$= (1.1143)e^{-(0.9408)^2(1.54)}0.9214)$$

$$= (1.1143)(0.2559)(0.9214) = 0.2627$$

Thus,

$$\begin{aligned} T(1.2\,\text{cm},\ 120\,\text{s}) &= (0.263)(T_i - T_\infty) + T_\infty \\ &= (0.263)(25^\circ\text{C} - 125^\circ\text{C}) + 125^\circ\text{C} \\ &= -26.3^\circ\text{C} + 125^\circ\text{C} = 98.7^\circ\text{C} \Longleftarrow \end{aligned}$$

(c) Table 21.4 shows that $J_1(\beta_1) = J_1(0.9408) = 0.4199$ (with interpolation). Equation 21.45 then gives

$$\begin{aligned} \frac{Q}{Q_i} &= 1 - \frac{2\theta_o}{\beta_1} J_1(\beta_1) \\ &= 1 - \frac{(2)(0.2851)}{0.9408}(0.4199) \\ &= 0.746 \end{aligned}$$

and with $Q = 245{,}000\,\text{J/m}$

$$Q = (0.746)Q_i = (0.746)(-245{,}000\,\text{J/min}) = -1.828 \times 10^5\,\text{J/min}$$

21.5.5 The Solid Sphere

Figure 21.12 shows a solid sphere of radius, r_o made of a material with thermal conductivity, k, density, ρ, specific heat, c, and thermal diffusivity, α. The sphere is initially at a uniform temperature, T_i. At time, $t = 0$, the sphere is exposed to an environment at temperature T_∞, which is less than T_i. The convective heat transfer coefficient between the surface of the sphere and the environment is h. Although convective cooling is considered, the final results are equally applicable to convective heating.

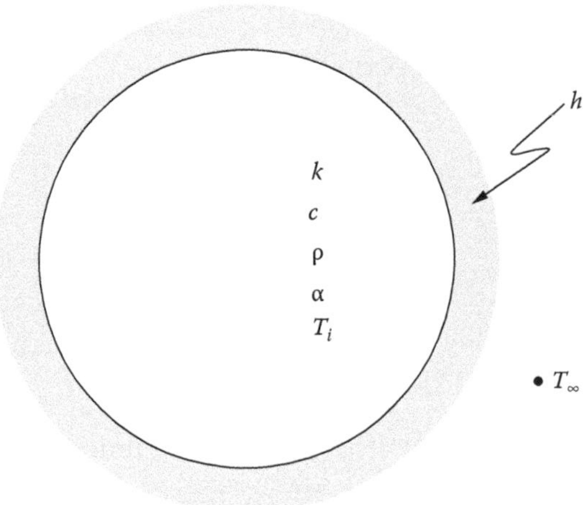

FIGURE 21.12
Solid sphere of radius, r_o.

The cooling of the sphere is described by the Fourier equation in the radial direction but in spherical coordinates

$$\frac{\partial^2 T}{\partial r^2} + \frac{2}{r}\frac{\partial T}{\partial r} = \frac{1}{\alpha}\frac{\partial T}{\partial t} \tag{21.46}$$

Here Equation 21.38a provides the initial condition

$$T(r, 0) = T_i \tag{21.38a}$$

and Equations 21.38b and 21.38c provide the boundary conditions

$$\frac{\partial T}{\partial r}(0, t) = 0 \tag{21.38b}$$

and

$$-k\frac{\partial T}{\partial r}(r_o, t) = h[T(r_o, t) - T_\infty] \tag{21.38c}$$

so that the solution to Equation 21.46 is given by

$$\theta = \frac{T(r, t) - T_\infty}{T_i - T_\infty} = 2\sum_{n=1}^{\infty} \frac{\sin \lambda_n r_o - \lambda_n r_o \cos \lambda_n r_o}{\lambda_n r_o - \sin \lambda_n r_o \cos \lambda_n r_o} \frac{\sin \lambda_n r}{\lambda_n r} e^{-\lambda_n^2 \alpha t} \tag{21.47}$$

where the λ_n s are the positive roots of

$$\tan \lambda_n r_o - \frac{\lambda_n r_o}{1 - \text{Bi}} = 0 \qquad (n = 1, 2, 3, \cdots, \infty) \tag{21.48}$$

and

$$\text{Bi} = \frac{h r_o}{k}$$

The instantaneous heat loss from the surface of the sphere is

$$\begin{aligned}\dot{Q} &= 8\pi k r_o (T_i - T_\infty) \\ &= \sum_{n=1}^{\infty} \frac{(\sin \lambda_n r_o - \lambda_n r_o \cos\lambda_n r_o)^2}{\lambda_n r_o(\lambda_n r_o - \sin \lambda_n r_o \cos \lambda_n r_o)} e^{-\lambda_n^2 \alpha t}\end{aligned} \tag{21.49}$$

and the cumulative heat loss over a period of time, t, is

$$Q = \int_0^t \dot{Q}\, dt$$

or

$$\begin{aligned}Q &= 8\pi\rho c(T_i - T_\infty) \\ &\sum_{n=1}^{\infty} \frac{(\sin \lambda_n r_o - \lambda_n r_o \cos \lambda_n r_o)^2}{\lambda_n^3(\lambda_n r_o - \sin \lambda_n r_o \cos \lambda_n r_o)} \left(1 - e^{-\lambda_n^2 \alpha t}\right)\end{aligned} \tag{21.50}$$

As in the case of a cylinder, θ may be expressed as a function of $R = r/r_o$, $\text{Fo} = \alpha t/r_o$ and $\text{Bi} = hr_o/k$, that is, $\theta = f(R, \text{Fo}, \text{Bi})$. Likewise $\theta_o = f(\text{Fo}, \text{Bi})$ and $Q/Q_i = f(\text{Fo}, \text{Bi})$ where

$$Q_i = \frac{4}{3}\pi r_o^3 \rho c(T_i - T_\infty)$$

is the initial internal energy of the solid above the environment.

Figure 21.13a shows θ_o as a function of Fo and 1/Bi. The ratio θ/θ_o as a function of 1/Bi for various values of R is plotted in Figure 21.13b. The ratio Q/Q_i is plotted as a function of Bi^2Fo for various values of Bi in Figure 21.13c.

Example 21.8 In an experimental set-up, a 4-cm-diameter copper ball ($k = 401\,\text{W/m-K}$, $c = 385\text{J/kg-K}$, and $\rho = 8933\,\text{kg/m}^3$) is heated in a furnace to a uniform temperature of 100°C and suddenly immersed in water at 25°C. A thermocouple placed at the center of the ball records a temperature of 44.5°C after 2 min of immersion. Use the appropriate Heisler chart to determine the convective heat transfer coefficient on the outer surface of the ball.

Solution

Assumptions and Specifications

(1) Unsteady-state conditions exist.
(2) Thermal properties do not vary with temperature.
(3) There is no radiation exchange with the surroundings.

We first calculate θ_o and Fo corresponding to 2 min or 120 s.

$$\theta_o = \frac{T(0,\ 120\,\text{s}) - T_\infty}{T_i - T_\infty} = \frac{44.5°\text{C} - 25°\text{C}}{100°\text{C} - 25°\text{C}} = 0.26$$

and

$$\text{Fo} = \frac{\alpha t}{r_o^2} = \frac{kt}{\rho c r_o^2}$$

so that

$$\text{Fo} = \frac{(401\,\text{W/m-K})(120\,\text{s})}{(8933\,\text{kg/m}^3)(385\,\text{J/kg-K})(0.02\,\text{m})^2} = 34.98$$

With $\theta_o = 0.26$ and Fo = 34.98, we read Figure 21.13a to obtain

$$\frac{1}{\text{Bi}} = \frac{k}{hr_o} = 80$$

which then gives

$$h = \frac{k}{\text{Bi}r_o} = \frac{401\,\text{W/m-K}}{80(0.02\,\text{m})} = 250.6\,\text{W/m}^2\text{-K} \Longleftarrow$$

21.5.6 One-Term Approximate Solutions

When Fo exceeds 0.20, the one-term approximate solutions are

$$\theta = \frac{T(r,\ t) - T_\infty}{T_i - T_\infty} = C_1 e^{-\beta_1^2 \text{Fo}} \frac{\sin(\beta_1 r/r_o)}{\beta_1 r/r_o} \tag{21.51}$$

$$\theta_o = \frac{T(0,\ t) - T_\infty}{T_i - T_\infty} = C_1 e^{-\beta_1^2 \text{Fo}} \tag{21.52}$$

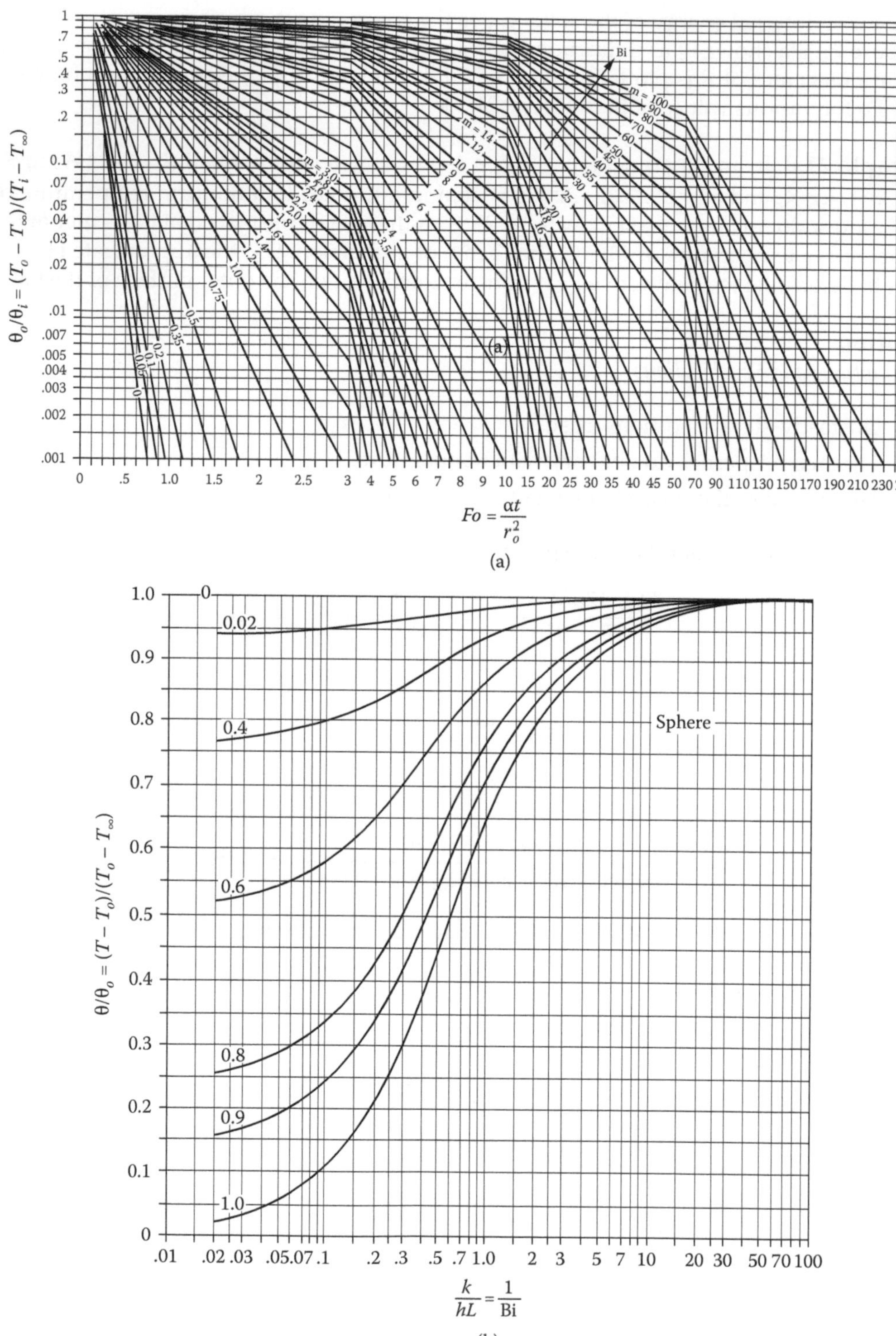

FIGURE 21.13
Charts for unsteady conduction in a solid sphere (a) center temperature, (b) temperature as a function of the center temperature, and (c) dimensionless heat loss. Figures 21.13a and 21.13b are reproduced from the original Heisler paper [Heisler (1947), courtesy of ASME] and Figure 21.13c is attributed to Gröeber et al. (1961).

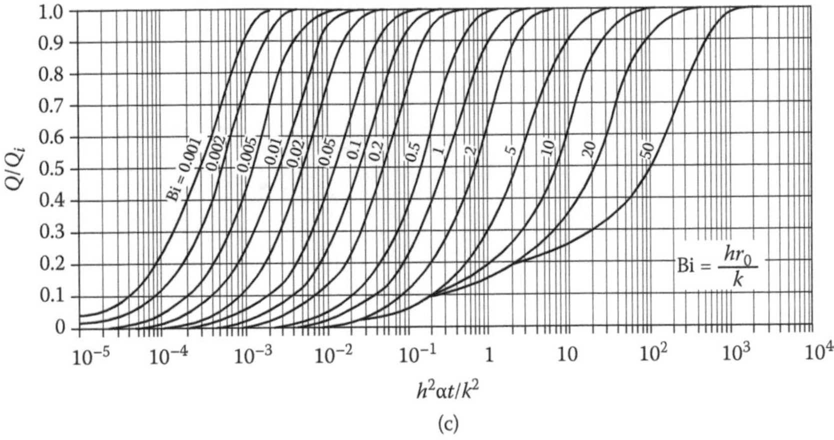

(c)

FIGURE 21.13
Continued

and

$$\frac{Q}{Q_i} = 1 - \frac{3\theta_o}{\beta_1^3}(\sin\beta_1 - \beta_1\cos\beta_1) \tag{21.53}$$

where $\beta_1 = \lambda_1 r_o$. The values of C_1 and β_1 for a range of values of Bi are provided in Table 21.5.

Example 21.9 A 4-cm-diameter lead ball ($k = 35.3\,\text{W/m-K}$, $c = 129\,\text{J/kg-K}$, and $\rho = 11{,}433\,\text{kg/m}^3$) at 300°C as in Figure 21.14, is suddenly placed in a water jet. At 30°C, the water jet produces a heat transfer coefficient of $1765\,\text{W/m}^2\text{-K}$. Use the one-term approximate solutions, determine (a) the center and the surface temperatures after ten seconds,

TABLE 21.5
Values of C_1 and β_1 Used in the One-Term Approximation to the Series Solution for Transient One-Dimensional Conduction in the Solid Sphere

Bi	β_1, rad	C_1	Bi	β_1, rad	C_1
0.01	0.1730	1.0030	0.80	1.4320	1.2236
0.02	0.2445	1.0060	0.90	1.5044	1.2488
0.03	0.2989	1.0090	1.00	1.5708	1.2732
0.04	0.3450	1.0120	2.00	2.0288	1.4793
0.05	0.3852	1.0149	3.00	2.2889	1.6227
0.06	0.4217	1.0179	4.00	2.4456	1.7201
0.07	0.4550	1.0209	5.00	2.5704	1.7870
0.08	0.4860	1.0239	6.00	2.6537	1.8338
0.09	0.5150	1.0268	7.00	2.7165	1.8674
0.10	0.5423	1.0298	8.00	2.7654	1.8921
0.15	0.6608	1.0445	9.00	2.8044	1.9106
0.20	0.7593	1.0592	10.0	2.8363	1.9249
0.25	0.8448	1.0737	21.0	2.9857	1.9781
0.30	0.9208	1.0880	30.0	3.0372	1.9898
0.40	1.0528	1.1164	40.0	3.0632	1.9942
0.50	1.1656	1.1441	50.0	3.0788	1.9962
0.60	1.2644	1.1713	100	3.1102	1.9990
0.70	1.3525	1.1978	∞	3.1415	2.0000

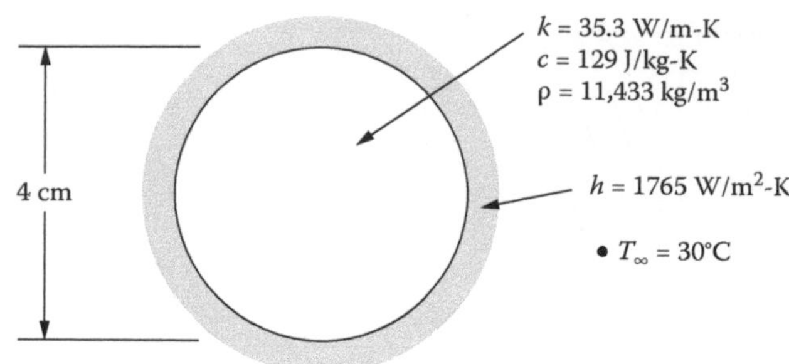

FIGURE 21.14
Configuration for Example 21.9.

(b) the initial energy of the ball above the water jet environment, and (c) the cumulative heat loss over a period of 10 s.

Solution

Assumptions and Specifications

(1) Unsteady-state conditions exist.
(2) Thermal properties do not vary with temperature.
(3) There is no radiation exchange with the surroundings.
(4) The one-term approximate solution is to be used.

(a) We first calculate Bi and Fo.

$$\text{Bi} = \frac{hr_o}{k} = \frac{(1765\,\text{W/m}^2\text{-K})(0.02\,\text{cm})}{35.3\,\text{W/m-K}} = 1.0$$

and

$$\text{Fo} = \frac{kt}{\rho c r_o^2} = \frac{(35.3\,\text{W/m-K})(10\,\text{s})}{(11,433\,\text{kg/m}^3)(129\,\text{J/kg-K})(0.02\,\text{m})^2} = 0.60$$

With Bi = 1.0, Table 21.5 gives for the sphere, $\beta_1 = 1.5708$ rad and $C_1 = 1.2732$. Solving Equation 21.52 for θ_o gives

$$\begin{aligned}\theta_o &= C_1 e^{-\beta_1^2 \text{Fo}}\\ &= (1.2732)e^{-(1.5708)^2(0.60)}\\ &= (1.2732)e^{-1.4804}\\ &= (1.2732)(0.2275) = 0.2897\end{aligned}$$

which then gives

$$\begin{aligned}T(0,\ 10\,\text{s}) &= (0.2897)(T_i - T_\infty) + T_\infty\\ &= (0.2897)(300^\circ\text{C} - 30^\circ\text{C}) + 30^\circ\text{C}\\ &= 78.2^\circ\text{C} + 30^\circ\text{C} = 108.2^\circ\text{C} \Longleftarrow\end{aligned}$$

The surface temperature may be obtained from Equation 21.51

$$\frac{T(r,\, t) - T_\infty}{T_i - T_\infty} = \theta_o \frac{\sin(\beta_1 r/r_o)}{\beta_1 r/r_o}$$

$$= (0.2897)\frac{\sin 1.5708}{1.5708} = 0.1844$$

which gives

$$T(0.02\,\text{m},\ 10\,\text{s}) = (0.1844)(T_i - T_\infty) + T_\infty$$

$$= (0.1844)(300°\text{C} - 30°\text{C}) + 30°\text{C}$$

$$= 49.8°\text{C} + 30°\text{C} = 79.8°\text{C} \Longleftarrow$$

(b) The initial energy of the ball above the water jet environment may be found as

$$Q_i = \frac{4\pi r_o^3}{3}\rho c(T_i - T_\infty)$$

$$= \frac{4\pi(0.02\,\text{m})^3}{3}(11{,}433\,\text{kg/m}^3)(129\,\text{J/kg-K})(300°\text{C} - 30°\text{C})$$

$$= 13{,}344\,\text{J} \Longleftarrow$$

Using Equation 21.53, we have

$$Q = Q_i\left[1 - \frac{3\theta_o}{\beta_1^3}(\sin\beta_1 - \beta_1\cos\beta_1)\right]$$

$$= (13{,}344\,\text{J})\left[1 - \frac{3(0.2897)}{(1.5708)^3}(\sin 1.5708 - 1.5708\cos 1.5708)\right]$$

$$= 10{,}352\,\text{J}$$

21.6 Summary

Certain transient heating or cooling problems can be handled by the **lumped capacitance model** in which the Biot modulus is less than 0.10

$$\text{Bi} = \frac{hL_c}{k} < 0.1$$

where L_c is the ratio of the volume of the solid to its surface area, $L_c = V/S$. In this case, from Equation 21.5, we have

$$\frac{T - T_\infty}{T_i - T_\infty} = e^{t/\tau} \tag{21.5}$$

where the time constant, τ, is defined by

$$\tau = \rho Vc/hS \tag{21.6}$$

In the event that internal heat generation, $\dot{E}_g$, is present, the temperature-time history is given by

$$\frac{\dot{E}_g - hS(T - T_\infty)}{\dot{E}_g - hS(T_i - T_\infty)} = e^{-t/\tau} \qquad (21.16)$$

The semi-infinite model considers the solid as extending to infinity in all dimensions except for one surface. For a constant surface temperature, T_o

$$T(x, t) = T_i + (T_o - T_i)\,\text{erfc}(\phi) \qquad (21.22)$$

where erfc is the complementary error function tabulated in Table 21.1 and where

$$\phi = \frac{x}{2(\alpha t)^{1/2}} \qquad (21.23)$$

The surface heat flux for a constant surface temperature is given by

$$\dot{q}_o = \frac{k(T_o - T_i)}{(\pi\alpha t)^{1/2}} \qquad (21.25)$$

The temperature for constant heat flux is given by Equation 21.27. For surface convection where the convection is specified, Equation 21.29 pertains.

The Heisler and Gröeber charts provide approximations to the unsteady-heat flow with convection present and are reproduced for the long, solid plane wall in Figure 21.7. They are based on a dimensionless spatial coordinate, $X = x/L$, and the Fourier and Biot moduli

$$\text{Fo} \equiv \frac{\alpha t}{L^2} \qquad \text{and} \qquad \text{Bi} \equiv \frac{hL}{k}$$

The charts give

$$\theta = f(X, \text{Fo}, \text{Bi}), \quad \theta_o = f(\text{Fo}, \text{Bi}), \quad \text{and} \quad \frac{Q}{Q_i} = f(\text{Fo}, \text{Bi})$$

For the long solid cylinder (Figure 21.10) and the solid sphere (Figure 21.13), $R = r/r_o$

$$\text{Fo} \equiv \frac{\alpha t}{r_o^2} \qquad \text{and} \qquad \text{Bi} \equiv \frac{hr_o}{k}$$

and the charts give

$$\theta = f(X, \text{Fo}, \text{Bi}), \quad \theta_o = f(\text{Fo}, \text{Bi}), \quad \text{and} \quad \frac{Q}{Q_i} = f(\text{Fo}, \text{Bi})$$

The classical solutions for the long, solid plane wall, the long solid cylinder and the solid sphere involve a series summation with each term in the series based upon a particular **eigenvalue.** One-term approximate solutions are obtainable and are given for θ, θ_o, and $\dot{Q}/\dot{Q}_i$ for cases where Fo exceeds 0.20. For the long, solid plane wall, Equations 21.34, 21.35, and 21.36 are applicable. For the long solid cylinder, Equations 21.43, 21.44, and 21.45 pertain; and for the solid sphere, the equations are Equations 21.51, 21.52, and 21.53.

21.7 Problems

The Lumped Capacitance Model

21.1: A plain carbon steel cube ($k = 60.5\,\text{W/m-K}$) 3 cm on a side is heated to 200°C and then allowed to cool in a stream of air at 20°C, which provides a convection coefficient of 50 W/m^2-K. Determine the Biot number and check to see whether the lumped capacitance model is valid.

21.2: Reconsider the cube in Problem 21.1. The cube has a density of 7833 kg/m^3 and a specific heat of 465J/kg-K. Determine (a) the time needed to cool the cube to 25°C and (b) the cumulative heat loss from the cube over this period of time.

21.3: In a laboratory experiment, the forced convection heat transfer coefficient is measured by heating a thermocouple equipped copper sphere with a Bunsen burner and then suddenly exposing the sphere to a stream of air. One such set-up uses a 25.4-cm-diameter copper sphere ($k = 401\,\text{W/m-K}$ and $c = 385\text{J/kg-K}$), which is heated to 320°C and allowed to cool in the air at 25°C. After 1 min of exposure to the air, the thermocouple records a temperature of 280°C. Determine the average heat transfer coefficient between the sphere and the airstream.

21.4: Consider the lumped capacitance cooling of a solid whose specific heat changes linearly with temperature according to

$$c = c_\infty[1 + \beta(T - T_\infty)]$$

where c_∞ is the specific heat at T_∞ and β is a constant. Show that the temperature-time history of the solid is given by

$$\ln\frac{T - T_\infty}{T_i - T_\infty} + \beta(T - T_\infty) = -\frac{hSt}{\rho c_\infty}$$

21.5: A steel ball 20 mm in diameter with an initial temperature of 800°C is drawn from a furnace. At time $t = 0$ it is suddenly placed in an oil bath at 20°C, which provides a heat transfer coefficient of 120 W/m^2-K. Assuming that the density of steel is $\rho = 8055\,\text{kg/m}^3$ and that

$$c = 500[1 + 3 \times 10^{-4}(T - 20°\text{C})]$$

with c in J/kg-K and T in °C, determine the time required for the ball to cool down to 300°C.

21.6: A 2-cm-thick AISI 302 stainless steel plate initially at a uniform temperature of 520°C is suddenly placed in a coolant bath at 25°C, which provides a convective heat transfer coefficient of 95 W/m^2-K. Assuming that $c = 480\text{J/kg-K}$, $\rho = 8055\,\text{kg/m}^3$, and $k = 15.1\,\text{W/m-K}$, for the plate, determine the Biot number to see if the lumped capacitance model is applicable and, if applicable, determine the time needed for the plate to cool down to 50°C.

21.7: Perform the lumped capacitance analysis of Section 21.2.1 to develop an expression for the temperature-time history of the solid when the heat transfer coefficient is temperature dependent in the form

$$h = h_i\left(\frac{T - T_\infty}{T_i - T_\infty}\right)^n$$

where h_i and n are constants.

21.8: A 2-cm-diameter aluminum ball ($c = 903 \text{J/kg-K}$, $\rho = 2702 \text{kg/m}^3$, and $k = 237 \text{W/m-K}$) initially at a uniform temperature of 90°C is suddenly exposed to the air at 25°C, which provides a heat transfer coefficient

$$h = 1.60(T - T_\infty)^{1/4} \text{ W/m}^2\text{-K}$$

where T is the ball temperature at any instant of time. Assuming that the lumped capacitance model is applicable, determine the temperature of the ball 5 min after exposure to the air.

The Effect of Internal Heat Generation

21.9: An electronic component mounted on a heat sink is cooled by the air at 20°C, which provides a heat transfer coefficient of $80 \text{W/m}^2\text{-K}$. When the component is turned on, it generates 150 W. Assume the entire assembly to be a lumped capacitance system with a surface area $S = 96 \text{cm}^2$ and $\rho VC = 50 \text{J/K}$. Given an initial temperature for the assembly of 20°C, determine (a) the temperature of the assembly after 4 min of operation and (b) the steady-state temperature of the assembly.

21.10: The initiation of a chemical reaction in a spherical vessel of 25-cm diameter gives rise to a heat generation of 800 W. The outer surface of the vessel is exposed to the air at 27°C, which provides a heat transfer coefficient of $15 \text{W/m}^2\text{-K}$. The vessel and its contents may be regarded as a lumped capacitance system with $\rho = 900 \text{kg/m}^3$ and $c = 2600 \text{J/kg-K}$ with an initial temperature of 27°C. Determine (a) the temperature of the vessel and its contents after 5 min of reaction and (b) the steady-state temperature of the vessel and its contents.

21.11: A 3-cm-diameter thinly coated steel ball heated electrically generates 80 W. When the power is turned on, the ball loses heat to its surroundings by convection and radiation. The convective environment is at 27°C, which provides a heat transfer coefficient of $8 \text{W/m}^2\text{-K}$ while the emissivity of the surface of the coating is $\epsilon = 0.85$. Assuming that the ball behaves as a spatially isothermal solid with $\rho = 8933 \text{kg/m}^3$ and $c = 385 \text{J/kg-K}$, determine the temperature of the ball at the instant where $dT/dr = 25 \text{K/s}$.

21.12: An electronic component with $\rho Vc = 75 \text{J/K}$ generates 10 W when energized. The component, which is initially at 25°C, is cooled by the air at 25°C. The temperature of the component reaches 32°C after 1 min of operation. The surface area of the component is $S = 40 \text{cm}^2$ and a lumped capacitance system may be assumed applicable. Determine the heat transfer coefficient provided by the air.

21.13: In Problem 21.12, how long will it take for the component to reach a temperature of 85°C?

21.14: In Problem 21.12, assume that the component is turned off after 4 min. Determine the time required for the temperature to fall back to 40°C.

21.15: In Problem 21.12, assume that the heat transfer coefficient provided on the exterior of the component is $80 \text{W/m}^2\text{-K}$. Determine (a) the temperature after 3 min and (b) the steady-state temperature.

21.16: A steel plate that is 12-cm thick ($c = 436 \text{J/kg-K}$, $\rho = 7775 \text{kg/m}^3$, and $k = 48 \text{W/m-K}$) at 627°C is plunged into an oil bath at 27°C. The heat transfer coefficient at the surface is $420 \text{W/m}^2\text{-K}$. Neglecting the edge effects, determine how long it will take for the plate to cool to 97°C.

The Semi-Infinite Solid

21.17: The surface at $x = 0\,\text{cm}$ of a semi-infinite solid initially at 85°C throughout is suddenly lowered to 30°C and kept at that value. A thermocouple located at $x = 25\,\text{cm}$ records a temperature of 53.56°C after 10 min. Determine the thermal diffusivity of the solid.

21.18: A plane carbon steel plate is initially at a uniform temperature of 27°C. At $t = 0$, the temperature at the surface ($x = 0\,\text{cm}$) is suddenly increased to 100°C and maintained at that value. The thermal conductivity and thermal diffusivity of the plate are 50.5 W/m-K and $1.70 \times 10^{-6}\,\text{m/s}^2$, respectively. Treat the plate as a semi-infinite solid and determine at time, $t = 60\,\text{s}$ (a) the temperature at $x = 5\,\text{cm}$ and (b) the surface heat flux.

21.19: A thick piece of material ($c = 900\text{J/kg-K}$, $\rho = 1800\,\text{kg/m}^3$, and $k = 3.2\,\text{W/m-K}$) is initially at a surface temperature of 60°C. The surface at $x = 0\,\text{cm}$ is suddenly reduced to 25°C and maitained at that value. Plot the temperature and heat flux as a function of time at $x = 5\,\text{cm}$.

21.20: A thick tungsten block ($c = 132\text{J/kg-K}$, $\rho = 19,300\,\text{kg/m}^3$, and $k = 174\,\text{W/m-K}$) is initially at a surface temperature of 60°C. The surface at $x = 0\,\text{cm}$ is suddenly exposed to a heat source that creates a constant heat flux of $800\,\text{W/m}^2$ at the surface. Determine (a) the temperature at a depth of 1 m after 50 min, (b) the surface temperature after 50 min, and (c) the cumulative heat supply over the 50-min period.

21.21: The thermal diffusivity of a material ($k = 75\,\text{W/m-K}$) is determined by imposing a known heat flux at the surface ($x = 0\,\text{cm}$) of a thick piece of the material and recording the temperature at a certain depth from the surface. In an experiment, a thermocouple registers a temperature of 54.4°C at $x = 5\,\text{cm}$ and $t = 2\,\text{min}$ when a heat flux of $5\,\text{kW/m}^2$ is established at the surface. Assuming the material to be at 50°C initially, determine the thermal diffusivity of the material.

21.22: Consider a thick slab of aluminum ($c = 903\text{J/kg-K}$, $\rho = 2702\,\text{kg/m}^3$ and $k = 237\,\text{W/m-K}$) which is initially at a uniform temperature of 20°C. When the surface at $x = 0\,\text{cm}$ is suddenly exposed to a constant radiant heat flux, and after 2 min, the temperature at a depth of 12 cm is measured at 44.8°C. Determine the radiant heat flux imposed at the surface.

21.23: A concrete wall ($c = 880\text{J/kg-K}$, $\rho = 2100\,\text{kg/m}^3$, and $k = 1.4\,\text{W/m-K}$) that is 30-mm thick is initially at a uniform temperature of 22°C. At time $t = 0$, the surface at $x = 0\,\text{cm}$ is subjected to a constant heat flux of $950\,\text{W/m}^2$. Assuming the wall to behave as a semi-infinite solid, determine the temperature at the two faces of the wall after 2 h of exposure to the heat flux.

21.24: Solve Problem 21.23 if, instead of a constant heat flux, the surface at $x = 0$ is suddenly exposed to a stream of hot fluid, which promotes a heat transfer coefficient of $80\,\text{W/m}^2\text{-K}$.

21.25: A hot, thick slab of steel ($c = 477\text{J/kg-K}$, $\rho = 7000\,\text{kg/m}^3$, and $k = 14.9\,\text{W/m-K}$) is suddenly exposed at its surface ($x = 0\,\text{cm}$) to a cold stream of water at 25°C, which provides a heat transfer coefficient of $240\,\text{W/m}^2\text{-K}$. The slab is assumed to be of infinite extent and a thermocouple installed at a depth of 3 cm from the surface records 48°C after 3 min. Determine the initial temperature of the slab.

21.26: A thick slab of material ($c = 850\text{J/kg-K}$, $\rho = 2000\,\text{kg/m}^3$, and $k = 2\,\text{W/m-K}$) is at uniform temperature of 20°C. The surface at $x = 0\,\text{cm}$ is suddenly brought into contact with a convective environment at a temperature of 90°C with a heat transfer coefficient of 200 W/m^2-K. Determine the surface heat flux after 5 min of exposure to the convective environment.

21.27: A thick wooden wall with $c = 1255$J/kg-K, $k = 0.16$ W/m-K, and $\rho = 720$ kg/m^3, initially at a uniform temperature of 22°C and is suddenly exposed at its surface at $x = 0$ cm to hot gases, which provide a heat transfer coefficient of 40 W/m^2-K. The exposed surface of the wall reaches a temperature of 180°C in 10 min. Determine the temperature of the hot gases.

21.28: A very thick concrete slab at 147°C is sprayed with a large quantity of water at 27°C. Determine how long it will take for a point 5 cm below the surface of the concrete to cool to 62°C. The heat transfer coefficient at the concrete surface is 270 W/m^2-K.

Finite-Sized Solids

21.29: An aluminum plate with $c = 903$J/kg-K, $\rho = 2702$ kg/m^3, and $k = 237$ W/m-K is 30-cm thick and is initially at a uniform temperature of 220°C. At time $t = 0$, the plate is suddenly placed in a water stream at 25°C, which promotes a heat transfer coefficient of 1000 W/m^2-K on both faces of the plate. Determine (a) the time required to cool the centerline of the plate to 80°C, (b) the surface temperature at the time determined in (a), and (c) the cumulative heat loss per unit area to the water over this time period.

21.30: A plastic slab ($c = 1480$J/kg-K, $\rho = 1150$ kg/m^3, and $k = 0.28$ W/m-K) is 3-cm thick and is initially at a uniform temperature of 75°C. At time $t = 0$, one surface of the slab is suddenly exposed to a cold stream of air at 20°C providing a heat transfer coefficient of 20 W/m^2-K. The other surface of the slab is insulated. Determine (a) the temperature of the cooled surface and (b) the temperature of the insulated surface after 45-min exposure to the cold air.

21.31: A 4-cm-thick bronze plate ($c = 343$J/kg-K, $\rho = 8666$ kg/m^3, and $k = 26$ W/m-K), which is initially at a uniform temperature of 20°C, is placed in an oven at 500°C that provides a heat transfer coefficient of 108 W/m^2-K on both faces of the plate. If a centerline temperature of 400°C is to be obtained, determine (a) how long the plate should stay in the oven and (b) how much energy in kilojoules is gained by the plate during this time.

21.32: For the plastic slab of Problem 21.30 determine the time for the slab to lose half of its original energy to the air.

21.33: A 10-cm-diameter stainless steel rod ($c = 480$J/kg-K, $\rho = 8055$ kg/m^3, and $k = 15.1$ W/m-K), initially at a uniform temperature of 800°C, is removed from a furnace and placed in an oil bath at 25°C at time $t = 0$. The rod loses heat by convection to the oil via a heat transfer coefficient of 55 W/m^2-K. Determine (a) the temperature at the center of the rod after 30 min, (b) the surface temperature after 30 min, and (c) the cumulative heat loss from the rod over 30 min.

21.34: A 8-cm-diameter stainless steel rod ($k = 10$ W/m-K and $\alpha = 4.2 \times 10^{-5}$ m^3/s) is initially at a uniform temperature of 200°C and is to be cooled convectively by a coolant at 25°C such that the center of the rod must reach 50°C in 5 min. Determine the heat transfer coefficient that the coolant must provide.

21.35: A long steel shaft ($k = 40$ W/m-K and $\alpha = 1.2 \times 10^{-5}$ m^3/s) of 5 cm radius, initially at a uniform temperature of 400°C, is removed from a furnace and placed in a coolant bath at 20°C. The heat transfer coefficient is 270 W/m^2-K. Determine the time required for a radial location at 3 cm to attain a temperature of 200°C.

21.36: A long bronze rod ($c = 343\text{J/kg-K}$, $\rho = 8666\,\text{kg/m}^3$, and $k = 25\,\text{W/m-K}$) of 6-cm diameter is initially at a uniform temperature of 300°C when it is suddenly placed in a coolant bath at 30°C. The convective heat transfer coefficient is $87\,\text{W/m}^2\text{-K}$. Determine (a) the time for the rod to lose 60 % of its initial energy to the coolant, (b) the energy lost during the time period found in (a), and (c) the temperature in the center of the rod at the time found in (a).

21.37: A solid stainless steel sphere ($k = 15.1\,\text{W/m-K}$ and $\alpha = 3.91 \times 10^{-6}\,\text{m}^3/\text{s}$) has a radius of 4 cm. From an initial uniform temperature of 250°C, it is suddenly exposed to the air at 25°C with a heat transfer coefficient of $20\,\text{W/m}^2\text{-K}$. Determine (a) the center temperature, (b) the temperature at a radius of 2.4 cm, and (c) the surface temperature after 30 min of exposure to the air.

21.38: A 12-cm-diameter sphere ($k = 15\,\text{W/m-K}$ and $\alpha = 3.8 \times 10^{-6}\,\text{m}^3/\text{s}$) is heated in a furnace to a uniform temperature of 80°C and then is suddenly placed in flowing water at 20°C. A thermocouple located in the center of the sphere records a temperature of 50°C after 5 min of exposure to the water. Determine the convective heat transfer coefficient created by the flowing water.

21.39: A 20-cm-diameter spherical pellet of lead ($k = 35.3\,\text{W/m-K}$ and $\alpha = 2.41 \times 10^{-5}\,\text{m}^3/\text{s}$) is heated to a uniform temperature of 400°C before it is suddenly allowed to quench in a bath of oil at 25°C. The convection heat transfer coefficient is $160\,\text{W/m}^2\text{-K}$. Determine (a) the time that it will take for the pellet to lose 40% of its energy to the oil and, for this time determine (b) the center temperature, (c) the temperature at a radius of 6 cm, and (d) the surface temperature.

One-Term Approximations

21.40: Use the one-term approximation for the long plane wall to solve Problem 21.29.

21.41: Use the one-term approximation for the long plane wall to solve Problem 21.30.

21.42: Use the one-term approximation for the long plane wall to solve Problem 21.31.

21.43: Calculate the Fourier number, $\text{Fo} = \alpha t / r_o^2$, for the rod in problem 21.33. If $\text{Fo} > 0.2$, use the one-term approximation for the long cylinder in a trial-and-error procedure to solve Problem 21.33.

21.44: Use the one-term approximation for the long cylinder to solve Problem 21.34.

21.45: Use the one-term approximation for the long cylinder to solve Problem 21.35.

21.46: Use the one-term approximation for the long cylinder to solve Problem 21.36.

21.47: Use the one-term approximation for the solid sphere to solve Problem 21.37.

21.48: A 12-cm-diameter sphere ($k = 15\,\text{W/m-K}$ and $\alpha = 3.8 \times 10^{-6}\,\text{m/s}^2$) is heated in a furnace to a uniform temperature of 80°C and then suddenly placed in flowing water at 20°C. The heat transfer coefficient between the sphere and the water is $60\,\text{W/m}^2\text{-K}$ and a thermocouple located in the center of the sphere records a temperature of 50°C. Using the one-term approximation, determine the time required for the sphere to obtain this temperature.

21.49: Suppose that the lead pellet considered in Problem 21.39 loses 50% of its energy to the oil after immersion at $t = 0$. Use the one-term approximation to determine the time required for this energy loss.

22

Forced Convection—Internal Flow

Chapter Objectives

- To evaluate axial temperature distributions for fluids flowing inside closed conduits with either constant wall temperature or constant wall heat flux.
- To generate applicable dimensionless parameters for internal forced convection using dimensional analysis.
- To catalog the empirical correlations for laminar, transition, and turbulent internal flow.
- To use the empirical expressions for h, along with the energy equation, to evaluate the heat transfer behavior of representative systems.

22.1 Introduction

Heat transfer by convection occurs through the exchange of thermal energy between a surface and an adjacent fluid. Fluid motion is frequently involved in practical energy exchange processes. In this chapter, we will build upon the knowledge acquired in earlier chapters involving fluid mechanics to include the situations in which temperature differences exist between a fluid and the surface it contacts.

Initially, we will treat **forced convection** leaving the subject of natural or free condition until Chapter 24. In this chapter, we will address the situation where the fluid, being either heated or cooled, is flowing in a closed conduit such as a pipe, tube, or duct. Fluid flow issues associated with internal flow were discussed in Chapter 16. Many of the concepts and much of the terminology introduced earlier will be used in this treatment without additional discussion.

The basic rate equation for convective heat transfer, as introduced earlier, is the **Newton rate equation**. This relationship is

$$\dot{Q} = hS(T_H - T_C) \tag{22.1}$$

where $\dot{Q}$ is the rate of heat transfer in watts, S is the surface area, T_H and T_C are the higher and cooler temperatures, respectively, and h is the convective heat transfer coefficient in W/m^2-K. Equation 22.1 is, in a practical sense, the definition of h, the heat transfer coefficient.

Much of the challenge in evaluating convection heat transfer lies in the evaluation of h, which involves the complexities of fluid flow in addition to temperature distribution effects. We will examine the means for evaluating h in a later section. First, we will look at the bigger picture involving energy exchange with internal flow.

22.2 Temperature Distributions with Internal Forced Convection

22.2.1 The Constant Wall Heat Flux Case

Flow in a portion of a circular conduit is shown in Figure 22.1 for the condition where heat is being added to the fluid. A first law analysis of the control volume for steady-state conditions yields the expression

$$\dot{Q}_{cv} + \dot{m}h_i - \dot{m}h_e = 0 \tag{22.2}$$

Here, $\dot{Q}_{cv}$ is the heat transferred from the wall to the fluid

$$\dot{Q}_{cv} = \frac{\dot{Q}}{S}\pi d\,(dx) = \dot{q}\,\pi d\,(dx)$$

where $\dot{Q}/S = \dot{q}$ is the heat flux, $\dot{m}h_i$ is the energy entering the control volume via fluid flow

$$\dot{m}h_i = \rho\frac{\pi d^2}{4}\hat{V}c_pT$$

and $\dot{m}h_e$ is the energy leaving the control volume by fluid flow

$$\dot{m}h_e = \rho\frac{\pi d^2}{4}\hat{V}c_p(T + dT)$$

Observe that, as a result of heat addition from the wall, the fluid temperature has increased by the amount, dT.

With the expressions for the individual terms substituted, Equation 22.2 becomes

$$\frac{\dot{Q}}{S}\pi d(dx) + \rho\frac{\pi d^2}{4}\hat{V}c_pT - \rho\frac{\pi d^2}{4}\hat{V}c_p(T + dT) = 0$$

and, after simplifying, this expression can be written as

$$\frac{\dot{Q}}{S} - \frac{d}{4}\rho\hat{V}c_p\frac{dT}{dx} = 0 \tag{22.3a}$$

or

$$\frac{dT}{dx} - \frac{4}{d}\frac{\dot{Q}/S}{\rho\hat{V}c_p} = 0 \tag{22.3b}$$

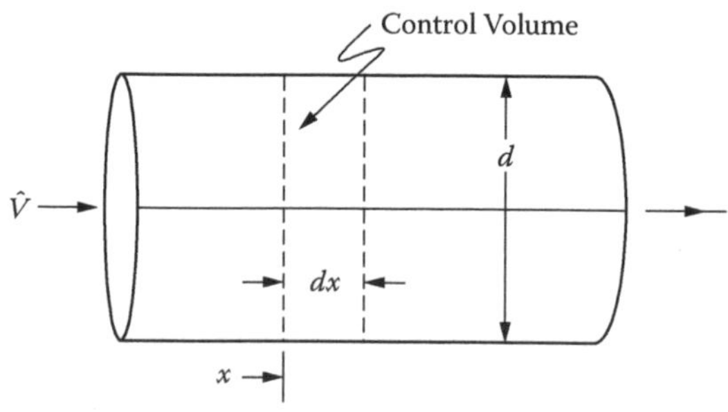

FIGURE 22.1
Control volume with fluid in forced convection.

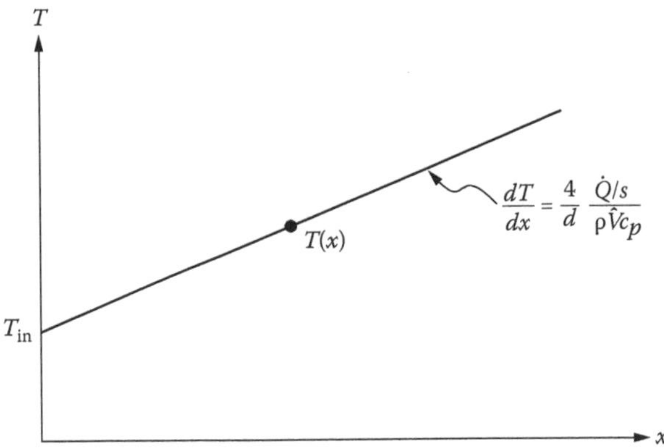

FIGURE 22.2
Axial temperature variation of a fluid being heated uniformly.

Equation 22.3b indicates that the rate of temperature increase by the fluid is proportional to the wall heat flux. If $\dot{Q}/S$ is constant, as would be true for a metal pipe wall subjected to a constant voltage source, the value of dT/dx would be a constant and the temperature variation of the fluid will be as shown in Figure 22.2.

Information of interest, in addition to the fluid temperature variation, $T(x)$, is the wall temperature. Equation 22.1 provides the means for evaluating the wall temperature, T_w (which, in this case, is T_H). The wall temperature is

$$T_w = T_{\text{fluid}} + \frac{\dot{Q}/S}{h} \tag{22.4}$$

This expression indicates that we must know the value of h in order to evaluate T_w. If h is constant, Equation 22.4 indicates that the wall temperature variation will be linear with a constant slope equal to that given by Equation 22.3b. A constant value of h is generally achieved after the fluid has progressed some distance from the entrance to the conduit. Typical wall temperature variation with axial position is shown in Figure 22.3, where the distance from the entrance to the point where the two temperature profiles become parallel is designated as the **thermal entrance length,** L_{eT}.

Example 22.1 Lubricating oil, to be used in an internal combustion engine test (Figure 22.4), is preheated by passing it through an electrically heated tube 1 cm in diameter and 50-cm long. The oil flows at a rate of 800 kg/h and enters at a temperature of 0°C. The density and specific heat of the oil are $\rho = 836\,\text{kg/m}^3$ and $c = 2.26\,\text{J/kg-K}$, respectively, and the heat transfer coefficient is 68 W/m^2-K. Determine (a) the amount of energy in watts that must be transferred to the oil to achieve a temperature of 20°C at the tube exit and (b) the maximum tube surface temperature.

Solution

Assumptions and Specifications

(1) The heat transfer process is in steady state.

(a) For the lubricating oil, a liquid, the specific heat is designated as c. The required heat flux can be determined from Equation 22.3a

$$\frac{\dot{Q}}{S} - \frac{d}{4}\rho\hat{V}c\frac{dT}{dx} = 0 \tag{22.3a}$$

T
T_{wall}
T_{fluid}
T_{in}
x
L_{eT}

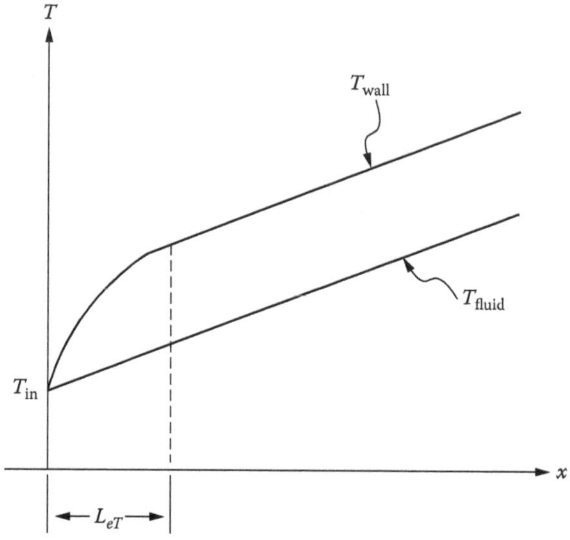

FIGURE 22.3
Axial temperature variation of a fluid and a uniformly heated wall.

This expression can be modified by noting that

$$\dot{m} = \rho A\hat{V} = \rho\hat{V}\frac{\pi d^2}{4}$$

which gives

$$\frac{\dot{Q}}{S} = \frac{d}{4}\left(\frac{4\dot{m}}{\pi d^2}\right)c\frac{dT}{dx} = \frac{\dot{m}}{\pi d}c\frac{dT}{dx}$$

Because the heat flux, $\dot{Q}/S$, is a constant, the temperature gradient, dT/dx, must be constant as well. Hence, we can write

$$\frac{dT}{dx} = \frac{\Delta T}{\Delta x}$$

and with this substitution, the heat flux expression is

$$\frac{\dot{Q}}{S} = \frac{\dot{m}}{\pi d}c\frac{\Delta T}{\Delta x}$$

h = 68 W/m²-K
Electrically Heated Tube
ṁ = 800 kg/h
T = 0°C
ρ = 836 kg/m³
c_p = 2.26 J/kg-K
1 cm
T = 20°C
50 cm

FIGURE 22.4
Configuration for Example 22.1.

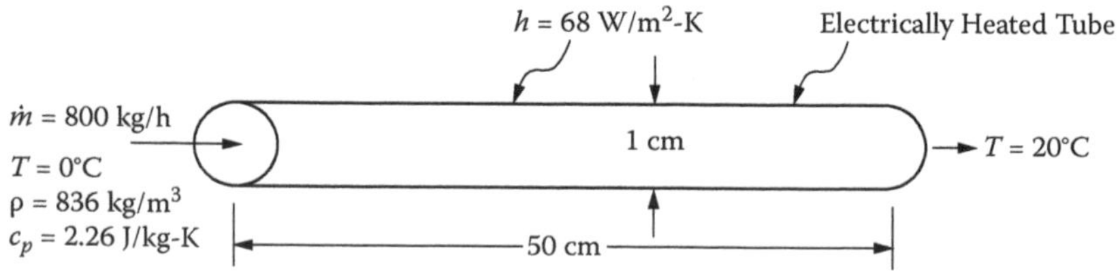

With numerical values inserted we have

$$\frac{\dot{Q}}{S} = \frac{(800\,\text{kg/h})(2.26\,\text{J/kg-K})(20\,\text{K})}{\pi(0.01\,\text{m})(0.50\,\text{m})(3600\,\text{s/h})} = 639\,\text{W/m}^2$$

The energy to the oil is thus,

$$\dot{Q} = (639\,\text{W/m}^2)\pi(0.01\,\text{m})(0.50\,\text{m}) = 10.0\,\text{W} \Longleftarrow$$

(b) Equation 22.4 will yield the maximum wall surface temperature. Because the maximum fluid temperature, at the exit, is 20°C, the maximum wall temperature is

$$T_{\text{w, max}} = T_{\text{fluid, max}} + \frac{\dot{Q}/S}{h}$$

$$= 20°\text{C} + \frac{639\,\text{W/m}^2}{68\,\text{W/m}^2\text{-K}} = 20°\text{C} + 9.40°\text{C} = 29.4°\text{C} \Longleftarrow$$

The foregoing example dealt with the simplest case, that of a uniform heat flux. When the wall heat is known to vary along the surface, that is $(\dot{Q}/S)(x)$ is a known function, Equations 22.3 and 22.4 still apply but their solutions will become more mathematically complex.

22.2.2 The Constant Wall Temperature Case

The other boundary condition of general interest is the case where the wall temperature variation, $T(x)$, is known. For the simplest case, where $T(x)$ is constant, the axial temperature variations of the wall and the fluid will be as indicated in Figure 22.5. In this case, the wall temperature is higher than the entering fluid temperature, T_1. The fluid temperature will increase due to energy transport from the wall, approaching the wall temperature for large values of x.

An analysis of this situation will involve the same control volume shown in Figure 22.1 with Equations 22.3 as the results. The difference between this case and the constant wall

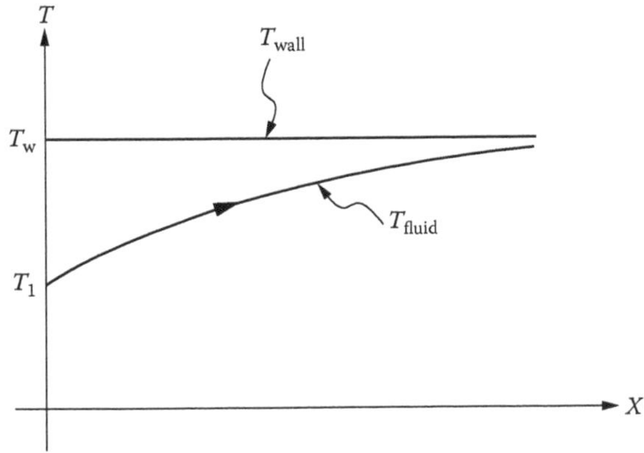

FIGURE 22.5
Temperature profiles for T_w = constant and $T_w > T_1$.

heat flux case is that we must now utilize Equation 22.1 for the heat flux in the form

$$\frac{\dot{Q}}{S} = h(T_w - T) \tag{22.5}$$

The combination of Equations 22.3b and 22.5 yields

$$\frac{dT}{dx} - \frac{4}{d}\frac{h(T_w - T)}{\rho\hat{V}c_p} = 0 \tag{22.6}$$

The solution to Equation 22.6 is made easier by defining

$$\theta = T_w - T$$

and because T_w is constant, we have

$$\frac{dT}{dx} = -\frac{d\theta}{dx}$$

With appropriate substitution, Equation 22.6 becomes

$$\frac{d\theta}{dx} + \frac{4}{d}\frac{h}{\rho\hat{V}c_p}\theta = 0 \tag{22.7}$$

For the boundary conditions

$$\theta(0) = \theta_1 \qquad \text{and} \qquad \theta(x) = \theta$$

Equation 22.7 can be solved by separation of the variables and integration:

$$\int_{\theta_1}^{\theta} \frac{d\theta}{\theta} + \frac{4}{d}\frac{h}{\rho\hat{V}c_p}\int_0^x dx = 0$$

$$\ln\frac{\theta}{\theta_1} + \frac{4}{d}\frac{h}{\rho\hat{V}c_p}x = 0$$

This yields the result

$$\frac{\theta}{\theta_1} = e^{-4hx/d\,\rho c_p\hat{V}} \tag{22.8}$$

or equivalently

$$\frac{T_w - T}{T_w - T_1} = e^{-4hx/d\,\rho c_p\hat{V}} \tag{22.9a}$$

The fluid temperature shown in Figure 22.5 is thus seen to approach the wall temperature in a decreasing exponential fashion.

At some length, $x = L$, where $T = T_2$, Equation 22.9a becomes

$$\frac{T_w - T_2}{T_w - T_1} = e^{-(4hL/d\,\rho c_p\hat{V})}$$

For the case of a liquid where c_p may be taken as c, and with $\dot{m} = \rho A\hat{V}$, the exponent becomes

$$\phi = -\frac{4hL}{d\,\rho c\hat{V}} = -\frac{4hLA}{d\,\rho Ac\hat{V}} = -\frac{4hLA}{d\,\dot{m}c}$$

But $A = \pi d^2/4$ so that $4A = \pi d^2$ and $S = \pi L d$. Hence,

$$\phi = -\frac{hS}{\dot{m}c}$$

and for the case of sized length, L, with a constant wall temperature, T_W, Equation 22.9a becomes

$$\frac{T_W - T_2}{T_W - T_1} = e^{-(hS/\dot{m}c)} \tag{22.9b}$$

Because the wall temperature was specified in this case, the quantity of heat transferred must be determined using a simple energy balance between the entrance and any downstream location. Thus,

$$\dot{Q} = \dot{m}c_p[T(x) - T_1] \tag{22.10a}$$

or

$$\dot{Q} = \dot{m}c_p\theta_1\left[1 - e^{-(4hx/d\,\rho c_p \hat{V})}\right] \tag{22.10b}$$

Example 22.2 For the same circular tube as in Example 22.1 with the lubricating oil entering at 0°C and flowing at a rate of 800 kg/h, the wall temperature is held constant at $T_w = 20$°C. Using the value of h given in Example 22.1 (68 W/m^2-K), determine (a) the exit temperature and (b) the total heat transfer.

Solution

Assumptions and Specifications

(1) The heat transfer process is in steady state.

(a) The exit temperature of the oil can be determined from Equation 22.9a

$$\frac{T_w - T}{T_w - T_1} = e^{-4hx/d\rho c\hat{V}} \tag{22.9a}$$

Because the mass flow rate is a given quantity, we can rewrite the right-hand side by substituting

$$\dot{m} = \rho A\hat{V} = \rho\hat{V}\frac{\pi d^2}{4}$$

and the expression becomes

$$\frac{T_w - T}{T_w - T_1} = e^{-\pi hd\,x/c\dot{m})}$$

With $T_w = 20$°C and $T_1 = 0$°C, the left-hand side becomes

$$\frac{T_w - T}{T_w - T_1} = \frac{20°\text{C} - T}{20°\text{C} - 0°\text{C}}$$

and for the right-hand side with $\dot{m} = 800/3600 = 0.222$ kg/s, we can evaluate the exponent.

$$-\frac{\pi d\,hx}{c\dot{m}} = -\frac{\pi(0.01\text{ m})(68\text{ W/m}^2\text{-K})(0.50\text{ m})}{(2.26\text{ J/kg-K})(0.222\text{ kg/s})} = -2.129$$

We can now solve for the exit temperature

$$\frac{20°\text{C} - T}{20°\text{C} - 0°\text{C}} = e^{-2.129} = 2.129$$

which yields

$$T = 20(1 - 0.119) = 20(0.881) = 17.6°\text{C} \Longleftarrow$$

(b) The energy transfer to the oil can be determined from an energy balance. Application of Equation 22.10a yields

$$\dot{Q} = \dot{m}c[T(x) - T_1]$$

$$= (0.222\,\text{kg/s})(2.26\,\text{J/kg-K})(17.6\,\text{K}) = 8.85\,\text{W} \Longleftarrow$$

The foregoing developments and examples have presumed a knowledge of the heat transfer coefficient, h. In the next section, we will discuss the means by which it is evaluated.

22.3 Convective Heat Transfer Coefficients

Equation 22.4 defines the heat transfer coefficient, h as the ratio of convective heat flux to the temperature difference, that is, the **driving force**, which produces an energy exchange between a surface and an adjacent fluid.

The **hydrodynamic boundary layer** considered in Chapter 16, is the region, close to the solid boundary, at which velocity gradients exist. When a temperature difference exists between a fluid and a surface, there is also a **thermal boundary layer** where, by definition, temperature gradients exist. It is within these boundary layers that the resistances to momentum and energy exchange occur.

One means of evaluating h is through an analysis of the hydrodynamic and thermal boundary layers using the fundamental conservation laws. The governing relationships that must be solved are higher-order nonlinear partial differential equations, which require considerable mathematical sophistication and/or massive numerical capability to achieve useful results. Such methods are beyond the scope of this text.

The equations that we will use for evaluating h are based upon empirical (experimental) results achieved by numerous researchers over the years. They have been thoroughly corroborated and are generally accepted within the technical community. Before presenting the working equations for h, we will examine the basis for using the dimensionless parameters that are involved. As discussed in Chapter 15, dimensionless analysis is a vital tool in analyzing and presenting experimental results.

Table 22.1 lists the variables that are important in evaluating h, the heat transfer coefficient. Also listed are the units of each variable as well as its dimensionless representation in terms of length, L, mass, M, time, T, temperature, Θ, and heat, Q. Note that two new fundamental dimensions, Θ and Q, have been added to those utilized previously.

Use of the Buckingham method introduced in Chapter 15 shows that the required number of dimensionless parameters is given, in this case, by

$$i = n - r = 8 - 4 = 4$$

The reader may verify that the rank of the dimensional matrix is 4.

TABLE 22.1

Fundamental Variables for Internal Forced Convection

Variable	Symbol	Units	Dimensions
Diameter	d	m	L
Length	L	m	L
Fluid density	ρ	kg/m^3	M/L^3
Fluid viscosity	μ	Pa-s	M/LT
Fluid specific heat	c_p	J/kg-K	$Q/M\Theta$
Fluid thermal conductivity	k	W/m-K	$Q/TL\Theta$
Velocity	$\hat{V}$	m/s	L/T
Heat transfer coefficient	h	W/m^2-K	$Q/TL^2\Theta$

We will choose $d,\ k,\ \mu$, and $\hat{V}$ as the core variables and, continuing with the Buckingham method we find that the four π-groups are

$$\pi_1 = d^a k^b \mu^c \hat{V}^d L$$

$$\pi_2 = d^a k^b \mu^c \hat{V}^d \rho$$

$$\pi_3 = d^a k^b \mu^c \hat{V}^d c_p$$

$$\pi_4 = d^a k^b \mu^c \hat{V}^d h$$

We see by inspection that

$$\pi_1 = \frac{L}{d} \tag{22.11a}$$

or in the preferred form, which is also clearly dimensionless,

$$\pi_1 = \frac{d}{L}$$

Writing π_2 in a dimensionless form gives us

$$\pi_2 = M^0 L^0 T^0 \Theta^0 Q^0 = L^a \left(\frac{Q}{TL\Theta}\right)^b \left(\frac{M}{LT}\right)^c \left(\frac{L}{T}\right)^d \frac{M}{L^3}$$

and equating exponents on $M,\ L,\ T,\ \Theta$, and Q yields

$$\begin{aligned} \mathrm{M}:&\quad 0 = c + 1 \\ \mathrm{L}:&\quad 0 = a - b - c + d - 3 \\ \mathrm{T}:&\quad 0 = -b - c - d \\ \Theta:&\quad 0 = -b \\ \mathrm{Q}:&\quad 0 = b \end{aligned}$$

The unknowns are easily evaluated as

$$a = 1 \qquad b = 0$$

$$c = -1 \qquad d = 1$$

and π_2 becomes

$$\pi_2 = \frac{d\hat{V}\rho}{\mu} \tag{22.11b}$$

which we recognize as the Reynolds number.

Solving for π_3 and π_4 in this manner yields

$$\pi_3 = \frac{\mu c_p}{k} \equiv \text{Pr} \qquad \text{(The Prandtl number)} \tag{22.11c}$$

and

$$\pi_4 = \frac{hd}{k} \equiv \text{Nu} \qquad \text{(The Nusselt number)} \tag{22.11d}$$

The last two groups are encountered for the first time. They are both fundamental in forced convection heat transfer.

The **Prandtl number**, Pr, expressed by Equation 22.11c, is seen to be a combination of fluid properties. Thus, it too can be considered to be a fluid property. Physically, it is the ratio of the diffusivities of momentum and heat, ν/α.

The **Nusselt number**, Nu, given by Equation 22.14, is the dimensionless dependent variable because it contains the heat transfer coefficient, h, which is the objective of our analysis.

In a dimensionless form we should expect to see correlations for forced convection in the form

$$\text{Nu} = f\left(\frac{d}{L}, \text{Re}, \text{Pr}\right) \tag{22.12}$$

As mentioned earlier, functional forms for the right-hand side of Equation 22.12 have been evaluated using experimental results.

We should note that the Nusselt number and the Reynolds number each include a length variable. Because the important length scale for internal flow is the diameter d, or the equivalent diameter if the conduit is not circular, we have currently written

$$\text{Nu} = \frac{hd}{k} \quad \text{and} \quad \text{Re} = \frac{d\hat{V}\rho}{\mu}$$

In subsequent chapters, the significant length variable may not be so obvious. To avoid confusion, we will use subscripts or Nu and Re to reflect the appropriate dimension on which they are based. Examples are

$$\text{Nu}_d = \frac{hd}{k}, \quad \text{Nu}_x = \frac{hx}{k}, \quad \text{and} \quad \text{Nu}_L = \frac{hL}{k}$$

and

$$\text{Re}_d = \frac{d\hat{V}\rho}{\mu}, \quad \text{Re}_x = \frac{x\hat{V}\rho}{\mu}, \quad \text{and} \quad \text{Re}_L = \frac{L\hat{V}\rho}{\mu}$$

For the present, it should be noted that for the region near the entrance, $0 < x < L_{\text{eT}}$, the heat transfer coefficient varies with axial location. For $x > L_{eT}$, the heat transfer coefficient is constant with a value that depends on the wall heating condition. For the case of constant wall heat flux, with $x > L_{\text{eT}}$

$$\text{Nu}_d = \frac{hd}{k} = 4.364 \tag{22.13}$$

and for the case of constant wall temperature with $x > L_{\text{eT}}$

$$\text{Nu}_d = \frac{hd}{k} = 3.658 \tag{22.14}$$

TABLE 22.2

Empirical Expressions for Internal Forced Convection

Flow Type	Re Range	Pr Range	L/d Range	Applicable Equation
Laminar	Re < 2300	Pr > 0.50	all	22.16
Transition	2300 < Re < 10^4	Pr > 0.50	all	22.17
Turbulent	Re > 10^4	0.70 < Pr < 160	$L/d > 10$	22.18

where for laminar flow

$$\frac{L_{eH}}{d} \approx 0.05\text{Re}_d \approx 0.05\frac{d\,\hat{V}\rho}{\mu}$$

and

$$\frac{L_{eT}}{d} \approx 0.05\text{Re}_d\text{Pr} \approx 0.05\frac{d\,\hat{V}\rho}{\mu}\text{Pr} \tag{22.15a}$$

For turbulent flow

$$L_{eH} = L_{eT} = 10d \tag{22.15b}$$

Table 22.2 lists the empirical expressions that can be used for evaluating Nu, or h, for various internal flow cases. The equations to be used for each case are provided in Sections 22.3.1 through 22.3.3.

22.3.1 Case 1—Laminar Flow

The widely used **Sieder-Tate equation** (1936) is

$$\text{Nu}_d = 1.86\text{Re}_d^{1/3}\text{Pr}^{1/3}\left(\frac{d}{L}\right)^{1/3}\left(\frac{\mu_b}{\mu_w}\right)^{0.14} \tag{22.16}$$

The ratio, μ_b/μ_w, involves the viscosity at the temperature of the **bulk fluid**, μ_b, and the viscosity at the wall temperature, μ_w. All other fluid properties are evaluated at the bulk fluid temperature.

The reader may recall that, for internal flow, the hydrodynamic entrance length is defined as the distance between the entrance to the downstream location where the flow becomes fully developed.

As noted earlier in the present chapter, a thermal boundary layer develops in internal flow when the fluid is heated or cooled by contact with the wall of the conduit. One should, thus, expect there to be a thermal entrance length analogous to the hydrodynamic case, and this is indeed true. The concept of a thermal entrance length is, however, not so obvious; a fluid will continue to be heated or cooled as it progresses through a conduit—even after the thermal entrance length has been exceeded.

Analytical solutions have been achieved for laminar flow cases. A classic solution exists for internal flow with the velocity profile fully developed (parabolic) when the wall temperature is suddenly raised and maintained constant for $x > L_{eT}$. This is the **Graetz problem** first solved in 1885.

Other variations include cases of both the velocity and temperature profiles developing simultaneously, with the wall condition being one of specified heat flux rather than temperature. The interested reader may consult Welty et al. (2008) or Kays and Crawford (1980) for a more detailed discussion.

As a practical matter, the Sieder-Tate equation, which is empirical, does a satisfactory job of predicting h for laminar internal flow, and its use is widely accepted within the heat transfer community.

22.3.2 Case 2—Transition Flow

The correlation of Hausen (1943) is recommended.

$$\mathrm{Nu}_d = 0.116\left[\mathrm{Re}_d^{2/3} - 125\right]\mathrm{Pr}^{1/3}\left(\frac{\mu_b}{\mu_w}\right)^{0.14}\left[1 + \left(\frac{d}{L}\right)^{2/3}\right] \tag{22.17}$$

Fluid properties are evaluated at the bulk fluid temperature in all cases except μ_w, which is evaluated at the wall temperature.

22.3.3 Case 3—Turbulent Flow

The Dittus-Boelter (1930) equation is

$$\mathrm{Nu}_d = 0.023\mathrm{Re}_d^{0.8}\mathrm{Pr}^n \tag{22.18}$$

where the exponent, n, is taken as

$$n = 0.40 \quad (T_w > T_b)$$

$$n = 0.30 \quad (T_w < T_b)$$

Fluid properties are evaluated at the bulk fluid temperature.

22.4 Applications of Internal Flow Forced Convection Correlations

Examples 22.3 through 22.5 illustrate the use of the capabilities developed in the preceding sections.

Example 22.3 Oil at 300°C is heated by steam condensing at 370°C on the outside of a steel pipe with an outside diameter (OD) of 2.67 cm and an inside diameter (ID) of 2.09 cm. The pipe is 2.5-m long and the mass flow rate through the pipe is 0.245 kg/s. Properties of the oil are as tabulated:

T, K	ρ, kg/m^3	c_p, J/kg-K	k, W/m-K	μ Pa-s
300	910	1.84×10^3	0.133	41.4×10^3
310	897	1.92×10^3	0.131	22.8×10^3
340	870	2.00×10^3	0.130	7.89×10^3
370	865	2.13×10^3	0.128	3.72×10^3

Determine (a) the temperature of the oil as it leaves the pipe and (b) the heat transferred.

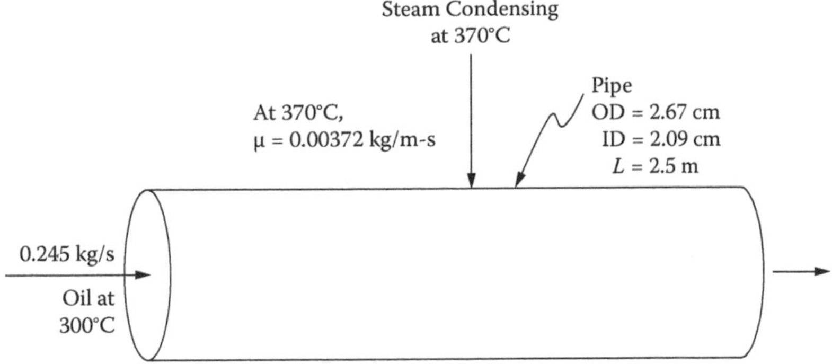

FIGURE 22.6
Configuration for Example 22.3.

Solution

Assumptions and Specifications

(1) Negligible temperature drop occurs across the tube wall so that we are dealing with a constant inside surface temperature of 370°C.

(2) The heat transfer process is in the steady state.

In this constant surface temperature situation Equation 22.9a applies and, because the fluid properties are temperature dependent and the exit temperature of the oil is unknown, a trial-and-error approach will be required. We will assume an exit temperature for the initial evaluation of fluid properties. This assumption will be checked and the solution repeated until sufficient agreement is achieved.

(a) The pipe is sketched in Figure 22.6. We will employ Equation 22.9a

$$\frac{T_w - T}{T_w - T_{\text{in}}} = e^{-4hx/d\,\rho c_p \hat{V}} \tag{22.9a}$$

We observe that the evaluation of the heat transfer coefficient, h, requires a knowledge of the flow regime which, in turn requires a knowledge of the Reynolds number. Since the mass flow rate, $\dot{m}$, is specified, the expression for Re is

$$\text{Re}_d = \frac{d\rho\hat{V}}{\mu} = \frac{\dot{m}d}{\mu A} = \frac{4\dot{m}d}{\mu\pi d^2} = \frac{4\dot{m}}{\pi d\mu}$$

Initially, an exiting fluid temperature of 310 K will be assumed and the corresponding average or bulk temperature of the oil is 305°C. At this temperature, the properties of the oil are

$$c = 1880\,\text{J/kg-K}, \quad \mu = 0.0321\,\text{Pa-s}, \quad \text{and} \quad k = 0.132\,\text{W/m-K}$$

and assuming that the pipe wall is at the temperature of the condensing steam, we have

$$\mu_w = 0.00372\,\text{Pa-s}$$

The Reynolds number is

$$\frac{d\,\hat{V}\rho}{\mu} = \frac{4\dot{m}}{\pi d\,\mu} = \frac{4(0.245\,\text{kg/s})}{\pi(0.0209\,\text{m})(0.0321\,\text{Pa-s})} = 465$$

This result indicates that the flow is laminar, which dictates that Equation 22.16 applies and we can write

$$\frac{hd}{k} = 1.86\text{Re}_d^{1/3}\text{Pr}^{1/3}\left(\frac{d}{L}\right)^{1/3}\left(\frac{\mu_b}{\mu_w}\right)^{0.14} \tag{22.16}$$

The Reynolds number has already been determined. The Prandtl number is evaluated as

$$\text{Pr} = \frac{\mu c}{k} = \frac{(0.0321\,\text{kg/m-s})(1880\,\text{J/kg-K})}{0.132\,\text{W/m-K}} = 457$$

Substituting numerical values into Equation 22.16 yields

$$h = \left(\frac{0.132\,\text{W/m-K}}{0.0209\,\text{m}}\right)(1.86)(465)^{1/3}(457)^{1/3}$$

$$\times\left(\frac{0.0209\,\text{m}}{2.5\,\text{m}}\right)^{1/3}\left(\frac{0.0321\,\text{kg/m-s}}{0.00372\,\text{kg/m-s}}\right)^{0.14}$$

or

$$h = 192.4\,\text{W/m}^2\text{-K}$$

With this result, we can solve Equation 22.9a, which may be modified with $\dot{m} = \rho A\hat{V}$ to the form

$$\frac{T_w - T}{T_w - T_1} = e^{-hL\pi d/\dot{m}c}$$

Inserting appropriate numerical values we obtain

$$\frac{hL\pi d}{\dot{m}c} = \frac{(192.4\,\text{W/m}^2\text{-K})(2.5\,\text{m})\pi(0.0209\,\text{m})}{(0.245\,\text{kg/s})(1880\,\text{J/kg-K})} = 0.0686$$

and, finally,

$$\frac{T - 370^\circ\text{C}}{300^\circ\text{C} - 370^\circ\text{C}} = e^{-0.0686} = 0.934$$

The temperature at the exit of the pipe is thus,

$$T = 370^\circ\text{C} - (0.934)(70^\circ\text{C}) = 370^\circ\text{C} - 65.4^\circ\text{C} = 304.6^\circ\text{C}$$

This result is lower than the assumed value of 310°C. At an adjusted value of the exit temperature of 304°C, we have a bulk temperature 302°C and adjusted properties of

$$c = 1858\,\text{J/kg-K}, \quad \mu = 0.0377\,\text{kg/m-s}, \quad \text{and} \quad k = 0.133\,\text{W/m-K}$$

The updated results are

$$\frac{d\,\hat{V}\rho}{\mu} = \frac{4\dot{m}}{\pi d\,\mu} = \frac{4(0.245\,\text{kg/s})}{\pi 0.0209\,\text{m})(0.0377\,\text{kg/m-s})} = 396$$

$$\text{Pr} = \frac{\mu c}{k} = \frac{(0.0377\,\text{kg/m-s})(1858\,\text{J/kg-K})}{0.133\,\text{W/m-K}} = 527$$

$$h = \left(\frac{0.133\,\text{W/m-K}}{0.0209\,\text{m}}\right)(1.86)(396)^{1/3}(527)^{1/3}$$

$$\times \left(\frac{0.0209\,\text{m}}{2.5\,\text{m}}\right)^{1/3}\left(\frac{0.0377\,\text{kg/m-s}}{0.00372\,\text{kg/m-s}}\right)^{0.14}$$

or

$$h = 197.1\,\text{W/m}^2\text{-K}$$

Then

$$\frac{hL\pi d}{\dot{m}c} = \frac{(197.1\,\text{W/m}^2\text{-K})(2.5\,\text{m})\pi(0.0209\,\text{m})}{(0.245\,\text{kg/s})(1858\,\text{J/kg-K})} = 0.0715$$

$$\frac{T - 370°\text{C}}{300°\text{C} - 370°\text{C}} = e^{-0.0715} = 0.931$$

and, the temperature at the exit of the pipe is

$$T = 370°\text{C} - (0.931)(70°\text{C}) = 370°\text{C} - 65.2°\text{C} = 304.8°\text{C} \Longleftarrow$$

This is sufficiently close to the initial assumption. Hence, another calculation is unnecessary.

(b) The heat received by the oil will be

$$\dot{Q} = \dot{m}c(T_{\text{exit}} - T_1)$$

$$= (0.245\,\text{kg/s})(1858\,\text{J/kg-K})(304.8°\text{C} - 300°\text{C})$$

$$= 2182\,\text{W} \Longleftarrow$$

Example 22.4 Hot water flows through a 1.6-cm-internal diameter pipe (Figure 22.7) at a velocity of 0.10 m/s. Its temperature at the pipe inlet is 80°C and the pipe wall temperature remains constant at 30°C. Determine (a) the temperature of the water at the exit from the pipe and (b) the quantity of heat lost in a waterline that is 1.55-m long.

Solution

Assumptions and Specifications

(1) The wall temperature is specified constant at 30°C.
(2) The heat transfer process is in the steady state.

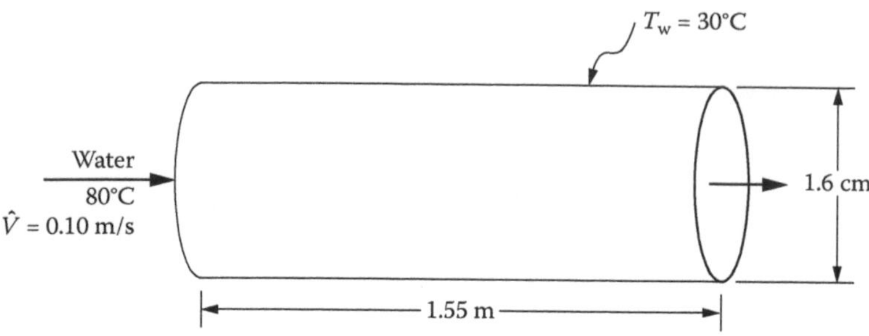

FIGURE 22.7
Configuration for Example 22.4.

Equation 22.9a applies in this case

$$\frac{T_w - T}{T_w - T_1} = e^{-4hx/d\,\rho c \hat{V}} \tag{22.9a}$$

A value for the heat transfer coefficient, h, is needed.

The first task is to evaluate the Reynolds number to determine which equation for h applies. To get a handle on Re, fluid properties at $T_b = 60°\text{C}$, which is consistent with an exit temperature of 40°C, will be used. Table A.14 in Appendix A shows that at 60°C (333 K), ν for water $= 0.483 \times 10^{-6}\,\text{m}^2/\text{s}$. The Reynolds number is thus determined as

$$\text{Re}_d = \frac{d\hat{V}}{\nu} = \frac{(0.016\,\text{m})(0.10\,\text{m/s})}{0.483 \times 10^{-6}\,\text{m}^2/\text{s}} = 3313$$

The water flow appears to be in the transition region that is Case 2 in Table 22.1. Equation 22.17 applies

$$\text{Nu}_d = 0.116[\text{Re}_d^{2/3} - 125]\text{Pr}^{1/3}\left(\frac{\mu_b}{\mu_w}\right)^{0.14}\left[1 + \left(\frac{d}{L}\right)^{2/3}\right] \tag{22.17}$$

For a first set of calculations, we will use $T_b = 60°\text{C}$. The properties of water obtained from Table A.14 are

$$\rho = 983.3\,\text{kg/m}^3 \qquad k = 0.654\,\text{W/m-K}$$

$$c = 4183\,\text{J/kg-K} \qquad \text{Pr} = 3.04$$

$$\mu = 475 \times 10^{-6}\,\text{Pa-s} \qquad \mu_w = 803 \times 10^{-6}\,\text{kg/m-s}$$

and after appropriate substitution, with

$$\frac{k}{d} = \frac{0.654\,\text{W/m-K}}{0.016\,\text{m}} = 40.88\,\text{W/m}^2\text{-K}$$

$$1 + \left(\frac{d}{L}\right)^{2/3} = 1 + \left(\frac{0.016\,\text{m}}{1.55\,\text{m}}\right)^{2/3} = 1 + 0.0474 = 1.0474$$

and

$$\left(\frac{\mu_b}{\mu_w}\right)^{0.14} = \left(\frac{475 \times 10^{-6}\,\text{kg/m-s}}{803 \times 10^{-6}\,\text{kg/m-s}}\right)^{0.14} = (0.592)^{0.14} = 0.929$$

we obtain

$$h = (40.88)(0.116)[(3313)^{2/3} - 125](3.04)^{1/3}(0.929)(1.0474)$$
$$= 650 \text{ W/m}^2\text{-K}$$

The argument of the exponential term in Equation 22.9a is evaluated as

$$\frac{4}{d}\frac{hL}{\rho c_p \hat{V}} = \frac{4}{0.016\,\text{m}}\frac{(650\,\text{W/m}^2\text{-K})(1.55\,\text{m})}{(983.3\,\text{kg/m}^3)(4183\,\text{J/kg-K})(0.10\,\text{kg/s})} = 0.612$$

which yields

$$\frac{T_2 - 30°\text{C}}{80°\text{C} - 30°\text{C}} = e^{-0.612} = 0.542$$

and

$$T_2 = 30°\text{C} + (0.542)(50°\text{C}) = 30°\text{C} + 27.1°\text{C} = 57.1°\text{C}$$

This value differs significantly from the assumed value of 40°C and we must redo the calculations.

With the exit temperature next assumed to be 56°C with $T_b = 68°\text{C}$, the fluid properties obtained from Table A.14 (with interpolation) are

$$\rho = 978.9\,\text{kg/m}^3 \qquad k = 0.662\,\text{W/m-K}$$
$$c_p = 4186\,\text{J/kg-K} \qquad \text{Pr} = 2.67$$
$$\mu = 421 \times 10^{-6}\,\text{Pa-s} \qquad \mu_w = 803 \times 10^{-6}\,\text{kg/m-s}$$

The Reynolds number will be

$$\text{Re}_d = \frac{d\rho\hat{V}}{\mu} = \frac{(0.016\,\text{m})(978.9\,\text{kg/m}^3)(0.10\,\text{m/s})}{421 \times 10^{-6}\,\text{Pa-s}} = 3720$$

After appropriate substitution, with

$$\frac{k}{d} = \frac{0.662\,\text{W/m-K}}{0.016\,\text{m}} = 41.38\,\text{W/m}^2\text{-K}$$

and

$$\left(\frac{\mu_b}{\mu_w}\right)^{0.14} = \left(\frac{421 \times 10^{-6}\,\text{kg/m-s}}{825 \times 10^{-6}\,\text{kg/m-s}}\right)^{0.14} = (0.510)^{0.14} = 0.910$$

we obtain

$$h = (41.38)(0.116)[(3720)^{2/3} - 125](2.67)^{1/3}(0.910)(1.0474)$$
$$= 730.4 \text{ W/m}^2\text{-K}$$

Then, with the exponential argument determined as

$$\frac{4}{d}\frac{h}{\rho c_p \hat{V}}L = \frac{4}{0.016\,\text{m}}\frac{(730.4\,\text{W/m}^2\text{-K})(1.55\,\text{m})}{(978.4\,\text{kg/m}^3)(4186\,\text{J/kg-K})(0.10\,\text{kg/s})} = 0.691$$

we have

$$\frac{T_2 - 30°\text{C}}{80°\text{C} - 30°\text{C}} = e^{-0.691} = 0.501$$

and

$$T_2 = 30°C + (0.501)(50°C) = 30°C + 25°C = 55°C \Longleftarrow$$

This value is sufficiently close to our last assumed value that we may consider it to be our desired result.

(b) With $\dot{m} = \rho A V$, the heat lost by the water will be

$$\dot{Q} = \rho A \hat{V} c_p (T_2 - T_1)$$

and with $T_2 - T_1 = 80°C - 55°C = 25°C$

$$\dot{Q} = (978.9\,\text{kg/m}^3)\left[\frac{\pi(0.016\,\text{m})^2}{4}\right](0.10\,\text{m/s})(4186\,\text{J/kg-K})(25°C)$$

$$= 2060\,\text{W} \Longleftarrow$$

Example 22.5 Air at atmospheric pressure and 10°C enters a 10-m-long rectangular duct (Figure 22.8) whose walls are maintained at 70°C by solar irradiation. The duct is 15-cm wide by 7.5-cm high. Determine the flow rate of the air if its exit temperature is to be 50°C.

Solution

Assumptions and Specifications

(1) The wall temperature is specified constant at 70°C.
(2) The air inlet and outlet temperatures are specified as 10°C and 50°C, respectively.
(3) The duct is specified as rectangular.
(4) The heat transfer process is in the steady state.

Initially, we will assume turbulent flow and this assumption will be validated as we proceed. The situation, as shown, is one of internal flow with a constant wall temperature. Equation 22.9a

$$\frac{T_w - T}{T_w - T_1} = e^{-4hx/d\,\rho c_p \hat{V}} \tag{22.9a}$$

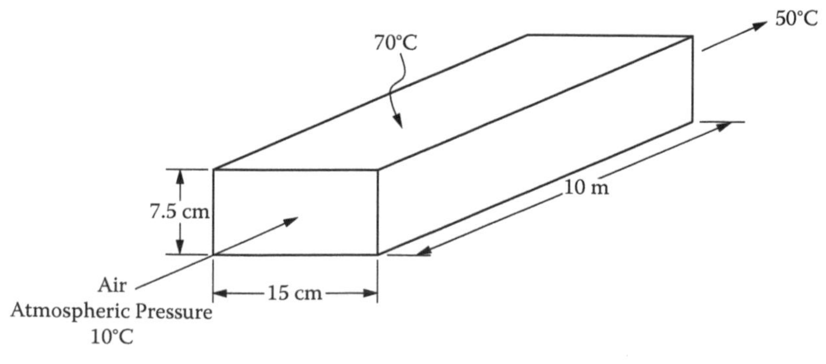

FIGURE 22.8
Configuration for Example 22.5.

is therefore applicable and we can evaluate the left-hand side by inserting the specified temperature values

$$\frac{T_w - T_2}{T_w - T_1} = \frac{70°\text{C} - 50°\text{C}}{70°\text{C} - 10°\text{C}} = \frac{20°\text{C}}{60°\text{C}} = 0.333$$

With this value, Equation 22.9a yields

$$e^{-4hx/d\,\rho c_p \hat{V}} = 0.333$$

which can be simplified to

$$-\frac{4}{d}\frac{hx}{\rho c_p \hat{V}} = \ln 0.333 = -1.099$$

or

$$h = 1.099\frac{d\rho c_p \hat{V}}{4x} \tag{a}$$

We note that a diameter is involved in this expression even though the duct has been specified as having a rectangular cross section. This is handled by employing the **equivalent diameter** that was introduced in Chapter 15. For this problem, the equivalent diameter is

$$d_e = \frac{4A}{P} = \frac{4(0.15\,\text{m})(0.075\,\text{m})}{2(0.15\,\text{m} + 0.075\,\text{m})} = 0.10\,\text{m}$$

Because turbulent flow was assumed, we choose Equation 22.18 (the Dittus-Boelter equation) as the correlating equation. For $T_w > T_b$, $n = 0.40$

$$\text{Nu}_d = \frac{hd}{k} = 0.023\text{Re}_d^{0.8}\text{Pr}^{0.4} = 0.023\left(\frac{d_e\hat{V}\rho}{\mu}\right)^{0.8}\text{Pr}^{0.4}$$

The properties, k, ρ, μ, and Pr are temperature dependent. At the average bulk temperature of 30°C (303 K), the property values, obtained from Table A.13 (with interpolation), are

$$k = 0.0264\,\text{W/m-K} \qquad \rho = 1.167\,\text{kg/m}^3$$
$$c_p = 1005.9\,\text{J/kg-K} \qquad \text{Pr} = 0.707$$
$$\mu = 1.86 \times 10^{-5}\,\text{Pa-s}$$

With appropriate substitution we obtain

$$h = 0.023\frac{k}{d_e}\left(\frac{d_e\hat{V}\rho}{\mu}\right)^{0.8}\text{Pr}^{0.4}$$

$$= 0.023\left(\frac{0.0264\,\text{W/m-K}}{0.10\,\text{m}}\right)\left[\frac{(0.10\,\text{m})\hat{V}(1.167\,\text{kg/m}^3)}{1.86 \times 10^{-5}\,\text{kg/m-s}}\right]^{0.8}(0.707)^{0.4}$$

$$= 5.77\hat{V}^{0.8}\,\text{W/m}^2\text{-K}$$

and from Equation (a), we have

$$h = 1.099\frac{d_e\rho c_p\hat{V}}{4x}$$

$$= 1.099\frac{(0.10\,\text{m})(1.167\,\text{kg/m}^3)(1005.9\,\text{J/kg-K})\hat{V}}{4(10\,\text{m})}$$

$$= 3.225\hat{V}\,\text{W/m}^2\text{-K}$$

The two expressions for h may be combined

$$5.77\hat{V}^{0.80} = 3.225\hat{V}$$

$$\hat{V}^{0.20} = 1.790$$

$$\hat{V} = 18.38\,\text{m/s}$$

and a check of the Reynolds number provides

$$\text{Re} = \frac{\rho\hat{V}d_e}{\mu} = \frac{(1.167\,\text{kg/m}^3)(18.38\,\text{m/s})(0.10\,\text{m})}{1.86\times10^{-5}\,\text{kg/m-s}} = 115{,}320$$

This indicates that the flow is well into the turbulent range. Hence, our preliminary assumption of turbulent flow was correct and the use of Equation 22.18 was appropriate.

The problem may now be completed by determining $\dot{m}$

$$\dot{m} = \rho A\hat{V} = (1.167\,\text{kg/m}^3)(0.15\,\text{m})(0.075\,\text{m})(18.38\,\text{kg/s}) = 0.241\,\text{kg/s} \Longleftarrow$$

22.5 Design Example 13

A copper tube of length, L, is to be welded to the underside of a flat plate (Figure 22.9) solar collector. The plate absorbs heat such that a uniform temperature of 85°C is maintained. The tube has an inner diameter of 1.25 cm and is to be designed so that it will heat $1.262\times10^{-5}\,\text{m}^3/\text{s}$ of water from 15°C to 75°C. Assuming laminar flow in the tube, determine the necessary length of the tube to accomplish the desired temperature rise of the water by comparing the results using two approaches: (a) assuming the fully developed value for the Nusselt number over the entire tube length and (b) using the Sieder-Tate correlation (Equation 22.16) to determine the Nusselt number.

Solution

Assumptions and Specifications

(1) Steady-state conditions exist.
(2) There is no temperature drop across the tube wall.
(3) The effect of bends in the tube is negligible.

The first step is to confirm that the flow in the tube is indeed laminar. The volumetric flow rate is

$$\dot{V} = 1.262\times10^{-5}\,\text{m}^3/\text{s}$$

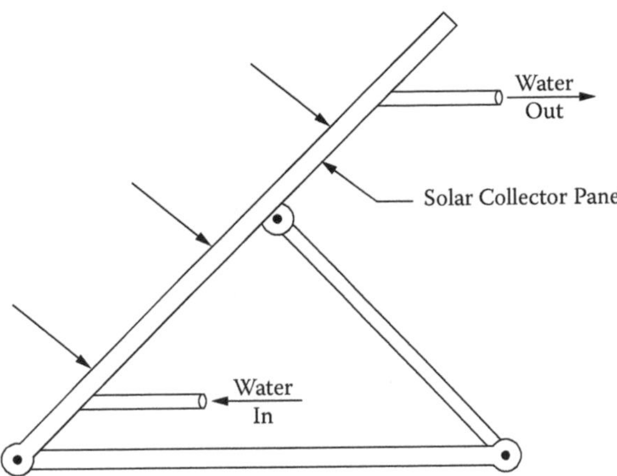

FIGURE 22.9
Configuration for Design Example 13.

and with

$$A = \frac{\pi}{4}d^2 = \frac{\pi}{4}(0.0125\,\text{m})^2 = 1.2272 \times 10^{-4}\,\text{m}^2$$

we have

$$\hat{V} = \frac{\dot{V}}{A} = \frac{1.262 \times 10^{-5}\,\text{m}^3/\text{s}}{1.2272 \times 10^{-4}\,\text{m}^2} = 0.1028\,\text{m/s}$$

With

$$T_b = \frac{T_1 + T_2}{2} = \frac{15°\text{C} + 75°\text{C}}{2} = 45°\text{C}$$

Table A.14 gives (with interpolation)

$$\rho = 990.3\,\text{kg/m}^3 \qquad \text{Pr} = 3.905$$
$$\mu = 0.596 \times 10^{-3}\,\text{kg/m-s} \qquad k = 0.6385\,\text{W/m-K}$$
$$c = 4.177\,\text{kJ/kg-K}$$

Thus, the Reynolds number is

$$\frac{d\,\hat{V}\rho}{\mu} = \frac{\rho\hat{V}d}{\mu} = \frac{(990.3\,\text{kg/m}^3)(0.1028\,\text{m/s})(0.0125\,\text{m})}{0.596 \times 10^{-3}\,\text{kg/m-s}} = 2135$$

This shows that the flow in the tubing is indeed laminar. Moreover,

$$\dot{m} = \rho A\hat{V} = (990.3\,\text{kg/m}^3)(1.2272 \times 10^{-4}\,\text{m}^2)(0.1028\,\text{m/s}) = 0.0125\,\text{kg/s}$$

Then,

$$e^{-\phi} = \frac{T_w - T_2}{T_w - T_1} = \frac{85°\text{C} - 75°\text{C}}{85°\text{C} - 15°\text{C}} = 0.1429$$

so that

$$-\phi = \frac{hS}{\dot{m}c} = \ln 0.1429 = -1.9456$$

or

$$\phi = 1.9456$$

This makes the hS product

$$hS = (1.9456)\,\dot{m}c = (1.9456)(0.0125\,\text{kg/s})(4177\,\text{J/kg-K}) = 101.6\,\text{W/K}$$

This hS product forms the underpinning for both parts (a) and (b) which now follow.
(a) Equation 22.14 gives the fully developed value for h/dk

$$\frac{hd}{k} = 3.658$$

Then

$$h = \frac{k}{d}\frac{hd}{k} = \left(\frac{0.6385\,\text{W/m-K}}{0.0125\,\text{m}}\right)3.658 = 186.9\,\text{W/m}^2\text{-K}$$

and to achieve an hS product of 101.6 W/K, we have

$$hS = 101.6\,\text{W/K} = \pi hd\,L = \pi(186.9\,\text{W/m}^2\text{-K})(0.0125\,\text{m})L$$

or

$$L = \frac{101.6\,\text{W/K}}{\pi(181.6\,\text{W/m}^2\text{-K})(0.0125\,\text{m})} = 14.25\,\text{m} \Longleftarrow$$

(b) The heat transfer coefficient via the Sieder-Tate correlation may be obtained from Equation 22.16.

$$h = \frac{k}{d}\frac{hd}{k} = 1.86\left(\frac{k}{d}\right)\left[\frac{d\hat{V}\rho}{\mu}\text{Pr}\left(\frac{d}{L}\right)\right]^{1/3}\left(\frac{\mu_b}{\mu_w}\right)^{0.14}$$

and with $S = \pi dL$, the hS product will be

$$hS = 101.6\,\text{W/K} = 1.86\pi\left(\frac{k}{d}\right)\left[\frac{d\hat{V}\rho}{\mu}\text{Pr}\left(\frac{d}{L}\right)\right]^{1/3}\left(\frac{\mu_b}{\mu_w}\right)^{0.14} dL$$

$$= 1.86\pi k\left(\frac{d\hat{V}\rho}{\mu}\text{Pr}\,d\right)^{1/3}\left(\frac{\mu_b}{\mu_w}\right)^{0.14} L^{2/3}$$

The length required to yield $hS = 101.6$ W/K will be

$$L = \left[\frac{101.6\,\text{W/K}}{1.86\pi k(\rho\hat{V}d/\mu)\text{Pr}\,d^{1/3}(\mu/\mu_w)^{0.14}}\right]^{3/2}$$

At $T_w = 85°\text{C}$, $\mu_w = 0.0.3385 \times 10^{-3}$ kg/m-s. Hence, with

$$\frac{d\hat{V}\rho}{\mu}\text{Pr}\,d = (2135)(3.905)(0.0125\,\text{m}) = 104.21\,\text{m}$$

and

$$\frac{\mu_b}{\mu_w} = \frac{0.596 \times 10^{-3}\,\text{kg/m-s}}{0.3385 \times 10^{-3}\,\text{kg/m-s}} = 1.7607$$

we obtain

$$L = \left[\frac{101.6}{1.86\pi(0.6385\,\text{W/m-K})(104.21\,\text{m})^{1/3}(1.7607)^{0.14}} \right]^{3/2} = 12.36\,\text{m} \Longleftarrow$$

The results of parts (a) and (b) indicate that entrance lengths are significant for this case. Use of the fully developed value for Nu gives a result for L, which is approximately 10% larger than the value obtained using the Sieder-Tate correlation. With an eye toward being conservative, we will choose the length of tubing to be

$$L = 14.25\,\text{m} \Longleftarrow$$

and we observe that, for fluids with appreciably larger values of Pr, the error in using the fully developed limit can be quite large (in excess of 200% in some cases).

We turn now to a design example involving a high-heat dissipating electronic component.

The traveling wave tube is an electron tube in which a stream of electrons interacts continuously (or repeatedly) with a guided electromagnetic wave moving substantially with it. The interaction occurs in such a way that there is a net transfer of energy from the stream of electrons to the guided electromagnetic wave. The tube is used as an amplifier or oscillator at frequencies in the microwave region.

An artist's conception of the anode (collector) end of a traveling wave tube is shown in Figure 22.10. Notice that the entire configuration is very small and, because the heat dissipation is substantial, the heat release in a traveling wave tube collector can approach that of a steam generator in a modern steam power plant.

The electrons are "collected" in the small cavity at the center of the configuration and, because of the high-heat release, the collector must be liquid cooled.

Several liquids are available and they all meet the requirements of high-dielectric strength, chemical inertness, thermal stability, effects of moisture, pour and flash points, flammability, toxicity, surface tension, and vapor pressure. Water is not suitable as a liquid coolant because of its low-dielectric strength.

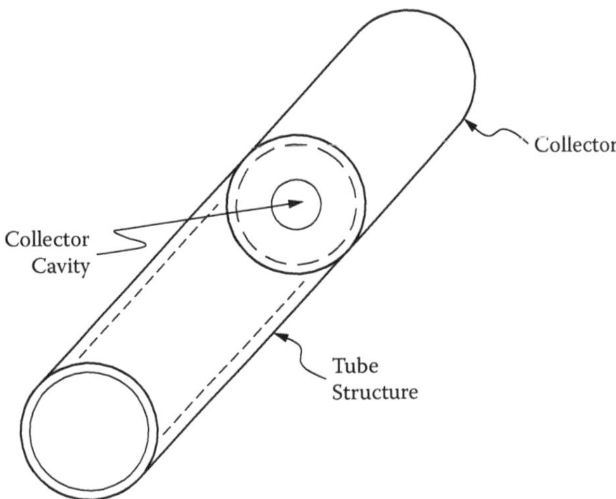

FIGURE 22.10
Artist's conception of the collector end of a traveling wave tube.

TABLE 5.1

Performance of a Liquid-Cooled System Dissipating 1500 W

Assume $\dot{m}$	kg/s	0.100	0.125	0.150	0.175	0.200
T_1	°C	77.0	77.0	77.0	77.0	77.0
Assume T_2	°C	84.6	83.0	81.8	81.0	80.6
T_b	°C	80.8	80.0	79.4	79.0	78.8
c at T_b	J/kg-K	2093	2090	2088	2087	2086
$\Delta T = \dot{Q}/\dot{m}c$	°C	7.2	5.7	4.8	4.1	3.6
T_2	°C	84.2	82.7	81.8	81.1	80.6
T_2 OK?		no	no	Yes	Yes	Yes
T_b	°C	80.6	79.9			
c at T_b	J/kg-K	2092	2089			
$\Delta T = \dot{Q}/\dot{m}c$	°C	7.2	5.7			
T_2	°C	84.2	82.7			
T_2 OK?		Yes	Yes			

The determination of the anode (collector) temperature using a liquid coolant requires an input from at least two of the disciplines treated in this book:

1. It begins with a first law consideration (Chapter 5) in which the range of liquid flows to accommodate the specified heat dissipation are established. This was performed as Design Example 1 and the results were summarized in Table 5.1, which is repeated here for the reader's convenience.
2. The surface temperature of the collector wall is then obtained using procedures found here in Chapter 22.

22.6 Design Example 14

A traveling wave tube collector is to be cooled in an application where the *uniform* dissipation in the collector is 1.5 kW (1500 W). Coolanol-45 (Monsanto Chemical Co.), which enters the collector at 77°C, is to be used. The collector material is copper ($k = 385$ W/m-K) and in order to prevent loss of vacuum due to "outgassing" of the copper material, the walls of the collector must be held to a maximum temperature of 150°C.

The details of the collector configuration are indicated in Figure 22.11 and the properties of Coolanol-45 are provided in Table 22.4. Because it is deemed risky to interpolate the dynamic viscosity data, a plot of dynamic viscosity as a function of temperature is provided in Figure 22.12.

Notice that 16 fins will be employed and the 1500 W may be considered as flowing radially outward from the collector cavity, with a diameter of 0.625 cm, to the coolant passage.

There are four fins per pass in this four-pass collector and relevant dimensions include:

$d_i = 2.54$ cm — $d_o = 5.715$ cm

fin height, $b = 1.875$ cm — fin thickness, $\delta = 2.381$ mm

$L = 8.255$ cm — thermal conductivity, $k = 385$ W/m-K

number of fins $= 16$ — fins per pass $= 4$

We will need to determine several pertinent parameters. These are the passage flow area, the passage wetted perimeter, the passage equivalent diameter, the finned surface area, and the base surface area.

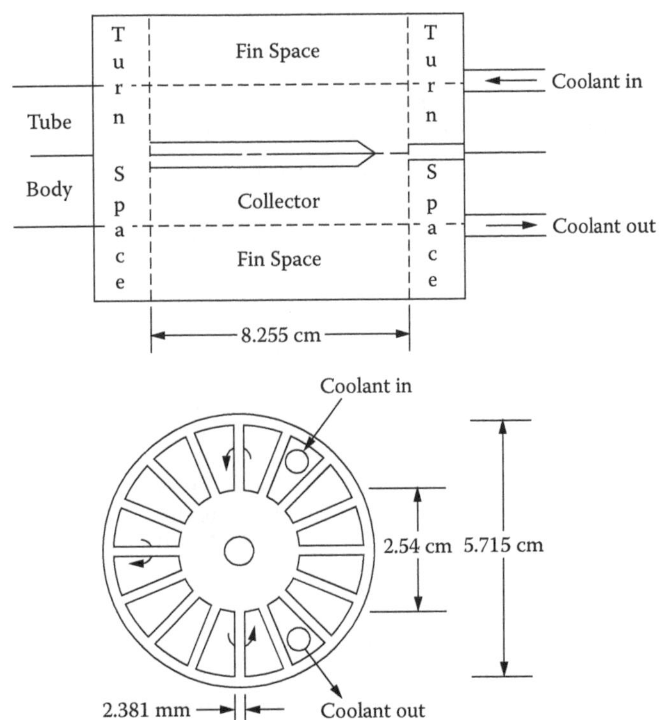

16 Fins 2.381-mm Thick–1.5875-cm High per Pass
Fins are Copper
4 Passes

FIGURE 22.11
Details of the collector of a traveling wave tube.

Strategy

1. A modification of Equation 20.21 will yield the maximum temperature allowed at the inner surface of the coolant passage.
2. The procedure is then straightforward

- For each flow rate, we determine the pertinent fluid properties, the Reynolds number, and the heat transfer coefficient. Because the flow in each case is laminar, Equation 22.16 is employed.
- After the heat transfer coefficient, h, is obtained, the fin performance parameter, m, is determined.

$$m = \left(\frac{2h}{k\delta}\right)^{1/2}$$

Then, with this value of m, the fin efficiency is obtained from Equation 20.67

$$\eta = \frac{\tanh mb}{mb}$$

TABLE 22.3

Summary of Calculations for the Surface Temperature of the Finned Passage in Figure 22.11: In All Cases, $T_1 = 77°C$

Assume $\dot{m}$	kg/s	0.100	0.125	0.150	0.175	0.200
T_2	°C	84.2	82.7	81.8	81.1	80.6
T_b	°C	80.6	79.9	79.4	79.0	78.8
c at T_b	J/kg-K	2092	2089	2088	2087	2086
$\mu \times 10^{-3}$ at T_b	kg/m-s	4.106	4.144	4.196	4.237	4.258
Pr at T_b		63.2	63.6	64.4	65.0	65.3
ρ at T_b	kg/m^3	858.1	858.7	858.9	859.2	859.3
$\mu \times 10^{-3}$ at 140°C	kg/m-s	2.0126	2.0126	2.1026	2.1026	2.1026
$\hat{V}$	m/s	0.3207	0.4005	0.4805	0.5604	0.6404
$\mathrm{Re}_d = \rho \hat{V} d_e / \mu$		563.8	698.7	827.4	956.0	1087.2
μ/μ_w		2.0267	2.0454	2.0711	2.0913	2.1017
$(\mu/\mu_w)^{0.14}$		1.1040	1.1054	1.1073	1.1088	1.1096
d_e/L		0.1019	0.1019	0.1019	0.1019	0.1019
$\mathrm{Re}_d \mathrm{Pr} d_e/L$		3631.2	4528.1	5430.0	6332.2	7234.6
$(\mathrm{Re}_d \mathrm{Pr} d_e/L)^{1/3}$		15.37	16.54	17.58	18.50	19.34
Nu_d		31.56	34.02	36.21	38.16	39.92
h	W/m^2-K	510.2	550.0	585.4	616.9	645.4
m^2	m^{-2}	1113.2	1200.0	1277.2	1346.0	1408.1
m	m^{-1}	33.36	34.64	35.74	36.69	37.53
mb		0.5297	0.5499	0.5673	0.5824	0.5957
$\tanh mb$		0.4841	0.5005	0.5134	0.5244	0.5340
η_f		0.916	0.910	0.905	0.900	0.896
S_b	m^2	0.00344	0.00344	0.00344	0.00344	0.00344
$\eta_f S_f$	m^2	0.03840	0.03816	0.03795	0.03775	0.03759
$S = S_b + \eta_f S_f$	m^2	0.04184	0.04160	0.04139	0.04129	0.04103
$\phi = hS/\dot{m}c$		0.1020	0.0876	0.0774	0.0697	0.0635
$F = e^{\phi}$		1.1074	1.0916	1.0805	1.0722	1.0656
FT_2	°C	93.2	90.3	88.4	87.0	85.9
$FT_2 - T_1$	°C	16.2	13.3	11.4	10.0	8.9
$F - 1$		0.1074	0.0916	0.0805	0.0722	0.0656
T_w	°C	151.2	145.0	141.4	137.9	135.5

TABLE 22.4

Properties of Coolanol-45: Data Extracted from Monsanto Bulletin AV-3 and Converted to SI Units

T °C	k W/m-K	$\mu \times 10^{-3}$ kg/m-s	$\nu \times 10^{-5}$ m^2/s	ρ kg/m^3	c kJ/kg-K	Pr
40	0.132	9.721	1.086	894.9	1.954	143.9
50	0.133	8.185	0.932	877.8	1.988	122.3
60	0.135	6.201	0.712	871.4	2.022	92.9
70	0.136	5.167	0.597	865.8	2.056	78.1
80	0.136	4.134	0.482	858.5	2.090	63.5
90	o.135	3.669	0.431	852.1	2.124	57.7
100	0.133	3.101	0.367	845.7	2.158	50.3
110	0.132	2.687	0.320	839.3	2.192	44.6
120	0.130	2.356	0.283	832.8	2.226	40.3
130	0.127	2.191	0.265	826.4	2.260	39.0
140	0.125	2.026	0.247	820.1	2.294	37.2

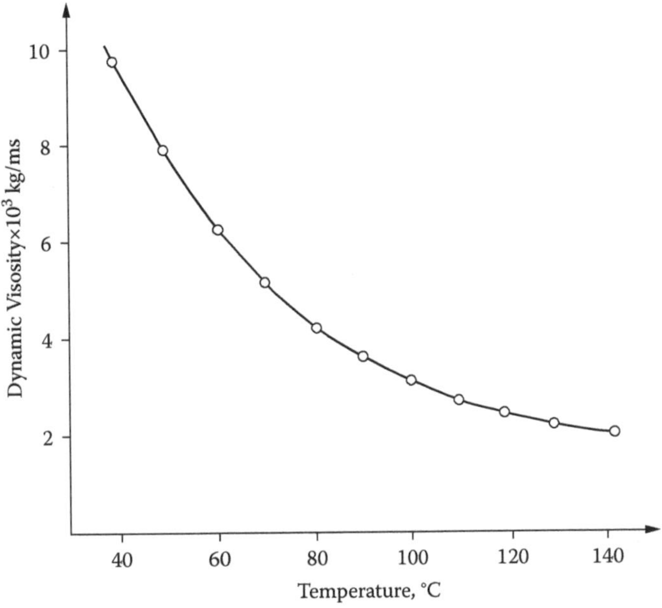

FIGURE 22.12
Dynamic viscosity of Coolanol-45 as a function of temperature.

- The total surface will be

$$S = S_b + \eta S_f$$

and then, in each case, using Equation 22.9b, we form the parameter, F, defined as

$$F = e^{-hS/\dot{m}c}$$

The temperature at the inner surface of the annular region is then found from a modification of Equation 22.9b

$$T_w = \frac{F T_2 - T_1}{F - 1}$$

3. The last step is the subtraction of the temperature drop through the copper collector from the collector cavity to the inside surface of the coolant passage.
4. The results for the five flow rates may be plotted to yield the solution.

Now that the strategy has been outlined, the first step is the determination of the temperature allowed at the inside surface of the coolant passage. The maximum temperature permitted in the collector is 150°C and it will exist at the walls of the collector cavity. A modification of Equation (20.30) will provide the maximum temperature allowed at the inner surface of the coolant passage. With d_c and T_c used to respectively designate the diameter and temperature of the cavity

$$T_c - T_o = \frac{\dot{Q}\ln d_o/d_c}{2\pi k L}$$

$$T_o = T_c - \frac{\dot{Q}\ln d_o/d_c}{2\pi k L}$$

$$= 150°\text{C} - \frac{(1500\,\text{W})(\ln 2.54\,\text{cm}/0.625\,\text{cm})}{2\pi(385\,\text{W/m-K})(0.08255\,\text{m})}$$

$$= 150°\text{C} - 10.4°\text{C} = 139.6°\text{C}$$

Supporting data for Table 22.3 include:
Flow Area

$$A = \frac{1}{4}\left[\frac{\pi}{4}(d_o^2 - d_i^2) - n_f b\delta\right]$$

$$= \frac{1}{4}\left\{\frac{\pi}{4}\left[(5.715\,\text{cm})^2 - (2.54\,\text{cm})^2\right] - (16\,\text{fins})(1.5875\,\text{cm})(0.2381\,\text{cm})\right\}$$

$$= \frac{1}{4}(20.585\,\text{cm}^2 - 6.048\,\text{cm}^2) = 3.6343\,\text{cm}^2$$

or

$$A = 3.6343 \times 10^{-4}\,\text{m}^2$$

Wetted Perimeter

$$P = \frac{1}{4}\left[\pi(d_o + d_i) - 2n_f\delta + 2n_f b\right]$$

$$= \frac{1}{4}\,[\pi(5.715\,\text{cm} + 2.540\,\text{cm}) - 2(16\,\text{fins})(0.2381\,\text{cm})$$

$$+2(16\,\text{fins})(1.5875\,\text{cm})]$$

$$= \frac{1}{4}(25.9338\,\text{cm} - 7.6192\,\text{cm} + 50.800\,\text{cm}) = 17.2787\,\text{cm}$$

or

$$P = 0.1728\,\text{m}$$

The Passage Equivalent Diameter

$$d_e = \frac{4A}{P} = \frac{4(3.6343 \times 10^{-4}\,\text{m}^2)}{0.1728\,\text{m}} = 8.4127 \times 10^{-3}\,\text{m}$$

The Surface Area of the Sixteen Fins

$$S_f = 2Ln_f b = 2(8.255\,\text{cm})(16\,\text{fins})(1.5875\,\text{cm}) = 419.34\,\text{cm}^2$$

or

$$S_f = 0.04193\,\text{m}^2$$

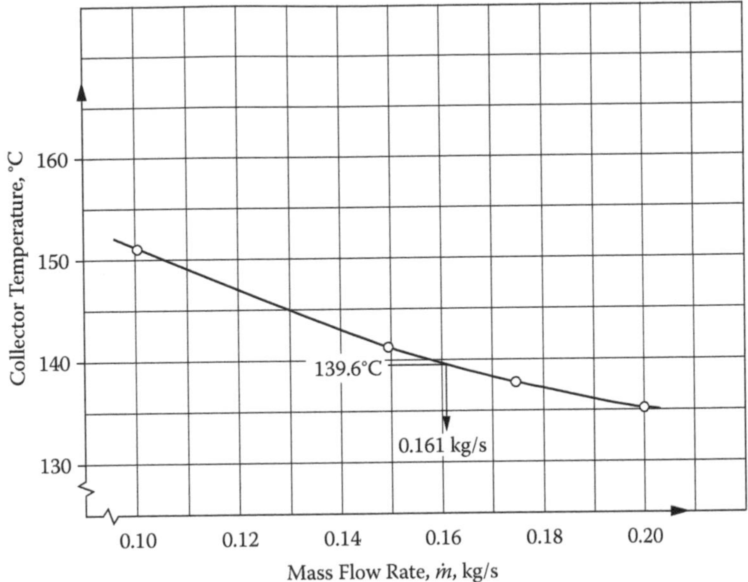

FIGURE 22.13
Collector temperature as a function of Coolanol-45 flow rate.

The Surface Area of the Inner Portion of the Coolant Passage

$$S_b = (\pi d_i - n_f \delta)L = [\pi(2.54\,\text{cm}) - (16\,\text{fins})(0.2381\,\text{cm})](8.255\,\text{cm})$$

$$= (7.9796\,\text{cm} - 3.8096\,\text{cm}))(8.255\,\text{cm}) = 34.4237\,\text{cm}^2$$

The actual calaculations are displayed in Table 22.3 and a plot of the collector surface temperature as a function of the Coolanol-45 mass flow rate is shown in Figure 22.13. From this plot, we are able to deduce that the mass flow rate of the Coolanol-45 will be approximately 0.161 kg/s.

22.7 Summary

This chapter has dealt with convective heat transfer when the fluid flow is confined within closed conduits such as tubes, pipes, or ducts.

A first law of thermodynamics application to internal flow provided the characteristic behavior of the fluid in terms of its axial temperature variation and total heat exchange for wall conditions of both constant wall temperature and constant heat input.

Dimensional analysis generated the parameters that apply to internal forced convection. These parameters are

- The Reynolds number, $\text{Re} = d\hat{V}\rho/\mu$
- The Prandtl number, $\text{Pr} = \mu c_p/k$
- The Nusselt number, $\text{Nu} = hd/k$

Empirical equations were introduced and their applications were demonstrated. The expressions that apply are

- For laminar flow (Re < 2300)

$$\frac{hd}{k} = 1.86\text{Re}_d^{1/3}\text{Pr}^{1/3}\left(\frac{d}{L}\right)^{1/3}\left(\frac{\mu}{\mu_w}\right)^{0.14} \tag{22.16}$$

- For transition flow (2300 < Re < 10^4)

$$\frac{hd}{k} = 0.116[\text{Re}_d^{2/3} - 125]\text{Pr}^{1/3}\left[1 + \left(\frac{d}{L}\right)^{2/3}\right]\left(\frac{\mu}{\mu_w}\right)^{0.14} \tag{22.17}$$

- For turbulent flow (Re > 10^4)

$$\frac{hd}{k} = 0.023\text{Re}_d^{0.80}\text{Pr}^n \tag{22.18b}$$

where $n = 0.4$ for $T_w > T_b$ and $n = 0.3$ for $T_w < T_b$.

22.8 Problems

Flow in Circular Tubes and Pipes

22.1: Lubricating oil (approximately SEA 50) flows in a 4-m-long pipe having an inner diameter of 2.50 cm at a velocity of 0.42 m/s. The oil has a mean bulk temperature of 100°C and the wall is at 40°C. Determine the heat transfer coefficient.

22.2: Air at 2.5 atm and a velocity of 6 m/s flows in 1.25-m pipe having an inner diameter of 2.50 cm. The air has a bulk temperature of 250°C and the wall is at 80°C. Determine the heat transfer coefficient.

22.3: Nitrogen at 1 atm flows in a 2-in-schedule 80 pipe at a flow rate of 12 m/s. The bulk temperature of the nitrogen is 27°C and the wall temperature is 77°C. Determine the heat transfer coefficient.

22.4: Rework Problem 22.3 if the fluid is carbon dioxide.

22.5: Rework Problem 22.3 if the fluid is air.

22.6: Water at a bulk temperature of 27°C enters a 1 in × 14 Birmingham wire gage tube at a velocity of 3.25 m/s. The tube is 1 m in length and the wall temperature is 47°C. Determine the heat transfer coefficient.

22.7: Rework Problem 22.6 for glycerin.

22.8: Rework Problem 22.6 for ethylene glycol.

22.9: Water at a velocity of 0.04 m/s flows through a 2-m-long tube having an inner diameter of 3 cm. The bulk temperature of the water is 27°C and the tube wall is at 80°C. Determine the heat transfer coefficient.

22.10: Air at 1 atm flows through a circular conduit that is 6.25 cm in inner diameter and 1.25-m long. The air has a bulk temperature of 27°C and the wall of the conduit is at 77°C. The mass rate of flow is $\dot{m} = 5.25$ kg/h. Determine the heat transfer coefficient.

22.11: Superheated steam at atmospheric pressure and a bulk temperature of 477°C enters a 2-m long 8-in-schedule 40 pipe at 2.375 m/s. The pipe wall temperature is 427°C. Determine the heat transfer coefficient.

22.12: Exhaust gases (which may be assumed to have the properties of air) enter the tubes of an air heater at 527°C and discharge to the atmosphere at 27°C. The air heater, which is vertical and 7.5-m high, contains tubes that have inner diameters of 7.5 cm. The exhaust gas flow rate is $\dot{m} = 0.0625$ kg/s. Determine the heat transfer coefficient.

22.13: Oxygen at 700 kPa and a flow rate of 3.50 m/s enters a pipe with an inner diameter of 1.905 cm at 17°C and leaves at 37°C. The pipe wall is at 77°C. Determine the heat transfer coefficient.

22.14: Ethylene glycol is heated from 17°C to 37°C in a tube whose inner diameter is 3.5 cm. The tube wall is held at 77°C and the velocity in the tube is 3.125 m/s. Estimate the length of the tube to ensure that h, the heat transfer coefficient, is 1325 W/m^2-K.

22.15: Air at 1 atm flows through a circular conduit at a volumetric flow rate of 1.125 m^3/s; the inner diameter of the conduit is 20 cm. The bulk temperature of the air is 15°C and the wall temperature is 77°C. Determine the heat transfer coefficient.

22.16: Lubricating oil (approximately SAE 50) with a bulk temperature of 160°C flows at a velocity of 0.32 m/s through a pipe with an inner diameter of 2.5 cm. The pipe wall temperature is 140°C and the tube length is 1 m. Determine the heat transfer coefficient.

22.17: Glycerin with a bulk temperature of 47°C flows at a velocity of 0.50 m/s through a pipe with an inner diameter of 5 cm. The pipe wall temperature is 27°C. Determine the heat transfer coefficient.

22.18: In a chilled water air-conditioning system, water enters and leaves a pipe having an inner diameter of 3.5 cm at 15°C and 5°C, respectively. The velocity of the water is 0.75 m/s, and the pipe wall temperature is 2°C. Determine the heat transfer coefficient.

Heat Transfer Coefficients: Constant Wall Temperature and Uniform Heat Flux

22.19: Water enters a 2-in-schedule 40 pipe at 80°C and a flow rate of 50,400 kg/h. The pipe wall temperature is constant at 140°C. Determine the length of pipe necessary to allow the water to leave at 120°C.

22.20: For the conditions of Problem 22.19, assume that the pipe wall is subjected to a uniform heat flux. Determine the length of pipe required to hold the maximum pipe wall temperature to 150°C.

22.21: Water at 57°C and a flow velocity of 0.02125 m/s enters a pipe having an inner diameter of 2.54 cm. The pipe is 2.75 m in length, and the wall temperature is constant at 86°C. Determine the exit water temperature.

22.22: Water at 17°C at a flow rate of 2.0 kg/s enters a pipe having an inner diameter of 2.54 cm. The pipe is 5.00 m in length and the wall temperature is constant at 125°C. Determine the exit water temperature.

22.23: Water at 15°C at a flow rate of 3.12 kg/s enters a pipe having an inner diameter of 4.00 cm. The exit water temperature is to be 25°C and the wall temperature is constant at 90°C. Determine the length of the pipe.

22.24: Air at 2.5 atm and a bulk temperature of 250°C is heated as it flows through a tube with a diameter of 3.81 cm with a velocity of 8 m/s. The heating is via a constant heat flux and the wall temperature is 25°C above the air temperature. Determine the heat flux.

22.25: Water enters a pipe with an inner diameter of 5.0 cm at 10°C and a flow rate of 2.75 kg/s. The pipe is 4.8 m in length and the wall temperature is constant over its entire length at 70°C. Determine the exit water temperature.

22.26: In Problem 22.25, the wall is subjected to a constant heat flux and all other conditions remain the same. Determine the maximum tube wall temperature.

22.27: Water enters a 2.00-in-schedule 40 pipe at 7°C at a flow rate of 2.625 kg/s. The wall temperature is constant over the entire length of the tube at 90°C and the water leaves at 23°C. Determine the length of the pipe.

22.28: Water at 15°C and a flow rate of 3.12 kg/s enters a pipe with a 4-cm-inner diameter. The wall temperature is constant over the entire length of the tube at 90°C and the water leaves at 35°C. Determine the length of the pipe.

The Equivalent Diameter

22.29: Determine the equivalent diameter of the trapezoidal passage shown in Figure P22.29.

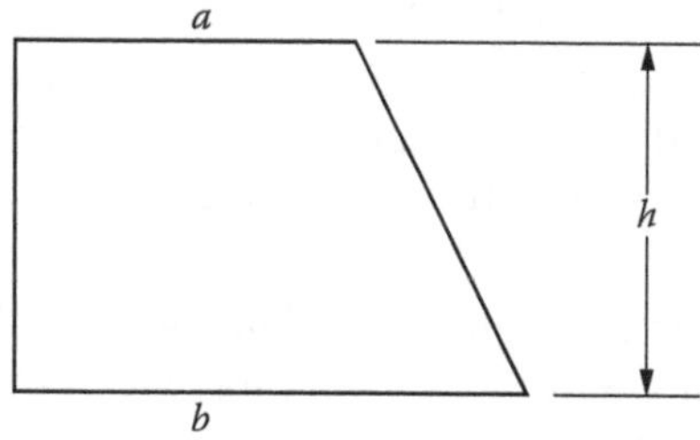

FIGURE P22.29

22.30: Determine the equivalent diameter of the trapezoidal passage shown in Figure P22.30.

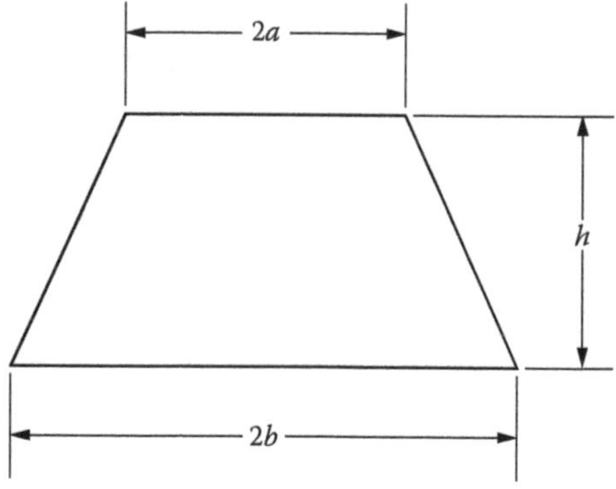

FIGURE P22.30

22.31: Determine the equivalent diameter of the triangular passage shown in Figure P22.31.

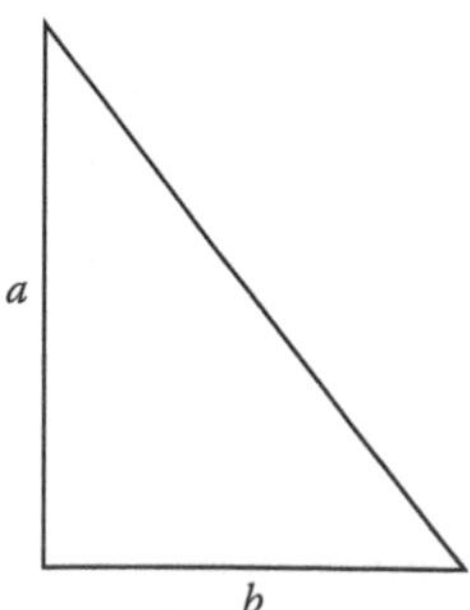

FIGURE P22.31

22.32: Determine the equivalent diameter of the isosceles right triangular passage shown in Figure P22.32.

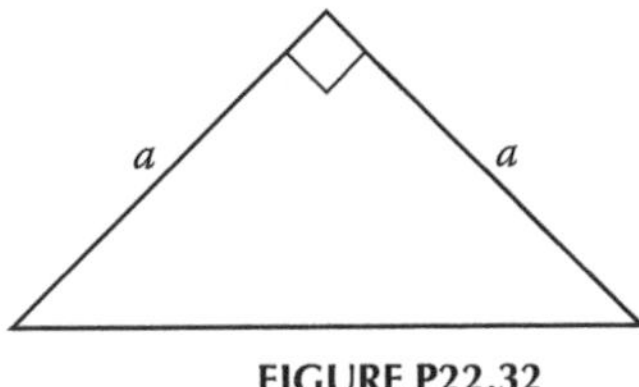

FIGURE P22.32

22.33: Determine the equivalent diameter of the equilateral triangular passage shown in Figure P22.33.

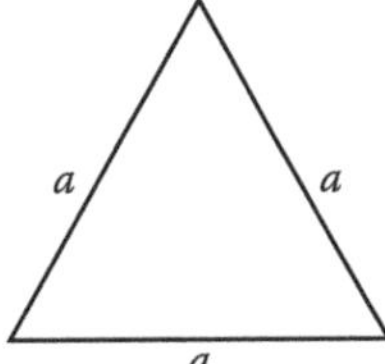

FIGURE P22.33

22.34: Determine the equivalent diameter of the circular passage shown in Figure P22.34.

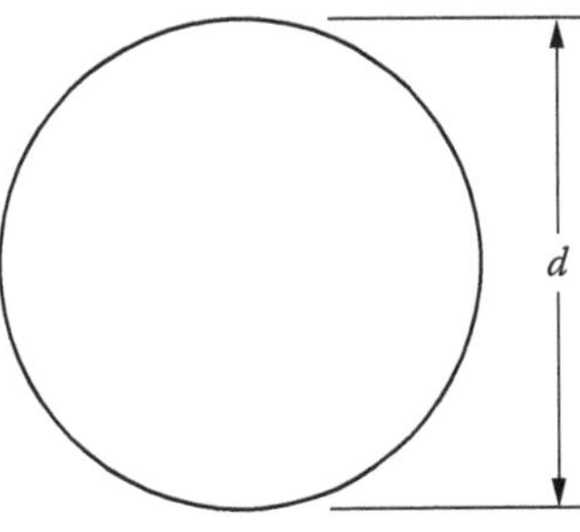

FIGURE P22.34

22.35: Determine the equivalent diameter of the rectangular passage shown in Figure P22.35.

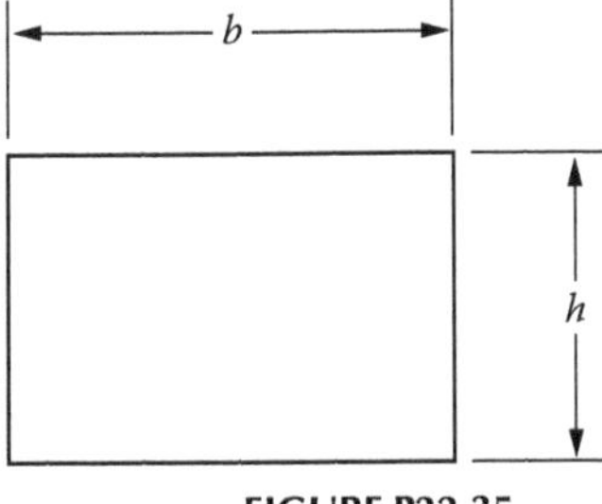

FIGURE P22.35

22.36: Determine the equivalent diameter of the annular passage shown in Figure P22.36.

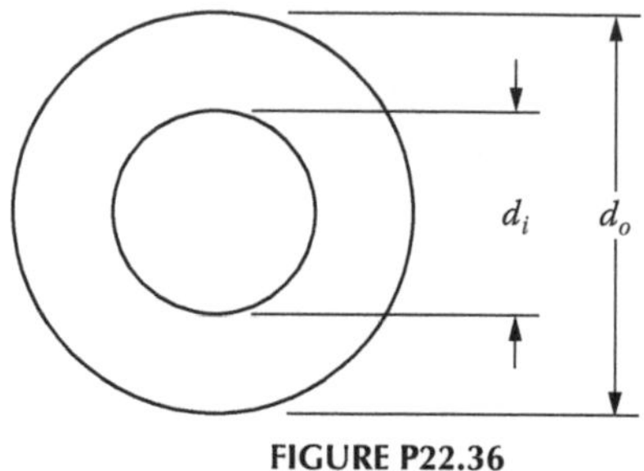

FIGURE P22.36

22.37: Figure P22.37 shows the interior of a circular tube containing four fins. Determine the equivalent diameter.

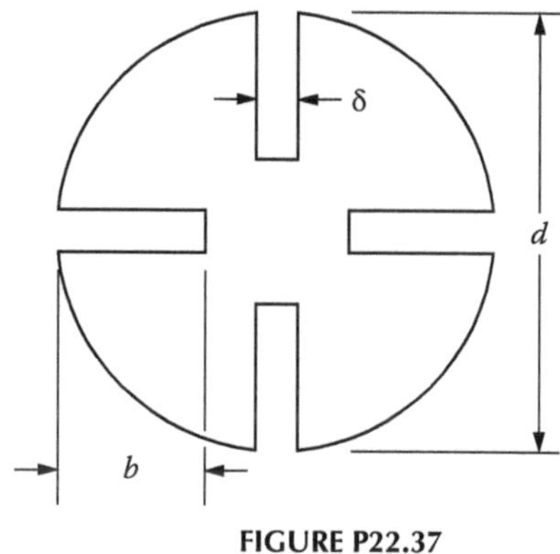

FIGURE P22.37

22.38: Determine the equivalent diameter of the finned annular passage shown in Figure P22.38.

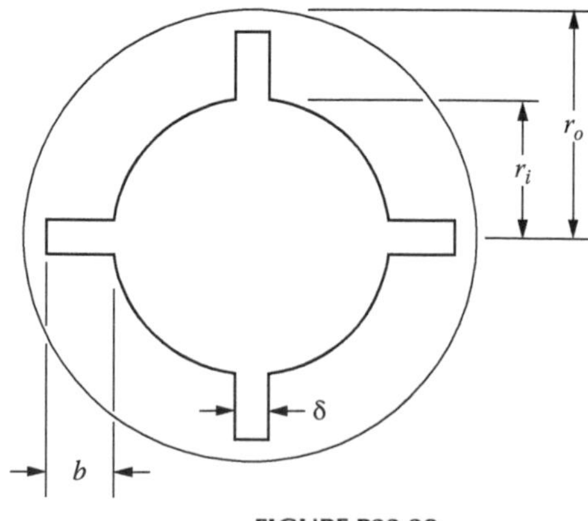

FIGURE P22.38

22.39: Determine the equivalent diameter of the octagonal passage shown in Figure P22.39.

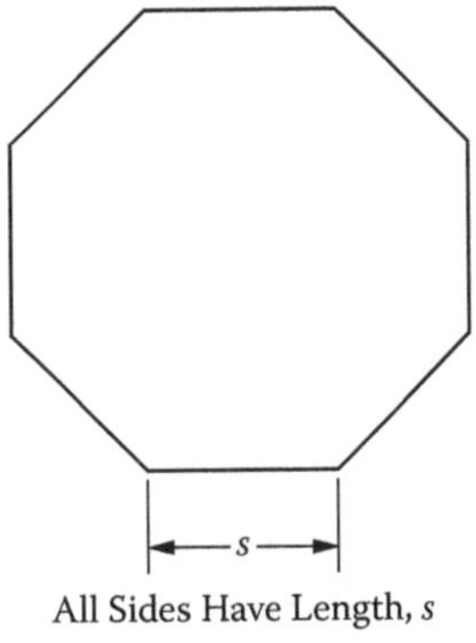

All Sides Have Length, *s*

FIGURE P22.39

Heat Transfer Coefficients in Passages Other than Circular

22.40: Lubricating oil (approximately SAE 50) flows at a velocity of 0.42 m/s in a 4-m-long rectangular duct that is 2 cm × 4 cm. The oil has a bulk temperature of 100°C, and the wall is at 40°C. Determine the heat transfer coefficient.

22.41: Lubricating oil (approximately SAE 50) flows at a velocity of 0.42 m/s in a 4-m-long equilateral triangular duct that is 2.00 cm on a side. The oil has a bulk temperature of 100°C, and the wall is at 40°C. Determine the heat transfer coefficient.

22.42: Air at 25 atm flows at a velocity of 6 m/s in a 1.25-m-long square duct that is 2.50 cm on a side. The air has a bulk temperature of 250°C, and the wall is at 80°C. Determine the heat transfer coefficient.

22.43: Air at 25 atm flows at a velocity of 6 m/s in a 1.25-m-long annular region (d_i = 2.50 cm, d_o = 4 cm). The air has a bulk temperature of 250°C, and the wall is at 80°C. Determine the heat transfer coefficient.

22.44: Rework Problem 22.43 for the case of air at 1 atm.

22.45: Nitrogen at 1 atm flows in a 1.5-m-long 2-in schedule 40 pipe that is equipped at the interior wall with four fins, each with a height of 0.75 cm and a thickness of 2.286 mm. The flow rate is 12 m/s, the bulk temperature of the nitrogen is 27°C, and the wall temperature is 77°C. Determine the heat transfer coefficient.

22.46: Nitrogen at 1 atm flows at a velocity of 12 m/s in the configuration of Figure P22.46. Pertinent dimensions are:

$$a_1 = 4 \text{ cm} \qquad b_1 = 1.5 \text{ cm}$$

$$a_2 = 3 \text{ cm} \qquad b_2 = 1.25 \text{ cm}$$

$$\delta = 40 \text{ cm}$$

The bulk temperature of the nitrogen is 27°C and the wall temperature is 77°C. Determine the heat transfer coefficient.

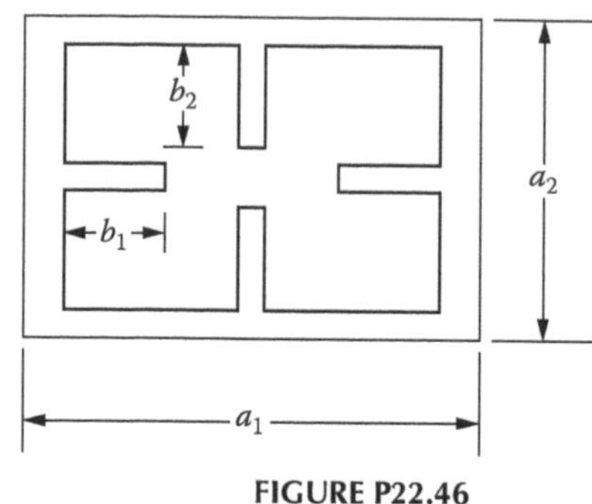

FIGURE P22.46

22.47: Superheated steam at atmospheric pressure and 450°C enters a square duct that is 20 cm on a side and 4-m long. The velocity of the steam is 1.375 m/s and the pipe wall temperature is 400°C. Determine the heat transfer coefficient.

22.48: Superheated steam at atmospheric pressure and 450°C enters a trapezoidal duct that has a height of 4 cm and bases of 20 and 24 cm and is 4-m long. The velocity of the steam is 1.375 m/s and the duct wall temperature is 400°C. Determine the heat transfer coefficient.

22.49: Water at a bulk temperature of 27°C flows in a square duct whose side dimension is 1 cm. The velocity is 3.25 m/s and the duct is 1 m in length. For a wall temperature of 47°C, determine the heat transfer coefficient.

22.50: Water at a bulk temperature of 27°C flows in the interior of an isosceles right triangular duct (45°-90°-45°) whose two equal sides measure 2 cm. The velocity is 3.25 m/s and the duct is 1 m in length. For a wall temperature of 47°C, determine the heat transfer coefficient.

22.51: Rework Problem 22.50 for water flowing at a velocity of 0.04 m/s in a rectangular duct that is 3 cm × 4 cm. All other conditions remain the same.

22.52: Rework Problem 22.50 for ethylene glycol in a rectangular duct that is 3 cm × 4 cm. The wall temperature is 77°C, and the duct length is 2 m. All other conditions remain the same.

22.53: Rework Problem 22.50 for glycerin in a rectangular duct that is 3 cm × 4 cm with a wall temperature of 47°C and a duct length of 2 m. All other conditions remain the same.

22.54: Rework Problem 22.50 for unused engine oil in a rectangular duct that is 3 cm × 4 cm. All other conditions remain the same except for the duct wall temperature, which is at 77°C.

22.55: Ethylene glycol is heated from 37°C to 27°C on the inside of a circular annular region. The annular region has an inner diameter, d_i, of 2.50 cm and outer diameter, d_o, of 4.50 cm. The metal wall is held at 77°C, and the velocity in the annulus is 2.125 m/s. Estimate the length of the annular region to ensure that h, the heat transfer coefficient, is 1325 W/m^2-K.

22.56: Rework Problem 22.55 for an annular region resembles a pair of concentric squares. The inside wall is 2 cm on a side and the outside wall is 4 cm on a side. All other conditions remain the same.

22.57: Rework Problem 22.55 for flows through an equilateral triangular passage that is $2\sqrt{3}$ cm on a side. All other conditions remain the same.

22.58: In a chilled water air-conditioning system, water enters and leaves a circular annular region ($d_i = 3$ cm and $d_o = 4.5$ cm) at 15°C and 5°C, respectively. The velocity of the water is 0.75 m/s and the annulus walls are at 2°C. Determine the heat transfer coefficient.

23

Forced Convection—External Flow

Chapter Objectives

- To examine the factors that influence heat transfer between a solid surface and an adjacent flowing fluid.
- To catalog the dimensionless empirical relations that allow the determination of the heat transfer coefficient, h, for external forced flow.
- To use the appropriate empirical expressions for h in evaluating the heat transfer behavior of representative systems.

23.1 Introduction

In this chapter, we continue to discuss and illustrate how convective heat transfer rates are determined using empirical correlations that have been generated and verified over many years of effort. Our focus will be on convective heat transfer between a solid surface and a fluid that flows over its outside surface, that is, **external flow.** External flows, as discussed in Chapter 16, are considerably more complex than the internal flows considered in the previous chapter.

Dimensional analysis techniques were employed in Chapter 22 to generate the dimensionless parameters that are useful in determining the heat transfer coefficient, h. This same technique would yield the same parameters for external flow that were generated earlier for internal flow. We thus state, without proof, that correlations for external flow convection are expected to have the form

$$\text{Nu}_L = C\text{Re}^a\text{Pr}^b \tag{23.1}$$

where

$$\text{Nu}_L = \frac{hL}{k}$$

is the Nusselt number

$$\text{Re}_L = \frac{L\hat{V}\rho}{\mu}$$

is the Reynolds number and

$$\text{Pr} = \frac{\mu c_p}{k}$$

is the Prandtl number.

The subscript, L, associated with Nu_L and Re_L, is the length scale that is appropriate for a given geometric case.

The constants, a, b, and C in Equation 23.1 have been experimentally determined for a large number of geometric and flow conditions. Subsequent sections will include discussions of several important cases.

As noted earlier, physical properties of interest are temperature dependent. In heat transfer processes involving external flow, conventional practice is to evaluate properties at the **film temperature**, T_f, which is the average of the wall and bulk fluid temperatures

$$T_f = \frac{T_w + T_\infty}{2} \tag{23.2}$$

23.2 Flow Parallel to a Plane Wall

23.2.1 Laminar Boundary Layer Flow

As noted in Chapter 16, the case of flow parallel to a flat surface has been solved analytically to yield operational expressions for boundary layer thickness, δ, and the coefficient of skin friction, $\overline{C}_f$. In Figure 23.1, the features of the hydrodynamic boundary layer are represented in a similar fashion as was illustrated in Figure 16.7.

When a temperature difference exists between the plate surface and the fluid, heat transfer will occur and the fluid layers near the wall will undergo changes in temperature. This process generates a **thermal boundary layer** that will grow as a function of x much like the hydrodynamic boundary layer. The thickness of the thermal boundary layer is designated as δ_t in Figure 23.1. The boundary layer region is exaggerated in Figure 23.1 for clarity; in reality it is quite thin. The fluid temperature varies from T_w to T_∞ for $0 < y < \delta_t$.

The trends indicated in Figure 23.1 apply to the case of steady, incompressible laminar flow along a plane surface. The physical situation is one for which an analytical solution exists. This is a classic problem in the fluid mechanics literature and is referred to as the **Blasius problem** in recognition of the original work of H. Blasius.[1]

The heat transfer counterpart to the Blasius problem has also been solved and is discussed in most heat transfer texts (Welty et al., 2008 and Incropera and Dewitt, 2002). The important relationships provided by solving the governing equations of mass conservation, the first law of thermodynamics, and Newton's second law of motion are the boundary layer thickness

$$\frac{\delta}{\delta_t} = \text{Pr}^{1/3} \tag{23.3}$$

or incorporating Equation 16.5

$$\delta_t = 5x\text{Re}_x^{-1/2}\text{Pr}^{-1/3} \tag{23.4}$$

and the Nusselt number

$$\text{Nu}_L = \frac{hL}{k} = 0.664\text{Re}_L^{1/2}\text{Pr}^{1/3} \tag{23.5}$$

Equation 23.5 applies for fluids with $\text{Pr} \geq 0.6$. This range includes all gases and most liquids except for the liquid metals such as mercury and molten sodium. A general

[1] P. R. H. Blasius (1883–1970) was one of Prandtl's students who provided an analytical solution to the boundary layer equations in 1908.

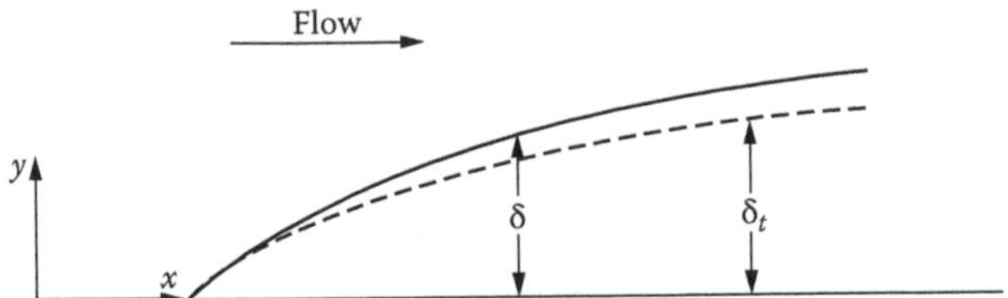

FIGURE 23.1
Laminar boundary layer growth for flow along a plane wall.

expression, which applies to any fluid in laminar flow over a flat plate, has been recommended by Churchill and Ozoe (1973):

$$\mathrm{Nu}_L = \frac{0.677\mathrm{Re}_L^{1/2}\mathrm{Pr}^{1/3}}{\left[1+\left(\frac{0.0468}{\mathrm{Pr}}\right)^{2/3}\right]^{1/4}} \tag{23.6}$$

Note that the thermal boundary layer thickness, δ_t, increases as $x^{1/2}$. The Nusselt number represented by Equation 23.6 is the mean value over the distance, L, measured from the leading edge of the plate.

The reader is reminded specifically of the Reynolds number dependence of boundary layer flow along a flat plate.

- The flow is laminar for

$$0 < \mathrm{Re}_x < 2 \times 10^5$$

- The flow is turbulent for

$$3 \times 10^6 < \mathrm{Re}_x$$

- And the transition region exists for

$$2 \times 10^5 < \mathrm{Re}_x < 3 \times 10^6$$

Example 23.1 Air at atmospheric pressure and a temperature of 280 K flows at a velocity of 3 m/s. A plane surface, 15 cm in width at a temperature of 400 K, is oriented parallel to the direction of flow (Figure 23.2). Determine (a) the length of the plate for the boundary layer flow to remain laminar, (b) the maximum thermal boundary layer thickness for a plane surface of this length, and (c) the rate of heat transfer between the plate surface and the air.

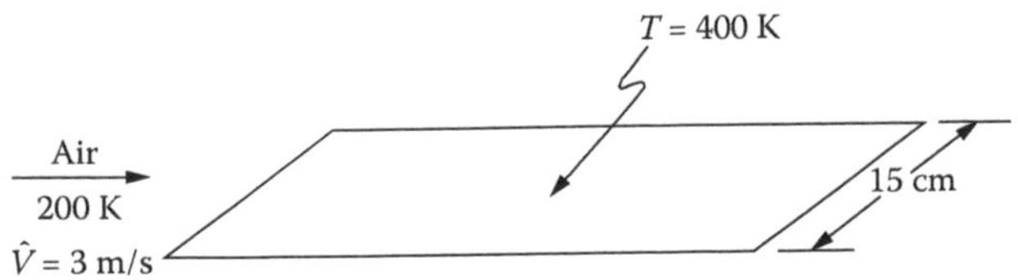

FIGURE 23.2
Configuration for Example 23.1.

Solution

Assumptions and Specifications

(1) Steady-state conditions exist.
(2) Radiation heat exchange with the surroundings is negligible.
(3) The bottom surface of the plate is adiabatic.

Fluid properties must be evaluated at the film temperature

$$T_f = \frac{T_w + T_\infty}{2} = \frac{400\,\text{K} + 280\,\text{K}}{2} = 340\,\text{K}$$

At 340 K, the properties of interest obtained from Table A.13 in Appendix A (with interpolation) are

$$k = 0.0292\,\text{W/m-K}, \qquad \nu = 1.975 \times 10^{-5}\,\text{m}^2/\text{s} \quad \text{and} \quad \text{Pr} = 0.699$$

(a) The laminar boundary layer will exist for $\text{Re} < 2 \times 10^5$. The corresponding length is given by

$$\text{Re}_L = \frac{L\hat{V}}{\nu} = 2 \times 10^5$$

which yields

$$L = \frac{\text{Re}_L \nu}{\hat{V}} = \frac{(2 \times 10^5)(1.975 \times 10^{-5}\,\text{m}^2/\text{s})}{3\,\text{m/s}} = 1.32\,\text{m} \Longleftarrow$$

(b) The thermal boundary layer thickness given by Equation 23.4 will reach its maximum value at $L = 1.32$ m, where $\text{Re}_x = 2 \times 10^5$ and the value of δ_t is determined as

$$\begin{aligned}\delta_t &= 5x\text{Re}_x^{-1/2}\text{Pr}^{-1/3}\\ &= 5(1.32\,\text{m})(2 \times 10^5)^{-1/2}(0.699)^{-1/3}\\ &= (6.60\,\text{m})(2.236 \times 10^{-3})(1.1268) = 0.0164\,\text{m} \qquad (1.64\,\text{cm}) \Longleftarrow\end{aligned}$$

(c) The heat transfer to a plate measuring 0.15 m × 1.32 m is determined from

$$\dot{Q} = hS(T_w - T_\infty)$$

where h can be evaluated using Equation 23.5

$$\frac{hL}{k} = 0.664\text{Re}_L^{1/2}\text{Pr}^{1/3}$$

or

$$h = \frac{k}{L}0.664\text{Re}_L^{1/2}\text{Pr}^{1/3}$$

Appropriate substitution yields

$$\begin{aligned}h &= \frac{0.0292\,\text{W/m-K}}{1.32\,\text{m}}(0.664)(2 \times 10^5)^{1/2}(0.699)^{1/3}\\ &= (0.0221\,\text{W/m}^2\text{-K})(0.664)(447.2)(0.888) = 5.85\ \text{W/m}^2\text{-K}\end{aligned}$$

and, finally,

$$\dot{Q} = hS(T_w - T_\infty)$$
$$= (5.85\,\text{W/m}^2\text{-K})(0.15\,\text{m})(1.32\,\text{m})(400\,\text{K} - 280\,\text{K}) = 139\,\text{W}$$

23.2.2 Turbulent Boundary Layer Flow

Flow in the boundary layer will be turbulent for $\text{Re}_x > 3 \times 10^6$. As mentioned in Section 16.4, the relevant fluid mechanics parameters are

$$\delta = 0.376x\text{Re}_L^{-1/5} \tag{16.10}$$

and

$$\bar{C}_f = 0.072\text{Re}_L^{-1/5} \tag{16.12}$$

The analysis leading to these relationships can be extended to include heat transfer phenomena, that is, behavior observed when a temperature difference exists between the surface and flowing fluid. The expressions that pertain to the boundary layer region are

$$\delta_t = 0.376x\text{Re}_x^{-1/5}\text{Pr}^{-1/3} \tag{23.7}$$

and

$$\text{Nu}_L = \frac{h_L L}{k} = 0.036\text{Re}_L^{4/5}\text{Pr}^{1/3} \tag{23.8}$$

which applies for $0.60 < \text{Pr} < 60$.

The natural question at this point is "How should one proceed to evaluate heat transfer for flow along a plane wall when more than one type of flow is present?"

A relatively simple approach to this situation is to presume two types of flow, laminar and turbulent with transition from one to the other occurring abruptly at some value of Re_x lying between 2×10^5 and 3×10^6. Quantitatively, this concept may be expressed as

$$h_L = \frac{1}{L}\left[\int_0^{x_c} h_{\text{lam}}dx + \int_{x_c}^{L} h_{\text{turb}}dx\right] \tag{23.9}$$

In one conventional choice for x_c, $\text{Re} = 5 \times 10^5$, and the result is

$$\text{Nu}_L = \frac{h_L L}{k} = \left[0.036\text{Re}_L^{4/5} - 871\right]\text{Pr}^{1/3} \tag{23.10}$$

which applies for

$$\text{Re}_{x_c} = 5 \times 10^5, \quad 0.60 < \text{Pr} < 60, \quad \text{and} \quad 5 \times 10^5 < \text{Re}_L < 10^8$$

Example 23.2 For air flowing along a plane surface as described in Example 23.1, determine the total heat transfer over a flat plate that is 10-m long.

Solution

Assumptions and Specifications

(1) Steady-state conditions exist.

(2) Radiation heat exchange with the surroundings is negligible.

(3) The bottom surface of the plate is adiabatic.

Recall that at $T_f = 340\,\text{K}$, the fluid properties of interest are

$$k = 0.0292\,\text{W/m-K}, \qquad \nu = 1.975 \times 10^{-5}\,\text{m}^2/\text{s} \quad \text{and} \quad \text{Pr} = 0.699$$

The Reynolds number, Re_L, for this situation is

$$\text{Re}_L = \frac{L\hat{V}}{\nu} = \frac{(10\,\text{m})(3\,\text{m/s})}{1.975 \times 10^{-5}\text{m}^2/\text{s}} = 1.519 \times 10^6$$

Equation 23.10 applies. Solving for h, we obtain

$$h = \frac{k}{L}\left[0.036\text{Re}_L^{4/5} - 871\right]\text{Pr}^{1/3}$$

$$= \frac{0.0292\,\text{W/m-K}}{10\,\text{m}}\left[0.036(1.519 \times 10^6)^{4/5} - 871\right](0.699)^{1/3}$$

$$= (0.00292\,\text{W/m}^2\text{-K})[0.036(88{,}154) - 871](0.8875) = 5.97\,\text{W/m}^2\text{-K}$$

The total heat transfer is then determined as

$$\dot{Q} = hS(T_w - T_\infty)$$

$$= (5.97\,\text{W/m}^2\text{-K})(0.15\,\text{m})(10\,\text{m})(400\,\text{K} - 280\,\text{K}) = 1075\,\text{W}$$

23.2.3 Additional Considerations

23.2.3.1 Constant Heat Flux Wall Condition

The foregoing discussion and example problems presumed a constant wall temperature. As presented in Chapter 20, the other type of boundary condition of interest is that of a prescribed heating rate or wall heat flux, $\dot{q} = \dot{Q}/S$.

When the wall heat flux is prescribed, the total heat transfer is easily calculated as

$$\dot{Q} = \dot{q}S \tag{23.11}$$

The other quantity of interest is the wall temperature that is related to the heat flux according to

$$\dot{Q} = hS(T_w - T_\infty)$$

or

$$T_w = T_\infty + \frac{\dot{Q}/S}{h}$$

or

$$T_w = T_\infty + \frac{\dot{Q}/S}{(k/L)\text{Nu}_L} \tag{23.12}$$

The evaluation of Nu_L for a constant wall heat flux is discussed in considerable detail by Kays and Crawford (1993). They conclude that for external flow, the Nusselt numbers for constant heat rate and constant wall temperature differ, quantitatively, by approximately 2%. In light of other inaccuracies in determining h it is concluded that Equation 23.12 may be used with any of the constant surface temperature expressions to determine the wall temperature.

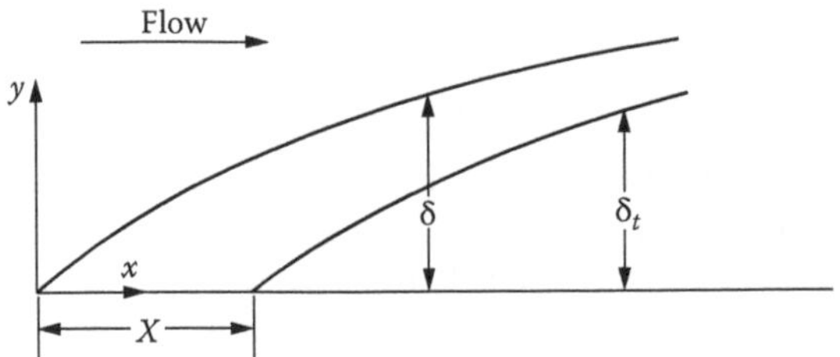

FIGURE 23.3
Two boundary layers with an unheated starting length.

23.2.3.2 Unheated Starting Length

All cases considered thus far presume that the wall surface is heated over its entire length. Figure 23.3 shows the two boundary layers for the case of a hydrodynamic boundary that begins at the leading edge, $x = 0$, but with the thermal boundary layer at some starting length, X, downstream from the leading edge. The wall boundary condition can be expressed as

$$T = T_\infty \qquad 0 < x < X$$
$$T = T_w \qquad X < x$$

for a prescribed wall temperature or

$$\dot{Q}/S = 0 \qquad 0 < x < X$$
$$\dot{Q}/S = q'' \qquad X < x$$

for a prescribed wall heat flux.

Solutions for unheated starting lengths are presented in a number of texts Kays and Crawford, 1993; Welty et al. 2008.

The analysis of this case involves techniques that are beyond the scope of this book and the results that follow are given without proof. The interested reader may consult the references cited for the complete development.

For laminar boundary layer flow, the local Nusselt number is given by

$$\mathrm{Nu}_x = \frac{\mathrm{Nu}_x\big|_{x=0}}{\left[1 - \left(\dfrac{X}{x}\right)^{3/4}\right]^{1/3}} \tag{23.13}$$

and for turbulent boundary layer flow

$$\mathrm{Nu}_x = \frac{\mathrm{Nu}_x\big|_{x=0}}{\left[1 - \left(\dfrac{X}{x}\right)^{9/10}\right]^{1/9}} \tag{23.14}$$

For a given situation the mean Nusselt number, Nu_L, must be evaluated by integrating the appropriate expression over the range $0 < x < L$.

23.3 External Flow over Bluff Bodies

Flow over nonstreamlined or **bluff** bodies involves the phenomenon of boundary layer separation and a series of complex surface phenomena that are not amenable to analysis. Common geometries in this category include the cylinder in cross flow, spheres, and a host of other cases where a surface causes significant disruption of flow. In this section, we will treat a number of representative cases.

23.3.1 The Cylinder in Cross Flow

External flow normal to the axis of a circular cylinder was discussed in some detail in Chapter 16. The drag coefficient, C_D, for this configuration was observed to vary in a complicated manner as a function of Re_d. Figure 16.5 shows C_D as a function of Re_d for both a cylinder and a sphere.

For a given value of Re_d, the phenomena of laminar-to-turbulent flow transition, boundary layer separation and other characteristics of flow around the cylindrical surface will cause the local Nusselt number to vary in an irregular manner. These variations are shown in Figure 23.4 for low Reynolds numbers and in Figure 23.5 for high values of Re_d. In engineering practice, we are more interested in mean values of Nu_d rather than the detailed local values shown in these figures.

A comprehensive compilation of experimental data relating Nu_d was presented by McAdams (1954). This classic work is shown in Figure 23.6. The variation of Nu_d as a function of Re_d in Figure 23.6 has been put into the form of Equation 23.1 by putting a

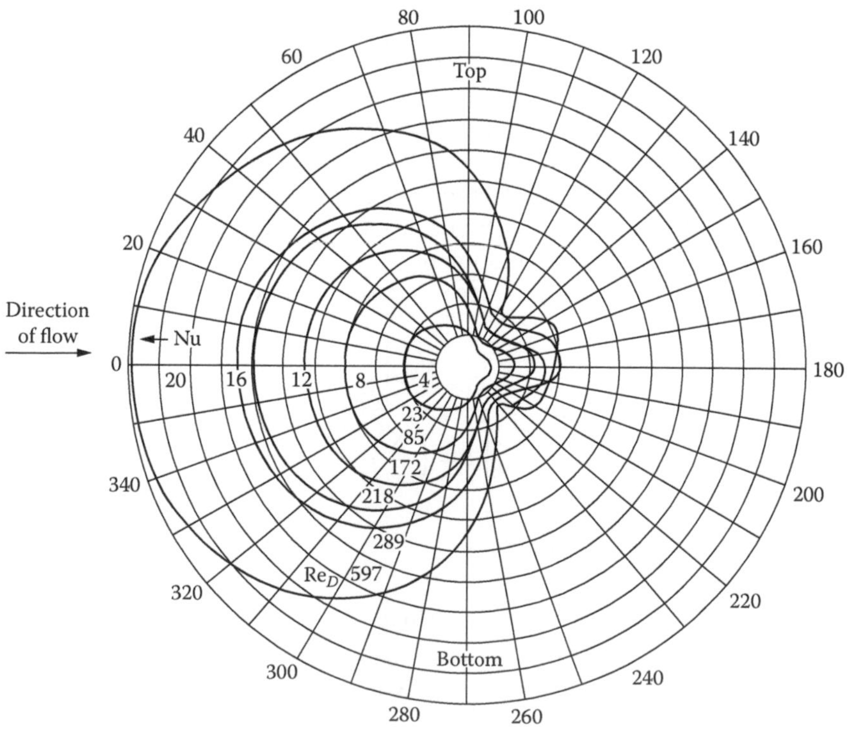

FIGURE 23.4
Local Nusselt numbers for flow normal to a circular cylinder at low Reynolds numbers.

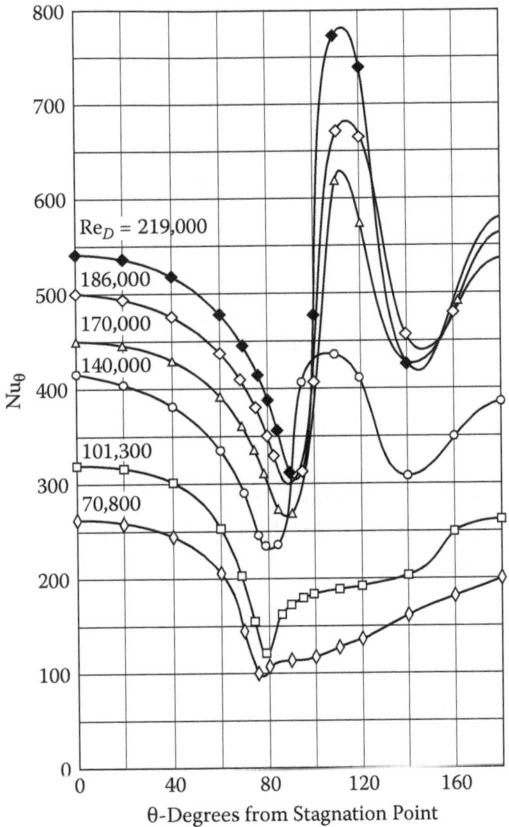

FIGURE 23.5
Local Nusselt numbers for flow normal to a circular cylinder at high Reynolds numbers.

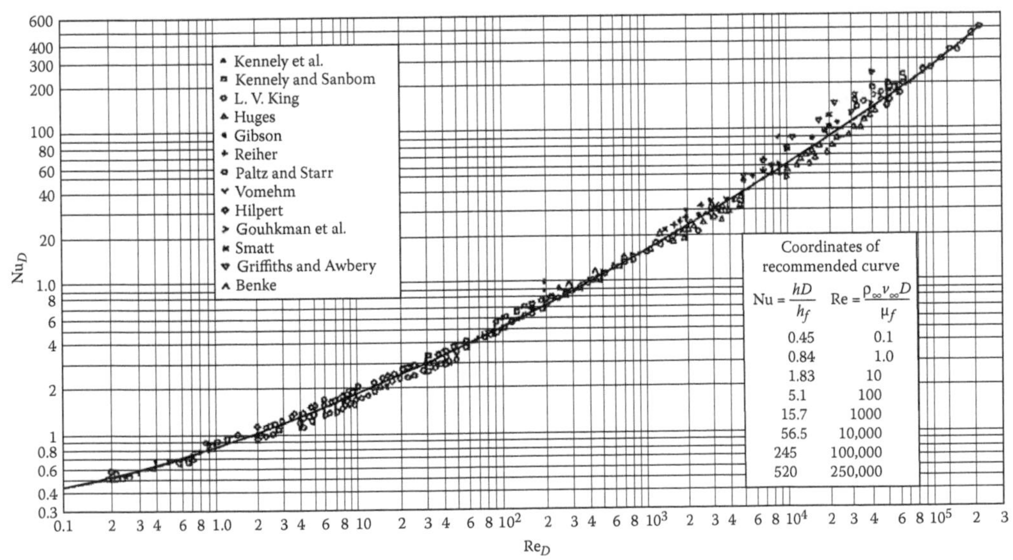

FIGURE 23.6
Nu_d as a functions of Re_d for flow normal to circular cylinders.

TABLE 23.1
Values of C and n for Use in Equation 23.15

Re_d	C	n
0.40–4.0	0.989	0.330
4.0–40	0.911	0.385
40–4000	0.683	0.466
$4000–4 \times 10^4$	0.143	0.618
$4 \times 10^4 - 4 \times 10^5$	0.027	0.805

solid line (best fit) piecewise for portions of the horizontal range of Re_d. The resulting set of empirical expressions have the form (Hilpert, 1933)

$$\mathrm{Nu}_d = C\mathrm{Re}_d^n\mathrm{Pr}^{1/3} \tag{23.15}$$

with the values of C and n given in Table 23.1.

A single correlation, applicable for all values of $\mathrm{Re}_d\mathrm{Pr} > 0.2$ has been recommended by Churchill and Bernstein (1977):

$$\mathrm{Nu}_d = 0.30 + \frac{0.62\mathrm{Re}_d^{1/2}\mathrm{Pr}^{1/3}}{\left[1 + \left(\frac{0.40}{\mathrm{Pr}}\right)^{2/3}\right]^{1/4}}\left[1 + \left(\frac{\mathrm{Re}_d}{282,000}\right)^{5/8}\right]^{4/5} \tag{23.16}$$

As noted, Equation 23.15, which utilizes the values of C and n in Table 23.1, has been available for many years. Equation 23.16 is considerably newer and generally considered to yield more accurate results. The following example problem will be worked using both approaches so that the results may be compared.

Example 23.3 An insulated steam pipe, 30 cm in diameter, extends between two buildings with 18 m of pipe exposed to the outside air (Figure 23.7). On a winter day, the pipe surface temperature is 140°C when 10°C air flows across the pipe surface at a velocity of 7.8 m/s. Determine (a) the heat loss from the steam pipe to the air and (b) the mass of steam condensed per hour resulting from this heat loss ($h_{fg} = 2147$ kJ/kg).

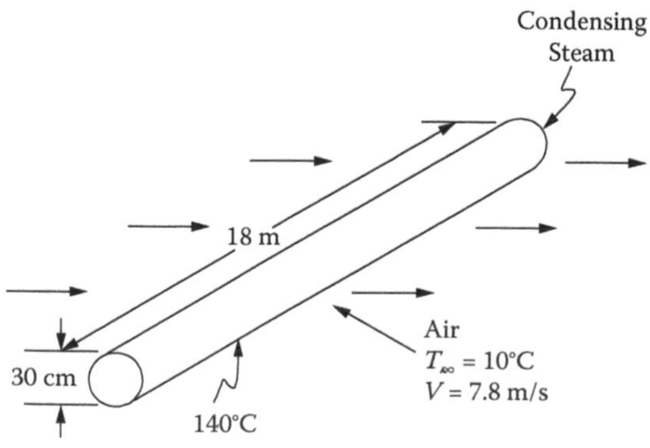

FIGURE 23.7
Configuration for Example 23.3.

Solution

Assumptions and Specifications

(1) Steady-state conditions exist.
(2) Radiation heat exchange with the surroundings is negligible.

(a) The film temperature for this situation is

$$T_f = \frac{140^\circ\text{C} + 10^\circ\text{C}}{2} = 75^\circ\text{C} \qquad (348\,\text{K})$$

Properties of air at this temperature are obtained from Table A.13 (with interpolation)

$$\rho = 1.0052\,\text{kg/m}^3$$
$$c_p = 1.0089\,\text{kJ/kg-K}$$
$$\nu = 20.56 \times 10^{-6}\,\text{m/s}^2$$
$$\text{Pr} = 0.697$$
$$k = 0.0298\,\text{W/m-K}$$

The value of Re_d is determined as

$$\text{Re}_d = \frac{\hat{V}d}{\nu} = \frac{(7.8\,\text{m/s})(0.30\,\text{m})}{2.056 \times 10^{-5}\,\text{m/s}^2} = 113{,}800$$

From Equation 23.15 and Table 23.1, we find the value of Nu_d that applies:

$$\text{Nu}_d = C\text{Re}_d^n\text{Pr}^{1/3}$$
$$= 0.027\text{Re}_d^{0.805}\text{Pr}^{1/3}$$

which yields

$$\text{Nu}_d = 0.027(113{,}800)^{0.805}(0.697)^{1/3}$$
$$= 0.027(11{,}755)(0.887) = 281.5$$

Alternatively, Equation 23.16

$$\text{Nu}_d = 0.30 + \frac{0.62\text{Re}_d^{1/2}\text{Pr}^{1/3}}{\left[1 + \left(\frac{0.40}{\text{Pr}}\right)^{2/3}\right]^{1/4}}\left[1 + \left(\frac{\text{Re}_d}{282{,}000}\right)^{5/8}\right]^{4/5} \qquad (23.16)$$

yields

$$\text{Nu}_d = 0.30 + \frac{0.62(113{,}800)^{1/2}(0.697)^{1/3}}{\left[1 + \left(\frac{0.40}{0.697}\right)^{2/3}\right]^{1/4}}\left[1 + \left(\frac{113{,}800}{282{,}000}\right)^{5/8}\right]^{4/5}$$

or

$$\text{Nu}_d = 0.30 + \frac{0.62(338.9)(0.887)}{[1 + 0.691]^{1/4}}[1 + 0.571]^{4/5} = 233.3$$

The difference, approximately 17%, is unfortunately typical of the uncertainties involved in convective heat transfer calculations. One making such calculations should be mindful of the approximate nature of any quantitative results.

Using the lower of the two Nu_d values calculated, the heat transfer coefficient is determined as

$$h = \frac{k}{d}\mathrm{Nu}_d$$

$$= \frac{0.0298\,\mathrm{W/m\text{-}K}}{0.30\ \mathrm{m}}(233.3) = 23.17\,\mathrm{W/m^2\text{-}K}$$

and the associated heat loss over 18 m is

$$\dot{Q} = hS(T_w - T_\infty)$$

$$= (23.17\,\mathrm{W/m^2\text{-}K})[\pi(0.30\,\mathrm{m})(18\,\mathrm{m})](140°\mathrm{C} - 10°\mathrm{C})$$

$$= 51{,}100\,\mathrm{W} \qquad (51.10\,\mathrm{kW}) \Longleftarrow$$

(b) The corresponding condensation rate of steam is given by $\dot{Q} = \dot{m}h_{fg}$ so that

$$\dot{m} = \frac{\dot{Q}}{h_{fg}} = \frac{51{,}100\,\mathrm{W}}{2147\,\mathrm{kJ/kg}} = 23.80\,\mathrm{kg/s} \Longleftarrow$$

23.3.2 Tube Bundles in Cross Flow

Considerable work has been directed toward evaluating heat transfer associated with arrays of cylinders exposed to a fluid in cross flow. The motivation to study these configurations is primarily due to the configuration of pipes and tubes in heat exchangers.

When many tubes are bundled together, the flow and heat transfer are influenced by the spacing and geometric arrangement of the tubes, as well as the factors discussed in Section 23.2.1.

Two common arrangements of tube banks are shown in Figure 23.8. Both in-line and staggered arrangements are shown along with other parameters used in specifying the spacing.

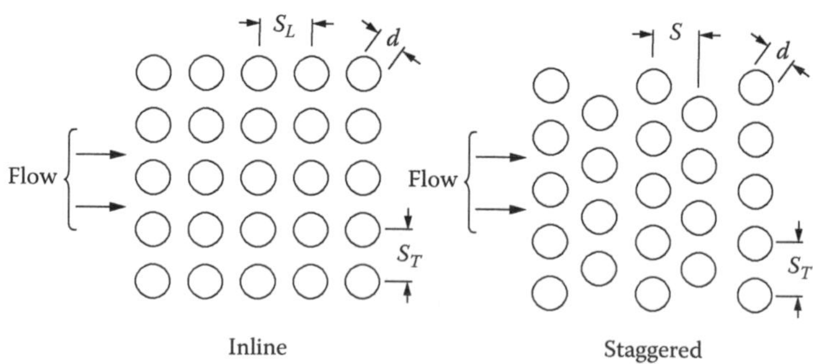

FIGURE 23.8
In-line (left) and staggered (right) configurations for tubes in a tube bank.

TABLE 23.2

Values of Constants C and m for Use in Equation 23.17

Configuration	$\mathrm{Re}_{d,\max}$	C	m
Aligned	10–100	0.80	0.40
Staggered	10–100	0.90	0.40
Aligned	100–1000	Use single Cylinder	
Staggered	100–1000	Use single Cylinder	
Aligned, ($S_T/S_L > 0.70$)	$1000–2 \times 10^5$	0.27	0.63
Staggered ($S_T/S_L < 2.00$)	$1000–2 \times 10^5$	$0.35(S_T/S_L)^{1/5}$	0.60
Staggered ($S_T/S_L > 2.00$)	$1000–2 \times 10^5$	0.40	0.60
Aligned	$2 \times 10^5–2 \times 10^6$	0.021	0.84
Staggered	$2 \times 10^5–2 \times 10^6$	0.022	0.84

Several correlations are available in the literature concerning this subject. The correlation suggested by Zukauskas (1972) is recommended and it has the form

$$\mathrm{Nu}_d = C\mathrm{Re}_{d,\max}^m \mathrm{Pr}^{0.36}\left(\frac{\mathrm{Pr}}{\mathrm{Pr_w}}\right)^{1/4} \tag{23.17}$$

where $\mathrm{Nu}_d = h_{\mathrm{avg}}d/k$

h_{avg} = average heat transfer coefficient for the tube bank

d = thc tube diameter

$\mathrm{Re}_{d,\max} = d\hat{V}_{\max}/\nu$

$\hat{V}_{\max}$ = the maximum velocity within the tube bank

$\mathrm{Pr_w}$ = the Prandtl number evaluated at T_{w}

C, m = constants presented in Table 23.2

Fluid properties other than $\mathrm{Pr_w}$ are evaluated at the mean bulk temperature.

The maximum flow velocity can be determined using Figure 23.8. For flow normal to in-line tube banks, the maximum flow velocity will occur through the minimum frontal flow area, $S_T - d$. With the free-stream velocity designated as $\hat{V}_\infty$, the maximum velocity designated as $\hat{V}_{\max}$, will be

$$\hat{V}_{\max} = \hat{V}_\infty\left[\frac{S_T}{S_T - d}\right] \tag{23.18a}$$

For flow normal to staggered tube banks, the picture is a little more complicated. Equation 23.18a will hold and may be used if the longitudinal spacing, S_L, is large. However, for close longitudinal spacing

$$\hat{V}_{\max} = \hat{V}_\infty\left[\frac{S_T/2}{[(S_T)^2 + S_L^2]^{1/2} - d}\right] \tag{23.18b}$$

In this case, the bracketed terms in Equation 23.18 must both be evaluated and the larger of the two values selected.

Example 23.4 An air preheater uses condensing steam at 120°C on the inside of a bank of tubes to heat combustion air that enters the tube bank at atmospheric pressure and 15°C.

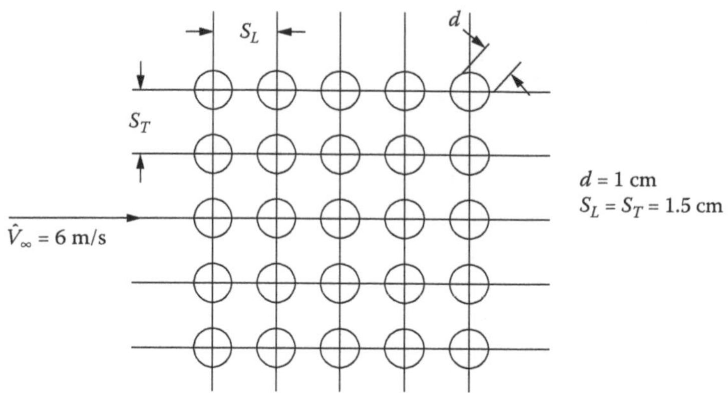

FIGURE 23.9
A portion of the tube bank used in Example 23.4.

The bank consists of an aligned array that is 15 tubes wide and 12 tubes deep. Each tube has an outside diameter of 10 mm and is 1.2-m long. The spacing parameters have the values $S_T = S_L = 15$ mm and the air velocity entering the tube bank is 6 m/s. Determine (a) the total heat transfer to the air and (b) the air temperature rise in the preheater.

Solution

Assumptions and Specifications

(1) Steady-state conditions exist.
(2) Radiation heat exchange with the surroundings is negligible.
(3) The heat transfer coefficient between the steam and the inner tube wall is high and the temperature drop across the tube wall is negligible.
(4) The tube bank is vertical.

A diagram of a portion of the tube bank is shown in Figure 23.9.

A stipulation on the use of Equation 23.17 is that fluid properties are evaluated at the average bulk temperature. Because we do not know the exiting temperature, a trial-and-error procedure is required.

We will assume that the air experiences a 30°C temperature rise and exits the tube bank at 45°C. The average bulk temperature is thus,

$$T_{b,\text{avg}} = \frac{15^\circ\text{C} + 45^\circ\text{C}}{2} = 30^\circ\text{C} \qquad (303\,\text{K})$$

Properties of air at this temperature may be obtained from Table A.13 (with interpolation)

$$\rho = 1.167\,\text{kg/m}^3 \qquad c_p = 1.006\,\text{kJ/kg-K}$$
$$\nu = 15.99 \times 10^{-6}\,\text{m/s}^2 \qquad \text{Pr} = 0.707$$
$$k = 0.0264\,\text{W/m-K}$$

and at $T_w = 120^\circ\text{C}$ (393 K)

$$\text{Pr}_w = 0.690$$

The solution for h includes the evaluation of Re_d which, in turn, requires the evaluation of $\hat{V}_{max}$.

For the aligned configuration of this problem, the minimum space between the tube rows, 5 mm, will be at a location where the maximum flow velocity, $\hat{V}_{max}$, will occur. Since $\hat{V}_\infty = 6\,\mathrm{m/s}$

$$\hat{V}_{max} = 6\,\mathrm{m/s}\left(\frac{S_T}{S_T - d}\right) = 6\,\mathrm{m/s}\left(\frac{1.5\,\mathrm{cm}}{1.5\,\mathrm{cm} - 1.0\,\mathrm{cm}}\right) = 18\,\mathrm{m/s}$$

We can now determine $\mathrm{Re}_{d,max}$ as

$$\mathrm{Re}_{d,max} = \frac{\hat{V}_{max} d}{\nu} = \frac{(18\,\mathrm{m/s})(0.010\,\mathrm{m})}{15.99 \times 10^{-6}\,\mathrm{m/s^2}} = 11{,}250$$

and then evaluate h via

$$h = \frac{k}{d}\left[C\mathrm{Re}_{d,max}^{m}\mathrm{Pr}^{0.36}\left(\frac{\mathrm{Pr}}{\mathrm{Pr_w}}\right)^{1/4}\right]$$

With $C = 0.27$ and $m = 0.63$ from Table 23.2 (the aligned arrangement), we have

$$\begin{aligned} h &= \frac{0.0264\,\mathrm{W/m\text{-}K}}{0.010\,\mathrm{m}}(0.27)(11{,}250)^{0.63}(0.707)^{0.36}\left(\frac{0.707}{0.690}\right)^{1/4} \\ &= 225.7\,\mathrm{W/m^2\text{-}K} \end{aligned}$$

The total surface area for $15 \times 12 = 180$ tubes will be

$$S = n\pi d L = 180\pi(0.010\text{ m})(1.2\,\mathrm{m}) = 6.786\,\mathrm{m^2}$$

The total heat transfer to the air is

$$\begin{aligned} \dot{Q} &= hS(T_w - T_\infty) \\ &= (225.7\,\mathrm{W/m^2\text{-}K})(6.786\,\mathrm{m^2})(120°\mathrm{C} - 30°\mathrm{C}) \\ &= 137{,}800\,\mathrm{W} \qquad (137.8\,\mathrm{kW}) \Longleftarrow \end{aligned}$$

This quantity of heat is related to the temperature rise of the air according to

$$\dot{Q} = \dot{m}c_p\Delta T_{air} = \rho A\hat{V}c_p\Delta T_{air}$$

The area in this expression is associated with the upstream velocity, $\hat{V}_\infty$. If we assume the space between the outside rows of the array and their adjacent walls to be $S_T/2$, the total width of the air duct will be

$$W = 15(0.015\,\text{m}) = 0.225\,\text{m}$$

and

$$A = WL = (0.225\,\text{m})(1.2\,\text{m}) = 0.27\,\text{m}^2$$

The temperature rise of the air can now be determined as

$$\begin{aligned}\Delta T_{\text{air}} &= \frac{\dot{Q}}{\rho A \hat{V} c_p} \\ &= \frac{137.8\,\text{kW}}{(1.167\,\text{kg/m}^3)(0.27\,\text{m}^2)(6\,\text{m/s})(1.006\,\text{kJ/kg-K})} \\ &= 72.5°\text{C} \Longleftarrow\end{aligned}$$

This result is higher than our original estimate; thus, an additional set of calculations must be made.

Assuming the exit air temperature to be 75°C, we evaluate the average bulk temperature to be

$$T_{b,\text{avg}} = \frac{15°\text{C} + 75°\text{C}}{2} = 45°\text{C} \qquad (318\,\text{K})$$

Properties of air at this temperature may be obtained from Table A.13 (with interpolation)

$$\rho = 1.113\,\text{kg/m}^3 \qquad c_p = 1.007\,\text{kJ/kg-K}$$

$$\nu = 17.51 \times 10^{-6}\,\text{m/s}^2 \qquad \text{Pr} = 0.704$$

$$k = 0.0276\,\text{W/m-K}$$

Repeating the calculations using these revised property values yields

$$\text{Re}_{d,\max} = \frac{\hat{V}_{\max} d}{\nu} = \frac{(18\,\text{m/s})(0.010\,\text{m})}{17.51 \times 10^{-6}\,\text{m/s}^2} = 10{,}280$$

$$h = \frac{k}{d}\left[C\text{Re}^m_{d,\max}\text{Pr}^{0.36}\left(\frac{\text{Pr}}{\text{Pr}_w}\right)^{1/4}\right]$$

With $C = 0.27$ and $m = 0.63$ from Table 23.2 (the aligned arrangement), we have

$$\begin{aligned}h &= \frac{0.0276\,\text{W/m-K}}{0.010\,\text{m}}(0.27)(10{,}280)^{0.63}(0.704)^{0.36}\left(\frac{0.704}{0.690}\right)^{1/4} \\ &= 222.4\,\text{W/m}^2\text{-K}\end{aligned}$$

$$\begin{aligned}\dot{Q} &= hS(T_w - T_\infty)\\ &= (222.4\,\text{W/m}^2\text{-K})(6.786\,\text{m}^2)(120°\text{C} - 45°\text{C})\\ &= 113{,}200\,\text{W} \qquad (113.2\,\text{kW}) \Longleftarrow\end{aligned}$$

$$\begin{aligned}\Delta T_{\text{air}} &= \frac{\dot{Q}}{\rho A\hat{V}c_p}\\ &= \frac{113.2\,\text{kW}}{(1.111\,\text{kg/m}^3)(0.27\,\text{m}^2)(6\,\text{m/s})(1.007\,\text{kJ/kg-K})}\\ &= 62.4°\text{C} \Longleftarrow\end{aligned}$$

An additional calculation is not necessary since this result is quite close to the assumed value. The results are

$$\dot{Q} = 113.2\ \text{kW} \Longleftarrow$$

$$\Delta T_{\text{air}} = 62.4°\text{C} \Longleftarrow$$

23.3.3 Single Spheres

Research on convection heat transfer with flow over single spheres has been conducted by several investigators. The relatively recent correlation of Whitaker (1972) is recommended. The Whitaker relationship is

$$\text{Nu}_d = 2.0 + \left[0.40\text{Re}_d^{1/2} - 0.06\text{Re}_d^{2/3}\right]\text{Pr}^{0.40}\left(\frac{\mu_\infty}{\mu_w}\right)^{1/4} \tag{23.19}$$

and is applicable for air and other fluids with

$$0.71 < \text{Pr} < 380$$
$$3.50 < \text{Re}_d < 7.6 \times 10^4$$
$$1 < \mu_\infty/\mu_w < 3.20$$

with all properties except μ_w evaluated at T_∞.

23.3.4 Bodies with Noncircular Cross Sections

Equation 23.1, presented in Section 23.1, represents the common form used in correlating experimental data for two-dimensional bluff bodies in cross flow. A modified form of this expression has been presented by Hilpert (1933).

$$\text{Nu}_d = C\text{Re}^a\text{Pr}^{1/4} \tag{23.20}$$

with values of the constants C and a for different cross-sectional geometries given in Table 23.3.

TABLE 23.3

Values of Constants C and n for use in Equation 23.20

(1) Air flow to corner of square rod

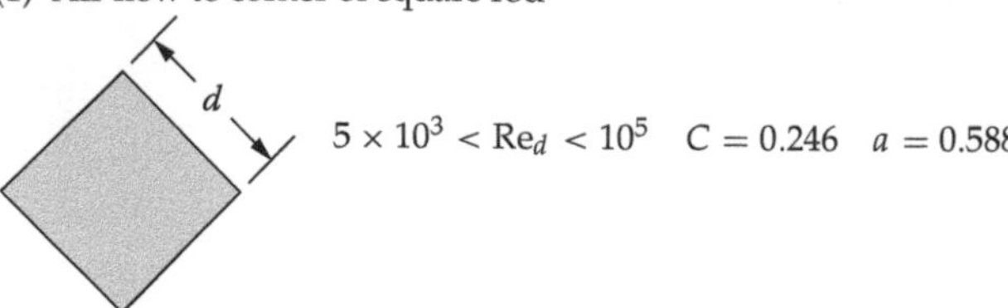

$5 \times 10^3 < \mathrm{Re}_d < 10^5$ $C = 0.246$ $a = 0.588$

(2) Air flow to face of square rod

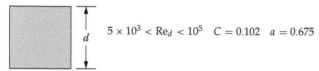

d $5 \times 10^3 < \mathrm{Re}_d < 10^5$ $C = 0.102$ $a = 0.675$

(3) Air flow to face of hexagonal rod

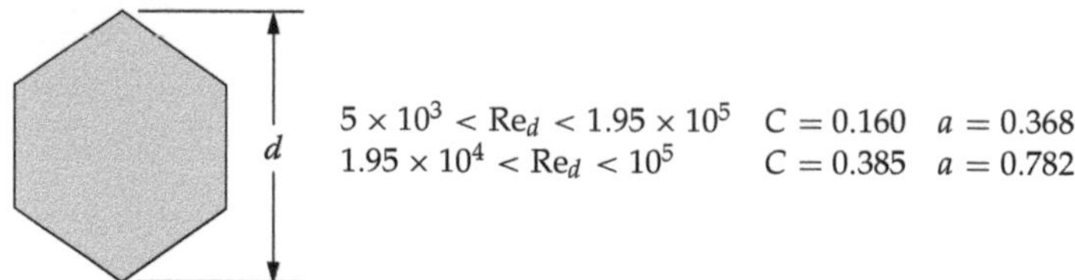

$5 \times 10^3 < \mathrm{Re}_d < 1.95 \times 10^5$ $C = 0.160$ $a = 0.368$

$1.95 \times 10^4 < \mathrm{Re}_d < 10^5$ $C = 0.385$ $a = 0.782$

(4) Air flow to corner of hexagonal rod

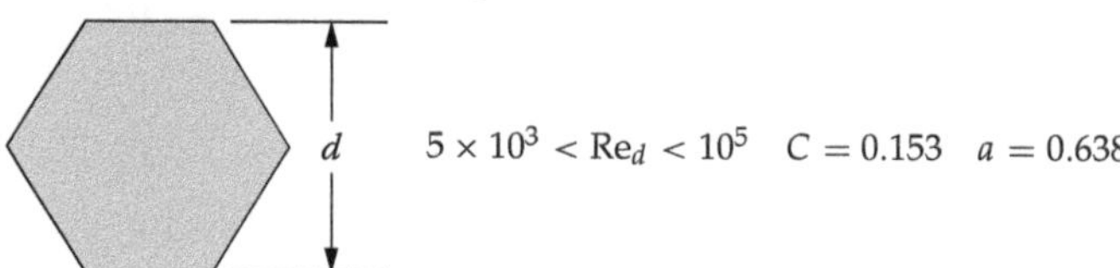

$5 \times 10^3 < \mathrm{Re}_d < 10^5$ $C = 0.153$ $a = 0.638$

(5) Air flow to normal to vertical strip

d $4 \times 10^3 < \mathrm{Re}_d < 1.5 \times 10^5$ $C = 0.228$ $a = 0.731$

For each of the shapes considered in Table 23.3, it is presumed that the length, that is, the axial dimension of the body, is very large relative to the characteristic cross-section dimension, d.

Attention now turns to Design Example 15, which involves the design of a duct heater.

23.4 Design Example 15

A bundle of 1-cm-diameter tubes is arranged in a staggered array to heat air. The duct width and tube length are both 1 m and the number of tube rows deep is to be determined. The bundle is 6 tubes high and the spacings are $S_T = 2$ cm and $S_L = 1$ cm. The height of the duct is 11.75 cm. Air enters the heater at 17°C at a mass flow rate of 0.50 kg/s and is to leave at 87°C. The heating is accomplished by condensing steam within the tubing at 150 kPa. Determine the number of tube rows deep to accomplish the heating.

Solution

Assumptions and Specifications

(1) Steady-state conditions exist.
(2) The temperature of the tube wall may be taken as the temperature of the condensing steam.
(3) Radiation heat exchange with the surroundings is negligible.

The average bulk temperature is

$$T_{b,\text{avg}} = \frac{17°\text{C} + 87°\text{C}}{2} = \frac{104°\text{C}}{2} = 52°\text{C} \qquad (325\,\text{K})$$

and properties of air at this temperature may be found in Table A.13 (with interpolation).

$$\rho = 1.0877\,\text{kg/m}^3 \qquad c_p = 1.0074\,\text{kJ/kg-K}$$

$$\nu = 18.23 \times 10^{-6}\text{m}^3 \qquad \text{Pr} = 0.703$$

$$k = 0.0281\,\text{W/m-K}$$

Table A.4 indicates that at 150 kPa, the saturation temperature is $T_w = 111.4°\text{C}$. The Prandtl number for air at this temperature is

$$\text{Pr}_w = 0.695$$

We must evaluate the bracketed terms in Equations 23.18. With $S_T = 2\,\text{cm}$ and $d = 1\,\text{cm}$, Equation 23.18a gives

$$\frac{S_T}{S_T - d} = \frac{2\,\text{cm}}{2\,\text{cm} - 1\,\text{cm}} = \frac{2\,\text{cm}}{1\,\text{cm}} = 2.0$$

and, in addition to S_T and d, with $S_L = 1\,\text{cm}$, Equation 23.18b yields

$$\frac{S_T/2}{[(S_T/2)^2 + S_L^2]^{1/2} - d} = \frac{1\,\text{cm}}{[(1\,\text{cm})^2 + (1\,\text{cm})^2]^{1/2} - 1\,\text{cm}}$$

$$= \frac{1\,\text{cm}}{\sqrt{2}\,\text{cm} - 1\,\text{cm}} = 2.414$$

The foregoing indicates that Equation 23.18b should be employed and that

$$\hat{V}_{\text{max}} = 2.414\hat{V}_\infty$$

The duct area normal to the flow will be

$$A = W(H - 6d) = (1.00\,\text{m})[0.1175\,\text{m} - 6(0.01\,\text{m})] = 0.0575\,\text{m}^2$$

and this makes the free-stream velocity

$$\hat{V}_\infty = \frac{\dot{m}}{\rho A} = \frac{0.50\,\mathrm{kg/s}}{(1.0877\,\mathrm{kg/m^3})(0.0575\,\mathrm{m^2})} = 7.99\,\mathrm{m/s}$$

and

$$\hat{V}_{\mathrm{max}} = 2.414\hat{V}_\infty = 2.414(7.99\,\mathrm{m/s}) = 19.30\,\mathrm{m/s}$$

We can now determine $\mathrm{Re}_{d,\mathrm{max}}$

$$\mathrm{Re}_{\mathrm{d,max}} = \frac{\hat{V}_{\mathrm{max}} d}{\nu} = \frac{(19.30\,\mathrm{m/s})(0.01\,\mathrm{m})}{18.23 \times 10^{-6}\mathrm{m^3}} = 10{,}586$$

This indicates that the flow is turbulent. We can evaluate the heat transfer coefficient, h, via Equation 23.17

$$h = \frac{k}{d}\left[C\mathrm{Re}_{d,\mathrm{max}}^{m}\mathrm{Pr}^{0.36}\left(\frac{\mathrm{Pr}}{\mathrm{Pr}_w}\right)^{1/4}\right]$$

With C and m taken from Table 23.3 (the staggered arrangement), we have with $S_T/S_L = 2\,\mathrm{cm}/1\,\mathrm{cm} = 2.00$

$$C = 0.35\left(\frac{S_T}{S_L}\right)^{1/5} = 0.35(2.00)^{1/5} = 0.40$$

and

$$m = 0.60$$

Hence,

$$\begin{aligned} h &= \frac{0.0281\,\mathrm{W/m\text{-}K}}{0.01\,\mathrm{m}}\left[(0.40)(10{,}586)^{0.60}(0.703)^{0.36}\left(\frac{0.703}{0.695}\right)^{1/4}\right] \\ &= 258.2\,\mathrm{W/m^2\text{-}K} \end{aligned}$$

The surface area per row will be designated as S_1

$$S_1 = 6\pi d L = 6\pi(0.01\,\mathrm{m})(1\,\mathrm{m}) = 0.1885\,\mathrm{m^2}$$

and the heat duty is

$$\begin{aligned} \dot{Q} &= \dot{m}c_p(T_2 - T_1) \\ &= (0.50\,\mathrm{kg/s})(1.0074\,\mathrm{kJ/kg\text{-}K})(87^\circ\mathrm{C} - 17^\circ\mathrm{C}) = 35.26\,\mathrm{kW} \end{aligned}$$

The total required surface area can be obtained from

$$-\frac{hS}{\dot{m}c_p} = \ln\frac{T_W - T}{T_W - T_1}$$

and with $T_1 = 17°C$ and $T = T_2 = 87°C$

$$-\frac{hS}{\dot{m}c_p} = \ln\frac{111.4°C - 87°C}{111.4°C - 17°C} = \ln\frac{24.4°C}{94.4°C} = -1.3530$$

This yields the total surface needed for the given duty

$$S = \frac{(1.3530)(0.50\,\text{kg/s})(1.0074\,\text{kJ/kg-K})}{258.2\,\text{W/m}^2\text{-K}} = 2.640\,\text{m}^2$$

and with $S_1 = 0.1885\,\text{m}^2$, the number of rows deep will be

$$n = \frac{S}{S_1} = \frac{2.640\,\text{m}^2}{0.1885\,\text{m}^2} = 14\,\text{rows} \Longleftarrow$$

23.5 Summary

Forced convection heat transfer for flow over the external surfaces of some common geometric configurations has been examined in this chapter.

A common form for expressing the Nusselt number for these configurations is

$$\text{Nu}_L = C\text{Re}^a\text{Pr}^b \tag{23.1}$$

where the constants, a, b, and C have been determined by numerous experiments over time.

The geometries considered and the empirical equations for the applicable Nusselt numbers are as follows:

- For flow parallel to a plane wall with $\text{Re}_L < 5 \times 10^5$

$$\text{Nu}_L = \frac{0.677\text{Re}_L^{1/2}\text{Pr}^{1/3}}{\left[1 + \left(\frac{0.0468}{\text{Pr}}\right)^{2/3}\right]^{1/4}} \tag{23.6}$$

- For flow parallel to plane surfaces with $3 \times 10^6 < \text{Re}_L$ and $0.60 < \text{Pr} < 60$

$$\text{Nu}_L = \frac{h_L L}{k} = 0.036\text{Re}_L^{4/5}\text{Pr}^{1/3} \tag{23.8}$$

- For flow parallel to plane surfaces with $5 \times 10^5 < \text{Re}_L < 10^7$, $0.60 < \text{Pr} < 60$ and $\text{Re}_{x,c}$

$$\text{Nu}_L = \frac{h_L L}{k} = \left[0.036\text{Re}_L^{4/5} - 871\right]\text{Pr}^{1/3} \tag{23.10}$$

- For a cylinder in cross flow

$$\mathrm{Nu}_d = 0.30 + \frac{0.62\mathrm{Re}_d^{1/2}\mathrm{Pr}^{1/3}}{\left[1+\left(\dfrac{0.40}{\mathrm{Pr}}\right)^{2/3}\right]^{1/4}}\left[1+\left(\frac{\mathrm{Re}_d}{282{,}000}\right)^{5/8}\right]^{4/5} \tag{23.16}$$

- For tube banks in cross flow

$$\mathrm{Nu}_d = C\mathrm{Re}_{d,\mathrm{max}}^{m}\mathrm{Pr}^{0.36}\left(\frac{\mathrm{Pr}}{\mathrm{Pr}_\mathrm{w}}\right)^{1/4} \tag{23.17}$$

where the values of C and m may be taken from Table 23.2.

- For spheres in cross flow

$$\mathrm{Nu}_d = 2.0 + [0.40\mathrm{Re}_d^{1/2} - 0.06\mathrm{Re}_d]^{2/3}\mathrm{Pr}^{0.40}\left(\frac{\mu_\infty}{\mu_\mathrm{w}}\right)^{1/4} \tag{23.19}$$

- For bodies of noncircular cross section

$$\mathrm{Nu}_d = C\mathrm{Re}^{a}\mathrm{Pr}^{1/4} \tag{23.20}$$

where the values of C and a are obtained from Table 23.3.

In all of the foregoing equations, except where specifically noted, fluid properties are evaluated at the film temperature defined as

$$T_f = \frac{T_\mathrm{w} + T_\infty}{2} \tag{23.2}$$

23.6 Problems

Flow over Flat Surfaces

23.1: Glycerin flows at 40°C and at a velocity of 3.96 m/s past a flat surface that is maintained at 10°C. The plate is 0.375-m long and 0.625-m wide. Determine the heat transferred to one side of the plate.

23.2: Air at atmospheric pressure and 20°C flows at 25 m/s past a flat surface that is kept throughout at 80°C. The surface is 12-m long. Determine the average heat transfer coefficient.

23.3: Air at atmospheric pressure and 15°C flows with a velocity of 6 m/s past a flat surface that is 56-cm long and is maintained at 75°C. Determine (a) the local heat transfer coefficient at locations of 14 cm, 28 cm, 42 cm, and 56 cm from the leading edge and (b) the average heat transfer coefficient for one side of the surface.

23.4: Water at 85°C flows with a velocity of 1.25 m/s past a flat surface maintained at 45°C. The total length of the surface is 15 cm. Determine (a) the local heat transfer coefficient at locations of 1 cm, 5 cm, 10 cm, and 15 cm from the leading edge and (b) the average heat transfer coefficient for one side of the surface.

23.5: Atmospheric air at 101.2 kPa and −15°C flows past a flat surface that is maintained at 40°C. The surface is 0.675-m long and 0.3125-m wide. Estimate the air velocity that is required to remove 80 W from one side of the surface.

23.6: A flat plate that is 1.25 cm × 0.5 m is towed, through unused engine oil, parallel to its long side, at a velocity of 1.85 m/s. The oil is at 65°C and the plate is maintained by an electrical heater at 89°C. Determine the power required to heat the plate.

23.7: Nitrogen at atmospheric pressure and 55°C flows with a velocity of 20 m/s past a flat surface held at 5°C. The surface is 87.5-cm long and 25-cm wide. Determine (a) the total heat transferred to one side of the surface and (b) the fraction of the heat transferred in the laminar portion of the boundary layer.

23.8: A flat plate has dimensions 7.5 cm × 45 cm and is placed in an airstream at atmospheric pressure and 10°C. The plate is maintained at 90°C, and the air flows with a velocity of 30 m/s. Determine the total heat transferred from one side of the plate when the leading edge is (a) the 7.5-cm side and (b) the 45-cm side.

23.9: A flat plate is 20-cm long and 10-cm wide. It is placed in an airstream at atmospheric pressure and 50°C. The plate is maintained at 100°C and the air flows at a Reynolds number, Re_L of 42,500. Determine (a) the total heat transferred from one side of the plate and (b) the total heat transferred if the velocity is doubled and the pressure is increased to 1 MPa.

23.10: Air flows with a velocity of 17.5 m/s at a temperature of 17°C past a flat surface that is 20-cm wide. Determine the heat transferred and the skin friction coefficient at a point 2.5 cm from the leading edge if the plate is maintained at (a) 27°C, (b) 37°C, and (c) 47°C.

23.11: A fluid flows at 4.8 m/s over a flat plate that is 18-cm long. The fluid temperature is 37°C and the plate temperature is 17°C. Determine the Reynolds number at the downstream end of the plate and state whether the flow is laminar, turbulent, or in transition for (a) air, (b) unused engine oil, (c) gaseous CO_2, and (d) water.

23.12: Ethylene glycol at 127°C flows over and is parallel to a flat plate held at 27°C at a velocity of 3.6 m/s. Determine the thickness of the hydrodynamic boundary layer at 0.325 cm from the leading edge.

23.13: An aluminum fin ($k = 190$ W/m-K) on the top of an electronic chassis is 7.5-cm high, 2.286-mm thick, and 0.50-m long. Air flows in the long direction at a velocity of 2.25 m/s and a temperature of 17°C. The base temperature of the fin is 87°C. Determine the total heat transferred from the fin to the air.

23.14: Air at a pressure of 300 kPa flows across a 0.875-m square flat plate at a velocity of 67.5 m/s. The plate is maintained at a temperature of 180°C and the free stream air temperature is 30°C. Determine the heat loss from one side of the plate.

23.15: The surface of a 1.6-m-long flat plate is maintained at 45°C and water at a temperature of 5°C and a velocity of 0.625 m/s flows across the plate in the length direction. Determine the heat transfer coefficient.

23.16: Air at a pressure of 100 kPa and a temperature of 47°C flows over a surface having a length of 0.25 m and a width of 0.125 m. The temperature of the plate is uniform at 97°C and the Reynolds number, based on the plate length, is $Re_L = 42,500$. Determine the rate of heat transfer from the plate to the air.

23.17: Rework Problem 23.16 for air at 1 MPa and a doubled air velocity.

23.18: Both sides of a flat plate 1-m long on a side are cooled by an airstream flowing with a velocity of 8 m/s at 20°C. The plate is held at 95°C. Determine the heat loss from the plate.

Flow across Single Cylinders

23.19: A wire that is 0.025 cm in diameter is placed normal to an airstream at atmospheric pressure and 17°C. When the wire is heated electrically to a surface temperature of 100°C, the measured heat loss from the wire is 40 W/m of length. Determine the velocity of the airstream.

23.20: A fluid at 17°C flows with a velocity of 12 m/s across a 6-cm-outside-diameter tube whose surface is maintained at 97°C. Determine the heat transfer rate per meter of length if the fluid is (a) air at atmospheric pressure, (b) water, and (c) unused engine oil.

23.21: A fluid at 37°C flows with a velocity of 10 m/s across a 5-cm-outside-diameter cylinder whose surface is maintained at 117°C. Determine the heat transfer rate per meter of length if the fluid is (a) air and (b) nitrogen.

23.22: Hot combustion gases, having the properties of air, flow at a pressure of 180 kPa and a temperature of 67°C normal to a 5-cm cylinder at a velocity of 16 m/s. The cylinder is held at 117°C. Determine the surface heat transfer coefficient.

23.23: Air at 27°C flows with a velocity of 32 m/s normal to a 2.5-cm-outer-diameter cylinder held at 110°C. Determine the heat transfer coefficient per unit of length for air at a pressure of (a) 1 atm, (b) 200 kPa, and (c) 400 kPa.

23.24: Air at 90°C and atmospheric pressure flows past a heated wire that is 0.15 cm in diameter. The air velocity is 6.25 m/s, and the surface temperature of the wire is 150°C. Determine the heat loss from the wire per meter of wire length.

23.25: Ethylene glycol at 47°C flows across a horizontal cylinder at a velocity of 3 m/s. The cylinder has a diameter of 3.2 cm and a length of 1.2 m and is maintained at 62°C. Determine the heat lost by the cylinder.

23.26: Nitrogen at 200 kPa and 7°C flows across a 5-cm-diameter horizontal cylinder at a velocity of 5 m/s. The cylinder is 1.25-m long and is maintained at 77°C. Determine the heat lost by the cylinder.

23.27: Air at atmospheric pressure and 12°C flows across a horizontal cylinder that is maintained at 52°C at a velocity of 20 m/s. The cylinder has a diameter of 4.5 cm and a length of 1 m. Determine the heat lost by the cylinder.

23.28: Air at 200 kPa and 7°C flows across a horizontal cylinder at a velocity of 25 m/s. The cylinder has a diameter of 18 cm and a length of 1 m and is maintained at 77°C. Determine the heat lost by the cylinder.

23.29: Air at 75 kPa and 22°C flows across a horizontal cylinder at a velocity of 16 m/s. The cylinder has a diameter of 4.8 cm, a length of 0.75 m, and a surface temperature that is maintained at 62°C. Determine the heat lost by the cylinder.

23.30: A heated 2.5-cm-diameter cylinder at 400 K is placed in an airstream at 100 kPa and 320 K. The air velocity is 32 m/s and the length of the cylinder is 0.875 m. Determine the heat lost by the cylinder.

23.31: Air at 100 kPa and 87°C flows across a heated 1.6-mm-diameter wire at a velocity of 6.25 m/s. The wire is maintained at 147°C, and is 0.75-m long. Determine the heat transferred from the wire.

23.32: Carbon dioxide at atmospheric pressure and 52°C flows across a horizontal wire at a velocity of 10 m/s. The wire has a diameter of 0.35 cm and is maintained at 152°C. Determine the heat transfer per unit length of wire.

23.33: Determine the heat transferred from 1 m of very thin wire (0.025 mm) maintained at 67°C if the fluid is (a) air at 100 kPa and 20°C flowing at 2.5 m/s and (b) water at 22°C flowing at 6.25 m/s.

23.34: Oil in an Arctic pipeline is heated at various stations along the line in order enhance its flow capability. The 50-cm-diameter pipeline carries hot oil (SAE 50) at 50°C and there is a strong wind at 100 kPa and −30°C blowing at 12.5 m/s. Assuming that the surface of the pipe is at 50°C, determine the heat lost by the pipe per kilometer of length.

23.35: A 4-cm-diameter cylinder is subjected to a cross flow of CO_2 at 100 kPa and 180°C. The cylinder is maintained at a constant temperature of 80°C and the CO_2 velocity is 36 m/s. Determine the heat transferred to the cylinder per meter of length.

23.36: Oxygen at 175 kPa and 20°C is forced across a horizontal cylinder at a velocity of 48 m/s. The cylinder has a diameter of 24 cm, a length of 5 m, and is maintained at 100°C. Determine the heat lost by the cylinder.

23.37: Carbon dioxide at 100 kPa and 30°C flows across a horizontal cylinder at a velocity of 36 m/s. The cylinder has a diameter of 0.625 cm, a length of 12 cm, and is maintained at 180°C. Determine the heat lost by the cylinder.

23.38: Carbon dioxide at 100 kPa and 300 K flows across a horizontal cylinder at a velocity of 48 m/s. The cylinder has a diameter of 18 cm, a length of 1 m, and is maintained at 400 K. Determine the heat lost by the cylinder.

Tube Banks in Cross Flow

23.39: A tube bank employs an in-line arrangement with $S_T = S_L = 3.2$ cm and tubes that are 2 cm in outside diameter. There are 10 rows of tubes that are held at a surface temperature of 85°C. Air at 100 kPa and 20°C flows normal to the tubes with a free-stream velocity of 6 m/s. The tube bank is ten rows deep and the tubes are 1.6-m long. Determine the amount of heat transferred.

23.40: Rework Problem 23.39 for a staggered arrangement.

23.41: A tube bank employs an in-line arrangement with $S_T = 3.6$ cm and $S_L = 3.2$ cm and tubes that are 2 cm in outside diameter. There are eight rows of tubes that are held at a surface temperature of 10°C. Air at 100 kPa and 35°C flows normal to the tubes with a free-stream velocity of 6.25 m/s. The tube bank is 10 rows deep and the tubes are 1.8-m long. Determine the amount of heat transferred.

23.42: Rework Problem 23.41 for a staggered arrangement.

23.43: A tube bank employs tubes with $S_T = S_L = 1.625$ cm and tubes that are 1.25 cm in outside diameter. There are eight rows of tubes that are held at a surface temperature of 85°C. Air at 100 kPa and 35°C flows normal to the tubes with a free-stream velocity of 1.25 m/s. The tube bank is eight rows deep and the tubes are 1.8-m long. Determine the amount of heat transferred for (a) an in-line arrangement.

23.44: Rework Problem 23.43 for a staggered arrangement. All other conditions remain the same.

23.45: Water at a velocity of 2 m/s and a bulk temperature of 20°C flows normal to an in-line tube bundle containing 1-cm-diameter tubes with a transverse spacing of $S_T = 2$ cm and a longitudinal spacing of $S_L = 1.40$ cm. The surface temperature of the tubes is 60°C. Estimate the heat transfer coefficient.

23.46: Rework Problem 23.45 for a staggered arrangement. All other conditions remain the same.

23.47: Hydrogen at 1 atmosphere and a bulk temperature of 127°C is cooled in a bundle of tubes in an in-line arrangement at a tube wall temperature of 27°C. The transverse tube spacing, S_T, is 3 cm, the longitudinal tube spacing, S_L, is 2 cm, and the tube diameter is 1.75 cm. For a-free stream velocity of 5 m/s, estimate the heat transfer coefficient.

23.48: Rework Problem 23.47 for a staggered arrangement. All other conditions remain the same.

23.49: Ethylene glycol flows across a rather large tube bundle containing 5-cm-outside-diameter tubes having a transverse spacing, S_T, of 7.50 cm and a longitudinal spacing of S_L, of = 4.75 cm. The glycol enters the bundle at 17°C and leaves at 37°C, and flows at a free-stream air velocity of 10 m/s. The tube surface temperature is 47°C. Estimate the heat transfer coefficient.

23.50: Rework Problem 23.49 for an in-line arrangement. All other conditions remain the same.

Flow over Spheres

23.51: Air at 15°C and 100 kPa flows at a velocity of 10 m/s over a sphere that has a diameter of 10 cm. Determine the heat transferred if the sphere is held at 30°C.

23.52: Water at a free-stream temperature of 37°C flows over a 4-mm-diameter sphere at a velocity of 5 m/s. The sphere is maintained at 97°C. Determine the heat transferred.

23.53: An electrically heated small sphere having a diameter of 8 mm has a surface temperature maintained at 200°C. The sphere is exposed to an airstream at 40°C and atmospheric pressure at a velocity of 18 m/s. Determine the power, in watts, that must be supplied to the sphere.

23.54: A heated sphere having a diameter of 3 cm has a surface temperature maintained at 97°C. The sphere is exposed to a water stream at 27°C flowing at a velocity of 3.75 m/s. Determine the heat lost by the sphere.

23.55: A spherical tank having a diameter of 0.825 m has its surface temperature maintained at 35°C. An airstream at 15°C and 100 kPa at a velocity of 6.25 m/s flows across the tank. Determine the heat lost by the tank.

23.56: Water at a free-stream temperature of 17°C flows over a 2.5-cm-diameter sphere at a velocity of 3.6 m/s. The sphere is maintained at 67°C. Determine the heat transferred.

23.57: Water at 27°C flows over a 2.75-cm-diameter sphere with a velocity of 4.8 m/s. The surface of the sphere is at 62°C. Determine the rate of heat transfer from the sphere.

23.58: Air at 27°C flows over a 1.25-cm-diameter sphere with a velocity of 20 m/s. The surface of the sphere is at 75°C. Determine the rate of heat transfer from the sphere.

23.59: A fluid at 30°C flows over a 2-cm-diameter sphere with a velocity of 3 m/s. The surface of the sphere is at 60°C. Determine the heat transfer from the sphere when the fluid is (a) water and (b) air at atmospheric pressure.

23.60: A 100-W lightbulb may be approximated by a sphere of 5-cm diameter. Air at 27°C and a pressure of 100 kPa flows over the bulb at a velocity of 0.625 m/s. The surface of the bulb is held at 150°C. Determine the rate of heat transfer from the bulb.

Flow over Bluff Bodies

23.61: Hot combustion gases, having the properties of air, flow at a pressure of 180 kPa and a temperature of 67°C normal to a square pipe that is 4 cm on a side. The flow velocity is 16 m/s. The pipe is held at 117°C. Determine the surface heat transfer coefficient.

23.62: Air at 100 kPa and a free-stream temperature of 27°C flows normal to a square rod such that the Reynolds number is 12,000. Compare the heat transfer from the rod with that of a cylinder having the same diameter as the side of the square.

23.63: Air at 100 kPa and 300 K flows normal to a 7.5-mm heated strip maintained at 580 K. The air velocity is such that the Reynolds number is 16,000. Determine the heat loss for a 60-cm-long strip.

23.64: Rework Problem 23.62 for the case of the airstream flowing normal to one corner of the square rod.

23.65: Air flows normal to one face of a hexagonal rod that is 1 cm on a side. The rod is maintained at 80°C and the airstream is at 100 kPa and 20°C. The rod is 40-cm long. Determine the heat transferred from the rod to the air.

23.66: Rework Problem 23.64 for the case of the airstream striking the hexagonal rod at one of the vertices.

24

Free or Natural Convection

Chapter Objectives

- To describe the phenomenon of natural convection.
- To define the Grashof, Rayleigh, and Elenbaas numbers.
- To provide a summary of working correlations for the coefficient of heat transfer in natural convection.
- To consider the effects of plate spacing in an array of heated vertical plates in natural convection.
- To examine natural convection in enclosures.

24.1 Introduction

In our discussion of forced convection in Chapters 22 and 23, we noted that the convecting fluid was **forced** through or past the heat transfer surface by an external mover such as a pump, fan, or blower. In **free** or **natural convection**, the fluid is set into motion by density differences between the confining surface and the bulk of the fluid. These density differences give rise to **buoyant forces** that circulate the convective fluid (either a liquid or a gas) and transfer heat between the surface and the fluid.

If we consider a heated vertical plate in cooler air, it is apparent that, because the density of air is inversely proportional to the temperature, there is a region of lower density air adjacent to the plate. This air will tend to rise, and the cooler air from the bulk of the adjacent fluid will replace it. Thus, there will be a continuous flow of air in a "channel" around the plate, and this flow, referred to as a **natural convection current**, leads to what is called **natural convection heat transfer**.

Temperature and velocity fields associated with **laminar natural convection** are typified by the data of Schmidt and Beckmann (1930) shown in Figures 24.1 and 24.2 for a hot plate immersed in initially quiescent air. At each x location, the vertical air velocity is seen to peak, at a modest distance from the plate and decrease toward zero at greater lateral distances. We may define the **velocity boundary layer thickness** as the value of the lateral distance at which the velocity has decreased to 1% of its peak value. We then note that the thickness of the velocity boundary layer grows monotonically and the peak velocity shifts further and further from the wall at decreasing distances from the leading edge.

In a similar fashion, we observe that the temperature is seen to grow along the plate and to lead to progressively lower temperature gradients in the fluid adjacent to the plate. This, in turn, leads to lower heat transfer rates from the surface of the plate to the air. At larger

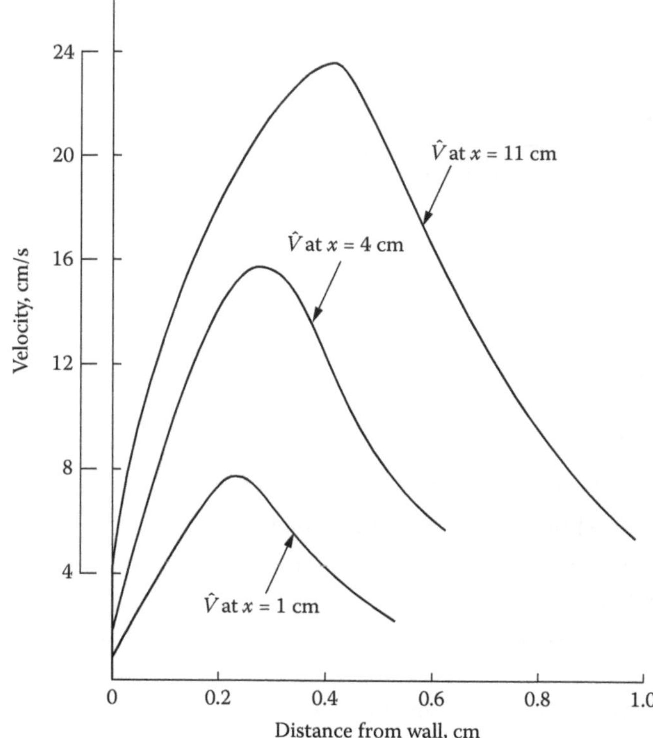

FIGURE 24.1
Velocity profiles in air adjacent to a vertical plate. The leading edge of the plate is at $x = 0$ and y is the distance normal to the plate. (After Schmidt and Beckmann, 1930.)

plate-to-air temperature differences and/or distances from the leading edge, turbulent flow is encountered and the temperature and velocity profiles attain an asymptotic, that is, x-invariant form.

24.2 Governing Parameters

It makes no difference whether the surface is hotter than the environment or the environment is hotter than the surface. The results will be identical. We will work with a hot surface and a cooler environment, and we first define the temperature difference between the surface at T_s, and the cooler surrounding environment at T_∞ as

$$\Delta T = T_s - T_\infty$$

so that the film temperature may be defined as

$$T_f = T_s - \frac{\Delta T}{2}$$

The fluid at the film temperature is related to the average temperature difference through the **volumetric coefficient of thermal expansion** taken as

$$\beta = -\frac{1}{\rho}\frac{\partial \rho}{\partial T}\bigg|_{P=C} \approx -\frac{1}{\rho}\frac{\Delta \rho}{\Delta T}\bigg|_{P=C} \qquad (\mathrm{K}^{-1}) \tag{24.1}$$

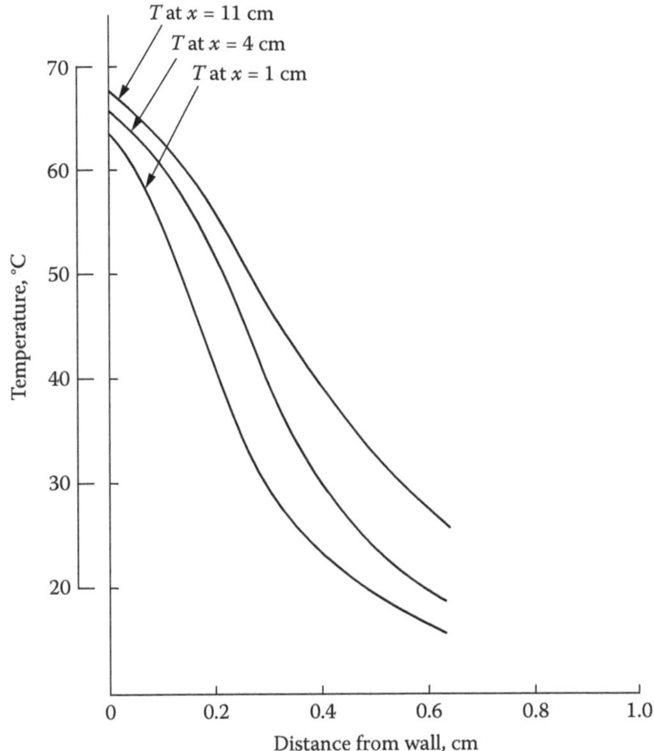

FIGURE 24.2
Temperature profiles in air adjacent to a vertical plate. The leading edge of the plate is at $x = 0$ and y is the distance normal to the plate. (After Schmidt and Beckmann, 1930.)

so that

$$|\Delta\rho| \approx \rho\beta\Delta T \qquad (\text{kg/m}^3) \tag{24.2}$$

For an ideal gas, the volumetric coefficient of thermal expansion can be evaluated by taking the reciprocal of the absolute temperature. For liquids, it is a tabulated property.

When Equation 24.2 is multiplied by g, we obtain the buoyant force per cubic meter of fluid

$$F_b = \rho g\beta\Delta T$$

and we see that the work done per cubic meter when this force acts through a height L will be

$$F_b L = \rho g\beta\Delta T L$$

If this work is converted into kinetic energy in the channel adjacent to the plate where the natural convection current flows, we have

$$\rho g\beta\Delta T L = \frac{\rho\hat{V}^2}{2}$$

and this produces a natural convection reference velocity of the form

$$\hat{V} = (g\beta L\Delta T)^{1/2} \tag{24.3}$$

It seems reasonable to relate the Nusselt representation (called the **Nusselt equation**) given by Equation 23.1

$$\text{Nu} = C\text{Re}^m\,\text{Pr}^n$$

to natural convection. To do this, we must use the velocity given by Equation 24.3 in the Reynolds number. With L taken as the significant length dimension, we write

$$\text{Re} = \frac{\rho L\hat{V}}{\mu} = \frac{\rho L(g\beta L\Delta T)^{1/2}}{\mu}$$

so that Nusselt's equation can be written with account taken of the square root of the velocity as

$$\text{Nu} = C\left[\frac{\rho^2 L^3 g\beta\Delta T}{\mu^2}\right]^{m/2}\left(\frac{c_p\mu}{k}\right)^n$$

The dimensionless term within the square brackets is known as the **Grashof number**

$$\text{Gr} \equiv \frac{\rho^2 g\beta L^3\Delta T}{\mu^2} \tag{24.4a}$$

which can also be defined using the kinematic viscosity as

$$\text{Gr} \equiv \frac{g\beta L^3\Delta T}{\nu^2} \tag{24.4b}$$

and both forms are sometimes referred to as the **natural convection modulus**. We are then able to write Nusselt's equation as

$$\text{Nu} = C\,\text{Gr}^{m/2}\,\text{Pr}^n$$

The results of exhaustive testing with gases and liquids indicate that $m/2$ and n are numerically equal so that we can write

$$\text{Nu} = C\left(\frac{\rho^2 c_p g\beta L^3\Delta T}{\mu k}\right)^n = C\,\text{Ra}^n \tag{24.5}$$

where the term within the parentheses is the dimensionless parameter known as the **Rayleigh number**

$$\text{Ra} \equiv \text{Gr}\,\text{Pr} = \frac{\rho^2 c_p g\beta L^3\Delta T}{\mu k} \tag{24.6}$$

We now turn to several correlations for the heat transfer coefficient in natural convection.

24.3 Working Correlations for Natural Convection

24.3.1 Introduction

We now list several natural convection correlations for various surface arrangements. All pertain to an isothermal surface and require that thermal properties be evaluated at the film temperature.

$$T_f = \frac{T_s + T_\infty}{2}$$

24.3.2 Plane Surfaces

24.3.2.1 Vertical Plates

McAdams (1954) gives, on the basis of the plate height, L

$$\frac{hL}{k} = 0.55\,\text{Ra}^{1/4} \qquad (10^3 < \text{Ra} < 10^9) \tag{24.7a}$$

and

$$\frac{hL}{k} = 0.13\,\text{Ra}^{1/3} \qquad (10^9 < \text{Ra}) \tag{24.7b}$$

Churchill and Chu (1975b) correlated the results of studies of several investigators and recommend on the basis of the plate length, L

$$\frac{hL}{k} = 0.68 + 0.670\,\text{Ra}^{1/4}\left[1 + \left(\frac{0.492}{\text{Pr}}\right)^{9/16}\right]^{-4/9} \qquad (0 < \text{Ra} < 10^9) \tag{24.8a}$$

and

$$\frac{hL}{k} = \left\{0.825 + 0.387\,\text{Ra}^{1/6}\left[1 + \left(\frac{0.492}{\text{Pr}}\right)^{9/16}\right]^{-8/27}\right\}^2 \qquad (10^9 < \text{Ra}) \tag{24.8b}$$

In Equations 24.7 and 24.8, all properties except β are obtained at the film temperature. The value of β for gases is obtained at T_∞ and for liquids it may be found at the film temperature. The range for the Prandtl number is $0 < \text{Pr} < \infty$.

Example 24.1 A 40-cm^2-vertical plate is insulated on one side and maintained at 140°C in atmospheric air at 20°C (Figure 24.3). Determine the rate of heat transfer.

Solution

Assumptions and Specifications

(1) Steady state exists.
(2) The air may be treated as an ideal gas.
(3) Buoyancy effects are included in the correlation employed.
(4) The ambient fluid is quiescent.
(5) Radiation effects are negligible.

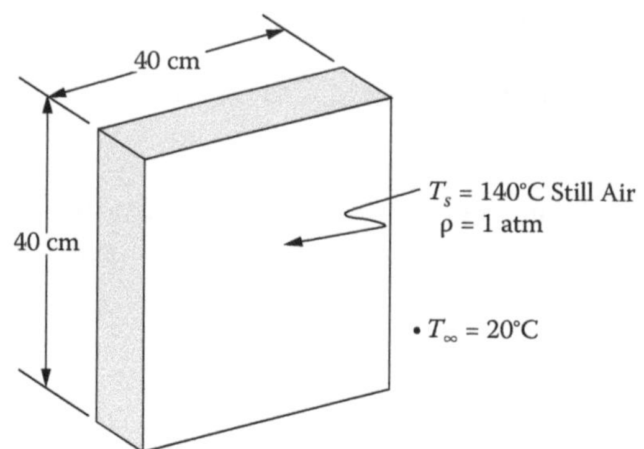

FIGURE 24.3
Configuration for Example 24.1.

(6) One face of the plate is insulated and the other dissipates.
(7) The acceleration of gravity is taken as $9.81\,\text{m/s}^2$.

This is a case of natural convection, and we will use one of Equation 24.8 depending on the value of the Rayleigh number. After the heat transfer coefficient has been found, the strategy will be to compute the rate of heat transfer from

$$\dot{Q} = hS\Delta T$$

The film temperature is

$$T_f = \frac{T_s + T_\infty}{2} = \frac{140°\text{C} + 20°\text{C}}{2} = 80°\text{C} \quad (353\,\text{K})$$

and Table A.13 in Appendix A can be used to find (with interpolation) at $T_f = 353\,\text{K}$

$$k = 0.0302\,\text{W/m-K}, \quad \nu = 21.07 \times 10^{-6}\,\text{m/s}^2, \quad \text{and} \quad \text{Pr} = 0.697$$

Thus,

$$\Delta T = T_s - T_\infty = 140°\text{C} - 20°\text{C} = 120°\text{C}$$

and at $T_\infty = 293\,\text{K}$

$$\beta = \frac{1}{293\,\text{K}} = 3.413 \times 10^{-3}\,\text{K}^{-1}$$

we determine the Rayleigh number with $L = 0.40\,\text{m}$

$$\begin{aligned}
\text{Ra} &= \text{Gr}\,\text{Pr} \\
&= \left(\frac{g\beta L^3 \Delta T}{\nu^2}\right)\text{Pr} \\
&= \frac{(9.81\,\text{m/s}^2)(3.413 \times 10^{-3}\,\text{K}^{-1})(0.40\,\text{m})^3(120°\text{C})(0.697)}{(21.07 \times 10^{-6}\,\text{m/s}^2)^2} \\
&= 4.037 \times 10^8
\end{aligned}$$

From the Churchill-Chu correlation, we see that for $\mathrm{Ra} = 4.037 \times 10^8$, Equation 23.8a should be used to determine the Nusselt number:

$$\mathrm{Nu} = 0.68 + 0.670\,\mathrm{Ra}^{1/4}\left[1 + \left(\frac{0.492}{\mathrm{Pr}}\right)^{9/16}\right]^{-4/9}$$

$$= 0.68 + 0.670(4.037 \times 10^8)^{1/4}\left[1 + \left(\frac{0.492}{0.697}\right)^{9/16}\right]^{-4/9}$$

$$= 0.68 + 0.670(141.75)[1 + (0.706)^{9/16}]^{-4/9}$$

$$= 0.68 + 94.97(1 + 0.8221)^{-4/9}$$

$$= 0.68 + 94,97(0.7659) = 73.42$$

Then, because $\mathrm{Nu} = hL/k$

$$h = 73.42\left(\frac{k}{L}\right) = 73.42\left(\frac{0.0302\,\mathrm{W/m\text{-}K}}{0.40\,\mathrm{m}}\right) = 5.54\,\mathrm{W/m^2\text{-}K}$$

and

$$\dot{Q} = hS\Delta T$$

$$= (5.54\,\mathrm{W/m^2\text{-}K})(0.40\,\mathrm{m})^2(120°\mathrm{C})$$

$$= 106.4\,\mathrm{W}$$

24.3.2.2 *Inclined Plates*

If the surface is inclined at no more than $\theta = 60°$ from the vertical and if Ra is less than 10^9, g is replaced with $g\cos\theta$ and Equation 24.8a is employed. If Ra exceeds 10^9, this modification is not employed.

24.3.3 Vertical Cylinders

For a vertical cylinder, Gebhart et al. (1988) showed that the equations for the vertical plate may be employed as long as

$$\frac{d}{L} > \frac{35}{\mathrm{Gr}_L^{1/4}} \tag{24.9}$$

where d is the cylinder diameter and the significant dimension in the Grashof number is L.

Example 24.2 A vertical heater 1-m high and 40 cm in diameter has a surface temperature of 35°C (Figure 24.4). The top surface of the cylinder does not dissipate, and the surrounding air is at 1 atm and 5°C. Determine the dissipation of the heater.

Solution

Assumptions and Specifications

(1) Steady state exists.

(2) The air may be treated as an ideal gas.

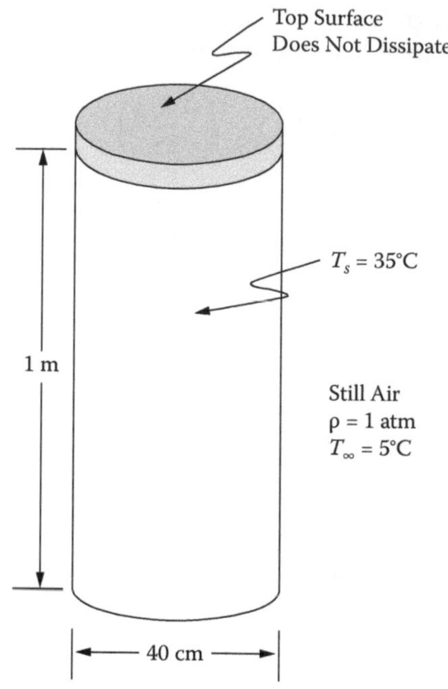

FIGURE 24.4
Configuration for Example 24.2.

(3) Buoyancy effects are included in the correlation employed.
(4) The ambient fluid is quiescent.
(5) Radiation effects are negligible.
(6) Lack of dissipation at the top is specified.
(7) The acceleration of gravity is taken as $9.81\,\text{m/s}^2$.

We must first determine the Grashof number to see whether the vertical plate correlation can be employed. Here, the temperature difference will be

$$\Delta T = T_s - T_\infty = 35°\text{C} - 5°\text{C} = 30°\text{C}$$

and the film temperature is

$$T_f = \frac{T_s + T_\infty}{2} = \frac{35°\text{C} + 5°\text{C}}{2} = 20°\text{C} \quad (293\,\text{K})$$

Thus, we can use Table A.13 to find (with interpolation) at $T_f = 293\,\text{K}$

$$k = 0.0257\,\text{W/m-K}, \quad \nu = 15.08 \times 10^{-6}\,\text{m/s}^2, \quad \text{and} \quad \text{Pr} = 0.710$$

Then, with $T_\infty = 278\,\text{K}$

$$\beta = \frac{1}{278\,\text{K}} = 3.597 \times 10^{-3}\,\text{K}^{-1}$$

we determine the Grashof number, which must be based on the cylinder height, $L = 1.0\,\text{m}$

$$\begin{aligned}
\text{Gr} &= \frac{g\beta L^3 \Delta T}{\nu^2} \\
&= \frac{(9.81\,\text{m/s}^2)(3.597 \times 10^{-3}\,\text{K}^{-1})(1.0\,\text{m})^3(30^\circ\text{C})}{(15.08 \times 10^{-6}\,\text{m/s}^2)^2} \\
&= 4.655 \times 10^9
\end{aligned}$$

Here

$$\frac{d}{L} = \frac{0.40\,\text{m}}{1.0\,\text{m}} = 0.40 > \frac{35}{\text{Gr}_L^{1/4}} = \frac{35}{(4.655 \times 10^9)^{1/4}} = 0.134$$

and the vertical plate correlations can be used. The Rayleigh number is

$$\text{Ra} = \text{Gr}\,\text{Pr} = (4.655 \times 10^9)(0.710) = 3.305 \times 10^9$$

and we have a choice of Equation 24.7b or 24.8b. This time choose the McAdams correlation of Equation 24.7b

$$\begin{aligned}
\text{Nu} &= \frac{hL}{k} = 0.13\,\text{Ra}^{1/3} \\
&= 0.13(3.305 \times 10^9)^{1/3} \\
&= 0.13(1490) = 193.64
\end{aligned}$$

Then

$$h = \text{Nu}\left(\frac{k}{L}\right) = 193.64\left(\frac{0.0257\,\text{W/m-K}}{1.0\,\text{m}}\right) = 4.98\,\text{W/m}^2\text{-K}$$

The surface area for the heater is the surface area of the lateral side:

$$S = \pi d_o L = \pi(0.40\,\text{m})(1.0\,\text{m}) = 1.2566\,\text{m}^2$$

and the power delivered by the heater is

$$\begin{aligned}
\dot{Q} &= hS\Delta T \\
&= (4.98\,\text{W/m}^2\text{-K})(1.2566\,\text{m}^2)(30^\circ\text{C}) \\
&= 187.7\,\text{W}
\end{aligned}$$

24.3.4 Horizontal Cylinders

For horizontal cylinders of sufficient length such that end effects may be neglected, McAdams (1954) gives, on the basis of the cylinder diameter, d

$$\frac{hd}{k} = 0.45\,\text{Ra}^{1/4} \qquad (10^3 < \text{Ra} < 10^9) \qquad (24.10\text{a})$$

and

$$\frac{hd}{k} = 0.11\,\mathrm{Ra}^{1/3} \qquad (10^9 < \mathrm{Ra}) \qquad (24.10b)$$

Churchill and Chu (1975a) recommend, on the basis of the cylinder diameter, d

$$\frac{hd}{k} = 0.60 + 0.518\,\mathrm{Ra}^{1/4}\left[1 + \left(\frac{0.559}{\mathrm{Pr}}\right)^{9/16}\right]^{-4/9} \qquad (\mathrm{Ra} < 10^9) \qquad (24.11a)$$

and

$$\frac{hd}{k} = \left\{0.60 + 0.387\,\mathrm{Ra}^{1/6}\left[1 + \left(\frac{0.559}{\mathrm{Pr}}\right)^{9/16}\right]^{-8/27}\right\}^2 \qquad (10^9 < \mathrm{Ra}) \qquad (24.11b)$$

Here too, all properties except β are obtained at the film temperature. The value of β for gases is obtained at the temperature of the surroundings; for liquids, it is obtained at the film temperature. The range for the Prandtl number is $0 < \mathrm{Pr} < \infty$.

Example 24.3 A 750-W electric heater is in the shape of a cylinder, 1.50 cm in diameter, and 75-cm long (Figure 24.5). The maximum allowable surface temperature of the heater is 96°C, and the heater is to be used to heat water at 24°C. Determine whether the heater must be derated.

Solution

Assumptions and Specifications

(1) Steady state exists.
(2) Buoyancy effects are included in the correlation employed.
(3) The ambient fluid is quiescent.
(4) Radiation effects are negligible.
(5) The ends of the heater are insulated.
(6) The acceleration of gravity is taken as $9.81\,\mathrm{m/s^2}$.

We will assume that the surface temperature is 96°C, and we will use this temperature to determine the heat dissipation. If the value so obtained exceeds 750 W, we will be drawn to the conclusion that the heater need not be derated.

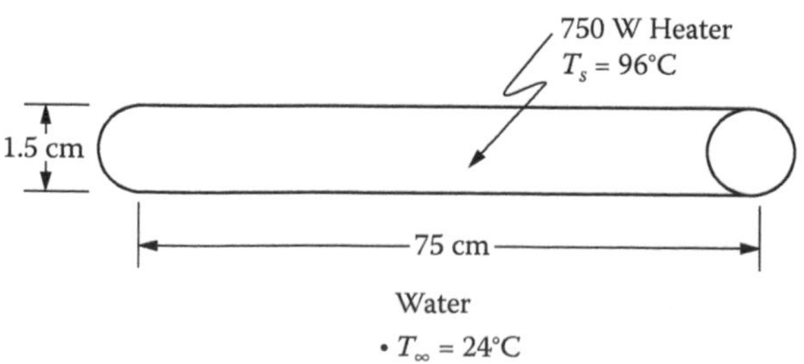

FIGURE 24.5
Configuration for Example 24.5.

Here, the temperature difference will be

$$\Delta T = T_s - T_\infty = 96°\text{C} - 24°\text{C} = 72°\text{C}$$

and the film temperature is

$$T_f = \frac{T_s + T_\infty}{2} = \frac{96°\text{C} + 24°\text{C}}{2} = 60°\text{C}$$

and Table A.14 can be used to find

$$k = 0.654\,\text{W/m-K} \qquad \nu = 0.4831 \times 10^{-6}\,\text{m/s}^2$$

$$\beta = 0.530 \times 10^{-3}\,\text{K}^{-1} \qquad \text{Pr} = 3.04$$

The Rayleigh number is determined with the diameter, d, as the significant dimension

$$\begin{aligned}\text{Ra} = \text{Gr}\,\text{Pr} &= \left(\frac{g\beta d^3 \Delta T}{\nu^2}\right)\text{Pr} \\ &= \frac{(9.81\,\text{m/s}^2)(0.530 \times 10^{-3}\,\text{K}^{-1})(0.015\,\text{m})^3(72°\text{C})(3.04)}{(0.4831 \times 10^{-6}\,\text{m/s}^2)^2} \\ &= 1.646 \times 10^7\end{aligned}$$

Equation 24.10a then gives

$$\begin{aligned}\text{Nu} = \frac{hd}{k} &= 0.45\,\text{Ra}^{1/4} \\ &= 0.45(1.646 \times 10^7)^{1/4} \\ &= 0.45(63.70) = 28.66\end{aligned}$$

and

$$h = \text{Nu}\left(\frac{k}{d}\right) = 28.66\left(\frac{0.654\,\text{W/m-K}}{0.015\,\text{m}}\right) = 1249.4\,\text{W/m}^2\text{-K}$$

The surface area for the heater is

$$S = \pi d_o L = \pi(0.015\,\text{m})(0.75\,\text{m}) = 0.0353\,\text{m}^2$$

and the power delivered by the heater when it is operating at its maximum allowable temperature is

$$\begin{aligned}\dot{Q} &= hS\Delta T \\ &= (1249.4\,\text{W/m}^2\text{-K})(0.0353\,\text{m}^2)(72°\text{C}) \\ &= 3180\,\text{W}\end{aligned}$$

Because 3180 W exceeds the rated value of 750 W by more than four times, the heater need not be derated.

24.3.5 Spheres

For free or natural convection around a sphere of diameter, d, with the Rayleigh number based on d, Churchill (1983) recommends for $\mathrm{Pr} > 0.700$ and $\mathrm{Ra} < 10^{11}$

$$\frac{hd}{k} = 2.00 + 0.589\,\mathrm{Ra}^{1/4}\left[1 + \left(\frac{0.469}{\mathrm{Pr}}\right)^{9/16}\right]^{-4/9} \tag{24.12}$$

where all properties except β are obtained at the film temperature. The value of β for gases is obtained at the temperature of the surroundings; for liquids, it is obtained at the film temperature. The range for the Prandtl number is $0 < \mathrm{Pr} < \infty$.

24.3.6 Horizontal Plates

For the vertical plate that has its significant dimension in line with the gravity vector, the natural convection currents flow parallel to the surface. However, for the horizontal plate, the significant dimension, whatever its magnitude, is normal to the direction of the flow, and we find a vast difference in the natural convection current patterns.

For example, with a horizontal heated plate facing downward, we see that to escape into the free stream, the dense air at the bottom surface must flow laterally to the edges of the plate. On the other hand, if the heated surface faces upward, the less dense air tends to rise but is impeded by the ever-present cooler air that flows downward.

In these cases, the correlations of McAdams (1954), Goldstein et al. (1973), and Lloyd and Moran (1974) are usually employed. These correlations are for hot plates facing upward or cold plates facing downward

$$\frac{hL}{k} = 0.54\,\mathrm{Ra}_L^{1/4} \qquad (2.6 \times 10^4 < \mathrm{Ra}_L < 3 \times 10^7) \tag{24.13a}$$

and

$$\frac{hL}{k} = 0.15\,\mathrm{Ra}_L^{1/3} \qquad (10^7 < \mathrm{Ra}_L < 3 \times 10^{10}) \tag{24.13b}$$

and for hot plates facing downward and cold plates facing upward

$$\frac{hL}{k} = 0.27\,\mathrm{Ra}_L^{1/4} \qquad (3 \times 10^5 < \mathrm{Ra}_L < 3 \times 10^{10}) \tag{24.13c}$$

In Equations 24.13, the significant dimension for the Rayleigh number is

$$L = \frac{\text{plate area}}{\text{plate perimeter}} \tag{24.14}$$

Example 24.4 A 0.50-m^2 plate with a surface temperature 120°C is suspended horizontally in a vat of still water at 20°C (Figure 24.6). Determine the heat loss from both faces of the plate.

Solution

Assumptions and Specifications

(1) Steady state exists.
(2) Buoyancy effects are included in the correlation employed.
(3) The ambient fluid is quiescent.

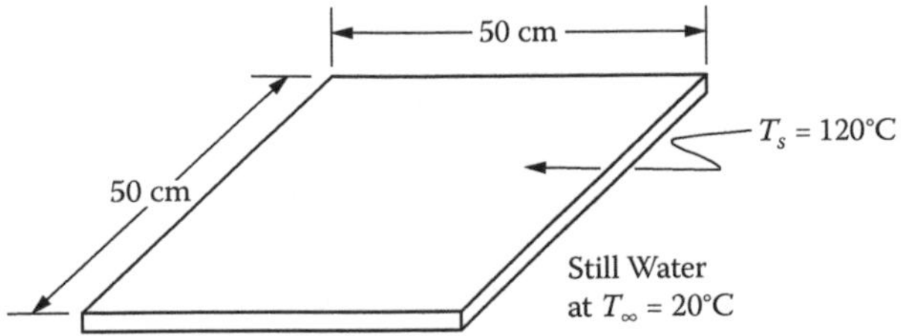

FIGURE 24.6
Configuration for Example 24.4.

(4) Radiation effects are negligible.
(5) The plate dissipates from both faces.
(6) The acceleration of gravity is taken as 9.81 m/s^2.

Still water and a horizontal surface are specified. Here we use Equation 24.14 for the significant dimension.

$$L = \frac{\text{plate area}}{\text{plate perimeter}} = \frac{L^2}{4L} = \frac{L}{4} = \frac{0.50\,\text{m}}{4} = 0.125\,\text{m}$$

In this case, the temperature difference will be

$$\Delta T = T_s - T_\infty = 120°\text{C} - 20°\text{C} = 100°\text{C}$$

and the film temperature is

$$T_f = \frac{T_s + T_\infty}{2} = \frac{120°\text{C} + 20°\text{C}}{2} = 70°\text{C}$$

and Table A.14 can be used to find

$$k = 0.664\,\text{W/m-K} \qquad \nu = 0.4173 \times 10^{-6}\,\text{m/s}^2$$

$$\beta = 0.590 \times 10^{-3}\,\text{K}^{-1} \qquad \text{Pr} = 2.573$$

The Rayleigh number is determined with $L = 0.125$ m as the significant dimension:

$$\text{Ra} = \text{Gr}\,\text{Pr} = \left(\frac{g\beta L^3 \Delta T}{\nu^2}\right)\text{Pr}$$

$$= \frac{(9.81\,\text{m/s}^2)(0.590 \times 10^{-3}\,\text{K}^{-1})(0.125\,\text{m})^3(100°\text{C})(2.573)}{(0.4173 \times 10^{-6}\,\text{m/s}^2)^2}$$

$$= 1.670 \times 10^{10}$$

We can use Equation 24.13b for the top surface and Equation 24.13c for the bottom surface. For the top surface

$$\text{Nu} = \frac{hL}{k} = 0.15\,\text{Ra}^{1/3}$$

$$= 0.15(1.670 \times 10^{10})^{1/3}$$

$$= 0.15(2556.4) = 383.4$$

and for the bottom surface

$$\text{Nu} = \frac{hL}{k} = 0.27\,\text{Ra}^{1/4}$$
$$= 0.27(1.670 \times 10^{10})^{1/4}$$
$$= 0.27(359.5) = 97.06$$

These values of the Nusselt number yield the heat transfer coefficients. For the top

$$h = 383.6\left(\frac{k}{L}\right) = 383.6\left(\frac{0.664\,\text{W/m-K}}{0.125\,\text{m}}\right) = 2036.7\,\text{W/m}^2\text{-K}$$

and for the bottom

$$h = 97.06\left(\frac{k}{L}\right) = 97.06\left(\frac{0.664\,\text{W/m-K}}{0.125\,\text{m}}\right) = 515.6\,\text{W/m}^2\text{-K}$$

The surface area for each face of the plate is

$$S = L^2 = (0.50\,\text{m})^2 = 0.25\,\text{m}^2$$

Thus, the rates of heat transfer will be

$$\dot{Q} = hS\Delta T = (2036.7\,\text{W/m}^2\text{-K})(0.25\,\text{m}^2)(100°\text{C}) = 50.92\,\text{kW} \Longleftarrow$$

for the top and

$$\dot{Q} = hS\Delta T = (515.6\,\text{W/m}^2\text{-K})(0.25\,\text{m}^2)(100°\text{C}) = 12.89\,\text{kW} \Longleftarrow$$

for the bottom.

We note that the dissipation from the top of the plate is about four times the dissipation from the bottom.

24.4 Natural Convection in Parallel Plate Channels

24.4.1 The Elenbaas Correlation

Elenbaas (1942) was apparently the first to document a detailed study of the thermofluid behavior of a vertical channel in natural convection. His experimental results for isothermal plates in air were later confirmed by others. A unified picture of the thermal transport in such channels emerged from these and complementary studies.

Figure 24.7 shows a gap between two isothermal plates. Individual thermal and velocity boundary layers are in evidence along each surface, and the rates of heat transfer approach those associated with laminar flow along isolated plates in infinite media. The development of the resulting boundary layers progresses toward the fully developed condition at the channel outlet.

Alternatively, for long narrow channels, the boundary layers merge near the entrance and we observe that fully developed flow prevails along much of the channel.

The analytic Nusselt number relations for the fully developed flow region and the isolated plate region can be expected to bound the Nusselt number values over the complete range of flow development. In the fully developed limit and for an isothermal channel, in the

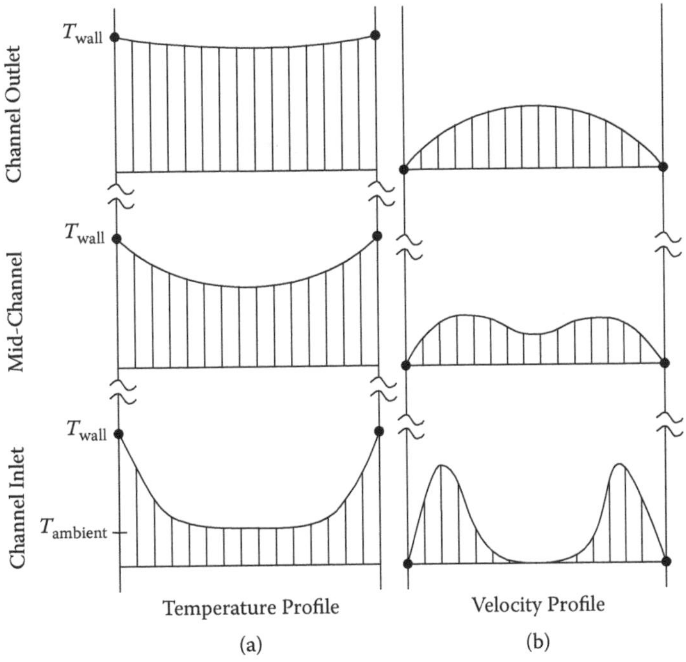

FIGURE 24.7
Developing temperature and velocity profiles for air in natural convection in a vertical channel: (a) the temperature profiles and (b) the velocity profiles. Broken lines designate isothermal channel boundaries.

fully developed limit, the gap-based Nusselt number

$$\mathrm{Nu}_z \equiv \frac{hz}{k}$$

is seen to depend on a **gap-based Rayleigh number** with z taken as the spacing between the plates

$$\mathrm{Ra}_z \equiv \frac{g\beta z^4 \Delta T}{L\nu^2}\,\mathrm{Pr}$$

In recognition of his seminal contributions to the field of natural convection in rectangular channels, the gap-based Rayleigh number is referred to as the **Elenbaas number**, El. Thus,

$$\mathrm{El} = \mathrm{Ra}_z = \frac{\rho^2 g\beta c_p z^4 \Delta T}{L\mu k} = \varphi z^4 \tag{24.15}$$

where φ is a thermal property parameter

$$\varphi = \frac{\rho^2 g\beta c_p \Delta T}{L\mu k} \tag{24.16}$$

Elenbaas used his understanding of these bounding relationships to establish the form of the relationship between the Elenbaas number and the fluid properties over the entire range of fluid properties. Application of this form to the data for relatively short isothermal plates in the air yielded the Elenbaas correlation

$$\mathrm{Nu}_z = \frac{hz}{k} = \frac{\mathrm{El}}{24}\left(1 - e^{-35/\mathrm{El}}\right)^{3/4} \tag{24.17}$$

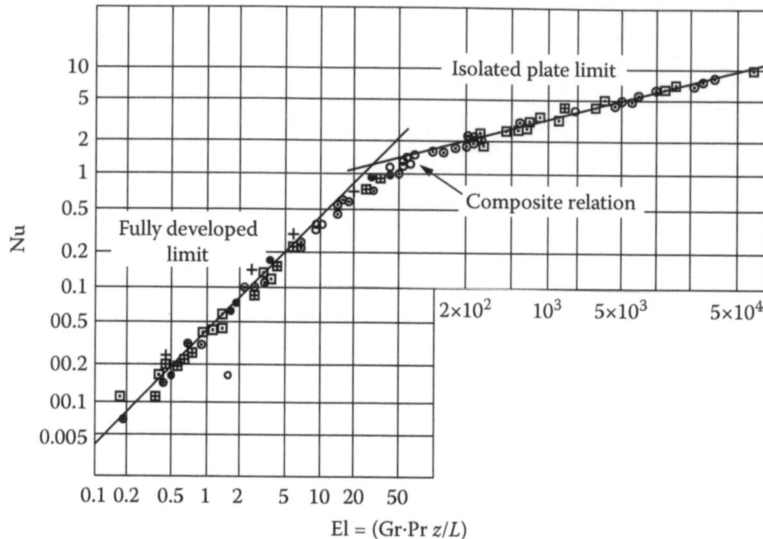

FIGURE 24.8
Nusselt number variation for symmetric isothermal plates. (Data points from Elenbaas, 1942.)

24.4.2 A Composite Relation

Churchill and Usagi (1972) have shown that when a function is known to vary smoothly between two limiting expressions that are, themselves, well defined and when solutions for intermediate values are difficult to obtain, an approximate composite relationship can be obtained by summing the two limiting expressions. Based on the Churchill-Usagi procedure, Bar-Cohen and Rohsenow (1984) provided a composite equation for isothermal plates

$$\mathrm{Nu}_z = \left[\frac{576}{(\mathrm{El})^2} + \frac{2.87}{(\mathrm{El})^{1/2}}\right]^{-1/2} \tag{24.18}$$

where Nu_z and El (including β) are based on properties evaluated at the average film temperature. The validity of this solution is borne out by the close proximity of the Elenbaas (1942) data points to the composite relationship of Equation 24.18 and the asymptotic equations at both limits as indicated in Figure 24.8.

Example 24.5 Determine the heat transfer coefficient for air in natural convection within a 15-cm-long vertical parallel plate channel (Figure 24.9). The ambient air temperature is 23°C, the channel walls are maintained at 77°C, and plate spacings of (a) 2 cm, (b) 0.50 cm, and (c) 0.20 cm are to be considered.

Solution

Assumptions and Specifications

(1) Steady state exists.
(2) The air may be treated as an ideal gas.
(3) Buoyancy effects are included in the correlation employed.
(4) The ambient fluid is quiescent.
(5) Radiation effects are negligible.
(6) The acceleration of gravity is taken as $9.81\,\mathrm{m/s^2}$.

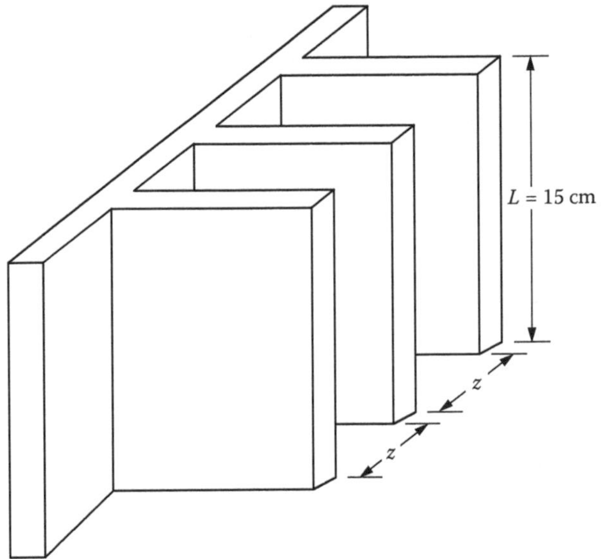

FIGURE 24.9
Configuration for Example 24.5.

We will calculate the Elenbaas number as a function of the plate spacing for all three spacings first. Here

$$\Delta T = T_s - T_\infty = 77^\circ\text{C} - 23^\circ\text{C} = 54^\circ\text{C}$$

and the film temperature is

$$T_f = \frac{T_s + T_\infty}{2} = \frac{77^\circ\text{C} + 23^\circ\text{C}}{2} = 50^\circ\text{C} \qquad (323\,\text{K})$$

Then with

$$\beta = \frac{1}{323\,\text{K}} = 3.096 \times 10^{-3}\,\text{K}^{-1}$$

Table A.13 provides (with interpolation)

$$k = 0.0279\,\text{W/m-K} \qquad \mu = 19.52 \times 10^{-6}\,\text{kg/m-s}$$

$$\rho = 1.0949\,\text{kg/m}^3 \qquad c_p = 1007.2\,\text{J/kg-K}$$

Here

$$\begin{aligned}\varphi &= \frac{\rho^2 g \beta c_p \Delta T}{\mu k L} \\ &= (1.0949\,\text{kg/m}^3)^2(9.81\,\text{m/s}^2) \\ &\quad \times \frac{(3.096 \times 10^{-3}\,\text{K}^{-1})(1007.2\,\text{J/kg-K})(54^\circ\text{C})}{(19.52 \times 10^{-6}\,\text{kg/m-s})(0.0279\,\text{W/m-K})(0.15\,\text{m})} \\ &= 2.424 \times 10^{10}\text{m}^{-4}\end{aligned}$$

and

$$\text{El} = \varphi z^4 = (2.424 \times 10^{10}\text{m}^{-4})z^4$$

(a) For a plate spacing of 2 cm

$$\text{El} = (2.424 \times 10^{10}\text{m}^{-4})(0.02\,\text{m})^4 = 3880$$

and

$$\text{Nu}_z = \left[\frac{576}{(\text{El})^2} + \frac{2.87}{(\text{El})^{1/2}}\right]^{-1/2} = \left[\frac{576}{(3880)^2} + \frac{2.87}{(3880)^{1/2}}\right]^{-1/2}$$

$$= (3.826 \times 10^{-5} + 0.0461)^{-1/2} = 4.656$$

Then

$$h = \text{Nu}_z\left(\frac{k}{z}\right) = 4.656\left(\frac{0.0279\,\text{W/m-K}}{0.02\,\text{m}}\right) = 6.50\,\text{W/m}^2\text{-K} \Longleftarrow$$

(b) For a plate spacing of 0.50 cm

$$\text{El} = (2.424 \times 10^{10}\text{m}^{-4})(0.005\,\text{m})^4 = 15.15$$

and

$$\text{Nu}_z = \left[\frac{576}{(\text{El})^2} + \frac{2.87}{(\text{El})^{1/2}}\right]^{-1/2} = \left[\frac{576}{(15.15)^2} + \frac{2.87}{(15.15)^{1/2}}\right]^{-1/2}$$

$$= (2.5096 + 0.7374)^{-1/2} = 0.5550$$

Then

$$h = \text{Nu}_z\left(\frac{k}{z}\right) = 0.5550\left(\frac{0.0279\,\text{W/m-K}}{0.005\,\text{m}}\right) = 3.10\,\text{W/m}^2\text{-K} \Longleftarrow$$

(c) For a plate spacing of 0.20 cm

$$\text{El} = (2.424 \times 10^{10}\text{m}^{-4})(0.002\,\text{m})^4 = 0.3880$$

and

$$\text{Nu}_z = \left[\frac{576}{(\text{El})^2} + \frac{2.87}{(\text{El})^{1/2}}\right]^{-1/2} = \left[\frac{576}{(0.3880)^2} + \frac{2.87}{(0.3880)^{1/2}}\right]^{-1/2}$$

$$= (3826.1 + 4.608)^{-1/2} = 0.0162$$

Then

$$h = \text{Nu}_z\left(\frac{k}{z}\right) = 0.0162\left(\frac{0.0279\,\text{W/m-K}}{0.002\,\text{m}}\right) = 0.225\,\text{W/m}^2\text{-K} \Longleftarrow$$

24.4.3 Optimum Plate Spacing

An examination of Figure 24.8 shows that the rate of heat transfer from each plate decreases as the plate spacing, z, is decreased. Because the total number of plates or the plate surface area increases with reduced spacing, we see that the total heat transfer may be maximized by finding the plate spacing at which the **total plate surface-heat transfer coefficient product** is maximized. Bar-Cohen and Rohsenow (1984) point out that Elenbaas (1942), based on his experimental results, determined that this optimum spacing for negligibly thick plates could be obtained by setting El = 46 leading to

$$z_{opt} = \frac{2.714}{\varphi^{1/4}} \tag{24.19}$$

where the thermal property parameter, φ, is defined in Equation 24.16.

We now turn to Design Example 16, which is a rather comprehensive problem involving the design of a heat sink for a package of electronic equipment.

24.5 Design Example 16

A vertical surface, $W = 4$ cm by $L = 4$ cm (Figure 24.10) must dissipate 2.25 W at 75°C by natural convection to the air at 45°C. Determine whether the surface without fins is sufficient to accommodate the required dissipation. Determine the number of fins with a height $b = 3$ cm, a length $L = 4$ cm, a thickness, $\delta = 2.286$ mm, and $k = 180$ W/m-K that will be required if the bare surface fails to provide the required dissipation.

Solution

Assumptions and Specifications

(1) Steady state exists.
(2) The air may be treated as an ideal gas.
(3) Buoyancy effects are included in the correlation employed.
(4) The ambient fluid is quiescent.
(5) Radiation effects are negligible.
(6) The acceleration of gravity is taken as 9.81 m/s^2.

The configuration is shown in Figure 24.10. In order to make sure that the fins, if used, are isothermal plates, we will aim for a fin efficiency of at least 0.990, thereby ensuring a negligible temperature drop from base to tip. Here, the film temperature is

$$T_f = \frac{T_s + T_\infty}{2} = \frac{75°\text{C} + 45°\text{C}}{2} = 60°\text{C} \qquad (333\text{ K})$$

and

$$\Delta T = T_s - T_\infty = 75°\text{C} - 45°\text{C} = 30°\text{C}$$

Then at $T_\infty = 45°\text{C}$ (318 K)

$$\beta = \frac{1}{318\text{ K}} = 3.148 \times 10^{-3}\text{ K}^{-1}$$

and Table A.13 provides (with interpolation)

$$k = 0.0287\text{ W/m-K}, \quad \nu = 19.04 \times 10^{-6}\text{ m/s}^2, \quad \text{and} \quad \text{Pr} = 0.701$$

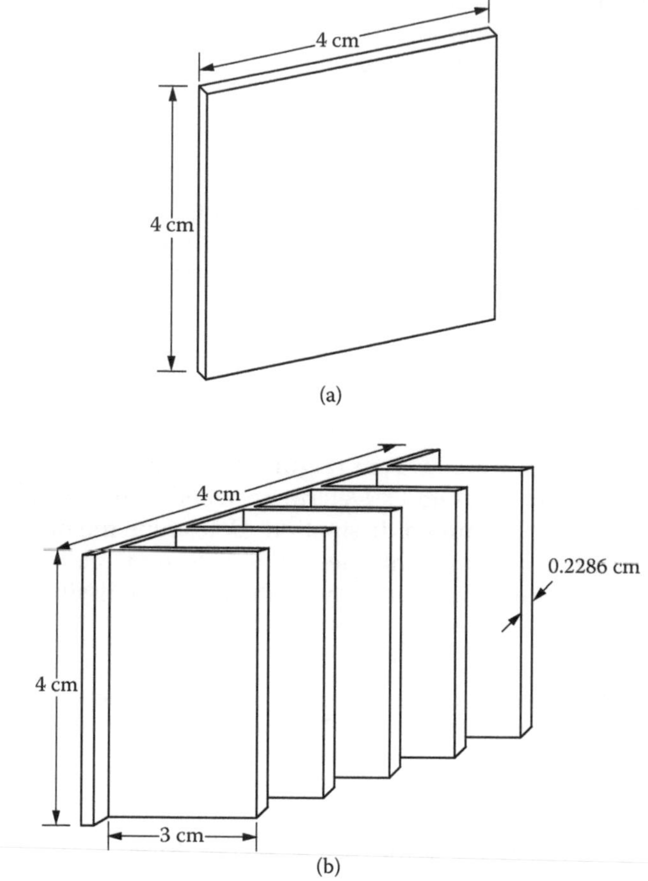

FIGURE 24.10
Configuration for Design Example 16: (a) a vertical plate and (b) the vertical plate equipped with longitudinal fins of rectangular profile.

The Rayleigh number is

$$\begin{aligned}
\mathrm{Ra} &= \mathrm{Gr}\,\mathrm{Pr} = \left(\frac{g\beta L^3 \Delta T}{\nu^2}\right)\mathrm{Pr} \\
&= \frac{(9.81\,\mathrm{m/s^2})(3.148\times10^{-3}\,\mathrm{K^{-1}})(0.04\,\mathrm{m})^3(30^\circ\mathrm{C})(0.701)}{(19.04\times10^{-6}\,\mathrm{m/s^2})^2} \\
&= 1.147\times10^5
\end{aligned}$$

We employ Equation 24.7a, which is applicable for the case of the vertical plate without the fins.

$$\mathrm{Nu} = 0.55\,\mathrm{Ra}^{1/4} = 0.55(1.147\times10^5)^{1/4} = 0.55(18.40) = 10.12$$

Then

$$h = \mathrm{Nu}\left(\frac{k}{L}\right) = 10.12\left(\frac{0.0287\,\mathrm{W/m\text{-}K}}{0.04\,\mathrm{m}}\right) = 7.26\,\mathrm{W/m^2\text{-}K} \Longleftarrow$$

and with

$$S = (0.04\,\text{m})^2 = 0.0016\,\text{m}^2$$

we obtain the heat dissipation as

$$\dot{Q} = hS\Delta T = (7.26\,\text{W/m}^2\text{-K})(0.0016\,\text{m}^2)(30^\circ\text{C}) = 0.349\,\text{W} < 2.25\,\text{W}$$

We will need to use the fins and when they are employed at the film temperature of 60°C (333 K), Table A.13 provides (with interpolation)

$$k = 0.0287\,\text{W/m-K} \qquad \mu = 19.97 \times 10^{-6}\,\text{m/s}^2$$

$$\rho = 1.059\,\text{kg/m}^3 \qquad c_p = 1007.9\,\text{J/kg-K}$$

and with β taken at the film temperature

$$\beta = \frac{1}{333\,\text{K}} = 3.003 \times 10^{-3}\,\text{K}^{-1}$$

Here

$$\begin{aligned}
\varphi &= \frac{\rho^2 g\beta c_p \Delta T}{\mu k L} \\
&= (1.059\,\text{kg/m}^3)^2(9.81\,\text{m/s}^2) \\
&\quad \times \frac{(3.003 \times 10^{-3}\,\text{K}^{-1})(1007.9\,\text{J/kg-K})(30^\circ\text{C})}{(19.97 \times 10^{-6}\,\text{m/s}^2)(0,0287\,\text{W/m-K})(0.04\,\text{m})} \\
&= 4.357 \times 10^{10}\text{m}^{-4}
\end{aligned}$$

The optimum clear space between the fins will then be

$$z_{\text{opt}} = \frac{2.714}{\varphi^{1/4}} = \frac{2.714}{456.89\,\text{m}^1} = 5.940 \times 10^{-3}\,\text{m} \quad (0.594\,\text{cm})$$

and the number of fins required is

$$n_f = \frac{W}{z_{\text{opt}} + \delta} = \frac{4.00\,\text{cm}}{0.5940\,\text{cm} + 0.2286\,\text{cm}} = \frac{4.00\,\text{cm}}{0.8226\,\text{cm}} = 4.86\,\text{fins}$$

We will call for five fins with the knowledge that the fin spacing will be slightly smaller than optimum. Moreover, we will space the fins so that there will be five equal spaces

$$z + \delta = \frac{4.00\,\text{cm}}{5\,\text{fins}} = 0.8000\,\text{cm}$$

so that

$$z = 0.8000\,\text{cm} - \delta = 0.8000\,\text{cm} - 0.2286\,\text{cm} = 0.5714\,\text{cm}$$

The Elenbaas number will be

$$\text{El} = \varphi z^4 = (4.357 \times 10^{10}\text{m}^{-4})(5.714 \times 10^{-3}\,\text{m})^4 = 46.45$$

and the Nusselt number will then be

$$\mathrm{Nu}_z = \left[\frac{576}{(\mathrm{El})^2} + \frac{2.87}{(\mathrm{El})^{1/2}}\right]^{-1/2} = \left[\frac{576}{(46.45)^2} + \frac{2.87}{(46.45)^{1/2}}\right]^{-1/2}$$
$$= (0.2670 + 0.4211)^{-1/2} = 1.2055$$

Then

$$h = \mathrm{Nu}\left(\frac{k}{z}\right) = 1.2055\left(\frac{0.0287\,\mathrm{W/m\text{-}K}}{5.714 \times 10^{-3}\,\mathrm{m}}\right) = 6.06\,\mathrm{W/m^2\text{-}K}$$

Before we make the surface computation, we will check to see if the fin efficiency is close to 0.990. Following the procedure illustrated in Section 20.7.3 with $L \gg \delta$, we have

$$m = \left(\frac{2h}{k\delta}\right)^{1/2} = \left[\frac{2(6.06\,\mathrm{W/m^2\text{-}K})}{(180\,\mathrm{W/m\text{-}K})(2.286 \times 10^{-3}\,\mathrm{m})}\right]^{1/2} = 5.427\,\mathrm{m}^{-1}$$

and

$$mb = (5.427\,\mathrm{m}^{-1})(0.03\,\mathrm{m}) = 0.1628$$

The fin efficiency is

$$\eta = \frac{\tanh mb}{mb} = \frac{\tanh 0.1628}{0.1628} = 0.991$$

This checks the assumption of the fin efficiency and allows us to accept the foregoing calculations.

The surface area of the base plate will be the area, WL, *minus* the footprints of the five fins:

$$S_b = L(W - n_f\delta)$$
$$= (4.00\,\mathrm{cm})[4.00\,\mathrm{cm} - 5(0.2286\,\mathrm{cm})]$$
$$= (4.00\,\mathrm{cm})[4.00\,\mathrm{cm} - 1.143\,\mathrm{cm})$$
$$= (4.00\,\mathrm{cm})(2.857\,\mathrm{cm}) = 11.428\,\mathrm{cm}^2$$

The surface area of the five fins is

$$S_f = 2n_f bL = 2(5)(3.00\,\mathrm{cm})(4.00\,\mathrm{cm}) = 120.0\,\mathrm{cm}^2$$

Thus, the total surface area for this application is

$$S = S_b + \eta S_f$$
$$= 11.428\,\mathrm{cm}^2 + 0.991(120\,\mathrm{cm}^2)$$
$$= 11.428\,\mathrm{cm}^2 + 118.92\,\mathrm{cm}^2 = 130.35\,\mathrm{cm}^2$$

and the heat transferred will be

$$\dot{Q} = hS\Delta T = (6.06\,\mathrm{W/m^2\text{-}K})(0.0130\,\mathrm{m}^2)(30^\circ\mathrm{C}) = 2.36\,\mathrm{W} > 2.25\,\mathrm{W} \Longleftarrow$$

This design is deemed to be satisfactory.

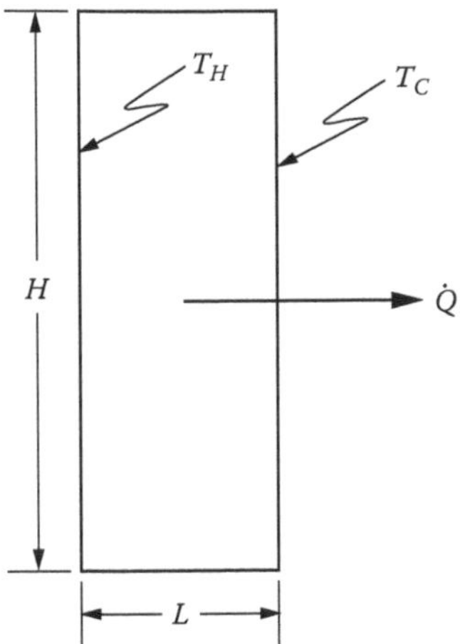

FIGURE 24.11
A pair of rectangular plates that form an enclosure oriented in the vertical direction.

24.6 Natural Convection in Enclosures

The **cavity** or **enclosure** shown in Figure 24.11 has a height, H, and a plate spacing, L. Here, the two opposing plates are held at different temperatures, T_H and T_C, where $T_H > T_C$. The dimension perpendicular to the plane of the paper, W, is presumed to be very large, that is $W \gg L$.

This configuration is of more than casual interest. For example, the air spaces between two vertical panes of glass or between two vertical walls in a structure are two important examples of such an enclosure. Moreover, the air space enclosed by the absorbing surfaces of a solar collector and its cover plate (glass or plastic) provides us with another important example. In the case of the solar collector, the enclosure is usually tilted at some angle, θ, from the vertical.

When considering enclosures, we define the **aspect ratio**, Λ, as

$$\Lambda = \frac{H}{L} \tag{24.20}$$

the Nusselt number with respect to L

$$\mathrm{Nu}_L = \frac{hL}{k} \tag{24.21}$$

the Grashof number with respect to L

$$\mathrm{Gr}_L = \frac{g\beta(T_H - T_C)L^3}{\nu^2} \tag{24.22}$$

and the Rayleigh number with respect to L

$$\mathrm{Ra}_L = \mathrm{Gr}_L\,\mathrm{Pr} = \frac{g\beta(T_H - T_C)L^3\,\mathrm{Pr}}{\nu^2} \tag{24.23}$$

In Equations 24.22 and 24.23, β is evaluated at the average of the two plate temperatures.

In the limiting case where there are no convection currents, heat transfer between the hot wall, at $T = T_H$, and the cooler one, at $T = T_C$, will be by conduction only. In such a case

$$h = \frac{k}{L}$$

and, consequently,

$$\text{Nu}_L = 1$$

This lower limit for Nu_L is often referred to as the **conduction limit.**

The **effective** or **apparent conductivity** for an enclosure gap, defined as

$$k_e \equiv \text{Nu}_L\, k \tag{24.24}$$

allows the rate of heat transfer to be expressed, using the Fourier law, in the form

$$\dot{Q} = \frac{k_e}{L} S(T_H - T_C) \tag{24.25}$$

24.6.1 Working Correlations

24.6.1.1 Vertical Rectangular Enclosures

For conditions where $2 < \Lambda \leq 10$, $\text{Pr} < 10^5$, and $10^5 < \text{Ra}_L < 10^{10}$, Catton (1978) recommends

$$\text{Nu}_L = 0.22 \left(\frac{\text{Pr Ra}_L}{0.22 + \text{Pr}} \right)^{0.28} \Lambda^{-1/4} \tag{24.26}$$

When $1 < \Lambda < 2$, $10^3 < \text{Pr} < 10^5$, and the quantity in the parentheses is greater than 10^3, the recommended expression is

$$\text{Nu}_L = 0.18 \left(\frac{\text{Pr Ra}_L}{0.22 + \text{Pr}} \right)^{0.29} \tag{24.27}$$

For larger aspect ratios, McGregor and Emory (1969) proposed

$$\text{Nu}_L = 0.42\, \text{Ra}_L^{1/4}\, \text{Pr}^{0.012} \Lambda^{-0.30} \tag{24.28}$$

for $10 < \Lambda < 40$, $1 < \text{Pr} < 2 \times 10^6$, and $10^4 < \text{Ra}_L < 10^7$, and

$$\text{Nu}_L = 0.046\, \text{Ra}_L^{1/3} \tag{24.29}$$

when $1 < \Lambda < 40$, $1 < \text{Pr} < 20$, and $10^6 < \text{Ra}_L < 10^9$.

24.6.1.2 Tilted Vertical Enclosures

Figure 24.12 shows an inclined vertical enclosure with the angle θ measured from the horizontal. Correlations are available for $0° < \theta < 70°$, but we present the correlation for $0° < \theta < 70°$ for the air-filled enclosure with $\Lambda < 12$ and for $0 < \text{Ra}_L < 10^5$ (Hollands et al., 1970). The maximum allowable value of the inclination angle, θ, is a function of the aspect ratio. Table 24.1 provides these values.

$$\text{Nu}_L = 1.00 + 1.44\varphi_1\varphi_2 + \varphi_3 \tag{24.30}$$

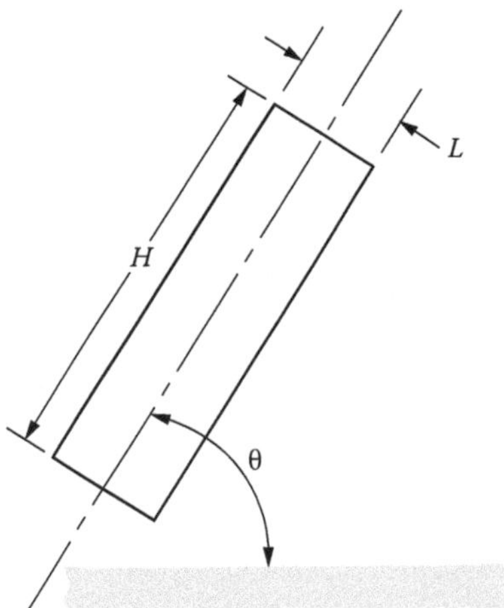

FIGURE 24.12
A tilted enclosure.

where

$$\varphi_1 = 1 - \frac{1708}{\mathrm{Ra}_L \cos\theta}$$

$$\varphi_2 = 1 - \frac{1708(\sin 1.8\theta)^{1.60}}{\mathrm{Ra}_L \cos\theta}$$

and

$$\varphi_3 = \left(\frac{\mathrm{Ra}_L \cos\theta}{5830}\right)^{1/3} - 1$$

If either φ_1 or φ_2 is negative, its value is set equal to zero.

24.6.2 Concentric Cylinders

For the case of the concentric horizontal cylinder arrangement shown in Figure 24.13a, the correlation of Raithby and Hollands (1975) provides the effective gap thermal conductivity, k_e, to be used in

$$\frac{\dot{Q}}{L} = \frac{2\pi k_e}{\ln d_o/d_i}(T_o - T_i) \tag{24.31}$$

TABLE 24.1

Maximum Allowable Value of the Angle of Inclination, θ

Λ = *H*/*L*	θ
1	25°
3	53°
6	60°
12	67°
>12	70°

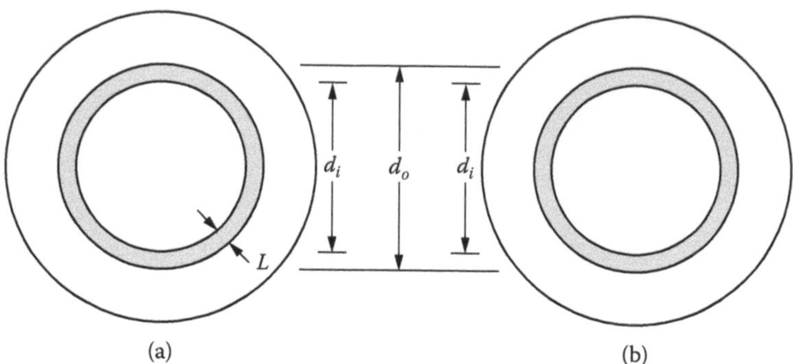

FIGURE 24.13
The gap formed by (a) concentric cylinders and (b) concentric spheres.

where the subscripts i and o refer, respectively, to the inner and outer cylinder. Here

$$k_e = 0.386k\left(\frac{\mathrm{Pr}}{0.861+\mathrm{Pr}}\right)^{1/4}\psi^{1/4} \tag{24.32}$$

where

$$\psi = \frac{\ln\left(\dfrac{d_o}{d_i}\right)^4 \mathrm{Ra}_L}{L^3(d_o^{-3/5}+d_i^{-3/5})^5}$$

and

$$L = \frac{d_o - d_i}{2}$$

is the gap width. Equation 24.32 may be used in the range $10^2 < \psi < 10^7$ and if $\psi < 100$, k_e is taken as $k_e = k$. Otherwise, k_e must be greater than k.

24.6.3 Concentric Spheres

Raithby and Hollands (1975) also considered the case of concentric spheres and proposed that

$$\dot{Q} = k_e\pi\frac{d_o d_i}{L}(T_o - T_i) \tag{24.33}$$

with k_e expressed as

$$k_e = 0.74k\left(\frac{\mathrm{Pr}}{0.861+\mathrm{Pr}}\right)^{1/4}\psi^{1/4} \tag{24.34}$$

and

$$\psi = \frac{L\,\mathrm{Ra}_L}{(d_o d_i)^4(d_o^{-7/5}+d_i^{-7/5})^5}$$

Equation 24.34 is valid in the range $10^2 < \psi < 10^7$.

24.7 Summary

The reader is cautioned that the following summary is simply a guide to the working correlations. Reference must always be made to the conditions of applicability, described in the text, that pertain to these correlations.

The dimensionless correlations to be employed in design and analysis of configurations involving natural convection heat transfer are based on the Grashof and Rayleigh numbers.

$$\text{Gr} \equiv \frac{\rho^2 g\beta L^3 \Delta T}{\mu^2} \tag{24.4a}$$

and

$$\text{Ra} = \text{Gr}\,\text{Pr}$$

The working correlations provided in the text are summarized in the following tabulation:

Configuration	Equation(s)
Vertical plates	24.7 and 24.8
Inclined vertical plates	24.8a
Vertical cylinders	24.9
Horizontal cylinders	24.10 and 24.11
Spheres	24.12
Hot plates facing upward	24.13a and 24.13b
Hot plates facing downward	24.13c
Vertical rectangular enclosures	24.26 to 24.29
Concentric cylinders	24.31
Concentric spheres	24.33

For parallel plate channels, the Elenbaas number is used with z as the gap spacing

$$\text{El} = \text{Ra}_z = \frac{g\beta z^4 \Delta T}{L\nu^2}\,\text{Pr}$$

and the Nusselt number is obtained from Equation 24.18.

24.8 Problems

Vertical Plates

24.1: Still water at 15°C is heated on both sides of an 18-cm × 18-cm-vertical plate held at 95°C. Use the McAdams correlation to predict the heat transferred.

24.2: Still water at 15°C is heated on both sides of an 18-cm × 18-cm-vertical plate held at 95°C. Use the Churchill-Chu correlation to predict the heat transferred.

24.3: Still air at atmospheric pressure and 15°C is heated on both sides of an 18-cm × 18-cm-vertical plate held at 95°C. Use the McAdams correlation to predict the heat transferred.

24.4: Still air at atmospheric pressure and 15°C is heated on both sides of an 18-cm × 18-cm-vertical plate held at 95°C. Use the Churchill-Chu correlation to predict the heat transferred.

24.5: Still air at 2 atm and 15°C is heated on both sides of an 18 cm × 18 cm vertical plate held at 95°C. Use the McAdams correlation to predict the heat transferred.

24.6: One side of a vertical plate that is 12-cm wide and 15-cm high is held at atmospheric pressure and a temperaure of 82°C in a still-air environment of 22°C. The emissivity of the plate is 0.76, the reverse side of the plate is insulated, and the enclosure is also at 22°C. Use the Churchill-Chu correlation to determine the heat transferred.

24.7: Suppose that in Problem 24.6, the heat loss is 22 W. Using the McAdams correlation, determine the surface temperature to the nearest degree if all other conditions remain the same.

24.8: Use the McAdams correlation to estimate the heat loss by natural convection from a vertical wall 2-m high and 3-m wide. At 37°C, the wall is exposed to a still-nitrogen environment at 1 atm at 17°C.

24.9: Use the Churchill-Chu correlation to estimate the heat loss by natural convection from a vertical wall 2-m high and 3-m wide held at 37°C. The wall is exposed to a still-nitrogen environment at 1 atm at 17°C.

24.10: Still air at 0.94 atm and 17°C receives heat by natural convection from the door of a dishwasher that is held at 37°C and is 60-cm wide and 80-cm high. Determine the heat loss using the McAdams correlation.

24.11: For a vertical plate at 1 atm and 94°C in a still-air environment at 60°C, determine the plate height needed to make the Rayleigh number equal to 10^9.

24.12: For a vertical plate at 0.15 atm and 94°C in a still-air environment at 60°C, determine the plate height needed to make the Rayleigh number equal to 10^9.

Vertical Cylinders and Tilted Vertical Plates

24.13: A vertical cylinder has a diameter of 8 cm and a height of 28 cm. The cylinder is held at 85°C in a still-air environment at atmospheric pressure and at 25°C. Neglect the heat transferred from the top of the cylinder and determine the heat dissipation.

24.14: In Problem 24.13, determine the minimum diameter that will permit the cylinder to be treated as a vertical plate.

24.15: A vertical cylinder that is 1.75-m high and 20 cm in diameter is held at 97°C. It is used as a heater in an unstirred tank of water at 23°C. The top of the cylinder is insulated. Determine whether the cylinder may be treated as a vertical plate and, if so, use the Churchill-Chu correlation to establish the rate of heat transfer.

24.16: A vertical cylinder that is 2-m high and 25 cm in diameter is held at 114°C. The cylinder is used in stagnant air at 40°C and 1 atm. Neglect heat transfer from the top of the cylinder and determine the rate of heat transfer to the air (a) without radiation and (b) with radiation if the emissivity of the surface of the cylinder is 0.825.

24.17: A vertical cylinder is 80-cm high. It is used in a large tank to heat glycerin held at a bulk temperature of 13°C. The top of the cylinder does not dissipate. Determine the heat transferred using the McAdams correlation.

24.18: Assume that an adult male can be approximated by a vertical cylinder that is 1.95-m high and 32 cm in diameter. With normal body temperature at 37°C and a still-air environment at 20°C, determine the heat transferred using the McAdams correlation.

24.19: The vertical plate in Problem 24.1 is tilted at an angle of 38° from the vertical, and all other conditions remain the same, use the McAdams correlation to predict the heat transferred.

24.20: The vertical plate in Problem 24.4 is tilted at an angle of 40° from the vertical and all other conditions remain the same, use the Churchill-Chu correlation to predict the heat transferred.

24.21: A 25-cm × 25-cm-vertical plate is held at 74°C and is inclined at an angle of 56° from the vertical. Use the McAdams correlation to predict the heat loss to a still-air environment at 0.625 atm and 20°C.

24.22: The plate in Problem 24.21 is tilted at an angle of 50° from the vertical and is exposed to oxygen at 20°C at 0.75 atm. Determine the heat loss.

Horizontal Cylinders

24.23: A 2.5-cm-diameter cylinder is placed horizontally in a pool of water at 20°C. The surface of the cylinder is maintained at 60°C and the cylinder is 1-m long. Use the McAdams correlation to predict the heat dissipated by the cylinder.

24.24: A 2.5-cm-diameter cylinder is placed horizontally in a pool of water at 20°C. The surface of the cylinder is maintained at 60°C and the cylinder is 1-m long. Use the Churchill-Chu correlation to predict the heat dissipated by the cylinder.

24.25: A 2.5-cm-diameter cylinder is placed horizontally in still air 1 atm and 20°C. The surface of the cylinder is maintained at 60°C and the cylinder is 1-m long. Use the McAdams correlation to predict the heat dissipated by the cylinder.

24.26: A 2.5-cm-diameter cylinder is placed horizontally in still air at 1 atm and 20°C. The surface of the cylinder is maintained at 60°C and the cylinder is 1-m long. Use the Churchill-Chu correlation to predict the heat dissipated by the cylinder.

24.27: A horizontal cylindrical heater with a diameter of 2.5 cm and a surface temperature of 117°C is placed in a pool of ethylene glycol at 37°C. For a length of 40 cm, determine the heat transferred using the McAdams correlation.

24.28: A horizontal cylindrical heater with a diameter of 2.5 cm and a surface temperature of 117°C is placed in a pool of ethylene glycol at 37°C. For a length of 40 cm, determine the heat transferred using the Churchill-Chu correlation.

24.29: A horizontal rod with a diameter of 1.75 cm is to be held at 129°C in still air at 1 atm and 25°C. The rod is 5-m long and has a resistivity of 180 $\mu\Omega\text{-cm}^2/\text{cm}$. Neglect radiation and using the McAdams correlation, determine the current required.

24.30: A horizontal rod with a diameter of 1.75 cm is to be held at 129°C in still air at 1 atm and 25°C. The rod is 5-m long, has a resistivity of 180 $\mu\Omega\text{-cm}^2/\text{cm}$, and an emissivity of 0.845. Assume that the radiation is to surroundings at 25°C, and using the McAdams correlation, determine the current required.

24.31: A horizontal steam pipe has a diameter of 6 cm and a surface temperature of 127°C. The pipe passes through a room at atmospheric pressure with walls and still air at 27°C. The pipe is 2-m long and the surface emissivity is 0.555. Using the McAdams correlation, determine the pipe heat dissipation.

24.32: A horizontal steam pipe has a diameter of 6 cm and a surface temperature of 127°C. The pipe passes through a room where the pressure is 0.80 atm with walls and still air at 27°C. The pipe is 2-m long and the surface emissivity is 0.555. Using the McAdams correlation, determine the pipe heat dissipation.

Spheres

24.33: A 5-cm-diameter sphere is suspended in still air at one atmosphere and 20°C. The surface of the sphere is at 80°C. Determine the heat loss.

24.34: A 5-cm-diameter sphere is suspended in water at 20°C and the surface of the sphere is at 80°C. Determine the heat loss.

24.35: The skin of a spherical gondola of a high-altitude balloon with a diameter of 4 m is to be held at 27°C in a still-air environment at 1 atm and 17°C. Neglecting radiation, determine the heat loss from the gondola.

24.36: The skin of a spherical gondola of a high-altitude balloon with a diameter of 4 m is to be held at 23°C in a still-air environment at 0.20 atm and 17°C. Neglecting radiation, determine the heat loss from the gondola.

24.37: A sphere that is 10 cm in diameter with a skin temperature of 127°C is placed in still-carbon dioxide at 1 atm and 27°C. Neglecting radiation, determine the heat loss from the sphere.

24.38: Suppose that the carbon dioxide in Problem 24.37 is at 5 atm, but all other conditions remain the same. Determine the heat loss from the sphere.

24.39: Determine the fraction of the dissipation that is due to radiation if the emissivity of the sphere in Problem 24.37 is 0.94.

24.40: Determine the diameter of the sphere necessary to raise the natural convection heat dissipation to 30 W if all other conditions in Problem 24.37 remain the same.

24.41: Determine the heat loss by natural convection if the sphere in Problem 24.37 is placed in a pool of ethylene glycol with all other conditions remaining the same.

24.42: Determine the heat loss by natural convection if the sphere in Problem 24.37 is placed in a pool of water with all other conditions remaining the same.

Horizontal Plates

24.43: A horizontal plate in the form of an isosceles right triangle whose equal sides are each 32-cm long is held at 70°C in still air at atmospheric pressure and 20°C. Determine the heat loss by natural convection from the upper surface of the plate.

24.44: A horizontal plate in the form of an isosceles right triangle whose equal sides are each 32-cm long is held at 70°C in water at 20°C. Determine the heat loss by natural convection from the upper surface of the plate.

24.45: The plate in Problem 24.3 is placed horizontally in a still-air environment at two atmopheres with all other conditions remaining the same. Determine the heat loss by natural convection from the upper surface of the plate.

24.46: The plate in Problem 24.3 is in a horizontal position at 1 atm and all other conditions remain the same. Determine the heat loss by natural convection from the upper surface of the plate.

24.47: A cast-iron plate is 28 cm × 36 cm is used as a grill in a restaurant. It is held horizontally at 82°C and sits in a still-air environment at 1 atm and 22°C. The emissivity of the cast-iron is 0.82 and the walls of the restaurant are at 17°C. Determine the total heat dissipation of the grill.

24.48: The top of a disk 40 cm in diameter is used to heat unused engine oil (SAE 50) that is at 4°C. The disk is placed face up at the bottom of the engine oil container. Determine the power required to maintain the surface temperature of the disk at 70°C.

24.49: A rectangular duct of height 40 cm and width 1 m carries conditioned air such that its surface temperature is 20°C. It is placed in a confined space containing still air at 70°C such that its length is 4 m. Assuming that the airflow is such that the 20°C temperature is maintained, determine the total natural convection heat gained for the duct.

24.50: A steel sheet that is 3 m × 3 m is removed from a furnace where the temperature is 400°C. It is placed in a still-air environment at 1 atm and 20°C. Taking 0.28 as the emissivity of the steel and considering the sheet as a horizontal surface facing upward, determine the rate of heat loss from the surface.

24.51: A cube that is 24 cm on a side is suspended by nonconducting strings in still air at 1 atm and 40°C. The surfaces of the cube are held at 94°C, radiation is to be neglected, and the top surface of the cube is absolutely horizontal. Determine the heat dissipation from the cube surface.

24.52: A cube that is 24 cm on a side is suspended by nonconducting strings in water at 40°C. The surfaces of the cube are held at 94°C, radiation is to be neglected, and the top surface of the cube is absolutely horizontal. Determine the heat dissipation from the cube surface.

Arrays of Vertical Plates and Fins

24.53: Two plates that are each 16 cm × 16 cm are located with a clear space of 1 cm between them in still air at 1 atm and 17°C. Each plate is at a temperature of 77°C. Determine the heat transfer by natural convection from the plates to the air.

24.54: Consider an aluminum ($k = 182$ W/m-K) longitudinal fin of uniform thickness with an adiabatic tip 16-cm high, 2.5-cm thick, and 50 cm in vertical length. The fin base temperature T_b is 120°C and the still-air environment at 1 atm is at 20°C. Consider the length dimension as the base of the fin and determine the rate of heat dissipation by natural convection from the surface of the fin.

24.55: Twenty-four vertical aluminum ($k = 200$ W/m-K) longitudinal fins with adiabatic tips, each with a height of 5 cm, a thickness of 2.286 mm, and a vertical length of 16 cm are placed over the rear wall of an electronic chassis that is 25-cm wide. Each fin base temperature, T_b, is 85°C, and the still-air environment at 1 atm is at 25°C. Determine the heat dissipation by natural convection by the fins.

24.56: Suppose that in Problem 24.55, it is required that the heat dissipation from the finned arrangement is 145 W. Determine how this can be accomplished if the base temperature of each fin remains at 85°C.

24.57: In Problem 24.53, make an estimate of the optimum spacing between the fins and determine, if, and by how much, the heat dissipation capability improves.

24.58: A vertical plate at the rear of an electronic chassis measures 36-cm wide × 18-cm high. Fins of aluminum ($k = 205$ W/m-K) that are 7.5-cm high, 2.286-mm thick, and 18-cm long can be used to enhance the heat transfer capability of the plate. The surrounding fluid is at 100 kPa and 22°C and the chassis temperature is 95°C. Determine the number of fins to be employed and the heat dissipation.

24.59: Eight vertical plates are placed x cm apart in the air at 100 kPa and 22°C. The plates are 16 cm × 16 cm and are held at 67°C. Determine the total heat flow from the plates when (a) $x = 2$ cm, (b) $x = 1.5$ cm, and (c) $x = 1$ cm.

24.60: The dissipation from a transformer is improved by equipping it with six vertical fins of aluminum ($k = 210$ W/m-K) that are each 8-cm high, 4-mm thick, and 48-cm long. The fin centerline-to-centerline spacing is 2 cm, and the surface temperature of the transformer is held at 77°C. The surrounding air pressure is at 100 kPa. Determine the air temperature if 160 W are to be dissipated.

24.61: Determine the heat transfer coefficient in natural convection within a 15-cm high × 7.5-cm wide parallel plate channel with a spacing of 2 cm. The air is at 100 kPa and 300 K, and the channel walls are maintained at 350 K.

Enclosures

24.62: A vertical enclosure consists of two plates that are 20-cm high at a plate spacing of 2 cm. The plate on the left is held at 85°C and the one on the right is held at 25°C. Determine the heat transfer coefficient for the gap.

24.63: Rework Problem 24.62 for a plate height of 35 cm at a gap spacing of 1.4 cm. All other conditions remain the same.

24.64: Rework Problem 24.62 for a tilted enclosure at 53.13°. All other conditions remain the same.

24.65: Two concentric cylinders form a horizontal annular region with $d_i = 2$ cm and $d_o =$ 4 cm. The temperatures of the inner and outer walls are 57°C and 17°C, respectively, and air at 100 kPa fills the annular region. Determine the heat flow for unit length.

24.66: Rework Problem 24.65 for an annular region with an unknown outer diameter but with heat flow across the gap of 16.75 W/m. All other conditions remain the same.

24.67: Rework Problem 24.65 for a pair of concentric spheres. All other conditions remain the same.

24.68: Rework Problem 24.62 for plate spacings of (a) 2.5 cm, (b) 4 cm, and (c) 5 cm. All other conditions remain the same.

24.69: Rework Problem 24.65 for outer cylinder diameters of (a) 4.5 cm, (b) 5 cm, and (c) 6.25 cm. All other conditions remain the same.

24.70: Rework Problem 24.69 for outer sphere diameters of (a) 4.5 cm, (b) 5 cm, and (c) 6.25 cm. All other conditions remain the same.

25

Heat Exchangers

Chapter Objectives

- To describe the types of heat exchangers and to delineate the differences between the recuperator, direct contact heat exchanger, and the regenerator.
- To provide the governing relationships for the unfinned heat exchanger.
- To consider the various forms of the overall heat transfer coefficient.
- To develop the concept of the logarithmic mean temperature difference.
- To give the details of the logarithmic mean temperature difference correction factor method of heat exchanger design.
- To give the details of the effectiveness N_{tu} method of heat exchanger design.
- To provide the governing relationships for finned heat exchangers.

25.1 Introduction

A heat exchanger is a device that transfers heat from one fluid to another or between a fluid and the environment. In the open type, or **direct contact** type of heat exchanger, there is no intervening surface between the fluids. In the closed type or **indirect contact** heat exchangers, our definition pertains to a device that is employed in the transfer of heat between two fluids or a surface and a fluid. Shah (1981) and Mayinger (1988) have categorized heat exchangers according to the following properties:

- Transfer processes
- Number of fluids
- Construction
- Heat transfer mechanisms
- Surface compactness
- Flow arrangement
- Number of fluid passes
- Type of surface

Recuperators (or closed-type heat exchangers) are heat exchangers in which heat transfer occurs between two fluid streams at different temperatures that are separated by a thin solid wall (a parting sheet or tube wall). Heat is transferred by convection from the hotter fluid to the wall surface and by convection from the wall surface to the cooler fluid. Figure 25.1 shows four examples of the recuperator that are surface-type exchangers. The recuperator is the only type of heat exchanger that we will consider in detail.

Regenerators are heat exchangers in which a hot fluid and a cold fluid alternately flow through the same surface at prescribed time intervals. The surface of the regenerator receives

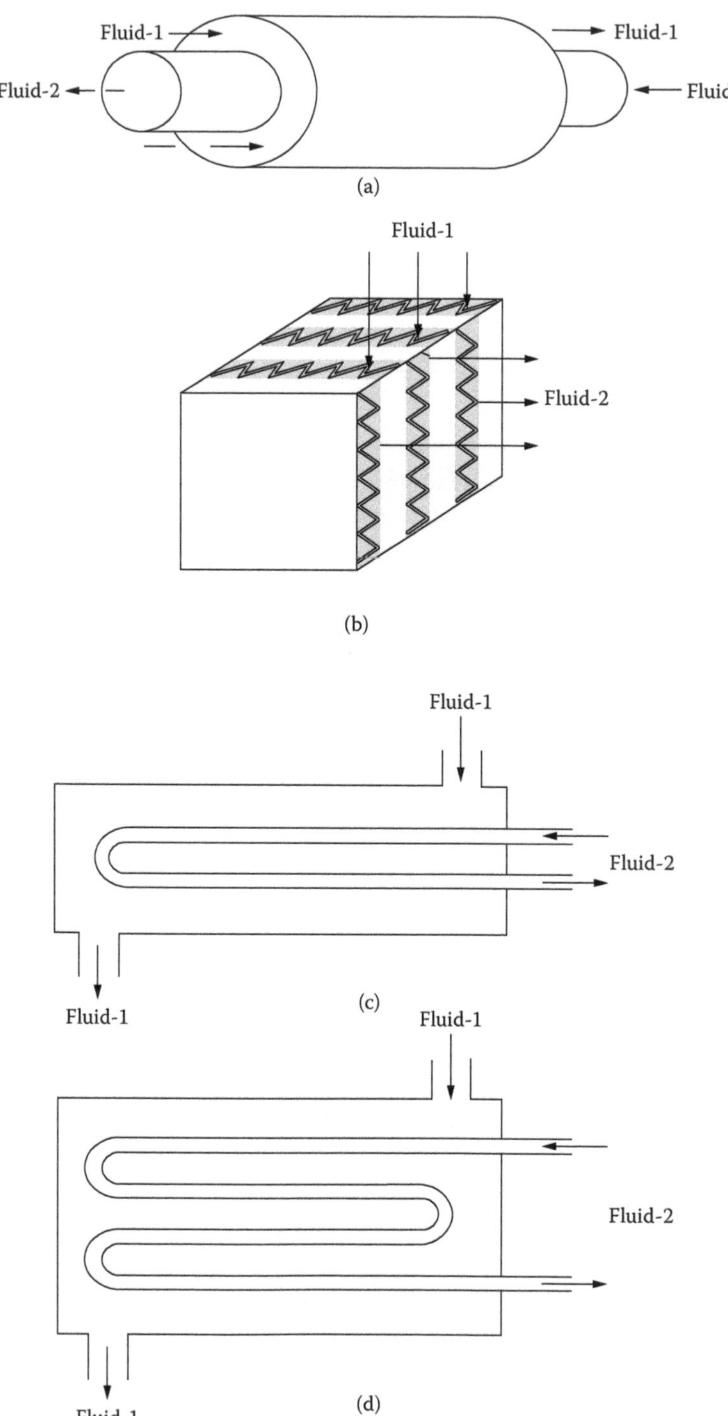

FIGURE 25.1
Examples of closed-type heat exchangers or recuperators. (a) The double-pipe counterflow heat exchanger, (b) the compact heat exchanger, (c) the shell and tube heat exchanger with one shell pass and two tube passes, and (d) the shell and tube heat exchanger with one shell pass and four tube passes.

heat by convection from the hot fluid and then releases heat by convection to the cold fluid. The process is transient, that is, the temperature of the surface (and the fluids themselves) vary with time during the heating and cooling of the common surface. The regenerator is also a surface-type heat exchanger.

In **direct contact** (or open type) heat exchangers, heat is transferred by partial or complete mixing of the hot and cold fluid streams. Hot and cold fluids that enter this type of exchanger separately leave together as a single mixed stream.

25.2 Governing Relationships

25.2.1 The Rate Equation

We will assume that there are two **process streams** in a heat exchanger, a hot stream flowing with a **capacity rate** $C_H = \dot{m}_H c_{p,H}$ and a cooler (or cold stream) flowing with a capacity rate $C_C = \dot{m}_C c_{p,C}$. Then, conservation of energy demands that the heat transferred between the streams be described by the enthalpy balance

$$\dot{Q} = C_H(T_1 - T_2) = C_C(t_2 - t_1) \tag{25.1}$$

where the subscripts 1 and 2 refer to the inlet and outlet of the exchanger and where the T's and t's are employed to indicate hot and cold fluid temperatures, respectively.

Equation 25.1 represents an ideal that must hold in the absence of losses and, while it describes the heat that will be transferred (the **duty** of the heat exchanger) for the case of prescribed flow and temperature conditions, it does not provide an indication of the size of the heat exchanger necessary to perform this duty. We find that the size of the exchanger is determined using what is called the **rate equation**

$$\dot{Q} = US\theta_m = U_H S_H \theta_m = U_C S_C \theta_m \tag{25.2}$$

where S_H and S_C are the surface areas on the hot and cold sides of the exchanger, U_H and U_C are the **overall heat transfer coefficients** referred to the hot and cold sides of the exchanger and θ_m is a driving temperature difference.

The entire heat exchange process can be represented by

$$\dot{Q} = U_H S_H \theta_m = U_C S_C \theta_m = C_H(T_1 - T_2) = C_C(t_2 - t_1) \tag{25.3}$$

which is merely a combination of Equations 25.1 and 25.2.

25.2.2 The Exchanger Surface Area

Consider the unfinned tube of length L shown in Figure 25.2 and observe that because of the tube wall thickness, δ_w, the inner diameter will be smaller than the outer diameter and the surface areas will be different

$$S_i = \pi d_i L \tag{25.4a}$$

and

$$S_o = \pi d_o L \tag{25.4b}$$

where $d_o = d_i + 2\delta_w$.

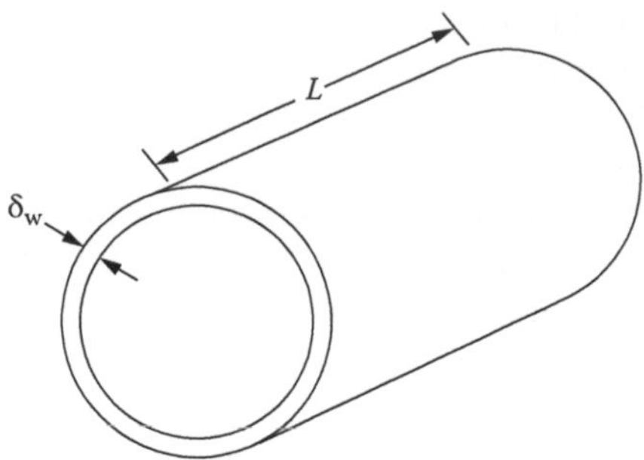

FIGURE 25.2
A heat exchanger tube of length, L.

25.2.3 The Overall Heat Transfer Coefficient

In a heat exchanger involving hot and cold streams, heat must flow, in turn, from the hot fluid to the cold fluid through as many as five thermal resistances:

- The hot-side convective layer resistance

$$R_H = \frac{1}{h_H S_H} \qquad \text{(K/W)} \tag{25.5}$$

- The hot-side **fouling** resistance due to an accumulation of foreign (and undesirable) material on the hot fluid exchanger surface.

$$R_{d,H} = \frac{1}{h_{d,H} S_H} \qquad \text{(K/W)} \tag{25.6}$$

  where the subscript d is intended to infer "dirt." Fouling is discussed in Section 25.2.5.

- The resistance of the exchanger material that has a finite thermal conductivity and may take on a value that is a function of the type of exchanger

$$R_m = \begin{cases} \dfrac{\delta_m}{k_m S_m} & \text{(K/W)} \quad \text{plane walls} \\[2ex] \dfrac{\ln(d_o/d_i)}{2\pi k_m L n_t} & \text{(K/W)} \quad \text{circular tubes} \end{cases} \tag{25.7}$$

  where δ_m is the thickness of the wall, S_m is the surface area of the wall, and n_t is the number of tubes.

- The cold-side fouling resistance

$$R_{d,C} = \frac{1}{h_{d,C} S_C} \qquad \text{(K/W)} \tag{25.8}$$

- The cold-side convective layer resistance

$$R_C = \frac{1}{h_C S_C} \qquad \text{(K/W)} \tag{25.9}$$

The resistances listed in Equations 25.5 through 25.9 are in "**series**" and the total resistance can be represented by $R_T = 1/US$

$$\frac{1}{US} = \frac{1}{h_H S_H} + \frac{1}{h_{d,H} S_H} + R_m + \frac{1}{h_{d,C} S_C} + \frac{1}{h_C S_C} \tag{25.10}$$

where, for the moment, U and S on the left side of Equation 25.10 are not assigned subscripts. We see that Equation 25.10 is perfectly general and may be put into specific terms depending on the selection of the reference surface, whether or not fouling is present and whether or not the conduction resistance need be considered. If Equation 25.10 is solved for U, the result is

$$U = \frac{1}{\dfrac{S}{h_H S_H} + \dfrac{S}{h_{d,H} S_H} + S R_m + \dfrac{S}{h_{d,C} S_C} + \dfrac{S}{h_C S_C}} \tag{25.11}$$

and if the thickness of the tube material is small or its thermal conductivity is high, the conduction resistance becomes negligible and

$$U = \frac{1}{\dfrac{S}{h_H S_H} + \dfrac{S}{h_{d,H} S_H} + \dfrac{S}{h_{d,C} S_C} + \dfrac{S}{h_C S_C}} \tag{25.12}$$

Several forms of Equation 25.12 are

- For a hot-side reference with fouling on both sides ($S = S_H$)

$$U_H = \frac{1}{\dfrac{1}{h_H} + \dfrac{1}{h_{d,H}} + \dfrac{1}{h_{d,C}}\dfrac{S_H}{S_C} + \dfrac{1}{h_C}\dfrac{S_H}{S_C}} \tag{25.13}$$

- For a cold-side reference with fouling on both sides ($S = S_C$)

$$U_C = \frac{1}{\dfrac{1}{h_H}\dfrac{S_C}{S_H} + \dfrac{1}{h_{d,H}}\dfrac{S_C}{S_H} + \dfrac{1}{h_{d,C}} + \dfrac{1}{h_C}} \tag{25.14}$$

- For a hot-side reference without fouling ($S = S_H$)

$$U_H = \frac{1}{\dfrac{1}{h_H} + \dfrac{1}{h_C}\dfrac{S_H}{S_C}} \tag{25.15}$$

- For a cold-side reference without fouling ($S = S_C$)

$$U_C = \frac{1}{\dfrac{1}{h_H}\dfrac{S_C}{S_H} + \dfrac{1}{h_C}} \tag{25.16}$$

A listing of some approximate overall heat transfer coefficients for various services is provided in Table 25.1.

Example 25.1 Hot water flows through a 3-in-schedule 80 pipe with an inside heat transfer coefficient of 1800 W/m^2-K. It is cooled by air with a natural convection heat transfer

TABLE 25.1

Approximate Overall Heat Transfer Coefficients for Some Process Services

Application	U, W/m²-K
Water to water	750–2500
Water to air	7.5–50
Water to lube oil	100–400
Water to organic vapor	250–1000
Water to heavy fuel oil	50–200
Heavy organics to heavy organics	40–200
Light organics to light organics	180–450
Gas to gas	8–30
Condensing steam to water	1000–6000
Condensing steam to air	50–200
Refrigerator evaporators	250–1000

coefficient of $8\,\text{W/m}^2\text{-K}$, and there is no fouling. Determine the heat lost per unit of length per degree Celsius temperature difference driving force based on (a) the outside surface and (b) the inside surface.

Solution

Assumptions and Specifications

(1) Steady-state conditions exist.
(2) There is negligible heat loss to and from the surroundings.
(3) Kinetic and potential energy changes are negligible.
(4) The tube wall thermal resistance is negligible.

We first note that the hot fluid (the water) flows through the pipe with an inside diameter, d_i. This leaves the cold fluid (the air) on the outside of the pipe with diameter, d_o. The ratio of the cold side to the hot-side surface areas will be

$$\frac{S_C}{S_H} = \frac{\pi d_o L}{\pi d_i L} = \frac{d_o}{d_i}$$

Table A.20 in Appendix A gives for the 3-in-schedule 80 pipe

$$d_o = 8.890\text{ cm} \qquad \text{and} \qquad d_i = 7.366\text{ cm}$$

so that

$$\frac{d_o}{d_i} = \frac{S_C}{S_H} = \frac{8.890\text{ cm}}{7.366\text{ cm}} = 1.207$$

and

$$\frac{d_i}{d_o} = \frac{1}{1.207} = 0.829$$

(a) Because the cold fluid (the air) is on the outside of the pipe, use Equation 25.16 with $S_C/S_H = d_o/d_i = 1.207$

$$U_C = \frac{1}{\dfrac{1}{h_H}\dfrac{S_C}{S_H} + \dfrac{1}{h_C}} = \frac{1}{\dfrac{1.207}{1800\,\text{W/m}^2\text{-K}} + \dfrac{1}{8\,\text{W/m}^2\text{-K}}}$$

or

$$U_C = \frac{1}{(6.705 \times 10^{-4}\,\mathrm{W/m^2\text{-}K})^{-1} + (0.125\,\mathrm{W/m^2\text{-}K})^{-1}} = 7.96\,\mathrm{W/m^2\text{-}K}$$

Then, for a surface area of

$$S_C = \pi d_o L = \pi(0.0889\,\mathrm{m})(1\,\mathrm{m}) = 0.2793\,\mathrm{m^2}$$

we have

$$\frac{\dot{Q}}{\theta_m} = U_C S_C = (7.96\,\mathrm{W/m^2\text{-}K})(0.2793\,\mathrm{m^2}) = 2.22\,\mathrm{W/K} \Longleftarrow$$

(b) With the hot fluid (the water) on the inside of the pipe, we use Equation 25.15 with $S_H/S_C = d_i/d_o = 0.829$

$$U_H = \frac{1}{\dfrac{1}{h_H} + \dfrac{1}{h_C}\dfrac{S_H}{S_C}} = \frac{1}{\dfrac{1}{1800\,\mathrm{W/m^2\text{-}K}} + \dfrac{0.829}{8\,\mathrm{W/m^2\text{-}K}}}$$

or

$$U_H = \frac{1}{(5.556 \times 10^{-4}\,\mathrm{W/m^2\text{-}K})^{-1} + (0.104\,\mathrm{W/m^2\text{-}K})^{-1}} = 9.60\,\mathrm{W/m^2\text{-}K}$$

Then, for a surface area of

$$S_H = \pi d_i L = \pi(0.0737\,\mathrm{m})(1.00\,\mathrm{m}) = 0.2314\,\mathrm{m^2}$$

we have

$$\frac{\dot{Q}}{\theta_m} = U_H S_H = (9.60\,\mathrm{W/m^2\text{-}K})(0.2314\,\mathrm{m^2}) = 2.22\,\mathrm{W/K} \Longleftarrow$$

25.2.4 The Logarithmic Mean Temperature Difference

We now focus our attention on the value of θ_m in Equation 25.2 and in Figure 25.3, which shows the temperature length profile for the two fluids in a parallel or cocurrent flow heat exchanger. It appears that various alternatives are available to us. We could use $\theta_m = T_1 - t_1$; however, this temperature difference is the maximum difference available, its use would be optimistic and would lead to erroneously high values of $\dot{Q}$. On the other hand, use of $\theta_m = T_2 - t_2$, which is the minimum temperature difference, would be pessimistic and lead to values of $\dot{Q}$ that are too low. Of course, we could take an arithmetic average, but this would need verification.

In working with Figure 25.3 we will assume that

- The overall coefficient of heat transfer, U, is constant throughout the entire heat exchanger.
- There is no heat lost to the surroundings.
- There is no longitudinal heat conduction in the direction of the flowing fluids.
- There is no phase change (or constant temperature zone) in either fluid.

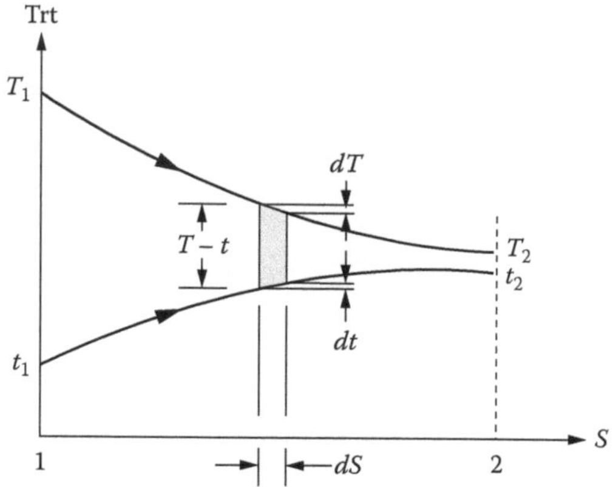

FIGURE 25.3
The hot and cold temperature profiles for the fluids in a parallel flow heat exchanger of length, L. The inlet and outlet end of the exchanger are designated as points 1 and 2, respectively. The profiles are plotted as a function of the surface area, S, and the differential element of surface area may noted. These are used to develop the LMTD for this type of exchanger.

- The specific heats, flow rates (and hence, the capacity rates) for both fluids are constant.
- The fluid temperatures at any cross section are characterized by the two temperatures, T and t.

For the differential element of surface area, dS, the heat flow, $d\dot{Q}$, can be related to the enthalpy change of each fluid

$$d\dot{Q} = -C_H dT = C_C dt \tag{25.17}$$

and the rate equation applied to the differential element shown in Figure 25.3 provides

$$d\dot{Q} = U(T - t)dS \tag{25.18}$$

From Equation 25.17 we have

$$dT = -\frac{d\dot{Q}}{C_H} \quad \text{and} \quad dt = \frac{d\dot{Q}}{C_C}$$

and a subtraction shows that

$$dT - dt = d(T - t) = -d\dot{Q}\left(\frac{1}{C_H} + \frac{1}{C_C}\right) \tag{25.19}$$

Equations 25.18 and 25.19 may then be combined to obtain

$$\frac{d(T - t)}{T - t} = -U\left(\frac{1}{C_H} + \frac{1}{C_C}\right) dS$$

and an integration between point 1 where $\Delta T_1 = T_1 - t_1$ and point 2 where $\Delta T_2 = T_2 - t_2$ will yield

$$\ln\left(\frac{T_2 - t_2}{T_1 - t_1}\right) = -US\left(\frac{1}{C_H} + \frac{1}{C_C}\right) \tag{25.20}$$

Now with

$$C_H = -\frac{\dot{Q}}{T_1 - T_2} \qquad \text{and} \qquad C_C = \frac{\dot{Q}}{t_2 - t_1}$$

Equation 25.20 may be displayed as

$$\dot{Q} = US\left[\frac{(T_2 - t_2) - (T_1 - t_1)}{\ln\left(\dfrac{T_2 - t_2}{T_1 - t_1}\right)}\right] \tag{25.21}$$

Then a comparison of Equations 25.2 and 25.21 gives θ_m for the parallel flow heat exchanger in Figure 25.3. Because of the natural logarithm in the denominator, θ_m is referred to as the **logarithmic mean temperature difference** and designated merely as LMTD.

$$\theta_m = \text{LMTD} = \frac{(T_2 - t_2) - (T_1 - t_1)}{\ln\left(\dfrac{T_2 - t_2}{T_1 - t_1}\right)} \tag{25.22}$$

Similar derivations for the four basic simple arrangements indicated in Figure 25.4 will provide the logarithmic mean temperature difference that can be written as

$$\theta_m = \text{LMTD} = \frac{\Delta T_1 - \Delta T_2}{\ln \dfrac{\Delta T_1}{\Delta T_2}} = \frac{\Delta T_2 - \Delta T_1}{\ln \dfrac{\Delta T_2}{\Delta T_1}} \tag{25.23}$$

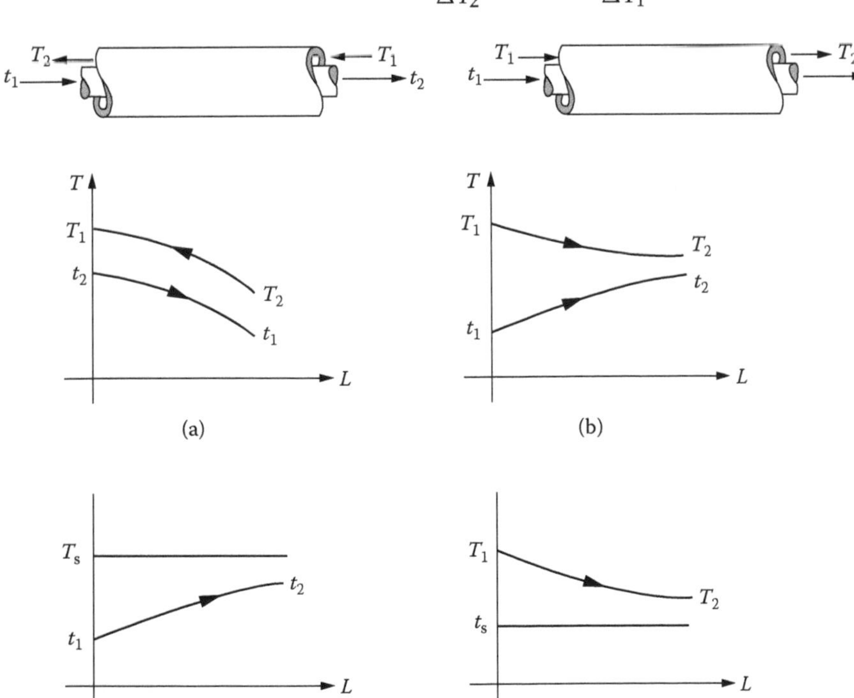

FIGURE 25.4
Temperature profiles for the four basic heat exchanger arrangements in which the LMTD may be employed. The counterflow heat exchanger, (b) the parallel or cocurrent flow heat exchanger, (c) the exchanger with a constant temperature source fluid and a rising temperature receiver fluid, and (d) the exchanger with a constant temperature receiver fluid and a falling temperature source fluid.

- For the **counterflow exchanger** where the fluids flow in opposite directions through the exchanger (Figure 25.4a)

$$\text{LMTD} = \frac{(T_1 - t_2) - (T_2 - t_1)}{\ln \dfrac{T_1 - t_2}{T_2 - t_1}} \quad (25.24)$$

- For the **cocurrent** or **parallel flow exchanger** where the fluids flow in the same direction through the exchanger (Figure 25.4b)

$$\text{LMTD} = \frac{(T_1 - t_1) - (T_2 - t_2)}{\ln \dfrac{T_1 - t_1}{T_2 - t_2}} \quad (25.22)$$

- For an exchanger that has a constant temperature source, $T_s = T_1 = T_2$, and a rising temperature receiver (Figure 25.4c)

$$\text{LMTD} = \frac{t_2 - t_1}{\ln \dfrac{T_s - t_1}{T_s - t_2}} \quad (25.25)$$

- For an exchanger that has a constant temperature receiver, $t_s = t_1 = t_2$, and a falling temperature source (Figure 25.4d)

$$\text{LMTD} = \frac{T_1 - T_2}{\ln \dfrac{T_1 - t_s}{T_2 - t_s}} \quad (25.26)$$

The logarithmic mean temperature difference must not be employed for arrangements other than those shown in Figure 25.4. The procedure for the case of cross flow and shell and tube exchangers is given in the next section.

Example 25.2 A process stream is cooled from 320°C to 160°C by another stream that is heated from 80°C to 140°C (Figure 25.5). Determine the LMTD for (a) counterflow and (b) cocurrent or parallel flow.

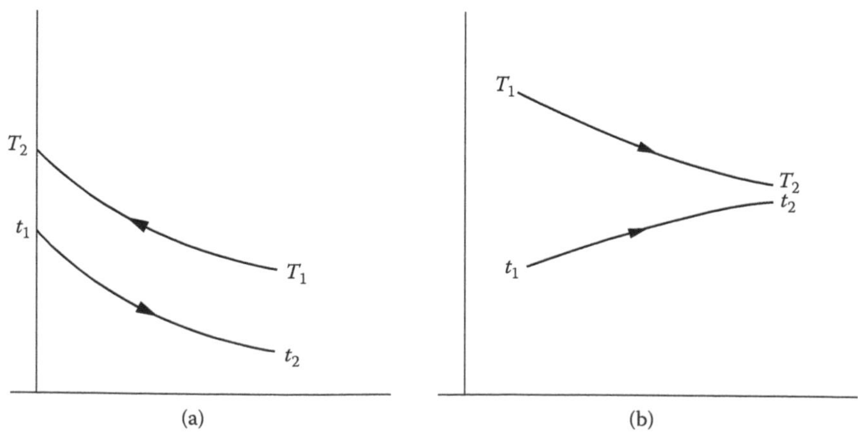

FIGURE 25.5
Temperature profiles for Example 25.2.

Solution

Assumptions and Specifications

(1) None.

Here, we are to find the LMTD for two flow arrangements with given temperatures at inlet and outlet for both streams. We have

$$T_1 = 320°\text{C} \quad \text{and} \quad T_2 = 160°\text{C}$$

and

$$t_1 = 80°\text{C} \quad \text{and} \quad t_2 = 140°\text{C}$$

(a) For counterflow, use Equation 25.24

$$\text{LMTD} = \frac{(T_1 - t_2) - (T_2 - t_1)}{\ln \dfrac{T_1 - T_2}{T_2 - t_1}}$$

$$= \frac{(320°\text{C} - 140°\text{C}) - (160°\text{C} - 80°\text{C})}{\ln \dfrac{320°\text{C} - 140°\text{C}}{160°\text{C} - 80°\text{C}}}$$

$$= \frac{180°\text{C} - 80°\text{C}}{\ln \dfrac{180°\text{C}}{80°\text{C}}} = \frac{100°\text{C}}{\ln 2.25} = 123.3°\text{C}$$

(b) For cocurrent or parallel flow, use Equation 25.22

$$\text{LMTD} = \frac{(T_1 - t_1) - (T_2 - t_2)}{\ln \dfrac{T_1 - t_1}{T_2 - t_2}}$$

$$= \frac{(320°\text{C} - 80°\text{C}) - (160°\text{C} - 140°\text{C})}{\ln \dfrac{320°\text{C} - 80°\text{C}}{160°\text{C} - 140°\text{C}}}$$

$$= \frac{240°\text{C} - 20°\text{C}}{\ln \dfrac{240°\text{C}}{20°\text{C}}} = \frac{220°\text{C}}{\ln 12} = 88.5°\text{C}$$

We can see that the parallel flow LMTD is much lower than the counterflow LMTD. In fact, it can be proven that the best possible LMTD for parallel flow reaches the limit of the counterflow LMTD. You might ask why, under these circumstances, is parallel flow ever used? The answer lies in the fact that cramped space in a proposed application may not permit the piping arrangement for counterflow and it may be desirable, in attempting to heat a highly viscous fluid, to provide a high-temperature difference at the front end to get the fluid flowing.

25.2.5 Fouling

25.2.5.1 Fouling Mechanisms

Somerscales and Knudsen (1981) have identified six categories of fouling:

1. *Particulate fouling:* The accumulation of solid particles suspended in the process stream on the heat transfer surfaces.

2. *Precipitation fouling:* The precipitation of dissolved substances carried in the process stream upon the heat transfer surfaces. **Scaling** occurs when precipitation occurs on heated rather than cooled surfaces.
3. *Chemical reaction fouling:* In certain cases, deposits on the heat transfer surfaces which are not, in themselves, reactants are formed by chemical reactions.
4. *Corrosion fouling:* In this type of fouling, the heat transfer surface reacts, at certain pH levels, to produce products that adhere to the heat transfer surfaces and, in turn, this may promote the attachment of additional fouling materials.
5. *Biological fouling:* Materials such as algae, bacteria, molds, seaweed, and barnacles carried in the process stream cause biological fouling of the heat transfer surfaces. Marine power plant condensers are notorious for their vulnerability to biological fouling.
6. *Freezing fouling:* Sometimes a liquid, or some of its higher melting point components, will deposit upon a subcooled heat transfer surface.

25.2.5.2 Fouling Factors

When a heat exchanger is placed in service, the heat transfer surfaces are, presumably, clean. With time, in some services in the power and process industries, the apparatus may undergo a decline in its ability to transfer heat. This is due to the accumulation of heat insulating substances on either or both of the heat transfer surfaces. The Tubular Exchanger Manufacturers Association (TEMA) published a table of **fouling factors** to assist the designer. Fouling resistances were tabulated and these were to be added to the film resistances ($1/S_i h_i$ and $1/h_o S_o$) of specific process streams. In this way, the operating period of all heat exchangers would be similar and ensure some desired period of continuous operation. The tables of fouling factors were intended as a crude guide toward the equalization of cumulative fouling in all fouling streams in the assembly.

Extracts from the table of fouling factors given by Chenoweth (1990), but in terms of $1/h_d$, are provided in Table 25.2.

Example 25.3 A heat exchanger employs 2-in OD × 12-BWG (Birmingham Wire Gage) tubes that have outer and inner diameters of 5.080 and 4.526 cm, respectively. The tubes are 4-m long and carry a hot process stream on the inside with $h_H = 800\,\text{W/m}^2\text{-K}$ and a cold fluid stream on the outside with $h_C = 500\,\text{W/m}^2\text{-K}$. Determine the overall heat transfer coefficient referred to the cold side when (a) the exchanger is clean and (b) when the exchanger is fouled to the extent that a fouling resistance of $0.0008\,(\text{W/m}^2\text{-K})^{-1}$ is present on the hot side.

Solution

Assumptions and Specifications

(1) Steady-state conditions exist.
(2) There is negligible heat loss to and from the surroundings.
(3) Kinetic and potential energy changes are negligible.
(4) The tube wall thermal resistance is negligible.

First, we need to determine S_H and S_C. The hot fluid is in the inside of the tubes

$$S_H = \pi d_i L$$

and hence,

$$S_C = \pi d_o L$$

TABLE 25.2

Some Fouling Resistances of Various Gas, Vapor, and Liquid Streams (Adapted from Chenoweth, 1990)

Fluid	Fouling Resistance $\frac{1}{h_d} \times 10^4$, m^2 K/W
Cooling tower water	1.75–3.50
Artificial spray pond water	1.75–3.50
Brackish water	3.50–5.25
Treated boiler feedwater	0.90
Seawater	1.75–3.50
No. 2 Fuel oil	3.50
Engine lube oil	1.75
Refrigerants (liquid)	1.75
Gasolene	3.50
Heavy fuel oil	5.25–12.50
Natural gas	1.75–3.50
Compressed air	1.75
Refrigerant vapor (oil bearing)	3.50
Steam (non-oil bearing)	9.00

Thus,

$$\frac{S_C}{S_H} = \frac{\pi d_o L}{\pi d_i L} = \frac{d_o}{d_i}$$

With the given values that can be verified in Table A.21

$$d_o = 5.080\,\text{cm} \qquad \text{and} \qquad d_i = 4.526\,\text{cm}$$

we have

$$\frac{d_o}{d_i} = \frac{S_C}{S_H} = \frac{5.080\,\text{cm}}{4.526\,\text{cm}} = 1.122$$

(a) Because the cold fluid is on the outside of the pipe, use Equation 25.16 with $S_C/S_H = d_o/d_i = 1.207$

$$U_C = \frac{1}{\frac{1}{h_H}\frac{S_C}{S_H} + \frac{1}{h_C}} = \frac{1}{\frac{1.122}{800\,\text{W/m}^2\text{-K}} + \frac{1}{500\,\text{W/m}^2\text{-K}}}$$

$$= \frac{1}{(1.403 \times 10^{-3}\,\text{W/m}^2\text{-K})^{-1} + (2 \times 10^{3}\,\text{W/m}^2\text{-K})^{-1}}$$

$$= 293.9\,\text{W/m}^2\text{-K} \Longleftarrow$$

(b) When a fouling factor of $1/h_{d,H} = 0.0008\,(\text{W/m}^2\,\text{K})^1$ is employed on the hot side, we use Equation 25.14

$$U_C = \frac{1}{\frac{1}{h_H}\frac{S_C}{S_H} + \frac{1}{h_{d,H}}\frac{S_C}{S_H} + \frac{1}{h_{d,C}} + \frac{1}{h_C}}$$

and with $1/h_{d,C} = 0$, we have

$$U_C = \frac{1}{\dfrac{1.122}{800\,\mathrm{W/m^2\text{-}K}} + (0.0008\,\mathrm{m^2\text{-}K/W})\,(1.122) + \dfrac{1}{500\,\mathrm{W/m^2\text{-}K}}}$$

$$= \frac{1}{[1.403 \times 10^{-3} + 8.976 \times 10^{-4} + 2 \times 10^{-3}]\,\mathrm{W/m^2\text{-}K}}$$

$$= 232.6\,\mathrm{W/m^2\text{-}K} \Longleftarrow$$

25.3 Heat Exchanger Analysis Methods

25.3.1 The Logarithmic Mean Temperature Difference Correction Factor Method

The logarithmic mean temperature difference developed in Section 25.2.4 is not applicable to shell and tube or cross-flow heat exchangers. In these configurations, the temperature parameter θ_m in Equations 25.2 and 25.3 is the true or effective mean temperature difference and is related to the counterflow logarithmic mean temperature difference

$$\mathrm{LMTD}_c = \frac{(T_1 - t_2) - (T_2 - t_1)}{\ln \dfrac{T_1 - T_2}{T_2 - t_1}} \tag{25.24}$$

via

$$\theta_m = F(\mathrm{LMTD}_c) \tag{25.27}$$

where F is a factor that is referred to as the **logarithmic mean temperature difference correction factor**. It is a function of two parameters, P and R where

$$P = \frac{t_2 - t_1}{T_1 - t_1} \tag{25.28a}$$

is defined as the **cold-side effectiveness** and

$$R = \frac{T_1 - T_2}{t_2 - t_1} = \frac{C_C}{C_H} \tag{25.28b}$$

is a **capacity rate ratio**.

The evaluation of the the logarithmic mean temperature difference correction factor for the multitude of multipass and cross-flow heat exchanger arrangements dates from the early 1930s: Nagle (1933), Underwood (1934), Fischer (1938), and Bowman et al. (1940). The correction factors are available in chart form such as those shown in Figures 25.6 through 25.9.

Example 25.4 In a water-to-water shell and tube heat exchanger, hot water is on the shell side and flows at 2.655 kg/s at an inlet temperature of 95°C. Cold water flows within the tubes at a rate of 4.00 kg/s between temperatures of 40°C and 60°C. The overall heat transfer coefficient referred to the hot side of the exchanger is 1250 W/m^2-K. The exchanger may be considered unfouled with tubes that are to be limited to a maximum length of 2.25 m. If

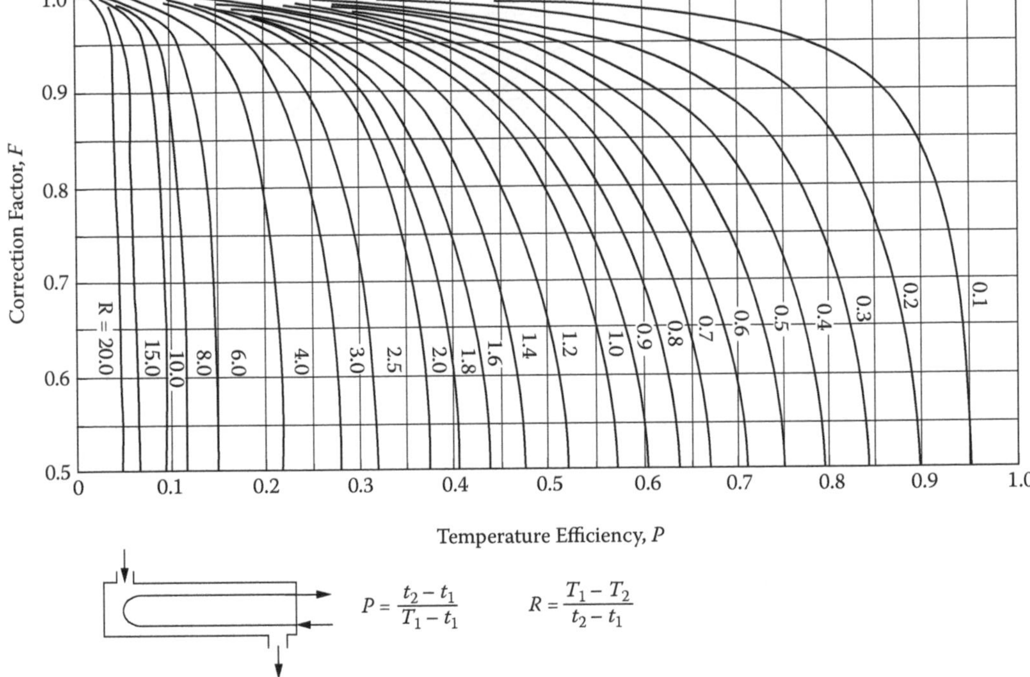

FIGURE 25.6
The logarithmic mean temperature difference correction factor F for a shell and tube heat exchanger with one shell pass and two tube passes. (From Bowman et al., 1940.)

the tubes are 3/4-in OD × 18 BWG and the velocity within the tubes is to be maintained as close as possible to 0.425 m/s, determine the number of tube passes and the number and length of the tubes.

Solution

Assumptions and Specifications

(1) Steady-state conditions exist.
(2) There is negligible heat loss to and from the surroundings.
(3) Kinetic and potential energy changes are negligible.
(4) The tube wall thermal resistance is negligible.
(5) The flow of the water in the tubes is fully developed.
(6) Fluid specific heats and the overall heat transfer coefficient are constant.
(7) Axial conduction along the exchanger tubing is negligible.

We first assume a single pass on the tube side and we will check this assumption against the maximum tube length of 2.25 m. The hot-side outlet temperature is obtained from a rearrangement of Equation 25.1

$$T_2 = T_1 - \frac{\dot{m}_C c_C (t_2 - t_1)}{\dot{m}_H c_H}$$

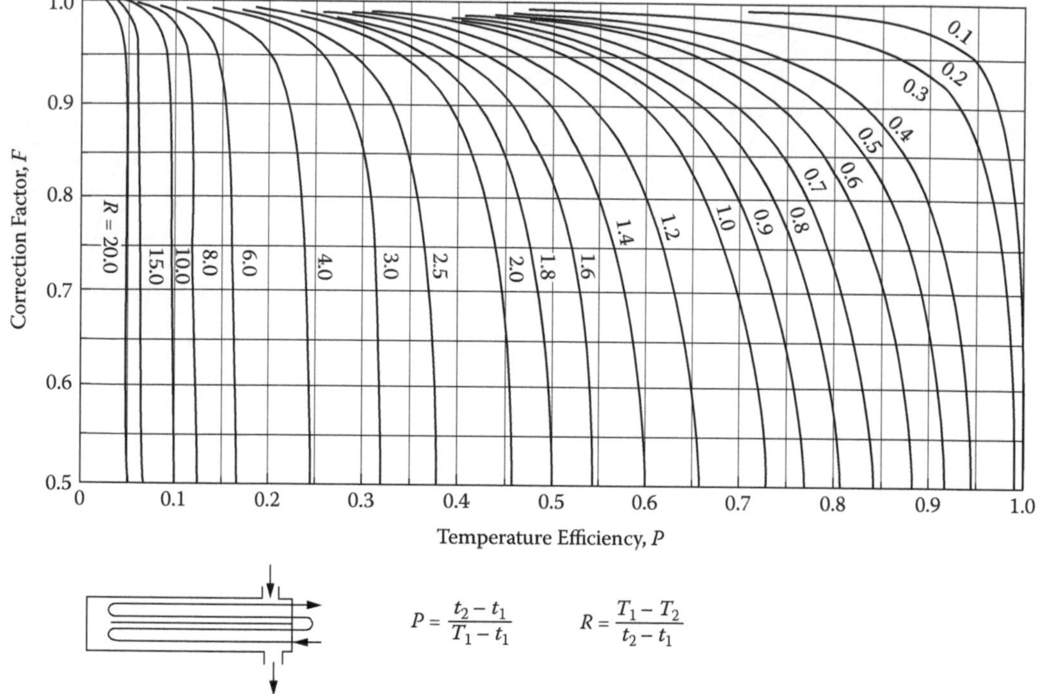

FIGURE 25.7
The logarithmic mean temperature difference correction factor F for a shell and tube heat exchanger with two shell passes and four tube passes. (From Bowman et al., 1940.)

Here, we retain the subscripts because of the variation of specific heat with the temperature of the water. The average cold water temperature is

$$t_{av} = \frac{t_1 + t_2}{2} = \frac{40°C + 60°C}{2} = 50°C$$

and Table A.14 gives

$$c_C = 4.178\,\text{kJ/kg-K}$$

When we assume $T_2 = 65°C$ (which will also require verification) we find that the average hot water temperature will be

$$T_{av} = \frac{T_1 + T_2}{2} = \frac{95°C + 65°C}{2} = 80°C$$

and Table A.14 gives

$$c_H = 4.197\,\text{kJ/kg K}$$

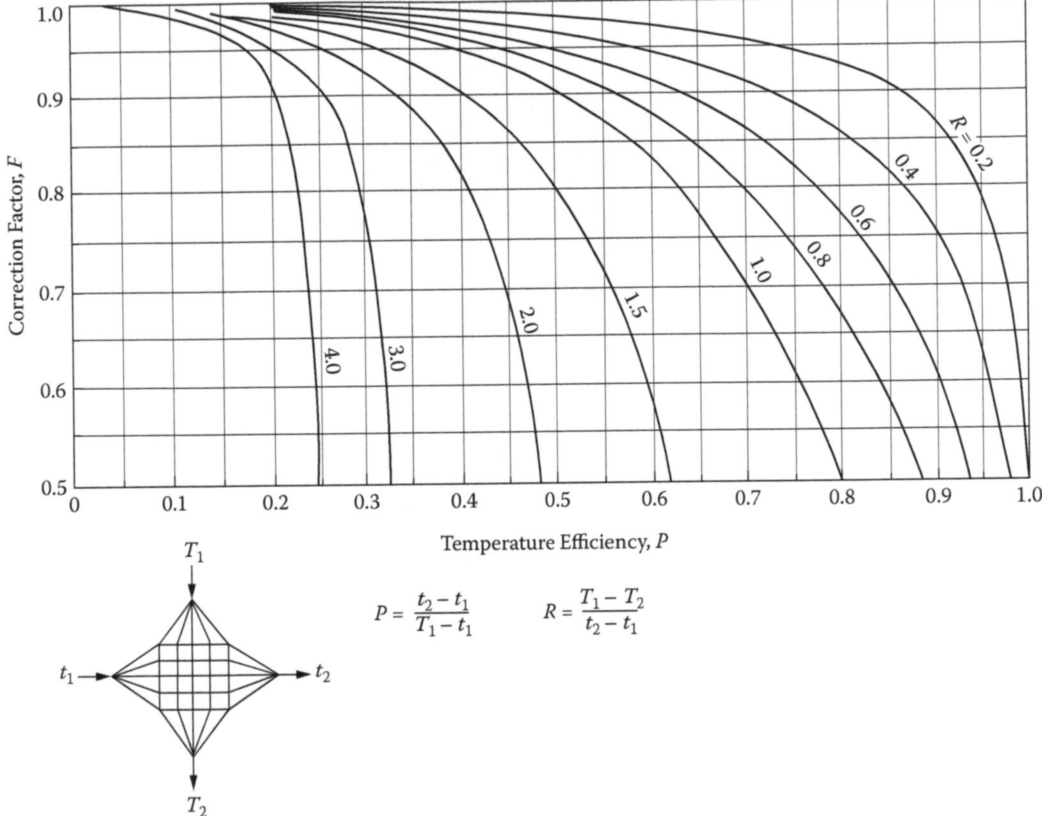

FIGURE 25.8
The logarithmic mean temperature difference correction factor F for a cross-flow heat exchanger with both fluids unmixed. (From Bowman et al., 1940.)

Thus,

$$
\begin{aligned}
T_2 &= T_1 - \frac{\dot{m}_C c_C (t_2 - t_1)}{\dot{m}_H c_H} \\
&= 95^\circ\text{C} - \frac{(4\,\text{kg/s})(4.178\,\text{kJ/kg-K})(60^\circ\text{C} - 40^\circ\text{C})}{(2.655\,\text{kg/s})(4.197\,\text{kJ/kg-K})} \\
&= 95^\circ\text{C} - 30^\circ\text{C} = 65^\circ\text{C}
\end{aligned}
$$

This is indeed the assumed value and permits the calculation procedure to continue.

The heat duty may also be obtained from Equation 25.1. Working with the cold side gives

$$
\begin{aligned}
\dot{Q} &= \dot{m}_C c_C (t_2 - t_1) \\
&= (4\,\text{kg/s})(4.178\,\text{kJ/kg-K})(60^\circ\text{C} - 40^\circ\text{C}) \\
&= 334.2\,\text{KW}
\end{aligned}
$$

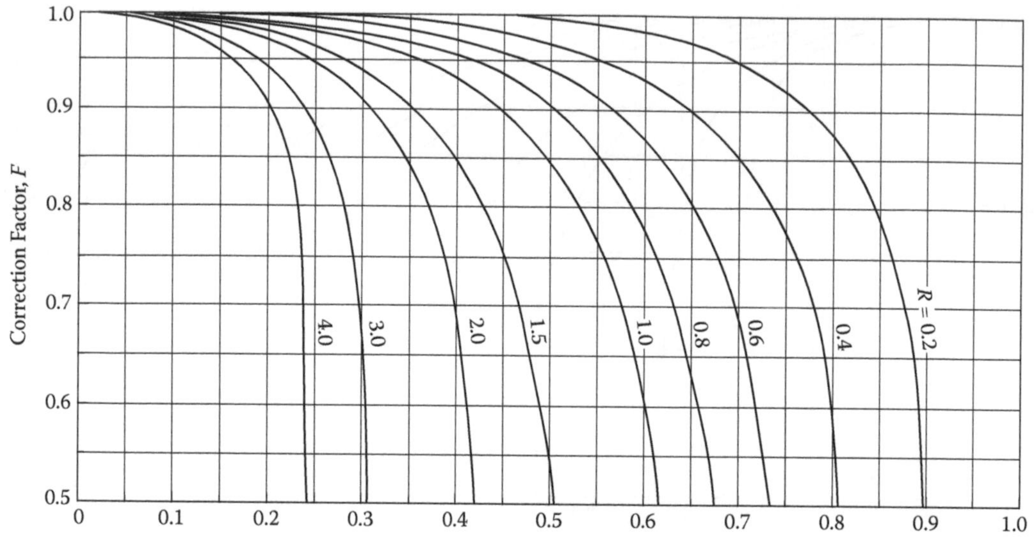

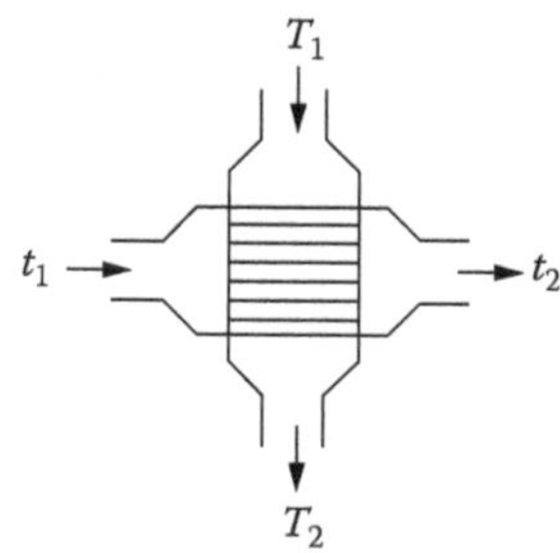

FIGURE 25.9
The logarithmic mean temperature difference correction factor F for a cross-flow heat exchanger with one fluid mixed and the other fluid unmixed. (From Bowman et al., 1940.)

For the one shell pass-one tube pass exchanger, we have true counterflow and we employ Equation 25.24 for the LMTD

$$\text{LMTD}_c = \frac{(T_1 - t_2) - (T_2 - t_1)}{\ln \dfrac{T_1 - T_2}{T_2 - t_1}} = \frac{(95^\circ\text{C} - 60^\circ\text{C}) - (65^\circ\text{C} - 40^\circ\text{C})}{\ln \dfrac{95^\circ\text{C} - 60^\circ\text{C}}{65^\circ\text{C} - 40^\circ\text{C}}}$$

or

$$\text{LMTD}_c = \frac{35^\circ\text{C} - 25^\circ\text{C}}{\ln \dfrac{35^\circ\text{C}}{25^\circ\text{C}}} = \frac{10}{\ln 1.40} = 29.72^\circ\text{C}$$

Because of the cold-side velocity requirement, we need to determine the total inside or cold-side cross-sectional area. At $t_{av} = 50^\circ$C, Table A.14 gives

$$\rho_C = 988.1\,\text{kg/m}^3$$

and then we have

$$A_C = \frac{\dot{m}_C}{\rho_C \hat{V}_C} = \frac{4.00\,\text{kg/s}}{(988.1\text{kg/m}^3)(0.425\,\text{m/s})} = 9.525 \times 10^{-3}\,\text{m}^2$$

Table A.21 shows that for a 3/4-in × 18-BWG tube

$$d_o = 1.905\,\text{cm} \quad \text{and} \quad d_i = 1.656\,\text{cm}$$

and it may be shown that

$$A_i = 2.154 \times 10^{-4}\,\text{m}^2 \quad \text{and} \quad S'_o = 0.05202\ \text{m}^2/\text{m}$$

where S'_o is the outside surface of the tube in m^2/m. To maintain a "come as close as possible to" velocity of 0.425 m/s, we have

$$n_t = \frac{A_C}{A_i} = \frac{9.525 \times 10^{-3}\,\text{m}^2}{2.154 \times 10^{-4}\,\text{m}^2} = 44.22 \qquad \text{(say 44 tubes)}$$

and with the tubes at their maximum length of 2.25 m, we have

$$S_H = n_t S'_o L = (44\,\text{tubes})(0.05202\,\text{m}^2/\text{m})(2.25\,\text{m}) = 5.150\ \text{m}^2$$

Equation 25.2 with θ_m = LMTD now gives

$$\dot{Q} = U_H S_H \text{LMTD} = (1250\ \text{W/m}^2\text{-K})(5.150\,\text{m}^2)(29.72°\text{C}) = 191.34\,\text{kW}$$

This falls far short of the required duty of 334.2 Kwa.

Because the single-pass heat exchanger does not perform the required duty, we must go to two passes still maintaining 44 tubes per pass in order to satisfy the water velocity requirement. But when we go to the two-pass exchanger, we must make a correction to the LMTD in accordance with Equation 25.27. This correction will require calculation of the values of P and R in accordance with Equation 25.28

$$P = \frac{t_2 - t_1}{T_1 - t_1} = \frac{60°\text{C} - 40°\text{C}}{95°\text{C} - 40°\text{C}} = \frac{20°\text{C}}{55°\text{C}} = 0.364$$

and

$$R = \frac{T_1 - T_2}{t_2 - t_1} = \frac{95°\text{C} - 65°\text{C}}{60°\text{C} - 40°\text{C}} = \frac{30°\text{C}}{20°\text{C}} = 1.50$$

Figure 25.6 then provides

$$F = 0.88$$

and with $\dot{Q}$ = 334.2 Kwa and $n_t = 2(44) = 88$ tubes

$$S_H = \frac{\dot{Q}}{U_H F(\text{LMTD}_c)} = \frac{334.2\,\text{kW}}{(1250\,\text{W/m}^2\text{-K})(0.88)(29.72°\text{C})} = 10.22\,\text{m}^2$$

The tube length will be

$$L = \frac{S_H}{\pi n_t d_o} = \frac{10.22\,\text{m}^2}{(88\,\text{tubes})(0.05202\ \text{m}^2/\text{m})} = 2.23\,\text{m} < 2.25\,\text{m}$$

Thus, there will be

2 passes ⟸
88 tubes ⟸
2.23 m long ⟸

25.3.2 The Effectiveness N_{tu} Method

25.3.2.1 Dimensionless Parameters

In the logarithmic mean temperature difference correction factor method, the parameter P, requires three temperatures for its computation. The inlet temperature of both the hot and cold streams is usually a given; when the cold-side outlet temperature is not known, however, a trial-and-error method is required to evaluate P. The trial-and-error procedure may be avoided by using the ϵ-N_{tu} method and, because of its suitability for computer aided design, the ϵ-N_{tu} method is gaining in popularity.

Kays and London (1984) have shown that the heat exchanger transfer equations may be written in a dimensionless form that results in three dimensionless groups:

1. The **capacity rate ratio**

$$C^* = \frac{C_{min}}{C_{max}} \tag{25.29}$$

Notice that this differs from the capacity rate ratio, R, used in the determination of the logarithmic mean temperature difference correction factor. Here, the capacity ratio, C^*, is *always* less than or equal to unity. It may take on the value $C^* = 0$ for the cases of the constant temperature source and rising temperature receiver fluid and the falling temperature source and constant temperature receiver fluid.

2. The exchanger **heat transfer effectiveness**

$$\epsilon = \frac{\dot{Q}}{\dot{Q}_{max}} \qquad (\epsilon < 1) \tag{25.30}$$

This is the ratio of the actual heat transferred to the maximum heat that could be transferred if the exchanger were a counterflow exchanger. We note that the limit of the effectiveness for an exchanger of infinite size would be unity.

3. The **number of transfer units**

$$N_{tu} \equiv \frac{US}{C_{min}} = \frac{1}{C_{min}} \int_S U\, dS \tag{25.31}$$

The number of transfer units is a measure of the size of the exchanger.

The actual heat transfer is given by the enthalpy balance of Equation 25.1. Observe that if

$$C_H > C_C \quad \text{then} \quad (T_1 - T_2) < (t_2 - t_1)$$

and if

$$C_C > C_H \quad \text{then} \quad (t_2 - t_1) < (T_1 - T_2)$$

the fluid that "might" experience the maximum temperature change, $T_1 - t_1$, is the fluid that has the minimum capacity rate. Thus, the maximum possible heat transfer can be expressed as

$$\dot{Q}_{max} = C_C(T_1 - t_1) \qquad (C_C < C_H) \tag{25.32a}$$

or

$$\dot{Q}_{max} = C_H(T_1 - t_1) \qquad (C_H < C_C) \tag{25.32b}$$

and either of these can be obtained with the counterflow exchanger. Therefore, the exchanger effectiveness can be written as

$$\epsilon = \frac{\dot{Q}}{\dot{Q}_{max}} = \frac{C_H(T_1 - T_2)}{C_{min}(T_1 - t_1)} = \frac{C_C(t_2 - t_1)}{C_{min}(T_1 - t_1)} \tag{25.33}$$

Observe that the value of ϵ will range between zero and unity and that for a given ϵ and $\dot{Q}_{max}$, the actual heat transfer in the exchanger will be

$$\dot{Q} = \epsilon C_{min}(T_1 - t_1) \tag{25.34}$$

Because

$$\epsilon = f(C^*, \mathrm{N_{tu}}, \text{flow arrangement})$$

each exchanger arrangement has its own effectiveness relationship. Figure 25.10 provides graphs for four such arrangements.

25.3.2.2 Specific Effectiveness ϵ-N_{tu} Relationships

Specific ϵ $\mathrm{N_{tu}}$ relationships along with their limiting values for seven flow arrangements, summarized from the work of Kays and London (1984) and Kakac et al. (1987) now follow:

1. **For counterflow**

$$\epsilon = \frac{1 - e^{-\mathrm{N_{tu}}(1-C^*)}}{1 - C^* e^{-\mathrm{N_{tu}}(1-C^*)}} \qquad (C^* < 1) \tag{25.35a}$$

$$\epsilon = \frac{\mathrm{N_{tu}}}{\mathrm{N_{tu}} + 1} \qquad (C^* = 1) \tag{25.35b}$$

and

$$\mathrm{N_{tu}} = \frac{1}{1 - C^*} \ln \frac{1 - \epsilon C^*}{1 - \epsilon} \tag{25.36}$$

2. **For cocurrent or parallel flow**

$$\epsilon = \frac{1 - e^{-\mathrm{N_{tu}}(1+C^*)}}{1 + C^*} \tag{25.37}$$

and

$$\mathrm{N_{tu}} = \frac{1}{1 + C^*} \ln \frac{1}{1 - (1 + C^*)\epsilon} \tag{25.38}$$

3. **For single-pass cross flow with both fluids unmixed**

$$\epsilon = 1 - e^{(\mathrm{N_{tu}})^{0.22}\phi / C^*} \tag{25.39}$$

where

$$\phi = e^{-C^*(\mathrm{N_{tu}})^{0.78}} - 1$$

Equation 25.39 cannot be rearranged to yield $\mathrm{N_{tu}} = f(\epsilon, C^*)$.

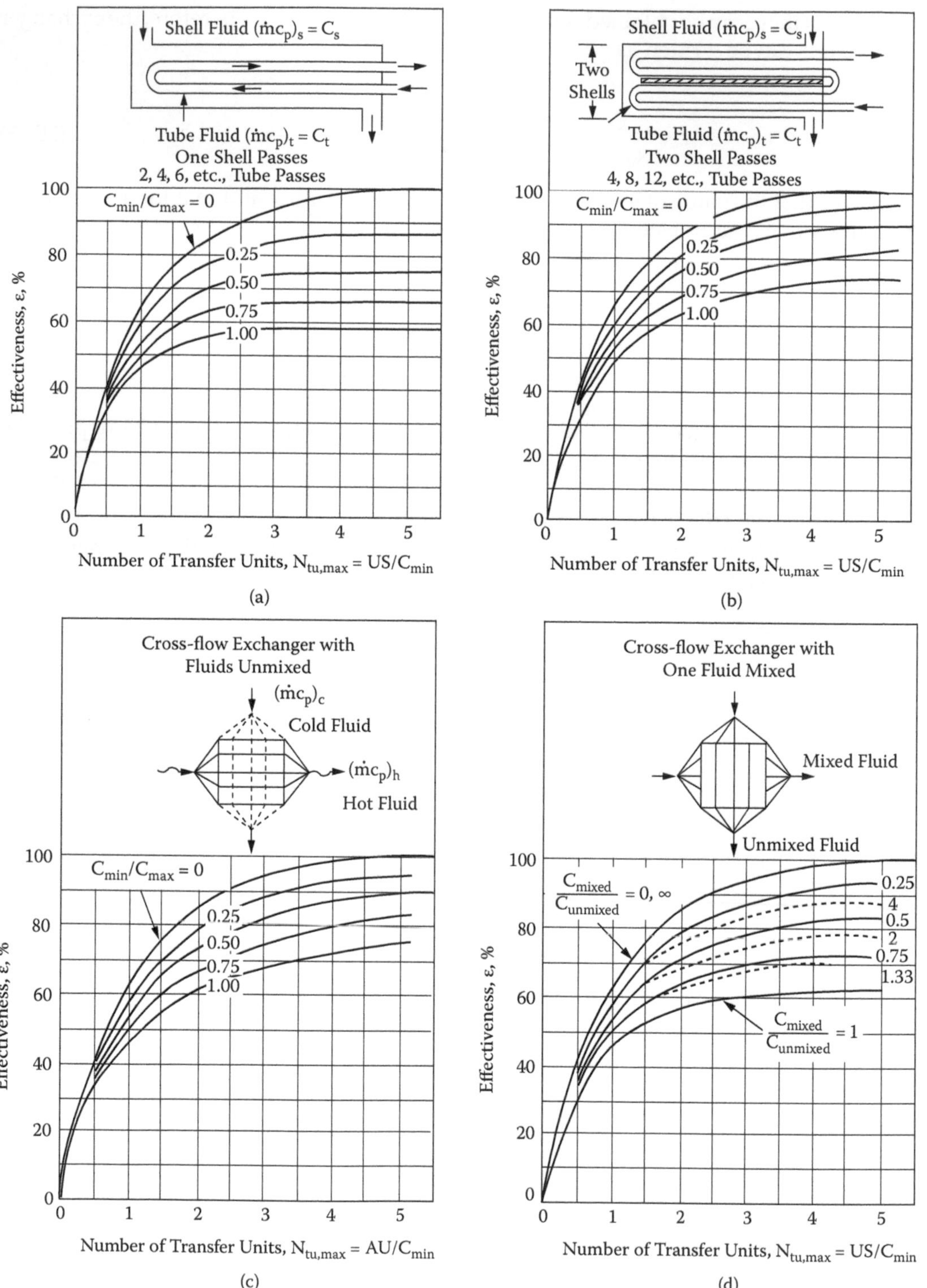

FIGURE 25.10
The effectiveness ϵ as a function of the number of transfer units, N_{tu}, for (a) the shell and tube exchanger with one shell pass and two tube passes, (b) the shell and tube exchanger with two shell passes and four tube passes, (c) the cross-flow heat exchanger with both fluids unmixed, and (d) the cross-flow heat exchanger with one fluid mixed and one fluid unmixed.

4. **For single-pass cross flow with both fluids mixed**

$$\epsilon = \left[\frac{1}{1-e^{-N_{tu}}} + \frac{C^*}{1-e^{-C^*N_{tu}}} - \frac{1}{N_{tu}}\right]^{-1} \tag{25.40}$$

Equation 25.40 cannot be rearranged to yield $N_{tu} = f(\epsilon, C^*)$.

5. **For single-pass cross flow with C_{max} mixed and C_{min} unmixed**

$$\epsilon = \frac{1}{C^*}(1-\phi) \tag{25.41}$$

where

$$\phi = e^{-C^*(1-e^{-N_{tu}})}$$

and

$$N_{tu} = -\ln\left[1 + \frac{1}{C^*}\ln((1-\epsilon C^*)\right] \tag{25.42}$$

6. **For single-pass cross flow with C_{min} mixed and C_{max} unmixed**

$$\epsilon = 1 - e^{\phi} \tag{25.43}$$

where

$$\phi = -\frac{1}{C^*}\left(1-e^{-C^*N_{tu}}\right)$$

and

$$N_{tu} = -\frac{1}{C^*}\ln[C^*\ln(1-\epsilon)+1] \tag{25.44}$$

7. **For the shell and tube heat exchanger with one shell pass and 2, 4, 6, . . ., tube passes**

$$\epsilon = \epsilon_1 = 2\left[1 + C^* + (1+C^{*2})^{1/2}\frac{1+e^{-N_{tu}(1+C^{*2})^{1/2}}}{1-e^{-N_{tu}(1+C^{*2})^{1/2}}}\right] \tag{25.45}$$

and

$$N_{tu} = \frac{2/\epsilon - (1+C^*)}{\epsilon(1+C^{*2})^{1/2}} \tag{25.46}$$

8. ***n* Shell passes and 2*n*, 4*n*, 6*n*, . . ., tube passes**

$$\epsilon = \left[\left(\frac{1-\epsilon_1 C^*}{1-\epsilon_1}\right)^n - 1\right]\left[\left(\frac{1-\epsilon_1 C^*}{1-\epsilon_1}\right)^n - C^*\right]^{-1} \tag{25.47}$$

where ϵ_1 is based on the N_{tu} per shell pass (N_{tu}/n) and

$$N_{tu,1} = \frac{\dfrac{2(F-C^*)}{F-1} - (1+C^*)}{(1+C^{*2})^{1/2}} \tag{25.48}$$

where

$$F = \left(\frac{\epsilon C^* - 1}{\epsilon - 1} \right)^{1/n}$$

with $\mathrm{N_{tu,1}}$ taken as the $\mathrm{N_{tu}}$ per shell pass ($\mathrm{N_{tu}} = n\mathrm{N_{tu,1}}$).

Example 25.5 Determine the surface area for a cross-flow heat exchanger that will cool 8 kg/s of methyl alcohol ($c = 2.470$ kJ/kg-K) from 96°C to 70°C using 6.75 kg/s of water available at 12°C. The water flows on the inside of 1-in OD × 16-BWG tubes. Consider that the methyl alcohol is mixed and the overall heat transfer coefficient based on the outside tube surface area is 600 W/m²-K.

Solution

Assumptions and Specifications

(1) Steady-state conditions exist.
(2) There is negligible heat loss to and from the surroundings.
(3) Kinetic and potential energy changes are negligible.
(4) The tube wall thermal resistance is negligible.
(5) The methyl alcohol is specified as mixed and the cooling water is assumed to be unmixed.
(6) Fluid specific heats and the overall heat transfer coefficient are constant.
(7) Axial conduction along the heat exchanger tubing is negligible.

We are given the specific heat of the methyl alcohol and assume that it is the value at the average temperature. Thus,

$$C_H = (8.00\,\text{kg/s})(2.470\,\text{kJ/kg-K}) = 19.76\,\text{kW/K}$$

and we can find the duty of the exchanger using Equation 25.1

$$\dot{Q} = C_H(T_1 - T_2) = (19.76\,\text{kW/K})(90°\text{C} - 70°\text{C}) = 395.2\,\text{kW}$$

We need the specific heat of the water and to find it, we assume that the water will leave the exchanger at $t_2 = 26°$C, which we will need to verify. Then with

$$t_{\text{av}} = \frac{t_1 + t_2}{2} = \frac{12°\text{C} + 26°\text{C}}{2} = 19°\text{C}$$

From Table A.14 we find

$$c_C = 4.191\,\text{kJ/kg K}$$

and

$$C_C = \dot{m}_C c_C = (6.75\,\text{kg/s})(4.191\,\text{kJ/kg-K}) = 28.29\,\text{kW/K}$$

We then use Equation 25.1 to verify t_2

$$t_2 = t_1 + \frac{\dot{Q}}{\dot{m}_C c_C} = 12°\text{C} + \frac{395.2\,\text{kW}}{(6.75\,\text{kg/s})(4.191\,\text{kJ/kg-K})}$$
$$= 12°\text{C} + 14.00°\text{C} = 26°\text{C}$$

Hence, we take $t_2 = 26°C$ and we find that C_C is correct at 28.29 kW/K.

The foregoing values of C_H and C_C show that $C_H = C_{min}$ and $C_C = C_{max}$ and via Equation 25.29

$$C^* = \frac{C_{min}}{C_{max}} = \frac{19.76\,\text{kW/K}}{28.29\,\text{kW/K}} = 0.699$$

It is indeed fortunate that $C_H = C_{min}$ because the reference value of the overall heat transfer coefficient corresponds to C_{min}. This eliminates the hassle of converting the overall heat transfer coefficient to the cold side of the exchanger.

With $C_H = C_{min}$, Equation 25.33 gives us the effectiveness

$$\epsilon = \frac{T_1 - T_2}{T_1 - t_1} = \frac{96°C - 70°C}{96°C - 12°C} = \frac{26°C}{84°C} = 0.310$$

Here, $C_H = C_{min}$ is mixed and the value of N_{tu} required to give $\epsilon = 0.310$ with $C^* = 0.699$ is determined from Equation 25.44

$$\begin{aligned} N_{tu} &= -\frac{1}{C^*} \ln[C^* \ln(1-\epsilon) + 1] \\ &= -\frac{1}{0.699} \ln[(0.699)\ln(1 - 0.310) + 1] \\ &= -(1.4306)\ln[(0.699)(\ln 0.690) + 1] \\ &= -(1.4306)\ln[-2694 + 1] \\ &= -(1.4306)(\ln 0.7406) \\ &= -(1.4306)(-0.3003) = 0.4296 \end{aligned}$$

We then find from Equation 25.31 that

$$S_H = \frac{N_{tu}C_H}{U_H} = \frac{(0.4296)(19.76\ \text{KW/K})}{0.600\,\text{kW/m}^2\text{-K}} = 14.15\,\text{m}^2 \Longleftarrow$$

Our attention now turns to a design example in which the heat transfer coefficient on both the hot and cold sides of the exchanger must be evaluated.

25.4 Design Example 17

A five-pass tubular air preheater with tubes having a maximum length of 7.0 m is to be used to heat 12 kg/s of air from 0°C to 204°C. Flue gas, having the properties of air, flows in the inside of the tubes at a rate of 12.93 kg/s and enters the tubes at 320°C.

The air heater is to contain a bank of 2-in × 12-BWG tubing that is 48 tubes wide and 14 tubes deep on transverse and in-line spacings of $S_T = S_L = 7.5$ cm. The five air passes and one flue gas pass will permit the heater to be designed on the basis of true counterflow. Fouling of the surfaces may be considered negligible.

Determine the total length of the tubing necessary to yield the 204°C outlet air temperature.

Solution

Assumptions and Specifications

(1) Steady-state conditions exist.
(2) There is negligible heat loss to and from the surroundings.
(3) Kinetic and potential energy changes are negligible.
(4) The tube wall thermal resistance is negligible.
(5) Fluid specific heats are constant.
(6) Axial conduction along the heat exchanger tubing is negligible.
(7) The air heater may be treated as an exchanger in counterflow.

The bulk temperature of the air is

$$t_b = \frac{t_1 + t_2}{2} = \frac{0^\circ\text{C} + 204^\circ\text{C}}{2} = 102^\circ\text{C} \quad (375\,\text{K})$$

and at this value, Table A.13 reveals (with interpolation) that

$$c_{\text{p,C}} = 1.0115\,\text{kJ/kg-K}$$

The duty of the air heater is therefore

$$\begin{aligned}\dot{Q} &= \dot{m}_C c_{\text{p,C}}(t_2 - t_1)\\ &= (12\,\text{kg/s})(1.0115\,\text{kJ/kg-K})(204^\circ\text{C} - 0^\circ\text{C})\\ &= 2.476\,\text{MW}\end{aligned}$$

Assume that the temperature of the flue gas at the exit is $T_2 = 134^\circ\text{C}$. This would yield an average bulk temperature for the flue gas of

$$T_b = \frac{T_1 + T_2}{2} = \frac{320^\circ\text{C} + 134^\circ\text{C}}{2} = 227^\circ\text{C} \quad (500\,\text{K})$$

and we find from Table A.13 that

$$c_{\text{p,H}} = 1.0295\,\text{kJ/kg-K}$$

provides an assumed heat duty of

$$\begin{aligned}\dot{Q} &= \dot{m}_h c_{\text{p,H}}(T_1 - T_2)\\ &= (12.93\,\text{kg/s})(1.0295\,\text{kJ/kg-K})(320^\circ\text{C} - 134^\circ\text{C})\\ &= 2.476\,\text{MW}\end{aligned}$$

The actual duty determined from the cold air side must match the assumed duty determined from the hot flue gas side. This means that the temperature of the air leaving the heater is indeed 134°C and demonstrates, once again, the value of the heat balance.

The thermal properties for both the flue gas and the air may be obtained from Table A.13:

Flue Gas Hot Side		Air Cold Side
500	T_b/t_b, K	375
0.0404	k, W/m-K	0.0318
1.0295	c_p, kJ/kg-K	1.0115
2.671	$\mu \times 10^5$, kg/m-s	2.180
0.680	Pr	0.693
0.7048	ρ, kg/m^3	0.9403

and we estimate Pr at the wall temperature as

$$\mathrm{Pr_w} = 0.685$$

where this should be verified.

Note that

$$C_C = \dot{m}_C c_{p,C} = (12\,\mathrm{kg/s})(1.0115\,\mathrm{kJ/kg\text{-}K}) = 12.1380\,\mathrm{W/K}$$

and

$$C_H = \dot{m}_H c_{p,H} = (12.93\,\mathrm{kg/s})(1.0295\,\mathrm{kJ/kg\text{-}K}) = 13.3114\,\mathrm{W/K}$$

which makes $C_C = C_{\min}$ and

$$C^* = \frac{12.1380\,\mathrm{W/K}}{13.3114\,\mathrm{W/K}} = 0.9118$$

Moreover, the required heat exchanger effectiveness will be, as given by Equation 25.23

$$\epsilon = \frac{C_C(t_2 - t_1)}{C_{\min}(T_1 - t_1)} = \frac{t_2 - t_1}{T_1 - t_1} = \frac{204°\mathrm{C} - 0°\mathrm{C}}{320°\mathrm{C} - 0°\mathrm{C}} = 0.6375$$

and the value of $\mathrm{N_{tu}}$ required to yield this effectiveness may be obtained from Equation 25.36

$$\begin{aligned}
\mathrm{N_{tu}} &= \frac{1}{1 - C^*} \ln \frac{1 - \epsilon C^*}{1 - \epsilon} \\
&= \frac{1}{1 - 0.9118} \ln \frac{1 - (0.6375)(0.9118)}{1 - 0.6375} \\
&= 11.344 \ln \frac{0.4187}{0.3625} \\
&= 11.344 \ln 1.1550 = (11.344)(0.1441) = 1.6349
\end{aligned}$$

The total number of tubes will be

$$n_t = (48)(14) = 672\,\text{tubes}$$

and Table A.21 reveals that the tubes have inner and outer diameters of

$$d_i = 4.526\,\mathrm{cm} \quad \text{and} \quad d_o = 5.080\,\mathrm{cm}$$

which makes the required $U_C S_C$ product $U_C S_C = 19.844$ w/k.

In the first trial, we will assume a total length of 6 m and for tubes spaced on 7.5 cm centers (both $S_T = S_L = 7.5$ cm), the air preheater width will be

$$W = (48)(0.075\,\mathrm{m}) = 3.60\,\mathrm{m}$$

The free-flow area on the tube (flue gas) side will be

$$A_H = \frac{\pi}{4} n_t d_i^2 = \frac{\pi}{4}(672)(0.04526\,\text{m})^2 = 1.0812\,\text{m}^2$$

and with L_p and L designated as the tube length per pass and the total tube length, respectively, we have the cold-side free duct area

$$A_C = L_p W = \left(\frac{L}{5}\right)(3.60\,\text{m}) = \left(\frac{6\,\text{m}}{5}\right)(3.60\,\text{m}) = 4.32\,\text{m}^2$$

The two heat transfer coefficients may now be determined. For the hot (flue gas) side

$$\hat{V}_H = \frac{\dot{m}_H}{\rho_H A_H} = \frac{12.93\,\text{kg/s}}{(0.7048\,\text{kg/m}^3)(1.0812\,\text{m}^2)} = 16.97\,\text{m/s}$$

and

$$\text{Re}_d = \frac{\rho_h \hat{V}_h d_i}{\mu_h} = \frac{(0.7048\,\text{kg/m}^3)(16.97\,\text{m/s})(0.04526\,\text{m})}{2.671 \times 10^{-5}\,\text{kg/m-s}} = 20{,}270$$

which shows that the flue gas on the inside of the tubes is in turbulent flow. We therefore employ Equation 22.18 with the exponent on Pr taken as 0.30 because $T_w < T_b$.

$$\text{Nu}_d = 0.023\text{Re}_d^{0.80}\text{Pr}^{0.30} = 0.023(20,270)^{0.80}(0.680)^{0.30} = 57.14$$

so that

$$h_H = \frac{k}{d}\text{Nu}_d = \left(\frac{0.0404\,\text{W/m-K}}{0.04526\,\text{m}}\right)(57.14) = 51.00\,\text{W/m}^2\text{-K}$$

For the cold (air) side, the free duct velocity is

$$\hat{V}_\infty = \frac{\dot{m}_c}{\rho_c A_c} = \frac{12.00\,\text{kg/s}}{(0.9403\,\text{kg/m}^3)(4.32\,\text{m}^2)} = 2.954\,\text{m/s}$$

and by Equation 23.18a

$$\hat{V}_{\text{max}} = \hat{V}_\infty\left(\frac{S_T}{S_T - d_o}\right) = (2.954\,\text{m/s})\left(\frac{7.5\,\text{cm}}{7.5\,\text{cm} - 5.080\,\text{cm}}\right) = 9.155\,\text{m/s}$$

Then

$$\text{Re}_{\text{d,max}} = \frac{\rho_c \hat{V}_{\text{max}} d_o}{\mu_c} = \frac{(0.9403\,\text{kg/m}^3)(9.155\,\text{m/s})(0.0508\,\text{m})}{2.18 \times 10^{-5}\,\text{kg/m-s}} = 20,060$$

and with C and m taken from Table 23.2

$$C = 0.27 \quad \text{and} \quad m = 0.63$$

we have from Equation 23.17

$$\text{Nu}_{\text{d,max}} = 0.27(\text{Re}_{\text{d,max}})^{0.63}(\text{Pr})^{0.36}\left(\frac{\text{Pr}}{\text{Pr}_\text{w}}\right)^{1/4}$$

$$= 0.27(20,060)^{0.63}(0.693)^{0.36}\left(\frac{0.693}{0.685}\right)^{1/4} = 121.83$$

and

$$h_C = \frac{k}{d}\text{Nu}_{\text{max}} = \left(\frac{0.0318\,\text{W/m-K}}{0.0508\,\text{m}}\right)(121.83) = 76.26\,\text{W/m}^2\text{-K}$$

With both h_H and h_C in hand, we may form the overall heat transfer coefficient. With $C_c = C_{\text{min}}$, we use Equation 25.10 with

$$\frac{S_C}{S_H} = \frac{\pi d_o L}{\pi d_i L} = \frac{d_o}{d_i} = \frac{5.080\,\text{cm}}{4.526\,\text{cm}} = 1.122$$

we have

$$U_C = \frac{1}{\dfrac{S_C/S_H}{h_H} + \dfrac{1}{h_C}} = \frac{1}{\dfrac{1.122}{51.00\,\text{W/m}^2\text{-K}} + \dfrac{1}{76.26\,\text{W/m}^2\text{-K}}}$$

$$= \frac{1}{0.0220\,\text{m}^2\text{-K/W} + 0.0131\,\text{m}^2\text{-K/W}} = \frac{1}{0.0351\,\text{m}^2\text{-K/W}}$$

$$= \frac{1}{0.0351\,\text{m}^2\text{-K/W}} = 28.48\,\text{W/m}^2\text{-K}$$

With

$$S_C = \pi n_t d_o L = \pi(672)(0.0508\,\text{m})(6.00\,\text{m}) = 643.47\,\text{m}^2$$

the $U_C S_C$ product will be

$$U_C = (28.48\,\text{W/m}^2\text{-K})(643.48\,\text{m}^2) = 18.325\,\text{kW/K}$$

which falls short of the required value of 19.829 kW/K.

The difference between the calculated and required values of the $U_C S_C$ product requires that a trial and error be employed. Table 25.3 summarizes the procedure. It may be noted that the computations for $L = 6\,\text{m}$ are repeated in the table.

TABLE 25.3

Summary of Trial-and-Error Procedure for Design Example 17

Assume L, m	**6.00**	**6.40**	**6.75**
S_C, m^2	643.47	664.23	718.14
L_p, m	1.20	1.28	1.34
A_C, m^2,	4.320	4.608	4.824
$\hat{V}_\infty$, m/s	2.954	2.769	2.568
$\hat{V}_{\text{max}}$, m/s	9.155	8.583	7.959
$\text{Re}_{d,\text{max}}$	20,060	18,800	17,440
$\text{Nu}_{d,\text{max}}$	121.83	116.95	111.54
h_C, W/m^2-K	76.26	73.21	69.83
h_H, W/m^2-K	51.00	51.00	51.00
$S_C/S_H h_H$, m^2-K/W	0.0220	0.0220	0.0220
$1/h_C$, m^2-K/kW	0.0131	0.0137	0.0143
Sum	0.0351	0.0357	0.0363
$U_C = 1/\text{sum}$, W/m^2-K	28.47	28.01	27.53
$U_C S_C$, kW/K	18.325	19.225	19.770
OK?	No	No	Yes
Remedy	Increase L	Increase L	Accept this result

Although slightly lower than required, we have accepted the result of $L = 6.70$ m because this is within any accuracy limitations.

25.5 Finned Heat Exchangers

Attention is now focused on the finned heat exchanger in which the surface area on the hot and/or cold side of the exchanger is augmented with fins. Three such exchangers are shown in Figure 25.11 and the performance of these exchangers is governed by the enthalpy balance in Equation 25.1

$$\dot{Q} = C_H(T_1 - T_2) = C_C(t_2 - t_1) \tag{25.1}$$

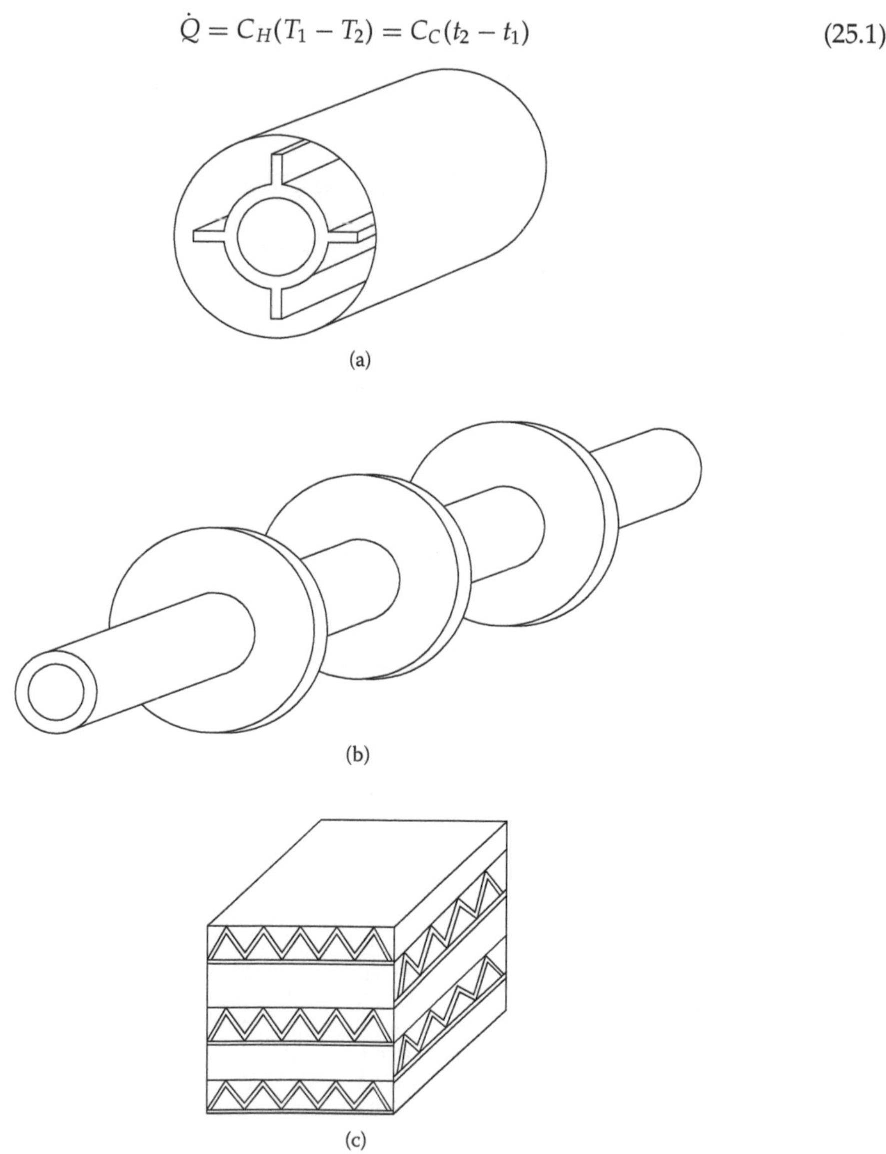

FIGURE 25.11
Examples of finned heat exchangers. (a) The double pipe with the outside surface of the inner pipe containing longitudinal fins of rectangular profile, (b) a single tube in a tubular heat exchanger with radial or annular fins of a constant cross section, and (c) the compact heat exchanger with rectangular plate fins on both hot and cold sides.

and the rate equation of Equation 25.2

$$\dot{Q} = US\theta_m = U_H S_H \theta_m = U_C S_C \theta_m \tag{25.2}$$

25.5.1 The Surface Area and the Overall Surface Efficiency

We first consider the outside surface of the inner tube or pipe in the double-pipe exchanger shown in Figure 25.11a. The heat exchange is between a fluid inside this inner pipe or tube and a fluid flowing in the annular region The outside surface of the inner tube or pipe contains n_f longitudinal fins, usually fins of rectangular profile of height b and thickness δ. The entire exchanger has a length L and the diameter of the inner pipe is d_o. The **base** or **prime** surface area for the outside surface of the inner pipe will be

$$S_b = (\pi d_o - n_f\delta)L \qquad \text{(m)} \tag{25.49a}$$

and because no heat flows from the tips of the fins, the finned surface is

$$S_f = 2n_f bL \qquad \text{(m)} \tag{25.49b}$$

The total surface area is

$$S_o = S_b + S_f = [\pi d_o - n_f\delta + 2n_f b]L \qquad \text{(m)} \tag{25.49c}$$

For the exchanger containing tubes with n_f radial (also referred to as **annular** or **circular**) fins a similar procedure yields

$$S_b = \pi(L - n_f\delta)d_o \qquad \text{(m)} \tag{25.50a}$$

where δ is the fin thickness and d_o is the outside diameter of the tube. Because no heat flows from the tips of the fins, the finned surface is

$$S_f = 2\frac{\pi}{4}(d_a^2 - d_o^2)n_f \qquad \text{(m)} \tag{25.50b}$$

where d_a and d_o are, respectively, the outer and inner diameter of the fins. The total surface is

$$S_o = S_b + S_f = \pi(L - n_f\delta)d_o + 2\frac{\pi}{4}(d_a^2 - d_o^2)n_f \quad \text{(m)} \tag{25.50c}$$

There are several types of surfaces that can be used in the compact heat exchanger shown in Figure 25.11c. Typical of these are the plain plate fin, louvered fin, strip fin, and wavy fin surfaces as described by Kays and London (1984). Each surface possesses its own geometric properties and heat transfer and flow friction data. The total surface is determined from the volume of the exchanger and the finned surface to total surface ratio is given.

We have discussed the efficiency of extended surface or fins in Chapter 20. There it was shown that the efficiency of the longitudinal fin of rectangular profile or a spine of constant cross section is given by

$$\eta = \frac{\tanh mb}{mb} \tag{20.67}$$

The graph of the efficiency of the annular or radial fin of the rectangular profile given as Figure 20.24 can be replaced by an approximate expression due to McQuiston and Tree (1972)

$$\eta = \frac{\tanh m\phi}{m\phi} \tag{25.51}$$

where, in both Equation 20.67 and Equation 25.51

$$m = \left(\frac{2h}{k\delta}\right)^{1/2}$$

and in Equation 25.51

$$\phi = r_b\left(\frac{1-\rho}{\rho}\right)\left(1 + 0.35\ln\frac{1}{\rho}\right)$$

where ρ is the radius ratio, $\rho = r_o/r_a \leq 1$.

We now develop an expression for the **overall surface efficiency**. With the total surface equal to the sum of the base or prime surface and the finned surface, we may consider the prime surface at an efficiency of unity and the finned surface at its own fin efficiency. Hence, in

$$\eta_{ov} S_o = (1.00) S_b + \eta_f S_f$$

we note that, with $S_b = S_o - S_f$, we will have

$$\eta_{ov} S_o = (1.00)(S_o - S_f) + \eta_f S_f$$

$$\eta_{ov} = \left(1 - \frac{S_f}{S_o}\right) + \eta_f \frac{S_f}{S_o}$$

or

$$\eta_{ov} = 1 - (1 - \eta_f)\frac{S_f}{S_o} \tag{25.52}$$

25.5.2 The Overall Heat Transfer Coefficient

Several relationships for the overall heat transfer coefficient have been listed in Section 25.2.3; all were for the unfinned exchanger. If the exchanger contains fins, then the expressions for the overall heat transfer coefficient must include the effect of the overall fin efficiency. In the equations for the overall heat transfer coefficient that follow, no provision is made for fouling on the side of the exchanger that contains the fins. Such a provision is somewhat beyond our scope and information on this topic can be found in Kraus et al. (2001).

- For the hot fluid flowing on the finned side of the exchanger with fouling on the unfinned cold side and no metal thermal resistance

$$U_H = \frac{1}{\dfrac{1}{h_H \eta_{ov,H}} + \dfrac{1}{h_{d,C}}\dfrac{S_H}{S_C} + \dfrac{1}{h_C}\dfrac{S_H}{S_C}} \tag{25.53}$$

- For the cold fluid flowing on the finned side of the exchanger with fouling on the unfinned hot side and no metal thermal resistance

$$U_C = \frac{1}{\dfrac{1}{h_H}\dfrac{S_C}{S_H} + \dfrac{1}{h_{d,H}}\dfrac{S_C}{S_H} + \dfrac{1}{h_C \eta_{ov,C}}} \tag{25.54}$$

- For the hot fluid flowing on the finned side of the exchanger with no fouling and no metal thermal resistance

$$U_H = \frac{1}{\dfrac{1}{h_H \eta_{ov,H}} + \dfrac{1}{h_C}\dfrac{S_H}{S_C}} \tag{25.55}$$

- For the cold fluid flowing on the finned side of the exchanger with no fouling and no metal thermal resistance

$$U_C = \frac{1}{\dfrac{1}{h_H}\dfrac{S_C}{S_H} + \dfrac{1}{h_C \eta_{ov,C}}} \tag{25.56}$$

- For the case of both sides of the exchanger finned with no fouling and no metal thermal resistance, we have for the hot-side reference

$$U_H = \frac{1}{\dfrac{1}{h_H \eta_{ov,H}} + + \dfrac{1}{h_C \eta_{ov,C}}\dfrac{S_H}{S_C}} \tag{25.57}$$

- For the case of both sides of the exchanger finned with no fouling and no metal thermal resistance, we have for cold-side reference

$$U_C = \frac{1}{\dfrac{1}{h_H \eta_{ov,H}}\dfrac{S_C}{S_H} + \dfrac{1}{h_C \eta_{ov,C}}} \tag{25.58}$$

We now turn to an example that will illustrate the concepts discussed in this section.

Example 25.6 A double-pipe heat exchanger with a finned annulus is to be used in a liquid-to-liquid application. The hot liquid with a capacity rate of 2.275 kW/K and an inlet temperature of $T_1 = 85°C$ flows in the annulus in counterflow with a heat transfer coefficient of 2500 W/m^2-K. The cold liquid with a capacity rate of 2.844 kW/K and an inlet temperature of $t_1 = 25°C$ flows in the inner pipe with a heat transfer coefficient of 2000 W/m^2-K.

The 3-in × schedule-40 inner pipe resides within a 5-in-schedule 40 pipe, and both pipes are 1.2-m long. The inner pipe is equipped with 8 longitudinal fins of rectangular profile, each 1.905-cm high and 0.3175-cm thick. The thermal conductivity of the inner pipe material is 42 W/m-K and fouling may be considered negligible. Determine the outlet temperatures of both fluids.

Solution

Assumptions and Specifications

(1) Steady-state conditions exist.
(2) There is negligible heat loss to and from the surroundings.
(3) Kinetic and potential energy changes are negligible.
(4) The tube wall thermal resistance is negligible.
(5) The Murray-Gardner assumptions pertaining to fins listed in Chapter 20 apply.
(6) The fins possess adiabatic tips.
(7) Fouling of the exchanger surfaces is specified as being negligible.

(8) Fluid specific heats and the overall heat transfer coefficient remain constant.
(9) Axial conduction along the piping is negligible.

For the 3-in × schedule-40 inner pipe, Table A.20 gives

$$d_o = 8.890\,\text{cm} \qquad \text{and} \qquad d_i = 7.793\,\text{cm}$$

We then make the observation that $C_H = 2.275\,\text{kW/K}$ and $C_C = 2.844\,\text{kW/K}$ so that $C_H = C_{min} = 2.275\,\text{KW/K}$. The hot fluid flows in the annulus and only the surface area of the inner pipe need be obtained. Thus, with $S_o = S_H$. The base surface is obtained from Equation 25.49a

$$\begin{aligned} S_b &= (\pi d_o - n_f\delta)L \\ &= [\pi(8.890\,\text{cm}) - 8(0.3175\,\text{cm})](120\,\text{cm}) \\ &= (27.929\,\text{cm} - 2.540\,\text{cm})(120\,\text{cm}) \\ &= 3046.7\,\text{cm}^2 \quad \text{or} \quad 0.3047\,\text{m}^2 \end{aligned}$$

and for the finned surface, Equation 25.49b gives

$$\begin{aligned} S_f &= 2n_f bL \\ &= 2(8)(1.905\,\text{cm})(120\,\text{cm}) \\ &= 3657.6\,\text{cm} \quad \text{or} \quad 0.3658\,\text{cm}^2 \end{aligned}$$

Then

$$S_o = S_H = S_b + S_f = 0.3047\,\text{cm}^2 + 0.3658\,\text{cm}^2 = 0.6705\,\text{cm}^2$$

and

$$\frac{S_f}{S_o} = \frac{0.3658\,\text{cm}^2}{0.6705\,\text{cm}^2} = 0.546$$

The fin efficiency for the hot side is given by Equation 20.67. We first find m

$$m = \left[\frac{2h}{k\delta}\right]^{1/2} = \left[\frac{2(2500\,\text{W/m}^2\text{-K})}{(42\,\text{W/m-K})(0.003175\,\text{m})}\right]^{1/2} = 193.64\,\text{m}^{-1}$$

Then

$$mb = (193.64\,\text{m}^{-1})(0.01905\,\text{m}) = 3.689$$

and Equation 20.67 gives

$$\eta_f = \frac{\tanh mb}{mb} = \frac{\tanh 3.689}{3.689} = \frac{0.9988}{3.689} = 0.271$$

Then we use Equation 25.52 to find the overall passage efficiency for the annulus:

$$\begin{aligned} \eta_{ov,H} &= 1 - (1 - \eta_f)\frac{S_f}{S_H} \\ &= 1 - (1 - 0.271)(0.546) = 1 - (0.729)(0.546) = 0.602 \end{aligned}$$

The surface area of the inside of the inner pipe will be needed for the determination of the overall heat transfer coefficient. Because the inside of the inner pipe carries the cold fluid, this will be designated as $S_i = S_C$

$$S_C = \pi d_i L = \pi(7.793\,\text{cm})(120\,\text{cm}) = 2937.8\,\text{cm}^2 \quad (0.2938\,\text{m}^2)$$

We then use Equation 25.52 to determine the overall heat transfer coefficient

$$\begin{aligned} U_H &= \frac{1}{\dfrac{1}{h_H \eta_{ov,H}} + \dfrac{1}{h_C}\dfrac{S_H}{S_C}} \\ &= \frac{1}{\dfrac{1}{(2500\,\text{W/m}^2\text{-K}))(0.602)} + \dfrac{1}{2000\,\text{W/m}^2\text{-K}}\dfrac{0.6705\,\text{m}^2}{0.2938\,\text{m}^2}} \\ &= \frac{1}{6.645 \times 10^{-4}\,\text{m}^2\text{-K/W} + 1.141 \times 10^{-3}\,\text{m}^2\text{-K/W}} \\ &= \frac{1}{1.806 \times 10^{-3}\,\text{m}^2\text{-K/W}} = 553.9\,\text{W/m}^2\text{-K} \end{aligned}$$

Then, with $C_H = C_{min} = 2.275\,\text{kW/K}$

$$N_{tu} = \frac{U_H S_H}{C_H} = \frac{(553.9\,\text{W/m}^2\,\text{K})(0.6705\,\text{m}^2)}{2.275\,\text{kW/K}} = 0.1632$$

and

$$C^* = \frac{C_{min}}{C_{max}} = \frac{2.275\,\text{kW/K}}{2.844\,\text{kW/K}} = 0.800$$

Equation 25.35a for the case of counterflow then gives

$$\begin{aligned} \epsilon &= \frac{1 - e^{-N_{tu}(1-C^*)}}{1 - C^* e^{-N_{tu}(1-C^*)}} = \frac{1 - e^{-(0.1632(1-0.800)}}{1 - (0.800)e^{-0.1632(1-0.800)}} \\ &= \frac{1 - e^{\,0.0326}}{1 - (0.800)e^{-0.0326}} = \frac{1 - 0.9679}{1 - (0.800)(0.9679)} \\ &= \frac{0.0321}{0.2257} = 0.142 \end{aligned}$$

Then from Equation 25.33 with $C_H = C_{min}$

$$\epsilon = \frac{T_1 - T_2}{T_1 - t_1}$$

$$\begin{aligned} T_1 - \epsilon(T_1 - T_1) &= 85^\circ\text{C} - (0.142)(85^\circ\text{C} - 25^\circ\text{C}) \\ &= 85^\circ\text{C} - (0.142)(60^\circ\text{C}) = 85^\circ\text{C} - 8.52^\circ\text{C} = 76.5^\circ\text{C} \Longleftarrow \end{aligned}$$

and because by definition with $C_H = C_{min}$

$$C^* = \frac{C_{min}}{C_{max}} = \frac{C_H}{C_C} = \frac{t_2 - t_1}{T_1 - T_2}$$

we find that

$$\begin{aligned} t_2 &= t_1 + C^*(T_1 - T_2) \\ &= 25°\text{C} + (0.800)(85°\text{C} - 76.5°\text{C}) \\ &= 25°\text{C} + (0.800)(8.5) = 31.8°\text{C} \Longleftarrow \end{aligned}$$

25.6 Summary

The duty of a heat exchanger is given by the enthalpy balance

$$\dot{Q} = C_H(T_1 - T_2) = C_c(t_2 - t_1) \tag{25.1}$$

and whether this duty can be accomplished by an exchanger of a given size depends on the rate equation

$$\dot{Q} = US\theta_m = U_H S_H \theta_m = U_C S_C \theta_m \tag{25.2}$$

In Equation 25.2, the inside and outside surfaces for an unfinned tubular heat exchanger are given by

$$S_i = \pi d_i L \qquad \text{and} \qquad S_o = \pi d_o L$$

The overall heat transfer coefficient involves as many as five resistances, some of which may not be present. A consideration of all of the resistances except the metal resistance leads to

$$U = \frac{1}{\dfrac{S}{h_H S_H} + \dfrac{S}{h_{d,H} S_H} + \dfrac{S}{h_{d,C} S_C} + \dfrac{S}{h_C S_C}} \tag{25.12}$$

with other more specific cases given by Equations 25.13 through 25.16.

For true counterflow, true cocurrent flow, or constant temperature source and receiver fluids, the logarithmic mean temperature difference can be evaluated from

$$\theta_m = \text{LMTD} = \frac{\Delta T_1 - \Delta T_2}{\ln \dfrac{\Delta T_1}{\Delta T_2}} = \frac{\Delta T_2 - \Delta T_1}{\ln \dfrac{\Delta T_2}{\Delta T_1}} \tag{25.23}$$

For arrangements other than true counterflow, true cocurrent flow, or constant temperature source and receiver fluids, the LMTD correction factor method or the ϵN_{tu} method can be employed. The LMTD correction factor, F, can be obtained from charts that consider two parameters

$$P = \frac{t_2 - t_1}{T_1 - t_1} \qquad \text{and} \qquad R = \frac{T_1 - T_2}{t_2 - t_1} = \frac{C_C}{C_H}$$

with $F = f(P, R)$ obtained from charts such as those shown in Figures 25.6 through 25.9. With the F factor so obtained, the rate equation becomes

$$\dot{Q} = USF(\text{LMTD}_c)$$

where LMTD_c is the logarithmic mean temperature difference for true counterflow.

The ϵN_{tu} method is based on three parameters

$$C^* = \frac{C_{min}}{C_{max}}, \quad \epsilon = \frac{\dot{Q}}{\dot{Q}_{max}}, \quad \text{and} \quad N_{tu} = \frac{US}{C_{min}}$$

The heat transfer effectiveness is a function of N_{tu} and C^*. Values have been provided for eight arrangements in Equations 25.35 through 25.48.

In dealing with finned surfaces, care must be used to consider the fin efficency. Most often, an overall passage efficiency is employed

$$\eta_{ov} = 1 - (1 - \eta_f)\frac{S_f}{S_o} \tag{25.52}$$

where S_o is the total surface and S_f is the finned surface. In the overall heat transfer coefficient, the individual heat transfer coefficients must be multiplied by their respective values of η_{ov}.

25.7 Problems

The Heat Balance

25.1: Ethylene glycol, flowing at a rate of 200 kg/s, enters a heat exchanger at 80°C and leaves at 40°C. The glycol is cooled by water, flowing at 160 kg/s, that enters at 15°C. Determine (a) the heat transferred in the exchanger and (b) the exit temperature of the water.

25.2: Rework Problem 25.1 with the glycol cooled by an airflow of 1425 kg/s of air at 20°C. All other conditions remain the same.

25.3: An air-cooled power plant condenser handles 3.2 kg/s of steam that has a quality of 98% when it enters and is saturated liquid when it leaves. The condenser pressure is 10 kPa and air enters the condenser at 20°C at a flow rate of 540 kg/s. Determine (a) the heat transferred in the condenser and (b) the exit temperature of the air.

25.4: The steam in Problem 25.3 is condensed by a more conventional flow of river water that enters at 17°C at a flow rate of 135 kg/s. Determine the exit temperature of the water.

25.5: A stream of lubricating oil (SAE 50) at 70°C is cooled by water. The oil flows at a rate of 24 kg/s; the water, flowing at a rate of 20 kg/s, enters the exchanger at 15°C and leaves at 23°C. Determine (a) the heat transferred in the exchanger and (b) the exit temperature of the oil.

25.6: Rework Problem 25.5 using air at a flow rate of 8 kg/s that enters at 15°C and leaves at 33°C. All other conditions remain the same.

25.7: In a gas-to-gas application, nitrogen is cooled from 80°C to 30°C by air, entering at 15°C. The flow rate of both the nitrogen and the air is 10 kg/s. Determine (a) the heat transferred in the exchanger and (b) the exit temperature of the air.

25.8: Chilled water entering a heat exchanger at 8°C is used to cool air from 40°C to 12°C. The flow rates of the water and the air are 8 and 15 kg/s, respectively. Determine (a) the heat transferred in the exchanger and (b) the exit temperature of the water.

25.9: Rework Problem 25.1 using glycerin. All other conditions remain the same.

25.10: Rework Problem 25.8 using carbon dioxide. All other conditions remain the same.

The Overall Heat Transfer Coefficient

25.11: A hot liquid flows in the interior of 2-in × 12-BWG tubing with a heat transfer coefficient of 1250 W/m^2-K. A cold fluid flows on the outside of the tubing with a heat transfer coefficient of 600 W/m^2-K. Fouling may be neglected. Determine the overall heat transfer coefficient based on (a) the hot side and (b) the cold side.

25.12: Rework Problem 25.11 with a fouling resistance on the hot side of $1/h_{d,H} = 0.0002$ m^2-K/W.

25.13: Rework Problem 25.11 with a fouling resistance on the cold side of $1/h_{d,C} = 0.0002$ m^2-K/W.

25.14: Rework Problem 25.11 with fouling resistances on both the hot and cold sides of $1/h_{d,H} = 1/h_{d,C} = 0.0002$ m^2-K/W.

25.15: A hot liquid flows in the interior of 2-in × 12-BWG tubing with a heat transfer coefficient of 800 W/m^2-K. A cold gas flows on the outside of the tubing with a heat transfer coefficient of 100 W/m^2-K. Fouling may be neglected. Determine the overall heat transfer coefficient based on (a) the hot side and (b) the cold side.

25.16: Rework Problem 25.15 with a fouling resistance on the hot side of $1/h_{d,H} = 0.0004$ m^2-K/W.

25.17: Rework Problem 25.15 with a fouling resistance on the cold side of $1/h_{d,C} = 0.0004$ m^2-K/W.

25.18: Rework Problem 25.15 with the same fouling resistances on both the hot and cold sides: $1/h_{d,H} = 1/h_{d,C} = 0.0004$ m^2-K/W.

25.19: A tubular air heater is constructed of low carbon steel tubing with a thermal conductivity k of 42 W/m-K. The tubes are arranged in a bank that is twelve tubes wide by ten tubes deep. The tubes are 2-m long and have an outside diameter of 7.62 cm and an inside diameter of 7.198 cm. The hot (inside) and cold (outside) heat transfer coefficients are 150 W/m^2-K and 100 W/m^2-K, respectively. Fouling may be considered negligible. Determine the overall heat transfer coefficient based on (a) the hot side, (b) the cold side, and (c) the hot side when the tube metal resistance is included.

25.20: Rework Problem 25.19 with a fouling resistance on the hot side of 0.0025 m^2-K/W.

25.21: Rework Problem 25.19 with a fouling resistance on the cold side of $1/h_{dC} = 0.0004$ m^2-K/W.

25.22: Rework Problem 25.19 with a fouling resistances on both the hot and cold sides of $1/h_{dH} = 1/h_{dC} = 0.0002$ m^2-K/W.

25.23: The tubing in Problem 25.19 is used as an air-cooled steam condenser. The steam is condensed on the inside of the tubing with a heat transfer coefficient of 8000 W/m^2-K. Air with a heat transfer coefficient of 100 W/m^2-K is used on the outside of the tubes. Fouling may be neglected. Determine the overall heat transfer coefficient based on the hot side.

25.24: Rework Problem 25.23 to determine the overall heat transfer coefficient based on the cold side.

25.25: Rework Problem 25.23 for fouling resistances on the hot and cold sides of $h_{d,H} = 1/0.0020$ m^2-K/W and $1/h_{d,C} = 0.0002$ m^2-K/W, respectively.

The Logarithmic Mean Temperature Difference

25.26: In a single-pass heat exchanger, the hot fluid enters at 300°C and leaves at 120°C. The cold fluid enters at 80°C and leaves at 140°C. Assume true counterflow and determine the LMTD.

25.27: In a single-pass counterflow heat exchanger, the hot fluid enters at 300°C and leaves at 140°C, and the cold fluid enters at 80°C and leaves at 120°C. Determine the LMTD.

25.28: In a single-pass cocurrent flow heat exchanger, the hot fluid enters at 300°C and leaves at 140°C, and the cold fluid enters at 80°C and leaves at 120°C. Determine the LMTD.

25.29: Steam at 5 kPa is condensed in a steam power plant in a condenser that takes cooling water from an adjacent river. The cooling water enters at 20°C and leaves at 24°C. Determine the LMTD.

25.30: Air enters the evaporator of an air-conditioning unit at 32°C and leaves at 22°C. The system employs refrigerant-134a at 2.8 bar in the evaporator coil. Determine the LMTD.

25.31: The surface of a metal plate used as a heat exchanger is held at a constant temperature. Air enters at 25°C and leaves at 60°C; the LMTD is 78°C. Determine the surface temperature.

25.32: In a counterflow heat exchanger, the hot stream is cooled from 340°C to 140°C by a fluid that is heated from 60°C to an unknown temperature, t_2. Determine t_2 for an LMTD of 135°C.

25.33: In a cocurrent flow heat exchanger, the cold fluid is heated from 40°C to 70°C. The hot fluid enters at 260°C and leaves at an unknown temperature, T_2. Determine T_2 for an LMTD of 86°C.

The LMTD Correction Factor Method

25.34: A shell-and-tube heat exchanger is used for heating 12 kg/s of oil ($c_C = 2.2$ kJ/kg-K) from 20°C to 38°C. The heat exchanger has one shell pass and two tube passes. Hot water ($c_H = 4.18$ kJ/kg-K) enters the shell at 75°C and leaves the shell at 55°C. The overall heat transfer coefficient based on the outside surface of the tubes is estimated to be 950 W/m^2-K. Determine (a) the corrected LMTD and (b) the required surface area in the exchanger.

25.35: Consider the heat exchanger in Problem 25.34. After 4 years of operation, the outlet of the oil reaches only 30°C instead of 38°C with everything else being the same. Determine the fouling resistance on the oil side of the exchanger.

25.36: A shell-and-tube heat exchanger is used to cool oil ($c_H = 2.2$ kJ/kg-K) from 110°C to 65°C. The heat exchanger has two shell passes and four tube passes. The coolant ($c_C = 4.20$ kJ/kg-K) enters the shell at 20°C and leaves the shell at 42°C. For an overall hot-side heat transfer coefficient of 1200 W/m^2-K and an oil flow of 11 kg/s, determine (a) the coolant mass flow rate and (b) the required surface area in the exchanger.

25.37: In an engine oil cooler, oil ($c_H = 2.08$ kJ/kg-K) enters at 58°C with a mass flow rate of 2 kg/s. The coolant ($c_C = 4.18$ kJ/kg-K) enters the cooler at 12°C and leaves at 24°C with a mass flow rate of 1.66 kg/s. Assuming an overall heat transfer coefficient of 2800 W/m^2-K, determine the outlet oil temperature and the heat transfer surface area for (a) a parallel flow and (b) a cross-flow arrangement with both fluids unmixed.

25.38: In a shell-and-tube heat exchanger, the hot fluid ($c_H = 3.798$ kJ/kg-K) flows at the rate of 7 kg/s through eighty 2-cm OD tubes. The hot fluid experiences a temperature drop of 20°C. The cold fluid ($c_C = 4.18$ kJ/kg-K) enters the shell at 12°C with a mass flow rate of 6.4 kg/s. The overall heat transfer coefficient based on the tube outside surface area is 680 W/m^2-K. The exchanger has one shell pass and two tube passes. Determine the length of each tube.

25.39: A finned tube cross-flow heat exchanger with both fluids unmixed is used to heat water ($c_C = 4.2$ kJ/kg-K) from 20°C to 75°C. The mass flow rate of the water is 2.5 kg/s. The hot stream ($c_H = 1.2$ kJ/kg-K) enters the heat exchanger at 280°C and leaves at 120°C. The overall heat transfer coefficient is 130 W/m^2-K. Determine (a) the mass flow rate of the hot stream and (b) the exchanger surface area.

25.40: A cross-flow heat exchanger is used to preheat combustion air to a boiler from 20°C to 275°C by using hot exhaust gases at 780°C. The overall heat transfer coefficient is 150 W/m^2-K. The air flowing on the unmixed side of the exchanger has a mass flow rate of 17.31 kg/s with a specific heat of $c_{p,C} = 1.0167$ kJ/kg-K. The exhaust gases flowing on the mixed side of the exchanger has a mass flow rate of 16.60 kg/s and a specific heat of $c_{p,H} = 1.1020$ kJ/kg-K. Determine the required heat transfer area.

25.41: A steam condenser consists of a bundle of brass tubes housed in a shell. Each tube has an inside diameter of 2 cm. The cooling water ($c_C = 4.18$ kJ/kg-K), which flows through the tubes enters the condenser at 20°C and leaves at 29°C with a mass flow rate of 1200 kg/s. Steam enters the condenser as saturated vapor at 15 kPa, condenses on the outside of the tube bundle and emerges from the condenser as saturated liquid at 15 kPa. The overall heat transfer coefficient based upon the outside tube surface area is 4800 W/m^2-K. Determine (a) the condensation rate of the steam in kg/s and (b) the number of tubes per pass for a two-tube pass arrangement with each tube 2.6-m long.

25.42: A steam condenser has brass tubes with a 1.9-cm inside diameter, a wall thickness of 4 mm, and a length of 3 m. The condenser is of a one shell pass-two tube pass design with 120 tubes per pass. Cooling water ($c_C = 4.18$ kJ/kg-K) enters the tubes at 20°C with a velocity of 1.5 m/s. The heat transfer coefficients on the inside and outside of the tubes are 6500 W/m^2-K and 10,000 W/m^2-K, respectively. The pressure of the condensing steam is 10 kPa. Neglecting any conduction or fouling resistances, determine (a) the temperature of the water leaving the condenser and (b) the rate of steam condensation.

25.43: In a water-to-water shell and tube heat exchanger containing one shell pass and two tube passes, hot water flows on the shell side at 3.2 kg/s with an inlet temperature of 110°C. Cold water flows inside the tubes at a rate of 4.5 kg/s with an inlet temperature of 25°C and an outlet temperature of 60°C. The overall heat transfer coefficient based on the outside surface area of the tubes is 1300 W/m^2-K. Determine (a) the hot water exit temperature and (b) the surface in the exchanger.

25.44: In a shell and tube heat exchanger, glycerin flows through the 2-in × 10-BWG tubes and is heated from 22°C to 36°C. Hot water enters the shell side at 85°C and exits at 50°C providing a heat transfer coefficient of 350 W/m^2-K. There are a total of 80 tubes, each carrying glycerin at 0.6 kg/s. The inside and outside tube diameters are 5.00 cm and 5.08 cm, respectively. The heat transfer coefficient on the inside of the tubes may be taken as 400 W/m^2-K and the specific heats of water and glycerin are

4.185 kJ/kg-K and 2.39 kJ/kg-K, respectively. Determine (a) the mass flow rate of the water, (b) the corrected LMTD, and (c) the tube length if each tube makes four passes.

25.45: Determine the heat transfer duty of the following heat exchanger arrangements if $T_1 = 150°C$, $T_2 = 90°C$, $t_1 = 25°C$, $t_2 = 75°C$, and $US = 130\,W/°C$.

(a) Parallel flow double pipe.
(b) Counterflow double pipe.
(c) One shell pass-two tube passes.
(d) Two shell passes-four tube passes.
(e) Cross flow—both fluids unmixed.
(f) Cross flow—one fluid mixed, the other fluid unmixed.

25.46: In a cross-flow exchanger with both fluids unmixed, the temperatures of the hot and cold fluids are $T_1 = 135°C$, $T_2 = 50°C$, $t_1 = 30°C$, $t_2 = 68°C$. The total hot-side heat transfer surface area provided by the heat exchanger is $10.5\,m^2$. The heat exchanger is to meet a transfer duty of 210 kW. Determine (a) the overall heat transfer coefficient referred to the hot side and (b) the ratio, C_H/C_C.

25.47: In a one shell pass-two tube pass heat exchanger, cooling water ($c_C = 4.18\,kJ/kg\text{-}K$) enters the shell side at 20°C and a flow rate of 7000 kg/h. Hot oil ($c_H = 2.095\,kJ/kg\text{-}K$) flows through the tubes, entering at 110°C and flowing at a rate of 14,000 kg/h. The total heat transfer area is $15.4\,m^2$ and the overall heat transfer coefficient is $450\,W/m^2\text{-}K$ with both referred to the hot side of the exchanger. Determine (a) the outlet temperatures of both the water and the oil and (b) the heat transfer duty of the exchanger.

25.48: Determine the hot-side heat transfer surface area for each of the following heat exchanger arrangements if $T_1 = 150°C$, $T_2 = 85°C$, $t_1 = 30°C$, $t_2 = 80°C$, $\dot{Q} = 180\,kW$, and $U_H = 350\,W/m^2\text{-}K$:

(a) Parallel flow double pipe.
(b) Counterflow double pipe.
(c) One shell pass-two tube passes.
(d) Two shell passes-four tube passes.
(e) Cross flow—both fluids unmixed.
(f) Cross flow—one fluid mixed, the other fluid unmixed.

25.49: Hot fluid flows through the tubes in a cross-flow heat exchanger in which the hot-side heat transfer surface is $16\,m^2$. The inlet and outlet temperatures of the hot fluid are 120°C and 65°C, respectively. The cold fluid, assumed to be mixed, enters the heat exchanger at 25°C and leaves at 65°C. The heat transfer duty is $1000\,W/m^2$. Determine the overall heat transfer coefficient.

25.50: In a cross-flow heat exchanger, hot exhaust gases ($c_{p,H} = 1.045\,kJ/kg\text{-}K$), assumed to be unmixed, flow over a bank of tubes at the rate of 6 kg/s. The inlet and outlet temperatures of the exhaust gases are 380°C and 180°C, respectively. Water enters through the tubes as saturated liquid at 0.60 MPa and emerges as saturated vapor at the same pressure. The overall hot-side heat transfer coefficient is $210\,W/m^2\text{-}K$. Determine (a) the surface area in the exchanger and (b) the steam generation rate.

25.51: Show that the logarithmic mean temperature difference correction factor, F, for a cocurrent (parallel) flow exchanger may be expressed as

$$F = \frac{R+1}{R-1} \frac{\ln \dfrac{1-P}{1-PR}}{\ln \dfrac{1}{1-P(R-1)}}$$

were P and R are as defined in the text.

25.52: In a steam condenser, saturated steam enters the shell at 10 kPa and condenses into saturated liquid at the same pressure. Water enters the tubes at 25°C and flows at the rate of 200 kg/s. The mass of steam condensed is 6 kg/s. Determine (a) the outlet temperature of the water, (b) the hot-side heat transfer area if the condenser is in a one shell-two tube pass arrangement with an overall hot-side heat transfer coefficient of $U_H = 2500\,\text{W/m}^2\text{-K}$.

25.53: Consider the steam condenser in Problem 25.52. With the mass flow rates and the inlet and outlet temperatures fixed, determine the mass of steam condensed if the pressure of the steam is 15 kPa.

The ϵ-N_{tu} Method

25.54: Show that for a cocurrent (parallel flow) exchanger, the ϵ-N_{tu} relationship is

$$\epsilon = \frac{1 - e^{-\text{N}_{\text{tu}}(1+C^*)}}{1+C^*}$$

where $C^* = C_{min}/C_{max}$.

25.55: Considering the expression derived in Problem 25.54, show that for an infinitely long, cocurrent (parallel) flow heat exchanger, the expression reduces to

$$\epsilon = \frac{1}{1+C^*}$$

25.56: Using the equations

$$\dot{Q} = C_H(T_1 - T_2) = C_C(t_2 - t_1)$$

and

$$\dot{Q} = US \frac{(T_1 - t_2) - (T_2 - t_1)}{\ln \dfrac{T_1 - t_2}{T_2 - t_1}}$$

derive the ϵ-N_{tu} relationship for the counterflow heat exchanger

$$\epsilon = \frac{1 - e^{-\text{N}_{\text{tu}}(1-C^*)}}{1 - C^* e^{-\text{N}_{\text{tu}}(1-C^*)}}$$

where $C^* = C_{min}/C_{max}$.

25.57: Show that, in the limit as $C^* \longrightarrow 1$, the ϵ-N_{tu} expression derived in Problem 25.56 reduces to

$$\epsilon = \frac{\text{N}_{\text{tu}}}{1+\text{N}_{\text{tu}}}$$

25.58: The temperature of the hot fluid in a heat exchanger drops from 75°C to 55°C while the cold fluid increases in temperature from 25°C to 35°C. Determine (a) the capacity rate ratio, C^* and (b) the number of transfer units, N_{tu}, if the exchanger is in counterflow, and (c) the number of transfer units, N_{tu}, if the exchanger is in cocurrent (parallel) flow.

25.59: The temperature of the hot fluid in a heat exchanger decreases from 68°C to 48°C while the cold fluid increases in temperature from 24°C to 36°C. Determine (a) the capacity rate ratio, C^*, (b) the number of transfer units, N_{tu}, if the exchanger is in cross flow with both fluids mixed, and (c) the number of transfer units, N_{tu}, if the exchanger is in cross flow with both fluids unmixed.

25.60: A double-pipe exchanger is used to cool oil ($c_H = 3.20\,\text{kJ/kg-K}$) from 110°C to 65°C. Water ($c_C = 4.20\,\text{kJ/kg-K}$) is used as the coolant entering at 20°C and leaving at 42°C. For an overall heat transfer coefficient of $U_H = 1000\,\text{W/m}^2\text{-K}$ and an oil flow of 1.125 kg/s, determine (a) the coolant flow rate, (b) the required surface for a counterflow arrangement, and (c) the required surface for a cocurrent (parallel) flow arrangement.

25.61: Rework Problem 25.37 using the ϵ-N_{tu} method of analysis to determine the heat transfer area.

25.62: Rework Problem 25.39 using the ϵ-N_{tu} method of analysis to determine the heat transfer area. The heat transfer coefficient is referred to the hot side.

25.63: Rework Problem 25.40 using the ϵ-N_{tu} method of analysis to determine the heat transfer area.

25.64: A double-pipe exchanger, operating in counterflow, has the same operating conditions as in Problem 25.34. The hot water is in the outer pipe and the cold water is in the inner pipe. The overall heat transfer coefficient is based on the hot side. Use the ϵ-N_{tu} analysis method to find the heat transfer area.

25.65: A double-pipe exchanger, operating in counterflow is used as a clean water chiller. Clean water enters the annulus at 80°C and leaves at 40°C at a flow rate of 8 kg/s. Cold water enters the inner pipe at 20°C at a flow rate of 10 kg/s. The overall heat transfer coefficient is $U_H = 3450\,\text{W/m}^2\text{-K}$. Determine (a) the duty of the exchanger and (b) the surface required.

25.66: The water chiller in Problem 25.65 has the conditions of Problem 25.65 but the configuration contains 360 tubes that are 2 in × 10 BWG and are 80-cm long. The hot water flows on the outside of the tubes with a heat transfer coefficient of $4200\,\text{W/m}^2\text{-K}$ and the cold water flows on the inside of the tubes with a heat transfer coefficient of $4800\,\text{W/m}^2\text{-K}$. It is typical of this cross-flow arrangement that both fluids are unmixed. Determine (a) the exchanger effectiveness and (b) the outlet temperature of the cold water.

26

Radiation Heat Transfer

Chapter Objectives

- To consider electromagnetic waves and the electromagnetic spectrum and to show that thermal radiation is a form of electromagnetic radiation.
- To consider the emission of radiant energy and the Stefan-Boltzmann and Wien displacement laws.
- To define the terminology peculiar to radiation heat transfer such as emissivity, absorptivity, reflectivity, and transmissivity.
- To describe what is meant by a blackbody and a gray body.
- To examine the directional characteristics of surface radiation.
- To provide a relationship for radiant heat interchange with perfect absorbers and emitters and for surfaces not in full view of each other.
- To modify the relationship for radiant heat interchange with nonperfect absorbers and emitters and for surfaces that are not in full view of each other.
- To present an electrothermal analog method for handling radiation inside enclosures.

26.1 The Electromagnetic Spectrum

Radiation of thermal energy is believed to be a specific form of radiation within the general phenomenon of electromagnetic radiation. As but one of numerous electromagnetic phenomena, thermal radiant energy travels at the speed of light: 2.9979×10^8 m/s.

The existence of radiation as a mode of heat transfer can be observed from everyday experience. Consider, for example, a warm body enclosed without physical contact inside a cooler enclosure under complete vacuum. The warm body will eventually attain the temperature of the surrounding enclosure without the aid of conduction or convection. This statement may appear intuitive, and we can easily imagine, as well, the approximation of the warm body suspended by nonconducting cords in the evacuated cooler enclosure.

All bodies continuously emit radiation. Figure 26.1 displays the electromagnetic spectrum showing a range of electromagnetic waves from long radio waves to the shorter wavelengths. The unit of wavelength used in this figure is the **micrometer** (μm) or 10^{-6} m,

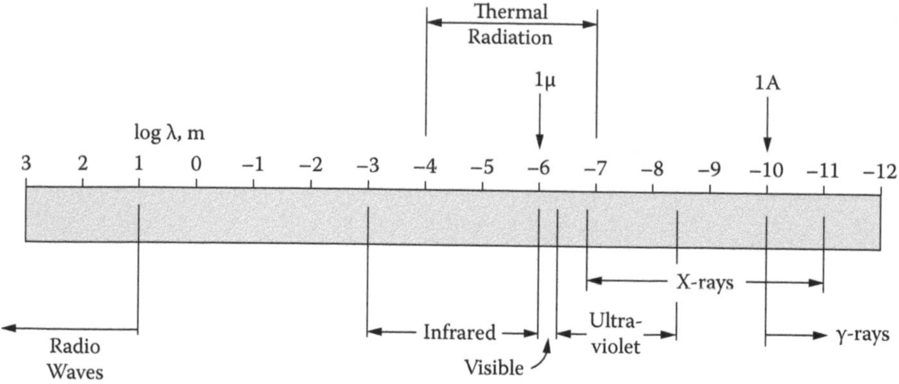

FIGURE 26.1
The electromagnetic spectrum.

and the relationship between the frequency of this radiation, ν, and the wavelength, λ, is

$$\lambda = \frac{c}{\nu} \tag{26.1}$$

where c is the speed of light.

In Figure 26.1 we may note the very narrow band of wavelengths designated as the **visible region**. This band from 0.38 to 0.76 μm is sensed by the optic nerve and is what we refer to as **light**. The radiation between 0.1 and 100 μm is termed **thermal radiation** and we note that thermal radiation includes portions of the **ultraviolet** and **infrared** portions of the electromagnetic spectrum.

26.2 Monochromatic Emissive Power

Monochromatic emission refers to electromagnetic radiation having wavelengths confined to an extremely narrow wavelength range. In the present context, monochromatic emission may also be defined as **hemispherical** or **spectral** emission.

26.2.1 The Black Surface or Blackbody

An ideal surface that absorbs all (and therefore reflects none) of the radiation falling upon it, regardless of the wavelength or the angle of incidence is defined as a **black surface** or **blackbody**. Because we "see" reflected light (in reality, electromagnetic radiation), the so-called blackbody, from which no light is reflected, will appear black. This is the idea behind the color black and we are quick to point out that surface color is not a requisite to the terminology. Because of the lacquer that it contains, black automobile paint does not yield a black surface. Moreover, freshly fallen snow, because of the diffuse nature of its reflection, may come very close to providing a black surface.

A black surface emits a spectrum of radiant energy and this is considered in Section 26.2.2.

26.2.2 Planck's Law

Max Planck introduced the quantum concept in 1900 and, with it, the idea that radiation is emitted, not in a continuous state, but in discrete amounts or **quanta**. Planck's law for the monochromatic emissive power of a surface is given as

$$E_{b\lambda}(\lambda, T) = \frac{2\pi C_1 c^2}{\lambda^5(e^{cC_1/C_2\lambda T} - 1)} \quad (\mathrm{W/m^2}) \tag{26.2}$$

where, in addition to c, which is the speed of light, the accepted values of the two constants are

$$C_1 = 6.624 \times 10^{-27}\ \mathrm{erg}$$

which is **Planck's constant** and

$$C_2 = 1.380 \times 10^{-16}\ \mathrm{erg/K}$$

which is **Boltzmann's constant.**

We may note that while Planck's law does indeed show that the monochromatic emissive power is a function of temperature and wavelength, it does not tell us how much radiation will occur at a given temperature. This is obtained from its integration over all wavelengths.

26.2.3 Wien's Displacement Law

Figure 26.2 indicates that there will be some value of wavelength for a particular absolute temperature that will yield a maximum value of $E_{b\lambda}(\lambda, T)$. A relationship between the maximum value of the product λT can be obtained at the point where $dE_{b\lambda}(\lambda, T)/d\lambda$ vanishes. First, however, we make the transformation

$$x = \frac{A}{\lambda} \qquad \left(A = \frac{C_1 c}{C_2 T}\right)$$

so that

$$\lambda = \frac{A}{x} \qquad \text{and} \qquad \frac{dx}{d\lambda} = -\frac{A}{\lambda^2}$$

Then

$$\frac{dE_{b\lambda}(\lambda, T)}{d\lambda} = \frac{dE_{b\lambda}(\lambda, T)}{dx}\frac{dx}{d\lambda} = \frac{d}{dx}\left[\frac{(2\pi C_1 c^2/A^5)x^5}{e^x - 1}\right]\left(-\frac{A}{\lambda^2}\right) = 0$$

and the differentiation and algebraic expansion of terms can proceed to give

$$5(e^x - 1) = xe^x$$

$$5\left(e^{C_1 c/C_2\lambda T} - 1\right) = \frac{C_1 c}{C_2 \lambda T} e^{C_1 c/C_2 \lambda T}$$

which will be satisfied when

$$\lambda T = 2897.8\ \mu\text{-K} \tag{26.3}$$

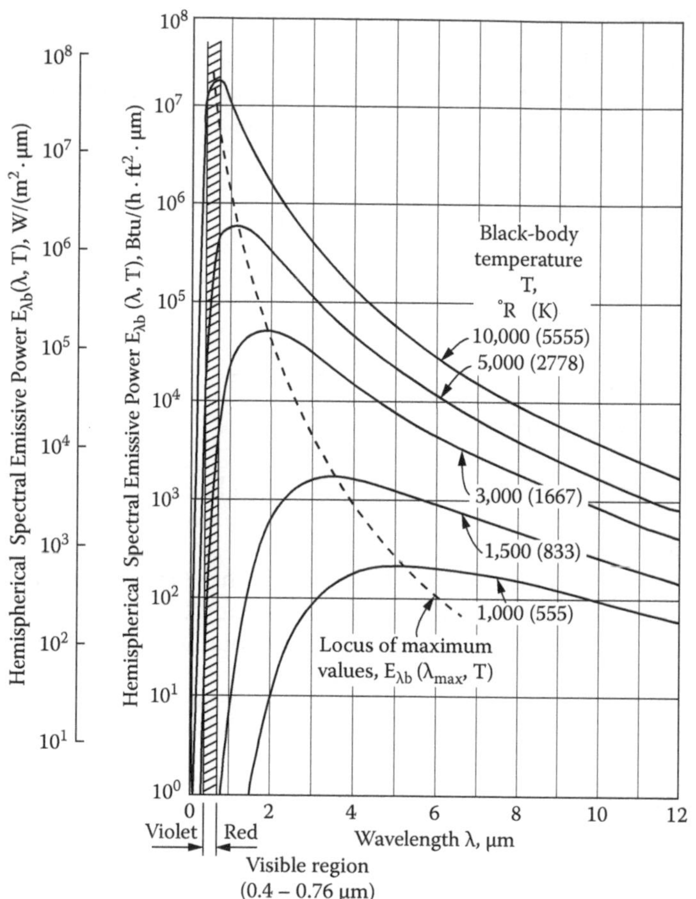

FIGURE 26.2
Monochromatic emissive power as a function of wavelength and temperature.

Equation 26.3 is **Wien's[1] displacement law** which provides the wavelength for the maximum monochromatic emissive power for the temperature specified.

26.2.4 The Stefan-Boltzmann Law

The total quantity of blackbody radiation emitted at a particular temperature can be obtained from an integration of Planck's law over all wavelengths

$$E_b = \int_0^\infty E_{b\lambda}(\lambda, T)d\lambda = \int_0^\infty 2\pi C_1 c^2 \frac{\lambda^{-5} d\lambda}{e^{C_1 c/C_2 \lambda T} - 1}$$

This time we make the transformation

$$z = \frac{A}{\lambda} = \frac{C_1 c/C_2}{T}$$

[1] W. Wien (1864–1928) was a German physicist who formulated the two Wien laws pertaining to radiation from blackbodies.

so that

$$\lambda = \frac{A}{z} \qquad \text{and} \qquad d\lambda = -\frac{A}{z^2}dz$$

With the substitution of the transformed variable we obtain

$$E_b = -2\pi C_1 c^2 \int_0^\infty \frac{z^3 dz}{A^4(e^z - 1)}$$

However, $1/(e^z - 1)$ can be written as an infinite series so that a term-by-term integration can be performed to yield, after a substitution of the limits

$$E_b = \frac{2\pi C_1 c^2}{A^4}\left(\frac{3!}{1^4} + \frac{3!}{2^4} + \frac{3!}{3^4} + \frac{3!}{4^4} + \cdots\right) = 2\pi c^2 (6.45)\left(\frac{C_2^4 T^4}{C_1^2 c^4}\right)$$

or

$$E_b = \frac{2\pi C_1 c^2 (6.45)}{A^4} = \left[\frac{2\pi (6.45) C_2^4}{C_1^3 c^2}\right] T^4$$

where the bracketed term, designated by σ, is a constant. Hence,

$$E_b = \sigma T^4 \tag{26.4}$$

which is the **Stefan-Boltzmann law.**

In Equation 26.4, the constant, σ, is universally designated as the **Stefan-Boltzmann constant** and has the value

$$\sigma = 5.67 \times 10^{-8}\ \text{W/m}^2\text{-K}^4$$

and we also observe that the Stefan-Boltzmann law can be written as

$$E_b = 5.67 \times 10^{-8} T^4 \tag{26.5}$$

We are frequently interested in the quantity of emission in a specific portion of the wavelength spectrum. This may be conveniently expressed by a fraction of the total emissive power and is designated by $F_{\lambda_1\text{-}\lambda_2}$

$$F_{\lambda_1\text{-}\lambda_2} = \frac{\int_{\lambda_1}^{\lambda_2} E_{b\lambda}(\lambda, T)d\lambda}{\int_0^\infty E_{b\lambda}(\lambda, T)d\lambda} = \frac{\int_{\lambda_1}^{\lambda_2} E_{b\lambda}(\lambda, T)d\lambda}{\sigma T^4}$$

or with both sides divided by σT^5

$$F_{\lambda_1\text{-}\lambda_2} = \frac{1}{\sigma T^5}\left[\int_0^{\lambda_2} E_{b\lambda}(\lambda T)d(\lambda, T) - \int_0^{\lambda_1} E_{b\lambda}(\lambda, T)d(\lambda T)\right] = F_{0\text{-}\lambda_2} - F_{0\text{-}\lambda_1}$$

We see that at a given temperature, the fraction of the emission between the two wavelengths of interest may be obtained from a subtraction of values. This subtraction process may be simplified if the variable of integration is changed to the wavelength-temperature product, λT

$$F_{\lambda_1\text{-}\lambda_2} = \int_0^{\lambda_2 T} \frac{E_{b\lambda}(d\lambda, T)}{\sigma T^5} d(\lambda T) - \int_0^{\lambda_1 T} \frac{E_{b\lambda}(\lambda T)}{\sigma T^5} d(\lambda T)$$

TABLE 26.1

Radiation Functions

λT µm-K	$F_{0\text{-}\lambda T}$	$\frac{E_{b\lambda}(\lambda T)}{\sigma T^5}$ (cm-K)$^{-1}$	λT µm	$F_{0\text{-}\lambda T}$	$\frac{E_{b\lambda}(\lambda, T)}{\sigma T^5}$ (cm-K)$^{-1}$
1000	0.0003	0.0732	6200	0.7541	0.7844
1200	0.0021	0.0855	6400	0.7692	0.7255
1400	0.0078	0.1646	6600	0.7832	0.6715
1600	0.0197	0.7825	6800	0.7961	0.6220
1800	0.0396	1.1797	7000	0.8081	0.5765
2000	0.0667	1.5499	7500	0.8344	0.4786
2200	0.1009	1.8521	8000	0.8562	0.3995
2400	0.1402	2.0695	8500	0.8736	0.3354
2600	0.1831	2.2028	9000	0.8900	0.2832
2800	0.2279	2.2623	9500	0.9030	0.2404
2897.8	0.2501	2.2688	10,000	0.9142	0.2052
3000	0.2732	2.2624	11,000	0.9318	0.1518
3200	0.3181	2.2175	12,000	0.9451	0.1145
3400	0.3617	2.1408	13,000	0.9551	0.0878
3600	0.4036	2.0429	14,000	0.9628	0.0684
3800	0.4434	1.9324	15,000	0.9689	0.0540
4000	0.4809	1.8157	16,000	0.9738	0.0432
4200	0.5160	1.6974	17,000	0.9777	0.0389
4400	0.5488	1.5807	18,000	0.9808	0.0285
4600	0.5793	1.4679	19,000	0.9834	0.0235
4800	0.6075	1.3601	20,000	0.9856	0.0196
5000	0.6337	1.2590	25,000	0.9922	0.0087
5200	0.6579	1.1640	30,000	0.9953	0.0044
5400	0.6803	1.0756	35,000	0.9970	0.0025
5600	0.7010	0.9938	40,000	0.9979	0.0015
5800	0.7201	0.9181	45,000	0.9985	0.0009
6000	0.7378	0.8485	50,000	0.9989	0.0004

or

$$F_{\lambda_1\text{-}\lambda_2} = F_{0\text{-}\lambda_2 T} - F_{0\text{-}\lambda_1 T} \tag{26.6}$$

Values of $F_{0\text{-}\lambda T}$ are given in Table 26.1.

Example 26.1 The surface of a blackbody is held at 800 K. Determine (a) the total emissive power, (b) the wavelength at which the maximum monochromatic emissive power occurs, (c) the magnitude of the monochromatic emissive power at this wavelength, and (d) the fraction of the total emission between 3 µm and 4 µm.

Solution

Assumptions and Specifications

(1) Steady-state conditions exist.
(2) The surface emits as a blackbody.
(3) Radiation is the only mode of heat transfer.

(a) At $T = 800$ K, Equation 26.4, which is the Stefan-Boltzmann law, gives

$$E_b = \sigma T^4 = (5.67 \times 10^{-8}\,\text{W/m}^2\text{-K}^4)(800\,\text{K})^4 = 23.22\,\text{kW/m}^2 \Longleftarrow$$

(b) Equation 26.3, which is the Wien displacement law, yields

$$\lambda = \frac{2897.8\,\mu\text{m-K}}{800\,\text{K}} = 3.622\,\mu\text{m} \Longleftarrow$$

(c) At 3.622 μm and 800 K where $\lambda T = 2897.8$ μm-K, Table 26.1 gives

$$E_{b\lambda}(\lambda, T)/\sigma T^5 = 2.2688, (\text{cm-K})^{-1}$$

Hence,

$$E_{b\lambda}(\lambda, T) = 2.2688 \times 10^{-4}\,(\text{cm-K}^{-1})\sigma T^5$$

$$E_{b\lambda}(3.622\,\mu\text{m}, 800\,\text{K}) = [2.2688 \times 10^{-4}(\text{cm-K}^{-1})](5.670 \times 10^{-8}\,\text{W/m}^2\text{-K}^4)(800\,\text{K})^5$$

$$= 4215.3\,\text{W/m}^2\text{-}\mu\text{m} \Longleftarrow$$

(d) The fraction of the total emission that occurs between $\lambda_1 = 3$ μm and $\lambda_2 = 6$ μm is obtained from Table 26.1. At $\lambda_1 T = (3\,\mu\text{m})(800\,\text{K}) = 2400$ μm-K, read

$$F_{0\text{-}2400} = 0.1402$$

and at $\lambda_2 T = (6\,\mu\text{m})(800\,\text{K}) = 4800$ μm-K, read

$$F_{0\text{-}4800} = 0.6075$$

Thus,

$$F_{6.0\text{-}3.0} = F_{0\text{-}4800} - F_{0\text{-}2400} = 0.6075 - 0.1402 = 0.4673 \qquad (46.73\%) \Longleftarrow$$

26.3 Radiation Properties and Kirchhoff's Law

26.3.1 Absorptivity, Reflectivity, and Transmissivity

Consider a system (or body) that possesses an element of surface area, dS, as shown in Figure 26.3a. We define the **irradiation**, G, as the total radiation heat flux that is intercepted by dS from all other surfaces that can "see" dS and we note that, in general, G may possess:

- An absorbed portion, $G_\alpha = \alpha G$, which takes place within the molecular layer immediately below the element of surface, dS.
- A reflected portion, $G_\rho = \rho G$, which is reflected back into the entire space that can be "seen" by dS. Here we note that the reflected radiation may leave dS in all directions (Figure 28.3b) with no direction preferred. This type of radiation is referred to as **diffuse** and it is the type of radiation most likely to occur with rough surfaces. On the other hand, if the radiation reflected from the surface is directional such that the angle of incidence is equal to the angle of reflection (Figure 26.3c), the radiation is termed **specular** and is most likely to occur on smooth polished surfaces. The development here considers only diffuse radiation.
- A transmitted portion, $G_\tau = \tau G$, which can pass through the body and exit at some surface remote from dS.

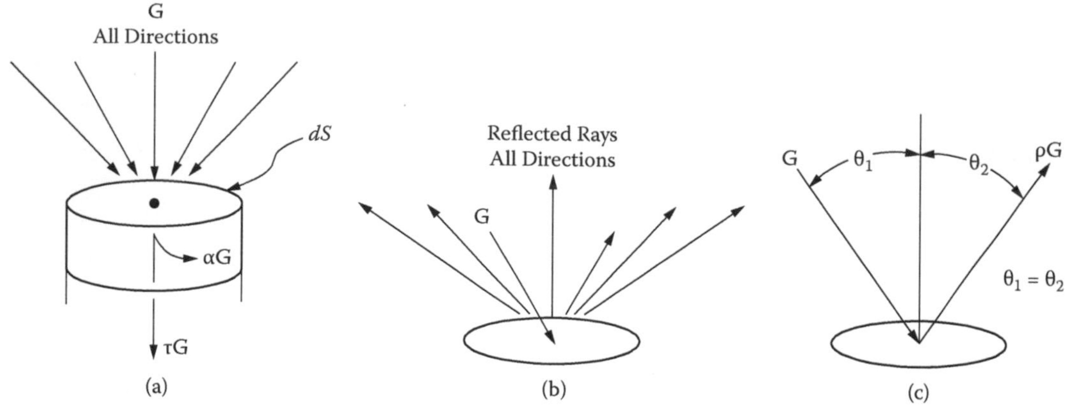

FIGURE 26.3
(a) A small areas dS subjected to a total irradiation G, (b) the reflection pattern for diffuse radiation, and (c) the reflection pattern for specular radiation.

Conservation of energy may be applied to the very thin disk-shaped control volume that encloses dS in Figure 26.3a to show that with $G_\alpha + G_\rho + G_\tau = G$

$$\alpha G + \rho G + \tau G = G$$

or

$$\alpha + \rho + \tau = 1 \tag{26.7}$$

where α, ρ, and τ are nonnegative dimensionless numbers that are system properties and we observe that none of them can exceed unity. Here α is the total **absorptivity**, ρ is the total **reflectivity**, and τ is the total **transmissivity** of the material. Because G is the irradiation from all directions over a hemispherical space above dS over all wavelengths, these properties are referred to as total **hemispherical** properties.

If the surface is **opaque**, $\tau = 0$ and Equation 26.7 reduces to

$$\alpha + \rho = 1 \qquad (\tau = 0) \tag{26.8}$$

If, in addition, the reflectivity is zero, the surface is black and the solid is said to be a **black body**. In this case,

$$\alpha = 1 \qquad (\tau = 0, \ \rho = 0) \tag{26.9}$$

26.3.2 Kirchhoff's Radiation Law

Consider Figure 26.4 and let the temperatures of the black enclosure and the small nonblack test body be, respectively, designated as T_b and T_1. Further, let blackbody radiation impinge on the smaller body so that the amount of heat absorbed by the smaller body is equal to $\alpha_1 E_b$. However, the smaller body will also emit radiation, say E_1, and we note that the net rate of radiant heat interchange between the enclosure and the small body will be

$$\dot{q} = \alpha_1 E_b - E_1$$

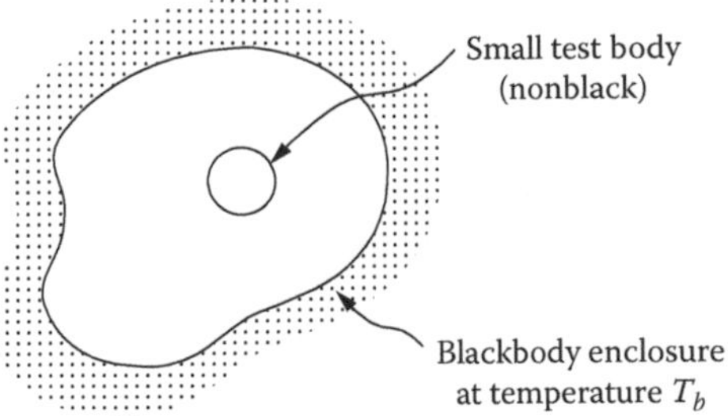

FIGURE 26.4
A small nonblack test body within a large black enclosure.

At thermal equilibrium when $T_1 = T_b$, there can be no radiant heat interchange and $\dot{q} = 0$. Hence,

$$\alpha_1 E_b - E_1 = 0 \quad \text{or} \quad E_b = \frac{E_1}{\alpha_1}$$

and we can derive a similar relationship for n small nonblack bodies in thermal equilibrium

$$E_b = \frac{E_1}{\alpha_1} = \frac{E_2}{\alpha_2} = \cdots = \frac{E_n}{\alpha_n} \tag{26.10}$$

The general relationship of Equation 26.10 is known as **Kirchhoff's law.**[2]

26.3.3 Emissivity

All blackbodies at the same temperature will emit radiation at the same rate. If the smaller body in Figure 26.4 is a blackbody, then $\alpha_1 = 1$ and $E_1 = E_b$. If, for the sake of argument, we propose that $E_1 > E_b$, the smaller body must be cooled by a transfer of heat from the lower-temperature smaller body to the higher-temperature blackbody. Clearly, this would be in violation of the Second Law of Thermodynamics, which prohibits heat from flowing of its own accord from a region at low temperature to one at high temperature. It is also obvious from the foregoing that, if the smaller body is a blackbody, at thermal equilibrium it will emit exactly the same amount of radiation as the blackbody in which it is enclosed.

The foregoing reasoning leads to three conclusions:

1. For thermal equilibrium, Kirchhoff's law shows that the ratio of the emissive power of the surface to its absorptivity is the same for all bodies.
2. Because the absorptivity can never exceed unity at a particular temperature, a blackbody has the maximum emissive power of any surface.
3. The blackbody may be considered as a perfect emitter as well as a perfect absorber of radiant energy.

In addition, because a perfect emitter or absorber does not exist, the concept of a blackbody is an idealized one. Indeed, the emissive characteristics of the so-called blackbody may be quite different from its absorptive characteristics. The emissive characteristics at a particular temperature are represented by the property known as the **emissivity,** ϵ, defined as the ratio of the actual rate of energy emission to the rate of energy emission of a blackbody

[2] G. R. Kirchhoff (1824–1887) was a German physicist who formulated the laws of electric currents and the electromotive forces in an electrical network and with R. W. Bunsen who developed the method of spectrum analysis.

at the same temperature. Thus,

$$E = \epsilon E_b$$

Consider Figure 26.4 once again and assume that the smaller body is not a blackbody. We see that for a black enclosure with energy emitted, E_b, the energy absorbed by the smaller body is

$$E_1 = \alpha_1 E_b$$

and that the energy emitted by the smaller body is

$$E_1 = \epsilon_1 E_b$$

as required by the definition of emissivity.

The net rate of heat interchange between the enclosure and small body is

$$\dot{q} = \alpha_1 E_b - \epsilon_1 E_b$$

and when the small body and enclosure are at the same temperature ($\dot{q} = 0$),

$$\alpha_1 E_b = \epsilon_1 E_b$$

from which we observe that the absorptivity equals the emissivity

$$\alpha_1 = \epsilon_1$$

26.3.4 An Approximation to a Blackbody

Although no body or surface is a perfect emitter or blackbody, such a body may be approximated by a small hole cut into a hollow cavity. Suppose that the cavity in Figure 26.5 has its interior wall at a constant temperature, that is, let the interior wall of the cavity be an isothermal surface. The wall, of course, is nonblack and has an emissivity, ϵ.

Now imagine that a small portion of the surface emits energy, ϵE_b, and that this energy escapes through the hole. Next, consider another portion of the surface that emits energy, ϵE_b, and let this energy be reflected once before it escapes through the hole. This once-reflected energy will be equal in magnitude to $\rho\epsilon E_b$. For another portion of the surface that emits energy that is reflected twice, the magnitude of the escaping energy will be $\rho^2\epsilon E_b$. We see that the total energy emitted from the hole will contain energy that has been directly

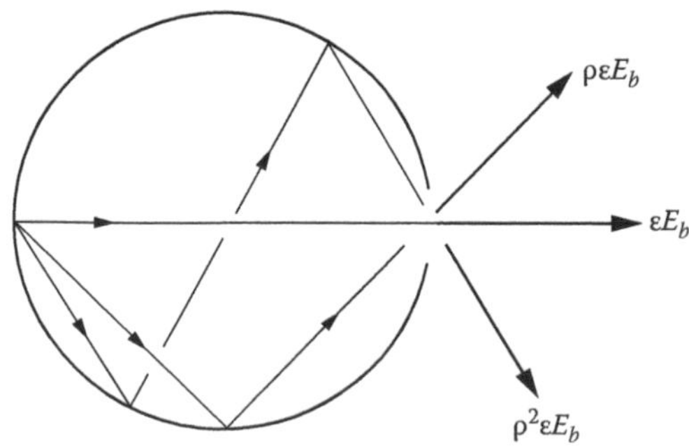

FIGURE 26.5
Cavity used to produce blackbody radiation.

radiated, reradiated once, twice, and so on. The total escaping energy will be equal to

$$E_T = \epsilon E_b + \rho\epsilon E_b + \rho^2\epsilon E_b + \rho^3\epsilon E_b + \cdots$$

or

$$E_T = (1 + \rho + \rho^2 + \rho^3 + \cdots)\epsilon E_b$$

However, we know that at any temperature,

$$\epsilon = \alpha = 1 - \rho$$

and we may take note of the infinite series

$$\frac{1}{1-\rho} = 1 + \rho + \rho^2 + \rho^3 + \cdots$$

Therefore, we see that the total energy can be represented by

$$E_T = \frac{\epsilon E_b}{1-\rho} = \frac{1-\rho}{1-\rho}E_b$$

or

$$E_T = E_b$$

This indicates that the energy emitted from the hole is essentially blackbody radiation.

26.4 Radiation Intensity and Lambert's Cosine Law

In Figure 26.6, we consider that dS_1 is a diffuse emitting surface that emits radiation in all directions. We will use the **intensity of radiation**, designated by I, as the quantity that describes how this radiation varies with respect to the azimuthal angle $d\psi$ and the zenith

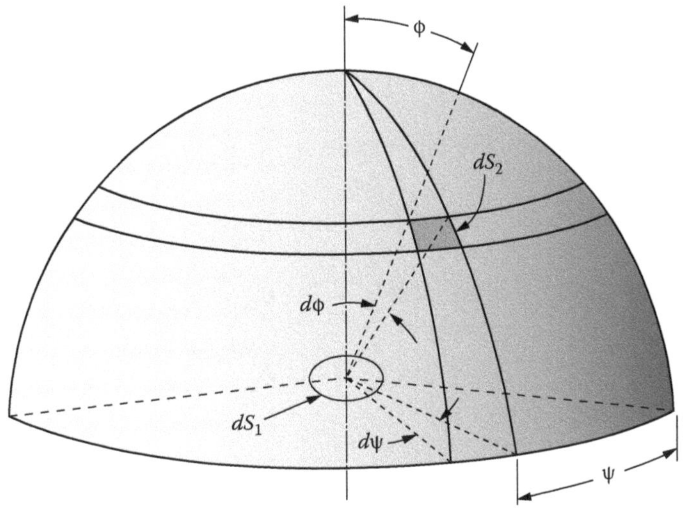

FIGURE 26.6
Configuration and nomenclature for the analysis of intensity of radiation.

angle $d\phi$ in this spherical coordinate system. When placed above the emitting surface, dS_1, the hemisphere shown will intercept all of the energy emitted by dS_1 but this emitted energy will only be "seen" without distortion from a point directly above dS_1. At the point dS_2, which is displaced by an angle of ϕ from the normal to dS_1, the element dS_1 appears as the projected area $dS_1 \cos\theta$.

We define the intensity of radiation I at some point in space due to the emission from area dS_1 as the radiant energy emitted per unit time per unit solid angle subtended at dS_1 per unit of emitting surface normal to the direction of a point in space from the emitting source. If dS_2 is considered at an angle ϕ from the normal, the solid angle subtended by dS_2 as viewed from dS_1 will be $d\omega = dS_2/r^2$ and the heat flow from dS_1 to dS_2 will be

$$d\dot{Q} = I\cos\phi dS_1 d\omega = I\cos\phi dS_1 \frac{dS_2}{r^2} \tag{26.11}$$

We note that $dS_1 \cos\phi$ is the effective or projected area of dS_1 as viewed from dS_2 and that

$$I = \frac{d\dot{Q}}{dS_1 \cos\phi d\omega} \qquad (\mathrm{W/m^2\text{-}sr}) \tag{26.12}$$

The total emissive power will be a summation of the intensity over the entire hemispherical surface. In the case of Figure 26.6, the radiation intercepted by the element dS_2 is integrated over the half-space represented by the hemisphere

$$E = \int_\omega \int_\phi I\cos\phi d\phi d\omega = \int_\omega \int_\phi I\cos\phi \frac{dS_2}{r^2} d\phi \tag{26.13}$$

With regard to Figure 26.6, an exercise in spherical coordinates will show that

$$d\omega = \frac{(r\sin\phi)(d\phi)(r d\psi)}{r^2}$$

or

$$d\omega = \sin\phi d\phi d\psi$$

so that, with this substituted for $d\omega$ in the integral in Equation 26.13, we obtain

$$E = \int_0^{2\pi} d\psi \int_0^{\pi/2} I\cos\phi\sin\phi d\phi$$

First, we integrate with respect to ψ to obtain

$$E = 2\pi I \int_0^{\pi/2} \cos\phi\sin\phi d\phi$$

and then, when we integrate with respect to ϕ, the result is

$$E = 2\pi I\left(\frac{1}{2}\right)\sin^2\phi\Big|_0^{\pi/2}$$

or

$$E = \pi I \tag{26.14}$$

Equation 26.14 is **Lambert's law** and shows that the total emissive power is equal to π times the radiation intensity.

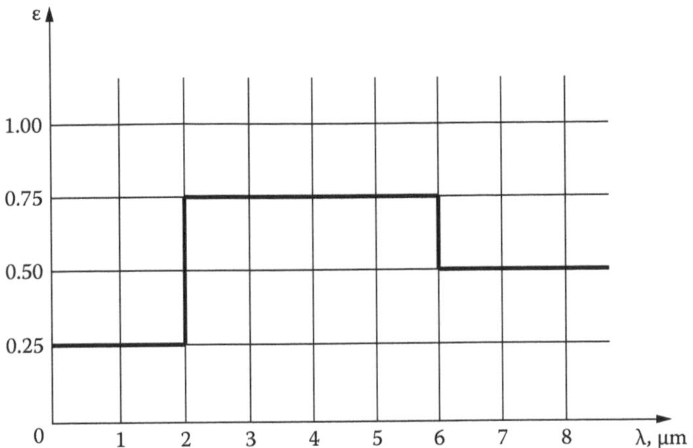

FIGURE 26.7
Emissivity for Example 26.2.

26.5 Monochromatic and Total Emissivity and Absorptivity

26.5.1 Emissivity

If $E_\lambda(\lambda T)$ is denoted as the **hemispherical monochromatic emissive power** of a non-black surface, then the **hemispherical monochromatic emissivity** is defined as the ratio of $E_\lambda(\lambda, T)$ to the hemispherical monochromatic emissive power of a black surface at the same temperature and wavelength

$$\epsilon_\lambda(\lambda, T) = \frac{E_\lambda(\lambda, T)}{E_{b\lambda}(\lambda, T)} \tag{26.15}$$

An even more global measure of the difference between a real surface and a black surface at the same temperature is the **total hemispherical emissivity,** $\epsilon(T)$

$$\epsilon(T) = \frac{E(T)}{E_b(T)} = \frac{1}{\sigma T^4}\int_0^\infty E_\lambda(\lambda, T)d\lambda = \frac{\int_0^\infty \epsilon_\lambda(\lambda, T)E_{b\lambda}(\lambda, T)d\lambda}{\sigma T^4} \tag{26.16}$$

Example 26.2 A diffuse emitter at 1800 K possesses a discontinuous hemispherical monochromatic emissivity of 0.25 from 0 μm to 2 μm, 0.75 from 2 μm to 6 μm and 0.50 above 6 μm. Determine (a) the total hemispherical emissivity and (b) the total emissive power.

Solution

Assumptions and Specifications

(1) Steady-state conditions exist.
(2) The surface is a diffuse emitter.
(3) Radiation is the only mode of heat transfer.

(a) The total hemispherical emissivity is given by Equation 26.16. Because the values of the monochromatic emissive power are constants within the three wavelength bands, we may write

$$\epsilon = \frac{\epsilon_1 \int_0^{\lambda_1 T} E_{b\lambda}(\lambda, T)}{\sigma T^4} + \frac{\epsilon_2 \int_{\lambda_1, T}^{\lambda_2, T} E_{b\lambda}(\lambda, T)}{\sigma T^4} + \frac{\epsilon_3 \int_{\lambda_2, T}^{\infty} E_{b\lambda}(\lambda, T)}{\sigma T^4}$$

or

$$\epsilon = \epsilon_1 F_{0\text{-}\lambda_1 T} + \epsilon_2 (F_{0\text{-}\lambda_2 T} - F_{0\text{-}\lambda_1 T}) + \epsilon_3 (F_{0\text{-}\lambda_3 T} - F_{0\text{-}\lambda_2 T})$$

With

$$\lambda_1 T = (2\,\mu\text{-m})(1800\,\text{K}) = 3600\,\mu\text{m-K}$$

$$\lambda_2 T = (6\,\mu\text{-m})(1800\,\text{K}) = 10{,}800\,\mu\text{m-K}$$

and $\lambda_3 T = \infty$, Table 26.1 gives

$$F_{0\text{-}3600} = 0.4036, \quad F_{0\text{-}10,800} = 0.9283, \quad \text{and} \quad F_{0\text{-}\infty} = 1.0000$$

Thus,

$$\begin{aligned}\epsilon &= 0.25(0.4036) + 0.75(0.9283 - 0.4036) + 0.50(1.0000 - 0.9283)\\ &= 0.1009 + 0.75(.5247) + 0.50(0.0717)\\ &= 0.1009 + 0.3935 + 0.0359 = 0.530 \Longleftarrow\end{aligned}$$

(b) The total emissive power is

$$\begin{aligned}E &= \epsilon \sigma T^4\\ &= 0.530(5.670 \times 10^{-8}\,\text{W/m}^2\text{-K}^4)(1800\,\text{K})^4\\ &= 315.6\,\text{kW/m}^2 \Longleftarrow\end{aligned}$$

26.5.2 Absorptivity

In dealing with the hemispherical monochromatic emissivity, we consider the hemispherical irradiation, G_λ, incident on a surface and define the **hemispherical monochromatic absorptivity** as the fraction of this incident irradiation that is absorbed, $G_{a\lambda}(\lambda, T)$

$$\alpha_\lambda(\lambda, T) = \frac{G_{a\lambda}(\lambda, T)}{G_\lambda(\lambda, T)} \tag{26.17}$$

Here we note that the absorptivity is also a function of both the wavelength and the temperature but that the wavelength to be employed is the wavelength of the incident irradiation.

The **total hemispherical absorptivity** will be

$$\alpha(T) = \frac{G_a(T)}{G(T)} = \frac{\int_0^\infty G_{a\lambda}(\lambda, T)d\lambda}{G(T)} \tag{26.18}$$

26.5.3 The Gray Surface or Gray Body

A **gray surface** or gray body with temperature, T, is a surface or body whose hemispherical monochromatic emissivity is independent of wavelength

$$\epsilon_\lambda(\lambda, T) = \epsilon_\lambda(T) \tag{26.19}$$

This infers that, in addition to satisfying the requirement of Equation 26.19, the gray surface or body is an opaque diffuse emitter, absorber, and reflector.

26.6 Heat Flow between Blackbodies

26.6.1 The Shape Factor

The Stefan-Boltzmann law applies to emission from a single surface. In this section, we consider the interchange of thermal energy between two surfaces such that not all of the radiation from one is intercepted by the other.

Consider Figure 26.8, which shows two surfaces, S_1 and S_2, separated by a nonabsorbing medium. The distance between the emitting surface, dS_1, and the receiving surface, dS_2, is designated by r. The rate of radiant energy interchange between dS_1 and dS_2 is

$$d\dot{Q}_{12} = I_1 \cos\theta_1 dS_1 d\omega_{12}$$

where $d\omega_{12}$ is the solid angle subtended by dS_2 with respect to dS_1. This is equal to the projected area of the receiving surface divided by the square of the distance between the surfaces

$$d\omega_{12} = \frac{dS_2 \cos\phi_2}{r^2}$$

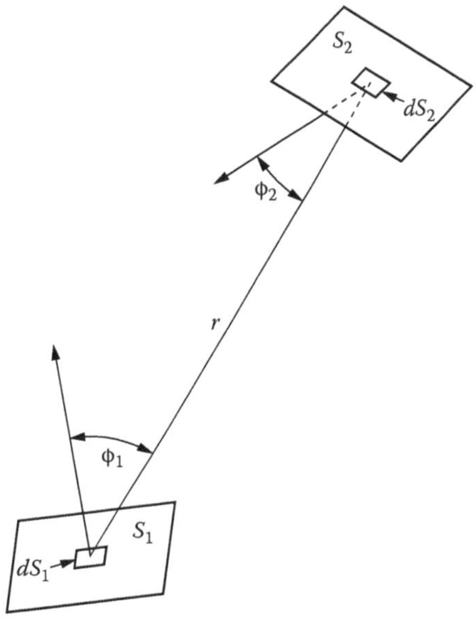

FIGURE 26.8
Geometric shape factor notation.

so that

$$d\dot{Q}_{12} = \frac{I_1 \cos\phi_1 \cos\phi_2 dS_1 dS_2}{r^2}$$

and because $I_1 = E_1/\pi$

$$d\dot{Q}_{12} = E_1 dS_1 \left(\frac{\cos\phi_1 \cos\phi_2 dS_2}{\pi r^2}\right) \tag{26.20a}$$

and we consider the term within the parentheses as the fraction of the total emission from dS_1 that is intercepted by dS_2.

It can be shown in a similar manner that

$$d\dot{Q}_{21} = E_2 dS_2 \left(\frac{\cos\phi_1 \cos\phi_2 dS_1}{\pi r^2}\right) \tag{26.20b}$$

so that the net rate of radiant energy interchange between the differential elements can be obtained from the difference of Equations 26.20a and 26.20b

$$d\dot{Q} = (E_1 - E_2)\left(\frac{\cos\phi_1 \cos\phi_2 dS_1 dS_2}{\pi r^2}\right)$$

or

$$\dot{Q}_{12} = (E_1 - E_2)\int_{S_1}\int_{S_2} \frac{\cos\phi_1 \cos\phi_2 dS_1 dS_2}{\pi r^2} \tag{26.21}$$

The double integral may be written as $S_1 F_{12}$, where F_{12} is called the **shape factor** (also referred to as the **view factor**, the **configuration factor**, or the **arrangement factor**) with respect to area S_1. The value of F_{12} may be obtained from

$$F_{12} = \frac{1}{S_1}\int_{S_1}\int_{S_2} \frac{\cos\phi_1 \cos\phi_2}{\pi r^2} dS_1 dS_2 \tag{26.22a}$$

and, in a similar fashion, we may define F_{21} as

$$F_{21} = \frac{1}{S_2}\int_{S_2}\int_{S_1} \frac{\cos\phi_1 \cos\phi_2}{\pi r^2} dS_2 dS_1 \tag{26.22b}$$

and with the shape factor defined in this manner, the heat interchange will be

$$\dot{Q}_{12} = F_{12} S_1 (E_1 - E_2)$$

or in the case of heat exchange from body 2 to body 1

$$\dot{Q}_{21} = F_{21} S_2 (E_2 - E_1)$$

In general,

$$\dot{Q} = S_i F_{ij} \Delta E \tag{26.23}$$

and the equality

$$S_1 F_{12} = S_2 F_{21} \tag{26.24}$$

is known as the **reciprocity theorem**.

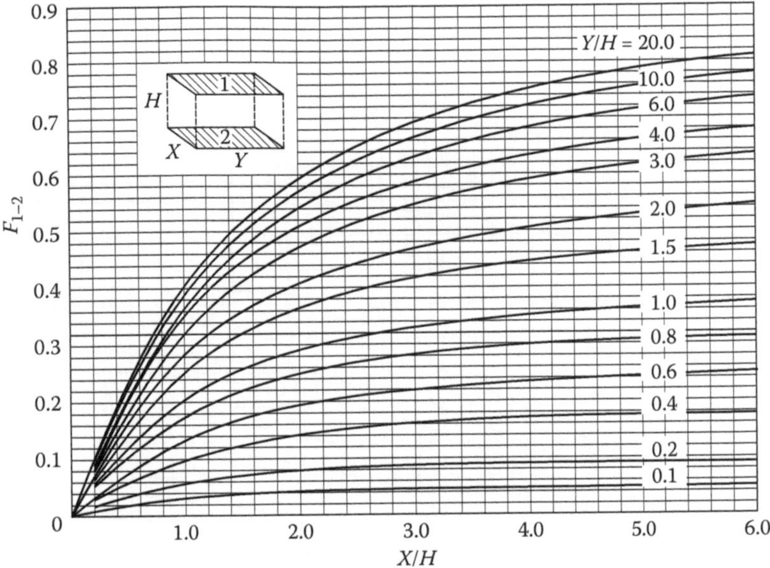

FIGURE 26.9
Shape factor between directly opposed rectangles.

The determination of the shape factor by means of the evaluation of the double integral is tedious even for the most simple configurations. Fortunately, however, the literature contains many references to specific configurations of practical interest (Howell, 1982; Mahan, 2002; Modest, 2003; Siegel and Howell, 2001). Figures 26.9 through 26.11 provide graphs of the shape factor for three common arrangements and relationships for nine shape factors are listed in Section 26.6.2.

Example 26.3 Determine the shape factor between a small area, dS_1 and a large, parallel area, S_2 of diameter, D. The areas are aligned on their centerlines and are R units apart.

Solution

Assumptions and Specifications

(1) Steady-state conditions exist.
(2) Both surfaces are diffuse.
(3) Radiation is the only mode of heat transfer.
(4) $S_2 \gg S_1$.

The configuration is displayed in Figure 26.12 where we note that the area dS_2 is in the shape of a ring with an area

$$dS_2 = x\, dx d\theta$$

and that

$$\phi_1 = \phi_2 = \phi, \quad r = (R^2 + x^2)^{1/2}, \quad \text{and} \quad \cos\phi = \frac{R}{(R^2 + x^2)^{1/2}}$$

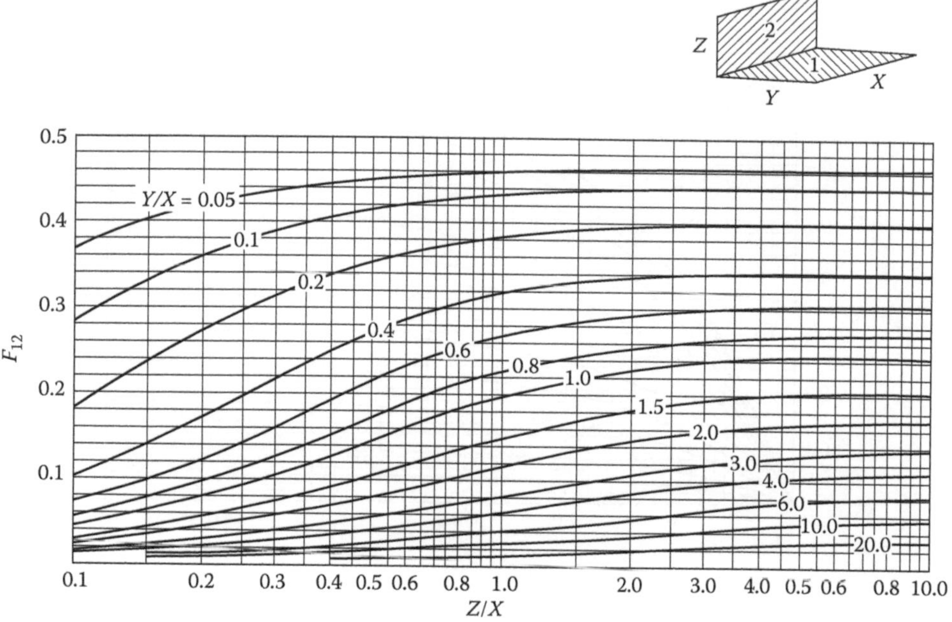

FIGURE 26.10
Shape factor between perpendicular rectangles with a common edge.

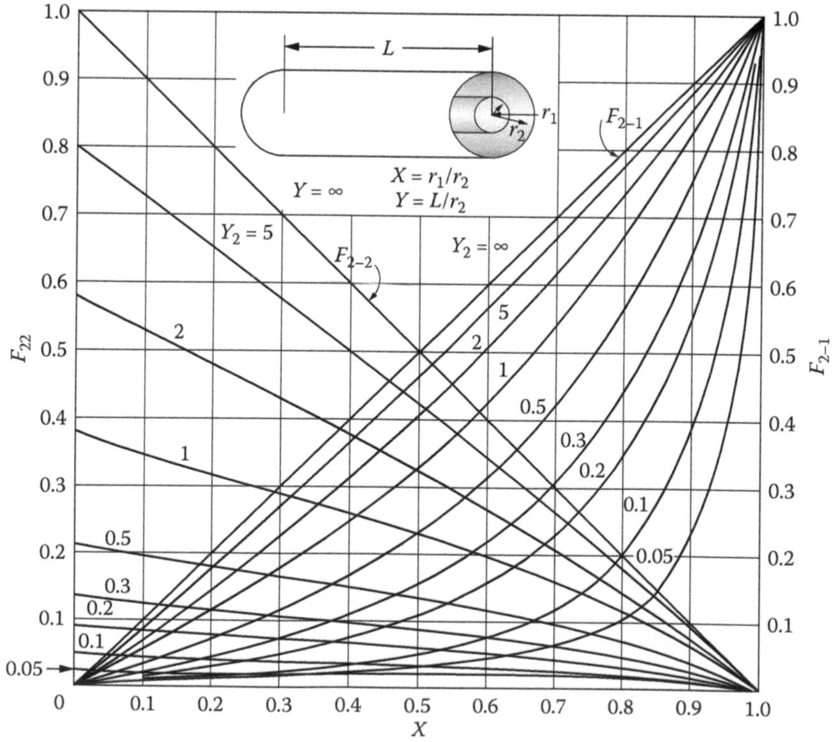

FIGURE 26.11
Shape factor between infinitely long concentric cylinders.

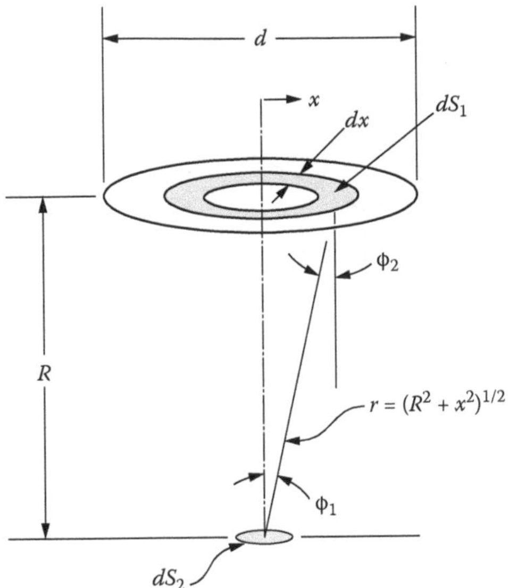

FIGURE 26.12
Radiation from a small area element to a disk for Example 26.3.

The shape factor will be given by Equation 26.22a

$$F_{12} = \frac{1}{S_1}\int_{S_1}\int_{S_2} \frac{\cos\phi_1 \cos\phi_2}{\pi r^2} dS_1 dS_2 \qquad (26.22a)$$

and with the foregoing substitutions

$$F_{12} = \int_{S_2} \frac{\cos^2\phi}{\pi r^2} dS_2 = \int_0^{d/2} \frac{2R^2 x dx}{(R^2+x^2)^{1/2}}$$

or

$$F_{12} = \frac{d^2}{4R^2 + d^2} \Longleftarrow$$

26.6.2 A Catalog of Simple Shape Factors in Two Dimensions

A tabulation of the relationships for the computation for nine two-dimensional shape factors extracted from Howell (1982) are given by Equations 26.25 to 26.33. By **two dimensional,** we mean that the dimension into the plane of the paper is infinite. The configurations for the nine cases are illustrated in Equations 26.24 through 26.33.

1. Two infinitely long plates of equal width joined at an edge

$$F_{12} = F_{21} = 1 - \sin\frac{\alpha}{2} \qquad \text{(Figure 26.13)} \qquad (26.25)$$

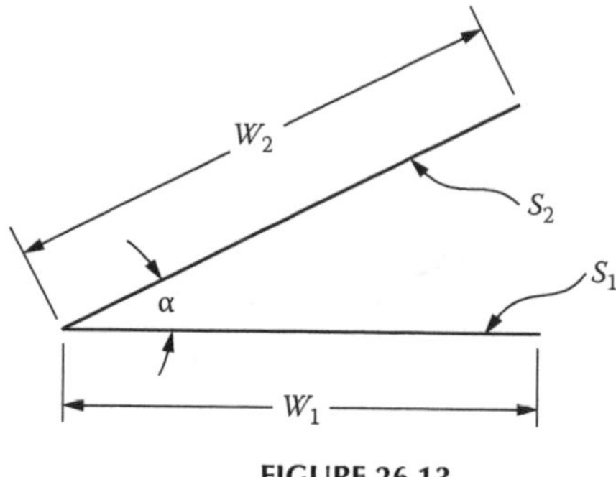

FIGURE 26.13

2. Two infinitely long plates of different widths, W_1 and W_2, joined at a common edge at an angle of 90°

$$F_{12} = \frac{1}{2}\left\{1 + \frac{W_1}{W_2} - \left[1 + \left(\frac{W_1}{W_2}\right)^2\right]^{1/2}\right\} \qquad \text{(Figure 26.14)} \qquad (26.26)$$

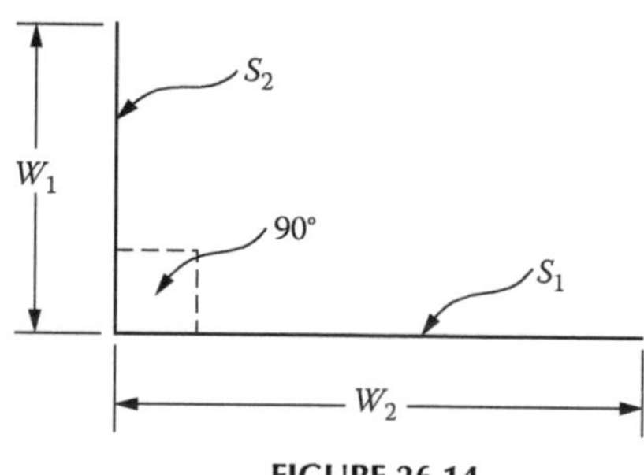

FIGURE 26.14

3. Triangular enclosure formed from three infinitely long plates of widths W_1, W_2, and W_3

$$F_{12} = \frac{W_1 + W_2 - W_3}{2W_1} \qquad \text{(Figure 26.15)} \qquad (26.27)$$

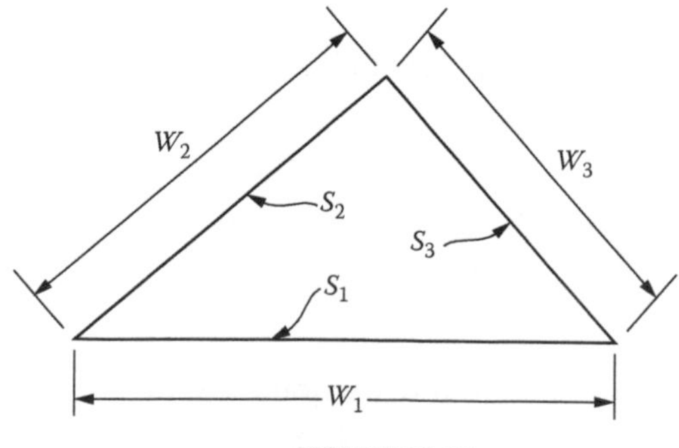

FIGURE 26.15

4. Disk of radius R and infinitesimal parallel disk positioned H units away on the disk centerline

$$F_{12} = \frac{R^2}{H^2 + R^2} \qquad \text{(Figure 26.16)} \tag{26.28}$$

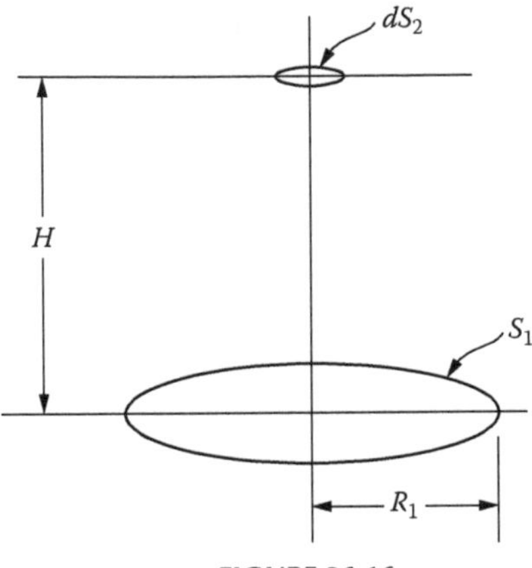

FIGURE 26.16

5. Two parallel disks of radii R_1 and R_2 positioned H units apart on the same centerline. With

$$x_1 = R_1/H, \quad x_2 = R_2/H, \quad \text{and} \quad X = 1 + \frac{1 + x_2^2}{x_1^2}$$

$$F_{12} = \frac{1}{2}\left\{X - \left[X^2 - 4\left(\frac{x_2}{x_1}\right)^2\right]^{1/2}\right\} \qquad \text{(Figure 26.17)} \tag{26.29}$$

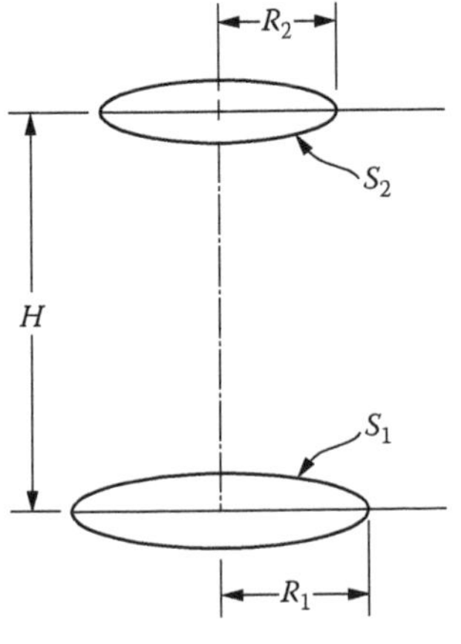

FIGURE 26.17

6. Infinitely long cylinder parallel to infinitely long plate with finite width, W

$$F_{12} = \frac{1}{2\pi}\left(\arctan\frac{L_1}{H} - \arctan\frac{L_2}{H}\right) \qquad \text{(Figure 26.18)} \qquad (26.30)$$

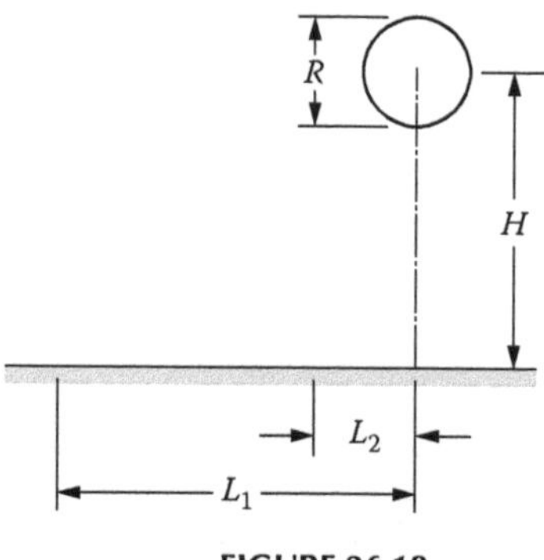

FIGURE 26.18

7. Two parallel and infinite cylinders of radius R with a clear spacing between them of L. With $X = 1 + L/2R$ and $Y = (X^2 - 1)^{1/2}$

$$F_{12} = F_{21} = \frac{1}{\pi}\left(Y + \arcsin\frac{1}{X} - X\right) \qquad \text{(Figure 26.19)} \qquad (26.31)$$

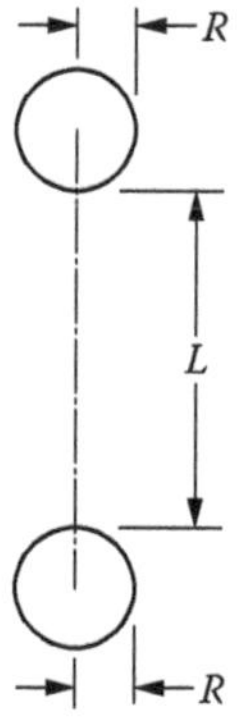

FIGURE 26.19

8. Row of equidistant infinitely long cylinders of diameter D with centerline spacing of W parallel to an infinite plane. With $x = D/W$

$$F_{12} = 1 - (1 - x^2)^{1/2} + x\arctan\left(\frac{1 - x^2}{x^2}\right)^{1/2} \qquad \text{(Figure 26.20)} \qquad (26.32)$$

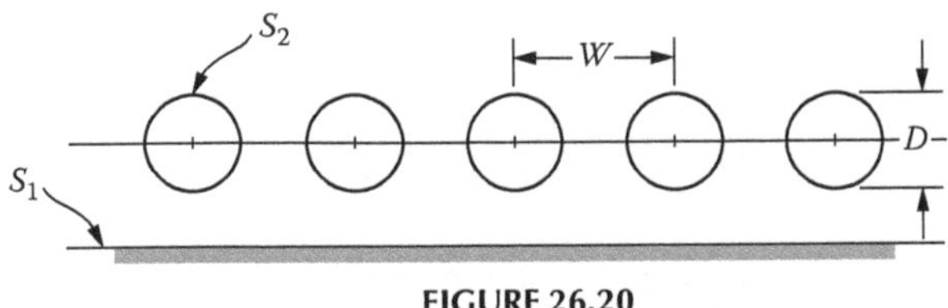

FIGURE 26.20

9. Sphere of radius R_1 positioned H units on the same centerline from a disc of radius R_2. With $X = R_2/H$

$$F_{12} = \frac{1}{2}\left[1 - (1 + X^2)^{-1/2}\right] \qquad \text{(Figure 26.21)} \tag{26.33}$$

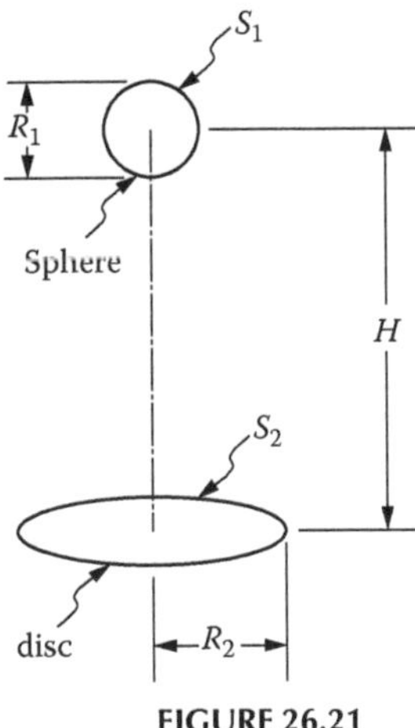

FIGURE 26.21

26.6.3 A Catalog of Simple Shape Factors in Three Dimensions

The relationships for three-dimensional shape factors with no restrictions placed on the dimension into the plane of the paper are provided in Equations 26.34 through 26.36.

1. Directly opposed rectangles of side dimensions X and Y located H units apart (Figure 26.9) with $x = X/H$ and $y = Y/H$

$$F_{12} = \frac{2}{\pi x y}\left[\Upsilon_1 + \Upsilon_2 + \Upsilon_3 - \Upsilon_4\right] \tag{26.34}$$

where

$$\Upsilon_1 = \ln\left[\frac{(1+x^2)(1+y^2)}{1+x^2+y^2}\right]^{1/2}$$

$$\Upsilon_2 = x(1+y^2)^{1/2}\arctan\frac{x}{(1+y^2)^{1/2}}$$

$$\Upsilon_3 = y(1+x^2)^{1/2}\arctan\frac{y}{(1+x^2)^{1/2}}$$

$$\Upsilon_4 = x\arctan x + y\arctan y$$

2. Perpendicular rectangles with a common edge (Figure 26.10) with $x = X/Z$ and $y = Y/Z$

$$F_{12} = \frac{1}{\pi y}\left[(\Upsilon_1 - \Upsilon_2 + \frac{1}{4}\ln(\Upsilon_3\Upsilon_4\Upsilon_5)\right] \quad (26.35)$$

where

$$\Upsilon_1 = y\arctan\frac{1}{y} + x\arctan\frac{1}{x}$$

$$\Upsilon_2 = (x^2+y^2)^{1/2}\arctan\frac{1}{(x^2+y^2)^{1/2}}$$

$$\Upsilon_3 = \frac{(1+x^2)(1+y^2)}{1+x^2+y^2}$$

$$\Upsilon_4 = \left[\frac{y^2(1+x^2+y^2)}{(1+y^2)(x^2+y^2)}\right]^{y^2}$$

$$\Upsilon_5 = \left[\frac{x^2(1+x^2+y^2)}{(1+x^2)(x^2+y^2)}\right]^{x^2}$$

3. Coaxial cylinders of finite length (Figure 26.11) with $x = r_2/r_1$, $y = L/r_1$, $A = y^2 + x^2 - 1$, and $B = y^2 - x^2 + 1$

$$F_{21} = \frac{1}{x} - \frac{1}{\pi x}\left[\Upsilon_1 - \frac{1}{2y}(\Upsilon_2 + \Upsilon_3)\right] \quad (26.36a)$$

where

$$\Upsilon_1 = \arccos\frac{B}{A}$$

$$\Upsilon_2 = (A^2 + 4A - 4x^2 + 4)^{1/2}\arccos\frac{B}{Ax}$$

$$\Upsilon_3 = B\arcsin\frac{1}{x} - \frac{\pi A}{2}$$

and

$$F_{22} = 1 - \frac{1}{x} + \frac{2}{\pi x}\Upsilon_4 - \frac{y}{2\pi x}(\Upsilon_5 - \Upsilon_6) \quad (26.36b)$$

where

$$\Upsilon_4 = \arctan \frac{2(x^2-1)^{1/2}}{y}$$

$$\Upsilon_5 = \frac{(4x^2+y^2)^{1/2}}{y} \arcsin \frac{4(x^2-1)+(y^2/x^2)(x^2-2)}{y^2+4(x^2-1)}$$

$$\Upsilon_6 = \arcsin \frac{x^2-2}{x^2} + \frac{\pi}{2}\left[\frac{(4x^2+y^2)^{1/2}}{y} - 1\right]$$

26.6.4 Properties of the Shape Factor

While the shape factor can be determined (often with great difficulty) from the relationships of Equation 26.22a,b, it is possible to deduce their value from one or more known factors that pertain to a related geometry. The use of this procedure is referred to as making use of **shape factor algebra** and is based on the **reciprocity, additivity**, and **enclosure** properties.

26.6.4.1 The Reciprocity Property

A modification of Equation 26.24 introduced in Section 26.6.1

$$S_1 F_{12} = S_2 F_{21} \tag{26.24}$$

is known as the reciprocity theorem and is the reciprocity property. It shows that, in a two-body system involving diffuse black surfaces, the heat emitted by the hot surface is equal to the heat absorbed by the cooler surface.

26.6.4.2 The Additivity Property

The additivity property considers the radiation interchange between a surface and another surface that may be subdivided into n subsurfaces

$$F_{12} = \sum_{i=1}^{n} F_{12,i} \tag{26.37}$$

If, for example, surface 2 is subdivided into three subsurfaces, S_{2a}, S_{2b}, and S_{2c}, the additivity property tells us that

$$F_{12} = F_{12a} + F_{12b} + F_{12c}$$

which shows that the shape factor F_{12} is the shape factor between surface 1 and *all* of the pieces of surface 2.

Equation 26.37 describes the additivity property.

26.6.4.3 The Enclosure Property

Suppose that a surface S_1, which may be concave so that it "sees" itself is completely surrounded by an enclosure containing n surfaces. Then the blackbody emission from surface 1, E_{b1}, will be intercepted by each surface, S_i and will be a fraction of E_{bi}. Hence,

$$E_{b1} S_1 = F_{11} S_1 E_{b1} + F_{12} S_1 E_{b1} + F_{13} S_1 E_{b1} + \cdots$$

and after we divide by the product $E_{b1} S_1$ we have

$$1 = F_{11} + F_{12} + F_{13} + \cdots$$

or

$$\sum_{j=1}^{n} F_{ij} = 1 \qquad (i = 1, 2, 3, \ldots, n) \tag{26.38}$$

Equation 26.38 describes the enclosure property and, in it, we note that the subscript, i, refers to the emitting surface and the subscript, j, accounts for all of the other surfaces that contribute to the enclosure. We reemphasize that if none of the surfaces "see" themselves, then $F_{11}, F_{22}, F_{33}, \ldots, F_{nn}$ will not be present.

A simple application of the reciprocity and enclosure properties is now provided in Example 26.4.

Example 26.4 Consider (a) a pair of infinite rectangular plates, S_1 and S_2 and (b) an infinite cylinder, S_1, within another infinite cylinder, S_2. In both cases, displayed in Figure 26.22, the end effects are negligible. Determine F_{12}, F_{21}, F_{11}, and F_{22}.

Solution

Assumptions and Specifications

(1) Steady-state conditions exist.
(2) All surfaces are diffuse.
(3) Radiation is the only mode of heat transfer.
(4) The plates and cylinders are of infinite extent.

(a) Because the plates are of infinite extent with $S_1 = S_2$ and since the end effects are negligible, by reciprocity

$$F_{12} = F_{21} = 1 \Longleftarrow$$

Then, because the plates don't "see" themselves

$$F_{11} = F_{22} = 0 \Longleftarrow$$

(b) All of the radiation from S_1 is intercepted by S_2. Hence,

$$F_{12} = 1 \Longleftarrow$$

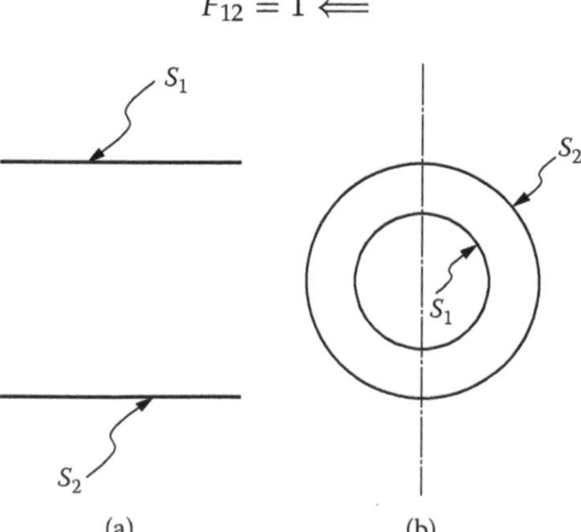

FIGURE 26.22
Configurations for Example 26.4: (a) an infinite pair of rectangular plates and (b) a pair of infinite cylinders.

and by reciprocity $S_1F_{12} = S_2F_{21}$ so that

$$F_{21} = \frac{S_1}{S_2}F_{12} = \frac{S_1}{S_2} \Longleftarrow$$

Because surface 1 does not see itself

$$F_{11} = 0 \Longleftarrow$$

and by the enclosure property

$$F_{22} = 1 - F_{21} = 1 - \frac{S_1}{S_2} \Longleftarrow$$

A straightforward application of the additivity property is provided in Example 26.5.

Example 26.5 Determine the shape factor F_{14} for the configuration of perpendicular rectangles indicated in Figure 26.23.

Solution

Assumptions and Specifications

(1) Steady-state conditions exist.
(2) All surfaces are diffuse.

With

$$S_5 = S_1 + S_2 \qquad \text{and} \qquad S_6 = S_3 + S_4$$

we have

$$S_5F_{54} = S_1F_{14} + S_2F_{24}$$

or

$$S_1F_{14} = S_5F_{54} - S_2F_{24} \tag{a}$$

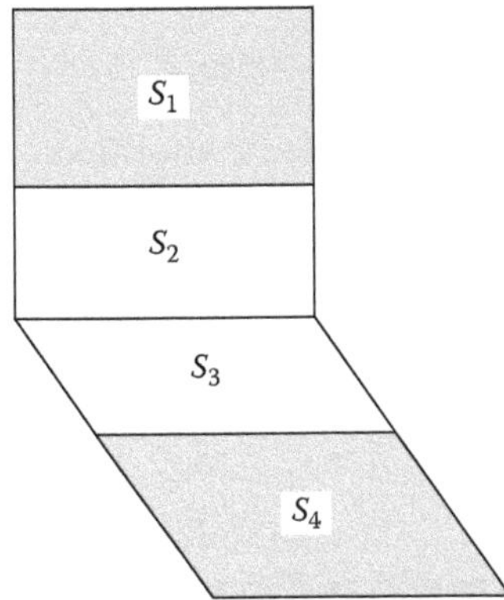

FIGURE 26.23
Configuration for Example 26.5.

Then

$$S_5F_{56} = S_5F_{54} + S_5F_{53}$$

or

$$S_5F_{54} = S_5F_{56} - S_5F_{53} \tag{b}$$

and

$$S_2F_{26} = S_2F_{24} + S_2F_{23}$$

or

$$S_2F_{24} = S_2F_{26} - S_2F_{23} \tag{c}$$

When we put (b) and (c) into (a), we obtain the result sought, which is in terms of shape factors that can be obtained from a graph, Equation 26.35 or software

$$F_{14} = \frac{1}{S_1}[S_5(F_{56} - F_{53}) - S_2(F_{26} - S_2F_{23})] \Longleftarrow$$

26.6.5 The Symmetry Property

If two surfaces, j and k, possess the same surface area and are symmetric about the surface, i, then $F_{ij} = F_{ik}$.

26.7 Heat Flow by Radiation between Two Bodies

26.7.1 Diffuse Blackbodies

For two diffuse blackbodies having surfaces, S_1 and S_2, the radiant heat exchange will be

$$E_{b1} - E_{b2} = \sigma(T_1^4 - T_2^4) \qquad (\text{W/m}^2)$$

where $T_1 > T_2$. This heat exchange may be multiplied by the surface area of either body with appropriate cognizance taken of the shape factor

$$\dot{Q}_{12} = \sigma S_1F_{12}(T_1^4 - T_2^4) = \sigma S_2F_{21}(T_1^4 - T_2^4) \quad (\text{W}) \tag{26.39a}$$

Moreover, if the surfaces are in full view of one another so that $F_{12} = F_{21} = 1$

$$\dot{Q}_{12} = \sigma S_1(T_1^4 - T_2^4) \qquad (\text{W}) \tag{26.39b}$$

26.7.2 Opaque Gray Bodies

If two gray surfaces are in full view of one another such as in the case of infinite parallel planes where $F_{12} = F_{21} = 1$, Equation 26.39b will not hold because the surfaces are not black. If the surfaces are opaque and gray, then

$$\rho = 1 - \alpha \qquad \text{and} \qquad \alpha = \epsilon$$

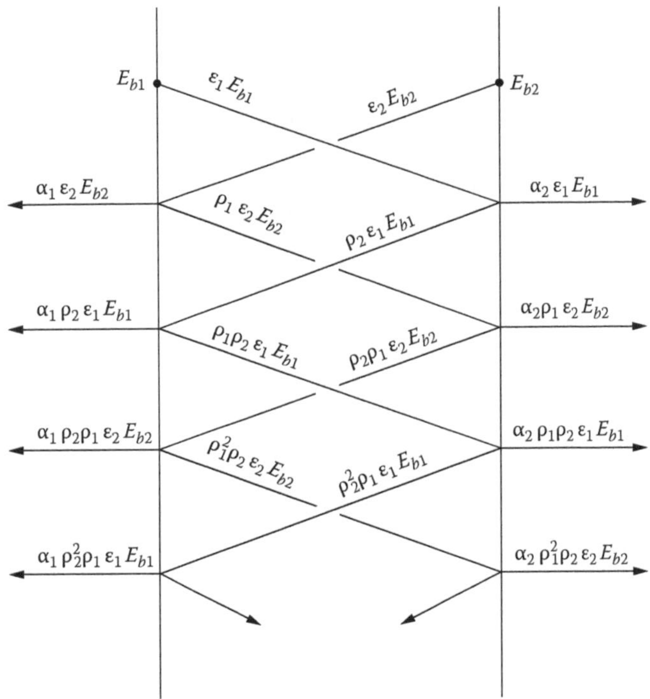

FIGURE 26.24
Reflection and absorption characteristics associated with two infinite parallel gray surfaces.

If surface 1 emits $\epsilon_1 E_{b1}$, surface 2 will absorb the fraction $\alpha_2 \epsilon_1 E_{b1}$ and reflect $\rho_2 \epsilon_1 E_{b2}$ back toward surface 1. Further reflections will occur and they are indicated in Figure 26.24 where a similar picture is shown for surface 2.

The net heat transferred per unit of surface 1 to surface 2 will be the original transmission $\epsilon_1 E_{b1}$ minus the fraction of $\epsilon_1 E_{b1}$ and of $\epsilon_2 E_{b2}$, which is ultimately absorbed by surface 1 after successive reflections. Hence, we have with $\rho_1 = 1 - \alpha_1 = 1 - \epsilon_1$ and $\rho_2 = 1 - \alpha_2 = 1 - \epsilon_2$

$$\begin{aligned}\dot{Q}_{12} = {} & S_1 \epsilon_1 E_{b1}[1 - \epsilon_1(1-\epsilon_2) - \epsilon_1(1-\epsilon_1)(1-\epsilon_2)^2 + \cdots] \\ & - S_2 \epsilon_2 E_{b2}[\epsilon_1 + \epsilon_1(1-\epsilon_1)(1-\epsilon_2) + \epsilon_1((1-\epsilon_1)^2(1-\epsilon^2)^2 + \cdots]\end{aligned}$$

and, if $S_1 = S_2 = S$, the two infinite series reduce such that

$$\dot{Q}_{12} = S(E_{b1} - E_{b2}) \left[\frac{1}{\epsilon_1} + \frac{1}{\epsilon_2} - 1 \right] \tag{26.40}$$

It is convenient to define a factor $\mathcal{F}_{12}$ (apparently first introduced by Hottel, 1954) to handle the fact that the surfaces are gray

$$\dot{Q}_{12} = \sigma S \mathcal{F}_{12} (T_1^4 - T_2^4) \tag{26.41}$$

where $\mathcal{F}_{12}$ can be considered as an **emissivity** factor. This factor is to be used for opaque gray configurations with surfaces in full view of one another in the same way as the shape factor is used for diffuse black surfaces. Indeed, for infinite parallel planes

$$\mathcal{F}_{12} = \frac{1}{\dfrac{1}{\epsilon_1} + \dfrac{1}{\epsilon_2} - 1} \tag{26.42}$$

and it can be shown that for a small body having surface, S_1, within a large enclosure, S_2

$$\mathcal{F}_{12} = \epsilon_1 \tag{26.43}$$

and for concentric cylinders or spheres, S_1 surrounded by S_2

$$\mathcal{F}_{12} = \frac{\epsilon_1}{1 + \dfrac{S_1}{S_2}\dfrac{\epsilon_1}{\epsilon_2}(1 - \epsilon_2)} \tag{26.44}$$

In this section, we have considered diffuse black surfaces (ideal emitters and absorbers) in geometries where not all of the radiant energy emitted by the source is intercepted by the receiver and gray opaque surfaces in full view of one another. Both of these considerations have pertained to a single source and a single receiver. The next two sections lead to an electrothermal analog due to Oppenheim (1956), which leads to expeditious calculations for multiple radiating and reradiating surfaces.

Example 26.6 A very thin radiation shield is to be placed between two opaque gray parallel planes of infinite extent as indicated in Figure 26.25. The shield has two surfaces denoted by the subscripts ℓ and r (for "left" and "right"). Determine the temperature of the shield if conduction through the shield is to be neglected and

$$T_1 = 1000\,\text{K} \qquad \epsilon_1 = 0.800 \qquad \epsilon_\ell = 0.625$$

$$T_2 = 400\,\text{K} \qquad \epsilon_2 = 0.750 \qquad \epsilon_r = 0.400$$

Solution

Assumptions and Specifications

(1) Steady-state conditions exist.

(2) All surfaces are opaque, gray, and diffuse.

(3) Radiation is the only mode of heat transfer.

(4) End effects are negligible.

In the steady state, the heat gained by the shield

$$\dot{Q}_{1\ell} = \sigma S \mathcal{F}_{1\ell}\left(T_1^4 - T_s^4\right)$$

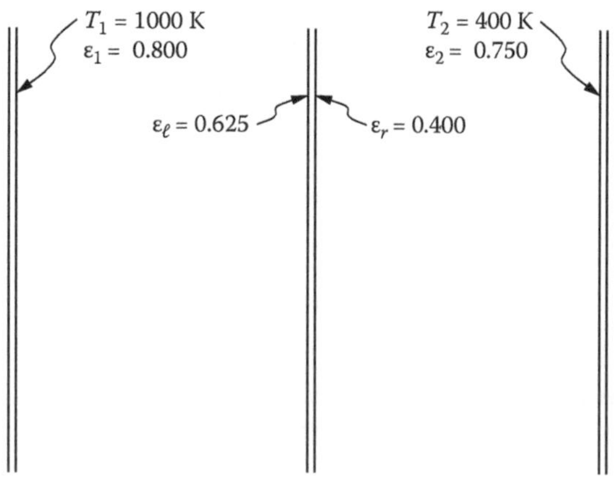

FIGURE 26.25
Configuration for Example 26.6.

is equal to the heat lost by the shield

$$\dot{Q}_{r2} = \sigma S \mathcal{F}_{r2}(T_s^4 - T_2^4)$$

where T_s is the shield temperature. A little algebra provides

$$\mathcal{F}_{1\ell}(T_1^4 - T_s^4) = \mathcal{F}_{r2}(T_s^4 - T_2^4)$$

$$\left(1 + \frac{\mathcal{F}_{r2}}{\mathcal{F}_{1\ell}}\right) T_s^4 = T_1^4 + \frac{\mathcal{F}_{r2}}{\mathcal{F}_{1\ell}} T_2^4$$

or

$$T_s = \left[\frac{T_1^4 + \dfrac{\mathcal{F}_{r2}}{\mathcal{F}_{1\ell}} T_2^4}{1 + \dfrac{\mathcal{F}_{r2}}{\mathcal{F}_{1\ell}}}\right]^{1/4}$$

Equation 26.40 can then be employed to give

$$\mathcal{F}_{1\ell} = 0.541, \quad \mathcal{F}_{r2} = 0.353, \quad \text{and} \quad \mathcal{F}_{r2}/\mathcal{F}_{1\ell} = 0.653$$

which leads to a shield temperature of 885.6 K $\Longleftarrow$

26.8 Radiosity and Irradiation

We have seen that the emissive power, whether total or monochromatic, refers only to the *original* emission from a surface. It does not include any additional components that are due to reflection or multiple reflections of any of the incident radiation.

Radiosity, denoted by J, is the term used to indicate all of the radiation leaving a surface (per unit time and per unit surface area). The total radiosity, J, may be obtained from the monochromatic radiosity, J_λ, via

$$J = \int_0^\lambda J_\lambda d\lambda$$

Moreover, we have seen in Section 26.3.1 that **irradiation** is the term used to denote the rate at which thermal radiation (again per unit time and per unit surface area) is incident upon a surface. This incident radiation, designated by G, may be the result of emissions and reflections from other surfaces. The total irradiation may be obtained from the monochromatic radiation, G_λ, from

$$G = \int_0^\lambda G_\lambda d\lambda$$

Both radiosity and irradiation are used in the next section, which employs a network method to discuss radiation within enclosures.

26.9 Radiation within Enclosures by a Network Method

For surfaces within an enclosure that obey Lambert's cosine law and reflect diffusively, a "network method" is quite useful for the evaluation of the multiple heat flow paths from surface to surface. We begin with the radiosity of the i surface

$$J_i = \rho_i G_i + \epsilon_i E_{bi} \tag{26.45}$$

where G_i is the irradiation, ρ_i and ϵ_i are the reflectivity and emissivity, and E_{bi} is the blackbody emissive power.

If both surface temperature and irradiation over the surface are uniform, then the net rate of heat loss or gain will be a function of the difference between the radiosity and the irradiation

$$\dot{Q}_i = S_i(J_i - G_i) \tag{26.46}$$

Moreover, if the surfaces are gray, we have the important limiting assumptions

$$\epsilon_i = \alpha_i \quad \text{and} \quad \rho_i = 1 - \epsilon_i$$

Equations 26.45 and 26.46 may be combined after we first observe from Equation 26.45 that

$$G_i = \frac{1}{\rho_i}(J_i - \epsilon_i E_{bi})$$

and hence,

$$\dot{Q}_i = S_i\left[J_i - \frac{1}{\rho_i}(J_i - \epsilon_i E_{bi})\right]$$

or

$$\dot{Q}_i = \frac{S_i}{\rho_i}\left[\epsilon_i E_{bi} + (\rho_i - 1)J_i\right] = \frac{S_i\epsilon_i}{1-\epsilon_i}(E_{bi} - J_i) \tag{26.47}$$

When we recast Equation 26.47 into the form

$$\dot{Q}_i = \frac{E_{bi} - J_i}{\dfrac{1-\epsilon_i}{S_i\epsilon_i}} \tag{26.48}$$

we see that an analogy may be drawn that shows $E_{bi} - J_i$ as a difference of potential, and the fraction $(1-\epsilon_i)/S_i\epsilon_i$ as a radiation thermal resistance. This resistance governs the flow of radiant energy, $\dot{Q}_i$, between a node point representing the blackbody emission, $E_{bi} = \sigma T_i^4$ and another node point representing the radiosity, J_i.

The direct radiative interchange between surface i and surface j is governed by the difference between the two radiosities, J_i and J_j, the shape factor, F_{ij}, and the surface, S_i, according to

$$\dot{Q}_{ij} = \frac{J_i - J_j}{1/S_i F_{ij}}$$

The reciprocity relationship, Equation 26.19, will allow this to be written as

$$\dot{Q}_{ij} = \frac{J_i - J_j}{1/S_i F_{ij}} = \frac{J_i - J_j}{1/S_j F_{ji}} \tag{26.49}$$

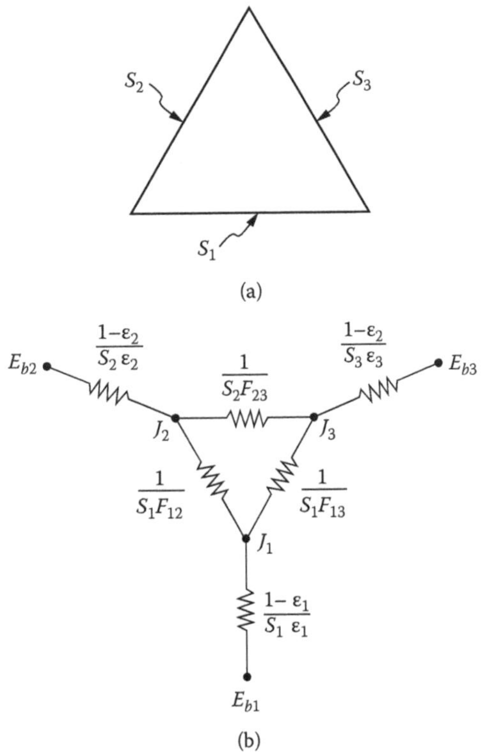

FIGURE 26.26
(a) An arrangement of three radiating surfaces and (b) the electrothermal analog network.

Here, the radiative thermal surface-to-surface resistance is $1/S_i F_{ij} = 1/S_j F_{ji}$. This resistance governs the flow of heat between two node points, each represented by a radiosity.

Equations 26.48 and 26.49 form the basis for the electrical analog network, which was first proposed by Oppenheim (1956). An enclosure with three surfaces is shown in Figure 26.26a and the electrothermal analog network is provided in Figure 26.26b. The nodes marked E_{b1}, E_{b2}, and E_{b3} represent heat sources and each of these heat sources may be related to a temperature, $E_{bi} = \sigma T_i^4$. The resistances $(1 - \epsilon_i)/S_i e_i$ "transfer" the E_{bi}'s to the surface radiosities, J_i, thereby accounting for the fact that the surfaces are not black but gray. The resistances, $1/S_i F_{ij} = 1/S_j F_{ji}$, link the radiosities and account for the actual radiative interchange between surface i and surface j.

Example 26.7 shows the set-up of the actual circuit. It should be kept in mind that the evaluation of the radiosities is only rarely required and that the object of the game is the determination of the surface temperatures.

Example 26.7 Use the network method to determine the temperature of the shield in Example 26.6.

Solution

Assumptions and Specifications

(1) Steady-state conditions exist.
(2) All surfaces are opaque, gray, and diffuse.
(3) Radiation is the only mode of heat transfer.
(4) End effects are negligible.

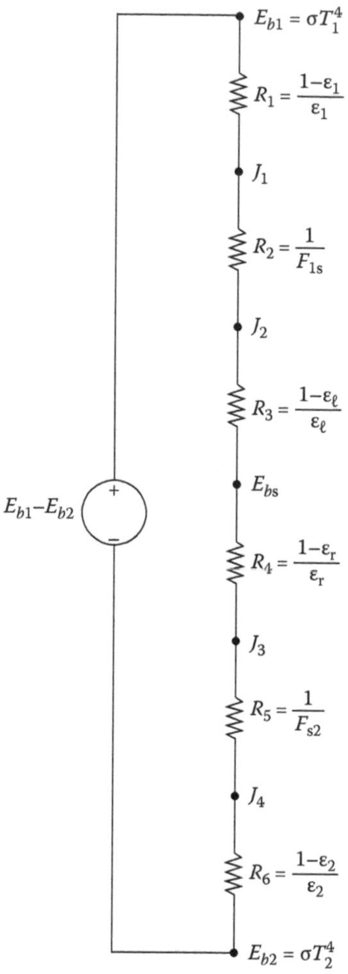

FIGURE 26.27
The electrothermal analog network for Example 26.7.

The analog circuit is displayed in Figure 26.27 where we note the designators for the various resistances. With the same data as in Example 26.6,

$$T_1 = 1000\,\text{K} \qquad \epsilon_1 = 0.800 \qquad \epsilon_\ell = 0.625$$

$$T_2 = 400\,\text{K} \qquad \epsilon_2 = 0.750 \qquad \epsilon_r = 0.400$$

we have for $S = 1\,\text{m}^2$ and both of the walls in full view of the shield

$$R_1 = \frac{1-\epsilon_1}{\epsilon_1} = \frac{1-0.800}{0.800} = 0.2500\,\text{m}^{-2}$$

$$R_2 = \frac{1}{F_{1s}} = \frac{1}{1} = 1.0000\,\text{m}^{-2}$$

$$R_3 = \frac{1-\epsilon_\ell}{\epsilon_\ell} = \frac{1-0.625}{0.625} = 0.6000\,\text{m}^{-2}$$

$$R_4 = \frac{1-\epsilon_r}{\epsilon_r} = \frac{1-0.400}{0.400} = 1.5000\,\text{m}^{-2}$$

$$R_5 = \frac{1}{F_{s2}} = \frac{1}{1} = 1.0000\,\text{m}^{-2}$$

and

$$R_6 = \frac{1-\epsilon_2}{\epsilon_2} = \frac{1-0.750}{0.750} = 0.3333\,\text{m}^{-2}$$

and the sum of these six resistances is

$$\sum R = 4.6833\,\text{m}^{-2}$$

The heat flow from wall 1 to wall 2 will be

$$\dot{Q}_{12} = \frac{E_{b1}-E_{b2}}{\sum R} = \frac{\sigma(T_1^4 - T_2^4)}{\sum R}$$

or

$$\begin{aligned}\dot{Q}_{12} &= \frac{(5.67\times10^{-8}\,\text{W/m}^2\text{-K}^4)[(1000\,\text{K})^4 - (400\,\text{K}^4)]}{4.6833\,\text{m}^{-2}}\\ &= (1.211\times10^{-8}\,\text{W/K}^4)[1.000\times10^{12}\text{K}^4 - 2{,}560\times10^{10}\text{K}^4]\\ &= (1.211\times10^{-8}\,\text{W/K}^4)(9.744\times10^{11}\,\text{K}^4)\\ &= 11.797\,\text{kW}\end{aligned}$$

We can obtain the shield temperature from the temperature drop through the first three resistances

$$\begin{aligned}E_{b1} - E_{bs} &= \dot{Q}_{12}(R_1 + R_2 + R_3)\\ &= (11.797\,\text{kW})(0.250\,\text{m}^{-2} + 1.000\,\text{m}^{-2} + 0.600\,\text{m}^{-2})\\ &= (11.797\,\text{kW})(1.850\,\text{m}^{-2})\\ &= 21.83\,\text{kW/m}^2\end{aligned}$$

Thus,

$$\begin{aligned}E_{bs} &= E_{b1} - 21.83\,\text{kW/m}^2\\ &= \sigma T_1^4 - 21.83\,\text{kW/m}^2\\ &= (5.67\times10^{-8}\,\text{W/m}^2\text{-K}^4)(1000\,\text{K})^4 - 21.83\,\text{kW/m}^2\\ &= 56.70\,\text{kW/m}^2 - 21.83\,\text{kW/m}^2\\ &= 34.87\,\text{kW/m}^2\end{aligned}$$

Then

$$T_s^4 = \frac{34.87\,\text{kW/m}^2}{\sigma} = \frac{34.87\,\text{kW/m}^2}{5.67\times10^{-8}\,\text{W/m}^2\text{-K}^4} = 6.151\times10^{11}\text{K}^4$$

and

$$T_s = 885.6\,\text{K} \Longleftarrow$$

This value checks the value obtained in Example 26.6 and we note that for those of the electrical persuasion, the voltage divider will work as long as due cognizance is taken of the fourth-power dependency of the emissive powers.

26.10 Summary

Thermal radiation is but one form of electromagnetic radiation. The narrow band of radiation between $0.36 < \lambda < 0.76\,\mu\text{m}$ is known as the visible region, whereas the band between $0.1\,\mu\text{m} < 100\,\mu\text{m}$ is generally considered as thermal radiation; and this range includes the infrared and a portion of the ultraviolet components of the electromagnetic spectrum.

A blackbody is an ideal surface that absorbs all and reflects none of the radiation that falls upon it. The monochromatic emissive power, which is a function of both wavelength and temperature, is given by Planck's law

$$E_{b\lambda}(\lambda, T) = \frac{2\pi C_1 c^2}{\lambda^5 (e^{cC_1/C_2\lambda T} - 1)} \qquad (\text{W/m}^2) \tag{26.2}$$

where

$$C_1 = 6.624 \times 10^{-27}\,\text{erg}$$

and

$$C_2 = 1.380 \times 10^{-16}\,\text{erg/K}$$

The wavelength of maximum monochromatic emissive power at a particular temperature is given by Wien's displacement law

$$\lambda T = 2897.8\,\mu\text{-K} \tag{26.3}$$

and the quantity of blackbody radiation at a particular temperature is given by the Stefan-Boltzmann law

$$E_b = \sigma T^4 \tag{26.4}$$

The fraction of emission falling between wavelengths of λ_1 and λ_2 is given by

$$F_{\lambda_1\text{-}\lambda_2} = F_{0\text{-}\lambda_2 T} - F_{0\text{-}\lambda_1 T} \tag{26.6}$$

with values tabulated as a function of λT in Table 26.1.

Three system properties that concern radiation and pertain to a surface are the absorptivity, α, the transmissivity, τ, and the reflectivity, ρ. These have a magnitude less than or equal to unity and are related by

$$\alpha + \tau + \rho = 1 \tag{26.7}$$

Kirchhoff's law states that for n small blackbodies in thermal equilibrium in a blackbody enclosure

$$E_b = \frac{E_1}{\alpha_1} = \frac{E_2}{\alpha_2} + \cdots + \frac{E_n}{\alpha_n}$$

The emissivity of a surface is related to the blackbody emissive power and is defined by

$$\epsilon = \frac{E}{E_b}$$

and for a small body emitting E_1 in a blackbody enclosure at the same temperature

$$\alpha_1 = \epsilon_1$$

Lambert's cosine law for the intensity of radiation is

$$E = \pi I \tag{26.16}$$

A gray body having a gray surface is a surface whose emissivity is independent of wavelength.

The shape factor (also referred to as the view factor, the configuration factor, or the arrangement factor) between two surfaces accounts for the fact that not all of the radiation emitted by one surface is intercepted by the other. Several shape factors are provided in Sections 26.6.1 and 26.6.2. Shape factors have properties that allow an analyst to employ shape factor algebra. These properties are the reciprocity property

$$S_1 F_{12} = S_2 F_{21} \tag{26.19}$$

the additive property

$$F_{12} = F_{12a} + F_{12b} + F_{12c}$$

where surface 2 is divided into n subsurfaces, the enclosure property

$$\sum_{j=1}^{n} F_{ij} = 1 \qquad (i = 1, 2, 3, \ldots, n) \tag{26.38}$$

and the symmetry property, which states that if two surfaces j and k possess the same area and are symmetric about a surface, i, then $F_{iy} = F_{ik}$.

Radiosity, denoted by J, is defined as the total radiation leaving a surface and irradiation, denoted by, G, is defined as the total radiation incident on a surface. An electrothermal analog method may be employed for situations where the surfaces are not in full view of one another and where the surfaces are not blackbodies. The analog is based on surface blackbody node points represented by $E_{b,i}$ and their associated radiosities, J_i. The connection between $E_{b,i}$ and J_i is governed by a resistance, $R = (1 - \epsilon_1)/S_i\epsilon_i$. The radiosities, J_i and J_j, are connected by a resistance, $R = 1/S_i F_{ij} = 1/S_j F_{ji}$.

26.11 Problems

Emissive Power and Spectral Characteristics

26.1: Determine the monochromatic emissive power for a black surface at a wavelength of 10 μm at temperatures of (a) 200 K, (b) 500 K, and (c) 1000 K.

26.2: In an experiment, the monochromatic emissive power of a blackbody is measured to be 5010 W/m^2-μm-K^4 at a wavelength of 1.5 μm. Determine (a) the temperature of the blackbody, (b) the wavelength at which the maximum monochromatic emissive power occurs, and (c) the maximum monochromatic emissive power.

26.3: Determine the peak wavelengths and the peak monochromatic emissive power of black surfaces at (a) 400 K, (b) 1200 K, (c) 3000 K, and (d) 3500 K.

26.4: Determine the radiation emitted, in W/m^2 between 4 and μm by black surfaces at (a) 400 K, (b) 800 K, and (c) 1200 K.

26.5: The fraction of the radiation emitted between 2 μm and a higher but unknown wavelength, λ, is 0.25 for a black surface at 800 K. Determine the unknown wavelength.

26.6: Determine the total emissive power of a blackbody at 600°C and 1200°C.

26.7: The so-called solar constant measured at the fringe of the earth's atmosphere located 93×10^6 miles from the center of the sun is 1394 W/m^2. Determine its value at (a) Venus (mean orbital radius of 36×10^6 miles) and (b) Pluto (mean orbital radius of 3670×10^6 miles).

26.8: Suppose that the filament of an ordinary 100-W lightbulb is at 2910 K and that the bulb can be approximated as a blackbody. Determine (a) the wavelength of maximum emission and (b) the fraction of emission in the visible region of the spectrum.

26.9: For a blackbody maintained at 227°C, determine (a) the total emissive power, (b) the wavelength for maximum emissive power, (c) the magnitude of the maximum emissive power, and (d) the fraction of the emission between wavelengths of 1.25 and 4.25 μm.

26.10: Rework Problem 26.9 for a blackbody temperature of 1027°C.

26.11: The radiation emitted by a blackbody at 620°C is 5408 W/m^2 between wavelengths of λ and $\lambda = \infty$. Determine (a) the wavelength λ and (b) the radiation emitted between $\lambda = 0$ and λ in part (a).

26.12: The inside surface of a large spherical enclosure is isothermal and maintained at 1200 K. Determine (a) the radiant heat flux, in W/m^2, from a small opening on the enclosure, (b) the fraction of this heat flux that is confined between 2 and 6 μm, and (c) the maximum monochromatic emissive power associated with the emission.

26.13: A small circular hole is to be drilled in the surface of a large, hollow, spherical enclosure maintained at 2000°C. It is required that 100 W emerge from the hole. Determine (a) the hole diameter, (b) the number of watts emitted in the visible range from 0.6 to 0.7 μm, (c) the ultraviolet range between 0.0 and 0.4 μm, and (d) the infrared range between 0.7 and 100 μm.

26.14: A 100-cm-diameter thin disk is maintained at a uniform temperature of 427°C. Both surfaces of the disk behave as black surfaces. Determine the total radiant power, in watts, emitted from the disk in the wavelength bands (a) 2–4 μm, (b) 6–8 μm, and (c) 10–20 μm.

26.15: Consider Planck's law

$$E_{b\lambda} = \frac{2\pi C_1 c^2}{\lambda^5 [e^{cC_1/C_2\lambda T} - 1]}$$

where C_1 and C_2 are constants and c is the speed of light. Show that for very large values of λT, $E_{b\lambda}$ may be approximated by

$$E_{b\lambda} = \frac{2\pi c C_1 c T}{C_2 \lambda^4}$$

and for very small values of λT the approximation will be

$$E_{b\lambda} = \frac{2\pi c^2 C_1}{\lambda^5 e^{cC_2/\lambda T}}$$

26.16: A substantial cavity with a small opening that is 0.0025 m^2 in area emits 8 W. Determine the wall temperature of the cavity.

26.17: Determine the wavelength of maximum emission for (a) the sun with an assumed temperature of 5790 K, (b) a lightbulb filament at 2910 K, (c) a surface at 1550 K, and (d) the human skin at 308 K.

26.18: For a blackbody surface at 1660 K, determine the fraction of the emission that occurs from (a) 0–1.25 μm, (b) 1.25–3.25 μm, (c) 3.25–6.50 μm, and (d) 6.50–2.5 μm.

26.19: The monochromatic emissive power of a surface varies with wavelength as $\epsilon = 0.125$ for 0–0.625 μm, $\epsilon = 0.515$ for 0.625–5.15 μm, $\epsilon = 0.700$ for 5.15–12.5 μm, and $\epsilon = 0.875$ for 12.5–∞ μm. Determine the total emissivity of the surface when the temperature is (a) 900 K, (b) 1800 K, and (c) 2700 K.

26.20: Assuming the sun to be a blackbody emitter at 5800 K, determine the fraction of the sun's emission that falls in the visible region of the spectrum.

26.21: A furnace that has black interior walls maintained at 1227°C contains a peephole with a diameter of 10 cm. The glass in the peephole has a transmissivity of 0.78 between 0 and 3.2 μm and 0.08 between 3.2 μm and ∞. Determine the heat lost through the peephole.

26.22: Determine the temperature required for a blackbody emitter that will allow 40% of the energy emitted to fall between and 12 μm.

Radiation Properties and Kirchhoff's Law

26.23: A horizontal opaque rectangular plate that is 15 cm × 20 cm receives radiation from its surroundings at the rate of 2000 W. Only one side of the plate is radiatively active and, the amount of radiation absorbed by the plate is 500 W. Determine (a) the irradiation, G in W/m^2, (b) the emissivity, ϵ, and (c) the reflectivity, ρ.

26.24: A hemispherical cavity of radius 2 cm is machined into a block. The radiant heat flux leaving the cavity is 35,000 W/m^2, and the cavity has an emissivity of 0.70. Determine the inside surface temperature of the cavity.

26.25: An opaque circular plate, 8 cm in diameter is irradiated by 10 W on its top surface having a reflectivity of 0.22. The top surface is at a uniform temperature of 120°C. Determine (a) the radiant heat flux in W/m^2 leaving the top surface and (b) the emissivity of the top surface.

26.26: In Problem 26.25, suppose that the top surface loses heat by convection to the surroundings at 20°C, which provides a heat transfer coefficient 0f 65 W/m^2-K. Determine the radiant heat flux in W/m^2.

26.27: A transparent plate is irradiated with 800 W of radiant energy. The amount of energy absorbed is 200 W. The reflectivity of the plate is 50% of its absorptivity. Determine the transmissivity of the plate.

26.28: The total radiant energy incident on a surface is 2400 W. The surface absorbs 825 W/m^2 and reflects 600 W/m^2. Determine the transmissivity.

26.29: A set of reflectivity measurements for a solid surface at 800°C shows $\rho = 0.12(0 < \lambda < 3.2\,\mu\text{m})$, $\rho = 0.335(3.2 < \lambda < 6.4\,\mu\text{m})$, $\rho = 0.615(6.4 < 8.9\,\mu\text{m})$, and $\rho = 0.98(8.9 < \lambda < \infty\,\mu\text{m})$. Determine the total emissive power.

26.30: The reflectivity and transmissivity of a gray diffuse surface are respectively one half and one third of its absorptivity. Determine the absorptivity of the surface.

26.31: A triangular opaque horizontal plate having an emissivity of 0.72 is maintained at a temperature of 700°C. Each side of the plate measures 15 cm. The irradiation on the plate is 5200 W/m^2. The plate also loses heat to the surrounding air at 300 K through a heat transfer coefficient of 35 W/m^2-K. Determine (a) the reflected radiation, (b) the total energy leaving the plate, and (c) the net heat transfer from the plate.

26.32: An inclined plate receives 1100 W/m^2 of radiant energy on its front surface. The back side of the plate is insulated. The plate has an absorptivity of 0.72 and an emissivity of 0.32. Determine (a) the steady temperature of the plate and (b) whether the plate can be considered a gray surface.

26.33: Reconsider the plate in Problem 26.32 under the following conditions: The plate also dissipates to the surroundings at 300 K by convection through a heat transfer coefficient of 27 W/m^2-K. Determine the steady-state temperature of the plate.

26.34: An opaque surface is initially at a uniform temperature of 500 K. The surface, having an absorptivity of 0.40 and an emissivity of 0.60, is suddenly exposed to a radiant energy flux of 12,000 W/m^2. Determine (a) whether the temperature increases or decreases with time and (b) how the picture changes, if at all, if the radiant energy flux is increased to 15,000 W/m^2.

Radiation Properties and Lambert's Cosine Law

26.35: Two small surfaces with areas $S_1 = 0.01\,\text{m}^2$ and $S_2 = 0.02\,\text{m}^2$ are oriented as shown in Figure P26.35. The surface S_1 acts as a black surface at 800 K. Determine the rate at which radiant energy from S_1 is intercepted by S_2.

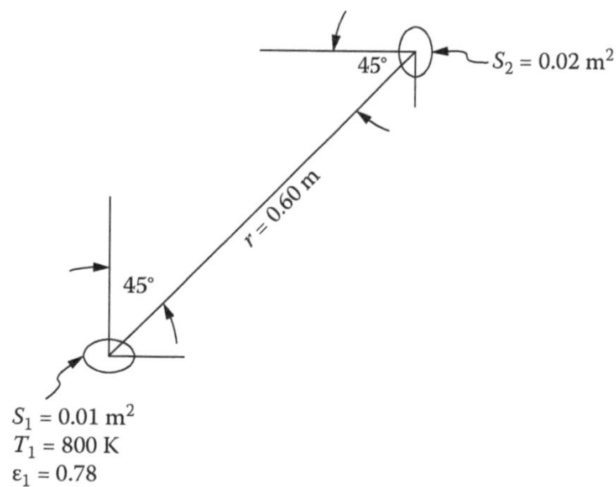

FIGURE P26.35

26.36: Determine the monochromatic intensity of radiation, $I_{b\lambda}$, in W/m^2-sr-μm at 5 μm for a black surface at (a) 1000 K, (b) 800 K, and (c) 500 K.

26.37: Consider three small surfaces with areas, $S_1 = 0.01\,\text{m}^2$, $S_2 = 0.015\,\text{m}^2$, and $S_3 = 0.02\,\text{m}^2$ oriented as shown in Figure P26.37. The emissive power of S_1 is 29,600 W/m^2 and the emission is diffuse. Determine the rate at which radiant energy from S_1 is intercepted by S_2 and S_3.

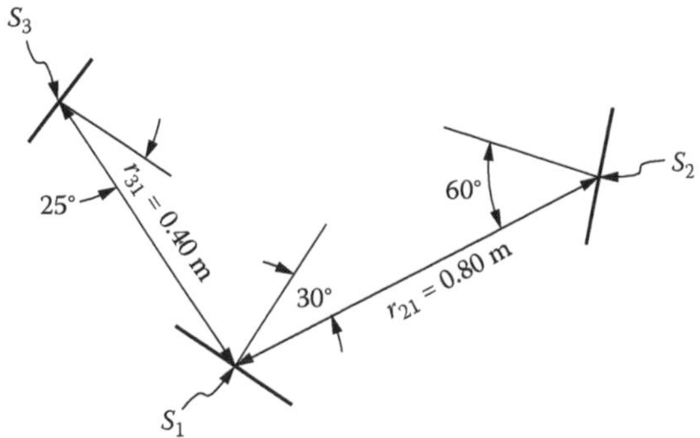

FIGURE P26.37

26.38: Determine the maximum monochromatic intensity of radiation $I_{b\lambda}$ for a black surface at 1000, 800, and 500 K. Use the results of Problem 26.36 to determine the ratio of monochromatic intensity of radiation to the maximum monochromatic intensity of radiation for each temperature.

26.39: Consider Figure P26.39 and observe that a small surface, $S_1 = 0.025\,\text{m}^2$, is maintained at 750 K and emits as a black surface. A second surface, $S_2 = 0.03\,\text{m}^2$, is to be oriented such that it captures 20% of the emission from S_1. Determine the angle of orientation, θ.

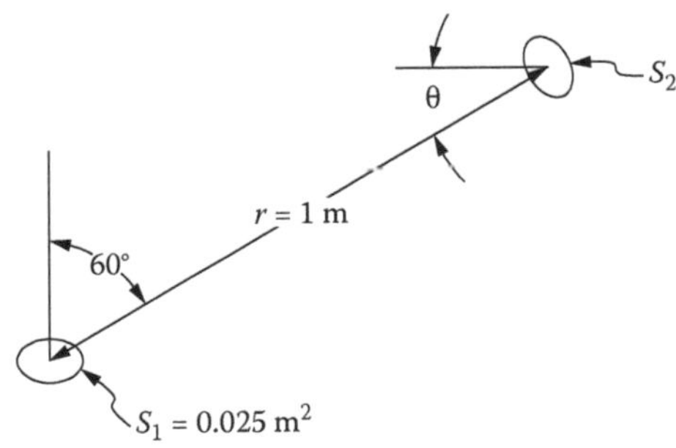

FIGURE P26.39

Monochromatic and Total Emissivity and Absorptivity

26.40: A diffuse emitter at 1500 K has a monochromatic emissivity ϵ_λ such that $\epsilon_\lambda = 0.20$ from 0 to 4 μm, $\epsilon_\lambda = 0.40$ from 4 to 10 μm, and $\epsilon_\lambda = 0.60$ from 10 to ∞. Determine (a) the total emissivity and (b) the total emissive power.

26.41: The top surface of a circular metallic disk of 1-cm diameter is maintained at 900 K while the bottom surface is perfectly insulated. The disk is placed in an environment at 100 K. The monochromatic emissivity of the disk is 0.35 between $\lambda = 0$ μm and $\lambda = 2$ μm and 0.20 between $\lambda = 2$ μm and $\lambda = \infty$. Determine the radiant energy (watts) emitted by the disk.

26.42: The monochromatic emissivity of a diffuse surface at 500 K is 0.65 between 0 and λ and 0.52 between λ and ∞; the total emissivity is 0.61. Determine the value of λ.

26.43: The monochromatic emissivity of a surface varies with wavelength as follows: $\epsilon_\lambda =$ 0.20 between $\lambda = 0$ and 1 μm, $\epsilon_\lambda = 0.50$ between $\lambda = 1$ and 6 μm, $\epsilon_\lambda = 0.70$ between $\lambda = 6$ and 15 μm, and $\epsilon_\lambda = 0.82$ between $\lambda = 15$ μm and ∞. Determine the total emissive power of the surface at a temperature of 1200 K.

26.44: Assuming the sun to be a blackbody at 5800 K, determine the absorptivity of the surface in Problem 26.43 to solar radiation.

26.45: The surface in Problem 26.42 receives an irradiation of 10,000 W/m^2 originating from a blackbody at 2500 K. Determine (a) the radiant heat flux in W/m^2 absorbed and (b) the reflected heat flux if the surface is opaque.

The Shape Factor

26.46: In the arrangement of surfaces shown in Figure P26.46, surface 1 is a disk of 1-m diameter, surface 2 is a hemisphere of 2-m diameter, and surface 3, an imaginary surface, is shown dashed. Determine the shape factors, F_{12}, F_{21}, F_{23}, F_{32}, and F_{22}.

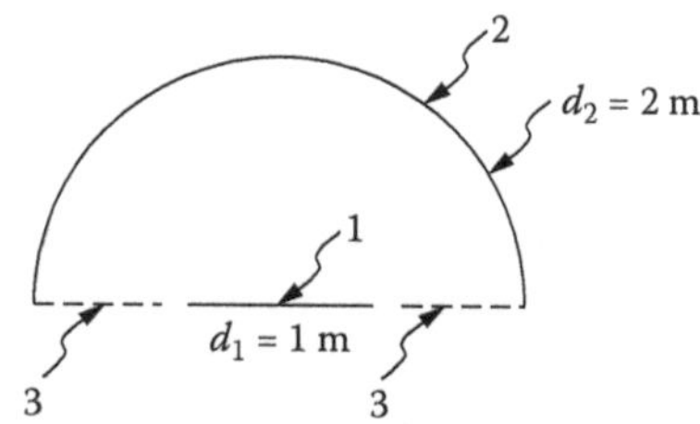

FIGURE P26.46

26.47: For the right circular cylinder of diameter d and length L, indicated in Figure P26.47, the end surfaces are designated as surfaces 1 and 2 and the lateral surface is designated as surface 3. If $d = L$, determine F_{12} and F_{13}.

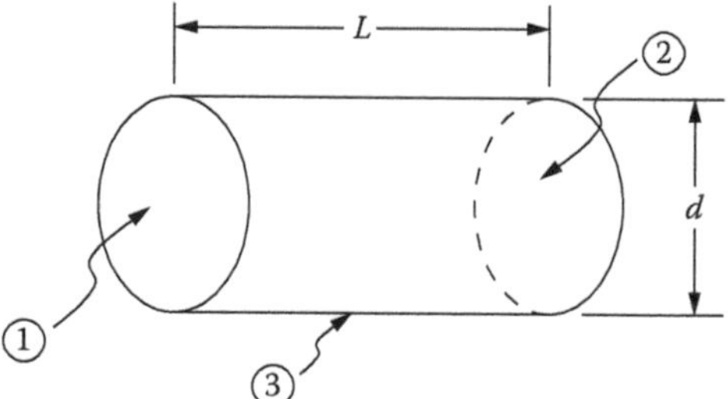

FIGURE P26.47

26.48: For a long duct with the cross section indicated in Figure P26.48, show that

$$F_{21} = \frac{1}{\left(\pi - \frac{\theta}{2}\right)}$$

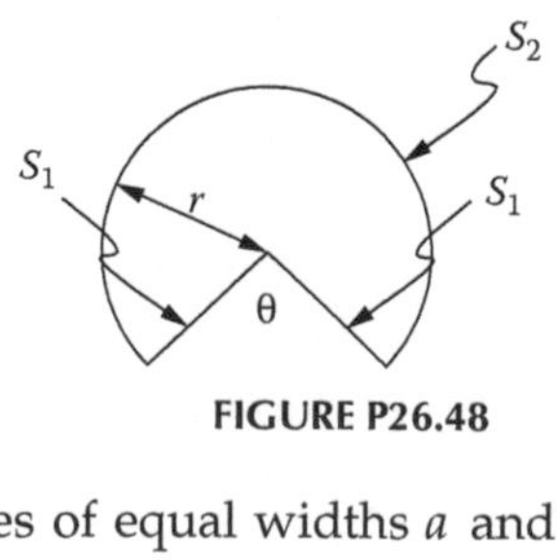

FIGURE P26.48

26.49: For inclined parallel plates of equal widths a and a common edge as indicated in Figure P26.49, prove that

$$F_{12} = 1 - \sin\frac{\theta}{2}$$

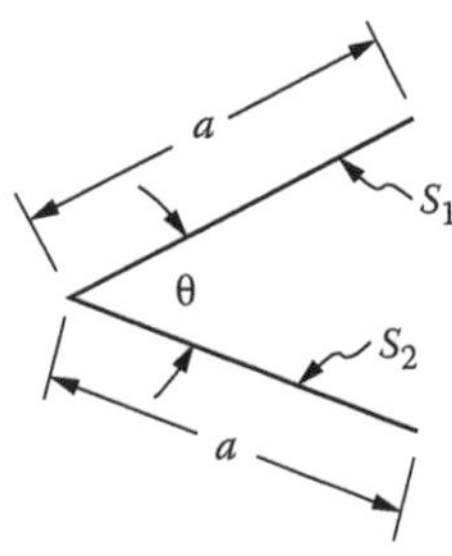

FIGURE P26.49

26.50: For the two-dimensional three-sided enclosure shown in Figure P26.50, show that the shape factor, F_{12}, is given by

$$F_{12} = \frac{a_1 + a_2 - a_3}{2a_1}$$

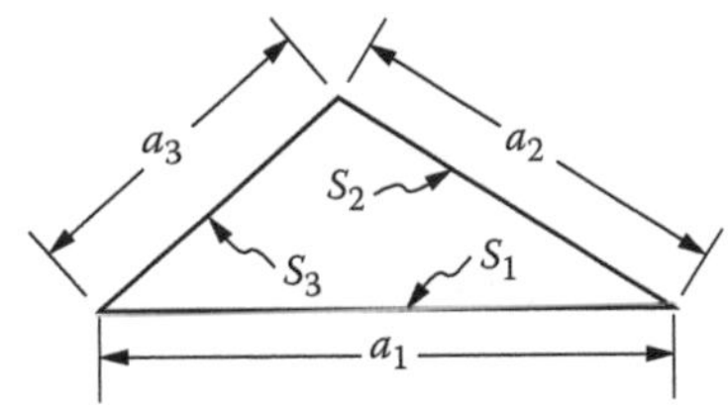

FIGURE P26.50

26.51: For the two-dimensional perpendicular arrangement shown in Figure P26.51, establish the relation

$$F_{12} = \frac{1 + \dfrac{a_2}{a_1} - \left[1 + \left(\dfrac{a_2}{a_1}\right)^2\right]^{1/2}}{2}$$

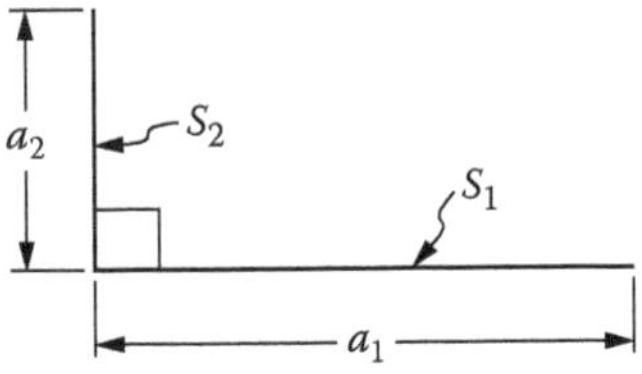

FIGURE P26.51

26.52: Consider the truncated cone of height h and top and bottom radii of r_1 and r_2, respectively, as shown in Figure P26.52. For the surface designations shown, determine the shape factors, F_{12}, F_{32}, and F_{22}.

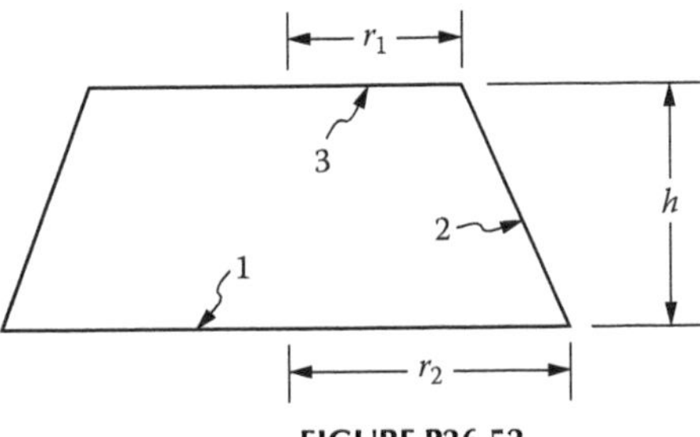

FIGURE P26.52

26.53: In Figure P26.53, determine the shape factors F_{14}, F_{41}, F_{32}, and F_{23}.

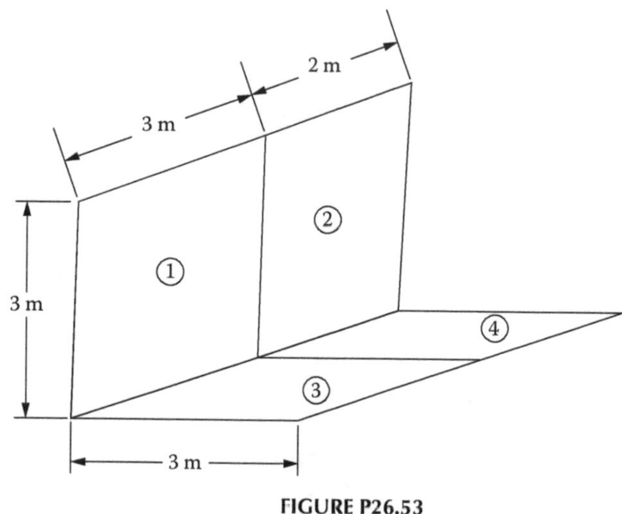

FIGURE P26.53

26.54: Determine the shape factors F_{12} and F_{21} for the configuration shown in Figure P26.54.

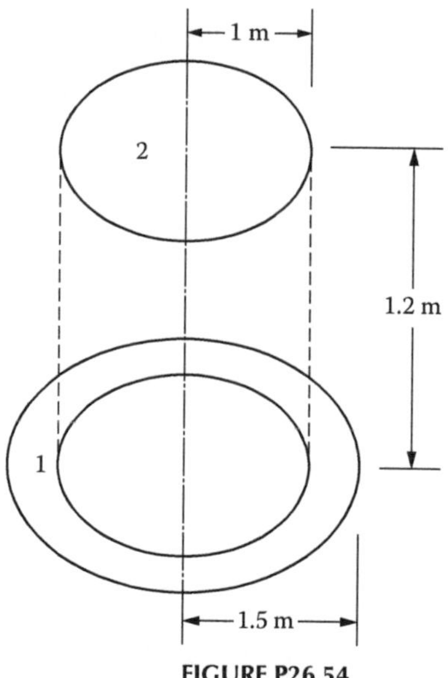

FIGURE P26.54

26.55: Consider the concentric cylinder arrangement shown in Figure P26.55 and derive the following expressions for the shape factors.

$$F_{12} = \frac{1}{2}(1 - F_{12} - F_{11})$$

$$F_{23} = \frac{1}{2}\left(1 - \frac{S_1}{S_2}F_{12}\right)$$

$$F_{33} = 1 - \frac{S_1 + S_2}{2S_3} + \frac{S_1}{2S_2}(2F_{12} + F_{11})$$

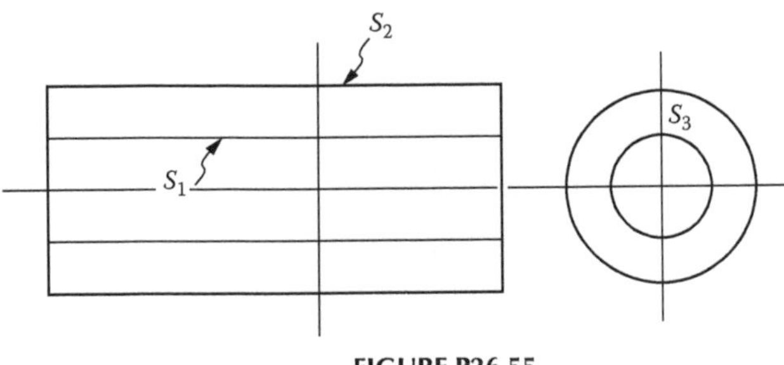

FIGURE P26.55

26.56: For the two-dimensional configuration shown in Figure P26.56, determine the shape factors F_{12}, F_{13}, and F_{14}.

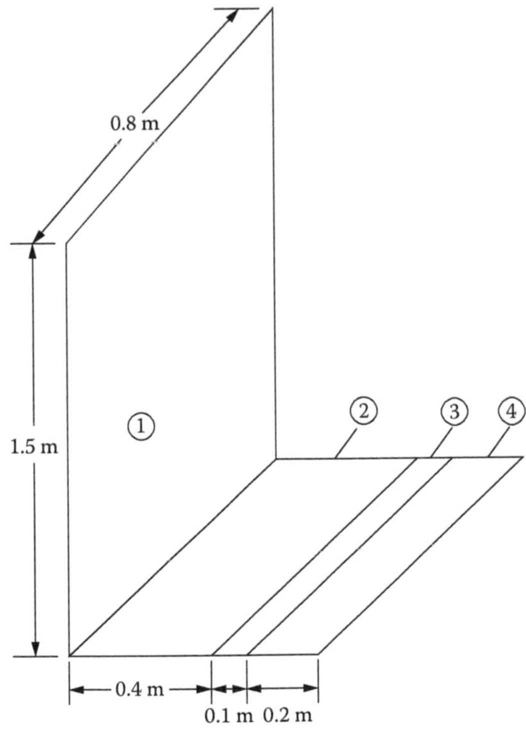

FIGURE P26.56

26.57: Determine the shape factors F_{12} and F_{21} for the two rectangles that lie in the mutually perpendicular planes indicated in Figure P26.57.

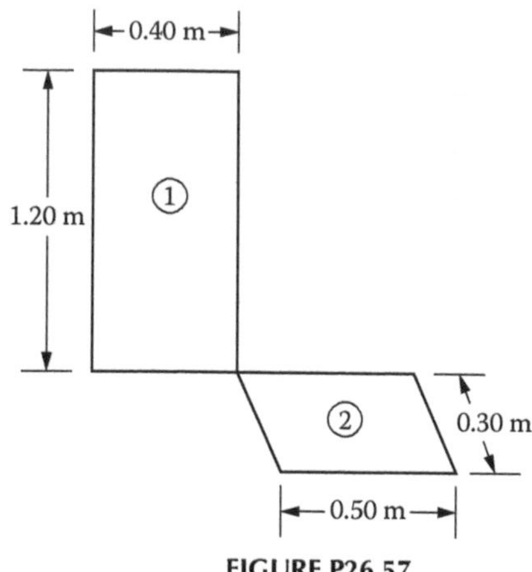

FIGURE P26.57

26.58: As indicated in Figure P26.58, the radiation from a cylindrical cavity of diameter 0.4 m and depth 0.8 m is measured by a circular sensor of 0.4 m placed parallel to the cavity opening at a distance of 0.3 m. Determine the fractions of the radiation received by the sensor (surface 3) from (a) the base of the cavity (surface 1) and (b) the lateral surfaces of the cavity (surface 2).

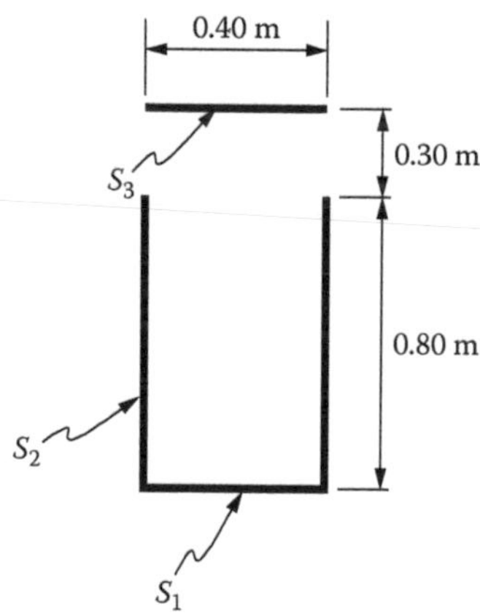

FIGURE P26.58

Heat Flow by Radiation between Two Bodies

26.59: Two black, infinitely long parallel surfaces at 1200 and 600 K face each other. Determine the net radiant heat exchange per unit surface area between the two surfaces.

26.60: Assume the surfaces in Figure P26.60 to be black, and determine the net radiant heat exchange, $\dot{Q}_{12}$, between the two aligned, parallel, rectangular surfaces shown.

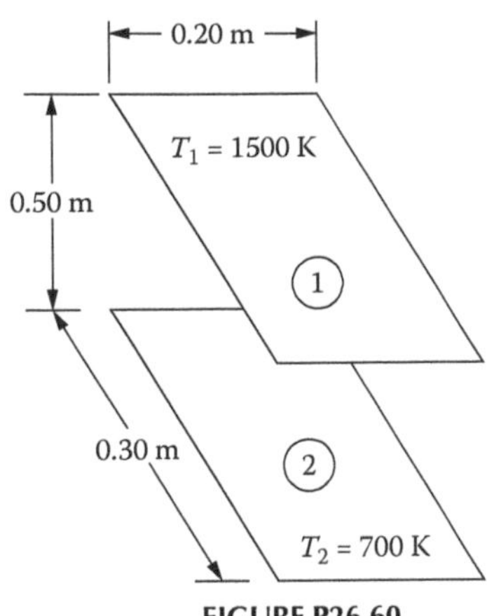

FIGURE P26.60

26.61: Refer again to Problem 26.56. Assume all surfaces to be black, and take as the surface temperatures $T_1 = 1600\,\text{K}$ and $T_2 = T_3 = T_4 = 500\,\text{K}$. Determine the net radiation heat transfers, $\dot{Q}_{12}$, $\dot{Q}_{13}$, and $\dot{Q}_{14}$.

26.62: For the configuration of Figure P26.54, determine $\dot{Q}_{12}$ when $T_1 = 500\,\text{K}$ and $T_2 = 1000\,\text{K}$ if both surfaces are black.

26.63: Consider the concentric cylinder arrangement of Figure P26.55. Assuming surfaces 1 and 2 to be black and at 1600 K and 600 K, respectively, determine $\dot{Q}_{12}$. Pertinent dimensions are $d_1 = 1\,\text{m}$, $d_2 = 2\,\text{m}$, and $L = 5\,\text{m}$.

26.64: Two gray, infinitely long, parallel surfaces at 1600 K and 400 K face each other. The emissivity of the hot surface is 0.90, while that of the cold surface is 0.50. Per unit of area, determine the net radiant energy heat exchange between the surfaces.

26.65: A thin radiation shield of emissivity 0.40 on both sides is inserted between the two gray surfaces described in Problem 26.64. Determine (a) the equilibrium temperature of the shield and (b) the percentage reduction in radiation heat exchange between the surfaces as a result of the shielding.

26.66: Two large gray parallel planes with $T_1 = 2000\,\text{K}$ and $\epsilon_1 = 0.82$ and $T_2 = 800\,\text{K}$ and $\epsilon_2 = 0.75$ face each other. The radiant heat exchange between the plates is to be reduced by 60% and the emissivity of the proposed radiation shield is the same on both sides. Determine (a) the emissivity of the shield and (b) the equilibrium temperature of the shield.

26.67: A thin radiation shield is placed between two parallel gray surfaces with $T_1 = 1400\,\text{K}$ and $\epsilon_1 = 0.90$ and $T_2 = 600\,\text{K}$ and $\epsilon_2 = 0.70$. The emissivities of the shield are 0.56 on the side facing the hot surface and 0.42 on the side facing the cooler surface. Determine the heat transfer per unit area (a) with the shield and (b) without the shield, and (c) determine the equilibrium temperature of the shield.

26.68: Consider two parallel gray surfaces of infinite extent with temperatures and emissivities of $T_1 = 1500\,\text{K}$ and $\epsilon_1 = 0.80$ and $T_2 = 600\,\text{K}$ and $\epsilon_2 = 0.70$. The two plates face each other and it is proposed that the heat transfer between the surfaces be reduced by a factor of 10 using radiation shields with emissivities of 0.40 on both sides. Determine the number of shields needed.

26.69: In a concentric cylinder arrangement, the outer surface of the inner cylinder (surface 1) has a diameter of 40 cm and the inner surface of the outer cylinder (surface 2) has a diameter of 80 cm. Temperatures and emissivities are $T_1 = 700\,\text{K}$ and $\epsilon_1 = 0.75$ and

$T_2 = 300\,\text{K}$ and $\epsilon_2 = 0.65$. Assuming that the cylinders are infinitely long, determine the rate of heat transfer between the hot and the cold surfaces.

26.70: For the arrangement in Problem 26.69, a thin cylindrical shield of 60-cm diameter and an emissivity of 0.25 is placed between the cylinders. Determine (a) the new rate of heat transfer between the two surfaces and (b) the equilibrium temperature of the shield.

26.71: For the concentric sphere arrangement indicated in Figure P26.71, show that the rate of heat transfer, $\dot{Q}_{12}$, is given by

$$\dot{Q}_{12} = \frac{4\pi r_1^2 \sigma \epsilon_1 (T_1^4 - T_2^4)}{\dfrac{1}{\epsilon_1} + \left(\dfrac{1}{\epsilon_2} - 1\right)\left(\dfrac{r_1}{r_2}\right)^2} \quad (T_1 > T_2)$$

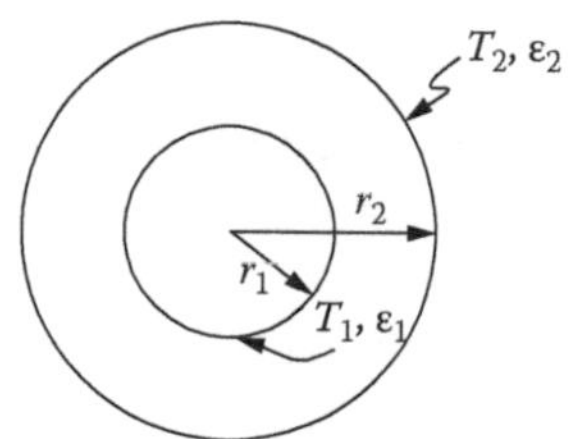

FIGURE P26.71

Radiosity, Irradiation, and Radiation Networks

26.72: Two gray parallel plates of infinite extent with $T_1 = 1000\,\text{K}$ and $\epsilon_1 = 0.75$ and $T_2 = 400\,\text{K}$ and $\epsilon_2 = 0.85$ radiate to each other. Determine (a) the radiosities, J_1 and J_2, (b) the irradiations, G_1 and G_2, and (c) the rate of heat transfer per unit area.

26.73: A circular disk of radius 25 cm sits centrally at the base of a hemisphere of radius 50 cm as shown in Figure P26.73. Determine (a) the radiosities, J_1 and J_2, (b) the irradiations G_1 and G_2, (c) the rates of heat transfer, $\dot{Q}_1$, and $\dot{Q}_2$, and (d) the heat transfer to the environment at 280 K.

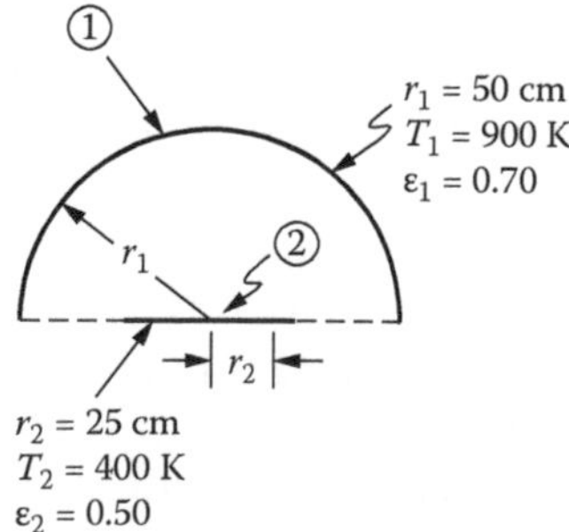

FIGURE P26.73

26.74: For the concentric arrangement in Problem 26.69 (see Figure P26.71) determine (a) the radiosities J_1 and J_2, (b) the irradiations G_1 and G_2, and (c) the net heat transfer $\dot{Q}_{12}$.

26.75: Consider the concentric sphere arrangement of Figure P26.71. With $T_1 = 800\,\text{K}$, $r_1 = 0.30\,\text{m}$, and $\epsilon_1 = 0.72$ and $T_2 = 300\,\text{K}$, $r_2 = 0.60\,\text{m}$, and $\epsilon_2 = 0.62$, determine (a) the radiosities J_1 and J_2, (b) the irradiations G_1 and G_2, and (c) $\dot{Q}_{12}$.

Radiation Networks

26.76: Two identical rectangular plates, each 30-cm wide and 50-cm high, face each other as shown in Figure P26.76. The plates sit in an environment at 0°C. Using the thermal network method, determine the net radiant heat exchange between the surfaces and the radiant heat loss from both plates to the environment areas.

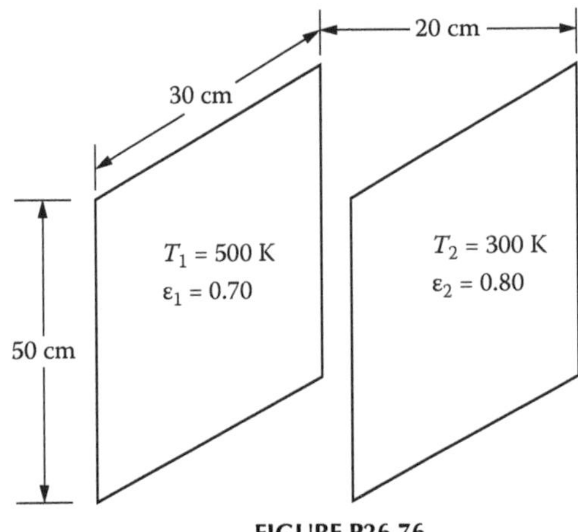

FIGURE P26.76

26.77: Consider the three-surface configuration shown in Figure P26.77. Draw the thermal network representing the configuration. The depth of all surfaces may be considered infinite and the surroundings may be considered to be at 0 K.

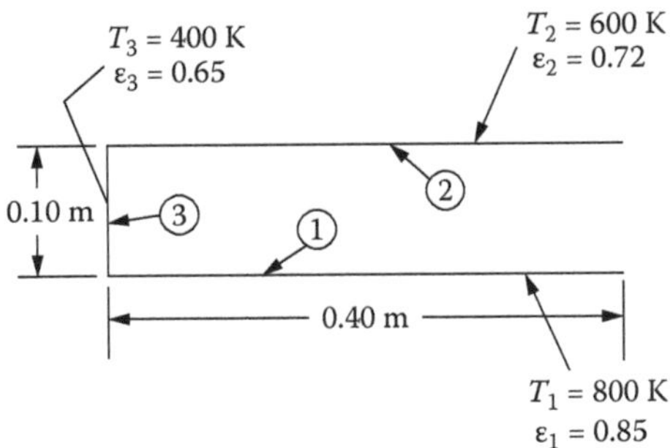

FIGURE P26.77

Appendix A

Tables and Charts

TABLE A.1

Values of the Gas Constant and Specific Heats for Several Gases

Gas	Molecular Weight, kg/kg mol	R at 300 K, kJ/kg-K	c_p at 300 K, kJ/kg-K	c_v at 300 K, kJ/kg-K	k^a
Air	28.97	0.287	1.005	0.718	1.400
Argon, A	39.94	0.208	0.515	0.310	1.661
Carbon dioxide, CO_2	44.01	0.189	0.846	0.653	1.296
Carbon monoxide, CO	28.01	0.297	1.041	0.745	1.400
Ethane, C_2H_6	30.07	0.276	1.767	1.495	1.182
Helium, He	4.003	2.078	5.234	3.146	1.664
Hydrogen, H_2	2.016	4.127	14.32	10.16	1.409
Methane, CH_4	16.04	0.519	2.227	1.735	1.284
Oxygen, O_2	32.00	0.260	0.917	0.653	1.404
Nitrogen, N_2	28.02	0.297	1.038	0.741	1.401
Propane, C_3H_8	44.09	0.188	1.692	1.507	1.123

[a] Dimensionless parameter = c_p/c_v.

Source: L. C. Nelson and E. F. Obert, Generalized Compressibility Charts. *Chem. Eng.*, 61, 203 (1954).

TABLE A.2

Specific Heats and $k = c_p/c_v$ for Six Gases as a Function of Absolute Temperature

	Air			N_2		
T, K	c_p, kJ/kg-K	c_v, kJ/kg-K	k	c_p, kJ/kg-K	c_v, kJ/kg-K	k
250	1.003	0.726	1.401	1.039	0.742	1.400
300	1.005	0.718	1.400	1.039	0.743	1.400
350	1.008	0.721	1.398	1.041	0.744	1.399
400	1.013	0.726	1.395	1.044	0.747	1.397
450	1.020	0.733	1.391	1.049	0.752	1.385
500	1.029	0.742	1.387	1.056	0.759	1.391
550	1.040	0.753	1.381	1.065	0.768	1.387
600	1.051	0.764	1.376	1.075	0.778	1.382
650	1.063	0.776	1.370	1.086	0.789	1.376
700	1.075	0.788	1.364	1.098	0.801	1.371
750	1.087	0.800	1.359	1.110	0.813	1.365
800	1.099	0.812	1.354	1.121	0.825	1.360
900	1.121	0.834	1.344	1.145	0.849	1.349
1000	1.142	0.855	1.336	1.167	0.870	1.341

(*continued*)

TABLE A.2 (Continued)

Specific Heats and $k = c_p/c_v$ for Six Gases as a Function of Absolute Temperature

	O_2			H_2		
T, K	c_p, kJ/kg-K	c_v, kJ/kg-K	*k*	c_p, kJ/kg-K	c_v, kJ/kg-K	*k*
250	0.913	0.653	1.398	14.051	9.927	1.416
300	0.915	0.658	1.395	14.307	10.183	1.405
350	0.928	0.668	1.389	14.427	10,302	1.400
400	0.941	0.681	1.382	14.476	10.352	1.398
450	0.956	0.696	1.373	14.501	10.377	1.398
500	0.972	0.712	1.365	14.513	10.389	1.397
550	0.988	0.728	1.358	14.530	10.405	1.396
600	1.003	0.743	1.350	14.546	10.422	1.396
650	1.017	0.758	1.343	14.571	10.457	1.395
700	1.031	0.771	1.337	14.604	10.480	1.394
750	1.043	0.783	1.332	14.645	10.521	1.392
800	1.054	0.794	1.327	14.695	10.570	1.390
900	1.074	0.814	1.319	14.822	10.698	1.385
1000	1.090	0.830	1.313	14.983	10.859	1.380

	CO			CO_2		
T, K	c_p, kJ/kg-K	C_v, kJ/kg-K	*k*	c_p, kJ/kg-K	c_v, kJ/kg-K	*k*
250	1.039	0.743	1.400	0.791	0.602	1.314
300	1.040	0.744	1.399	0.846	0.657	1.288
350	1.043	0.746	1.398	0.895	0.706	1.268
400	1.047	0.751	1.395	0.939	0.750	1.252
450	1.054	0.757	1.392	0.978	0.790	1.239
500	1.063	0.767	1.387	1.014	0.825	1.229
550	1.075	0.778	1.382	1.046	0.857	1.220
600	1.087	0.790	1.376	1.075	0.886	1.213
650	1.100	0.803	1.370	1.102	0.913	1.207
700	1.113	0.816	1.364	1.126	0.937	1.202
750	1.126	0.829	1.358	1.148	0.969	1.197
800	1.139	0.842	1.353	1.169	0.980	1.193
900	1.163	0.866	1.343	1.204	1.015	1.186
1000	1.185	0.868	1.335	1.234	1.045	1.181

Source: U.S. National Bureau of Standards, Tables of Thermal Properties of Gases, Circular 564, 1955.

TABLE A.3

Thermodynamic Properties of Steam: Temperature Table

		Specific Volume, m^3/kg			Internal Energy, kJ/kg			Enthalpy, kJ/kg			Entropy, kJ/kg-K		
T_{sat}, °C	P_{sat}, kPa	v_f	v_{fg}	v_g	u_f	u_{fg}	u_g	h_f	h_{fg}	h_g	s_f	s_{fg}	s_g
0	0.61	0.001000	206.13	206.13	0.00	2,373.9	2,373.9	0.0	2,500.0	2,500.0	−0.0012	9.1590	9.1578
5	0.87	0.001000	147.20	147.20	21.04	2,361.1	2,382.1	21.0	2,489.6	2,510.6	0.0757	8.9510	9.0267
10	1.23	0.001000	106.36	106.36	42.02	2,347.8	2,389.8	42.0	2,478.4	2,520.4	0.1509	8.7511	8.9020
15	1.71	0.001001	78.036	78.037	62.95	2,333.7	2,396.7	63.0	2,466.8	2,529.7	0.2244	8.5582	8.7827
20	2.34	0.001002	57.801	57.802	83.86	2,319.9	2,403.7	83.9	2,455.0	2,538.9	0.2965	8.3718	8.6684
25	3.17	0.001003	43.446	43.447	104.75	2,305.5	2,410.3	104.8	2,443.1	2,547.9	0.3672	8,1919	8.5591
30	4.24	0.001004	32.907	32.908	125.63	2,291.6	2,417.2	125.6	2,431.2	2,556.8	0.4367	8.0180	8.4546
35	5.62	0.001006	25.250	25.251	146.50	2,277.3	2,423.8	146.5	2,419.2	2,565.7	0.5049	7.8496	8.3545
40	7.37	0.001008	19.536	19.537	167.37	2,263.2	2,430.6	167.4	2,407.3	2,574.6	0.5720	7.6864	8.2584
45	9.58	0.001010	15.262	15.263	188.24	2,249.1	2,437.3	188.3	2,395.3	2,583.5	0.6381	7.5281	8.1662
50	12.33	0.001012	12.046	12.047	209.12	2,234.7	2,443.8	209.1	2,383.2	2,592.3	0.7031	7.3745	8.0776
55	15.74	0.001014	9.5771	9.5781	230.01	2,220.4	2,450.4	230.0	2,371.1	2,601.1	0.7672	7.2253	7.9925
60	19.92	0.001017	7.6776	7.6786	250.91	2,206.0	2,456.9	250.9	2,358.9	2,609.8	0.8303	7.0804	7.9107
65	25.00	0.001020	6.1996	6.2006	271.83	2,191.6	2,463.4	271.9	2,346.6	2,618.4	0.8926	6.9394	7.8320
70	31.15	0.001023	5.0452	5.0462	292.76	2,177.0	2,469.7	292.8	2,334.2	2,626.9	0.9540	6.8023	7.7563
75	38.54	0.001026	4.1328	4.1338	313.70	2,162.3	2,476.0	313.7	2,321.6	2,635.4	1.0146	6.6687	7.6834
80	47.35	0.001029	3.4074	3.4085	334.67	2,147.6	2,482.3	334.7	2,309.8	2,643.7	1.0744	6.5387	7.6131
85	57.80	0.001032	2.8276	2.8286	355.65	2,132.8	2,488.4	355.7	2,296.2	2,651.9	1.1335	6.4118	7.5453
90	70.10	0.001036	2.3604	2.3614	376.66	2,117.8	2,494.5	376.7	2,283.3	2,660.0	1.1917	6.2881	7.4798
95	84.52	0.001039	1.9806	1.9817	397.69	2,102.8	2,500.5	397.8	2,270.2	2,668.0	1.2493	6.1673	7.4166
100	101.32	0.001043	1.6689	1.6699	418.75	2,087.9	2,506.6	418.9	2,257.0	2,675.8	1.3062	6.0492	7.3554
105	120.80	0.001047	1.4142	1.4152	439.83	2,072.8	2,512.6	440.0	2,243.6	2,683.6	1.3624	5.9338	7.2962
110	143.27	0.001051	1.2063	1.2074	460.95	2,057.2	2,518.2	461.1	2,230.0	2,691.1	1.4179	5.8209	7.2388
115	169.07	0.001056	1.0350	1.0361	482.10	2,041.3	2,523.4	482.3	2,216.3	2,698.6	1.4728	5.7105	7.1833
120	198.55	0.001060	0.89100	0.8921	503.28	2,025.4	2,528.7	503.5	2,202.3	2,705.8	1.5271	5.6023	7.1293
125	232.11	0.001065	0.76938	0.7704	524.51	2,009.6	2,534.1	524.8	2,188.2	2,712.9	1.5807	5.4962	7.0770
130	270.15	0.001070	0.66702	0.6681	545.78	1,993.6	2,539.4	546.1	2,173.8	2,719.9	1.6338	5.3922	7.0261
135	313.09	0.001075	0.58074	0.5818	567.09	1,977.3	2,544.4	567.4	2,159.2	2,726.6	1.6864	5.2902	6.9766
140	361.39	0.001080	0.50739	0.5085	588.46	1,960.9	2,549.3	588.8	2,144.3	2,733.1	1.7384	5.1900	6.9284
145	415.53	0.001085	0.44462	0.4457	609.88	1,944.3	2,554.2	610.3	2,129.1	2,739.4	1.7899	5.0916	6.8815
150	475.99	0.001091	0.39100	0.3921	631.35	1,927.5	2,558.8	631.9	2,113.6	2,745.5	1.8409	4.9948	6.8358

(*continued*)

TABLE A.3 (Continued)

Thermodynamic Properties of Steam: Temperature Table

		Specific Volume, m^3/kg			Internal Energy, kJ/kg			Enthalpy, kJ/kg			Entropy, kJ/kg-K		
T_{sat}, °C	P_{sat}, kPa	v_f	v_{fg}	v_g	u_f	u_{fg}	u_g	h_f	h_{fg}	h_g	s_f	s_{fg}	s_g
155	543.30	0.001096	0.34514	0.3462	652.89	1,910.3	2,563.2	653.5	2,097.8	2,751.3	1.8915	4.8996	6.7911
160	618.00	0.001102	0.30566	0.3068	674.50	1,892.8	2,567.3	675.2	2,081.7	2,756.9	1.9416	4.8059	6.7475
165	700.68	0.001108	0.27131	0.2724	696.18	1,875.1	2,571.3	697.0	2,065.2	2,762.2	1.9912	4.7135	6.7048
170	791.86	0.001114	0.24141	0.2425	717.93	1,857.2	2,575.2	718.8	2,048.4	2,767.2	2.0405	4.6224	6.6630
175	892.20	0.001121	0.21538	0.2165	739.77	1,839.0	2,578.8	740.8	2,031.2	2,772.0	2.0894	4.5325	6.6220
180	1,002.3	0.001127	0.19266	0.1938	761.69	1,820.5	2,582.1	762.8	2,013.6	2,776.4	2.1380	4.4437	6.5817
185	1,122.9	0.001134	0.17272	0.1739	783.70	1,803.6	2,585.3	785.0	1,995.5	2,780.5	2.1862	4.3559	6.5421
190	1,254.5	0.001141	0.15513	0.1563	805.80	1,782.4	2,588.3	807.2	1,977.1	2,784.3	2.2341	4.2691	6.5032
195	1,398.0	0.001148	0.13964	0.1408	828.01	1,762.9	2,590.9	829.6	1,958.1	2,787.8	2.2817	4.1834	6.4651
200	1,553.9	0.001156	0.12597	0.1271	850.32	1,743.0	2,593.3	852.1	1,938.8	2,790.9	2.3290	4.0986	6.4276
205	1,723.1	0.001164	0.11386	0.1150	872.74	1,722.7	2,595.4	874.7	1,918.9	2,793.6	2.3761	4.0147	6.3908
210	1,906.3	0.001172	0.10307	0.1042	895.28	1,702.0	2,597.3	897.5	1,898.5	2,796.0	2.4230	3.9314	6.3544
215	2,104.3	0.001180	0.09345	0.0946	917.94	1,681.0	2,598.9	920.4	1,877.6	2,798.0	2.4696	3.8485	6.3181
220	2,317.8	0.001189	0.08486	0.0860	940.73	1,659.5	2,600.2	943.5	1,856.2	2,799.7	2.5161	3.7661	6.2821
225	2,547.8	0.001198	0.07716	0.0784	963.66	1,637.6	2,601.3	966.7	1,834.2	2,800.9	2.5623	3.6841	6.2464
230	2,795.0	0.001208	0.07022	0.0714	986.73	1,615.4	2,602.2	990.1	1,811.7	2,801.8	2.6084	3.6025	6.2109
235	3,060.3	0.001218	0.06400	0.0652	1,010.0	1,592,7	2,602.7	1,033.7	1,788.6	2,802.3	2.6544	3.5213	6.1757
240	3,344.7	0.001228	0.05851	0.0597	1,033.6	1,569.1	2,602.5	1,037.5	1,764.8	2,802.3	2.7002	3.4404	6.1406
245	3,649.0	0.001239	0.05353	0.0548	1,056.9	1,545.2	2,602.1	1,061.4	1,740.5	2,801.9	2.7460	3.3597	6.1057
250	3,974.2	0.001250	0.04893	0.0502	1,080.7	1,521.0	2,601.7	1,085.6	1,715.5	2,801.2	2.7917	3.2792	6.0708
255	4,321.3	0.001262	0.04471	0.0460	1,104.6	1,496.7	2,601.3	1,110.1	1,689.9	2,800.0	2,8373	3.1986	6.0359
260	4,691.2	0.001275	0.04086	0.0421	1,128.8	1,471.9	2,600.7	1,134.8	1,663.5	2,798.3	2.8829	3.1180	6.0009
265	5,085.0	0.001288	0.03738	0.0387	1,153.2	1,446.4	2,599.6	1,159.8	1,636.5	2,796.3	2.9286	3.0372	5.9657
270	5,503.8	0.001302	0.03424	0.0355	1,177.9	1,420.3	2,598.1	1,185.1	1,608.7	2,793.7	2.9743	2.9560	5.9303
275	5,948.6	0.001317	0.03139	0.0327	1,202.8	1,393.4	2,596.3	1,210.7	1,580.1	2,790.8	3.0200	2.8745	5.8945

280	6,420.5	0.001333	0.02878	0.0301	1,228.1	1,366.0	2,594.0	1,236.6	1,550.8	2,787.4	3.0660	2.7924	5.8584
285	6,920.8	0.001349	0.02639	0.0277	1,253.7	1,337.9	2,591.6	1,263.0	1,520.6	2,783.6	3.1121	2.7097	5.8218
290	7,450.6	0.001366	0.02418	0.0255	1,279.6	1,297.7	2,577.3	1,289.8	1,477.9	2,767.7	3.1585	2.6262	5.7847
295	8,011.1	0.001385	0.02214	0.0235	1,306.0	1,265.5	2,571.5	1,317.1	1,442.8	2,759.9	3.2052	2.5417	5.7469
300	8,603.7	0.001404	0.02025	0.0217	1,332.8	1,232.0	2,564.8	1,344.9	1,406.2	2,751.1	3.2523	2.4560	5.7083
305	9,214.4	0.001425	0.01850	0.0199	1,360.2	1,197.5	2,557.6	1,373.3	1,367.9	2,741.2	3.3000	2.3688	5.6687
310	9,869.4	0.001447	0.01688	0.0183	1,388.0	1,161.1	2,549.2	1,402.3	1,327.8	2,730.1	3.3483	2.2797	5.6279
315	10,561.	0.001470	0.01538	0.0169	1,416.5	1,123.0	2,539.6	1,432.1	1,285.5	2,717.6	3.3973	2.1884	5.5858
320	11,289.	0.001499	0.01398	0.0155	1,445.7	1,083.0	2,528.7	1,462.6	1,240.9	2,703.5	3.4473	2.0947	5.5420
325	12,056.	0.001528	0.01267	0.0142	1,475.5	1,040.9	2,516.4	1,494.0	1,193.6	2,687.3	3.4984	1.9979	5.4962
330	12,862.	0.001561	0.01143	0.0130	1,506.2	996.3	2,502.5	1,526.3	1,143.3	2,669.6	3.5507	1.8973	5.4480
335	13,712.	0.001598	0.01026	0.0119	1,537.8	949.0	2,486.8	1,559.7	1,089.6	2,649.3	3.6045	1.7922	5.3967
340	14,605.	0.001639	0.00914	0.0108	1,570.4	898.7	2,469.0	1,594.3	1,032.2	2,626.5	3.6601	1.6820	5.3420
345	15,545.	0.001686	0.00808	0.0098	1,606.3	842.1	2,448.4	1,632.5	967.7	2,600.2	3.7176	1.5658	5.2834
350	16,535.	0.001741	0.00706	0.0088	1,643.0	780.5	2,423.5	1,671.8	897.2	2,569.0	3.7775	1.4416	5.2191
355	17,577.	0.001808	0.00605	0.0079	1,682.1	710.9	2,393.0	1,713.9	817.3	2,531.2	3.8400	1.3054	5.1454
360	18,675.	0.001896	0.00504	0.0069	1,726.2	629.5	2,355.7	1,761.6	723.7	2,485.3	3.9056	1.1531	5.0587
365	19,833.	0.002016	0.00400	0.0060	1,777.9	531.0	2,308.9	1,817.8	610.3	2,428.1	3.9746	0.9822	4.9569
370	21,054.	0.002225	0.00274	0.0050	1,843.3	394.1	2,237.3	1,890.1	451.9	2,342.0	4.0476	0.7555	4.8030
374.4	22,090.	0.00315	0.00000	0.00315	2,029.6	0.0	2,029.6	2,099.3	0.0	2,099.3	4.4298	0.0	4.4298

Source: Properties obtained from software, STEAMCALC. John Wiley & Sons, New York, 1983.

TABLE A.4

Thermodynamic Properties of Steam: Pressure Table

		Specific Volume, m³/kg			Internal Energy, kJ/kg			Enthalpy, kJ/kg			Entropy, kJ/kg-K		
P_{sat}, kPa	T_{sat}, °C	v_f	v_{fg}	v_g	u_f	u_{fg}	u_g	h_f	h_{fg}	h_g	s_f	s_{fg}	s_g
1.00	7.0	0.001000	129.08	129.08	29.40	2,356.1	2,385.5	29.4	2,485.2	2,514.6	0.1058	8.8704	8.9763
1.50	13.0	0.001001	88.067	88.068	54.68	2,339.3	2,394.0	54.7	2,471.4	2,526.3	0.1956	8.6337	8.8292
2.00	17.5	0.001001	67.073	67.074	73.41	2,326.8	2,400.2	73.4	2,460.9	2,534.3	0.2607	8.4642	8.7249
2.50	21.1	0.001002	54.290	54.291	88.41	2,316.7	2,405.1	88.4	2,452.4	2,540.8	0.3120	8.3322	8.6442
3.00	24.1	0.001003	45.751	45.752	100.96	2,308.0	2,409.0	101.0	2,445.3	2,546.2	0.3545	8.2240	8.5786
3.50	26.7	0.001003	39.483	39.484	111.81	2,300.9	2,412.7	111.8	2,439.1	2,550.9	0.3908	8.1324	8.5233
4.00	29.0	0.001004	34.779	34.780	121.37	2,294.5	2,415.9	121.4	2,433.6	2,555.0	0.4226	8.0529	8.4756
4.50	31.0	0.001005	31.128	31.129	129.95	2,288.6	2,418.6	130.0	2,428.7	2,558.7	0.4509	7.9827	8.4336
5.00	32.9	0.001005	28.194	28.195	137.73	2,283.3	2,421.0	137.7	2,424.3	2,562.0	0.4764	7.9197	8.3961
5.50	34.6	0.001006	25.773	25.774	144.86	2,278.4	2,423.3	144.9	2,420.2	2,565.0	0.4996	7.8626	8.3622
6.00	36.2	0.001006	23.742	23.743	151.45	2,274.0	2,425.4	151.5	2,416.4	2,567.9	0.5209	7.8104	8.3313
6.50	37.7	0.001007	22.013	22.014	157.58	2,269.8	2,427.4	157.6	2,412.9	2,570.5	0.5407	7.7623	8.3030
7.00	39.0	0.001007	20.522	20.523	163.31	2,266.0	2,429.3	163.3	2,409.6	2,572.9	0.5590	7.7177	8.2768
7.50	40.3	0.001008	19.225	19.226	168.70	2,262.3	2,431.0	168.7	2,406.5	2,575.2	0.5763	7.6762	8.2524
8.00	41.5	0.001008	18.086	18.087	173.79	2,258.9	2,432.7	173.8	2,403.6	2,577.4	0.5924	7.6372	8.2296
8.50	42.7	0.001009	17.080	17.081	178.61	2,255.6	2,434.2	178.6	2,400.8	2,579.4	0.6077	7.6006	8.2083
9.00	43.8	0.001009	16.185	16.186	183.19	2,252.5	2,435.7	183.2	2,398.2	2,581.4	0.6222	7.5660	8.1881
9.50	44.8	0.001010	15.383	15.384	187.56	2,249.5	2,437.1	187.6	2,395.7	2,583.2	0.6359	7.5332	8.1691
10.00	45.8	0.001010	14.660	14.661	191.74	2,246.7	2,438.4	191.7	2,393.3	2,585.0	0.6490	7.5021	8.1511
15.00	54.0	0.001014	10.020	10.021	225.83	2,223.2	2,449.0	225.8	2,373.5	2,599.4	0.7544	7.2548	8.0092
20.00	60.1	0.001017	7.6483	7.6493	251.28	2,205.7	2,457.0	251.3	2,358.7	2,610.0	0.8314	7.0779	7.9093
25.00	65.0	0.001020	6.2015	6.2025	271.80	2,191.6	2,463.3	271.8	2,346.6	2,618.4	0.8925	6.9396	7.8321
30.00	69.1	0.001022	5.2277	5.2287	289.09	2,379.5	2,468.6	289.1	2,336.3	2,625.5	0.9433	6.8260	7.7693
35.00	72.7	0.001024	4.5249	4.5259	304.11	2,169.0	2,473.1	304.1	2,327.4	2,631.5	0.9869	6.7295	7.7164
40.00	75.9	0.001026	3.9918	3.9929	317.42	2,159.7	2,477.1	317.5	2,319.4	2,636.8	1.0253	6.6455	7.6707
45.00	78.7	0.001028	3.5744	3.5755	329.40	2,151.3	2,480.7	329.4	2,312.2	2,641.6	1.0595	6.5710	7.6305
50.00	81.3	0.001030	3.2389	3.2398	340.31	2,143.6	2,483.9	340.4	2,305.5	2,645.9	1.0904	6.5042	7.5946
60.00	86.0	0.001033	2.7305	2.7316	359.66	2,129.9	2,489.6	359.7	2,293.7	2,653.5	1.1446	6.3880	7.5326
70.00	90.0	0.001036	2.3638	2.3648	376.49	2,117.9	2,494.4	376.6	2,283.4	2,659.9	1.1913	6.2891	7.4804
80.00	93.5	0.001038	2.0859	2.0869	391.43	2,107.2	2,498.7	391.5	2,274.1	2,665.6	1.2323	6.2029	7.4352
90.00	96.7	0.001041	1.8667	1.8678	404.90	2,097.7	2,502.6	405.0	2,265.7	2,670.7	1.2689	6.1265	7.3954
100.00	99.6	0.001043	1.6898	1.6908	417.20	2,089.0	2,506.2	417.3	2,258.0	2,675.3	1.3020	6.0578	7.3598
101.32	100.0	0.001043	1.66895	1.6700	418.74	2,087.9	2,506.6	418.8	2,257.0	2,675.8	1.3062	6.0493	7.3554

125.00	106.0	0.001048	1.36965	1.3707	444.01	2,069.7	2,513.7	444.1	2,240.9	2,685.1	1.3734	5.9113	7.2847
150.00	111.4	0.001053	1.15612	1.1572	466.74	2,052.9	2,519.6	466.9	2,226.3	2,693.2	1.4330	5.7904	7.2234
175.00	116.1	0.001057	1.00248	1.0035	486.58	2,037.9	2,524.5	486.8	2,213.4	2,700.1	1.4844	5.6873	7.1717
200.00	120.2	0.001060	0.88498	0.8860	504.25	2,024.7	2,529.0	504.5	2,201.7	2,706.2	1.5295	5.5974	7.1269
225.00	124.0	0.001064	0.79229	0.7934	520.22	2,012.8	2,533.0	520.5	2,191.1	2,711.5	1.5700	5.5175	7.0874
250.00	127.4	0.001067	0.71751	0.7186	534.82	2,001.9	2,536.7	535.1	2,181.2	2,716.3	1.6066	5.4455	7.0521
275.00	130.6	0.001070	0.65602	0.6571	548.30	1,991.7	2,540.0	548.6	2,172.1	2,720.7	1.6401	5.3800	7.0201
300.00	133.5	0.001073	0.60457	0.6056	560.83	1,982.1	2,542.9	561.2	2,163.5	2,724.6	1.6710	5.3199	6.9910
325.00	136.3	0.001076	0.56082	0.5619	572.57	1,973.1	2,545.7	572.9	2,155.4	2,728.3	1.6998	5.2643	6.9641
350.00	138.9	0.001079	0.52305	0.5241	583.60	1,964.6	2,548.2	584.0	2,147.7	2,731.6	1.7266	5.2126	6.9392
375.00	141.3	0.001081	0.49007	0.4911	594.03	1,956.6	2,550.6	594.4	2,140.3	2,734.8	1.7519	5.1642	6.9161
400.00	143.6	0.001084	0.46105	0.4621	603.93	1,948.9	2,552.8	604.4	2,133.3	2,737.7	1.7757	5.1187	6.8944
425.00	145.8	0.001086	0.43534	0.4364	613.36	1,941.6	2,554.9	613.8	2,126.6	2,740.4	1.7982	5.0758	6.8740
450.00	147.9	0.001088	0.41242	0.4135	622.35	1,934.5	2,556.9	622.8	2,120.1	2,743.0	1.8196	5.0352	6.8548
475.00	149.9	0.001091	0.39188	0.3930	630.97	1,927.7	2,558.7	631.5	2,113.9	2,745.4	1.8400	4.9966	6.8366
500.00	151.8	0.001093	0.37336	0.3745	639.24	1,921.2	2,560.4	639.8	2,107.9	2,747.6	1.8595	4.9598	6.8193
550.00	155.5	0.001097	0.34129	0.3424	654.86	1,908.7	2,563.5	655.5	2,096.4	2,751.8	1.8960	4.8910	6.7871
600.00	158.8	0.001101	0.31443	0.3155	669.42	1,896.9	2,566.3	670.1	2,085.5	2,755.6	1.9298	4.8278	6.7576
650.00	162.0	0.001104	0.29151	0.2926	683.06	1,885.8	2,568.8	683.8	2,075.2	2,759.0	1.9613	4.7692	6.7305
700.00	164.9	0.001108	0.27168	0.2728	695.93	1,875.3	2,571.2	696.7	2,065.4	2,762.1	1.9907	4.7146	6.7053
750.00	167.7	0.001111	0.25439	0.2555	708.11	1,865.3	2,573.4	708.9	2,056.0	2,765.0	2.0183	4.6634	6.6817
800.00	170.4	0.001115	0.23919	0.2403	719.68	1,855.7	2,575.4	720.6	2,047.0	2,767.6	2.0445	4.6152	6.6597
850.00	172.9	0.001118	0.22572	0.2268	730.71	1,846.5	2,577.2	731.7	2,038.4	2,770.0	2.0692	4.5696	6.6388
900.00	175.3	0.001121	0.21372	0.2148	741.27	1,837.7	2,578.9	742.3	2,030,0	2.772.3	2.0928	4.5264	6.6192
950.00	177.7	0.001124	0.20295	0.2041	751.39	1,829.1	2,580.5	752.5	2,021.9	2,774.3	2.1152	4.4853	6.6005
1,000	179.9	0.001127	0.19322	0.1943	761.11	1,820.8	2,581.9	762.2	2,014.0	2,776.3	2.1367	4.4460	6.5827
1,100	184.1	0.001133	0.17631	0.1774	779.52	1,805.1	2,584.6	780.8	1,999.0	2,779.8	2.1771	4.3725	6.5495
1,200	187.9	0.001138	0.16209	0.1632	796.71	1,790.2	2,586.9	798.1	1,984.7	2,782.8	2.2145	4.3047	6.5191
1,300	191.6	0.001143	0.14998	0.1511	812.87	1,776.1	2,589.0	814.4	1,971.1	2,785.4	2.2493	4.2417	6.4910
1,400	195.0	0.001148	0.13956	0.1407	828.13	1,762.7	2,590.8	829.7	1,958.0	2,787.8	2.2820	4.1829	6.4649
1,500	198.3	0.001153	0.13050	0.1317	842.61	1,749.8	2,592.4	844.3	1,945.5	2,789.8	2.3127	4.1277	6.4405
1,600	201.4	0.001158	0.12254	0.1237	856.39	1,737.3	2,593.7	858.2	1,933.4	2,791.7	2.3418	4.0757	6.4176
1,700	204.3	0.001163	0.11549	0.1167	869.56	1,725.4	2,595.0	871.5	1,921.7	2,793.3	2.3695	4.0265	6.3960
1,800	207.1	0.001167	0.10918	0.1104	882.18	1,713.9	2,596.1	884.3	1,910.4	2,794.7	2.3958	3.9797	6.3755
1,900	209.8	0.001172	0.10351	0.1047	894.30	1,702.7	2,597.0	896.5	1,899.4	2,795.9	2.4210	3.9350	6.3559
2,000	212.4	0.001176	0.09839	0.0996	905.97	1,691.9	2,597.9	908.3	1,888.7	2,797.0	2.4450	3.8922	6.3372

(continued)

TABLE A.4 (Continued)

Thermodynamic Properties of Steam: Pressure Table

P_{sat}, kPa	T_{sat}, °C	Specific Volume, m³/kg			Internal Energy, kJ/kg			Enthalpy, kJ/kg			Entropy, kJ/kg-K		
		v_f	v_{fg}	v_g	u_f	u_{fg}	u_g	h_f	h_{fg}	h_g	s_f	s_{fg}	s_g
2,250	218.4	0.001186	0.08751	0.0887	933.41	1,666.2	2,599.6	936.1	1,863.1	2,799.2	2.5012	3.7925	6.2936
2,500	223.9	0.001196	0.07873	0.0799	958.75	1,642.1	2,600.9	961.7	1,839.0	2,800.7	2.5525	3.7015	6.2540
2,750	229.1	0.001206	0.07147	0.0727	982.37	1,619.4	2,601.8	985.7	1,816.0	2,801,7	2.5997	3.6178	6.2176
3,000	233.8	0.001215	0.06538	0.0666	1,004.5	1,597.9	2,602.4	1,008.2	1,794.0	2,802.2	2.6437	3.5402	6.1839
3,250	238.3	0.001225	0.06028	0.0615	1,025.4	1,577.0	2,602.5	1,029.4	1,772.9	2,802.3	2.6848	3.4676	6.1524
3,500	242.5	0.001234	0.05594	0.0572	1,045.3	1,556.8	2,602.1	1,049.6	1,752.6	2,802.2	2.7234	3.3995	6.1229
3,750	246.5	0.001242	0.05208	0.0533	1,064.2	1,537.6	2,601.8	1,068.8	1,732.9	2,801.7	2.7600	3.3350	6.0950
4,000	250.3	0.001251	0.04864	0.0499	1,082.2	1,519.3	2,601.5	1,087.2	1,713.9	2,801.1	2.7947	3.2739	6.0686
5,000	263.9	0.001285	0.03811	0.0394	1,147.9	1,451.9	2,599.8	1,154.3	1,642.4	2,796.7	2.9186	3.0548	5.9734
6,000	275.5	0.001319	0.03109	0.0324	1,205.6	1,390.4	2,596.0	1,213.5	1,576.9	2,790.5	3.0251	2.8655	5.8906
7,000	285.8	0.001352	0.02603	0.0274	1,257.8	1,333.5	2,591.3	1,267.2	1,515.7	2,782.9	3.1194	2.6965	5.8159
8,000	295.0	0.001385	0.02214	0.0235	1,305.9	1,265.8	2,571.7	1,317.0	1,442.9	2,759.9	3.2050	2.5421	5.7471
9,000	303.3	0.001418	0.01907	0.0205	1,351.0	1,209.3	2,560.3	1,363.7	1,380.9	2,744.7	3.2840	2.3981	5.6821
10,000	311.0	0.001452	0.01658	0.0180	1,393.7	1,153.8	2,547.5	1,408.2	1,319.5	2,727.7	3.3580	2.2616	5.6196
11,000	318.1	0.001489	0.01450	0.0160	1,434.6	1,098.6	2,533.1	1,451.0	1,258.0	2,709.0	3.4283	2.1305	5.5588
12,000	324.7	0.001527	0.01273	0.0143	1,474.0	1,043.3	2,517.3	1,492.4	1,196.0	2,688.4	3.4958	2.0028	5.4986
13,000	331.0	0.001568	0.01119	0.0128	1,512.4	987.6	2,499.9	1,532.8	1,133.1	2,665.8	3.5611	1.8771	5.4382
14,000	336.9	0.001612	0.00984	0.0114	1,549.8	931.1	2,480.9	1,572.4	1,068.8	2,641.2	3.6249	1.7520	5.3769
15,000	342.4	0.001661	0.00862	0.0103	1,586.6	873.4	2,460.0	1,611.5	1,002.8	2,614.3	3.6877	1.6265	5.3142
16,000	347.7	0.001715	0.00752	0.0092	1,626.1	810.1	2,436.2	1,653.6	930.4	2,584.0	3.7498	1.4996	5.2494
17,000	352.3	0.001769	0.00660	0.0084	1,660.2	750.3	2,410.5	1,690.3	862.5	2,552.8	3.8054	1.3819	5.1872
18,000	357.0	0.001839	0.00566	0.0075	1,698.6	680.8	2,379.4	1,731.7	782.6	2,514.3	3.8652	1.2481	5.1134
19,000	361.4	0.001926	0.00475	0.0067	1,740.1	603.6	2,343.6	1,776.7	693.9	2,470.5	3.9249	1.1063	5.0312
20,000	365.7	0.002037	0.00384	0.0059	1,785.8	515.1	2,300.9	1,826.6	591.9	2,418.5	3.9846	0.9568	4.9414
21,000	369.8	0.002208	0.00281	0.0050	1,839.7	401.7	2,241.4	1,886.0	460.8	2,346.8	4.0443	0.7681	4.8124
22,000	373.7	0.002623	0.00114	0.0038	1,944.6	174.0	2,118.6	2,002.3	199.0	2,201.3	4.1042	0.4563	4.5605
22,090	374.4	0.00315	0.00000	0.00315	2,029.6	0.0	2,029.6	2,099.3	0.0	2,099.3	4.4298	0.0	4.4298

Source: Properties obtained from software, STEAMCALC. John Wiley & Sons, New York, 1983.

TABLE A.5

Thermodynamic Properties of Steam: Superheated Vapor Table

P, kPa	*T*, °C	*v*, m³/kg	*u*, kJ/kg	*h*, kJ/kg	*s*, kJ/kg-K
10 (T_{sat} = 45.8°C)					
	100	17.196	2516.2	2688.1	8.4498
	150	19.513	2588.2	2783.3	8.6893
	200	21.826	2661.2	2879.5	8.9040
	250	24.136	2735.5	2976.8	9.0996
	300	26.446	2811.2	3075.6	9.2799
	350	28.754	2888.3	3175.9	9.4476
	400	31.063	2967.1	3277.7	9.6048
	450	33.371	3047.4	3381.1	9.7530
	500	35.679	3129.4	3486.2	9.8935
	550	37.987	3213.1	3593.0	10.027
	600	40.295	3298.5	3701.5	10.155
	650	42.603	3385.7	3811.7	10.278
	700	44.911	3474.5	3923.7	10.396
	750	47.219	3565.2	4037.4	10.510
	800	49.526	3657.6	4152.9	10.620
	850	51.834	3751.8	4270.2	10.727
50 (T_{sat} = 81.3°C)					
	100	3.4182	2512.0	2682.9	7.6959
	150	3.8894	2586.0	2780.5	7.9413
	200	4.3561	2659.8	2877.6	8.1583
	250	4.8206	2734.5	2975.6	8.3551
	300	5.2840	2810.5	3074.7	8.5360
	350	5.7468	2887.8	3175.1	8.7040
	400	6.2092	2966.6	3277.1	8.8614
	450	6.6715	3047.1	3380.6	9.0098
	500	7.1336	3129.1	3485.8	9.1504
	550	7.5956	3212.9	3592.6	9.2843
	600	8.0575	3298.3	3701.2	9.4123
	650	8.5193	3385.5	3811.4	9.5351
	700	8.9811	3474.4	3923.4	9.6532
	750	9.4428	3565.0	4037.2	9.7672
	800	9.9045	3657.5	4152.7	9.8775
	850	10.366	3751.7	4270.0	9.9843
100 (T_{sat} = 99.6°C)					
	100	1.6956	2506.4	2676.0	7.3610
	150	1.9363	2583.1	2776.8	7.6146
	200	2.1724	2658.1	2875.3	7.8347
	250	2.4062	2733.3	2974.0	8.0329
	300	2.6388	2809.6	3073.5	8.2146
	350	2.8708	2887.1	3174.2	8.3831
	400	3.1025	2966.1	3276.4	8.5407
	450	3.3340	3046.6	3380.0	8.6893
	500	3.5654	3128.8	3485.3	8.8300
	550	3.7966	3212.5	3592.2	8.9640
	600	4.0277	3298.0	3700.8	9.0921
	650	4.2588	3385.2	3811.1	9.2149
	700	4.4898	3474.1	3923.1	9.3331
	750	4.7208	3564.8	4036.9	9.4471
	800	4.9518	3657.3	4152.4	9.5574
	850	5.1827	3751.5	4269.8	9.6643
101.32 (T_{sat} = 100.0°C)					
	150	1.9109	2583.1	2776.7	7.6084
	200	2.1439	2658.0	2875.3	7.8286
	250	2.3747	2733.3	2973.9	8.0268
	300	2.6043	2809.6	3073.4	8.2085
	350	2.8334	2887.1	3174.2	8.3770
	400	3.0621	2966.1	3276.3	8.5347
	450	3.2905	3046.6	3380.0	8.6832
	500	3.5189	3128.7	3485.3	8.8240
	550	3.7471	3212.5	3592.2	8.9580
	600	3.9752	3298.0	3700.8	9.0860
	650	4.2033	3385.2	3811.1	9.2089
	700	4.4313	3474.1	3923.1	9.3271
	750	4.6593	3564.8	4036.9	9.4411
	800	4.8872	3657.3	4152.4	9.5513
	850	5.1152	3751.5	4269.8	9.6582
200 (T_{sat} = 120.2°C)					
	150	0.9596	2577.2	2769.1	7.2804
	200	1.0804	2654.5	2870.6	7.5072
	250	1.1989	2730.9	2970.7	7.7084
	300	1.3162	2807.8	3071.1	7.8916
	350	1.4329	2885.8	3172.4	8.0610
	400	1.5492	2965.0	3274.9	8.2192
	450	1.6653	3045.7	3378.8	8.3682
	500	1.7813	3128.0	3484.3	8.5092
	550	1.8971	3211.9	3591.3	8.6434
	600	2.0129	3297.4	3700.0	8.7716
	650	2.1286	3384.7	3810.4	8.8945
	700	2.2442	3473.7	3922.5	9.0128
	750	2.3598	3564.4	4036.4	9.1268
	800	2.4754	3656.9	4152.0	9.2372
	850	2.5909	3751.1	4269.3	9.3441
300 (T_{sat} = 133.5°C)					
	150	0.6338	2570.8	2760.9	7.0779
	200	0.7164	2650.8	2865.7	7.3122
	250	0.7965	2728.5	2967.4	7.5165
	300	0.8753	2806.1	3068.7	7.7014
	350	0.9535	2884.4	3170.5	7.8717
	400	1.0314	2963.9	3273.4	8.0305
	450	1.1091	3044.8	3377.6	8.1798
	500	1.1866	3127.2	3483.2	8.3211
	550	1.2639	3211.2	3590.4	8.4554
	600	1.3413	3296.9	3699.2	8.5838
	650	1.4185	3384.2	3809.7	8.7068
	700	1.4957	3473.2	3921.9	8.8252
	750	1.5728	3564.0	4035.8	8.9393
	800	1.6499	3656.5	4151.5	9.0497
	850	1.7270	3750.8	4268.9	9.1566

(continued)

TABLE A.5 (Continued)

Thermodynamic Properties of Steam: Superheated Vapor Table

P, kPa	*T*, °C	*v*, m³/kg	*u*, kJ/kg	*h*, kJ/kg	*s*, kJ/kg-K
400					
(T_{sat} =143.6°C)					
	150	0.4707	2563.9	2752.2	6.9287
	200	0.5343	2647.0	2860.7	7.1712
	250	0.5952	2726.0	2964.1	7.3789
	300	0.6549	2804.3	3066.2	7.5655
	350	0.7139	2883.1	3168.6	7.7367
	400	0.7725	2962.9	3271.9	7.8961
	450	0.8309	3044.0	3376.3	8.0458
	500	0.8892	3126.5	3482.2	8.1873
	550	0.9474	3210.6	3589.5	8.3219
	600	1.0054	3296.3	3698.5	8.4504
	650	1.0635	3383.7	3809.0	8.5735
	700	1.1214	3472.7	3921.3	8.6919
	750	1.1793	3563.5	4035.3	8.8062
	800	1.2372	3656.1	4151.0	8.9166
	850	1.2951	3750.4	4268.4	9.0236
600					
(T_{sat} =158.8°C)					
	200	0.3521	2639.0	2850.2	6.9669
	250	0.3939	2720.8	2957.2	7.1819
	300	0.4344	2800.6	3061.3	7.3719
	350	0.4742	2880.3	3164.8	7.5451
	400	0.5136	2960.7	3268.9	7.7057
	450	0.5528	3042.2	3373.8	7.8562
	500	0.5919	3125.0	3480.1	7.9982
	550	0.6308	3209.3	3587.7	8.1332
	600	0.6696	3295.1	3696.9	8.2619
	650	0.7084	3382.6	3807.7	8.3853
	700	0.7471	3471.8	3920.1	8.5039
	750	0.7858	3562.7	4034.2	8.6182
	800	0.8245	3655.3	4150.0	8.7287
	850	0.8631	3749.7	4267.6	8.8358
800					
(T_{sat} =170.4°C)					
	200	0.2608	2630.4	2839.1	6.8156
	250	0.2932	2715.5	2950.1	7.0388
	300	0.3241	2796.9	3056,2	7.2326
	350	0.3544	2877.5	3161.0	7.4079
	400	0.3842	2958.5	3265.8	7.5697
	450	0.4137	3040.4	3371.4	7.7209
	500	0.4432	3123.5	3478.0	7.8635
	550	0.4725	3208.0	3586.0	7.9988
	600	0.5017	3294.0	3695.4	8.1278
	650	0.5309	3381.6	3806.3	8.2514
	700	0.5600	3470.9	3918.9	8.3702
	750	0.5891	3561.9	4033.1	8.4846
	800	0.6181	3654.6	4149.1	8.5953
	850	0.6471	3749.0	4266.7	8.7024

P, kPa	*T*, °C	*v*, m³/kg	*u*, kJ/kg	*h*, kJ/kg	*s*, kJ/kg-K
1000					
(T_{sat} =179.9°C)					
	200	0.2059	2621.4	2827.3	6.6930
	250	0.2328	2710.0	2942.8	6.9251
	300	0.2580	2793.1	3051.1	7.1229
	350	0.2824	2874.7	3157.1	7.3003
	400	0.3065	2956.3	3262.7	7.4633
	450	0.3303	3038.5	3368.8	7.6154
	500	0.3540	3121.9	3475.9	7.7585
	550	0.3775	3206.6	3584.2	7.8942
	600	0.4010	3292.8	3693.8	8.0235
	650	0.4244	3380.6	3805.0	8.1473
	700	0.4477	3470.0	3917.7	8.2662
	750	0.4710	3561.0	4032.0	8.3808
	800	0.4943	3653.8	4148.1	8.4916
	850	0.5175	3748.3	4265.8	8.5988
1500					
(T_{sat} =198.3°C)					
	250	0.15199	2695.4	2923.4	6.7093
	300	0.16971	2783.3	3037.8	6.9183
	350	0.18654	2867.4	3147.2	7.1014
	400	0.20292	2950.6	3255.0	7.2677
	450	0.21906	3034.0	3362.5	7.4219
	500	0.23503	3118.1	3470.6	7.5664
	550	0.25089	3203.3	3579.7	7.7030
	600	0.26666	3289.9	3689.9	7.8331
	650	0.28237	3378.0	3801.5	7.9574
	700	0.29803	3467.6	3914.7	8.0767
	750	0.31364	3558.9	4029.4	8.1917
	800	0.32921	3651.8	4145.7	8.3027
	850	0.34475	3746.5	4263.6	8.4101
2000					
(T_{sat} = 212.4°C)					
	250	0.11145	2679.5	2902.4	6.5451
	300	0.12550	2772.9	3023.9	6.7671
	350	0.13856	2860.0	3137.1	6.9565
	400	0.15113	2944.8	3247.1	7.1263
	450	0.16343	3029.3	3356.1	7.2826
	500	0.17556	3114.2	3465.3	7.4286
	550	0.18757	3200.0	3575.1	7.5662
	600	0.19950	3287.0	3686.0	7.6970
	650	0.21137	3375.4	3798.1	7.8218
	700	0.22318	3465.3	3911.6	7.9416
	750	0.23494	3556.8	4026.7	8.0569
	800	0.24667	3649.9	4143.2	8.1681
	850	0.25836	3744.7	4261.5	8.2758

TABLE A.5 (Continued)

Thermodynamic Properties of Steam: Superheated Vapor Table

P, kPa	T, °C	v, m³/kg	u, kJ/kg	h, kJ/kg	s, kJ/kg-K	P, kPa	T, °C	v, m³/kg	u, kJ/kg	h, kJ/kg	s, kJ/kg-K
2500						5000					
(T_{sat} = 223.9°C)						(T_{sat} = 263.9°C)					
	250	0.08699	2662.2	2879.7	6.4076		300	0.045302	2698.2	2924.7	6.2085
	300	0.09893	2762.0	3009.3	6.6446		350	0.051943	2809.9	3069.6	6.4512
	350	0.10975	2852.2	3126.6	6.8409		400	0.057792	2907.7	3196.6	6.6474
	400	0.12004	2938.9	3239.1	7.0145		450	0.063252	3000.0	3316.2	6.8188
	450	0.13005	3024.6	3349.7	7.1730		500	0.068495	3090.0	3432.5	6.9743
	500	0.13987	3110.3	3459.9	7.3205		550	0.073603	3179.5	3547.5	7.1184
	550	0.14958	3196.6	3570.6	7.4592		600	0.078617	3269.2	3662.3	7.2538
	600	0.15921	3284.1	3682.1	7.5906		650	0.083560	3359.6	3777.4	7.3820
	650	0.16876	3372.8	3794.7	7.7161		700	0.088447	3451.1	3893.4	7.5044
	700	0.17827	3462.9	3908.6	7.8362		750	0.093289	3543.9	4010.4	7.6216
	750	0.18772	3554.6	4024.0	7.9518		800	0.098094	3638.2	4128.7	7.7345
	800	0.19714	3648.0	4140.8	8.0633		850	0.102867	3734.0	4248.3	7.8435
	850	0.20653	3742.9	4259.3	8.1712						
3000						6000					
(T_{sat} = 233.8°C)						(T_{sat} = 275.6°C)					
	250	0.07055	2643.2	2854.9	6.2855		300	0.036146	2667.1	2884.0	6.0669
	300	0.08116	2750.6	2994.1	6.5399		350	0.042223	2790.9	3044.2	6.3354
	350	0.09053	2844.3	3115.9	6.7437		400	0.047380	2894.3	3178.6	6.5429
	400	0.09931	2932.9	3230.9	6.9213		450	0.052104	2989.7	3302.3	6.7202
	450	0.10779	3019.8	3343.1	7.0822		500	0.056592	3081.7	3421.3	6.8793
	500	0.11608	3106.3	3454.5	7.2312		550	0.060937	3172.5	3538.1	7.0258
	550	0.12426	3193.2	3566.0	7.3709		600	0.065185	3263.1	3654.2	7.1627
	600	0.13234	3281.1	3678.1	7.5031		650	0.069360	3354.3	3770.4	7.2922
	650	0.14036	3370.2	3791.3	7.6291		700	0.073479	3446.4	3887.2	7.4154
	700	0.14832	3460.6	3905.6	7.7497		750	0.077552	3539.6	4004.9	7.5333
	750	0.15624	3552.5	4021.2	7.8656		800	0.081588	3634.3	4123.8	7.6467
	800	0.16412	3646.0	4138.4	7.9774		850	0.085592	3730.4	4243.9	7.7562
	850	0.17197	3741.1	4257.1	8.0855						
4000						7000					
(T_{sat} = 250.3°C)						(T_{sat} = 285.8°C)					
	300	0.058835	2725.8	2961.2	6.3622		300	0.029459	2631.9	2838.1	5.9299
	350	0.066448	2827.6	3093.4	6.5835		350	0.035234	2770.6	3017.2	6.2301
	400	0.073377	2920.6	3214.1	6.7699		400	0.039922	2880.4	3159.8	6.4504
	450	0.079959	3010.0	3329.8	6.9358		450	0.044132	2979.1	3288.1	6.6343
	500	0.086343	3098.2	3443.6	7.0879		500	0.048087	3073.2	3409.8	6.7971
	550	0.092599	3186.4	3556.8	7.2298		550	0.051890	3165.4	3528.6	6.9460
	600	0.098764	3275.2	3670.2	7.3636		600	0.055591	3257.0	3646.1	7.0846
	650	0.10486	3364.9	3784.3	7.4907		650	0.059218	3348.9	3763.4	7.2153
	700	0.11090	3455.9	3899.5	7.6121		700	0.062788	3441.6	3881.1	7.3394
	750	0.11690	3548.2	4015.8	7.7287		750	0.066312	3535.3	3999.5	7.4580
	800	0.12285	3642.1	4133.5	7.8411		800	0.069798	3630.3	4118.9	7.5720
	850	0.12878	3737.6	4252.7	7.9496		850	0.073253	3726.7	4239.5	7.6818

(continued)

TABLE A.5 (Continued)

Thermodynamic Properties of Steam: Superheated Vapor Table

P, kPa	*T*, °C	*v*, m³/kg	*u*, kJ/kg	*h*, kJ/kg	*s*, kJ/kg-K	*P*, kPa	*T*, °C	*v*, m³/kg	*u*, kJ/kg	*h*, kJ/kg	*s*, kJ/kg-K
8000						25,000					
(T_{sat} =295.0°C)											
	300	0.024265	2592.0	2786.2	5.7926		500	0.011128	2894.0	3172.2	5.9731
	350	0.029949	2748.7	2988.3	6.1316		550	0.012721	3024.0	3342.1	6.1861
	400	0.034311	2865.8	3140.3	6.3665		600	0.014126	3139.0	3492.2	6.3632
	450	0.038145	2968.3	3273.5	6.5574		650	0.015416	3247.0	3632.4	6.5194
	500	0.041705	3064.6	3398.2	6.7243		700	0.016631	3351.6	3767.4	6.6618
	550	0.045103	3158.2	3519.0	6.8757		750	0.017790	3454.6	3899.4	6.7941
	600	0.048395	3250.8	3638.0	7.0160		800	0.018907	3557.2	4029.8	6.9186
	650	0.051612	3343.5	3756.4	7.1479		850	0.019990	3659.9	4159.6	7.0368
	700	0.054770	3436.8	3874.9	7.2729						
	750	0.057883	3531.0	3994.0	7.3922						
	800	0.060957	3626.4	4114.0	7.5068						
	850	0.064000	3723.1	4235.1	7.6170						
10,000						30,000					
(T_{sat} =311.0°C)											
	350	0.022422	2699.5	2923.7	5.9448		500	0.008681	2833.1	3093.6	5.8080
	400	0.026409	2834.8	3098.9	6.2156		550	0.010166	2980.2	3285.2	6.0484
	450	0.029743	2945.8	3243.2	6.4226		600	0.011437	3103.9	3447.0	6.2393
	500	0.032760	3046.9	3374.5	6.5982		650	0.012582	3217.3	3594.7	6.4039
	550	0.035598	3143.6	3499.6	6.7549		700	0.013647	3325.6	3735.0	6.5519
	600	0.038320	3238.4	3621.6	6.8988		750	0.014655	3431.4	3871.1	6.6882
	650	0.040964	3332.6	3742.2	7.0332		800	0.015619	3536.2	4004.8	6.8158
	700	0.043547	3427.0	3862.5	7.1601		850	0.016549	3640.7	4137.2	6.9364
	750	0.046083	3522.2	3983.0	7.2809						
	800	0.048581	3618.4	4104.2	7.3965						
	850	0.051047	3715.8	4226.3	7.5077						
15,000						35,000					
(T_{sat} = 342.4°C)											
	350	0.011460	2536.0	2707.9	5.4667		600	0.009520	3068.0	3401.2	6.1278
	400	0.015662	2742.8	2977.8	5.8845		650	0.010562	3187.1	3556.7	6.3011
	450	0.018452	2884.1	3160.8	6.1472		700	0.011521	3299.2	3702.5	6.4548
	500	0.020796	3000.2	3312.1	6.3496		750	0.012420	3408.0	3842.6	6.5953
	550	0.022909	3105.7	3449.3	6.5216		800	0.013275	3515.0	3979.6	6.7260
	600	0.024884	3206.3	3579.6	6.6753		850	0.014095	3621.4	4114.7	6.8491
	650	0.026768	3304.8	3706.3	6.8164						
	700	0.028587	3402.4	3831.2	6.9482						
	750	0.030356	3500.1	3955.4	7.0727						
	800	0.032086	3598.3	4079.6	7.1912						
	850	0.033783	3697.4	4204.2	7.3046						
20,000						40,000					
(T_{sat} = 365.7°C)											
	400	0.009948	2622.8	2821.8	5.5595		650	0.009053	3156.5	3518.7	6.2077
	450	0.012707	2812.4	3066.5	5.9110		700	0.009930	3272.6	3669.9	6.3672
	500	0.014772	2949.4	3244.8	6.1497		750	0.010748	3384.2	3814.2	6.5118
	550	0.016548	3065.8	3396.8	6.3402		800	0.011521	3493.6	3954.4	6.6456
	600	0.018162	3173.2	3536.4	6.5049		850	0.012259	3601.8	4092.2	6.7712
	650	0.019672	3276.2	3669.7	6.6533						
	700	0.021112	3377.2	3799.5	6.7093						
	750	0.022499	3477.5	3927.5	6.9186						
	800	0.023845	3577.9	4054.8	7.0400						
	850	0.025159	3678.7	4181.9	7.1558						

Source: Properties obtained from software, STEAMCALC. John Wiley & Sons, New York, 1983.

TABLE A.6

Thermodynamic Properties of Refrigerant R-134a: Temperature Table

		v, m^3/kg; u, kJ/kg; h, kJ/kg; s, kJ/kg-K								
		Specific Volume		Internal Energy		Enthalpy			Entropy	
Temp., °C T	Press., bar P	Sat. Liquid $v_f \times 10^3$	Sat. Vapor v_g	Sat. Liquid u_f	Sat. Vapor u_g	Sat. Liquid h_f	Evap. h_{fg}	Sat. Vapor h_g	Sat. Liquid s_f	Sat. Vapor s_g
−40	0.5164	0.7055	0.3569	−0.04	204.45	0.00	222.88	222.88	0.0000	0.9560
−36	0.6332	0.7113	0.2947	4.68	206.73	4.73	220.67	225.40	0.0201	0.9506
−32	0.7704	0.7172	0.2451	9.47	209.01	9.52	218.37	227.90	0.0401	0.9456
−28	0.9305	0.7233	0.2052	14.31	211.29	14.37	216.01	230.38	0.0600	0.9411
−26	1.0199	0.7265	0.1882	16.75	212.43	16.82	214.80	231.62	0.0699	0.9390
−24	1.1160	0.7296	0.1728	19.21	213.57	19.29	213.57	232.85	0.0798	0.9370
−22	1.2192	0.7328	0.1590	21.68	214.70	21.77	212.32	234.08	0.0897	0.9351
−20	1.3299	0.7361	0.1464	24.17	215.84	24.26	211.05	235.31	0.0996	0.9332
−18	1.4483	0.7395	0.1350	26.67	216.97	26.77	209.76	236.53	0.1094	0.9315
−16	1.5748	0.7428	0.1247	29.18	218.10	29.30	208.45	237.74	0.1192	0.9298
−12	1.8540	0.7498	0.1068	34.25	220.36	34.39	205.77	240.15	0.1388	0.9267
−8	2.1704	0.7569	0.0919	39.38	222.60	39.54	203.00	242.54	0.1583	0.9239
−4	2.5274	0.7644	0.0794	44.56	224.84	44.75	200.15	244.90	0.1777	0.9213
0	2.9282	0.7721	0.0689	49.79	227.06	50.02	197.21	247.23	0.1970	0.9190
4	3.3765	0.7801	0.0600	55.08	229.27	55.35	194.19	249.53	0.2162	0.9169
8	3.8756	0.7884	0.0525	60.43	231.46	60.73	191.07	251.80	0.2354	0.9150
12	4.4294	0.7971	0.0460	65.83	233.63	66.18	187.85	254.03	0.2545	0.9132
16	5.0416	0.8062	0.0405	71.29	235.78	71.69	184.52	256.22	0.2735	0.9116
20	5.7160	0.8157	0.0358	76.80	237.91	77.26	181.09	258.36	0.2924	0.9302
24	6.4566	0.8257	0.0317	82.37	240.01	82.90	177.55	260.45	0.3113	0.9089
26	6.8530	0.8309	0.0298	85.18	241.05	85.75	175.73	261.48	0.3208	0.9082
28	7.2675	0.8362	0.0281	88.00	242.08	88.61	173.89	262.50	0.3302	0.9076
30	7.7006	0.8417	0.0265	90.84	243.10	91.49	172.00	263.50	0.3396	0.9070
32	8.1528	0.8473	0.0250	93.70	244.12	94.39	170.09	264.48	0.3490	0.9064
34	8.6247	0.8530	0.0236	96.58	245.12	97.31	168.14	265.45	0.3584	0.9058
36	9.1168	0.8590	0.0223	99.47	246.11	100.25	166.15	266.40	0.3678	0.9053
38	9.6298	0.8651	0.0210	102.38	247.09	103.21	164.12	267.33	0.3772	0.9047
40	10.164	0.8714	0.0199	105.30	248.06	106.19	162.05	268.24	0.3866	0.9041
42	10.720	0.8780	0.0188	108.25	249.02	109.19	159.04	269.14	0.3960	0.9035
44	11.299	0.8847	0.0177	111.22	249.96	112.22	157.79	270.01	0.4054	0.9030
48	12.526	0.8989	0.0159	117.22	251.79	118.35	153.33	271.68	0.4243	0.9017
52	13.851	0.9142	0.0142	123.31	253.55	124.58	148.66	273.24	0.4432	0.9004
56	15.278	0.9308	0.0127	129.51	255.23	130.93	143.75	274.68	0.4622	0.8990
60	16.813	0.9488	0.0114	135.82	256.81	137.42	138.57	275.99	0.4814	0.8973
70	21.162	1.0027	0.0086	152.22	260.15	154.34	124.08	278.43	0.5302	0.8918
80	26.324	1.0766	0.0064	169.88	262.14	172.71	106.41	279.12	0.5814	0.8827
90	32.435	1.1949	0.0046	189.82	261.34	193.69	82.63	276.32	0.6380	0.8655
100	39.742	1.5443	0.0027	218.60	248.49	224.74	34.40	259.13	0.7196	0.8117

Source: Properties generated from D. P. Wilson and R. S. Basu, Thermodynamic Properties of a New Stratospherically Safe Working Fluid—Refrigerant 134a. *ASHRAE Trans.*, (94) 2, 2095–2118 (1988).

TABLE A.7

Thermodynamic Properties of Refrigerant R-134a: Pressure Table

v, m^3/kg; *u*, kJ/kg; *h*, kJ/kg; s, kJ/kg-K										
		Specific Volume		Internal Energy		Enthalpy			Entropy	
Press., bar P	Temp., °C T	Sat. Liquid $v_f \times 10^3$	Sat. Vapor v_g	Sat. Liquid u_f	Sat. Vapor u_g	Sat. Liquid h_f	Evap. h_{fg}	Sat. Vapor h_g	Sat. Liquid s_f	Sat. Vapor s_g
0.6	−37.07	0.7097	0.3100	3.41	206.12	3.46	221.27	224.72	0.0147	0.9520
0.8	−31.21	0.7184	0.2366	10.41	209.46	10.47	217.92	228.39	0.0440	0.9447
1.0	−26.43	0.7258	0.1917	16.22	212.18	16.29	215.06	231.35	0.0678	0.9395
1.2	−22.36	0.7323	0.1614	21.23	214.50	21.32	212.54	233.86	0.0879	0.9354
1.4	−18.80	0.7381	0.1395	25.66	216.52	25.77	210.27	236.04	0.1055	0.9322
1.6	−15.62	0.7435	0.1229	29.66	218.32	29.78	208.19	237.97	0.1211	0.9295
1.8	−12.73	0.7485	0.1098	33.31	219.94	33.45	206.26	239.71	0.1352	0.9273
2.0	−10.09	0.7532	0.0993	36.69	221.43	36.84	204.46	241.30	0.1481	0.9253
2.4	−5.37	0.7618	0.0834	42.77	224.07	42.95	201.14	244.09	0.1710	0.9222
2.8	−1.23	0.7697	0.0719	48.18	226.38	48.39	198.33	246.52	0.1911	0.9197
3.2	2.48	0.7770	0.0632	53.06	228.43	53.31	195.35	248.66	0,2089	0.9177
3.6	5.84	0.7839	0.0564	57.54	230.28	57.82	192.76	250.58	0.2251	0.9166
4.0	8.93	0.7904	0.0509	61.69	231.97	62.00	190.32	252.32	0.2399	0.9145
5.0	15.74	0.8056	0.0409	70.93	235.64	71.33	184.74	256.07	0.2723	0.9117
6.0	21.58	0.8196	0.0341	78.99	238.74	79.48	179.71	259.19	0.2999	0.9097
7.0	26.72	0.8328	0.0292	86.19	241.42	86.78	175.07	261.85	0.3242	0.9080
8.0	31.33	0.8454	0.0255	92.75	243.78	93.42	170.73	264.15	0.3459	0.9066
9.0	35.53	0.8576	0.0226	98.79	245.88	99.56	166.62	266.18	0.3656	0.9054
10.0	39.39	0.8695	0.0202	104.42	247.77	105.29	162.68	267.97	0.3838	0.9043
12.0	46.32	0.8928	0.0166	114.69	251.03	115.76	155.23	270.99	0.4164	0.9023
14.0	52.43	0.9159	0.0140	123.98	253.74	125.26	148.14	273.40	0.4453	0.9003
16.0	57.92	0.9392	0.0121	132.52	256.00	134.02	141.31	275.33	0.4714	0.8982
18.0	62.91	0.9631	0.0105	140.49	257.88	142.22	134.60	276.83	0.4954	0.8959
20.0	67.49	0.9878	0.0093	148.02	259.41	149.99	127.95	277.94	0.5178	0.8934
25.0	77.59	1.0562	0.0069	165.48	261.84	168.12	111.06	279.17	0.5687	0.8854
30.0	86.22	1.1416	0.0053	181.88	262.16	185.30	92.71	278.01	0.6156	0.8735

Source: Properties generated from D. P. Wilson and R. S. Basu, Thermodynamic Properties of a New Stratospherically Safe Working Fluid—Refrigerant 134a. *ASHRAE Trans.*, (94) 2, 2095–2118 (1988).

TABLE A.8

Thermodynamic Properties of Refrigerant R-134a: Superheated Vapor Table

$T,°C; v, m^3/kg; u, kJ/kg; h, kJ/kg; s, kJ/kg\text{-}K$								
T	v	u	h	s	v	u	h	s
	0.6 bar (0.060 MPa) (T_{sat} = −37.07°C)				**1.0 bar (0.10 MPa) (T_{sat} = −26.43°C)**			
Sat.	0.31003	206.12	224.72	0.9520	0.19170	212.18	231.35	0.9395
−20	0.33536	217.86	237.98	1.0062	0.19770	216.77	236.54	0.9602
−10	0.34992	224.97	245.96	1.0371	0.20686	224.01	244.70	0.9918
0	0.36433	232.24	254.10	1.0675	0.21587	231.41	252.99	1.0227
10	0.37861	239.69	262.41	1.0973	0.22473	238.96	261.43	1.0531
20	0.39279	247.32	270.89	1.1267	0.23349	246.67	270.02	1.0829
30	0.40688	255.12	279.53	1.1557	0.24216	254.54	278.76	1.1122
40	0.42091	263.10	288.35	1.1844	0.25076	262.58	287.66	1.1411
50	0.43487	271.25	297.34	1.2126	0.25930	270.79	296.72	1.1696
60	0.44879	279.58	306.51	1.2405	0.26779	279.16	305.94	1.1977
70	0.46266	288.08	315.84	1.2681	0.27623	287.70	315.32	1.2254
80	0.47650	296.75	325.34	1.2954	0.28464	296.40	324.87	1.2528
90	0.49031	305.58	335.00	1.3224	0.29302	305.27	334.57	1.2799
	1.4 bars (0.14 MPa) (T_{sat} = −18.80°C)				**1.8 bars (0.18 MPa) (T_{sat} = −12.73°C)**			
Sat.	0.13945	216.52	236.04	0.9322	0.10983	219.94	239.71	0.9273
−10	0.14549	223.03	243.40	0.9606	0.11135	222.02	242.06	0.9362
0	0.15219	230.55	251.86	0.9922	0.11678	229.67	250.69	0.9684
10	0.15875	238.21	260.43	1.0230	0.12207	237.44	259.41	0.9998
20	0.16520	246.01	269.13	1.0532	0.12723	245.33	268.23	1.0304
30	0.17155	253.96	277.97	1.0828	0.13230	253.36	277.17	1.0604
40	0.17783	262.06	286.96	1.1120	0.13730	261.53	286.24	1.0898
50	0.18404	270.32	296.09	1.1407	0.14222	269.85	295.45	1.1187
60	0.19020	278.74	305.37	1.1690	0.14710	278.31	304.79	1.1472
70	0.19633	287.32	314.80	1.1969	0.15193	286.93	314.28	1.1753
80	0.20241	296.06	324.39	1.2244	0.15672	295.71	323.92	1.2030
90	0.20846	304.95	334.14	1.2516	0.16148	304.63	333.70	1.2303
	2.0 bars (0.20 MPa) (T_{sat} = −10.09°C)				**2.4 bars (0.24 MPa) (T_{sat} = −5.37°C)**			
Sat.	0.09933	221.43	241.30	0.9253	0.08343	224.07	244.09	0.9222
−10	0.09938	221.50	241.38	0.9256				
0	0.10438	229.23	250.10	0.9582	0.08574	228.31	248.89	0.9399
10	0.10922	237.05	258.89	0.9898	0.08993	236.26	257.84	0.9721
20	0.11394	244.99	267.78	1.0206	0.09399	244.30	266.85	1.0034
30	0.11856	253.06	276.77	1.0508	0.09794	252.45	275.95	1.0339
40	0.12311	261.26	285.88	1.0804	0.10181	260.72	285.16	1.0637
50	0.12758	269.61	295.12	1.1094	0.10562	269.12	294.47	1.0930
60	0.13201	278.10	304.50	1.1380	0.10937	277.67	303.91	1.1218
70	0.13639	286.74	314.02	1.1661	0.11307	286.35	313.49	1.1501
80	0.14073	295.53	323.68	1.1939	0.11674	295.18	323.19	1.1780
90	0.14504	304.47	333.48	1.2212	0.12037	304.15	333.04	1.2055

(continued)

TABLE A.8 (Continued)

Thermodynamic Properties of Refrigerant R-134a: Superheated Vapor Table

T,°C; v, m^3/kg; u, kJ/kg; h, kJ/kg; s, kJ/kg-K

T	v	u	h	s	v	u	h	s
	2.8 bars (0.28 MPa) (T_{sat} = −1.23°C)				**3.2 bars (0.32 MPa) (T_{sat} = 2.48°C)**			
Sat.	0.07193	226.38	246.52	0.9197	0.06322	228.43	248.66	0.9177
0	0.07240	227.37	247.64	0.9238				
10	0.07613	235.44	256.76	0.9566	0.06576	234.61	255.65	0.9427
20	0.07972	243.59	265.91	0.9883	0.06901	242.87	264.95	0.9749
30	0.08320	251.83	275.12	1.0192	0.07214	251.19	274.28	1.0062
40	0.08660	260.17	284.42	1.0494	0.07518	259.61	283.67	1.0367
50	0.08992	268.64	293.81	1.0789	0.07815	268.14	293.15	1.0665
60	0.09319	277.23	303.32	1.1079	0.08106	276.79	302.72	1.0957
70	0.09641	285.96	312.95	1.1364	0.08392	285.56	312.41	1.1243
80	0.09960	294.82	322.71	1.1644	0.08674	294.46	322.22	1.1525
90	0.10275	303.83	332.60	1.1920	0.08953	303.50	332.15	1.1802
100	0.10587	312.98	342.62	1.2193	0.09229	312.68	342.21	1.2076
	4.0 bars (0.40 MPa) (T_{sat} = 8.93°C)				**5.0 bars (0.50 MPa) (T_{sat} = 15.74°C)**			
Sat.	0.05089	231.97	252.32	0.9145	0.04086	235.64	256.07	0.9117
10	0.05119	232.87	253.35	0.9182				
20	0.05397	241.37	262.96	0.9515	0.04188	239.40	260.34	0.9264
30	0.05662	249.89	272.54	0.9837	0.04416	248.20	270.28	0.9597
40	0.05917	258.47	282.14	1.0148	0.04633	256.99	280.16	0.9918
50	0.06164	267.13	291.79	1.0452	0.04842	265.83	290.04	1.0229
60	0.06405	275.89	301.51	1.0748	0.05043	274.73	299.95	1.0531
70	0.06641	284.75	311.32	1.1038	0.05240	283.72	309.92	1.0825
80	0.06873	293.73	321.23	1.1322	0.05432	292.80	319.96	1.1114
90	0.07102	302.84	331.25	1.1602	0.05620	302.00	330.10	1.1397
100	0.07327	312.07	341.38	1.1878	0.05805	311.31	340.33	1.1675
110	0.07550	321.44	351.64	1.2149	0.05988	320.74	350.68	1.1949
	6.0 bars (0.60 MPa) (T_{sat} = 21.58°C)				**7.0 bars (0.70 MPa) (T_{sat} = 26.72°C)**			
Sat.	0.03408	238.74	259.19	0.9097	0.02918	241.42	261.85	0.9080
30	0.03581	246.41	267.89	0.9388	0.02979	244.51	265.37	0.9197
40	0.03774	255.45	278.09	0.9719	0.03157	253.83	275.93	0.9539
50	0.03958	264.48	288.23	1.0037	0.03324	263.08	286.35	0.9867
60	0.04134	273.54	298.35	1.0346	0.03482	272.31	296.69	1.0182
70	0.04304	282.66	308.48	1.0645	0.03634	281.57	307.01	1.0487
80	0.04469	291.86	318.67	1.0938	0.03781	290.88	317.35	1.0784
90	0.04631	301.14	328.93	1.1225	0.03924	300.27	327.74	1.1074
100	0.04790	310.53	339.27	1.1505	0.04064	309.74	338.19	1.1358
110	0.04946	320.03	349.70	1.1781	0.04201	319.31	348.71	1.1637
120	0.05099	329.64	360.24	1.2053	0.04335	328.98	359.33	1.1910
130	0.05251	339.93	370.88	1.2320	0.04468	338.76	370.04	1.2179

TABLE A.8 (Continued)

Thermodynamic Properties of Refrigerant R-134a: Superheated Vapor Table

T, °C; v, m^3/kg; u, kJ/kg; h, kJ/kg; s, kJ/kg-K								
T	v	u	h	s	v	u	h	s
	8.0 bars (0.80 MPa) (T_{sat} = 31.33°C)				**9.0 bars (0.90 MPa) (T_{sat} = 35.53°C)**			
Sat.	0.02547	243.78	264.15	0.9066	0.02255	245.88	266.18	0.9054
40	0.02691	252.13	273.66	0.9374	0.02325	250.32	271.25	0.9217
50	0.02846	261.62	284.39	0.9711	0.02472	260.09	282.34	0.9566
60	0.02992	271.04	294.98	1.0034	0.02609	269.72	293.21	0.9897
70	0.03131	280.45	305.50	1.0345	0.02738	279.30	303.94	1.0214
80	0.03264	289.89	316.00	1.0647	0.02861	288.87	314.62	1.0521
90	0.03393	299.37	326.52	1.0940	0.02980	298.46	325.28′	1.0819
100	0.03519	308.93	337.08	1.1227	0.03095	308.11	335.96	1.1109
110	0.03642	318.57	347.71	1.1508	0.03207	317.82	346.68	1.1392
120	0.03762	328.31	358.40	1.1784	0.03316	327.62	357.47	1.1670
130	0.03881	338.14	369.19	1.2055	0.03423	337.52	368.33	1.1943
140	0.03997	348.09	380.07	1.2321	0.03529	347.51	379.27	1.2211
	10.0 bars (1.00 MPa) (T_{sat} = 39.39°C)				**12.0 bars (1.20 MPa) (T_{sat} = 46.32°C)**			
Sat.	0.02020	247.77	267.97	0.9043	0.01663	251.03	270.99	0.9023
40	0.02029	248.39	268.68	0.9066				
50	0.02171	258.48	280.19	0.9428	0.01712	254.98	275.52	0.9164
60	0.02301	268.35	291.36	0.9768	0.01835	265.42	287.44	0.9527
70	0.02423	278.11	302.34	1.0093	0.01947	275.59	298.96	0.9868
80	0.02538	287.82	313.20	1.0405	0.02051	285.62	310.24	1.0192
90	0.02649	297.53	324.01	1.0707	0.02150	295.59	321.39	1.0503
100	0.02755	307.27	334.82	1.1000	0.02244	305.54	332.47	1.0804
110	0.02858	317.06	345.65	1.1286	0.02335	315.50	343.52	1.1096
120	0.02959	326.93	356.52	1.1567	0.02423	325.51	354.58	1.1381
130	0.03058	336.88	367.46	1.1841	0.02508	335.58	365.68	1.1660
140	0.03154	346.92	378.46	1.2111	0.02592	345.73	376.83	1.1933
	14.0 bars (1.40 MPa) (T_{sat} = 52.43°C)				**16.0 bars (1.60 MPa) (T_{sat} = 57.92°C)**			
Sat.	0.01405	253.74	273.40	0.9003	0.01208	256.00	275.33	0.8982
60	0.01495	262.17	283.10	0.9297	0.01233	258.48	278.20	0.9069
70	0.01603	272.87	295.31	0.9658	0.01340	269.89	291.33	0.9457
80	0.01701	283.29	307.10	0.9997	0.01435	280.78	303.74	0.9813
90	0.01792	293.55	318.63	1.0319	0.01521	291.39	315.72	1.0148
100	0.01878	303.73	330.02	1.0628	0.01601	301.84	327.46	1.0467
110	0.01960	313.88	341.32	1.0927	0.01677	312.20	339.04	1.0773
120	0.02039	324.05	352.59	1.1218	0.01750	322.53	350.53	1.1069
130	0.02115	334.25	363.86	1.1501	0.01820	332.87	361.99	1.1357
140	0.02189	344.50	375.15	1.1777	0.01887	343.24	373.44	1.1638
150	0.02262	354.82	386.49	1.2048	0.01953	353.66	384.91	1.1912
160	0.02333	365.22	397.89	1.2315	0.02017	364.15	369.43	1.2181

Source: Properties generated from D. P. Wilson and R. S. Basu, Thermodynamic Properties of a New Stratospherically Safe Working Fluid—Refrigerant 134a. *ASHRAE Trans.*, (94) 2, 2095–2118 (1988).

TABLE A.9

Critical Constants

Gas	Critical Temperature, K	Critical Pressure, bar	Critical Compressibility, Z_c	Molecular Weight, M
Acetylene	309	62.8	0.274	26.04
Air	133	37.7	0.284	28.97
Ammonia	406	112.8	0.242	17.04
Argon	151	48.6	0.290	39.94
Benzene	563	49.3	0.274	78.11
Butane	425	38.0	0.274	58.12
Carbon dioxide	304	73.9	0.276	44.01
Carbon monoxide	133	35.0	0.294	28.01
Ethane	305	48.8	0.285	30.01
Ethylene	283	63.8	0.270	28.05
Helium	5.2	2.3	0.300	4.003
Hydrogen	33.2	13.0	0.304	2.016
Methane	191	46.4	0.290	16.04
Nitrogen	126	33.9	0.291	28.01
Octane	569	24.9	0.258	114.22
Oxygen	154	50.5	0.290	32.00
Propane	370	42.7	0.276	44.09
Propylene	365	46.2	0.276	42.08
Refrigerant 12	385	41.2	0.278	120.92
Refrigerant 134a	374	40.7	0.260	102.03
Water	647.3	220.9	0.233	18.02

Source: L. C. Nelson and E. F. Obert, Generalized Compressibility Charts. *Chem. Eng.* 61, 203 (1954).

TABLE A.10

Constants for the van der Waals Equation of State[a]

Gas	a_r bar-(m^3/kg mol))^2	b_r m^3/kg mol
Acetylene	4.410	0.0510
Air	1.368	0.0367
Ammonia	4.223	0.0373
Benzene	18.63	0.1181
Butane	13.860	0.1162
Carbon dioxide	3.647	0.0428
Carbon monoxide	1.474	0.0395
Ethane	5.575	0.0650
Ethylene	4.563	0.0574
Helium	0.0341	0.0234
Hydrogen	0.2476	0.0265
Methane	2.293	0.0428
Nitrogen	1.366	0.0386
Oxygen	1.369	0.0317
Propane	9.349	0.0901
Refrigerant 134a	10.05	0.0957
Water	5.531	0.0305

[a] Determined from the generalized behavior of gases represented by the derivatives $\partial P/\partial \bar{v} = 0$ and $\partial^2 P/\partial \bar{v}^2 = 0$.

TABLE A.11

Constants for the Redlich-Kwong Equation of State[a]

Gas	a_r bar-(m^3/kg mol)^2 K$^{1/2}$	b_r m^3/kg mol
Carbon dioxide	64.64	0.02969
Carbon monoxide	17.26	0.02743
Methane	32.19	0.02969
Nitrogen	15.59	0.002681
Oxygen	17.38	0.02199
Propane	183.070	0.06269
Refrigerant 134a	197.1	0.06634
Water	142.64	0.02110

[a] Determined from the generalized behavior of gases represented by the derivatives $\partial P/\partial \bar{v} = 0$ and $\partial^2 P/\partial \bar{v}^2 = 0$.

TABLE A.12

Thermodynamic Properties of Air

T, K	P_r	v_r	u, kJ/kg	h, kJ/kg	s°, kJ/kg-K
230	0.5477	1205.0	164.00	230.02	1.4356
240	0.6355	1084.0	171.13	240.02	1.4782
250	0.7329	979.0	178.28	250.05	1.1925
260	0.8405	887.8	185.45	260.09	1.5585
270	0.9590	808.0	192.60	270.11	1.5963
280	1.0889	738.0	199.75	280.13	1.6328
285	1.1584	706.1	203.33	285.14	1.6506
290	1.2311	676.1	206.91	290.16	1.6680
295	1.3068	647.9	210.49	295.17	1.6852
300	1.3860	621.2	214.07	300.19	1.7020
305	1.4686	596.0	217.67	305.22	1.7186
310	1.5546	572.3	221.25	310.24	1.7350
315	1.6442	549.8	224.85	315.27	1.7511
320	1.7375	528.6	228.42	320.29	1.7669
325	1.8345	508.4	232.02	325.31	1.7825
330	1.9352	489.4	235.61	330.34	1.7978
340	2.149	454.1	242.82	340.42	1.8279
350	2.379	422.2	250.02	350.49	1.8571
360	2.626	393.4	257.24	360.58	1.8854
370	2.892	367.2	264.46	370.67	1.9131
380	3.176	343.4	271.69	380.77	1.9400
390	3.481	321.5	278.93	390.88	1.9663
400	3.806	301.6	286.16	400.98	1.9919
410	4.153	283.3	293.43	411.32	2.0170
420	4.522	266.6	300.69	421.26	2.0414
430	4.915	251.1	307.99	431.43	2.0653
440	5.332	236.8	315.30	441.61	2.0887
450	5.775	223.6	322.82	451.80	2.1161
460	6.245	211.4	329.97	462.02	2.1341
470	6.742	200.1	337.32	472.24	2.1560
480	7.268	189.5	344.70	482.49	2.1776
490	7.824	179.7	352.08	492.74	2.1988
500	8.411	170.6	359.49	503.02	2.2195
510	9.031	162.1	366.92	513.32	2.2399
520	9.684	154.1	374.36	523.63	2.2600
530	10.37	146.7	381.84	533.98	2.2797
540	11.10	139.7	389.34	544.35	2.2991
550	11.86	133.1	396.86	555.74	2.3181
560	12.66	127.0	404.42	565.17	2.3369
570	13.50	121.2	411.97	575.59	2.3553
580	14.38	115.7	419.55	586.04	2.3735
590	15.31	110.6	427.15	596.57	2.3914
600	16.28	105.8	434.78	607.02	2.4090
610	17.30	101.2	442.42	617.53	2.4264

TABLE A.12 (Continued)

Thermodynamic Properties of Air

T, K	P_r	v_r	u, kJ/kg	h, kJ/kg	$s°$, kJ/kg-K
620	18.36	96.92	450.09	628.07	2.4436
630	19.84	92.84	457.78	638.63	2.4605
640	20.64	88.99	465.50	649.22	2.4772
650	21.86	85.34	473.25	659.84	2.4936
660	23.13	81.89	481.01	670.47	2.5099
670	24.46	78.61	488.81	681.14	2.5259
680	25.85	75.50	496.62	691.82	2.5418
690	27.29	72.56	504.45	702.51	2.5573
700	28.80	69.76	512.33	713.27	2.5728
720	32.02	64.53	528.14	734.82	2.6032
740	35.50	59.82	544.02	756.44	2.6328
760	39.27	55.54	560.01	778.18	2.6618
780	43.35	51.64	576.12	800.03	2.6901
800	47.75	48.08	592.30	821.95	2.7179
820	52.59	44.84	608.59	843.98	2.7450
840	57.60	41.85	624.95	866.08	2.7717
860	63.09	39.12	641.40	888.27	2.7978
880	68.98	36.61	657.95	910.56	2.8234
900	75.29	34.31	674.58	932.93	2.8459
920	82.05	32.18	691.28	955.38	2.8732
940	89.28	30.22	708.08	977.92	2.8975
960	97.00	28.40	725.02	1000.55	2.9213
980	105.2	26.73	741.98	1023.55	2.9497
1000	114.0	25.17	758.94	1046.04	2.9677
1020	123.4	23.72	776.10	1068.89	2.9903
1040	133.3	22.39	793.36	1091.85	3.0126
1060	143.9	21.14	810.62	1114.86	3.0345
1080	155.2	19.98	827.88	1137.89	3.0561
1100	167.1	18.896	845.33	1161.07	3.0773
1120	179.7	17.886	862.79	1184.28	3.0983
1140	193.1	16.946	880.35	1207.57	3.1188
1160	207.2	16.064	897.91	1230.92	3.1392
1180	222.2	15.241	915.57	1254.34	3.1592
1200	238.0	14.470	933.33	1277.79	3.1789
1220	254.7	13.747	951.09	1301.31	3.1983
1240	272.2	13.069	968.95	1324.93	3.2175
1260	290.8	12.435	986.90	1348.55	3.2364
1280	310.4	11.835	1004.76	1372.24	3.2551
1300	330.9	11.275	1022.82	1395.97	3.2735
1320	352.5	10.747	1040.88	1429.76	3.2916
1340	375.3	10.247	1058.94	1443.60	3.3096
1360	399.1	9.780	1077.10	1467.49	3.3272
1380	424.2	9.337	1095.26	1491.44	3.3447
1400	450.5	8.919	1113.52	1515.42	3.3620
1420	478.0	8.526	1131.77	1539.44	3.3790

(*continued*)

TABLE A.12 (Continued)

Thermodynamic Properties of Air

T, K	P_r	v_r	u, kJ/kg	h, kJ/kg	s°, kJ/kg-K
1440	506.9	8.153	1150.13	1563.51	3.3959
1460	537.1	7.801	1168.49	1587.63	3.4125
1480	568.8	7.468	1186.95	1611.79	3.4289
1500	601.9	7.152	1205.41	1635.97	3.4452
1520	636.5	6.854	1223.87	1660.23	3.4612
1540	672.8	6.569	1242.43	1684.51	3.4771
1560	710.5	6.301	1260.99	1708.82	3.4928
1580	750.0	6.046	1279.65	1733.17	3.5083
1600	791.2	5.804	1298.30	1757.57	3.5236
1620	834.1	5.574	1316.96	1782.00	3.5388
1640	878.9	5.355	1335.72	1806.76	3.5538
1660	925.6	5.147	1354.48	1839.96	3.5687
1680	974.6	4.949	1373.24	1855.50	3.5834
1700	1025	4.761	1392.7	1880.1	3.5979
1750	1161	4.328	1439.8	1941.6	3.6336
1800	1310	3.944	1487.2	2003.2	3.6684
1850	1475	3.601	1534.9	2065.3	3.7013
1900	1655	3.295	1582.6	2127.4	3.7354
1950	1852	3.022	1630.6	2189.7	3.7667
2000	2068	2.776	1678.7	2251.1	3.7994
2050	2303	2.555	1726.8	2314.6	3.8303
2100	2559	2.356	1775.3	2377.7	3.8605
2150	2837	2.175	1823.8	2440.3	3.8901
2200	3138	2.012	1872.4	2503.2	3.9191
2250	3464	1.864	1921.3	2566.4	3.9474

Source: J. H. Keenan and J. Kaye, *Gas Tables*. John Wiley & Sons, New York, 1945.

TABLE A.13

Thermal Properties of Air at Atmospheric Pressure

T, K	ρ, kg/m^3	c_p, kJ/kg-K	μ, kg/m-s $\times 10^5$	v, m^2/s $\times 10^6$	k, W/m-K	α, m^2/s $\times 10^4$	Pr
100	3.6010	1.0266	0.6924	1.923	0.0092	0.0250	0.770
150	2.3675	1.0099	1.0283	4.343	0.0137	0.0575	0.753
200	1.7684	1.0061	1.3289	7.490	0.0181	0.3017	0.739
250	1.4128	1.0053	1.5990	11.31	0.0228	0.1568	0.722
300	1.1774	1.0057	1.8462	15.69	0.0262	0.2216	0.708
350	0.9980	1.0090	2.075	20.76	0.0300	0.2983	0.697
400	0.8826	1.0140	2.286	25.90	0.0337	0.3760	0.689
450	0.7823	1.0207	2.484	31.71	0.0371	0.4220	0.683
500	0.7048	1.0295	2.671	37.90	0.0404	0.5564	0.680
550	0.6423	1.0392	2.848	44.34	0.0436	0.6532	0.680
600	0.5879	1.0551	3.018	51.34	0.0466	0.7512	0.680
650	0.5430	1.0635	3.177	58.51	0.0495	0.8578	0.682
700	0.5030	1.0752	3.332	66.25	0.0523	0.9672	0.684
750	0.4709	1.0856	3.481	73.91	0.0551	1.0774	0.686
800	0.4405	1.0978	3.625	82.29	0.0578	1.1951	0.689
900	0.3925	1.1212	3.899	99.30	0.0629	1.4271	0.696
1000	0.3524	1.1417	4.152	117.8	0.0675	1.6679	0.702
1100	0.3204	1.160	4.440	138.6	0.0732	1.969	0.704
1200	0.2947	1.179	4.690	159.1	0.0782	2.251	0.707
1300	0.2707	1.197	4.930	182.1	0.0837	2.583	0.705
1400	0.2515	1.214	5.170	205.5	0.0891	2.920	0.705
1500	0.2355	1.230	5.400	229.1	0.0946	3.262	0.705
1600	0.2211	1.248	5.630	254.5	0.1000	3.609	0.705
1800	0.1970	1.287	6.070	308.1	0.1110	4.379	0.704
2000	0.1762	1.338	6.50	369.0	0.1240	5.260	0.700

Source: U.S. National Bureau of Standards, Tables of Thermal Properties of Gases, Circular 564, 1955.

TABLE A.14

Physical Properties of Water at Atmospheric Pressure

T,°C	ρ, kg/m^3	c, kJ/kg-K	μ, kg/m-s $\times 10^3$	v, m^2/s $\times 10^6$	k, W/m-K	α, m^2/s $\times 10^6$	Pr	β, K^{-1} $\times 10^3$
0	1000.0	4.194	1.790	1.790	0.566	0.1350	13.26	0.000
10	1000.0	4.202	1.310	1.310	0.585	0.1392	9.41	0.100
20	998.0	4.190	1.010	1.012	0.602	0.1440	7.03	0.200
30	996.0	4.179	0.803	0.8062	0.619	0.1487	5.42	0.290
40	992.6	4.177	0.656	0.6609	0.633	0.1527	4.33	0.380
50	988.1	4.178	0.536	0.5435	0.644	0.1560	3.48	0.460
60	983.3	4.183	0.475	0.4831	0.654	0.1590	3.04	0.530
70	977.8	4.187	0.408	0.4173	0.664	0.1622	2.573	0.590
80	971.8	4.197	0.359	0.3694	0.671	0.1645	2.245	0.640
90	965.4	4.206	0.318	0.3294	0.676	0.1665	1.979	0.690
100	958.6	4.219	0.283	0.2952	0.682	0.1686	1.751	0.740
120	943.4	4.251	0.229	0.2427	0.085	0.1708	1.421	0.860
140	926.4	4.294	0.196	0.2116	0.685	0.1722	1.229	0.960
160	907.8	4.350	0.174	0.1917	0.681	0.1725	1.111	1.080
180	887.3	4.420	0.154	0.1736	0.676	0.1724	1.007	1.200
200	865.0	4.506	0.139	0.1607	0.667	0.1711	0.939	1.380
220	840.8	4.616	0.127	0.1511	0.654	0.1685	0.896	1.540
240	813.7	4.761	0.116	0.1426	0.638	0.1647	0.866	1.700
260	783.7	4.955	0.107	0.1365	0.616	0.1586	0.861	2.140
280	750.8	5.224	0.098	0.1305	0.581	0.1481	0.881	2.390
300	712.3	5.610	0.091	0.1278	0.530	0.1326	0.963	2.880

Source: J. P. Todd and H. B. Ellis, *Applied Heat Transfer*. Harper & Row, New York, 1982.

TABLE A.15

Physical Properties of Gases at Atmospheric Pressure

T, K	ρ, kg/m^3	c_p, kJ/kg-K	k, W/m-K	μ, kg/m-s $\times 10^6$	ν, m^2/s $\times 10^6$	Pr
N_2						
200	1.7108	1.0429	0.0182	12.95	7.57	0.747
300	1.1421	1.0408	0.0262	17.84	15.63	0.713
400	0.8538	1.0459	0.0334	21.96	25.74	0.691
500	0.6824	1.0555	0.0398	25.70	27.66	0.684
600	0.5687	1.0756	0.0458	29.11	51.19	0.686
700	0.4934	1.0969	0.0512	32.13	65.13	0.691
800	0.4277	1.1225	0.0561	34.84	81.46	0.700
O_2						
200	1.9959	0.9131	0.0182	14.85	7.593	0.745
250	1.5618	0.9157	0.0226	17.87	11.45	0.725
300	1.3007	0.9203	0.0268	20.63	15.86	0.709
350	1.1133	0.9291	0.0307	23.16	20.80	0.702
400	0.9755	0.9420	0.0346	25.54	26.18	0.695
450	0.8682	0.9567	0.0383	27.77	31.99	0.684
500	0.7801	0.9722	0.0417	29.91	38.34	0.697
550	0.7096	0.9881	0.0452	31.07	45.05	0.700
CO_2						
250	2.1657	0.804	0.0129	12.59	5.813	0.793
300	1.7973	0.871	0.0166	14.96	8.321	0.770
350	1.5362	0.900	0.0205	17.21	11.19	0.747
400	0.3424	0.942	0.0246	19.32	14.39	0.738
450	1.1918	0.980	0.0290	21.34	17.90	0.721
500	1.0732	1.013	0.0335	23.26	21.67	0.702
550	0.9739	1.047	0.0382	25.08	25.74	0.685
600	0.8938	1.076	0.0483	26.83	30.02	0.686
CO						
300	1.139	1.042	0.0253	17.84	15.67	0.737
350	0.974	1.043	0.0288	20.09	20.62	0.728
400	0.854	1.048	0.0323	22.19	25.99	0.722
450	0.758	1.055	0.0356	24.18	31.88	0.718
500	0.682	1.064	0.0386	26.06	38.19	0.718
550	0.620	1.076	0.0416	27.89	44.97	0.721
600	0.569	1.088	0.0444	29.60	52.06	0.724

TABLE A.15 (Continued)

Physical Properties of Gases at Atmospheric Pressure

T, K	ρ, kg/m^3	c_p, kJ/kg-K	k, W/m-K	μ, kg/m-s $\times 10^6$	ν, m^2/s $\times 10^6$	Pr
H_2						
150	0.1637	12.602	0.0981	5.595	34.18	0.718
200	0.1227	13.540	0.1282	6.813	55.53	0.719
250	0.0982	14.069	0.1561	7.919	80.64	0.713
300	0.0819	14.314	0.182	8.963	109.5	0.706
350	0.0702	14.436	0.208	9.954	141.9	0.697
400	0.0614	14.491	0.228	10.864	177.1	0.690
450	0.0546	14.499	0.251	11.779	215.6	0.682
500	0.0492	14.507	0.272	12.636	257.0	0.675
600	0.0409	14.537	0.315	14.285	349.7	0.664
700	0.0349	14.574	0.351	15.89	455.1	0.659
800	0.0306	14.675	0.384	17.40	569.0	0.664
900	0.0272	14.821	0.412	18.78	690.0	0.676
He						
144	0.3379	5.200	0.0928	12.55	37.11	0.700
200	0.2435	5.200	0.1177	16.66	64.38	0.694
255	0.1906	5.200	0.1357	18.17	95.50	0.700
368	0.1328	5.200	0.1691	23.05	173.6	0.710
477	0.1020	5.200	0.197	27.50	269.3	0.720
589	0.0828	5.200	0.225	31.13	375.8	0.720
700	0.0703	5.200	0.251	34.75	494.2	0.720
800	0.0602	5.200	0.275	38.17	634.1	0.720
Steam						
400	0.5542	2.014	0.0261	13.44	24.2	1.040
450	0.4902	1.980	0.0299	15.25	31.1	1.010
500	0.4405	1.985	0.0339	17.04	38.6	0.996
550	0.4005	1.997	0.0379	18.84	47.0	0.991
600	0.3652	2.026	0.0422	20.67	56.6	0.986
650	0.3380	2.056	0.0464	22.47	66.4	0.995
700	0.3140	2.085	0.0505	24.26	77.2	1.000
750	0.2931	2.119	0.0549	26.04	88.8	1.005
800	0.2739	2.152	0.0592	27.86	102.0	1.010
850	0.2579	2.186	0.0637	29.69	115.2	1.019

Source: Converted to SI units from E. R. G. Eckert and R. M. Drake, *Heat and Mass Transfer*, 2nd ed. McGraw-Hill, New York, 1959.

TABLE A.16

Physical Properties of Liquids at Atmospheric Pressure

T, K	ρ, kg/m^3	c, kJ/kg-K	k, W/m-K	μ, kg/m-s $\times 10^2$	ν, m^2/s $\times 10^6$	Pr	β, K^{-1} $\times 10^3$
Glycerin							
273	1276.0	2.261	0.282	1060	8310	85,000	0.47
280	1271.9	2.298	0.284	534	4200	43,200	0.47
290	1265.8	2.367	0.286	185	1460	15,300	0.48
300	1259.9	2.427	0.286	79.9	634	6780	0.48
310	1253.9	2.490	0.286	35.2	281	3060	0.49
320	1247.2	2.564	0.287	21.0	168	1870	0.50
Ethylene Glycol							
273	1130.8	2.294	0.242	6.51	57.6	617	0.65
280	1125.8	2.323	0.244	4.20	37.3	400	0.65
290	1118.8	2.368	0.248	2.47	22.1	236	0.65
300	1114.4	2.415	0.252	1.57	14.1	151	0.65
310	1103.7	2.505	0.255	1.07	9.65	103	0.65
320	1096.2	2.549	0.258	0.757	6.91	73.5	0.65
330	1089.5	2.592	0.260	0.561	5.15	55.0	0.65
340	1083.8	2.637	0.261	0.431	3.98	42.8	0.65
350	1079.0	2.682	0.261	0.342	3.17	34.6	0.65
Engine Oil (unused)							
280	895.3	1.827	0.144	217.0	2430	27,500	0.70
290	890.0	1.868	0.145	99.9	1120	12,900	0.70
300	884.1	1.909	0.145	48.6	550	6400	0.70
310	877.9	1.951	0.145	25.3	288	3400	0.70
320	871.8	1.993	0.143	14.1	161	1965	0.70
330	865.8	2.035	0.141	8.36	96.6	1205	0.70
340	859.9	2.076	0.139	5.31	61.7	793	0.70
350	853.9	2.118	0.138	3.56	41.7	596	0.70
Lubricating Oil ($\approx$ SAE 50)							
273	899.12	1.796	0.147	384.8	4280	47,100	
293	888.23	1.880	0.145	79.94	900	10,400	
313	876.05	1.964	0.144	21.03	240	2870	
333	864.04	2.047	0.140	7.25	83.9	1050	
353	852.02	2.131	0.138	3.20	37.5	490	
373	840.01	2.219	0.137	1.71	20.3	276	
393	828.96	2.307	0.135	1.03	12.4	175	
413	816.94	2.395	0.133	0.65	8.0	116	
433	805.89	2.483	0.132	0.45	5.6	84	
CO_2							
223	703.69	4.463	0.547	3.06	0.435	2.60	
233	691.68	4.467	0.547	2.81	0.406	2.28	
243	679.24	4.476	0.549	2.63	0.387	2.15	
253	666.69	4.509	0.547	2.54	0.381	2.09	
263	653.55	4.564	0.543	2.47	0.378	2.07	
273	640.10	4.635	0.540	2.39	0.373	2.05	
283	626.16	4.714	0.531	2.30	0.368	2.04	
293	611.75	4.798	0.521	2.20	0.359	2.02	
303	596.37	4.890	0.507	2.07	0.349	2,01	
313	580.99	4.999	0.493	1.98	0.340	2.00	
323	564.33	5.116	0.476	1.86	0.330	1.99	

Source: Converted to SI units from E. R. G. Eckert and R. M. Drake, *Heat and Mass Transfer*, 2nd ed., McGraw-Hiil, New York, 1959.

TABLE A.17

Thermal Properties of Metals at 20°C

Metal	ρ, kg/m³	c, kJ/kg-K	α, m²/s ×10⁵	k, W/m-K 20°C	100°C	200°C	300°C
Aluminum							
Pure	2,707	0.896	8.418	204	206	215	228
1% Cu	2,659	0.867	5.933	137	144	152	161
22% Si	2,627	0.854	7.172	161	168	175	178
1% Mn	2,707	0.892	7.311	177	189	204	
Brass, 70% Cu, 30% Zn	8,522	0.385	3.412	111	128	144	147
Copper	8,954	0.383	11.234	386	379	374	369
Iron	7,897	0.452	2.034	73	67	62	55
Wrought iron	7,849	0.460	1.626	59	57	52	48
Lead	11,373	0.130	2.343	35	33.4	31.5	29.8
Magnesium	1,746	1.013	9.708	171	168	163	157
Molybdenum	10,220	0.251	3.605	66	62	74	83
Nickel	8,906	0.446	2.266	90	83	73	64
Silver	10,525	0.234	16.563	407	415	374	362
Steel							
0.5% C	7,833	0.465	1.474	54	52	48	45
1.0% C	7,801	0.473	1.172	43	43	42	40
1.0% Cr	7,865	0.460	1.665	61	55	52	47
5.0% Cr	7,833	0.460	1.110	40	38	36	36
Tin	7,304	0.227	3.884	64	59	57	55
Tungsten	19,350	0.134	6.271	163	151	142	133
Zinc	7,144	0.384	4.106	112	109	106	100

Source: Converted to SI units from E. R. G. Eckert and R. M. Drake, *Heat and Mass Transfer*, 2nd ed. McGraw-Hill, New York, 1959.

TABLE A.18

Thermal Properties of Nonmetals

Material	T,°C	ρ, kg/m³	c, J/kg-K	k, W/m-K	$\alpha \times 10^6$ m²/s
Asbestos	20	383	816	0.113	0.036
Bakelite	30	1200	1600	0.23	0.012
Brick					
Common	20	1800	840	0.38–0.52	0.028–0.034
Masonry	20	1760	837	0.638	0.0406
Clay	20	1545	880	1.26	0.101
Coal, anthracite	20	1370	1260	0.238	0.013–0.015
Concrete, dry	20	2300	837	1.80	0.094
Corkboard	20	150	1880	0.042	0.015–0.044
Glass, window	20	2700	840	0.78	0.034
Glass wool	20	200	670	0.040	0.028
Ice	0	910	1930	2.22	0.126
Paper	30	900	1200	0.12	0.011
Plexiglas	20	1200	1500	0.20	0.0164
Rock wool	20	160	1200	0.040	0.011
Rubber, hard	20	2250	2009	0.163	0.0162
Soil, dry	20	1500	1842	0.35	0.138
Teflon	30	2300	1050	0.35	0.112–0.119
Wood					
Oak	20	609-801	2390	0.17–0.21	0.0111–0.1021
Pine	20	116–421	2720	0.15	0.0124

Source: Converted to SI units from E. R. G. Eckert and R. M. Drake, *Heat and Mass Transfer*, 2nd ed. McGraw-Hill, New York, 1959.

TABLE A.19

Normal Total Emissivities of Several Metals

Material	T,°C	ε
Aluminum		
Polished	100	0.095
24ST	232–504	0.16–0.20
Brass, polished	100	0.06
Chromium, polished	100	0.075
Copper, polished	100	0.052
Gold, polished	227–627	0.018–0.35
Iron		
Roughly polished	100	0.17
Cast, polished	200	0.21
Wrought, polished	38–250	0.28
Lead, unoxidized	127–227	0.057–0.075
Magnesium oxide	277–827	0.55–0.20
Mercury	0–100	0.09–0.12
Molybdenum, polished	100	0.071
Nickel, polished	100	0.072
Silver, polished	100	0.052
Steel		
Stainless, polished	100	0.074
Type 301	24	0.21
Type 316	24	0.28
Type 347	24	0.39
Tin, bright	25	0.43
Tungsten, polished	100	0.066
Zinc		
Commercial, polished	227–327	0.045–0.053
Galvanized sheet	100	21

Source: From a compilation by H. C. Hottel in W. H. McAdams, *Heat Transmission*, 3rd ed. McGraw-Hill, New York, 1954.

TABLE A.20

Standard Pipe Sizes

Nominal Size, in.	Schedule Number	Outside Diameter, cm	Outside Surface m^2/m	Inside Diameter, cm
1/2	40	2.134	0.06704	1.580
	80			1.387
1	40	3.340	0.1049	2.664
	80			2.431
2	40	6.033	0.1895	5.250
	80			4.925
3	40	8.890	0.2793	7.793
	80			7.366
4	40	11.43	0.3591	10.226
	80			9.718
5	40	14.13	0.4439	12.82
	80			12.23
6	40	16.83	0.5287	15.41
	80			14.63
8	40	21.91	0.6683	20.27
	80			19.37
10	40	27.31	0.8580	25.45
	80			24.29
12	40	32.29	0.10176	30.32
	80			28.90

Source: NAVCO Piping Catalog.

TABLE A.21

Standard Tubing Sizes

Nominal Size, in.	BWG	Outside Diameter, cm	Outside Surface, m^2/m	Inside Diameter, cm
5/8	18	1.270	0.03990	1.021
	16			0.940
	14			0.848
1/2	18	1.588	0.04989	1.339
	16			1.257
	14			1.166
3/4	18	1.905	0.05985	1.656
	16			1.575
	14			1.483
1	18	2.540	0.07980	2.291
	16			2.210
	14			2.118
	12			1.986
	10			1.589
1 1/4	18	3.175	0.09975	2.926
	16			2.845
	14			2.753
	12			2.621
	10			2.494
1 1/2	13	3.810	0.1197	3.327
	12			3.256
	11			3.200
	10			3.129
	8			2.972
2	13	5.080	0.1596	4.597
	12			4.526
	11			4.470
	10			4.399
	8			4.242

Source: NAVCO Piping Catalog.

TABLE A.22

Properties of the Standard Atmosphere

Altitude, h meters	T, K K	P Pa	ρ kg/m^3	$\mu \times 10^5$ kg/m-s
0	288.2	1.0133×10^5	1.225	1.789
500	284.9	0.95461×10^5	1.167	1.774
1000	281.7	0.89876×10^5	1.111	1.758
1500	278.4	0.84560×10^5	1.058	1.742
2000	275.2	0.79501×10^5	1.007	1.726
2500	271.9	0.74692×10^5	0.9570	1.710
3000	268.7	0.70121×10^5	0.9093	1.694
3500	265.4	0.65780×10^5	0.8634	1.678
4000	262.2	0.61660×10^5	0.8194	1.661
4500	258.9	0.57753×10^5	0.7770	1.645
5000	255.7	0.54048×10^5	0.7364	1.628
5500	252.4	050539×10^5	0.6975	1.612
6000	249.2	0.47218×10^5	0.6601	1.595
6500	245.9	0.44075×10^5	0.6243	1.578
7000	242.7	0.41105×10^5	0.5900	1.561
7500	239.5	0.38300×10^5	0.5572	1.544
8000	236.2	0.35652×10^5	0.5258	1.527
8500	233.0	0.33154×10^5	0.4958	1.510
9000	229.7	0.30801×10^5	0.4671	1.493
9500	226.5	0.28585×10^5	0.4397	1.475
10,000	223.3	0.26500×10^5	0.4135	1.458
11,000	216.8	0.22700×10^5	0.3648	1.422
12,000	216.7	0.19399×10^5	0.3119	1.422
13,000	216.7	0.16580×10^5	0.2666	1.422
14,000	216.7	0.14170×10^5	0.2279	1.422
15,000	216.7	0.12112×10^5	0.1948	1.422
16,000	216.7	0.10353×10^5	0.1665	1.422
17,000	216.7	8.8597×10^3	0.1423	1.422
18,000	216.7	5.5293×10^3	0.12l7	1.422
19,000	216.7	2.5492×10^3	0.1040	1.422
20,000	216.7	5.5293×10^3	0.08891	1.422
25,000	221.5	2.5492×10^3	0.04008	1.448

Source: Extracted from NACA-TN-1428

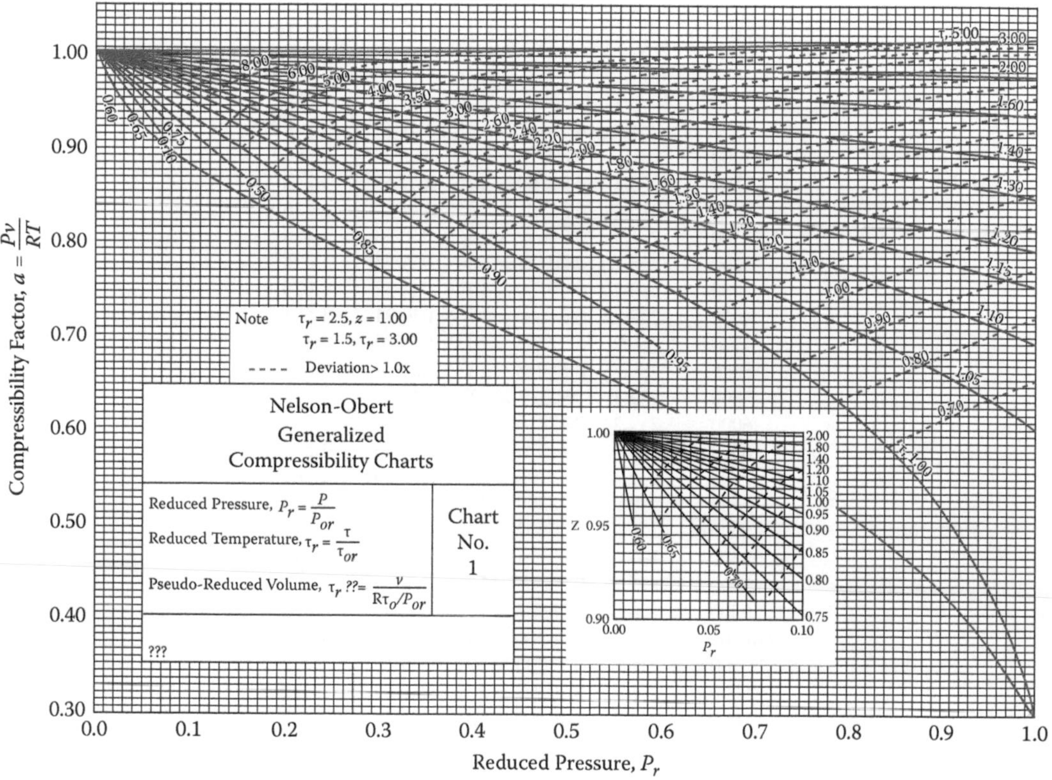

FIGURE A.1
Compressibility chart for $p_r < 1.0$. (From L. C. Nelson and E. F. Obert, Compressibility Charts. *ASME Trans.*, 76, 1057 ff, 1954.)

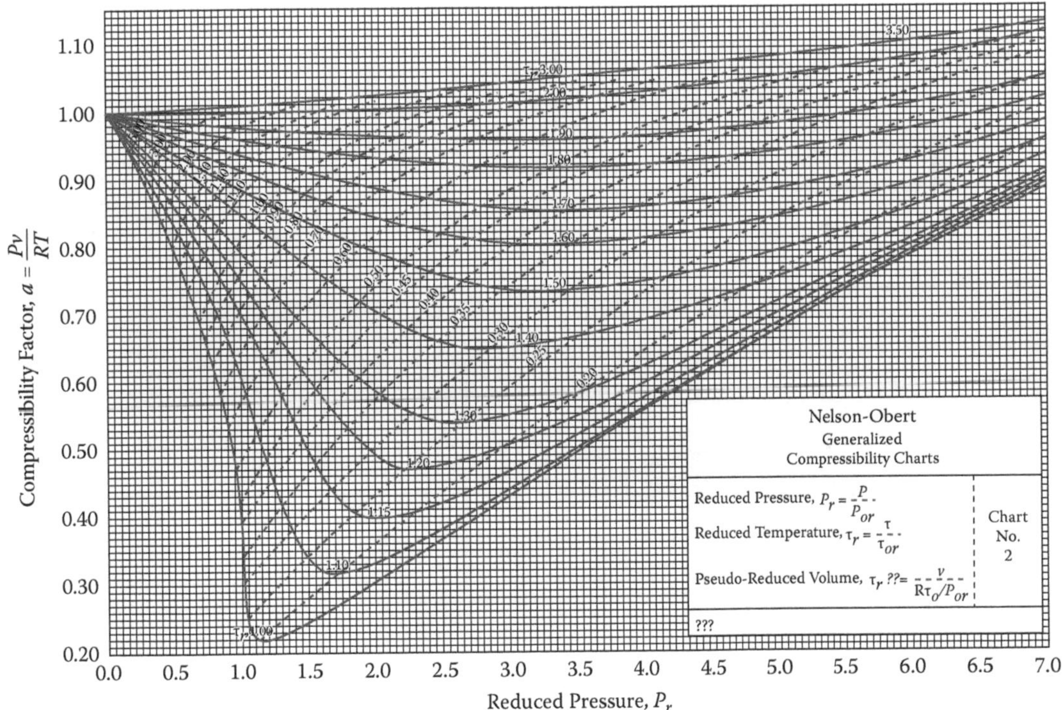

FIGURE A.2
Compressibility chart for 0.50 < p_r < 7.0. (From L. C. Nelson and E. F. Obert, Compressibility Charts. *ASME Trans.*, 76, 1057 ff, 1954.)

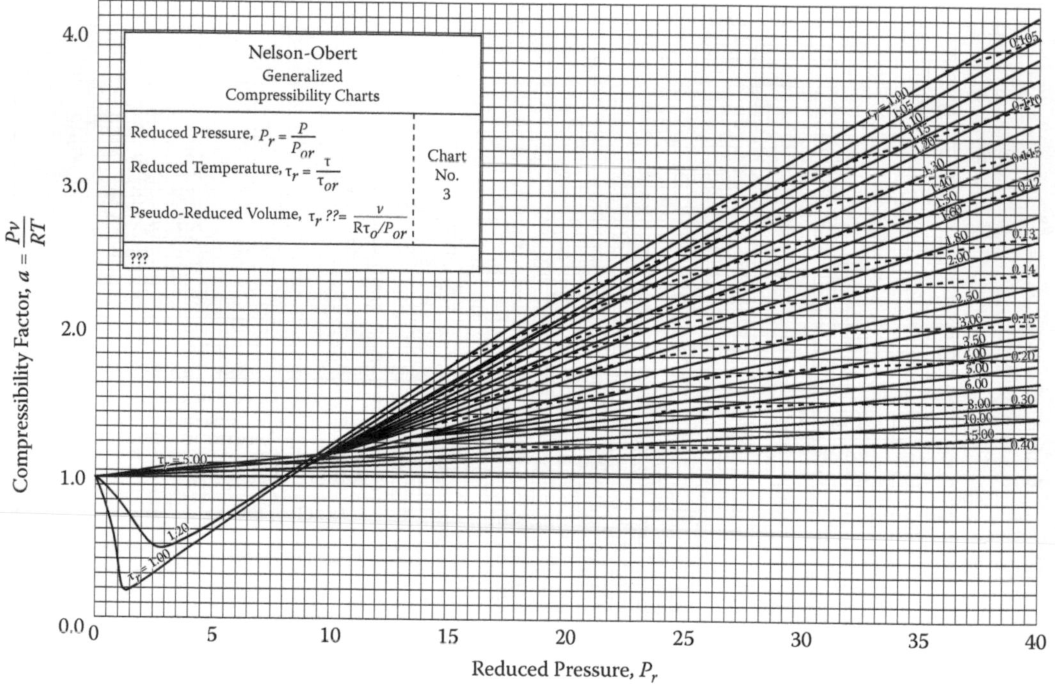

FIGURE A.3
Compressibility chart for $0 < p_r < 40$. (From L. C. Nelson and E. F. Obert, Compressibility Charts. *ASME Trans.*, 76, 1057 ff, 1954.)

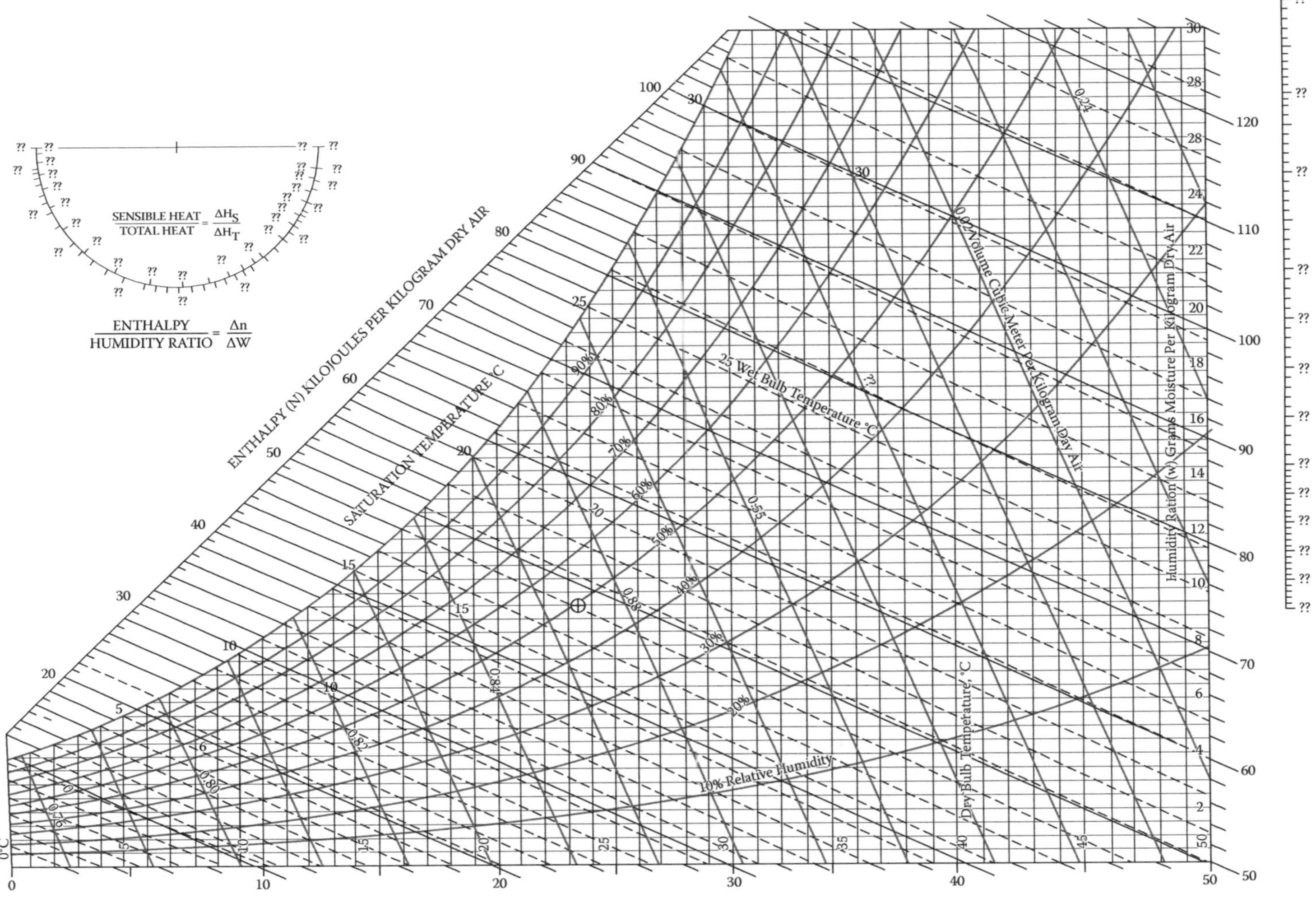

FIGURE A.4
Psychrometric chart for air at atmospheric pressure. (Courtesy of ASHRAE.)

Appendix B

Summary of Differential Vector Operations in Three Coordinate Systems

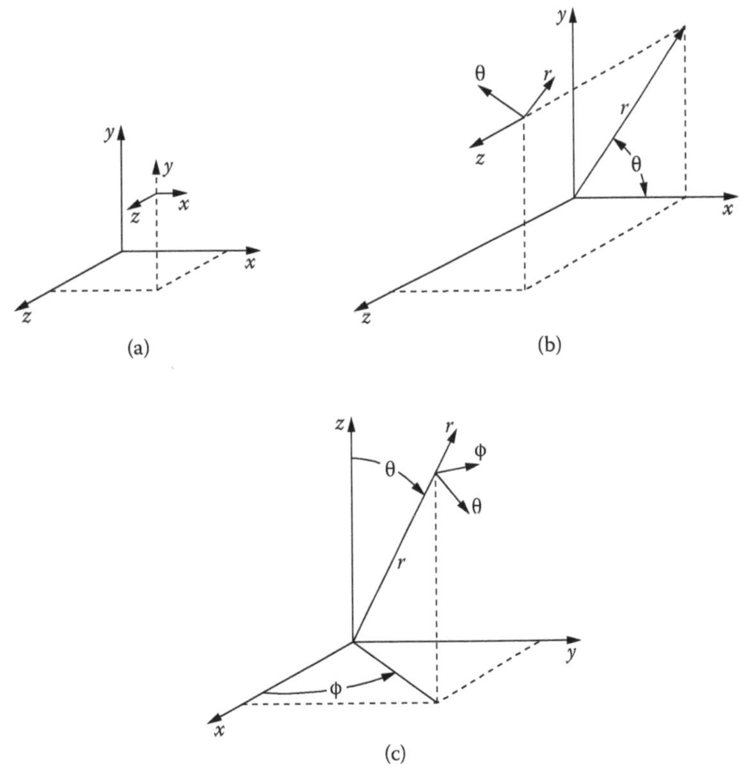

FIGURE B.1
Coordinate systems: (a) Cartesian, (b) cylindrical, and (c) spherical.

B.1 Cartesian Coordinates

Gradient

$$\nabla P = \frac{\partial P}{\partial x}\mathbf{x} + \frac{\partial P}{\partial y}\mathbf{y} + \frac{\partial P}{\partial z}\mathbf{z}$$

Divergence

$$\nabla \cdot \mathbf{v} = \frac{\partial v_x}{\partial x} + \frac{\partial v_y}{\partial y} + \frac{\partial v_z}{\partial z}$$

Curl

$$\nabla \times \mathbf{v} = \begin{Bmatrix} \left(\frac{\partial v_z}{\partial y} - \frac{\partial v_y}{\partial z}\right)\mathbf{x} \\ \left(\frac{\partial v_x}{\partial z} - \frac{\partial v_z}{\partial x}\right)\mathbf{y} \\ \left(\frac{\partial v_y}{\partial x} - \frac{\partial v_x}{\partial_y}\right)\mathbf{z} \end{Bmatrix}$$

Laplacian of a scalar

$$\nabla^2 T = \frac{\partial^2 T}{\partial x^2} + \frac{\partial^2 T}{\partial y^2} + \frac{\partial^2 T}{\partial z^2}$$

B.2 Cylindrical Coordinates

Gradient

$$\nabla P = \frac{\partial P}{\partial r}\mathbf{r} + \frac{1}{r}\frac{\partial P}{\partial \theta}\boldsymbol{\theta} + \frac{\partial P}{\partial z}\mathbf{z}$$

Divergence

$$\nabla \cdot \mathbf{v} = \frac{1}{r}\frac{\partial}{\partial r}(r v_r) + \frac{1}{r}\frac{\partial v_\theta}{\partial \theta} + \frac{\partial v_z}{\partial z}$$

Curl

$$\nabla \times \mathbf{v} = \begin{Bmatrix} \left(\frac{1}{r}\frac{\partial v_z}{\partial \theta} - \frac{\partial v_\theta}{\partial z}\right)\mathbf{r} \\ \left(\frac{\partial v_r}{\partial z} - \frac{\partial v_z}{\partial r}\right)\boldsymbol{\theta} \\ \left\{\frac{1}{r}\left[\frac{\partial}{\partial r}(r v_\theta) - \frac{\partial v_r}{\partial \theta}\right]\right\}\mathbf{z} \end{Bmatrix}$$

Laplacian of a scalar

$$\nabla^2 T = \frac{1}{r}\frac{\partial}{\partial r}\left(r\frac{\partial T}{\partial r}\right) + \frac{1}{r^2}\frac{\partial^2 T}{\partial \theta^2} + \frac{\partial^2 T}{\partial z^2}$$

B.3 Spherical Coordinates

Gradient

$$\nabla P = \frac{\partial P}{\partial r}\mathbf{r} + \frac{1}{r}\frac{\partial P}{\partial \theta}\boldsymbol{\theta} + \frac{1}{r\sin\theta}\frac{\partial P}{\partial \phi}\phi$$

Divergence

$$\nabla \cdot \mathbf{v} = \frac{1}{r^2}\frac{\partial}{\partial r}(r^2 v_r) + \frac{1}{r\sin\theta}\frac{\partial}{\partial\theta}(v_\theta \sin\theta) + \frac{1}{r\sin\theta}\frac{\partial v_\phi}{\partial\phi}$$

Curl

$$\nabla \times \mathbf{v} = \left\{ \begin{array}{l} \dfrac{1}{r\sin\theta}\left[\dfrac{\partial}{\partial\theta}(v_\phi \sin\theta) - \dfrac{\partial v_\theta}{\partial\phi}\right]\mathbf{r} \\ \left[\dfrac{1}{r\sin\theta}\dfrac{\partial v_r}{\partial\phi} - \dfrac{1}{r}\dfrac{\partial}{\partial r}(r v_\phi)\right]\boldsymbol{\theta} \\ \dfrac{1}{r}\left[\dfrac{\partial}{\partial r}(r v_\theta) - \dfrac{\partial v_r}{\partial\theta}\right]\boldsymbol{\phi} \end{array} \right\}$$

Laplacian of a scalar

$$\nabla^2 T = \frac{1}{r^2}\frac{\partial}{\partial r}\left(r^2\frac{\partial T}{\partial r}\right) + \frac{1}{r^2\sin\theta}\frac{\partial}{\partial\theta}\left(\sin\theta\frac{\partial T}{\partial\theta}\right) + \frac{1}{r^2\sin^2\theta}\frac{\partial^2 T}{\partial\phi^2}$$

References and Additional Readings

References

Achenbach, E. (1968). Distribution of Local Pressure and Skin Friction Around a Circular Cylinder in Cross-Flow up to Re = 5×10^5. *J. Fluid Mech.*, 34, Part 4.

Adamson, A. W. (1960). *Physical Chemistry of Surfaces*. Interscience, New York.

American Society of Heating, Refrigerating and Air-Conditioning Engineers. (1981). *ASHRAE Handbook of Fundamentals*. ASHRAE, Atlanta.

American Society of Heating, Refrigerating and Air-Conditioning Engineers. (1994). *Handbook of Refrigeration*. ASHRAE, Atlanta.

American Society of Mechanical Engineers. (1967). *Steam Tables*. ASME, New York.

American Society of Mechanical Engineers. (1976). ASME Orientation and Guide for Use of SI Units, Guide no. SI-1, 7th ed. ASME, New York.

Bar-Cohen, A., and Rohsenow, W. M. (1984). Thermally Optimum Spacing of Vertical Natural Convection Cooled Parallel Plates. *J. Heat Trans.*, 106, 116–122.

Bejan, A., and Kraus, A. D. (2003). *Handbook of Heat Transfer*. John Wiley & Sons, New York, Chap. 4.

Blasius, H. (1908). Grenzschichten in Flüssigkeiten mit kleiner Reibung, *Z. Math. Phys. Sci.*, 1.

Boelter, L. M. K., Cherry, H., Johnson, H. A., and Martinelli, R. C. (1965). *Heat Transfer Notes*. McGraw-Hill, New York.

Bowman, R. A., Mueller, A. C., and Nagle, W. M. (1940). Mean Temperature Difference in Design. *Trans. ASME*, 62, 283–294.

Buckingham, E. (1914). On Physically Similar Systems: Illustrations of the Use of Dimensional Equations. *Phys. Rev.*, 4(4), 345–376.

Catton, I. (1978). Natural Convection in Enclosures. *Proc. Int. Heat Trans. Conf.*, Toronto, Canada, 6, 13–31.

Chenoweth, J. M. (1990). Final Report of the HTRI/TEMA Joint Committee to Review the Fouling Section of the TEMA Standards. *Heat Trans. Eng.*, 11(1), 73–107.

Churchill, S. W. (1983). Free Convection Around Immersed Bodies in *Heat Exchanger Design Handbook*, E. Schlunder, ed., Section 2.5.7. Hemisphere, New York.

Churchill, S. W., and Bernstein, M. (1977). A Correlating Equation for Forced Convection from Gases and Liquids to a Circular Cylinder in Cross Flow. *J. Heat Trans.*, 99(2), 300–306.

Churchill, S. W., and Chu, H. H. S. (1975a). Correlating Equations for Laminar and Turbulent Free Convection from a Horizontal Cylinder. *Int. J. Heat Mass Trans.*, 18, 1049–1054.

Churchill, S. W., and Chu, H. H. S. (1975b). Correlating Equations for Laminar and Turbulent Free Convection from a Vertical Plate. *Int. J. Heat Mass Trans.*, 18, 1323–1328.

Churchill, S. W., and Ozoe, H. (1973). Correlations for Laminar Forced Convection with Uniform Heating in Flow Over a Flat Plate and in Developing and Fully Developing Flow in a Tube. *J. Heat Trans.*, 95(1), 78–84.

Churchill, S. W., and Usagi, R. (1972). A General Expression for the Correlation of Rates of Heat Transfer and Other Phenomena. *AIChe J.*, 18(6), 1121–1138.

Colebrook, C. F. (1939). Turbulent Flow in Pipes with Particular Reference to the Transition Between the Smooth and Rough Pipe Laws, *J. Inst. Civil Eng.*, 11, 133–136.

Comstock, J. P., ed. (1967). *Principles of Naval Architecture*. Society of Naval Architects and Marine Engineers, New York.

Crane. (1957). Flow of Fluids Through Valves, Fittings and Pipe. Crane Company Technical Paper 410, Chicago.

Dittus, F. W., and Boelter, L. M. K. (1930). *Univ. Calif. (Berkeley) Publ. Eng.*, 2, 443.

Elenbaas, W. (1942). Heat Distribution of Parallel Plates by Free Convection. *Physica*, 9(11), 665–671.

Fischer, F. K. (1938). Mean Temperature Difference Correction in Multipass Exchangers. *Ind. Eng. Chem.*, 30(4), 377–383.

Fourier, J. B. J. (1822). *The Analytic Theory of Heat*, translated by A. Freeman. Cambridge University Press, Cambridge.

Gardner, K. A. (1945). Efficiency of Extended Surfaces. *Trans. ASME*, 67, 621–628.

Gebhart, B., Jaluria, Y., Mahajan, Y. L., and Sammakia, B. (1988). *Buoyancy Induced Flows and Transport*. Hemisphere, Washington, DC.

Goldstein, R. J., Sparrow, E. M., and Jones, D. C. (1973). Natural Convection Mass Transfer Adjacent to Horizontal Plates. *Int. J. Heat Mass Trans.*, 16, 1025–1032.

Gröeber, H. (1925). The Heating and Cooling of Simple Geometric Bodies. *Z. Ver. Dtsch. Ing.*, 69, 705–711.

Gurney, H. P., and Lurie, J. (1923). Charts for Estimating Temperature Distributions in Heating and Cooling Solid Shapes. *Ind. Eng. Chem.*, 15, 1170–1178.

Hausen, H. (1943). Darstellung des Wärmeauberganges in Rohren durch verallgemeinerte Potenzbeziehungen. *Z. Ver. Dtsch. Ing.*, 4, 91.

Heisler, M. P. (1947). Temperature Charts for Induction and Constant Temperature Heating. *Trans. ASME*, 69, 227.

Hilpert, R. (1933). Wärmeabgabe von Geheizen Drähten und Rohren. *Forsch. Geb. Ing.*, 4, 215–224.

Hollands, K. G. T. , Unny, S. E., Raithby, G. D., and Konicek, L. (1976). Free Convective Heat Transfer Across Inclined Air Layers. *J. Heat Trans.*, 98, 189–196.

Hottel, H. (1954). "Radiant Heat Transmission," in W. H. McAdams, ed., *Heat Transmission*, 3rd ed. McGraw-Hill, New York.

Howarth, L. (1938). On the Solution of the Laminar Boundary Layer Equations. *Proc. R. Soc. London*, AI64.

Howell, J. R. (1982). *A Catalog of Radiation Configuration Factors*. McGraw-Hill, New York.

Kakac, S. Shah, R. K., and Aung, W. (1987). *Handbook of Single Phase Heat Transfer*. John Wiley & Sons, New York.

Kays, W. M., and London, A. L. (1984). *Compact Heat Exchangers*, 3rd ed. McGraw-Hill, New York.

Keenan, J. H., Keyes, F. G., Hill, P. G., and Moore, J. G. (1969). *Steam Tables*. John Wiley & Sons, New York.

Kraus, A. D., Aziz, A., and Welty, J. R. (2001). *Extended Surface Heat Transfer*. John Wiley & Sons, New York.

Lloyd, J. R., and Moran, W. R. (1974). Natural Convection Adjacent to Horizontal Surfaces. ASME Paper 74WA/HT-66. American Society of Mechanical Engineers, New York.

MacGregor, R. K., and Emery, A. P. (1969), Free Convection Through Vertical Plane Layers: Moderate and High Temperature Fluids. *J. Heat Trans.*, 91, 391–396.

Mayinger, F. (1988). Classification and Applications of Two-Phase Flow Heat Exchangers, in *Two-Phase Flow Heat Exchangers*, S. Kakac, A. E. Bergles, and E. O. Fernandes, eds. Kluwer Academic, Dordrecht, The Netherlands.

McAdams, W. H. (1954). *Heat Transmission*, 3rd ed. McGraw-Hill, New York.

McQuiston, F. C., and Tree, D. R. (1972). Optimum Space Envelopes of the Finned Tube Heat Transfer Surface. *Trans. ASHRAE*, 78, 144–148.

Moody, L. F. (1944). Friction Factors for Pipe Flow. *Trans. ASME*, 66, 681–684.

Munson, B. R., Young, D. F, and Okishi, T. K. (1994). *Fundamentals of Fluid Mechanics*, 2nd ed. John Wiley & Sons, New York.

Murray, W. M. (1938). Heat Transfer Through an Annular Disk or Fin of Uniform Thickness. *Trans. ASME, J. Appl. Mech.*, 60, A78–A81.

Nagle, W. M. (1933). Mean Temperature Difference in Multipass Exchangers. *Ind. Eng. Chem.*, 25, 604–609.

Oppenheim, A. K. (1956). Radiation Analysis by the Network Method. *Trans. ASME*, 78, 725–732.

Planck, M. (1959). *The Theory of Heat Radiation*. Dover, New York.

Raithby, G. D., and Hollands, K. G. T. (1975). A General Method of Obtaining Approximate Solutions to Laminar and Turbulent Free Convection Problems, in *Advances in Heat Transfer*, T. F. Irvine and J. P. Hartnett, eds. Academic Press, New York, pp. 265–315.

Schmidt, E., and Beckmann, W. (1930). Das Temperatur vor einer wärmeab-gebenden senkrechten Platte bei natürlicher Convection. *Tech. Mech. Thermodyn.*, 1(10), 341–349; 1(11), 391–160.

Schneider, P. J. (1965). *Temperature Response Charts*. John Wiley & Sons, New York.

Sieder, E. N., and Tate, G. E. (1936). Heat Transfer and Pressure Drop of Liquids in Tubes. *Ind. Eng. Chem.*, 28, 1429.

Shah, R. K. (1981). Classification of Heat Exchangers, in *Heat Exchangers, Thermal–Hydraulic Fundamentals and Design*, S. Kakac, A. E. Bergles, and F. Mayinger, eds. Hemisphere, New York.

Streeter, V. L. ed. (1961). *Handbook of Fluid Dynamics*. McGraw-Hill, New York.

Underwood, A. J. V. (1934). The Calculation of Mean Temperature Difference in Multipass Heat Exchangers. *J. Inst. Petrol. Technol.*, 20, 145–158.

U.S. Standard Atmosphere (1962). Washington, DC: Government Printing Office.

U.S. Standard Atmosphere (1976). Washington, DC: Government Printing Office.

Whitaker, S. (1972). Forced Convection Heat Transfer Correlations for Flow in Pipes, Past Flat Plates, Single Cylinders, Single Spheres and Flow in Packed Beds and Tube Bundles. *AIChE J.*, 18, 361–371.

Zhukauskas, A. (1972). Heat Transfer from Tubes in Cross Flow, in *Advances in Heat Transfer*, vol. 8, J. P. Hartnett and T. F. Irvine, eds. Academic Press, New York.

Additional Readings

Anderson, D. A., Tannehill, J. C., and Pletcher, R. H. (1984). *Computational Fluid Mechanics and Heat Transfer*. McGraw-Hill, New York.

Batchelor, G. K. (1967). *An Introduction to Fluid Mechanics*. Cambridge University Press, London.

Becker, H. A. (1976). *Dimensionless Parameters*. Halstead Press (Wiley), New York.

Bejan, A. (1993). *Heat Transfer*. John Wiley & Sons, New York.

Benedict, R. P. (1984). *Fundamentals of Temperature, Pressure, and Flow Measurement*. John Wiley & Sons, New York.

Black, W. Z., and Hartley, J. G. (1996). *Thermodynamics*, 3rd ed. Prentice-Hall, Upper Saddle River, NJ.

Blevens, R. D. (1984). *Applied Fluid Mechanics Handbook*. Van Nostrand Reinhold, New York.

Bridgeman, P. W. (1922). *Dimensional Analysis*. Yale University Press, New Haven, CT.

Cengel, Y. A. (2003). *Heat Transfer: A Practical Approach*. 2nd ed. McGraw-Hill, New York.

Cengel, Y. A., and Boles, M. A. (1989). *Thermodynamics: An Engineering Approach*. McGraw-Hill, New York.

Currie, I. G. (1974). *Fundamental Mechanics of Fluids*. McGraw-Hill, New York.

Duncan, W. J. (1953). *Physical Similarity and Dimensional Analysis; An Elementary Treatise*. Edward Arnold, London.

Fox, R. W., and McDonald, A. C. (1999). *Introduction to Fluid Mechanics*, 5th ed. John Wiley & Sons, New York.

Gilmer, T. C. (1970). *Modern Ship Design*. 5, United States Naval Institute, Annapolis, MD, Chap. 5.

Goldstein, S. (1938). *Modern Developments in Fluid Mechanics*. Oxford University Press, London.

Happel, J. (1965). *Low Reynolds Number Hydrodynamics*. Prentice-Hall, Englewood Cliffs, NJ.

Hoerner, S. F. (1965). *Fluid-Dynamic Drag*. Published by the author. Library of Congress No. 64-1966.

Holman, J. P. (2002). *Heat Transfer*, 9th ed. McGraw-Hill, New York.

Howell, J. R., and Buckius, R. O. (1992). *Fundamentals of Engineering Thermodynamics*, 2nd ed. McGraw-Hill, New York.

Huntley, H. E. (1952). *Dimensional Analysis*, Macdonald, London.

Incropera, F. P., and DeWitt, D. P. (2003). *Introduction to Heat Transfer*, 5th ed. John Wiley & Sons, New York.

Ipsen, D. C. (1960). *Units, Dimensions and Dimensionless Numbers*. McGraw-Hill, New York.

Isaacson, E. de St. Q., and Isaacson, M. de St. Q. (1975). *Dimensional Methods in Engineering and Physics*. John Wiley & Sons, New York.

Jepson, R. W. (1976). *Analysis of Flow in Pipe Networks*. Ann Arbor Science Publishers, Ann Arbor, MI.

Kays, W. M., and Crawford, M. E. (1980). *Convective Heat and Mass Transfer*. McGraw-Hill, New York.

Kline, S. J. (1965). *Similitude and Approximation Theory*. McGraw-Hill, New York.

Kraus, A. D. (2001). *Matrices for Engineers*. Oxford University Press, New York.

Kreider, J. F. (1985). *Principles of Fluid Mechanics*. Allyn & Bacon, Boston.

Kuethe, A. M., and Chow, C. Y. (1986). *Foundations of Aerodynamics, Bases of Aerodynamics Design*, 4th ed. John Wiley & Sons, New York.

Langhaar, H. L. (1951). *Dimensional Analysis and the Theory of Models*. John Wiley & Sons, New York.

Mahan, J. R. (2002). *Radiation Heat Transfer*. John Wiley & Sons, New York.

Mills, A. F. (1999). *Heat Transfer*, 2nd ed. Prentice-Hall, Upper Saddle River, NJ.

Milne-Thomson, L. M. (1968). *Theoretical Hydrodynamics*. Macmillan, New York.

Modest, M. (1993). *Radiative Heat Transfer*. McGraw-Hill, New York.

Moran, J. (1984). *An Introduction to Theoretical and Computational Aerodynamics*. John Wiley & Sons, New York.

Moran, M. J., and Shapiro, H. N. (2004). *Fundamentals of Engineering Thermodynamics*, 5th ed. John Wiley & Sons, New York.

Munson, B. R., Young, D. F., and Okiishi, T. (1998). *Fundamentals of Fluid Mechanics*, 3rd ed. John Wiley & Sons, New York.

Murphy, G. (1950). *Similitude in Engineering*. Ronald Press, New York.

Ozisik, M. N. (1985). *Heat Transfer: A Basic Approach*, McGraw-Hill, New York.

Panton, R. L. (1984). *Incompressible Flow*. Wiley-Interscience, New York.

Potter, C., and Wiggert, D. C. (1997). *Mechanics of Fluids*, 2nd ed. Prentice-Hall, Upper Saddle River, NJ.

Reiner, M. (1969). *Deformation, Strain and Flow: An Elementary Introduction to Rheology*, 3rd ed. Lewis, London.

Roberson, J. A., and Grove, C. L. (1997). *Engineering Fluid Mechanics*. John Wiley & Sons, New York.

Rosenhead, L. (1963). *Laminar Boundary Layers*. Oxford University Press, London.

Schlicting, H. (1960). *Boundary Layer Theory*. McGraw-Hill, New York.

Schuring, D. J. (1977). *Scale Models in Engineering*. Pergamon Press, New York.

Sedov, L. I. (1959). *Similarity and Dimensional Methods in Mechanics*. Academic Press, New York.

Siegel, R. and Howell, J. R. (2001). *Thermal Radiation Heat Transfer*, 4th ed. Taylor & Francis, New York.

Sharp, J. J. (1981). *Hydraulic Modeling*. Butterworth, London.

Somerscales, E. F. C., and Knudsen, J. G. (1981). *Fouling of Heat Transfer Equipment*, Hemisphere, New York.

Sonntag, R. E., Borgnakke, C., and VanWylen, G. J. (2003). *Fundamentals of Thermodynamics*, 6th ed. John Wiley & Sons, New York.

Sovran, G., ed. (1978). *Aerodynamic Drag Mechanisms of Bluff Bodies and Road Vehicles*. Plenum Press, New York.

Taylor, E. S. (1974). *Dimensional Analysis for Engineers*. Clarendon Press, Oxford.

Wark, K. Jr., and Richards, D. E. (1999). *Thermodynamics*, 6th ed. McGraw-Hill, New York.

Welty, J. R., Wicks, C. E., Wilson, R. E., and Rorrer, G. L. (2008). *Fundamentals of Momentum Heat and Mass Transfer*, 5th ed. John Wiley & Sons, New York.

White, F. (1974). *Viscous Fluid Flow*. McGraw-Hill, New York.

White, F. M. (1994). *Fluid Mechanics*, 3rd ed. McGraw-Hill, New York.

White, F. M. (1999). *Fluid Mechanics*, 4th ed. McGraw-Hill, New York.

Yalin, M. S. (1971). *Theory of Hydraulic Models*. Macmillan, London.

Nomenclature

Roman Letter Symbols

A area, m^2
AFR air fuel ratio, kg air/kg fuel
$\overline{AFR}$ molal air fuel ratio, kgmol air/kgmol fuel
a acceleration, m/s^2
constant in Redlich-Kwong equation of state, bar-$(m^3/kgmol)^2$-$K^{1/2}$
constant in van der Waals equation of state, bar-$(m^3/kgmol)^2$
B bulk modulus of elasticity, K^{-1}
Bi Biot number, dimensionless
b constant in Redlich-Kwong equation of state, $m^3/kgmol$
constant in van der Waals equation of state, $m^3/kgmol$
fin height, m
C a constant, dimensionless
capacitance, Farads
capacity rate, W/K
C_H hot-side capacity rate, W/K
C_C cold-side capacity rate, W/K
C^* capacity rate ratio, dimensionless
C_D drag coefficient, dimensionless
C_f coefficient of skin friction, dimensionless
c speed of sound, m/s
speed of light, 2.99×10^8, m/s
specific heat, dimensionless
c_p specific heat at constant pressure, J/kg-K
c_v specific heat at constant volume, J/kg-K
C_n polytropic specific heat, J/kg-K
D diameter, m diffusion coefficient, m^2/s
d diameter, m
d_e equivalent diameter, m
E energy, J
modulus of elasticity, N/m^2
E radiant energy emissive power, W/m^2
E_b blackbody emissive power, W/m^2-μm
E_g heat generation, W
E_λ hemispherical monochromatic emissive power, W/m^2-μm
El Elenbaas number, dimensionless
Eu Euler number, dimensionless
e specific energy, J/kg
pipe or tube wall roughness, m
F emission fraction, dimensionless
force, N
logarithmic mean temperature difference correction factor, dimensionless
shape, view or arrangement factor, dimensionless
F_c compression force, N
F_n normal force, N
F_t tension force, N
F_x x-component of force, N
F_y y-component of force, N
F_z z-component of force, N
Fo Fourier modulus, dimensionless
f_f Fanning friction factor, dimensionless
G gravimetric fraction, dimensionless total irradiation, W/m^2
G_λ hemispherical monochromatic irradiation, W/m^2
Gr Grashof number, dimensionless
g acceleration of gravity, m/s^2

g_c proportionality constant, 32.174, lb_m-ft/s^2-lb_f
H enthalpy, J
height, m
hp horsepower
h specific enthalpy, J/kg
heat transfer coefficient, W/m^2-K
head, m
h_L head lost, m
I current, amperes
intensity of radiation, W/m^2-sr
moment of inertia, m^4
I_o moment of inertia about centroidal axis, m^4
i instantaneous electric current, amperes
J Bessel function of first kind, dimensionless
total radiosity, W/m^2
J_λ monochromatic radiosity, W/m^2-μm
KE kinetic energy, J
k Boltzmann constant
ratio of specific heats, dimensionless
thermal conductivity, W/m-K
mass transfer coefficient, m/s
ke specific kinetic energy, J/kg
k_g spring constant, kg/m
L liters, m^3
length, m
inductance, H
L_e entrance length, m
LMTD logarithmic mean temperature difference, K
M molecular weight, kg/kgmol
Mach number, dimensionless
moment, N-m
m mass, kg
fin performance factor, m^{-1}
$\dot{m}$ mass flow rate, kg/s
Nu Nusselt number, dimensionless
$\mathbf{N_{tu}}$ number of transfer units. dimensionless
n number of moles, dimensionless
number of tubes, dimensionless
number of fins, dimensionless
rotational speed, rpm
P pressure, N/m^2 (Pa)
power, W
fin perimeter, m
thermodynamic probability, dimensionless
P_{cr} critical pressure, N/m^2 (Pa)
p_r reduced pressure, dimensionless
PE potential energy, J
Pr Prandtl number, dimensionless
p number of dependent intrinsic variables, dimensionless
p_r reference pressure, N/m^2 (Pa)
pe specific internal energy, J/kg
Q heat, J
$\dot{Q}$ heat flow, W
q instantaneous charge, Coulomb
q_i heat generated per unit volume, W/m^3
$\dot{q}$ heat flux, W/m^2
R gas constant, J/kg-K
radius of gyration, m
radiation resistance, m^{-2}
R_d fouling resistance, K/W
$\bar{R}$ universal gas constant, 8314 J/kgmol-K
Ra Rayleigh number, dimensionless
Re Reynolds number, dimensionless
r radius, m
radial coordinate, m
compression ratio, dimensionless
r_e cutoff ratio, dimensionless
r_c critical radius, m
r_p pressure ratio, dimensionless
S surface area, m^2
entropy, J/K
SG specific gravity, dimensionless
s specific entropy, J/kg-K
T Temperature, units vary
tensile force, N
torque N-m
T_{as} adiabatic saturation temperature, K
T_c critical temperature, K
T_∞ free stream or surrounding temperature, K
t time, sec
exchanger cold-side temperature, units vary
U internal energy, J
overall heat transfer coefficient, W/m^2-K
u specific internal energy, J/kg
V volume, m^3
voltage, volts

$\hat{V}$	velocity, m/s
$\hat{V}_\infty$	free-stream velocity, m/s
$\dot{V}$	volumetric flow rate, m^3/s
v	specific volume, m^3/kg
$\bar{v}$	molal specific volume, $m^3/kgmol$
W	weight, N
	work, J
	width, m
$\dot{W}$	power, W
W_e	electrical work, J
W_p	paddle wheel work, J
W_s	spring work, J
x	length coordinate, m
	quality, kg vapor/kg mixture, percent
Y	volumetric fraction, dimensionless
y	length coordinate, m
	moisture fraction, kg vapor/kg mixture, percent
	centroidal axis location, m
Z	compressibility factor, dimensionless
z	length coordinate, m
	intrinsic variable, dimensionless
	plate spacing, m

Greek Letter Symbols

α	angular acceleration, rad/s^2
	absorbtivity, dimensionless
	angle, degree or radian
	thermal diffusivity, m^2/s
β	refrigerator coefficient of performance, dimensionless
	volumetric coefficient of expansion, K^{-1}
	angle, degree or radian
	aircraft attitude angle, degree or radian
γ	heat pump coefficient of performance, dimensionless
	specific weight, N/m^3
Δ	change in, dimensionless gap thickness, m
ΔT	time interval, s
δ	inexact differential, dimensionless
	fin thickness, m
	boundary layer thickness, m
	static deflection, m
δW	differential work, J
δQ	differential heat, J
ϵ	emissivity, dimensionless
	heat exchanger effectiveness, dimensionless
	fin effectiveness, dimensionless
η	efficiency, dimensionless
	fin efficiency, dimensionless
θ	temperature difference, K
	angular displacement, radians
	wetting angle, degrees or radians
θ_m	mean temperature difference, K
Λ	aspect ratio, dimensionless
λ	wavelength, μm
	eigenvalue, dimensionless
μ	dynamic viscosity, kg/m-s
v	kinematic viscosity, m^2/s
ρ	density, kg/m^3
	reflectivity, dimensionless
ρ_e	electrical resistivity, ω-m
σ	Stefan-Boltzmann constant, 5.67×10^{-8}, W/m^2-K^4
	normal stress component, N/m^2
	surface tension, N/m
τ	torque, N-m
	shear stress, N/m^2
	tangential shear component, N/m^2
	transmissivity, dimensionless
	time constant, s
γ	percent throretical air
ϕ	an angle, degrees or radians
ψ	wet bulb depression, K
Ω	electrical resistance, Ohm
ω	angular velocity, rad/s
	specific humidity, kg water vapor/kg dry air
	solid angle, steradians

Roman Subscripts

avg indicates average condition
atm indicates atmospheric pressure
C indicates Carnot
indicates cold or cold side
indicates compressor
c indicates critical condition
indicates compression
cv indicates control volume
cyc indicates cycle
da indicates dry air
dirt indicates dirt or fouling
eng indicates engine
e indicates exit or exiting condition
indicates equivalent
flow indicates flow
f indicates saturated liquid condition
fg indicates difference between saturated liquid and saturated vapor
gage indicates gage pressure
gen indicates generator or generated value
g indicates saturated vapor
H indicates hot or hot side
i indicates in or inlet condition
indicates inner or inside
id indicates ideal
in indicates input
irrev indicates irreversible
liq indicates liquid
L indicates low or low side
max indicates a maximum
min indicates a minimum
mix indicates mixture
nonflow indicates nonflow condition
n indicates normal component
o indicates original or nominal value
indicates outer or outside
opt indicates optimum condition
out indicates output
proj indicates projected
P indicates pump
ref indicates reference condition
rev indicates reversible
r indicates reduced condition
sat indicates saturated condition
sink indicates sink
source indicates source
t indicates tension
total indicates total
trans indicates translational
v indicates vapor
vap indicates vapor
vac indicates vacuum
wv indicates water vapor

Index

N

O

P

Q

R

V

W

Z

Milton Keynes UK
Ingram Content Group UK Ltd.
UKHW051857071024
449327UK00025B/1996